Foodborne pathogens

Related titles:

Emerging foodborne pathogens
(ISBN 978-1-85573-963-5)
Developments such as the increasing globalisation of the food industry, constant innovations in technologies and products, and changes in the susceptibility of populations to disease, have all highlighted the problem of emerging pathogens. Pathogens may be defined as emerging in a number of ways. They can be newly-discovered (e.g. through more sensitive analytical methods), linked for the first time to disease in humans, or first associated with a particular food. A pathogen may also be defined as 'emerging' when significant new strains emerge from an existing pathogen, or if the incidence of a pathogen increases. Designed for microbiologists and QA staff in the food industry, and food safety scientists working in governments and academia, this collection discusses ways of identifying emerging pathogens and includes chapters on individual pathogens, their epidemiology, methods of detection, and means of control.

Handbook of hygiene control in the food industry
(ISBN 978-1-85573-957-4)
Complementing the highly successful *Hygiene in food processing*, this book reviews recent research on improving hygiene in food processing. Part I considers research on contamination risks such as biofilms and how they can be assessed. Part II reviews ways of improving hygienic design of buildings and equipment, including clean room technology. The final part discusses ways of improving hygiene practice and management.

Modelling microorganisms in food
(ISBN 978-1-84569-006-9)
While predictive microbiology has made a major contribution to food safety, there remain many uncertainties, e.g. growing evidence that traditional microbial inactivation models do not always fit the experimental data and an awareness that bacteria of one population do not behave homogeneously, that they may interact and behave differently in different food systems. These problems are all the more important because of the growing interest in minimal processing techniques that operate closer to death, survival and growth boundaries and thus require a greater precision from models. Edited by leading authorities, this collection reviews current developments in quantitative microbiology. Part I discusses best practice in constructing quantitative models and Part II looks at specific areas in new approaches to modelling microbial behaviour.

Details of these books and a complete list of Woodhead's titles can be obtained by:
- visiting our web site at www.woodheadpublishing.com
- contacting Customer Services (e-mail: sales@woodheadpublishing.com; fax: +44 (0) 1223 893694; tel.: +44 (0) 1223 891358 ext.130; address: Woodhead Publishing Limited, Abington Hall, Granta Park, Great Abington, Cambridge CB21 6AH, UK)

Foodborne pathogens

Hazards, risk analysis and control

Second edition

Edited by
Clive de W. Blackburn and Peter J. McClure

CRC Press
Boca Raton Boston New York Washington, DC

WOODHEAD PUBLISHING LIMITED
Oxford Cambridge New Delhi

Published by Woodhead Publishing Limited, Abington Hall, Granta Park,
Great Abington, Cambridge CB21 6AH, UK
www.woodheadpublishing.com

Woodhead Publishing India Private Limited, G-2, Vardaan House, 7/28 Ansari Road,
Daryaganj, New Delhi – 110002, India

Published in North America by CRC Press LLC, 6000 Broken Sound Parkway, NW,
Suite 300, Boca Raton, FL 33487, USA

First published 2009, Woodhead Publishing Limited and CRC Press LLC

British Library Cataloguing in Publication Data
A catalogue record for this book is available from the British Library.

Library of Congress Cataloging in Publication Data
A catalog record for this book is available from the Library of Congress.

Woodhead Publishing ISBN 978-1-84569-362-6 (book)
Woodhead Publishing ISBN 978-1-84569-633-7 (e-book)
CRC Press ISBN 978-1-4398-0768-2
CRC Press order number N10071

The publishers' policy is to use permanent paper from mills that operate a sustainable forestry policy, and which has been manufactured from pulp which is processed using acid-free and elemental chlorine-free practices. Furthermore, the publishers ensure that the text paper and cover board used have met acceptable environmental accreditation standards.

Typeset by Replika Press Pvt Ltd, India
Printed by TJ International, Padstow, Cornwall, UK

Cover image: Electron micrograph of *Salmonella* kindly provided by Bill Cooley, Veterinary Laboratories Agency, Addlestone, Surrey, UK

Contents

Part II Bacterial hazards

Contributor contact details

(* = main contact)

Chapter 1
Dr Clive de W. Blackburn* and Dr Peter J. McClure
Unilever
Colworth Science Park
Sharnbrook
Bedford
MK44 1LQ
UK

E-mail: Clive.Blackburn@unilever.com
peter.mcclure@unilever.com

Chapter 2
Dr Roy Betts
Campden BRI
Chipping Campden
Gloucestershire
GL55 6LD
UK

E-mail: betts@campden.co.uk

Dr Clive Blackburn*
Group Leader – Drinks Design
Centre of Excellence Drinks
Unilever
Colworth Science Park
Sharnbrook
Bedford
MK44 1LQ
UK

E-mail: Clive.Blackburn@unilever.com

Chapter 3
Dr J. David Legan*
Kraft Foods
Kraft Foods Technology Center
801 Waukegan Road
Glenview
IL 60025
USA

E-mail: david.legan@kraft.com

Dr Martin B. Cole
National Center for Food Safety and Technology
6502 South Archer Road
Summit-Argo
IL 60501
USA

E-mail: cole@iit.edu

Dr Cynthia M. Stewart
Silliker Inc
Silliker Food Science Center
160 Armory Drive
South Holland
IL 60473
USA

E-mail: cynthia.stewart@silliker.com

Chapter 4

Dr Tom Ross* and Professor Tom A. McMeekin
Food Safety Centre
Tasmanian Institute of Agricultural Reasearch
Private Bag 54
University of Tasmania
Hobart 7001 Tasmania
Australia

E-mail: Tom.Ross@utas.edu.au
Tom.McMeekin@utas.edu.au

Chapter 5

Dr Les Smoot*
Director, Nestlé Quality Assurance Center
Nestlé USA
6625 Eiterman Road
Dublin
OH 43017
USA

E-mail: Leslie.Smoot@us.nestle.com

Dr Jean-Louis Cordier
Food Safety Manager
Nestlé Nutrition
Avenue Reller 22
1800 Vevey
Switzerland

E-mail: Jean-Louis.Cordier@nestle.com

Chapter 6

Dr Geraldine Duffy
Ashtown Food Research Centre
Teagasc
Ashtown
Dublin 15
Ireland

E-mail: geraldine.duffy@teagasc.ie

Chapter 7

Ralph Early
Principal Lecturer in Food Science and Moral Philosophy
Harper Adams University College
Newport
Shropshire
TF10 8NB
UK

E-mail: rearly@harper-adams.ac.uk
00483560@harper-adams.ac.uk

Chapter 8

Dr Birte Fonnesbech Vogel
Technical University of Denmark
National Institute of Aquatic Resources
Søltofts Plads, Building 221
DK-2800 Kgs. Lyngby
Denmark

E-mail: bfv@aqua.dtu.dk

Chapter 9
Dr W. J. Doré
Shellfish Microbiology
Marine Institute
Oranmore
Co. Galway
Ireland

E-mail: bill.dore@marine.ie

Chapters 10 and 12
Dr John T. Holah
Campden BRI
Station Road
Chipping Campden
Gloucestershire
GL55 6LD
UK

E-mail: j.holah@campden.co.uk

Chapter 11
Dr A. P. M. Hasting
Tony Hasting Consulting
37 Church Lane
Sharnbrook
Bedford
MK44 1HT
UK

E-mail: tony.hasting@virgin.net

Chapter 13
Professor Martyn Brown
mhb Consulting
46 High Street
Harrold
Bedfordshire
MK43 7DQ
UK

E-mail: martynbrown46@btinternet.com

Chapter 14
Tony Mayes*
Safety & Environmental Assurance Centre
Unilever Colworth
Sharnbrook
Bedford
MK44 1LQ
UK

E-mail: tony.mayes@unilever.com

Sara Mortimore
Vice President Quality and Regulatory Affairs
Land O'Lakes Inc.
P O Box 64101
MS 2350
St Paul
MN 55164-0101
USA

E-mail: semortimore@landolakes.com

Chapter 15
Professor Chris Griffith* and Dr Elizabeth Redmond
Food Research and Consultancy Unit
Cardiff School of Health Sciences
University of Wales Institute Cardiff
Western Avenue
Cardiff
CF5 2YB
Wales
UK

E-mail: cgriffith@uwic.ac.uk

Chapter 16
Professor G. W. Gould
Visiting Professor
University of Leeds
Leeds
LS2 9JT
UK

17 Dove Road
Bedford
MK41 7AA
UK

E-mail: gould7aa@btinternet.com

Chapters 17, 18 and 19
Dr Chris Bell
Independent Consultant Microbiologist
UK

Alec Kyriakides*
Head of Product Quality
Safety & Supplier Performance
Sainsbury's Supermarkets Ltd
UK

E-mail: alec.kyriakides@sainsburys.co.uk

Chapter 20
Dr Clive de W. Blackburn* and Dr Peter J. McClure
Unilever
Colworth Science Park
Sharnbrook
Bedford
MK44 1LQ
UK

E-mail: Clive.Blackburn@unilever.com
peter.mcclure@unilever.com

Chapter 21
Professor Steve Forsythe*
Nottingham Trent University
Nottingham
N91 4BU
UK

E-mail: stephen.forsythe@ntu.ac.uk

Dr Jane Sutherland and Dr Alan Varnam (deceased)
University of North London
London
UK

Chapter 22
Professor Martin Adams
Division of Microbial Sciences
Faculty of Health and Medical Sciences
University of Surrey
Guildford
Surrey GU2 7XH
UK

E-mail: m.adams@surrey.ac.uk

Chapter 23

Paul A. Gibbs, BSc, PhD, FIFST, Ch Sci.
Leatherhead Food International
Randalls Road
Leatherhead
Surrey
KT22 7RY
UK

E-mail: pgibbs@LeatherheadFood.com

Escola Superior de Biotecnologia
Universidade Católica Portuguesa
4200-072 Porto
Portugal

E-mail: pgibbs@esb.ucp.pt

Chapter 24

Dr Clive de W. Blackburn* and Dr Peter J. McClure
Unilever
Colworth Science Park
Sharnbrook
Bedford
MK44 1LQ
UK

E-mail: Clive.Blackburn@unilever.com
peter.mcclure@unilever.com

Chapter 25

Dr Kirsten Mattison,* Dr Sabah Bidawid and Dr Jeff Farber
Health Canada
251 Sir Frederick Banting Driveway
PL2204E
Ottawa
ON K1A 0K9
Canada

E-mail: Kirsten_Mattison@hc-sc.gc.ca;
jeff_farber@hc-sc.gc.ca

Chapter 26

Professor H. V. Smith*
Scottish Parasite Diagnostic Laboratory
Stobhill Hospital
Glasgow
G21 3UW
Scotland
UK

E-mail: huw.smith@northglasgow.scot.nhs.uk

Dr Roger Evans
Department of Microbiology
Raigmore Hospital
Inverness
IV2 3UJ
Scotland
UK

E-mail: roger.evans@haht.scot.nhs.uk

Chapter 27

Dr K. Darwin Murrell*
Adjunct Professor
Department of Veterinary Disease Biology
Faculty of Life Science
Copenhagen University
Gronnegardsvey 15
1879 Frederiksberg C
Kobenhavn
Denmark

E-mail: kdmurrell@comcast.net

Professor D. W. T. Crompton
Faculty of Biomedical and Life Sciences
Graham Kerr Building
University of Glasgow
Glasgow
G12 8QQ
Scotland
UK

Chapter 28

Visiting professor Maurice Moss
Faculty of Health and Medical Sciences
University of Surrey
Guildford
Surrey
GU2 7XH
UK

E-mail: m.moss@surrey.ac.uk

Chapter 29

Dr Mansel W. Griffiths
Chair in Dairy Microbiology and Director, Canadian Research Institute for Food Safety
Department of Food Science
University of Guelph
Guelph, ON
Canada
N1G 2W1

E-mail: mgriffit@uoguelph.ca

Chapter 30

Paul Brown
7815 Exeter Road
Bethesda
MD 20814
USA

E-mail: paulwbrown@comcast.net

Chapter 31

Paw Dalgaard* and Jette Emborg
Technical University of Denmark
National Institute of Aquatic Resources
Department of Seafood Research
Building 221, Søltofts Plads
DK-2800 Kgs. Lyngby
Denmark

E-mail: pad@aqua.dtu.dk

Chapter 32

Dr Erwin Duizer and Dr Marion Koopmans
Laboratory for Infectious Diseases and Perinatal Screening
National Institute for Public Health and the Environment (RIVM)
P O Box 1
3720 BA Bilthoven
The Netherlands

E-mail: Erwin.Duizer@rivm.nl

Part I

Risk assessment and management in the food chain

1

Introduction

C. de W. Blackburn and P. J. McClure, Unilever, UK

Abstract: Foodborne disease continues to be a common and serious threat to public health all over the world, and both industrialised and developing countries suffer large numbers of illnesses. This introductory chapter sets the scene for the rest of this book on foodborne pathogens by considering the trends in foodborne disease in the context of a world of societal, demographic, economic and climatic change.

Key words: foodborne disease, emerging foodborne disease, epidemiology, demographics, climate change.

1.1 Trends in foodborne disease

Foodborne disease continues to be a common and serious threat to public health all over the world and is a major cause of morbidity. Both industrialised and developing countries suffer large numbers of illnesses, costing billions of dollars to society every year. In the USA alone, the estimated costs related to foodborne disease in 2001 were between $10 and $83 billion annually (FDA, 2001). Most foodborne illnesses are mild, and are associated with acute gastrointestinal symptoms such as diarrhoea and vomiting. However, sometimes foodborne disease is much more serious; it can be life-threatening, particularly in children in developing countries, and infection can also be followed by chronic sequelae or disability. In many countries where information on foodborne diseases is collected, the total number of cases has been increasing over the past 20–30 years (Käferstein *et al.*, 1997). In the UK, figures rose from just under 10 000 cases recorded in 1977, to more than 90 000 cases in 1998. Most European countries reported a doubling of salmonellosis cases between 1985 and 1992 (FAO/WHO, 1995). In the UK, for instance, the Steering Group on the Microbiological Safety of Food (SGMSF) found 12 846 infections (23 per 100 000) in 1981 compared with 33 600 (58 per 100 000) in 1994 (SGMSF, 1995). In North America, there has been a notable increase in infections caused by *Salmonella* Enteritidis since the late 1980s (St Louis *et al.*, 1988; Levy *et al.*, 1996). Overall, *Salmonella* and *Campylobacter* continue to dominate bacterial foodborne diseases, with incidences of 38.2 and 51.6 cases per 100 000 population, respectively, in Europe in 2005 (EFSA, 2006).

In recent years, the increased awareness of food safety, changes in regulatory and educational measures and changes in practice in food production have led to decreases in incidence of particular foodborne diseases in some regions. For example, in the year 2000, salmonellosis in the UK was at its lowest level since 1985, with a 54 % decrease in the number of reported cases compared with the previous year. A review of gastrointestinal disease outbreaks in the UK between 1993 and 2003 by Hughes *et al.* (2007) reported a marked decline in the number of *Salmonella* outbreaks during this period, substantially reducing the burden of morbidity and mortality associated with foodborne disease. A decrease in the number of cases of salmonellosis has also been observed recently in the US (Olsen *et al.*, 2001). These particular decreases in salmonellosis have been attributed to vaccination programmes for poultry and other changes that have been implemented in these regions. Decreases in the number of cases of listeriosis have also been observed in the UK and the USA in the past 20 years although there has been some recent concern raised in Europe due to an increase in cases of listeriosis in the elderly. *Campylobacter* spp. are the largest cause of sporadic bacterial gastroenteritis in the world but, in some countries, cases of campylobacteriosis appear to have peaked. There has been a decrease in the incidence rate of campylobacteriosis in the UK and the USA, and it has been suggested that the decrease in the USA is due to the introduction of the Hazard Analysis Critical Control Points system (HACCP) (FoodNet, 2006).

1.2 Incidence of foodborne disease

Accurate estimates of the yearly incidence of foodborne disease are difficult and sometimes impossible, depending on the reporting systems in different countries. Foodborne disease statistics in some European countries and the Americas, where reporting systems are better than some other regions, are still dominated by cases of salmonellosis and campylobacteriosis. In other regions, however, foodborne disease statistics tend to rely on outbreak information only and, in some cases, other organisms are identified as leading causes of illness. For example, in Taiwan, 74 % of outbreaks in 1994 were caused by bacterial pathogens of which *Vibrio parahaemolyticus* (56.7 %), *Staphylococcus aureus*

(20.3 %), *Bacillus cereus* (14.9 %) and salmonellas (8.1 %) were the major agents identified (Pan *et al.*, 1996). In a study of diarrhoeal disease in south eastern China between 1986 and 1987, the overall incidence of diarrhoeal illness was 730 episodes per 1000 population (Kangchuan *et al.*, 1991). The most commonly isolated organisms in order of frequency of occurrence were enterotoxigenic *Escherichia coli*, *Shigella* species, enteropathogenic *E. coli*, *Campylobacter jejuni*, vibrios and enteroinvasive *E. coli*. These organisms and *Entamoeba histolytica* are typical causes of diarrhoea in developing countries (DuPont, 1995). It is important to remember that foods will only be one of a number of possible sources of infection in these cases, but the lack of good epidemiological data in these regions leads to the role of food being poorly acknowledged (Käferstein *et al.*, 1997). Even massive outbreaks of illness, such as an outbreak associated with *E. coli* O157:H7 in China in 2001 affecting 20 000 individuals and causing 1777 deaths in Jiangsu and Anhui provinces, fail to be widely reported.

In the USA, it has been estimated that foodborne diseases cause approximately 76 million illnesses, 325 000 hospitalisations and 5000 deaths each year, with known pathogens accounting for 14 million illnesses, 60 000 hospitalisations and 1800 deaths (Mead *et al.*, 1999). In this study, estimates were made using data from a number of sources, including outbreak-related cases and passive and active surveillance systems. The organisms identified as causing the largest number of foodborne-related cases of illness were Norwalk-like viruses, followed by campylobacters, salmonellas, *Clostridium perfringens*, *Giardia lamblia*, staphylococci, *E. coli* and *Toxoplasma gondii*, respectively. Norovirus is estimated to be the most common cause of diarrhoea in people of all age groups although the proportion of infections attributable to food remains unknown (Koopmans and Duizer, 2004). Incidence of foodborne disease in different countries is often difficult to compare because of different reporting systems.

1.3 Foodborne disease surveillance

Quantifying numbers of reported foodborne illnesses and identifying emerging pathogens and elements that increase the risk of disease can all be achieved with surveillance systems. These systems can be considered passive, e.g. laboratory-based reporting of specific pathogens, illnesses reported by physicians, outbreak investigations, or active, such as use of special epidemiological studies to determine more realistic levels of morbidity of foodborne disease. Information from such systems is used to determine the priorities for food safety actions and prevention and control programmes, including development of new or modified policies and procedures, monitoring efficacy of programmes, identifying new hazards, educating and training those involved in food manufacturing, handling and preparation, including consumers. Each surveillance system has its drawbacks and strengths and focuses on different aspects of foodborne disease investigation. Many systems are not able to determine the true incidence of foodborne illness, because of the reporting systems used and various other reasons, such as under-reporting due to methodologies unable to determine the actual causes of outbreaks. It is estimated that for every case of salmonellosis reported to the Centers for Disease Control and Prevention in the USA, between 20 and 100 cases go unreported (Tauxe, 1991).

Experts on foodborne disease estimate that most cases of foodborne illness in the USA originate from foods prepared in the home (Knabel, 1995). Surveys of consumer practices and perceptions (Altekruse *et al.*, 1995; Fein *et al.*, 1995) tend to demonstrate that awareness of the major causes of foodborne illness such as salmonellas and campylobacters is extremely poor and emphasise the need for and importance of effective education and training. Equally worrying is a recent study (Jones *et al.*, 2008) that used analytical epidemiology to study management risk factors in the catering industry in the UK reporting that case businesses (i.e. those causing outbreaks) were neither less likely to have staff with formal training nor less likely to use HACCP. The authors also concluded that food hygiene inspection scores were not useful in predicting which businesses were associated with outbreaks. Other studies also show that greater knowledge/understanding does not necessarily result in implementation of good food safety practices and awareness of HACCP is not independently associated with outbreak status in restaurants in the USA.

Different surveillance systems are reviewed in a series of studies published by Guzewich, Bryan and Todd (Bryan *et al.*, 1997a, b; Guzewich *et al.*, 1997; Todd *et al.*, 1997). It is critical that surveillance systems share common information across national boundaries and where possible exchange information on outbreaks of foodborne disease utilising the power and capability of modern telecommunication facilities. An example of such a system is Enter-net, which is used to provide early recognition and subsequent comprehensive investigation of outbreaks of salmonellosis and vero cytotoxin-producing *E. coli* O157 in Europe (Fisher and Gill, 2001). International networks such as Enter-net are important tools considering the large-scale production of food and globalisation of food trade. Development in nucleic acid-based techniques has had a major impact on disease surveillance, enabling rapid pathogen detection and characterisation. This has resulted in 'sporadic cases' being linked, sometimes over large geographical areas, and identified as outbreaks, with a common source of infection. The 'new outbreak scenario' resulting from low-level contamination of widely distributed food products or ingredients is described in detail by Tauxe (1997) and attributed to changes in the way food is produced and distributed. A recent review of surveillance systems used in different regions is provided by Todd (2006) and examples of countries and regions that have consolidated and improved their activities for surveillance and control of foodborne diseases are discussed. The support of national and international systems by agencies such as the World Health Organisation will help developed and developing nations to gain a better understanding of the true burden of illness in these regions and also on a global scale.

1.4 Emerging foodborne disease and changing patterns in epidemiology

In recent years, the epidemiology of foodborne diseases has been changing as new pathogens have emerged. 'Emerging diseases' are described as those that have increased in prevalence in recent decades or are likely to do so in the near future (Altekruse and Swerdlow, 1996), so it is not necessary for an emerging pathogen to be newly evolved. Foodborne diseases that are regarded as emerging include illness caused by enterohaemorrhagic (EHEC) *E. coli* O157:H7, *Campylobacter jejuni* and antibiotic-resistant *Salmonella*. In some cases, disease has been associated with food vehicles only relatively recently. Examples of these pathogens include *Listeria monocytogenes*, *Cryptosporidium parvum* and *Cyclospora cayetanensis*. Many of these foodborne pathogens have a non-human animal reservoir, and are termed zoonoses, but they do not necessarily cause disease in the animal. Previously, animal or carcass inspection was used as a method of preventing zoonotic diseases being transferred through food, but this can no longer be relied upon. The various factors that impact on presence of pathogens on meat in Europe have been recently reviewed by Nørrung and Buncic (2008). This study assessed the different stages in the food chain including the pre-harvest phase, harvest and processing stages.

Both *E. coli* O157:H7 and antibiotic-resistant *Salmonella* are found in cattle and are examples of relatively newly evolved pathogens. According to Whittam (1996), *E. coli* O157:H7 probably evolved from an enteropathogenic O55:H7 ancestor through horizontal gene transfer and recombination, and a stepwise evolutionary model has been proposed by Feng *et al.* (1998). When outbreaks of vero cytotoxigenic *E. coli* associated illness were first identified, many were associated with undercooked beef, such as burgers. More recently, the list of food vehicles associated with EHEC is becoming longer and longer, and an increasing number of infections are being linked to fresh produce such as vegetables and fruit. Enterohaemorrhagic *E. coli* O157:H7 and other EHEC have changed the 'rule book' in a number of ways, primarily because they are able to cause illness in relatively low numbers and infection can result in very serious illness or even death. Some foods that were traditionally regarded as 'safe', such as apple juice and fermented meats, have caused haemorrhagic colitis and more serious illness associated with EHEC. Growth of the organism in foods is not necessary to cause infection, so any contamination, even at very low levels, may have serious consequences. Worryingly, a new sub-clone of a second group of EHEC (primarily comprising O26:H11 and O11:H8 serotypes) has emerged in Europe and this clone shares the same prominent virulence factors of O157:H7 and is common in the bovine reservoir (Donnenberg and Whittam, 2001). These organisms, and others like them, may well emerge as important foodborne pathogens in the future.

Genetic promiscuity is facilitated by a range of genetic elements including plasmids, transposons, conjugative transposons and bacterophages. The ability to evolve through horizontal gene transfer and acquire 'foreign' DNA has resulted in novel phenotypes and genotypes emerging, and this is causing confusion amongst some microbiologists who prefer to group organisms according to one or two characteristics. For example, there are six pathotypes of diarrhoeagenic *E. coli*, characterised usually on the basis of disease caused and also on presence of mainly non-overlapping virulence factors. There is, however, an increasing number of studies describing *E. coli* isolates associated with diarrhoeal disease that possess previously unreported combinations of virulence factors.

In some cases, virulence factors are encoded on large DNA regions termed pathogenicity islands, and these are shared amongst different pathogenic organisms, contributing to microbial evolution (Hacker *et al.*, 1997). Some genetic elements, such as bacteriophages, as well as being important vectors for transmission of virulence genes, also serve as important precursors for the expression of bacterial virulence, as shown in *Vibrio cholerae* (Boyd *et al.*, 2001). It has also been shown that some members of the Enterobacteriaceae carry defects in the *mutS* gene, which directs DNA-repair processes (LeClerc *et al.*, 1996). The formation of deletions also plays a major role in so-called 'genome plasticity' and can contribute to development of organisms with improved functionality (Stragier *et al.*, 1989). Fortunately, new techniques determining DNA sequences specific to particular regions should allow investigation of evolutionary mechanisms that allow development of new pathogens, and will facilitate identification and characterisation of these organisms. Such techniques have recently been used to show that old lineages of *E. coli* have acquired the same virulence factors in parallel, indicating that natural selection has favoured an ordered acquisition of genes and the progressive build-up of molecular mechanisms that increase virulence (Reid *et al.*, 2000). These new techniques and approaches, such as sequencing, mean that there will soon be an abundance of data and the focus will then need to shift to solving the problem of interpreting these data and visualising them.

Adaptation to particular environments appears to have played a part in the emergence of *S.* Typhimurium DT 104 and other antibiotic-resistant salmonellas. A number of researchers have proposed that the use of antibiotics in human health, agriculture and aquaculture has resulted in the selection of *Salmonella* strains, notably DT 104, that are resistant to multiple antibiotics. There is increasing evidence that 'stresses' imposed by an organism's environment can modulate and enhance virulence, providing there is a potential driving force promoting adaptive mutations that may serve to select strains that are even more virulent (Archer, 1996). Factors associated with demographics, consumer trends and changes in food production have also been put forward as possible contributors to the emergence of new pathogens that have appeared in different areas of the globe simultaneously. Many of these shifts have magnified the potential impact of a single source of infection.

1.4.1 Demographics

An increasing world population places increased pressure on global food production and the question 'will supply meet expected demand?', especially in developing countries, cannot be answered with any certainty (Doos and Shaw, 1999). Fuelled by urbanisation and higher incomes, there are likely to be changes in the pattern of food consumption. For example, there is likely to be a major increase in consumption of meat in the developing world and this will place more pressure on animal production systems (van der Zijpp, 1999).

Demographic changes occurring in industrialised nations have resulted in an increase in the proportion of the population with heightened susceptibility to severe foodborne infections. Growing segments of the population have immune impairment as a consequence of infection with HIV, ageing or underlying chronic disease (Slutsker *et al.*, 1998).

Increasing migration is leading to ethnic diversity within countries which in turn is leading to changing dietary patterns and increased globalisation of food choices (Jensen, 2006). Immigration has also contributed to the epidemiology of foodborne disease, as some reports of foodborne illnesses involve transmission through foods consumed primarily by immigrant groups, an example of this being the increase in parasite infections in the USA (Slutsker *et al.*, 1998).

1.4.2 Consumer trends

There are several consumer trends that may have an impact on foodborne disease. Shifting preferences due to new information about the links between diet and health (Jensen, 2006) and the trend towards 'more natural' and 'fresh' food with less preservation and processing are contributing to new demands for foods. This has manifested itself in increasing consumption of fresh fruits and vegetables, and the number of outbreaks associated with these types of foods has also increased. Numerous studies have described the health and nutritional benefits of fresh fruits and vegetables, and various regulatory agencies have emphasised the need for consumption of fresh produce by recommending regular daily servings in the diet. This has led to a dramatic increase in sales of fresh-cut produce in many countries and also, unfortunately, to an increase in foodborne disease linked to these products. These foods are particularly vulnerable as potential foodborne disease vehicles because they are easily contaminated during growing and harvesting, and reliable intervention methods, such as heating, cannot be used because of their effects on the organoleptic quality of the foods. These products now account for 12 % of disease cases and 6 % of outbreaks in the USA (CDCP, 2005), with *Salmonella* and *E. coli* O157 being the most common aetiological agents for produce-related outbreaks (Mukherjee *et al.*, 2006). Anecdotes about the health properties of raw foods may also be interfering with health messages about the risks associated with eating some raw or lightly cooked foods (Slutsker *et al.*, 1998).

Another consumer trend is the increase in the percentage of spending on food eaten away from home. This places greater importance on the safe operation of catering establishments for the control of foodborne disease. By the 1990s, for example, foodborne outbreaks that occurred outside the home accounted for almost 80 % of all reported outbreaks in the USA (Slutsker *et al.*, 1998). It is also suggested that this situation is compounded by a decrease in home food hygiene instruction, particularly in light of other important health concerns tackled in schools, e.g. substance abuse, HIV infection and obesity (Slutsker *et al.*, 1998).

International travel has increased dramatically during the last century. Travellers may become infected with foodborne pathogens that are uncommon in their nation of residence and may transmit the pathogen further when they return home. International travel is also one of the drivers for an increasing demand for international foods in local markets, and this in turn fuels the international trade in foods.

1.4.3 Trends in food production

Technological changes in production, processing and distribution, growth of large-scale retailing, changes in product availability, as well as expansion of trade world wide, have contributed to a rapidly changing market for food products (Jensen, 2006).

There is a trend towards global sourcing of raw materials and processing in large, centralised facilities and distribution of product over large geographical areas using longer and more complex supply chains. As a result, there are now many more potential points at which pathogens, including those that might otherwise not have been considered, can be introduced into the supply chain and spread within a country and across regional and national borders. However, on the positive side this provides large companies with the opportunity to significantly reduce foodborne disease globally by focusing their resources on identifying hazards, assessing risks and implementing effective preventative measures.

Consumer-driven demands for demonstrable reductions in the 'carbon footprint' associated with food manufacture may lead to pressure on food producers, manufacturers and retailers in a number of areas that may have an impact on food safety, e.g. reducing transportation costs thereby limiting sourcing and manufacture options; reduction in packaging leading to a higher risk of loss of product integrity; and increased reuse of water during manufacture. Pressure to use land to cultivate biofuel crops may also have implications for the safety of the food chain as sourcing options, and potentially standards of safety, are reduced.

In contrast to this centralisation of food processing there is a trend for increasing numbers of localised catering and food preparation operations, driven by consumer demand for out-of-home food consumption. This leads to greater challenges in order to provide a unified standard of food safety in this food sector. For example, the most frequently reported setting for outbreaks where food workers have been implicated in the spread of foodborne disease is restaurants (Todd *et al.*, 2007a).

As the global demand for food grows there will be an increasing need for intensification of agriculture and this will have dramatic impacts on the diversity, composition and functioning of the world's ecosystems (Tilman, 1999). The likelihood for the proliferation of human pathogens in more intensive and centralised forms of animal and crop production, and potential contamination of water supplies, will be greater and will require effective management. The current status of livestock production systems, drivers of global livestock production and major trends in production of livestock and how these trends link to food safety are covered in a recent review by Steinfeld *et al.* (2006).

1.4.4 Climate change

There is near unanimous scientific consensus that greenhouse gas emissions generated by human activity will change Earth's climate and this may adversely affect human health (McMichael *et al.*, 2006). Climate change is likely to manifest itself in greater variability in, and changes to, mean values of: temperature, precipitation, humidity and wind patterns. As a consequence increases in the frequency and severity of extreme weather conditions are likely, together with sea-level rise and environmental degradation, which in turn are likely to adversely affect crop, livestock and fisheries yields placing pressure on the food chain. Transmission of infectious disease is determined by many factors including climatic conditions and there is evidence that increased temperatures may lead to more foodborne disease, especially salmonellosis, and waterborne infection, such as cholera (McMichael *et al.*, 2006).

1.5 Control of foodborne disease

The complexity of the global food market means that the control of foodborne disease is a joint responsibility and requires action at all levels from the individual to international groups, and at all parts of the supply chain from the farm to the fast-food restaurant. The tools used and approaches taken to ensure control require different emphasis, depending on a number of factors such as where food materials have come from, how they have been processed and handled and how they are stored. The risk of foodborne illness can be reduced by using existing technologies, such as pasteurisation and refrigeration, and by adopting some simple precautions such as avoiding cross-contamination by separation of raw and cooked foods and employing good hygienic practices. These key principles of food safety are captured in the 'Five Keys' for safer food, developed by the WHO:

- keep clean;
- separate raw and cooked;
- cook thoroughly;
- keep food at safe temperatures;
- use safe water and raw materials.

For many foodborne diseases, multiple choices for prevention are available, and the best answer may be to apply several steps simultaneously, for example measures both to eliminate organisms during the food process and to reduce the likelihood of the organisms being present in the first place. A better understanding of how pathogens persist in animal reservoirs (such as farm herds) is also critical to successful long-term prevention. In the past, the central challenge of foodborne disease lay in preventing contamination of human food with sewage of animal manure. In the future, prevention of foodborne disease will increasingly depend on controlling contamination of feed and water consumed by the animals themselves (Tauxe, 1997).

Decision making to ensure that the most effective controls are implemented is key to making the most efficient use of risk management options. 'Fit-for-purpose' microbiological risk assessment (MRA) and MRA-based approaches and tools are increasingly being employed to give a holistic view of complex food safety challenges and enabling risk-based decision making by both governmental risk managers and food safety managers in industry (Lammerding, 2007).

Although the onus is on prevention of foodborne disease, valuable lessons can be learned by reviewing food poisoning statistics and incidents. This in turn can provide a focus for effective control measures to help reduce food poisoning (Bryan, 1988; Todd *et al.*, 2007b). It is, however, notable that outbreaks continue to occur for the same reasons, such as poor implementation of HACCP, marketing products that are not 'safe by design', breakdowns in manufacturing, use of inappropriate and unhygienic handling practices. These are basic failures that can be easily avoided. Ranking the factors that contributed to outbreaks of foodborne diseases can indicate trends and also differences in the different foodborne pathogens reflecting their association with raw material and physiological properties. Increasing cooperation between countries to collect and share high-quality outbreak data will help to provide a more complete picture of the main sources of foodborne disease and to be able to monitor changes over time (EFSA, 2006). However, unless these learnings are translated into material and messages that can change behaviours, then collecting and sharing the information may have limited use. It seems that improvements in risk communication and training in food safety are needed, to influence and change behaviours that currently contribute to the prevalence of foodborne disease that occurs across the globe.

1.6 Rationale for this book

Ultimately, the control of foodborne pathogens requires the understanding of a number of factors including the knowledge of possible hazards, their likely occurrence in different products, their physiological properties, the risks they pose to the consumer and the availability and effectiveness of different preventative/intervention measures. The aim of this book is to help provide this understanding.

Whilst there are good reference texts for the microbiologist on foodborne pathogens, there are less that relate current research to practical strategies for hazard identification, risk assessment and control. This text takes this more applied approach. It is designed both for the microbiologist and the non-specialist, particularly those whose role involves the safety of food processing operations.

Part I looks at general techniques in assessing and managing microbiological hazards. After a review of analytical methods and their application, there are chapters on modelling pathogen behaviour, carrying out risk assessments as the essential foundation for effective food safety management, and the significance of emerging foodborne pathogens. The following chapters then look at good management practice at key stages in the supply chain, starting with primary production and ending with the consumer. In between there are chapters on hygienic plant and equipment design, sanitation and safe process design and operation. These provide the foundation for what makes for effective HACCP systems implementation.

This discussion of pathogen control then provides a context for Part II which starts with an overview of preservation principles and new technologies and then looks at what this means in practice for major pathogens such as pathogenic *E. coli*, *Salmonella*, *Listeria monocytogenes*, *Campylobacter* and pathogenic *Clostridium* and *Bacillus* species. Each chapter discusses pathogen characteristics, detection methods and control procedures. Part III then looks at non-bacterial hazards such as toxigenic fungi, viruses, protozoa, helminths and prions, as well as emerging potential hazards such as *Mycobacterium avium* subspecies *paratuberculosis*.

1.7 References

Altekruse S F and Swerdlow D L (1996) The changing epidemiology of foodborne diseases, *The American Journal of the Medical Sciences*, **311**, 23–9.

Altekruse S F, Street D A, Fein S B and Levy A S (1995) Consumer knowledge of foodborne microbial hazards and food-handling practices, *Journal of Food Protection*, **59**, 287–94.

Archer D L (1996) Preservation microbiology and safety: evidence that stress enhances virulence and triggers adaptive mutations, *Trends in Food Science and Technology*, **7**, 91–5.

Boyd F E, Davis B M and Hochhut B (2001) Bacteriophage-bacteriophage interactions in the evolution of pathogenic bacteria, *Trends in Microbiology*, **9**, 137–44.

Bryan F L (1998) Risks of practices, procedure and processes that lead to outbreaks of foodborne diseases, *Journal of Food Protection*, **51**, 663–73.

Bryan F L, Guzewich J J and Todd E C D (1997a) Surveillance of foodborne disease. Part 2. Summary and presentation of descriptive data and epidemiologic patterns: their values and limitations, *Journal of Food Protection*, **60**, 567–78.

Bryan F L, Guzewich J J and Todd E C D (1997b) Surveillance of foodborne disease. Part 3. Summary and presentation of data on vehicles and contributory factors; their value and limitations, *Journal of Food Protection*, **60**, 701–14.

CDCP (2005) *U.S. Foodborne Disease Outbreaks*, Foodborne Outbreak Response and Surveillance Unit, Centers for Disease Control and Prevention, Atlanta, GA, available at: http://www.cdc.gov/foodborneoutbreaks/, accessed November 2008.

Donnenberg M S and Whittam T S (2001) Pathogenesis and evolution of virulence in enteropathogenic and enterohemorrhagic *Escherichia coli*, *Journal of Clinical Investigation*, **107**, 539–48.

Doos B R and Shaw R (1999) Can we predict the future food production? A sensitivity analysis, *Global Environmental Change – Human and Policy Dimensions*, **9**, 261–83.

DuPont H (1995) Diarrhoeal diseases in the developing world, *Infectious Disease Clinics of North America*, **9**, 313–24.

EFSA (2006) The community summary report on trends and sources of zoonoses, zoonotic agents and antimicrobial resistance and foodborne outbreaks in the European Union in 2005, European Food Safety Authority, *The EFSA Journal*, **94**, 1–288.

FAO/WHO (1995) FAO/WHO Collaborating Centre for Research and Training in Food Hygiene and Zoonoses, *WHO surveillance programme for control of food-borne infections and intoxication's in Europe. Sixth Report: 1990–1992*, Federal Institute for Health Protection of Consumers and Veterinary Medicine, Berlin.

FDA (2001) *Food code*, Food and Drug Administration, Washington, DC, available at; http://www.cfsan.fda.gov/~dms/fc01-toc.html, accessed November, 2008.

Fein S B, Lin C T J and Levy A S (1995) Foodborne illness: perceptions, experience, and preventative behaviours in the United States, *Journal of Food Protection*, **58**, 1405–11.

Feng P, Lampel K A, Karch H and Whittam T S (1998) Genotypic and phenotypic changes in the emergence of *Escherichia coli* O157:H7, *Journal of Infectious Diseases*, **177**, 1750–3.

Fisher I S T and Gill O N (2001) International surveillance networks and principles of collaboration, *Eurosurveillance*, **6**, 17–21.

FoodNet (2006) *FoodNet surveillance report for 2004 (final report)*, Centers for Disease Control and Prevention, Atlanta, GA.

Guzewich J J, Bryan F L and Todd E C D (1997) Surveillance of foodborne disease. Part 1. Purposes and types of surveillance systems and networks, *Journal of Food Protection*, **60**, 555–66.

Hacker J, Blum-Oehler G, Mühidorfer I and Tschäpe H (1997) Pathogenicity islands of virulent bacteria: structure, function and impact on microbial evolution, *Molecular Microbiology*, **23**, 1089–97.

Hughes C, Gillespie I A, O'Brien S J and The Breakdowns in Food Safety Group (2007) Foodborne transmission of infectious intestinal disease in England and Wales, 1992–2003, *Food Control*, **18**, 766–72.

Jensen H J (2006) Changes in seafood consumer preference patterns and associated changes in risk exposure, *Marine Pollution Bulletin*, **53**, 591–8.

Jones S L, Parry S M, O'Brien S J and Palmer S R (2008) Are staff management practices and inspection risk ratings associated with foodborne disease outbreaks in the catering industry in England and Wales?, *Journal of Food Protection*, **71**, 550–7.

Käferstein F K, Motarjemi Y and Bettcher D W, (1997) Foodborne disease control: a transnational challenge, *Emerging Infectious Diseases*, **3**, 503–10.

Kangchuan C, Chensui L, Qingxin Q, Ningmei Z, Guokui Z, Gongli C, Yijun X, Yiejie L and Shifu Z (1991) The epidemiology of diarrhoeal diseases in southeastern China, *Journal of Diarrhoeal Disease Research*, **2**, 94–9.

Knabel S J (1995) Foodborne illness: role of home food handling practices, *Food Technology*, **49**, 119–31.

Koopmans M and Duizer E (2004) Foodborne viruses: an emerging problem, *International Journal of Food Microbiology*, **90**, 23–41.

Lammerding A (2007) *Using microbiological risk assessment (MRA) in food safety management*, ILSI Europe Report Series, International Life Sciences Institute, Brussels.

LeClerc J E, Li B G, Payne W L and Cebula T A (1996) High mutation frequencies among *Escherichia coli* and *Salmonella* pathogens, *Science*, **274**, 1208–11.

Levy M, Fletcher M and Moody M (1996) Outbreaks of *Salmonella* serotype *enteridis* infection associated with consumption of raw shell eggs: United States 1994–5, *Morbidity and Mortality Weekly Report*, **45**, 737–42.

McMichael A J, Woodruff R E and Hales S (2006) Climate change and human health: present and future risks, *Lancet*, **367**, 859–69.

Mead P S, Slutsker L, Dietz V, McCaig L F, Bresee J S, Shapiro C, Griffin P M and Tauxe R V (1999) Food-related illness and death in the United States, *Emerging Infectious Disease*, **5**, 607–25.

Mukherjee A, Speh D, Jones A T, Buesing K M and Diez-Gonzalez F (2006) Longitudinal microbiological survey of fresh produce grown by farmers in the upper Midwest, *Journal of Food Protection*, **69**, 1928–36.

Nørrung B and Buncic S (2008) Microbial safety of meat in the European Union, *Meat Science*, **78**, 14–24.

Olsen S J, Bishop R, Brenner F W, Roels T H, Bean N, Tauxe R V and Slutsker L (2001) The changing epidemiology of *Salmonella*: Trends in serotypes isolated from humans in the United States, 1987–1997, *Journal of Infectious Diseases*, **183**, 753–61.

Pan T-M, Chiou C-S, Hsu S-Y, Huang H-C, Wang T-K, Chiu S-I, Yea H-L and Lee C-L (1996) Food-borne disease outbreaks in Taiwan, 1994, *Journal of Formosan Medical Association*, **95**, 417–20.

Reid S D, Herbelin C J, Bumbaugh A C, Selander R K and Whittam T S (2000) Parallel evolution of virulence in pathogenic *Escherichia coli*, *Nature*, **406**, 64–7.

Slutsker L, Altekruse S F and Swerdlow D L (1998) Foodborne diseases: emerging pathogens and trends, *Infectious Disease Clinics of North America*, **12**, 199–216.

SGMSF (1995) *Annual Report 1994*, Steering Group on the Microbiological Safety of Food, Ministry of Agriculture, Fisheries and Food, London.

Steinfeld H, Wassenaar T and Jutzi S (2006) Livestock production systems in developing countries: status, drivers, trends, *Revue Scientifique et Technique de L'Office International des Epizooties*, **25**, 505–16.

St Louis M E, Peck S H S and Bowering D (1988) The emergence of grade A eggs as a major source of *Salmonella enteritidis* infection, *Journal of American Medical Association*, **259**, 2103–7.

Stragier P, Kunkel B, Kroos L and Losick R (1989) Chromosomal rearrangements generating a composite gene for the developmental transcription factor, *Science*, **243**, 507–12.
Tauxe R V (1991) *Salmonella*: A postmodern pathogen, *Journal of Food Protection*, **54**, 563–8.
Tauxe R V (1997) Emerging foodborne diseases: an evolving public health challenge, *Emerging Infectious Diseases*, **3**, 425–34.
Tilman D (1999) Global environmental impacts of agricultural expansion: the need for sustainable and efficient practices, *Proceedings of the National Academy of Sciences of the United States of America*, **96**, 5995–6000.
Todd E C D (2006) Challenges to global surveillance of disease patterns, *Marine Pollution Bulletin*, **53**, 569–78.
Todd E C D, Guzewich J J and Bryan F L (1997) Surveillance of foodborne disease. Part 4. Dissemination and uses of surveillance data, *Journal of Food Protection*, **60**, 715–23.
Todd E C D, Greig J D, Bartleson C A and Michaels B S (2007a) Outbreaks where food workers have been implicated in the spread of foodborne disease. Part 2. Description of outbreaks by size, severity and setting, *Journal of Food Protection*, **70**, 1975–93.
Todd E C D, Greig J D, Bartleson C A and Michaels B S (2007b) Outbreaks where food workers have been implicated in the spread of foodborne disease. Part 3. Factors contributing to outbreaks and description of outbreak categories, *Journal of Food Protection*, **70**, 2199–217.
van der Zijpp A J (1999) Animal food production: the perspective of human consumption, production, trade and disease control, *Livestock Production Science*, **59**, 199–206.
Whittam T S (1996) Genetic variation and evolutionary processes in natural populations of *Escherichia coli*, in F C Neidhardt, R Curtiss III, J L Ingraham, E C C Lin, K B Low, B Magasanik, W S Reznikoff, M Riley, M Schaechter and H E Umbarger (eds), *Escherichia coli and Salmonella: Cellular and Molecular Biology*, 2nd edn, ASM Press, Washington, DC, 2708–20.

2

Detecting pathogens in food

R. Betts, Campden BRI, UK, and C. de W. Blackburn, Unilever, UK

Abstract: The detection and enumeration of microorganisms either in foods or on food contact surfaces form an integral part of any quality control or quality assurance plan. This chapter considers the application of microbiological methods in the identification of hazards, the assessment of risk and hazard control, as well as providing a comprehensive overview of the principles behind both conventional and rapid and automated methods.

Key words: rapid and automated methods, conventional cultural methods, electrical methods, adenosine triphosphate (ATP) bioluminescence, microscopy methods, direct epifluorescent filter technique (DEFT); flow cytometry, immunological methods, nucleic acid hybridisation, nucleic acid amplification techniques.

2.1 Introduction

The detection and enumeration of microorganisms either in foods or on food contact surfaces form an integral part of any quality control or quality assurance plan. Microbiological tests done on foods can be divided into two types: (i) quantitative or enumerative, in which a group of microorganisms in the sample is counted and the result expressed as the number of organisms present per unit weight of sample; or (ii) qualitative or presence/absence, in which the requirement is simply to detect whether a particular organism is present or absent in a known weight of sample.

The basis of methods used for the testing of microorganisms in foods is very well established, and relies on the incorporation of a food sample into a nutrient medium in which microorganisms can replicate thus resulting in a visual indication of growth. Such methods are simple, adaptable, convenient and generally inexpensive. However, they have two drawbacks: firstly, the tests rely on the growth of organisms in media, which can take many days and result in a long test elapse time; and secondly, the methods are manually oriented and are thus labour intensive.

Over recent years, there has been considerable research into rapid and automated microbiological methods. The aim of this work has been to reduce the test elapse time by using methods other than growth to detect and/or count microorganisms and to decrease the level of manual input into tests by automating methods as much as possible. These rapid and automated methods have gained some acceptance and application within the food industry.

There has been an inexorable move from quality control (QC) to a quality assurance (QA) approach in the control of microbiological hazards in food, with the wide adoption of Hazard Analysis Critical Control Point (HACCP), HACCP-based approaches and pre-requisite programmes (PRPs) as preventative management systems. Microbiological testing is now becoming integrated within these preventative management systems, and it can have a number of roles in monitoring, validation and verification. In addition, microbiological testing may be required to demonstrate compliance with microbiological criteria (whether standards, guidelines or specifications) and in the investigation of a suspected breakdown of process control.

This chapter considers the application of microbiological methods in the identification of hazards, the assessment of risk and hazard control, as well as providing a comprehensive overview of the principles behind both conventional and rapid and automated methods.

2.2 Quality control and quality assurance

QC and QA are two different approaches to delivering safety; both systems share tools, but the emphasis is very different (Table 2.1). Both approaches are legitimate but they need totally different organisations, structures, skills, resource and ways of working (Kilsby, 2001).

QC is a reactive approach influenced by the pressures in the external world. In a QC organisation the emphasis is on measurement, which needs to be robust and statistically relevant, and the focus is on legal and commercial issues. In contrast, QA is a preventative approach driven by the company's

Table 2.1 A comparison of quality assurance and quality control

Factor	QA	QC
Approach	Preventative	Reactive
Reliance for delivering safety	Central standards and processes	Measurement
Focus	Consumer	Legal and commercial issues

Source: From Kilsby (2001).

internal standards. The emphasis is on operational procedures, which must be robust and regularly reviewed, and the focus is on the consumer.

There are several problems associated with relying on testing for product safety assurance (van Schothorst and Jongeneel, 1994). In order to apply any statistical interpretation to the results, the contaminant should be distributed homogeneously through the batch. As microbiological hazards are usually heterogeneously distributed this means that there is often a major discrepancy between the microbiological status of the batch and the microbial test results (ICMSF, 1986). Even if the microbial distribution is homogeneous, it still may be prohibitive to test a sufficient number of sample units for all the relevant hazards to obtain meaningful information. Microbiological testing detects only the effects and neither identifies nor controls the causes.

As a consequence, there has been an inexorable move from QC to QA in the management of microbiological hazards in food, with the focus on preventative control measures rather than finished product testing. Although microbiological analysis has subsequently borne the brunt of much denigration, it still has a vital role to play as part of a QA framework, albeit with a shift in application and emphasis.

2.3 Role of microbiology methods

Microbiological methods, and specifically those for pathogens, can play a number of important roles within a food safety framework. This section has a focus from the perspective of those involved in food production and preparation, but the role of microbiological testing from the perspective of local authorities in the UK has been the subject of a recent review (Tebbutt, 2007).

2.3.1 Hazard Analysis Critical Control Point (HACCP)

Owing to the difficulty of assuring microbiological safety through testing alone there is now widespread adoption of the quality assurance approach using the HACCP system. Successful implementation of a fully validated HACCP study means that the supposed reliance on microbiological testing, with all its sampling limitations, is relinquished, and this should enable a significant reduction in the volume of testing. Some in the food industry went so far as to surmise that microbiological testing would become obsolete (Struijk, 1996). In reality, however, microbiological testing has continued, albeit with a shift in application and emphasis and accompanying changes in the role of the microbiologist (Kvenberg and Schwalm, 2000).

It is recognised that microbiological testing can serve several functions within HACCP (de Boer and Beumer, 1999) and that the extent and scope of microbial testing is likely to vary with differences in facilities and equipment, the scales of processes and the types of products involved (Brown *et al.*,

2000). It has also been pointed out that although in spite of meticulous adherence to HACCP-based good practices occasional human, instrumental or operational hiatuses can and will occur (Struijk, 1996). Microbiological methods are still required for trouble shooting and forensic investigation in order to identify the cause of the contamination and rectify it.

HACCP has increasingly been developed, particularly within and for larger food businesses over the past 20 years or so, but European legislation in force from January 2006 (EC, 2004) now requires such approaches for all food production and processing businesses (post primary production).

Hazard analysis

The HACCP process comprises seven principles (see Chapter 14), which are further broken down into series of stages. The first principle is to conduct a hazard analysis, and the use of microbiological tests may be required by the HACCP team to gather relevant data. This may involve determining the incidence of pathogens or indicator organisms in raw materials, the efficacy of equipment cleaning procedures, the presence of pathogens (e.g. *Listeria*) in the environment and microbial loads in foods and on equipment (Stier, 1993).

Published sources of microbiological data, including epidemiological and surveillance data, together with knowledge gained through commercial experience can provide the HACCP team with relevant information on the likely hazards associated with the product and process. However, when existing data are lacking, microbiological testing is often needed (Kvenberg and Schwalm, 2000). This may involve determining the incidence of pathogens or indicator organisms in raw materials, the presence of pathogens (e.g. *Listeria monocytogenes*) in the environment, and microbial loads in foods and on equipment (Stier, 1993). Here the links with HACCP PRPs are important.

The use of molecular characterisation techniques has further increased the microbiologist's armoury and epidemiological tracking of strains can provide a more in-depth knowledge of the food manufacturing process. This may enable the determination of sites of cross-contamination, or sites where strains appear and disappear, thus pinpointing the positions contributing to the final flora of the product, permitting more precise identification of critical control points (CCPs) (Dodd, 1994).

Validation of the technical accuracy of the hazard analysis and effectiveness of the preventative measures is important before the HACCP study is finalised and implemented. Examples where microbiology methods may be used for validation include pre-operation checks of cleaning and sanitising, screening of sensitive raw materials, challenge testing and monitoring of critical sites for microbiological build-up during processing (Hall, 1994).

Critical control points (CCPs)

The HACCP process requires the establishment of systems to monitor all identified CCPs. In most cases, it is not feasible to use microbial testing to monitor CCPs due to the long analytical time, low method sensitivity and the heterogeneous nature of most microbial contamination. However, a notable exception to this is the application of ATP bioluminescence for checking the cleaning of equipment. Results from these methods can be obtained in only a few minutes, which allows sufficient time for equipment to be recleaned before production begins if they are found to be contaminated. Although care in the application of these methods is required to prevent being lulled into a false sense of security (Stier, 1993), the methodology can have a beneficial impact in demonstrating to staff responsible for cleaning the importance of their role.

In the context of HACCP, microbiological criteria play a role in the monitoring of CCPs in food processing and distribution (Hall, 1994), and both conventional and rapid methods have a role to play in the checking of raw materials and monitoring of supplies. The receipt of raw materials within defined microbiological specifications is often identified as a CCP. As a consequence, preventative measures are likely to focus on the supplier's own microbiology assurance procedures and may include a Certificate of Analysis for selected contaminants, with the use of 'in-house' laboratory testing to confirm acceptability and when selecting new suppliers.

Microbial testing can play a major role in the validation of CCPs to demonstrate their effectiveness (van Schothorst, 1998; Blackburn, 2000; Kvenberg and Schwalm, 2000). For safe product design a defined reduction of target microorganisms may be required, delivered either in one CCP or over a series of process steps. Quantitative data may be required to demonstrate that the process can deliver the defined level of microbial kill or that the end-product meets the specification for safety (EC, 2004, 2005). This is particularly true if unconventional or unique control measures and/or critical limits are used. For example, the use of microbiological testing to enumerate *Campylobacter* has demonstrated the reduction (1–2 $\log_{10}$ cfu/ml of carcass rinse) achievable using a variety of antimicrobial processes applied during the processing of broiler chickens (Oyarzabal, 2005). In some cases experimental trials to document the adequacy of a control measure may involve laboratory challenge testing or in-plant challenge testing, using surrogate microorganisms (Scott, 2005).

Microbial methods, particularly molecular characterisation ones, can be useful in answering questions that may arise as part of the HACCP validation exercise. For example, if a hazardous organism appears in a product at a point in the production line beyond the CCP designed to control it, does this mean failure of the CCP, or does it indicate post-process contamination (Dodd, 1994).

HACCP verification and review

Part of the HACCP process involves establishing procedures for verification to confirm that the HACCP system is working effectively. Once a HACCP plan is operational, finished product testing can be one of the means by which its successful implementation is verified. In addition, microbiological data can provide valuable sources of information for trend analysis and statistical process control. In theory, a well-functioning HACCP plan should only require occasional testing as part of the verification process. However, sometimes local legislation, customer requirements or the company's own standards demand a higher level of testing (Stier, 1993).

Microbial testing for verification purposes may involve pathogen testing, although quantitative indicators can provide a much more effective tool for verifying that HACCP is properly implemented (Swanson and Anderson, 2000). The choice of appropriate indicators is product- and process-specific. For example, testing for coliform bacteria provides an effective verification technique for pasteurised milk and water potability. However, in certain applications, finished product testing for even indicator organisms provides no meaningful data, e.g. in the case of canned products.

Once production starts and, depending on the product, the distribution system is filled in readiness for product launch, there is the opportunity to ensure that the HACCP plan is fully implemented. Verification activities are likely to be at their most intense at the early stages and a high level of monitoring (e.g. finished product testing) may be appropriate. After product launch, monitoring of consumer complaints plays a key role to verify that the product design is meeting consumer expectation.

In the USA, to verify that PRP/HACCP systems are effective in controlling pathogen contamination of raw meat and poultry products, product-specific *Salmonella* performance standards must be met by slaughter establishments and establishments producing raw ground products (Rose *et al.*, 2002). Test results showed that *Salmonella* prevalence in most product categories was lower after the implementation of PRP/HACCP than in pre-PRP/HACCP baseline studies.

HACCP is a living system and therefore review of the HACCP plan is an important aspect in ensuring that it remains fully valid and implemented. A formal review should be triggered if there is a change to the product or process but, if this is not the case, then it should be reviewed at regular intervals, e.g. annually. In these reviews, it may be decided that microbiological data are required to assess the significance of a new hazard or to ensure that the CCPs can still control the existing hazards in the light of any proposed changes to the product or process.

2.3.2 Pre-requisite programmes (PRPs)

It is generally agreed that the most successful implementation of HACCP is done within an environment of well-managed PRPs (see Chapter 14). Although definitions vary, the concept of PRPs does not differ significantly from what may be termed good manufacturing practices (GMPs). GMP is concerned with the general (i.e. non-product-specific) policies, practices, procedures, processes and other precautions that are required to consistently yield safe, suitable foods of uniform quality. With regard to catering/food service businesses, good catering practice (GCP) is analogous to GMP (see Chapter 15). Good hygienic practice (GHP) is the part of GMP that is concerned with the precautions needed to ensure appropriate hygiene and as such tends to focus on the pre-requisites required for HACCP.

Microbiological testing has been shown to have a role to play within the PRPs targeted at primary production. For example, *Salmonella* testing has been used to show that much pre-harvest *Salmonella enterica* infection in pigs occurs immediately before slaughter, during lairage in the contaminated abattoir holding pens (Rostagno *et al.*, 2005). Subsequently, a potential intervention strategy to reduce the prevalence of *S. enterica*-positive pigs at slaughter, which consisted of resting pigs prior to slaughter in their transport vehicle instead of in the abattoir holding pen, was evaluated. In addition, programmes for monitoring *Salmonella* in the pork production chain have begun in several European countries, and this has included the use of serological assays for the presence of antibodies against *Salmonella*. The aim is to give farms a particular status, reflecting a certain level of prevalence of *Salmonella*, in order to make decisions on the implementation of certain hygiene measures and slaughtering logistics (Achterberg *et al.*, 2005).

GMP/GHP systems have been found to be effective provided that they are: well-documented with standard operating procedures (SOPs); fully implemented; and include monitoring records and verification procedures (Kvenberg and Schwalm, 2000). From a manufacturing perspective, there are several key sources of microbial contamination of a product that require control: raw materials, equipment, process/production environment and people. The extent to which microbial testing plays a role, and the degree of sampling required, should reflect the category of risk associated with the particular raw material, area or operation. For example, a 'high-risk' raw material that is added to a product post-pasteurisation may require more testing to verify compliance with a specification than one added before pasteurisation. Also, the food contact surfaces and air quality in a 'high-care hygiene' area may also require a higher level of sampling.

The food production/processing environment can be a source of general contamination. Many surfaces not directly in contact with food may harbour microorganisms, e.g. non-food-contact equipment surfaces, walls, floors, drains, overhead structures. These microorganisms can then be transferred to the food in the air via water droplets and dust. Sampling of this environment can provide information on the likely presence and incidence of pathogens, their distribution in relation to processing lines and thus the risk of product contamination (Cordier, 2002). This allows preventative measures to be established in the framework of GHP, such as layout of processing lines and zoning within the factory.

Sampling the cleaning equipment is a very useful index of what is actually present in a production environment, because cleaning 'collects' dirt and bacteria from all parts of the factory, e.g. floor mops, brushes, vacuum cleaners (Fraser, 2002). In a similar way, sampling of drains also gives a better chance of determining whether a particular pathogen is present in the production environment, e.g. *Listeria monocytogenes*. This can often be a better approach than sampling finished products. In addition, other wet areas such as sinks, taps, cleaning cloths and brushes and boot-washing baths should also be checked routinely. Aerosols can be created from such areas and contaminants can find their way into products on the manufacturing line. Testing for indicator organisms generally gives the most useful information on environmental hygiene, an exception to this being the testing for *L. monocytogenes* in high-hygiene environments.

2.3.3 Trouble shooting and 'forensic' investigation

It has been pointed out that, in spite of meticulous adherence to HACCP-based good practices, occasional human, instrumental or operational hiatuses can and will occur (Struijk, 1996). Microbiological testing may be required for trouble shooting and 'forensic' investigation in order to identify the cause of the problem and rectify it. Usually, the first action required is to identify and control the affected product. Microbiological testing may be appropriate to determine, or confirm, whether there is a microbiological problem and, if so, whether it is a safety or spoilage/quality incident. In combination with a review of the process records, particularly at CCPs, and any historical microbiological test data, it may be necessary to instigate a sampling and testing plan to determine how much product is affected. As speed is often critical, rapid test methods can play an important role (Stier, 1993). Once this information has been obtained, decisions can be made regarding segregation, blocking, recall and salvaging of affected batches, and the status of further production.

Microbiological analysis is often required to determine the cause or source of the problem, and the type and extent of testing required will vary enormously, depending on the situation. Rapid techniques like ATP bioluminescence can be useful trouble-shooting tools to quickly identify problem areas. Tests ranging from indicator organisms, through specific pathogen detection methods, to genetic finger-printing of strains may also be appropriate.

Following this immediate action, an assessment of the integrity of the HACCP plan is required. It has to be determined whether the HACCP has failed due to its validity or its implementation. Here again, microbiological analysis may have a role to play in any subsequent review and re-validation.

2.3.4 Microbiological criteria

A microbiological criterion for food defines the acceptability of a product or a food lot, based on the absence or presence, or number of microorganisms including parasites, and/or quantity of their toxins/metabolites, per unit(s) of mass, volume, area or lot (CAC, 1997; ICMSF, 2002). Microbiological criteria are essentially of three types: standards, guidelines and specifications (IFST, 1999). A standard is a microbiological criterion contained in a law or regulation where compliance is mandatory and is usually focused on safety, although process hygiene criteria have been defined (EC, 2005). A specification is a microbiological criterion applied to raw materials, ingredients or the end-product, which is used in a purchase agreement. A guideline is a microbiological criterion applied at any stage of food processing and retailing which indicates the microbiological condition of the sample and aids in identifying situations requiring attention for food safety or quality reasons.

There are many reasons for setting and applying different types of microbiological criteria (e.g. determining acceptability of raw materials/ingredients, intermediates and finished products; supporting validation and verification of HACCP and PRPs) and hence pathogen criteria have an important role to play (CCFRA, 2007).

2.3.5 Risk assessment

In a food safety context, the formalised meaning of risk assessment has evolved primarily from the CAC definitions (FAO/WHO, 1995), where risk assessment is the primary science-based part of risk analysis, dealing specifically with condensing scientific data to an assessment of the human health risk related to the specific foodborne hazard, and risk analysis according to these definitions also comprises risk management and risk communication (Schlundt, 2000). Microbiological risk assessment approaches have been utilised primarily by regulatory bodies and researchers in order to determine the best and/or most effective risk management options, but the food and beverage industry is beginning to apply these risk assessment approaches in order to help better manage microbial pathogens (Membré *et al.*, 2005, 2006; Syposs *et al.*, 2005). Risk assessment approaches have particular value when a risk management decision is required related to a critical and complex food safety issue where there may be a high degree of uncertainty and variability in the relevant information and data. For example, risk assessment outputs have enabled decisions to be made about heat process optimisation (Membré *et al.*, 2006) and shelf-life determination (Membré *et al.*, 2005). Risk assessment approaches are not without disadvantages in that they can be time consuming and require a great deal of detailed knowledge, as well as considerable skill, to implement.

One important area within the food industry where methodology is raising its profile is quantitative risk assessment. Risk assessment is very much tied in with microbiological data, and microbiological examinations of samples of ingredients and end-products may be necessary. Risk assessment methods can identify gaps in our knowledge that are crucial to providing better estimates of risk, and this may in fact lead to an increase in the level of microbiological testing (de Boer and Beumer, 1999). Assessing the risk posed by a 'new' or 'emerging' organism may also highlight deficiencies in current methodology requiring the need for method development.

2.4 Applying microbiological testing

2.4.1 To test or not to test?

There are many reasons for doing microbiological tests, as described in the preceding text. Testing to meet microbiological criteria may be enforced by legislation (e.g. microbiological standards) or customers (e.g. microbiological specifications); or identified as part of HACCP or PRPs (e.g. microbiological guidelines). However, one of the most useful questions to ask before deciding which type of test method is most suitable is 'what is the purpose of the test and how will the result be used?' If there is no clear justification for the test or a use for the result, with identified actions depending on the outcome of the test, then the testing is likely to be of no value. Asking this question will help to target testing more effectively and lead to the implementation of the most appropriate methods and more effective use of test results.

2.4.2 Choice of microbiological test

The numbers of pathogenic microorganisms in most raw materials and food products are usually low and so pathogen tests may provide little information of use for the implementation and maintenance of GMP and HACCP systems. Instead, the enumeration of so-called 'indicator organisms' has an important role. Indicator organisms are single species or groups that are indicative of the possible presence of pathogens. Although there is not necessarily a relationship between indicator and pathogen numbers, it can be generally assumed that the likely numbers of a pathogen will be less than those of the organisms indicative of its presence. It can also be assumed that reducing the numbers of indicator organisms will produce a similar reduction in the numbers of any corresponding pathogen (Brown *et al.*, 2000). For the same reasons indicator organisms can also provide a measure of post-pasteurisation contamination that might lead to pathogen contamination.

Microbiology methods can differ widely in their comparative advantages and disadvantages. The role of a particular test will help to determine the relative importance of possible selection criteria (Table 2.2). For example, tests used for HACCP validation and verification may not need to be particularly rapid, but accuracy may be more critical. Alternatively, speed is more likely to be important for methods used to monitor CCPs. For products with a short shelf-life, rapidity of test result may be an important factor, but when maximising the volume of material sampled is crucial, sample throughput and low cost/test may be higher on the priority list. In recent years, a plethora of rapid test kits has become available that, to a greater or lesser extent,

Table 2.2 Factors that may influence the choice of microbiological method

Factor	Considerations
Performance	Sensitivity, specificity, accuracy, precision, reproducibility, repeatability
Time	Total test time (presumptive/confirmed results), 'hands-on' time, time constraints
Ease of use	Complexity, automation, robustness, training requirement, sample throughput, result interpretation
Standardisation	Validation, accreditation, international acceptance
Cost	Cost/test, capital outlay/equipment running cost, labour costs

have helped to expedite, simplify, miniaturise and automate methodology. The drive for standardisation, validation and international acceptance of methods, with regard to good laboratory practice and accreditation, means that this is often a constraint on method selection.

2.4.3 Interpretation of test results

The importance of reliable laboratory management and test methodology is highlighted by the fact that 'false-positive', 'false-negative' and erroneous results can occur for a number of reasons, including: insufficiently experienced personnel; laboratory contamination or cross-contamination from controls; contamination from packaging; the use of an unvalidated method; insufficient controls; variation in methodology between different laboratories; identification errors and misinterpretation (e.g. product residues/particles appearing similar to microbial colonies).

Inaccurate results can lead to poor management decisions, ranging from unnecessary blocking, disposal or recall of product to failure to identify a problem leading to a food poisoning outbreak. Thus, microbiological testing should be used to complement a preventative approach to food safety, rather than being relied upon as the only approach to managing microbiological hazards.

2.5 Sampling

Although this chapter deals with the methodologies employed to test foods, it is important for the microbiologist to consider sampling. No matter how good a method is, if the sample has not been taken correctly and is not representative of the batch of food that it has been taken from, then the test result is meaningless.

It is useful to devise a sampling plan in which results are interpreted from a number of analyses, rather than a single result. It is now common for microbiologists to use two or three class sampling plans, in which the number of individual samples to be tested from one batch are specified, together with the microbiological limits that apply. These types of sampling plan are fully described elsewhere (ICMSF, 1986, 2002).

Once a sampling plan has been devised then a representative portion must be taken for analysis. In order to do this the microbiologist must understand the food product and its microbiology in some detail. Many chilled products will not be homogeneous mixtures but will be made up of layers or sections: a good example would be a prepared sandwich. It must be decided if the microbiological result is needed for the whole sandwich (i.e. bread and filling), or just the bread, or just the filling; indeed in some cases one part of a mixed filling may need to be tested. When this has been decided then the sample for analyses can be taken, using the appropriate aseptic technique and sterile sampling implements (Kyriakides *et al.*, 1996). The sampling procedure having been developed, the microbiologist will have confidence that samples taken are representative of the foods being tested and test methods can be used with confidence.

2.6 Conventional microbiological techniques

As outlined in the introduction, conventional microbiological techniques are based on the established method of incorporating food samples into nutrient media and incubating for a period of time to allow the microorganisms to grow. The detection or counting method is then a simple visual assessment of growth. These methods are thus technically simple and relatively inexpensive, requiring no complex instrumentation. The methods are, however, very adaptable, allowing the enumeration of different groups of microorganisms.

Before testing, the food sample must be converted into a liquid form in order to allow mixing with the growth medium. This is usually done by accurately weighing the sample into a sterile container and adding a known volume of sterile diluent (the sample to diluent ratio is usually 1:10); this mixture is then homogenised using a homogeniser (e.g. stomacher or pulsifier) that breaks the sample apart, releasing any organisms into the diluent. The correct choice of diluent is important. If the organisms in the sample are stressed by incorrect pH or low osmotic strength, then they could be injured or killed, thus affecting the final result obtained from the microbiological test. The diluent must be well buffered at a pH suitable for the food being tested and be osmotically balanced. When testing some foods (e.g. dried products) which may contain highly stressed microorganisms, then a suitable recovery period may be required before the test commences, in order to ensure cells are not killed during the initial phase of the test procedure (Davis and Jones, 1997).

Injured pathogenic microorganisms in food may still pose a consumer risk and therefore it is important that methods reliant on microbial growth provide sufficient resuscitation. Microbial injury in food and the development of recovery methods has been the subject of a recent review (Wu, 2008).

2.6.1 Conventional quantitative procedures

The enumeration of organisms in samples is generally done by using plate count, or most probable number (MPN) methods. The former are the most widely used, whilst the latter tend to be used only for certain organisms (e.g. *Escherichia coli*) or groups (e.g. coliforms).

Plate count method

The plate count method is based on the deposition of the sample, in or on an agar layer in a Petri dish. Individual organisms or small groups of organisms will occupy a discrete site in the agar, and on incubation will grow to form discrete colonies that are counted visually. Various types of agar media can be used in this form to enumerate different types of microorganisms. The use of a non-selective nutrient medium that is incubated at 30 °C aerobically will result in a total viable count or mesophilic aerobic count. By changing the conditions of incubation to anaerobic, a total anaerobe count will be obtained. Altering the incubation temperature will result in changes in the type of organism capable of growth, thus showing some of the flexibility in the conventional agar approach. If there is a requirement to enumerate a specific type of organism from the sample, then in most cases the composition of the medium will need to be adjusted to allow only that particular organism to grow. There are three approaches used in media design that allow a specific medium to be produced: the elective, selective and differential procedures.

Elective procedures refer to the inclusion in the medium of reagents, or the use of growth conditions, that encourage the development of the target organisms, but do not inhibit the growth of other microorganisms. Such reagents may be sugars, amino acids or other growth factors. Selective procedures refer to the inclusion of reagents or the use of growth conditions that inhibit the development of non-target microorganisms. It should be noted that, in many cases, selective agents will also have a negative effect on the growth of the target microorganism, but this will be less great than the effect on non-target cells. Examples of selective procedures would be the inclusion of antibiotics in a medium or the use of anaerobic growth conditions. Finally, differential procedures allow organisms to be distinguished from each other by the reactions that their colonies cause in the medium. An example would be the inclusion of a pH indicator in a medium to differentiate acid-producing organisms. In most cases, media will utilise a multiple approach system, containing elective, selective and differential components in order to ensure that the user can identify and count the target organism.

The types of agar currently available are far too numerous to list. For details of these, the manuals of media manufacturing companies (e.g. Oxoid, LabM, Difco, Merck) should be consulted.

Most probable number (MPN) method
The second enumerative procedure, the MPN method, allows the estimation of the number of viable organisms in a sample based on probability statistics. The estimate is obtained by preparing decimal (tenfold) dilutions of a sample, and transferring sub-samples of each dilution to (usually) three tubes of a broth medium. These tubes are incubated, and those that show any growth (turbidity) are recorded and compared to a standard table of results (ICMSF, 1986) that indicate the contamination level of the product.

As indicated earlier, this method is used only for particular types of test and tends to be more labour- and materials-intensive than plate count methods. In addition, the confidence limits are large even if many replicates are studied at each dilution level. Thus the method tends to be less accurate than plate counting methods but has the advantage of greater sensitivity.

An automated MPN system (TEMPO®, bioMérieux, France) has been recently developed for a number of groups of microorganisms relevant to foods. A recent study (Paulsen *et al.*, 2008) reporting enumeration of Enterobacteriaceae and comparing TEMPO® with other methods including ISO method 21528-2 (ISO-VRBG) (ISO, 2004) showed that the automated MPN system yielded results comparable to those of the confirmed Enterobacteriaceae ISO-VRBG method but required only 24 h of analysis time.

2.6.2 Conventional qualitative procedures

Qualitative procedures are used when a count of the number of organisms in a sample is not required and only their presence or absence needs to be determined. Generally such methods are used to test for potentially pathogenic microorganisms such as *Salmonella* spp., *Listeria* spp., *Yersinia* spp. and *Campylobacter* spp. The technique requires an accurately weighed sample (usually 25 g) to be homogenised in a primary enrichment broth and incubated for a stated time at a known temperature. In some cases, a sample of the primary enrichment may require transfer to a secondary enrichment broth and further incubation. The final enrichment is usually then streaked out onto a selective agar plate that allows the growth of the organisms under test. The long enrichment procedure is used because the sample may contain very low levels of the test organism in the presence of high numbers of background microorganisms. Also, in processed foods the target organisms themselves may be in an injured state. Thus the enrichment methods allow the resuscitation of injured cells followed by their selective growth in the presence of high numbers of competing organisms.

The organism under test is usually indistinguishable in a broth culture, so the broth must be streaked onto a selective/differential agar plate. The microorganisms can then be identified by their colonial appearance. Colonies formed on the agar that are typical of the microorganism under test are described as presumptive colonies. In order to confirm that the colonies are composed of the test organism, further biochemical and serological tests are usually performed on pure cultures of the organism. This usually requires colonies from primary isolation plates being restreaked to ensure purity. The purified colonies are then tested biochemically by culturing in media that will indicate whether the organism produces particular enzymes or utilises certain sugars.

At present a number of companies market miniaturised biochemical test systems that allow rapid or automated biochemical tests to be quickly and easily set up by microbiologists. Serological tests are done on pure cultures of some isolated organisms, e.g. *Salmonella*, using commercially available antisera.

2.7 Rapid and automated methods

The general interest in alternative microbiological methods has been stimulated in part by the increased output of food production sites. This has resulted in the following:

1. greater numbers of samples being stored prior to positive release – a reduction in analysis time would reduce storage and warehousing costs;
2. a greater sample throughput being required in laboratories – the only way that this can be achieved is by increased laboratory size and staff levels, or by using more rapid and automated methods;
3. a requirement for a longer shelf-life in the chilled foods sector – a reduction in analysis time could expedite product release thus increasing the shelf-life of the product;
4. the increased application of HACCP procedures – rapid methods can be used in HACCP verification procedures.

There are a number of different techniques referred to as rapid methods and most have little in common either with each other or with the conventional procedures that they replace. The methods can generally be divided into quantitative and qualitative tests, the former giving a measurement of the number of organisms in a sample, the latter indicating only presence or absence. Laboratories considering the use of rapid methods for routine testing must carefully consider their own requirements before purchasing such a system. Every new method will be unique, giving a slightly different result, in a different timescale with varying levels of automation and sample throughput. In addition, some methods may work poorly with certain types of food or may not be able to detect the specific organism or group that is required. All of these points must be considered before a method is adopted by a laboratory. It is also of importance to ensure that staff using new methods are aware of the principles of operation of the techniques and thus have the ability to troubleshoot if the method clearly shows erroneous results.

2.7.1 Electrical methods

The enumeration of microorganisms in solution can be achieved by one of two electrical methods, one measuring particle numbers and size, the other monitoring metabolic activity.

Particle counting

The counting and sizing of particles can be done with the 'Coulter' principle, using instruments such as the Coulter Counter® (Beckman Coulter, USA). The method is based on passing a current between two electrodes placed on either side of a small aperture. As particles or cells suspended in an electrolyte are drawn through the aperture they displace their own volume of electrolyte solution, causing a drop in d.c. conductance that is dependent on cell size. These changes in conductance are detected by the instrument and can be presented as a series of voltage pulses, the height of each pulse being proportional to the volume of the particle, and the number of pulses equivalent to the number of particles.

The technique has been used extensively in research laboratories for experiments that require the determination of cell sizes or distribution. It has found use in the area of clinical microbiology where screening for bacteria is required (Alexander *et al.*, 1981). In food microbiology, however, little use has been made of the method. There are reports of the detection of cell numbers in milk (Dijkman *et al.*, 1969) and yeast estimation in beer (MaCrae, 1964), but little other work has been published. Any use of particle counting for food microbiology would probably be restricted to non-viscous liquid samples or particle-free fluids, since very small amounts of sample debris could cause significant interference, and cause aperture blockage.

Metabolic activity

Stewart (1899) first reported the use of electrical measurement to monitor microbial growth. This author used conductivity measurements to monitor the putrefaction of blood and concluded that the electrical changes were caused by ions formed by the bacterial decomposition of blood constituents. After this initial report a number of workers examined the use of electrical measurement to monitor the growth of microorganisms. Most of the work was successful; however, the technique was not widely adopted until reliable instrumentation capable of monitoring the electrical changes in microbial cultures became available.

There are currently four instruments commercially available for the detection of organisms by electrical measurement. The Malthus System (IDG, UK) based on the work of Richards *et al.* (1978) monitors conductance changes occurring in growth media as does the Rabit® System (Don Whitley Scientific, UK), whilst the Bactometer® (bioMeriéux, UK) and the Batrac (SyLab, Austria) (Bankes, 1991) can monitor both conductance and capacitance signals. All of the instruments have similar basic components: (i) an incubator system to hold samples at a constant temperature during the test; (ii) a monitoring unit that measures the conductance and/or capacitance of every cell at regular frequent intervals (usually every 6 min); and (iii) a computer-based data handling system that presents the results in usable format.

The detection of microbial growth using electrical systems is based on the measurement of ionic changes occurring in media, caused by the metabolism of microorganisms. The changes caused by microbial metabolism and the detailed electrochemistry that is involved in these systems have been previously described in some depth (Eden and Eden, 1984; Easter and Gibson, 1989; Bolton and Gibson, 1994). The principle underlying the system is that as bacteria grow and metabolise in a medium, the conductivity of that medium will change. The electrical changes caused by low numbers of bacteria are impossible to detect using currently available instrumentation, approximately 10^6 organisms/ml must be present before a detectable change is registered. This is known as the *threshold of detection*, and the time taken to reach this point is the *detection time*.

In order to use electrical systems to enumerate organisms in foods, the sample must initially be homogenised. The growth well or tube of the instrument containing medium is inoculated with the homogenised sample and connected to the monitoring unit within the incubation chamber or bath. The electrical properties of the growth medium are recorded throughout the incubation period. The sample container is usually in the form of a glass or plastic tube or cell, in which a pair of electrodes is sited. The tube is filled with a suitable microbial growth medium, and a homogenised food sample is added. The electrical changes occurring in the growth medium during microbial metabolism are monitored via the electrodes and recorded by the instrument.

As microorganisms grow and metabolise they create new end-products in the medium. In general, uncharged or weakly charged substrates are transformed into highly charged end-products (Eden and Eden, 1984), and thus the conductance of the medium increases. The growth of some organisms such as yeasts does not result in large increases in conductance. This is possibly due to the fact that these organisms do not produce ionised metabolites and this can result in a decrease in conductivity during growth.

When an impedance instrument is in use, the electrical resistance of the growth medium is recorded automatically at regular intervals (e.g. 6 min) throughout the incubation period. When a change in the electrical parameter being monitored is detected, then the elapsed time since the test was started is calculated by a computer; this is usually displayed as the detection time. The complete curve of electrical parameter changes with time (Fig. 2.1) is similar to a bacterial growth curve, being sigmoidal and having three stages: (i) the inactive stage, where any electrical changes are below the threshold limit of detection of the instrument; (ii) the active stage, where rapid electrical changes occur; and (iii) the stationary or decline stage, that occurs at the end of the active stage and indicates a deceleration in electrical changes.

The electrical response curve should not be interpreted as being similar

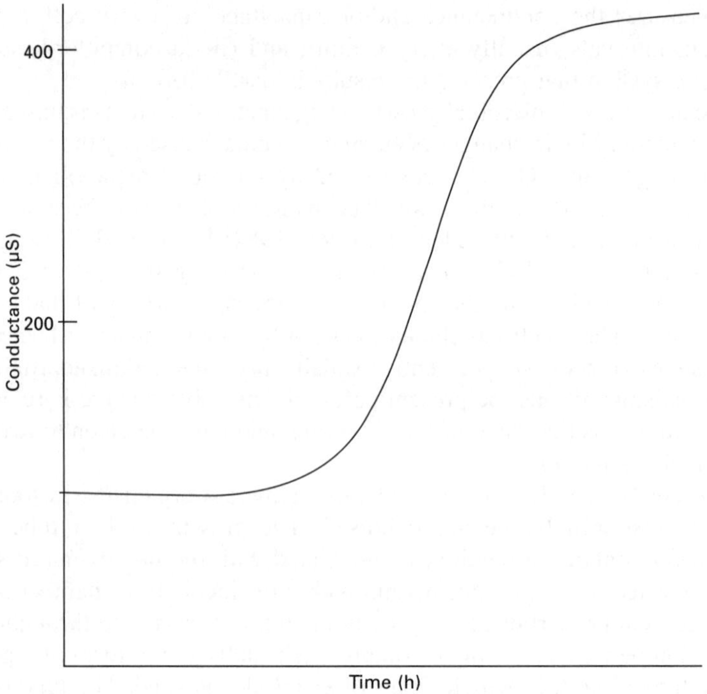

Fig. 2.1 A conductance curve generated by the growth of bacteria in a suitable medium.

to a microbial growth curve. It is accepted (Easter and Gibson, 1989) that the lag and logarithmic phases of microbial growth occur in the inactive and active stages of the electrical response curve, up to and beyond the detection threshold of the instrument. The logarithmic and stationary phases of bacterial growth occur during the active and decline stages of electrical response curves.

In order to use detection time data generated from electrical instruments to assess the microbiological quality of a food sample, calibrations must be done. The calibration consists of testing samples using both a conventional plating test and an electrical test. The results are presented graphically with the conventional result on the *y*-axis and the detection time on the *x*-axis (Fig. 2.2). The result is a negative line with data covering 4 to 5 log cycles of organisms and a correlation coefficient greater than 0.85 (Easter and Gibson, 1989). Calibrations must be done for every sample type to be tested using electrical methods; different samples will contain varying types of microbial flora with differing rates of growth. This can greatly affect electrical detection time and lead to incorrect results unless correct calibrations have been done.

So far, the use of electrical instruments for total microbial assessment has been described. These systems, however, are based on the use of a

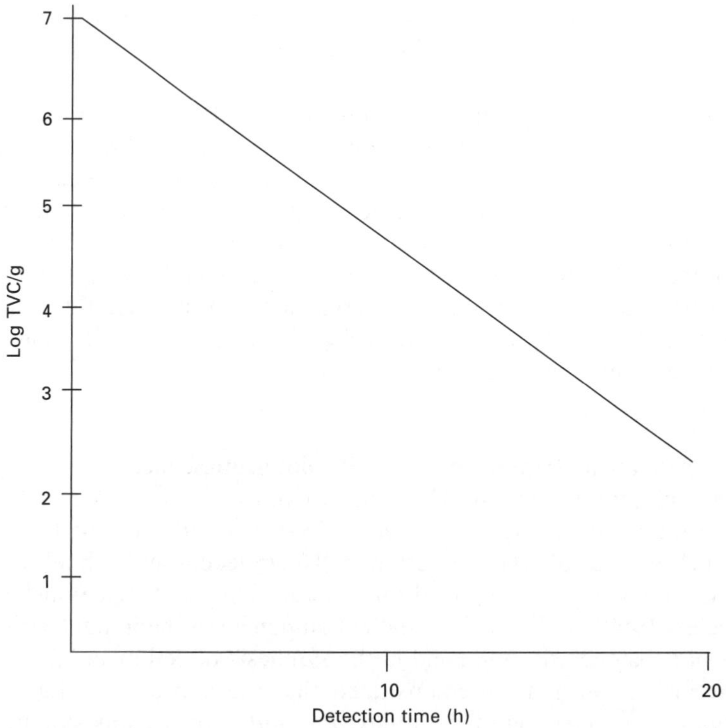

Fig. 2.2 Calibration curve showing changes in conductance detection time with bacterial total viable count (TVC).

growth medium and it is thus possible, using media engineering, to develop methods for the enumeration or detection of specific organisms or groups of organisms. Many examples of the use of electrical measurement for the detection/enumeration of specific organisms have been published; these include: Enterobacteriaceae (Petitt, 1989; Cousins and Marlatt, 1990), *Pseudomonas* (Banks *et al.*, 1989), *Yersinia enterocolitica* (Walker, 1989), yeasts (Connolly *et al.*, 1988), *E. coli* (Druggan *et al.*, 1993) and *Campylobacter* (Bolton and Powell, 1993). In the future, the number of types of organisms capable of being detected will undoubtedly increase. Considerable research is currently being done on media for the detection of *Listeria*, and media for other organisms will follow.

Most of the electrical methods described above involve the use of direct measurement, i.e. the electrical changes are monitored by electrodes immersed in the culture medium. Some authors have indicated the potential for indirect conductance measurement (Owens *et al.*, 1989) for the detection of microorganisms. This method involves the growth medium being in a separate compartment to the electrode within the culture cell. The liquid surrounding the electrode is a gas absorbent, e.g. potassium hydroxide for carbon dioxide. The growth medium is inoculated with the sample and, as the microorganisms grow, gas is released. This is absorbed by the liquid surrounding the electrode, causing a change in conductivity, which can be detected.

This technique may solve the problem caused by microorganisms that produce only small conductance changes in conventional direct conductance cells. These organisms, e.g. many yeast species, are very difficult to detect using conventional direct conductance methods, but detection is made easy by the use of indirect conductance monitoring (Betts, 1993). The increased use of indirect methods in the future could considerably enhance the ability of electrical systems to detect microorganisms that produce little electrical change in direct systems, thus increasing the number of applications of the technique within the food industry.

2.7.2 Adenosine triphosphate (ATP) bioluminescence

The non-biological synthesis of ATP in the extracellular environment has been demonstrated (Ponnamperuma *et al.*, 1963), but it is universally accepted that such sources of ATP are very rare (Huernnekens and Whiteley, 1960). ATP is a high-energy compound found in all living cells (Huernnekens and Whiteley, 1960), and it is an essential component in the initial biochemical steps of substrate utilisation and in the synthesis of cell material.

McElroy (1947) first demonstrated that the emission of light in the bioluminescent reaction of the firefly, *Photinus pyralis*, was stimulated by ATP. The procedure for the determination of ATP concentrations utilising crude firefly extracts was described by McElroy and Streffier (1949) and has since been used in many fields as a sensitive and accurate measure of ATP. The light-yielding reaction is catalysed by the enzyme luciferase, this being the enzyme found in fireflies causing luminescence. Luciferase takes part in the following reaction:

$$\text{Luciferase} + \text{Luciferin} + \text{ATP} \rightarrow \text{Mg}_2 + \text{Luciferase} - \text{Luciferin} - \text{AMP} + \text{PP}$$

The complex is then oxidised:

$$\text{Luciferase} - \text{Luciferin} - \text{AMP} + \text{O}_2 \rightarrow (\text{Luciferase} - \text{Luciferin} - \text{AMP} = \text{O}) + \text{H}_2\text{O}$$

The oxidised complex is in an excited stage, and as it returns to its ground stage a photon of light is released:

$$\text{Luciferase} - \text{Luciferin} - \text{AMP} = \text{O} \rightarrow (\text{Luciferase} - \text{Luciferin} - \text{AMP} = \text{O}) + \text{Light}$$

The light-yielding reaction is efficient, producing a single photon of light for every luciferin molecule oxidised and thus every ATP molecule used (Seliger and McElroy, 1960).

Levin *et al.* (1964) first described the use of the firefly bioluminescence assay of ATP for detecting the presence of viable microorganisms. Since this initial report considerable work has been done on the detection of viable organisms in environmental samples using a bioluminescence technique (Stalker, 1984). As all viable organisms contain ATP, it could be considered simple to use a bioluminescence method to rapidly enumerate microorganisms. Research, however, has shown that the amount of ATP in different microbial cells varies depending on species, nutrient level, stress level and stage of growth (Stalker, 1984; Stannard, 1989). Thus, when using bioluminescence it is important to consider:

(1) the type of microorganism being analysed; generally, vegetative bacteria will contain 1 fg of ATP/cell (Karl, 1980), yeasts will contain ten times this value (Stannard, 1989), whilst spores will contain no ATP (Sharpe *et al.*, 1970);
(2) whether the cells have been subjected to stress, such as nutrient depletion, chilling or pH change – in these cases a short resuscitation may be required prior to testing;
(3) whether the cells are in a relatively ATP-free environment, such as a growth medium, or are contained within a complex matrix, like food, that will have very high background ATP levels.

When testing food samples one of the greatest problems is that noted in (3) above. All foods will contain ATP, and the levels present in the food will generally be much higher than those found in microorganisms within the food. Data from Sharpe *et al.* (1970) indicated that the ratio of food ATP to bacterial ATP ranges from 40 000:1 in ice cream to 15:1 in milk. Thus, to be able to use ATP analysis as a rapid test for foodborne microorganisms, methods for the separation of microbial ATP were developed. The techniques that have been investigated fall into two categories: either to physically separate microorganisms from other sources of ATP, or to use specific extractants to remove and destroy non-microbial ATP. Filtration methods have been successfully used to separate microorganisms from drinks (LaRocco *et al.*, 1985; Littel and LaRocco, 1986) and brewery samples (Hysert *et al.*, 1976). These methods are, however, difficult to apply to particulate-containing solutions as filters rapidly become blocked. A potential way around this problem has been investigated by some workers and utilises a double filtration system/scheme (Littel *et al.*, 1986), the first filter removing food debris but allowing microorganisms through, the second filter trapping microorganisms prior to lysis and bioluminescent analysis. Other workers (Baumgart *et al.*, 1980; Stannard and Wood, 1983) have utilised ion exchange resins to trap selectively either food debris or microorganisms before bioluminescent tests were done.

The use of selective chemical extraction to separate microbial and non-microbial ATP has been extensively tested for both milk (Bossuyt, 1981) and meat (Billte and Reuter, 1985) and found to be successful. In general, this technique involves the lysis of somatic (food) cells followed by destruction of the released ATP with an apyrase (ATPase) enzyme. A more powerful extraction reagent can then be used to lyse microbial cells, which can then be tested with luciferase, thus enabling the detection of microbial ATP only.

There are a number of commercially available instruments aimed specifically at the detection of microbial ATP; Lumac (Netherlands), Foss Electric (Denmark), Bio Orbit (Finland) and Biotrace (UK) all produce systems, including separation methods, specifically designed to detect microorganisms in foods. Generally, all of the systems perform well and have similar specifications, including a minimum detection threshold of 10^4 bacteria (10^3 yeasts) and analysis times of under 1 h.

In addition to testing food samples for total viable microorganisms, there have been a number of reports concerning potential alternative uses of ATP bioluminescence within the food industry. The application of ATP analysis to rapid hygiene testing has been considered (Holah, 1989), both as a method of rapidly assessing microbiological contamination and as a procedure for measuring total surface cleanliness. It is in the latter area that ATP measurement can give a unique result. As described earlier, almost all foods contain very high levels of ATP; thus food debris left on a production line could be detected in minutes using a bioluminescence method, allowing a very rapid check of hygienic status to be done. The use of ATP bioluminescence to monitor surface hygiene has now been widely adopted by industry. The availability of relatively inexpensive, portable, easy to use luminometers has now enabled numerous food producers to implement rapid hygiene testing procedures that are ideal for HACCP monitoring applications where surface hygiene is a critical control point. Reports suggest (Griffiths, 1995) that all companies surveyed that regularly use ATP hygiene monitoring techniques note improvements in cleanliness after initiation of the procedure. Such ATP-based test systems can be applied to most types of food processing plant, food service and retail establishments and even assessing the cleanliness of transportation vehicles such as tankers.

One area that ATP bioluminescence has not yet been able to address has been the detection of specific microorganisms. It may be possible to use selective enrichment media for particular microorganisms in order to allow selective growth prior to ATP analysis. This approach would, however, considerably increase analysis time and some false high counts would be expected. The use of specific lysis agents that release ATP only from the cells being analysed has been investigated (Stannard, 1989) and shown to be successful. The number of these specific reagents is, however, small and thus the method is of only limited use. Perhaps the most promising method developed for the detection of specific organisms is the use of genetically engineered bacteriophages (Ulitzur and Kuhn, 1987; Schutzbank *et al.*, 1989; Ulitzur *et al.*, 1989).

Bacteriophages are viruses that infect bacteria. Screening of bacteriophages has shown that some are very specific, infecting only a particular type of bacteria. Workers have shown it is possible to add into the bacteriophage the genetic information that causes the production of bacterial luciferase. Thus, when a bacteriophage infects its specific host bacterium, the latter produces luciferase and becomes luminescent. This method requires careful selection of the bacteriophage in order to ensure that false-positive or false-negative results do not occur; it does, however, indicate that, in the future, luminescence-based methods could be used for the rapid detection of specific microorganisms (Stewart, 1990).

In conclusion, the use of ATP bioluminescence in the food industry has been developed to a stage at which it can be reliably used as a rapid test for viable microorganisms, as long as an effective separation technique for microbial ATP is used. Its potential use in rapid hygiene testing has been realised and the technique is being used within the industry. Work has also shown that luminescence can allow the rapid detection of specific microorganisms but such a system would need to be commercialised before widespread use within the food industry.

2.7.3 Microscopy methods

Microscopy is a well-established and simple technique for the enumeration of microorganisms. One of the first descriptions of its use was for rapidly counting bacteria in films of milk stained with the dye methylene blue (Breed and Brew, 1916). One of the main advantages of microscope methods is the speed with which individual analyses can be done; however, this must be balanced against the high manual workload and the potential for operator fatigue caused by constant microscopic counting.

The use of fluorescent stains, instead of conventional coloured compounds, allows cells to be more easily counted and thus these stains have been the subject of considerable research. Microbial ecologists first made use of such compounds to visualise and count microorganisms in natural waters (Francisco *et al.*, 1973; Jones and Simon, 1975). Hobbies *et al.* (1977) first described the use of Nuclepore® polycarbonate membrane filters (Whatman plc, UK) to capture microorganisms before fluorescent staining, while enumeration was considered in depth by Pettipher *et al.* (1980), the method developed by the latter author being known as the direct epifluorescent filter technique (DEFT).

The DEFT is a labour-intensive manual procedure and this has led to research into automated fluorescence microscope methods that offer both automated sample preparation and high sample throughput. The first fully automated instrument based on fluorescence microscopy was the BactoScan™ (Foss Electric, Denmark), which was developed to count bacteria in milk and urine. Milk samples placed in the instrument are chemically treated to lyse somatic cells and dissolve casein micelles. Bacteria are then separated by continuous centrifugation in a dextran/sucrose gradient. Microorganisms recovered from the gradient are incubated with a protease to remove residual protein, then stained with acridine orange and applied as a thin film to a disc rotating under a microscope. The fluorescent light from the microscope image is converted into electrical impulses and recorded. The BactoScan™ has been used widely for raw milk testing in continental Europe, and correlations with conventional methods have reportedly been good (Kaereby and Asmussen, 1989). The technique does, however, have a poor sensitivity (approximately 5×10^4 cells/ml) and this negates its use on samples with lower bacterial counts.

An instrument-based fluorescence counting method, in which samples were spread onto a thin plastic tape, was developed for the food industry. The instrument (Autotrak) deposited samples onto the tape, which was then passed through staining and washing solutions, before travelling under a fluorescence microscope. The light pulses from the stained microorganisms were then enumerated by a photomultiplier unit. Tests on food samples using this instrument (Betts and Bankes, 1988) indicated that the debris from food samples interfered with the staining and counting procedure and gave results that were significantly higher than corresponding total viable counts.

Perhaps the most recent development in fluorescence microscope techniques to be used within the food industry for rapid counting is flow cytometry. In this technique the stained sample is passed under a fluorescence microscope system as a liquid in a flow cell. Light pulses caused by the light hitting a stained particle are transported to a photomultiplier unit and counted. This technique is automated, rapid and potentially very versatile. Of the microscope methods discussed here, the DEFT has perhaps the widest usage, while flow cytometry could offer significant advantages in the future. These two procedures will therefore be discussed in more detail.

Direct epifluorescent filter technique (DEFT)

The DEFT was developed for rapidly counting the numbers of bacteria in raw milk samples (Pettipher *et al.*, 1980; Pettipher and Rodrigues, 1982). The method is based on the pre-treatment of a milk sample in the presence of a proteolytic enzyme and surfactant at 50 °C, followed by a membrane filtration step that captures the microorganisms. The pre-treatment is designed to lyse somatic cells and solubilise fats that would otherwise block the membrane filter. After filtration the membrane is stained with the fluorescent nucleic acid binding dye acridine orange, then rinsed and mounted on a microscope slide. The membrane is then viewed with an epifluorescent microscope. This illuminates the membrane with ultraviolet light, causing the stain to emit visible light that can be seen through the microscope. As the stain binds to nucleic acids it is concentrated within microbial cells by binding to DNA and RNA molecules; thus any organisms on the membrane can be easily visualised and counted. The complete pre-treatment and counting procedure can take as little as 30 minutes.

Although the original DEFT was able to give a very rapid count, it was very labour-intensive, as all of the pre-treatment and counting were done manually. This led to a very poor daily sample throughput for the method.

The development of semi-automated counting methods based on image analysis (Pettipher and Rodrigues, 1982) overcame some of the problems of manual counting and thus allowed the technique to be more user-friendly.

The early work on the uses of DEFT for enumerating cells in raw milk was followed with examinations of other types of foods. It was quickly recognised that the good correlations between DEFT count and conventional total viable counts that were obtained with raw milk samples did not occur when heat-treated milks were examined (Pettipher and Rodrigues, 1981). Originally this was considered to be due to heat-induced staining changes occurring in Gram-positive cocci (Pettipher and Rodrigues, 1981); however, more recent work (Back and Kroll, 1991) has shown that similar changes occur in both Gram-positives and Gram-negatives. Similar staining phenomena have also been observed with heat-treated yeasts (Rodrigues and Kroll, 1986) and in irradiated foods (Betts *et al.*, 1988). Thus the use of DEFT as a rapid indication of total viable count is mainly confined to raw foods.

The type of food with which DEFT can be used has been expanded since the early work with raw milk. Reports have covered the use of the method with frozen meats and vegetables (Rodrigues and Kroll, 1989), raw meats (Shaw *et al.*, 1987), alcoholic beverages (Cootes and Johnson, 1980; Shaw, 1989), tomato paste (Pettipher *et al.*, 1985), confectionery (Pettipher, 1987) and dried foods (Oppong and Snudden, 1988) and with hygiene testing (Holah *et al.*, 1988). In addition, some workers (Rodrigues and Kroll, 1988) have suggested that the method could be modified to detect and count specific groups of organisms.

In conclusion, the DEFT is a very rapid method for the enumeration of total viable microorganisms in raw foods and has been used with success within the industry. The problems of the method are a lack of specificity and an inability to give a good estimate of viable microbial numbers in processed foods. The former could be solved by the use of short selective growth stages or fluorescently labelled antibodies; however, these solutions would have time and cost implications. The problem with processed foods can be eliminated only if alternative staining systems that mark viable cells are examined; preliminary work (Betts *et al.*, 1989) has shown this approach to be successful, and the production and commercialisation of fluorescent viability stains could advance the technique. At present the high manual input and low sample throughput of DEFT procedures has limited the use of the procedure in the food industry.

Flow cytometry

Flow cytometry is a technique based on the rapid measurement of cells as they flow in a liquid stream past a sensing point (Carter and Meyer, 1990). The cells under investigation are inoculated into the centre of a stream of fluid (known as the sheath fluid). This constrains them to pass individually past the sensor and enables measurements to be made on each particle in turn, rather than average values for the whole population. The sensing point consists of a beam of light (either ultraviolet or laser) that is aimed at the sample flow and one or more detectors that measure light scatter or fluorescence as the particles pass under the light beam. The increasing use of flow cytometry in research laboratories has largely been due to the development of the reliable instrumentation and the numerous staining systems. The stains that can be used with flow cytometers allow a variety of measurements to be made. Fluorescent probes based on enzyme activity, nucleic acid content, membrane potential and pH have all been examined, while the use of antibody-conjugated fluorescent dyes confers specificity to the system.

Flow cytometers have been used to study a range of eukaryotic and prokaryotic microorganisms. Work with eukaryotes has included the examination of pathogenic amoeba (Muldrow *et al.*, 1982) and yeast cultures (Hutter and Eipel, 1979), while bacterial studies have included the growth of *Escherichia coli* (Steen *et al.*, 1982), enumeration of cells in bacterial cultures (Pinder *et al.*, 1990) and the detection of *Legionella* spp. in cooling tower waters (Tyndall *et al.*, 1985).

Flow cytometric methods for the food industry have been developed and have been reviewed by Veckert *et al.* (1995). Donnelly and Baigent (1986) explored the use of fluorescently labelled antibodies to detect *Listeria monocytogenes* in milk, and obtained encouraging results. The method used by these authors relied on the selective enrichment of the organisms for 24 hours, followed by staining with fluorescein isothiocyanate labelled polyvalent *Listeria* antibodies. The stained cells were then passed through a flow cytometer, and the *L. monocytogenes* detected. The author suggested that the system could be used with other types of food. A similar approach was used by McClelland and Pinder (1994) to detect *Salmonella* Typhimurium in dairy products.

Patchett *et al.* (1991) investigated the use of a Skatron Argus flow cytometer (Skatron AS, Norway) to enumerate bacteria in pure cultures and foods. The results obtained with pure cultures showed that flow cytometer counts correlated well with plate counts down to 10^3 cells/g. With foods, however, conflicting results were obtained. Application of the technique to meat samples gave a good correlation with plate counts and enabled enumeration down to 10^5 cells/g. Results for milk and paté were poorer, the sensitivity of the system for paté being 10^6 cells/ml, whilst cells inoculated into milk were not detected at levels in excess of 10^7 ml. The poor sensitivity of this flow cytometer with foods was thought to be due to interference of the counting system caused by food debris and it was suggested that the application of separation methods to partition microbial cells from food debris would overcome the problem.

Perhaps the most successful application of flow cytometric methods to food products has been the use of a D-Count® (AES Chemunex, France) system to detect contaminating yeast in dairy and fruit products (Bankes *et al.*, 1991). The procedure used with this system calls for an incubation of the product for 16–20 hours followed by centrifugation to separate and concentrate the cells. The stain is then added, and a sample is passed through the flow cytometer for analysis. An evaluation of the system by Pettipher (1991), using soft drinks inoculated with yeasts, showed that it was reliable and user-friendly. The results obtained indicated that cytometer counts correlated well with DEFT counts; however, the author did not report how the system compared to plate counts.

Investigations of the D-Count® system by Bankes *et al.* (1991) utilised a range of dairy and fruit-based products inoculated with yeast. Results indicated that yeast levels as low as 1 cell/25 g could be detected in 24 hours in dairy products. In fruit juices a similar sensitivity was reported: however, a 48-hour period was required to ensure that this was achieved. The system was found to be robust and easy to use. The D-Count® system has now been adapted to detect bacterial cells as well as yeasts, and applications are available for fermenter biomass and enumeration of total flora in vegetables. The D-Count® system has been fully evaluated in a factory environment (Dumain *et al.*, 1990) testing fermented dairy products. These authors report a very good correlation between cytometer count and plate count ($r = 0.98$), results being obtained in 24 hours, thus providing a time saving of three days over classical methods.

Flow cytometry has also been successfully applied in the water industry for the detection of *Cryptosporidium* oocysts in environmental samples. Combined with fluorescence activated cell sorting (FACS) and epifluorecence microscopy, this technology allows detection of low numbers, and more recent developments using two-colour immunofluorescence assays mean that very specific tests are now possible (Quintero-Betancourt *et al.*, 2002) and inclusion of commercially-available antibodies allows separation of *Cryptosporidium* oocysts and *Giardia* cysts in water samples (Hsu *et al.*, 2005).

In conclusion, flow cytometry can provide a rapid and sensitive method for the rapid enumeration of microorganisms. The success of the system depends on the development and use of (a) suitable staining systems, and (b) protocols for the separation of microorganisms from food debris that would otherwise interfere with the detection system. In the future a flow cytometer fitted with a number of light detection systems could allow the analysis of samples for many parameters at once, thus considerably simplifying testing regimes.

Solid phase cytometry

A relatively new cytometric technique has been developed by AES Chemunex (France) based on solid phase cytometry. In this procedure samples are passed through a membrane filter which captures contaminating microorganisms. A stain is then applied to the filter to fluorescently mark metabolically active microbial cells. After staining, the membrane is then transferred to a ChemScan® RDI instrument (AES Chemunex, France), which scans the whole membrane with a laser, counting fluorescing cells. The complete procedure takes around 90 minutes to perform and can detect single cells in the filtered sample. The ChemScan® RDI solid phase cytometry system is an extremely powerful tool for rapidly counting low levels of organisms. It is ideally suited to the analysis of waters or other clear filterable fluids, and specific labelling techniques could be used to detect particular organisms of interest. Foods containing particulate materials could, however, be problematic as organisms would need to be separated from the food material before filtration and analysis.

2.7.4 Immunological methods

Antibodies and antigens

Immunological methods are based on the specific binding reaction that occurs between an antibody and the antigen to which it is directed. *Antibodies* are protein molecules that are produced by animal white blood cells, in response to contact with a substance causing an immune response. The area to which an antibody attaches on a target molecule is known as the *antigen*. Antigens used in immunochemical methods are of two types. The first occurs when the analyte is of low molecular weight and thus does not stimulate an immune response on its own; these substances are described as haptens and must be bound to a larger carrier molecule to elicit an immune response and cause antibody production. The second type of antigen is immunogenic and is able to elicit an immune response on its own.

Two types of antibody can be employed in immunological tests. These are known as *monoclonal* and *polyclonal* antibodies. *Polyclonal* antibodies are produced if large molecules such as proteins or whole bacterial cells are used to stimulate an immune response in an animal. The many antigenic sites result in numerous different antibodies being produced to the molecule or cell. *Monoclonal* antibodies are produced by tissue culture techniques and are derived from a single white blood cell; thus they are directed towards a single antigenic site. The binding of an antigen is highly specific. Immunological methods can therefore be used to detect particular specific microorganisms or proteins (e.g. toxins). In many cases, when using these methods a label is attached to the antibody, so that binding can be visualised more easily when it occurs.

Labels

The labels that can be used with antibodies are of many types and include radiolabels, fluorescent agents, luminescent chemicals and enzymes; in addition agglutination reactions can be used to detect the binding of antibody to antigen. Radioisotopes have been extensively used as labels, mainly because of the great sensitivity that can be achieved with these systems. They do, however, have some disadvantages, the main one being the hazardous nature of the reagents. This would negate their use in anything other than specialist laboratories, and certainly their use within the food industry would be questioned.

Fluorescent labels have been widely used to study microorganisms, the most frequently used reagent being fluorescein. However, others such as rhodamine and umbelliferone have also been utilised. The simplest use of fluorescent antibodies is in microscopic assays. Recent advances in this approach have been the use of flow cytometry for multiparameter flow analysis of stained preparations, and the development of enzyme-linked immunofluorescent assays (ELIFA), some of which have been automated.

Luminescent labels have been investigated as an alternative to the potentially hazardous radiolabels (Kricka and Whitehead, 1984). The labels can be either chemiluminescent or bioluminescent, and have the advantage over radiolabels that they are easy to handle and measure using simple equipment, while maintaining a similar sensitivity (Rose and Stringer, 1989). A number of research papers have reported the successful use of immunoluminometric assays (Lohneis *et al.*, 1987); however, none has yet been commercialised.

Antibodies have been used for the detection of antigens in precipitation and agglutination reactions. These assays tend to be more difficult to quantify than other forms of immunoassay and usually have only a qualitative application. The assays are quick and easy to perform and require little in the way of equipment.

A number of agglutination reactions have been commercialised by manufacturers and have been successfully used within the food industry. These methods have tended to be used for the confirmation of microbial identity, rather than for the detection of the target organisms. They offer a relatively fast test time, are easy to use and usually require no specialist equipment, thus making ideal test systems for use in routine testing laboratories.

Several latex agglutination test kits are available for the confirmation of *Salmonella* from foods. These include the Oxoid *Salmonella* Latex Kit (Oxoid, Basingstoke, UK) designed to be used with the Oxoid Rapid *Salmonella* Test Kit (Holbrook *et al.*, 1989); the Microscreen *Salmonella* Latex Slide Agglutination Test (Mercia Diagnostics Ltd, UK); the Wellcolex® Colour *Salmonella* Test (Remel Inc, USA) (Hadfield *et al.*, 1987a, b); and the Spectate® *Salmonella* test (Rhône Poulenc Diagnostics Ltd, UK) (Clark *et al.*, 1989). The latter two kits use mixtures of coloured latex particles that allow not only detection but also serogrouping of *Salmonella*. Latex agglutination test kits are also available for *Campylobacter* (Microscreen, Mercia Diagnostics Ltd), *Staphylococcus aureus* (Staphaurex®, Remel Inc), *Shigella* (Wellcolex® Colour *Shigella* Test, Remel Inc) and *Escherichia coli* O157:H7 (Oxoid). Agglutination kits have also been developed for the detection of microbial toxins, e.g. Oxoid Staphylococcal Enterotoxin Reverse Passive Latex Agglutination Test (Bankes and Rose, 1989; Rose *et al.*, 1989).

Enzyme immunoassays have been extensively investigated as rapid detection methods for foodborne microorganisms. They have the advantage of specificity conferred by the use of a specific antibody, coupled with coloured or fluorescent end-points that are easy to detect either visually or with a spectrophotometer or fluorimeter. Most commercially-available enzyme immunoassays use an antibody sandwich method in order initially to capture and then to detect specific microbial cells or toxins. The kits are supplied with two types of antibody: capture antibody and conjugated antibody. The capture antibody is attached to a solid support surface such as a microtitre plate well. An enriched food sample can be added to the well and the antigens from any target cells present will bind to the antibodies. The well is washed out, removing food debris and unbound microorganisms. The enzyme-conjugated antibody can then be added to the well. This will bind to the target cell, forming an antibody sandwich. Unbound antibodies can be washed from the well and the enzyme substrate added. The substrate will be converted by any enzyme present from a colourless form, into a coloured product. A typical microplate enzyme immunoassay takes between two and three hours to perform and will indicate the presumptive presence of the target bacterial cells. Thus positive samples should always be confirmed by biochemical or serological methods.

There are a number of commercially-available enzyme immunoassay test kits for the detection of *Listeria*, *Salmonella*, *Escherichia coli* O157, staphylococcal enterotoxins and *Bacillus* diarrhoeal toxin, from food samples. The sensitivity of these systems is approximately 10^6 cells/ml, so that a suitable enrichment procedure must be used before analysis using the assay. Thus, results can be obtained in two to three days, rather than the three to five days required for a conventional test procedure.

Over recent years a number of highly automated immunoassays have been developed; these add to the benefit of the rapid test result, by reducing the level of manual input required to do the test. Automation of enzyme immunoassays has taken a number of forms; a number of manufacturers market instruments which simply automate standard microplate enzyme-linked immunosorbent assays (ELISAs). These instruments hold reagent bottles and use a robotic pipetting arm, which dispenses the different reagents required in the correct sequence. Automated washing and reading completes the assay with little manual input needed. At least two manufacturers have designed immunoassay kits around an automated instrument, to produce very novel systems.

The Vidas® system (bioMérieux, UK) uses a test strip, containing all of the reagents necessary to do an ELISA test: the first well of the strip is inoculated with an enriched food sample, and placed into the Vidas instrument, together with a pipette tip internally coated with capture antibody. The instrument then uses the pipette tip to transfer the test sample into the other cells in the strip containing various reagents needed to carry out the ELISA test. All of the transfers are completely automatic, as is the reading of the final test result. Vidas ELISA tests are available for a range of organisms including *Salmonella*, *Listeria*, *Listeria monocytogenes*, *E. coli* O157, *Campylobacter* and staphylococcal enterotoxin. Evaluations of a number of these methods have been done (Bobbitt and Betts, 1993; Blackburn *et al.*, 1994) and indicated that results were at least equivalent to conventional test methods.

The EiaFoss (Foss Electric, Denmark) is another fully automated ELISA system. In this case the instrument transfers all of the reagents into sample containing tubes, in which all of the reactions occur. The EiaFoss procedure is novel as it uses antibody-coated magnetic beads as a solid phase. During the assay these beads are immobilised using a magnet mounted below the sample tube. EiaFoss test kits are available for *Salmonella*, *Listeria*, *E. coli* O157 and *Campylobacter*, and evaluations have indicated that these methods operate well (Jones and Betts, 1994).

The newest immunoassay procedure that has been developed into a commercial format is arguably the simplest to use. Immunochromatography operates on a dipstick, composed of an absorbent filter material which contains coloured particles coated with antibodies to a specific organism. The particles are on the base of the dipstick and when dipped into a microbiological enrichment broth, they move up the filter as the liquid is moved by capillary action. At a defined point along the filter material lies a line of immobilised specific antibodies. In the presence of the target organism, binding of that organism to the coloured particles will occur. This cell/particle conjugate moves up the filter dipstick by capillary action until it meets the immobilised antibodies where it will stick. The build-up of coloured particles results in a clearly visible coloured line, indicating a positive test result.

A number of commercial kits based around this procedure, including the Oxoid *Listeria* Rapid Test (Jones *et al.*, 1995a) and the Celsis Lumac Pathstik (Celsis International plc, USA) (Jones *et al.*, 1995b), have been developed and appear to give good results. The immunochromatography techniques require an enrichment in the same way as other immunoassays; they do not, however, require any equipment or instrumentation and, once the dipstick is inoculated, need only minutes to indicate a positive or negative result.

Immunoassay conclusions

In conclusion, immunological methods have been extensively researched and developed. There is now a range of systems that allow the rapid detection of the specific organism to which they are directed. Numerous evaluations of commercially available immuno-based methods have indicated that the results generally correlate well with conventional microbiological methods. Enzyme immunoassays in particular appear to offer a simple way of reducing analysis times by one or two days; automation or miniaturisation of these kits has reduced the amount of person time required to do the test and simplified the manual procedures considerably.

The main problem with the immunological systems is their low sensitivity. The minimum number of organisms required in an enzyme immunoassay system to obtain a positive result is approximately 10^5/ml. As the food microbiologist will want to analyse for the presence or absence of a single target organism in 25 g of food, an enrichment phase is always necessary.

The inclusion of enrichment will always add 24–48 hours to the total analysis time.

2.7.5 Nucleic acid hybridisation

Nucleic acids

The specific characteristics of any organism depend on the particular sequence of the nucleic acids contained in its genome. The nucleic acids themselves are made up of a chain of units each consisting of a sugar (deoxyribose or ribose, depending on whether the nucleic acid is DNA or RNA), a phosphorus-containing group and one of four organic purine or pyrimidine bases. DNA is constructed from two of these chains arranged in a double helix and held together by bonds between the organic bases. The bases specifically bind adenine to thymine and guanine to cytosine. It is the sequence of bases that make different organisms unique.

The development of nucleic acid probes

Nucleic acid probes are small segments of single-stranded nucleic acid that can be used to detect specific genetic sequences in test samples. Probes can be developed against DNA or RNA sequences. The attraction of the use of gene probes in the problem of microbial detection is that a probe consisting of only 20 nucleotide sequences is unique and can be used to identify an organism accurately (Gutteridge and Arnott, 1989).

In order to be able to detect the binding of a nucleic acid probe to DNA or RNA from a target organism, it must be attached to a label of some sort that can easily be detected. Early work was done with radioisotope labels such as phosphorus (^{32}P) that could be detected by autoradiography or scintillation counting. Radiolabels, however, have inherent handling, safety and disposal problems that make them unsuitable for use in food laboratories doing routine testing. Thus the acceptance of widespread use of nucleic acid probes required the development of alternative labels.

A considerable amount of work has been done on the labelling of probes with an avidin–biotin link system. This is based on a very high binding specificity between avidin and biotin. The probe sequence of nucleic acid is labelled with biotin and reacted with target DNA. Avidin is then added, linked to a suitable detector, e.g. avidin-alkaline phosphatase, and binding is detected by the formation of a coloured product from a colourless substrate. These alternative labeling systems proved that non-radiolabelled probes could be used for the detection of microorganisms. However, the system was much less sensitive than isotopic procedures, requiring as much as a 100-fold increase in cell numbers for detection to occur, compared with isotope labels.

In order to develop non-isotopic probes with a sensitivity approaching that of isotope labels, it was necessary to consider alternative probe targets within cells. Probes directed toward cell DNA attach to only a few sites on the chromosome of the target cell. By considering areas of cell nucleic acid that are present in relatively high copy number in each cell and directing probes toward these sites, it is possible to increase the sensitivity of non-isotopic probes considerably. Work on increasing probe sensitivity centred on the use of RNA as a target. RNA is a single-stranded nucleic acid that is present in a number of forms in cells. In one form it is found within parts of the cell protein synthesis system called ribosomes. Such RNA is known as ribosomal RNA (rRNA), and is present in very high copy numbers within cells. By directing nucleic acid probes to ribosomal RNA it is possible to increase the sensitivity of the assay system considerably.

Probes for organisms in food

Nucleic acid hybridisation procedures for the detection of pathogenic bacteria in foods have been described for *Salmonella* spp. (Fitts, 1985; Curiale *et al.*, 1986), *Listeria* spp. (Klinger *et al.*, 1988; Klinger and Johnson, 1988), *Yersinia enterocolitica* (Hill *et al.*, 1983b; Jagow and Hill, 1986), *Listeria monocytogenes* (Datta *et al.*, 1988), enterotoxigenic *Escherichia coli* (Hill *et al.*, 1983a, 1986), *Vibrio vulnificus* (Morris *et al.*, 1987), enterotoxigenic *Staphylococcus aureus* (Notermans *et al.*, 1988), *Clostridium perfringens* (Wernars and Notermans, 1990) and *Clostridium botulinum* (Wernars and Notermans, 1990).

The first commercially available nucleic-probe-based assay system for food analysis was introduced by Gene-Trak Systems (USA) in 1985 (Fitts, 1985). This test used *Salmonella*-specific DNA probes directed against chromosomal DNA to detect *Salmonella* in enriched food samples. The format of the test involved hybridisation between target DNA bound to a membrane filter and phosphorus 32-labelled probes. The total analysis time for the test was 40–44 hours of sample enrichment in non-selective and selective media, followed by the hybridisation procedure lasting 4–5 hours. Thus the total analysis time was approximately 48 hours. The *Salmonella* test was evaluated in collaborative studies in the USA and appeared to be at least equivalent to standard culture methods (Flowers *et al.*, 1987). Gene-Trak also produced a hybridisation assay for *Listeria* spp., based on a similar format (Klinger and Johnson, 1988).

The Gene-Trak® probe kits gained acceptance within the USA and a number of laboratories began using them. In Europe, however, there was a reluctance among food laboratories to use radioisotopes within the laboratory. In addition, ^{32}P has a short half-life, which caused difficulties when transporting kits to distant sites. In 1988 Gene-Trak began marketing non-isotopically labelled probes for *Salmonella*, *Listeria* and *Escherichia coli*. The detection system for the probes was colorimetric. In order to overcome the reduction in sensitivity caused by the use of non-isotopic labels, the target nucleic acid within the cell was ribosomal RNA. This nucleic acid is present in an estimated 500–20 000 copies per cell.

The colorimetric hybridisation assay is based on a liquid hybridisation reaction between the target rRNA and two separate DNA oligonucleotide probes (the capture probe and the reporter probe) that are specific for the organism of interest. The capture probe molecules are extended enzymatically with a polymer of approximately 100 deoxyadenosine monophosphate residues. The reporter probe molecules are labelled chemically with the hapten fluorescein.

Following a suitable enrichment of the food under investigation, a test sample is transferred to a tube and the organisms lysed, releasing rRNA targets. The capture and detector probes are then added and hybridisation is allowed to proceed. If target rRNA is present in the sample, hybridisation takes place between the probes and the target 16s rRNA. The solution containing the target probe complex is then brought into contact with a solid support dipstick, containing bound deoxythymidine homopolymer, under conditions that will allow hybridisation between the poly-deoxyadenosine polymer of the capture probe and the poly-deoxythymidine on the dipstick. Unhybridised nucleic acids and cellular debris are then washed away, leaving the captured DNA–RNA complex attached to the surface of the dipstick. The bound fluoresceinated reporter probe is detected by the addition of an antifluorescein antibody conjugated to the enzyme horseradish peroxidase. Subsequent addition of a chromogenic substrate for the enzyme results in colour development that can be measured spectrophotometrically.

Results of the colorimetric assays (Mozola *et al.*, 1991) have indicated a good comparison between the probe methods and conventional cultural procedures for both *Salmonella* and *Listeria*. The sensitivity of the kits appeared to be between 10^5 and 10^6 target organisms/ml, and thus the enrichment procedure is a critical step in the methodology. Since the introduction of the three kits previously mentioned, Gene-Trak have marketed systems for *Staphylococcus aureus*, *Campylobacter* spp. and *Yersinia enterocolitica*, although the latter assay is not currently available.

Commercially available nucleic acid probes for the confirmation of *Campylobacter*, *Staphylococcus aureus* and *Listeria* are available from Gen-Probe (Gen-Probe Inc., USA). These kits are based on a single-stranded DNA probe that is complementary to the ribosomal RNA of the target organism. After the ribosomal RNA is released from the organism, the labelled DNA probe combines with it to form a stable DNA:RNA hybrid. The hybridised probe can be detected by its luminescence.

The assay method used is termed a hybridisation protection assay and is based on the use of a chemiluminescent acridinium ester. This ester reacts with hydrogen peroxide under basic conditions to produce light that can be measured in a luminometer. The acridinium esters are covalently attached to the synthetic DNA probes through an alkylamine arm. The assay format is based on differential chemical hydrolysis of the ester bond. Hydrolysis of the bond renders the acridinium permanently non-chemiluminescent. When the DNA probe which the ester is attached to hybridises to the target RNA, the acridinium is protected from hydrolysis and can thus be rendered luminescent. The test kits for *Campylobacter* and *Listeria* utilise a freeze-dried probe reagent. The *Campylobacter* probe reacts with *C. jejuni*, *C. coli* and *C. pylori*; the *Listeria* probe reacts with *L. monocytogenes*. In both cases a full cultural enrichment protocol is necessary prior to using the probe for confirmation testing.

An evaluation of the *L. monocytogenes* probe kit (Bobbit and Betts, 1991) indicated that it was totally specific for the target organism. The sensitivity required approximately 10^6 *L. monocytogenes* to be present in order for a positive response to be obtained. The kit appeared to offer a fast reliable culture confirmation test and had the potential to be used directly on enrichment broth, thus reducing test times even further.

Probes – the future

The development and use of probes in the food industry have advanced little in recent years. The kits that are currently available show great promise but are not as widely used as immunoassays. Microbiologists must always consider the usefulness of analysing the genetic information within cells, for example genes coding for toxins could be detected, even when not expressed, and screening methods could be devised for pathogenicity plasmids, such as that in *Yersinia enterocolitica*. It may be, however, that the advances in molecular biology mean that the best way to test for such information is by using nucleic acid amplification methods such as the polymerase chain reaction (PCR).

Nucleic acid amplification techniques

In recent years, several genetic amplification techniques have been developed and refined. The methods usually rely on the biochemical amplification of cellular nucleic acid and can result in a 10^7-fold amplification in two to three hours. The very rapid increase in target that can be gained with nucleic acid amplification methods makes them ideal candidates for development of very rapid microbial detection systems. A number of amplification methods have been developed and applied to the detection of microorganisms:

- polymerase chain reaction (PCR) and variations, including nested PCR, reverse transcriptase (RT) PCR and multiplex PCR;
- Q Beta Replicase;
- ligase amplification reaction (LAR);
- transcript amplification system (TAS), also known as Self-Sustained Sequence Replication (3SR) or nucleic acid sequence-based amplification (NASBA).

Of these amplification methods only PCR has been commercialised as a kit-based procedure for the detection of foodborne microorganisms. Much research has been done with NASBA and there are a number of research papers outlining its use for detecting food pathogens but, as yet, no commercially available kits are on the market.

Polymerase chain reaction (PCR)
PCR is a method used for the repeated *in vitro* enzymic synthesis of specific DNA sequences. The method uses two short oligonucleotide primers that hybridise to opposite strands of a DNA molecule and flank the region of interest in the target DNA. PCR proceeds via series of repeated cycles, involving DNA denaturation, primer annealing and primer extension by the action of DNA polymerase. The three stages of each cycle are controlled by changing the temperature of the reaction, as each stage will occur only at particular defined temperatures. These temperature changes are accomplished by using a specialised instrument known as a thermocycler. The products of primer extension from one cycle act as templates for the next cycle, thus the number of target DNA copies doubles at every cycle.

Reverse transcriptase polymerase chain reaction
This involves the use of an RNA target for the PCR reaction. The PCR must work on a DNA molecule; thus initially reverse transcriptase is used to produce copy DNA (cDNA). The latter is then used in a conventional PCR reaction. The RT-PCR reaction is particularly applicable to certain microbiological tests. Some foodborne viruses contain RNA as their genetic material; thus RT-PCR must be used if amplification and thus detection of these viruses is necessary. A second use of RT-PCR is in the detection of viable microorganisms. One of the problems associated with PCR is its great sensitivity and ability to amplify very low concentrations of a target nucleic acid. Thus, if using PCR to detect the presence or absence of a certain microorganism in a food, PCR could 'detect' the organism, even if it had been previously rendered inactive by a suitable food process. This could result in a false-positive detection. A way to overcome this problem is to use an RT-PCR targeted against cellular messenger RNA, which is only produced by active cells and once produced has a short half-life. Thus a detection of specific mRNA by an RT-PCR procedure is indicative of the presence of a viable microorganism.

Nucleic acid sequence-based amplification (NASBA)
NASBA is a multi-enzyme, multicycle amplification procedure, requiring more enzymes and reagents than standard PCR. It does, however, have the advantage of being isothermal; therefore all stages of the reaction occur at a single temperature and a thermocycler is not required. Various research papers have been published which use NASBA to detect foodborne pathogens (e.g. Uyttendaele *et al.*, 1996); however, the procedure has yet to be commercialised.

Commercial polymerase chain reaction-based kits
Currently there are three manufacturers producing kits based on PCR for the detection of foodborne microorganisms. BAX® (DuPont Qualicon, USA) utilises tableted reagents and a conventional thermocycler, gel electrophoresis-

based approach. Positive samples are visualised as bands on an electrophoresis gel. BAX® kits are available for *Salmonella* (Bennett *et al.*, 1998), *Listeria* genus, *Listeria monocytogenes*, *E. coli* O157:H7; the tests for *Salmonella* and *E. coli* O157 have been through an Association of Official Analytical Chemists Research Institute (AOACRI) testing procedure and have gained AOACRI Performance Tested Status.

The second of the commercially available PCR kits is the Probelia™ kit (Sanofi, France); this uses conventional PCR followed by an immunoassay and colorimetric detection system. Kits are available for *Salmonella* and *Listeria*.

The final commercial PCR system is the Perkin Elmer TaqMan® system (Roche Molecular Systems Inc, USA). This uses a novel probe system incorporating a TaqMan Label. This is non-fluorescent in its native form, but once the probe is bound between the primers of the PCR reaction, it can be acted upon by the DNA polymerase enzyme used in PCR to yield a fluorescent end-product. This fluorescence is detected by a specific fluorescence detection system. TaqMan kits are available for *Salmonella* and under development for *Listeria* and *E. coli* O157. Perhaps one of the most interesting future aspects of TaqMan is its potential to quantify an analyte. Currently PCR-based systems are all based on presence/absence determinations. TaqMan procedures and instrumentation can give information on actual numbers. Therefore the potential for using PCR for rapidly counting microorganisms could now be achieved.

Separation and concentration of microorganisms from foods

In recent years there has been considerable interest in the potential for separating microorganisms from food materials and subsequently concentrating them to yield a higher number per unit volume. The reason for this interest is that many of the currently-available rapid test methods have a defined sensitivity, examples are: 10^4/ml for ATP luminescence, 10^6/ml for electrical measurement, 10^5–10^6/ml for immunoassay and DNA probes and approximately 10^3/ml for current PCR-based kits. These sensitivity levels mean that a growth period is usually required before the rapid method can be applied, and this growth period may significantly increase the total test time.

One way in which this problem can be addressed is by separating and concentrating microorganisms from the foods, in order to present them to the analytical procedure in a higher concentration. An additional advantage is that the microbial cells may be removed from the food matrix, which in some cases may contain materials which interfere with the test itself. A simple example of the use of concentration is in the analysis of clear fluids (water, clear soft drinks, wines, beers, etc.). Here contamination levels are usually very low, thus large volumes are membrane filtered to concentrate the microorganisms onto a small area. These captured organisms can then be analysed. A thorough review of separation concentration methods has been given by Betts (1994). They broadly fall into five categories:

1. filtration
2. centrifugation
3. phase separation
4. electrophoresis
5. immuno-methods.

Of these categories only one has reached commercialisation for use in solid foods, that is the immuno-methods. Immunomagnetic separation relies on coating small magnetic particles with specific antibodies for a known cell. The coated particles can be added into a food suspension or enrichment and, if present, target cells will attach to the antibodies on the particles.

Application of a magnetic field retains the particles and attached cells allowing food debris and excess liquid to be poured away, thus separating the cells from the food matrix and concentrating them. This type of system has been commercialised by Dynal (Norway), LabM (England) and Denka (Japan), an automated system incorporating the procedure is produced by Foss Electric (EiaFoss™). The various companies produce kits for *Salmonella*, *Listeria*, *E. coli* O157, other verocytotoxin producing *E. coli* and *Campylobacter*. Immunomagnetic separation systems for detecting the presence of *E. coli* O157 have been very widely used and become accepted standard reference methods in many parts of the world.

Identification and characterisation of microorganisms

Once an organism has been isolated from a food product it is often necessary to identify it; this is particularly relevant if the organism is considered to be a pathogen. Traditionally, identification methods have involved biochemical or immunological analyses of purified organisms. With the major advances now taken in molecular biology, it is possible to identify organisms by reference to their DNA structure. The sensitivity of DNA-based methods will in fact allow identification to a level below that of species (generally referred to as characterisation or sub-typing). Sub-typing is a powerful new tool that can be used by food microbiologists not just to name an organism, but also to find out its origin. Therefore it is possible in some cases to isolate an organism in a finished product and then, through a structured series of tests, find whether its origin was a particular raw material, the environment within a production area or a poorly cleaned piece of equipment.

A number of DNA-based analysis techniques have been developed that allow sub-typing, and many of these have been reviewed by Betts *et al.* (1995). There is, however, only one technique that has been fully automated and made available to food microbiologists on a large scale, and that is ribotyping through use of the Qualicon RiboPrinter® (Dupont Qualicon, USA). This fully automated instrument accepts isolated purified colonies of bacteria, and produces DNA band images (RiboPrint patterns) that are automatically compared to a database to allow identification and characterisation. The technique has been used successfully within the food industry to identify contaminants, indicate the sources and routes of contamination and check for culture authenticity (Betts, 1998).

2.8 Future trends

The food industry has the responsibility to produce safe and wholesome food and providing this assurance is the ultimate microbiological goal. A microbiology test that could analyse a batch of food non-destructively, on-line and with the required accuracy, sensitivity and specificity is the 'Holy Grail' and would provide this assurance. However, our current technical capabilities fall well short of this ideal situation.

Conventional microbiological methods have remained little changed for many decades. Microbiologists generally continue to use lengthy enrichment and agar-growth-based methods to enumerate, detect and identify organisms in samples. As the technology of food production and distribution has developed, there has been an increasing requirement to obtain microbiological results in shorter time periods.

The rapid growth of the chilled foods market, producing relatively short shelf-life products, has led this move into rapid and automated methods, as the use of such systems allows: (i) testing of raw materials before use; (ii) monitoring of the hygiene of the production line in real time; and (iii) testing of final products over a reduced time period. All of these points will lead to better quality food products with an increased shelf-life.

All of the methods considered in this chapter are currently in use in Quality Control laboratories within the food industry. Some (e.g. electrical methods) have been developed, established and used for a considerable time period, while others (e.g. PCR), are a much more recent development. The future of all of these methods is good; they are now being accepted as standard and routine, rather than novel. Some users are beginning to see the benefits of linking different rapid methods together to gain an even greater test rapidity, e.g. using an enzyme immunoassay to detect the presence of *Listeria* spp., then using a species-specific nucleic acid probe to confirm the presence or absence of *L. monocytogenes*.

One of the problems of many of the rapid methods is a lack of sensitivity. This does in many cases mean that lengthy enrichments are required prior to using rapid methods. Research on methods for the separation and concentration of microorganisms from food samples would enable microorganisms to be removed from the background of food debris and concentrated, thus removing the need for long incubation procedures. The developments in DNA-based methods for both detection and identification/characterisation have given new tools to the food microbiologist; there is no doubt that these developments will continue in the future giving significant analytical possibilities that are currently difficult to imagine.

The new 'omics' technologies are likely to provide opportunities to achieve goals that were previously beyond reach. In particular, DNA microarray technology has the potential to generate data that could be used to improve the safety of our food supply (Al-Khaldi *et al.*, 2002). By simultaneously examining the presence and expression of all genes of a specific microorganism at a given point in time, it is possible to study microbial ecology related to food safety, e.g. the relationships between virulence genes and fitness genes (Al-Khaldi *et al.*, 2002). By understanding growth potential and survival characteristics, it is conceivable that the technology could be applied in process setting based on the microbial characteristics of the raw material contaminants, and safe shelf-life setting based on knowledge of the degree of injury of any survivors (Brul *et al.*, 2002). The technology is also being used to understand how pathogenic microorganisms overcome lag phase, with a view to being able to extend safe shelf-life. By identifying new cellular pathways and hence new antimicrobial targets, existing preservation systems may be optimised or new preservatives may be identified. By hybridising genomic DNA to microarrays, a system known as genomotyping (Lucchini *et al.*, 2001), large sets of bacterial strains can be characterised and compared, thus making the technique applicable to epidemiology and in tracing the sources of bacterial contamination (van der Vossen *et al.*, 2005).

In summary, methods have an important place in our armoury against the threats posed by microorganisms in food. Owing to the diversity of applications and user requirements and the shift from QC to QA, methodology still plays a key role in assuring food safety and new methods still have the potential to bring benefits.

2.9 References and further reading

Achterberg R, Maneschijn-Bonsing J, Bloemraad R, Swanenburg M and Maassen K (2005) Detecting *Salmonella* antibodies in pork, *New Food*, **4** 56–8.

Al-Khaldi S F, Martin S A, Rasooly A and Evans J D (2002) DNA microarray technology used for studying foodborne pathogens and microbial habitats: minireview, *Journal of AOAC International*, **85** 906–10.

Alexander M K, Khan M S and Dow C S (1981) Rapid screening for bacteriuria using a particle counter, pulse-height analyser and computer, *Journal of Clinical Pathology*, **34** 194–8.

Back J P and Kroll R G (1991) The differential fluorescence of bacteria stained with acridine orange and the effect of heat, *Journal of Applied Bacteriology*, **71** 51–8.

Bankes P (1991) An evaluation of the Bactrac 4100, *Campden Food & Drink Research Association Technical Memorandum 628*, Campden Food & Drink Research Association, Chipping Campden.

Bankes P and Rose S A (1989) Rapid detection of staphylococcal enterotoxins in foods with a modification of the reversed passive latex agglutination assay, *Journal of Applied Bacteriology*, **67** 395–9.

Bankes P, Rowe D and Betts R P (1991) *The Rapid Detection of Yeast Spoilage Using the Chemflow System*, Technical Memorandum 621, Campden Food & Drink Research Association, Chipping Campden.

Banks J G, Rossiter L M and Clark A E (1989) Selective detection of *Pseudomonas* in foods by a conductance technique, in A Balows, R C Tilton and A Turano (eds) *Rapid Methods and Automation in Microbiology and Immunology*, Florence 1987, Brixia Academic Press, Brescia, 725–7.

Baumgart J, Frickle K and Kuy C (1980) Quick determination of surface bacterial content of fresh meat using a bioluminescence method to determine adenosine triphosphate, *Fleischwirtschaft*, **60** 266–70.

Bennett A R, Greenwood D, Tennant C, Banks J G and Betts R P (1998) Rapid and definitive detection of *Salmonella* in foods by PCR, *Letters in Applied Microbiology*, **26** 437–41.

Betts R P (1993) Rapid electrical methods for the detection and enumeration of food spoilage yeasts, *International Biodeterioration and Biodegradation*, **32** 19–32.

Betts R P (1994) The separation and rapid detection of microorganisms, in R C Spencer, E P Wright and S W B Newsom (eds) *Rapid Methods and Automation in Microbiology and Immunology*, Intercept Press, Andover, 107–20.

Betts R P (1998) Foodborne bacteria in the spotlight, *Laboratory News*, September p. A6–7.

Betts R P and Bankes P (1988) *An Evaluation of the Autotrak – A rapid method for the enumeration of microorganisms*, Technical Memorandum 491, Campden & Chorleywood Food Research Association, Chipping Campden.

Betts R P, Bankes P, Farr L and Stringer M F (1988) The detection of irradiated foods using the Direct Epifluorescent Filter Technique, *Journal of Applied Bacteriology*, **64** 329–35.

Betts R P, Bankes P and Banks J G (1989) Rapid enumeration of viable microorganisms by staining and direct microscopy, *Letters in Applied Microbiology*, **9** 199–202.

Betts R P, Bankes P and Green J (1991) An evaluation of the Tecra enzyme immunoassay for the detection of *Listeria* in foods, in M R A Morgan, C J Smith and P A Williams (eds) *Food Safety and Quality Assurance: Applications of Immunoassays Systems*, Elsevier, London, 283–98.

Betts R P, Stringer M, Banks J G and Dennis C (1995) Molecular methods in food microbiology, *Food Australia*, **47** 319–22.

Billte M and Reuter G (1985) The bioluminescence technique as a rapid method for the determination of the microflora of meat, *International Journal of Food Microbiology*, **2** 371–81.

Blackburn C de W (2000) Modelling shelf-life, in *The Stability and Shelf-Life of Food*, D Kilcast and P Subramaniam, (eds) Woodhead, Cambridge, 55–78.

Blackburn C de W and Stannard C J (1989) Immunological detection methods for *Salmonella* in foods, in C L Stannard, S B Petitt and F A Skinner (eds) *Rapid Microbiological Methods for Foods, Beverages and Pharmaceuticals*, Society for Applied Bacteriology Technical Series 25, Blackwell Scientific, Oxford, 249–63.

Blackburn C de W, Curtis L M, Humpheson L and Petitt S (1994) Evaluation of the Vitek Immunodiagnostic Assay System (VIDAS) for the detection of *Salmonella* in foods, *Letters in Applied Microbiology*, **19** 32–36.

Bobbitt J A and Betts R P (1991) *Evaluation of Accuprobe Culture Confirmation Test for Listeria monocytogenes*, Technical Memorandum 630, Campden Food & Drink Research Association, Chipping Campden.

Bobbitt J A and Betts R P (1993) *Evaluation of the VIDAS Listeria System for the Detection of Listeria in Foods*, Technical Memorandum 674, Campden & Chorleywood Food Research Association, Chipping Campden.

Bolton F J and Gibson D M (1994) Automated electrical techniques in microbiological analysis, in P Patel (ed.) *Rapid Analysis Techniques in Food Microbiology*, Blackie Academic, London, 131–69.

Bolton F J and Powell S J (1993) Rapid methods for the detection of *Salmonella* and *Campylobacter* in meat products, *European Food and Drink Review*, Autumn 73–81.

Bossuyt R (1981) Bacteriological quality determination of raw milk by an ATP assay technique, *Milchwissenschaft*, **36** 257–60.
Breed R S and Brew J D (1916) Counting bacteria by means of the microscope, *Technical Bulletin 49*, Agricultural Experimental Station, Albany, New York.
Brown M H, Gill C O, Hollingsworth J, Nickelson I I R, Seward S, Sheridan J J, Stevenson T, Sumner J L, Theno D M, Usborne W R and Zink D (2000) The role of microbiological testing in systems for assuring the safety of beef, *International Journal of Food Microbiology*, **62** 7–16.
Brul S, Klis F M, Oomes S J C M, Montijn R C, Schuren F H J, Coote P and Hellingwerf K J (2002) Detailed process design based on genomics of survivors of food preservation processes, *Trends in Food Science and Technology*, **13** 325–33.
CAC (1997) *Principles for the Establishment and Application of Microbiological Criteria for Foods*, Codex Alimentarius Commission, Brussels, available at: http://www.fao.org/docrep/W6419E/w6419e04.htm.
Candlish A (1992) Rapid testing methods for *Salmonella* using immunodiagnostic kits, presented at *Complying with Food Legislation through Immunoassay*, February, Glasgow.
Carter N P and Meyer E W (1990) Introduction to the principles of flow cytometry, in M G Ormerod (ed.) *Flow Cytometry, A Practical Approach*, IRL Press, Oxford, 1–28.
CCFRA (2007) *Establishment and use of microbiological criteria (standards, specifications and guidelines) for foods*, Guideline No. 52, Campden & Chorleywood Food Research Association, Chipping Campden.
Clark C, Candlish A A G and Steell W (1989) Detection of *Salmonella* in foods using a novel coloured latex test. *Food and Agricultural Immunology*, **1** 3–9.
Clayden J A, Alcock S J and Stringer M F (1987) Enzyme linked immunoabsorbant assays for the detection of *Salmonella* in foods, in J M Grange, A Fox and N L Morgan (eds) *Immunological Techniques in Microbiology*, Society for Applied Bacteriology Technical Series 24, Blackwell Scientific, Oxford, 217–30.
Connolly P, Lewis S J and Corry J E L (1988) A medium for the detection of yeasts using a conductimetric method, *International Journal of Food Microbiology*, **7** 31–40.
Coombes P (1990) Detecting *Salmonella* by conductance. *Food Technology International Europe*, No. 4229, 241–6.
Cootes R L and Johnson R (1980) A fluorescent staining technique for determination of viable and non-viable yeast and bacteria in wines, *Food Technology in Australia*, **32** 522–4.
Cordier J-L (2002) Sampling and testing for pathogens essential in safe food manufacture, *New Food*, **2** 37–40.
Cousins D L and Marlatt F (1990) An evaluation of a conductance method for the enumeration of Enterobacteriaceae in milk, *Journal of Food Protection*, **53** 568–70.
Curiale M S, Flowers R S, Mozola M A and Smith A E (1986) A commercial DNA probe based diagnostic for the detection of *Salmonella* in food samples, in L S Lerman (ed.) *DNA Probes: Applications in Genetic and Infectious Disease and Cancer*, Cold Spring Harbour Laboratory, New York, 143–8.
Curiale M S, Klatt M L, Robinson B J and Beck L T (1990a) Comparison of colorimetric monoclonal enzyme immunoassay screening methods for detection of *Salmonella* in foods, *Journal of the Association of Official Analytical Chemists*, **73** 43–50.
Curiale M S, McIver D, Weathersby S and Planer C (1990b) Detection of *Salmonella* and other Enterobacteriaceae by commercial deoxyribonucleic acid hybridization and enzyme immunoassay kits, *Journal of Food Protection*, **53** 1037–46.
Datta A R, Wentz B A, Shook D and Trucksess M W (1988) Synthetic oligonucleotide probes for detection of *Listeria monocytogenes*, *Applied and Environmental Microbiology*, **54** 2933–7.
Davis S C and Jones K L (1997) *A Study of the Recovery of Microorganisms from Flour*, R&D Report No. 51, Campden & Chorleywood Food Research Association, Chipping Campden.

De Boer E and Beumer R R (1999) Methodology for detection and typing of foodborne microorganisms, *International Journal of Food Microbiology*, **50** 119–30.

Dijkman A J, Schipper C J, Booy C J and Posthumus G (1969) The estimation of the number of cells in farm milk, *Netherlands Milk and Dairy Journal*, **23** 168–81.

Dodd C E R (1994) The application of molecular typing techniques to HACCP, *Trends in Food Science and Technology*, **5** 160–4.

Donnelly C W and Baigent G T (1986) Method for flow cytometric detection of *Listeria monocytogenes* in milk, *Applied and Environmental Microbiology*, **52** 689–95.

Druggan P, Forsythe S J and Silley P (1993) Indirect impedance for microbial screening in the food and beverage industries, in R G Kroll and A Gilmour (eds) *New Techniques in Food and Beverage Microbiology*, SAB Technical Series 31, Blackwell, London, 115–30.

Dumain P P, Desnouveauz R, Bloc'H L, Leconte C, Fuhrmann B, De Colombel E, Plessis M C and Valery S (1990) Use of flow cytometry for yeast and mould detection in process control of fermented milk products: The Chemflow System – A Factory Study, *Biotech Forum Europe*, **7** 224–9.

Easter M C and Gibson D M (1985) Rapid and automated detection of *Salmonella* by electrical measurement, *Journal of Hygiene, Cambridge*, **94** 245–62.

Easter M C and Gibson D M (1989) Detection of microorganism by electrical measurements, in M R Adams and C F A Hope (eds) *Rapid Methods in Food Microbiology. Progress in Industrial Microbiology Vol. 26*, Elsevier, Amsterdam, 57–100.

EC (2004) EC Regulation (EC) No 852/2004 of the European Parliament and of the Council of 29 April 2004 on the hygiene of foodstuffs, *Official Journal of the European Union*, **L139**, 30 April, 1–54.

EC (2005) EC Commission Regulation (EC) No 2073/2005 of 15 November 2005 on microbiological criteria for foodstuffs, *Official Journal of the European Union*, **L338**, 22 December, 1–26.

Eden R and Eden G (1984) *Impedance Microbiology*, Research Studies Press. Letchworth.

FAO/WHO, (1995) Application of risk analysis to food standards issues, in *Report of the Joint FAO/WHO Expert Consultation*, Geneva, Switzerland, World Health Organisation, 13–17.

Fitts R (1985) Development of a DNA hybridisation test for the presence of *Salmonella* in foods, *Food Technology*, **39**, 95–102.

Flowers R S, Klatt M J and Keelan S L (1988) Visual immunoassay for the detection of *Salmonella* in foods. A collaborative study, *Journal of the Association of Official Analytical Chemists*, **71** 973–80.

Flowers R S, Mozola M A, Curiale M S, Gabis D A and Silliker J H (1987) Comparative study of DNA hybridization method and the conventional cultural procedure for detection of *Salmonella* in foods, *Journal of Food Science*, **52** 842–5.

Francisco D D, Mah R A and Rabin A C (1973) Acridine orange epifluorescent technique for counting bacteria in natural waters, *Transactions of the American Microscopy Society*, **92** 416–21.

Fraser E (2002) Environmental monitoring in the food industry, *New Food*, **5** 9–14.

Griffiths M W (1995) Bioluminescence and the food industry, *Journal of Rapid Methods and Automation in Microbiology*, **4** 65–75.

Gutteridge C S and Arnott M L (1989) Rapid methods: an over the horizon view, in M R Adams and C F A Hope (eds) *Rapid Methods in Microbiology. Progress in Industrial Microbiology Vol. 26*, Elsevier, Amsterdam, 297–319.

Hadfield S G, Lane A and McIllmurray M B (1987a) A novel coloured latex test for the detection and identification of more than one antigen, *Journal of Immunological Methods*, **97** 153–8.

Hadfield S G, Jovy N F and McIllmurray M B (1987b) The application of a novel coloured latex test to the detection of *Salmonella*, in J M Grange, A Fox and N L Morgan (eds) *Immunological Techniques in Microbiology*, Society for Applied Bacteriology Technical Series 24, Blackwell Scientific, Oxford, 145–52.

Hall P A (1994) Scope of rapid microbiological methods in modern food production, in: P D Patel (ed.) *Rapid Analysis Techniques in Food Microbiology*, Blackie Academic, Glasgow, 255–67.

Hill W E, Madden J M, McCardell B A, Shah D B, Jagow J A, Payne W L and Boutin B K (1983a) Foodborne enterotoxigenic *Esch. coli*: detection and enumeration by DNA colony hybridisation, *Applied and Environmental Microbiology*, **45** 1324–1330.

Hill W E, Payne W L and Aulisio C C G (1983b) Detection and enumeration of virulent *Yersinia enterocolitica* in food by colony hybridisation, *Applied and Environmental Microbiology*, **46** 636–41.

Hill W E, Wentz B A, Jagow J A, Payne W L and Zon G (1986) DNA colony hybridization method using synthetic oligonucleotides to detect enterotoxigenic *Esch. coli:* a collaborative study, *Journal of the Association of Official Analytical Chemists*, **69** 531–6.

Hobbies J F, Daley R J and Jasper S (1977) Use of nucleoport filters for counting bacteria by fluorescence microscopy, *Applied and Environmental Microbiology*, **33** 1225–8.

Holah J T (1989) Monitoring the hygienic status of surfaces, in *Hygiene – The issues for the 90s*, Symposium Proceedings, Campden & Chorleywood Food Research Association, Chipping Campden.

Holah J T, Betts R P and Thorpe R H (1988) The use of direct epifluorescent microscopy (DEM) and the direct epifluorescent filter technique (DEFT) to assess microbial populations on food contact surfaces, *Journal of Applied Bacteriology*, **65** 215–21.

Holbrook R, Anderson J M, Baird–Parker A C and Stuchbury S H (1989) Comparative evaluation of the Oxoid *Salmonella* Rapid Test with three other rapid *Salmonella* methods, *Letters in Applied Microbiology*, **9** 161–4.

Hsu B M, Wu N M, Jang H D, Shih F C, Wan M T and Kung C M (2005) Using the flow cytometry to quantify the *Giardia* cysts and *Cryptosporidium* oocysts in water samples, *Environmental Monitoring and Assessment*, **104** 155–62.

Huernnekens F M and Whiteley H R (1960) in M Florkin and H. S. Mason (eds) *Comparative Biochemistry*, Vol. 1, Academic Press, New York, 129.

Hughes D, Sutherland P S, Kelly G and Davey G R (1987) Comparison of the Tecra *Salmonella* tests against cultural methods for the detection of *Salmonella* in foods, *Food Technology in Australia*, **39** 446–54.

Hutter K J and Eipel H E (1979) Rapid determination of the purity of yeast cultures by immunofluorescence and flow cytometry, *Journal of the Institute of Brewing*, **85** 21–2.

Hysert D W, Lovecses F and Morrison N M (1976) A firefly bioluminescence ATP assay method for rapid detection and enumeration of brewery micro-organisms, *Journal of the American Society of Brewing Chemistry*, **34** 145–50.

ICMSF (1986) *Microorganisms in Foods 2, Sampling for microbiological analysis: Principles and specific applications*, International Commission on Microbiological Specifications for Foods, Blackwell Scientific, Oxford.

ICMSF (2002) *Microorganisms in Foods 7, Microbiological Testing in Food Safety Management*. International Commission on Microbiological Specifications for Foods, Kluwer Academic/Plenum, London, UK.

IFST (1999; 3rd ed in preparation), *Development and Use of Microbiological Criteria for Foods*, 2nd ed, Institute of Food Science and Technology, London.

ISO (2004), *ISO 21528:-2:2004 Microbiology of food and animal feeding stuffs – Horizontal methods for the detection and enumeration of Enterobacteriaceae – Part 2: Colony-count method*, International Organization for Standardization, Geneva.

Jagow J and Hill W E (1986) Enumeration by DNA dolony hybridisation of virulent *Y. enterocolitica* colonies in artificially contaminated food, *Applied and Environmental Microbiology*, **51** 411–43.

Jones J G and Simon B M (1975) An investigation of errors in direct counts of aquatic bacteria by epifluorescence microscopy with reference to a new method of dyeing membrane filters, *Journal of Applied Bacteriology*, **39** 1–13.

Jones K L and Betts R P (1994) *The EIAFOSS System for Rapid Screening of* Salmonella *from Foods*, Technical Memorandum 709, Campden & Chorleywood Food Research Association, Chipping Campden.

Jones K L, MacPhee S, Turner A and Betts R P (1995a) *An Evaluation of the Oxoid Listeria Rapid Test Incorporating Clearview for the Detection of* Listeria *from Foods*, R&D Report No. 19, Campden & Chorleywood Food Research Association, Chipping Campden.

Jones K L, MacPhee S, Turner A and Betts R P (1995b) *An Evaluation of Pathstick for the Detection of* Salmonella *from Foods*, R&D Report No. 11, Campden & Chorleywood Food Research Association, Chipping Campden.

Kaereby F and Asmussen B (1989) Bactoscan – Five years of experiences, in A Balows, R C Tilton and A Turano (eds) *Rapid Methods and Automation in Microbiology and Immunology*, Florence 1987, Brixia Academic Press, Brescia, 739–45.

Karl D M (1980) Cellular nucleotide measurements and applications in microbial ecology, *Microbiological Review*, **44** 739–96.

Kilsby D (2001) Personal communication.

Klinger J D and Johnson A R (1988) A rapid nucleic acid hybridisation assay for *Listeria* in foods, *Food Technology*, **42** 66–70.

Klinger J D, Johnson A, Croan D, Flynn P, Whippie K, Kimball M, Lawne J and Curiale M (1988) Comparative studies of a nucleic acid hybridisation assay for *Listeria* in foods, *Journal of the Association of Official Analytical Chemists*, **73** 669–73.

Kricka L J and Whitehead T P (1984) Luminescent immunoassays: new labels for an established technique, *Diagnostic Medicine*, May 1–8.

Kvenberg J E and Schwalm D J (2000) Use of microbial data for hazard analysis and critical control point verification – Food and Drug Administration perspective, *Journal of Food Protection*, **63** 810–14.

Kyriakides A, Bell C and Jones K L (eds) (1996) *A Code of Practice for Microbiology Laboratories Handling Food Samples*, Guideline No. 9, Campden & Chorleywood Food Research Association, Chipping Campden.

LaRocco K A, Galligan P, Little K J and Spurgash A (1985) A rapid bioluminescent ATP method for determining yeast concentration in a carbonated beverage, *Food Technology*, **39** 49–52.

Levin G V, Glendenning J R, Chapelle E W, Heim A H and Roceck E (1964) A rapid method for the detection of microorganisms by ATP assay: its possible application in cancer and virus studies, *Bioscience*, **14** 37–8.

Littel K J and LaRocco K A (1986) ATP screening method for presumptive detection of microbiologically contaminated carbonated beverages, *Journal of Food Science*, **51** 474–6.

Littel K J, Pikelis S and Spurgash A (1986) Bioluminescent ATP assay for rapid estimation of microbial numbers in fresh meat, *Journal of Food Protection*, **49** 18–22.

Lohneis M, Jaschke K H and Terplan G (1987) An immunoluminometric assay (ILMA) for the detection of staphylococcal enterotoxins, *International Journal of Food Microbiology*, **5** 117–27.

Lucchini S, Thompson A and Hinton J C D (2001) Microarrays for microbiologists, *Microbiology*, **147** 1403–14.

McClelland R G and Pinder A C (1994) Detection of *Salmonella typhimurium* in dairy products with flow cytometry and monoclonal antibodies *Applied and Environmental Microbiology*, **60** (12) 4255–62.

MaCrae R M (1964) Rapid yeast estimations, in *Proceedings of European Brewing Convention Brussels 1963*, Elsevier, Amsterdam, 510–12.

McElroy W D (1947) The energy source for bioluminescence in an isolated system, *Proceedings of the National Academy of Science, USA*, **33** 342–5.

McElroy W D and Streffier B L (1949) Factors influencing the response of the bioluminescent reaction to adenosine triphosphate, *Archives of Biochemistry*, **22** 420–33.

Mattingly J A, Butman B T, Plank M, Durham R J and Robinson B J (1988) Rapid monoclonal antibody based enzyme-linked immunosorbant assay for the detection of *Listeria* in food products, *Journal of the Association of Official Analytical Chemists*, **71** 679–81.

Membré J-M, Johnston M D, Bassett J, Naaktgeboren G, Blackburn C de W and Gorris L G M (2005) Application of modelling techniques in the food industry: determination of shelf-life for chilled foods, in M L A T M Hertog and B M Nicolaï (eds), *III International Symposium on Applications of Modelling as an Innovative Technology in the Agri-Food Chain; MODEL-IT, ISHS Acta Horticulture 674*, Leuven, Belgium, 407–14.

Membré J-M, Amézquita A, Bassett J, Giavedoni P, Blackburn C de W and Gorris L G M (2006) A probabilistic modeling approach in thermal inactivation: estimation of post-process *Bacillus cereus* spore prevalence and concentration, *Journal of Food Protection*, **69** 118–29.

Morris J G, Wright A C, Roberts D M, Wood P K, Simpson L M and Oliver J D (1987) Identification of environmental *Vibrio vulnificus* isolates with a DNA probe for the cytotoxin-hemolysin gene, *Applied and Environmental Microbiology*, **53** 193–5.

Mozola M, Halbert D, Chan S, Hsu H Y, Johnson A, King W, Wilson S, Betts R P, Bankes P and Banks J G (1991) Detection of foodborne bacterial pathogens using a colorimetric DNA hybridisation method, in J M Grange, A Fox and N L Morgan (eds) *Genetic Manipulation Techniques and Applications*, Society for Applied Bacteriology Technical Series No. 28, Blackwell Scientific, Oxford, 203–16.

Muldrow L L, Tyndall R L and Fliermans C B (1982) Application of flow cytometry to studies of free-living amoebae, *Applied and Environmental Microbiology*, **44** 1258–69.

Neaves P, Waddell M J and Prentice G A (1988) A medium for the detection of Lancefield group D cocci in skimmed milk powder by measurement of conductivity changes, *Journal of Applied Bacteriology*, **65** 437–48.

Notermans S, Heuvelman K J and Wernars K (1988) Synthetic enterotoxin B, DNA probes for the detection of enterotoxigenic *Staphylococcus aureus*, *Applied and Environmental Microbiology*, **54** 531–3.

Oppong D and Snudden B H (1988) Comparison of acridine orange staining using fluorescence microscopy with traditional methods for microbiological examination of selected dry food products, *Journal of Food Protection*, **51** 485–8.

Owens J D, Thomas D S, Thompson P S and Timmerman J W (1989) Indirect conductimetry: a novel approach to the conductimetric enumeration of microbial populations, *Letters in Applied Microbiology*, **9** 245–50.

Oyarzabal O (2005) Reduction of *Campylobacter* spp. by commercial antimicrobials applied during the processing of broiler chickens: a review from the United States perspective, *Journal of Food Protection*, **68** 1752–60.

Patchett R A, Back J P, Pinder A C and Kroll R G (1991) Enumeration of bacteria in pure cultures and foods using a commercial flow cytometer, *Food Microbiology*, **8** 119–25.

Paulsen P, Borgetti C, Schopf E and Smulders F J M (2008) Enumeration of Enterobacteriaceae in various foods with a new automated most-probable-number method compared with petrifilm and international organization for standardization procedures, *Journal of Food Protection*, **71** 376–9.

Petitt S B (1989) A conductance screen for Enterobacteriaceae in foods, in C J Stannard, S B Petitt and F A Skinner (eds) *Rapid Microbiological Methods for Foods Beverages and Pharmaceuticals*, Blackwell Scientific, Oxford, 131–42.

Pettipher G L (1987) Detection of low numbers of osmophilic yeasts in crème fondant within 24 h using a pre-incubated DEFT count, *Letters in Applied Microbiology*, **4** 95–8.

Pettipher G L (1991) Preliminary evaluation of flow cytometry for the detection of yeasts in soft drinks, *Letters in Applied Microbiology*, **12** 109–12.

Pettipher G L and Rodrigues U M (1981) Rapid enumeration of bacteria in heat treatment milk and milk products using a membrane filtration-epifluorescence microscopy technique, *Journal of Applied Bacteriology*, **50** 157–66.

Pettipher G L and Rodrigues U M (1982) Semi-automated counting of bacteria and somatic cells in milk using epifluorescence microscopy and television image analysis, *Journal of Applied Bacteriology*, **53** 323–9.

Pettipher G L, Mansell R, McKinnon C H and Cousins C M (1980) Rapid membrane filtration-epifluorescence microscopy technique for the direct enumeration of bacteria in raw milk, *Applied and Environmental Microbiology*, **39** 423–9.

Pettipher G L, Williams R A and Gutteridge C S (1985) An evaluation of possible alternative methods to the Howard Mould Count, *Letters in Applied Microbiology*, **1** 49–51.

Phillips A P and Martin K L (1988) Limitations of flow cytometry for the specific detection of bacteria in mixed populations, *Journal of Immunological Methods*, **106** 109–17.

Pinder A C, Purdy P W, Poulter S A G and Clark D C (1990) Validation of flow cytometry for rapid enumeration of bacterial concentrations in pure cultures, *Journal of Applied Bacteriology*, **69** 92–100.

Ponnamperuma C, Sagan C and Mariner R (1963) Synthesis of adenosine triphosphate under possible primitive earth conditions, *Nature*, **199** 222–6.

Quintero-Betancourt W, Peele E R and Rose J B (2002) *Cryptosporidium parvum* and *Cyclospora cayetanensis*: a review of laboratory methods for detection of these waterborne parasites, *Journal of Microbiological Methods*, **49** 209–24.

Richards J C S, Jason A C, Hobbs G, Gibson D M and Christie R H (1978) Electronic measurement of bacterial growth, *Journal of Physics, E: Scientific Instruments*, **11** 560–8.

Rodrigues U M and Kroll R G (1986) Use of the direct epifluorescent filter technique for the enumeration of yeasts, *Journal of Applied Bacteriology*, **61** 139–44.

Rodrigues U M and Kroll R G (1988) Rapid selective enumeration of bacteria in foods using a microcolony epifluorescence microscopy technique. *Journal of Applied Bacteriology*, **64** 65–78.

Rodrigues U M and Kroll R G (1989) Microcolony epifluorescence microscopy for selective enumeration of injured bacteria in frozen and heat treated foods, *Applied and Environmental Microbiology*, **55** 778–87.

Rose S A and Stringer M F (1989) Immunological methods, in M R Adams and C F A Hope (eds) *Rapid Methods in Food Microbiology. Progress in Industrial Microbiology Vol. 26*, Elsevier, Amsterdam, 121–67.

Rose S A, Bankes P and Stringer M F (1989) Detection of staphylococcal enterotoxins in dairy products by the reversed passive latex agglutination (Setrpla) kit, *International Journal of Food Microbiology*, **8** 65–72.

Rose B E, Hill W E, Umholtz R, Ransom G M and James W O (2002) Testing of *Salmonella* in raw meat and poultry products collected at federally inspected establishments in the United States, 1998 through 2000, *Journal of Food Protection*, **65** 937–47.

Rostagno M H, Hurd H S and McKean J D (2005) Resting pigs on transport trailers as an intervention strategy to reduce *Salmonella enterica* prevalence at slaughter, *Journal of Food Protection*, **68** 1720–23.

Schlundt J (2000) Comparison of microbiological risk assessment studies published, *International Journal of Food Microbiology*, **58** 197–202.

Schutzbank T E, Tevere V and Cupo A (1989) The use of bioluminescent bacteriophages for the detection and identification of *Salmonella* in foods, in A Balows, R C Tilcon and A Turano (eds) *Rapid Methods and Automation in Microbiology and Immunology*, Florence 1987, Brixia Academic Press, Brescia, 241–51.

Scott V N (2005) How does industry validate elements of HACCP plans?, *Food Control*, **16** 497–503.

Seliger H and McElroy M D (1960) Spectral emission and quantum yield of firefly bioluminescence, *Archives of Biochemistry and Biophysics*, **88** 136–41.

Sharpe A N, Woodrow M N and Jackson A K (1970) Adenosine triphosphate levels in foods contaminated by bacteria, *Journal of Applied Bacteriology*, **33** 758–67.

Shaw S, (1989) Rapid methods and quality assurance in the brewing industry, in M R Adams and C F A Hope (eds) *Rapid Methods in Food Microbiology. Progress in Industrial Microbiology Vol. 26*, Elsevier, Amsterdam, 273–96.

Shaw B G, Harding C D, Hudson W H and Farr L (1987) Rapid estimation of microbial numbers in meat and poultry by the direct epifluorescent filter technique, *Journal of Food Protection*, **50** 652–7.

Skarstad K, Steen H B and Boye E (1983) Cell cycle parameters of growing *Escherichia coli*. B/r studied by flow cytometry, *Journal of Bacteriology*, **154** 652–6.

Skjerve E and Olsvik O (1991) Immunomagnetic separation of *Salmonella* from foods, *International Journal of Food Microbiology*, **14** 11–18.

Stalker R M (1984) *Bioluminescent ATP Analysis for the Rapid Enumeration of Bacteria – Principles, Practice, Potential Areas of Inaccuracy and Prospects for the Food Industry*, Technical Bulletin 57, Campden & Chorleywood Food Research Association, Chipping Campden.

Stannard C J (1989) ATP estimation, in M R Adams and C F A Hope (eds) *Rapid Methods in Food Microbiology, Progress in Industrial Microbiology Vol. 26*, Elsevier, Amsterdam, 1–18.

Stannard C J and Wood J M (1983) The rapid estimation of microbial contamination of raw meat by measurement of adenosine triphosphate (ATP), *Journal of Applied Bacteriology*, **55** 429–38.

Steen H B and Boye E (1980) *Escherichia coli* growth studied by dual parameter flow cytophotometry, *Journal of Bacteriology*, **145** 145–50.

Steen H B, Boye E, Skarstad K, Bloom B, Godal T and Mustafa S (1982) Applications of flow cytometry on bacteria: cell cycle kinetics, drug effects and quantification of antibody binding, *Cytometry*, **2** 249–56.

Stewart G N (1899) The change produced by the growth of bacteria in the molecular concentration and electrical conductivity of culture media, *Journal of Experimental Medicine*, **4** 23S–24S.

Stewart G S A B (1990) A review: *In vivo* bioluminescence: new potentials for microbiology, *Letters in Applied Microbiology*, **10** 1–9.

Stier R F (1993) Development and confirmation of CCPs using rapid microbiological tests, *Journal of Rapid Methods and Automation in Microbiology*, **2** 17–26.

Struijk C B (1996) The Hamlet option in food microbiology: to analyze or not to analyze food specimens as marketed once HACCP implemented, *Acta Alimentaria*, **25** 57–72.

Swanson K M J and Anderson J E (2000) Industry perspectives on the use of microbial data for hazard analysis and critical control point validation and verification, *Journal of Food Protection*, **63** 815–8.

Syposs Z, Reichart O and Mészáros L (2005) Microbiological risk assessment in the beverage industry, *Food Control*, **16** 515–21.

Tebbutt G M (2007) Does microbiological testing of foods and the food environment have a role in the control of foodborne disease in England and Wales?, *Journal of Applied Microbiology*, **102** 883–91.

Tyndall R L, Hand R E Jr, Mann R C, Evan C and Jernigan R (1985) Application of flow cytometry to detection and characterisation of *Legionella* spp., *Applied and Environmental Microbiology*, **49** 852–7.

Ulitzur S and Kuhn J (1987) Introduction of lux genes into bacteria, a new approach for specific determination of bacteria and their antibiotic susceptibility, in J Scholmerich, A Andreesen, A Kapp, M Earnst and E G Woods (eds), *Proceedings of Xth International Symposium on Bioluminescence and Chemoluminescence*, Freiburg, Germany 1986, Wiley, Bristol, 463–72.

Ulitzur S, Suissa M and Kuhn J C (1989) A new approach for the specific determination of bacteria and their antimicrobial susceptibilities, in A Balows, R C Tilton and A Turano (eds) *Rapid Methods and Automation in Microbiology and Immunology*, Florence 1987, Brixia Academic Press, Brescia, 235–40.

Uyttendaele M, Schukkink A, van Gemen B and Debevere J (1996) Comparison of the nucleic acid amplifications system NASBA and agar isolation for the detection of pathogenic campylobacters in naturally contaminated poultry, *Journal of Food Protection*, **59** 683–9.

Vanderlinde P B and Grau F H (1991) Detection of *Listeria* spp. in meat and environmental samples by an enzyme-linked immunosorbant assay (ELISA), *Journal of Food Protection*, **54** 230–1.

van der Vossen J, Schuren F and Montijn R (2005) Genomics in microbial food quality and safety, *New Food*, **8** 74–7.

van Schothorst M (1998) Principles for the establishment of microbiology food safety objectives and related control measures, *Food Control*, **9** 379–84.

van Shothorst M and Jongeneel S (1994) Line monitoring, HACCP and food safety, *Food Control*, **5** 107–10.

Veckert J, Breeuwer P, Abee T, Stephens P and Nebe Von Caron G (1995) Flow cytometry applications in physiological study and detection of foodborne organisms, *International Journal of Food Microbiology*, **28** 317–26.

Vermunt A E M, Franken A A J M and Beumer R R (1992) Isolation of salmonellas by immunomagnetic separation, *Journal of Applied Bacteriology*, **72** 112–8.

Walker S J (1989) Development of an impedimetric medium for the detection of *Yersinia enterocolitica* in pasteurised milk, in IDF, *Modern Microbiological Methods for Dairy Products*, proceedings of International seminar, Santander, International Dairy Federation, Brussels.

Walker S J, Archer P and Appleyard J (1990) Comparison of *Listeria*-Tek ELISA kit with cultural procedures for the detection of *Listeria* species in foods, *Food Microbiology*, **7** 335–42.

Wernars K and Notermans S (1990) Gene probes for detection of foodborne pathogens, in A J L Macario and E C de Macario (eds) *Gene Probes for Bacteria*, Academic Press, London, 355–88.

Wolcott M J (1991) DNA-based rapid methods for the detection of foodborne pathogens, *Journal of Food Protection*, **54** 387–401.

Wu V C H (2008) A review of microbial injury and recovery methods in food, *Food Microbiology*, **25** 735–44.

3

Modeling the growth, survival and death of microbial pathogens in foods

J. D. Legan, Kraft Foods, USA, C. M. Stewart, Silliker Inc, USA, and M. B. Cole, National Center for Food Safety and Technology, USA

Abstract: This chapter aims to provide a practical guide to microbiological predictive modeling, particularly of pathogen growth, survival and death. It will cover the principles involved in creating useful predictive microbiological models including experimental design, execution and data analysis. Types of models include kinetic growth, survival and death models and end-point 'time-to-growth' or 'time-to-inactivation' models, also known as 'boundary models'. Each of these types is covered. Applications of models and anticipated developments are also discussed.

Key words: predictive microbiology, kinetic growth models, growth boundary models survival models, inactivation models.

3.1 Introduction

Simple mathematical models have been used in food microbiology since the early 1920s (Bigelow *et al.*, 1920; Esty and Meyer., 1922) and were designed to assure safety of canned foods by calculating the process time needed to produce 'commercial sterility'. The Bigelow *et al.* (1920) studies reported data on the heat penetration into cans filled with various ingredients and included equations that can be used to calculate the time required for the center of the can to reach the desired temperatures. Based on these thermal processing data, they then estimated lethality rates for bacterial sporeformers and reported equivalent processes at different temperatures from these estimated lethality rates. Esty and Meyer (1922) reported 'end-point' data (time at temperature between samples having surviving spores and samples having no survivors or 'destruction') and a series of times at five different temperatures are reported as maximum heat resistance values. From this work, these early 'models' were used to define the processing conditions needed to assure destruction of *Clostridium botulinum* spores leading to the so-called 'botulinum cook', or minimum process lethality required to destroy 10^{12} spores of *C. botulinum* in 'low-acid' foods (those with pH greater than 4.5).

Mathematical modeling in food microbiology began to extend beyond process-setting into bacterial growth prediction but gained ground slowly at first. McMeekin (2007) attributes Scott (1937) with being the first to clearly express the concept that the type and rate of spoilage could be predicted from knowledge of the growth rates of relevant organisms at different storage temperatures. Although it took some time for the concept to catch on, the potential of models to support improvements in food safety and product-development efficiency led to a considerable expansion of activity in the 1980s when large government-funded research programs in the USA, UK and elsewhere developed models of microbial growth, survival and death in response to product or environmental factors including temperature, a_w and pH (Buchanan and Phillips, 1990; Gibson *et al.*, 1988; McClure *et al.*, 1994a; McClure *et al.*, 1997; Sutherland and Bayliss, 1994). The discipline is now established with a number of reference books (McMeekin *et al.*, 1993; McKellar and Lu, 2004; Peleg, 2006; Brul *et al.*, 2007) and electronic resources (see Sources of further information and advice) and a range expanding beyond bacteria to fungi (Gibson *et al.*, 1994; Pardo *et al.*, 2006; Stewart, 2006), viruses (Deboosere *et al.*, 2004; Chen *et al.*, 2004; Isbarn *et al.*, 2007; Thomas and Swayne, 2007; Grove *et al.*, submitted for publication 2009) and parasites (Kim *et al.*, 2007).

This chapter aims to provide a practical guide to microbiological predictive modeling, to give the reader an appreciation of the principles involved in creating useful models and help them to identify and apply soundly based models.

3.2 Approaches to modeling

3.2.1 Principles

By 'model' we mean an equation that describes or predicts the growth, survival or death of microorganisms in foods. Unlike models developed in the food engineering discipline, in food microbiology, predictive models are generally empirical, relating the microbial responses to the levels of the controlling factors. Also known as 'black box', descriptive, inductive or observational models, they tell us nothing about the physiological mechanisms or biological, chemical or physical principles that drive the responses and should not be used outside the range of the data used to create them. Hence, the intended range of use is a critical consideration in experimental design. The range may be decided from our knowledge of the conditions found in relevant product categories and/or knowledge of the likely microbial responses, e.g. minimum a_w or pH for growth of the microorganism of interest. Microbial responses to single factors are well-documented (ICMSF, 1996). Responses to factors acting together are less well characterized, although an increase in published and electronically accessible models since the mid-1990s provides the opportunity to explore the effect of multiple factors.

A simplifying common approach is to experiment in liquid laboratory media that are homogeneous and less complex than the foods that we are ultimately interested in. This restricts the range of controlling factors to relatively few, such as a_w, pH, temperature, acid concentration, salt concentration or preservative concentration. As long as appropriate factors are chosen they will predict the greatest part of the microbial response. Normally we select the three or four factors most relevant to the intended use of the model based on previous experience with the microorganism of interest. Experimentation with more factors is possible but rapidly becomes more complex and expensive as the number of factors increases. Where previous experience is limited, as with a new or emerging pathogen, screening experiments can help to identify the range of conditions to test and estimate the response. Growth models created in these simplified systems tend to be conservative, that is, to predict faster growth than is observed in real foods. This can build in a wide safety margin around predictions at the cost of unnecessarily restricting product shelf-life. Models generated in a matrix that closely simulates a food product of interest may give more accurate predictions of the behavior of that particular food type. The cost of this increase in prediction accuracy is greater experimental complexity and a model with a restricted range of applications.

In modeling experiments, our objective is to obtain high-quality data. The range of conditions must be covered in sufficient detail to see the full range of responses to the conditions of interest with at least some replication to minimize the impact of variability in the responses. This variability is usually highest in conditions closest to the limits permitting growth or causing death. This implies that considerable attention should be paid to these 'marginal' conditions. Unfortunately for experimenters the 'marginal' conditions are usually the hardest to work in! Several authors have given detailed descriptions of experimental design for modeling in food microbiology (Davies, 1993; McMeekin *et al.*, 1993; Ratkowsky, 1993; Rasch, 2004; van Boekel and Zwietering, 2007). Guidelines for data collection and storage are given by Kilsby and Walker (1990) and Walker and Jones (1993).

3.2.2 Types of models

One way to consider the different types of models is in terms of their use to assess the increase of a hazard (e.g. growth models) or the reduction of a hazard (inactivation or survival models). In this way, models can be classified based on the nature of the microbial response.

Growth models are those where at least part of the range of conditions permits growth to occur.

- **Kinetic growth models** are usually based on growth curves over a range of conditions and can allow growth rates to be predicted. These models can also predict the time between the starting condition and the final condition of interest, e.g. hours for a 3-$\log_{10}$ increase in numbers. If desired, they can predict a complete growth curve for the conditions of interest.
- **Growth boundary models** describe the limits of a set of conditions that permit or do not permit growth of the microorganism. They predict time to growth based on data collected as a qualitative response, i.e. growth or no-growth, at intervals over a specified time period. Boundary models do not predict growth curves but can be used to identify no-growth conditions for the microorganism of interest, and often predict the responses over time periods of weeks or months.
- **Probability** or **probabilistic models** describe the likelihood of a particular response being observed or the time until an event occurs. Such models are appropriate when replicated binary responses (i.e. either 'growth' or 'no-growth') are available and a model for the probability of growth at a defined point in time is desired.

Death or **inactivation models** are designed to predict for conditions where a lethal process is deliberately applied and microbial death is relatively rapid.

- **Kinetic inactivation models** are based on inactivation curves over a range of conditions and can allow death rates to be calculated. Most were developed for thermal processes, but there are some for other deliberately lethal conditions including high-pressure processing (Chen and Hoover, 2004; Klotz *et al.*, 2007).
- **Time-to-inactivation models** describe the hold time needed in a process of interest in order to reduce the level of a specified organism from a specified initial level to 'not detectable' even by enrichment methods and are based on measured time to no survivors from different initial levels (Bull *et al.*, 2005).

Survival models relate to transitional conditions between growth and death. Typically death occurs relatively slowly in conditions where growth is prevented but no deliberately lethal treatment is applied (e.g. at ambient temperature and low a_w).

3.2.3 Definitions

In this chapter we use the following terms:

- **Factor** – An independent variable such as temperature, pH, etc., that takes more than one value.

- **Treatment** – A unique combination of factors and their levels. For example pH 6.5 and 25 °C.
- **Response** – The thing we are measuring and modeling, e.g. growth rate or viable count, sometimes known as a dependent variable.
- **Parameter** – A term in a model that is applied to the value of a constant to obtain the prediction. For example in $\ln N = \ln N_0 + \mu t$: N is the dependent variable (cell numbers), N_0 is the initial value of N, μ is a parameter (in this case growth rate) and t is the independent variable (in this case time).

3.2.4 Traceability of measurements

When using models we tend to assume that the results are independent of the laboratory in which the model was created. This may never be true, but will be most closely approached when the accuracy of all measurements is known. This is not simple to achieve because all measuring devices (thermometers, balances, etc.) and control equipment (incubators, waterbaths, etc.) introduce errors. Formal quality management systems include those based on the ISO 9000 series, ISO Guide 25 or the principles of Good Laboratory Practice (Wood *et al.*, 1998). They help to assure the accuracy of measurements through a chain of evidence linking laboratory measurements to national or international standards. At the very least, all measurements of weight, volume, temperature, pH and a_w (or relative humidity) should be made using equipment with calibration traceable to the relevant national standard.

Temperature of incubation should ideally be monitored continuously. Where this is not feasible some indication of the range of temperatures experienced during incubation should be obtained, e.g. using a maximum and minimum thermometer.

Time can be measured using clocks or timers with traceable calibration and this is encouraged. However, modern electronic clocks are generally so accurate that any errors attributable to the clock would be insignificant compared to other sampling and recording errors. When the interval between observations is long and the record is of time to an event (e.g. time for optical density to exceed some threshold value) it is important to also record the last time at which the event was *not* observed. This is likely to be more of a concern in boundary modeling than in kinetic modeling.

3.3 Kinetic growth models

3.3.1 Experimental design

When designing experiments we must first consider the intended use of the model. We can never consider all the factors involved in food composition so we must select those that are most important for controlling microbial growth in the foods of interest. Significant factors are commonly a_w, pH and temperature and there is good evidence that these account for the largest part of the microbial response in many foods. Careful consideration should be given both to relevant factor levels and to how they are achieved.

Salt (NaCl) is the humectant most commonly used to control a_w because it is cheap, convenient and used in preparing a wide range of foods. However, different growth responses can be observed at a single a_w when it is controlled by different humectants (Slade and Levine, 1988; Stewart *et al.*, 2002). For practical purposes, a_w is frequently good enough to guide decision making but by itself does not accurately predict the limits of growth of any microorganism. The minimum a_w for growth of proteolytic *C. botulinum*, for example, is given as 0.96 with NaCl as the humectant and 0.93 with glycerol (Lund and Peck, 2000). There is a growing understanding of the influence of the humectant on the physical properties of the system and the physiological effects that together influence the microbial growth response. For example, NaCl is an ionic humectant that has specific effects on membrane transport systems but does not form an aqueous glass. It can place both osmotic and ionic stresses on bacterial cells. Glycerol is a non-ionic humectant that passively permeates the cell membrane and has no direct effect on ion transport systems. It forms an aqueous glass that reduces the molecular mobility of the system and affects the amount of osmotic stress placed on the bacterial cells (Slade and Levine, 1988). Each glass-forming solute has a characteristic glass transition temperature (T_g). The role of T_g in control of microbial growth is controversial, but it is sensible to approach differences in humectants with caution when designing models to support safety-critical decisions (Stewart *et al.*, 2001).

The acidulant used to adjust pH can also influence the microbial response. The simplest approach is to use hydrochloric acid, which minimizes the influence of undissociated organic acids and ignores the effects of pH buffering that can be seen in foods. Models based on HCl as the acidulant tend to predict faster growth than models based on organic acids. Where organic acids are significant components of the foods of interest, for example in fermented foods and pickles, it is important to be able to determine the concentration of undissociated acid.

Other factors that may be important include the type and concentration of any added preservatives (e.g. nitrite, sorbate, benzoate, propionate, nisin), the composition of the atmosphere around the product and the structure of the food. Since the number of experimental treatments rises rapidly with the number of factors, we may need to recognize a trade-off between the number of factors and the number of factor levels. Alternatively, we could substitute an approach such as growth boundary modeling that is more readily automated using, for example, automated turbidimetry or we could use a reduced experimental design and trade the ability to quantify all interactions for a smaller number of experimental treatments.

Models predicting the growth of pathogens should be ‘fail safe’; they should not over-predict the time to the final condition of interest. For this reason it is more efficient to experiment with cocktails of representative strains than single strains. A cocktail helps to capture the variability between strains and increases the chance of detecting the fastest growth at different points in the experimental matrix, to create a 'leading-edge' model without repeating the experiment many times using different strains. A representative cocktail is likely to contain a blend of isolates from relevant food-poisoning outbreaks, well-characterized experimental strains and isolates from relevant foods. Strains should, as far as possible, be reasonably typical of those likely to be encountered in the foods that comprise the intended use of the model. The value of models is significantly reduced if they become unrealistically conservative as can result from using strains with unusually fast growth rates, or extreme tolerance to one or more of the controlling factors.

Inoculum size and condition are also important. For growth studies, inoculation to give an initial concentration of 10^2–10^3 cfu/ml is ideal because it allows counts to be measured during the lag phase but reduces the risk of unrealistically raising the probability of growth occurring as can happen with high inoculation levels. The culture history in terms of growth, storage conditions, etc. can significantly affect the length of the lag phase so preparation of the inoculum must be standardized to minimize its influence on variability between repeat experiments. The method of inoculum preparation should tend to minimize the length of the lag phase.

Some pathogens may produce toxins even when growth is very limited. For example, it is commonly considered that *C. botulinum* may form toxin in association with only 1–2 $\log_{10}$ increase in numbers. For such pathogens, the methods used to measure growth must be sensitive enough to detect all relevant amounts of growth. Toxin tests on media from marginal conditions at the end of experimentation can be used to confirm 'no-growth' observations.

A number of formal statistical designs may be used to help design microbiological modeling experiments (Davies, 1993; McMeekin *et al.*, 1993; Ratkowsky, 1993; Rasch, 2004; van Boekel and Zwietering, 2007). Even so, for the microbiologist beginning a modeling study there is no substitute for involving a good statistician or mathematician at all stages of experimental design, experimentation and data analysis. Some principles to consider include the following.

Levels of factors

We are usually interested in modeling second-order behavior, i.e. quantifying any curvature in the response. For this we need at least three different levels spanning the range of relevant conditions and more treatments than degrees of freedom for the model to be built. For simple three-factor models, e.g. pH, a_w, temperature, a basic matrix of five levels of each factor is better. A full factorial design would then have 5^3, or 125, different treatments requiring growth curves to be measured. In practice, considerably fewer (perhaps 70–100) will be sufficient because some of the original 125 treatments will involve combinations of factors that do not support growth. By dropping certain treatment conditions, the design becomes 'unbalanced' and the ability to independently gauge the effect of each factor is lost. Since the main objective here is to develop predictive models, this is not an important concern. Use of a reduced design such as a fractional factorial or a central composite design may considerably reduce the number of treatments required. In these cases the reduction in treatments comes at the cost of reduced resolving power in the design or an increased risk that the final model will not predict well.

For most factors, equal linear spacing of levels is initially preferred. However, geometric spacing may be preferred when this leads to more uniform changes between levels in the size of the microbial response. For example, equal spacing in the level of hydrogen ion concentration ($[H^+]$) may be better than equal intervals of pH (which give logarithmic changes in $[H^+]$). Also the use of a zero level for a factor requires careful consideration, e.g. for a preservative the levels might be 0, 500 and 1000 ppm. The use of a zero level is potentially risky because there may be a discontinuity in the response as the level of a factor goes from presence (however small) to complete absence. When subject matter expertise indicates that such a discontinuity is unlikely, the experimenter should feel free to use zero level designs.

Placement of treatments is also important. The microbial responses change most rapidly, and the variance is highest, as the limits for growth are approached. Placing treatments closer together in these conditions is valuable.

Replication

Replication is used to improve the estimate of the response through averaging. Classical training in microbiology emphasizes repetition of experiments (treatment replicates) and replication of data points (measurement replicates) because microbiological data can have high variability, but tends to follow a one-factor-at-a-time approach. When using designed experiments for model-building we have the opportunity to reduce the replication of individual points without losing much information. The principle is to build in independent tests of individual results by comparing them with the results from similar conditions. Just as individual points on a curve contribute to defining the shape of the curve, individual treatments in a matrix contribute to defining the overall shape of the response described by the model. Hence the first emphasis is on populating the experimental matrix as thoroughly as possible, even at the expense of limited repeats. The second emphasis is on independent repeats of selected treatments in the matrix where variability is known to be high, usually those in the most marginal growth conditions. The repeats use media made on different occasions and are run on different days, preferably by different experimenters. Sample replicates, i.e. multiple samples from each treatment, are also useful but make the smallest contribution to the overall performance of the model. The common practice of counting colonies on duplicate plates to reduce measurement error can itself contribute some information on variability, particularly if duplicate plates are identified and specifically recorded when the difference between them falls outside a pre-determined range. These steps are all taken on the understanding that the model will be validated against data produced in foods (ideally in a different laboratory) to give as independent a test as can be arranged of the performance of the model. None of these comments is intended to imply that replication is not valuable, merely that, as with other elements of experimental design, there can be a trade-off between costs and benefits.

Growth curves

A good kinetic growth model requires high-quality growth curves across the range of conditions of interest. Good growth curves typically have 10 or more data points, though the placement of points to identify regions of rapid change can be more important than the number of points (Fig. 3.1). For good estimates of lag phase duration and growth rate during the logarithmic phase data points are needed close to the point of inflection between these two phases. Similarly, if the final population density is of interest, points are needed close to the point of inflection between the logarithmic growth and stationary phases. There is no simple rule for the number of growth curves required. A model based on three controlling factors could need 70 to 100 curves but, with appropriate experimental design, fewer may be sufficient.

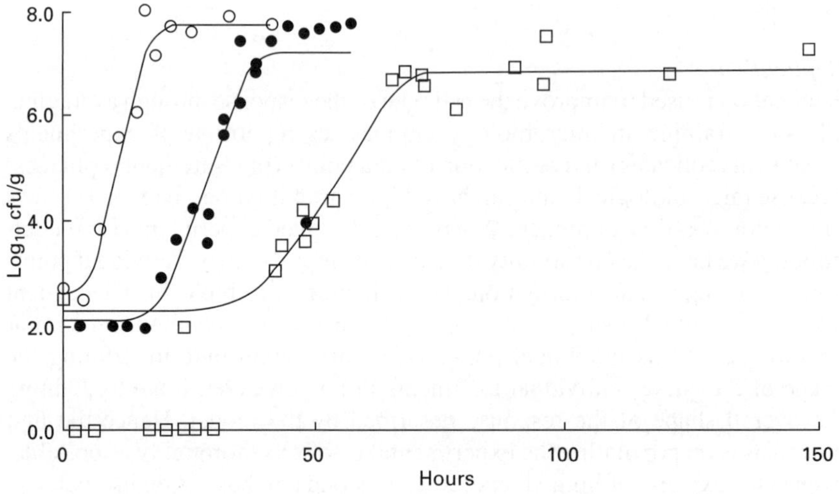

Fig. 3.1 Growth curves for *Bacillus subtilis* under different conditions showing fitted curves from the model of Baranyi and Roberts (1994) giving good descriptions of growth when based on high-quality colony count data in sufficient amounts. The data were accessed from ComBase and fitted using the DMFit option. Growth was in nutrient broth at 34 °C with: pH 7.0, a_w 0.960, ○; pH 6.2, a_w 0.960, ●; and pH 5.9, a_w 0.932, □. Points shown at 0 $\log_{10}$ cfu/g represent counts below the limit of detection and were not included in curve fitting.

3.3.2 Experimentation

Screening experiments

To produce large numbers of growth curves is laborious but worthwhile when a high proportion of the experimental data can be used for model building. For many of the factors controlling the growth of microorganisms the limits of growth given in the literature are defined in terms of only a single factor. When two or more factors are studied together, the limits can differ radically from those based on single factors. It is also hard to find good estimates of the lag phase at non-optimal values even of single factors. Both of these limitations make the planning of modeling studies harder. One solution is to run screening experiments over a wide range of the factors of interest in the intended combinations. Laboratory instruments based on automated readings of turbidity, conductance or capacitance, metabolite production or other markers of microbial growth can be particularly valuable when the data can be plotted as 'pseudo growth curves'. These results can indicate where to place growth curves and how frequently to take samples for counting. However, such instruments are costly. Simple visual inspection of flasks or tubes for turbidity, indicating that growth has occurred, repeated at intervals, can exclude 'no-growth' conditions and indicate a range of lag times to expect under different non-optimal growth conditions.

Order of experiments

When working in broth systems it is probably most efficient to begin working at conditions near the optimum for growth where work progresses quickly. Good growth curves can be obtained in only a few 'all night' experiments and the results can be used to refine the plans for work with less optimal combinations of factors. As experiments work out towards the limits for growth it may become apparent that the position of growth curves should be modified to obtain the best definition of the rapidly changing responses in the most marginal conditions.

This selective ordering of treatments ignores the usual notions of randomization. For reasons outlined above, such restricted randomization can be tolerated but we must be especially careful to ensure that extraneous factors do not affect the experimental results.

When working with 'model food' systems the need to formulate and prepare 'products' assumes greater importance, and access to suitable facilities may dictate a less flexible approach. As always, we need to balance design considerations with practicality and understand the impact of the decisions that we make.

Stability of treatments

Treatments may change during the course of an experiment, particularly when they permit growth. For example, as microbes ferment sugars in the medium, this causes the pH to drop and the a_w to rise. Physical and chemical processes such as evaporation and oxidation may also take place whether growth occurs or not. Hence, it is good practice to make measurements of the experimental variables during and at the end of an experiment, in addition to during the set-up, to help to plan follow-up studies and interpret unexpected results.

3.3.3 Data analysis

To generate the model(s) that relate the growth responses to the experimental treatments, the first step is to plot and inspect the growth curves for each experimental treatment. Reject those where no growth was seen or where the growth curves are clearly inadequate because of known laboratory errors. Continue with the remaining high-quality datasets. Fit growth curves for individual treatments, extract the values of the parameters that describe each curve and then fit the model that relates the changing values of the growth parameters to the levels of the experimental factors. This is often called the two-step approach. Selection of growth curves for modeling does not imply arbitrary discarding of data. For example, significant variability in the observed response, often termed 'noisy data', will lead to a model having higher error associated with its predictions. However, the variability may itself be an important feature of the response and an estimate of that variability can be valuable.

With access to appropriate statistical software, it is possible to fit a response to directly define the microbial count in terms of the independent variables and time (Geysen *et al.*, 2005). This may be called a one-step approach. One-step fitting can be particularly helpful when the data set contains incomplete growth curves because it captures more of the information in the data than the two-step approach. However it is more demanding both of software and expertise than the two-step approach.

Handling variance

To obtain the best fit of a model to the data the variance should be independent of the value of the response variable (in this case the microbial count). However, in microbial count data the variance tends to increase with the count. Transforming the data by taking the logarithm of the counts normalizes the variance, i.e. makes it independent of the value of the count. Kilsby and Walker (1990) discussed some considerations relevant to the most appropriate counting technique.

Fitting growth curves

Several different mathematical models give reasonable descriptions of typical, sigmoid microbial growth curves (i.e. those with lag, log and stationary phases). Common models (Table 3.1) include the modified Gompertz equation, the Baranyi equation and the logistic equation (Gibson *et al.*, 1988; Baranyi *et al.*, 1993). Combinations of linear models for exponential, lag and exponential or lag, exponential and stationary phases have also been

Table 3.1 Commonly-used models that describe microbial growth curves

Model	Equation	Where:
Exponential	$\ln(n) = \ln(n_0) + \mu t$	n = count/g; n_0 = count/g when t = zero; t = time (hours) μ = specific growth rate (per hour)
Lag-exponential	$\ln(n) = \ln(n_0)$, for $t < \lambda$ $\ln(n) = \ln(n_0) + \mu(t - \lambda)$, for $t \geq \lambda$	n = count/g; n_0 = count/g when t = zero; t = time (hours) μ = specific growth rate (per hour); λ = lag time (hours)
Modified Gompertz	$L(t) = A + C \exp\{-\exp[-B(t - M)]\}$	$L(t) = \log_{10}$ bacterial count at time t; B = the relative maximum growth rate (/h); M = the time at which maximum growth rate occurs (h); A = the lower asymptotic $\log_{10}$ bacterial count as t decreases indefinitely; C = the difference between A and the upper asymptotic $\log_{10}$ bacterial count as t increases indefinitely. Also: Lag = $M - (1/B)$ Growth rate = $(B*C*\ln(10))/\exp(1)$ Peak population density Generation time = $\log_{10}(2)\exp(1)/(B*C)$ = $\log_{10}(A + C)$
Baranyi	$y(t) = y_0 + \mu_{max}A_n(t) + \ln\left\{1 + \frac{\exp[\mu_{max}A_n(t)] - 1}{\exp(y_{max} - y_0)}\right\}$	μ_{max} = maximum specific growth rate; $y(t)$ = ln population concentration at time t; y_0 = ln initial population density; y_{max} = ln maximum population density; A_n is an adjustment function as defined by Baranyi *et al.* (1993), considered to account for the physiological state of the cells. Its role is to define the lag phase.
Jones and Walker	$N = N_0 2^{[G(t)-M(t)]}\ t \geq 0$	$G(t) = A\{1-[1 + (t/B) + \frac{1}{2}(t/B)^2 + \frac{1}{6}(t/B)^3]\exp(-t/B)\}$ $M(t) = \exp[(t - D)/C] - \exp[-(t - D)/C] - \exp(-D/C) + \exp(D/C)$ Also: A, B, C and D are constants, t represents time and N_0, the count at time 0, and N are actual counts not log counts.
Logistic	$\ln(n) = \ln(n_0) + \frac{a}{1 + \exp(b - ct)}$	n = count/g; n_0 = count/g when t = zero; t = time (hours); a, b, and c are all fit parameters; t = time (hours).

used (Buchanan *et al.*, 1997a). This 'three-phase linear model' attracted some initial controversy (Baranyi, 1997; Garthright, 1997) and it has not been used extensively despite its relative simplicity (McKellar and Lu, 2004). Appropriate statistical software is used to find the parameters of the chosen equation that give the best fit to the data. For the Gompertz, Baranyi and logistic equations, this involves non-linear regression analysis. The fitness of any particular approach is partly dependent on the objective underlying model building. Van Gerwen and Zwietering (1998) discussed some advantages of different approaches when the objective is to develop models for risk assessment purposes.

More than one curve-fitting approach may give acceptable models for a particular dataset. Selection of the best model is to some degree subjective and the choice may be simply pragmatic based on model validation, a view of the principles underlying the different modeling approaches or other relevant criteria.

Modeling the response of growth parameters to levels of experimental variables

The process here is to use regression modeling, available in many statistical analysis packages, to relate changes in the growth parameters to the levels of the experimental variables. A common strategy uses a polynomial model, though alternative models are available (Table 3.2). A polynomial model has the form:

$$\text{Response} = C_0 + C_1V_1 + C_2V_2 + C_3V_1V_2 + C_4(V_1)^2 + C_5(V_2)^2$$

where all *C*s are constants and all *V*s are variables. Start with a general model that includes terms for all the variables' main effects (V_1, V_2, V_3, etc.) their interactions, e.g. ($V_1 \times V_2$), and quadratics e.g. $(V_1)^2$. Main effects are known as first-order terms. Interactions and quadratics are known as second-order terms. Not all of the terms are equally influential. Regression coefficient estimates for each term in the model can be used to judge the relative importance of each term. The least influential terms are progressively removed and the regression modeling procedure is re-run iteratively until the most parsimonious model, that has the fewest possible terms that accurately matches the observations, is achieved. The principle of parsimony is observed because there is uncertainty associated with the estimate of each term. A parsimonious model minimizes the influence of this uncertainty whilst retaining most of the predictive power of the underlying factors. At each iteration it is useful to plot the experimental observations against the surface predicted by the model and look for outlying observations that reduce the overall goodness of fit and 'leverage points' that have a disproportionate influence on the model because of their position. It is sensible to discard outliers if, upon re-examination, they are shown to be due to some identifiable laboratory error.

The modeling process can sometimes be done with less iteration and an

Table 3.2 Some examples of secondary models for growth parameters. Other examples are given in, e.g. McMeekin *et al.* (1993) and van Gerwen and Zwietering (1998)

Model type	Equation	Where:	Comments
Polynomial	Response = $C_0 + C_1V_1 + C_2V_2 + C_3V_1V_2 + C_4(V_1)^2 + C_5(V_2)^2$	all Cs are constants and all Vs are variables.	Straightforward to apply by multiple linear regression. No knowledge of process needed. But no theoretical foundation, no biological meaning to parameters and no insights on mechanisms.
Square root	$\sqrt{\mu} = b(T - T_{min}) \times \sqrt{(V_1 - V_{1\,min})} \times \sqrt{(V_2 - V_{2\,min})}$	μ is specific growth rate (/h); T is temperature in degrees kelvin; V_1 and V_2 are independent variables; and b is a fit parameter.	Parameters claimed to be biologically interpretable and can calculate relative effects of each variable. But non-linear regression involved if pH and/or a_w are included. No theoretical foundation. Parameters extrapolated to growth limits.
Arrhenius-type	$\mu = A\exp(-E_a/RT)$	μ is specific growth rate (/h); A is the 'collision factor'; E_a is the 'activation energy' of the system; R is the universal gas constant (8.31 J/mol/K); T is temperature in degrees kelvin.	Theoretically grounded for chemical reactions involving collisions between simple models but purely empirical for biological systems. Describes temperature response only but has been modified to include other factors (e.g. a_w, Davey, 1989).

improvement in the overall goodness of fit if some of the inputs can be held constant. For example, we might use the mean initial $\log_{10}$ cfu/ml from all experimental treatments as the starting value for all growth curves. This makes sense because differences between treatments in the starting count are more likely to be a function of inoculation error than of any influence of the experimental factors. In this example we lose essentially nothing and gain some simplicity. Another example would be to fix the population density at stationary phase. Here the price of simplicity might be to overlook real differences in the final population density, which might be acceptable if we are mainly interested in the length of lag phase and the growth rate.

Diagnostic tests (mathematical tests) for model evaluation

A range of mathematical and visual diagnostic tests can be used to help evaluate models (Baranyi *et al.*, 1999). A good start is to plot observed growth against the growth curves predicted by the model for the same conditions. Inspection of the plots quickly establishes how well the model predicts across the range of conditions. Plots of observed versus predicted values for length of lag, growth rate and time to a defined increase in numbers for all the conditions used in building the model help to illustrate its overall performance. Normal probability plots of residuals test the performance of the model across the whole range of conditions. Regression statistics such as the R^2 value also help to quantify how well the model describes the variance in the data. Since R^2 varies inversely with the number of points and the variability in the data, values in the region of 0.8–0.9 (or even lower) are not unusual for models, unlike the 0.99 or so that we are accustomed to seeing for individual curves.

Another method for model selection is to apply each model to a separate, unmodeled hold-out sample. A hold-out sample is one that was included in the original design and run within the model-building experiment for the purpose of comparison with the model but not included in the statistical analysis for model building. Rather it is compared with a prediction from the model for the relevant conditions. For each sample, calculate the mean squared prediction error by computing a prediction and then averaging the squared prediction error. The model with the smallest squared prediction error is deemed the best.

Model acceptance: examination for biological sense

Once the model has been generated and the statistical diagnostic tests have shown that it gives a good description of the data it should be examined for 'biological sense'. This examination tests that predictions from the model behave as expected on the basis of microbiological experience. It is often easiest to assess biological sense using contour or surface plots of predictions for a matrix of conditions similar to that used to create the model (Fig. 3.2).

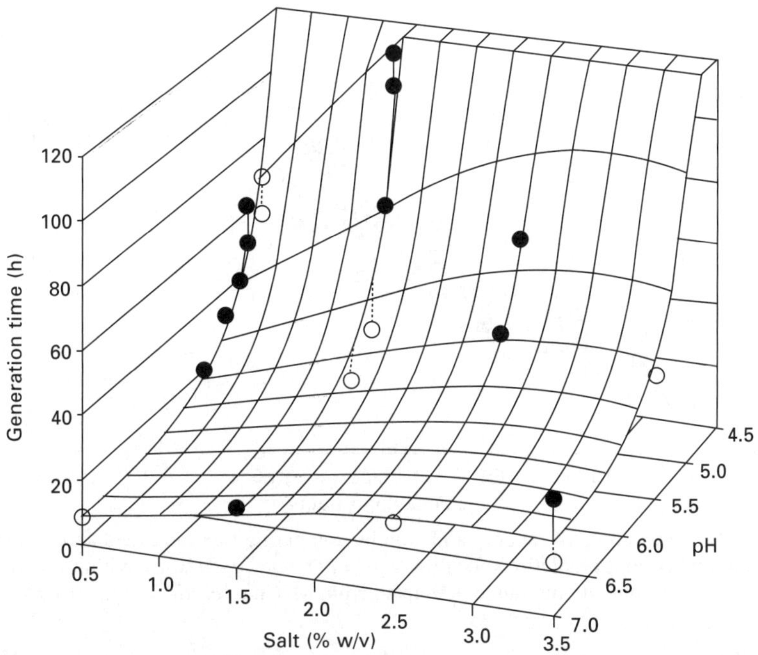

Fig. 3.2 Quadratic response surface for generation time of *Aeromonas hydrophila* with respect to pH and salt concentration (adapted from McClure *et al.*, 1994b). Open points lie below the surface and solid points are above it. This illustrates a model that makes 'biological sense' with generation times becoming longer as conditions become more inhibitory.

Model acceptance: validation

The final step in building confidence in the model is to validate its predictions against observed growth responses in relevant foods. In some cases it is possible to extract validation data from the literature. Unfortunately, data in the literature are often too incomplete to use and we must resort to experimentation. In most cases we need analytical measurements of the relevant factors (salt, pH, a_w, etc.) in the food. Then we compare actual growth data points with predicted growth curves (Fig. 3.3). It is less critical to catch the points of inflection than it is with the model building experiments and 4–6 well-spaced points per curve can be enough. However, more points may allow growth parameters to be derived from the food data, which may facilitate comparison with the model predictions. Good agreement between predicted and observed responses helps to build confidence in the model. The comparison is often shown as a plot of observed against predicted values (Fig. 3.4) in which the responses observed in foods should be no faster than those predicted by the model for maximum confidence. Data should be obtained for all food types (e.g. meat, poultry, fish, dairy, cereal, eggs) for which we intend to use the model to support decision making.

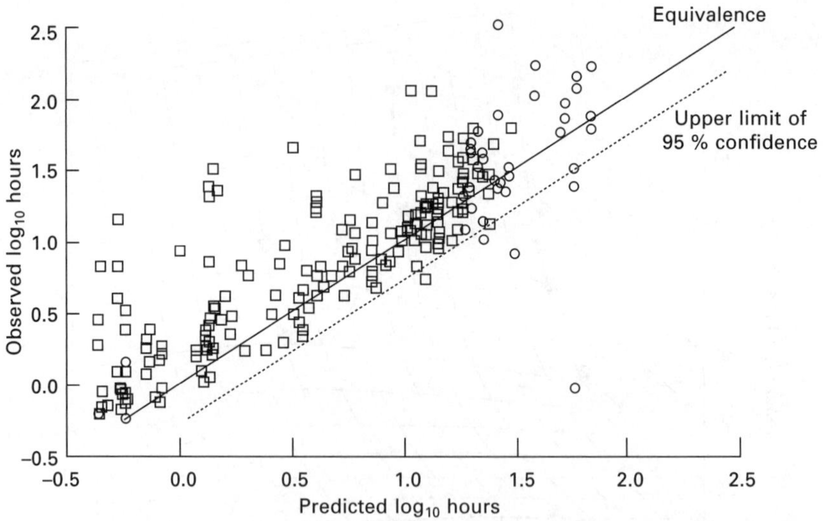

Fig. 3.3 An example of model validation by comparing literature data across a range of conditions with predictions for growth of *L. monocytogenes*. □ Within the range; ○ outside the range. (Adapted from McClure *et al*., 1994b).

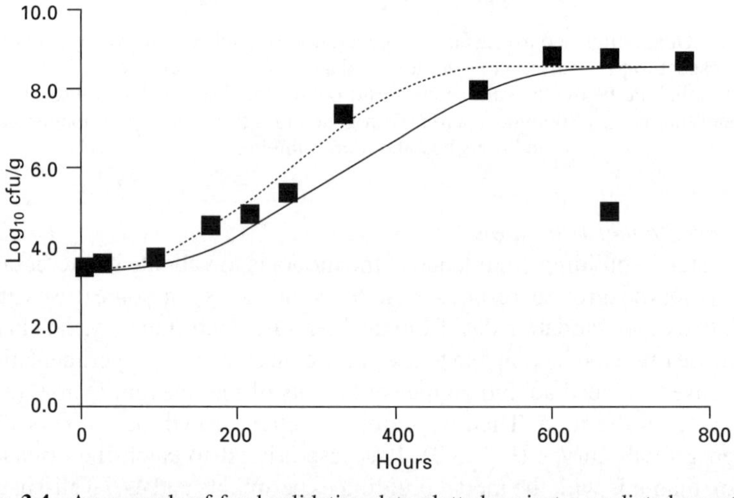

Fig. 3.4 An example of food validation data plotted against a predicted growth curve. Predictions and observations were for *L. monocytogenes* in quiche at pH 5.8 and a_w 0.995. The predictions were from Growth Predictor at 4 °C (solid line) and 5 °C (broken line) and the data were from ComBase. (Figure reprinted from Legan, 2007)

Perhaps the greatest difficulty in working with foods is that they are not naturally sterile. It might be possible to sterilize the foods, e.g. by irradiation, or we might use selective media to count just the species of interest. In this case, we need to be sure that the recovery is comparable with that on the media used during model building. We should also consider whether to count all microorganisms of the relevant species, or only those strains used to inoculate the food. If the latter, and if the incidence of the microorganism in the food of interest is significant, it may be necessary to use antibiotic-resistant (Curtis *et al.*, 1995) or other marked variants of the model building strains. It is important to be sure that the variants behave similarly to the parent strains.

Model acceptance: domain of validity
Predictions from models have high degrees of confidence only when interpolating within the conditions used to build them. This range includes only those conditions where good experimental data were obtained and may be narrower than the experimental design matrix. Establishing the range of interpolation helps to define the useable range of the model (Baranyi, 1999).

Fluctuating conditions
Real foods normally experience a range of temperatures throughout their shelf-life (non-isothermal conditions). Predicting growth in these circumstances is somewhat more involved than predicting for constant conditions but can be achieved by time/temperature integration or other approaches (Baranyi *et al.*, 1995; Corradini and Peleg, 2005).

Combining datasets
It is rarely appropriate to combine models. However, when there are several different datasets for the microorganism of interest, these may be combined and remodeled together. This allows models to be strengthened by using data produced in different laboratories. The range of existing models can also be extended by including additional data. In these circumstances it is particularly important to be aware of the domain of validity.

3.4 Growth boundary models

3.4.1 Experimental design

Many of the design considerations such as intended use of the model, choice of experimental variables and their levels, method of establishing levels, etc. are similar to those for kinetic growth models and have already been discussed. There are, however, some important differences.

The goal of a boundary model is primarily to determine the borderline conditions that prevent growth of the target microorganism without specific inactivation processes, like heat, being applied. To develop this type of model we need many replicates equally spaced about the design matrix, with approximately 50 % of the conditions allowing growth and 50 % not allowing growth. This gives us the best opportunity to find the position of the boundary and important advantages in data analysis. We need many replicates because the area of greatest interest contains the most marginal conditions for growth. In these conditions, the microorganisms are highly stressed and the variability in their growth response is at its highest.

In many disciplines, central composite designs are used for developing response surface models. For microbiological modeling these designs should be approached with some caution. Central composite designs place most replication in the center of the design space. Since replication serves to estimate variability and biological systems tend to exhibit greatest variability near the margins for growth, the central composite design performs best if centered in marginal growth conditions. When working in model foods, or with more than three independent variables, we may choose to use a central composite or other reduced design. If working in laboratory media, particularly if experiments can be automated, constraints on the number of experimental treatments are less rigid and full factorial experiments are preferred. In a factorial experiment, every factor level is run with every other factor level. For regions in the design space in which conditions are so harsh that no growth is expected, or conditions are so favorable that rapid growth is assured, treatments may be omitted without adversely affecting the model, provided that the proportion of growth to no-growth conditions remains approximately equal. Omitting treatments may lead to the experimental design becoming unbalanced. This could be important in a screening experiment because the ability to independently estimate the importance of each factor declines as the design becomes more unbalanced. When experimenting primarily for the purpose of developing predictive models it is less of a problem because, presumably, we already know which factors are important.

As always, randomization should be used to minimize the impact of external, uncontrollable factors on the outcome of the experiment. When complete randomization is not possible the experimenter should randomize wherever practical and watch for external factors that might influence outcomes.

Experimental design considerations include:

- create the experimental design so that about 50 % of the conditions allow and 50 % do not allow growth;
- for second-order models run 3–5 equally spaced levels of each factor;
- run 3–5 sample replicates at each treatment condition to estimate variability.

If using an automated method, a full factorial design is ideal and can be handled with relative ease. With careful planning, media can be made over a period of several days by one person and refrigerated as appropriate until all the treatments are ready. Inoculation and dispensing is then done in a single day ready to begin incubation.

3.4.2 Experimentation

Experimentation for boundary modeling in laboratory media often makes use of automated equipment that detects growth by changes in optical density, conductance or impedance or production of metabolites. Since we are only interested in if, when and under what conditions growth occurs (or not), specific quantification of growth is not needed. However, careful consideration must be given to the appropriate growth 'threshold value' (e.g. the optical density level) defining the borderline between growth and no-growth. The threshold should be the lowest realistic value of the measured response that is clearly greater than the noise seen over time at no-growth conditions. If the threshold is in doubt, more than one value can be selected and models created based on each. Comparison of the models may resolve the selection of the response threshold. Note that the automated method might correlate better with total microbial biomass than with colony count, particularly when changes in environmental stresses cause a change in the size or shape of the cells.

Automated methods may only be sensitive to the presence of 10^5 cfu/ml or higher and this may influence the inoculum level. Ideally we would still inoculate at a level of 10^2–10^3 cfu/ml to avoid inducing growth through the presence of an unrealistically high initial level. The difference in time to detect growth between an automated method and one based on plate counting will be relatively small because the length of the lag phase has much more effect on the time to any threshold population density than the growth rate. However, when working with toxin-forming bacteria we may choose to use a higher inoculum to ensure that the amount of growth needed for detection is less than that associated with toxin formation.

If working directly in foods, counting might be the best way to detect growth. The considerations on threshold level outlined above still apply, though experience suggests that a 1 $\log_{10}$ cfu/g increase in count is likely to be appropriate.

Order of experimentation

In boundary modeling the emphasis is on simply detecting growth, without the need to quantify it. Hence the balance of advantage favors setting up 'marginal' and 'no-growth' treatments first because these treatments will run for the longest time (possibly several months). Those conditions in which growth is expected to be relatively quick can be set up last because they only need monitoring until growth is detected. When automated methods are used it may be most effective to simply start all treatments at the same time. To minimize the chance for systematic error, the order of treatments should be randomized as far as practicable during preparation, dispensing, inoculation and reading and position in the laboratory equipment, if relevant.

Stability of treatments

As boundary-modeling experiments can span several months, we must ensure that the initial conditions of interest do not change over time solely through an uncontrolled interaction with the laboratory environment. For example, gain or loss of moisture can significantly change the a_w (or relative humidity, RH) of the system and its ability to support growth. The RH around the treatment can be stabilized by incubating in a sealed container over saturated salt slurries that give the same RH as the medium (Weast *et al.*, 1984). Changes resulting from microbial activity may, however, be an important part of the mechanism leading to growth initiation and we would not wish to stabilize the system at the expense of growth that would naturally occur in a food. For example, maintaining the initial pH over time is typically neither possible nor practical, even in buffered media. In fact, allowing a change in pH due to growth of the microorganism more closely mimics what happens in a food product.

Data analysis: initial inspection of data

We inspect the raw data for quality and consistency, using plots or tabulations to show patterns in the data, and identify the time to growth for each treatment. This determination can be made 'by eye' with small datasets. For large datasets it may be more efficient to identify a suitable interpolation method to determine time to growth for the majority of observations. Interpolation using a geometric time interval is theoretically more accurate than using linear time intervals, but differences in any particular circumstances may be small.

Modeling growth parameters: survival analysis

The final dataset ready for boundary modeling usually contains results where growth was not observed before the end of the experiment; usually presented as 'time to growth > t_{max}', where t_{max} is the maximum duration of the experiment. Such results are called 'censored data' because they tell us nothing about what might have happened if the experiment had continued. Least squares regression analysis cannot handle censored data. Instead, we need a statistical package that is capable of 'survival analysis', also known as generalized regression, using maximum likelihood estimation methods. A number of packages are available including SAS (SAS Institute Inc., Cary, NC), Minitab (Mintab Inc., State College, PA) and Statistica (Statsoft Inc., Tulsa, OK), but not all offer the same range of capability.

The SAS LIFEREG procedure is one survival analysis tool that has often been used for microbiological modeling. By default, the procedure fits a model to the log of the dependent variable (in this case, time). The resulting model can easily be transformed to a regular time scale. The result is a regression equation of the form:

$$\ln(\text{time to growth}) = C_0 + C_1V_1 + C_2V_2 + C_3V_1V_2 + C_4(V_1)^2 + C_5(V_2)^2$$

where all *C*s are constants and all *V*s are variables.

LIFEREG outputs a table of regression coefficient estimates and approximate chi-squared distribution *p*-values for each factor in the model. The relative importance of each factor can be judged by the *p*-value: factors with small

p-values are most influential and predictive of the time to growth (TTG; Table 3.3). Starting with a general model that includes all the variables' main effects (V_1, V_2, V_3, etc.) their interactions, e.g. ($V_1 \times V_2$), and quadratic terms, e.g. ${V_1}^2$, those factors that do not have a significant effect are progressively excluded through a series of iterations and the data re-analyzed. The final model is parsimonious, i.e. it gives the best description of the data in the fewest possible terms. As with kinetic modeling, relating the data to the model at each iteration and looking for outliers and leverage points can help both to improve the quality of the final model and identify any limitations on its application.

LIFEREG allows the user to specify the error distribution to account for the variation in TTG not explained by the regression model. Several distributions, including Weibull, lognormal and loglogistic, typically give good fits to survival data. All should be considered to identify the one giving the best fit. Since these are not all from the same class of distributions, it is not possible to formally test for goodness of fit using likelihood ratio tests. However, comparison of the sample log likelihood can informally be used and that with the largest log likelihood should be considered to give the best fit. Jenkins *et al.* (2000), Stewart *et al.* (2001) and Legan *et al.* (2004) give more detailed descriptions of boundary modeling using SAS LIFEREG.

Logistic regression modeling

Ratkowsky and Ross (1995) proposed an approach which advanced kinetic modeling of bacterial growth as limited by various growth controlling factors (i.e. pH, temperature, additives, etc.) by enabling data for no-growth to be incorporated into the model as well as data for growth and to deal with the variability in that response. The logistic regression model was developed in

Table 3.3 Example of parameters derived from a SAS LIFEREG output table for a time to growth boundary model for *Staphylococcus aureus* based on relative humidity (rh), pH and calcium propionate (cal) (Stewart *et al.*, 2001). In the first iteration of modeling, the quadratic term cal^2 was excluded because it was not significant (p = 0.6247) whereas all the other p values were < 0.0001 (see Pr > ChiSq column)

Variable	Degrees of freedom	Estimate	Standard error	Chi square	Pr > ChiSq
Intercept	1	2.2496	0.0319	4983.5007	< 0.0001
rh	1	–3.7032	0.0562	4338.8734	< 0.0001
ph	1	–1.4881	0.0327	2069.5765	< 0.0001
cal	1	0.4176	0.0301	192.5908	< 0.0001
rh^2	1	1.5717	0.0474	1101.7569	< 0.0001
ph^2	1	0.3430	0.0220	243.5212	< 0.0001
rh*ph	1	0.8145	0.0300	746.1667	< 0.0001
rh*cal	1	–0.1305	0.0302	18.7282	< 0.0001
ph*cal	1	–0.2209	0.0208	112.2248	< 0.0001
Scale	1	0.2000	0.0082		

the form of an expression containing the growth limiting factors suggested by a kinetic model, while response at a given combination of factors may either be presence/absence (i.e. growth/no-growth) or probabilistic (i.e. *r* successes in *n* trials). This growth/no-growth boundary model adopts a probabilistic approach using logistic regression to develop secondary models to define this interface. With this approach, either data obtained from kinetic experiments may be used to develop the model or, alternatively, data derived from experiments designed to determine whether growth is possible or not may be utilized. Ratkowsky and Ross (1995), Tienungoon *et al.* (2000), Vermeulen *et al.* (2007) and Gysemans *et al.* (2007) give more detailed descriptions of boundary modeling using logistic regression modeling. Gysemans *et al.* (2007) also compared two types of boundary modeling: (1) the linear logistic regression model; and (2) the non-linear logistic regression model derived from a square root-type kinetic model.

Model acceptance: validation and biological sense

The principles of model acceptance are similar to those discussed under kinetic modeling, though the validation plot can look more complex because it must take account of the censored data (Fig. 3.5).

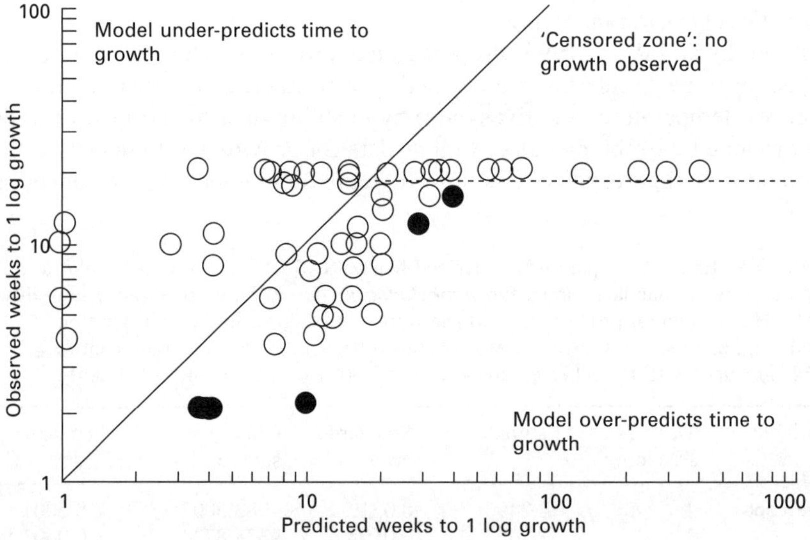

Fig. 3.5 Example of validation of a growth boundary model by comparing observed weeks to growth in independent experiments with predictions from a model. Notice the effect of censoring, where predicted weeks to growth continue to increase indefinitely but observed 'no-growth' is defined by the upper limit of the experimental time (in this case 18 weeks). Practically this amounts to good agreement since both observations and predictions show no growth at the upper limit of experimental time. Open points lie within the range and solid points outside it. (Data from Legan *et al.*, 2004).

There is an additional consideration for toxin-producing microorganisms. If toxin formation could occur with less growth than the difference between the inoculum level and the sensitivity of the technique used to determine growth, the boundary conditions should be confirmed by either plate counts, direct microscopic counts or toxin assays. This confirmation must be done before using the model to support safety-critical decisions.

Diagnostic tests (mathematical tests)
According to Box (1979) 'All models are wrong but some are useful' and diagnostic testing helps to select the best (most useful) model. One way to assess model quality is to examine plots of model predictions and residuals. Residuals are the difference between model prediction and the observed TTG. Useful plots include predicted TTG vs observed TTG, residual vs observed TTG, residuals vs individual factor levels and a probability plot of the residuals. If patterns are detected in the residual plots, the model might be improved by addition/deletion of factors or scale transformation. Such plots indicate whether the data might be better fit with a different error distribution. Most regression books provide detailed discussions of residual diagnostics and possible remedies. Another useful metric of model quality is the sample log likelihood. This is automatically calculated by most survival analysis software. Depending on the distributions being considered, we can do a generalized likelihood ratio test to compare models (Allison, 1995). Generally, models with relatively large sample likelihoods are preferred.

3.5 Kinetic death models

3.5.1 Experimental design

As with growth modeling, we must consider the intended use of the model and select the factors that most influence the rate of microbial death in the food categories and processes of interest. In particular, we must select realistic levels for factors that can either protect microorganisms from the lethal agent or increase their sensitivity to it. Naturally, spores will be more resistant than vegetative cells.

We must be open-minded and place points to determine the true nature of the microbial response to the lethal agent, whether log-linear or not. Ideally this involves 10–12 points, at least 6–7 $\log_{10}$ reduction in population, implying an inoculation level of 10^8–10^9 cfu/ml, though fewer points can be adequate provided that a good population decline is measured. A zero-time point is essential and time intervals increasing geometrically between samplings can be beneficial.

To develop realistic yet 'fail-safe' models, strains with above average, but not abnormal, resistance to the lethal agent are preferred. Generally experiments to generate death data should be done using single strains. Use of strain cocktails can lead to complex curves that are difficult to interpret.

In some cases it is easier to kill bacteria than to destroy their toxins. In applying models to help to design safe processes it is important to keep this in mind. For example, little is known about the effectiveness of novel preservation processes on destruction of microbial toxins.

3.5.2 Experimentation

In death modeling, the response (reduction in viable microbial count) occurs over a relatively short time, typically seconds to minutes. Hence errors in measuring the exposure time can be significant and the lethal agent is usually applied, and removed, rapidly to allow the exposure time to be accurately determined. Sometimes, the microbial response to the rate of change of the lethal agent is the factor of interest (Stephens *et al.*, 1997a). In these cases, we should be aware that slow inactivation may permit adaptation to occur and a change in resistance to develop. In any case, we must manage culture history so as to ensure that resistance of the microorganism is at least equal to that in the food or process of concern. We must also ensure that we are able to detect all survivors, including those that may be injured by a sub-lethal exposure to the lethal agent. A number of models of viral inactivation have appeared since the first edition of this book was published (Chen *et al.*, 2004; Deboosere *et al.*, 2004; Isbarn *et al.*, 2007; Thomas and Swayne, 2007; Grove *et al.*, 2009). Recovery methods present an interesting challenge when the microorganism of interest is not able to grow independently but requires a host cell to support it. Modeling has also been applied to simultaneously compare inactivation of protozoan parasites in water supplies with formation of bromate and decay of ozone (Kim *et al.*, 2007). In this case a major advantage of modeling is the potential to optimize the process for more than one variable.

3.5.3 Data analysis

As with growth modeling, the approach to discovering the model(s) that relate the death responses to the experimental treatments is to fit death curves for individual treatments, find the values of the parameters that describe each curve and fit the model that relates the values of the fitted parameters to the levels of the experimental factors.

The first step is to plot and inspect the inactivation curves for each experimental treatment. Reject those where the curves are clearly inadequate because of insufficient points, known laboratory errors, etc. Continue with the remaining high-quality datasets.

If microbial death data are approached with an open mind, several different approaches can give reasonable descriptions of microbial inactivation curves and Table 3.4 shows examples. The traditional approach using log-linear models of microbial death began in the 1920s (Bigelow *et al.*, 1920; Ball, 1923) when a linear assumption significantly simplified the work involved.

Table 3.4 Examples of commonly used models that describe microbial death curves

Model	Equation	Where:
Exponential	$\log(n) = \log(n_0) - kt$	n = count/g at time t t = time (usually in seconds) k = a rate constant Also: $1/k = D$ = time needed for one-log reduction in count/g Note: assumes log-linear death kinetics.
Logistic	$\ln(n) = \ln(n_0) + \dfrac{a}{1 + \exp(b - ct)}$	n = count/g; n_0 = count/g when t = zero; t = time (hours); a, b, and c are all fit parameters Note: fits shoulders and tails but 'expects' survival curves to be sigmoid.
Cumulative Weibull	$s = \exp(-bt^n)$ or, $\log(s) = -bt^n$	s = surviving fraction (n/n_0) at time t b is the scale parameter n is the shape parameter It is generally preferable to work with the second, semi-logarithmic form to capture detail in the tail of the curve. Note: this model fits shoulders or tails, but not both.
Gompertz	$L(t) = A + C \exp\{-\exp[-B(t - M)]\}$	$L(t) = \log_{10}$ bacterial count at time t; B = the relative maximum death rate (/h); M = the time at which maximum death rate occurs (h); A = the upper asymptotic $\log_{10}$ bacterial count as t decreases indefinitely; C = the difference between A and the lower asymptotic $\log_{10}$ bacterial count as t increases indefinitely. Note: fits shoulders and tails but 'expects' survival curves to be sigmoid.

In these days of inexpensive computers and statistical software, non-linear curve fitting avoids the need to prejudge the microbial response and allows more objective analysis of the data. The drawbacks of using non-linear models to make predictions outside the range of the experimental data are quite apparent (Fig. 3.6). The fact that extrapolation of linear models can be equally flawed is often overlooked.

Fitting death curves

There are many non-linear models that can be used to describe microbial death curves and examples are shown in Table 3.4. Van Gerwen and Zwietering (1998) discussed some advantages of the Gompertz and logistic equations that were initially popular. Peleg and Cole (1998) discussed the use of the cumulative Weibull function ($\log s = -bt^n$). This is particularly flexible because it can describe curves with upwards ($n < 1$) or downwards

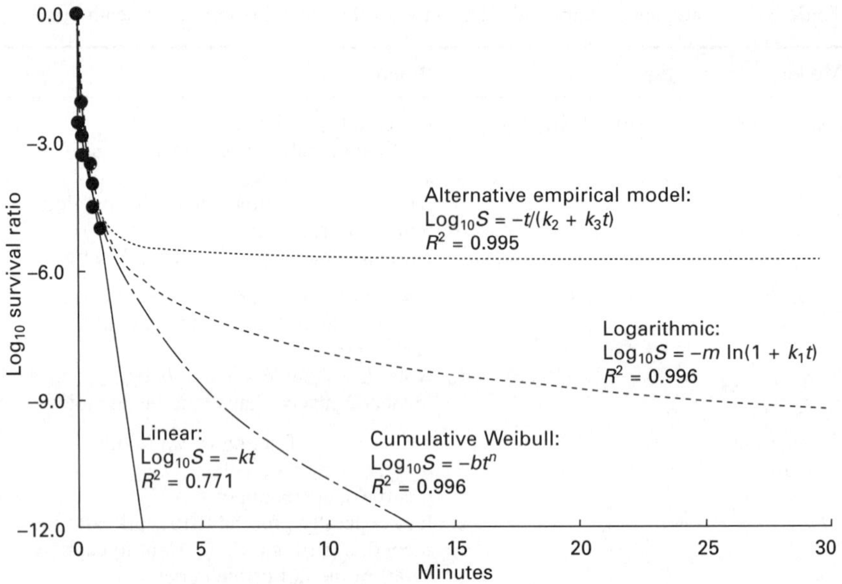

Fig. 3.6 Experimental death curves of *Clostridium botulinum* spores at 119 °C (Anderson *et al.*, 1996) fitted with different models, adapted from Peleg and Penchina (2000). The curves are all similar when interpolating between points but give drastically different values when extrapolated to a 12-log reduction in numbers. In the equations S is survival ratio, t is time, b, k, k_1, k_2, k_3, m and n are fit parameters and ln indicates natural logarithm.

($n > 1$) concavity and accommodate log-linear death curves simply as the special case where $n = 1$. The Weibull approach seems to have increased in popularity in the early 2000s, probably as a result of this flexibility combined with relative simplicity.

More than one curve-fitting approach may give acceptable models of a particular dataset. Selection of the best model is rather subjective and the choice may be based on model validation, the principles underlying the different modeling approaches or other relevant criteria. It is most important that the selected model clearly describes the observed data and is not merely forced to fit, or fitted to an arbitrarily selected subset of the data. Unfortunately, many research papers present only derived *D*- and/or *z*-values and make comparisons between conditions based on these quantities. This has two damaging consequences:

- without evidence that the chosen model gives an adequate description of the data the validity of any comparisons are in doubt,
- without the underlying data others cannot compare different modeling approaches and opportunities for future improvement are lost.

The opportunity for authors to deposit data with ComBase (see Sources of further information and advice) offers the possibility for much improvement in future.

Modeling death parameters

The principles are essentially the same as when modeling growth parameters. The simplest approach is to fit a polynomial regression model that relates the parameters describing the individual death curves to the factors (pH, a_w, etc.) that define the experimental treatments. However, other models have been used for this stage and examples are given in Table 3.5.

Varying conditions

An important application for death models is for evaluation of food process lethality. A characteristic of the 'kill-step' in most food processes is that the intensity of the lethal agent varies with time, for example due to heating and cooling times. To calculate microbial survival under such conditions we must integrate the momentary lethal effects throughout the process. Standard techniques for performing this calculation based on log-linear kinetic models are well documented (Texeira, 1992; Toledo, 1998). Though widely accepted, these approaches are technically invalid when survival curves are not log-linear. Approaches for calculating microbial survival in varying conditions have been proposed by Peleg and Penchina (2000) and others (Körmendy and Körmendy; 1997; Gil *et al.*, 2006). The Peleg and Penchina approach has been shown to be relatively easy to use and effective (Mattick *et al.*, 2001; Peleg *et al.*, 2001; Peleg *et al.*, 2003; Peleg and Normand, 2004; Corradini *et al.*, 2005, 2006) and certainly is worthy of further investigation. Additionally, models have been developed by Peleg (2002) and Corradini and Peleg (2003) which describe microbial inactivation in water treated with a dissipating chemical disinfecting agent, such as chlorine, ozone and similar agents.

Validation of death models

The most critical test of the usefulness of a lethality model is how well it can predict the outcome of an independent test of performance. As with growth modeling, validation can be done by comparing observations against predicted death curves using 4–6 observations per curve and a range of curves across the model design-space. Even with a validated model, it is wise to independently validate any process established using it. Process validation is described in detail in Keener (2006) and CAC (2006).

3.6 Time to inactivation models

The nature of kinetic models is that they are based on reductions in countable numbers of microbes. Non-linearity is primarily an issue when upwards concavity, or tailing, raises the possibility that a few individual cells, but too few to count, might survive a treatment that we expect to be lethal when considering anticipated log reductions. For pathogens with low infectious

Table 3.5 Some examples of secondary models for death parameters. Other examples are given in, e.g. van Gerwen and Zwietering (1998)

Model type	Equation	Where:	Comments
Polynomial	Response = $C_0 + C_1V_1 + C_2V_2 + C_3V_1V_2 + C_4(V_1)^2 + C_5(V_2)^2$	all *C*s are constants and all *V*s are variables.	Straightforward to apply by multiple linear regression. No knowledge of process needed. But no theoretical foundation, no biological meaning to parameters and no insights on mechanisms.
Arrhenius-type	$k = A \exp(-E_a/RT)$	k is death rate constant; A is the 'collision factor'; E_a is the 'activation energy' of the system; R is the universal gas constant (8.31 J/mol/K); T is temperature in degrees kelvin.	Parameters have no biological meaning. Has been adapted to include other parameters (e.g. pH) but does not predict limiting values for variables (Davey *et al.*, 1995).
z concept	$z = \dfrac{T_2 - T_1}{\mathrm{Log}D_{T_1} - \mathrm{Log}D_{T_2}}$	z is temperature for a 10-fold change in D value; D is time for 10-fold reduction in survivors; T_1 is lower temperature; T_2 is higher temperature.	Uses linear fitting by eye or by regression. Assumes a first-order death kinetic.

doses or pathogens in foods where there is a possibility of regrowth we may want greater assurance that we have eliminated the pathogen of concern.

One solution to the problem of non-linear death curves seen with some new technologies, such as high-pressure processing, is to avoid them altogether. Models of time to inactivation (no survivors recoverable by enrichment) from different inoculation levels sidestep the kinetics by modeling the end-point of interest (Bull *et al.*, 2005; Hannaford *et al.*, 2009). Conceptually this is similar to boundary models for growth, where the end-point is the minimum detectable amount of growth, but here the end-point is no recoverable survivors.

3.6.1 Experimental design

The overall design is a full factorial or a reduced design such as a fractional factorial with at least three and preferably five or more replicate samples of each treatment.

3.6.2 Experimentation

Several different inoculation levels between low, around 10^2 cfu/g, and high, 10^6 cfu/g or more, are exposed to the lethal temperature, pressure, etc., for different exposure times. Following treatment, samples are recovered by enrichment, and the numbers of positive and negative samples are recorded for each treatment and exposure time. Normally some reduction in the proportion of positive samples is seen as the exposure time increases until a point is reached where all replicate samples are negative, indicating no survivors. This end-point is confirmed when the next higher time also shows all replicate samples negative. The time to inactivation lies between the last exposure with at least one positive sample and the first exposure with no survivors in any replicate. If desired, experiments can be repeated at additional times between the last positive and first negative time in order to narrow down the estimate of the time to inactivation.

3.6.3 Data analysis

In time-to-inactivation modeling with a deliberately applied lethal treatment it is normally possible to find the time to inactivation (end-point of no survivors) for all treatments. With all end-points known there are no censored observations and statistical analysis is by multiple regression modeling of observed time to inactivation against the values of the independent variables for each treatment. If, for any reason, there are treatments for which the time to inactivation cannot be determined within the time or budget available for experimentation the longest time for which each treatment was exposed to the lethal process without achieving inactivation should be recorded. An option then is to treat the longest time recorded for the incomplete treatments as censored observations and use generalized regression techniques such as SAS LIFEREG as described for growth boundary modeling. The resulting model has the ability to predict time to inactivation from an inoculation level within the range of inoculation levels in the experimental design but no ability to predict lethality curves.

3.6.4 Model validation

As with other models, validation is by comparing the outcome of independent tests against model predictions. Practically, we should test times a little more and a little less than the predicted time to inactivation. We want to see no survivors in a little longer than the predicted time to inactivation and expect to see survivors in a little less than the predicted time to inactivation (though, if we see no survivors here, that indicates that the model is conservative).

If, instead, we test exactly at the predicted time to inactivation then we would expect a good model to be wrong close to 100 % of the time; over-predicting and under-predicting slightly in roughly equal proportions. However, since we cannot know the size of the over- and under-predictions we have no way to tell a good model from a poor one by this means.

3.7 Survival models

Survival models are generally concerned with non-thermal inactivation under relatively mild conditions, often just outside the limits for growth. Some pathogens can remain viable for many months or even years in no-growth conditions. An outbreak of food poisoning caused by *Salmonella* Napoli in chocolate occurred in 1982 after such an occurrence and the infectious dose was judged to be as few as 50 microorganisms per patient (1.6 cfu/g of chocolate; Greenwood and Hooper, 1983). Survival models can help us to understand the limits of survival in different food systems. They are less common than growth or death models because of some of the practical constraints on their creation, but there are examples in the literature (Buchanan *et al.*, 1994; Little *et al.*, 1994; Whiting *et al.*, 1996; Zaritzky *et al.*, 1999; Vora *et al.*, 2003; Hew *et al.*, 2006). Corradini and Peleg (2006) developed models to simulate the transition of microbial growth to inactivation and vice-versa, for examples when there were survivors after a heat treatment, and then went on to develop Weibullian models to describe microbial injury and subsequent mortality due to refrigeration/freezing processes, heat treatments, chemical disinfection and high-pressure processing (Corradini and Peleg, 2007).

3.7.1 Experimental design

Most of the design considerations are similar to those already discussed under growth and death models. The conditions of interest often include some that support growth whilst others lead to death. Accordingly the inoculation level is usually moderate (10^4–10^6 cfu/ml) to allow changes in either direction to be followed. Choice of single strains or cocktails is harder here because of the range of responses that we may see. Use of single strains simplifies interpretation of the inactivation kinetics but puts a particular emphasis on selecting strains with representative characteristics. Cocktails may allow us to compensate for variability between strains at the risk of survival curves that are hard to interpret.

3.7.2 Laboratory work

Automated methods such as automated turbidimetry, conductance or capacitance measurement, etc. are not effective for following population death. This means that laboratory work for survival models involves colony counting and is similar to that for kinetic growth models, with similar resource constraints and design considerations.

3.7.3 Data analysis

The approach to data analysis for survival models follows the principles already outlined, though the selection of models for curve fitting may be somewhat different (Jones and Walker, 1993).

3.8 Applications of models

Creation of models is an intellectually challenging and satisfying activity, but the greatest satisfaction comes from using the model for some practical purpose. Getting the most out of models in practical applications requires a little care, and some of the considerations were discussed by Legan (2007). Some circumstances where models can help with practical problems are illustrated below.

3.8.1 Product and process design

Approaches to providing more natural, higher-quality manufactured foods include milder heat treatments, reduced salt content, lower acidity, fewer preservatives, etc. These improve taste and texture but provide less microbiological protection. Accordingly, product and process design must maximize quality while minimizing risk. Our ability to approach the boundary between safe and hazardous conditions with confidence reflects both how well the process is controlled and how well the position of the boundary is known. When process parameters are well defined and controlled so that process variability is small, our ability to safely modify process conditions is limited by how well the boundary of the hazard conditions is known. Predictive models allow product and process designers to:

- explore the product and process parameters that allow growth, survival or death of the microorganisms of concern (Juneja *et al.*, 2003; Membré *et al.*, 2006);
- determine more accurately the limits between safe and hazardous processing conditions; define effective lethal processes; establish safe cooling times after cooking a bulk product; optimize the formula to achieve the desired shelf-life in the intended distribution chain (Stewart *et al.*, 2003; Hew *et al.*, 2006) and identify the maximum achievable shelf-life in the intended distribution conditions when the preferred shelf-life is not accessible.

In this respect, predictive models are more useful than traditional challenge tests, though challenge tests remain an ideal way to validate a product and process design obtained through modeling.

The majority of the microbiological models in the literature were derived from experiments in laboratory media. Media-based models are often very conservative and may give predictions for safe conditions that may be detrimental to product quality or manufacturing costs, or even restrict access to formulations giving pH, a_w, etc. that would be safe in foods. Although more costly to produce, models from experimentation in food systems may be more applicable to a manufacturer's specific product line as a result of formulation, processing or structural factors not applicable to microbiological media. Fat content, for example, can make a large difference to potential microbiological risk: a high fat content can have a 'protective' effect in thermal inactivation of microorganisms; some preservatives preferentially migrate to the fat phase and become unavailable for microbial inhibition; and high levels of fat can change the rate of oxygen diffusion through a food product.

An early example of creating predictive models in a complex food system was published by Tanaka (1982) and Tanaka *et al.* (1986) who comprehensively studied the factors of NaCl, phosphates, moisture and pH in preventing *C. botulinum* outgrowth in safe and stable processed cheese spreads. More recently, ter Steeg and co-workers developed predictive models on the time to 100-fold increase in proteolytic *C. botulinum* in high moisture pasteurized processed cheese spreads (ter Steeg *et al.*, 1995; ter Steeg and Cuppers, 1995). A few additional examples of models derived from collecting experimental data in model food systems include models for:

- high-pressure inactivation of *Salmonella* in orange juice (Bull *et al.*, 2005);
- non-thermal inactivation of *L. monocytogenes* in four types of European fermented sausages with lactic acid bacteria or bacteriocins in the product formulations (Drosinos *et al.*, 2006);
- survival of *E. coli*, *L. monocytogenes* and *Salmonella* in Mexican-style chorizos (Hew *et al.*, 2006);

- the behavior of *L. monocytogenes* in ice cream during static and dynamic storage conditions (Gougouli *et al.*, 2008).

Where available, relevant models from food-based systems can greatly facilitate product and/or process design.

3.8.2 Hygienic design of equipment

Numerical simulation of flow and temperature combined with microbial growth models can be used to predict microbial numbers in closed processes and give guidance on hygienic design of equipment. For example, if equipment contains dead ends, models can be used to see if microorganisms can grow in the dead end and then recontaminate the flow.

3.8.3 Hazard Analysis Critical Control Point (HACCP)

Significant progress has been made in developing the use of predictive modeling for risk assessment and HACCP (Notermans *et al.*, 1994). Predictive models can help to decide if a particular microorganism should be a concern for a particular food product in the risk analysis. They can also help to determine the Critical Control Points (CCPs) in HACCP studies by providing information on the likely growth or survival of the microbiological hazards at each point in the process. Once the CCPs have been identified, predictive models can help to define the critical limits at the CCPs by exploring the microbial responses to different critical limits. Models can help to calculate the parameters of any corrective action necessary (e.g. temperature and time for repasteurization; Walker and Jones, 1994). Models can also be used to determine parameters for new or intended processes by comparing the predicted potential for microbial growth with that for existing processes (Armitage, 1997).

3.8.4 Risk assessment

Predictive microbial models are essential tools in quantitative risk assessment. They can be used in:

- hazard identification, to determine whether a particular pathogen can grow in a product or process;
- disease characterization, e.g. to understand the behavior of newly recognized pathogens;
- dose–response assessment to assess the risk of infection (Foegeding, 1997);
- exposure assessment to determine the probability of consuming a biological agent and the amount consumed.

Microbial risk assessment is generally more complex than chemical risk assessment as the microbial hazard has the potential both to increase (through growth) or decrease (through inactivation or dilution) throughout the production and distribution of a food. Predictive microbial models make a significant contribution to risk assessment through their ability to explore new conditions rapidly and quantitatively (Buchanan and Whiting, 1996; Baker *et al.*, 1997; Whiting and Buchanan, 1997; Cassin *et al.*, 1998; ICMSF, 1998; Marks *et al.*, 1998; Juneja *et al.*, 2003; Drosinos *et al.*, 2006) and can be powerful tools to help microbiologists and others make risk management decisions. Together with acceptance sampling schemes of quantifiable performance (Legan *et al.*, 2001) they will transform the way in which the microbiological safety of foods is managed.

3.8.5 Validation of control measures to meet food safety objectives

A key provision of the World Trade Organization Sanitary and Phytosanitary (SPS) Agreement is that countries must provide risk assessments to facilitate resolution of trading disputes that involve food safety issues. The risk assessments help to quantify the risks faced by consumers and the levels of assurance in the exporting and importing countries (ICMSF, 1998). This is a significant advance on the previous requirement for risk to be 'as low as reasonably possible' because technological capabilities vary between countries and even between companies within the same country. The perception of 'reasonable' also differs between countries with 'acceptable risk' being culturally defined. Developments in quantitative risk assessment and validated microbial models that can accurately predict the growth, death or survival of the major foodborne pathogens have made it possible to link the exposure assessment of a pathogen to likely public health outcomes (Buchanan *et al.*, 1997b; Whiting and Buchanan, 1997).

The International Commission on Microbiological Specifications for Foods (ICMSF) has proposed a scheme for managing microbiological risks for foods in international trade in which the Food Safety Objective (FSO) is a functional link between risk assessment and risk management. The FSO is defined as 'a statement of the frequency or maximum concentration of a microbiological hazard in a food considered acceptable for consumer protection' (van Schothorst, 1998) and allows the equivalence of different control measures to be established. This scheme is described in detail, with examples in ICMSF book 7 (ICMSF, 2002). This framework is gaining ground internationally for establishing processes and food safety systems.

An excellent example of how microbiological modeling, risk assessment and risk management utilizing the ICMSF concepts has been published by Drosinos *et al.* (2006). The authors developed non-thermal inactivation models for *L. monocytogenes* in European fermented sausages formulated with lactic acid bacteria or bacteriocins and then conducted a case study for risk assessment using @Risk 4.5 (Palisade Decision Corp., New York, NY) to predict the simulated *L. monocytogenes* concentration in control and formulated sausages at each consumption period. Based on derived listeriosis incidence, they conducted and reported on an exercise to illustrate how risk can be managed based on the establishment of appropriate level of protection, food safety objectives, performance criteria, performance objectives, process criteria or product criteria and implementation of control measures.

3.9 Future trends

Access to models and modeling resources has improved noticeably since the first edition of this book was published, largely as a result of much material moving 'on-line', and we believe and hope that this trend will continue. Resources such as the Pathogen Modeling Information Portal (PMIP) from the USDA (see Sources of further information and advice) will fulfill an important role in education and will help to make it easier to find the many individual resources through a single access point. Information systems have become powerful tools in managing food safety and, as McMeekin *et al.* (2006) concluded, '…as information systems become more sophisticated (e.g. RFID technology) and are combined with increasingly extensive microbial databases (e.g. *ComBase*), the prospect for continual real-time monitoring will take food safety management to new levels of precision and flexibility'. Many of the resources identified in Sources of further information and advice will allow easier and broader exploration of the application of pathogen models which will reinforce the improvement in acceptance and use of models that we have seen since the first edition of this book.

Soundly-based, appropriately validated pathogen models are moving out of the laboratory and into commercial applications. The discussion has moved from whether models can be reliable in principle to whether a particular model is reliable for a particular application. The tendency to excessive conservatism of models derived from experiments in laboratory media will drive an increase in the number and quality of models created through experimentation in food systems, even though experiments in laboratory media can be simpler, less costly, and lead to models that can be applied over a broad range of food product categories. Additionally, there appears to be a trend towards model validation studies (in food products) being conducted either subsequently to the publication of predictive models derived from laboratory media (Vora *et al.*, 2003; Francois *et al.*, 2006) or for studies to include product validation data in their initial publications (Lindström *et al.*, 2003; Legan *et al.*, 2004; Bull *et al.*, 2005). Both food-based models and an increase in published validation of models in food systems will contribute to greater practical application of models for product and process development.

Since the late 1990s there has been increasing commercialization of non-thermal pasteurization techniques, particularly high-pressure processing, fuelled by consumer desire for foods that are more fresh-like yet still convenient and safe. Commercialization of non-thermal technologies has been slow, partly because of the high capital cost of the equipment, but also because there is no 'processing continuum' that allows industry to establish processes confidently based on the product type and microorganism of concern. To establish this confidence there is a need to obtain systematic inactivation kinetics data or time to inactivation data. Unfortunately, we cannot apply thermal inactivation kinetics to other processes; not only are the kinetics of inactivation different between heat and e.g. pressure but the microorganism of primary concern may change too, depending on the relative resistance of different pathogens to heat and the non-thermal process of interest.

An interesting feature of non-thermal process technologies is that kinetic inactivation data can be hard to fit to traditional first-order log-linear death curves. Non-linear modeling approaches that may be based on an assumption of biovariability such that, for any population of cells or spores, there will be a distribution of sensitivities have been successfully employed to model observed data (Cole *et al.*, 1993; Little *et al.*, 1994; Anderson *et al.*, 1996; Peleg and Cole, 1998). These approaches will undoubtedly see greater use in future and it will become increasingly important to understand the microbial risk implications of non-first-order inactivation kinetics. Chen and Hoover (2003, 2004) have compared the use of the log-logistic, Gompertz and Weibull models for describing high-pressure inactivation of *L. monocytogenes* in milk and found that the Weibull model gave the best predictions with their non-linear inactivation data. The practical downside to non-linear models of decreasing lethality with increasing exposure to the lethal agent, commonly known as tailing, is that it can be hard to create sufficient assurance of desired lethality. The time-to-inactivation approach introduced by Bull *et al.* (2005) makes the shape of the inactivation curve irrelevant. Although logistically more complex than kinetic inactivation studies the benefits for non-thermal processes are clear. We expect to see more models of this type, at least until the best way to apply non-linear kinetic models to food-safety applications is more clear.

Small numbers of cells do not always behave as anticipated based on studies of cell populations (classical microbial culture methods). This divergence in behavior between large and small populations is based on the natural biovariability between individual cells and has been termed quantum or quantal microbiology (Bridson, 1997, Bridson and Gould, 2000). Studies with small inocula have been used to model the lag time for vegetative bacteria (reviewed by Swinnen *et al.*, 2004) and the germination time and/or outgrowth time of bacterial spores (Billon *et al.*, 1997; Stringer *et al.*, 2005). Additionally, the effect of inoculum size on the ability of bacteria to grow or not in a particular model or food system has also been studied with regards to modeling or validating growth boundaries (Masana and Baranyi, 2000; Vora *et al.*, 2003). Many single-cell approaches including image analysis (Billon *et al.*, 1997), flow cytometry (Ueckert *et al.*, 1995), automated turbidimetry (Stephens *et al.*, 1997b), combination of phase-contrast microscopy and automated turbidity (Stringer *et al.*, 2005) will see increased application, particularly as adoption of new processing technologies demands new risk-based approaches to decision making.

To this point, we have discussed only empirical models. Such models are adequate for many practical purposes but provide no secure basis for extrapolation outside the range of the experimental data. Mechanistic, or deterministic, models, in contrast, are built upon a theoretical understanding of the system. They have the potential to give more accurate predictions than empirical models and provide a better basis for extrapolation outside the range of the experimental data, though this is still not a trivial matter because the mechanism itself may change, or prediction errors may become very large (Box *et al.*, 1978). Many 'quasi-mechanistic' models have been developed (Bazin and Prosser, 1992; McMeekin *et al.*, 1993; Ross, 1999) and have proved useful for developing and testing hypotheses. However, in all cases the 'key enzyme' is unknown and a model whose parameters cannot be determined experimentally cannot be considered truly mechanistic (Heitzer *et al.*, 1991). Despite much progress, the question of how growth (and death) kinetics are related to the physiology of microorganisms is generally not well understood. A rare example of a truly mechanistic model linking these elements is the work of Cayley *et al.* (1992) that relates the growth rate of *Escherichia coli* K12 under osmotic stress to the intracellular accumulation of betaine and proline and the thermodynamics of osmoprotection.

Box *et al.* (1978) commented that a mechanistic approach is justified whenever a basic understanding of the system is essential to progress or when the state of the art is sufficiently advanced to make a useful mechanistic model easily available. Clearly the latter is not yet true in microbiology, though the former would certainly be helpful. Cole (1991) observed that: 'researchers in the field of predictive microbiology are striving to develop models for microbial growth and death based upon an understanding of cell variability and physiology and that could be used to extrapolate to other conditions'. We remain confident that microbiology will eventually see a useful collection of mechanistic models though given the complexity of the systems involved it might take many years.

3.10 Sources of further information and advice

3.10.1 Models and databases

- http://www.purac.com/purac_com/a5348511153c582f5bd69fd6bd64bb49.php#suppression Brief information on models available from Purac to calculate inhibition of growth of *Listeria monocytogenes* using potassium lactate and sodium diacetate, including a link to request the model(s) on compact disc (accessed November 2008).

- http://portal.arserrc.gov/ Pathogen Modeling Information Portal from USDA, links to USDA pathogen models online, US food safety regulations and other information; designed to support small food manufacturers (accessed November 2008).
- http://www-unix.oit.umass.edu/~aew2000/GrowthAndSurvival/GrowthAndSurvival.html Links to spreadsheets for generating non-isothermal survival and growth curves provided by Dr Micha Peleg (accessed November 2008).
- http://www.dfu.min.dk/micro/sssp/Home/Home.aspx Links to download Seafood Spoilage and Safety Predictor from the Danish Institute for Fisheries Research and to use the online (English only) version (accessed November 2008).
- http://www.ifr.ac.uk/safety/growthpredictor Source for UK Growth Predictor and Perfringens Predictor software (accessed November 2008).
- http://www.ifr.ac.uk/safety/DMfit/default.html DMFit: an add-in for Microsoft Excel 5 or above; it fits sigmoid microbial growth curves using the model of Baranyi and Roberts (1994) (accessed November 2008).
- http://www.ifr.ac.uk/microfit/ MicroFit: a tool developed by the UK Institute for Food Research to allow product developers to extract microbial growth parameters from challenge test data. (accessed November 2008).
- http://ars.usda.gov/Services/docs.htm?docid=6786 Source for USDA pathogen modeling program and associated notes (accessed November 2008).
- http://www.symprevius.net/ Website of French national predictive microbiology program; accessible in French and English (click on the rotating flag to change language); registered users can access a database of over 3000 growth and survival curves and a simulation module (accessed November 2008).
- http://wyndmoor.arserrc.gov/combase/ Browser for ComBase: Combined Database for Predictive Microbiology, a database provided jointly by the United Kingdom Institute for Food Research and the USDA Eastern Regional Research Center; it contains data on growth, survival and death of many foodborne pathogens (accessed November, 2008).

3.10.2 Risk assessment

- http://www.foodrisk.org Food safety risk analysis site provided by JIFSAN (Joint Institute for Food Safety and Nutrition) with links to databases holding pathogen information including epidemiological data, disease incidence, etc. and lists of modeling and food safety tools (accessed November 2008).

3.10.3 Other information

- http://www.icmsf.iit.edu/publications/sampling_plans.html Source for download of a spreadsheet to calculate the concentration of microorganisms controlled by ICMSF sampling plans as described by Legan *et al.* (2001) (accessed November 2008).
- http://www.icmsf.iit.edu/main/articles_papers.html A simplified guide to using FSOs and performance objectives can be downloaded from this site (accessed November 2008).
- http://www.pathogencombat.com/ A multi-partner EU funded research program (accessed November 2008).

3.10.4 Additional reviews of predictive modeling

A collection of papers entitled 'Models for safety, quality, and competitiveness of the food processing sector' was published in *Comprehensive Reviews in Food Science and Food Safety* (2008) as a follow-up of a pre-Institute of Food Technologists workshop. Of particular interest to readers of this chapter include articles by Black and Davidson (Use of modeling to enhance the microbiological safety of the food system), van Boekel (Kinetic modeling of food quality: a critical review) and Marks (Status of microbial modeling in food process models). Another review, by McMeekin (2007), manages to cover the past, present and future of predictive modeling and quantitative microbiology, to discuss the food safety and economic benefits realized and to comment on the future sociological challenges to further improvements in food safety, and all in only ten pages!

3.11 Acknowledgements

The authors gratefully acknowledge several years of successful and enjoyable collaboration, including co-authorship of the first edition of this chapter, with Dr Mark Vandeven, now at Colgate Palmolive, and are indebted to him for his advice during preparation of this second edition.

3.12 References

Allison PD (1995) *Survival Analysis Using the SAS System: A Practical Guide*, Cary, NC, SAS Institute, Inc.

Anderson WA, McClure PJ, Baird-Parker AC and Cole MB (1996) The application of a log-logistic model to describe the thermal inactivation of *Clostridium botulinum* 213B at temperatures below 121.1°C, *Journal of Applied Bacteriology*, **80**, 283–90.

Armitage NH (1997) Use of predictive microbiology in meat hygiene regulatory activity, *International Journal of Food Microbiology*, **36**, 103–109.

Baker A, Ebel E, Hogue A, McDowell R, Morales R, Schlosser W and Whiting RC (1997)

Parameter values for a risk assessment *of Salmonella enteritidis* in shell eggs and egg products, Washington, DC, USDA Food Safety and Inspection Service.

Ball CO (1923) Thermal process time for canned foods, *Bulletin of the National Research Council*, vol. 7 part 1, number **37**, 1–76.

Baranyi J (1997) Simple is good as long as it is enough. (Models for bacterial growth curves), *Food Microbiology*, **14**, 391–94.

Baranyi J (1999) Creating predictive models, in TA Roberts (ed.) *COST 914 Predictive Modelling of Microbial Growth and Survival in Foods*, Luxembourg, Office for Official Publications of the European Communities, 65–7.

Baranyi J and Roberts TA (1994) A dynamic approach to predicting bacterial growth in food, *International Journal of Food Microbiology*, **23**, 277–94.

Baranyi J, Roberts TA and McClure P (1993) A non-autonomous differential equation to model bacterial growth, *Food Microbiology*, **10**, 43–59.

Baranyi J, Robinson TP, Kaloti A and Mackey BM (1995) Predicting growth of *Brochothrix thermosphacta* at changing temperature, *International Journal of Food Microbiology*, **27**, 61–75.

Baranyi J, Pin C and Ross T (1999) Validating and comparing predictive models, *International Journal of Food Microbiology*, **48**, 159–166.

Bazin MJ and Prosser JI (1992) Modelling microbial ecosystems, *Journal of Applied Bacteriology Supplement*, **73**, 89S–95S.

Bigelow WD, Bohart GS, Richardson AG and Ball CO (1920) Heat penetration in processing canned foods, *National Canners Association Bulletin*, **16L**, 128.

Billon CMP, McKirgan CJ, McClure PJ and Adair C (1997) The effect of temperature on the germination of single spores of *Clostridium botulinum* 62A, *Journal of Applied Microbiology*, **82**(1), 48–56.

Black DG and Davidson PM (2008) Use of modeling to enhance the microbiological safety of the food system, *Comprehensive Reviews in Food Science and Food Safety*, **7**, 159–67.

Box GEP (1979) Robustness in the strategy of scientific model building, in RL Launer and GN Wilkinson (eds) *Robustness in Statistics*, New York, Academic Press.

Box GEP, Hunter WG and Hunter JS (1978) *Statistics for Experimenters. An Introduction to Design, Data Analysis and Model Building*, New York, Wiley.

Bridson EY (1997) Are we in a time warp? in M. Sussman (ed.) *Cultural Concepts, Proceedings of a day of thought and contemplation*, London, Oxoid.

Bridson EY and Gould GW (2000) Quantal microbiology, *Letters in Applied Microbiology*, **30**, 95–98.

Brul S, van Gerwen S and Zwietering M (2007) *Modelling Microorganisms in Food*, Cambridge, Woodhead.

Buchanan RL and Phillips JG (1990) Response Surface Model for predicting the effects of temperature, pH, sodium nitrite concentrations and atmosphere on the growth of *Listeria monocytogenes, Journal of Food Protection*, **53**, 370–6.

Buchanan RL and Whiting RC (1996) Risk assessment and predictive microbiology, *Journal of Food Protection*, **59** (Supplement), 31–6.

Buchanan RL, Golden MH, Whiting RC, Philips JG and Smith JL (1994) Non-thermal inactivation models for *Listeria monocytogenes. Journal of Food Science*, **59**, 179–88.

Buchanan RL, Whiting C and Damert WC (1997a) When is simple good enough: a comparison of the Gompertz, Baranyi and three-phase linear models for fitting bacterial growth curves, *Food Microbiology*, **14**, 313–26.

Buchanan RL, Damert WG, Whiting RC and van Schothorst M (1997b) Use of epidemiologic and food survey data to estimate a purposefully conservative dose-response relationship for *Listeria monocytogenes* levels and incidence of listeriosis, *Journal of Food Protection*, **60**(8), 918–22.

Bull MK, Szabo EA, Cole MB and Stewart CM (2005) Validation of process criteria for high pressure pasteurization of orange juice using predictive models, *Journal of Food Protection*, **68**(5), 949–54.

Cassin MH, Lammerding AM, Todd ECD, Ross W and McColl RS (1998) Quantitative risk assessment for *Escherichia coli* 0157:H7 in ground beef hamburgers, *International Journal of Food Microbiology*, **41**, 21–44.

Cayley S, Lewis BA and Record TM (1992) Origins of the osmoprotective properties of betaine and proline in *Escherichia coli* K-2, *Journal of Bacteriology*, **174**, 1586–95.

Chen H and Hoover DG (2003) Modeling the combined effect of high hydrostatic pressure and mild heat on the inactivation kinetics of *Listeria monocytogenes* Scott A in whole milk, *Innovative Food Science & Emerging Technologies*, **4**(1), 25–34.

Chen H and Hoover DG (2004) Use of Weibull model to describe and predict pressure inactivation of *Listeria monocytogenes* Scott A in whole milk, *Innovative Food Science & Emerging Technologies*, **5**(3), 269–76.

Chen H, Joerger RD, Kingsley DH and Hoover DG (2004) Pressure inactivation kinetics of phage λ cI 857, *Journal of Food Protection*, **67**(3), 505–11.

CAC (2006) *Proposed draft guidelines for the validation of food safety control measures*, CX/FH 06/38/8, Rome/Geneva, Food and Agriculture Organization of the United Nations/World Health Organization.

Cole MB (1991) Predictive modelling – yes it is! *Letters in Applied Microbiology*, **13**, 218–19.

Cole MB, Davies KW, Munro G, Holyoak CD and Kilsby DC (1993) A vitalistic model to describe thermal inactivation of *L. monocytogenes*, *Journal of Industrial Microbiology*, **12**, 232–9.

Corradini MG and Peleg M (2003) A model of microbial survival curves in water treated with a volatile disinfectant, *Journal of Applied Microbiology*, **95**, 1268–76.

Corradini MG and Peleg M (2005) Estimating nonisothermal bacterial growth in foods from isothermal experimental data, *Journal of Applied Microbiology*, **99**(1), 187–200.

Corradini MG and Peleg M (2006) On modeling and simulating transitions between microbial growth and inactivation or vice versa, *International Journal of Food Microbiology*, **108**, 22–35.

Corradini MG and Peleg M (2007) A Weibullian model for microbial injury and mortality, *International Journal for Food Microbiology*, **119**, 319–28.

Corradini MG, Normand MD and Peleg M (2005) Calculating the efficacy of heat sterilization processes, *Journal of Food Engineering*, **67**, 59–69.

Corradini MG, Normand MD and Peleg M (2006) Expressing the equivalence of non-isothermal and isothermal heat sterilization processes, *Journal of the Science of Food and Agriculture*, **86**, 785–92.

Curtis LM, Patrick M and Blackburn C de W (1995) Survival of *Campylobacter jejuni* in foods and comparison with a predictive model, *Letters in Applied Microbiology*, **21**, 194–7.

Davey KR (1989) A predictive model for combined temperature and water-activity on microbial growth during the growth phase, *Journal of Applied Bacteriology*, **67**, 483–8.

Davey KR, Hall RF and Thomas CJ (1995) Experimental and model studies of the combined effect of temperature and pH on the thermal sterilization of vegetative bacteria in liquid, *Food and Bioproducts Processing*, **73**, 127–32.

Davies KW (1993) Design of experiments for predictive microbial modelling, *Journal of Industrial Microbiology*, **12**(3–5), 295–300.

Deboosere N, Legeay O, Caudrelier Y and Lange M (2004) Modelling effect of physical and chemical parameters on heat inactivation kinetics of hepatitis A virus in a fruit model system, *International Journal of Food Microbiology*, **93**, 73–85.

Drosinos EH, Mataragas M, Vesković-Moračanin S, Gasparik-Reichardt J, Hadžiosmanović M and Alagić D (2006) Quantifying nonthermal inactivation of *Listeria monocytogenes* in European fermented sausages using bacteriocinogenic lactic acid bacteria or their bacteriocins: a case study for risk assessment, *Journal of Food Protection*, **69**(11), 2648–63.

Esty JR and Meyer KF (1922) The heat resistance of spores of *Clostridium botulinum* and allied anaerobes, *Journal of Infectious Diseases*, **31**, 650–663.

Foegeding PM (1997) Driving predictive modeling on a risk assessment path for enhanced food safety, *International Journal of Food Microbiology*, **36**, 87–95.

Francois K, Devlieghere F, Uyttendaele M, Standaert AR, Geeraerd AH, Nadal P, Van Impe JF and Debevere J (2006) Single cell variability of *L. monocytogenes* grown on liver pâté and cooked ham at 7 °C: comparing challenge test data to predictive simulations, *Journal of Applied Microbiology*, **100**(4), 800–12.

Garthright WE (1997) The three-phase linear model of bacterial growth: a response, *Food Microbiology*, **14**, 395–97.

Geysen S, Verlinden BE, Geeraerd AH, Van Impe JF, Michiels CW and Nicolaï BM (2005) Predictive modelling and validation of *Listeria innocua* growth at superatmospheric oxygen and carbon dioxide concentrations, *International Journal of Food Microbiology*, **105**, 333–45.

Gibson AM, Bratchell N and Roberts TA (1988) Predicting microbial growth: growth response of salmonellae in a laboratory medium as affected by pH, sodium chloride and storage temperature, *International Journal of Food Microbiology*, **6**, 155–78.

Gibson AM, Baranyi J, Pitt JI, Eyles MJ and Roberts TA (1994) Predicting fungal growth: the effect of water activity on *Aspergillus flavus* and related species, *International Journal of Food Microbiology*, **23**(3–4), 419–31.

Gil MM, Brandão TRS and Silva CLM (2006) A modified Gompertz model to predict microbial inactivation under time-varying temperature conditions, *Journal of Food Engineering*, **76**, 89–94.

Gougouli M, Angelidis AS and Koutsoumanis K (2008) A study on the kinetic behavior of *Listeria monocytogenes* in ice cream stored under static and dynamic chilling and freezing conditions, *Journal of Dairy Science*, **91**, 523–30.

Greenwood MH and Hooper WL (1983) Chocolate bars contaminated with *Salmonella napoli*: an infectivity study, *British Medical Journal*, **286**(6375), 1394.

Grove SF, Lee A, Stewart CM and Ross T (2009) Development of a high pressure processing model for hepatitis A virus (submitted for publication).

Gysemans KPM, Beraerts K, Vermeulen A, Geeraerd AH, Debevere J, Devlieghere F and Van Impe JF (2007) Exploring the performance of logistic regression model types on growth/no growth data of *Listeria monocytogenes*, *International Journal of Food Microbiology*, **114**, 316–31.

Hannaford AM, Huang Y, Stewart CM, Kalinowksi RM and Legan JD (2009) Modeling time to inactivation of *Listeria monocytogenes* in response to high pressure, sodium chloride and sodium lactate, (submitted for publication).

Heitzer A, Kohler HE, Reichert P and Hamer G (1991) Utility of phenomenological models for describing temperature dependence of bacterial growth, *Applied and Environmental Microbiology*, **57**, 2656–65.

Hew CM, Hajmeer MN, Farver TB, Riemann HP, Glover JM and Cliver DO (2006) Pathogen survival in chorizos: ecological factor, *Journal of Food Protection*, **69**(5), 1087–95.

ICMSF (1996) *Microorganisms in Foods 5: Characteristics of Microbial Pathogens*, International Commission on Microbiological Specifications for Foods, London, Blackie Academic and Professional.

ICMSF (1998) Potential application of risk assessment techniques to microbiological issues related to international trade in food and food products, International Commission on Microbiological Specifications for Foods, *Journal of Food Protection*, **61**(8) 1075–86.

ICMSF (2002) *Microorganisms in Foods 7: Microbiological Testing in Food Safety*

Management, International Commission on Microbiological Specifications for Foods, New York, Kluwer Academic/Plenum.

Isbarn S, Buckow R, Himmelreich A, Lehmacher A and Heinz V (2007) Inactivation of avian influenza virus by heat and high hydrostatic pressure, *Journal of Food Protection*, **70**, 667–73.

Jenkins P, Poulos PG, Cole MB, Vandeven MH and Legan JD (2000) The boundary for growth of *Zygosaccharomyces bailii* in acidified products described by models for time to growth and probability of growth, *Journal of Food Protection*, **63**(2), 222–30.

Jones JE and Walker SJ (1993) Advances in modeling microbial growth, *Journal of Industrial Microbiology*, **12**, 200–5.

Juneja VK, Marks HM and Huang L (2003) Growth and heat resistance kinetic variation among various isolates of *Salmonella* and its application to risk assessment, *Risk Analysis*, **23**(1), 199–213.

Keener L (2006) Hurdling new technology challenges: Investing in process validation of novel technologies, *Food Safety Magazine*, **12**(1), 37–8, 40, 53–5.

Kilsby DC and Walker SJ (1990) *Predictive modelling of microorganisms in foods. Protocols document for production and recording of data*, Chipping Campden, Campden Food & Drink Research Association.

Kim JH, Elovitz MS, von Gunten U, Shukairy HM and Mariñas BJ (2007) Modeling *Cryptosporidium parvum* oocyst inactivation and bromate in a flow-through ozone contactor treating natural water, *Water Research*, **41**(2), 467–75.

Klotz B, Pyle DL and Mackey BM (2007) New mathematical modeling approach for predicting microbial inactivation by high hydrostatic pressure, *Applied and Environmental Microbiology*, **73**(8), 2468–78.

Körmendy I and Körmendy L (1997) Considerations for calculating heat inactivation processes when semilogarithmic thermal inactivation models are non-linear, *Journal of Food Engineering*, **34**, 33–40.

Legan JD (2007) Application of models and other quantitative microbiology tools in predictive microbiology, in S Brul, S van Gerwen and M Zwietering (eds) *Modelling Microorganisms in Food*, Cambridge, Woodhead, 82–110.

Legan JD, Vandeven MH, Dahms S and Cole MB (2001) Determining the concentration of micro-organisms controlled by attributes sampling plans, *Food Control*, **12**, 137–47.

Legan JD, Seman DL, Milkowski AL, Hirschey JA and Vandeven MH (2004) Modeling the growth boundary of *Listeria monocytogenes* in ready-to-eat cooked meat products as a function of the product salt, moisture, potassium lactate, and sodium diacetate concentrations, *Journal of Food Protection*, **67**(10), 2195–204.

Lindström M, Nevas M, Hielm S, Lähteenmäki L, Peck MW and Korkeala H (2003) Thermal inactivation of nonproteolytic *Clostridium botulinum* type E spores in model fish media and in vacuum-packaged hot-smoked fish products, *Applied and Environmental Microbiology*, **69**(7), 4029–36.

Little CL, Adams MR, Anderson WA and Cole MB (1994) Application of a log-logistic model to describe the survival of *Yersinia enterocolitica* at sub-optimal pH and temperature, *International Journal of Food Microbiology*, **22**, 63–71.

Lund BM and Peck MW (2000) *Clostridium botulinum*, in BM Lund, TC Baird-Parker and GW Gould (eds), *The Microbiological Safety and Quality of Food, Volume 2*, Gaithersburg, MD, Aspen 1057–109.

Marks BP (2008) Status of microbial modeling in food process models. *Comprehensive Reviews in Food Science and Food Safety*, **7**, 138–43.

Marks HM, Coleman ME, Lin CT and Roberts T (1998) Topics in microbial risk assessment: dynamic flow tree process, *Risk Analysis*, **18**, 309–28.

Masana MO and Baranyi J (2000) Growth/no growth interface of *Brochothrix thermosphacta* as a function of pH and water activity, *Food Microbiology*, **17**, 485–93.

Mattick KL, Legan JD, Humphrey TJ and Peleg M (2001) Calculating *Salmonella* survival during non-isothermal heat inactivation, *Journal of Food Protection*, **64**(5), 606–13.

McClure PJ, Blackburn C de W, Cole MB, Curtis PS, Jones JE, Legan JD, Ogden ID, Peck MW, Roberts TA, Sutherland JP and Walker SJ (1994a) Modelling the growth, survival and death of microoorganisms in foods: the UK Food Micromodel approach, *International Journal of Food Microbiology*, **23**, 265–75.

McClure PJ, Cole MB and Davies KW (1994b) An example of the stages in the development of a predictive mathematical model for microbial growth: the effects of NaCl, pH and temperature on the growth of *Aeromonas hydrophila*. *International Journal of Food Microbiology*, **23**, 359–75.

McClure PJ, Beaumont AL, Sutherland JP and Roberts TA (1997) Predictive modelling of growth of *Listeria monocytogenes*: the effects on growth of NaCl, pH, storage temperature and $NaNO_2$, *International Journal of Food Microbiology*, **34**, 221–32.

McKellar RC and Lu X (2004) *Modeling Microbial Responses in Food*, Boca Raton, FL, CRC Press.

McMeekin T (2007) Predictive microbiology: quantitative science delivering quantifiable benefits to the meat industry and other food industries, *Meat Science*, **17**, 17–27.

McMeekin T, Olley JN, Ross T and Ratkowsky DA (1993) *Predictive Microbiology: Theory and Application*, New York, Wiley.

McMeekin TA, Baranyi J, Bowman J, Dalgaard P, Kirk M, Ross T, Schmid S and Zwietering MH (2006) Information systems in food safety management, *International Journal of Food Microbiology*, **112**, 181–94.

Membré J-M, Amézquita A, Bassett J, Giavedoni P, Blackburn C de W and Gorris LGM (2006) A probabilistic modeling approach in thermal inactivation: estimation of postprocess *Bacillus cereus* spore prevalence and concentration, *Journal of Food Protection*, **69** (1), 118–129.

Notermans S, Gallhoff G, Zwietering MH and Mead GC (1994) The HACCP concept: specification of criteria using quantitative risk assessment, *Food Microbiology*, **11**, 397–408.

Pardo E, Marin S, Ramos AJ and Sanchis V (2006) Ecophysiology of ochratoxigenic *Aspergillus ochraceus* and *Penicillium verrucosum* isolates, Predictive models for fungal spoilage prevention a review, *Food Additives and Contaminants*, **23**(4), 398–410.

Peleg M (2002) Modeling and simulation of microbial survival during treatments with a dissipating lethal chemical agent, *Food Research International*, **35**, 327–36.

Peleg M (2006) *Advanced Quantitative Microbiology for Foods and Biosystems: Models for Predicting Growth and Inactivation*, Boca Raton, FL, CRC Press.

Peleg M and Cole MB (1998) Re-interpretation of microbial survival curves, *Critical Reviews in Food Science*, **38**, 353–80.

Peleg M and Normand MD (2004) Calculating microbial survival parameters and predicting survival curves from non-isothermal inactivation data, *Critical Reviews in Food Science and Nutrition*, **44**, 409–18.

Peleg M and Penchina CM (2000) Modeling microbial survival during exposure to a lethal agent with varying intensity, *CRC Critical Reviews in Food Science*, **40**, 159–72.

Peleg M, Penchina CM and Cole MB (2001) Estimation of the survival curve of *Listeria monocytogenes* during non-isothermal heat treatments, *Food Research International*, **34**, 383–8.

Peleg M, Normand MD and Campanella OH (2003) Estimating microbial inactivation parameters from survival curves obtained under varying conditions – the linear case, *Bulletin of Mathematical Biology*, **65**, 219–34.

Rasch M (2004) Experimental design and data collection, in RC McKellar and X Lu (eds) *Modeling Microbial Responses in Food*, Boca Raton, FL, CRC Press, 1–20.

Ratkowsky DA (1993) Principles of nonlinear regression modelling, *Journal of Industrial Microbiology*, **12**(3–5), 195–9.

Ratkowsky DA and Ross T (1995) Modelling the bacterial growth/no growth interface, *Letters in Applied Microbiology*, **20**, 29–33.

Ross T (1999) Assessment of a theoretical model for the effects of temperature on bacterial growth rate, in *Predictive microbiology applied to chilled food preservation. Proceedings of conference no 1997/2 of Commission C2 (16–18 June 1997) Quimper, France*, Luxembourg, Office for Official Publications of the European Communities, 64–71.

Scott WJ (1937) The growth of microorganisms on ox muscle. II. The influence of temperature, *Journal of the Council of Scientific and Industrial Research, Australia*, **9**, 266–77.

Slade L and Levine H (1988) Non-equilibrium behavior of small carbohydrate-water systems, *Pure and Applied Chemistry*, **60**, 1841–64.

Stephens PJ, Cole MB and Jones MV (1997a) Effect of heating rate on the thermal inactivation of *Listeria monocytogenes*, *Journal of Applied Bacteriology*, **77**, 702–8.

Stephens PJ, Joynson JA, Davies KW, Holbrook R, Lappin-Scott HM and Humphrey TJ (1997b) The use of an automated growth analyser to measure recovery times of single heat-injured *Salmonella* cells, *Journal of Applied Microbiology*, **83**, 445–55.

Stewart CM (2006) Use of microbiological predictive models to estimate product safety and shelf-life, *AACC International World Grains Summit*, 17–20 September, San Francisco, CA.

Stewart CM, Cole MB, Legan JD, Slade L, Vandeven MH and Schaffner DW (2001) Modeling the growth boundary of *Staphylococcus aureus* for risk assessment purposes, *Journal of Food Protection*, **64**(1), 51–7.

Stewart CM, Cole MB, Legan JD, Slade L, Vandeven MH and Schaffner DW (2002) *Staphylococcus aureus* growth boundaries: moving towards more mechanistic predictive models based on solute specific effects, *Applied & Environmental Microbiology*, **68**, 1864–71.

Stewart CM, Cole MB and Schaffner DW (2003) Managing the risk of staphylococcal food poisoning from cream-filled baked goods to meet a food safety objective, *Journal of Food Protection*, **66**(7), 1310–25.

Stringer SC, Webb MD, George SM, Pin C and Peck MW (2005) Heterogeneity of times required for germination and outgrowth from single spores of nonproteolytic *Clostridium botulinum*, *Applied and Environmental Microbiology*, **71**(9), 4998-5003.

Sutherland JP and Bayliss AJ (1994) Predictive modelling of growth of *Yersinia enterocolitica*: the effects of temperature, pH and sodium chloride, *International Journal of Food Microbiology*, **21**, 197–215.

Swinnen IA, Bernaerts K, Dens EJ, Geeraerd AH and Van Impe JF (2004) Predictive modeling of the microbial lag phase: a review, *International Journal of Food Microbiology*, **94**(2) 137–59.

Tanaka N (1982) Challenge of pasteurized processed cheese spreads with *Clostridium botulinum* using in-process and post-process inoculation, *Journal of Food Protection*, **45**(11), 1044–50.

Tanaka N, Traisman E, Plantinga P, Finn L, Flom W and Guggisberg J (1986) Evaluation of factors involved in antibotulinal properties of pasteurized process cheese spreads, *Journal of Food Protection*, **49**(7), 526–31.

ter Steeg PF and Cuppers HGAM (1995) Growth of proteolytic *Clostridium botulinum* in process cheese products: II. Predictive modeling, *Journal of Food Protection*, **58**(10) 1100–08.

ter Steeg PF, Cuppers HGAM, Hellemons JC and Rijke G (1995) Growth of proteolytic *Clostridum botulinum* in process cheese products: I. Data acquisition for modeling the influence of pH, sodium chloride, emulsifying salts, fat dry basis, and temperature, *Journal of Food Protection*, **58**(10), 1091–99.

Texeira A (1992) Thermal processing calculations, in DR Heldman and LB Lund (eds) *Handbook of Food Engineering*, New York, Marcel Dekker, 563–619.

Thomas C and Swayne DE (2007) Thermal inactivation of H5N1 high pathogenicity avian influenza virus in naturally infected chicken meat, *Journal of Food Protection*, **70**(3), 674–80.

Tienungoon S, Ratkowsky DA, McMeekin TA and Ross T (2000) Growth limits of *Listeria monocytogenes* as a function of temperature, pH, NaCl, and lactic acid, *Applied and Environmental Microbiology*, **66**(11), 4979–87.

Toledo R (1998) *Fundamentals of Food Process Engineering*, Gaithersburg, MD, Aspen.

Ueckert J, Breeuwer P, Abee T, Stephens P, Nebe-von-Caron G, ter Steeg PF (1995) Flow cytometry applications in physiological study and detection of foodborne microorganisms, *International Journal of Food Microbiology*, **28**(2), 317–26.

van Boekel MAJS and Zwietering MH (2007) Experimental design, data processing and model fitting in predictive microbiology, in S Brul, S van Gerwen and M Zwietering (eds) *Modelling Microorganisms in Foods*, Cambridge, Woodhead, 22–43.

van Boekel MAJS (2008) Kinetic modeling for food quality: a critical review, *Comprehensive Reviews in Food Science and Food Safety*, **7**, 144–58.

van Gerwen SJ and Zwietering M (1998) Growth and inactivation models to be used in quantitative risk assessment, *Journal of Food Protection*, **61**(11) 1541–9.

van Schothorst M (1998) Principles for the establishment of microbiological food safety objectives and related control measures, *Food Control*, **9**(6), 379–84.

Vermeulen A, Gysemans KPM, Beraerts K, Geeraerd AH, Van Impe JF, Debevere J and Devlieghere F (2007) Influence of pH, water activity and acetic acid concentration on *Listeria monocytogenes* at 7 °C: Data collection for the development of a growth/no growth model, *International Journal of Food Microbiology*, **114**, 332–41.

Vora P, Senecal A and Schaffner DW (2003) Survival of *Staphylococcus aureus* in intermediate moisture foods is highly variable, *Risk Anaylsis*, **23**(1), 229–36.

Walker SJ and Jones JE (1993) Protocols for data generation for predictive modelling, *Journal of Industrial Microbiology*, **12**(3–5), 273–6.

Walker S and Jones J (1994) Microbiology modelling and safety assessment, in A Turner (ed.), *Food Technology International Europe*, London, Sterling Publications, 25–9.

Weast RC, Astle MJ and Beyer WH (eds) (1984) *CRC Handbook of Chemistry and Physics*, 65th ed, p. E–42, Boca Raton, FL, CRC Press.

Whiting RC and Buchanan RL (1997) Development of a quantitative risk assessment model for *Salmonella enteritidis* in pasteurized liquid eggs, *International Journal of Food Microbiology*, **36**(2/3), 111–25.

Whiting RC, Sackitey S, Calderone S, Morely K and Phillips JG (1996) Model for the survival of *Staphylococcus aureus* in nongrowth environments, *International Journal of Food Microbiology*, **31**, 231–43.

Wood R, Nilsson A and Wallin H (1998) *Quality in the Food Analysis Laboratory*, Cambridge, RSC, 16–35.

Zaritzky N, Contreras E and Giannuzzi L (1999) Modelling of the aerobic growth and decline of *Staphylococcus aureus* under different conditions of water activity, pH and potassium sorbate concentration. In *Predictive microbiology applied to chilled food preservation. Proceedings of conference no 1997/2 of Commission C2 (16–18 June 1997) Quimper, France*, Office for Official Publications of the European Communities, Luxembourg, 120–27.

4

Risk assessment and pathogen management

T. Ross and T. A. McMeekin, University of Tasmania, Australia

Abstract: The commercial preparation of safe foods is widely considered to be best managed by 'pre-requisite programs' in combination with the Hazard Analysis Critical Control Point (HACCP) strategy. Formal risk assessment approaches have latterly been promoted and developed by national governments and international organisations for the purposes of establishing objective and fair rules for international trade in foods and for setting food safety management priorities and consequent regulatory actions. These initiatives are likely to affect the management of microbial food safety in industry. This chapter briefly describes the principles of and approaches to microbial risk assessment (MRA). The terminology of MRA, and its origins, are described and its application to managing the risk of foodborne pathogens and their toxins, including its integration with HACCP, are considered. It is suggested that the application of the tools and techniques of MRA in the food industry can assist in elucidating optimal food safety control options, identifying CCPs and specifying their limits and appropriate corrective actions. Thus, rather than replacing HACCP, MRA will help to optimise the HACCP systems of the food industry and individual food businesses to improve product safety and overall public health. Additionally, MRA can facilitate innovation in the food industry because it provides an agreed method for demonstration of the food safety equivalence of alternative processing technologies and formulations. Currently MRA is a high-level, resource- and time-intensive activity. Increasingly, the approaches of MRA are being applied to narrower food safety decisions, involving the development of 'fit-for-purpose' approaches to MRA. Examples are presented of the use of MRA approaches to support decision making by food businesses.

Key words: predictive microbiology, hazard, HACCP, food safety objective, stochastic simulation, modelling.

4.1 Introduction

4.1.1 Food safety management

Management of food safety by end-product testing is time consuming, expensive and often invasive or destructive of the material tested. Moreover, for many food:pathogen combinations, very large numbers of samples are required to provide assurance that a batch of the food is not contaminated at a level higher than that deemed to be consistent with an Appropriate Level of Protection[1] (ALOP) of consumers.

Rather, the preparation of safe foods is widely considered to be best managed by 'pre-requisite programs', e.g. good manufacturing practice (GMP), good agricultural practice (GAP), good hygienic practice (GHP), etc. in combination with the Hazard Analysis Critical Control Point (HACCP) strategy. Identification and analysis of hazard potential is implicit in the HACCP system yet, while there are many hazards and many scenarios imaginable by which those hazards could arise in the food, procedures within HACCP for differentiation of trivial or unlikely hazards from those that pose a serious threat to public health are seldom discussed. Correct identification of the CCPs is essential to enable the business to focus its resources at those steps critical to food safety (Gaze *et al.*, 2002).

Currently, consumer desire for minimally processed, yet *safe*, foods creates an additional paradoxical challenge for the food industry, not least because food must remain 'affordable'. Meeting these competing demands poses a challenge to the food industry, and regulators, alike. Thus, it is important to be able to assess and *quantify* the potential for pathogen contamination of foods and the consequent impact on public health, and to allocate resources and responsibility for pathogen management and/or control where they will have greatest effect in assuring public health.

The process of formal risk assessment can assist in these broader objectives, as well as helping to elucidate optimal control options, to identify CCPs and to specify their limits and appropriate corrective actions to optimise the HACCP systems of the food industry and individual food businesses.

4.1.2 Risk assessment

'Risk assessment' is one element of a formal and objective approach to the understanding of risks associated with specific activities. The overall framework has been termed 'risk analysis', and is considered also to include the elements 'risk communication', 'risk assessment' and 'risk management'. These terms are described more fully in Section 4.2.2.

Risk assessment techniques have been used for decades in many fields of commerce and technology including the insurance industry, financial market

[1]An ALOP is the 'expression of the level of protection in relation to food safety that is currently achieved' (FAO/WHO, 2006).

analysis, budgeting for large construction projects, threats to human and environmental health from industrial developments, and injury risks from mechanical failure of equipment or machinery. More recently the techniques have been applied to biological risks related to quarantine issues, to drinking water safety and, most recently, to the microbiological safety of foods.

Microbiological food safety risk assessment approaches have, to date, been promoted and formalised largely by national governments and international organisations for the purposes of establishing objective and fair rules for international trade in foods and to assist poorer nations along the path to economic development. National governments are adopting formal risk assessment methods for setting (objective) food safety regulations and for prioritisation of programmes to protect public health. Equally, these initiatives and approaches offer benefits to the food industry.

This chapter describes the principles of microbial (food safety) risk assessment (MRA) and considers its application to managing the risk of foodborne pathogens and their toxins[2], including its integration with HACCP. Potential applications of the tools and techniques of MRA in the food industry are discussed.

4.2 Overview of microbial food safety risk assessment

4.2.1 Formal risk assessment applied to foods

As noted above, formal risk assessment approaches have been used in a wide range of fields for decades. Extension of established risk assessment principles to microbial food safety management seems to have been in response to several needs including:

- the need to use available resources more efficiently to minimise food safety risks: this response is, in part, due to perceived failures in existing systems and also the perception by some scientists that, despite the great amount of effort expended to show that certain chemicals in foods *might* be toxic or carcinogenic, microbial hazards *do* make millions of people ill, and kill thousands every year, even in 'developed' nations (CAST, 1994; Mead *et al.*, 1999; Todd, 2008);
- the need for governments to justify the benefit, compared to the cost, of introduction of new food safety regulations;
- the need for objective, 'science-based', rules for international trade in foods.

A significant factor leading to the rapid increase of quantitative risk assessment methods in all fields was the availability of software for personal

[2] For simplicity, throughout the remainder of this chapter the term 'pathogen' will be used to denote pathogens and/or their toxins.

computers that enabled sophisticated probability, or 'stochastic', modelling techniques to be performed by users without high-level mathematical and programming skills. This brought quantitative risk assessment within reach of many fields of study (Morgan, 1993). Another factor that facilitated the introduction of risk analysis of foodborne pathogens was the development of predictive microbiology, a tool necessary for the 'exposure assessment' (see Section 4.2.2) stage in risk assessment (Buchanan and Whiting, 1996; Elliot, 1996; Foegeding, 1997; van Gerwen and Zwietering, 1998; Marks, 2008).

Despite the promotion of risk assessment methodology by various national governments and international trade and political organisations (NRC, 1983; FAO/WHO, 1996; Kindred, 1996; Buchanan, 1997; Craun *et al.*, 1996; McNab *et al.*, 1997; PCCRARM, 1997; ILSI, 2000; Lammerding, 2007), there are no universally agreed methods nor terminology for microbial food safety risk assessment. The Codex Alimentarius Commission[3] ('Codex'), however, have documented principles for risk assessment within microbial food safety management (CAC, 1999) and, at the request of Codex, the World Health Organisation (WHO) and Food and Agriculture Organisation (FAO) are developing and documenting approaches for microbial food safety risk assessment through a series of expert consultations, collectively entitled the Joint FAO/WHO Expert Meetings on Microbiological Risk Assessment (JEMRA). More information about JEMRA and the reports of those meetings can be obtained, by download, from the WHO or FAO websites: (http://www.who.int/foodsafety/micro/jemra/en/index.html and http://www.fao.org/ag/agn/agns/jemra_index_en.asp).

For conduct of valid microbiological food safety risk assessment Codex (CAC, 1999), emphasised various needs to be met including:

- a sound scientific basis;
- the explicit consideration of the dynamics of microbiological growth, survival and death in foods;
- consideration of the complexity of the interaction (including sequelae) between human and agent following consumption as well as the potential for further spread;
- transparency and documentation concerning methods of analysis, assumptions made, data used;
- estimation, or description, of uncertainty in the risk assessment and its consequences for the estimate of the risk;

[3] The Codex Alimentarius Commission was created in 1963 by FAO and WHO to develop food standards, guidelines and related texts such as Codes of Practice under the Joint FAO/WHO Food Standards Programme. Codex standards and guidelines are developed by committees of the Commission, which are open to all member countries. The objectives of the programme are to protect the health of the consumers, to ensure fair international food trade practices and to promote coordination of all food standards work undertaken by international governmental and non-governmental organisations.

- (functional) separation between risk assessment and risk management personnel.

The Codex Guidelines also recommend the use of a structured approach that includes the elements 'hazard identification', 'hazard characterization' (sometimes called 'dose–response assessment'), 'exposure assessment', and 'risk characterization'. These terms are explained in the following section.

4.2.2 Definitions and terminology

Risks and hazards

In common usage, the words 'risk' and 'hazard' or 'risky' and 'hazardous' are often used interchangeably. In the formal discipline of risk assessment, however, 'risk' and 'hazard' have specific meanings. Codex defines a hazard as: 'a biological, chemical, or physical agent in, or a condition of, food with the potential to cause an adverse health effect', and risk as: 'a function of the probability of an adverse health effect and the severity of that effect, consequential to a hazard(s) in food'.

In simple terms, a *hazard* is considered to be the possibility of an undesirable effect, e.g. that a food might contain viable pathogens at the time it is eaten and cause illness. *Risk*, however, is the combination of the fact that a hazard exists, the *likelihood* of it happening, and the *severity* of the consequences if it does. Thus, *the concept of risk differentiates trivial and serious hazards*. For example, a hazard that is very unlikely to occur and that, even if it did, would cause only mild discomfort, is relatively trivial in comparison to a hazard that is likely to be present, or a hazard that even if unlikely would cause serious harm to a consumer, and which would require active management and control.

Risk analysis

Codex describes risk analysis as a process involving the following three aspects:

- risk assessment;
- risk management;
- risk communication.

One interpretation of the interplay between these three aspects of risk analysis is shown in Fig. 4.1. Whereas *risk assessment* is the scientific evaluation of the probability of occurrence and severity of adverse health effects resulting from human exposure to a foodborne hazard, *risk management* is concerned with how to use the information from the risk assessment process to manage the risk to acceptable levels and in a manner that is acceptable to those affected by the risk, termed 'stakeholders'. Thus, risk managers may also need to consider, e.g.:

- what is technologically achievable;

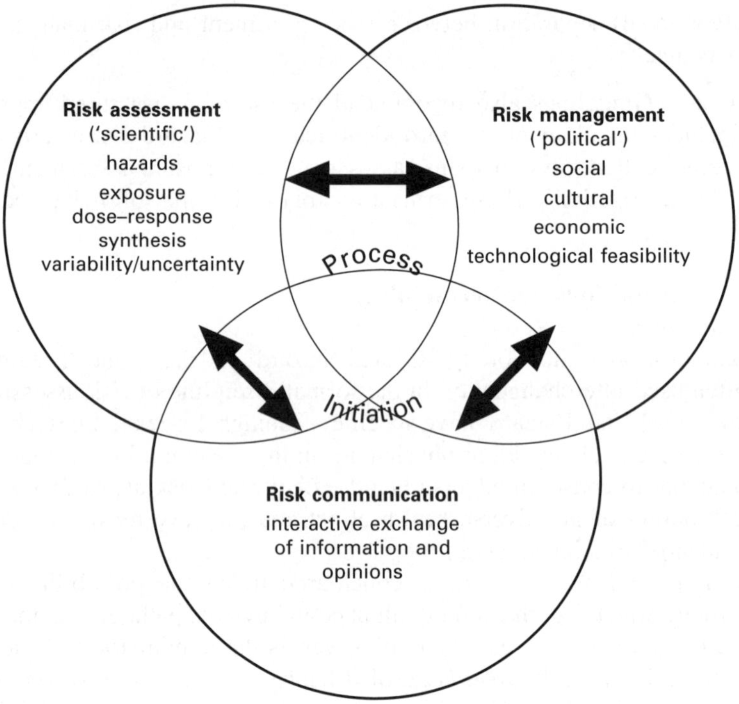

Fig. 4.1 A schematic representation of the interplay between risk management, risk communication and risk assessment. The risk assessment may be initiated from any source, but its conduct will typically be under the control of a risk manager who will co-ordinate the process, oversee exchange of information and turn the results of the assessment into a plan of action. The diagram also emphasises that the risk analysis process is not sequential, but interactive and iterative. (Modified from McNab *et al.*, 1997)

- what is sociologically acceptable;
- the cost, compared to the benefit, of proposed actions to control the risk, etc.

To determine the 'best' risk management strategies, risk managers need to understand the perspectives of stakeholders, e.g. their willingness to accept some risk or to accept some responsibility for control of the risk, so as to reduce cost or to increase choice. Equally, for this to occur, stakeholders may need to be informed about the source of the risk and options, and limitations, for its control. Thus, risk communication involves the exchange of information and opinions between risk managers, risk assessors and stakeholders both to understand the risk, the perceptions of risk, and acceptable and unacceptable options for risk management and to communicate the basis of risk management decisions and designated responsibilities for risk management among all stakeholders.

Risk assessment

As noted above, risk assessment is considered to include the elements of 'hazard identification', 'dose–response assessment', 'exposure assessment' and 'risk characterisation'. These terms are described briefly below.

Hazard identification involves the presentation and assessment of the evidence for a causal relationship between a pathogenic agent, an illness and a food as a vector of that illness. In MRAs the hazard is often unambiguous, e.g. in comparison to the risks posed by chemical toxins. Many microbiological hazards are already known and the relationship between human illness, the pathogen and a particular type of food as a vehicle is also well known from clear epidemiological and medical evidence, e.g. in the scientific and medical literature. There are also a number of decision support 'tools' to assist in determining whether a pathogen is, or could be, an important hazard in a given food/food process combination. These include various semi-quantitative scoring systems, decision trees, and expert systems (see e.g. ICMSF, 1996; Notermans and Mead, 1996; Todd and Harwig, 1996; van Gerwen *et al.*, 1997; van Schothorst, 1997), such as the one shown in Fig. 4.2 that may also be used for hazard analysis as part of HACCP plan development. Decision trees enable the experience of others to be shared and can assist in decision making by presenting a structured series of questions relevant to the decision being made. In essence, the structured approach of risk assessment offers the same assistance for more complex decision processes.

Exposure assessment seeks to estimate:

1. how often consumers become exposed to a hazardous agent in a food or group of foods, and to determine the dose ingested;
2. how heavily contaminated that food is;
3. how much of the food is eaten, e.g. per eating occasion, or over a period of time, etc.

The information to answer the first two questions above is rarely available directly and, usually, must be inferred from knowledge of the contamination level and frequency at some earlier time in the history of the product (e.g. from surveys or company quality analysis records for microbial loads, or incidence of contamination at the point of production) using predictive microbiology models. It requires knowledge of the changes that the product has undergone since the point of contamination, whether through processing, storage, transport or preparation for eating. This includes quantification of inactivation, concentration, dilution, recontamination or growth that will affect the frequency and concentration of pathogens in foods and their ingredients. Data on the duration and environmental conditions (e.g. temperature, product formulation, etc.) during processes and handling, and how this affects the fate of pathogens in foods, is required to assess exposure.

Translation of this knowledge into an estimate of the numbers of pathogens in the product at the time of consumption is also required. This task is the province of predictive microbiology, which was discussed in Chapter 3.

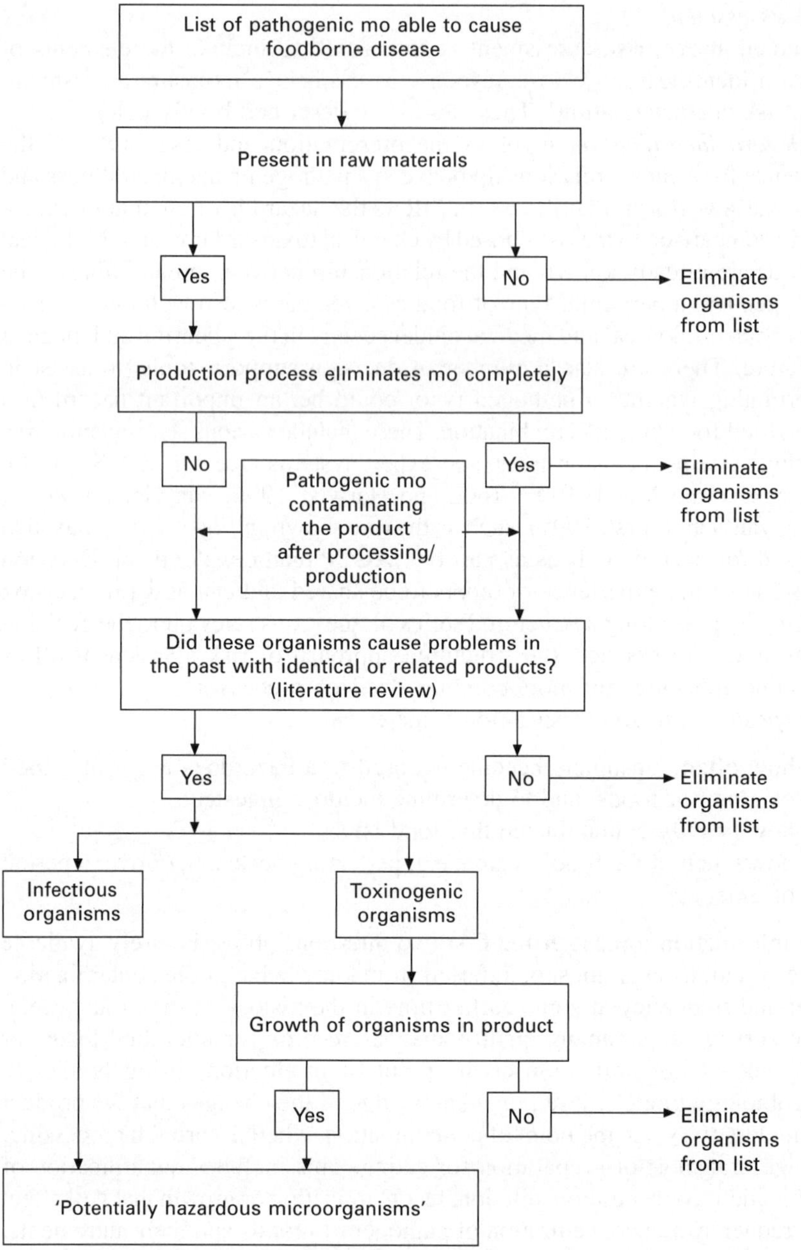

Fig. 4.2 A decision tree to aid the identification of microbial hazards in finished foods. (Reproduced from Notermans and Mead, 1996). mo = microorganisms.

Many commentators have noted the synergy between predictive microbiology and quantitative microbial risk assessment (Buchanan and Whiting, 1996, 1998; Foegeding, 1997; Lammerding, 1997; McNab, 1998; van Gerwen and Zwietering, 1998; Lammerding and Fazil, 2000; McMeekin *et al.*, 2000; Ross and McMeekin, 2003; Marks, 2008). Nauta (2005) introduced to MRA the concept of mathematical 'modules' that could be used for modelling different processes in food production, processing and distribution capable of affecting contamination frequency and concentration. He envisaged that the modules would encompass microbial growth and inactivation, mixing and partitions, and also include the effects of cell clustering, and that they could be combined in different ways to model more complex food processing operations and supply chains. Other studies (see Section 4.3.3) have focused on development of techniques and data for modelling of cross-contamination.

Hazard characterisation, sometimes also called 'dose–response characterisation' or 'dose–response assessment', attempts to relate the probability and severity of illness to the dose (i.e. total number) of the pathogen ingested. For foodborne pathogens, the dose is usually calculated from the concentration of the hazard in the product at the time of eating multiplied by the quantity of the food consumed in a single meal or serving.

There is a paucity of published dose–response data, a situation that is compounded by its relevance (or lack of it) to normal human populations. The few human challenge studies that have been performed have usually involved healthy adult males (typically prison inmates or soldiers) who would be expected to have higher intrinsic resistance than the young, old, pregnant or immunocompromised (YOPI) groups. Resulting from a series of detailed risk assessments, developed through JEMRA, dose–response data for *Listeria monocytogenes*, *Salmonella*, *Campylobacter* and *Vibrio* spp. have recently been reviewed and, in some cases, dose–response models developed. Dose–response information for enteric pathogens was summarised by Teunis *et al.* (1996) and dose–response relationships were derived by 'expert elicitation' for 13 foodborne pathogens by Martin *et al.* (1995). The latter authors concede that the information collected from the expert elicitation process cannot substitute for the scientific data needed to accurately estimate dose–response relationships and their variance. Martin *et al.* (1995) considered, nonetheless, that 'given the difficulty of collecting experimental data on harmful pathogens, this information can provide a characterisation of scientific judgements of relative risk to the population from these microbial pathogens'. Strachan *et al.* (2005) developed a dose–response model for *E. coli* O157:H7 from foodborne and environmental outbreak data, and Teunis *et al.* (2005) extended an existing dose–response model for *Campylobacter jejuni* by incorporation of data from an outbreak. Buchanan *et al.* (2000) and FAO/WHO (2003) provide detailed discussion of the approaches to, and problems of, preparing dose–response models for microbial pathogens in food and water.

Risk characterisation is the synthesis of the exposure assessment and dose–response characterisation, i.e. the combination of all the data gathered and their interrelationships to generate an estimate of the risk of human illness upon consumption of the specified food. The methods that can be used vary from semi-quantitative scoring matrices to fully quantitative, stochastic simulation models (see Section 4.3). Additionally, the risk can be expressed in various ways, such as:

- the risk of illness per serving of the food;
- change in risk *relative to the status quo*, e.g. due to some proposed risk management action, or new process;
- the estimated number of infections in a defined population due to the presence of the hazard in a specific food, or group of foods;
- the overall impact of the disease, allowing for differences in severity of infection from various microbial hazards, e.g. summarised as 'disability adjusted life years (DALY)' (Sassi, 2006). The DALY is primarily a measure of disease burden and is a useful method for including estimates of disease severity, rather than disease incidence, into a risk assessment and to enable comparison of risk from different hazards.

4.3 Approaches to microbial food safety risk assessment

The main purpose of risk assessment is to provide information and insights to support decisions. In MRA, those decisions can take a multitude of forms, e.g. which foods/food processes represent the greatest risk and, hence, require more regulatory attention; which sector of the food supply chain has greatest responsibility for the safety of a particular product; whether education of consumers is a more effective risk management strategy than requiring 'zero tolerance'; whether to invest in a particular food processing technology to increase product integrity, etc.

The nature of the risk management question often has a strong influence on the risk assessment approach taken. One of the main lessons learnt from risk assessments undertaken to date is the need for clear definition of the risk assessment question by the risk manager so that the risk assessor can develop the most efficient risk assessment approach to provide useful advice. Thus, while the Codex model emphasises functional separation between risk assessors and risk managers to maintain objectivity, newer paradigms also emphasise the need for ongoing communication between the risk manager and risk assessor to ensure that the risk assessment developed does address the needs of the risk manager for the risk question. An expert consultation on the interaction between assessors and managers of microbiological hazards in food concluded that 'communication [between risk managers and risk assessors] has to occur frequently and iteratively while striving to ensure scientific integrity and achieve freedom from bias in risk assessments' (FAO/WHO, 2000).

For example, to rank the risk of salmonellosis from different foods will require a different approach and different types of data than to estimate the number of consumers who will contract salmonellosis from all poultry products. In a *L. monocytogenes* risk assessment conducted by US authorities (CFSAN/FSIS, 2003), the risk management question was to establish the relative contribution to listeriosis in the USA from 20 different categories of food. In this case definition of a dose–response model was not completely necessary and, in each iteration of the model, the dose–response model was 'calibrated' to match the observed incidence of listeriosis in the USA, i.e. in effect a different dose–response function was used for each iteration of the model. In the FAO/WHO (2004) risk assessment of *L. monocytogenes* in ready-to-eat foods, some risk management questions could be answered without recourse to a detailed exposure assessment because they centred on the dose–response relationship. Conversely, in many cases industry-based microbiological risk assessment questions will focus on methods to limit exposure to consumers, or to satisfy some regulatory criterion, and will not need detailed hazard characterisation.

Lammerding (2007) notes that some food safety management questions can be answered without recourse to risk assessment methods at all, stating that 'MRA is not necessarily the approach needed to address all food safety management questions', citing as an example the use of epidemiological data, e.g. to gauge relative disease incidence from different foodborne pathogens.

Risk assessment can be very time- and labour-intensive, and the risk assessment approach should also be consistent with the scale of the problem to be resolved or decision to be made. For example, Lammerding (2007) observed that, from an industry perspective, 'the larger the market opportunity or economical benefit or the more complex the aspect of food safety is, the more relevant it may be to use MRA in new product or process design'. In other situations, appropriate answers may be obtained by more modest approaches than the holistic farm-to-fork approach that is currently most prevalent in the published literature. The scale of the risk assessment should be commensurate with the magnitude of the risk and the scope of the MRA sufficient to resolve the risk management question(s), within the resources (time, expertise, financial) available. Definition and documentation of scope and resources limitations are identified by Codex (CAC, 1999) as an important part of an MRA. This parallels the first steps in HACCP plan development which requires definition of the specific product, including composition, processes that will be applied, distribution conditions, required shelf-life, expected conditions of use by the end-user and information about the target consumer group for the product (Gaze *et al.*, 2002). A decision tree for determining whether a risk assessment is necessary or feasible is shown in Fig. 4.3.

Various approaches to risk assessment are described, and examples identified, in this section. Lammerding (2007) also provides a good overview of different approaches to the estimation of risk, and their suitability for different types of risk assessment question. Whatever the approach, the key

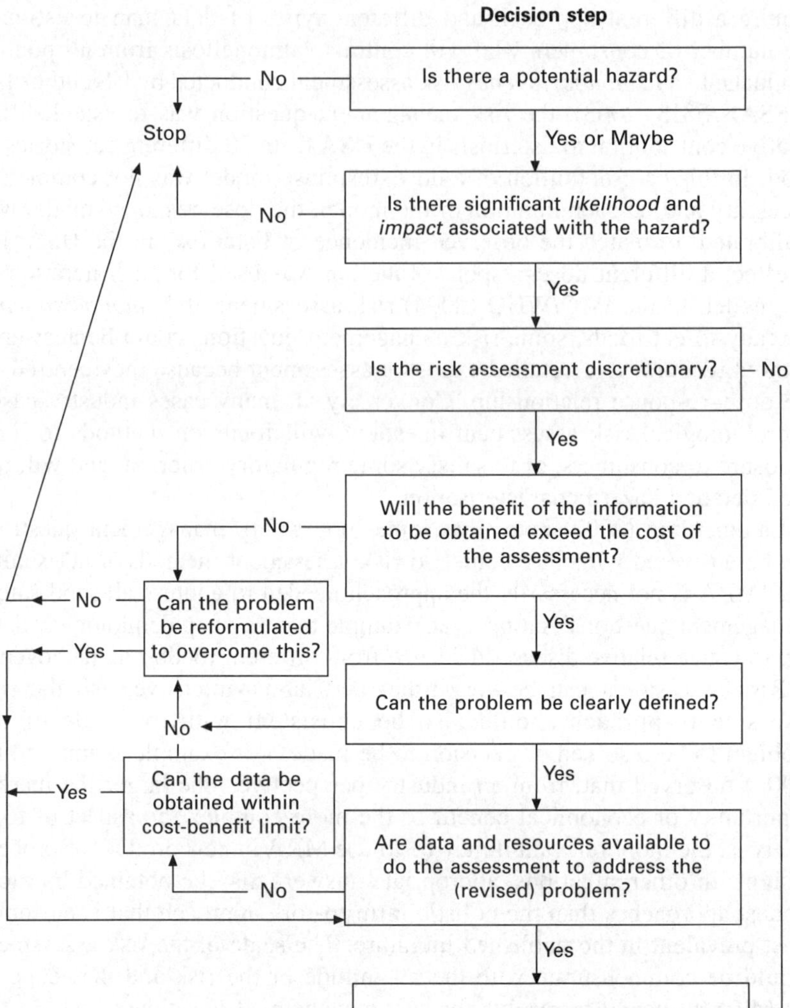

Fig. 4.3 A decision tree for microbial risk assessment projects. The same decision tree process can be applied for the first stage of the HACCP process.

principles of reliable risk assessment are that it is systematic, structured, objective, transparently documented and presented, and that it considers the influence of variability and uncertainty on the risk estimate, to provide support to the decision making process.

4.3.1 Conceptual model

Hazards can arise at any stage during the harvest, processing, distribution and preparation of a food. Furthermore, levels of microbial hazards, in particular, are likely to be affected by subsequent handling steps. The system under analysis is a continuum, and hazards at one point in the chain cannot be considered in isolation from the system as a whole. To assess risk it is necessary to *understand* where in the supply chain the risks arise, how the risk changes, and the interactions between risk-affecting factors.

The food harvest/processing/distribution system that delivers foods to consumers can be described in a number of ways, but it is often easiest to start the process of description and characterisation of the system using diagrams such as flow charts to show the sources and flow of ingredients into products, the handling and processing of product, and the origin of hazards and the relationships and operations that can change the risk during the life of the product. 'Influence diagrams' are another way of representing these relationships, and are inherent in the architecture of some stochastic simulation modelling software packages which are discussed further below. Fault trees and event trees (Kumamoto and Henley, 1996) can aid critical thinking in this phase, by highlighting the factors that are likely to be important inputs to the final risk, and that need to be included as explicit steps in the risk assessment. Again, these steps parallel those in the development of HACCP plans. Failure Mode and Effect Analysis (FMEA; see also Section 4.4.1) can help to identify critically important pathways of contamination and mishandling that could endanger customer safety (as well as business reputation). FMEA has been applied to study microbial food safety risk, e.g. industrial salmon processing operations (Arvanitoyannis and Varzarkas, 2008).

This description of the process has been termed a 'conceptual model'. For microbial food safety risk assessment, a formulation of the conceptual model that traces the evolution, magnitude and likelihood of risk in the product from harvest to consumption was developed and described as a 'process risk model' (Cassin *et al.*, 1998).

4.3.2 Approaches to estimation of risk

Qualitative schemes

There are a number of types of 'qualitative' approaches to risk assessment. The most widely observed are risk matrices, or simple qualitative scoring schemes, in which a series of descriptors of hazard severity and likelihood of occurrence are chosen and combined to generate an estimate of risk. A relatively simple example of a risk matrix is shown in Fig. 4.4. In that approach, the user selects a category of hazard severity, and a category of hazard likelihood. The combination chosen is assigned a risk category, which has additional text to provide interpretation of the magnitude of risk. Obvious limitations of this approach are the subjectivity of selection of the hazard severity and hazard likelihood categories, unless very specific instructions are given on the choice of categories as part of the matrix system. Usually, such descriptions have to be mathematically based to be able to be used objectively. In that case the tool then becomes, effectively, semi-quantitative (see below).

		RISK ESTIMATE			
Likelihood	Highly likely	Low	Moderate	High	High
	Likely	Negligible	Low	High	High
	Unlikely	Negligible	Low	Moderate	High
	Highly unlikely	Negligible	Negligible	Low	Moderate
		Marginal	Minor	Intermediate	Major
		CONSEQUENCES			

Fig. 4.4 An example of a simple risk assessment matrix from OGTR (2005) that considers both likelihood and severity of outcome using a series of qualitative descriptors. In this scheme a negligible risk is considered to be insubstantial with no need to invoke actions for mitigation. A low risk is considered to be minimal but may require additional actions, beyond normal practices, to mitigate the risk. A moderate risk would cause marked concern and will require mitigations that need to be demonstrated to be effective. A high risk is considered to be unacceptable, unless mitigations are highly feasible and effective.

Truly qualitative approaches to risk *characterisation* are also possible. This type of qualitative risk assessment will ideally be based on numerical data for exposure assessment and hazard characterisation (FAO/WHO, in press). However, rather than analysing the data mathematically, the risk assessment will use descriptive words or phrases that cannot be directly related to specific quantitative measures of risk. Nonetheless, there will need to be clear description and discussion of the evidence and the logic used to reach conclusions based on that evidence. Qualitative risk assessment is not, however, simply a literature review or description of all of the available information about a risk issue: it must also arrive at some conclusion about the probabilities of outcomes for a baseline risk and/or any reduction strategies that have been proposed (FAO/WHO, in press). Qualitative risk characterisation is well discussed and exemplified in Wooldridge (2008) and FAO/WHO (in press).

Semi-quantitative

Semi-quantitative risk assessment is an approach that is often used for prioritising risks from diverse sources, by focusing on major determinants of risk. It assists in management of the entire 'risk portfolio'. Thus it is an approach to 'risk profiling'[4]. From the results of a semi-quantitative risk

[4] Risk profiling: a risk profile is intended to provide contextual and background information relevant to a food/hazard combination so that risk managers can make decisions and, if necessary, take further action. Risk profiles are useful to regulators when developing requirements for risk-based food control programmes, and for those in the food industry to help understand the microorganisms they need to control in food processes and their associated public health significance.

assessment, the risk manager might decide to undertake a detailed, quantitative, risk assessment on specific, high-risk, foods/processes.

Similar to the risk matrix approach, several schemes (e.g. Corlett and Pierson 1992; NACMCF, 1992) suggest a two-step process for risk assessment in the context of HACCP. In those approaches hazard characteristics and risk-contributing factors are formulated into statements, and the number of positive answers to the questions determines the risk ranking. Huss *et al.* (2000) adopted and adapted this approach to develop a scheme for assessment of foodborne disease risk, as determined by the number of positive answers to the following questions.

1. Is there epidemiological evidence that the particular type of food product has been associated with foodborne disease many times, or with very serious disease?
2. Does the production process *not* include any CCPs for at least one identified hazard?
3. Is the product subject to potentially harmful recontamination after processing and before packaging?
4. Is there substantial potential for abusive handling in distribution or in consumer handling that could render the product harmful when consumed?
5. Is there potential for growth of pathogens in the product?
6. Is there *no* terminal heat process after packing or during preparation in the home?

If the product/pathogen combination is positive for four or more of the above questions the risk is considered to be high. Less than four positive characteristics is considered to be a low risk. The ICMSF tables (ICMSF, 1986, 2002) for determining the required rigour of a food safety sampling plan, i.e. the '15 cases', also represent a simple risk matrix that considers hazard severity and the likelihood of pathogen increase or decrease between the time of manufacture and consumption.

There are a number of deficiencies in the Huss *et al.* (2000) approach when used to assess risk. First, there is no differentiation of likelihood and severity. Second, the scheme does not discriminate well between different levels and sources of risk. A further difficulty in answering the question about disease severity is that severity varies with host susceptibility factors and, potentially, with dose ingested. The schemes are also implicitly geared toward bacterial pathogens that cause infections or produce toxins. This can lead to inconsistencies. For example, temperature abuse is only relevant if a pathogen could grow in the product to an infectious or toxic dose level, but is irrelevant to the risk if the answer to the next question ('is there potential for growth?') is negative. Other risks, e.g. due to algal toxins or viruses, are not well predicted by the above scheme. Each factor is considered to have equal importance, which can also lead to inconsistencies. For example, for the risk due to *L. monocytogenes* in cold smoked salmon, the score would be: – + + + + +, suggesting that this is a very hazardous product even though there is practically no epidemiological evidence to implicate this product/pathogen combination (FAO, 1999). Conversely, enteric viruses in oysters are a well-documented source of disease. Using the above scheme, the score would be: + + – – – +, suggesting that this is less of a risk. Overall, such schemes are really intended to determine whether a perceived microbial hazard is worthy of consideration in a HACCP plan, rather than determining whether an unacceptable risk exists or not.

Building on approaches such as those described above, Ross and Sumner (2002) developed a risk calculation tool[5] to aid determination of relative risks from various product/pathogen/processing combinations. The tool can be used by those without extensive experience in risk modelling to provide a first estimate of relative risk and for food safety risk management prioritisation. The model is intended to be simple, generic and robust, to include all factors that could affect food safety risks, and to be consistent with the formal risk assessment approach described earlier. In reality, reliable use of the model requires a relatively high level of knowledge about the pathogen and the process. The model is presented and automated in spreadsheet software in which the user selects descriptive answers to a series of questions on the spreadsheet. Many of the questions are similar to those in the Huss *et al.* (2000) scheme. A reason for using spreadsheet software as the platform is that the software is widely available thereby increasing accessibility to the model. The underlying conceptual model translates these descriptors into numbers and, using mathematical relationships and logical tests that define the conceptual model, into a range of risk estimates for the food:hazard combination and population considered. Some factors, such as processing or cooking, are modelled to completely eliminate the risk and have been assigned a value of zero. The model also recognises, however, that even if a process completely eliminates the risk, re-contamination may occur and reintroduce the risk. A particular advantage of spreadsheet software is that it enables logical tests to be part of the risk calculations. The spreadsheet, while providing estimates of risk, also helps to focus attention on the interplay of factors that contribute to the risk of foodborne disease, and can be used to explore the effect of different risk reduction strategies.

The scheme is a useful training tool and an example of an approach, rather than a definitive model. While it can be criticised on several grounds, it contains all elements required to estimate risk from foods and can be modified to suit the specific question of the risk assessor or risk manager. However, the equations and logical tests in the model are accessible by users and can be modified to suit other food safety risk assessment questions. Tools such as these can help managers to think about how risks arise, and change,

---

[5] The spreadsheet, designated 'Risk Ranger', can be downloaded from: http://www.foodsafetycentre.com.au/riskranger.php

and to decide where interventions might be applied with most likelihood of success.

While stochastic simulation modelling approaches (see below) are the option currently preferred by most food safety risk assessors, some workers who have attempted fully quantitative risk assessments have been frustrated by the lack of data to support that approach[6]. The process also usually requires considerable human resources and expertise, which may not be available in all cases, particularly for food businesses but even for some developing nations. As such, there is active interest in simpler methods for risk assessment. This, in part, has led to the introduction of the concept of the risk profile, described earlier, as a pre-cursor to full risk assessments. Risk assessment methods should be appropriate for the scale of the resources available and the scale of decision to be made. Lammerding (2007) provides discussion of different approaches to risk assessment as appropriate to different sorts of questions and tasks.

Semi-quantitative risk assessment methods offer the advantage of generating generic estimates of risk that can be quickly derived and that can be used to prioritise risks from different sources. They have often been used in this way, e.g. the risk ranking spreadsheet has been used to prioritise risks in the seafood industry (Sumner and Ross, 2002; Sumner *et al.*, 2004), the Australian meat industry (Pointon *et al.*, 2006) and, in a form subtly modified to accommodate egg-specific data, used to produce a profile of safety of eggs in Australia (Daughtry *et al.*, 2005). Risk matrices have been used in many applications including, e.g. in import risk analyses (Bartley *et al.*, 2006) using the risk matrix scheme of Biosecurity Australia (2002) which considers six levels each of 'likelihood' and 'consequences' to produce six categories of risk.

Despite the apparently much greater emphasis on quantitative risk assessment methods, both Codex (CAC, 1999) and OIE (1999) consider that qualitative and quantitative risk assessments have equal validity. Many commentators consider, however, that the quality and reliability of inferences derived from risk assessment are much greater when fully quantitative 'stochastic' or 'probabilisitic' models are used. Cox (2008) presents a detailed analysis and criticism of the reliability of risk matrices and strongly advocates fully quantitative approaches.

Quantitative approaches
In principle, the entire system being analysed – and the relationships between all variables that affect the risk of foodborne illness – can also be described

[6] In response to these needs various organizations have sought to co-ordinate collection and collation of relevant data to support microbiological risk assessments. The International Life Sciences Institute (Walls, 2007) provide useful background and discussion of the data needs of MRA and propose a framework for the identification of relevant data needed for decision support and means of making such data more useful and accessible.

mathematically, i.e. using algebraic notations and equations and numbers rather than words or diagrams. By substituting data, or values based on expert opinion, for the variables in the model the equations describing the origin and amount of the pathogen in the food, and the factors that impinge upon it, can be solved to yield a numerical estimate of the risk. For repeated execution of the model and calculation of risk under different scenarios, i.e. 'what-if scenario' analysis, it is often convenient to express the model in spreadsheet software to allow automation of the calculations. The risk spreadsheet model described above is a simple example.

Typically, the factors in a food production, processing and distribution system that affect risk do not have single, fixed, values but are characterised by a range of possible values. When estimating risk, a decision has to be made regarding the value of the variables to be used for estimation of the risk.

One approach is to describe the range of possible values by a single representative value, e.g. the average. In this case, the model is said to be deterministic. The alternative is to include the effect of all possible values and combinations of values for the various risk-affecting factors. This approach is termed 'stochastic' or 'probabilistic'. Both approaches are considered below.

Deterministic models

When adopting deterministic methods, the most obvious approach is to characterise the variable by its most common value (mode), or its average value. Thus, the mathematical model would produce an estimate of the risk characterised by the most commonly occurring scenario. However, this might ignore important, but unusual, circumstances. For example, exposure to low levels (e.g. < 100 cfu/g) of *L. monocytogenes* in ready-to-eat foods is common (CFSAN/FSIS, 2003; FAO/WHO, 2004). For the average person, the likelihood of serious illness from this exposure would be predicted to be virtually nil. It is known, however, that the risk to susceptible members of the population is considerably higher, both in terms of the likelihood of infection from low doses, and also the severity of that infection. For example, epidemiological data presented in FAO/WHO (2004) suggests that transplant recipients are between 500 and 2500 times more likely to acquire foodborne listeriosis than an otherwise healthy consumer. Equally, it is observed that foodborne disease outbreaks usually result when several sets of unusual circumstances occur simultaneously (Buchanan and Whiting, 1998). Thus, a risk assessment in which only average values were used to represent the variables could severely under-estimate the true risk. Accordingly, some risk assessors have characterised risk-affecting variables by values that encompass more of the possible scenarios, e.g. 90th or 95th percentiles, but this can lead to a problem described as 'compounding conservatism' (Cassin *et al.*, 1996), in which the combination of a sequence of conservative estimates leads to an overall assessment which is *too* conservative, i.e. which greatly *over-estimates* the risk.

Variability and uncertainty

The use of single-value estimates of variables determining the probability and severity of an adverse event cannot give a complete assessment of risk (Whiting and Buchanan, 1997; Cox, 2008). Risk managers who use the results of the risk assessment will also need to have an understanding of the level of certainty in the model predictions, e.g. to understand the range of possible outcomes and the likelihood of any particular outcome.

Confidence in MRA model outputs can be affected by two sources, uncertainty and variability, and explicit consideration of the consequences of uncertainty and variability in MRA is recommended. Uncertainty refers to information that is required for completion of the assessment but that is not available and has to be assumed, or inferred, e.g. simplification of complex processes into mathematical models; small sets of scenarios are generalised to all scenarios of importance; best guesses of ranges of values where primary data are simply not available. Identification of critical data needs can facilitate decisions by leading to research that specifically fills that need and clarifies the risk management decision.

Variability is an inherent property of some systems. There is natural variability (heterogeneity) among the constituents of a population, whether people or numbers of pathogens in a sample, in strain virulence, in the susceptibility of consumers and their eating patterns, etc.

To provide a measure of prediction confidence, many risk assessors advocated stochastic modelling techniques for risk assessment that take into account variability to reflect both the range and likelihood of possible outcomes arising from all possible combinations of risk-affecting factors. Stochastic models can also be used to model the importance of uncertainty. Until the advent of stochastic simulation modelling software, most public health risk assessments involved the assumption and combination of a series of conservative, average and worst-case values to derive a point estimate that was presumed to be conservative and protective of public health.

'Stochastic' or 'probabilistic' modelling

As noted above, in reality the values of factors that affect microbial food safety risks can take a range of values, some of which are more likely to occur than others, i.e. they form a *distribution*. The Normal distribution is a well-known example. Distributions are all characterised by a range and a most probable value, but they can have a variety of shapes, e.g. whether they are symmetrical around the mode, whether the distribution is low and flat around the mode, or strongly peaked at the mode, etc. Many can be described by a unique mathematical equation, and those mathematical equations can be used instead of discrete values for the variables in the mathematical model. The risk is then calculated using the model with distributions, rather than discrete values, representing those variables.

The answer obtained by solving a stochastic mathematical model is called the *explicit* solution. The explicit solution is itself a distribution of values, which is based on all the possible combinations of circumstances contributing to the risk and thus shows the range of possible outcomes, as well as the probability of each of those outcomes. Each combination of factors can be considered a 'scenario' (i.e. one set of circumstances affecting the presence and load of the hazard in the food), and the overall outcome of the model is based on a 'scenario set'. The explicit solution offers a complete 'picture' of the range of consequences and likelihood of all scenarios, and provides much more insight than a calculation based on average values. In most cases, however, the calculations required to reach the explicit solution become so complicated so quickly that they could not be solved for anything but the simplest models.

Instead, software is used to analyse complex systems or processes for which explicit mathematical models do not exist or are difficult, if not impossible, to solve. Stochastic simulation software – e.g. @Risk (Palisade Corporation, Ithaca, NY, http://www.palisade.com), Crystal Ball (Oracle Corporation, Denver, CO, http://www.oracle.com/crystalball/index.html), Analytica (Lumina Decision Systems Inc, Los Gatos, CA, http://www.lumina.com), etc. – automates calculation of possible combinations of factors by calculating the answer many times sequentially. Each time is called an *iteration* and represents one possible scenario and outcome. Typically tens, or hundreds, of thousands of iterations are performed and the results are calculated and collated. At each iteration a value is selected from each variable range (at random but according to the probability distribution describing that variable), and the outcome is evaluated for that set of circumstances. All of those values are collated to generate a distribution of possible outcomes, i.e. there is a range of outcomes, some of which are predicted to occur more often than others. This technique is called Monte-Carlo simulation. Stochastic simulation models provide much more insight about the possible outcomes than estimates based on sets of single values. As such, they can lead to better understanding of the risk and potential risk management decisions.

Stochastic simulation modelling tools offer other analytical functions that enable identification of steps where most 'risk' arises in a process or system, or assessing which risk management actions would have the most effect on reducing risk to consumers. As it goes through the iterations of the model the software records the relationship between the magnitude of each input value and the magnitude of the output. If an input value usually increases in parallel with the risk estimate, or usually moves in an opposite 'direction', it suggests that the input strongly affects the output, and is thus a critical factor in determining the risk. This process is called sensitivity, or 'importance' analysis, and is a powerful tool within simulation software for identifying those variables in the model that most affect the risk, and identifying them as potential targets for process control and risk management. (It should be noted, however, that sensitivity of the model can be affected by the range of values that a variable is allowed to take in the model, i.e. a factor may be judged unimportant simply because it does not vary widely. For example, temperature will be very important to the growth of pathogens and hence to the predicted risk but if temperature is always controlled within narrow limits it may not be an important variable in that situation.)

'Scenario analysis' can identify combinations of risk-affecting factors that cause certain outcomes. This can be used to identify particularly favourable, or unfavourable, combinations of factors, and can aid in development of risk management strategies.

Stochastic simulation modelling software has made it possible to model very complex and variable systems relatively easily. However, while it is easy to develop spreadsheet models, it is also possible to introduce errors that are not immediately obvious, i.e. to develop models that are mathematically or logically incorrect: to use stochastic simulation modelling software *correctly* its operation and limits must be understood. A common problem is to fail to include in the model relationships between variables, so that combinations of conditions that could never occur in practice are modelled as scenarios and the results of these unrealistic scenarios included in the risk estimate that is generated. For example, the range of storage times and storage temperatures for a food could be described independently by separate distributions. If the relationship between these factors was not explicit in the model, the model could generate predictions based on long storage times and high temperatures even though such situations are unlikely to occur because higher temperatures are usually associated with shortened shelf-life for perishable foods. This is a potential problem in all quantitative risk assessment (Morgan, 1993; Vose, 2008), not only food safety risk assessment. A full discussion of potential pitfalls in simulation modelling is beyond the scope of this chapter but Morgan (1993), Burmaster and Anderson (1994), EPA (1997) and Vose (2008) provide discussion and guidelines for simulation modelling in risk assessment. When used appropriately, simulation software can provide a way to identify and rank factors that contribute to risks, quantify levels of risk, and help to identify strategies and information needed to control or minimise risk.

4.3.3 Examples of quantitative microbial food safety risk assessments

Whiting and Buchanan (1997) first presented an example of a stochastic risk modelling approach in the food microbiology literature to assess the risk from *Salmonella* Enteritidis in liquid pasteurised eggs. Cassin *et al.* (1998) provided a more detailed analysis for public health risk from enterohaemorrhagic *E. coli* (EHEC) in beefburgers in North America. There have been numerous risk assessments of *E. coli* O157:H7 in meat products, five of which have been reviewed and compared by Duffy *et al.* (2006). Other examples are Lindqvist and Westöö (2000) for *L. monocytogenes* in cold-smoked fish, Alban *et al.* (2002) for *Salmonella* Typhimurium DT104 in Danish dry-cured pork sausages, Hartnett *et al.* (2001) and Rosenquist *et al.* (2003) for

Campylobacter in chicken meat, Hope *et al.* (2002) for *Salmonella* risks from shell eggs and egg products, Sanaa *et al.* (2004) for *L. monocytogenes* in French raw-milk soft cheeses, Salvat and Fravalo (2004) for *L. monocytogenes* in processed pork products in Europe, McNamara *et al.* (2007) for pork-borne salmonellosis in the USA, Yamamoto *et al.* (2008) for *Vibrio parahaemolyticus* in bloody clams in southern Thailand, etc. In addition, there are numerous other microbial food safety risk assessments that have been undertaken and reported by government, and other, organisations (e.g. CFSAN/FSIS, 2003; FSANZ, 2006; Nauta *et al.*, 2005; AFRC, 2008; EFSA, 2008). FAO/WHO's JEMRA have developed detailed risk assessments for:

- *Salmonella* in eggs and broiler chickens;
- *L. monocytogenes* in ready-to-eat foods;
- *Vibrio* spp. in seafoods;
- *Campylobacter* spp. in broiler chickens;
- *Cronobacter* (formerly Enterobacter) and other micro-organisms in powdered infant formula.

The risk assessments described above all adopt a 'big picture' view of risks across food supply and processing chains, and with a focus on public health outcomes. In most cases they represent the work of large teams of experts from many disciplines, and often many person-years of time spent in finding, collating and analysing the data to generate the estimates of risk and presenting the results of the analyses, the assumptions and the data and knowledge upon which they are based.

In general, such farm-to-fork risk assessments have been produced by government regulatory agencies or academics presenting examples of risk assessment methods and benefits, or who have been sponsored by industry organisations to undertake industry-relevant risk assessment tasks. The literature is replete with expositions and commentaries on the benefits of formal risk assessment approaches for 'farm-to-fork' food safety management, i.e. for prioritisation of food safety management needs (e.g. attribution of foods most responsible for particular types of foodborne infections, or cost–benefit analyses to evaluate proposed food safety regulation or programs), for identification of risk mitigation options and for appropriate allocation of responsibility for food safety management actions among different industry sectors along the farm-to-fork chain.

Many commentators have drawn attention to the potential of risk assessment methods to assist food businesses to improve their operations, and the complementarity of risk assessment and hazard analysis as a major step in the development of HACCP plans is evident. Membré *et al.* (2006) used stochastic modelling techniques to estimate the survival of both mesophilic and psychrotrophic strains of the pathogenic spore-former, *B. cereus*, and the numbers of each type expected in pouches of REPFED (refrigerated processed foods of extended durability) products, after thermal processing. This type of analysis can be used to design minimum thermal processes to optimise product quality without compromising safety. Membré and her colleagues also used stochastic modelling approaches to estimate shelf-lives for chilled foods (Membré *et al*., 2005), where shelf-life was determined by the extent of growth of *B. cereus* from spores surviving thermal processing. Earlier Rasmussen *et al.* (2002) had also developed a process risk model for the shelf-life of vacuum-packed Atlantic salmon that included influences from the point of slaughter to distribution and retail display. The model was developed to determine steps in the 'farm-to-retail' chain where improvements could be made to increase the shelf-life of the product.

Stochastic modelling approaches are being applied to improve understanding of different food processing operations and also mechanisms of contamination. Jordan *et al.* (1999) developed a stochastic model to assess the quantity of foodborne pathogens such as *Salmonella* and EHECs deposited on cattle carcasses under different pre-slaughter management regimens. Those authors concluded that the model showed promise as an inexpensive method for evaluating pathogen control strategies including those that could form part of a HACCP system. Schaffner (2004) proposed the use of stochastic modelling to explore mechanisms and magnitudes of cross-contamination of foods with pathogens by various processes and began to develop methods for that approach. Lindqvist and Lindblad (2008) developed a stochastic risk assessment model to assess the role of consumer handling of raw chickens, particularly that leading to cross-contamination, in human campylobacteriosis in Sweden. Aziza *et al.* (2006) developed a stochastic model to explore the potential for cross-contamination during smearing of washed rind cheeses, with a view to improving control of this step in the manufacture of that style of cheese. Hill *et al.* (2008) implemented a stochastic model of transmission of *Salmonella* between pigs in a herd to inform risk managers about the benefits of implementation of a monitoring program proposed to reduce *Salmonella* levels in pigs at slaughter. The model predicted that the most effective control strategies would be those that reduce between-pen transmission. Cassin *et al.* (1998) also used their farm-to-fork model for *E. coli* O157:H7 contamination of beefburgers to explore the effectiveness of alternative risk mitigation actions. The proposed actions included improvements in temperature control during retail, reducing the prevalence at slaughter of cattle shedding high levels of the pathogen and a communication program to encourage consumers to more thoroughly cook their beefburgers. Under the assumptions of the various 'what-if' scenarios reduction of growth of *E. coli* O157:H7 at retail, through improved temperature control, was shown to achieve the greatest risk reduction.

4.4 Industry use of microbial food safety risk assessment

As noted above, microbial food safety risk assessment was introduced to meet the needs of governments and nations for setting rules for international trade in food and for public health protection. MRA is still generally considered to be a high-level, governmental activity, and it is unlikely that any but the largest food businesses would undertake a fully quantitative MRA. Risk assessments initiated by industry organisations will often have more direct relevance to individual businesses, but many food businesses may be able to benefit from elements of governmental food safety risk assessments and approaches. For example, most food businesses will have implemented HACCP. HACCP is primarily a risk management system, but its effectiveness can be improved by adoption of elements of risk assessment. HACCP, despite being an essentially quantitative management approach characterised by identification of CCPs and specification of their safe operating limits, is often applied qualitatively (Notermans and Mead, 1996; Buchanan and Whiting, 1998; Gaze *et al.*, 2002). Since the general approach in HACCP system development is to err on the side of safety this qualitative approach often leads to HACCP plans with an excessive number of CCPs or to overly conservative application of lethal treatments at CCPs (Buchanan and Whiting, 1998). A risk-based approach can improve the effectiveness of HACCP plans by providing an objective basis for focusing on hazards that represent unacceptable *risks* and discounting hazards that are unlikely to occur or to cause harm to consumers.

Some of the tools used in risk assessment, and the data presented and insights gained from government-sponsored risk assessments, will provide invaluable information for businesses to establish and optimise their own food safety management plans. Risk profiles, such as those developed by the New Zealand Food Safety Authority (see: http://www.nzfsa.govt.nz/science/risk-profiles/), offer similar benefits.

Additionally, 'government' risk assessments are likely to affect individual food businesses through implementation of food safety regulations derived from them, for example, through the articulation of food safety objectives, performance objectives and performance criteria (discussed below) that are derived from MRAs.

'Farm-to-fork' risk assessments, such as those undertaken by governments and other competent authorities, attempt to characterise a supply chain for a commodity. Consequently, they are usually highly generic representations of the risks from that food. A food business, however, must develop pathogen management strategies that apply specifically to its stage and its operations within the commercial food chain. Nonetheless, some food processors will have to consider how the product will be handled after processing to ensure that if pathogens could be present in the product they do not reach dangerous levels prior to consumption. Exposure assessment methods, thus, should prove valuable for food manufacturers to be able to determine the acceptable levels of contamination at the point of production, that will satisfy ALOPs and related regulatory criteria, but taking into account the effect of handling of the product by transporters, sellers and consumers after it leaves the manufacturer's control. Because 'acceptable levels' would be expected to be established by regulatory authorities, and not subject to negotiation, industry risk assessments will often devolve to exposure assessments.

The adoption of risk-based approaches to microbial food safety management also provides opportunities to the food industry to be innovative. Risk assessment methods can be used to demonstrate equivalence of different food safety programs and technologies and can be used by the food industry to gain acceptance for alternative products and processing technologies. MRA could also be used by industry to undertake pre-market evaluations of complex product innovations, or to evaluate the consequences of changed material sources or changed formulations, etc.

This section considers the use of risk assessment methods by industry, particularly in helping to optimise HACCP plans. Useful discussions of the application of risk assessment methods by the food industry can also be found in Buchanan and Whiting (1998), van Gerwen and Gorris (2004), Brown (2002a, b) and CCFRA (2007).

4.4.1 Synthesising Hazard Analysis Critical Control Point (HACCP) and microbiological risk assessment

Formal risk assessment is a method for synthesising and structuring knowledge so that hazards and their potential consequences can be evaluated and compared. In some cases the creation of a conceptual model (see Section 4.3.1) may not generate new knowledge, but simply provide a tool to aid structured thinking and communication. If properly formulated, i.e. with sufficient differentiation and detail of risk-affecting steps, the risk model can help to identify alternative, or proposed, risk management actions that most reduce food safety risk by using the model to simulate the effect of those actions on the risk estimate. For a government food safety manager, this may involve identification of a specific stage in the farm-to-fork supply chain where control is most effective, e.g. prevention of infected animals from entering the food supply chain, requirement for a pathogen inactivation step during processing, temperature control during transport, limitation of shelf-life, etc. For a food business, those actions are more likely to relate to steps under the control of the food business, i.e. steps that could be CCPs. Van Schothorst (1998) succinctly describes the potential benefits of risk assessment techniques and the accompanying ability to relate microbial control to public health benefits including identification of critical control points, establishment of critical limits and determination of disposition of product produced during periods of loss of control of CCPs. This is discussed further in Section 4.4.2. The risk assessment approach also offers benefits during the 'hazard analysis' phase of risk assessment.

Use of microbial food safety risk assessment (MRA) to identify industry risk management options

According to FAO/WHO (2006) the use of 'what-if scenarios' for various control measures in risk assessments has already proven to be an effective means of evaluating different risk management options. By altering the model's structure, inputs and/or assumptions to represent those putative risk management actions, and evaluating the predicted impact on risk, the industry risk manager can begin formulating which of the potential strategies seem practical and feasible to implement. For example, in an industry-initiated assessment of the risk of *L. monocytogenes* in Australian processed meats, a variety of risk reduction options emerged including distributing the products frozen, reduction of product shelf-life, implementation of an in-pack listeriocidal treatment, reduction in contamination levels due to improved plant hygiene and reformulation of the product to reduce *L. monocytogenes* growth rate. Neither frozen distribution nor reduction of shelf-life was feasible due to consumer expectations and for commercial reasons. The remaining options were evaluated using a detailed, stochastic risk assessment model (Ross *et al.*, 2009). While in-pack treatment was shown to be effective, the structure of the Australian processed meat industry (i.e. many geographically dispersed small to medium enterprises) precluded adoption of technologies such as irradiation or high-pressure processing for economic (capital cost) reasons. Product reformulation, using organic acid salts, was predicted to be the most effective, feasible and acceptable risk mitigation strategy and has been adopted by many producers.

As noted above, food producers must also consider the microbiological fate of their product during distribution, sale and consumer handling under normal and even slightly abusive conditions so as to set shelf-life criteria and microbiological specifications for their end-product. The ability to estimate the consequences of these post-processing steps is essentially an exposure assessment. In general, government MRAs provide a high level of detail concerning microbial events that occur along the food chain and valuable information about the complex dynamics of pathogens during food processing. As such, the information presented in 'government' risk assessments will provide a ready source of information about the ecology of pathogens in foods, and also useful predictive microbiology models and modelling approaches. Discussions of the applications of predictive microbiology in the context of food safety risk and HACCP have been presented (Ross and McMeekin, 1995; van Gerwen and Zwietering, 1998; Soboleva *et al.*, 2000; Ross and McMeekin, 2003). Rasmussen *et al.* (2002), cited earlier, provides an example of stochastic exposure assessment modelling to support an individual food business.

As noted earlier, not all food safety management decisions require risk assessment approaches: use of epidemiological data was invoked as an example that risk management decisions of public health officials can be made without recourse to a full MRA. Similarly, in food processing businesses some food safety management decisions can be supported using predictive microbiology approaches (see e.g. Zwietering and Hasting, 1997a, b), particularly for identification of CCPs, specification of their critical limits and corrective actions in the event of a loss of control. Incorporation of stochastic modelling approaches in this type of analysis has also been demonstrated (Soboleva *et al.*, 2000).

In many cases a numerical estimate of risk won't be needed to inform risk management decisions. Often, risk managers require only a measure of risk compared to some other known level of risk to prioritise management strategies and actions. For example: 'does a 1 °C reduction in the average temperature of the chill chain reduce the growth of *L. monocytogenes* more than reducing the shelf-life of susceptible ready-to-eat foods by 30 %?'

Hazard analysis

Hazard analysis in HACCP involves developing and evaluating a list of all possible hazards. As discussed above, the idea of 'risk' can assist in the hazard analysis phase of HACCP to distinguish important hazards requiring control from those that are 'unrealistic'. The main issue is the need to discriminate between hazards that are unlikely to occur or to lead to human illness from those that pose a credible hazard. Notermans *et al.* (1995) give specific recommendations for identification of *relevant* microbial hazards in foods which involves creating a list of all known foodborne pathogens and elimination from that list of:

1. those organisms that have never been reported in the raw materials;
2. those that are likely to be completely eliminated by processing, unless they could be reintroduced by post-processing contamination;
3. those that have never caused a foodborne illness involving the same or a similar food;
4. toxigenic organisms unable to grow in the product where growth is required to produce sufficient toxin to cause illness.

All remaining organisms are considered as credible hazards requiring control.

Failure Mode and Effect Analysis (FMEA, see also Section 4.3.1) is an engineering 'quality assurance' system that systematically considers a production/assembly process to identify how things 'could go wrong', and the causes and the effects on product performance if they did. Effective control measures are then designed and implemented to ensure that failures are prevented from occurring. FMEA was originally introduced by the US military and adopted into the USA's aerospace program, and was the basis for the development of HACCP within that program but, whereas FMEA is more focused on the likelihood and consequences of failure of components, HACCP is concerned with failure of *processes*. An essential element of FMEA, that seems to have been lost in HACCP, involves rating failures in a manner that is risk-based, by considering:

1. the likelihood of a failure;
2. the severity of the consequence of failure; but also
3. the likelihood of that failure not being detected (prior to use).

A score is given to each of these elements from 0 to 10 corresponding to: 'extremely unlikely' to 'inevitable' for failure; 'insignificant' to 'catastrophic' for severity; and 'absolutely certain to be detected' to 'certain not to be detected' for failure detection. The product of the three scores, called the 'risk priority number', is used to rank the overall level of importance of that failure 'mode'. The concepts embodied in FMEA could be applied to good effect in hazard analysis during HACCP to discriminate credible, from trivial, hazards.

Identification and specification of CCPs

The identification and correct specification of CCPs is often one of the most difficult aspects of preparation of a HACCP plan, partly because the real critical limits (i.e. those corresponding to acceptable levels of consumer protection) are not well known or defined, or are not readily translated into measurable attributes of the product at the point of production. In practice, however, regulatory limits will often be imposed. From these, CCPs and critical limits can be deduced for specific products/processes.

Where the microbiological criteria apply to the point of production, the tools and methods of MRA, particularly predictive microbiological models and stochastic simulation modelling, will enable industry to establish critical limits that encompass variability in microbiological quality of ingredients and processing parameters to be able to optimise processes to achieve those regulatory levels with a specified level of confidence. Where criteria apply at the point of consumption, the same tools can be used to conduct an exposure assessment to determine levels at the point of production that will enable the point-of-consumption criterion to be achieved given the specific formulation of the product, its planned use and specified shelf-life, and expected changes in microbiology during the distribution, sale and consumer handling of the product (as discussed above).

Predictive models coupled with process specifications provide quantitation of potential for microbial increase or decrease during processing. Knowledge of the microbiological quality of ingredients will also be necessary. Coupled with stochastic simulation models, the effects of variability and uncertainty can be considered as well.

As discussed earlier, the tools of sensitivity analysis and scenario analyses can also be used to help to identify CCPs and their critical limits. Equally, such models could be used to systematically explore the consequences of loss of control at various steps and, from that, to specify corrective actions for different types and levels of loss of control. Under risk-based regulatory systems, the focus is on meeting an acceptable level of safety, rather than prescribing how that level should be achieved. The MRA approaches described (Section 4.3), particularly for exposure assessment, provide a means to demonstrate food safety equivalence of innovative processes and, thus, to foster innovation.

4.4.2 Linking appropriate levels of protection (ALOPs) to HACCP

To date, the specification of food safety regulations has largely been based on what food businesses employing good practices and available technology can achieve. Criteria were then based on the concept of ALARA which stands for 'as low as reasonably achievable' (Gorris *et al.*, 2006). Such implicit levels of controls were translated into practical risk management strategies for the food industry including prerequisite programs (e.g. GMP, GAP, GHP) and the development of food safety management systems such as HACCP.

As described in this chapter, as food supply chains become more complex and more 'rationalised' (i.e. into fewer, larger, producers), and as failures of the system leading to foodborne disease outbreaks are more evident, governments are becoming more actively involved in specifying the degree of consumer protection that is expected and required, i.e. specification of ALOPs. The introduction of the idea of 'risk' applied to microbial food safety led to the ability of governments and other competent authorities to specify levels of risk to public health that are considered 'acceptable' or 'tolerable' to provide an acceptable level of protection of consumers. The concept of food safety objectives (FSO) was proposed and described by Codex and ICMSF (ICMSF, 2002; CAC, 2005) as a tool to meet a food safety public health goal. A FSO specifies the maximum permissible level or frequency of a microbiological hazard in a food at the moment of consumption. A FSO could be statement of intent, of some desirable upper level of contamination, but is not necessarily expressed in a way that can be readily assayed, particularly because the result of the assay would not be available until after the food is consumed. Currently there is no example of a FSO that has been specified by a competent authority.

While risk assessments are generic and describe typical or representative food chains, processing technologies and contamination levels, individual businesses are responsible for ensuring the safety of their own products. Thus, it was recognised that for industry to be able to produce foods that satisfy the FSO, there must also be some way to translate the FSO into practical, measurable, product attributes that relate to the product at some earlier point in the food supply chain, and for it to be translated in a way that can be built into existing industry food safety management systems such as HACCP. To meet this need, the concepts of performance objectives (PO) and performance criteria (PC) have been introduced to provide practical guidance to industry. Thus, maximum hazard levels at other (earlier) points along the food chain that ensure that the food at the point of consumption satisfies the FSO are called POs. A PO is defined as: 'the maximum frequency and/or concentration of a hazard in a food at a specified step in the food chain before consumption that provides or contributes to an FSO or ALOP, as appropriate'. Guidance to industry on how to achieve performance objectives is given in PC, defined as 'the effect, in terms of frequency and/or the concentration of a hazard in a food that must be achieved by the application of one or more control measures to provide or contribute to a PO or an FSO'. A 'control measure' is any action or activity that can be used to prevent or eliminate a food safety hazard or to reduce it to an acceptable level. Thus, CCPs, in the sense used in HACCP, are control measures.

Industry will be engaged in this new risk-based food safety paradigm in two ways. The first involves its role as a stakeholder and contributor to the process of risk analysis, e.g. in the risk assessment phase by providing data and knowledge concerning hazard levels in raw materials, details of processing conditions, hazard levels in finished products, product composition and time and temperature details of distribution systems, etc., so that the risk assessment accurately represents reality. Industry can also provide input to the risk management phase by providing advice to regulatory risk managers about what is technologically achievable, costs of proposed risk management options, and translation of FSOs into practical POs and PCs. The second role is that of having to comply with the specified FSOs. While PCs and POs specify recommended food safety management criteria to satisfy the FSO, when stipulated by competent authorities they might be regarded as 'default' or 'safe-harbour' levels. FSOs were proposed to enable risk-based ALOPs to be translated into practical guidance, e.g. default values, that relate to products at the point of production. According to FAO/WHO (2006):

> The establishment of limits provides distinct advantages to both risk managers and the food industry by clearly articulating the level of control expected. When done on an industry wide basis, this establishes a 'level playing field' among companies in the industry. If done with a focus on the level of control required, rather than on a specific technology or practice, such limits can provide enhanced flexibility as to the approaches and technologies used to achieve the required level of control.

The FSO specifies the desired outcomes at the point of consumption, and might be regarded as a-benchmark of equivalence (Gorris *et al.*, 2006). As noted in Section 4.4.1, industry could set its own POs or PCs to satisfy those FSOs set by competent authorities. Performance criteria set by individual food processing businesses must be based on knowledge of the level of the pathogen in the raw materials used and the specified performance objective for that food at that stage in the food supply chain, bearing in mind possible changes in hazard levels at subsequent stages in the chain prior to consumption, so that the food satisfies the FSO.

The International Commission for the Microbiological Specifications of Foods (ICMSF) equation

The PO concept recognises that several actions or processes could act together to provide the required level of pathogen management at a particular stage in the food supply chain so that the FSO is satisfied. To specify the rigour of a process or combination of processes to achieve a PO it is, in principle, necessary only to know the microbiological quality of the incoming material and the PO required. This simple concept has been expressed by ICMSF (2002) in the following equation:

$$H_0 - \sum R + \sum I \leq \text{PO}$$

where H_0 is the level of the hazard(s) in the raw incoming material, $\sum R$ is the sum of the effect of processes in that step that reduce hazard levels, $\sum I$ is the sum of the effect of processes in that step that increase hazard levels, and PO is the Performance Objective for that step. If the step considered is the final step in the food supply chain, i.e. just prior to consumption, then the PO is the FSO. The equation can be applied sequentially for various steps in the chain, so that the PO for one step becomes H_0 for a subsequent step:

$$H_{0\text{-}1} - \sum R_1 + \sum I_1 \leq \text{PO}_1\ (= H_{0\text{-}2}) \quad \text{[Stage 1]}$$

$$\text{H}_{0\text{-}2} - \sum R_2 + \sum I_2 \leq \text{PO}_2\ (= H_{0\text{-}3}) \quad \text{[Stage 2]}$$

$$\text{H}_{0\text{-}3} - \sum R_3 + \sum I_3 \leq \text{PO}_3\ (= H_{0\text{-}4}) \quad \text{[Stage 3]}$$

$$\ldots$$

$$H_{0\text{-}n} - \sum R_n + \sum I_n \leq \text{PO}_n\ (= \text{FSO}) \quad \text{[Final stage]}$$

In this way, the effect of all steps in the food supply chain can be considered as part of the overall risk management activities in place to satisfy the FSO. Thus, for the whole chain;

$$H_0 - \sum R_{1\rightarrow n} + \sum I_{1\rightarrow n} \leq \text{FSO}$$

The equation is not fully mathematically correct: while it works for exponential processes such as microbial growth or microbial inactivation, it is not logically correct when arithmetic processes (e.g. cross-contamination) are considered. Nonetheless, the concepts embodied in the equation will be useful to industry to begin to design food safety systems for their part of the food supply chain, and for all businesses in the supply chain to understand their respective roles in overall management of the safety of the product.

Most importantly, this simple equation provides a simple conceptual approach to translate an FSO into POs for different stages of the food supply chain by applying the equation deterministically and working 'backwards'. The approaches and tools described in Section 4.4.1 apply equally here. While it was stated in Section 4.3.2 that stochastic modelling is preferred by many risk assessors, the accumulation of variability means that the process cannot be run 'backwards', i.e. one cannot infer a PO that is a distribution from an FSO that is a distribution because of all the sources of variation between these stages. This limitation is discussed in greater detail in Lammerding (2007). In principle, however, a PO distribution might be deduced by scenario analysis.

4.4.3 Conclusions

The new paradigm of risk-based food safety management approaches, and their translation into food safety regulations, is currently unlikely to have major impacts on the food industry because, in general, food safety management systems, such as HACCP and pre-requisite programs, are well developed and new food safety objectives will be implemented via these systems as outlined above. Nonetheless, as FSOs and POs are implemented, there will be implications for industry, such as the need to provide evidence of compliance with FSOs, but there are also opportunities, e.g. for innovation by demonstration of equivalence. Gorris *et al.* (2006) succinctly describe how industry can become a genuine stakeholder in this MRA-based regulatory environment. They state that industry will need to:

1. invest in skills and capability to understand FSOs and related concepts;
2. assess whether food safety systems currently used comply with FSOs;
3. understand how measures can be established (and validated) that improve the current systems to meet the FSO, where necessary;
4. understand what the legal status is of the concepts and what the industry situation is with regard to liability;

noting that these requirements may be beyond the means of small-to-medium enterprises and may require leadership from industry sector or trade organisations, and also governments, to enable the benefits to both consumers and industry to be achieved.

4.5 Future trends

Since the advent of HACCP, food safety management has focused on control of hazards, but the accessibility of risk assessment tools and technologies, in combination with predictive microbiology, has enabled exploration of risk-based approaches to microbial food safety management. This opportunity has been taken up by governments to facilitate international trade in foods and as a means to optimise strategies to reduce the incidence of foodborne disease. Development of optimal pathogen management strategies requires knowledge of the pathogen, the consumer, how the food becomes a vehicle for disease transmission and the differentiation of risks and hazards. Industry has also begun to explore the benefits that MRA may offer. Given the complexity of the food supply chain, and its influence on the frequency of contamination and number of pathogens that may be in foods, tools for synthesis of this information into an objective and reliable estimate of consumer health risk are also needed. The risk assessment paradigm provides a sound basis for organising relevant information and data to be able to understand and communicate where risks arise, and how they are altered, in the human commercial food chain.

There are challenges, however, for the implementation of risk-based approaches to food safety regulation, for example, the specification of an ALOP and, even more fundamentally, the development of consensus concerning appropriate methods for estimating risk, or for demonstrating equivalence of food safety programs *between* nations. As greater experience is obtained, it is likely that criteria for the conduct of 'valid' MRAs that satisfy the needs of different types of risk management questions will be developed. Equally, techniques to evaluate the reliability or validity of MRA models will also have to be developed before results of risk assessments are translated into management decisions that may have an effect on stakeholders. To date, results of risk assessments have been applied within nations, and have usually involved extensive stakeholder consultation and communication, in some cases leading to provision of new data or knowledge that has led to re-evaluation of the risk estimates. Within industry, the ability of insights gained from risk assessments to enhance HACCP effectiveness has already been demonstrated. Approaches to facilitate the translation of FSOs derived from MRAs into practical guidance for the food industry, and for integration into existing industry food safety management systems such as HACCP, will also be more fully developed and utilised: POs and PCs, and the ICMSF equation, begin to address these needs.

Methods for microbial food safety risk assessment are still evolving, and will continue to do so to meet the challenges and variety of needs of risk assessment. In addition to the need for robust mathematical modelling techniques to synthesise the information, there are data and knowledge challenges, e.g. including insufficient data on pathogen ecophysiology in foods and during food processing operations; data for human host response to different levels of pathogens and biotoxins. Initiatives such as data and software repositories, however, can help to make risk assessment a progressive, cumulative and co-operative process. A collation of links to relevant web-based resources is available at: http://www.foodrisk.org/resource_types/tools/index.cfm.

As experience with MRA has grown, it has also become evident that the risk assessment approach taken has to be 'fit-for-purpose' because resources for MRA are not unlimited. Not all risk management questions require a fully quantitative, stochastic, risk estimate and simpler risk assessment approaches for different types of risk questions will need to be developed to make the benefits of risk assessment more broadly available. Alternatively, software that does involve high-level stochastic models that are generic for a wide range of MRA questions, but that guide users through the process with 'user-friendly' interface and extensive data 'built-in', may be able to be developed. This will make the benefits of the risk assessment approach for providing support for decisions available to a wider range of users concerned with pathogen management and for a wider range of situations.

In essence, a risk assessment is never completed because new data become available, systems can change and new pathogens can emerge, so that the model must continue to be developed and refined. Uncertainty is reduced as more knowledge is incorporated into the model and confidence in the reliability of the model's predictions is increased. Also, models can be created as a series of connected modules so that once a stage or process in the food chain is modelled its structure or data or both can be reused in other risk assessments. Thus, a risk model itself, as an expression of an ongoing MRA, can be seen as a 'living' and structured repository of data and knowledge that can assist understanding of the complex commercial food chain and that can facilitate objective microbial food safety management decisions.

4.6 Sources of further information and advice

The preceding sections have identified many reviews and commentaries on the use of risk assessment methods and HACCP in developing pathogen management plans. Many additional resources are available from government and private internet sites around the world, and are readily located using internet search engines. Some sites of particular interest are presented below.

- The USA Joint Institute for Food Safety and Applied Nutrition provides a large range of resources related to microbial food safety risk assessment at the FoodRisk.org website:
  http://www.foodrisk.org
  including a range of downloadable tools at:
  http://www.foodrisk.org/resource_types/tools/index.cfm
- Risk analysis and HACCP information can be found at:
  http://haccpalliance.org/alliance/foodsafety.html
- US government microbiological food safety risk assessments are available for download at:
  http://www.health.gov.au/internet/main/publishing.nsf/content/cda-pubs-cdi-cdicur.htm
- Food and Agriculture Organization and World Health Organization microbial food safety risk assessment activities and major reports are documented at:
  http://www.who.int/foodsafety/micro/jemra/en/index.html
  or
  http://www.fao.org/ag/agn/agns/jemra_index_en.asp
- The European International Life Sciences Institute has a Food Safety Risk Assessment in Food Microbiology taskforce and has produced various reports that can be downloaded from:
  http://europe.ilsi.org/activities/taskforces/riskassessment/RiskAnalysisMicrobiology.htm
- The European Food Safety Authority also has produced a series of reports and expert opinions including those related to risks from specific foodborne microbial pathogens, which can be accessed at:

http://www.efsa.eu.int/EFSA/efsa_locale-1178620753812_ScientificOpinionPublicationReport.htm
- The Centre for Disease Control and Prevention (USA) web site http://www.cdc.gov/ provides much information on the incidence of foodborne disease, and links to other sites.

Other communicable disease resources can be found at:

- Centre for Food Safety and Applied Nutrition (USA):
  http://vm.cfsan.fda.gov/list.html
- and Communicable Diseases – Australia
  http://www.health.gov.au/internet/main/publishing.nsf/content/cda-pubs-cdi-cdicur.htm
  provides links to many international communicable disease sites.
- Risk world
  http://www.riskworld.com/
  provides many links to risk analysis related sites.

Useful texts include:

Brown, M. and Stringer, M. (eds) (2002) *Microbiological Risk Assessment in Food Processing*, Woodhead, Cambridge.

Gorris, L.G.M, Bassett, J. and Membré, J.-M. (2006) Food safety objectives and related concepts: the role of the food industry, in Y. Motarjemi and M. Adams (eds), *Emerging Foodborne Pathogens*, Woodhead, Cambridge, 153–78.

Lammerding, A. (2007) Using microbiological risk assessment (MRA) in food safety management, *ILSI Europe Report Series*, ILSI Europe, Brussels, 1–36.

Vose, D. (2008) *Risk Analysis: a Quantitative Guide*, *(3rd Edn)*. Wiley, New York.

Volumes 36 (2–3) and 58 (3) in 1997 and 2000 respectively of *International Journal of Food Microbiology* featured exclusively papers concerning microbial food safety risk assessment and its applications as does the *Journal of Food Protection*, 1996 Supplement.

4.7 References

AFRC (2008) E. coli *O157:H7 in beefburgers produced in the Republic of Ireland: A quantitative microbial risk assessment*, Ashtown Food Research Centre, Dublin available at: http://www.teagasc.ie/ashtown/research/foodsafety/fs_quantitative_micro_risk_assess.htm accessed November 2008.

Alban L, Olsen AM, Nielsen B, Sorenson R and Jessen B (2002) Qualitative and quantitative risk assessment for human salmonellosis due to multi-resistant *Salmonella* Typhimurium DT104 from consumption of Danish dry-cured pork sausages, *Preventive Veterinary Medicine*, **52** 251–65.

Arvanitoyannis ID and Varzarkas TH (2008) Application of ISO22000 and Failure Mode and Effect Analysis (FMEA) for Industrial Processing of Salmon: A case study, *Critical Reviews in Food Science and Nutrition*, **48** 411–29.

Aziza F, Mettler E, Daudin JJ and Sanaa M (2006) Stochastic, compartmental, and dynamic modeling of cross-contamination during mechanical smearing of cheeses, *Risk Analysis*, **26** 731–45.

Bartley DM, Bondad-Reantaso MG and Subasinghe RH (2006) A risk analysis framework for aquatic animal health management in marine stock enhancement programmes, *Fisheries Research*, **80** 28–36.

Biosecurity Australia (2002) *Import risk analysis of non-viable bivalve molluscs*, Agriculture, Fisheries and Forestry, Australia. Legislative Services, Info Products, Department of Finance and Administration, GPO Box 1920, Canberra, ACT, Australia, 2601.

Brown M (2002a) Exposure assessment, in M Brown and M Stringer (eds) *Microbiological Risk Assessment in Food Processing*, Woodhead, Cambridge, 100–26.

Brown M (2002b) Hazard identification in M Brown and M Stringer (eds) *Microbiological Risk Assessment in Food Processing*, Woodhead, Cambridge, 64–76.

Buchanan RL (1997) National Advisory Committee on Microbiological Criteria for Foods: 'Principles of risk assessment for illnesses caused by foodborne biological agents', *Journal of Food Protection*, **60** 1417–19.

Buchanan RL and Whiting RC (1996) Risk assessment and predictive microbiology, *Journal of Food Protection*, Supplement 31–6.

Buchanan RL and Whiting RC (1998) Risk assessment – a means for linking HACCP plans and public health, *Journal of Food Protection*, 61, 1531–4.

Buchanan RL, Smith JL and Long W (2000) Microbial risk assessment: dose-response relations and risk characterization, *International Journal of Food Microbiology*, **58** 159–72.

Burmaster DE and Anderson PD (1994) Principles of good practice for the use of Monte-Carlo techniques in human health and ecological risk assessments, *Risk Analysis*, **14** 477–81.

CAC (1999) *Principles and guidelines for the application of microbiological risk assessment*, GL-30, Food and Agriculture Organization of the United Nations/World Health Organization, Rome/Geneva, available at: www.codexalimentarius.net/download/standards/357/CXG_030e.pdf, accessed December 2008.

CAC (2005) *Proposed draft principles and guidelines for the conduct of Microbiological Risk Management (MRM) (at Step 5 of the Procedure)*, ALINORM 05/28/13, Appendix III, Food and Agriculture Organization of the United Nations, Rome.

Cassin MH, Paoli GM, McColl RS and Lammerding AM (1996) Hazard assessment of *Listeria monocytogenes* in the processing of bovine milk – Comment, *Journal of Food Protection*, **59** 341–2.

Cassin MH, Lammerding AM, Todd ECD, Ross W and McColl RS (1998) Quantitative risk assessment for *Escherichia coli* O157:H7 in ground beef hamburgers, *International Journal of Food Microbiology*, **41** 21–44.

CAST (1994) *Foodborne pathogens: risk and consequences*, Task Force Report No. 122, Council for Agricultural Science and Technology, Ames, IO.

CCFRA (2007) *Industrial microbiological risk assessment – a practical guide*, 2nd edn, CCFRA Guideline No. 28, Campden & Chorleywood Food Research Association, Chipping Campden.

CFSAN/FSIS (2003) *Quantitative Assessment of the Relative Risk to Public Health from Foodborne* Listeria monocytogenes *Among Selected Categories of Ready-to-Eat Foods*, Center for Food Safety and Applied Nutrition/Food Safety Inspection Service, College Park, MD/Washington DC.

Corlett DA and Pierson MD (1992) Hazard analysis and assignment of risk categories, in MD Pierson and DA Corlett, Jr (eds), *HACCP: Principles and Applications*, New York, Van Nostrand Reinhold, 29–38.

Cox LA (2008) What's wrong with risk matrices? *Risk Analysis*, **28** 497–512.

Craun G, Dufour A, Eisenberg J, Foran J, Gauntt C, Gerba C, Haas C, Highsmith A, Irbe R, Julkunen P, Juranek D, LeChevallier M, Levine M, Macler B, Murphy P,

Payment P, Pfaender F, Regli S, Roberson A, Rose J, Schaub S, Schiff G, Seed J, Smith C, Sobsey M, Spear R and Walls I (International Life Sciences Institute, North America – Risk Science Institute Pathogen Risk Assessment Working Group) (1996) A conceptual framework to assess the risk of human disease following exposure to pathogens, *Risk Analysis*, **16**(6) 841–8.

Daughtry B, Sumner J, Hooper G, Thomas C, Grimes T, Horn R, Moses A and Pointon A (2005) *National Food Safety Risk Profile of Eggs and Egg Products Part 1: Final Report and Part 2: Attachments. A Report For The Australian Egg Corporation Limited*, Australian Egg Corporation Limited, North Sydney, NSW, Australia, available at: http://www.aecl.org/index.asp?pageid=428, accessed November 2008.

Duffy G, Cummins E, Nally P, O'Brien S and Butler F (2006) A review of quantitative microbial risk assessment in the management of *Escherichia coli* O157:H7 on beef, *Meat Science*, **74**, 76–88.

EFSA (2008) A quantitative microbiological risk assessment on *Salmonella* in meat: Source attribution for human salmonellosis from meat. Scientific Opinion of the Panel on Biological Hazards, European Food Safety Authority, *The EFSA Journal*, **625** 1–32.

Elliott PH (1996) Predictive microbiology and HACCP, *Journal of Food Protection*, Supplement: 48–53.

EPA (1997) *Guiding principles for Monte Carlo analysis* (EPA/630/R-91/001), United States Environmental Protection Agency, Washington, DC.

FAO (1999) *FAO Expert Consultation on the Trade Impact of Listeria monocytogenes in Fish Products*, FAO Fisheries Report no. 564, Food and Agriculture Organization of the United Nations, Rome.

FAO/WHO (1996) *Report of the twenty-ninth session of the Codex Committee on Food Hygiene*, ALINORM 97/13A, Washington, DC, 21–25 October, Food and Agriculture Organization of the United Nations World Health Organization, Rome Geneva.

FAO/WHO (2000) *The Interaction between Assessors and Managers of Microbiological Hazards in Food*, Report of a WHO Expert Consultation in collaboration with The Institute for Hygiene and Food Safety of the Federal Dairy Research Center and The Food and Agriculture Organization of the United Nations, Kiel, Germany 21–23 March, available at: http://www.fao.org/ag/agn/agns/jemra_riskmanagement_en.asp, accessed November 2008.

FAO/WHO (2003) *Hazard characterization for pathogens in food and water: Guidelines. Microbiological Risk Assessment Series 3*, Food and Agriculture Organization of the United Nations/World Health Organization, Rome Geneva.

FAO/WHO (2004) *Risk assessment of Listeria monocytogenes in ready-to-eat foods*, MRA Series 4, Food and Agriculture Organization of the United Nations/World Health Organization, Rome/Geneva, available at: http://www.who.int/foodsafety/publications/micro/en/mra4.pdf, accessed December 2008.

FAO/WHO (2006) *The use of microbiological risk assessment outputs to develop practical risk management strategies: metrics to improve food safety*, Report of an Expert meeting, Kiel, Germany, April, Food and Agriculture Organization of the United Nations World Health Organization, Rome Geneva, available at: ftp://ftp.fao.org/ag/agn/food/kiel.pdf, accessed November 2008.

FAO/WHO (in press), *FAO/WHO Guidelines on Risk Characterization of Microbiological Hazards in Food*, Food and Agriculture Organisation of the United Nations/World Health Organisation, Rome Geneva.

Foegeding PM (1997) Driving predictive modelling on a risk assessment path for enhanced food safety, *International Journal of Food Microbiology*, **36** 87–95.

FSANZ (2006) *Scientific Assessment of the Public Health and Safety of Poultry Meat in Australia*, Food Standards Australia New Zealand, Canberra, available at: http://www.foodstandards.gov.au/_srcfiles/Poultry%20Meat%20in%20Australia.pdf, accessed December 2008.

Gaze R, Betts R and Stringer, MF (2002) HACCP systems and microbiological risk assessment, in M Brown and M Stringer (eds), *Microbiological Risk Assessment in Food processing*, Woodhead, Cambridge, 248–65.

Gorris LGM, Bassett J and Membré J-M (2006) Food safety objectives and related concepts: the role of the food industry, in Y Motarjemi and M Adams (eds), *Emerging Foodborne Pathogens*, Woodhead, Cambridge, 153–8.

Hartnett E, Kelly L, Newell D, Newell D, Wooldridge M and Gettinby G (2001) A quantitative risk assessment for the occurrence of *Campylobacter* in chickens at the point of slaughter, *Epidemiology and Infection*, **127** 195–206.

Hill AA, Snary EL, Arnold ME, Alban L and Cook AJC (2008) Dynamics of *Salmonella* transmission on a British pig grower-finisher farm: a stochastic model, *Epidemiology and Infection*, **13** 320–33.

Hope BK, Baker AR, Edel ED, Hogue AT, Schlosser WD, Whiting R, McDowell RM and Morales RA (2002) An overview of the *Salmonella enteritidis* risk assessment for shell eggs and egg products, *Risk Analysis*, **22** 203–18.

Huss HH, Reilly A and Ben Embarek PK (2000) Prevention and control of hazards in seafood, *Food Control*, **11**(2) 149–56.

ICMSF (1986) *Microorganisms in Foods 2. Sampling for Microbiological Analysis: Principles and Specific Applications*, 2nd edn, International Commission for the Microbiological Specifications for Foods, University of Toronto Press, Toronto.

ICMSF (1996) *Microorganisms in Foods 5. Microbiological Specifications of Food Pathogens*, International Commission for the Microbiological Specifications for Foods, Blackie Academic and Professional, London.

ICMSF (2002) *Microorganisms in Foods, Microbiological Testing in Food Safety Management vol. 7*, International Commission for the Microbiological Specifications for Foods, Kluwer Academic/Plenum, New York.

ILSI (2000) *Revised Framework for Microbial Risk Assessment. An ILSI Risk Science Research Institute Workshop Report*, ILSI Risk Science Research Institute, Washington, DC.

Jordan D, McEwen SA, Lammerding AM, McNab WB and Wilson JB (1999) A simulation model for studying the role of pre-slaughter factors on the exposure of beef carcasses to human microbial hazards, *Preventive Veterinary Medicine*, **41**, 37–54.

Kindred TP (1996) Risk analysis and its application in FSIS, *Journal of Food Protection*, Supplement 24–30.

Kumamoto H and Henley E (1996) *Probabilistic risk assessment and management for Engineers and Scientists*, 2nd edn, IEEE Press, New York.

Lammerding AM (1997) An overview of microbial food safety risk assessment, *Journal of Food Protection*, **60**(11) 1420–5.

Lammerding A (2007) Using microbiological risk assessment (MRA) in food safety management, *ILSI Europe Report Series*, 1–36.

Lammerding AL and Fazil A (2000) Hazard identification and exposure assessment for microbial food safety risk assessment, *International Journal of Food Microbiology*, **58**(3) 147–57.

Lindqvist R and Lindblad M (2008) Quantitative risk assessment of thermophilic *Campylobacter* spp. and cross-contamination during handling of raw broiler chickens evaluating strategies at the producer level to reduce human campylobacteriosis in Sweden, *International Journal of Food Microbiology*, **121** 41–52.

Lindqvist R and Westöö A (2000) Quantitative risk assessment for *Listeria monocytogenes* in smoked or gravad salmon and rainbow trout in Sweden, *International Journal of Food Microbiology*, **58**(3) 147–57.

Marks BP (2008) Status of microbial modeling in food process models, *Comprehensive Reviews in Food Science and Food Safety*, **7** 137–43.

Martin SA, Wallsten TS and Beaulieu ND (1995) Assessing the risk of microbial pathogens: application of judgement-encoding methodology, *Journal of Food Protection*, **58** 289–95.

McMeekin TA, Presser K, Ratkowsky D, Ross T, Salter M and Tienungoon S (2000) Quantifying the hurdle concept by modelling the bacterial growth/no growth interface, *International Journal of Food Microbiology*, **55** 93–8.

McNab WB (1998) A general framework illustrating an approach to quantitative microbial food safety risk assessment, *Journal of Food Protection*, **61** 1216–28.

McNab WB, Alves DM and Lammerding AM (1997) *A general framework for food safety risk assessment (Document #2)*, Ontario Ministry of Agriculture, Food and Rural Affairs, Guelph, Ontario.

McNamara PE, Miller GY, Liu W and Barber DA (2007) A farm-to-fork stochastic simulation model of pork-borne salmonellosis in humans: lessons for risk ranking, *Agribusiness*, **23** 157–72.

Mead PS, Slutsker L, Dietz V, McCaig LF, Bresee JS, Shapiro C, Griffin PM and Tauxe RV (1999) Food-related illness and death in the United States, *Emerging Infectious Diseases*, 5, 607–25.

Membré JM, Johnston MD, Bassett J, Naaktgeboren G, Blackburn CD and Gorris LGM (2005) Application of modelling techniques in the food industry: determination of shelf-life for chilled foods, *Acta Horticulturae*, **674** 407–14.

Membré JM, Amezquita A, Bassett J, Giavedoni P, Blackburn C de W and Gorris LGM (2006) A probabilistic modeling approach in thermal inactivation: Estimation of postprocess *Bacillus cereus* spore prevalence and concentration, *Journal of Food Protection*, **69** 118–29.

Morgan MG (1993) Risk analysis and management, *Scientific American*, **269** 32–41.

NRC (1983) *Risk Assessment in the Federal Government: Managing the Process*, National Research Council, National Academy Press, Washington, DC.

NACMCF (1992) Hazard analysis and critical control point system, The National Advisory Committee on Microbiological Criteria for Foods, *International Journal of Food Microbiology*, **16** 1–23.

Nauta MJ (2005) Microbiological risk assessment models for partitioning and mixing during food handling, *International Journal of Food Microbiology*, **100** 311–22.

Nauta MJ, Jacobs-Reitsma WF, Evers EG, van Pelt W and Havelaar AH (2005) *Risk assessment of* Campylobacter *in the Netherlands via broiler meat and other routes*, RIVM report 250911006/2005, available at: http://demo.openrepository.com/rivm/handle/10029/7248 accessed November 2008.

Notermans S, Gallhoff G, Zwietering MH and Mead CG (1995) The HACCP concept: specification of criteria using quantitative risk assessment, *Food Microbiology*, **12** 81–90.

Notermans S and Mead GC (1996) Incorporation of elements of quantitative risk analysis in the HACCP system, *International Journal of Food Microbiology*, **30** 157–73.

OGTR (2005) *Risk Analysis Framework (January 2005)*, Office of the Gene Technology Regulator, Canberra, ACT, Australia, available at: http://www.ogtr.gov.au/internet/ogtr/publishing.nsf/Content/raf-3/$FILE/raffinal2.2.pdf, accessed December 2008.

OIE (1999) *International animal health code: mammals, birds and bees*, 8th edn Office International des Epizooties, Paris.

Pointon A, Jenson I, Jordan D, Vanderlinde P, Slade J and Sumner J (2006) A risk profile of the Australian red meat industry: approach and management, *Food Control*, **17** 712–18.

PCCRARM (1997) *Framework for Environmental Health Risk Management*, The Presidential/Congressional Commission on Risk Assessment and Risk Management, available from: http://www.riskworld.com/riskcommission/Default.html, accessed December 2008.

Rasmussen SKJ, Ross T, Olley J and McMeekin T (2002) A process risk model for the shelf life of Atlantic salmon fillets, *International Journal of Food Microbiology*, **73** 47–60.

Rosenquist H, Neilsen NL, Sommer HM, Norrung B and Christensen BB (2003)

Quantitative risk assessment of human campylobacteriosis associated with thermophilic *Campylobacter* species inc chickens, *International Journal of Food Microbiology*, **83** 87–103.

Ross T and McMeekin TA (1995) Predictive microbiology and HACCP, in AM Pearson and TR Dutson (eds) *Hazard Analysis Critical Control Point (HACCP) in Meat Poultry and Seafoods*, Blackie Academic and Professional, London, 330–57.

Ross T and McMeekin TA (2003) Modeling microbial growth within food safety risk assessments, *Risk Analysis*, **23** 179–97.

Ross T and Sumner JL (2002) A simple, spreadsheet-based, food safety risk assessment tool, *International Journal of Food Microbiology*, **77** 39–53.

Ross T, Rasmussen S, Fazil A, Paoli G and Sumner J (2009) Quantitative risk assessment of *Listeria monocytogenes* in ready-to-eat meats in Australia, *International Journal of Food Microbiology* doi: 10.1016/j.ijfoodmicro.2009.02.007.

Sanaa M, Coroller L and Cerf O (2004) Risk assessment of listeriosis linked to the consumption of two soft cheeses made from raw milk: Camembert of Normandy and Brie of Meaux, *Risk Analysis*, **24**, 389–99.

Salvat G and Fravalo P (2004) Risk assessment strategies for Europe: Integrated safety strategy or final product control: Example of Listeria monocytogenes in processed products from pork meat industry, *Deutsche Tierarztliche Wochenschrift*, **111** 331–4.

Sassi F (2006) Calculating QALYs, comparing QALY and DALY calculations, *Health Policy and Planning*, **21** 402–8.

Schaffner DW (2004) Mathematical frameworks for modelling *Listeria* cross-contamination in food-processing plants, *Journal of Food Science*, **69** R155–59.

Soboleva TK, Pleasants AB and le Roux G (2000) Predictive microbiology and food safety, *International Journal of Food Microbiology*, **57** 183–92.

Strachan NJC, Doyle MP, Kasuga F, Rotariu O and Ogden ID (2005) Dose response modelling of *Escherichia coli* O157 incorporating data from foodborne and environmental outbreaks, *International Journal of Food Microbiology*, **103** 35–47.

Sumner JL and Ross T (2002) A semi-quantitative seafood safety risk assessment, *International Journal of Food Microbiology*, **77** 55–9.

Sumner J, Ross T and Ababouch L (2004) *Application of risk assessment in the fish industry. FAO Fisheries Technical Paper, No. 442*, Food and Agriculture Organisation of the United Nations, Rome.

Teunis PFM, van der Heijden OG, van der Geissen JWB and Havelaar AH (1996) *The dose-response relation in human volunteers for gastro-intestinal pathogens*, report 2845500002, National Institute of Public Health and Environment, Bilthoven, The Netherlands.

Teunis P, van den Brandhof W, Nauta M, Wagenaar J, van den Kerkhof H and van Pelt W (2005) A reconsideration of the *Campylobacter* dose-response relation, *Epidemiology and Infection*, **133** 583–92.

Todd ECD and Harwig J (1996) Microbial risk analysis of food in Canada, *Journal of Food Protection*, Supplement 10–18.

Todd ECD (2008) Trends in foodborne disease, *CAB Reviews: Perspectives in Agriculture, Veterinary Science, Nutrition and Natural Resources*, **3** 15 pp.

van Gerwen SJC, de Wit JC, Notermans S and Zwietering MH (1997) An identification procedure for foodborne microbial hazards, *International Journal of Food Microbiology*, **38** 1–15.

van Gerwen SJC and Gorris LGM (2004) Application of elements of microbiological risk assessment in the food industry via a tiered approach, *Journal of Food Protection*, **67** 256–64.

van Gerwen SJC and Zwietering MH (1998) Growth and inactivation models to be used in quantitative risk assessments, *Journal of Food Protection*, **61** 1541–9.

van Schothorst M (1997) Practical approaches to risk assessment, *Journal of Food Protection*, **60**(11), 1439–43.

van Schothorst M (1998) Principles for the establishment of microbiological food safety objectives and related control, *Food Control*, **9**(6) 379–84.

Vose D (2008) *Risk analysis: a quantitative guide* 3rd edn, Wiley, New York.

Walls I (2007) Framework for identification and collection of data useful for risk assessments of microbial foodborne or waterborne hazards: A report from the International Life Sciences Institute committee on data collection for microbial risk assessment, *Journal of Food Protection*, **70** 1744–51.

Whiting RC and Buchanan RL (1997) Development of a quantitative risk assessment model for *Salmonella enteritidis* in pasteurised liquid eggs, *International Journal of Food Microbiology*, **36**(2–3) 111–25.

Wooldridge M (2008) Qualitative risk assessment, in DW Schaffner, (ed.), *Microbial Risk Analysis of Foods*, ASM Press, Washington, DC, 1–28.

Yamamoto A, Iwahori J, Vuddhakul V, Charernjiratragul W, Vose D, Osaka K, Shigematsu M, Toyofuku H, Yamamoto S, Nishibuchi M and Kasuga F (2008) Quantitative modeling for risk assessment of *Vibrio parahaemolyticus* in bloody clams in southern Thailand, *International Journal of Food Microbiology*, **124** 70–78.

Zwietering MH and Hasting APM (1997a) Modelling the hygienic processing of foods – a global process overview, *Food and Bioproducts Processing*, **75** 159–67.

Zwietering MH and Hasting APM (1997b) Modelling the hygienic processing of foods—influence of individual process stages, *Food and Bioproducts Processing*, **75** 168–73.

5

Emerging foodborne pathogens and the food industry

L. Smoot, Nestlé USA, USA, J-L. Cordier, Nestlé Nutrition, Switzerland

Abstract: Implementation of effective food safety control measures requires a thorough understanding of the biological hazards relevant for the products that are being manufactured. This is not only true for well-known pathogens which may over time have adapted to control measures currently in place, but it is also important for emerging pathogens which may become a threat. This knowledge will then allow for an assessment of whether existing control measures are effective or whether modifications are necessary. This chapter will discuss the factors contributing to the emergence of new pathogens and the sources of information used in recognizing and defining their emergence. More importantly, the management options leading to the application of effective control measures for these food safety hazards will also be reviewed.

Key words: emerging foodborne pathogens, biological hazards, food safety management, Hazard Analysis Critical Control Points (HACCP), good hygiene practices (GHP).

5.1 Introduction

In order, as a food manufacturer, to implement effective control measures, it is important to have a thorough understanding of the pathogens relevant for the products that are being manufactured. This is, of course, true for well-established pathogens for which changes in behavior or sources may require adaptation and modifications of already implemented control measures. It is, however, also important for emerging ones which may become a threat, and it is necessary to assess whether existing control measures are effective or whether changes are required.

The description of new foodborne and waterborne pathogens has a long history. The first descriptions of most of those which are well-known today as well as their sources can be traced back through textbooks and publications. First descriptions of illness related to microbial pathogens such as *Vibrio cholerae*, *Salmonella* spp., *Clostridium botulinum* or the parasite *Trichinella spiralis* go back to the second half of the 19th century (Kaper *et al*., 1995; Poppe, 1999; Caya *et al*., 2004; Pozio and Zarlenga, 2005).

Others have been identified and described at regular intervals during the last decades: staphylococcal food poisoning in the 1910s (Bennett and Monday, 2003), *Bacillus cereus* (Rajkowski and Bennett, 2003), *Cl. perfringens* (Brynestad and Granum, 2002) and *V. parahaemolyticus* (Yeung and Boor, 2004) between 1940 and 1950, *Listeria monocytogenes* (Swaminathan and Gerner-Smidt, 2007) and the parasite *Anisakis simplex* (Audicana and Kennedy, 2008) between 1950 and 1960, *Campylobacter jejuni* (Engberg, 2006), *Yersinia enterocolitica* (Nesbakken, 2005), Norwalk virus (Estes *et al*., 2006) and *Giardia* spp. (Huang and White, 2005) between 1970 and 1980.

Over the last 20–25 years several infectious agents have been newly described or have been newly associated with foodborne outbreaks. *Escherichia coli* O157:H7, for example, was first recognized as a human pathogen in 1982 and identified as a cause of bloody diarrhoea. It has been associated with the consumption of insufficiently cooked hamburgers (Riley *et al*., 1983) and is at the origin of very severe symptoms such as haemolytic uremic syndrome (HUS), mainly in children. Since then, numerous outbreaks have been reported involving not only different meat products but also raw milk or products such as fresh cheese and curd manufactured thereof; unpasteurized fruit juices prepared with fruits such as apples having been exposed to bovine manure and thus likely to harbor the pathogen; sprouts grown from contaminated seeds; and up to the very recent global outbreaks caused by spinach exported from the USA and consumed raw (Espié *et al*., 2006; Doyle and Erickson, 2008; Heaton and Jones, 2008; Vojdani *et al*., 2008).

Although first cases of infections due to yellow-pigmented Enterobacteriaceae were described in the late 1960s, it is only since the mid 1980's that *Cronobacter* (formerly *Enterobacter*) *sakazakii* has been recognized as a new species and has emerged as an opportunistic pathogen (Bowen and Braden, 2008). Contrary to *E. coli* O157:H7, which has been associated with several types of food and caused infections in different age groups, *Cr. sakazakii* has caused rare but severe diseases only in premature babies and infants up to 6 months. Whether older infants between 6 and 12 months are susceptible is still a point of discussion and, while sporadic unpublished cases have been discussed within Codex Alimentarius, a definitive conclusion has not been drawn. This question and the need to introduce criteria for this organism for follow-up formulae consumed by infants between 6 and 12 months has been discussed in a 3rd expert meeting organized by FAO/WHO in July 2008, the publication of the report being expected by late 2008/early 2009 (consult the website of either FAO or WHO for the final report). *Cronobacter sakazakii* has been associated with reconstituted infant formulae, caused either by the use of powdered formulae showing low intrinsic contamination or by extrinsic contamination of the feeds during their preparation in bottle kitchens in the hospitals (FAO/WHO, 2004, 2006), followed, in several cases, by mishandling such as prolonged storage under inadequate conditions allowing for multiplication. It is interesting to note that in several recently reported cases, infants had been fed with sterile ready-to-feed formulae or mother's milk pointing to contamination during handling of the feeds from an environmental source in the hospital (Stoll *et al.*, 2004; Conde Aguirre *et al.*, 2007; Ray *et al.*, 2007). Recent taxonomic investigations have allowed demonstration of specific features and characteristics of isolates, and a new classification as *Cronobacter sakazakii* has been proposed (Iversen *et al.*, 2007).

Outbreaks due to viruses are usually associated with unprocessed foods such as seafood, produce or berries and contaminated water while industrially manufactured foods have rarely been identified as sources of viral infections. Several groups of viruses have emerged as significant causes of foodborne and waterborne infections. Of these, norovirus and hepatitis A are currently the most important in terms of the number of attributed cases or outbreaks. It has in fact been shown that noroviruses are the single most common cause of gastroenteritis in patients of all age groups (Fiore, 2004; Koopmans and Duizer, 2004; Carter, 2005; Vasickova *et al.*, 2005; Logan and O'Sullivan, 2008).

The outbreaks of bovine spongiform encephalopathy (BSE) in bovines in the UK and then in different European countries in the 1990s, as well as the discovery that a variant of Creutzfeld Jacob disease (vCJD) in teenagers and young adults is caused by the exposure to the agent (see Chapter 30) can certainly be considered as an emerging public health threat (Bradley *et al.*, 2006; Collee *et al*, 2006; Grist, 2007; Ducrot *et al.*, 2008). This emergence is likely to be linked to changes in husbandry and processing technologies, in particular the dry rendering of animal carcasses along with the more widespread use of mechanically recovered meat. Both are leading to an increased risk of introduction of tissues contaminated with particles of Transmissible Spongiform Encephalopathy (TSE) in the food chain (Cooper and Bird, 2002; Caramelli *et al.*, 2006; Doherr, 2006).

5.2 Factors contributing to the emergence of new pathogens

An updated literature survey published by Woolhouse and Gowtage-Sequeira (2005) identified 1407 recognized human pathogen species with 177 (13 %) being considered as emerging or re-emerging. While not restricted to foodborne or waterborne pathogens, this survey is nevertheless of interest as the authors discuss the 10 main factors believed to trigger this increase, in order of importance:

1. changes in land use or agricultural practices;
2. changes in human demographics and society;
3. changes in population health status (e.g. HIV, malnutrition);
4. hospital and medical procedures;
5. pathogen evolution (e.g. antibiotic resistance, increased virulence);
6. contamination of food or water sources;
7. international travel;
8. failure of public health programs;
9. international trade;
10. climate change.

With the exception of (4) and to some extent of (8), the different factors listed are certainly valid as well for foodborne and waterborne pathogens. This survey confirms similar discussions published by different authors (IFT, 2000, Blancou *et al.*, 2005; Pompe *et al.* 2005). In a significant number of cases of diarrhoea no infective agent can be recovered suggesting the occurrence of enteropathogenic agents yet to be discovered. Janda and Abbott (2006) have investigated the situation for five separate Gram-negative bacteria with respect to different criteria, ranking their potential ability to cause disease. While they concluded that new groups such as *Providencia alcalifaciens* or *Bacteroides fragilis* could be considered as enteropathogens, there was little evidence for others such *Hafnia alvei* to be considered as such. This type of approach may be useful in order to more systematically investigate other groups and to determine their role as emerging enteric pathogens.

The carriage of foodborne pathogens in animals is favored by different factors associated with increased densities of production animals. This is linked with modern and intensive breeding methods. Such intensive breeding techniques are not only applied to cattle, pigs or poultry but also in aquaculture for the production of fish, shrimp, mussels or oysters. They go along with the use of cheap sources of feed such as silage, rendered animal products, animal waste, plant- and animal-based fats which have been shown to harbor pathogens such as *Salmonella* spp., *E. coli* O157:H7, *L. monocytogenes*, *C. jejuni* as well as causative agents of BSE (Crump *et al.*, 2002; Sapkota *et al.*, 2007). In addition, supplementation of antibiotics in a number of the feeds used is also contributing to the build-up of antibiotic-resistant microbial populations including pathogens which then may enter the food chain.

Intensive breeding contributes to the spread of pathogens throughout herds and flocks and has not only been discussed for well-known pathogens but also for some of those considered as potentially emerging pathogens such as Avian Influenza Virus or *Mycobacterium paratuberculosis* (McKenna *et al.*, 2006; Alexander, 2007; Capua and Marangon, 2007; Benedictus *et al.*, 2008).

Fecal shedding leads then also to the contamination of the surrounding environment, including surface waters. Agricultural practices such as the land application of manure or biosolids contribute further to the spread of both enteric bacterial pathogens, and viruses and parasites of human or animal origin (Gerba *et al.*, 2002; Gerba and Smith, 2005). The use of contaminated manure or irrigation water has led to numerous documented outbreaks. Examples are fruits from tropical countries, tomatoes, berries, sprouts or leafy produce such as salads and very recently spinach. They have been at the origin of outbreaks involving well-established pathogens such as *Salmonella* spp. or *E. coli* O157:H7 as well as unusual ones such as protozoan parasites, and they need to be considered in food safety management systems (Beuchat, 2006; Matthews, 2006; Bassett and McClure, 2008)

Changes in the demographic structure of the populations also have an impact on the number of cases of waterborne and foodborne diseases. The absolute number of cases is increasing as a function of the world's growing population, in particular in developing countries. This leads also to increasing difficulties in the supply of safe food and in particular of safe water in more and more regions of the world (McMichael and Butler, 2005; Dixon *et al.*, 2007; Fewtrell *et al.*, 2007; Hrudey and Hrudey, 2007).

Regarding the hosts, there has been an increase in the number of people belonging to sensitive sub-populations over the last decade (IFT, 2000), and this trend will certainly persist in the future. These sub-populations show a higher susceptibility to food or waterborne illnesses caused by well-known pathogens but also from microorganisms not usually pathogenic to healthy consumers. They encompass pregnant women, more fragile groups such as prematures, infants up to a few months and elderly persons over 75 years of age, persons with reduced immune defenses such as AIDS patients or patients receiving anti-carcinogenic or immunosuppressive medications as well as patients suffering from chronic diseases such as diabetes or inflammatory bowel diseases (Kendall *et al.*, 2003, 2006; Koehler *et al.*, 2006; Lopez *et al.*, 2006; Millar and Moore, 2006). Under such conditions fragile sub-populations may become susceptible to established pathogens through unusual food sources and they are, as a consequence, considered as emerging pathogens by the authors describing the cases. An example is a recent outbreak of *B. cereus* attributed to the consumption of contaminated tea by immunocompromised children (El Saleeby *et al.*, 2004).

In developed countries these sub-populations represent up to 20 % or more of the total population (Morris and Potter, 1997) with an increasing tendency. Although reliable statistical data are not available from all countries, increases in the size of such sub-populations, in particular AIDS patients, are also occurring in developing countries. These countries and regions are in addition frequently affected by malnutrition, an important factor contributing to an increased susceptibility of hosts to infections.

While pathogenicity characterizes the ability of a microorganism to cause illness, virulence can be considered as the degree of pathogenicity. For example, *E. coli* encompasses a wide array of very different types. They range from completely avirulent strains which are part of the normal intestinal flora, to strains causing mild symptoms such as diarrhea, up to highly virulent strains of *E. coli* O157:H7 which are at the origin of illnesses with severe symptoms (Kaper *et al.*, 2004; Manning *et al.*, 2008).

The role of genetic diversity in adaptation and evolution of microorganisms has been discussed in detail by several authors (Wren, 2000; Woolhouse *et al.*, 2005; Raskin *et al.*, 2006). Changes in genome and emergence of more virulent clones causing more severe diseases have been described for very different foodborne pathogens, for example for *Salmonella* spp. (Morales *et al.*, 2007), *L. monocytogenes* (Orsi *et al.*, 2007), *B. cereus* (Cardazzo *et al.*, 2008) or *T. gondii* (Khan *et al.*, 2007). Other phenomena leading to an evolution in the virulence of certain microorganisms have been described. An example is the vertical or horizontal acquisition by otherwise avirulent strains of genetic elements such as pathogenicity islands endowing them with pathogenic characteristics (Hochhut *et al.*, 2005; Gal-Mor and Finley, 2006; Kelly *et al.*, 2008). The occurrence of such transfer events has been identified and discussed also for well-established pathogens relevant for foods, for example in Gram-positives such as *Staph. aureus* (Novick and Subedi, 2007) or Gram-negatives such as *Salmonella* spp. (Foley and Lynne, 2008) and entero-haemorrhagic *E. coli* (Caprioli *et al.*, 2005). Such mechanisms may also be relevant in the case of sporadic outbreaks due to microorganisms not usually considered as pathogens, i.e. opportunistic pathogens such as strains of different species of Enterobacteriaceae (Cordier, 2006).

Intracellular communication among bacteria is triggered by small molecules called autoinducers, such as acylated homoserine lactones (AHLs), which become effective as soon as a microbial population reaches a certain threshold level. This phenomenon has been described as quorum sensing and is thought to play a significant role in virulence (Waters and Bassler, 2005; Hu and Ehrlich, 2008). Quorum sensing has been discussed by several authors in relation with food spoilage (Rasch *et al.*, 2005) as well as for foodborne pathogens such as *Cl. botulinum* (Zhao *et al.*, 2006), *Salmonella* Typhimurium (Choi *et al.*, 2007). While several authors have investigated the role of quorum sensing on mechanisms such as virulence and biofilm formation including in food processing eveironments (Van Houdt *et al.*, 2003; Smith *et al.*, 2004; Brandl, 2006; Lindsay and von Holy, 2006; Jayaraman and Wood, 2008) others have focused on specific emerging pathogens such as *Helicobacter pylori* (Rader *et al.*, 2007) or Gram-negative enteric microorganisms such as *Burkholderia* spp. or *Serratia* spp (Ulrich *et al.*, 2004; Van Houdt *et al.*, 2007). Research to block such a communication in order to protect hosts is currently ongoing and is gaining in importance.

Another cause of the increase in bacterial zoonoses is linked to the occurrence of pathogens resistant to antibiotics. This occurrence has been reported for several microorganisms including pathogens such as *Salmonella* spp. for decades. It has, however, been showing an increasing trend during recent years, and the emergence of strains showing multiple resistance to several antibiotics has been described repeatedly (Velge *et al.*, 2005; Brenner *et al.*, 2006; Larson, 2007). This trend is attributed to the widespread use and misuse of antibiotics in both human and veterinary medicine. The use of antimicrobial drugs is often considered to be a major driver of antimicrobial resistance since the largest quantities of antibiotics used worldwide, even taking into consideration those normally used to treat humans, are used for veterinary applications. This frequently results in sub-therapeutic exposures leading to the occurrence of resistant pathogens which spread through the food chain (Silbergeld *et al.*, 2008). The risk of such an exposure has been evaluated by different expert groups, but no general conclusions could be drawn (Phillips *et al.*, 2004; Snary *et al.*, 2004). In Europe, for example, it was therefore recommended that a full risk assessment for specific food-pathogen combinations be performed (EFSA, 2006, 2008).

In terms of the impact of processing technologies, the example of BSE has already been discussed in the first part of the introduction. Food manufacturers are responding to the consumer's desire for minimally processed foods or foods containing less salt, less sugar or fewer preservatives. New technologies such as high-pressure processing or combinations of milder processing steps can lead to the modification of microbial populations in the food matrices not normally observed with traditional technologies. This is due to differences in the behaviour of different microorganisms and their responses to stress and consequent modifications to their survival and recovery. This impact has been discussed in general terms (Skovgaard, 2007) for specific categories of food such as meat (Duffy *et al.*, 2008) or for specific microorganisms such as *L. monocytogenes* (Gandhi and Chikindas, 2007).

Globalization leads to a minimization of geographic barriers to well-known or emerging pathogens. Global sourcing can allow the movement of microbial pathogens but also of parasites or viruses from areas in which they are indigenous to places were they have not yet been identified. Lack of familiarity with some of these 'exotic' pathogens may thus also lead to their misidentification and/or the application of inappropriate control measures. The availability of ethnic foods and foods imported from remote countries is increasing as well as food catering or street vending. This goes along with the occurrence and spread of different pathogens but in particular of viruses or parasites (McMichael, 2004; Blancou *et al.*, 2005; Flint *et al.*, 2005; Sakaguchi, 2005).

The impact of the evolving climate is influenced by different events such as heavy rainfalls and associated floods leading to a more rapid spread of vectors of infectious diseases. Increasing water temperatures lead to the occurrence of blooms of various planktonic species such as cyanobacteria or dinoflagellates which are at the origin of outbreaks linked with seafood contaminated with their toxins. The general climatic impact on food and waterborne diseases has been discussed in detail by Hunter (2003). In an effort to better understand the consequences and predict future trends, investigations into the impact of climatic parameters on different pathogens have also been conducted, in particular for parasites (Polley, 2005) but also for well-known pathogens such as *Campylobacter* spp. (Kovats *et al.*, 2005).

5.3 How to identify emerging pathogens: sources of information

Regular review and surveillance of published information on food and waterborne pathogens is certainly crucial to the food industry. It represents an important tool enabling food manufacturers to constantly keep abreast of information on relevant pathogens and also on new and emerging issues. This information should serve as an impetus to regularly review one's own situation and, where deemed necessary, to adapt implemented control and management measures in light of new learning and conclusions.

A first element to consider is the availability of published information related to food and waterborne outbreaks. The most easily accessible sources of information are websites reporting recalls and withdrawals of contaminated food products. While such alerts are frequently limited to well-known pathogens, they can prove useful in identifying new or uncommon foods or raw materials and ingredients as vectors. Examples are the occurrence of *S.* Agona or of *Cl. botulinum* in herbal tea, both causing unusual outbreaks of salmonellosis and infant botulism in Germany and Spain respectively (Koch *et al.*, 2005; Bianco *et al.*, 2007).

Such information is of particular importance to ensure that potential hazards are not overlooked when performing a hazard analysis in HACCP studies. Such sites are usually maintained by national Public Health Authorities such as www.food.gov.uk/enforcement/alerts in the UK; in the USA the site www.recalls.gov provides links to the sites of both the Food and Drug Administration (FDA) and the US Department of Agriculture (USDA); in France the Institut de Veille Sanitaire (InVS; www.invs.sante.fr) also includes food-related alerts. In addition there are national or supranational scientific organizations such as the Center for Science in the Public Interest (CSPI, www.cspinet.org/foodsafety) which maintains databases on outbreaks or the European Centre for Disease Prevention and Control (ECDC, www.ecdc.europa.net) posting weekly compilations of recalls, withdrawals and alerts within the European Union as well as providing links to bulletins issued by several European national health authorities. Other examples are food safety websites such as the ones maintained by private companies, e.g. http://www.foodhaccp.com/ or http://www.ecolab.com/Initiatives/FoodSafety/.

The situation is more difficult to deal with when completely new microorganisms (at least in the field of foods) are implicated or in case of very sporadic and infrequent outbreaks. In such cases the detailed information may not become available until long after the cases have occurred.

Information obtained from such websites needs nevertheless to be complemented by information posted by national, regional or international networks dedicated to the prevention and control of zoonoses and foodborne diseases. These networks are usually devoted to and specialize in the collection and exchange of epidemiological data on outbreaks and cases including molecular profiles of isolates from such events. While access to part of the data is restricted to members, public information is, however, still useful. Examples of such networks which have frequently been developed from national initiatives into regional or even international ones are SalmNet, EnterNet, PulseNet, MedVetNet (Fisher, 1999, 2004). While the surveillance and typing schemes are devoted to known pathogens, thus covering about 60 % of the known foodborne pathogens, in numerous cases the etiological agent is neither isolated nor identified. According to Mead *et al.* (1999), in the USA more than 80 % of the 76 million cases of foodborne illness are of 'unknown origin'. This will of course trigger new research to identify the vector; particular attention has, for example, recently been focused on parasites or viruses as causative agents for foodborne illnesses (Pozio, 2003; Koopmans and Duizer, 2004). Recently, the Divine-Net has been created as a project under the European Health and Consumer Protection Directorate General to foster national and transnational surveillance of outbreaks due to foodborne viruses and in particular noroviruses (Kroneman *et al.*, 2008).

Surveillance activities involve the structured and systematic collection of data on outbreaks and their evaluation and dissemination (Swaminathan *et al.*, 2001). Molecular subtyping schemes such as pulsed field gel electrophoresis (PFGE) or ribotyping provide a powerful tool in the investigation of outbreaks and in the establishment of links between apparently unrelated sporadic cases (Prager and Tschäpe, 2003). Recent examples are the investigation of two consecutive outbreaks due to infant formulae contaminated with *Salmonella* strains showing distinct molecular characteristics (Brouard *et al.*, 2007), a small outbreak in Switzerland due to a soft cheese contaminated with particular pulsovars of *L. monocytogenes* (Bille *et al.*, 2006), multistate outbreaks in the USA due to new clonal types of *L. monocytogenes* (Kathariou *et al.*, 2006), an outbreak in The Netherlands due to steak tartare contaminated with *E. coli* O157 (Doorduyn *et al.*, 2006), an outbreak in France linked with infant formulae contaminated with specific types of *Cr. sakazakii* (Desenclos *et al.*, 2007), the investigation of outbreaks in France and Canada, respectively, caused by foods contaminated with *Staph. aureus* (Kérouanton *et al.*, 2007) or *Cl. botulinum* (Leclair *et al.*, 2006).

The interconnection or expansion of existing schemes such as SalmNet, EnterNet and PulseNet collecting relevant information results in increased efficiency. The accumulation of information will help in determining the extent of the diseases and in identifying effective and appropriate control measures (Borgdorff and Motarjemi, 1997; ICMSF, 2006).

With respect to new pathogens, such surveillance schemes also provide an increasing chance of detecting strains which are not normally part of the routine analytical activities but which will draw particular attention in case of a sudden and unusual accumulation of cases of illnesses. Such patterns of illness will more than likely trigger specific investigations in order to determine the cause(s). In certain cases this will also allow identification of unusual or new pathogens such as *E. coli* O111 in Australia (Paton *et al.*,

1996), which has since then been identified in other outbreaks, or *E. coli* O39 in the USA (Hedberg *et al.*, 1997).

It is essential for food manufacturers to extend the gathering and collection of data and information beyond simple epidemiological data and statistics. The information collected needs to encompass detailed case studies and investigations into the cases and sources of contamination and into the fate and behavior of the pathogens involved. It is also important to understand as thoroughly as possible their ecology both in foods and matrices and in the environment, but in particular the processing environment. This knowledge is key in designing, reviewing or improving, where necessary, implemented hygiene control measures.

In the past, relevant details of the investigations of outbreaks or individual cases became available in peer-reviewed publications only 1–2 years later, in some extreme cases only several years later. Today, epidemiological investigations and updates of the progress on investigations are becoming much more rapidly available.

Another element to consider in the design and implementation of management options are the requirements or recommendations issued by Public Health Authorities at national or international levels. For example, over the last few years, microbiological risk analysis has become more widely applied by authorities and organizations in different countries. In terms of disease, risk is a measure of the probability of occurrence of a particular illness caused by a specific pathogen and the severity of that illness. Microbiological risk analysis is based, as defined by Codex Alimentarius (CAC, 1999), on the following three components: risk assessment, risk management and risk communication. It will increasingly become an essential part of the establishment of science-based policies.

When used to define regulations and policies, the risk assessment reflects the expected impact of a particular pathogen/food combination. It also provides indications as to which control measures are possible and evaluates their potential efficacy in eliminating or minimizing the public health burden. In addition, the systematic and structured assessment provides an idea of the level of urgency and controversy related to the assessed issue. The risk assessment, the scientific and technical part of the risk analysis, has four main components which are discussed in more detail in Chapter 4 of this book.

(1) The hazard identification involves the identification of the pathogen and its nature, its known or potential health effects as well as the individual risks;
(2) The exposure assessment describes the exposure pathways and estimates the likely frequency and level of consumption of the food contaminated with the pathogen under consideration as well as other likely sources of the pathogen.
(3) The hazard characterization explores the relationship between the

exposure level and the adverse effect caused by the pathogen, taking into consideration both the frequency and severity.

(4) The risk characterization identifies the probability of a population or subpopulation experiencing an adverse health impact from exposure to the particular food or category of foods in case of contamination. It also describes the variability and uncertainty of the risk and determines gaps in the assessment.

The outcome of such risk assessments is used to define a policy to adequately protect consumers from illness or injury caused by foods. It is used to establish an appropriate level of protection (ALOP), a concept introduced in the WTO/SPS agreements. The ALOP represents a level of protection deemed appropriate by the country or authority establishing a sanitary or phytosanitary measure to protect human, animal or plant life or health within a territory. The ALOP has been defined as the number of cases of illness per year and per 100 000 persons in a given population caused by a certain microorganism in a food considered to be appropriate, acceptable or tolerable. An ALOP is thus typically measured in terms of a probability of disease or a number of cases per year. An example is the Food Safety Initiative in the USA in which current values for major pathogens as well as target values are given with the goal being a reduction of the health burden by 50 % by 2010 (FDA, 2000). However, such values cannot readily be used by governments to determine requirements in terms of food safety or by manufacturers to define appropriate measures to reach the established requirements. For this reason the ALOP needs to be translated into a desired level of safety and thus to a level of a hazard in a food. This target is referred to as food safety objective (FSO) and defined as the maximum frequency and/or concentration of a hazard in a food at the time of consumption that provides or contributes to the appropriate level of health protection (ALOP) (ICMSF 2002; CAC, 2004; Stringer, 2005).

For certain products it may also be necessary to define specifications for the level of a hazard at an intermediate point in the food chain. For this purpose the term performance objective (PO) has recently been defined by Codex Alimentarius (CAC, 2004). This allows the diversity of food products to be taken into account and varying requirements for products at different points in the food chain to be established. In the case of shelf-stable products, for example, where no multiplication occurs during distribution, the FSO and PO are identical, while in the case of foodstuffs requiring cooking before consumption, the FSO > PO. As in the case of the FSO, the PO represents a maximum frequency and/or concentration of a hazard in the food at a specific point in the food chain, e.g. at the factory or at retail level. A detailed discussion on FSOs is provided in Chapter 4 of this book.

The FSO and PO thus represent quantifiable elements and serve as an interface between governmental and industrial food safety management systems. These values indeed quantitatively express the levels of pathogens in a food considered as acceptable by the authorities and which have to be achieved through preventative control measures during processing and further handling. A FSO can also be used to derive microbiological criteria which are then used to determine the acceptability of a specific lot with respect to quality and safety requirements. Microbiological criteria are, however, not identical to the FSO as outlined by the ICMSF (2002).

The conceptual formula (Eq. 5.1) developed by the ICMSF (2002) provides an approach to determine whether an FSO or a PO can be met as well as to understand which factors may contribute to achieving this goal or, on the contrary, may jeopardize the safety of a product:

$$H_0 - \sum R + \sum I \leq \text{FSO or PO} \qquad [5.1]$$

where H_0 represents the initial load of a specific pathogen, $\sum R$ represents the total reduction achieved during the whole process, $\sum I$ represents the total increase observed during the whole process and which is composed by growth or recontamination and FSO and PO represent the hazard levels at the moment of consumption and in the food chain, respectively. All values are expressed as log units.

This conceptual formula can be used to describe the fate of a specific pathogen in each step of the food chain, for example during primary processing, manufacturing of products, distribution or during preparation, such as in food service establishments. The effect of one or more control measures applied to achieve the required FSO or PO, i.e. a frequency or concentration of a pathogen in a determined food, is characterised by $\sum R$ and $\sum I$ and is expressed as a Performance Criterion (PC); the parameters at single processing steps are termed as Process Criteria (e.g. a time/temperature combination at a sterilization step) (CAC, 2004; Stringer, 2005).

So far, most of the risk assessments performed have been focused on well-established pathogens and foods frequently involved in foodborne outbreaks. Examples are the risk assessments for *L. monocytogenes* in selected ready-to-eat foods (FDA, 2003) or for *Salmonella* Enteritidis in eggs (FSIS, 1998) published in the past few years. Other risk assessments or risk profiles on less common combinations of pathogens and foods such as *L. monocytogenes* and ice cream have been recently published by the Food and Environmental Hygiene Department in Hong Kong (FEHD, 2001) and by the New Zealand Food Safety Authority (NZFSA, 2003) or for *B. cereus* and rice (NZFSA, 2004). These documents provide good summaries of existing knowledge and detailed information on current opinions with regard to these potential issues. This is important to assess whether the measures currently implemented by food manufacturers are appropriate or whether they need to be strengthened. In the case of the ice cream, for example, both authorities agreed to a very low risk related to this category of products. This element can be considered when reviewing current preventive measures applied in processing facilities which seem therefore adequate in achieving the appropriate protection. In the case of *B. cereus*, the document represents a very good basis for discussions on microbiological criteria demonstrating that a certain tolerance will most likely not affect public health.

It is certainly not the purpose of this chapter to discuss published risk assessments and their impact on food safety management systems implemented by food manufacturers. The websites of national Public Health Authorities as mentioned in the first section of this chapter, websites of the Joint FAO/WHO Expert Meetings on Microbiological Risk Assessments (JEMRA), the EFSA as well as sites such as the one maintained by Joint Institute for Food Safety and Applied Nutrition in collaboration with the University of Maryland and the Food and Drug Administration (http://www.foodrisk.org/) are good resources to keep abreast of existing or newly published risk profiles or risk assessments.

What is the situation concerning emerging pathogens? The situation concerning emerging or re-emerging pathogens is of course much more difficult to handle because detailed knowledge is frequently lacking. It is certainly not possible to perform risk assessments or even risk profiles in the case of individual sporadic cases. However, as the number of cases increases and the information on the causative agents and the food vehicle(s) becomes available, opinions and strategies may be developed. When epidemiological evidence indicates that a hazard is not under control or a new one is emerging, expert panels may identify ways to increase consumer protection. These expert panels, usually composed of participants from different disciplines and haring differing perspectives, will evaluate the risks based on the available scientific data and propose possible mitigation options if deemed necessary. Such panels may also be involved in addressing possible issues caused by changes in food processing technologies, food packaging or distribution systems.

Information is, however, not completely missing and certain newly recognized potential waterborne pathogens such as *Cryptosporidium* have recently been addressed in a risk assessment (Pouillot *et al.*, 2002). Government and international bodies have, for example, made extensive use of expert panels in assessing the situation with respect to *Cronobacter sakazakii* (FAO/WHO, 2004, 2006) and the opinion of the EFSA with respect to *Cr. sakazakii* (EFSA, 2004) or the relationship between Enterobacteriaceae and *Cr. sakazakii* or *Salmonella* in the case of infant formulae (EFSA, 2007). The report and recommendations issued by the FAO/WHO following the two expert consultations have been used to progress with the review of the Code of Hygiene for the Manufacture of Infant Formulae as well as for the revision of the microbiological criteria established by the Codex Alimentarius about 30 years ago (CAC, 1979). Based on the two assessments, recommendations in terms of management during processing but also during preparation in hospitals have been included in the revised draft code (CAC, 2007).

On different occasions industry expert panels have considered the factors leading to foodborne disease due to, at the time, emerging pathogens and developed recommendations as to appropriate control measures. Examples are guidance documents to control *E. coli* O157 in dry fermented sausages (Blue Ribbon Task Force, 1996) through an appropriate combination of control measures or to prevent post-process contamination of meat products by *L. monocytogenes* (Tompkin *et al.*, 1999; Tompkin, 2002). While these documents provide guidance on precise control measures, other reports, such as for example on *M. paratuberculosis* issued by the International Dairy Federation (IDF, 2001), are only providing a summary of the knowledge related to the organism, its ecology, detection and behaviour. Such reports, however, are useful in identifying gaps in knowledge and thus the need for further research.

5.4 Management options

As can be seen from the previous section, information and data related to emerging or re-emerging pathogens is abundant. However, the conditions and interactions of parameters leading to the emergence of new pathogens or to the re-emergence of well-known ones involved in outbreaks caused by products and types of foods so far recognized as safe are not always easy to interpret. Numerous questions can be raised when considering the need to implement control measures and to define the appropriate practices.

- Should otherwise harmless microorganisms causing one or few isolated outbreaks over a period of several years or even decades be considered as emerging pathogens?
- Should such sporadic events be considered as exceptional and due to an unfortunate combination of different factors such as microorganism, food and host and as unlikely to occur again?
- Should such events on the contrary, trigger the immediate implementation of specific control measures or should such measures only be considered in the case of recurring cases over a certain period of time?
- Should microorganisms such as *Mycobacterium paratuberculosis* (Waddell *et al.*, 2008), *Helicobacter pylori* (Quaglia *et al.*, 2008), *Cl. difficile* (Rupnik, 2007) or Avian Influenza Virus (Swayne and Thomas, 2008) which are discussed and perceived as potential foodborne human pathogens by several authors, but where definitive conclusions are still missing, be considered as hazards and thus be addressed in food safety management systems?
- Why are some of the pathogens such as *C. jejuni* or *L. monocytogenes* still considered as 'emerging pathogens' in recent publications (Oldfield, 2001; Bielecki, 2003; Rottman and Gaillard, 2003; Skovgaard, 2007)? Is this related to a real emergence or re-emergence or to a question of perception?

Several elements can help food manufacturers in answering these or similar questions and assist in the decision making process.

5.4.1 Control measures during food manufacture

At the level of the food manufacturers, the basic management tools to ensure the commercialization of safe products are the well-known good manufacturing and good hygiene practices (GMP/GHP) and Hazard Analysis and Critical Control Point (HACCP) principles. These management tools have proven effective in controlling well-known pathogens. Outbreaks which unfortunately still occur are due not to the inappropriateness of the tools but commonly to a poor understanding of issues and the poor implementation of GMP/GHPs and, to some extent, of HACCP or due to accidents. The factors contributing to outbreaks summarised by Hobbs and Roberts (1987), i.e. preparation of food too far in advance, storage at ambient temperature, inadequate cooling, inadequate reheating, contaminated processed foods, consumption of raw foods, etc. are still at the top of the lists established today. The main causes identified 20 years ago are still relevant today.

Both GHP/GMP and HACCP are, in principle, also well-adapted to control emerging and re-emerging pathogens. To be effective, however, these tools must be maintained and updated as new information on such hazards becomes available. The occurrence of new pathogens or the identification of new types of foods as vectors for the transmission of known agents requires revision of the control measures according to these new findings and their adaptation where necessary. This is in no way different to the required revisions of HACCP in case of changes in recipes for products or in processing conditions which, in principle, should help in avoiding issues.

A new element to consider in food safety management systems is the release and application of new ISO Standards and in particular ISO 22 000 on the requirements of food safety management systems of organizations along the food chain (ISO, 2005). ISO 22 000 will trigger the revision of existing control measures by requiring a more quantitative evaluation of existing control measures. While GMP/GHP still represent the basic general control measures necessary to manufacture safe foods, the specific control measures need to be subdivided into Operational Pre-Requisite Programs (OPRP) and HACCP (CCP) depending on their effectiveness in controlling identified hazards.

The question can be raised as to whether certain outbreaks due to such emerging or re-emerging pathogens could have been avoided through a systematic review of existing knowledge and experience or through a comparison of analogies in existing processes or processing steps. In the case of the outbreak due to berries contaminated with parasites, for example, the basic hygiene recommendations usually made to travellers (i.e. not to eat this type of fruit) in tropical countries have been 'bypassed' as a consequence of the importation into a country with a more temperate climate. The outbreaks related to *Salmonella*-contaminated mangoes have certain analogies with outbreaks related to contaminated canned foods and those where post-process contamination due to penetration of contaminated cooling and cleaning water through micro-leaks has been shown. Could this experience not have been used to design and implement appropriate control measures to ensure the safety of the water? *L. monocytogenes* has been recognised for years as the cause of post-process contamination in soft-cheese manufacturing or in refrigerated meats such as pâté. Why was this knowledge not taken into account when considering GMP/GHP in the case of sweet butter production, which has been at the origin of an outbreak (Lyytikainen *et al.*, 2000)? Knowledge and interpretation of the details of previous outbreaks related to refrigerated products, of the microbial ecology of *L. monocytogenes* in dairy processing environments and the principle(s) of post-process contamination may have enabled anticipation of the problem. Could the outbreak due to the consumption of garlic contaminated with *Cl. botulinum* mentioned earlier in this chapter not have been prevented? Although unusual, similar cases had already been described almost ten years earlier (St Louis *et al.* 1988; Morse *et al.*, 1990); the information was available and could have been used to design appropriate control measures. Other examples are recent cases of salmonellosis or recalls due to contaminated chocolate or peanut butter – a review of the literature reveals that such outbreaks have been recurrent over the last 20–30 years and with similar sources and causes.

Good manufacturing and good hygiene practices represent the basic operating and environmental conditions that are necessary for the production of safe, wholesome foods. These general conditions and practices, many of which are described and specified in international or national regulations/guidelines (e.g. CAC, 1997), are in general well-known and widely applied. In recognition of specific and particular issues related to individual products or product categories, more specific documents have been issued by international public health organizations or by professional organizations. Examples are the codes for the hygienic manufacture of low-acid canned products issued by the Codex Alimentarius (CAC, 1993) or the Code of Hygiene for milk and dairy products of the Codex Alimentarius (CAC, 2004) or Guidelines or Codes of Hygiene for the manufacture of chilled culinary products (CFA, 2006), of cocoa, chocolate and confectionery products (IOCCC, 1993) or of dry milk as well as other dairy products (IDF, 1991, 1994). These specific documents highlight particularly important control measures such as the roasting of cocoa beans or the heating conditions of different culinary products as well as requirements for the zoning within a processing facility to prevent post-process contaminations or recommendations on special cleaning procedures.

The HACCP system is a management tool which is applied to control specific hazards, i.e. to prevent, eliminate or reduce them to an acceptable level. It was designed to enhance the safety of foods produced for the American space foods program in the 1970s (Bauman, 1992). Numerous publications have since been devoted to HACCP and details can be found in different documents (e.g. ICMSF, 1988; CAC, 1997; Mortimore and Wallace, 1998; ILSI, 2004).

Examples of the evolution of control measures as a function of new information and learning experiences are provided for *Salmonella*, which are today well-known pathogens but which have been considered as emerging or re-emerging pathogens at one stage. The cases described should serve as an example for the necessary steps in the control of new emerging pathogens. Similar examples could be provided for other pathogens such as *L. monocytogenes*.

To our knowledge, no specific studies have been performed in the past on the occurrence and distribution of *Salmonella* in dry milk powder or infant formulae factories. The knowledge has in fact been gathered following outbreaks such as those reported in the mid-1970s in Australia (Forsyth *et al.*, 2003) or the Farley's case in 1985 in the UK (Rowe *et al.*, 1987). These cases have clearly shown the importance of post-process contamination after the drying step and the need to eradicate *Salmonella* from these areas in the production facilities.

In 1985 a study undertaken by the USDA on milk powder production showed that numerous milk powder plants were still not designed to ensure the containment of unknown contamination and in particular of *Salmonella* (Mettler, 1989). During this period several Codes of Practices have been issued by organisations such as the Codex Alimentarius (CAC, 1979) and the International Dairy Federation (IDF, 1991).

In addition to general requirements on the design of the factory and the layout of processing lines, most of these documents focus on preventing the ingress of *Salmonella* in the production premises. In particular, areas close to the processing line are critical. In the case of *Salmonella* presence, practices to avoid its spread throughout the whole plant but to prevent its eventual establishment in a more persistent manner are described. Recommendations also underline the importance of raw material and ingredient processing and handling as well as on the prevention of post-process contamination of the line and processing environment.

Today the incidence of *Salmonella* in dry milk powders as well as in infant formulae which are manufactured on similar processing lines can be considered as rare, despite the reporting of sporadic outbreaks. These outbreaks can be considered, as indicated above, as accidents or errors in the application of the preventive measures. Similar improvements in the performance of processing lines, i.e. a significant reduction in the incidence of Enterobacteriaceae and *Cr. sakazakii* through the implementation of very stringent hygiene measures during the manufacture of infant formulae, have been described in detail by Cordier (2008).

In the case of chocolate, *Salmonella* emerged as a pathogen in the late 1960s. The occurrence of several outbreaks, which have been summarised in ICMSF (2005) has triggered the development of specific guidelines to improve the hygienic control measures implemented by manufacturers. Particular emphasis has been put on the roasting and zoning of the subsequent processing steps in order to avoid post-processing recontamination (IOCCC 1991, 1993). For almost 15 years no further outbreaks have been registered until 2001/2002 when chocolate contaminated with *S. enterica* serovar Oranienburg was at the origin of a large outbreak (Werber *et al.*, 2005), showing that failures in the well-known preventive measures are still possible.

A question which can be raised in view of the current developments in food safety management, in particular the increasing importance of risk assessments, is related to the benefits of these developments with regard to emerging or re-emerging pathogens. Microbiological risk assessment is a tool for authorities in developing health protection policies. However, the principles and approach as well as certain elements such as the exposure assessment can be used by food manufacturers to review and strengthen their preventive measures. In particular the conceptual equation (5.1) proposed by ICMSF (2002) will contribute to performing a more systematic and structured assessment of the performance of individual processes. Although no FSOs and POs have been established so far, the equation can nevertheless be of use, for example, in benchmarking different products or newly developed products in comparison to existing categories showing a good (historical) safety record.

The fate of microorganisms and in particular of pathogens can be, and in fact is, well described using the different elements (H_0, ΣR and ΣI) of this conceptual formula. The systematic description of the elements of the product–pathogen–pathway (PPP) should allow the outcome of the different steps to be described and identification of the elements making the greatest contribution to the safety or weaknesses representing a potential for food safety issues and also to identify gaps in knowledge. While the elements H_0 and ΣR are usually well understood, ΣI represents, in fact, a combination of increase due to growth and increase due to recontamination. Numerous data are available on the growth behaviour of pathogens as a function of intrinsic and extrinsic parameters (e.g. ICMSF, 1996) and several growth models such the Pathogen Modelling Program or Growth Predictor are available, some directly on the World Wide Web. However, gaps still exist, in particular for specific hurdles and for different new processing technologies. More research will be needed to close these gaps.

In food processing, however, a major issue is the post-process contamination following kill steps, and this will lead to the presence of, frequently, low or very low levels of pathogens. For products supporting growth, this represents an additional risk for multiplication to unacceptable levels. Post-process contamination has been identified as a frequent cause of outbreaks as summarised by Reij *et al.* (2004). Its impact is, however, frequently overlooked and thus appropriate preventive measures are not always implemented or effective. In order to design and implement such control measures, it is important to understand the microbial ecology of these microorganisms and their behaviour in different niches in processing facilities, from their presence or ingress into critical areas to their establishment as a permanent population.

The importance of these routes of contamination as well as pathogen monitoring to verify the effectiveness of preventive measures has been underlined by the ICMSF (2002).

5.5 Future trends

A literature review performed by Taylor *et al.* (2001) has identified 1415 species of infectious agents known to be pathogenic to humans, 868 or 61 % being considered as zoonotic, i.e. they can be transmitted between humans and animals. 175 pathogenic species are associated with diseases considered to be emerging, 75 % (132) being zoonotic. The authors estimated that zoonotic pathogens are twice as likely to be associated with emerging diseases as non-zoonotic pathogens. Although no precise estimation of the role of food and water has been made, it is likely that a number of these 'emerging' pathogens will be transmitted through food vectors. While certain of them will behave similarly to known ones, important differences may exist for others. It will be important to understand these differences in order to adapt the control measures or to develop new ones.

In view of such reports and forecasts, it is important to underline the importance of careful monitoring of the occurrence of outbreaks due to new pathogens or to new food vehicles and of an appropriate interpretation of the situation with respect to prevention. Existing preventative measures (GMP/GHP and HACCP) are, in principle, effective in preventing outbreaks and in controlling pathogens, including those recognized as emerging agents for human foodborne illness. It is, however, important to consider the need for modifications of these measures to address specific emerging pathogens. The design and implementation of such modifications require a very good knowledge of the organisms, i.e. appropriate analytical and typing methods to understand their occurrence and incidence in raw materials and ingredients, their behaviour during processing and in foods as well as their ecology in food processing environments. This will frequently require investments in terms of research that can last for weeks, even up to years, depending upon the microorganism and the complexity of the questions addressed. However, these data and knowledge will allow food manufacturers to review and improve the preventive control measures already implemented.

5.6 References

Alexander DJ (2007) An overview of the epidemiology of avian influenza, *Vaccine*, **25** 5637–44.

Audicana MT and Kennedy MW (2008) *Anisakis simplex*: from obscure infectious worm to inducer of immune hypersensitivity, *Clin Microbiol Rev*, **21** 360–79.

Bassett J and McClure P (2008) A risk assessment approach for fresh fruits, *J Appl Microbiol*, **104** 925–43.

Bauman HE (1992) Introduction to HACCP, in M.D. Pierson and D.A. Corlett (eds), *HACCP Principles and Applications*, Van Nostrand Reinhold, New York, 1–5.

Benedictus A, Mitchell RM, Linde-Widmann M, Sweeney R, Fyock T, Schukken YH and Whitlock RH (2008) Transmission parameters of *Mycobacterium avium* subspecies *paratuberculosis* infections in a dairy herd going through a control program, *Prev Vet Med*, **83** 215–27.

Bennett RW and Monday SR (2003) *Staphylococcus aureus*, in M.D. Miliotis and J.W. Bier (eds), '*Internatioanl Handbook of Foodborne Pathogens*', Marcel Dekker, New York, 41–59.

Beuchat LR (2006) Vectors and conditions for preharvest contamination of fruits and vegetables with pathogens capable of causing enteric diseases, *Br Food J*, **108** 38–53.

Bianco ML, Lúquez C, de Jong LIT and Fernandez RA (2007) Presence of *Clostridium botulinum* spores in *Matricaria chamomilla* (chamomile) and its relationship with infant botulism, *Int J Food Microbiol*, **121** 357–60.

Bielecki J (2003) Emerging food pathogens and bacterial toxins, *Acta Microbiol Pol.*, **52** (Supplement), 17–22.

Bille J, Blanc DS, Schmid H, Boubaker K, Baumgartner A, Siegrist HH, Tritten ML, Lienhard R, Berner D, Anderau R, Treboux M, Ducommun JM, Maliverni R, Genn D, Erard P and Waespi U (2006) Outbreak of human listeriosis associated with Tomme cheese in Nordwest Switzerland, 2005, *Euro Surveill*, **11** 9–13.

Blancou J, Chomel BB, Belotto A and Meslin FX (2005) Emerging or re-emerging bacterial zoonoses: factors of emergence, surveillance and control, *Vet Res*, **36** 507–22.

Blue Ribbon Task Force (1996) *Dry fermented sausage and E. coli O157:H7*, Report no 11–316, National Cattlemen's Beef Association, Chicago, IL.

Borgdorff MW and Motarjemi Y (1997) Surveillance of foodborne diseases: what are the options? *World Health Stat Q*, **50** 12–23.

Bowen AB and Braden CR (2008) Enterobacter Sakazakii disease and epidemiology, in J.M. Farber and S.J. Forsythe (eds), '*Enterobacter Sakazakii*', ASM Press, Washington, DC, 101–26.

Bradley R, Collee JG and Liberski PP (2006) Variant CJD (vCJD) and Bovine Spongiform Encephalopathy (BSE): 10 and 20 years on: part 1, *Folia Neuropathol*, **44** 93–101.

Brandl MT (2006) Fitness of human enteric pathogens on plants and implications for food safety, *Ann Rev Phytophathol*, **44** 367–92.

Brenner GM, Butaye P, Cloeckaert A and Schwarz S (2006) Genes and mutations conferring antimicrobial resistance in *Salmonella*: an update, *Microbes Infect*, **8**, 1898–914.

Brouard C, Espie E, Weill FX, Kerouanton A, Brisabois A, Forgue AM, Vaillant V and de Valk H (2007) Two consecutive large outbreaks of *Salmonella enterica* serotype Agona infections in infants linked to the consumption of powdered infant formula, *Infect Dis J*, **26** 148–52.

Brynestad S and Granum PE (2002) *Clostridium perfringens* and foodborne infections, *Int J Food Microbiol*, **74** 195–202.

CAC (1979) *Recommended International Code of Hygienic Practice for Foods for Infants and Children*, CAC/RCP 2§-1979, Food and Agriculture Organization of the United Nations, Rome.

CAC (1993) *Recommended International Code of Hygienic Practice for Low and Acidified Low Acid Canned Foods*, CAC/RCP23-1979, Rev. 2 (1993), Food and Agriculture Organization of the United Nations, Rome.

CAC (1997) *Recommended International Code of Practice – General Principles of Food Hygiene*, CAC/RCP1-1969, Rev. 3-1997, Amd (1999), Food and Agriculture Organization of the United Nations, Rome.

CAC (1999) *Principles and Guidelines for the Conduct of Microbiological Risk Assessment*, CAC/GL-30, Food and Agriculture Organization of the United Nations, Rome.

CAC (2004) *Code of Hygienic Practice for Milk and Milk Products*, CAC/RCP 57-2004, Food and Agriculture Organization of the United Nations, Rome.

CAC (2007) *Proposed Draft Code of Hygienic Practice for Powdered Formulae for Infants and Young Children*, 39th Session of the Codex Committee on Food Hygiene, Oct 30–Nov 4, 2007, New Delhi, India.

Caprioli A, Morabito S, Brugère H and Oswald E (2005) Enterohaemorrhagic *Escherichia coli*: emerging issues on virulence and modes of transmission, *Vet Res*, **36** 289–311.

Capua I and Marangon S (2007) Control and prevention of avian influenza in an evolving scenario, *Vaccine*, **25** 5645–52.

Caramelli M, Ru G, Acutis P and Forloni G (2006) Prion diseases: current understanding of epidemiology and pathogenesis, and therapeutic advances, *CNS Drug*, **20** 15–28.

Cardazzo B, Negrisolo E, Carraro L, Alberghini L, Patarnello T and Giaccone V (2008) Multiple locus sequence typing and analysis of toxin genes in *Bacillus cereus* foodborne isolates, *Appl Envision Microbiol*, **74** 850–60.

Carter MJ (2005) Enterically infecting viruses: pathogenicity, transmission and significance for food and waterborne infection, *J Appl Microbiol*, **98** 1354–80.

Caya JG, Agni R and Miller JE (2004) *Clostridium botulinum* and the clinical laboratorian. A detailed review of botulism, including biological warfare ramification of botulinum toxin, *Arch Pathol Lab Med*, **128** 653–62.

CFA (2006) *CFA Best PracticeGuidelines for the Production of Chilled Foods*, 4th edn, Chilled Food Association, Peterborough.

Choi J, Shin D and Ryu S (2007) Implication of quorum sensing in *Salmonella enterica* serovar Typhimurium virulence: the luxS gene is necessary for expression of genes in pathogenicity island, *Infect Immun*, **75** 4885–90.

Collee JG, Bradley R and Liberski PP (2006) Variant CJD (vCJD) and Bovine Spongiform Encephalopathy (BSE): 10 and 20 years on: part 2, *Folia Neuropathol*, **44** 102–10.

Conde-Aguirre A, Perez-Legorburu A, Echaniz-Urcelay I, Hernando-Zarate Z and Arrate-Zugazabeitia JK (2007) Sepsis por *Enterobacter sakazakii*, *Anales de Pediatria*, **66** 196–7.

Cooper JD and Bird SM (2002) UK dietary exposure to BSE in beef mechanically recovered meat: by birth cohort and gender, *J Cancer Epidemiol Prevt*, **7** 59–70.

Cordier JL (2006) *Enterobacteriaceae*, in Y Motarjemi and M Adams (eds), *Emerging Foodborne Pathogens*, Woodhead, Cambridge, 450–75.

Cordier JL (2008) Production of powdered infant formulae and microbiological control measures, in J.M. Farber and S.J. Forsythe (eds), *Enterobacter sakazakii*, ASM Press, Washington, DC, 145–86.

Crump JA, Griffin PM and Angulo FJ (2002) Bacterial contamination of animal feed and its relationship to human foodborne illness, *Clin Infect Dis*, **35** 859–65.

Desenclos JC, Vaillant V, Delarocque Astagneau E, Campèse C, Che D, Coignard B, Bonmarin I, Lévy Bruhl D and de Valk H (2007) Les principes de l'investigation d'une épidémie dans une finalité de santé publique, *Médecine*, **37** 77–94.

Dixon J, Omwega AM, Friel S, Burns C, Donati K and Carlisle R (2007) The health equity dimensions of urban food systems, *J Urban Health*, **84** 118–29.

Doherr MG (2006) Brief review on the epidemiology of transmissible spongiform encephalopathies (TSE), *Vaccine*, **25** 5619–24.

Doorduyn Y, de Jager CM, van der Zwaluw WK, Friesema IHM, Heuvelink AE, de Boer E, Wanner WJB and van Duynhoven YTHP (2006) Shiga toxin producing *Escherichia coli* (STEC) O157 outbreak, The Netherlands, *Euro Surveill*, **11** 6636–40.

Doyle MP and Erickson MC (2008) Summer meeting 2007 – the problems with fresh produce: an overview, *J Appl Microbiol*, **105** 317–30.

Ducrot C, Arnold M, de Koeijer A, Heim D and Calavas D (2008) Review on the epidemiology and dynamics of BSE epidemics, *Vet Res*, **39** 2–18.

Duffy G, Lynch OA and Cagney C (2008) Tracking emerging zoonotic pathogens from farm to fork, *Meat Sci*, **78** 34–42.

EFSA (2004) Opinion of the Scientific Panel on biological hazards (BIOHAZ) related to the microbiological risks in infant formulae and follow-on formulae, European Food Safety Authority, *EFSA J*, **113** 1–35.

EFSA (2006) Review of the Community report on zoonoses, European Food Safety Authority, *EFSA J*, **403** 1–62.

EFSA (2007) Review of the opinion on microbiological risks in infant formulae and follow-on formulae with regard to *Enterobacteriaceae* as indicators, European Food Safety Authority, *EFSA J*, **444** 1–14.

EFSA (2008) Scientific Opinion of the Panel on Biological Hazards on a request from the European Food Safety Authority on foodborne antimicrobial resistance as a biological hazard, European Food Safety Authority, *The EFSA Journal*, **765** 1–87, available at: http://www.efsa.europa.eu/cs/BlobServer/Scientific_Opinion/biohaz_op_ej765_antimicrobial_resistance_en.pdf?ssbinary=true, accessed November 2008.

El Saleeby CM, Howard SC, Hayden RT and McCullers JA (2004) Association between tea ingestion and invasive Bacillus cereus infection among children with cancer, *Clin Infect Dis*, **39** 1536–9.

Engberg J (2006) Contribution to the epidemiology of *Campylobacter* infections – A review of clinical and microbiological studies, *Dan Med Bull*, **53** 361–89.

Espié F, Vaillant V, Mariani-Kurkdjian P, Grimont F, Martin-Scahller R, de Valk H and Vernozy-Rozand C (2006) *Escherichia coli* O157 outbreak associated with fresh unpasteurized goat's cheese, *Epidemiol Infect*, **134** 143–6.

Estes MK, Verkataram PBV and Atmar RL (2006) Noroviruses everywhere: has something changed? Gastrointestinal infections, *Current Opinion Infect Dis*, **19** 467–74.

FAO/WHO (2004) *Enterobacter sakazakii and other microorganisms in powdered infant formula: meeting report*, MRA Series 6, Food and Agriculture Organization of the United Nations/World Health Organization, Rome/Geneva, available at: http://www.who.int/foodsafety/publications/micro/es.pdf, accessed November 2008.

FAO/WHO (2006) *Enterobacter sakazakii and Salmonella in powdered infant formula: Meeting report*, MRA Series 10, Food and Agriculture Organization of the United Nations/World Health Organization, Rome/Geneva, available at: http://www.who.int/foodsafety/publications/micro/mra10.pdf, accessed November 2008.

FDA (2000) *Healthy People 2010*, http://www.healthypeople.gov/document/tableofcontents.htm#volume1, accessed November 2008.

FDA (2003) *Quantitative Assessment of Relative Risk to Public Health from Foodborne Listeria monocytogenes Among Selected Categories of Ready-to Eat Foods*, Food and Drug Administration, Washington DC, available at: http://www.foodsafety.gov/~dms/lmr2-toc.html, accessed November 2008.

FEHD (2001) *Microbiological Risk Assessment of Ice Cream*, Food and Environmental Hygiene Department, Hong Kong, available at: http://www.fehd.gov.hk/safefood/report/icecream/report.html#con, accessed November 2008.

Fewtrell R, Kaufmann D, Kay W, Enanoria L, Haller J and Colford J (2007) Water, sanitation, and hygiene interventions to reduce diarrhea in less developed countries: a systematic review and meta-analysis, *Lancet Inf Dis*, **5** 42–52.

Fiore AE (2004) Hepatitis A transmitted by food, *Clin Infect Dis*, **38** 705–15.

Fisher IS (1999) The Enternet international surveillance network – how it works, *Euro Surveill*, **4** 52–5.

Fisher IS (2004) International trends in salmonella serotypes 1998-2003 – a surveillance report from the Enternet international surveillance network, *Euro Surveill*, **9** 45–7.

Flint JA, Van Duynhoven YT, Angulo FJ, DeLong SM, Braun P, Kirk M, Scallan E, Fitzgerald M, Adak GK, Sockett P, Ellis A, Hall G, Gargouri N, Walke H, Braam P (2005) Estimating the burden of acute gastroenteritis, foodborne disease, and pathogens commonly transmitted by food: an international review, *Clin Inf Dis*, **41** 698–704.

Foley SL and Lynne AM (2008) Food animal-associated *Salmonella* challenges: pathogenicity and antimicrobial resistance, *J Anim Sci*, **86** 173–87.

Forsyth JR, Bennett NM, Hogben S, Hutchinson EM, Rouch G, Tan A and Taplin J (2003) The year of the *Salmonella* seekers-1977, *Aust N Z J Public Health*, **27** 385–9.

FSIS (1998) *Salmonella Enteritidis Risk Assessment – Shell eggs and egg products*,

Food Safety and Inspection Service, US Department of Agriculture, Washington, DC, available at: http:// www.fsis.usda.gov/ophs/risk, accessed November 2008.
Gal-Mor O and Finlay BB (2006) Pathogenicity islands: a molecular toolbox for bacterial virulence, *Cellular Microbiol*, **8** 1707–19.
Gandhi M and Chikindas ML (2007) *Listeria*: A foodborne pathogen that knows how to survive, *Int J Food Microbiol*, **113** 1–15.
Gerba CP and Smith JEJ (2005) Sources of pathogenic microorganisms and their fate during land application of wastes, *J Environ Qual*, **34** 42–8.
Gerba CP, Pepper IL and Whitehead LF (2002) A risk assessment of emerging pathogens of concern in the land application of biosolids, *Water Sci Technol*, **46**, 225–30.
Grist EPM (2007) vCJD and the unbounded legacy of BSE, in M.J. Stonebrook (ed.), *Creutzfeld-Jacob Disease – New Research*, Nova Publishers, New York, 127–46.
Heaton JC and Jones K (2008) Microbial contamination of fruit and vegetables and the behaviour of enteropathogens in the phyllosphere: a review, *J Appl Microbiol*, **104** 613–26.
Hedberg CW, Savarino SJ, Besser JM, Paulus CJ, Thelen VM, Myers LJ, Cameron DN, Barrett TJ, Kaper JB and Osterholm MT (1997) An outbreak of foodborne illness caused by *Escherichia coli* O39:NM, an agent not fitting into the existing scheme for classifying diarrheogenic *E. coli*, *J Infect Dis*, **176** 1625–8.
Hobbs BC and Roberts D (1987) *Food Poisoning and Food Hygiene*, 5th edn, Arnold, London.
Hochhut B, Dobrindt U and Hacker J (2005) Pathogenicity islands and their role in bacterial virulence and survival, in W. Russell and H. Herwald (eds), *Concepts in Bacterial Virulence*, Contributions to Microbiology, Karger, Basel, 234–54.
Hrudey SE and Hrudey EJ (2007) Published case studies of waterborne disease outbreaks – evidence of a recurrent threat, *Water Environ Res*, **79** 233–45.
Hu FZ and Ehrlich GD (2008) Population level virulence factors amongst pathogenic bacteria: relation to infection outcome, *Future Microbiol*, **3** 31–42.
Huang DB and White AC (2005) An updated review on *Cryptosporidium* and *Giardia*, *Gastroenterol Clin North Am*, **35**, 291–314.
Hunter PR (2003) Climate change and waterborne and vector-borne disease, *J Appl Microbiol*, **94** Suppl. 37S–46S.
ICMSF (1988) *HACCP in Microbiological Safety and Quality*, International Commission on Microbiological Specifications for Foods, Blackwell Scientific Publications, Oxford.
ICMSF (1996) *Microorganisms in Foods 5, Microbiological Specifications of Food Pathogens*, International Commission on Microbiological Specifications for Foods, Blackie Academic and Professional, London.
ICMSF (2002) *Microorganisms in Foods 7, Microbiological Testing in Food Safety Management*, International Commission on Microbiological Specifications for Foods, Kluwer Academic/Plenum, New York.
ICMSF (2005) *Microorganisms in Foods 6, 2nd edn, Microbial Ecology of Food Commodities*, International Commission on Microbiological Specifications for Foods, Kluwer Academic/Plenum, New York.
ICMSF (2006) Use of epidemiologic data to measure the impact of food safety control programs, International Commission on Microbiological Specifications for Foods, *Food Control*, **17** 825–37.
IDF (1991) *IDF Recommendations for the Hygienic Manufacture of Spray Dried Milk Powders*, Bulletin no. 267, International Dairy Federation, Brussels.
IDF (1994) *Recommendations for the Hygienic Manufacture of Milk and Milk Based Products*, Bulletin no. 292, International Dairy Federation, Brussels.
IDF (2001) *Mycobacterium Paratuberculosis*, IDF Bulletin no. 362, International Dairy Federation, Brussels.
IFT (2000) *IFT Expert Report on Emerging Microbiological Food Safety Issues –*

Implications for Control in the 21st Century, Institute of Food Technologists, Chicago, IL.

ILSI Europe (2004) *A simple Guide to Understanding and Applying the Hazard Analysis Critical Control Point Concept*, 3rd revised edn, International Life Sciences Institute, Brussels.

IOCCC (1991) *The IOCCC Code of Good Manufacturing Practice. Specific GMP for the Cocoa, Chocolate and Confectionery Industry*, International Office of Cocoa, Chocolate and Confectionery, Brussels.

IOCCC (1993) *Hazard Analysis of Critical Control Points (HACCP) System in Chocolate*, International Office of Cocoa, Chocolate and Confectionery, Brussels.

ISO (2005) *Food Safety Management Systems – Requirements for any Organization in the Food Chain*, International Organization for Standardization, Paris.

Iversen C, Lehner A, Mulane N, Marugg J, Fanning S, Stephan R and Joosten H (2007) Identification of *Cronobacter* spp. (*Enterobacter sakazakii*), *J Clin Microbiol*, **45** 3814–6.

Janda JM and Abbott SL (2006) New Gram-negative enteropathogens: fact or fancy? *Rev Med Microbiol*, **17** 27–37.

Jayaraman A and Wood TK (2008) Bacterial quorum sensing: signals, circuits and implications for biofilms and disease, *Ann Rev Biomed Eng*, **10** 145–67.

Kaper JB, Morris JG and Levine MM (1995) Cholera, *Clin Microbiol Rev*, **8** 48–86.

Kaper JB, Nataro JP and Mobley HL (2004) Pathogenic *Escherichia coli*, *Nat Rev Microbio*, **2** 123–40.

Kathariou S, Graves L, Buchrieser C, Glaser P, Siletzky RM and Swaminathan B (2006) Involvement of closely related strains of a new clonal group of *Listeria monocytogenes* in the 1998-99 and 2002 multistate outbreaks of foodborne listeriosis in the United States, *Foodborne Pathogens Dis*, **1**, 292–302.

Khan A, Fux B, Su C, Dubey JP, Darde ML, Ajioka JW, Rosenthal BM and Sibley LD (2007) Recent transcontinental sweep of *Toxoplasma gondii* driven by a single monomorphic chromosome, *Proc Natl Acad Sci*, **194** 14872–7.

Kelly BG, Vespermann A and Bolton DJ (2008) Horizontal gene transfer of virulence determinants in selected bacterial foodborne patogens, *Food Chem Toxicol*, (Epub ahead of print).

Kendall PA, Hillers VV and Medeiros LC (2006) Food safety guidance for older adults, *Clin Infect Dis*, **42** 1298–304.

Kendall P, Medeiros LC, Hillers V, Chen G and DiMascola S (2003) Food handling behaviours of special importance for pregnant women, infants and young children, the elderly, and immune-compromised people, *J Am Diet Assoc*, **103** 1646–9.

Kérouanton A, Hennekinne JA, Letertre C, Petit L, Chesneau O, Brisabois A and De Buyser ML (2007) Characterization of *Staphylococcus aureus* strains associated with food poisoning outbreaks in France, *Int J Food Microbiol*, **115** 369–75.

Koch J, Schrauder A, Alpers K, Werber D, Frank C, Prager R, Rabsch W, Broll S, Feil F, Roggentin P, Bockemühl J, Tschäpe H, Ammon A and Stark K (2005) *Salmonella* Agona outbreak from contaminated aniseed, Germany, *Emerg Infect Disect*, **11**, 1124–7.

Koehler KM, Lasky T, Fein SB, DeLong SM, Hawkins MA, Rabatsky-Her T, Ray SM, Shiferaw B, Swanson E and Vugia DJ (2006) Population-based incidence of infection with selected bacterial enteric pathogens in children younger than five years of age, 1996–1998, *Pediatr Infect Dis J*, **25** 129–34.

Koopmans M and Duizer E (2004) Foodborne viruses: an emerging problem, *Int J Food Microbiol*, **90** 23–41.

Kovats RS, Edwards SJ, Charron D, Cowden J, D'Souza RM, Ebi KL, Gauci C, Gerner-Smidt P, Hajat S, Hales S, Hernandez-Pezzi G, Kriz B, Kutsar K, McKeown P, Mellou K, Menne B, O'Brien S, van Pelt W, Schmid H (2005) Climate variability and *Campylobacter* infection: an international study, *Int J Biometeorol*, **49** 207–14.

Kroneman A, Harris J, Vennema H, Duizer E, van Duynhoven Y, Gray J, Iturriza M,

Böttiger B, Falkenhorst G, Johnsen C, von Bonsdorff H, Maunula L, Kuusi M, Pothier P, Gallay A, Schreier E, Kocj J, Szücs G, Reuter G, Krisztalovics K, Lynch M, McKeown P, Foley B, Coughlan S, Ruggeri FM, Di Bartolo L, Vainio K, Isakbaeva E, Poljsak-Prijatelj M, Hocevar-Grom A, Bosch A, Buesa J, Sanchez Fauquier A, Hernandéz Pezzi G, Hedlund K-O, Koopmans M (2008) Data quality of 5 years of central norovirus outbreak reporting in the European Network for food borne viruses, *J Public Health*, **30** 82–90.

Larson E (2007) Community factors in the development of antibiotic resistance, *Ann Rev Public Health*, **28** 435–47.

Leclair D, Pagotto F, Farber JM, Cadieux B and Austin JW (2006) Comparison of DNA fingerprinting methods for use in investigation of type E botulism outbreaks in the Canadian Arctic, *J Clin Microbiol*, **44** 1635–44.

Lindsay D and von Holy A (2006) Bacterial biofilms within the clinical setting: what healthcare professionals should know, *J Hosp Infect*, **64** 313–25.

Logan C and O'Sullivan N (2008) Detection of viral agents of gastroenteritis: Norovirus, Saprovirus and Aatrovirus, *Future Virol*, **3** 61–70.

Lopez A, Mathers C, Ezzati M, Jamison D and Murray C (2006) Global and regional burden of disease and risk factors, 2001: systematic analysis of population health data, *Lancet*, **367** 1747–57.

Lyytikainen O, Autio T, Maijala R, Ruutu P, Honkanen-Buzalski T, Miettinen M, Hatakka M, Mikkola J, Anttila V J, Johansson T, Rantala L, Aalto T, Korkeala H and Siitonen A (2000) An outbreak of *Listeria monocytogenes* serotype 3a infections from butter in Finland, *J Infect Dis*, **181** 1838–41.

Manning SD, Motiwala AS, Springman AC, Oi W, Lacher DW, Ouellette LM, Mladonicky JM, Somsel P, Rudrik JT, Dietrich SE, Zhang W, Swaminathan B, Alland D and Whittam TS (2008) Variation in virulence among clades of *Escherichia coli* O157:H7 associated with disease outbreaks, *Proc Natl Acad Sci USA*, **105** 4868–73.

Matthews KR (2006) *Microbiology of Fresh Produce*, ASM Press, Washington, DC.

McKenna SL, Keefe GP, Tiwari A, Van Leeuwen J and Barkema HW (2006) Johne's disease in Canada part II: disease impacts, risk factors, and control programs for dairy producers, *Can Vet J*, **47** 1089–99.

McMichael AJ (2004) Impact of climatic and other environmental changes on food production and population health in the coming decades, in *Globalization of food systems in developing countries: impact on food security and nutrition*, FAO Food and Nutrition Paper 83, Food and Agriculture Organization of the United Nations, Rome.

McMichael AJ and Butler CJD (2005) The effect of environmental change on food production, human nutrition and health, *Asia Pac J Clin Nutr*, **14** 39–47.

Mead PS, Slutsker L, Dietz V, McCaig LF, Bresee JS, Shapiro C, Griffin PM and Tauxe RV (1999) Food-related illness and death in the United States, *Emerg Infect Dis*, **5** 607–25.

Mettler AE (1989) Pathogens in milk powders – Have we learnt the lessons? *J Soc Dairy Technol*, **17** 101–6.

Millar BC and Moore JE (2006) Emerging pathogens in infectious diseases: definitions, causes and trends *Rev Med Microbiol*, **17** 101–6.

Morales CA, Musgrove M, Humphrey TJ, Cates C, Gast R and Guard-Bouldin J (2007) Pathotyping of *Salmonella enterica* by analysis of single nucleotide polymorphisms on cyaA and flanking 23S ribosomal sequences, *Environ Microbiol*, **9** 1047–59.

Morris JG and Potter M (1997) Emergence of new pathogens as a function of changes in host susceptibility, *Emerg Infect Dis*, **3** 435–41.

Morse DL, Pickard LK, Guzewich JJ, Devine BD and Shayegani M (1990) Garlic-in oil associated botulism: episode leads to product modification, *Am J Public Health*, **80** 1372–3.

Mortimore S and Wallace C (1998) *HACCP – A Practical Approach*, Aspen, Gaithersbuns, MD.

Nesbakken T (2005) *Yersinia enterocolitica*, in P.M. Fratamico, A.K. Bhunia and J.L. Smith (eds), *Foodborne Pathogens – Microbiology and Molecular Biology*, Caister, Norwich, 223–49.

Novick RP and Subedi A (2007) The SaPls: mobile pathogenicity islands of *Staphylococcus*, *Chem Immunol Allergy*, **93** 42–57.

NZFSA (2003) *Risk profile: Listeria monocytogenes in ice cream*, New Zealand Food Safety Authority, Wellington, available at: http://www.nzfsa.govt.nz/science/risk-profiles/lmono-in-ice-cream.pdf, accessed November 2008.

NZFSA (2004) *Risk profile: Bacillus spp. in rice*, New Zealand Food Safety Authority, Wellington, available at: http://www.nzfsa.govt.nz/science-technology/risk-profiles/bacillus-in-rice-1.pdf, accessed November 2008.

Oldfield EC (2001) Emerging foodborne pathogens: keeping your patients and your families safe, *Rev Gastroenterol Disord*, **1**, 177–86.

Orsi RH, Ripoll DR, Yeung M, Nightingale KK and Wiedmann M (2007) Recombination and positive selection contribute to evolution of *Listeria monocytogenes* inlA, *Microbiol*, **153** 2666–78.

Paton AW, Ratcliff RM, Doyle RM, Seymour-Murray J, Davos D, Lanser JA and Paton JC (1996) Molecular microbiological investigation of an outbreak of hemolytic-uremic syndrome caused by dry fermented sausage contaminated with Shiga-like toxin-producing *Escherichia coli*, *J Clin Microbiol*, **34** 1622–7.

Phillips I, Casewell M, Cox T, De Groot B, Friis C, Jones R, Nightingale C, Preston R and Wadell J (2004) Does the use of antibiotics in food animals pose a risk to human health? A critical review of published data, *J Antimicrobial Chemoth*, **53** 28–52.

Polley L (2005) Navigating parasite web and parasite flow. Emerging and re-emerging parasitic zoonoses of wildlife origin, *Int J Parasitol*, **35** 1279–94.

Pompe S, Simon J, Wiedemann PM and Tannert C (2005) Future trends and challenges in pathogenomics, *EMBO Rep*, **6** 600–5.

Poppe C (1999) Epidemiology of *Salmonella enterica* serovar Enteritidis, in *Salmonella enterica serovar Enteritidis* in humans and animals A.M. Saeed, R.K. Gast, M.E. Potter and P.G. Wall, (eds), Iowa State University Press, Ames, IA, 3–28.

Pouillot R, Beaudeau P, Denis JB and Derouin F (2004) A quantitative risk assessment of waterborne cryptosporidiosis in France using second-order Monte Carlo simulation, *Risk Analysis*, **24** 1–17.

Pozio E (2003) Foodborne and waterborne parasites, *Acta Microbiol Pol*, **52** (Supplement), 83–96.

Pozio E and Zarlenga DS (2005) Recent advances on the taxonomy, systematics and epidemiology of *Trichinella, Int J Parasitol*, **35** 1191–204.

Prager R and Tschäpe H (2003) Genetic fingerprinting (PFGE) of bacterial isolates for their molecular epidemiology, *Berl Munch Tierarztl Wochenschr*, **116** 474–81.

Quaglia NC, Dambrosio A, Normanno G, Parisi A, Patrono R, Ranieri G, Rella A and Celano GV (2008) High occurrence of *Helicobacter pylori* in raw goat, sheep and cow milk inferred by *glm*M gene: A risk of food-borne infection? *Int J Food Microbiol*, **124** 43–7.

Rader BA, Campagna SR, Semmelback MF, Bassler BL and Guillemin K (2007) The quorum sensing molecule autoinducer 2 regulates motility and flagellar morphogenesis in *Helicobacter pylori*, *J Bacteriol*, **189** 6109–17.

Rasch M, Nadersen JB, Nielsen KF, Flodgaard LR, Christensen H, Givskov M and Gram L (2005) Involvement of bacterial quorum sensing signals in spoilage of bean sprouts, *Appl Environ Microbiol*, **71** 3321–30.

Raskin D, Seshadri R, Pukatzki S and Mekalan J (2006) Bacterial genomics and pathogen evolution, *Cell*, **124** 703–14.

Ray P, Das A, Gautam V, Jain N, Narang A and Sharma M (2007) *Enterobacter sakazakii* in infants: Novel phenomenon in India, *Indian J Med Microbiol*, **25** 408–10.

Rajkowski KT and Bennett RW (2003) *Bacillus cereus*, in M.D. Miliotis and J.W. Bier (eds), *International Handbook of Foodborne Pathogens*, Marcel Dekker, New York, 27–40.

Reij MW, Den Aantrekker ED and ILSI Europe Risk Analysis in Microbiology Task Force (2004) Recontamination as a source of pathogens in processed foods, *Int J Food Microbiol*, **91** 1–11.

Riley LW, Remis RS, Helgerson SD, McGee HB, Wells JG, Davis BR, Hebert RJ, Olcott ES, Johnson LM, Hargrett NT, Blake PA and Cohen ML (1983) Hemorrhagic colitis associated with a rare *Escherichia coli* serotype, *N Engl J Med*, **308** 681–5.

Rottman M and Gaillard JL (2003) New foodborne infections, *Rev Prat*, **53** 1055–62.

Rowe B, Begg NT, Hutchinson DN, Dawkins HC, Gilbert RJ, Jacob M, Hales BH, Rae FA and Jepson M (1987) *Salmonella ealing* infections associated with consumption of infant dried milk, *Lancet*, **2** 900–3.

Rupnik M (2007) Is *Clostridium difficile*-associated infection a potentially zoonotic and foodborne disease? *Clin Microbiol Infect*, **13** 457–9.

Sakaguchi A (2005) Emerging and reemerging diseases and globalization, *J Health Politics Law*, **30** 1163–78.

Sapkota AR, Lefferts LY, McKenzie S and Walker P (2007) What do we feed to food-production animals? A review of animal feed ingredients and their potential impacts on human health, *Environ Health Perspect*, **115** 663–70.

Silbergeld EK, Graham J and Price LB (2008) Industrial food animal production, antimicrobial resistance, and human health, *Ann Rev Public Health*, **29** 151–69.

Snary EL, Kelly LA, Davison HC, Teale CJ and Wooldridge M (2004) Antimicrobial resistance: a microbial risk assessment *J Antimicrob Chemother*, **53** 906–17.

Skovgaard N (2007) New trends in emerging pathogens, *Int J Food Microbiol*, **120** 217–24.

Smith JL, Fratamico PM and Novak JS (2004) Quorum sensing: a primer for food microbiologists, *J Food Prot*, **67** 1053–70.

St. Louis ME, Peck SH, Bowering D, Morgan GB, Blatherwick J and Banerjee S (1988) Botulism from chopped garlic delayed recognition of a major outbreak, *Annals Int Med*, **108** 363–8.

Stoll B, Hansen N, Fanaroff A and Lemons J (2004) *Enterobacter sakazakii* is a rare cause of neonatal speticemia or meningitis in vlbw infants, *J Pediatrics*, **144** 821–3.

Stringer M (2005) Summary report – role in microbiological food safety management, *Food*, **16** 775–94.

Swayne DE and Thomas C (2008) Trade and food safety aspects for avian influenza viruses, in D.E. Swayne (ed.), *Avian Influenza*, Blackwell Publishing, Ames, IA, 499–512.

Swaminathan B, Barrett TJ, Hunter SB and Tauxe RV (2001) Pulsenet: the molecular subtyping network for foodborne bacterial disease surveillance, United States, *Emerg Infect Dis*, **7** 382–9.

Swaminathan B and Gerner-Smidt P (2007) The epidemiology of human listeriosis, *Microbes and Infect*, **9** 1236–43.

Taylor LH, Latham SM and Woolhouse MEJ (2001) Risk factors for human disease emergence, *Phil Trans R Soc Lond B*, **356** 983–9.

Tompkin RB (2002) Control of *Listeria monocytogenes* in the food-processing environment, *J Food Prot*, **65** 709–25.

Tompkin RB, Scott VN, Bernard DT, Sveum WH and Gombas KS (1999) Guidelines to prevent post-processing contamination from *Listeria monocytogenes*, *Dairy Food Environ Sanit*, **19** 551–62.

Ulrich RL, DeShazer D, Hines HB and Jeddeloh JA (2004) Quorum sensing: a transcriptional regulatory system involved in the pathogenicity of *Burkholderia mallei*, *Infect Immun*, **72** 6589–96.

Van Houdt R, Aertsen A, Jansen A, Quintana AL and Michiels CW (2003) Biofilm formation and cell-to-cell signalling in Gram-negative bacteria isolated from a food processing environment, *J Appl Microbiol*, **96** 177–84.

Van Houdt R, Givskov M and Michiels CW (2007) Quorum sensing in *Serratia*, *FEMS Microbiol Rev*, **31** 407–24.

Vasickova P, Dvorska L, Lorencova A and Pavlik I (2005) Viruses as a cause of foodborne diseases: a review of the literature, *Vet Med Czech*, **50** 89–104.

Velge P, Cloeckaert A and Barrow P (2005) Emergence of *Salmonella* epidemics: the problem related to *Salmonella enterica* serotype Enteritidis and multiple antibiotic resistance in other major serotypes, *Vet Res*, **36** 267–88.

Vojdani JD, Beuchat LR and Tauxe RV (2008) Juice-associated outbreaks of human illness in the United States, 1995 through 2005, *J Food Prot*, **71** 356–64.

Waddell LA, Rajic A, Sargeant J, Harris J, Amezcua R, Downey L, Read S and McEwen SA (2008) The zoonotic potential of *Mycobacterium avium* spp. *paratuberculosis*: a systematic review, *Can J Public Health*, **99** 145–55.

Waters CM and Bassler BL (2005) Quorum sensing: cell-to-cell communication in bacteria, *Annu Rev Cell Dev Biol*, **21** 319–46.

Werber D, Dreesman J, Feil F, van Treeck U, Fell G, Ethelberg S, Hauri AM, Roggentin P, Prager R, Fisher IST, Behnke SC, Bartelt E, Weise E, Ellis A, Siitonen A, Andersson Y, Tschäpe H, Kramer MH and Ammon A (2005) International outbreak of *Salmonella* Oranienburg due to German chocolate, *BMC Infect Dis*, **5** 7–17.

Woolhouse MEJ and Gowtage-Sequeir S (2005) Host range and emerging and reemerging pathogens *Emerg Infect Dis*, **11** 1842–7.

Woolhouse MEJ, Haydon DT and Antia R (2005) Emerging pathogens: the epidemiology and evolution of species jump *Trends Ecol Evol*, **20** 238–44.

Wren BW (2000) Microbial genome analysis: insight into virulence, host adaptation and evolution, *Nature Rev*, **1** 29–39.

Yeung PSM and Boor KJ (2004) Epidemiology, pathogenesis, and prevention of foodborne *Vibrio parahaemolyticus* infections, *Foodborne Path Dis*, **1** 74–88.

Zhao L, Montville TJ and Schaffner DW (2006) Evidence for quorum sensing in *Clostridium botulinum* 56A, *Lett Appl Microbiol*, **42** 54–8.

6

Pathogen control in primary production: meat, dairy and eggs

G. Duffy, Ashtown Food Research Centre, Teagasc, Ireland

Abstract: The eradication of pathogens from livestock and the farming environment is not yet an achievable goal; however, risk reduction measures can be implemented on the farm to minimise the spread of pathogens within a herd or flock, thus reducing the potential for their transmission into the human food chain. Control measures taken can range from good farming practices and general hygiene-based measures to targeted approaches which are specifically designed to reduce or eliminate carriage and shedding of key pathogenic bacteria by food animals or poultry. This chapter outlines the main microbial hazards associated with ruminant animals, pigs and poultry and the general and targeted approaches taken to control pathogens in these reservoirs.

Key words: farm, control, pathogens.

6.1 Introduction

Foodborne zoonotic diseases are of major public health concern in addition to causing huge economic losses for the agricultural-food sector and wider society. It is well documented that potentially harmful pathogens are shed in animal faeces and that human illness can occur through consumption of contaminated raw and processed meats or other derived foods of animal origin. Infection can also occur as a result of direct contact with faecal material, and this risk may be enhanced when animal waste and effluents from farming operations including manures and slurries are applied as fertiliser to land used for crop, silage production or animals grazing. Recreational use of land contaminated with animal faeces or handling/petting farm animals can also lead to human illness. Pathogens shed in faeces may persist in the farm environment for extended periods ranging from several weeks to many months, posing a risk for further transmission on the farm, the water chain, the fresh food chain and the wider environment. To address this risk there are many efforts focused on control of pathogen shedding by production animals and poultry and the approaches being taken are summarised in this chapter.

6.2 Identifying and assessing hazards and risks

6.2.1 Bovine animals

There is large variability across the world in the manner that beef and dairy cattle are produced. They may be raised in situations ranging from small herds on family holdings to very large feed-lot systems containing thousands of animals. They may be housed indoors in pens or sheds or outdoors on grass or in pens. Diets can vary from grass, hay or silage-based to grain, barley, soybean or other concentrate diets. The variations in scale of production and management systems mean that many different factors can impact on the carriage and shedding of microbial pathogens in the faeces of bovine animals. The most notorious group of bacterial pathogens associated with bovine animals are verocytotoxigenic *E. coli* (VTEC) in particular *E. coli* O157:H7, but *Salmonella* spp., *Listeria monocytogenes*, *Campylobacter* and others have also been recovered from bovine faeces in addition to parasitic agents such as *Cryptosporidium parvum* and *Giardia* (Table 6.1).

The shedding pattern for *E. coli* O157 in a herd is sporadic with epidemic periods of shedding interspersed with periods of non-shedding. Colonisation of cattle is short-term (1–2 months) and shedding of *E. coli* O157:H7 by the animal during this period appears to be transient. Within a herd there can be a small number of persistent high shedders or super shedders with reported concentrations of the pathogen in the faeces of up to 6.7×10^5 and 1.6×10^6 cfu g^{-1} (Matthews *et al.*, 2006). In addition, it is usual that only a small number of animals in the herd are shedders.

Studies on horizontal transmission of *E. coli* O157:H7 in cattle housed together have shown that the pathogen can be transmitted from animal to animal following ingestion of the pathogen at relatively low levels and that animal grooming and the hide are important sources of transmission (McGee *et al.*, 2004).

Pathogens can be readily transmitted within a cattle herd, and potential sources of contamination and cross-contamination include water troughs, feed, rodents, insects, bird droppings and the housing environment. Hoar *et al.* (2001) examined risk factors associated with beef cattle shedding pathogens of potential zoonotic concern including *Campylobacter* sp., *Giardia* sp. and *C. parvum*. *Campylobacter* sp. was positively associated with the number of female animals on the farm. For *Giardia* sp. none of

Table 6.1 Prevalence of selected pathogens in bovine faeces in different regions

Country	Pathogen	Type of cattle (%)	Number positive	Reference
Switzerland	*E. coli* O157	Dairy cattle	23/500 (4.6 %)	Kuhnert *et al.*, 2005
USA	*E. coli* O157	Cull dairy cattle	21/1026 (2.1 %)	Dodson and LeJeune, 2005
USA	*E. coli* O157	Feed-lot cattle	1087/10662 (10.2 %)	Sargeant *et al.*, 2003
England/Wales	*E. coli* O157	Dairy, suckler and fattening herds	219/4663 (4.7 %)	Paiba *et al.*, 2002
Netherlands	*E. coli* O157	Dairy cattle on farms	75/1152 (7 %)	Heuvelink *et al.*, 1998
Norway	*E. coli* O157	Heifers and milking cows	6/1970 (0.3 %)	Vold *et al.*, 1998
Canada	*E. coli* O157	Dairy cattle	12/1478 (0.8 %)	Wilson *et al.*, 1996
USA	*Salmonella*	Feed-lot cattle	4997 (5.5 %)	Fedorka-Cray *et al.*, 1998
USA	*Salmonella*	Beef cattle	509/1681 (30.3 %)	Kunze *et al.*, 2008
USA	*Salmonella*	Beef cows and calves	70/5049 (1.4 %)	Dargatz *et al.*, 2000
USA	*L. monocytogenes*	Beef cattle	255/825 (31 %)	Ho *et al.*, 2007
USA	*Campylobacter*	Dairy cattle	735/1435 (51.2 %)	Englen *et al.*, 2007
Finland	*Campylobacter*	Faeces at slaughter	296/952 (31.1 %)	Hakkinen *et al.*, 2007
USA	*C. parvum*	Beef cattle	1.1 %	Hoar *et al.*, 2001
Spain	*C. parvum*	Beef cattle	6.38 %	Villacorta *et al.*, 1991
Spain	*Giardia duodenalis*	Beef cattle	27/101 (26 %)	Castro-Hermida *et al.*, 2007

the management factors examined was significantly associated with the proportion in a herd testing positive. For *C. parvum* the length of calving season and whether any procedures were performed on the calves in the first two days of life were positively associated with the proportion that tested positive.

A number of studies have been conducted to identify the risk factors associated with transmission of pathogens on beef farms. Gunn *et al.* (2007) investigated factors associated with the prevalence of verocytotoxin producing *E. coli* O157 shedding in Scottish beef cattle. They showed an increased probability of a group containing a shedding animal associated with larger numbers of finishing cattle and the farm being classed as a dairy unit stocking beef animals. Farms that spread slurry on grazing land were more likely to have shedding animals, while those that spread manure were at a lower risk. Groups with older animals were less likely to be identified as positive. There were higher mean levels of shedding associated with animals being housed rather than at pasture, and this effect was stronger in groups which had recently had a change in housing or diet.

Pangloli *et al.*, 2008 investigated the risk factors impacting on the occurrence of *Salmonella* in dairy cows and the farm environment. Major sources of *Salmonella* on the dairy farm were the air in the milking parlor, bird droppings, feeds, water, calf bedding, soils and insects. *Salmonella* ribotyping indicated that most serovars came from different sources, but some might have originated from a common source and transmitted from site-to-site on the farm. These data provide some important information on key animal and environmental sampling sites needed to initiate on-farm management programs for control of this important foodborne pathogen.

6.2.2 Ovine animals

Sheep and goats generally have access to a wide environment for grazing and are thus exposed to a broad range of environmental factors and pathogens. They are reported to asymptomatically carry and shed a range of pathogens including *E. coli* O157, *Salmonella* and *L. monocytogenes*. *E. coli* O157 has been reported at levels of 1.7 % in sheep in the UK (Paiba *et al.*, 2002) and 7.3 % of dairy sheep in Spain (Oporto *et al.*, 2008).

6.2.3 Pigs

Pigs may be produced in situations ranging from enclosed pens in close contact with other pigs to outdoor environments including organic systems. Pigs are most commonly associated with the carriage and shedding of *Salmonella*, *Yersinia* and *Campylobacter*. The reported prevalence of these pathogens in porcine faeces from a number of different geographical studies is shown in Table 6.2.

Table 6.2 Prevalence of selected pathogens in porcine faeces in different regions

Country	Pathogen	Type of pig	Number positive	Reference
Hungary	*S. enterica*	Farrow to finishing	21.5 %	Biksi *et al.*, 2007
Japan	*Salmonella*	Pigs on farm	3.3 %	Futagawa-Saito *et al.*, 2007
USA	*Salmonella*	Finishing pigs: conventional production	4.2 %	Gebreyes *et al.*, 2006
USA	*Salmonella*	Finishing pigs: antimicrobial-free production	15.2 %	Gebreyes *et al.*, 2006
Germany	*Campylobacter*	Finishing pigs	64.7 %	Wehebrink *et al.*, 2008
Germany	*Yersinia*	Finishing pigs	13.3 %	Wehebrink *et al.*, 2008
Finland	*Yersinia enterocolitica*	Finishing pigs	17.7 %	Asplund *et al.*, 1990

6.2.4 Broiler and layer poultry flocks

In the developed world, the majority of broilers and layer hens are produced in intensive production systems (houses or barns) though some are raised outdoors roaming freely in the farm environment. The main bacterial pathogens associated with poultry are *Salmonella* and *Campylobacter.* The prevalence of *Campylobacter* has been reported at levels ranging from 42 % (McDowell *et al.*, 2008) to 100 % in broiler farms (Parisi *et al.*, 2007). A Canadian study reported that the prevalence of *Salmonella*-positive flocks was 50 % (95 % CI: 37, 64) for chickens and 54 % (95 % CI: 39, 70) for turkeys, respectively (Arsenault *et al.*, 2007). A Dutch study (Van de Giessen *et al.*, 2006) reported that the prevalence of *Salmonella* spp. in laying-hen flocks decreased from 21.1 % in 1999 to 13.4 % in 2002. In the UK after rising in the early 1980s and early 90s the number of recorded human cases of *Salmonella enterica* subsp. *enterica* in the UK fell in the late 1990s and early 2000 concomitant with the introduction of vaccination of egg-laying hens against serovar Enteritidis (Cogan and Humphrey, 2003). Pathogens can spread very rapidly within flocks and houses, with breaches in biosecurity being a major issue in operations where 'all in/all out' production is not in practice.

To gain an understanding of the risk factors which influence the contamination and the spread of *Salmonella* in laying hens in Belgium, Namata *et al.*, 2008 intregated data from a 2005 baseline study on the prevalence of *Salmonella* in laying flocks of *Gallus gallus* in Belgium. The main risk factor identified was rearing flocks in cages compared to barns and free-range systems. The results also showed a significantly higher risk for *Salmonella* for a one week increase in flocks' age as well as with a unit increase in the size of the flock.

6.2.5 Eggs

Salmonella is the main pathogen of concern in relation to eggs. Eggs may be contaminated from environmental faecal contamination, however, stringent procedures for cleaning and inspecting eggs have made salmonellosis caused by external, faecal contamination of egg shells rare. Most cases of infection are internal, occurring when the ovaries of apparently healthy hens are contaminated with *Salmonella* Enteritidis which contaminates the eggs before the shells are formed. In a flock, only a small number of hens seem to be infected at any given time, and an infected hen can lay many normal eggs while only occasionally laying an egg contaminated with the *Salmonella* bacterium. In the UK, *Salmonella* was estimated to be present in around one box in every 30 (3.3 %) imported into the UK (FSA, 2006).

6.3 Managing and controlling hazards and risks with ruminant animals

6.3.1 General controls

Pathogens can survive for many months in the farm environment, including animal waste, feed and farm surfaces and water (Fukushima *et al.*, 1999; Maule, 2000), providing a reservoir of contamination for both humans and animals who come in contact with this material. General measures to control the spread of pathogens on the farm are outlined in farm quality assurance schemes and good agricultural practices. These guidelines, often issued at national level, cover general issues such as management and housing of stock, pest control, management of feed and water supply, management of animal waste and disease prevention and control. The introduction of full on-farm Hazard Analysis Critical Control Point (HACCP) systems which attempt to identify, monitor and control all hazards on the farm is hugely challenging due to the diverse nature of the farm environment, and it has been argued that it is unrealistic to expect that primary producers should be able to reduce or eliminate microbial contamination on the farm. Concerns over the complexity of developing a site-specific HACCP plan on a farm, the reams of paperwork required to maintain the system and the need for a third-party audit to verify its effectiveness all indicate that HACCP is unsuitable for farm operations. However, on-farm HACCP has been shown to work when it is kept simple in terms of the key elements to control, the corrective actions to take and the records to keep.

Public awareness of the potential for pathogen transmission through environmental contact particularly in rural settings is important. Illness in farm families, extended families and visitors to farms/open farms has resulted from direct contact with livestock or their faeces. Very serious illness has also resulted from drinking raw milk or consuming unpasteurised dairy products. Washing hands after contact with animals or animal faeces, especially before eating and drinking, is a simple control measure. Children may need to be supervised and farm visitors need to know the importance of hand washing in the prevention of infection. As a result of a number of VTEC outbreaks among children visiting open farms, strict regulations regarding facilities and supervision are now implemented in many countries. Farmers should endeavour to present animals for slaughter with the minimum amount of soil and faecal contamination on their hides/fleeces as these are recognised as the major source of carcass contamination during slaughter. Some countries have introduced clean cattle and sheep policies to specifically address this issue. Farmers should follow the appropriate regulations regarding the application of animal waste to land to minimise the risk of contamination of either water sources or produce.

6.3.2 Specific control measures

A number of specific control measures have been developed to minimise the shedding of pathogens by ruminant animals particularly in the pre-slaughter period. The majority of this research has focused on reducing the shedding of *E. coli* O157:H7 by cattle. Some of these agents act by actively inhibiting the pathogen while some inhibit colonisation of the host animal by the pathogen. They are reviewed in the section below.

Dietary manipulation

A number of research studies have focused attention on the impacts of dietary regimes to minimise shedding of *E. coli* O157 by cattle in the preslaughter period. Berg *et al.*, (2004) evaluated the prevalence and magnitude of faecal of *E. coli* O157:H7 excretion in commercial feed-lot cattle fed a barley-based finishing ration versus cattle fed a corn-based diet. Average *E. coli* O157:H7 prevalence was 2.4 % in barley-fed cattle and 1.3 % in the corn-fed cattle ($P < 0.05$), and the average magnitude of fecal *E. coli* O157:H7 excretion was 3.3 log cfu g^{-1} in the barley-fed cattle and 3.0 log cfu g^{-1} in the corn-fed cattle ($P < 0.01$). However, another study (Fegan *et al.*, 2004) reported no significant difference in prevalence or numbers of *E. coli* O157 collected from grass-fed (pasture) and lot-fed (feed-lot) cattle. Overall the results gained from dietary manipulation are inconclusive and further research is needed on their potential commercial use.

Seaweed extract

Tasco 14™ is a seaweed extract that has been trialled to reduce *E. coli* O157 shedding in cattle (Braden *et al.*, 2004). While successful at reducing the pathogen it had adverse effects on animal performance levels indicating that further research is needed before it could be commercially applied.

Antibiotics

The impact of administering antibiotics to bovine animals either directly in milk replacers or indirectly in their drinking water on the shedding of *E. coli* O157:H7 has been studied. Alali *et al.*, 2004 compared the concentration and duration of faecal shedding of *E. coli* O157:H7 between calves fed milk replacer with or without antibiotic (oxytetracycline and neomycin) supplementation. A comparison of the duration of faecal shedding between treated and untreated calves showed no significant difference between groups. McAllister *et al.*, 2006 also revealed no evidence that dietary inclusion of monensin or tylosin, alone or in combination, increased faecal shedding of *E. coli* O157:H7 or its persistence in the environment.

Naylor *et al.* (2007) reported that the direct application to the bovine rectal mucosa of two therapeutic agents, polymyxin B or chlorhexidine, greatly reduced *E. coli* O157 shedding levels in the immediate post-treatment period. The most efficacious therapeutic agent, chlorhexidine, was compared in orally and rectally challenged calves. The treatment eliminated high-level shedding and reduced low-level shedding by killing bacteria at the terminal rectum. However, the continuing increase in antimicrobial resistance among bacteria and their transmission to humans mitigates against the use of antibiotics as a viable commercial approach to controlling *E. coli* O157:H7.

Sodium chlorate

Sodium chlorate ($NaClO_3$) is an oxidising agent which is mostly used to produce chlorine dioxide for bleaching paper pulp, but is also used as a herbicide and to prepare other chlorates. Sodium chlorate is bactericidal, containing the intracellular enzyme nitrate reductase. When bacteria which anaerobically respire on nitrate are exposed to chlorate, nitrate is converted to nitrite and chlorate to the cytotoxic chlorite. Sodium chlorate administered intraruminally in water or feed has been shown to significantly reduce *E. coli* O157 levels in cattle and in sheep (Callaway *et al.*, 2002, 2003). In cattle, chlorate treatment reduced *E. coli* O157:H7 counts throughout the intestinal tract but did not alter total culturable anaerobic bacterial counts or the ruminal fermentation pattern, indicating its potential use as a pre-harvest treatment, but further research and approval from an appropriate regulatory authority are needed before commercial application.

Disinfectants

Cetylpyridinium chloride (CPC) is a cationic quaternary ammonium compound which kills bacteria and other microorganisms. Boselivac *et al.* (2004) have assessed the efficacy of CPC as an antimicrobial intervention on beef cattle hides. Two high-pressure CPC washes lowered aerobic plate counts and Enterobacteriaceae counts by 4 log cfu/100 cm^2 while two medium-pressure CPC washes were only slightly less effective. These results indicate that under the proper conditions, CPC may be effective for reducing microbial populations and *E. coli* O157 on cattle hides. Further study is warranted to determine if this effect will result in reduction of hide-to-carcass contamination during processing. In general, hide decontamination processes are applied at the slaughter level and will thus be addressed in other sections of this book.

Vaccines

Vaccination has been investigated as a potential method to control pathogens in ruminant food animals. The animal's own immune system is triggered to produce antigens against particular pathogens thereby reducing pathogen loads.

A number of vaccines have been proposed for control of *E. coli* O157, some of which target intimin, an outer membrane protein encoded by the *eae* gene and required for intestinal colonisation and attaching and effacing activity of *E. coli* O157:H7. Testing of the $Intimin_{O157}$ vaccine using neonatal piglets as a challenge model demonstrated that it protected piglets from colonisation with the pathogen (Dean-Nystrom *et al.*, 2002) and was a potentially viable vaccine against *E. coli* O157:H7 in cattle. Other studies looked at the use of proteins (Tir, EspA and EspB) which are part of a type III secretion system involved in intestinal colonisation of *E. coli* O157 as possible vaccines (Potter *et al.*, 2004). Faecal shedding of the pathogen by cattle immunised with these vaccines was significantly reduced. This included the number of animals in the group shedding the pathogen and the duration of shedding. However, a subsequent field trial of the vaccine in nine feed-lots showed no significant association between vaccination and pen prevalence of faecal *E. coli* O157:H7 (Van Donkersgoed *et al.*, 2005). A number of possible reasons were suggested for this, including different preparation of the vaccine, different vaccination strategies and different pen sizes. Further research is ongoing in this area.

Bacteriocins

Bacteriocins are ribosomally synthesised antimicrobial peptides or proteins produced by bacteria that kill or inhibit the growth of other bacteria, either in the same species (narrow spectrum), or across genera (broad spectrum) (Cotter *et al.*, 2005). Bacteriocins can be administered to animals by mixing dried or wet cultures with feed or drinking water.

A number of studies have looked at the potential role of colicins (bacteriocins produced by *E. coli*) to reduce *E. coli* O157 in cattle populations (Schamberger *et al.*, 2004; Diez-Gonzalez, 2007). In calves fed with a colicin-producing *E. coli*, levels of *E. coli* O157:H7 were reduced by 1.8 $\log_{10}$ cfu g^{-1} over 24 days (Schamberger *et al.*, 2004). A control approach based on the expression of colicins by bacteria which are normally present in the animal rumen would decrease the chances of transferring colicin genes to other potentially pathogenic *E. coli* strains, such as *E. coli* O157:H7 (McCormick *et al.*, 1999). Research continues in this area.

Bacteriophage

Bacteriophage are viruses that infect and kill bacterial cells by reproducing within bacteria and disrupting the host metabolic pathways, causing the bacterium to lyse. Their action is thus host-specific. A number of studies suggest that application of a cocktail of bacteriophage yields more successful results in control of *E. coli* than bacteriophage administered singly (Bach *et al.*, 2002). Waddell and colleagues (2000) administered a cocktail of bacteriophage to control shedding of *E. coli* O157:H7 in calves and showed that shedding of the pathogen was reduced to 6–8 days compared to 6–14 days in the control calves. Sheng *et al.* (2006) investigated the application of bacteriophage (KH1 and SH1) applied orally and rectally to sheep and cattle, respectively. No reduction in intestinal carriage of *E. coli* O157:H7 in sheep was observed when KH1 was administered orally. When an equal mixture of KH1 and SH1 were administered rectally to cattle, it reduced the numbers of *E. coli* O157:H7, but did not eliminate the pathogen from the cattle. In another study a single oral dose of *E. coli* O157:H7 specific bacteriophage (CEV1) applied to sheep reduced shedding of the pathogen by 2 $\log_{10}$ cfu g^{-1} compared to levels in the control animals (Raya *et al.*, 2006). Bacteriophage are now close to commercial application.

Competitive exclusion
Competitive exclusion (CE) is based on the administration of non-pathogenic bacterial culture to promote microbial competition and thus reduce colonisation or decrease populations of pathogenic bacteria in the gastrointestinal tract (Callaway *et al.*, 2004). A CE culture should ideally be composed of species normally resident in the animal intestinal microflora. The use of CE cultures in cattle to eliminate *E. coli* O157:H7 from the rumen has been shown to be successful and has potential for further commercial development (Brashears *et al.*, 2003a, b; Zhao *et al.*, 2003; Younts-Dahl *et al.*, 2004). Much of the activity in this area is still at the research stage although they are getting closer to commercial application.

6.4 Managing and controlling hazards and risks with pigs

6.4.1 General controls

General farm management practices for control of pathogens in pigs include all-in/all-out policies in pens to avoid contamination between new younger pigs and older pigs (Dahl *et al.*, 1997). Sick animals or poor thrivers should not be housed with healthy animals. In addition to this, adequate down-time is needed for cleaning, washing and disinfecting, before moving a new batch of pigs into a house or pen. In general, *Salmonella* levels increase as pigs move closer to finishing age; therefore pigs should move in a one-way flow from farrowing/weaner area to the growing/finishing area. If people are moving back and forth between pens, separate footwear and/or properly maintained disinfectant footbaths should be in operation. Preferably, there should be no movement of equipment, and dedicated equipment should be provided for each section. Rodent and pest control programmes are essential and all water and feeding troughs should be positioned to avoid faecal contamination.

Hurd *et al.*, 2008 conducted a risk-based analysis of the Danish Control Programme for *Salmonella* in pork. This Danish pork *Salmonella* control program initiated in 1993 involved improved efforts at slaughter hygiene (post-harvest) and on-farm (pre-harvest) surveillance and control. The retrospective simulations showed that, except for the first few years (1994–1998), the on-farm program had minimal impact in reducing the number of positive carcasses and pork-attributable human cases (PAHC). Most of the reductions in PAHC up to 2003 were, according to this analysis, due to various improvements in abattoir processes. Prospective simulations showed that minimal reductions in human health risk (PAHC) could be achieved with on-farm programs alone. Carcass decontamination was shown as the most effective means of reducing human risk, reducing PAHC to about 10 % of the simulated 2004 level. Miller *et al.* (2005) also reported that for the on-farm strategies of vaccination and meal feeding, the cost:benefit ratio was less than 1 indicating they are not likely to be profitable from a socio-economic perspective.

6.4.2 Targeted approaches

Some efforts to develop specific control measures for pathogen manipulation in pigs have focused on reducing shedding of *Salmonella* by pigs using competitive exclusion and vaccines. CE cultures have been administered to pigs to eliminate *Salmonella* from the gastrointestinal tract and have shown some success with potential for further commercial development (Genovese *et al.*, 2003). Roesler *et al.* (2004) investigated the use of a metabolic drift mutant of *S.* Typhimurium as an oral vaccination for piglets that were subsequently challenged with a highly virulent *S.* Typhimurium strain. Vaccinated animals were shown to shed substantially smaller amounts of the challenge strain and for a shorter period of time.

6.5 Managing and controlling hazards and risks with poultry

6.5.1 General controls

As described above for ruminant animals and pigs, general hygiene and biosecurity measures can impact on levels of *Salmonella* and *Campylobacter* in flocks. Guerin *et al.* (2007) reported that factors associated with an increased risk of *Campylobacter* were increasing median flock size on the farm ($p \leq 0.001$), spreading manure on the farm ($p = 0.004$–0.035), and increasing the number of broiler houses on the farm ($p = 0.008$–0.038). Protective factors included the use of official (municipal) ($p = 0.004$–0.051) or official treated ($p = 0.006$–0.032) water compared to the use of non-official untreated water, storing manure on the farm ($p = 0.025$–0.029), and the presence of other domestic livestock on the farm ($p = 0.004$–0.028). A Belgian study on the risk factors associated with *Salmonella* in laying hens indicated that the main risk factors were rearing flocks in cages compared to barns and free-range systems. The results also showed a significantly higher risk for *Salmonella* for a one week increase in flocks' age as well as with a unit increase in the size of the flock (Namata *et al.*, 2008)

6.5.2 Targeted controls

Competitive exclusion (CE)

CE cultures have been examined for their role in reducing *Salmonella* and *Campylobacter* carriage in poultry for many years, from the work of Nurmi and Rantala (1973) and Mead *et al.* (1989) to more recent studies (Chen and Stern, 2001; La Ragione and Woodward, 2003; Wagner, 2006; Zhang *et al.*, 2007a, b). The use of CE cultures, including commercial products in poultry has been reported to reduce the colonisation of poultry with *Salmonella* spp. by up to 70 % or by 7–9 $\log_{10}$ cycles (Hoszowski and Truszczynski, 1997; Davies and Breslin, 2003). Reductions of between 3 and 100 % *Campylobacter* spp. colonisation on poultry have been reported by Schoeni and Wong (1994).

Bacteriocins

Bacteriocins have been shown to have direct antimicrobial activity, *in vitro* against *Campylobacter* (Schoeni and Doyle, 1992; Morency *et al.*, 2001; Svetoch *et al.*, 2005), while a purified bacteriocin produced by *Paenibacillus polymyxa* microencapsulated and administered in feed was reported to reduce caecal *Campylobacter* colonisation in young broiler chickens and turkeys (Stern *et al.*, 2005; Cole *et al.*, 2006). Treatment with the bacteriocin eliminated detectable *Campylobacter* concentrations in turkey caecal contents (detection limit, 1×10^2 cfu g^{-1}) compared to controls (1.0×10^6 cfu g^{-1} of caecal contents) (Cole *et al.*, 2006). In chicken caecal samples, administration of bacteriocin yielded reductions of $10^{4.6}$–$10^{6.3}$ cfu g^{-1} for *C. jejuni*. These studies reported significant reductions of both intestinal levels and frequency of chicken and turkey colonisation by *C. jejuni*, suggesting that this on-farm application would be an alternative to chemical disinfection of contaminated carcasses (Stern *et al.*, 2005).

Vaccines

There are a number of successful and commercially-available vaccines targeting *Salmonella* in poultry. Sub-unit vaccines i.e. vaccines which only contain individual proteins, were shown to reduce colonisation of *S.* Enteriditis in the chicken intestinal mucosa following challenge (Khan *et al.*, 2003). Vaccination can play an important role, as was confirmed by EFSA opinion (EFSA, 2007) which stated that the vaccination of poultry can be an additional measure to increase the resistance of birds against *Salmonella* exposure and decrease the shedding. The EU Regulation on requirements for the use of specific control methods for the control of *Salmonella* in poultry stipulates that from 2008 all Member States with *Salmonella* Enteritidis (responsible for 90 % of outbreaks in humans by eating eggs) prevalence above 10 % must vaccinate laying hen flocks.

There are presently no commercially-available vaccines against *Campylobacter* in poultry. The development of such vaccines is hampered by a number of factors including the antigenic variety of strains and the lack of knowledge of antigens which induce a protective immune response.

Rice *et al.* (1997) administered formalin-inactivated *C. jejuni* with and without *E. coli* heat-labile toxin and, after challenge, found that lower numbers of *C. jejuni* were isolated from the caecum. However, Cawthraw *et al.* (1998) also used formalin-inactivated *C. jejuni* but did not show reduced caecal colonisation by the pathogen on subsequent challenge. A number of studies have looked at using flagellin, which is involved in colonisation of *Campylobacter* in the chicken gut. Widders *et al.* (1998) reported that birds immunised twice intraperitoneally with killed *C. jejuni* and purified flagellin showed a significant reduction in caecal colonisation, whereas birds that received a second immunisation orally or birds immunised twice with flagellin alone showed no significant reductions in caecal colonisation of *C. jejuni*. Another study created a fusion protein where flagellin was fused to the B-sub-unit of the labile toxin of *E. coli* and administered as a vaccine to chickens that were subsequently challenged with *C. jejuni*. It was found that there was significantly less colonisation in comparison to the control birds (Khoury and Meinersmann, 1995). A cocktail of attenuated live *C. jejuni* strains has also been used, but subsequent challenge with the parent strain did not result in reduced colonisation compared to control birds (Ziprin *et al.*, 2002). These studies clearly indicate that there is a long way to go in designing an efficient *Campylobacter* vaccine.

Bacteriophage
Wagenaar *et al.* (2005) applied a bacteriophage (strains 71 and 69) with wide host range against *C. jejuni* strains to control *C. jejuni* colonisation in broiler chickens. They observed a 3 $\log_{10}$ cfu g^{-1} reduction in *C. jejuni* counts initially in the therapeutic group; however, counts stabilised after five days to 1 $\log_{10}$ cfu g^{-1} lower than the control group. They concluded that this bacteriophage treatment could be an alternative method for reducing *C. jejuni* colonisation in broiler chickens. Loc Carrillo *et al.* (2005) reported a reduction in *C. jejuni* counts of between 0.5 and 5 $\log_{10}$ cfu g^{-1} when broiler chickens were orally administered bacteriophage CP8 and CP34 in an antacid suspension to reduce *C. jejuni* colonisation in broiler chickens.

6.6 Managing and controlling hazards and risks with eggs

General hygiene practices to prevent faecal contamination of egg shells and careful control of storage temperatures are the best approaches to prevent the transfer of contamination from the external shell into the egg. Although *Salmonella* deposition inside yolks is uncommon in naturally contaminated eggs, migration through the vitelline membrane into the nutrient-rich yolk contents could enable rapid bacterial multiplication. Gast *et al.* (2007) used an *in vitro* egg contamination model to assess the ability of small numbers of *S.* Enteritidis strains ($n = 4$) and *S.* Heidelberg strains ($n = 4$) to penetrate the vitelline membrane and multiply inside yolks during 36 h of storage at either 20 or 30 °C. After inoculation onto the exterior surface of the vitelline membrane, all eight *Salmonella* strains penetrated to the yolk contents (at a mean frequency of 45.1 %), and most strains grew to significantly higher levels (with a mean $\log_{10}$ bacterial concentration of 2.2 cfu/mL) during incubation at 30 °C. Significant differences in penetration frequency and yolk multiplication were observed between individual strains and between serotypes (*S.* Enteritidis > *S.* Heidelberg for both parameters). Penetration and multiplication were significantly less frequent during incubation at 20 °C. These results demonstrate that controlling ambient temperatures during pre-refrigeration storage may be an important adjunct to prompt refrigeration for limiting *Salmonella* growth in eggs and thereby for preventing egg-transmitted human illness.

Targeted approaches to reduce infected flocks and thus internally contaminated eggs are routine monitoring for specific *Salmonella* serovars and subsequent culling of infected flocks or vaccination as described above.

6.7 Future strategies and regulatory issues

There are many novel approaches to pathogen control at primary production level as described above. The limitations and potential for future development of the targeted strategies are discussed below as well as pertinent regulatory issues.

6.7.1 Competitive exclusion (CE)

While CE has been shown to work in several animal and poultry species, the benefits have not been consistent although the variations may be in part attributed to differences in the host animals, cultures or experimental designs. It can also be a slow process to bring a CE from laboratory trial stage to commercial use as any CE culture in which the microbial isolates are unknown must be regulated and approved for use in the same manner as an animal drug. This is essential to determine if the CE has any impact on animal health or if there is a risk of transfer of undesirable bacteria (or genetic elements therein) to humans (Wagner, 2006). Ongoing research in this area aims to characterise the bacteria present in CE products which will lead to safe assurance of use of the product and will allow optimised potential application.

6.7.2 Antibiotics

The administration of antibiotics for non-clinical purposes to animals has been banned in the EU since 2006 due to public health risks associated with development, selection and spread of antimicrobial resistance so this approach will not be feasible in Europe. In addition, if poultry is treated by antibiotics, detection of the *Salmonella* is difficult so an infection may be hidden but not eliminated from the flock.

6.7.3 Bacteriocins

The extensive use of bacteriocins has been hindered by a number of factors. One of the main limitations is the narrow spectrum of activity of most bacteriocins, although the specificity of a bacteriocin can be advantageous for applications in which a single bacterial strain of species is targeted without disrupting other microbial populations. Research continues to identify alternative bacteriocins which may extend bacteriocin applications either through their use alone or in combination with other antimicrobials or hurdle technologies. Another potential problem associated with using bacteriocins in foods is the development of resistance to these agents by the target bacteria. Resistance can occur naturally, and it has been reported especially with regards to Class IIa bacteriocins (Naghmouchi *et al.*, 2007). Consequently, studies have examined the possibility of generating multiple bacteriocin producers to limit the potential of bacteriocin-resistance populations. Some bacteriocin producing bacteria may also spontaneously lose their ability to produce bacteriocin(s) due to genetic instability (Riley and Wertz, 2002) and can also become ineffective if the cell membrane of a target organism changes in response to a particular environmental condition. Although there are some limitations for the use of bacteriocins, the ongoing study of existing bacteriocins as well as the identification of new bacteriocins will enhance the potential of these agents in many different applications.

6.7.4 Bacteriophage

Bacteriophage also have the capability to transfer unfavorable genes from one bacterium to another; these could be virulence genes or antibiotic resistance genes (Alisky *et al.*, 1998; Figueroa-Bossi and Bossi, 1999; Mirold *et al.*, 2001). This is a major concern and much effort must be applied to characterise and determine the potential of a bacteriophage to transfer virulence factors and antibiotic resistance genes to commensal or pathogenic bacteria before it is commercialised.

The emergence of host resistance to bacteriophage due to DNA mutations has been reported in pathogens in pre-harvest environments (Sklar and Joerger, 2001). A cocktail of bacteriophage may settle the issues of host resistance (Barrow and Soothill, 1997; Tanji *et al.*, 2004). Desirably, bacteriophage with a broad host spectrum are favourable for biocontrol as there are many sub-types in each pathogen species that may possess different cell surface receptors.

6.7.5 Vaccines

Overall, the results from vaccine trials to reduce the pathogen load in food animals have been mixed and results appear to be very specific to the host used. Arguments also exist regarding the type of vaccine to be applied, i.e. live, killed or sub-unit vaccines. In the case of *Salmonella* live vaccines are seen to be more efficacious, but concerns exist, such as public acceptability and safety. Killed vaccines are considered safer but appear be less efficient at prompting a protective immune response. Sub-unit vaccines require an in-depth understanding of the pathogen's interaction with its host, in order to choose an effective antigen. While vaccines against *Salmonella* are approved and in use for poultry, those for *E. coli* O157:H7 reduction in cattle are not approved for use. Overall, the development of effective vaccines against foodborne pathogens for animals still faces many hurdles.

6.8 Sources of further information and advice

Animal Production Food Safety, World Organisation for Animal Health, Paris, available at: http://www.oie.int/eng/secu_sanitaire/en_introduction.htm, accessed December 2008.

Bach S. J., McAllister T. A., Veira D. M., Gannon V. P. J. and Holley R. A. (2002) Transmission and control of *Escherichia coli* O157:H7: a review, *Can J Anim Sci*, **82** 475–90.

Callaway T. R., Anderson R. C., Edrington T. S., Genovese K. J., Harvey R. B., Poole T. L., Nisbet D. J. (2004) Recent pre-harvest supplementation strategies to reduce carriage and shedding of zoonotic enteric bacterial pathogens in food animals, *Anim Health Res Rev*, **5**(1) 35–47.

EFSA (2007) Opinion of the Scientific Panel on Animal Health and Welfare (AHAW) related with the vaccination against avian influenza of H5 and H7 subtypes in domestic poultry and captive birds, European Food Safety Authority, *The EFSA Journal* (2007) **489**, 1–64, available at: http://www.efsa.europa.eu/cs/BlobServer/Scientific_Opinion/ahaw_op_ej489_AI_Vaccination_en.pdf?ssbinary=true, accessed December 2008.

Fries R (2002) Reducing *Salmonella* transfer during industrial poultry meat production, *World Poultry Science J*, **58** 527–40.

Industry Best Practices for Reducing E. coli O157:H7 at Pre-Harvest, the US National Cattlemen's Beef Association, Centennial, CO, available at: http://www.bifsco.org/BestPractices.aspx, accessed November 2008.

Keener K. M., Bashor M. P., Curtis P. A., Sheldon B. W. and Kathariou S. (2004) Comprehensive review of *Campylobacter* and poultry processing, *Comprehens Rev Food Sci Food Safe*, **3** 105–6.

LeJeune J. T., Wetzel A. N. (2007) Preharvest control of *Escherichia coli* O157 in cattle, *J Anim Sci*, **85**(13 Suppl) E73–80.

Sargeant J. M., Amezcua M. R., Rajic A., Waddell L. (2007) Pre-harvest interventions to reduce the shedding of *E. coli* O157 in the faeces of weaned domestic ruminants: a systematic review, *Zoonoses Public Health*, **54**(6–7) 260–77.

Stanley K. and Jones K. (2003) Cattle and sheep farms as reservoirs of *Campylobacter*, *J Appl Microbiol*, **94**, Suppl 104S–113S.

Sofos J. N. (ed.) (2005) *Improving the Safety of Fresh Meat*, Woodhouse, Cambridge.

USDA (2006) *Pre-Harvest Security Guidelines and Checklist*, United States Department of Agriculture, Washington, DC, available at: http://www.usda.gov/documents/PreHarvestSecurity_final.pdf, accessed November 2008.

Vanselow B. A., Krause D. O. and McSweeney C. S. (2001) The Shiga toxin-producing *Escherichia coli*, their ruminant hosts, and potential on-farm interventions: a review, *Australian Journal of Agricultural Research*, **56**(3) 219–44.

Wegener H. C., Hald T., Lo Fo Wong D., Madsen M., Korsgaard H., Bager F., Gerner-Smidt P. and Mølbak K. (2003) *Salmonella* Control Programs in Denmark, *Emerging Infectious Diseases*, **9** 774–80.

6.9 References

Alali WQ, Sargeant JM, Nagaraja TG and DeBey BM (2004) Effect of antibiotics in milk replacer on fecal shedding of *Escherichia coli* O157:H7 in calves, *J Anim Sci* **82**(7) 2148–52.

Alisky J, Iczkowski K, Rapoport, A and Troitsky N (1998) Bacteriophages show promise as antimicrobial agents, *J Infection*, **36** 5–15.

Arsenault J, Letellier A, Quessy S, Morin JP and Boulianne M (2007) Prevalence and risk factors for *Salmonella* and *Campylobacter* spp. carcass contamination in turkeys slaughtered in Quebec, Canada, *J Food Prot*, **70**(6) 1350–9.

Asplund K, Tuovinen V, Veijalainen P and Hirn J (1990) The prevalence of *Yersinia enterocolitica* 0:3 in Finnish pigs and pork, *Acta Vet Scand*, **31**(1), 39–43.

Bach SJ, McAllister TA, Viera DM, Gannon VPJ and Holley RA (2002) Transmission and control of *Escherichia coli* O157:H7, A review, *Can J Anim Sci*, **82** 475–90.

Barrow PA and Soothill JS (1997) Bacteriophage therapy and prophylaxis: rediscovery and renewed assessment of potential, *Trends Microbiol*, **5**, 268–71.

Berg J, McAllister T, Bach S, Stilborn R, Hancock D and LeJeune J (2004) *Escherichia coli* O157:H7 excretion by commercial feedlot cattle fed either barley- or corn-based finishing diets, *J Food Prot*, **67**(4) 666–71.

Biksi I, Lorincz M, Molnár B, Kecskés T, Takács N, Mirt D, Cizek A, Pejsak Z, Martineau GP, Sevin JL and Szenci O (2007) Prevalence of selected enteropathogenic bacteria in Hungarian finishing pigs, *Acta Vet Hung*, **55**(2) 219–27.

Bosilevac JM, Arthur TM, Wheeler TL, Shackelford SD, Rossman M, Reagan JO and Koohmaraie M (2004) Prevalence of Escherichia coli O157 and levels of aerobic bacteria and Enterobacteriaceae are reduced when hides are washed and treated with cetylpyridinium chloride at a commercial beef processing plant, *J Food Prot*, **67**(4), 646–50.

Braden KW, Blanton JR, Allen VG, Pond KR and Miller MF (2004) *Ascophyllum nodosum* supplementation: a preharvest intervention for reducing *Escherichia coli* and *Salmonella* in feedlot steers, *J Food Prot*, **67**(9) 1824–8.

Brashears MM, Jaroni D and Trimble J (2003a) Isolation, selection, and characterization of lactic acid bacteria for a competitive exclusion product to reduce shedding of *Escherichia coli* O157:H7 in cattle, *J Food Prot*, **66** 355–63.

Brashears MM, Galyean ML, Mann JE, Killinger-Mann K and Lonergan G (2003b) Prevalence of *Escherichia coli* O157:H7 and performance by beef feedlot cattle given *Lactobacillus* direct-fed microbials, *J Food Prot*, **66**(5) 748–54.

Callaway TR, Anderson RC, Genovese KJ, Poole TL, Anderson TJ, Byrd JA, Kubena LF and Nisbet DJ (2002) Sodium chlorate supplementation reduces *E. coli* O157:H7 populations in cattle, *J Anim Sci*, **80**(6) 1683–9.

Callaway TR, Edrington TS, Anderson RC, Genovese KJ, Poole TL, Elder RO, Byrd JA, Bischoff KM and Nisbet DJ (2003) *Escherichia coli* O157:H7 populations in sheep can be reduced by chlorate supplementation, *J Food Prot*, **66**(2) 194–9.

Callaway TR, Anderson RC, Edrington TS, Genovese KJ, Harvey RB, Poole TL and Nisbet DJ (2004) Recent pre-harvest supplementation strategies to reduce carriage and shedding of zoonotic enteric bacterial pathogens in food animals, *Anim Health Res Rev*, **5** 35–47.

Castro-Hermida JA, Almeida A, González-Warleta M, Correia da Costa JM, Rumbo-Lorenzo C and Mezo M (2007) Occurrence of *Cryptosporidium parvum* and *Giardia duodenalis* in healthy adult domestic ruminants, *Parasitol Res*, **101**(5) 1443–8.

Cawthraw S, Ayling R, Nuijten P, Wassenaar T and Newell DG (1998) Prior infection, but not a killed vaccine, reduces colonization of chickens by *Campylobacter jejuni*, in AJ Lastavica, DG Newell and EE Lastovica (eds) *Proceedings of the 9th International Workshop*, Xxiv + 1–627, Cape Town, Institute of Child Health, 364–72.

Chen H-C and Stern NJ (2001) Competitive exclusion of heterologous *Campylobacter* spp. in chicks, *Appl Environ – Microbiol*, **67** 848–51.

Cogan TA and Humphrey TJ (2003) The rise and fall of *Salmonella* Enteritidis in the UK. *J Appl Microbiol*, **94** (Supplement) 114S–119S.

Cole K, Farnell MB, Donoghue AM, Stern NJ and Evetoch EANE (2006) Bacteriocins reduce *Campylobacter* colonization and alter gut morphology in turkey poults, *Poul Sci*, **85**, 1570–5.

Cotter PD, Hill C and Ross RP (2005) Bacteriocins: developing innate immunity for food, *Nat Rev Microbiol*, **3** 777–88.

Dahl J, Wingstrand A, Nielsen B and Baggesen D L (1997) Elimination of *Salmonella typhumurium* infection by the strategic movement of pigs, *Vet Rec*, **140** 679–81.

Dargatz DA, Fedorka-Cray PJ, Ladely SR, Kopral CA, Ferris KE and Headrick ML (2000) Prevalence and antimicrobial susceptibility of Salmonella spp. isolates from US cattle in feedlots in 1999 and 2000, *J Appl Microbiol*, **95**(4), 753–61.

Davies RH and Breslin MF (2003) Observations on the distribution and persistence of *Salmonella enterica* serovar Enteritidis phage type 29 on a cage layer farm before and after the use of competitive exclusion treatment, *Brit Poult Sci*, **44** 551–7.

Dean-Nystrom EA, Gansheroff LJ, Mills M, Moon HW and O'Brien AD (2002) Vaccination of pregnant dams with intimin(O157) protects suckling piglets from *Escherichia coli* O157:H7 infection, *Infect Immun*, **70** 2414–18.

Diez-Gonzalez F (2007) Applications of bacteriocins in livestock, *Curr Issues Intest Microbiol*, **8** 15–24.

Dodson K and LeJeune J (2005) *Escherichia coli* O157:H7, *Campylobacter jejuni*, and *Salmonella* prevalence in cull dairy cows marketed in northeastern Ohio, *J Food Prot*, **68**, 927–31.

Englen MD, Hill AE, Dargatz DA, Ladely SR and Fedorka-Cray PJ (2007) Prevalence and antimicrobial resistance of *Campylobacter* in US dairy cattle, *J Appl Microbiol*, **102**(6), 1570–7.

EFSA (2007) Opinion of the Scientific Panel on Animal Health and Welfare (AHAW) related with the vaccination against avian influenza of H5 and H7 subtypes in domestic poultry and captive birds, European Food Safety Authority, *The EFSA Journal* (2007) **489** 1–64, available at: http://www.efsa.europa.eu/cs/BlobServer/Scientific_Opinion/ahaw_op_ej489_AI_Vaccination_en.pdf?ssbinary=true, accessed December 2008.

Fedorka-Cray PJ, Dargatz DA, Thomas LA and Gray JT (1998) Survey of *Salmonella* serotypes in feedlot cattle, *J Food Prot*, **61**(5), 525–30.

Fegan N, Vanderlinde P, Higgs G and Desmarchelier P (2004) The prevalence and concentration of *Escherichia coli* O157 in faeces of cattle from different production systems at slaughter, *J Appl Microbiol*, **97**(2) 362–70.

Figueroa-Bossi N and Bossi L (1999) Inducible prophages contribute to *Salmonella* virulence in mice, *Mol Microbiol*, **33**, 167–76.

FSA (2006) *FSA surveys non-UK eggs for salmonella*, Food Standards Agency, London, available at: http://www.food.gov.uk/news/newsarchive/2006/nov/eggs, accessed November 2008.

Fukushima H, Hoshina K and Gomyoda M (1999) Long-term survival of shiga toxin-producing *Escherichia coli* O26, O111, and O157 in bovine feces, *Appl Environ Microbiol*, **65**(11) 5177–81.

Futagawa-Saito K, Hiratsuka S, Kamibeppu M, Hirosawa T, Oyabu K and Fukuyasu T (2007) *Salmonella* in healthy pigs: prevalence, serotype diversity and antimicrobial resistance observed during 1998–1999 and 2004–2005 in Japan, *Epidemiol Infect*, **26** 1–6.

Gast RK, Guraya R, Guard-Bouldin J and Holt PS (2007) *In vitro* penetration of egg yolks by *Salmonella* Enteritidis and *Salmonella* Heidelberg strains during thirty-six-hour ambient temperature storage, *Poult Sci* **86**(7), 1431–5

Gebreyes WA, Thakur S and Morrow WE (2006) Comparison of prevalence, antimicrobial resistance, and occurrence of multidrug-resistant *Salmonella* in antimicrobial-free and conventional pig production *J Food Prot*, **69**(4) 743–8.

Genovese KJ, Anderson RC, Harvey RB, Callaway TR, Poole TL, Edrington TS, Fedorka-Cray PJ and Nisbet DJ (2003) Competitive exclusion of *Salmonella* from the cut of neonatal and weaned pigs, *J Food Prot*, **66**, 1353–9.

Guerin MT, Martin W, Reiersen J, Berke O, McEwen SA, Bisaillon JR and Lowman R (2007) A farm-level study of risk factors associated with the colonization of broiler flocks with *Campylobacter* spp. in Iceland, 2001–2004, *Acta Vet Scand*, **10**(49/1), 18.

Gunn GJ, McKendrick IJ, Ternent HE, Thomson-Carter F, Foster G and Synge BA (2007) An investigation of factors associated with the prevalence of verocytotoxin producing *Escherichia coli* O157 shedding in Scottish beef cattle, *Vet J*, **174**(3) 554–64.

Hakkinen M, Heiska H and Hänninen ML (2007) Prevalence of *Campylobacter* spp. in cattle in Finland and antimicrobial susceptibilities of bovine *Campylobacter jejuni* strains, *Appl Environ Microbiol*, **73**(10) 3232–8.

Heuvelink AE, van den Biggelaar FLAM, de Boer E, Herbes WJG, Melchers WJG and Huis in 't Veld (1998) Isolation and characterization of verocytotoxin-producing *Escherichia coli* O157 strains from Dutch cattle and sheep, *J Clin Microbiol*, **36** 878–82.

Ho AJ, Ivanek R, Gröhn YT, Nightingale KK and Wiedmann M (2007) *Listeria monocytogenes* fecal shedding in dairy cattle shows high levels of day-to-day variation and includes outbreaks and sporadic cases of shedding of specific *L. monocytogenes* subtypes, *Prev Vet Med*, **16**(80/4), 287–305.

Hoar BR, Atwill ER, Elmi C and Farver TB (2001) An examination of risk factors associated with beef cattle shedding pathogens of potential zoonotic, *Epidemiol Infect*, **127** 147–55.

Hoszowski A and Truszczynski M (1997) Prevention of *Salmonella typhimurium* caecal colonisation by different preparations for competitive exclusion, *Comp Immun Microbial Infect Dis*, **20**, 111–17.

Hurd HS, Enøe C, Sørensen L, Wachman H, Corns SM, Bryden KM and Grenier M (2008) Risk-based analysis of the Danish pork *Salmonella* program: past and future, *Risk Anal*, **28**(2), 341–51.

Khan MI, Fadl AA and Venkitanarayanan KS (2003) Reducing colonization of *Salmonella* Enteritidis in chicken by targeting outer membrane proteins, *J Appl Microbiol*, **95**, 142–5.

Khoury CA and Meinersmann RJ (1995) A genetic hybrid of the *Campylobacter jejuni flaA* gene with LT-B of *Escherichia coli* and assessment of the efficacy of the hybrid protein as an oral chicken vaccine, *Avian Dis*, **39**, 812–20.

Kuhnert P, Dubosson CR, Roesch M, Homfeld E, Doherr MG and Blum JW (2005) Prevalence and risk-factor analysis of Shiga toxigenic *Escherichia coli* in faecal samples of organically and conventionally farmed dairy cattle, *Vet Microbiol*, **109**(1–2), 37–45.

Kunze DJ, Loneragan GH, Platt TM, Miller MF, Besser TE, Koohmaraie M, Stephens T and Brashears MM (2008) *Salmonella* enterica burden in harvest-ready cattle populations from the southern high plains of the United States, *Appl Environ Microbiol*, **74**(2), 345–51.

La Ragione RM and Woodward MJ (2003) Competitive exclusion by *Bacillus subtilis* spores of *Salmonella enterica* serotype Enteritidis and *Clostridium perfringens* in young chickens, *Vet Microbiol*, **94**, 245–56.

Loc Carrillo C, Atterbury RJ, el-Shibiny A, Connerton PL, Dillon E, Scott A and Connerton IF (2005) Bacteriophage therapy to reduce *Campylobacter jejuni* colonization of broiler chickens, *Appl Environ Microbiol*, **71**(11), 6554–63.

Matthews L, McKenrick IJ, Ternent H, Gunn GJ, Synge B and Woolhouse MEJ (2006) Super-shedding cattle and the transmission dynamics of *Escherichia coli* O157, *Epidemiol Infect*, **134**, 131–42.

Maule A (2000) Survival of verocytotoxigenic *Escherichia coli* O157 in soil, water and on surfaces, *Symposium Series for Society of Applied Microbiology*, **29**, 71S–78S.

McAllister TA, Bach SJ, Stanford K and Callaway TR (2006) Shedding of *Escherichia coli* O157:H7 by cattle fed diets containing monensin or tylosin, *J Food Prot*, **69**(9) 2075–83.

McCormick JK, Klaenhammer TR and Stiles ME (1999) Colicin V can be produced by lactic acid bacteria, *Lett Appl Microbiol*, **29** 37–41.

McDowell SW, Menzies FD, McBride SH, Oza AN, McKenna JP, Gordon AW and Neill SD (2008) *Campylobacter* spp. in conventional broiler flocks in Northern Ireland: epidemiology and risk factors, *Prev Vet Med*, **84** 261–76.

McGee P, Scott L, Sheridan JJ, Earley B and Leonard N (2004) Horizontal transmission of *Escherichia coli* O157:H7 during cattle housing, *J Food Prot*, **67** 2651–6.

Mead GC, Barrow PA, Hinton MA, Humbert F, Impey CS, Lahellec RW, Muklder AW, Stavric S and Stern NJ (1989) Recommended assay for treatment of chicks to prevent *Salmonella* colonisation by competitive inhibition, *J Food Prot*, **52** 500–2.

Miller GY, Liu X, McNamara PE and Barber DA (2005) Influence of *Salmonella* in pigs preharvest and during pork processing on human health costs and risks from pork, *J Food Prot*, **68**(9) 1788–98.

Mirold S, Rabsch W, Tschape H and Hardt WD (2001) Transfer of the *Salmonella* type III effector *sopE* between unrelated phage families, *J Mol Biol*, **312** 7–16.

Morency H, Mota-Meira M, LaPointe G, Lacroix C and Lavoie M C (2001) Comparison of the activity spectra against pathogens of bacterial strains producing a mutacin or a lantibiotic, *Can J Microbiol*, **47**, 322–31.

Naghmouchi K, Kheadr E, Lacroix C and IF (2007) Class I/IIa bacteriocin cross-resistance phenomenon in *Listeria monocytogenes*, *Food Microbiol*, **24**, 718–27.

Namata H, Méroc E, Aerts M, Faes C, Abrahantes JC, Imberechts H and Mintiens K (2008) *Salmonella* in Belgian laying hens: an identification of risk factors, *Prev Vet Med*, **17**(83/3/4), 323–36.

Naylor SW, Nart P, Sales J, Flockhart A, Gally DL and Low JC (2007) Impact of the direct application of therapeutic agents to the terminal recta of experimentally colonized calves on *Escherichia coli* O157:H7 shedding, *Appl Environ Microbiol*, **73**(5), 1493–500.

Nurmi E and Rantala M (1973) New aspects of *Salmonella* infection in broiler production, *Nature* **241** 210–11.

Oporto B, Esteban JI, Aduriz G, Juste RA and Hurtado A (2008) *Escherichia coli* O157:H7 and non-O157 shiga toxin-producing *E. coli* in healthy cattle, sheep and swine herds in Northern Spain, *Zoonoses Public Health*, **55**(2), 73–81.

Paiba GA, Gibbens JC, Pascoe SJ, Wilesmith JW, Kidd SA, Byrne C, Ryan JB, Smith RP, McLaren M, Futter RJ, Kay AC, Jones YE, Chappell SA, Willshaw GA and Cheasty

T (2002) Faecal carriage of verocytotoxin-producing *Escherichia coli* O157 in cattle and sheep at slaughter in Great Britain, *Vet Rec*, **150**(19), 593–8.

Pangloli P, Dje Y, Ahmed O, Doane CA, Oliver SP and Draughon FA (2008) Seasonal incidence and molecular characterization of *Salmonella* from dairy cows, calves, and farm environment, *Foodborne Pathog Dis*, **5**(1), 87–96.

Parisi A, Lanzilotta SG, Addante N, Normanno G, Di Modugno G, Dambrosio A and Montagna CO (2007) Prevalence, molecular characterization and antimicrobial resistance of thermophilic campylobacter isolates from cattle, hens, broilers and broiler meat in south-eastern Italy, *Vet Res Commun*, **31**(1) 113–23.

Potter AA, Klashinsky S, Li Y, Frey E, Townsend H, Rogan D, Erickson G, Hinkley S, Klopfenstein T, Moxley RA, Smith DR and Finlay BB (2004) Decreased shedding of *Escherichia coli* O157:H7 by cattle following vaccination with type III secreted proteins, *Vaccine*, **22**, 362–9.

Raya RR, Varey P, Oot RA, Dyen MR, Callaway TR, Edrington TS, Kutter EM and Brabban AD (2006) Isolation and characterization of a new T-even bacteriophage, CEV1, and determination of its potential to reduce *Escherichia coli* O157:H7 levels in sheep, *Appl Environ Microbiol*, **72**, 6405–10.

Rice BE, Rollins DM, Mallinson ET, Carr L and Joseph SW (1997) *Campylobacter jejuni* in broiler chickens: colonization and humoral immunity following oral vaccination and experimental infection, *Vaccine*, **15** 1922–32.

Riley MA and Wertz JE (2002) Bacteriocins: Evolution, ecology and application, *Ann Rev Microbiol*, **56**, 117–37.

Roesler U, Marg H, Schroder I, Mauer S, Arnold T, Lehmann J, Truyen U and Hensel A (2004) Oral vaccination of pigs with an invasive *gyrA-cpxA-rpoB Salmonella* Typhimurium mutant, *Vaccine*, **23**, 595–603.

Sargeant JM, Sanderson MW, Smith RA and Griffin DD (2003) *Escherichia coli* O157 in feedlot cattle feces and water in four major feeder-cattle states in the USA, *Prev Vet Med*, **61**, 127–35.

Schamberger GP, Phillips RL, Jacobs JL and Diez-Gonzalez F (2004) Reduction of *Escherichia coli* O157:H7 populations in cattle by feeding colicin E7-producing *E. coli*, *Appl Environ Microbiol*, **70**, 6053–60.

Schoeni JL and Wong ACL (1994) Inhibition of *Campylobacter jejuni* colonization in chicks by defined competitive exclusion bacteria, *Appl Environ Microbiol*, **60**, 1191–7.

Schoeni JL and Doyle MP (1992) Reduction of *Campylobacter jejuni* colonization of chicks by cecum-colonizing bacteria producing anti-*C. jejuni* metabolites, *Appl Environ Microbiol*, **58** 664–70.

Sheng H, Knecht HJ, Kudva IT and Hovde CJ (2006) Application of bacteriophages to control intestinal *Escherichia coli* O157:H7 levels in ruminants, *Appl Environ Microbial*, **72** 5359–66.

Sklar IB and Joerger RD (2001) Attempts to utilize bacteriophage to combat *Salmonella enterica* serovar Enteritidis infection in chickens, *J Food Safety*, **21**, 15–29.

Stern NJ, Svetoch EA, Eruslanov BV, Kovalev YN, Volodina LI, Perelygin VV, Mitsevich EV, Mitsevich IP and Levchuk VP (2005) *Paenibacillus polymyxa* purified bacteriocin to control *Campylobacter jejuni* in chickens, *J Food Prot*, **68**, 1450–3.

Svetoch EA, Stern NJ, Eruslanov BV, Kovalev YN, Volodina LI, Perelygin VV, Mitsevich EV and Mitsevich, IP, Pokhilenko VD, Borzenkov VN, Levchuk VP, Svetoch OE and Kudriavtseva TY (2005) Isolation of *Bacillus circulans* and *Paenibacillus polymyxa* strains inhibitory to *Campylobacter jejuni* and characterization of associated bacteriocins, *J Food Prot*, **68** 11–17.

Tanji Y, Shimada T, Yoichi M, Miyanaga K, Hori K and Unno H (2004) Toward rational control of *Escherichia coli* O157:H7 by a phage cocktail, *Appl Microbiol Biotechnol*, **64**, 270–4.

van de Giessen AW, Bouwknegt M, Dam-Deisz WD, van Pelt W, Wannet WJ and Visser G (2006) Surveillance of *Salmonella* spp. and *Campylobacter* spp. in poultry production flocks in The Netherlands, *Epidemiol Infect*, **134**(6), 1266–75.

Van Donkersgoed J, Hancock D, Rogan D and Potter AA (2005) *Escherichia coli* O157:H7 vaccine field trial in 9 feedlots in Alberta and Saskatchewan, *Can Vet J*, **46** 724–8.

Villacorta I, Ares-Mazas E and Lorenzo MJ (1991) *Cryptosporidium parvum* in cattle, sheep and pigs in Galicia (N.W. Spain) *Vet Parasitol* **38** 249–52.

Vold L, Johansen BK, Kruse H, Skjerve E and Wasteson Y (1998) Occurrence of shigatoxinogenic *Escherichia coli* O157 in Norwegian cattle herds, *Epidemiol Infect*, **120**, 21–8.

Waddell T, Mazzocco A, Johnson RP, Pacan J, Campbell S, Perets A, MacKinnon J, Holtslander B, Pope C and Gyles CL (2000) Control of *Escherichia coli* O157:H7 infection in calves by bacteriophages, Presentation, *4th International Symposium and Workshop on Shiga toxin (Verocytotoxin)- Producing Escherichia coli Infections*, 29 October–2nd November, Kyoto, Japan.

Wagenaar JA, Van Bergen MA, Mueller MA, Wassenaar TM and Carlton RM (2005) Phage therapy reduces *Campylobacter jejuni* colonization in broilers, *Vet Microbiol*, **109**, 275–83.

Wagner RD (2006) Efficacy and food safety consideration of poultry competitive exclusion products, *Mol Nutr Food Res*, **50**, 1061–71.

Wehebrink T, Kemper N, grosse Beilage E, Krieter J (2008) Prevalence of *Campylobacter* spp. and *Yersinia* spp. in the pig production, *Berl Munch Tierarztl Wochenschr*, **121**(1/2), 27–32.

Widders PR, Thomas LM, Long KA, Tokhi MA, Panaccio M and Apos E (1998) The specificity of antibody in chickens immunised to reduce intestinal colonisation with *Campylobacter jejuni*, *Vet Microbiol*, **64**, 39–50.

Wilson JB, Clarke RC, Renwick SA, Rahn K, Johnson RP, Karmali MA, Lior H, Alves D, Gyles CL, Sandhu KS, McEwen SA and Spika JS (1996) Vero cytotoxigenic *Escherichia coli* infection in dairy farm families, *J Infect Dis*, **174**, 1021–7.

Younts-Dahl SM, Galyean ML, Lonergan GH, Elam N and Brashears M (2004) Dietary supplementation with *Lactobacillus* and priopionibacterium based direct fed microbials and prevalence of *Escherichia coli* O157 in beef feedlot cattle and on hides at harvest, *J Food Prot*, **67**(5), 889–93.

Zhang G, Ma L and Doyle MP (2007a) Salmonellae reduction in poultry by competitive exclusion bacteria *Lactobacillus salivarius* and *Streptococcus cristatus*, *J Food Prot*, **70**, 874–8.

Zhang G, Ma L and Doyle MP (2007b) Potential competitive exclusion bacteria from poultry inhibitory to *Campylobacter jejuni* and *Salmonella*, *J Food Prot*, **70**, 867–73.

Zhao T, Tkalcic S, Doyle MP, Harmon BG, Brown CA and Zhao P (2003) Pathogenicity of Enterohaemorrhagic *Escherichia coli* in neonatal calves and evaluation of fecal shedding by treatment with probiotic *Escherichia coli*, *J Food Prot*, **66** 924–30.

Ziprin RL, Hume ME, Young CR and Harvey RB (2002) Inoculation of chicks with viable non-colonizing strains of *Campylobacter jejuni*: evaluation of protection against a colonizing strain, *Curr Microbiol*, **44**, 221–3.

7

Pathogen control in primary production: crop foods

R. Early, Harper Adams University College, UK

Abstract: Food crops form an essential part of the diet of humans yet, during their production, harvesting, post-harvest handling and storage they can be subject to contamination by microbial pathogens with the potential to cause foodborne illness. Consequently, strategies are needed and technologies must be applied to ensure the food safety of food crops destined either for consumption as minimally-processed products, e.g. fresh produce, or for use as components in manufactured food products, e.g. cereals. This chapter reviews the pathogens that are commonly associated with food crops and their implications for food safety, and it considers the strategies and methods that can be employed to manage food safety in crop production. Attention is given to the use of good agricultural practice (GAP), Hazard Analysis Critical Control Point (HACCP) and farm assurance.

Key words: crops, food crops, cereals, fresh produce, foodborne illness, pathogens, mycotoxins, good agricultural practice (GAP), Hazard Analysis Critical Control Point (HACCP), farm assurance.

7.1 Introduction

The focus of this chapter is set at the level of agricultural food production and concerns aspects of food crop production and post-harvest management. It discusses systems approaches to the control of pathogens in food crops and specifically good agricultural practice (GAP) and the Hazard Analysis Critical Control Point (HACCP) system of food safety management. It also concerns the development of farm assurance as a means of controlling food quality and safety at farm level.

The need for farmers to exercise control over foodborne hazards has been emphasised since the 1980s by various food safety crises arising at farm level. One of the most significant was the BSE (Bovine Spongiform Encephalopathy) crisis in the UK in the 1990s which clearly illustrated that what happens at one end of the food chain – in this case at the level of the farm inputs sector with animal feed manufacturers recycling the remains of cattle and sheep into cattle feed – could have disastrous effects at the other end of the chain. Farmers are now encouraged, and in some cases required, by customers to use GAP and HACCP for the benefit of food safety. In the UK, farm assurance has been implemented in direct response to the BSE crisis to provide a systematic approach to both quality and food safety management on farms. Of course BSE is not an issue for food crop production. In crop production a number of possible hazards require control, including bacterial pathogens and fungal mycotoxins, and both GAP and HACCP have their part to play in ensuring control. This chapter is concerned principally with bacterial pathogens and fungal mycotoxins in food crops, but it should be recognised that other forms of hazard are also important to food crop safety, e.g. viruses. Chemical hazards, e.g. pesticide residues, and physical hazards, e.g. metal and glass, are also important for control, but they are not dealt with in detail here as they fall outside the scope of the chapter. Throughout this chapter the term 'farmer' is used to describe those whose occupation concerns the production of agricultural food materials, including those who might think of themselves more accurately as growers. For convenience no distinction is made between the terms 'farmer' and 'grower' and no prejudice is intended by the use of the former term rather than the latter.

7.2 Quality and safety in the food chain

When consumers buy food products they take it for granted that they will be of the right quality and safe to eat. The very act of food purchase implies trust in the vendor. Who would intentionally buy food that is not expected to live up to quality and safety expectations, and who would buy from a vendor that cannot be trusted to meet these expectations? Today most food in developed countries is sold to consumers by the multiple food retailers or supermarkets and by food service businesses. Consumers place their trust in these organisations to fulfil expectations of food quality and safety. However, the supermarkets and food service businesses represent only one end of the food supply chain. They are the tip of the iceberg that is the food supply system. Consumer trust may be invested in the food retail and food service sectors, but the assurance of food quality and specifically food safety must be provided by every organisation within the food supply system. Failure at any point in the system could have catastrophic consequences. If consumers are to be protected from foodborne harm the management of food safety must therefore extend from agricultural food production right to the point of consumption.

7.2.1 Defining food quality and food safety

The food processors that buy agricultural raw materials for processing into food products which meet consumer expectations set exacting demands on farmers today. The power in the food chain that was once held by farmers has, during the past 40 years or so, moved into the hands of the supermarkets who function as the representatives of consumers. The food processors must meet the demands of the supermarkets for food products that are of the right quality and safe to eat if they are to remain in business. They must meet the supermarkets' product specifications and, in turn, farmers must satisfy the food processors by supplying produce that meets their specifications. In the broad sense of the term, farmers must provide quality produce. But what do we mean by the word 'quality'?

Quality has been defined by ISO (2000a) as '*The degree to which a set of inherent characteristics fulfils requirements*'. Another definition is given by Crosby (1984) who says that '*Quality is conformance to requirements, not goodness or elegance*'. Interpretation of these definitions establishes that if a product is to be a quality product, then it must comply with customer requirements. It must meet the specification. This is as true for the agricultural raw materials used by food processors to make grocery products, as it is of the grocery products supplied to supermarkets for sale to consumers. Farmers, food processors and the supermarkets must all understand quality in objective terms if quality food products are to be produced, and in this they are different from consumers. Consumers judge food quality subjectively. Their perception of quality relates to the satisfaction delivered by a food product, generally in terms of the hedonistic pleasure that it delivers. To consumers the sensory or organoleptic properties (appearance, flavour, aroma, texture, sound) of a food product are what counts, as well as freedom from foodborne harm, e.g. foodborne illness. Although the terms 'quality' and 'food safety' are often used as separate concepts, food safety is part of quality. It is defined by the Codex Committee on Food Hygiene (CCFH) (2001) as the '*Assurance that food will not cause harm to the consumer when it is prepared and/or eaten according to its intended use*'. A food that is not of the required quality may be safe to eat, but one that is not safe to eat is automatically not of the required quality.

7.2.2 The systems approach to food quality and safety management

In the production of agricultural food materials and the processing of food products, farmers and food processors must recognise that if quality cannot be measured it cannot be controlled. Consequently, they must define quality (and food safety as part of quality) in terms of parameters which can be measured and for which methods of control can be established. The quality parameters of agricultural food materials and food products are normally defined in specifications. The controls that must be exercised to ensure that products meet specifications are normally established within quality management systems. Quality management is defined by ISO (2000a) as the '*Coordinated activities* [used] *to direct and control an organization with regard to quality.*' It concerns the systematic control of quality planning and operational activities required to meet the objectives of a business. As part of their quality management systems food processors operate quality assurance systems that ensure, through process control, that customer requirements are met. Quality assurance is defined by ISO (2000a) as '*Part of quality management focussed on providing confidence that quality requirements will be fulfilled.*' Some food processing businesses, particularly in mainland Europe and the rest of the world, develop their quality management systems against ISO 9001:2008, the international standard for quality management (ISO, 2008). This is a generic quality system standard published by the International Organisation for Standardisation (see Section 7.17.2). Food businesses that supply the retail sector in the UK develop their systems in compliance with the British Retail Consortium's Global Standard for Food Safety (see Section 7.17.2). Whilst some farmers, mainly in mainland Europe but not so commonly in the UK, have developed quality management systems registered to ISO 9001 for their enterprises, for many such a venture would entertain too many complexities and may not yield adequate benefits. For most farmers the management of quality can be achieved through the application of Good Agricultural Practice, as discussed in Section 7.9. In the UK the majority of farmers are now members of farm assurance schemes, which equate to and are interpretations of GAP, as discussed in Section 7.14. Both GAP and farm assurance can and should incorporate appropriate methods for food safety control and often HACCP-based food safety management systems are implemented, as discussed in Section 7.11.

7.2.3 The obligation to produce safe food

All food businesses have moral and legal responsibilities to produce and/or sell safe foods. These obligations bear on those who produce agricultural food materials and those who process them into food products. Most significantly they bear on those who sell food products to consumers. In some instances the production of agricultural food materials entails the presence of foodborne hazards, e.g. bacterial pathogens, that must be destroyed in later processing. In others, particularly in the case of minimally-processed food crops, e.g. leafy salad crops and some soft fruit, foodborne hazards such as bacterial pathogens will not be destroyed by processing. In the European Union (EU) the requirement for food businesses to produce safe food is embodied in Regulation (EC) No 852/2004 of the European Parliament and of the Council of 29 April 2004 on the hygiene of foodstuffs (EC, 2004).

In the UK the requirement to produce safe food is also stated in the Food Safety Act 1990, which defines the offences and the defence in the event of prosecution, as well as the penalties for breaking the law and the powers of the enforcement authorities. With the introduction of the Food Safety

Act in the UK the legal obligation to produce safe food was crystallised for food retailers and processors. It became evident that both food processors and retailers could be accountable for food safety problems that occurred further down the food supply chain, e.g. at farm level. Section 8 of the Act states essentially that any person who produces or makes available for sale food which is unsafe has broken the law. Section 14 of the Act states '*Any person who sells to the purchaser's prejudice any food which is not of the nature or substance or quality demanded by the purchaser shall be guilty of an offence.*' Atwood (2000) says that 'nature' relates to food sold that is different from that demanded by the purchaser, e.g. fruit that is not of the type required, while the word 'substance' relates to the composition of food being incompatible with that demanded and 'quality' concerns food that is not of the required commercial quality. In the event that a food business is prosecuted under the Food Safety Act 1990 a defence, commonly termed the 'due diligence' defence, is given in Section 21, which states that '*it shall be a defence for the person charged to prove that he took all reasonable precautions and exercised all due diligence to avoid the commission of the offence by himself or by a person under his control.*' The term 'reasonable precautions' is interpreted to mean the establishment of a food safety management system, while 'due diligence' is taken to mean using the system effectively.

To ensure compliance with the law and the ability to mount a due diligence defence, many food businesses in the UK have implemented HACCP-based food safety management systems. Significantly, HACCP is regarded as the best way to prevent the inheritance of food safety problems from suppliers and, if such problems occur, to claim due diligence and pass responsibility back to suppliers. The concept that, for example, a food retailer could become accountable for food safety problems caused by suppliers has increased the retailers' (the supermarkets') concern about the quality of produce passing up the chain. This has caused them to take greater control over their supply chains, of which more will be said later. For the moment it is sufficient to recognise that both food processors and retailers can be held accountable for food safety problems not directly of their making and perhaps which occurred even at farm level. Food law does not yet apply to farmers exactly as it applies to food processors and retailers. However, to assist food processors and retailers to comply with the law, farmers should employ agricultural practices that minimise the contamination of produce with pathogens and, in the case of some produce, such contamination must be prevented. To ensure control, farmers are best advised to take a systems approach to quality assurance and process control. This concept shall be developed throughout this chapter.

7.3 Crops as foodstuffs for humans

Plants form a critical part of the diet of mankind. Almost all of the carbohydrate we consume is derived from plants, either directly or as the result of processing and extraction, e.g. flour milling or the extraction and refining of sugar from sugar beet (*Beta vulgaris*) or sugar cane (*Saccharum officinarum*). Even honey is derived from plant nectar which contains 40–50 % sucrose hydrolysed to glucose and fructose as it is carried by the bee (Butler, 1954). Plants provide nearly three-quarters of the protein in the human diet and overall about 90 % of mankind's food (Fernández-Armesto, 2001). Much of the meat consumed by mankind depends on plants for its production as many meat animals convert plants of low nutrition quality which cannot be utilised by the human digestive system, e.g. grass, into materials of high nutrition quality, e.g. muscle as beef steak. In western societies many plant species are grown specifically for feeding to food animals, even some of high nutrition quality such as maize (corn) used to feed animals kept in beef-lot systems.

It is probable that the evolution of human beings was directly influenced by the search for plant foods. Milton (1987) suggests that with the migration of early members of the genus *Homo* from tropical regions, where high-quality plant foods rich in carbohydrate and fibre could be found in abundance, to regions where lower-quality plant foods were to be found, the development of social co-ordination and language would have provided advantages in the collection of food. While meat would have formed part of the diet of our early hominid ancestors, plant foods and particularly those containing carbohydrate would have been important for the energy they were able to provide. Even today the few hunter-gatherer peoples left in the world, such as the Hadza of Tanzania, consume a large proportion of plant material in their diet as a source of energy and other important nutrients. Plants which produce seed would have been important as sources of carbohydrates and, to a certain extent proteins and lipids, for our ancestors. Mazoyer and Roudart (2006) suggest that the planned growing of seed-bearing plants – specifically cereals – facilitated the provision of a diet which was of higher quality and quantity than that obtained by predation. This was coupled with the development of sedentary social groupings during the Neolithic agricultural revolution, and it is likely that success in the growing of such plants through a form of proto-agriculture initiated the development of human society. Mazoyer and Roudart propose that the growing of seed plants by humans automatically selected for improvements in the quality of seed, especially the carbohydrate content. Large seeds rich in sugars germinate earlier than those with a proportionally larger germ rich in protein and fat, and mature and produce seed earlier. The harvest performed only once selects for these seeds and successive growings results in crops of seed which are larger in size and higher in carbohydrate content.

The importance of seed-bearing plants to the diet of mankind should not be under-estimated. Heiser (1990) states that of greater than 20 000 species of flowering plants (the angiosperms) some 3000 have been used by humans for food and 200 or so have been domesticated, of which around a dozen form our primary source of food. Of the flowering plants, those which are of the greatest importance are the grasses of the family Gramineae which includes wheat, rice, maize and sorghum that form the staple plant foods for most of the world's population. The production of plants as food, or as sources of feed for animals, or sources of materials to be used in food manufacture, underpins the economies of most countries of the world. Table 7.1 gives crop production statistics for the main plant foods of the world and Table 7.2 shows the growth in world cereal production in recent years. The production statistics demonstrate the importance of wheat (*Triticum* spp.) and rice (*Oryza* spp.) as judged by the quantities produced. The former is of

Table 7.1 World crop production in 2004 ('000 tonnes)

Cereals	
Wheat	629 873
Rice	608 368
Coarse grains	1 032 199
Barley	153 830
Maize	724 515
Rye	17 650
Oats	25 828
Millett	27 767
Sorghum	57 924
Root and tuber crops	
Potatoes	330 125
Cassava	203 018
Pulses	
Beans (dry)	18 335
Broad beans (dry)	4 434
Oilseeds, oil nuts and kernels	
Soybeans	206 408
Groundnuts	36 421
Sunflower seed	26 466
Rapeseed	46 171
Cottonseed	43 334
Sugar cane and sugar beets	
Sugar cane	1 328 217
Sugar beets	248 616
Beverages	
Coffee	7 782
Cocoa beans	3 881
Tea	3 342

Source: Adapted from *FAO Statistical Yearbook 2005–2006* (FAO, 2007): www.fao.org/es.ess/yearbook/vol_1_1/index.asp.

Table 7.2 World cereal production ('000 tonnes)

1979–1981	1989–1991	1999–2001	2003	2004
1 573 227	1 903 961	2 084 615	2 085 774	2 270 360

Source: Adapted from *FAO Statistical Yearbook 2005–2006* (FAO, 2007); www.fao.org/es/ess/yearbook/vol_1_1/index.asp

significance mainly to societies in Europe, North America and the antipodes, while the latter forms the basis of most diets in Asia and the Pacific Rim and the Indian subcontinent. In other parts of the world wheat and rice are found in diets in varying proportions, while in many countries of the African continent sorghum (*Sorghum bicolor*) and millet (*Penisetum glaucum*) form staples. Although cereals are of great importance for many peoples, other carbohydrate rich plants such as the white potato (*Solanum tuberosum*) and cassava (*Manihot esculenta*: also known as manioc and yucca) form key parts of diets and food cultures. Potato consumption is widespread in South America, especially along the Andes, and in North America and Europe where it is commonly consumed within meals and as a material processed into a variety of snack foods. Cassava is important in South America and parts of Africa, where it may have been introduced from South America in the 15th century, if not the 16th century (Karasch, 2000). In Indonesia cassava ranks second to rice. Ko (1982) states that in parts of West Africa cassava (*Manihot utilissima*) is fermented to produce a material which can be fried or eaten without cooking. The reduction of pH during fermentation results in the hydrolysis of cyanogenic glucoside and the release of gaseous hydrocyanic gas, thus making the product safe to eat.

The fact that cassava has to be prepared carefully to eliminate toxic compounds illustrates that not all plant foods are intrinsically safe to eat. In the case of cassava and, for example, red kidney beans (*Phaseolus vulgaris*), thorough cooking is necessary to ensure safety. Red kidney beans, and other species of beans, contain toxic haemagglutinins that would cause foodborne illness if eaten raw or undercooked (McLauchlin and Little, 2007). In contrast, the safety of some plant foods has been improved by selective breeding, e.g. potato varieties which contain high levels of the alkaloids solanine and chaconine (found in green potatoes) and can give rise to foodborne illness, have been bred to eliminate these compounds and, at the same time, maintain or improve eating qualities. Similarly, oilseed rape (*Brassica napus*) has been selectively bred to reduce the content of both erucic acid and glucosinolates (Gunstone, 2008) which have toxic effects. Relatively few of the plants used as food by humans present problems of intrinsic toxicity, but all can be affected by extrinsic food safety factors the control of which is the focus of this chapter. It has been the intention here to give only a glance at the human use of plants as food and the influence of plant breeding to create improvements. It is important to recognise that mankind's relationship with plant foods has been a journey of exploration which began with the identification of edible plants and continues with work to improve in areas such as disease and pest resistance, yield, quality and food safety. Part of the journey inevitably concerns methods and systems of production and, in this context, systems for the management of food safety now exercise a priority.

7.4 Microbial food safety and food crops

The quality and safety of food crops are of paramount importance to consumers and, therefore, to regulatory authorities. Given the value placed on fruit and vegetables as components of a healthy diet, and also the growth in global trade in these foodstuffs, particular emphasis is now being placed by organisations, including the FAO (Piñeiro and Takeuchi, 2008), on ways to improve both the quality and safety of fresh fruit and vegetables. In this chapter there is neither space nor reason to present a consideration of the production and processing of any specific food crop. It is important, however, that we understand the scope of microbial food safety issues in crop production, and the points in production where such issues are likely to arise. Methods for the production of plant foods vary considerably and are determined in many ways by the crops involved. For instance, very many aspects of the production of coffee are different from those involved in the production of palm kernel oil, and both are different from the production of wheat for milling. In simplistic terms similarities exist in the production of all three and we can envisage the production of all crops to consist of three phases: crop production, harvesting (including post-harvest handling) and storage. For the purposes of this chapter, this tripartite division defines the scope of microbial food safety issues in crop production.

A variety of pathogen-related food safety issues can arise in the growing of crops for food. Some are the result of natural processes while others are a consequence of the agricultural interventions undertaken in crop production. During growth, plants are susceptible to infection by microorganisms. For example, naturally occurring fungal infections affect a wide range of plant species and can lead to the destruction of crops, the reduction of yields and, importantly, the formation of mycotoxins which render crops unsafe for consumption. As far as possible, agronomic strategies should be designed to limit such problems. Protecting growing cereal crops from water stress and heat stress, e.g. keeping them well irrigated, can restrict the onset of fungal infection and mycotoxin production. Insect pest damage can bring with it cellular damage to plants and access to plant tissues and nutrients by fungal spores, leading to growth and mycotoxin production. The judicious use of insecticides may be important in preventing pest infestation, and the use of fungicides may be necessary to prevent or deal with fungal infections. Additionally, stresses caused to crops during growth by, for example, excessive crop densities, competition from weeds and nutrient deficiencies in the soil can also increase the possibility of fungal infections becoming established. The use of appropriate crop densities, the control of weeds and the management of soil nutrition all become part of the strategic plan to protect against fungal infection. When pesticides (a collective term covering insecticides, herbicides, fungicides and rodenticides) are used it should be remembered that the agrochemical treatments can in themselves create a food safety issue. The treatment of crops with pesticides means the intentional application of toxic compounds to crops destined for consumption. Such treatments demand that food safety procedures are observed in relation to, for example, application rates, appropriate intervals between application and harvest, adherence to maximum residue levels (MRLs), etc.

Alongside concerns related to the fungal infection of food crops, care must be taken to avoid the contamination of crops with bacterial pathogens, viruses and protozoal parasites. Fresh fruit and vegetables are not normally thought of as vehicles for pathogenic microorganisms and sources of foodborne illness. However, an increasing number of outbreaks associated with this product group are being reported. Many studies of the microbiological quality of fresh produce have been reported in the literature. Johannessen *et al.* (2002) report that low levels of *Listeria monocytogenes* and *Staphylococcus aureus* were isolated from fresh produce in Norway. Isolates of *Enterococcus faecium* and *Enterococcus faecalis*, as well as *L. monocytogenes*, were taken from fresh produce by Johnston *et al.* (2006). Penteado and Leitão (2004) report outbreaks of salmonellosis associated with cut cantaloupe melons and watermelons. Branquinho Bordini *et al* (2007) report the occurrence and growth of *Salmonella* contaminating mangoes. Monaghan (2006) identifies *S. enterica* serovar Thompson (given as *S. thompson*) associated with salad rocket (*Eruca vesicaria* subsp. *sativa*) and *Salmonella enterica* serovar Newport (given as *Salmonella newport*) associated with head lettuce (*Lactuca sativa*) as causes of outbreaks of foodborne illness in Europe involving minimally-processed crop produce. Beretti and Stuart (2008) refer to outbreaks of *E. coli* O157:H7 associated with the leafy greens, lettuce, escarole, endive, spring mix, spinach, cabbage, kale, argula and chard. Stuart *et al.* (2006) record three deaths and more than 200 people infected by *E. coli* O157:H7 associated with spinach from California's Salinas Valley. Croci *et al.* (2002) report that fresh produce has often been implicated as the source of foodborne illness involving human viral infections, including hepatitis A. McCabe-Sellers and Beattie (2004) express concern about emerging pathogens and, for instance, *Yersinia pseudotuberculosis* causing foodborne illness in association with iceberg lettuce.

Pathogen-related food safety issues can arise during harvesting and post-harvest handling, and it is to be expected that the root causes of some cases of foodborne illness associated with crops, particularly fresh produce, will be much higher up the food chain in food processing and food service. However, any incident of foodborne illness associated with crop products will impact on the reputation of the industry as a whole. The type of food safety issue and its magnitude depends significantly on the crop species concerned and the way it is used. The harvesting of cereals demands the use of combine harvesters and grain trailers that have been cleaned properly before use. Grain trailers can represent a particular problem when they have dual functions and are used at other times for carrying materials which may present a source of contamination, e.g. manure which is itself contaminated with bacterial pathogens, viruses and fungal spores. In some instances the contamination of a crop by bacterial pathogens gives rise to lesser concern as later food processing, e.g. heat treatment, eliminates contaminant organisms and safeguards consumers. Potatoes may be contaminated by the infective organisms *E. coli* O157:H7 and *L. monocytogenes*, for example, but processing into crisps or oven ready chips involving heat treatments will result in their destruction and the risks to consumers will be virtually eliminated. Even with cooking in the home where there is the possibility of cross-contamination between raw potatoes and cooked potatoes there is negligible risk. In contrast, the potential for consumers to be infected by pathogens associated with salad crops is much greater as these crops are generally minimally processed. If the crops are grown inappropriately or managed and handled wrongly during harvesting and post-harvest handling they can be irreversibly contaminated.

In crop production, food safety starts in the field. As has already been observed, the prevention of fungal contamination in cereal and oil seed production, as well as, for example, in grape production, is essential to safeguarding against mycotoxin contamination of the product. Fungicides may offer a part solution and the use of irrigation water is important for protection against fungal invasion, for the reasons mentioned above. Whatever the crop, water used to promote crop growth must be of an appropriate microbiological quality. This is especially important if the crop is destined for consumption as a minimally-processed product. If, for instance, irrigation water is faecally contaminated this can lead to contamination by enteric pathogenic microorganisms. Irrigation water can become contaminated in various ways. Farmyard manure carrying bacterial pathogens, viruses and parasites can enter water directly by animals grazing at the margins of ponds and reservoirs, and also as a result of run-off from fields where manure has been spread, or from leaking slurry lagoons, etc. Leaking irrigation distribution pipes can draw in soil and microbial contaminants. In the production of fresh produce, contamination can be limited by the use of surface irrigation rather than spray irrigation. However, in harvesting, potable water must be used for practices such as hydrocooling salad crops to remove field heat, in order to prevent contamination with pathogenic organisms. Similarly, only potable water should be used in cleaning root crops to remove soil and stones, or with vegetables and salad crops to remove insect larvae. When crops are harvested by hand, cross-contamination with faecal pathogens must be prevented by ensuring that workers maintain acceptable personal hygiene standards.

Apart from concerns about the influence of water used in crop production on the food safety of produce, concerns are also raised about the survival of pathogenic organisms in soil and their potential to contaminate crop products. This has particular relevance for the fresh produce sector and the safety of minimally-processed products. The survival of pathogens in contaminated soil, and in association with crops grown in contaminated soil, is of interest to food safety authorities and purchasers of produce, e.g. supermarkets. Solomon *et al.* (2006) report work concerning the survival of *E. coli* O157:H7 and *S. enterica* on growing crops. Cools *et al.* (2001) assess the survival of *E. coli* and *Enterococcus* spp. from pig slurry in different soils. Gessel *et al.* (2004) examine the persistence of zoonotic pathogens in soil also treated with pig manure, while Boes *et al.* (2005) are concerned with *E. coli* and *S.* Typhimurium in slurry applied to clay soil on a pig farm. Islam *et al.* (2005) report that manure and water are able to contaminate vegetables pre-harvest with *E. coli* O157:H7 with equivalent persistence over a period of weeks. Nicholson *et al.* (2005) demonstrate that *E. coli* O157:H7, *Salmonella* and *Campylobacter* can survive for up to a month in soil treated with livestock manure and *Listeria* can survive for longer. Interpreting these latter authors, it is important to promote elevated temperatures of 55 °C or more in compost heaps and to allow a sufficient time for pathogenic organisms from manures to die off in the soil. Apart from concerns about soil and water acting as sources of microbial contamination in crop production, the potential exists for some invertebrates that inhabit soils to act as vectors in the pre-harvest contamination of fruits and vegetables. This is illustrated by Kenney *et al.* (2006), who report the migration to manure and manure compost of *Caenorhabditis elegans*, a free-living bacterivorous nematode, with the potential to contaminate fruits and vegetables with *Salmonella*, and Anderson *et al.* (2006), who report the shedding of foodborne pathogens by this nematode in compost-amended and unamended soils.

In addition to concerns about the potential for crops to be contaminated with pathogens during growing, harvesting and post-harvest handling, the potential exists for pathogen-related food safety problems to arise during storage. The storage of food crops is a complex subject, with different crops stored for different reasons. Some are stored for long periods of time to control market supply and stabilise prices, e.g. cereals and potatoes. Others are stored until capacity is available to process them, e.g. sugar beet. In some cases particular storage conditions are used initially to suspend maturation and then changed to initiate ripening. Immature climacteric fruits, for example, may be harvested and held in store to be ripened later under conditions that regulate chemical and physical changes that would normally occur independently, to bring the product to an acceptable state of ripeness at the required time (Thompson, 1996). Some crops are stored for only very brief periods of time. Many salad crops pass quickly from harvesting through processing and packing into the retail supply chain with the longest period of storage occurring in distribution and retail.

Whatever the storage circumstances of a food crop, it must be managed appropriately to prevent or minimise contamination by microbial pathogens. In this, different crops have different requirements. Cereals, nuts and beans, for instance, should be dried to a water activity (a_w) below 0.6 to inhibit the germination and growth of fungal spores. They should be stored under clean dry conditions in storage silos or other containers which may be treated with fungicides and insecticides to prevent, respectively, fungal growth and contamination by spores and insect contamination and damage. Various gasses including ethyl formate, methyl bromide and phosphine have been used to eliminate insect infestations in cereals (and legumes) (Bell, 2000a). Methyl bromide is described by Bell (2000b) as the agent that offers the widest range of applications, but it is due to be phased out in 2015 because of its ozone-depleting characteristics. Phosphine is given as an effective alternative to methyl bromide for the fumigation and disinfestation of grain as well as cocoa, coffee, dried fruit, nuts and bagged rice, but this is under regulatory review in the USA and Europe. In the future, the fumigation and disinfestation of grains and other seed crops, as well as facilities, e.g. storage silos etc., may depend on compounds such as sulphuryl fluoride, carbon dioxide used at high pressure where possible, carbonyl sulphide and ethyl formate, the use of which are briefly reviewed by Bell (2000b).

Clearly, if insect infestation does occur in stored crops insect respiration will raise the relative humidity in containers, leading to an increase in the a_w of stored produce and probably the germination and growth of fungal spores. Likewise, storage containers should be water-tight to prevent the moistening of produce and an increase in a_w that may give rise to fungal growth. There should also be protection against pest infestation, e.g. rodents and birds which may carry enteric and other pathogens and contaminate produce. The storage of fruit and vegetables must be managed appropriately to prevent damage to produce resulting in the loss of quality and contamination which may compromise food safety. Thompson (1996) recognises that stored vegetables may be subject to parasitic and saprophytic bacterial and fungal infections, whereas fruits are not normally infected by bacteria but are subject to attack by fungi. Although bacterial and fungal spoilage organisms are an economic issue, those that represent food safety hazards are of concern here and are discussed below. From the perspective of food safety, fungal infections that may give rise to mycotoxin production are of particular interest. As with the storage of cereals, nuts, beans, etc. fruit and vegetables must be protected in store from contamination by pests, e.g. rodents, as, in some cases, fruits may be eaten directly by consumers without washing. The presence of bacterial pathogens and viruses on the surface of fruit following contact by rodents has the potential to cause foodborne illness.

When cereals and other seed crops are infested with insects or contaminated with mycotoxin producing fungi, or when fruit and vegetables are contaminated with pathogenic microorganisms, the food businesses whose job it is to produce safe, marketable products from the crops face particular problems.

In some instances the nature or levels of contamination may be such that the produce must be disposed of appropriately. Often, however, suitable methods and technologies may be employed to disinfest or disinfect the crop produce. The use of methyl bromide and phosphine for the disinfestation of cereals has already been mentioned. Methyl bromide is also used for the disinfestation of other crops, e.g. packed table grapes (Leesch *et al.*, 2008). Heat treatments raising the temperature of grain storage bins above 50 °C have been used successfully for disinfestation purposes (Tilley *et al.*, 2007). The elimination of grain mites (Acari) and *Sitophilus granarius* infestation in wheat using high temperature/short time (HTST) heat treatments has been successful, as has the elimination of *Prostephanus truncatus* in maize (Mourier and Poulsen, 2000). Wheat has also been successfully disinfested using microwave energy (Vadivambal *et al.*, 2007). The use of chlorinated water has been a standard in the disinfection of fresh-cut vegetables, but concerns about the production of trihalomethanes and other carcinogenic compounds is driving the search for alternatives. Ozone gas offers a potential substitute and has been shown to be effective in the treatment of cherry tomatoes contaminated with *S.* Enteritidis (Daş *et al.*, 2006), shredded lettuce contaminated with *Shigella sonnei* (Selma *et al.*, 2007) and fresh-cut cantaloupe contaminated with *Salmonella* (Selma *et al.*, 2008). Neutral electrolysed water (NEW) has been proposed as an effective alternative to the use of chlorinated water in the disinfection of minimally-processed lettuce (Rico *et al.*, 2008) and minimally-processed vegetables (Abadias *et al.*, 2008a). The use of ultraviolet energy to destroy pathogens on fresh produce has been advocated in conjunction with GAP and good manufacturing practices (GMP) (Yaun *et al.*, 2004). Of course, irradiation technology has a significant potential to increase the safety of crop produce. Bidawid *et al.* (2000) demonstrate the use of gamma irradiation to inactivate hepatitis A in fruits and vegetables. Molins *et al.* (2001) propose the use of irradiation as a 'cold pasteurization' for ensuring the food safety of raw foods and as a critical control point (CCP) in HACCP-based food safety management systems.

While various technologies may be used to disinfest or disinfect crop produce, this means, essentially, applying a cure for a problem that has occurred when the focus, as far as possible, should be on the prevention of crop infestation by pests and contamination by microbial pathogens in the first instance. Some measures have already been discussed that might be considered to be preventative measures. Other preventative measures include soil solarisation, which is advocated (Stapleton, 2000) as a means of disinfesting seedbeds of pests and as a substitute to methyl bromide and other synthetic chemical toxicants. In the growing of fresh produce, fields may be fenced off to exclude feral animals and the possible contamination of crops with faecally-borne microbial pathogens for example. Non-crop vegetation may be removed to eliminate its attraction to insects and also, therefore, rodents and other small animals which consume insects and value the protection that vegetation cover provides. Pest baits may be utilised in fields to control rodents, while ponds and reservoirs may be fenced off adequately to exclude amphibians. Measures should also be taken to ensure that waterfowl cannot gain access to stored water and that other birds are not attracted to field margins by food sources, e.g. non-crop vegetation and places to roost. Whatever the measures taken to protect crops from contamination in the field, conflict may exist between crop production objectives and biodiversity objectives which may need careful resolution, as discussed by Stuart *et al.* (2006) and Beretti and Stuart (2008).

7.5 Fungal pathogens and mycotoxins

Fungi are a group of organisms whose body typically consists of a tangled mass of hyphae forming the mycelium or vegetative body. In some small groups the vegetative body is an amoeba-like cell or pseudoplasmodium. Fungi lack chlorophyll and are not photosynthetic. They include yeasts, moulds, mildews and mushrooms. There may be as many as 250 000 fungal species of which around 200 are considered to be human pathogens. Most are either parasitic or saprophytic. Jay (1996) lists 19 genera of foodborne moulds. Some are classed as spoilage organisms and some are considered as beneficial, e.g. yeasts in brewing and breadmaking, *Penicillium roqueforti* used to make blue cheeses, and a small number are of food safety concern because they produce mycotoxins. Bender and Bender (1995) state that mycotoxins are '*toxins produced by fungi (moulds), especially* Aspergillus flavus *under tropical conditions and* Penicillium *and* Fusarium *species under temperate conditions*'.

The potential for mycotoxins to cause human disease is recognised worldwide and many governments have established legal limits for their presence in food. Such limits are essentially speculative and balance the concern for human health with the maintenance of economic infrastructures built on food crop supply systems. Berg (2003) describes the Codex Alimentarius and EU processes for establishing maximum limits for mycotoxins in foods. Although knowledge about the ability of mycotoxins to cause disease in humans is still being accumulated, facts about the aetiology of mycotoxicosis in humans have come from the study of the occasional outbreak of disease in populations where an association is made with food contaminated by mycotoxins, and by laboratory studies using animals. It is clear that many foods of plant origin can allow mycotoxins to enter the human food chain and the contamination by mycotoxins of cereals, fruits, nuts, seeds and spices, and the products made from them, has been linked to a variety of diseases. The FAO (2001) identifies seven mycotoxin-producing fungal species of concern to food safety worldwide. To this list (see Table 7.3) are added *Penicillium expansum* and other species as producers of the mycotoxin, patulin, and *Claviceps purpurea*, the cause of ergotism. This is perhaps now only of historical interest as Galvano *et al.* (2005) say that the

Table 7.3 Mycotoxins of importance to food safety and the fungal species associated with their production

Mycotoxins	Fungal species
Aflatoxins	
Aflatoxins B1, B2	*Aspergillus flavus*
Aflatoxins B1, B2, G1, G2	*Aspergillus parasiticus*
Trichothecenes	
T-2 toxin	*Fusarium sporotrichioides*
Deoxynivalenol	*Fusarium graminearum*
Zearalenone	
Zearalenone	*Fusarium graminearum*
Fumonisins	
Fumonisin B1	*Fusarium moniliforme* (*verticillioides*)
Ochratoxin	
Ochratoxin A	*Aspergillus ochraceus*
Ochratoxin A	*Penicillium verrucosum*
Patulin	
Patulin	*Penicillium expansum* (also *Aspergillus* spp. & *Byssochlamys* spp.)
Ergot	
Ergotamine alkaloids	*Claviceps purpurea*

Source: Adapted from FAO (2001).

last incidence of ergotism reported in the literature dates from the 1970s. A detailed review of mycotoxins is given by Bennett and Klich (2003) and by Murphy *et al.* (2006). Mello *et al.* (1999) review the implications of *Fusarium* mycotoxins for animal health, welfare and productivity. The point at which the contamination of food crops by mycotoxins occurs is dependent on a number of factors, including the mould species involved and its optimum growing conditions (temperature and water activity), local climatic conditions and whether or not the crop involved is growing or has been harvested. For instance, organisms of the genus *Fusarium* produce a number of mycotoxins in growing cereal crops of maize (*Zea mays* L.), wheat (*Triticum aestivum*), barley (*Hordeum vulgare*), rice (*Oryza sativa*) and others, particularly in temperate climates. In contrast and under similar climatic conditions, stored cereals are more commonly contaminated by mycotoxins produced by *Penicillium* and *Aspergillus* species.

7.5.1 Aflatoxins

Aflatoxins are a group of mycotoxins which cause disease in humans and animals, e.g. farm livestock. They are the toxic metabolites of specific fungal species and affect both crop produce destined for direct consumption by humans and feedstuffs intended for food animal production. Aflatoxins have received much scientific attention, perhaps more than other mycotoxins. They are demonstrated carcinogens with high potency in laboratory animals and cause acute toxicological effects in humans and animals in the form of aflatoxicosis, which is principally a hepatic disease. Their effects are enhanced in the presence of hepatitis B and C virus. Aflatoxins are teratogenic and have immunosupressive action which can lead to bacterial infection in affected humans and animals. They reduce productivity in affected animals and can cause embryo toxicity. In 1959 the death of many tens of thousands of turkeys on farms in the East of England was initially thought to be caused by infectious disease, but was a consequence of the fungal contamination of turkey feed and the production of aflatoxin. The causative organism was found to be the fungal species *Aspergillus flavus* which grew on groundnuts used in the feed (Moss, 1996). Other species which produce aflatoxins are *A. parasiticus*, *A. nomius* and *A. niger*. *Aspergillus flavus* and *A. parasiticus* are more commonly associated with crop produce from tropical and sub-tropical regions, where temperatures and relative humidities favour growth.

The term 'aflatoxin' is derived from *A. flavus*. Aflatoxins consist of approximately 20 related secondary metabolites but only aflatoxins B1, B2, G1 and G2 are normally found in food (Imanaka *et al.*, 2007). Two metabolic products labelled M1 and M2 have been isolated from milk produced by animals fed with aflatoxin-contaminated materials, and it is clear that aflatoxins can be found in foodstuffs of animal origin, e.g. meat, milk and eggs. Pettersson (2004) explains that the degradation of aflatoxin B1 in the rumen of dairy cows yields a compound that enters the blood stream to be metabolised into the relatively stable aflatoxin M1 in the liver. This is then carried in the blood to pass into the cow's milk, urine and bile. The occurrence of aflatoxins in milk in Northern Italy in 2004–2005 and its reduction by surveillance and changes to agricultural practices is reported by Decastelli *et al.*, (2007).

Viridis *et al.* (2007) discuss the presence of aflatoxin M1 in goats' milk and cheese, while Tekinşen and Uçar (2007) report the same aflatoxin at levels that constitute a human health risk in butter and cream in Turkey. Surveys of aflatoxins in foodstuffs demonstrate that they are commonly associated with maize, groundnuts and cottonseed. They are also found in rice, nuts, e.g. almonds, figs, spices and cheese. Romagnoli *et al.* (2007) report the analysis of spices, aromatic herbs, herb-teas and medicinal plants marketed in Italy, where chilli was found to be the most frequently contaminated spice, but aromatic herbs, herb-teas and medicinal plants were not contaminated even when originating from tropical countries. Aflatoxins can be produced in growing crops. They are most likely to be found in produce which is badly managed during harvest and in storage. Produce must be managed appropriately to prevent contamination and kept dry to prevent the germination of spores and aflatoxin production.

7.5.2 Trichothecenes

Fusarium is a filamentous fungus that is widely distributed in plants and soils. The genus contains over 20 species, of which 14 are significant to crop producers because of the diseases they cause. Most *Fusarium* species are common in tropical and sub-tropical regions with some found in temperate zones. *Fusarium* causes crop disease in a wide range of commodity crops, from cereals to melon, pepper, potato and tomato. The *Fusarium* species *F. avenaceum*, *F. culmorum*, *F. graminearum* and *F. poae* affect barley and malt, oats, rye and wheat, causing canker, ear blight, glume, root rot, brown foot rots and wilts. *Fusarium* also affects rice and in maize it is responsible for pink ear rot. It causes wilt in bananas, when it is known as Panama disease, and is one of the most devastating diseases of this crop. It is important as an agent of biodegradation, surviving from one season to the next on decaying plant material, and is able to survive for up to 16 years in soil. *Fusarium* is of importance to food safety in crop production and particularly to the production of cereals as it produces a number of mycotoxins. *Fusarium* mycotoxins have been isolated from a wide variety of cereals, including maize and barley (Jin-Cheol *et al.*, 1993). It is one of several genera to produce mycotoxins known as trichothecenes and the principal genus of concern with relation to this group of mycotoxins. The trichothecenes are acutely toxic compounds linked to a number of diseases. They form the largest group of fusariotoxins. The trichothecenes are classified as four sub-groups. Galvano *et al.* (2005) state that the type A trichothecenes include T-2 toxin, HT-2 toxin, neosolaniol (NEO) and diacetoxyscripenol (DAS), while the type B group contains deoxynivalenol (DON), also known as vomitoxin, and its derivatives, as well as nivalenol (NIV) and fusarenon-X (FUS-X). The type A trichothecenes are the metabolic products of mainly *F. sporotrichioides* and also *F. poae*. The type B trichothecenes are produced by *F. culmorum* and *F. graminearum*. The occurrence of *Fusarium* mycotoxins in food in the European Community and implications for the diet is reported by the European Commission (EC, 2003), and a summary concerning malting barleys and malts is given by Booer *et al.* (2007).

7.5.3 Fumonisins

Other than the trichothecenes, *Fusarium* produces a group of mycotoxins known as fumonisins with fumonisin B1 (FB1) considered to be the most toxic. Fumonisin B1 is found in maize as the production of this toxin is favoured by warmer climates, and it commonly occurs along with other mycotoxins. Chu and Li (1994) report its occurrence with aflatoxin and trichothecene mycotoxins in maize collected in China. Fumonisins are recognised carcinogens in laboratory animals and have been found as contaminants of human food where rates of oesophageal cancer are high. They are responsible for hepatotoxic and nephratoxic effects in many animals. Jackson and Jablonski (2004) explain that fumonisins were isolated from *F. moniliforme (verticillioides)*, following research into equine leucoencephalomalacia, a cerebral disease in horses, and an investigation into oesophageal cancer in South African communities. *Fusarium moniliforme* (*verticillioides*) is able to produce more than ten mycotoxins, including fumonisins and fusarin C, the latter being mutagenic (Moss, 1996). *Fusarium moniliforme* has been shown to produce fusarin C on maize, other cereals and soybean (Bacon *et al.*, 1989).

7.5.4 Zearalenone

The species *F. graminearum* produces the mycotoxin zearalenone, which is structurally unrelated to fumonisins and trichothecenes. It is an oestrogenic mycotoxin and raises concern about hormonal imbalances in children and normal sexual development (Moss, 1996). Zearalenone is associated with cereals, including wheat and barley and, along with other mycotoxins, has been isolated from maize (Abbas *et al.*, 1988). It has also been found as a contaminant of sugar beet fibre from stored sugar beet, amongst other mycotoxins produced by *Fusarium* species (Bosch and Mirocha, 1992).

7.5.5 Ochratoxin A

Ochratoxin A (OTA) is a mycotoxin produced by *P. verrucosum* and *A. ochraceus* as well as other species of *Penicillium* and *Aspergillus*. The *Penicillium* species are mainly responsible for OTA production in temperate regions, while the *Aspergillus* species are more important in tropical and sub-tropical regions (Nicholson, 2004). Ochratoxin A has been found in a wide range of crop produce including cereals, grapes, coffee, cocoa and foodstuffs made from contaminated produce including beer, wine and vinegar (Battilani and Pietri, 2004: Galvano *et al.*, 2005). The mycotoxin is linked to the human diseases of Balkan endemic nephropathy and testicular cancer (Galvano *et al.*, 2005).

7.5.6 Patulin

Of the various mycotoxins associated with fruit juice, patulin is the most important. It is produced mainly by *P. expansum* in apple juice and other apple-based products (Scudamore, 2004) but also by *Aspergillus* and *Byssochlamys* species (Jouany and Diaz, 2005). Patulin has been of interest because of its wide-spectrum antimicrobial activity. Concern over its toxicity has arisen from incidents including the poisoning of cattle by barley malt sprouts contaminated by *A. clavatus*, poisoning from apples infected by *P. expansum* and silage contaminated by *Byssochlamys* spp. (Moss, 1996). Laboratory studies have shown that patulin acts as a neurotoxin and causes pathological changes to the viscera (FAO, 2001), but it is not considered to be as dangerous a mycotoxin as most others. The fermentation of apple juice to cider destroys patulin.

7.5.7 Ergot alkaloids

Ergot is the common name for fungal disease caused by organism *Claviceps purpurea.* It is a common disease of grasses and cereal crops and has a long and interesting history as an agent associated with foodborne illness linked to food crops. It occurs in many cereal grains and most often rye. In Europe from the Middle Ages to the time of the Enlightenment, ergot was responsible for some 132 poisoning epidemics and it has been associated with the Salem witch trials (Küster, 2000) in 17th century New England. When consumed ergot alkaloids cause a number of symptoms including vasoconstriction, gangrene and haemorrhagia and it is a neurotoxic agent (Aaronson, 2000).

7.6 Bacterial pathogens

Bacteria are a complex heterogeneous group of single-celled organisms classified as prokaryotes. They are widely distributed throughout the environment. Some are associated with man and animals, while others predominate in soils or in association with plants and plant materials. Most bacteria are heterotrophic and saprophytic. Of the tens of thousands of species of bacteria, some are pathogenic as disease-causing organisms. A relatively small number are classified as foodborne organisms causing foodborne illness and an even smaller number are of concern to food safety in relation to food crops and crop products, but some cause serious disease and even death, as discussed below. When incidents of foodborne illness occur involving bacteria and food crops and crop products, commonly the source of the causative organism can be traced to faecally-contaminated water or soil. The critical importance of preventing the bacterial contamination of produce through the adequate control of water, manures and municipal waste is recognised by the FDA (1998) in guidance given on minimising food safety hazards in fresh fruit and vegetables. The need to establish controls particularly with respect to manure recontamination is stated by De Waal (2003) who reports that illness related to contaminated food in the USA is estimated at 76 million a year, with 325 000 hospitalisations and 5000 deaths, and that foodborne illness outbreaks normally linked to foodborne hazards associated with animals are increasingly being traced to fruit and vegetables. De Waal urges the use of HACCP to ensure food safety. Also of importance to the safety of crop produce is the prevention of cross-contamination during harvesting and post-harvest handling and processing, e.g. from contaminated equipment, or as a result of the unacceptable personal hygiene practices of staff.

7.6.1 *Listeria monocytogenes*

Listeria monocytogenes is an infective organism, one of six in the genus *Listeria.* Within the genus, it is the pathogen of concern for humans and is represented by 13 serovars. The organism is considered to be a ubiquitous environmental organism found in soils, sewage, water, animal faeces and decaying vegetation. It grows at $a_w > 0.92$, in the pH range 4.39–9.6, in the temperature range –0.4–45 °C, and in salt solutions of >10% (see Chapter 19). The disease listeriosis was first thought to affect only animals but, in the 1980s, following outbreaks in Europe and North America, it became apparent that it affects humans as well. *L. monocytogenes* most commonly causes disease in vulnerable humans; the unborn, neonates (infants under 28 days), the elderly and the immunocompromised. It causes abortion, infection of the central nervous system and septicaemia. It has been associated with soft cheese, pâtés, sliced meats, smoked fish and vegetable products, e.g. coleslaw, cut fruits, pickled olives and salted mushrooms (McLauchlin and Little, 2007).

7.6.2 Enterohaemorrhagic producing *E. coli*

Although the genus *Escherichia* is the most widely studied of all bacterial genera, recognition that some *E. coli* can be infective pathogens in humans did not occur until 1971 (Jay, 1996). *E. coli* is an organism of normal intestinal microflora in man and animals. Consequently, it is used by the food industry as an indicator organism suggesting the presence of faecal contamination. The US Food and Drug Administration (FDA, 2007) advises that four classes of enterovirulent *E. coli* (termed EEC) are recognised as causing gastroenteritis in humans. The enterohaemorrhagic (EHEC) pathotype of EEC most commonly represented by *E. coli* O157:H7 can occur in the intestines of animals and humans along with other *E. coli*. It can therefore be found in faeces. The EHEC strains of *E. coli* produce shiga-like toxins (similar to those produced by *Shigella* spp.) known as verocytotoxins (VTs) which can severely damage the epithelium of the intestine when infection occurs. The consequences of infection include damage to the kidneys (haemolytic uraemic syndrome or HUS) and death. Children are especially vulnerable. The toxin-producing *E. coli* are of concern to food safety, particularly *E. coli* O157:H7 which has been responsible for many cases of foodborne illness worldwide. *E. coli* O157:H7 has a very low infective dose of between 10 and 100 cfu (colony forming units). It is a mesophilic organism with growth ceasing at around 42 °C, is acid tolerant, resistant to dehydration and is killed by pasteurisation. *Escherichia coli* O157:H7 has commonly been associated with foodborne illness cases involving undercooked hamburgers and cross-contamination from raw meats to cooked meats. Its low infective dose is of concern to the production of minimally-processed crop products, e.g. leafy salad products. Improperly composted manures and faecally-contaminated irrigation water are possible sources of contamination.

7.6.3 *Salmonella* spp.

Salmonella is of particular concern to the safety of food crops and particularly fresh fruits and vegetables. Surveys of fresh produce grown in the USA

(FDA, 2000) and imported fresh produce (FDA, 1999) demonstrated low levels of contamination with either *Salmonella* or *Shigella*. *Salmonella* has been implicated in many outbreaks of foodborne illness associated with fresh produce. Matthews (2006) states that in the USA outbreaks of foodborne illness associated with fresh produce grew from 0.7 % of all outbreaks in the early 1970s to 6 % in the 1990s, which is similar to the level of 5.5 % in England and Wales between 1992 and 2000. Matthews also states that nearly 60 % of illnesses associated with fresh produce are due to either *Salmonella* or Norovirus and that the produce most frequently implicated are salads, sprouts, melons, lettuce and berries. Microbiological surveys of fresh produce often report finding *Salmonella* and other pathogens. Abadias *et al.* (2008b) report finding *Salmonella* and *L. monocytogenes* in fresh, minimally-processed fruit and vegetables and sprouts sampled from retail establishments.

Like escherichiae and shigellae, organisms of the genus *Salmonella* belong to the family Enterobacteriaceae. The classification of salmonellas is complex and, as Heyndrickx *et al.* (2005) state, there is need for the clarification of nomenclature. Those who wish to understand this topic in detail are advised to refer to Heyndrickx *et al.* (2005), as well as Tindall *et al.* (2005) and Brenner *et al.* (2000). The genus *Salmonella* contains more than 2463 serovars, most of which are associated with disease in humans and animals. DNA and electrophoretic techniques have been significant to the identification of *Salmonella* serovars (Jay, 1996; Adams and Moss, 2008). Heyndrickx *et al.* (2005) explain that the species *S. enterica*, *S. bongori* and *S. subterranea* have been described, and that *S. enterica* is divided into six subspecies: *S. enterica* subsp. *enterica*, *S. enterica* subsp. *houtenae*, *S. enterica* subsp. *arizonae*, *S. enterica* subsp. *diarizonae*, *S. enterica* subsp. *indica* and *S. enterica* subsp. *salamae*. The serovars that mainly affect man and other mammals belong to the subspecies *enterica*. The development of *Salmonella* classification and nomenclature has meant, for example, that the organism once known as *S. typhimurium* is now properly known as *S. enterica* subsp. *enterica* serovar Typhimurium. The abbreviated names *Salmonella* serovar Typhimurium and *Salmonella* Typhimurium are acceptable.

Salmonellas are widely distributed in the environment and are found predominantly in the intestinal tracts of animals, including birds and reptiles, and man. Foodborne illness occurs when enough cells to cause intestinal infection are consumed. Symptoms include nausea, abdominal cramps, fever, vomiting and diarrhoea. Salmonellae are mesophilic and grow typically over the range 6–45 °C, with optimum growth temperatures around 37 °C, and in the pH range 4.0–9.0, at a_w above 0.94 (Jay, 1996). They do not tolerate high levels of salt or nitrite and are destroyed by pasteurisation. The infective doses of salmonellae are usually relatively high at thousands to millions of cfu, although a number of incidents of foodborne illness caused by *Salmonella* associated with chocolate and cocoa products have occurred (Craven *et al.*, 1975; D'Aoust, 1977; Gill *et al.*, 1983; Greenwood and Hooper,

1983; Cordier, 1994) where the infective doses were quite low. Incidents of foodborne illness involving *Salmonella* frequently concern high-risk foods (meat, milk and egg products) and, for exmple, inadequate heat processing or post-heat treatment cross-contamination. As with *E. coli*, the potential for food crops to become contaminated by *Salmonella* from faecal sources must be of concern to food safety management. Hygiene control in relation to workers and equipment, the control of manures and use of potable irrigation water, and the prevention of contamination of crops by feral animals and birds is important, particularly in relation to minimally-processed produce.

7.6.4 *Shigella* spp.

The genus *Shigella* contains the four species, *Sh. dysenteriae*, *Sh. flexneri*, *Sh. boydii* and *Sh. sonnei*, of which the first listed is a common cause of dysentery at a very low infection dose of some 10 cfu. *Shigella* species grow in the range 10–48 °C and best at neutral pH. They differ from *Salmonella* and *E. coli* in that they are found only in the intestines of humans. Consequently, and given the low infective dose level, the prevention of *Shigella* contamination in food processing must focus on the personal hygiene practices of staff, e.g. those handling crop produce destined to be eaten raw and with little or no processing.

7.6.5 *Campylobacter jejuni*

Campylobacter jejuni subspecies *jejuni* is one of 14 species of *Campylobacter* and the most significant to food safety. *Campylobacter jejuni* is a mesophilic organism that can grow at 42 °C, although it is destroyed by pasteurisation. Its principal reservoir is the intestines of animals and it has a low infective dose level of a few hundred cfu. It is a cause of acute enteritis, with fever, stomach cramps and bloody diarrhoea. In a very small number of cases the consequences of infection are reactive arthritis, IBS (irritable bowel syndrome) and a form of paralysis known as Guillain-Barré syndrome. Food safety management with respect to *C. jejuni* must be concerned with controls to prevent the faecal contamination of crops and/or elimination of the organism by appropriate post-harvest heat treatments.

7.7 Viral and parasitic pathogens

7.7.1 Gastrointestinal viruses

Gastrointestinal viruses can represent hazards to consumers of crop products, particularly when such products have been contaminated during handling and preparation. Fruit, vegetables and salad products have been implicated in outbreaks of foodborne illness involving viruses transmitted by humans.

Of the viruses that may be implicated in foodborne illness involving crop products, the Noroviruses are of significant importance. This is a group of viruses carried by people. Though transmission is often person-to-person via the faecal–oral route, it is possible for the faecal contamination of a product because of poor personal hygienic practices, for example, to give rise to infection. Some 10–100 viral particles are needed for infection and the onset period is around 24 hours. Symptoms include vomiting, mild fever and diarrhoea. Infection is self-limiting and lasts for around two days. The control of gastrointestinal viruses in the production of foodstuffs made from crop products will focus on the processing of minimally-processed products, e.g. fruit salads, coleslaw, etc.

7.7.2 Gastrointestinal parasites

Protozoan organisms of the genera *Cryptosporidium*, *Cyclospora* and *Giardia* are of concern to the production of some crops and some foodstuffs made from crop products. Most cases of cryptosporidiosis are caused by *Cr. parvum* and *Cr. hominis* although not all species of the genus affect humans. The same is true of the genera *Cyclospora* and *Giardia*, with *Cy. cayetanensis* and *G. duodenalis* the cause of cyclosporiasis and giardiasis in humans. Symptoms of infection include nausea, bloating, severe abdominal cramps and diarrhoea, and may last for a number of weeks. These protozoans are unable to reproduce in the environment and must maintain complex life cycles involving the infection of hosts in order to multiply. *Cryptosporidium parvum* is able to infect both humans and farm and wild animals. The means by which these parasites infect humans is the faecal–oral route involving human-to-human contact or via products contaminated by humans. In the case of *Cr. parvum* infection is through the consumption of faecally-contaminated drinking water or food processed with such water. Oocysts of *Cryptosporidium* are resistant to chlorine and can survive the chlorination of water processed for drinking and food manufacture. The oocysts of *Cryptosporidium*, *Cyclospora* and *Giardia* are sensitive to drying and freezing and will be killed by pasteurisation. Cryptosporidiosis has been associated with a number of foodstuffs, including apple juice made from apples taken from ground where cattle had grazed (McLauchlin and Little, 2007). The control of these parasites in the production of crop foods must necessarily focus on avoiding the faecal contamination of produce and ensuring the use of uncontaminated water in irrigation and food processing operations.

7.8 Food safety management in crop production

The management of food safety in crop production must be based on a reasoned understanding of which hazards are most likely to be problematical and where in the sequence of the production process they are most likely to occur. As stated in Section 7.4, for practical purposes the production process can be divided into the stages of (i) growing, (ii) harvesting (including post-harvest handling) and (iii) storage. The methods used to control hazards at each stage will depend on the crop itself, the way it is grown, the way it is harvested and handled and the way it is stored after harvesting, prior to post-harvest processing. The methods of control will also depend on the hazards believed to be likely at each stage from production to post-harvest processing. The HACCP system for food safety management provides a structured and reasoned approach to the control of hazards in food production and processing. In relation to crop production it should be appreciated that HACCP does not provide the only method of food safety management and alternative approaches may be more appropriate. Indeed, the concept that HACCP forms only part of the plan and the provision of processes required for food safety as well as regulatory compliance is made by Adams (2002) who says that as well as HACCP there must be additional controls. In crop production GAP can be used to manage aspects of food safety in growing, harvesting and storage without the need to apply HACCP principles.

7.8.1 The status of HACCP

HACCP has its roots in the early days of the US manned space programme when it was developed to ensure the production of defect-free foods for astronauts. It is now the method of food safety management recommended by the Codex Alimentarius Commission within the Food and Agriculture Organisation and World Health Organisation of the United Nations, and many other expert bodies, e.g. the National Advisory Committee for Microbiological Criteria for Foods (NACMCF) in the USA, as well as many governments. The requirement in EU Regulation (EC) no. 852/2004 that food businesses use HACCP and its importance to the concept of the due diligence defence under the UK's Food Safety Act 1990 were discussed in Section 7.2.3. The use of HACCP enables food businesses to avoid food safety incidents that may lead to prosecution, but it can yield other benefits. HACCP should be thought of as a form of insurance designed to prevent consumers from being harmed by food products. It can protect food businesses against prosecution, damage to reputation, litigation for compensation and closure. It can also provide a mechanism by which process controls can be improved and product losses can be reduced.

7.8.2 HACCP in agriculture

Although HACCP was originally developed for use by the food industry, the development of the food chain from farm to consumer and the occurrence of food safety issues originating on farms have led to interest in the application of HACCP at farm level. Sofos (2002) identifies a range of bacterial pathogens associated with food animal production and that the application of HACCP at farm level is likely to be encouraged to assist abattoirs in meeting their regulatory requirements. While particular focus may be given to the use of HACCP to control bacterial pathogens associated with animal production, this approach to food safety management also has significant utility in the control of hazards associated with the production of food crops. A key factor that has stimulated the application of HACCP at farm level has been the general recognition by national food safety agencies that the control of foodborne hazards originating on farms is critical to improving the safety of food products throughout the food supply chain.

While the importance of HACCP as an approach to food safety management in animal and crop production will increase, it is not the only method of food safety management of value to farmers. Indeed, the mechanics of HACCP can limit its applicability to some agricultural enterprises and activities. Also, the formalities of HACCP system implementation and maintenance can make this approach to food safety management inappropriate in some instances. For example, on farms where the agricultural processes are relatively simple, HACCP may add unnecessary complications. Where later processing will clearly eliminate hazards from agricultural produce, the cost of implementing and operating HACCP systems may, with judgement, be avoided. If it is decided that the use of HACCP is unnecessary, then alternative methods will need to be used to ensure that agricultural processes are managed appropriately to control produce with respect to food safety. GAP can be used to manage food safety issues where HACCP is inappropriate and, alone, GAP can facilitate the production of crops of high microbiological quality (Kokkinakis *et al.*, 2007). GAP should also be used when HACCP is used, as it provides much of the foundation on which effective HACCP systems applied at farm level are built: indeed, GAP aids the implementation of HACCP. However, the scope of GAP is wider than matters of food safety and it covers many other elements of agricultural practice.

7.9 Good agricultural practice (GAP)

The concept of GAP has been developing for more than a decade and is still evolving today. Da Cruz *et al.* (2006a) recognise GAP alongside GMP and HACCP as a quality assurance requirement in [crop] produce processing. GAP compares with GMP as employed in the food manufacturing industry, but the principles on which it is based draw significantly from integrated pest management (IPM) and integrated crop management (ICM). In essence GAP equates to integrated farm management (IFM) as advocated by LEAF (2000), and in some interpretations it embodies forms of values-driven agriculture of the kind proposed by the Food Ethics Council (2001). GAP recognises the irreplaceability of land as the medium from which most food is derived and the criticality of environmental sustainability to food production in the future. In this respect, it moves contemporary farming and growing towards a more enlightened position of respect for the environment than, for instance, is prevalent in the scientific reductionists' view that most problems in agricultural food production can be solved by recourse to artificial fertilisers and pesticides. GAP properly conceived resonates with notions of holistic agriculture encompassing the land, biodiversity, the welfare of farm animals and those who work the land, and broader society and the economy. Whatever the philosophical dimensions of GAP, its utility depends on its applicability and practicability for those who use it, i.e. farmers. For this reason various organisations are involved in the development and specification of GAP which can make for different interpretations and incoherencies in approach. In the UK for instance, GAP is manifested as a range of farm assurance schemes dealing with different kinds of agricultural enterprise and produce groups, but essentially many principles of good practice are transferable from one kind of enterprise to another just as the principles of quality management are transferable from one business to another. Indeed, GAP, and farm assurance for that matter, should be understood as quality management and quality assurance applied at farm level, as shall be discussed in Section 7.14.

Of the various organisations that have established principles for GAP, the FAO (2007) proposes a non-prescriptive approach to deal with the environmental, economic and social sustainability of farm production and post-production processes to produce safe, quality food. The FAO (2003) recognises that GAP encompasses 'three pillars of sustainability' in the form of economic viability, environmentally sound practice and socially acceptable practice. Food safety and quality are embedded within the three pillars. The FAO's framework for GAP recognises ten elements of agricultural practice that require the application of good management practice. These are: soil; water; crop and fodder production; crop protection; animal production; animal health and welfare; harvest and on-farm processing and storage; energy and waste management; human welfare, health and safety; and wildlife and landscape. Other organisations have been active in establishing a structured and coherent approach to GAP, including the GlobalGAP organisation. This is a private sector body that has developed from an initiative of food and produce retailers belonging to the Euro-Retailer Produce Working Group (EUREP) which established the EurepGAP standards, now the GlobalGAP standards. GlobalGAP sets voluntary standards for the certification of agricultural produce worldwide which will be discussed later.

The deconstruction of variants of GAP yields what is essentially a set of criteria which define good management practice in the operation of land-based farming and growing enterprises. Clearly different criteria have different levels of emphasis or relevance, depending on the nature of an enterprise. For instance, criteria concerning animal production and welfare are of no importance to businesses engaged solely in crop production. They would necessarily be incorporated into GAP standards applied to the farming of animals. When applied to the production of food crops, GAP should take a logical form elaborated as a set of best practice management principles, as follows:

1. **Management** – GAP requires the effective management of agricultural enterprises including the planning of operations, the implementation of management techniques for production activities, management to enhance the farm's value and production of food, and planning to sustain or increase the farm's contribution to society and the economy, without damaging the land and landscape or eliminating wildlife.
2. **Staff competence and training** – Business success and the achievement of objectives is dependent on the quality of management and leadership. The responsibility lies with managers to ensure that staff are capable of fulfilling their roles. This is recognised in GAP which requires that staff are adequately trained for their jobs, that training and development requirements are identified and updated as needed, and that staff work safely and in compliance with legal requirements and safe working periods. This is particularly important where specialist operations and activities are concerned.
3. **Hygiene and food safety** – In the production of foodstuffs food safety must come first. GAP requires the implementation of procedures to ensure the production of agricultural produce that presents no harm to consumers. Such procedures will involve the control of, for instance, fungal infections of growing crops and stored produce that could lead to mycotoxin contamination. It will also encompass the control of agrochemicals, e.g. observing pesticide application intervals and MRLs to prevent possible chemical toxicity to consumers. Additionally, it will concern the prevention of post-harvest contamination, e.g. the contamination of stored produce by pests and feral animals which could lead to foodborne illness caused by bacterial or viral infection. In this respect, it is also important that general hygiene management is managed within the GAP framework. So, for instance, the hygienic maintenance of equipment, e.g. irrigation equipment, harvesting and post-harvest processing machinery, and facilities, e.g. storage containers and buildings, must ensure that residues of organic material and other detritus do not accumulate and lead potentially to increased levels (or presence) of bacterial pathogens or viruses in the production and processing environment, with the potential to contaminate crop produce.
4. **Quality** – Apart from food safety, other aspects of produce quality are of importance to customers and consumers. Within the scope of GAP, procedures will be implemented to ensure that produce meets customers' specifications and that production practices comply with customer requirements and society's expectations. This latter point is important, as increasingly consumers incorporate in their definition of quality factors that go beyond product quality itself, e.g. environmental

sustainability. Such an interpretation of quality is made when consumers purchase organic produce. Here issues of food safety, e.g. the perceived harmful effects of pesticide residues, are pertinent, but also of importance is the perceived effect of agricultural practices on the sustainability of the land and associated biodiversity.

5. **Traceability** – ISO (2000a) defines traceability as: '*The ability to trace the history, application, or location of that which is under consideration*'. The need to trace a product back from the marketplace to the point of origin is of importance if a failure in quality or food safety is recognised, or questions are raised about the authenticity and provenance of a product. The ability to trace a product allows the root causes of failures to be identified and corrective actions to be implemented. Traceability facilitates product recall. GAP requires the establishment of systems of traceability and, importantly, systems of identification, as the ability to trace a product is contingent on the ability to identify it. Ideally, systems and procedures for identification and traceability will allow a scope of traceability that extends from 'field to fork'. EU Regulation 178/2002 (EC, 2002) in Article 18 states requirements for the implementation of systems for traceability from farm to consumer.
6. **Integrated crop management** – GAP incorporates principles of ICM to minimise and limit the adverse effects of crop production on the sustainability of the land and on biodiversity. Such good practices as are encompassed by GAP include the selection of crops, cultivars and varieties according to the constraints made by specific locations; the selection of crops for pest and disease resistance and tolerance of climatic conditions; crop rotations to provide disease breaks; the tactical use of pesticides; evaluating the need to use agrochemicals and weighing the advantages of use against the disadvantages and harms.
7. **Water protection** – The shortage of water is likely to be one of the greatest challenges faced by future generations, for without water food cannot be produced. GAP expresses the need to protect and preserve water by preserving and recycling water, using water efficiently, preventing soil erosion, run-off and leaching and protecting water supplies from contaminants.
8. **Soil fertility** – Alongside the preservation of water, preventing the damage and loss of soils is also a major problem facing future generations. GAP requires that measures are taken to maintain soil fertility, soil body and the capacity to hold water. It also requires that soil erosion, run-off and leaching are prevented.
9. **Pollution control, by-product and waste management** – GAP requires the implementation of controls for farm pollutants, by-products and wastes. Under GAP the intention should be to reduce off-farm inputs, e.g. to minimise and, where possible, eliminate the use of artificial fertilisers and pesticides. Plans should also be implemented to minimise

the production and handling of waste, to recycle organic wastes and to minimise the use of non-recyclable materials. Corrective action procedures should be implemented to rectify and prevent the recurrence of pollution incidents, e.g. pesticide or fertiliser spillage, leakage from silage, etc. It is also important to control adequately accumulations of manure and slurry as run-off resulting from poor management, for example, or rain may have the potential to flood into fields of growing crops, carrying with it pathogenic bacteria and viruses.

10. **Energy management** – The effective use of energy and reducing energy demands are of increasing importance to GAP. Ways to use energy more effectively include constructing buildings and facilities so as to conserve energy and using energy-efficient machinery. An aim must be to reduce off-farm energy inputs by using energy-saving methods of cultivation, adopting sustainable energy technologies and exploiting local and regional markets to reduce transport energy consumption in the provision of off-farm inputs when they are needed.
11. **Record keeping** – The systems and procedures implemented as good agricultural practice demonstrate the intention to use GAP according to standards such as those defined and administered by, for example, GlobalGAP. Records must be kept of the practice of GAP for the benefit of providing a source of information for problem solving and improvement. Records are essential to the function of audit and are needed for both internal and external audit purposes. Records should be kept of, for example, farm planning and production, the purchase of equipment and materials, equipment maintenance, the use of materials such as pesticides and fertilisers, as well as of staff training and competencies, and failures and associated corrective actions.
12. **Amenity, landscape and the protection of biodiversity** – As GAP evolves and concerns increase about the impact of human activities on the sustainability of the planet, elements of GAP will inevitably address with more forcefulness issues relating to amenity, landscape and biodiversity. Within GAP farm planning should operate to prevent or minimise damage to the landscape and biodiversity, as well as to enhance the landscape and success and security of natural flora and fauna. In this, practices such as selecting and organising agricultural operations to minimise adverse effects on the land and biodiversity become important, e.g. managing field margins and wetlands to encourage wildlife, and creating and maintaining areas of wild habitat to support biodiversity.

The management principles defined in GAP standards should be interpreted for application to a given agricultural enterprise. Using a GAP standard as a guide and source of reference, a farmer must decide how issues of food safety management, quality management and traceability will be managed in relation to the types of crops and produce being grown, the particular circumstances of the farm and the nature of customers' requirements. In working to GAP standards the aim should be to establish a planned and coherent approach to farm management that produces safe, quality produce which, at the same time, maintains the land as a sustainable resource for the production of food with minimal adverse effects on biodiversity. As GAP develops it can be seen to embody concepts of agricultural and food ethics, particularly those related to the ethical principles of beneficence, non-maleficence, autonomy and justice. Acting to produce good-quality, wholesome food for consumers is acting from the principle of beneficence. Acting to sustain the land as a source of food for future generations while also protecting biodiversity is acting from the principle of non-maleficence. Some approaches to GAP incorporate the need to ensure that those who work the land are fairly paid, have access to affordable housing and are not detached from local society for socioeconomic reasons. This resonates with the ethical principles of autonomy and justice and echoes the objectives of Fair Trade schemes.

7.10 GAP and food safety management

The management of food safety in crop production does not necessitate the use of HACCP. In some instances GAP may be sufficient. Pressure may, however, be brought to bear on farmers by customers, e.g. supermarkets, to use HACCP if, in the customer's view, HACCP is (i) necessary to protect consumers, and (ii) the customer's reputation can be safeguarded by the evidence of due diligence that HACCP provides. The decision whether or not to use HACCP will be influenced by a number of factors and must take into account the complexity of the production processes, the possibility of hazards arising, their source and their nature, and the ability of the enterprise to implement and maintain HACCP systems. A key consideration must be the ability to control foodborne hazards arising in agricultural production managed by GAP rather than HACCP. For instance, animal manures may be used to fertilise crops and the potential exists for crops to become contaminated by bacterial pathogens, e.g. *Salmonella* spp., enterohaemoraghic *E. coli*. The question arises: Should the management of manure be governed by GAP or HACCP? Chambers (1999) advocates that the treatment and control of animal wastes should be managed under GAP and not HACCP. Thus, manure may be well rotted over time and achieve the temperatures required to destroy microbial pathogens as part of the GAP procedures of a farm, thereby avoiding the complexities that HACCP brings. Likewise, irrigation water has the potential to become contaminated with faecal material which, in turn, can contaminate crops, e.g. leafy salad crops and soft fruit, with bacterial pathogens. This should also be managed under GAP and not HACCP (Early, 2002). Another factor influencing the decision not to use HACCP may be the extent to which hazards are eliminated in later, post-harvest processing. It is known that fungi have the potential to produce mycotoxins when growing on sugar beet (Bosch *et al.*, 1992) and that this has food safety implications for the feeding of sugar beet fibre to food animals (Bosch and Mirocha, 1992). The practice of clamping of sugar beet for protection against frost has the potential to allow fungal growth and mycotoxin formation. The chances of mycotoxins from sugar beet passing in significant quantities into refined sugar are very small. It is appropriate therefore to use GAP rather than HACCP to preserve the quality of sugar beet post-harvest and prior to delivery to a sugar factory. Whatever the crop, a farmer must use knowledge, judgement and an understanding of the produce and associated hazards, as well as an understanding of customer and market needs, to guide decision making in the choice between GAP and HACCP.

7.10.1 GAP and HACCP pre-requisite programmes

While GAP may in some instances be used to manage matters of food safety instead of HACCP, HACCP itself relies on GAP to create the environment in which HACCP systems can function effectively. GAP provides an underpinning structure on which to build HACCP systems, indeed da Cruz *et al.* (2006b) state that the GAP programme is a pre-requisite for HACCP implemented at field level. Elements of GAP form what are known as HACCP pre-requisites: procedures and activities necessary to the implementation and operation of HACCP systems. ISO (2005) defines prerequisite programmes as the '*basic conditions and activities that are necessary to maintain a hygienic environment throughout the food chain suitable for the production, handling and provision of safe end products and safe food for human consumption*'. In a food processing factory there is the need, for example, to achieve and maintain a hygienic production environment and to ensure that staff are trained in personal hygienic practices and follow them accordingly. Failures in environmental hygiene or staff hygienic practices can impact adversely on the operation of HACCP systems, giving rise to hazards or causing systematic failures. They need managing, but to manage them as part of a HACCP system would be complex and costly. It would also divert attention from the objective of HACCP: controlling the hazards that have been identified as essential for ensuring food safety. Creating and sustaining a hygienic production environment and training staff in good hygienic practice, and enforcing staff hygiene policies, are functions of GMP, which, in food factories, underpins HACCP. They are activities that are more appropriately classified as HACCP pre-requisite programmes or PRPs, as they must be in place before HACCP systems are developed and implemented.

In the context of agricultural food production, GAP is the equivalent of the food manufacturing industry's GMP as described by IFST (2006). Aspects of GAP fulfil the role of pre-requisite programmes as they establish the basic conditions and activities necessary to the effective implementation and operation of HACCP systems at farm level. Decisions about what to make a PRP in crop production will be determined by the crop itself and the systems and methods of production and post-harvest management. An example of a PRP can be seen in the management of crop sprayers used to apply pesticides. These toxic compounds need to be applied in the correct concentration and at the right rate to be effective. If they are applied at too low a concentration or rate they will not be effective and money will be wasted. Conversely, if they are applied at too high a concentration or rate, money will also be wasted and, importantly, they may represent a potential hazard to consumers in the form of residues above MRLs. While some aspects of pesticide application can be defined as part of HACCP, e.g. ensuring that the correct pesticides are used and made up to the right concentrations, the maintenance of spraying equipment is an activity that should be managed under GAP. If crop spraying is to be carried out effectively the crop sprayer must be maintained and calibrated properly as there is little point in spraying with a badly maintained sprayer or one that is out of calibration. If mycotoxins and levels of pesticides on crops above MRLs are recognised as potential hazards for consumers they should be controlled by HACCP, but the maintenance and calibration of the crop sprayer which is instrumental to effective hazard control should be established as a PRP under GAP. Examples of other crop production activities that may be established as PRPs include maintaining farm facilities in an adequate condition for the handling of food materials; maintaining storage facilities (grain silos, bins, etc.) in a clean and hygienic condition; managing staff hygienic practices in the manual harvesting and handling of minimally-processed produce to prevent contamination, etc.

7.11 Hazard Analysis Critical Control Point system (HACCP)

The status of HACCP as the method of food safety management recommended by various international and national authorities, as well as its inclusion in law, was discussed earlier. The importance of HACCP to the quality of food products, and the ability of food businesses to produce safe food according to the requirements of customers, has been recognised by the International Organisation for Standardisation which has published the standard ISO 22000:2005, *Food safety management systems – Requirements for any organisation in the food chain*, to provide a model for the development of HACCP-based food safety management systems. This standard is compatible with the quality system standard ISO 9001:2008, *Quality management system – Requirements*, and can be used to guide the design, development, implementation, operation and improvement of food safety management systems by any food business at any point in the food chain. It is stated that ISO 22000 is intended to function as a standard which can be applied to the management of food safety 'from field to fork' and this reflects the growing interest in the need to control foodborne hazards, including pathogens, at farm level.

7.11.1 HACCP at farm level

Although HACCP was originally devised to manage food safety in food manufacturing, its application now encompasses catering and food service operations, where it is beneficial to the food safety management of high-risk foods, e.g. milk and meat-based products, and also foods of generally lower risk, e.g. fresh produce and processed vegetables (Kokkinakis and Fragkiadakis, 2007). Since the mid-1990s, it has increasingly been applied to food safety management at farm level. This is particularly a response to food safety issues that have their root cause in the agricultural inputs sector and on farms, e.g. BSE and *E. coli* O157:H7 associated with beef production as discussed earlier, and which have brought a particular focus on the need to control the safety of agricultural produce passing up the food chain. With respect to the control of foodborne hazards in crop production, it is now recognised that microbial pathogens and hazards such as pesticide residues can be controlled by the application of HACCP. Consequently, many crop production standards and crop related farm assurance schemes incorporate HACCP as a key requirement.

7.11.2 Foodborne hazards

Many hazards of microbial origin of relevance to food crops have been identified in this chapter. The term 'hazard' is defined (CCFH, 2001) as '*A biological, chemical or physical agent in, or condition of, food with the potential to cause an adverse health effect*'. Biological hazards of relevance to crop production are mainly: pathogenic bacteria that cause infective foodborne illness, e.g. *E. coli* O157:H7, *Campylobacter* spp., *L. monocytogenes*, *Salmonella*, *Shigella*, *Vibrio cholerae*, *Yersinia enterocolitica*; toxigenic fungi, e.g. *A. flavus*, *A. clavatus*, *Fusarium* spp., *P. verrucosum*; foodborne illness viruses, e.g., Norovirus, hepatitis A; protozoan parasites, e.g. *Cr. parvum*, *Toxoplasma gondii*. The intoxicating bacteria that cause foodborne illness, e.g., *Staphylococcus aureus*, *Clostridium botulinum*, are much less likely to be of significance to food safety in crop production, as they tend to be problematical in processed food production. The same is so for allergenic materials, e.g. nuts and wheat gluten. Most foodborne illness in the world is bacterial in origin, although the biological hazards of most concern to crop production are bacterial and fungal pathogens. The bacterial pathogens associated with food crops may be infective or intoxicating. *Listeria monocytogenes* is an environmental organism found in soils which may contaminate root and salad crops. *E. coli* O157, *Campylobacter* spp. and *Salmonella* spp., for example, are associated with farm animals (as well as some feral animals) and may contaminate crops via inadequately composted manures or faecally contaminated irrigation water. *Shigella* spp. and other pathogens such as *E. coli* O157 and *Salmonella* spp. may be carried by humans who contaminate crops through handling. Protozoan parasites may contaminate crops, e.g. leafy salad crops, as a result of the faecal contamination of irrigation water by grazing livestock or direct contamination by feral animals. Viruses may contaminate fresh produce through handling and poor hygiene. Farmers need to know of the sources of biological hazard that may affect crops and factor them into their food safety management systems.

Many chemical hazards are associated with agricultural food production and have relevance to crop production, but their importance to this chapter is limited and the examples below are listed for completeness. Chemical hazards include: naturally occurring environmental contaminants, e.g. heavy metals; industrial contaminants, e.g. dioxins, PCBs; pesticide (insecticide, herbicide and fungicide) residues in fruit, vegetables and cereals, etc.; nitrates (mainly leafy crops) and other fertiliser residues; pesticides and nitrates in drinking water; residues of veterinary medicines and zootechnical substances in animal products, e.g. meat, milk, eggs; contaminants arising from the handling, storage and processing of foodstuffs, e.g. grain treatment compounds, machine lubricants, cleaning agents, rodenticides and other pest control poisons; contaminants arising from food packaging, e.g. plasticisers and other packaging material additives, adhesives, inks, metals leached from cans. As with chemical hazards, physical hazards are also of academic interest to this chapter but they should be evaluated when developing HACCP systems. Physical hazards may be categorised as: slicing hazards – sharp glass fragments, sharp plastic fragments, wood splinters, sharp metal filings and swarf; dental hazards – glass particles, pieces of wood, pieces of hard plastic, stones, metal fragments and parts, e.g. nuts, washers; choking hazards – wood, stones, metal fragments, string, nuts, e.g. peanuts. Most crop production, harvesting and post-harvest processing and storage activities will either prevent the physical contamination of crop produce or remove physical contaminants. For instance, the screening of cereals within a combine harvester should remove stones and other physical materials picked up with the crop, while in the milling of cereals metal detection and screening should remove ferrous and other contaminants from grain, e.g. bits of combine harvesters. In sugar beet processing water is used to wash away soil and separate stones from beet. The same is so in the cleaning of potatoes for packing or processing. In the harvesting of salad crops and vegetables the management systems employed should ensure the careful control of glass and, for example, wooden packing containers and pallets, to prevent fragments of glass or wood from contaminating produce.

Biological hazards and hazards in the form of bacterial pathogens probably represent the key source of food safety focus for crop producers, in part because the symptoms of bacterial foodborne illness are manifested rapidly when compared to other forms of foodborne hazard. In comparison, disease from mycotoxin poisoning or the consumption of pesticide residues takes longer to become apparent. Indeed, a link between the cause of disease and the disease itself may be extremely difficult, or impossible, to establish. The evidence of mycotoxin poisoning manifested as, for example, oesophageal cancer requires a sufficiency of epidemiological data linked to fungally-contaminated food supplies. The evidence of disease associated with pesticide residues is likely to be more difficult to establish, not least because exposure to pesticides by consumers in industrialised countries is very low and pesticide products are continually being reformulated and replaced. For many pesticide products a sufficient body of epidemiological data may never accumulate to establish a link between a given product and a disease in consumers. Kishi (2005) states that this is not the case for migrant or seasonal agricultural workers in industrialised countries, or for farm workers in developing countries, both of whom are the most highly-exposed populations and are rarely the subjects of research into the health-related effects of pesticide exposure. Agricultural workers may receive significant exposure and, at times in developing countries, be exposed to highly toxic products that have been withdrawn from use in industrialised countries, but which are still manufactured and exported to developing countries where regulation is less restrictive. The use of a number of different pesticides on a crop complicates the picture. Individually a pesticide may represent a low risk of causing disease in consumers as a result of the consumption of residues over time, or indeed to farm workers through exposure in use. However, Kishi (2005) expresses concern about the synergistic effects of using many pesticides in concert and the possible increased risk to the health of farm workers.

Questions must therefore be asked about the effects of pesticide synergy on the health of consumers, but establishing with any confidence the degree to which consumers are at risk is likely to be complicated if not impossible. Pesticide manufacturers have a duty of care to ensure that their products do not represent a risk to consumers and agricultural workers when used properly, and to support through the provision of trustworthy data and information epidemiological and environmental studies into the effects of pesticides. Farmers use in good faith the proprietary compounds made available to them for crop production. From their perspective the use of pesticides achieves an economic benefit as a crop is produced with reduced damage for example from insects and fungi. There is also a benefit to the consumer as farmers are able to produce high-quality foodstuffs. The farmer must weigh the use of pesticides against the risks associated with not using them, e.g. crops contaminated with mycotoxins as a result of fungal infestation which could make the crop unsafe, and manage the production system accordingly. Through the judicious use of HACCP the farmer can recognise the possible hazards associated with crop production and assess the risk of harm to consumers of either not using pesticides, or of using pesticides. Importantly, when pesticides must be used HACCP allows the implementation of controls that should safeguard the health of consumers, agricultural workers and the farmer's business.

7.11.3 The seven principles of HACCP

The HACCP concept is based on the seven principles of HACCP which provide a methodology for the development of food safety management systems. The seven principles of HACCP as defined by CCFH (2001) are:

Principle 1: Conduct a hazard analysis.
Principle 2: Determine the critical control points (CCPs).
Principle 3: Establish critical limit(s).
Principle 4: Establish a system to monitor control of the CCP(s).
Principle 5: Establish the corrective action to be taken when monitoring indicates that a particular CCP is not under control.
Principle 6: Establish procedures for verification to confirm the HACCP system is working effectively.
Principle 7: Establish documentation concerning all procedures and records appropriate to these principles and their application.

Application of the seven principles of HACCP is achieved through a logical sequence of activities in which a HACCP study is carried out to develop a HACCP plan which is implemented as a HACCP system. A HACCP plan is defined (CCFH, 2001) as '*A document prepared in accordance with the principles of HACCP to ensure control of hazards which are significant for food safety in the segment of the food chain under consideration*'. A HACCP system is defined (CCFH, 2001) as '*A system which identifies, evaluates and controls hazards which are significant for food safety*'.

Conventionally, HACCP systems are developed in relation to a given food product and the process by which it is produced. At times generic HACCP plans are proposed for certain food products or types of product, but they risk overlooking hazards which are intrinsic to a particular process. Two producers of leafy salad crops, for example, may operate very similar production processes, but this does not mean that the hazards of concern at one enterprise will be the same as those at the other. This said, when compared with the complexities of many food manufacturing businesses the processes used in crop production tend to be relatively simple. The principles used to define food safety control in a generic HACCP plan may therefore be applied to similar agricultural enterprises with some conviction irrespective of the differences between the enterprises. It would, however, be negligent not to urge caution in the use of generic HACCP plans.

7.12 Developing a food safety management system

A HACCP study is carried out to provide an understanding of the product itself and the process by which it is produced. The information gained during the study is evaluated to design and develop the HACCP plan which identifies the hazards associated with the product and process and establishes the required controls. The HACCP system is implemented, operated and maintained according to the requirements of the HACCP plan. A HACCP study is usually carried out according to a 12-stage process.

7.12.1 Stage 1: assemble the HACCP team (and define the scope and terms of reference of the study)

It is usual to carry out a HACCP study with a HACCP team. In food crop production this may not be possible as there may not be enough people available in an agricultural enterprise to constitute a team. This does not mean that HACCP systems cannot be developed and used. The ideal HACCP team will be multidisciplinary and consist of people with expertise that covers knowledge of the product and production process as well as food safety management and HACCP. In food manufacturing a HACCP team usually contains people with technical, production and engineering expertise. In crop production on a farm it may be that only one person is available to undertake the HACCP study, and clearly that person should be conversant in the theory and practice of HACCP. They should also have knowledge of the product and production process as well as an appreciation of the food safety hazards associated with crops and crop production. When a person (or people) with the right knowledge and experience is not available this may be rectified through training with attendance on an appropriate HACCP course. Once a person has grasped the fundamentals of HACCP they should be able to make an effective attempt at the development of a farm-based HACCP system and, in so doing, develop their experience and confidence in food safety management at farm level. At times questions arise that cannot easily be answered and in such instances external sources of information may be sought. Consultants can be used to carry out or supplement HACCP studies, but this will be at a cost. Alternative sources of advice may be university academics and government advisers, as well as experts within farm inputs businesses, e.g. those supplying pesticides and fertilisers, etc. Additionally, professional and industry organisations can be a useful source of information. For example, the Chilled Food Association publication, *Microbiological Guidance for Growers* (CFA, 2007), advises on HACCP implementation in the field relating to fresh produce and combinable crops.

Once it has been established who will carry out the HACCP study, the scope and terms of reference of the study must be set. The scope of the HACCP study confines the HACCP system. It dictates the start- and end-points. Given that a HACCP system applied to crop production may cover a range of activities carried out discretely and at different times of the year it can be sensible to break a study into separate components or study units. This can make the study simpler and reduce the chance of something critical being missed. On completion of a series of studies the HACCP systems covering the entire process can be linked to form the overall system. For instance, seed preparation and propagation may form one HACCP study, crop management may form another and harvesting and post-harvest crop management may form yet another. The terms of reference of the study establish which hazards will be considered. For instance, it may be convenient or more resource efficient to select only biological, or chemical or physical hazards, or a specific form of hazard from one of the categories (e.g. microbiological hazards), rather than tackle all three categories at once.

7.12.2 Stage 2: describe the product

In Stage 2 of the HACCP study a complete description of the food product is developed to inform the hazard analysis stage (Stage 6) of the study. In food manufacture the description of the product involves gathering information about the ingredients used to make the product and information about the product itself and the way it is packaged, stored, transported and distributed. The information is used to establish facts of relevance to the identification and evaluation of hazards, such as intrinsic preservation factors, e.g. pH, salt-in-moisture content and low a_w, and their effects on the survival and growth of bacterial pathogens. Also of interest are extrinsic preservation factors, e.g. heat treatments and hygienic or aseptic packaging processes affecting food safety, and the need to store food under refrigeration to keep it safe.

In crop production the description of the product will need to consider the crop itself and the nature of the product with respect to factors that may give rise to hazards or which can influence the prevention of hazards. Agricultural practices that can be interpreted as inputs to the product should be considered for their potential to introduce hazards to the product, e.g. inadequately composted manures or faecally-contaminated irrigation water as a source of bacterial pathogens. Likewise, the application of pesticides equates to the addition of 'ingredients' of a kind that have the potential to be hazardous if not properly controlled. There are parallels here with the addition of preservatives to manufactured foods which are considered safe when used in very small amounts, but which could be toxic when used to excess. The growth of the crop should be assessed for its potential to allow the development of hazards, e.g. the fungal infections due to water stress and insect attack giving rise to mycotoxins. The harvest and post-harvest management stages of production also have to be considered for their potential to allow hazards to develop, e.g. inadequately dried cereals or nuts allowing fungal growth and mycotoxin development. A variety of information sources may need to be consulted when describing the product, including textbooks, scientific journals, food research organisations, consultants and academics, pesticide manufacturers, national and local government organisations with responsibility for food safety, and sources on the Internet. A useful source of information about foodborne pathogens and foodborne disease is the US Food and Drug Administration's 'Bad Bug Book' which can be found at (http://www.cfsan.fda.gov/~mow/intro.html).

7.12.3 Stage 3: identify the intended use of the product

It is important to consider the intended use of the product and the ways in which use by a processor or consumer could give rise to hazards. For example, could a soft fruit or salad crop eaten with little or no washing or domestic processing prior to consumption compromise food safety? Leafy salad crops or soft fruit such as peaches or nectarines may be consumed without washing. If they had been irrigated with faecally-contaminated water the potential may exist for consumers to be poisoned by *E. coli* O157 which has a very low infective dose level of 10–100 cfu, or some other foodborne pathogen. Although the focus of this chapter is pathogen control in relation to food crops, a similar line of thinking must also take into account sources of hazard such as pesticide residues, some of which may be used to control microbial pathogens.

As well as considering the intended use of the product, consideration should be given to any intrinsic characteristics which might be harmful to sensitive consumer groups (children, the elderly, people who suffer from allergies, people with depressed immune systems, etc.). During a HACCP study in food manufacture concern would be given to consumers who suffer from nut allergy and to children who might choke on peanuts when the products involved contain nuts or peanuts, or when they are made in factories that handle nuts. At farm level such concerns may be academic as most crop produce is taken into food processing and manufacturing, where issues concerning sensitive groups become the concern of the food manufacturer and food retailer.

7.12.4 Stage 4: construct a flow diagram

In Stages 2 and 3 of the HACCP study information should be gathered that helps to identify hazards associated with the product and the way it is used. In Stage 4 attention should be turned to the production process and information is gathered for use in identifying hazards that may arise in the production process or as a result of its operation or an associated activity. A process flow diagram of the crop production process should be prepared (see example in Fig. 7.1) which identifies each step in the process. A flow diagram is '*A systematic representation of the sequence of steps or operations used in the production or manufacture of a particular food item*' (CCFH, 2001). A step is defined (CCFH, 2001) as '*A point, procedure, operation or stage in the food production chain including raw materials, from primary production to final consumption*'. The process flow diagram should identify the steps involved in the crop production process, including, for instance, seed treatment and propagation, field or site preparation, fertiliser applications, planting, growing, irrigation, pesticide applications, harvesting, post-harvest handling and post-harvest pre-treatments, e.g. cleaning, trimming, as well as storage and transport, etc. The process control parameters applied at each production stage should be recorded on the process flow diagram, as should the inputs to and outputs from the process. Inputs to crop production include seed, seed treatment agents, irrigation water, manure, fertilisers and pesticides, etc. In harvesting and post-harvest operations inputs will include, for example, water used in hydrocooling to remove field heat and washing to remove soil and contaminants, etc. The product itself is the principal output, but others may include product rejected in grading due to damage or deterioration, waste botanical material, e.g. from trimming and other preparation processes, soil from washing operations, etc., all of which, given the right circumstances, could give rise to microbial and other hazards. The scope of the HACCP study

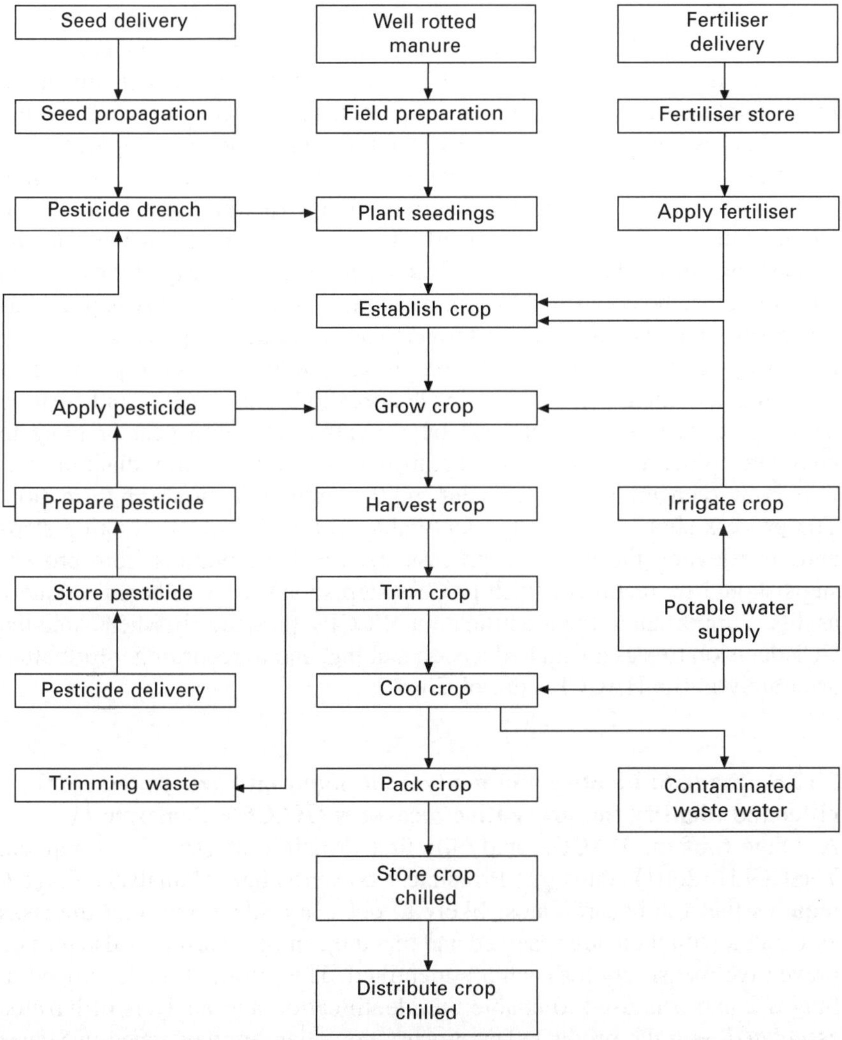

Fig. 7.1 Generalised flow diagram for field crop production. Note: a working flow diagram would present a greater level of detail giving process parameters, and with each step of the process numbered, etc.

will constrain the scope of the process flow diagram which should be logically and systematically structured. A competently prepared process flow diagram should provide a level of detail that allows evaluation and the identification of hazards without constant reference to additional information.

7.12.5 Stage 5: confirm the flow diagram

The process flow diagram may be developed through the application of knowledge about the production process itself and by reference to information and data concerning the process or the operations contained within it. Sources of information will include specifications for seeds, fertilisers and pesticides, information about crop handling conditions required to control contamination for example by microbial pathogens, procedures for cleaning and making ready harvesting, post-harvest crop handling and storage equipment, etc. Preparation of the process flow diagram against a theoretical understanding of the process can mean that the diagram is not a true representation of what happens. Items that have been overlooked or unauthorised changes to the process may mean that the diagram is inaccurate. The validity of the diagram must therefore be confirmed by comparing it with what actually happens, as it happens. This is commonly referred to as 'walking the process', but it is not always easy or immediate in execution in relation to crop production, given that a complete production cycle needs to be examined. Confirmation of the process flow diagram may take months or even a year or more to complete. When developing and confirming the process flow diagram it is important to ensure clarity in layout and the communication of information. The process should be clearly identifiable as a series of interlinking steps and, as relevant, the inputs to process steps and the outputs from process steps should be recorded. Each process step should be numbered uniquely as this is important to the identification of CCPs, particularly when using the CCP decision tree as an aid to decision making, and to recording information accurately in the HACCP Control Chart.

7.12.6 Stage 6: identify and analyse all potential hazards, assess the risks and identify the preventive measures (HACCP Principle 1)

At Stage 6 of the HACCP study the first Principle of HACCP is applied. The CCFH (2001) states that Principle 1 concerns hazard analysis. Stage 6 requires that the hazards most likely to occur are identified, that the risks associated with them are assessed and that the control measures (also termed preventive measures) for each are identified. The information developed in Stages 2 and 3 is used to enable the identification and analysis of hazards associated with the product. The process flow diagram developed in Stages 4 and 5 enables the identification and analysis of hazards associated with the process. In the process of hazard analysis the possible hazards should be listed and a value for risk should be determined for each. Risk is calculated as the product of the probability of occurrence and the severity of consequences for consumers. Risk assessment is a valuable part of the hazard analysis. It allows the rejection from consideration of hazards that are unlikely to occur so that resources can be focused on those that must be controlled. The concept of reasonable precautions in the UK's Food Safety Act 1990 allows for the identification and control of only the hazards over which it would be reasonable to expect control to be exercised. Hazards that are hardly likely to arise can be rejected by the hazard analysis, as there is little point in building into HACCP systems the need to control all hazards however remote the possibility of their occurrence. To do so would reduce the level of attention given to hazards that are very important for control and would waste resources. When it has been decided which hazards must be controlled then the method of control for each needs to be established. This is a hazard-specific exercise as the method of control for one hazard may be different to that for another, even for hazards within the same category.

Hazard analysis in a study where the terms of reference have been limited to pathogens affecting a food crop should focus on the identification of, for example, bacterial pathogens associated with manure and irrigation water and/or the growth of fungal organisms on crops and stored materials, with the development of mycotoxins. Control methods are likely to centre on the prevention of hazards, rather than their elimination or reduction to an acceptable level. Control methods will include, for example, the use of properly composted manures and potable irrigation water to prevent the contamination of crops by bacterial pathogens, the use of fungicides to prevent the fungal contamination of growing crops and some stored crops, and the reduction of water activity in stored crops such as grain to stop fungal growth. In some instances refrigeration may also be used as a control method, for example, to prevent the growth to unacceptable levels of very low levels of pathogenic bacteria that may be associated with fresh produce. When establishing control measures for hazards care must be taken not to incorporate in the HACCP plan controls which should be dealt with under GAP. For instance, as already established the composting of manures to eliminate bacterial pathogens is best managed under GAP. When composted manures are used to fertilise crops the HACCP plan may recognise the potential for contamination by bacterial pathogens e.g. EHEC or *Salmonella* spp. GAP will produce properly composted manures suitable for use as fertiliser. HACCP should ensure that only such manures are used. Some authorities advocate not using manures in crop production which eliminates from the HACCP study the need to consider bacterial pathogens associated with manure. If manure is used as a fertiliser, e.g. for reasons of cost or to maintain the sustainability of soil – its biota and ability to hold water and to produce food, etc. – then due consideration must be given to controls that fall under GAP and HACCP.

7.12.7 Stage 7: determine the Critical Control Points (CCPs) (HACCP Principle 2)

Stage 7 of the HACCP Study requires that CCPs are identified for each hazard. A CCP is defined (CCFH, 2001) as '*A step at which control can be applied and is essential to prevent or eliminate a food safety hazard or reduce it to an acceptable level*'. Control (or preventive) methods are applied at CCPs. The hazards identified in Stage 6 must be assessed in relation to each step in the production process to decide whether a particular step constitutes a CCP. Experience and judgement are used to decide whether a process step is a CCP. When a process step is being evaluated for its potential to be a

CCP for a given hazard, if it is clear that (i) the step can exercise control over the hazard, and (ii) there is no subsequent step where the hazard can be prevented or eliminated or reduced to an acceptable level, then the process step can objectively be identified as the CCP. As an aid to the identification of CCPs the CCP decision tree (Fig. 7.2) can be very useful, particularly for those who have little experience in HACCP and require support in decision

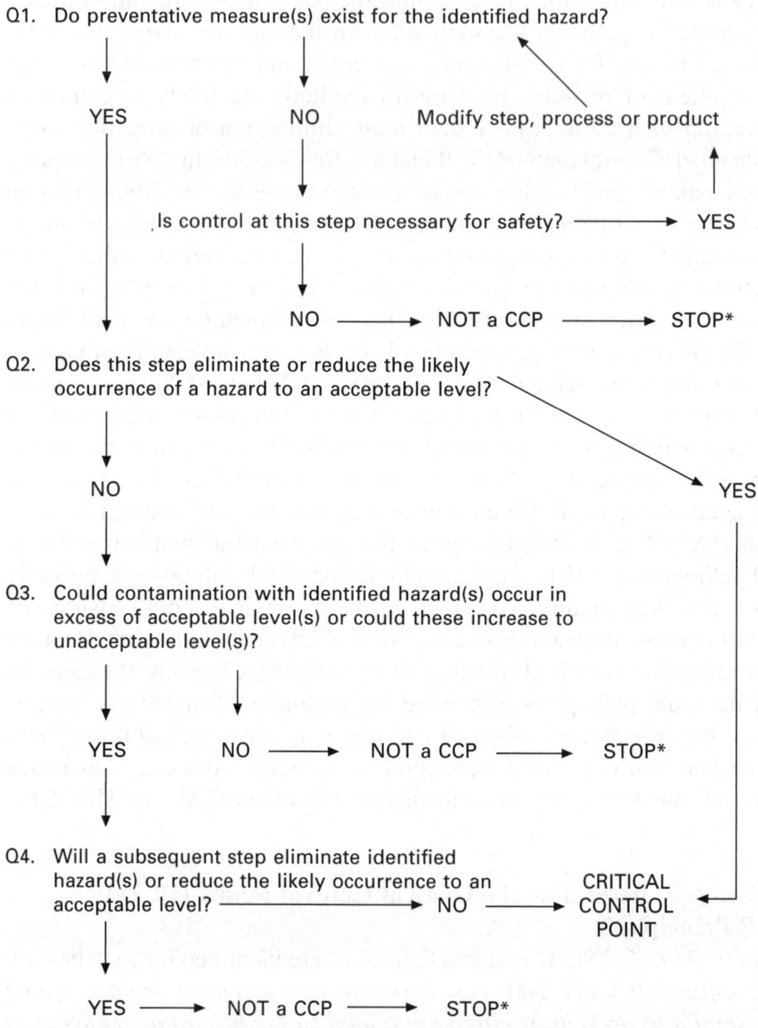

Fig. 7.2 The CCP decision tree. (Source: FLAIR, undated, HACCP User Guide, Concerted Action No. 7, Food Linked Agro Industrial Research, 191 rue de Vaurigand - 75015, Paris). *Proceed to next step in the described process. Note: For each hazard identified answer each question in relation to each step of the production process.

making. Care must be taken when using the CCP decision tree to ask each question in the decision tree of each step in the production process with respect to each hazard identified in Stage 6. The value of the CCP decision tree is that it causes HACCP practitioners to reject those process steps where application of a control method for a given hazard is either absurd or unsuitable and to recognise precisely where the control method should be applied to ensure food safety. It is important that no more CCPs are created for a HACCP system than are necessary to achieve food safety. A common error in the application of HACCP methodology is to recognise too many CCPs. The root cause of such error is usually found in an inadequate risk assessment and the inability to reject from inclusion in the HACCP study the hazards that are unlikely to arise. The inclusion of unnecessary CCPs will add to the complexity and costs of maintaining the HACCP system and may risk overburdening the staff who operate the system, with the consequence that work errors occur leading to failures in food safety management. An intelligent interpretation of HACCP is required if the creation of false CCPs is to be avoided.

7.12.8 Stage 8: establish critical limits for each CCP (HACCP Principle 3)

Control methods are applied at CCPs. When CCPs have been identified it is necessary to establish operating requirements for the control methods and specifically the critical limits to which CCPs should adhere. A critical limit is '*A criterion which separates acceptability from unacceptability*' (CCFH, 2001). In practice it is a process parameter that separates safe from unsafe and beyond the limit food will potentially be unsafe. Critical limits may concern quantitative values based on physical measurements such as time, temperature, pH, a_w, concentration, etc.. Alternatively, they may be based on qualitative assessments proven to demonstrate food safety, e.g. the absence of glass in a product as confirmed by a glass audit which shows that no glass item in a premises has been broken. When applying pesticides to crops, values for concentration and application rates may be set as quantitative critical limits. So may the intervals between application and harvesting and legally defined MRLs. It is important not to confuse critical limits with operational control limits. In some instances operational limits are set for the management of a process, e.g. a target and tolerance for the operation of a process may be set for process control purposes, but the values for target and tolerance may not be the same as that set for the critical limit. Such differentiation is seen when heat treatments are used to kill pathogenic microorganisms. Target temperatures are set with tolerances that maintain processes at an acceptable level apart from critical limits which define temperatures below which pathogens will not be destroyed, thereby creating a safety net.

7.12.9 Stage 9: establish a monitoring system for each CCP (HACCP Principle 4)

It is essential to know that the control methods applied at CCPs are effective and that hazards are being controlled. Stage 9 of the HACCP study concerns the implementation of methods for monitoring the effectiveness of control methods applied at CCPs. Monitor is defined (CCFH, 2001) as '*The act of conducting a planned sequence of observations or measurements of control parameters to assess whether a CCP is under control*'. Monitoring methods should be established that are simple to apply and which quickly confirm the status of CCPs, i.e. that they remain in control or control has been lost. The methods may be quantitative or qualitative and may be linked to the nature of control measures, e.g. if heat treatment is the method of control then temperature measurement is likely to be used to monitor the control. Methods of monitoring in crop production include checks on the usage of pesticides which indicate that appropriate dilutions have been made; the measurement of moisture content in grain which indicates that grain drying has been effective and that water activity has been lowered below that required for the growth of fungi; the measurement of pesticide residues in crops to ensure MRLs are not exceeded, etc. Monitoring requirements should be identified in HACCP plans, as should the frequency of monitoring and the identification of who is responsible for carrying out monitoring activities.

7.12.10 Stage 10: establish corrective action procedures (HACCP Principle 5)

At times monitoring may reveal that control at a CCP has been lost. In such circumstances it is essential that (i) control is restored immediately to the implicated CCP to ensure the continued production of safe food, and (ii) any product associated with the period during which the CCP was out of control is identified and itself controlled to prevent potentially unsafe food from entering the food supply chain. Corrective action procedures should be established at Stage 10 of the HACCP study for each of the CCPs identified in the HACCP plan. It is necessary to plan corrective action in advance – even though it may never be implemented if control is never lost – to ensure that the right actions are taken without delay when control is lost, thereby safeguarding consumers. Any product that has been identified as potentially unsafe because the loss of control at a CCP has been recognised should be assessed, e.g. by appropriate laboratory testing, to establish its food safety status. However caution should be exercised in the interpretation of results. Traditional quality control methods used for the food safety assessment of food products are based on the random sampling of product, and the principle of acceptance or rejection based on analytical results. Even when using statistical sampling plans to obtain what is considered to be a representative sample, it is possible that a hazard may be missed because food products are frequently heterogeneous. A hazard may not be evenly distributed throughout a product. The sampling and testing of a potentially unsafe food product may not confirm with confidence the safety of the product and it is generally better to dispose of such product rather than releasing it to the marketplace. In contrast to traditional quality control which relies on product inspection to confirm compliance with specifications, the HACCP approach to food safety management is a preventive approach which emphasises process control to ensure that hazards are prevented or eliminated or reduced to an acceptable level. Because control methods are applied to all parts of the product the issue of heterogeneity and hazards not being detected is overcome. It is important that the staff whose job includes corrective action are clearly identified in the HACCP plan and are given both responsibility and authority to undertake the tasks associated with any corrective action that may need to be implemented in relation to any aspect of the HACCP system. The responsibility for corrective action should also include confirming that control has been restored to CCPs.

7.12.11 Stage 11: Establish verification procedures (HACCP Principle 6)

Prior to the implementation of a HACCP plan as a HACCP system it is essential to confirm that the plan is complete and will fulfil its purpose when implemented. Following the implementation of a HACCP system it is essential to confirm that the system is operating effectively. Stage 11 of the HACCP study requires that procedures are established for the validation of the HACCP plan and verification of the HACCP system. Validation concerns answering the question: Will the system work when we put it into practice? (ILSI, 1999) and is defined (CCFH, 2001) as '*Obtaining evidence that the HACCP plan is* [likely to be] *effective*'. Validation involves making an assessment of the scientific and technical content of the HACCP plan. In relation to crop production firstly the elements of the plan should be checked for completeness to ensure that there are no missing elements or gaps in information. There should be a logical flow through the plan and associated documentation, from the identification of hazards that need to be controlled to the corrective action associated with the loss of control over given hazards. The validation should confirm that the decisions and assumptions made during the study are sound. So, for example, when named pathogenic bacteria (e.g. *Salmonella*, EHEC, etc.) sourced from the faeces of feral animals are identified as the hazards that are likely to contaminate a salad crop and the need to fence fields to prevent the movement of animals through the crop is identified as a control measure, then the rationale and justifications for making decisions about the hazards and means of control should be clear and properly recorded. Validation should ensure that the HACCP plan is confirmed as adequate to create a workable and effective food safety management system. A series of validation activities which confirm the adequacy of the HACCP plan in relation to the 7 Principles of HACCP are recommended by ILSI (1999) and have been adapted as follows.

Principle 1: hazard analysis
Confirm that (i) those undertaking the HACCP study are appropriately qualified and experienced, (ii) the process flow diagram is suitable for the purposes of the study, and (iii) significant hazards and control measures have been identified. The qualification and experience of HACCP practitioners can be problematical for farmers. It may therefore be necessary to recognise that although those who undertook the HACCP study may have an appropriate theoretical knowledge of the practice of HACCP, time will be needed to gain experience. Also, it is easy to believe *prima facie* that all possible hazards must be controlled, but in crop production, as in all other areas of food production, absolute food safety is not possible. Validation should show that a rational case has been made about which hazards must be controlled and which can be rejected from consideration because the risks are so low.

Principle 2: identify CCPs
Confirm that (i) for each significant hazard CCPs have been identified where control measures can be applied, and (ii) the CCPs are at suitable stages of the process (e.g. the last point in a process where a hazard can be controlled). Note that control methods appropriate to the control of hazards in crop production may not provide as great a degree of confidence as control methods applied in food processing factories. Even in the case of crops grown under glass with hydroponic irrigation and nutrition, the levels of control may not be equivalent to those possible in food manufacturing. The validation of HACCP plans in crop production must take into consideration the limitations of individual control methods and confirm, as appropriate, the use of combinations of control methods to achieve overall food safety. In this respect the approach is similar to the 'hurdle' concept used to prevent microbial growth in food preservation. This concept recognises that a single food safety factor may be insufficient in a particular food system to prevent the growth of pathogens, but that growth may be inhibited by a suitable combination of inhibiting factors operating at sub-inhibitory levels (Fonseca, 2006; McLauchlin and Little, 2007). So, the use of fencing to exclude feral animals from salad crops may reduce the likelihood of faecal contamination and thus the presence of pathogenic bacteria on a crop, but it may not have the power to prevent contamination entirely (birds will still over-fly the crop with the potential of contaminating it). It is important therefore to ensure that when it is harvested the crop is cooled to prevent the growth of contaminant organisms and that it is later washed in a sanitising solution, e.g. chlorinated, cold water with a pH of less than 3 to increase the availability of chlorine. Together, the fencing, cooling and sanitising work in concert to ensure food safety.

Principle 3: critical limits
Confirm that (i) critical limits are established for each significant hazard, and (ii) critical limits are appropriate to the control measures applied. It is

important in validation to confirm the basis for establishing critical limits and, therefore, their acceptability. Those based on physical measurements are usually convincing, but some may of necessity be subjective. For example, in grain storage dry grain kept in a clean, dry grain silo, or other suitable container, is necessary to the prevention of mould growth and the possible formation of mycotoxins. The moisture content of stored grain can easily be measured using an appropriate calibrated instrument, but how is one to assess the cleanliness of a grain silo? In most cases this will be done visually and, therefore, subjectively: but, subjective, visual assessment may be quite satisfactory.

Principle 4: monitoring
Confirm that (i) monitoring methods are able to demonstrate the effectiveness of control measures, and (ii) appropriate procedures exist for the calibration of monitoring methods. The issue of subjectivity also bears on the validation of monitoring in relation to HACCP plans created for crop production purposes. Ideally, monitoring activities will be based on physical measurements. Washing fresh-cut produce in chlorinated water can reduce microbial loading by 90–99 % (Bhagwat, 1995). The use of chlorinated water to sanitise produce may represent a control measure and the measurement of available chlorine in the water may, therefore, provide an objective, physical method of monitoring. In contrast, if the application of fungicide to a crop is deemed a method of control, the associated monitoring procedure may be somewhat subjective and consist of little more than checks on the usage of fungicide by the sprayer operator by noting the depletion of fungicide stock levels. The sprayer operator has to be trusted to use the required quantity of fungicide when preparing a solution for spraying.

Principle 5: corrective action
Confirm that (i) corrective action procedures have been established for each CCP, (ii) procedures exist to prevent non-conforming product from entering the food supply chain, and (iii) those responsible for taking and verifying corrective action, and for approving the disposition of non-conforming product, have been identified. The validation of Principle 5 can present certain difficulties in crop production when, for instance, all of the decision making power and authority in the business rests with one person, i.e. the farmer. Indeed, it may be that the farmer carries out the validation. It is important to ensure that the HACCP plan can be implemented and operated effectively without the undue influence of vested interests. The implementation of corrective action can mean the loss of money when produce deemed potentially unsafe has to be disposed of. Validation should ensure that not only are the corrective action procedures and associated procedures complete and, in principle, fit for use, but also that corrective action will not be inhibited by conflicts of interest in the event that monitoring highlights the failure to control hazards.

Principle 6: verification
Confirm that procedures and a plan for the validation and verification of the HACCP system have been established. The procedures and plan for validation should exist as they should be used to guide the validation process. When validating the plan for verification of the HACCP system it should be recognised that the verification process may, in some instances, take a long time, because of the nature of crop production cycles. It may also be necessary to allow some latitude in the timing of verification activities, because of seasonal demands on personnel making it difficult to release people to carry out verification duties when, for example, crops have to be harvested.

Principle 7: documentation
Confirm that documentation describing the entire HACCP system has been established and that the records required to support the system have been created.

The outcome of validation should be that a HACCP plan which is complete and suitable for implementation has been produced and that, when implemented as the HACCP system, it should be effective in protecting consumers. The process of validation is essentially a systems audit. It can be done with the auditing techniques used in quality systems auditing.

Verification can only be carried out on an operational HACCP system. Prior to implementation of the HACCP plan as the HACCP system, and as part of the HACCP study, the procedures for verification should be established. Verification concerns answering the question: Are we doing what we planned to do? (ILSI, 1999) and is defined (CCFH, 2001) as '*The application of methods, procedures, tests and other evaluations, in addition to monitoring, to determine compliance with the HACCP plan*'. Verification is a check that what the HACCP plan says will be done, is actually being done. It should confirm that the HACCP system has been implemented in compliance with the plan. Verification is an audit process and should be undertaken as often as necessary to confirm that the HACCP system is working correctly. Verification should aim at confirming that:

1 The HACCP plan remains suitable for its purposes and has been implemented effectively as the HACCP system.
2 The HACCP system records have been appropriately designed and are being used appropriately to record the required information.
3 The CCPs are functioning correctly through the application of control measures and in compliance with critical limits.
4 The monitoring activities are demonstrating that CCPs remain under control and are detecting deviations in control and the breach of critical limits.
5 The disposition of product is known.
6 The corrective action procedures are effective and capable of ensuring

the return of control to CCPs and the identification, segregation and control of potentially unsafe product.

7 The verification procedures themselves are implemented properly and used effectively.
8 HACCP system documentation has been established properly and is adequate for its purpose.
9 HACCP system records have been established, are used properly and retained for appropriate periods of time.

The verification of HACCP systems used in crop production cannot always be undertaken quickly as crop production cycles may make it necessary to extend verification activities over a year or more. Also, it can be important to take into account deviations from intended production processes caused for example by the weather, pests, disease, etc. when analysing verification results to ensure that conclusions about the adequacy of the HACCP plan and its implementation are valid. Following verification, modifications and improvements can be made to the HACCP plan, as required.

7.12.12 Stage 12: establish documentation and record keeping requirements (HACCP Principle 7)

The development, operation and improvement of a capable food safety management system demands the implementation of system documentation and records. A documented food safety management system will be auditable and can be improved. It can also provide evidence to customers and enforcement agencies of a business's competence to provide safe food. Within the context of the UK's Food Safety Act 1990 the use of a documented HACCP system can demonstrate due diligence. The HACCP plan defines the structure and operation of the HACCP system and provides the forms used to record that food safety management activities have been undertaken. Documents and records are also necessary for HACCP system improvement. The process of verification should check that requirements stated in the documents, or plan, are fulfilled. The records provide the principal source of objective evidence that the requirements have been met. The review of verification results should lead to HACCP system improvement. A key output of the HACCP study should be the HACCP control chart (an example is given in Table 7.4) which is central to the existence and operation of the HACCP system. Other documents that can form parts of the HACCP plan include, for example, pesticide specifications, the process flow diagram, crop production procedures, CCP control and monitoring procedures, preventive and corrective action procedures, and verification procedures. HACCP system records should include those concerning CCP control and monitoring, preventive and corrective action, the control of non-conforming product, HACCP system validation and verification, and HACCP plan revision and improvement.

Table 7.4 Example HACCP control chart for field crop production

Process step	Step no.	CCP no.	Hazard	Control measure	Critical limit(s)	Monitoring		Corrective action
						Procedure	Frequency	Procedure
Fertiliser delivery	1							
Fertiliser in store	2							
Seed delivery	3	1	Unacceptable pesticide residues in seed.	Seed meets specification.	As stated in specification.	Check certificate of analysis on delivery.	Each delivery.	Reject delivery. Review supplier.
Seed propagation	4							
Pesticide drench	5	2	Microbial pathogens in drench water, e.g. *E. coli* O157, *Salmonella*.	Use potable water.	Absence of named pathogens.	Agree micro-biological specification for water with supplier.	Annually.	Agree measures for improvement with supplier.
Field preparation	6	3	Contamination with pathogens from unrotted manure.	Check history of manure use on site.	No unrotted manure used in past two years.	Confirm site history.	Prior to site preparation.	Use site only if free from manure deposits or choose another site.
Fertiliser application (P & K) during field preparation	7							
Planting of seedlings	8							

Irrigation during crop establishment and growth	9	4	Microbial pathogens in water, e.g. *E. coli* O157, *Salmonella*.	Use clean water.	Absence of named pathogens.	Agree micro-biological specification for water with upplier.	Prior to use.	Agree measures for improvement with supplier or use another source.
Fertiliser (N) application during growth	10							
Pesticide preparation	11	5	Pesticide(s) prepared at excessive concentration giving unacceptable residue levels.	Use calibrated equipment for preparation.	As defined by manufacturer for named pesticide(s).	Observe preparation procedures complied with and check usage of pesticides against stocks.	Periodically when pesticide(s) used.	Calibrate equipment, revise procedures, or test crop for excess residue, according to nature of failure.
		6	Microbial pathogens in water.	Use clean water.	Absence of pathogens.	Confirm water quality with supplier.	Prior to use.	Agree measures for improvement with supplier or use another source.
Pesticide application	12	7	Pesticide(s) applied at excessive concentration giving unacceptable residue levels.	Apply with appropriate and calibrated equipment. Use trained staff for job.	As defined by manufacturer for named pesticide(s) and/or legally defined limits.	Check pesticide usage and observe application procedure.	Periodically when pesticide(s) used.	Calibrate equipment, revise procedures, or test crop for excess residue, according to nature of failure.
Pesticide post-application period	13	8	Pesticide(s) remain in crop at unacceptable residue levels.	Observe pre-harvest interval. Do not harvest until interval	Pre-harvest interval as advised by manufacturer	Record dates of pesticide applications and observe	Each crop.	Test crop for excess residue and reject if levels exceed

Table 7.4 (Cont'd)

Process step	Step no.	CCP no.	Hazard	Control measure	Critical limit(s)	Monitoring		Corrective action
						Procedure	Frequency	Procedure
				elapses.	for named pesticide(s).	interval to harvest.		requirements.
Harvesting (glass control)	14	9	Contamination with wood from machinery.	Glass policy – only use glass when needed and care taken when glass involved.	No glass contamination of product.	Check all glass for damage.	Daily.	Segregate and check implicated product before approving for use.
Harvesting (wood control)	15	10	Contamination with wood from packaging.	Care taken when wood is involved.	No wood contamination of product.	Check packaging materials for damage.	Daily.	Segregate and check implicated product before approving for use.
Harvesting (staff control)	16	11	Contamination with microbial pathogens from staff.	Good personal hygiene practised by staff.	Staff adhering to personal hygiene policy.	Observation and supervision of staff.	Continuous.	Appropriate managementof staff breaking the rules.
Storage	17	12	Growth of microbial pathogens on produce.	Select temperature and humidity suitable to prevent growth.	Adequate temperature and humidity to maintain product quality, but unsuitable for microbial growth.	Check storage temperature and humidity.	Daily.	Segregate and check implicated product before approving for use. Rectify temperature and humidity.
Transport	18	13	Contamination with microbial	Use only approved	Vehicles clean, hygienic and	Check records of vehicle	Daily.	Agree measures for improvement

		pathogens from transport vehicles.	vehicles and hauliers. Check vehicles before use.	fit for use.	inspection.		with haulier or use another approved haulier.
	14	Growth of microbial pathogens.	Check temperature and humidity suitable to prevent growth prior to despatch.	Temperature and humidity suitable to maintain product quality, but unsuitable for microbial growth.	Check records of vehicle temperature and humidity assessment.	Daily.	Review control procedures. Agree measures for improvement with haulier or use another approved haulier.

Notes:
1. The focus of the chapter has been pathogens in food crop production, but this table identifies all possible sources of hazard for illustrative purposes.
2. The HACCP control chart is not as detailed as would be possible because specific crops and pesticides are not identified.
3. The responsibility for monitoring and corrective action as listed in the 'monitoring' and 'corrective action' columns would normally be identified to a person by job title.
4. It is recommended by some authoroties that the HACCP control chart list only the process steps which correlate with CCPs. In this chart all process steps are given.

7.13 Implementing and maintaining HACCP systems

A HACCP plan must be implemented effectively and the HACCP system must be maintained if food safety is to be achieved. An eight-step approach to implementation is advocated by Mortimore and Wallace (2001) which is adapted by Early (2002) as a ten-step process which emphasises the implementation of preventive measures, or confirmation of their adequacy if they already exist.

1. Determine the approach to implementation – Implement the HACCP system as a complete system or in smaller, more manageable units.
2. Agree the activities to be undertaken and the timetable – Identify implementation activities and those responsible for them. Define the timetable for completion. As appropriate, use project management techniques, e.g. Gantt charts.
3. Confirm the existence of adequate control measures or implement them – If control measures do not exist as part of operational procedures ensure they are implemented.
4. Conduct training in the operation of control measures or confirm adequate operation exists – Confirm that preventive measures are effective and train staff in their use.
5. Set up CCP monitoring methods – Establish methods for monitoring the control of CCPs.
6. Conduct training in CCP monitoring – Ensure staff competence in CCP monitoring activities through training.
7. Complete 'once-only' activities – Carry out activities required to put in place everything needed to complete the HACCP system, e.g. write procedures, create records, implement document and record control systems, train staff, etc.
8. Confirm the monitoring systems are in place – Make sure that the monitoring systems are implemented and are operating correctly.
9. Confirm implementation is complete and operate the HACCP system – Begin operation of the HACCP system following confirmation that implementation activities have been completed.
10. Audit to confirm adequate implementation – Use audit to confirm that the HACCP system has been implemented correctly. Allow one (or more) complete crop production cycle(s) to achieve complete confidence in the system.

The HACCP plan should be reviewed annually to confirm that it meets food safety requirements. When changes are made to crop production methods the HACCP plan should be revised and the system modified accordingly. The HACCP system should be audited at least annually to confirm that it continues to meet food safety requirements. Finally, the implementation of HACCP systems by farmers will present many challenges, not least those related to possessing or having access to an adequate knowledge of HACCP and food safety issues, and having the resources to implement and maintain such systems. A review of the technical barriers to HACCP, some of which will be faced by farmers, is given by Panisello and Quantick (2001).

7.14 Farm assurance

Farm assurance has developed in direct response to concerns about the safety of food products entering the food supply chain. The concept of the food supply chain extending from farm to consumer was reinforced during the last decade of the 20th century, although the term food supply system is more accurate (Tansey and Worsley, 1995). Various factors have increased understanding of the chain as a structure which links farms to consumers. Of these the most important has been the occurrence in many countries of a range of highly publicised food safety scares. During the 1980s various countries experienced serious food safety incidents caused by *L. monocytogenes* and *E. coli* O157 and both pathogens were associated with agricultural food materials, i.e. milk and meat. The former has been linked to soft cheeses and pâté, the latter to undercooked hamburgers and cross-contamination between raw and cooked meat products. In 1996/97 an outbreak of *E. coli* O157 poisoning in Scotland involving meat products infected 496 people (Pennington, 1997) and killed 20 (SCIEH, 1997). This incident added to the growing concern about the safety of British farm produce already intensified by the BSE crisis. Government health authorities, farming organisations, food manufacturers and the supermarkets in the UK began to consider with urgency the problem of controlling food safety hazards arising at farm level. Farm assurance was seen as the solution. Prior to the exposure of BSE as a human health hazard in 1996, farm assurance had already been proposed by some supermarkets in the UK as a method of assuring the quality and safety of agricultural food products. The UK's major supermarket chains were increasingly establishing vertically integrated supply chains reaching back to individual farms, and even beyond to the agricultural inputs sector. This gave increased control over the quality of purchased produce and the prices charged by suppliers. It was also a way for supermarkets to prevent competitors from gaining access to their suppliers. Importantly, it shifted responsibility for food safety from the supermarkets to the suppliers under the terms of the 'due diligence' defence in the Food Safety Act 1990. This was particularly important from the perspective of the supermarkets given, for example, the growth in fresh produce sales, including many product lines subjected only to minimal processing or in some cases no processing at all. But what is farm assurance?

7.14.1 Farm assurance: quality assurance at farm level

In the early 1990s the National Farmers' Union of England and Wales (NFU) stated that farm assurance is a system for ensuring that customers and final consumers are confident in UK farm produce (Early, 1998). This stated a desirable outcome of farm assurance, but not what farm assurance is. The term 'farm assurance' has since been defined differently by many organisations with an interest in the subject. Whatever the definition, farm assurance is essentially the application of quality assurance principles at farm level. The objective of farm assurance is to ensure that farm produce meets the quality and food safety requirements of customers (food processors and supermarkets) and the food marketplace (consumers). Farmers today are quality assurance managers who function to ensure that their produce consistently meets product specifications and that customers are completely satisfied.

The principles of quality assurance are manifested through a quality assurance system which, in basic terms, should ensure that:

1 customers' requirements are properly understood and agreed with the customer;
2 ability to supply the customer is confirmed;
3 resources needed to supply the customer are provided;
4 resources are managed appropriately to produce the product the customer requires;
5 product that meets the customer's requirements is supplied to the customer.

At farm level the quality assurance system is the farm assurance system and it should be designed and managed such that the processes of the enterprise are adequately controlled. The concept of process control underpins quality assurance and the quality assurance system effects process control. The process model (Fig. 7.3) expresses the concept that provided the business process itself is properly defined and controlled, and that the inputs to the process are also properly defined and controlled, then the probability will be high that the outputs will be right first time. This applies as much to growing cereals or pineapples as it does to canning and sterilising meat or making beer. Farmers have not traditionally perceived their roles as being quality assurance managers and process control managers, but that is what they are today. The greater integration of the food supply chain coupled to increased concerns about food safety issues arising at farm level have necessitated changes in their role.

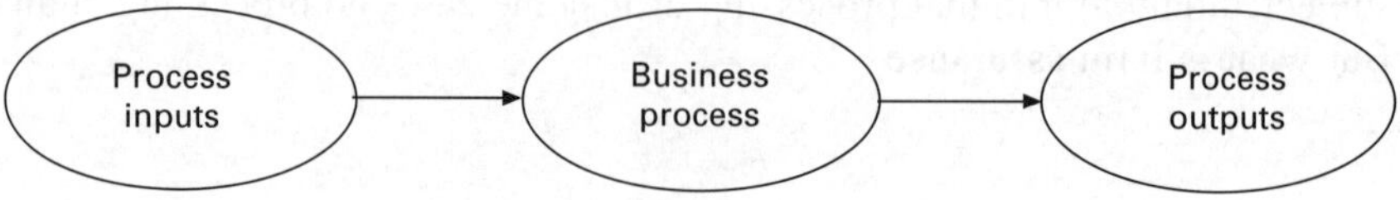

Fig. 7.3 The process model.

7.14.2 Farm assurance: UK developments

The farm assurance sector in the UK operates a range of farm assurance schemes which are effectively farm assurance standards. The farm assurance schemes define requirements for good agricultural practice and the competent management of agricultural enterprises applied to the production of livestock or the growing of crops. Essentially, farmers use the schemes to model their quality assurance systems and methods of process control to ensure that their produce complies with customers' specifications. Most of the UK's farm assurance schemes were established by independent organisations representing particular agricultural products or product groupings. For example, the National Dairy Farm Assurance scheme was developed to encompass milk production while the Farm Assured British Beef and Lamb scheme covered both beef and lamb production. The development of farm assurance in the UK presents a muddled history and the inadequate application of intelligent reasoning to the subject of quality assurance applied at farm level. Initially there was a distinct lack of leadership from the UK Government and the Ministry of Agriculture, Fisheries and Food (now replaced by the Department of Environment, Food and Rural Affairs (DEFRA)) as well as the bodies that represent farmers, e.g. the NFU. During the early 1990s very many independent farm assurance schemes were established, with a number covering the same types of agricultural produce and, essentially, having the same scope and objectives. A high degree of nationalism was also seen and still exists, as some schemes apply only to agricultural foodstuffs produced in England or Wales or Scotland. Fundamentally, the UK missed the opportunity to establish a single farm assurance standard applying to the production of all agricultural foodstuffs supplemented by intelligent interpretations for different types of produce. If the same principles of quality assurance and process control can be applied to the manufacture of bread or pork pies or soft drinks, as well as electronic products and motor cars for that matter, why can't they be applied via a single standard with appropriate interpretations to all agricultural enterprises?

That the UK approach to the development of farm assurance was unstructured and ill-considered has been evidenced by the establishment of Assured Food Standards (AFS). This is an independent organisation that was created to manage the 'Red Tractor' mark, which is the logo used by approved produce packers and food processors to demonstrate that food products are farm assured, i.e. that they have been produced, and verified to have been produced, under the authority of an approved farm assurance scheme. It is stated on the AFS website (see Section 7.17.2) that AFS belongs to the food chain; which raises certain ontological questions, as the food chain is a concept and it is difficult to understand how a concept can own anything. It is said that the interests of the key links in the UK's food chain – being the NFU, the Ulster Farmers' Union, the Meat & Livestock Commission, Dairy UK and the British Retail Consortium – are represented by AFS and that the observers to AFS include DEFRA and the UK's Food & Drink Federation. AFS's role is to establish standards that apply to agricultural produce as well as various of the links within the food chain, e.g. the haulage of agricultural produce and livestock markets, and to manage the independent certification bodies that audit farmers against the requirements of the standards. AFS also defines the rules for use of the Red Tractor logo and licenses the food packers and processors who use it. AFS owns a number of farm assurance schemes, identified as the Red Tractor schemes which are: the Assured Combinable Crops Scheme; Assured Produce; Assured Chicken Production; Assured British Pigs; Assured Dairy Farms; and Assured British Meat. It also manages a number of schemes: Farm Assured Welsh Livestock; Northern Ireland Farm Quality Assurance Scheme; Quality Meat Scotland; and Genesis Quality Assurance. In addition to the farm assurance schemes established by AFS or under the authority of AFS, there are many others in the UK, including, for illustrative purposes, the LEAF (Linking Environment And Farming) scheme based on principles of Integrated Farm Management used in Germany, the organic produce schemes operated by the Soil Association, and the Nature's Choice scheme established by Tesco, the supermarket chain.

An appraisal of farm assurance schemes will reveal that they usually embody the principles of quality assurance in structures that essentially describe GAP in the production of agricultural foodstuffs. Farm assurance schemes may vary in their scope and content because they have been written to apply to different types of agricultural produce. However, whatever their orientation with respect to the agricultural produce to which they relate, they typically express the core principles of quality assurance employed in the context of farm assurance. Farm assurance standards generally express requirements for:

1 **System organisation and management** – concerning the implementation and management of the farm assurance system and for document and record control, and maintenance of the system through auditing.
2 **Food safety** – concerning the production of agricultural food products which meet the food safety requirements of customers and are suitable for release to the food chain.
3 **Food quality** – concerning the production of agricultural food products which comply with the requirements of customer and market specifications.
4 **Product realisation and process control** – concerning the methods of production and the control of production processes. In relation to crop production, requirements will concern, e.g. seed selection and the use of disease-resistant cultivars, seed treatment, seed bed preparation, the judicious use of pesticides and fertilisers, the use of uncontaminated irrigation water, the use of manures, etc. (Farm assurance standards concerned with animal production will centre here on issues of animal welfare, appropriate housing, the use of veterinary medicines, etc.)
5 **Product protection** – concerning the management and protection of produce after production, i.e. post-harvest crop management covering

storage and transport designed and managed to prevent damage to product or deterioration, e.g. contamination by fungi and mycotoxin production in stored grain. (Farm assurance standards concerned with animal production will centre here on issues of animal transportation and holding at markets or prior to slaughter, etc.)

6 **Traceability** – concerning the identification and traceability of agricultural produce as well as the inputs to the production processes, enabling, for example, the tracing of product back from the customer to the farm where it was produced and records of production and inputs to production such as seed and pesticides, etc.
7 **Environmental protection** – concerning the implementation of methods and practices aimed at preventing or minimising damage to the environment (including biodiversity) through agriculture, the management of the land to achieve agricultural sustainability and, in some schemes, the enhancement of the land as a resource and amenity.

The benefits to farmers of working to farm assurance schemes should be:

1 The schemes direct the development and use of a structured approach to quality assurance at farm level, which should increase the probability that produce meets customer specifications.
2 Compliance with the requirements of schemes, and the demonstration of competence to produce agricultural products to specification, is verified by independent certification bodies through the process of audit, which is recognised nationally and, in some instances, internationally.
3 Certification to a farm assurance scheme allows produce to be sold to certain customers and into certain market sectors.
4 The implementation of quality assurance at farm level through the use of farm assurance schemes should enable more effective cost management and the reduction of failure costs.

Not all farmers consider that farm assurance schemes work sufficiently to their advantage. Many consider that the schemes establish a form of licensing process by which approved farmers are allowed to access given markets. Many also believe farm assurance brings to the farmer an unnecessary level of bureaucracy. Contact with farmers indicates that many are concerned that farm assurance auditors are variable in their interpretation of the requirements of farm assurance schemes. Of the many farm assurance schemes operated in the UK two of most significance to crop production are the Assured Combinable Crops Scheme (ACCS) and the Assured Produce (AP) scheme.

7.14.3 Assured combinable crops scheme (ACCS)

The Assured Combinable Crops Scheme (ACCS) is owned by AFS and covers combinable crops, including barley, oats, wheat, oilseeds (including linseed, oilseed rape, sunflower seed) and crops with high protein including beans and peas. The ACCS is open to any producer of combinable crops in England

and Wales. ACCS state in its website (see Section 7.17.2) that the scheme was established to enable farmers to 'avoid a multitude of individual and differing schemes being imposed' and to enable the traceability and assurance required by customers. As with most UK farm assurance schemes, the ACCS is revised regularly. To gain a current understanding of the scheme the reader is advised to examine the ACCS website as listed in Section 7.17.2.

7.14.4 Assured Produce (AP) scheme

The Assured Produce (AP) scheme is also owned by AFS. It covers the production of 'produce' which includes fruit, salad crops and vegetables. It promotes the use of integrated crop management (ICM) which is concerned with achieving both environmental and economic sustainability. The AP scheme aims at ensuring the safety and integrity of produce and, consequently, maintaining consumers' confidence. The AP standards take the form of (1) Generic Standards which state requirements against which third party assessment is undertaken, and (2) Guidance Notes on the use of the standards. Crop Protocols are defined to communicate commercially acceptable best practice in crop production including methodology in integrated pest management, disease management and crop management systems, as well as ways to minimise pesticide residues. AP states in its website (see Section 7.17.2) that Integrated Crop Management (ICM) is a 'whole farm' system concerned with managing crops and respecting the environment, which seeks to ensure the sustainable use of natural resources. ICM, it is stated, includes: **Soil management** – covering structures, cultivations and establishment systems; **Crop protection** – covering monitoring, diagnostics, thresholds and application; **Cropping & variety** – covering rotations, resistant varieties and seed selection; **Integrated conservation** – covering field margins, beetle banks, etc.; **Crop nutrition** – covering soil and tissue analysis and thresholds; **Management** – covering training, energy usage, planning and organisation. To gain a current understanding of the AP scheme the reader is advised to examine the website as listed in Section 7.17.2.

7.14.5 Farm assurance scheme verification

When using a farm assurance scheme to model a quality assurance system, a farmer will need to carry out what is effectively an internal quality system audit to confirm compliance with the scheme or standard. The term 'audit' in relation to quality systems has been variously defined, but a definition that has particular explanatory power is that by ISO (1990) which states that an audit is '*A systematic and independent examination to determine whether quality activities and related results comply with planned arrangements and whether these arrangements are implemented effectively and are suitable to achieve objectives*'. A farmer carrying out an internal audit of compliance with a farm assurance standard may not be able to be entirely independent and the danger exists that the results may be biased. If farm produce is to be sold to customers under the authority of a farm assurance scheme then confirmation that the produce has been produced according to the requirements of the scheme must be achieved by independent audit. With the development of farm assurance in the UK, various organisations have been established (see Section 7.17.2) which provide independent, third party auditing services to farm assurance scheme operators and their members, the farmers. These certification bodies are so called because they provide certification services, i.e. they audit farmers against the requirements of farm assurance schemes and, provided the requirements are met, they issue certificates of compliance. In the UK the certification bodies are themselves accredited for the services they provide by UKAS (United Kingdom Accreditation Service) against the European Standard *EN 45011:1989 General criteria for certification bodies operating product certification.*

7.14.6 Farm assurance: international developments

Although the UK's entry into farm assurance preceded that of most other countries, in some instances international developments in farm assurance are occurring which will provide a more reasoned and coherent approach than that so far taken in the UK. In particular the GlobalGAP organisation (see Section 7.17.2) aims at establishing a single standard for GAP applicable to all global agriculture, with interpretations made for different types of agricultural produce. With foresight and leadership such an approach could have been taken in the UK, but it was not. The origins of GlobalGAP are found in the EurepGAP organisation which was established as 'The Global Partnership for Safe and Sustainable Agriculture'. The organisation started in 1997 as an initiative of retailers belonging to the Euro-Retailer Produce Working Group (EUREP). The name EurepGAP is thus a contraction of Euro-Retailer Produce Working Group Good Agricultural Practice. EurepGAP's mission was '*To develop widely accepted standards and procedures for the global certification of Good Agricultural Practices*', with the standards established as normative documents accredited to international certification criteria. The standard and certification system was developed by Technical and Standards Committees working in each product sector and the committees have 50:50 retailer and producer representation. The creation of EurepGAP was driven by food safety scares, such as the occurrence of BSE in the UK, and consumer concerns about pesticide residues in foods and the introduction of GM foods. At the outset it was recognised that consumers now ask more complex questions about food and that there is a need to ensure that consumers can be confident in the safety of foodstuffs and the sustainability of production methods. During the last decade EurepGAP has established itself and is now transformed into GlobalGAP which administers product certification against its standard in many countries. Indeed, GlobalGAP product certification is carried out by over 100 certification bodies in more than 80 countries. The

GAP standard is defined as three 'product bases' covering crops, livestock and aquaculture (GlobalGAP, 2007). The crops base includes fruit and vegetables, flowers and ornamentals, combinable crops, green coffee, tea and cotton. In addition, GlobalGAP operates a Plant Propagation Material Standard. Farmers working to GlobalGAP's standard are subject to annual announced and unannounced audits.

In the USA the SQF (Safe Quality Food) Program was developed in response to the need to control hazards at farm level and also in food processing. Two SQF codes have been developed. The SQF 1000 code was developed in 1997 to provide a food safety and quality management certification programme for primary producers, or farmers. Alongside it, the SQF 2000 code was developed with the same food safety and quality intentions, but its scope of application is the food industry. Both codes are essentially HACCP-based supplier assurance codes of practice. In 2003 the Food Marketing Institute (see Section 7.17.2) acquired the rights to the SQF Program and established the SQF Institute to manage the programme. The SQF Program has been recognised as a global food safety system by the seven global food retailers: Ahold, Carrefour, Delhaize, Metro, Migros, Tesco and Wal-Mart (SQF, undated). For application to the agricultural production of food materials the SQF 1000 code merges GAP with HACCP systems based on the Codex Alimentarius Commission HACCP Guidelines. When using the code farmers essentially implement quality and food safety management systems which integrate GAP and HACCP. Compliance with the code is demonstrated by independent, third party audit which leads to certification. The certification bodies are licensed by the SQF Institute and must be accredited under the requirements of ISO/IEC Guide 65:1996 *General requirements for certification bodies offering certification*. The controls and process of third party certification are the same as those established for farm assurance systems in the UK and by GlobalGAP for its European and international activities. In 2007 GlobalGAP and the SQF Institute agreed to work towards harmonising the GlobalGAP and SQF 1000 farm standards by developing a combined audit checklist (Freshplaza, 2008). With GlobalGAP and the SQF Institute spanning the world with their collaborative GAP- and HACCP-based approaches to quality and food safety management at farm level, where that leaves the UK with its multiplicity of farm assurance schemes is anyone's guess. Given that Tesco, the UK's largest supermarket business, has approved the SQF codes, it will be interesting to observe the degree to which in the future Tesco demands that its UK farm-based suppliers use the SQF 1000 code (or GlobalGAP standard) instead of UK-based farm assurance schemes.

7.14.7 Implications of GAP and farm assurance standards for pathogen related food safety

Food safety is fundamental to the rationale and objectives of the standards established by GlobalGAP, the SQF Foundation and the UK's farm assurance bodies. Irrespective of all other aspects of quality, if agricultural produce is not safe for use and consumption then nothing else matters. The standards outlined above define good agricultural practice which, amongst other things, is concerned with the implementation of practices and methods that will ensure food safety. Thus, standards covering the production of combinable crops are concerned with, for example, the control of *Fusarium* spp. in the growing crop and the prevention of mycotoxin contamination. Combinable crops standards are also concerned with the appropriate post-harvest control of grain, such as drying to prevent mould growth and mycotoxin contamination, as well as hygienic storage to prevent pest infestation and contamination, e.g. faecal contamination by rodents with the possibility of spread of enteric pathogens. Standards applied to the production of non-combinable oilseeds will require similar controls with respect to the prevention of food safety problems caused by microbial pathogens. Standards used in the production of fruit and vegetables will also be concerned about the potential for the contamination of produce by microbial pathogens, e.g. as a result of the use of contaminated water for irrigation, cooling or washing, and they will express requirements for the control of pesticide residues. Apart from the benefits of working to the schemes or standards identified in Section 7.14.2, their use offers farmers distinct benefits in relation to being able to manage food safety. Few farmers would describe themselves as experts in food safety or food safety management. In many ways the lack of such expertise is compensated for by the use of farm assurance standards, as the documents provide guidance that directs the development and implementation of appropriate food safety controls. As long as farmers faithfully adhere to the requirements of the standards they should then be confident that food safety requirements will be met.

7.15 Future trends

Looking to the future we are likely to see developments in the application of GAP and HACCP used for food safety control at farm level, as well as the further evolution of farm assurance standards. One possible development may be the rationalisation of terminology involved in the systems approach to crop management. First there was integrated pest management which was followed by integrated crop management and then integrated farm management, which contains integrated pest management and integrated crop management. Additionally, integrated farm management equates to good agricultural practice, which is also farm assurance: and if the equation is not precise at least all of these 'territories' overlap in various ways. Confused? For the sake of avoiding confusion and to limit the proliferation of different terminologies all describing related or overlapping concepts, it would seem that the time is right for the revision of terminology and the adoption of universally agreed terms and definitions. Leadership in the standardisation of systems and terminology could come from the International Organisation for Standardisation. In particular ISO might consider developing an international standard for the application of quality assurance and process control at farm level, thus to standardise the approach to GAP (or farm assurance). Such a development might be undertaken in conjunction with GlobalGAP and the SQF Foundation. When viewed purely as a process of rationalisation the concept of a single internationally agreed standard for farm assurance may be unappealing to some. This may be particularly so for those organisations that have already made significant investment in the development of their own standards. However, issues of global climate change, fossil fuel shortage, reduced land availability for food production and global food shortage should bring a different and less parochial light to bear on the subject of farm assurance. In the future there is likely to be the need to produce as much food as possible with reduced levels of off-farm inputs, to the highest standards of quality and safety possible, and ensuring at the same time the sustainability of farm land for food production. Farmers will need clear guidance in the principles and practices of good agricultural management for food production in the face of such constraints. An international standard for quality assurance and process control based on GAP could prove to be invaluable to ensuring that existing consumers continue to be provided with quality and safe food, and that the consumers not yet born will have farm land that is still fit for the production of food.

As pressure grows for the control of foodborne pathogens at farm level, as well as other foodborne hazards arising at farm level, we are likely to see the increased use of HACCP in agricultural food production. Approaches to the interpretation of the seven principles of HACCP for use on farms are likely to develop still further, though it should be recognised that HACCP is not in every case appropriate for the control of food safety at farm level. Sperber (2005) recognises limitations in the use of HACCP to control hazards 'from farm to table' and that many actions taken in the management of food safety hazards at farm level concern establishing HACCP pre-requisites rather than critical control points. It should be recognised that in many instances the control of hazards that arise at farm level and particularly microbial pathogens cannot be undertaken on the farm. Some can only be dealt with in food processing, e.g. by use of appropriate heat treatments to destroy microorganisms. The ability to implement HACCP fully is limited by circumstances. In the future we may see farmers using ISO 22000 to develop food safety management systems that fully incorporate the concept of pre-requisite programmes, thus to integrate the systems with GAP. Where it is not possible to implement CCPs at farm level, the execution of pre-requisite programmes under ISO 22000 and as part of GAP may result in an acceptable level of control if not the prevention or elimination of hazards.

The increased use of HACCP at farm level either directly or embodied within food safety management systems developed against ISO 22000 is likely to encourage developments in the training of staff to use HACCP at farm level.

In the future we may see co-ordinated approaches to training development, such as the creation of a HACCP training standard for use in relation to food safety management at farm level. We may also see a revision of the seven principles of HACCP as the eight principles of HACCP, by delineating the requirement for HACCP plan validation as a separate principle.

7.16 Conclusions

This chapter has discussed aspects of pathogen control in the primary production of food crops. Its focus has been on the use of management systems for the control of food safety and specifically good agricultural practice and the Hazard Analysis Critical Control Point system for food safety management. It has outlined farm assurance as a form of GAP. Throughout the chapter the key hazards of bacterial pathogens sourced from inadequately composted animal manures or contaminated irrigation water, and mycotoxins resulting from the fungal contamination of growing and stored crops, have been identified. Use by farmers of the food safety management systems advocated by this chapter will ensure the control of these and other hazards that may be associated with food crops.

7.17 Sources of further information and advice

7.17.1 Further reading

Alford D (2007) *Pests of Fruit Crops*, 2nd edn, Oxford, Blackwell Publishers.

Barkai-Golan R and Pastor N (eds) (2008) *Mycotoxins in Fruits and Vegetables*, London, Academic Press.

Barug D, Bhatnagar D, van Egmond HP, van der Kamp JW, van Osenbruggen WA and Visconti A (eds) (2006) *The Mycotoxin Factbook: Food and Feed Topics*, Wageningen, Wageningen Academic.

do Nascimento Nunes MC (2008) *Color Atlas of Postharvest Quality of Fruits and Vegetables*, Oxford, Wiley-Blackwell.

Gollob P, Farrell G and Orchard JE (eds) (2002) *Crop Post-Harvest Science and Technology Vol. 1: Principles and Practice*, Oxford, Wiley-Blackwell.

Hagg M, Ahvenainen R and Evers AM (1999) *Agri-food Quality II: Quality Management of Fruit and Vegetables*, Cambridge, Woodhead.

Hodges R and Graham Farrell G (2004) *Crop Post-Harvest: Science and Technology Vol. 2: Principles and Practice*, Oxford, Wiley-Blackwell.

Hui YH, Barta J, Cano MP, Gusek TW, Sidhu J and Sinha N (2006) *Handbook of Fruit and Vegetable Processing*, Oxford, Blackwell.

Jongen W (2005) *Improving the Safety of Fresh Fruit and Vegetables*, Cambridge, Woodhead.

Magan N and Olsen M (eds) (2004) *Mycotoxins in Food: Detection and Control*, Cambridge, Woodhead.
Matthews KR (ed.) (2006) *Microbiology of Fresh Produce*, Washington, DC, ASM Press.
Mortimore S and Wallace C (2001) *HACCP*, Oxford, Blackwell Science.
Motarjemi Y (2006) *Emerging Foodborne Pathogens*, Cambridge, Woodhead.
Sinha KK and Bhatnagar D (1998) *Mycotoxins in Agriculture and Food Safety*, New York, Marcel Dekker.
Thompson K (2003) *Fruit and Vegetables: Harvesting, Handling and Storage*, 2nd edn, Oxford, Blackwell.

7.17.2 Organisations of interest

Standards organisations

- **Assured Combinable Crops**: ACCS Secretariat, Unit 4b, Highway Farm, Horsley Road, Downside, Cobham, Surrey, KT11 3JZ, UK (www.assuredcrops.co.uk).
- **Assured Food Standards**: 4th Floor, Kings Building, 16 Smith Square, London SW1P 3JJ, UK (www.redtractor.org.uk).
- **Assured Produce**: Unit 4b, Highway Farm, Horsley Road, Downside, Cobham, Surrey, KT11 3JZ, UK (www.assuredproduce.co.uk).
- **British Retail Consortium:** 21 Dartmouth Street, London, SW1H 9BP, UK (www.brc.org.uk).
- **Chilled Food Association**: 90 Lincoln Road, Peterborough, PE1 2SP, UK (www.chilledfood.org).
- **GlobalGAP**: GLOBALGAP Secretariat, c/o FoodPLUS GmbH, P.O. Box 19 02 09, 50499 Cologne, Germany (www.globalgap.org)
- **International Organisation for Standardisation (ISO)**: 1, ch. de la Voie-Creuse, Case postale 56, CH-1211 Geneva 20, Switzerland (www.iso.org).
- **Soil Association**: South Plaza, Marlborough Street, Bristol BS1 3NX, UK (www.soilassociation.org/farmassurance).
- **SQF Institute**: 2345 Crystal Drive, Suite 800, Arlington VA, 22202 USA (www.sqfi.com).

Certification bodies

- **BSI Management Systems**: 389 Chiswick High Road, London, W4 4AL, UK (www.bsi-emea.com).
- **CMi plc**: Long Hanborough Oxford OX29 8SJ, UK (www.cmi-plc.com).
- **Genesis Quality Assurance**: Ryknield House, Alrewas, Burton on Trent, DE13 7AB, UK (www.genesisqa.com).
- **SAI Global**: SAI Global has offices in many countries. The postal address for assurance in the UK is: SAI Global, PO Box 44, Winterhill

House, Snowdon Drive, Milton Keynes, MK6 1AX, UK (www.saiglobal.com)

Organisations for advice and information

- **Campden BRI**: Chipping Campden, Gloucestershire, GL55 6LD, UK (www.campden.co.uk).
- **Food Marketing Institute**: 2345 Crystal Drive, Suite 800, Arlington, VA 22202, USA (www.fmi.org).
- **Harper Adams University College**: Newport, Shropshire, TF10 8NB, UK (www.harper-adams.ac.uk).
- **Pesticides Safety Directorate**: Mallard House, Kings Pool, 3 Peasholme Green, York, YO1 7PX, UK (www.pesticides.gov.uk).
- **US Food and Drug Administration (FDA)**: 5600 Fishers Lane, Rockville MD, USA (www.fda.gov).

7.18 References

Aaronson, S. (2000) Fungi, in, K.F. Kiple and K. Coneè Ornelas (eds), *The Cambridge World History of Food*, Cambridge, Cambridge University Press, 313–36.

Abadias, M., Usall, J., Olivera, M., Alegre, I. and Viñas, I. (2008a) Efficacy of neutral electrolyzed water (NEW) for reducing microbial contamination on minimally-processed vegetables. *International Journal of Food Microbiology*, **123**, 151–8.

Abadias, M., Usall, J., Anguera, M., Solsona, C. and Viñas, I. (2008b) Microbiological quality of fresh, minimally-processed fruit and vegetables, and sprouts from retail establishments, *International Journal of Food Microbiology*, **123**, 121–9.

Abbas, H.K., Mirocha, C.J., Meronuck, R.A., Pokorny, J.D., Gould, S.L. and Kommedahl, T. (1988) Mycotoxins and *Fusarium* spp. associated with infected ears of corn in Minnesota. *Applied and Environmental Microbiology*, **54**(8), 1930–3.

Adams, C.E. (2002) Hazard analysis and critical control point – original 'spin', *Food Control*, **13**(6/7), 355–8.

Adams, M.R. and Moss, M.O. (2008) *Food Microbiology*, 3rd edn, Cambridge: RSC Publishing.

Anderson, G.L., Kenney, S.J., Millner, P.D., Beuchat, L.R. and Williams, P.L. (2006) Shedding of foodborne pathogens by *Caenorhabditis elegans* in compost-amended and unamended soil, *Food Microbiology*, **23**, 146–53.

Atwood, B. (2000) *Butterworth's Food Law*, 2nd edn, London, Butterworth's.

Bacon, C.W., Marijanovic, D.R., Norred, W.P. and Hinton, D.M. (1989) Production of fusarin C on corn and soybean by *Fusarium moniliforme*, *Applied and Environmental Microbiology*, **55**(11), 2745–8.

Battilani, P. and Pietri, A. (2004) Risk assessment and management in practice: ochratoxin in grapes and wine, in N. Magan and M. Olsen, (eds), *Mycotoxins in Foods*, Cambridge, Woodhead 244–61.

Bell, C.H. (2000a) Pest control: insects and mites, in H.L.M. Lelieveld, M.A. Mostert, J. Holah and B. White (eds), *Hygiene in Food Processing*, Cambridge, Woodhead, 335–79.

Bell, C.H. (2000b) Fumigation in the 21st century, *Crop Protection*, **19**, 563–9.

Bender, A.E. and Bender, D.A. (1995) *A Dictionary of Food and Nutrition*, Oxford, Oxford University Press.

Bennett, J.W. and Klich, M. (2003) Mycotoxins, *Clinical Microbiology Reviews*, **16**(3), 497–516.

Beretti, M. and Stuart, D. (2008) Food safety and environmental quality impose conflicting demands on Central Coast growers, *California Agriculture*, **62**(2), 68–73.

Berg, T. (2003) How to establish international limits for mycotoxins in food and feed? *Food Control*, **14**(4), 219–24.

Bhagwat, A.A. (1995) Microbial safety of fresh-cut produce: Where are we now? in K.R. Matthews (ed.) *Microbiology of fresh produce*, Washington, DC, ASM Press.

Bidawid, S., Farber, J.M. and Sattar, S.A. (2000) Inactivation of hepatitis A (HAV) in fruits and vegetables by gamma irradiation, *International Journal of Food Microbiology*, **57**, 91–7.

Boes, J., Alban, L., Bagger, J., Møgelmose, V., Baggesen, D.L. and Olsen, J.E. (2005) Survival of *Escherichia coli* and *Salmonella* Typhimurium in slurry applied to clay soil on a Danish swine farm, *Preventive Veterinary Medicine*, **69**, 213–28.

Booer, C., Faulkner, A., Muller, R., Slaiding, I. and Baxter, D. (2007) The Incidence of some *Fusarium* mycotoxins in malting barleys and malts, *Food Science and Technology*, **21**(4), 36–7.

Bosch, U. and Mirocha, C.J. (1992) Toxin production by *Fusarium* species from sugar beets and natural occurrence of zearalenone in beets and beet fibre, *Applied and Environmental Microbiology*, **58**(10), 3233–9.

Bosch, U., Mirocha, C.J. and Wen, Y. (1992) Production of zearalenone, moniliformin and trichothecenes in intact sugar beets under laboratory conditions, *Mycopathologia*, **19**(3), 167–73.

Branquinho Bordini, M.E., Ristori, C.A., Jakabi, M. and Gelli, D.S. (2007) Incidence, internalization and behaviour of *Salmonella* in mangoes, var. Tommy Atkins, *Food Control*, **18**, 1002–7.

Brenner, F.W., Villar, R.G., Angulo, F.J., Tauxe, R. and Swaminathan, B. (2000) *Salmonella* nomenclature, *Journal of Clinical Microbiology*, **38**(7), 2465–7.

Butler, C.G. (1954) *The World of the Honeybee*, London, Collins.

CCFH (2001) Hazard Analysis Critical Control Point (HACCP) System and Guidelines for its Application. Annex to CAC/RCP–1 (1969), Rev. 3 (1997). Codex Committee on Food Hygiene, in *Codex Alimentarius Commission Food Hygiene Basic Texts*, Rome, Food and Agriculture Organization of the United Nations.

CFA (2007) *Microbiological Guidance for Growers*, 2nd edn, Peterborough, Chilled Food Association.

Chambers, B.J. (1999) *Good Agricultural Practice for Fresh Produce*, Proceedings of the International Conference on Fresh-Cut Produce. 9–10 September, Chipping Campden, Campden & Chorleywood Food Research Association.

Chu, F.S. and Li, G.Y. (1994) Simultaneous occurrence of fumonisin B1 and other mycotoxins in mouldy corn collected from the People's Republic of China in regions with high incidences of esophogeal cancer, *Applied and Environmental Microbiology*, **60**(3), 847–52.

Cools, D., Merckx, R., Vlassak, K. and Verhaegen, J. (2001) Survival of *E. coli* and *Enterococcus* spp. derived from pig slurry in soils of different texture, *Applied Soil Ecology*, **17**, 53–62.

Cordier, J.-L. (1994) HACCP in the chocolate industry, *Food Control*, **5**(3), 171–5.

Craven, P.C., Baine, W.B., Mackel, D.C., Barker, W.H., Gangarosa, E.J., Goldfiled, M., Rosenfield, H., Altman, R., Lachapelle, G., Davies, J.W. and Swanson, R.C. (1975) International outbreak of *Salmonella eastbourne* infection traced to contaminated chocolate, *Lancet*, **305**(7910), 788–92.

Croci, L., de Medici, D., Scalfaro, C., Fiore, A. and Toti, L. (2002) The survival of hepatitis A virus in fresh produce, *International Journal of Food Microbiology*, **73**, 29–34.

Crosby, P.B. (1984) *Quality Without Tears*, New York, McGraw-Hill.

da Cruz, A.G., Cenci, S.A. and Maia, M.C.A. (2006a) Quality assurance requirements in produce processing, *Trends in Food Science and Technology*, **17**, 406–11.

da Cruz, A.G., Cenci, S.A. and Maia, M.C.A. (2006b) Good agricultural practices in a Brazilian produce plant, *Food Control*, **17**, 781–8.

D'Aoust, J.Y. (1977) *Salmonella* and the chocolate industry: A review, *Journal of Food Protection*, **40**, 718–27.

Daş, E., Gürakan, G.C. and Bayindirli, A. (2006) Effect of controlled atmosphere storage, modified atmosphere packaging and gaseous ozone treatment on the survival of *Salmonella* Enteritidis on cherry tomatoes, *Food Microbiology*, **23**, 430–8.

Decastelli, L., Lai, J., Gramaglia, M., Monaco, A., Nachtmann, C., Oldano, F., Ruffier, M., Sezian, A. and Bandirola, C. (2007) Aflatoxins occurrence in milk and feed in northern Italy during 2004–2005, *Food Control*, **18**(10), 1263–6.

De Waal, C.S. (2003) Safe food from a consumer perspective, *Food Control*, **14**(2), 75–9.

Early, R. (1998) Farm assurance – benefit or burden? *Journal of the Royal Agricultural Society of England*, **159**, 32–43.

Early, R. (2002) Use of HACCP in fruit and vegetable production, in W. Jongen (ed.), *Fruit and Vegetable Processing*, Cambridge, Woodhead, 91–118.

EC (2002) Regulation (EC) No 178/2002 laying down the general principles and requirements of food law. Establishing the European Food Safety Authority and laying down procedures in matters of food safety, *Official Journal of the European Union*, **L31**, 28 January 1–24.

EC (2003) *Final report of SCOOP task 3.2.10. Collection of occurrence data of Fusarium toxins in food and assessment of dietary intake by the population of EU Member States*, Brussels, European Commission, DG Health and Consumer Protection.

EC (2004) Corrigendum to Regulation (EC) No 852/2004 of the European Parliament and of the Council of 29 April 2004 on the hygiene of foodstuffs, *Official Journal of the European Union*, **L226**, 25 June, 3–21.

FAO (2001) *Manual on the Application of the HACCP System in Mycotoxin Prevention and Control*, Rome, Food and Agriculture Organization of the United Nations.

FAO (2003) *Report on the Expert Consultation on Good Agricultural Practices (GAP) Approach, Rome, Italy, 10–12 November 2003*, Rome, Food and Agriculture Organisation of the United Nations.

FAO (2007) *Good Agricultural Practice*. Rome, Food and Agriculture Organisation of the United Nations, available at: www.fao.org/prods/GAP/home/principles_11_en.htm, accessed November 2008.

FDA (1998) *Guide to Minimize Microbial Food Safety Hazards for Fresh Fruit and Vegetables*, Washington, DC, US Food and Drug Administration.

FDA (1999) *FDA Survey of Imported Fresh Produce*. FY 1999 Field Assignment, Washington, DC, US Food and Drug Administration, available at: www.cfsan.fda.gov/~dms/prodsur6.html, accessed November, 2008.

FDA (2000) *FDA Survey of Domestic Fresh Produce*, FY 2000/2001 Field Assignment, Washington, DC, US Food and Drug Administration, available at: www.cfsan.fda.gov/~dms/prodsu10.html.

FDA (2007) *Foodborne Pathogenic Microorganisms and Natural Toxins Handbook*, Washington, DC, US Food and Drug Administration, Centre for Food Safety and Applied Nutrition, available at: www.cfsan.fda.gov/~mow/intro.html, accessed November, 2008.

Fernández-Armesto, F. (2001) *Food: A History*, London, Macmillan.

Fonseca, J.M. (2006) Postharvest handling and processing: sources of microorganisms and impact of sanitising procedures in K.R. Matthews (ed.). *Microbiology of Fresh Produce*, Washington, DC, ASM Press, 85–120.

Food Ethics Council (2001) *After FMD: Aiming for a Values-driven Agriculture*, Southwell, Food Ethics Council.

Freshplaza (2008) GLOBALGAP and FMI/SQF to work in partnership on audit checklist and standards for growers, available at: www.freshplaza.com/news_detail.asp?id=11585, accessed November, 2008.

Galvano, F., Ritieni, A., Piva G. and Pietri, A. (2005) Mycotoxins in the human food chain, in D.E. Diaz (ed.) *The Mycotoxin Blue Book* Nottingham, Nottingham University Press, 187–224.

Gessel, P.D., Hansen, N.C., Goyal, S.M., Johnston, L.J. and Webb, J. (2004) Persistence of zoonotic pathogens in surface soil treated with different rates of liquid pig manure, *Applied Soil Ecology*, **25**, 237–43.

Gill, O.N., Bartlett, C.L.R., Sockett, P.N., Vaile, M.S.B., Rowe, B., Gilbert, R.J., Dulake, C., Murrell, H.C. and Salmaso, S. (1983) Outbreak of *Salmonella napoli* infection caused by contaminated chocolate bars, *Lancet*, **321**(8324), 574–7.

GlobalGAP (2007) *General Regulations Integrated Farm Assurance, Version 3.0-2 September 2007*, Köln, GLOBALGAP.

Greenwood, M.H. and Hooper, W.L. (1983) Chocolate bars contaminated with *Salmonella napoli*: an infectivity study, *British Medical Journal*, **286**, 1394.

Gunstone, F.D. (2008) *Oils and Fats for the Food Industry*, Oxford, Blackwell Science.

Heiser, B. (1990) *Seeds to Civilization: the Story of Food*, Cambridge, MA, Harvard University Press.

Heyndrickx, M., Pasmans, F, Ducatelle, R. Decostere, A. and Haesebrouck, F. (2005) Recent changes in *Salmonella* nomenclature: the need for clarification, *Veterinary Journal*, **170**, 275–7.

IFST (2006) *Food and Drink: Good Manufacturing Practice*, 5th edn, London, Institute of Food Science and Technology (UK).

ILSI (1999) *Validation and Verification of HACCP*, Brussels, International Life Sciences Institute.

Imanaka, B.T., de Menezes, H.C., Vicente, E., Leite, R.S.F. and Taniwaki, M.H. (2007) Aflatoxigenic fungi and aflatoxin occurrence in sultanas and dried figs commercialized in Brazil, *Food Control*, **18**(5), 454–7.

Islam, M., Doyle, M.P, Phatak, S.C., Millner, P. and Jiang, X. (2005) Survival of *Escherichia coli* O157:H7 in soil and on carrots and onions grown in fields treated with contaminated manure composts and irrigation water, *Food Microbiology*, **22**, 63–70.

ISO (1990) *ISO 10011-1:1990 Guidelines for auditing quality systems – Part 1: Auditing*. Geneva, International Organisation for Standardization.

ISO (2000a) *ISO 9000:2000 Quality management systems – Fundamentals, and vocabulary*, Geneva, International Organisation for Standardization.

ISO (2005) *ISO 22000:2005. Food safety management systems – Requirements, for any organization in the food chain*, Geneva, International Organisation for Standardization.

ISO (2008) ISO 9001:2008 *Quality management systems – Requirements*, Geneva, International Organisation for Standardization.

Jackson. L. and Jablonski, J. (2004). Fumonisins, in N. Magan and M. Olsen (eds), *Mycotoxins in Foods*, Cambridge, Woodhead, 367–405.

Jay, J.M. (1996) *Modern Food Microbiology*, 5th edn, New York, Chapman and Hall.

Jin-Cheol, K., Hyo-Jung, K., Dong-Hyan, L., Yin-Won, L. and Yoshizawa, T. (1993) Natural occurrence of fusarium mycotoxins (trichothecenes and zearalenone) in barley and corn in Korea, *Applied and Environmental Microbiology*, **59**(11), 3789–802.

Johannessen, G.S., Loncarevic, S. and Kruse, H. (2002) Bacteriological analysis of fresh produce in Norway, *International Journal of Food Microbiology*, **77**, 199–204.

Johnston, L.M., Jaykus, L.-A., Moll, D. Anciso, J. Mora, B. and Moe, C.L. (2006) A field study of the microbiological quality of fresh produce of domestic and Mexican origin, *International Journal of Food Microbiology*, **112**, 83–95.

Jouany, J-P. and Diaz, D.E. (2005) Effects of mycotoxins in ruminants, in D.E. Diaz (ed.), *The Mycotoxin Blue Book*, Nottingham, Nottingham University Press, 295–321.

Karasch, M. (2000) Manioc, in K.E. Kiple and K. Coneè Ornelas (eds), *Cambridge World History of Food: Volume one*, Cambridge, Cambridge University Press, 181–7.

Kenney, S.J., Anderson, G.L., Williams, P.L., Millner, P.D. and Beuchat, L.R. (2006) Migration of *Caenorhabditis elegans* to manure and manure compost and potential for transport of *Salmonella newport* to fruits and vegetables, *International Journal of Food Microbiology*, **106**, 61–8.

Kishi, M. (2005) The health impacts of pesticides: What do we now know? in J. Pretty (ed.), *The Pesticide Detox*, London, Earthscan.

Ko, S.D. (1982) Indigenous fermented foods, in A.H. Rose (ed.), *Fermented Foods*, London, Academic Press.

Kokkinakis, E.N. and Fragkiadakis, G.A. (2007) HACCP effect on microbiological quality of minimally processed vegetables: a survey in six mass-catering establishments, *International Journal of Food Science and Technology*, **42**(1), 18–23.

Kokkinakis, E.N., Boskou, G., Fragkiadakis, G.A., Kokkinaki, A. and Lapidakis, N. (2007) Microbiological quality of tomatoes and peppers produced under good agricultural practices protocol AGRO 2–1 and 2–2 in Crete, Greece, *Food Control*, **18**, 1538–46.

Küster, H. (2000) Rye, in K.F. Kiple and K. Coneè Ornelas (eds), *The Cambridge World History of Food*, Cambridge, Cambridge University Press, 149–52.

LEAF (2000) *The Leaf Handbook for Integrated Farm Management*, Stoneleigh, LEAF.

Leesch, J.G., Smilanick, J.L. and Tebbets, J.S. (2008) Methyl bromide fumigation of packed table grapes: effect of shipping box on gas concentrations and phytotoxicity, *Postharvest Biology and Technology*, **49**, 283–6.

Matthews, K.R. (2006) Microorganisms associated with fruits and vegetables, in K.R. Matthews (ed.). *Microbiology of Fresh Produce*, Washington, DC, ASM Press, 1–20.

Mazoyer, M. and Roudart, L. (2006) *A History of World Agriculture*, translated by J. H. Membrez, London, Earthscan.

McCabe-Sellers, B.J. and Beattie, S.E. (2004) Food safety: emerging trends in foodborne illness surveillance and prevention, *Journal of the American Dietetic Association*, **104**, 1708–17.

McLauchlin, J. and Little, C. (eds) (2007) *Hobbs' Food Poisoning and Food Hygiene*. London, 7th edn, Hodder Arnold.

Mello, J.P.F., Placinta, C.M. and Macdonald, A.M.C. (1999) *Fusarium* mycotoxins: a review of global implications for animal health, welfare and productivity *Animal Feed Science and Technology*, **80**, 183–205.

Milton, K. (1987) Primate diets and gut morphology: implications for hominid evolution, in M. Harris and E.B. Ross (eds) *Food and Evolution: Toward a Theory of Human Food Habits*. Philadelphia, PA, Temple University Press, 93–116.

Molins, R.A., Motarjemi, Y. and Käferstein, F.K. (2001) Irradiation: a critical control point in ensuring the microbiological safety of raw food, *Food Control*, **12**, 347–56.

Monaghan, J.M. (2006) United Kingdom and European approach to fresh produce food safety and security, *HortTechnology*, **16**, 548–693.

Mortimore, S. and Wallace, C. (2001) *HACCP*, Oxford, Blackwell Science.

Moss, M.O. (1996) Mycotoxic fungi, in A.R. Eley (ed.), *Microbial Food Poisoning*, London, Chapman and Hall, 75–93.

Mourier, H. and Poulsen, K.P. (2000) Control of insects and mites in grain using a high temperature/short time (HTST) technique *Stored Products Research*, **36**, 309–18.

Murphy, P.A., Hendrich, S., Landgren, C. and Bryant, C.M. (2006) Food mycotoxins: an update, *Journal of Food Science*, **71**(5), 51–65.

Nicholson, P. (2004) Rapid detection of mycotoxigenic fungi in plants, in N. Magan and M. Olsen (edn) *Mycotoxins in Foods*, Cambridge, Woodhead, 111–36.

Nicholson, F.A., Groves, S.J. and Chambers, B.J. (2005) Pathogen survival during livestock manure storage and following land application, *Bioresource Technology*, **96**, 135–43.

Panisello, P.J. and Quantick, P.C. (2001) Technical barriers to Hazard Analysis Critical Control Point (HACCP), *Food Control*, **12**, 165–73.

Pennington (1997) *The Pennington Group Report*, Edinburgh, The Stationery Office.

Penteado, A.L. and Leitão, M.F.F. (2004) Growth of *Salmonella enteritidis* in melon, watermelon and papaya pulp stored at different times and temperatures, *Food Control*, **15**, 369–73.

Pettersson, H. (2004) Controlling mycotoxins in animal feeds, in N. Magan and M. Olsen (eds), *Mycotoxins in Foods*, Cambridge, Woodhead, 262–304.

Piñeiro, M. and Takeuchi, M.T. (2008) An FAO programme for improving the quality and safety of fresh fruits and vegetables, *Food Science and Technology*, **22**(2), 50–3.

Rico, D., Martín-Diana, A.B., Barry-Ryan, C., Frías, J.M., Henehan, G.T.M. and Barat, J.M. (2008) Use of neutral electrolyzed water (NEW) for quality maintenance and shelf-life extension of minimally processed lettuce, *Innovative Food Science and Emerging Technologies*, **9**, 37–48.

Romagnoli, B., Menna, V., Gruppioni, N. and Bergamin, C. (2007) Aflatoxins in spices, aromatic herbs, herb-teas and medicinal plants marketed in Italy, *Food Control*, **18**(6), 697–701.

SCIEH (1997) Scottish Centre for Infection and Environmental Health (SCIEH) Weekly report, 03 June 1997, **31**, No. 97/22.

Scudamore, K.A. (2004) Control of mycotoxins: secondary processing, in N. Magan and M. Olsen (eds), *Mycotoxins in Foods*, Cambridge, Woodhead, 224–43.

Selma, M.V., Beltrán, D., Allende, A., Chacón-Vera, E. and Gil, M.I. (2007) Elimination by ozone of *Shigella sonnei* in shredded lettuce in water, *Food Microbiology*, **24**, 492–9.

Selma, M.V., Ibáñez, A.M., Cantwell, M. and Suslow, T. (2008) Reduction by gaseous ozone of *Salmonella* and microbial flora associated with fresh-cut cantaloupe, *Food Microbiology*, **25**, 558–65.

Sofos, J.N. (2002) Approaches to pre-harvest food safety assurance, in F.J.M Smulders and J.D. Collins (eds), *Food Safety Assurance in the Pre-harvest Phase*, Vol. 1, Wageningen, Wageningen Academic, 23–48.

Solomon, E.B., Brandl, M.T. and Mandrell, R.E. (2006) Biology of foodborne pathogens on produce, in K.R. Matthews (ed.). *Microbiology of Fresh Produce*. Washington, DC, ASM Press, 55–84.

Sperber, W.H. (2005) HACCP does not work from farm to table, *Food Control*, **16**(6), 511–14.

SQF (undated) *The SQF Program: A brief guide*, Arlington, VA, SQF Institute.

Stapleton, J.J. (2000) Soil solarization in various agricultural production systems, *Crop Protection*, **19**, 837–41.

Stuart, D., Shennan, C. and Brown, M. (2006) Food safety versus environmental protection on the Central California coast: exploring the science behind the apparent conflict, *Research Brief 10*, University of California, Santa Cruz: Centre for Agroecology and Sustainable Food Systems.

Tansey, G. and Worsley, T. (1995) *The Food System*, London, Earthscan.

Tekinşen, K.K. and Uçar, G. (2007) Aflatoxin M1 levels in butter and cream consumed in Turkey, *Food Control*, **19**(1), 27–30.

Thompson, A.K. (1996) *Postharvest Technology of Fruit and Vegetables*, Oxford, Blackwell Science.

Tilley, D.R., Casada, M.E. and Arthur, F.H. (2007) Heat treatment for disinfestation of empty grain storage bins, *Stored Products Research*, **43**, 221–8.

Tindall, B.J., Grimont, P.A.D., Garrity, G.M. and Euzéby, J.P. (2005) Nomenclature and taxonomy of the genus *Salmonella*, *International Journal of Systematic and Evolutionary Microbiology*, **55**, 521–4.

Vadivambal, R., Jayas, D.S. and White, N.D.G. (2007) Wheat disinfestation using microwave energy, *Stored Products Research*, **43**, 508–14.

Viridis, S., Corgiolu, G., Scarano, C., Pilo A.L. and De Santis, E.P.L. (2007) Occurrence of Aflatoxin M1 in tank bulk goat milk and ripened goat cheese, *Food Control*, **19**(1), 44–9.

Yaun, B.R., Sumner, S.S., Eifet, J.D. and Marcy, J.E. (2004) Inhibition of pathogens on fresh produce by ultraviolet energy, *International Journal of Food Microbiology*, **90**, 1–8.

8

Pathogen control in primary production: fisheries and aquaculture

B. Fonnesbech Vogel, Technical University of Denmark, Denmark

Abstract: The true incidence of disease transmitted by foods including seafood, is not known. However, based on reported cases, there is substantial evidence that fish and fish products can cause several types of foodborne diseases. The present chapter reviews in detail the potential biological hazards related to seafood, including aquatic biotoxins, parasites and pathogenic bacteria and histamine fish poisoning. Furthermore the requirement to manage and control these hazards in primary production is discussed.

Key words: seafood, foodborne disease, pathogenic bacteria, parasite, biological hazards.

8.1 Introduction

Seafood products are produced from fish, mollusca and crustaceans from both wild populations and aquaculture. More than 20 000 species of fish, crustacean shellfish, mollusca and cephalopods have been identified and they are caught or harvested over a wide range of conditions, from freshwater lakes to oceans with 3–4 % salt and from the cold Arctic waters to the warm waters of Southeast Asia (Gram and Huss, 2000). In this chapter seafood denotes fin fish and shellfish. Shellfish cover bivalve molluscan shellfish, gastropods and crustacean shellfish. The molluscan shellfish are dealt with in Chapter 9.

Fin fish, crustacea and molluscs may carry human pathogenic bacteria that can cause foodborne disease. They may harbour potential human pathogens from their natural environment and may also be contaminated during further processing with one or more of the indigenous pathogens (bacteria naturally present in the aquatic environment) or bacteria from the human/animal reservoir. The latter will be present as a result of contamination with human or animal faeces or otherwise introduced into the aquatic environment. The levels of potential human pathogens are normally quite low, and for several of the pathogens it is unlikely that the numbers naturally present in uncooked seafood can cause disease. In most cases, disease only occurs if the pathogenic bacteria multiply to levels able to cause food intoxication or foodborne bacterial infection or when infection occurs at low numbers. An exception is the concentration of pathogens that may occur in bivalve molluscs due to filtration (see Chapter 9).

The true incidence of disease transmitted by foods, including seafood, is not known. However, based on reported cases, there is substantial evidence that fish and fish products can cause several types of foodborne diseases (Huss, 1994, ICMSF, 2005). The present chapter discusses some of the causes of seafood-borne disease and the requirement to control the hazards in primary production.

8.2 Identifying and assessing hazards and risk

The microflora of cold-blooded animals is to some extent a reflection of the microflora in their environment. Hence, the microflora of fish from colder waters is likely to be psychrotrophic in nature and very different from warm-blooded animals. However, during further processing, seafood processed in the modern fish industry is exposed to many of the same potential risks in terms of contamination as any other food product.

8.2.1 Microflora of wild and farmed aquatic animals

Microorganisms are found on the skin, on the gills and in the intestine, whereas the flesh is sterile if the animal is healthy. However, the circulatory system of some shellfish is not 'closed'. For example, the haemolymph of crabs commonly contains marine bacteria, such as vibrios, sometimes at high levels. Numbers of bacteria vary and 10^2–10^5 cfu/cm^2 are found on skin, 10^3–10^7 cfu/g on the gills and, depending on the intake of food, between 10 and more than 10^8 cfu/g in the intestines. The level of bacteria on animals from warmer waters is probably slightly higher than on animals from colder waters (Liston, 1980; Cahill, 1990). The level of human pathogenic bacteria in the general environment and in fish is generally low (Table 8.1); however, it is important to realise that several potentially human pathogenic bacteria are a natural part of the aquatic environment. Most of the genera of pathogenic bacteria also contain non-pathogenic strains, i.e. most strains of *Vibrio parahaemolyticus* isolated from the environment or seafood are not pathogenic to humans (Nishibuchi and Kaper, 1995; Su and Liu, 2007) and similarly *Vibrio cholerae* consist of more than 200 serotypes, of which only two (O1

Table 8.1 Human pathogenic bacteria associated with seafood-borne disease. The bacteria may be indigenous to the environment and naturally present on fish or originate from the animal/human reservoir, which are present due to faecal contamination

General habitat	Bacteria	Primary habitat	Quantitative levels	Of concern in	
				primary raw or undercooked products	processed products
Aquatic environment	*Clostridium botulinum* Non-proteolytic types B, E, F	Temperate and Arctic aquatic environment	Generally low (< 0.1 spores/g fish) but up to 5.3 spores/g fish has been recorded	–	+
	V. parahaemolyticus	Ubiquitous in warm (> 15 °C) seawater environment	Up to 10^2–10^3 cfu/g in shellfish; up to 10^4–10^8 cfu/g in intestines of shellfish-eating fish	+	(+)
	V. vulnificus			+	–
	V. cholerae			+	–
	Aeromonas hydrophila	Aquatic environment	Generally low, but up to 10^4 cfu/ml in seawater, 10^7 cfu/ml in sewage and 10^6 cfu/g in raw seafood	+	–
	Plesiomonas shigelloides	Warm aquatic environment	Generally low	+	–
General environment	*Clostridium botulinum;* Proteolytic types A, B	Soil	Generally low(< 0.01 spore/g soil)	–	+
	Clostridium perfringens	Soil (type A), animals (types B, C, D, E)	10^3–10^4 cfu/g soil	–	+
	Listeria monocytogenes	Ubiquitous in general environment like soil, decaying vegetation. Often occurring in fish processing plants	Generally low < 100 cfu/g in freshly produced fish products	–	+
	Pathogenic *Bacillus* spp.	Ubiquitous in general environment (soil,	10^1–10^3 cfu/g or ml raw, processed food	–	+

		natural waters, vegetation)			
Human/animal reservoir	*Salmonella* spp.	Intestines of warm blooded animals/humans	Levels in symptomatic and asymptomatic carriers vary; sporadic, low levels, but may accumulate in molluscan shellfish	+	–
	Shigella spp.	Intestines of warm blooded animals/humans	Sporadic, low levels, but may accumulate in molluscan shellfish	+	–
	E. coli	Intestines of warm blooded animals/humans	Sporadic, low levels, but may accumulate in molluscan shellfish	+	–
	C. jejuni spp. and other mesophilic campylobacter	Birds, insects, intestines of warm blooded animals	Sporadic, low levels, but may accumulate in molluscan shellfish	+	–
	Staphylococcus aureus	Outer surface (skin) and mucus membranes (nose)	Transient, but present on 50 % of population. Generally < 100 cfu/cm^2	–	+

Source: Based on Huss (1997) and Huss *et al.* (2004).

and O139) are involved in cases of cholera (Kaper *et al.*, 1995; FAO/WHO, 2005a). For some species such as *Listeria monocytogenes* it has been shown that the pathogenic potential is different among strains but no general method is available to distinguish between virulence of different strains, (Wiedmann *et al.*, 1997). *L. monocytogenes* and *Clostridium botulinum* are found in the general environment and can hence be detected on fish and shellfish (Table 8.1). Fish caught in coastal areas may carry pathogenic bacteria from the animal/human reservoir due to contamination of this aquatic environment as a result of water run-off from the surrounding land (Table 8.1) (Huss, 1997). For example *Salmonella* spp., *Shigella* spp., and *E. coli* are associated with faecal contamination of seafood (Table 8.1) (Nilsson and Gram, 2002). Furthermore contamination can occur during production, i.e. contamination with *Staphylococcus aureus*, *Salmonella* spp. or other pathogens may occur during manual handling of cooked products, such as shrimps and crustaceans (Nilsson and Gram, 2002). In the same way, unless processing equipment in fish slaughterhouses is adequately maintained from a sanitary perspective, contamination problems due to plant-persistent bacteria like *L. monocytogenes* may occur. *L. monocytogenes* is commonly isolated from food production plants, including fish slaughterhouses and fish smokehouses where some sub-types may be able to persist for months or many years (Rørvik *et al.*, 1995; Fonnesbech Vogel *et al.*, 2001; Wulff *et al.*, 2006).

The incidence of *L. monocytogenes* in raw fish is typically low, whereas the contamination on the processing equipment in slaughter and cold-smoking fish plants can be quite high. Molecular sub-typing of strains from raw fish, processing environments and packed products has suggested that contamination during processing rather than contamination directly from raw fish is the principal cause of product contamination (Autio *et al.*, 1999; Fonnesbech Vogel *et al.*, 2001; Thimothe *et al.*, 2004; Wulff *et al.*, 2006).

The microflora of fish and crustaceans reared in aquaculture does not, in principle, differ from wild aquatic animals other than that they are closer to land and human activities that may influence their environment and microflora (Dalgaard, 2006). Aquaculture has the potential to be a highly-controlled and hygienic operation at the initial harvesting and handling step. The microbial flora present on aquacultured fish is affected by water quality and feed composition. Water quality can range from that in raceways or tanks fed by microbe-free spring water, well water, ozone-treated water and ultraviolet light-treated water, to that in mud ponds purposely enriched with untreated human or animal waste. Hence regardless of whether a fish has been wild-caught or under which conditions it is farm-raised, it may carry various types of pathogenic bacteria.

8.2.2 Statistics on seafood-borne diseases

The exact numbers of outbreaks and cases of foodborne human diseases resulting from seafood consumption are uncertain. It has been estimated that only about 1 % of the actual cases of foodborne diseases are reported (Huss *et al.*, 2004). In many countries there is no obligation to report on foodborne disease and in countries that do have a reporting system there is substantial under-reporting. Furthermore in many cases it is difficult to identify both the contaminated food item and the responsible pathogenic agent.

The US Center for Science in the Public Interest (CSPI) maintains a database of foodborne illness outbreaks categorised by food vehicle, compiled from sources including the Center for Disease Control and Prevention, state health department and scientific journals. Between 1990 and 2005, the foods most commonly linked to outbreaks with identified vehicles were seafood (n = 1053), produce and produce dishes (n = 713), poultry and poultry dishes (n = 580) and beef and beef dishes (n = 506) (Fig. 8.1). Although seafood was associated with a high number of outbreaks, relatively few people were involved in each outbreak and the number of cases (10 415) was much lower than the number of cases caused by produce, meat or poultry (CSPI, 2007).

Of the 984 seafood-borne disease outbreaks (1990–2004) the major part of outbreaks (63 %) were linked to fish such as tuna, grouper and mahi mahi. Sixteen per cent were linked to molluscan shellfish including oysters, clams and mussels. Seafood dishes such as crab cakes and tuna burgers caused 14 % of the outbreaks and the remaining 7 % were linked to other seafood, such as shrimp and lobster (CSPI, 2006). The majority of seafood outbreaks were caused by toxins, either naturally-occurring aquatic toxins such as ciguatera or scombrotoxin (histamine fish poisoning (HFP)). Outbreaks caused by shellfish typically involved *Vibrio* bacteria and Norovirus (Fig. 8.2). Also *Salmonella* accounted for a considerable proportion of outbreaks.

From other investigations it also appears that HFP is the greatest cause of seafood-associated disease (Lipp and Rose, 1997; Wallace *et al.*, 1999; Olsen *et al.*, 2000). The disease is generally mild and self-resolving and, even though

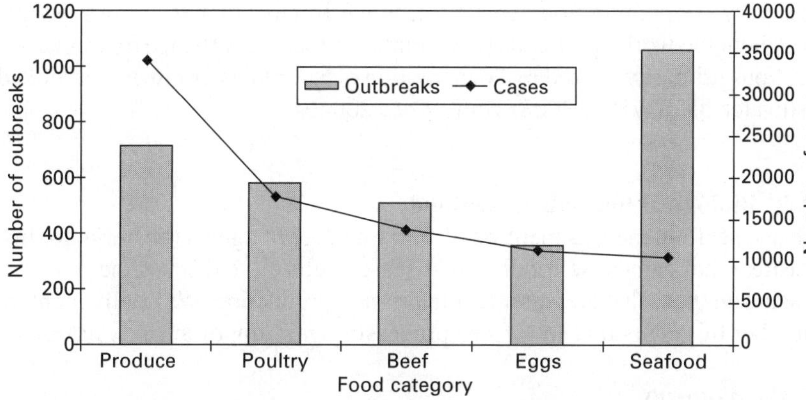

Fig. 8.1 Cases linked to outbreaks, 1990–2005, in the USA by the Center for Science in the Public Interest. Outbreak Alert! Database, 2007.

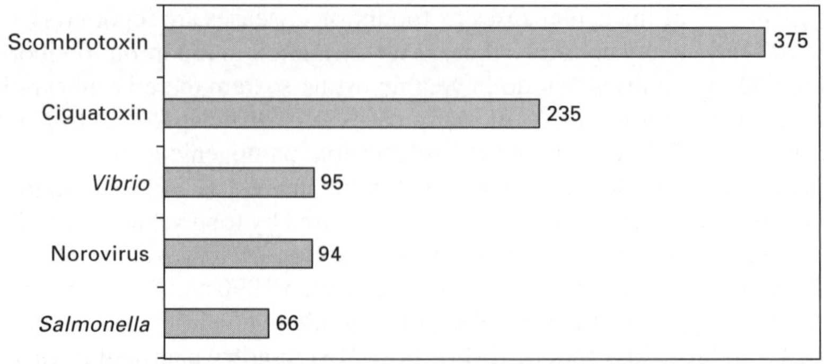

Fig. 8.2 Leading seafood pathogens, 1990–2005, with numbers of outbreaks in the USA by the Center for Science in the Public Interest. Outbreak Alert! Database, 2007.

seafood ranked fourth on the list of products which caused foodborne disease between 1993 and 1997 in the USA, seafood-borne disease did not, as opposed to other diseases caused by foods, result in death (Olsen *et al.*, 2000).

The statistics from England and Wales reveal a similar pattern, with scombrotoxin (HFP), virus and *Salmonella* as the major agents involved in human seafood-borne diseases between 1992 and 2003 (Hughes *et al.*, 2007). This situation is probably similar for many countries in Europe although outbreak statistics are incomplete.

Vibrio parahaemolyticus causes 20–30 % of food poisoning cases in Japan and is identified as a common cause of seafood-borne illness in many Asian countries (Alam *et al.*, 2002; Deepanjali *et al.*, 2005; Su and Liu, 2007). In contrast to Asian countries, *V. parahaemolyticus* infections are rarely reported in European countries, but sporadic cases occur in USA, particularly caused by contamination of *V. parahaemolyticus* in oysters (FDA, 2001; Su and Liu, 2007).

Seafood-borne parasitic infection rarely occurs in the USA and Europe, but is very common in Southeast Asia. It is estimated that 40 million people throughout the world, particularly in Southeast Asia, are affected by trematode infections primarily caused by consumption of raw plants or raw/undercooked freshwater fish (WHO, 2002; Butt *et al.*, 2004b).

8.2.3 Biological hazards in seafood

Fish and seafood may, as mentioned, carry a range of agents (bacteria, toxins, parasites) that can cause foodborne disease. Below, we discuss these agents focusing on those that are important in primary production. We briefly mention a number of agents that in further processing gain importance as hazards.

Aquatic biotoxins

Most of the so-called aquatic toxins are produced by species of naturally-occurring marine algae (phytoplankton) and may accumulate in the flesh of fish, crustaceans or molluscs. The toxins are not toxic to the aquatic animal and cannot be detected by sensory assessment. There are many types of aquatic toxins with ciguatoxin, tetrodotoxin and saxitoxin being accumulated in fin fish. Paralytic shellfish toxin, diarrheic shellfish toxin and others accumulate in shellfish and are discussed in Chapter 9. In the developed world, toxic seafood poisonings are more common than seafood-borne parasitic infections, such as fish tapeworms and liver flukes (Diaz, 2004).

Ciguatera fish poisoning (CFP) is one of the most common foodborne illnesses related to fish consumption. The reported incidence of CFP ranges between 10 000 and 50 000 cases per year, but this is probably much lower than the actual number of cases due to under-reporting of a self-limiting disorder and misdiagnosis as fish allergy or mild decompression sickness (Diaz, 2004). It is caused by consumption of fish that have become toxic by feeding on toxic dinoflagellates or toxic herbivore fish. It is usually large (greater than 10 kg) predacious coral reef fish such as kingfish, barracuda, red snapper, amberjack, mackerel and grouper that causes disease (Withers 1982; Karalis *et al.*, 2000). CFP is endemic in tropical and sub-tropical waters of the Pacific Ocean, Indian Ocean, Gulf of Mexico and Caribbean Sea (Diaz, 2004).

Tetrodotoxin fish poisoning (TFP). Tetrodotoxin is a neurotoxin found in species of *Tetraodontiae* such as all puffer fish, marine sunfish and porcupine fish found throughout the Atlantic, Pacific and Indian Oceans. Although puffer fish contain tetrodotoxin at a high concentration mainly in liver, the underlying mechanism remains to be elucidated (Matsumoto *et al.*, 2007). Tetrodotoxin has been assumed to be endogenous; however, it is hypothesised that two routes are implicated; symbiosis or parasitism with tetrodotoxin-producing bacteria and the food chain (Matsumoto *et al.*, 2007). It is assumed that tetrodotoxin in fish comes directly from the feed. The toxin is believed to be produced by bacteria, absorbed on or precipitated with plankton and transmitted by the food chain to fish (Huss *et al.*, 2004). TFP has traditionally been a problem in Japan and Okinawa where raw fugu fish is considered a delicacy and TFP has frequently occurred (Diaz, 2004; Huss *et al.*, 2004). Fugu fish is the Japanese word for pufferfish. Sale of fish belonging to this genus is forbidden altogether in the European Union (EC, 2004).

Parasites

Parasites cover several groups of organisms that can cause infectious diseases in man. Parasites able to infect humans include nematodes, trematodes, cestodes and protozoa (Table 8.2). The last which include *Cryptosporidium* and *Giardia* have their natural reservoir in the mammal gastrointestinal tract but are typically transmitted with contaminated water (Fayer, 2001; Butt *et al.*, 2004b). They have not, so far, been associated with seafood-borne disease, but their waterborne association indicates that they represent a risk to be considered in seafood.

Table 8.2 Pathogenic parasites transmitted by seafood

Type of parasite	Genera/species	Geographical distribution
Nematodes	*Anisakis simplex*	Worldwide
	Pseudoterranova decipiens	Worldwide
	Gnathostoma	Thailand, Japan and Mexico
	Capillaria spp.	The Philippines and Thailand
Trematodes	*Clonorchis* sinensis	China, Korea, Taiwan and Vietnam
	Opisthorchis spp.	The Far East and Eastern Europe
	Paragonimus spp.	The Far East, The Americas and Africa
	Heteropyes spp.	Egypt, the Middle East and the Far East
	Metagonimus spp.	The Far East, Siberia, Musharia, The Balkan states and Spain
Cestodes	*Diphyllopbothrium* spp.	Worldwide
Protozoa	*Cryptosporidium* spp.	Worldwide
	Giardia spp.	Worldwide, more prevalent in warm climates

Source: Based on Nuss *et al.* (2004) and modified by http://www.dpd.cdc.gov, accessed November 2008.

Nematodes, trematodes and cestodes are responsible for a substantial number of seafood-associated infections. It is estimated than more than 40 million people are infected with foodborne trematodes worldwide (WHO, 2002). Most of these infections are associated with consumption of raw or undercooked seafood. Heat inactivation is an effective method for eliminating the risk of parasitic disease from seafood. For internal temperature a minimum of 63 °C for 15 s should be enough to inactivate parasites (FDA, 2001). Freezing is also an acceptable method for parasite inactivation and the US regulations stipulate that 15 h at –35 °C or seven days at – 20 °C will be effective (FDA, 2001). The highest prevalence of these infections is in southeast and east Asia (Butt *et al.*, 2004b). Parasites are very common in fish, but most do not cause disease in their natural host or in humans. However, more than 50 species of helminth parasites from fish and shellfish are known to cause disease in humans (Huss *et al.*, 2004).

All the parasitic helminths have complicated life cycles. They do not spread directly from fish to fish but must pass through a number of intermediate hosts in their development. Very often sea-snails or crustaceans are involved as first intermediate host and marine fish as second intermediate host, while the sexually mature parasite is found in mammals as the final host (Huss *et al.*, 2004). In most cases infections of humans are caused by consumption of intermediate hosts, thereby disrupting the natural life cycle. As an example the complete life cycle of *Anisakis simplex* is shown in Fig. 8.3.

The most common nematode associated with seafood-borne disease is *Anisakis simplex*, followed by *Pseudoterranova desipiens* (Herreras *et al.*, 2000). Large outbreaks of human anisakiasis have been reported from countries such as Japan and Korea with a high consumption of raw or undercooked seafood, but anisakiasis also occurs in Europe (the Netherlands) and the USA (Herreras *et al.*, 2000; Butt *et al.*, 2004b). Anisakiasis is a gastrointestinal parasitosis caused by the larval stages of anisakid nematodes. Humans become infected by eating fish containing live third stage larvae and are accidental hosts, since transfer to humans results in a non-complete life cycle for the parasite (Fig. 8.3). Many patients will recover spontaneously, but some may require endoscopic removal of the worms.

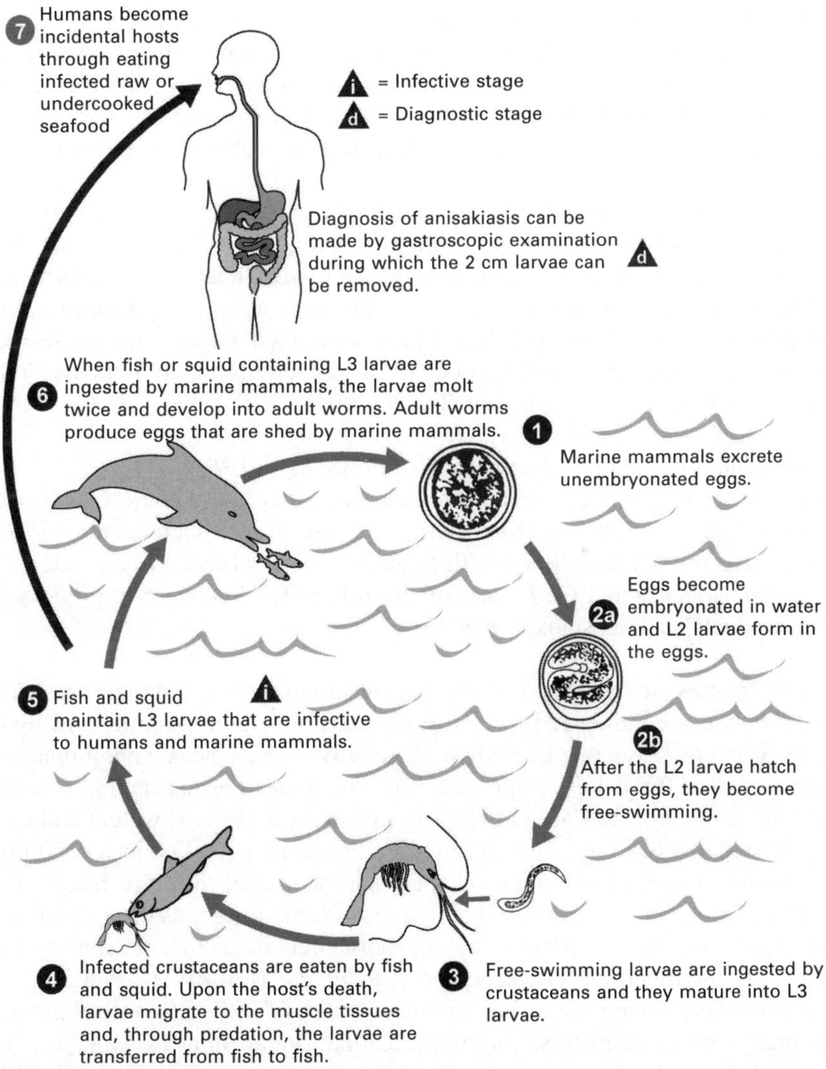

Fig. 8.3 Life cycle of *Anisakis simplex*. (Adapted from http://www. dpd.cdc.gov/dpdx/HTML/Anisakiasis.htm) accessed November 2008).

There are three main groups of fish-borne trematodes infecting humans: the liver flukes, the lung flukes and the intestinal flukes. *Clonorchis sinensis,* known as the Chinese liver fluke, is highly prevalent in China, Korea, Taiwan, Vietnam and Japan with more than 5 million people believed to be infected in China alone (Dixon and Flohr, 1997). Lung flukes like *Paragonimus* spp. are located in the lungs of humans and a number of domestic and wild carnivorous animals (dogs, cats, leopards, tigers) (Butt *et al.*, 2004b; Huss *et al.*, 2004). Intestinal flukes live in the intestines of the final host as the name indicates. The most important species are *Heterophyes* and *Metagonimus* spp. that are acquired by eating raw, marinated or improperly cooked fish.

Cestodes associated with seafood consumption are mostly *Diphyllobothrium* spp. *Diphyllobothrium latum* is the largest human tape worm and can reach more than 10 m in length. It resides in the small intestines of fish. Cases are generally reported from areas where fish are eaten raw, marinated or undercooked (Butt *et al.*, 2004b).

About 40 parasitic protozoans are known to be infectious to humans. The most important of those are primarily being transmitted via water like *Cryptosporidium.* This is the leading cause of diarrhoea due to protozoal infections worldwide (Butt *et al.*, 2004b). The only stage of *Cryptosporidium* outside the body is the oocyst stage which is excreted in faeces from infected humans and animals. Finding oocysts in the environment indicates the presence of faecal contamination (Fayer, 2001).

Pathogenic bacteria and histamine fish poisoning (HFP)

As mentioned, a number of bacteria indigenous to the aquatic or general environment can cause seafood-borne disease. Some, such as the *Vibrio* spp., are primarily associated with fresh bivalves and fish. Others, such as *L. monocytogenes* and *Cl. botulinum* are primarily of concern in processed fish products such as smoked fish.

***Vibrio* species** are indigenous to the aquatic environment, and their presence and numbers are influenced by factors such as temperature, salinity and algal density, but not associated with human activity such as faecal contamination (Huss *et al.*, 2004). *Vibrio* spp. are very common in estuarine and coastal environments, but they are particularly prevalent in tropical waters and can be isolated in temperate zones during the summer months (Feldhusen, 2000). Since the marine environment is their natural niche, they are frequently isolated from fish, crustaceans and bivalves. Within the genus that comprises 34 species, the most important seafood-borne human pathogenic species are *V. parahaemolyticus*, *V. cholerae* and *V. vulnificus* (FAO/WHO, 2005b).

Vibrio parahaemolyticus is a common cause of seafood-borne illnesses in many Asian countries, including China, Japan and Taiwan, and is recognised as the leading cause of human gastroenteritis associated with seafood consumption in the USA (DePaola *et al.*, 2003; Su and Liu, 2007). In contrast, *V. parahaemolyticus* infections are rarely reported in European countries. However, sporadic outbreaks have been reported in Spain and France (Martinez-Urtaza *et al.*, 2005). In Western cultures, raw molluscs and cooked crustaceans are the most common food source of *V. parahaemolyticus*, while in Asian cultures finfish is also a common source, as it is often eaten raw (Lake *et al.*, 2003). Consumption of raw or undercooked seafood contaminated with high numbers of *V. parahaemolyticus* causes an acute gastroenteritis characterised by diarrhoea, abdominal cramps, vomiting, headache and nausea (Gooch *et al.*, 2001). The infectious dose for *V. parahaemolyticus* is thought to be high and growth in seafood is required for disease to occur. The probability of illness is relatively low (< 0.001 %) for consumption of 10 000 *V. parahaemolyticus* cells/serving (equivalent to about 50 cells/gram oysters) (FDA, 2005). However, findings during the 1998 outbreaks caused by oyster consumption in the USA suggest a possible low infectious dose of the pathogenic strains involved in the outbreaks and that the FDA guidance may not be sufficient to protect consumers from *V. parahaemolyticus* infection associated with raw oyster consumption (Su and Liu, 2007). Most strains of *V. parahaemolyticus* isolated from the environment or seafood are not pathogenic (Nishibuchi and Kaper, 1995; DePaola *et al.*, 2003). Clinical strains of *V. parahaemolyticus* are differentiated from environmental strains by their ability to produce the enzymes thermolytic direct hemolysin (TDH) or a TDH-related hemolysin (TRH) which are recognised as the major virulence factors (DePaola *et al.*, 2003; Su and Liu, 2007).

Vibrio vulnificus illness has one of the highest mortality rates of any foodborne disease and has emerged as a food safety issue in a number of countries and regions including Europe, Japan, New Zealand, Republic of Korea and the USA (FAO/WHO, 2005b). Biotype 1 of this bacterium can be lethal to humans and Biotype 2 is known to be pathogenic for Asian and European eels and may be an opportunistic pathogen for humans (Feldhusen, 2000). Biotype 3 has been associated with seafood mediated bacteremia (Bisharat *et al.*, 1999). Foodborne *V. vulnificus* infection is clearly associated with underlying medical conditions (Strom and Paranjpye, 2000). Two distinct syndromes are recognised – septicaemia with high mortality and severe necrotising wound infections from exposure to contaminated water or handling of fish (Butt *et al.*, 2004a). Primary septicaemia is almost exclusively caused by consumption of raw bivalve molluscs such as oysters. *Vibrio vulnificus* naturally inhabits warm estuarine environments and the presence is strongly correlated with water temperature and the disease is therefore more common in the summer months. Infection with *V. vulnificus* is a safety issue in the Gulf Coast area of the USA, but is not common in Europe (Huss *et al.*, 2004; FAO/WHO, 2005b).

Choleragenic *V. cholerae* is an important pathogen in many developing countries, where it can cause devastating disease and economic burdens (FAO/WHO, 2005a). Choleragenic *V. cholerae* serotypes O1 and O139 produce the cholera toxin and are the only serotypes associated with epidemic and pandemic cholera. Cholera is a gastrointestinal disease with severe watery diarrhoea that results in dehydration and can be fatal. Some *V. cholerae* non-O1 and non-O139 have caused milder gastrointestinal disease, but they are not associated with the epidemic disease (Kaper *et al.*, 1995). *Vibrio cholerae* non-O1 and non-O139 are as other *Vibrio* spp. ubiquitous in marine and estuarine waters. The primary cause of choleragenic *V. cholerae* is the faeces of persons acutely infected with the organism and thus it reaches water most often through sewage, but outbreaks of cholera have also been associated with consumption of seafood including oysters, crabs and shrimp being contaminated with cholerae-containing water (FAO/WHO, 2005a). Inadequate sanitation and lack of safe water are the major causes of cholera epidemics therefore improved sanitation is the key to solving the problem (Huss *et al.*, 2004).

Clostridium botulinum is a spore-forming bacterium that produces neurotoxins (BoNT types A–G). Disease is caused by neurotoxin produced in the food product. Botulinum toxin is one of the most potent of all poisons and as little as 30–100 ng has been estimated to cause death (Lund and Peck, 2000). The disease can vary from a mild illness to a serious disease, which may be fatal within 24 hours (Huss *et al.*, 2004). *Clostridium botulinum* strains are divided into four groups (I–IV) based on their metabolism and pathogenesis, where groups I and II include human pathogenic strains. Group I consists of proteolytic strains producing types A, B and F toxins and group II of non-proteolytic strains producing types B, E and F toxins (Eklund, 1992; Merivirta *et al.*, 2006). The proteolytic types are heat-resistant, mesophilic, NaCl-tolerant and have the general environment as the natural habitat. In contrast, the non-proteolytic types are heat-sensitive, psychrotolerant, NaCl-sensitive and are indigeneous to the aquatic environment (Huss *et al.*, 2004). *Clostridium botulinum* group I occurs frequently in soil, but is also found on aquatic animals, primarily in warm waters. In contrast, Group II is ubiquitous in aquatic environments and has been isolated from water, ocean sediments, the intestinal tracts of fish and the gills and viscera of crabs and other shellfish (Huss, 1980). Spores of *Cl. botulinum* type E have been found in newly caught or processed seafood. The presence of spores or low levels of vegetative cells does not constitute a hazard, however; of concern is the ability of *Cl. botulinum* type E to grow and produce toxin at temperatures as low as 3.0–3.3 °C or in up to 5 % NaCl (water phase salt, WPS). The organism is strictly anaerobic and sensitive to oxygen (Gram, 2001). Toxin production at these limits, however, generally requires that all other growth conditions are near optimal. As a consequence of its capabilities to grow at low temperatures and its resistance to salt, it may, if anoxic conditions are created, germinate and produce toxins in many food products with extended chilled shelf-lives and cause foodborne botulism (Gram, 2001). Lightly preserved fish products, particularly fermented fish, have been linked to cases of botulism (Lund and Peck, 2000). In most cases the products have been under-processed with less than 1 % NaCl, vacuum-packed and temperature-abused before consumption (Eklund, 1992). Hot-smoked fish has led to several outbreaks. In contrast cold-smoked fish has not. However, the risk of toxin formation in cold-smoked fish must be taken seriously, because there is no heating step that is able to eliminate the spores during processing of cold-smoked fish (Gram, 2001).

Listeria monocytogenes can cause listeriosis, which is an infectious disease, primarily transmitted to humans via the food chain. It is a ubiquitous bacterium found in environments such as soil, water, decaying vegetation, silage and pasture grasses. It occurs naturally in many raw foods, and has a unique ability to establish itself in the processing plants and thereby contaminate food products during processing.

Although indigenous to the general environment, the number of positive environmental samples is low, between 0 and 6 % (Frances *et al.*, 1991; MacGowan *et al.*, 1994; Hansen *et al.*, 2006). This correlates with the findings that the bacterium survived poorly in both freshwater and seawater in a lab-setup (Hansen *et al.*, 2006). The bacterium can be present in raw fish, albeit typically at a low prevalence ranging from 0–10 % (Jemmi and Keusch, 1994; Autio *et al.*, 1999; Johansson *et al.*, 1999; Miettinen and Wirtanen, 2005; Wulff *et al.*, 2006).

Listeria monocytogenes is resistant to high sodium chloride concentrations (10 %), has a lower pH limit for growth between 4.4 and 4.6 (George *et al.*, 1988), and grows up to pH 9.2 (Petran and Zottola, 1989). Optimal growth temperature is 30–37 °C, but the bacterium can grow at temperatures as low as –1.5 °C (Hudson *et al.*, 1994) and up to 45 °C (Seeliger and Jones, 1986). It is believed that low levels of the pathogen are not likely to cause disease and that growth must occur in the food product for infectious levels to be reached (FSIS/FDA, 2003; FAO/WHO, 2004). *Listeria* tolerates the preserving parameters (low temperature, salt and vacuum packaging) of some ready-to-eat products well, and inoculation trials have shown that it may multiply to considerable numbers at refrigeration temperatures. As these products are not heat treated by the consumer, there is no elimination step and thus the products must be considered high-risk products with respect to *Listeria*. This includes a range of seafood, such as cold-smoked fish (Huss *et al.*, 2000). Although *L. monocytogenes* has been found in a variety of fishery products and these products have been considered potential sources of listeriosis, only a few confirmed cases have been established. Three investigations link fish products directly with human listeriosis. The fish products were smoked rainbow trout (Ericsson *et al.*, 1997), smoked mussels (Brett *et al.*, 1998) and cold-smoked trout (Autio *et al.*, 1999).

***Salmonella* species** are among the most important causes of gastrointestinal disease worldwide, and in the USA, approximately 7 % of all *Salmonella* outbreaks are linked to seafood (Bean *et al.*, 1997; Olsen *et al.*, 2001). Compared to other types of food, seafood has caused relatively few, and no major, outbreaks of salmonellosis, but in many seafood importing countries occurrence of *Salmonella* in seafood has been a major reason to reject products at port of entry (Feldhusen, 2000). The incidence of *Salmonella* in seafood is highest in the central Pacific and African countries and lowest in Europe and North America (Heinitz *et al.*, 2000). The natural reservoir is the gastrointestinal tract of humans and animals. Therefore the presence of *Salmonella* in seafood has been seen as a sign of poor standards of process hygiene and sanitation. However, recent studies have revealed that *Salmonella* could be part of the natural flora of cultured shrimp and shrimp culture environments (Bhaskar and Sachindra, 2006). The survival rate of *Salmonella* in aquatic environments can be high (Shabarinath *et al.*, 2007). The organism can be found in tropical estuaries and coastal water and, accordingly, on fish and shellfish from such waters (Shabarinath *et al.*, 2007). The presence of *Salmonella* in water depends on many factors. The integrated fish farming in some areas of Southeast Asia and Asia where poultry manure is sometimes used as fertiliser for ponds may add to the *Salmonella* contamination (Huss *et al.*, 2004). However, the use of chicken manure does not always lead to presence of *Salmonella* (Dalsgaard *et al.*, 1995). Until recently, *S. enterica* serovar Typhimurium was the most common sereovar causing human salmonellosis. However, in Southeast Asia, *S. enterica* serovar Weltevreden has been reported as a frequent and increasing cause of human infection (Ponce *et al.*, 2008). It is the predominant serovar in Malaysia, Thailand and Vietnam and is the most frequent serovar in imported seafood samples in the USA (Heinitz *et al.*, 2000; Ponce *et al.*, 2008).

Histamine fish poisoning (HFP) is a mild disease with a variety of allergy-like symptoms that occurs rapidly after intake of seafood containing above 500–1000 ppm of histamine (Dalgaard, 2006). The symptoms include rash, nausea, headache and sometimes diarrhea (Emborg and Dalgaard, 2006). HFP occurs worldwide and is perhaps the most common form of toxicity caused by the ingestion of fish. Many HFP incidents probably go unreported because of the mildness of the disease and misdiagnosis as fish allergy. Products become toxic when specific histidine decarboxylase producing bacteria, at some stage between catch and consumption, have been able to grow to high concentrations and to form significant amounts of histamine (Lehane and Olley, 2000; Emborg and Dalgaard, 2006). Not all seafoods cause HFP because a high content of free histidine and high activity of microbial histidine decaboxylases are pre-requisites for histamine production in toxic concentrations (Dalgaard, 2006). Fish species involved in HFP include tuna, swordfish, mahi-mahi, bluefish, bonito, mackerel, yellowtail, marlin and anchovy (Lehane and Olley, 2000; Dalgaard, 2006). In contrast Atlantic salmon (*Salmo salar*) has a low concentration of free histidine in the muscle tissue and cold-smoked salmon has not been associated with HFP (Emborg and Dalgaard, 2006). Relatively few bacteria species are prolific histamine-producers able to form more than 1000 ppm histamine. The most important mesophilic strains include *Morganella morganii*, *Hafnia alvei*,

Raoultella planticola, *Enterobacter aerogenes* and *Proteus vulgaris*. Also psychrotolerant bacteria such as *M. psychrotolerans* and *Photobacterium phosphoreum* can produce histamine (Emborg *et al.*, 2005, 2006). The identification of psychrotolerant bacteria responsible for histamine formation is rather new information and important as these bacteria may cause a significant proportion of the reported HFP incidents (Kanki *et al.*, 2004). To control histamine formation in seafood chilling and storage time are the most important factors. Freezing can inactivate histamine-producing bacteria, but histidine decarboxylase is active during frozen storage and therefore products of dubious freshness should not be frozen (Dalgaard, 2006).

8.3 Managing and controlling hazards and risk

Good Manufacturing Practices (GMP) and Hazard Analysis Critical Control Points (HACCP) programmes in particular are important for improving the safety of seafood. HACCP programmes should be designed specifically to address food safety hazards of particular products and to eliminate or reduce them to an acceptable level. The process from catching to processing should be performed by GMP including good hygiene and prevention of cross-contamination.

When all hazards are considered, fresh and frozen fish products are a low-risk category. Most fish are cooked before consumption and cooking will kill the bacteria present. Furthermore thorough cooking prevents parasitic diseases. The risk of bacterial infection from eating raw fish is also low, because there is no accumulation of pathogens in the live (healthy) fish. Furthermore the sensory quality needs to be very high, which means that the product has to be fresh and consequently the associated temperature control and shelf-life setting will also give pathogens less opportunity to multiply.

As mentioned above, aquatic toxins (ciguatera) and histamine are the major hazards in fresh fish. Only if fish or crustacea are consumed raw or undercooked do parasites and some of the indigenous bacteria, in particular *V. parahaemolyticus* and bacteria from the human/animal reservoir such as *Salmonella* spp., become potential risks. Control options for hazards are listed in Table 8.3.

8.3.1 Control of aquatic biotoxins

Ciguatera and algal intoxications are handled through control of where the fish has been caught or cultured. Fishing from areas during an algal bloom should be avoided because of risk of ciguatera and algal intoxications in fish. The present approach to control is embargoing potentially toxic species, especially large fish caught near known high-risk ('hot spot') areas and discouraging commercial and sport fishing in these areas (ICMSF, 1998, 2005; OzFoodNet,

Table 8.3 Control of finfish and crustacea of marine and freshwater origin with regard to aquatic biotoxins, parasites, pathogenic bacteria and histamine fish poisoning

Significant hazards[a]		Control measures
Aquatic biotoxins	Ciguatera	Avoid fish from certain areas or areas with algal blooms Surveying catching waters for algal blooms
	Tetrodotoxin	Avoid eating fish of the order *Tetraodontiformers* Fugu-fish should only be prepared by a trained chef
Parasites		Freeze to inactivate parasites (if consumed raw) Thorough cooking will kill vegetative pathogens
Pathogenic bacteria		Avoid catching or harvesting in contaminated waters (for raw consumption) Control growth by time and temperature Thorough cooking will destroy vegetative bacteria in crustacea
Histamine fish poisoning		Control growth by time and temperature Avoid temperature abuse

[a]In particular circumstances, other hazards may need to be considered.
Source: Based on ICMSF (2005).

2004). Ciguatoxin is not inactivated by cooking, freezing or stomach acid and cannot be detected by sight, smell or taste (Diaz, 2004).

In all fish that belong to the order *Tetraodontiformers* presence of tretrodotoxin is possible and therefore the EU has prohibited import of fish from this family (EEC, 1991; ICMSF, 2005). Tetrodotoxin is distributed throughout the skin and muscle and is concentrated in liver and bile, and in ovaries and roe during spawning. Without access to intensive care, case fatality rates for tetrodotoxin poisoning can exceed 60 %, especially in Japan and Okinawa, where raw fugu fish is considered a delicacy. Preventing TFP consists of avoiding inherently toxic fish, especially during spring spawning, and furthermore fugu fish should only be prepared by trained and licensed fugu fish preparers (Diaz, 2004).

8.3.2 Control of parasites

All parasites of concern are transmitted to humans by eating raw or lightly prepared fish products. Infection can be prevented by heat treatment or frozen storage and therefore, according to the EU, all fish that are to be consumed raw or almost raw must be subjected to a freezing process (–20 °C for at least 24 h) and this also applies to fish products heated to a temperature less than 60 °C for 24 h (i.e. hot-smoked) (EEC, 1991; Huss *et al.*, 2004). Visual inspection can be used to determine presence/absence of parasites and is used as a quality check also for fish intended for freezing (ICMSF, 2005). Eating habits are deeply rooted in a culture and therefore difficult to change and seafood has been eaten raw for centuries, particularly in Asian countries. Measures can be taken during harvesting, processing or post-processing to reduce the risk of infection with parasites. Before harvesting, measures to be taken may include determining the type and size of fish, their feeding habits and their environment. After capture, fish should be rapidly chilled or gutted to prevent parasites migrating from the gut to the flesh (Butt *et al.*, 2004b).

For a full discussion of control of virus please see Chapter 9.

8.3.3 Control of pathogenic bacteria and histamine fish poisoning

The usual level of naturally-occurring pathogenic bacteria on non-molluscan seafood is so low that there is little or no risk of causing disease unless growth and multiplication can occur. To be a hazard, the toxin producers like *Cl. botulinum* need to multiply in the food in order to produce their toxins, and the infective types of bacteria need to multiply to larger numbers to cause disease (Huss, 1997). However, some bacteria cause illness even at low levels, e.g. *E. coli* O157 and campylobacters. They are rarely associated with seafood-borne disease, but waterborne outbreaks have occurred (Sharma *et al.*, 2003).

Fish are caught by net, hook and line or traps in bodies of water. After capture, fish must be protected from spoilage to the extent possible during transport to the processing plant to ensure both microbiological quality and safety. All fish should be cooled as soon as possible after capture to low temperatures, preferablly to below 2 °C. Raw fish are spoiled by psychrotrophic microorganisms that grow well in temperatures between 0 and 30 °C, whereas most of the pathogenic bacteria are mesophiles, which grow well between 8 and 42 °C. Thus in general seafood refrigerated below 5 °C will not support growth of pathogens (ICMSF, 2005). Exceptions to this are *Cl. botulinum* type E, which can grow at temperatures as low as 3.0–3.3 °C and *L. monocytogenes*, *Aeromonas hydrophila* and *Yersinia enterocolitica*, which can grow at temperatures below 5 °C. However, many of the pathogens are poor competitors and in the temperature range where both groups grow, typically psychrotrophic bacteria have a faster growth rate than the pathogenic bacteria and spoil the fish before the pathogen has reached dangerous levels. Exceptions to this exist, i.e. *V. parahaemolyticus* can grow in competition with spoilage bacteria (ICMSF, 2005).

Crabs and lobsters are trapped in baited pots or cages and transported live directly to processing plants or holding tanks. The primary control for crabs and lobsters is to keep them alive until cooking, because live muscle tissue remains sterile. Therefore dead animals should be discarded. Once cooked, crabs and lobsters are either distributed refrigerated or frozen (ICMSF, 2005). Unlike crabs and lobsters, shrimps die quickly after capture and cannot be kept alive. Because shrimps are harvested by trawling they should be washed with fresh seawater to remove mud and sediment. Shrimps are cooked headed or not and distributed whole with shell or peeled and either distributed refrigerated or frozen (ICMSF, 2005). Peeling is either done by peeling machines or by hand and needs to be done carefully to avoid contamination, and furthermore the product must be kept chilled. A sufficiently cooked product is free of vegetative microorganisms and therefore very sensitive to contamination either by spoilage or pathogenic bacteria, because they may be able to multiply to high levels in a short time without competitive bacteria.

Rapid chilling after capture and maintenance of cold storage is particularly important for fish with high levels of endogenous free histidine and is necessary to prevent development of histamine. Toxic concentrations of histamine in seafood are produced by bacteria and high concentrations of strongly histamine-producing bacteria are required. Therefore, it is possible to control histamine formation by limiting the contamination and by reducing the growth of these bacteria (Dalgaard *et al.*, 2008). Although some psychrotolerant histamine-producing bacteria may produce biogenic amines at low temperatures, rapid chilling and maintenance of low temperature are the primary means for preventing biogenic amines (histamine) production at levels able to cause HFP.

8.4 Future trends

In the developed part of the world increased seafood consumption is desirable for health-promoting effects. Several national authorities recommend increased seafood consumption. In Denmark, the USA and in the UK two portions of fish per week is recommended (Andersen *et al.*, 2003; FDA, 2004; IOM, 2006).

Despite decades of research and efforts by the seafood sector and national and international authorities HFP remains the most common disease associated with seafood. Probably this will not change in the near future. However, the recent discovery of psychrotolerant bacteria, as major producers of histamine in seafood, provides new options to reduce HFP (Dalgaard *et al.*, 2008).

Ciguatera poisoning is still a serious threat to public health and fisheries development along tropical and sub-tropical shorelines. Predictions are for more CFP outbreaks as increased fish consumption drives worldwide trade, particularly exports from fish-rich tropical island nations. However, harmful algal blooms are followed by authorities (http://www.cdc.gov/hab/) and fishing from these areas can be avoided. Furthermore a commercial test has been developed (The Cigua-Check® test kit, ToxiTec Inc. Honolulu, Hawai) to allow persons to test sport caught fish for ciguatoxins.

Aquaculture production in marine and particularly in freshwater is one of the fastest growing food-producing sectors in the world, with an annual increase of 6.1 % from 2000 to 2002 and it is estimated that this increase will continue during the next decade. The total world aquaculture production in 2002 was 51.4 million tons. In comparison, capture fisheries has remained at approximately 90 million tons since the mid-1990s (FAO, 2004). Farming of fish has obvious advantages in terms of ability to control chemical composition of the fillet, growth rate, harvesting time, etc. (Gram and Huss, 2000). However, disease is one of the major constraints on the aquaculture sector as fish in intensive rearing experience stress and rapid spread of infection (Gram and Ringø, 2005). When vaccination is impossible (vaccination of fish smaller than 35 g is not recommended) antibiotics are widely used for controlling bacterial fish disease (Gram and Ringø, 2005). This causes concern because of the risk of developing antibiotic-resistant bacteria that, via the food, can be transferred to humans (Sørum, 1999; Gram and Ringø, 2005). It is desirable to reduce the amount of antibiotics by disease prevention, i.e. by using probiotics. This is a major area needing further investigation.

The developing countries have surpassed the developed countries in terms of amount of fish supplied. Asia contributes more than 90 % to the world's aquaculture production (Bondad-Reantaso *et al.*, 2005). This emphasises the need for an understanding of the microbiology and safety of fish and crustacea caught or harvested in warmer, tropical waters (Gram and Huss, 2000). Foodborne trematodes are emerging as a serious public health problem, especially in Southeast Asia but also in Latin America, in part due to a combination of increased aquaculture production, often under unsanitary conditions, and increased consumption of raw and lightly-processed freshwater fish and fishery products (WHO, 2002).

Fish and shellfish are processed using almost every known food preservation technique; however, the trend in most parts of the world is toward more fresh and less preserved products (Gram and Huss, 2000; Nilsson and Gram, 2002). Milder preservation strategies such as reduced use of NaCl and chemical additives in combination with refrigeration may represent new food systems with respect to health risks (Nilsson and Gram, 2002). However, if growth of psychrotrophic pathogens can be controlled these products will not result in more seafood-borne diseases. Raw fish like sushi is increasingly consumed in Europe, but this has not yet resulted in outbreaks caused by *Vibrio* spp. or to our knowledge other problems related to sushi consumption.

8.5 Sources of further information and advice

For a thorough discussion of microorganisms and their control in fish and fish product, the reader is referred to ICMSF (2005). In Huss *et al.* (2004) a thorough discussion of all the agents causing seafood-borne disease and their importance in fish products is also given. Furthermore, this book gives a detailed description of the requirements for the implementation of GMP and of the HACCP system and of the monitoring programmes to control biotoxins, microorganisms and chemical pollutants. The application of HACCP to various fish/seafood products is also described by Tzouros and Arvanitoyannis (2000).

Reviews of specific subjects like parasites and histamine formation in seafood can be found in Butt *et al.* (2004b) and Dalgaard *et al.* (2008), respectively. For valuable information on risk assessment of *V. parahaemolyticus*, *V. vulnificus* and *V. cholerae* in seafood the reader is referred to FDA (2001), FAO/WHO (2005b) and FAO/WHO (2005a), respectively. Furthermore Su and Liu (2007) have summarised information on *V. parahaemolyticus* and its role in seafood-related foodborne disease.

8.6 References

Alam M J, Tomochika K I, Miyoshi S I and Shinoda S (2002) Environmental investigation of potentially pathogenic *Vibrio parahaemolyticus* in the Seto-Inland Sea, Japan, *FEMS Microbiology Letters*, **208**(1), 83–7.

Andersen J K, Büchert A, Koch B, Ladefoged O, Leth T, Licht D and Ovesen L (2003), *Helhedssyn på fisk og fiskevarer*, Fødevaredirektoratet, Søborg.

Autio T, Hielm S, Miettinen M, Sjoberg A M, Aarnisalo K, Bjorkroth J, Mattila-Sandholm T and Korkeala H (1999) Sources of *Listeria monocytogenes* contamination in a cold-smoked rainbow trout processing plant detected by pulsed-field gel electrophoresis typing, *Applied and Environmental Microbiology*, **65**(1), 150–5.

Bean N H, Goulding J S, Daniels M T and Angulo F J (1997), Surveillance for foodborne disease outbreaks – United States, 1988–1992, *Journal of Food Protection*, **60**(10), 1265–86.

Bhaskar N and Sachindra N M (2006) Bacteria of public health significance associated with cultured tropical shrimp and related safety issues: A review, *Journal of Food Science and Technology-Mysore*, **43**(3), 228–38.

Bisharat N, Agmon V, Finkelstein R, Raz R, Ben-Dror G, Lerner L, Soboh S, Colodner R, Cameron D N, Wykstra D L, Swerdlow D L and Farmer J J (1999) Clinical, epidemiological, and microbiological features of *Vibrio vulnificus* biogroup 3 causing outbreaks of wound infection and bacteraemia in Israel, *Lancet*, **354**(9188), 1421–4.

Bondad-Reantaso M G, Subasinghe R P, Arthur J R, Ogawa K, Chinabut S, Adlard R, Tan Z and Shariff M (2005) Disease and health management in Asian aquaculture, *Veterinary Parasitology*, **132**(3–4), 249–72.

Brett M S, Short P and McLauchlin J (1998) A small outbreak of listeriosis associated with smoked mussels, *International Journal of Food Microbiology*, **43**(3), 223–9.

Butt A A, Aldridge K E and Sanders C V (2004a) Infections related to the ingestion of seafood Part I: viral and bacterial infections, *Lancet Infectious Diseases*, **4**(4), 201–12.

Butt A A, Aldridge K E and Sanders C V (2004b) Infections related to the ingestion of seafood. Part II: parasitic infections and food safety, *Lancet Infectious Diseases*, **4**(5), 294–300.

Cahill M M (1990) Bacterial-flora of fishes – a review, *Microbial Ecology*, **19**(1), 21–41.

CSPI (2006) *Outbreak Alert*, Washington, DC, Center for Science in the Public Interest, available at: http://www.cspinet.org/new/pdf/outbreakalert2005.pdf, accessed November 2008.

CSPI (2007) *Outbreak Alert Database*, Washington, DC, Center for Science in the Public Interest, available at: http://www.cspinet.org/foodsafety/outbreak/pathogen.php, accessed November 2008.

Dalgaard P (2006) Microbiology of marine muscle foods, in Y. Hui (ed.) *Handbook of Food Science, Technology and Engineering*, CRC Press, Boca Raton, FL, 1–20.

Dalgaard P, Emborg J, Kjølby j, Sørensen N D and Ballin N Z (2008) Histamine and biogenic amines – formation and importance in seafood, in T. Børresen (ed.) *Improving Seafood Products for the Consumer*, Woodhead, Cambridge, 292–324.

Dalsgaard A, Huss H H, Kittikun A and Larsen J L (1995) Prevalence of *Vibrio cholerae* and *Salmonella* in a major shrimp production area in Thailand, *International Journal of Food Microbiology*, **28**(1), 101–13.

Deepanjali A, Kumar H S, Karunasagar I and Karunasagar I (2005) Seasonal variation in abundance of total and pathogenic *Vibrio parahaemolyticus* bacteria in oysters along the southwest coast of India, *Applied and Environmental Microbiology*, **71**(7), 3575–80.

DePaola A, Ulaszek J, Kaysner C A, Tenge B J, Nordstrom J L, Wells J, Puhr N and Gendel S M (2003) Molecular, serological, and virulence characteristics of *Vibrio parahaemolyticus* isolated from environmental, food, and clinical sources in north America and Asia, *Applied and Environmental Microbiology*, **69**(7), 3999–4005.

Diaz J (2004) Is fish consumption safe?, *Journal of the Louisiana State Medical Society*, **156**, 42–9.

Dixon B and Flohr R (1997) Fish- and shellfish-borne trematode infections in Canada, *Southeast Asian Journal of Tropical Medicine and Public Health*, **28**, 58–64.

EC (2004) Regulation (EC) No. 853/2004 of the European Parliament and of the Council of 29 April 2004 laying down specific hygiene rules for the hygiene of foodstuffs, *Official Journal of the European Union*, **L139**, 30 April, 55–205 (see App III, Sec VIII).

EEC (1991) Council directive 91/493/EEC of 22th July 1991 laying down the health conditions for the production and the placing on the market off fishery products. *Official Journal of the European Union*, **L268**, 24 September, 15–34.

Eklund M W (1992) Control in fishery products, in A. H. W. Hauschild and K. L. Dodds (eds.) Clostridium botulinum: *Ecology and Control in Foods*, Marcel and Dekker, New York, 209–32.

Emborg J and Dalgaard P (2006) Formation of histamine and biogenic amines in cold-smoked tuna: An investigation of psychrotolerant bacteria from samples implicated in cases of histamine fish poisoning, *Journal of Food Protection*, **69**(4), 897–906.

Emborg J, Laursen B G and Dalgaard P (2005) Significant histamine formation in tuna (*Thunnus albacares*) at 2 degrees C – effect of vacuum- and modified atmosphere-packaging on psychrotolerant bacteria, *International Journal of Food Microbiology*, **101**(3), 263–79.

Emborg J, Dalgaard P and Ahrens P (2006) *Morganella psychrotolerans* sp nov., a histamine-producing bacterium isolated from various seafood, *International Journal of Systematic and Evolutionary Microbiology*, **56**, 2473–9.

Ericsson H, Eklow A, Danielsson T M, Loncarevic S, Mentzing L O, Persson I, Unnerstad H and Tham W (1997) An outbreak of listeriosis suspected to have been caused by rainbow trout, *Journal of Clinical Microbiology*, **35**, 2904–7.

FAO (2004) *The State of World Fisheries and Aquaculture*, Food and Agriculture Organization of the United Nations, Rome.

FAO/WHO (2004) *Risk assessment of* Listeria monocytogenes *in ready-to-eat foods – Interpretative summary*, Food and Agriculture Organization of the United Nations/World Health Organization, Rome/Geneva.

FAO/WHO (2005a) *Risk assessment of choleragenic* Vibrio cholerae *O1 and O139 in warm-water shrimp in international trade*, Food and Agriculture Organization of the United Nations/World Health Organization, Rome/Geneva.

FAO/WHO (2005b) *Risk assessment of* Vibrio vulnificus *in raw oysters*, Food and Agriculture Organization of the United Nations/World Health Organization, Rome/Geneva.

Fayer J J (2001) Waterborne and foodborne protozoa, in *Foodborne Disease Handbook*, 2nd edn, Y H Hui, S A Sattar, K D Murrell, W K Nip and P S Stanfield (eds.), Marcel Dekker, New York and Basel, 289–321.

FDA (2001) *Fish and Fishery Products Hazards and Controls Guide*, 3rd edn, Food and Drug Administration, Washington, DC.

FDA (2004) *Advice on Fish Consumption: Benefits and Risks*, Food Standards Agency, Scientific Advisory Committee on Nutrition, London.

FDA (2005) *Quantitative Risk Assessment on the Public Health Impact of Pathogenic* Vibrio parahaemolyticus *in Raw Oysters*, CFSAN, Food and Drug Administration, Washington, DC.

Feldhusen F (2000) The role of seafood in bacterial foodborne diseases, *Microbes and Infection*, **2**, 1651–60.

Fonnesbech Vogel B, Huss H H, Ojeniyi B, Ahrens P and Gram L (2001) Elucidation of *Listeria monocytogenes* contamination routes in cold-smoked salmon processing plants detected by DNA-based typing methods, *Applied and Environmental Microbiology*, **67**(6), 2586–95.

Frances N, Hornby H and Hunter P R (1991) The isolation of *Listeria* species from fresh-water sites in Cheshire and North-Wales, *Epidemiology and Infection*, **107**(1), 235–8.

FSIS/FDA (2003) *Interpretative Summary: Quantitative assessment of the relative risk to public health from foodborne* Listeria monocytogenes *among selected categories of ready-to-eat foods*, Food Safety and Inspection Services/Food and Drug Administration, Washington, DC.

George S M, Lund B M and Brocklehurst T F (1988) The effect of pH and temperature on initiation of growth of *Listeria monocytogenes*, *Letters in Applied Microbiology*, **6**, 153–6.

Gooch J A, DePaola A, Kaysner C A and Marshall D L (2001) Evaluation of two direct plating methods using nonradioactive probes for enumeration, of *Vibrio parahaemolyticus* in oysters, *Applied and Environmental Microbiology*, **67**(2), 721–4.

Gram L (2001) Potential hazards in cold-smoked fish: *Clostridium botulinum* type E, *Journal of Food Science*, **66** (supplement), 1082–7.

Gram L and Huss H H (2000) Fresh and processed fish and shellfish, in B. M. Lund, T. C. Baird-Parker, and G. W. Gould (eds.) *The Microbiological Safety and Quality of Food*, Aspen, Gaithersburg, MD, 472–506.

Gram L and Ringø E (2005) Prospects of fish probiotics, in W. H. Holzapfel and P. J. Naughton (eds), *Microbial Ecology in Growing Animals*, Elsevier, Amsterdam 379–417.

Hansen C H, Vogel B F and Gram L (2006) Prevalence and survival of *Listeria monocytogenes* in Danish aquatic and fish processing environments, *Journal of Food Protection*, **69**(9), 2113–22.

Heinitz M L, Ruble R d, Wagner D E and Tatini S R (2000) Incidence of *Salmonella* in fish and seafood, *Journal of Food Protection*, **63**, 579–92.

Herreras M V, Aznar F J, Balbuena J A and Raga J A (2000) Anisakid larvae in the musculature of the Argentinean hake, *Merluccius hubbsi*, *Journal of Food Protection*, **63**(8), 1141–3.

Hudson J A, Mott S J and Penney N (1994) Growth of *Listeria monocytogenes*, *Aeromonas hydrophila*, and *Yersinia enterocolitica* on vacuum and saturated carbon dioxide controlled atmosphere-packaged sliced roast beef, *Journal of Food Protection*, **57**, 204–8.

Hughes C, Gillespie I A and O'Brien S J (2007) Foodborne transmission of infectious intestinal disease in England and Wales, 1992–2003, *Food Control*, **18**(7), 766–72.

Huss H H (1980) Distribution of *Clostridium botulinum*, *Applied and Environmental Microbiology*, **39**(4), 764–9.

Huss H H (1994) *Assurance of Seafood Quality*, FAO Fishery Technical Paper No. 334, Food and Agriculture Organization of the United Nations, Rome.

Huss H H (1997) Control of indigenous pathogenic bacteria in seafood, *Food Control*, **8**(2), 91–8.

Huss H H, Jorgensen L V and Vogel B F (2000) Control options for *Listeria monocytogenes* in seafoods, *International Journal of Food Microbiology*, **62**(3), 267–74.

Huss H H, Ababouch L and Gram L (2004) *Assessment and Management of Seafood Safety and Quality*, FAO Fisheries Technical Paper No. 444, Food and Agriculture Organization of the United Nations, Rome.

ICMSF (1998) Fish and fish products, in *Microorganisms in Foods 6*, International Commission on Microbiological Specifications for Foods, Blackie Academic and Professional, London, 130–89.

ICMSF (2005) Fish and fish products, in *Microorganisms in Foods 6*, 2nd edn, International Commission on Microbiological Specifications for Foods, Kluwer Academic/Plenum Publishers, New York, 174–249.

IOM (2006) *Seafood Choices: Balancing Benefits and Risks*, Institute of Medicine, Washington, DC.

Jemmi T and Keusch A (1994) Occurrence of *Listeria monocytogenes* in freshwater fish farms and fish-smoking plants, *Food Microbiology*, **11**, 309–16.

Johansson T, Rantala L, Palmu L and Honkanen-Buzalski T (1999) Occurrence and typing of *Listeria monocytogenes* strains in retail vacuum-packed fish products and in a production plant, *International Journal of Food Microbiology*, **47**(1–2), 111–19.

Kanki M, Yoda T, Ishibashi M and Tsukamoto T (2004) *Photobacterium phosphoreum* caused a histamine fish poisoning incident, *International Journal of Food Microbiology*, **92**(1), 79–87.

Kaper J B, Morris G and Levine M M (1995) Cholera (Vol 8, Pg 51, 1995), *Clinical Microbiology Reviews*, **8**(2), 316.

Karalis T, Gupta L, Chu M, Campbell B A, Capra M F and Maywood P A (2000) Three clusters of ciguatera poisoning: clinical manifestations and public health implications, *Medical Journal of Australia*, **172**(4), 160–2.

Lake R, Hudson A, and Cressey P (2003) *Risk Profile:* Vibrio parahaemolyticus *in Seafood*, Institute of Environmental Science and Research Limited, Wellington.

Lehane L and Olley J (2000) Histamine fish poisoning revisited, *International Journal of Food Microbiology*, **58**(1–2), 1–37.

Lipp E K and Rose J B (1997) The role of seafood in foodborne diseases in the United States of America, *Revue Scientifique et Technique de l Office International des Epizooties*, **16**(2), 620–40.

Liston J (1980) Microbiology in fishery science, in J Connell (ed.), *Advances in Fish Science and Technology*, Fishing News Book Ltd, London, 138–57.

Lund B M and Peck M W (2000) *Clostridium botulinum*, in B. M. Lund, T. C. Baird-Parker, and G. W. Gould (eds.), *The Microbiological Safety and Quality of Foods*, Aspen, Gaithersburg, MD, 1057–109.

MacGowan A P, Bowker K, McLauchlin J, Bennett P M and Reeves D S (1994) The occurrence and seasonal-changes in the isolation of *Listeria* spp in shop bought food stuffs, human feces, sewage and soil from urban sources, *International Journal of Food Microbiology*, **21**(4), 325–34.

Martinez-Urtaza J, Simental L, Velasco D, DePaola A, Ishibashi M, Nakaguchi Y, Nishibuchi M, Carrera-Flores D, Rey-Alvarez C and Pousa A (2005) Pandemic *Vibrio parahalemolyticus* O3 : K6, Europe, *Emerging Infectious Diseases*, **11**(8), 1319–20.

Matsumoto T, Nagashima Y, Kusuhara H, Sugiyama Y, Ishizaki S, Shimakura K and Shiomi K (2007) Involvement of carrier-mediated transport system in uptake of tetrodotoxin into liver tissue slices of puffer fish *Takifugu rubripes*, *Toxicon*, **50**(2), 173–9.

Merivirta L O, Lindstrom M, Bjokroth K J and Korkeala H J (2006) The prevalence of *Clostridium botulinum* in European river lamprey (*Lampetra fluviatilis*) in Finland, *International Journal of Food Microbiology*, **109**(3), 234–7.

Miettinen H and Wirtanen G (2005) Prevalence and location of *Listeria monocytogenes* in farmed rainbow trout, *International Journal of Food Microbiology*, **104**(2), 135–43.

Nilsson L and Gram L (2002) Improving the control of pathogens in fish products, in A. H. Bremner (ed.), *Safety and Quality Issues in Fish Processing*, Woodhead Cambridge, 54–84.

Nishibuchi M and Kaper J B (1995) Thermostable direct hemolysin gene of *Vibrio parahaemolyticus* – a virulence gene acquired by a marine bacterium, *Infection and Immunity*, **63**(6), 2093–9.

Olsen S J, Mackinnon L C, Goulding J S, Bean N H and Slutsker L (2000) Surveillance for food-borne-disease outbreaks United States, 1993–1997, *Morbidity and Mortality Weekly Report*, **49**, 1–64.

Olsen S J, Bishop R, Brenner F W, Roels T H, Bean N, Tauxe R V and Slutsker L (2001) The changing epidemiology of *Salmonella*: Trends in serotypes isolated from humans in the United States, 1987–1997, *Journal of Infectious Diseases*, **183**(5), 753–61.

OzFoodNet (2004) *Foodborne disease investigation across Australia: Annual report of the OzFoodNet network, 2003*, available at: http://www.health.gov.au/internet/main/Publishing.nsf/Content/cda-pubs-2004-cdi2803-pdf-cnt.htm/$FILE/cdi2803h.pdf, accessed November 2008.

Petran R L and Zottola E A (1989) A study of factors affecting growth and recovery of *Listeria monocytogenes* Scott A., *Journal of Food Science*, **54**, 458–60.

Ponce E, Khan A A, Cheng C M, Summage-West C and Cerniglia C E (2008) Prevalence and characterization of *Salmonella enterica* serovar Weltevreden from imported seafood, *Food Microbiology*, **25**(1), 29–35.

Rørvik L M, Caugant D A and Yndestad M (1995) Contamination pattern of *Listeria monocytogenes* and other *Listeria* spp. in a salmon slaughterhouse and smoked salmon processing plant, *International Journal of Food Microbiology*, **25**, 19–27.

Seeliger H P R and Jones D (1986) Listeria, in P H A Sneath, N S Mair, M E Sharpe and J G Holt (eds), *Bergey's manual of Systematic Bacteriology: Volume 2*, Williams and Wilkins, Baltimore, MD, 1235–45.

Shabarinath S, Kumar H S, Khushiramani R, Karunasagar I and Karunasagar I (2007) Detection and characterization of *Salmonella* associated with tropical seafood, *International Journal of Food Microbiology*, **114**(2), 227–33.

Sharma S, Sachdeva P and Virdi J S (2003) Emerging water-borne pathogens, *Appl. Microbiol. Biotechnol.*, **61**, 424–8.

Sørum H (1999) Antibiotic resistance in aquaculture, *Acta Veterinamia Scandinavica*, **92** (Supplement), 29–36.

Strom M S and Paranjpye R N (2000) Epidemiology and pathogenesis of *Vibrio vulnificus*, *Microbes and Infection*, **2**(2), 177–88.

Su Y C and Liu C C (2007) *Vibrio parahaemolyticus*: a concern of seafood safety, *Food Microbiology*, **24**(6), 549–58.

Thimothe J, Nightingale K K, Gall K L, Scott V N and Wiedmann M (2004) Tracking of *Listeria monocytogenes* in smoked fish processing plants, *Journal of Food Protection*, **67**, 328–41.

Tzouros N E and Arvanitoyannis I S (2000) Implementation of hazard analysis critical control point system to the fish/seafood industry, *Food Reviews International*, **16**, 273–325.

Wallace B J, Guzewich J J, Cambridge M, Altekruse S and Morse D L (1999) Seafood-associated disease outbreaks in New York, 1980–1994, *American Journal of Preventive Medicine*, **17**(1), 48–54.

WHO (2002) *Foodborne Diseases Emerging*, Fact sheet No. 124, World Health Organization, Geneva.

Wiedmann M, Bruce J L, Keating C, Johnson A E, McDonough P L and Batt C A (1997) Ribotypes and virulence gene polymorphisms suggest three distinct *Listeria monocytogenes* lineages with differences in pathogenic potential, *Infection and Immunity*, **65**(7), 2707–16.

Withers N W (1982) Ciguatera fish poisoning, *Annual Review of Medicine*, **33**, 97–111.

Wulff G, Gram L, Ahrens P and Vogel B F (2006) One group of genetically similar *Listeria monocytogenes* strains frequently dominate and persist in several fish slaughter- and smokehouses, *Applied and Environmental Microbiology*, **72**, 4313–22.

9

Pathogen control in primary production: bivalve shellfish

W. J. Doré, Marine Institute, Ireland

Abstract: Bivalve molluscan shellfish such as oysters, mussels, cockles and clams can accumulate and concentrate human pathogenic bacteria and viruses from their surrounding environment. Such shellfish can cause illness when consumed raw or lightly cooked. This chapter identifies the major pathogens responsible for illness associated with bivalve shellfish consumption. The current control measures employed to reduce these risks are highlighted and their effectiveness assessed.

Key words: bivalve shellfish, human viruses, vibrios, shellfish treatment, sewage contamination.

9.1 Introduction

Bivalve molluscan shellfish such as oysters, mussels, cockles and clams have been used as a food source by humans for thousands of years. Today, commercial exploitation of cultivated and fished bivalve shellfish is widely practised and represents a significant worldwide trade. Bivalve shellfish feed by filtering large volumes of water and accumulating food particles from their surrounding environment. During this process bivalves can also accumulate microorganisms present in the water column. When the growing environment contains human pathogens, shellfish can accumulate and concentrate those pathogens, and such shellfish represent a health risk when consumed raw or only lightly cooked (Rippey, 1994). A significant public health risk is associated with bivalve shellfish grown in faecally-contaminated environments, particularly those contaminated by the introduction of human sewage (Lees, 2000). Sewage contains a wide range of human pathogens including a number of viruses and bacteria which can survive both during the sewage treatment process and in marine environments. The risks associated with sewage-contaminated bivalve shellfish have been recognized for many years and have been documented in a number of large-scale illness outbreaks (Halliday *et al.*, 1991; Christensen *et al.*, 1998; Le Guyader *et al.*, 2006b). In addition, bivalve shellfish contaminated with naturally-occurring marine pathogens can also cause illness in consumers (Drake *et al.*, 2007). The ability of bivalve shellfish to concentrate microbiological contaminants through filter feeding and a widespread consumer preference for eating shellfish products raw or only lightly cooked are the fundamental reasons for the high level of illness associated with their consumption.

9.2 Identifying and assessing hazards and risk

Human pathogens found in bivalve shellfish may occur naturally in the marine environment or may be introduced as result of faecal contamination originating from human sewage. Faecal contamination of bivalve shellfish harvesting areas from animal waste also represents a potential hazard but is rarely implicated in illness.

9.2.1 Sewage contaminated bivalve shellfish

Human faecal pollution in the marine environment resulting in the microbiological contamination of bivalve molluscan shellfish has long been recognized as a significant threat to human health. Typhoid fever associated with the consumption of raw oysters contaminated with *Salmonella* Typhi was documented in the late 1800s. The first link to viral illness associated with sewage-contaminated shellfish was made relatively more recently (Roos, 1956). Today, viral illness, particularly infectious hepatitis caused by hepatitis A virus (HAV) and gastroenteritis caused by norovirus (NoV) are recognized as the most significant infectious risk associated with the consumption of sewage-contaminated shellfish (Lees, 2000). It is worth noting that the largest recorded outbreak of foodborne illness was associated with the consumption of clams in Shanghai in 1988 when almost 300 000 people contracted hepatitis A (Halliday *et al.*, 1991). Despite the extensive implementation of sanitary controls to prevent the health risk associated with sewage-contaminated bivalve shellfish worldwide, outbreaks of viral illness continue to occur throughout the world (Christensen *et al.*, 1998; Berg *et al.*, 2000; Doyle *et al.*, 2004; Kageyama *et al.*, 2004).

Viral illness associated with shellfish consumption is directly linked with sewage-contamination at source in the harvesting area. Several large-scale outbreaks have particularly been associated with shellfisheries contaminated with untreated sewage resulting from high rainfall events which cause sewage to bypass the treatment process (Doyle *et al.*, 2004) or other pollution events such as direct discharge of untreated sewage from boats (Berg *et al.*, 2000). In recent studies viral contamination in shellfish resulting from sewage overflows has been directly demonstrated (Keaveney *et al.*, 2006). Contamination of bivalve shellfish through the introduction of untreated sewage in to shellfisheries, particularly during epidemics of illness in the community when virus levels are elevated, represents a major risk factor for causing illness in consumers.

9.2.2 Principle viruses associated with shellfish consumption

The two most commonly identified viral illnesses associated with bivalve shellfish consumption are infectious hepatitis caused by HAV and gastroenteritis caused by NoVs (Lees, 2000).

HAV is classified within the enterovirus group of the Picornaviridae family. The virus is excreted in faeces of infected people and can produce clinical disease by person-to-person spread or via the faecal–oral route. Susceptible individuals consuming contaminated water or foods are also at risk. Cooked meats and sandwiches, fruits and fruit juices, milk and milk products, vegetables, salads, shellfish and iced drinks are commonly implicated in outbreaks. Water, shellfish and salads are the most frequent sources of food-associated infections (Carter, 2005). HAV infections generally cause a mild illness characterized by sudden onset of fever, malaise, nausea, anorexia and abdominal discomfort, followed several days later by jaundice. The infectious dose is unknown but is presumed to be 10–100 virus particles (Mele *et al.*, 1990). The incubation period for hepatitis A, which varies from 10 to 50 days (mean 30 days), is dependent upon the number of infectious particles consumed. Many infections with HAV do not result in clinical disease, especially in children. When disease does occur, it is usually mild and recovery is complete in 1–2 weeks. Occasionally, the symptoms are severe and convalescence can take several months. Patients suffer from feeling chronically tired during convalescence, and their inability to work can cause considerable financial loss. Death can occur but is very rare and usually associated with the elderly.

Norovirus is one of the four genera of the calicivirus that make up the Caliciviridae family. NoV causes acute gastroenteritis which is usually a mild self-limiting infection lasting 12–24 h. Symptoms often include diarrhoea, nausea, vomiting and abdominal pain; more rarely reported are fever and headache. Illness is caused by infection of the intestinal mucosa and destruction of absorptive cells at the top of the intestinal villi and follows an incubation period of 24–60 h. Some infections are more severe and hospitalization of elderly and debilitated individuals has been required on rare occasions. In countries where molecular and serological tests are used for diagnostic purposes NoVs have been identified as the major cause of epidemic gastroenteritis. This includes the UK, the Netherlands, Japan and the USA.

Transmission of NoVs is via the faecal-oral route, either directly or via contaminated food or water. The virus is highly infectious requiring a low infectious dose, maybe < 10 virus particles. Person-to-person transmission is common, especially in closed communities, and large outbreaks of NoV-induced gastroenteritis have been reported in hospitals, nursing homes, cruise ships and military settings. Infections in Northern Europe demonstrate a strong seasonal distribution with infections peaking in the winter months (November through to March) and the illness is often known as winter vomiting disease.

The immunological response to NoV infection is complex and still relatively poorly understood. Infection by NoVs has been shown to result in short-term immunity with little cross-protection between different strains. However, studies have demonstrated that pre-existing antibody levels in volunteers before exposure to NoVs do not correlate with resistance to infection. In fact, paradoxically, increased antibody levels appear to correlate with increased risk of infection. It has been hypothesized that detection of antibody levels indicates previous infection without necessarily conferring immunity and suggests the host is susceptible. Conversely it is proposed that absence of antibodies may be indicative of no past infection and a pre-disposed resistance which is not necessarily due to immune response. Recently it has been suggested that susceptibility and resistance may be associated with an individual's blood group. Individuals with an O antigen phenotype were more likely to be infected with NoV than individuals with a blood group B antigen (Hutson *et al.*, 2004). The recent use of immunological assays based on recombinant virus-like particles in epidemiological studies has demonstrated that the age of antibody acquisition and illness is much lower than previously believed. It appears that infection can occur early in life and repeated re-infection can occur throughout life in susceptible individuals.

9.2.3 Naturally-occurring pathogens in bivalve shellfish

As well as illness associated with the consumption of sewage-contaminated bivalve shellfish, illness can also result following the consumption of shellfish containing naturally occurring bacteria. *Vibrio vulnificus* and *V. parahaemolyticus* in particular are recognized as a significant public health problem associated with the consumption of raw shellfish.

Vibrios are Gram-negative bacteria which are associated with estuarine and coastal waters. Under ideal conditions the doubling time of vibrios can be as short as 10 min, but this will be considerably longer in seafood stored at ambient temperatures and in seawater. Higher seawater temperature is a major risk factor for the presence of *V. vulnificus* and *V. parahaemolyticus* in marine waters and subsequent uptake in by bivalve shellfish (Kaspar and Tamplin, 1993). *Vibrio vulnificus* and *V. parahaemolyticus* are halophilic bacteria with an absolute requirement for the presence of Na^+ for growth.

The infectious dose of human pathogenic marine vibrios may be high and vibrios are generally present at relatively low levels in harvested shellfish. Therefore post-harvest multiplication is usually required to reach levels which present a risk to consumers (Huss *et al.*, 2004). Temperature control post-harvest is therefore considered to be the major control for prevention of illness.

Illness associated with *V. vulnificus* infections can occur in individuals consuming contaminated oysters or through exposure of open wounds to seawater where *V. vulnificus* is present. Illness associated with the consumption of raw oysters usually presents as a primary septicaemia and is defined as systemic illness characterized by fever and shock with or without gastrointestinal symptoms. *Vibrio vulnificus* causes significant mortality in persons with pre-existing conditions including liver disease, diabetes mellitus and immunodeficiency. Infections have primarily been reported in the USA and more specifically associated with oysters harvested form the Gulf of Mexico (Lee *et al.*, 2008), although illness has also been demonstrated in other countries. Infections are associated with sporadic cases and do not manifest in outbreak situations.

Gastroenteritis in humans caused by *V. parahaemolyticus* infections is exclusively associated with the consumption of raw or inadequately cooked seafoods (Huss *et al.*, 2004). Symptoms are diarrhoea, vomiting, nausea, headache and fever. The incubation period ranges from 8–72 h with symptoms generally lasting for 48–72 h. The exact virulence mechanism is not known; however, the link between pathogenicity of *V. parahaemolyticus* strains and their ability to form thermostable direct haemolysin (TDH) or TDH-related haemolysin (TRH) is recognized. TDH-positive strains cause haemolysis of red blood cells, and this phenomenon is known as the Kanagawa reaction. The proportion of environmental strains which produce TDH has been reported to be low (Nishibuchi and Kaper, 1995). For example only 0.2–3.2 % of environmental isolates of *V. parahaemolyticus* in the USA contain the *tdh* gene (FDA, 2001). Therefore the presence of *V. parahaemolyticus* in bivalve shellfish does not in itself infer a pathogenic risk without further confirmation of the presence of pathogenicity markers.

9.2.4 Relative incidence of illness

The true incidence of viral illness associated with shellfish consumption is of course impossible to estimate. The most complete reviews of data on foodborne outbreaks of illness exist in the UK and the USA. In the USA between 1993 and 1997 shellfish, both molluscan and crustacean, were responsible for 1.7 % (number 47) of all reported foodborne outbreaks and resulted in 1868 cases of illness which accounted for 2.2 % of all foodborne illnesses (Olsen *et al.*, 2000). The aetiological agent was identified in approximately 50 % of these outbreaks. Further data over an eight-year period from 1990 to 1998, looking only at outbreaks in the USA where the aetiological agent responsible for illness was identified, indicated that bivalve molluscan shellfish were responsible for 66 outbreaks and 3281 cases of illness (CSPI, 2001). NoV was identified as the aetiological agent responsible for 15 outbreaks (23 %) and 2175 cases (66 %). During the same period *V. parahaemolyticus* was identified as the cause of 18 outbreaks (27 %) and 733 cases (22 %). In the UK between 1992 and 1999, 1425 outbreaks of infectious intestinal disease from all foodstuffs were recorded. Molluscan bivalve shellfish were confirmed as being responsible for 54 of these outbreaks (Gillespie *et al.*, 2001). Viruses were positively identified as being responsible for 21 outbreaks. However, it is worth noting that in 31 of the outbreaks associated with bivalve mollusc consumption the aetiological agent was not accounted for. It is probable that these outbreaks were viral in nature but that a lack of sensitive methods for detecting viruses in shellfish during this period failed to identify the causative agent.

The incidence of *V. parahaemolyticus* varies geographically and is related to warmer seawater temperatures. In Europe few reports of *V. parahaemolyticus* infection have been associated with shellfish consumption. However, there have been suggestions that such illness may be increasing (Lee *et al.*, 2008). In the USA several large outbreaks of gastroenteritis caused by *V. parahaemolyticus* infection associated with raw oyster consumption have been recorded (Drake *et al.*, 2007), and the virus constitutes a significant public health risk. In Japan *V. parahaemolyticus* infection associated with the consumption of raw seafoods including bivalve shellfish is a major health problem and outbreaks have been increasing since 1996. Significantly increased *V. parahaemolyticus* infections have been reported worldwide since the late 1990s, and this has been associated with the emergence of pandemic O3:K6 strains (Drake *et al.*, 2007).

Vibrio vulnificus infections usually present themselves as sporadic cases and are not associated with outbreak situations. Despite this, significant numbers of illness associated with oyster consumption are reported in the USA. Between 1988 and 2006, CDC received reports of more than 900 *V. vulnificus* infections from the Gulf Coast states, where most cases occur. Presently an average of 32 *V. vulnificus* culture confirmed primary septicaemia cases are reported to the CDC annually (Drake *et al.*, 2007). In 2007, infections caused by *V. vulnificus* and other *Vibrio* species became nationally notifiable in the USA.

9.3 Managing and controlling hazards and risks

The hazards associated with the consumption of bivalve shellfish have been recognized for a long time, and extensive regulatory controls aimed at preventing illness have been introduced around the world including Europe, the USA, Australia and Asia. While the details of regulatory requirements vary in different parts of the world, the underlying principles of control remain the same. In particular, to control the risk associated with sewage-contaminated bivalve shellfish, current regulations adopt the approach of assessing the level of contamination in harvesting areas by categorizing them. Depending on the classification of the harvesting area, a minimum level of post-harvest treatment is required before consumption. During this process, the harvest area risk assessment is generally based on a sanitary survey of the harvesting area and routine bacterial monitoring, usually *E. coli* or faecal coliforms, of either the shellfish flesh or overlying water.

In Europe shellfish harvesting areas are classified as category A, B or C based on *E. coli* levels found in shellfish flesh (EC, 2004) and in the USA areas are classified as Approved, Conditionally Approved, Restricted or Conditionally Restricted on the basis of total or faecal coliform testing in water (FDA, 2005). In both systems each level of classification requires a minimum level of post-harvest treatment in order for shellfish to meet end-product standards and be considered safe for consumers. In some countries, for example New Zealand, production areas are closed for harvesting on the basis of heavy rainfall. Production areas are reopened after a short period and bacterial levels in shellfish returning to normal. This step is a useful additional control, but generally such closures are not sufficiently long to allow viruses to be eliminated if contamination occurs during the pollution event. As the presence of *V. vulnificus* and *V. parahaemolyticus* in the marine environment is not related to faecal contamination, a sanitary assessment of harvesting areas does not provide information on the risk of illness associated with these pathogens (Tamplin *et al.*, 1982).

A number of post-harvest options are practised to treat shellfish to make them fit for human consumption. The preferred option for treatment depends on the extent of the initial contamination, consumer preferences in the intended market and logistical considerations.

9.3.1 Temperature control

Both *V. parahaemolyticus* and *V. vulnificus* levels in freshly harvested bivalve shellfish are generally present in concentrations below the infectious dose required to cause illness. Further multiplication is required post-harvest to present a risk to consumers (Huss *et al.*, 2004). Therefore temperature control of bivalve shellfish between harvesting and consumption is a major control measure to prevent illness. Refrigeration controls the multiplication of *V. parahaemolyticus* and *V. vulnificus* in oysters. For example, *V. vulnificus* has been demonstrated as incapable of multiplication in oysters stored below 13 °C (Cook and Ruple, 1992). However, if temperature of harvested oysters is not immediately controlled, growth of vibrios can occur rapidly. For example, levels of *V. vulnificus* in freshly harvested oysters have been demonstrated to increase by almost 2 log units in 14 h if stored at ambient temperatures (Cook, 1997). Therefore, in the USA stringent storage conditions are employed for oysters harvested from at-risk harvesting areas. These conditions are determined on a risk assessment basis and vary depending on the water temperature where oysters were harvested (Drake *et al.*, 2007). Freezing has been shown to significantly reduce the level of *V. parahaemolyticus* and *V. vulnificus*. Cook and Ruple (1992) demonstrated a 4–5 $\log_{10}$ reduction of *V. vulnificus* in oysters stored at –40 °C for three weeks. A combination of freezing and storage has been approved for use a suitable control measure in the USA (Drake *et al.*, 2007).

9.3.2 Heat treatment and cooking

Commercial heat treatment of bivalve shellfish is applied to a number of shellfish species which are not consumed raw or lightly cooked. Heat treatment studies in the UK have demonstrated that heating shellfish to 90 °C for 90 seconds completely inactivates viral pathogens present in bivalve shellfish (Appleton, 1994). This cooking regime forms the basis for approved commercial heat treatment processes in European regulations (EC, 2004) and is used extensively to treat bivalve shellfish such as cockles. There have been no recorded incidents of illness associated with shellfish where this process has been correctly applied.

In the UK, fewer outbreaks of illness are associated with bivalve shellfish traditionally cooked after sale, such as mussels, compared to those more often consumed raw, such as oysters. This indicates that protection is also provided by home and restaurant cooking. However, reports in the UK have shown occasional outbreaks of NoV-associated gastroenteritis associated with the consumption of mussels which are traditionally consumed lightly cooked (CDSC, 1993, 1998). Cases of gastroenteritis have been associated with shellfish which had been grilled, fried, steamed or stewed (Mcdonnell *et al.*, 1995). Home or restaurant cooking cannot be considered to provide complete consumer protection against the risk of viral infection.

Most cooking procedures will destroy *V. parahaemolyticus* and *V. vulnificus* and constitute an effective control where practised. Commercial heat-shock treatments are also effective at controlling the risks associated with vibrios, and application of these is practised in the USA. The process involves submerging chilled oysters in water at a temperature of 67 °C for approximately 5 min. This process has been found to reduce levels of *V. vulnificus* by 2–4 $\log_{10}$ in oysters.

9.3.3 Depuration

Depuration or controlled purification is the process of reducing levels of microbiological contaminants in live bivalve shellfish (Richards, 1988). Shellfish are placed in a controlled clean water environment, generally onshore seawater tanks, where they continue to filter-feed and purge themselves of contaminants. Depuration has been practised for over a century in both the USA and Europe to produce shellfish free from sewage-derived bacteria.

Depuration has been shown to be effective at eliminating sewage-derived bacteria from even highly-contaminated shellfish. For example, studies have demonstrated that levels of *E. coli* in excess of 18 000 MPN $100g^{-1}$ can be eliminated during depuration within 24–48 hours (Doré and Lees, 1995). In contrast, viral elimination during depuration is known to be relatively inefficient and numerous outbreaks of viral illness have been associated with the consumption of depurated shellfish shown to be free of bacterial contaminants (Murphy and Grohmann, 1980; Chalmers and McMillan, 1995; Perret and Kudesia, 1995). Laboratory studies have demonstrated extended virus persistence over sewage-derived bacteria during depuration (Power and Collins, 1989; Doré and Lees, 1995; Muniain-Mujika *et al.*, 2002; Henshilwood *et al.*, 2003). Although viruses are not completely eliminated during depuration they are reduced, and the widespread application undoubtedly plays a role in reducing the incidence of viral illness associated with shellfish consumption.

The mechanism for viral persistence in shellfish remains unclear. Early studies suggested that because of their size viruses are readily sequestered into tissues outside of the shellfish alimentary tract and were therefore refractory to the elimination process during depuration. However, subsequent studies have demonstrated that the major site of viral retention following depuration is the digestive gland (Doré and Lees, 1995). More recent work has suggested that NoVs and maybe other human viruses bind to specific receptors present in oyster tissues (Le Guyader *et al.*, 2006a). Temperature has been shown to have a major effect on the rate of virus depuration in shellfish. In general, increase in temperature has been shown to increase the rate of virus depuration during laboratory investigations (Power and Collins, 1989; Doré *et al.*, 1998; Henshilwood *et al.*, 2003). Increased removal of viruses at higher temperatures could be due to either increased elimination of viruses because of higher rates of filter feeding by shellfish or greater inactivation of viruses due to the increased metabolic activity of the shellfish.

Marine vibrios are not readily removed during depuration (Lee *et al.*, 2008). A number of studies have demonstrated that no significant reduction of *V. parahaemolyticus* and *V. vulnificus* occurs during depuration for periods of 24–48 hours as commonly practised (Drake *et al.*, 2007). The mechanism for persistence of vibrios in shellfish is unclear. However, their resistance to the activity of bivalve haemolymph compared with other bacteria has been proposed as a key factor (Pruzzo *et al.*, 2005). In fact, studies have demonstrated that levels of vibrio in bivalve shellfish actually increased when depuration was conducted at elevated temperatures because vibrios were able to multiply in shellfish tissues at those temperatures (Tamplin and Capers, 1992). Depuration cannot be considered a suitable control measure for reducing the risk of vibrio infections associated with bivalve shellfish harvested from at-risk areas.

9.3.4 Relaying

During relaying, contaminated shellfish are moved to microbiologically clean environments to remove bacterial and viral contaminants in the natural marine setting. It is an alternative to depuration whereby shellfish can be purified for longer than during depuration but in a less controlled manner. As shellfish can be held in the natural environment for longer periods than in tanks, this is also a suitable method for treating more heavily-polluted shellfish. Typically relaying should be carried out for a period of at least one month.

Although relaying can be an effective treatment to remove sewage-derived bacteria, limited data are available on virus removal during relaying. One study has indicated that NoVs may be eliminated following relaying for 4–6 weeks combined with subsequent depuration (Doré *et al.*, 1998). However, virus persistence was demonstrated in excess of this period in low temperatures. Water temperature and quality appear to be critical parameters in determining the effectiveness of virus removal.

9.3.5 High hydrostatic pressure treatment

High hydrostatic pressure (HHP) processing has emerged in recent years as promising technology for pasteurizing food products without the application of heat. HHP is currently used commercially in the USA by oyster processors, principally because it facilitates the shell shucking (opening) process and in order to extend the shelf-life of products due to the reduction of spoilage bacteria (He *et al.*, 2002). Experimental work has also demonstrated that *V. parahaemolyticus* and *V. vulnificus* levels in oysters are reduced by the application of HHP technology. Recent studies have demonstrated that *V. parahaemolyticus* and *V. vulnificus* are susceptible to HHP treatment at pressures between 200 and 300 MPa (He *et al.*, 2002).

The potential of using HHP treatment to reduce virus levels in bivalve shellfish has also been investigated. It has been reported that HAV levels in oysters can be significantly reduced by HHP treatment. In one study, reductions of greater than 3 $\log_{10}$ HAV were recorded at 400 MPa for one minute (Calci *et al.*, 2005). More recently, using HHP treatment at just 300 MPa for 600 seconds produced a > 1 $\log_{10}$ reduction in HAV levels (Grove *et al.*, 2008). Similarly, using murine NoV as a surrogate for human NoV, workers have demonstrated that NoVs in laboratory media can be significantly reduced by HHP (Kingsley *et al.*, 2007). Additional studies combined HHP treatment at 450 MPa with heat treatment at 75 °C for two minutes at ambient temperature and were able to demonstrate a 7 $\log_{10}$ reduction of feline calicivirus in culture media (Buckow *et al.*, 2008). The same workers have developed a predictive model for inactivation of feline calicivirus by heat and HHP treatment. Further work is required to demonstrate similar effects in shellfish.

HHP represents a promising technology for post-harvest treatment to reduce levels of both pathogenic viruses and bacteria in bivalve shellfish post-harvest. However, concerns have been raised about the effect on the organoleptic quality of oysters treated with higher-pressure treatments which may be required for viruses, and this requires further investigation. As yet HHP processing is not recognized as an approved treatment for contaminated bivalve shellfish in European regulations.

In general, despite the high level of regulatory controls implemented worldwide to prevent infectious illness associated with shellfish consumption, viral- and vibrio-related illness continue to be associated with the consumption of raw shellfish and represent a significant public health problem. This continued incidence of viral and bacterial illness would indicate that there is a systemic and underlying failure of the controls applied to these hazards rather than the application of poor handling and management procedures during production and processing.

9.4 Future trends

There is a clear recognition that worldwide the current regulatory controls for preventing viral illness associated with sewage-contaminated bivalve shellfish do not fully protect public health and require improvement. The use of bacterial monitoring of harvesting areas and treated shellfish to assess the virus risk is inappropriate. Historically, bacterial indicator organisms have been used because it has not been possible to routinely detect NoVs and HAV in shellfish. In recent years, significant advances in molecular detection techniques for viruses in shellfish have been made and a number of usable methods have become available. In particular real-time polymerase chain reaction (PCR) procedures that allow virus levels to be quantified have been developed (Jothikumar *et al.*, 2005; Costafreda *et al.*, 2006). This offers, for the first time, the prospect of virus detection methods being employed to determine the sanitary quality of bivalve molluscan shellfish. The methods also provide the opportunity for virus standards for shellfish to be determined in an international setting. However, there is currently a wide range of methods being used in laboratories throughout the world. Therefore there is a fundamental requirement for standardizsation of virus detection methods for shellfish. In Europe, significant progress is being made to validate and standardize real-time PCR methods for the detection of NoV and HAV in shellfish, as well as other foods, through the Committee for European Standardization (CEN). These methods will be submitted as standard methods to the International Standards Organization (ISO). Standardized virus detection methods will provide an important tool to determine the sanitary quality of shellfish production areas and the effectiveness of subsequent treatment methods.

The recently developed virus detection methods have already been applied in field studies providing an increasing insight into environmental conditions leading to viral contamination of shellfish and subsequent persistence during treatment (Pommepuy *et al.*, 2008). As a result there is a growing recognition that proactive real-time management of virus risk in shellfisheries may provide a more productive control than the existing reactive monitoring programmes currently in place. This approach will rely on site-specific early identification or prediction of high-risk periods such as sewage overflows due to high rainfall during community epidemics of viral illness and the timely implementation of appropriate control or treatment measures. Critically the real-time PCR procedures will allow virus levels in shellfish to be monitored to determine the effectiveness of the additional controls and treatments.

The number of *V. parahaemolyticus* infections has been rising worldwide since the late 1990s due to the emergence of pandemic O3:K6 strains. This trend may increase further as a result of global warming. Global warming is now an accepted fact and the recent figures indicate that the earth's temperatures may rise by as much 6.4 °C by the year 2100 (IPCC, 2007). This has potential implications for the incidence of food poisoning in general and for incidents associated with bivalve shellfish consumption in particular. Already, increasing seawater temperatures have been linked to an outbreak of gastroenteritis caused by *V. parahaemolyticus* associated with the consumption of Alaskan raw oysters in waters previously considered too cold to support the growth of this organism (McLaughlin *et al.*, 2005). Increasing temperatures also pose difficulties for controlling product temperature at all stages of the supply chain. Temperature control is critical in preventing the proliferation of pathogenic vibrios. Global warming is also linked to increasing storm events. Storm events are in turn linked to the overflow of untreated sewage into aquatic environments where sewage treatment plant infrastructure cannot cope with surges in influent volumes. Such events can lead to increased viral contamination of shellfisheries and subsequent outbreaks associated with shellfish consumption.

In recent years there has been a significant increase in the international trade of fishery products including bivalve shellfish. This is due in part to a general global rise in consumer income, changing demographics and shifting preference in the light of links between health and diet (Jensen, 2006). This globalization of trade in bivalve shellfish may present a significant increase in the risks associated with the consumption of bivalve shellfish. Increasing international trade in shellfish often involves the importation of shellfish into more developed countries with sophisticated food safety controls from countries with less well-developed food safety programmes. To control these risks extensive regulations governing the import of shellfish already exist in many countries. However, these controls may be placed under increasing pressure as consumer demand for shellfish and other fishery products develops. Clearly the threats to public health resulting from increased shellfish trade and global warming require careful consideration, and developing a robust understanding of the processes involved in shellfish-borne illnesses will be important in combating the risks.

9.5 Sources of further information and advice

The following websites provide up-to-date information related to shellfish safety.

- European Community Reference Laboratory for monitoring bacteriological and viral contamination of bivalve shellfish: http://crlcefas.org.
- US Food and Drug Administration, National Shellfish Sanitation Program Guide for the Control of Molluscan Shellfish: http://vm.cfsan.fda.gov/~ear/nss2-toc.html, accessed November 2008.

The following articles provide full reviews in their respective areas of shellfish safety:

Lees DN (2000) Viruses and bivalve shellfish. *International Journal of Food Microbiology* **59**, 81–116.

Drake SL, DePaola A and Jaykus LA (2007) An overview of the *Vibrio vulnificus* and *Vibrio parahaemolyticus*, Comprehensive Reviews in Food Science and Food Safety, **6**, 120–44.

Huss HH, Ababouch L and Gram L (2004) *Assessment and management of seafood safety and quality*, FAO Fisheries Technical Paper 444, Food and Agriculture Organization of the United Nations, Rome.

9.6 References

Appleton H (1994) Norwalk virus and the small round viruses causing foodborne gastroenteritis, in Y. Hui, J. Gorham, K. Murrell and D.O. Cliver (eds), *Foodborne Disease Handbook*, Marcell Decker, New York, 77–97.

Berg DE, Kohn MA, Farley TA and McFarland LM (2000) Multi-state outbreaks of acute gastroenteritis traced to fecal-contaminated oysters harvested in Louisiana, *Journal of Infectious Diseases*, **181** (Supplement 2), S381–S386.

Buckow R, Isbarn S, Knorr D, Heinz V and Lehmacher A (2008) Predictive model for inactivation of feline calicvirus, a norovirus surrogate, by heat and high hydrostatic pressure, *Applied and Environmental Microbiology*, **74**, 1030–8.

Calci KR, Meade GK, Tezloff RC and Kingsley DH (2005) High-pressure inactivation of hepatitis A virus within oysters, *Applied and Environmental Microbiology*, **71**, 339–43.

Carter MJ (2005) Enterically infecting viruses: pathogenicity, transmission and significance for food and waterborne infection, *Journal of Applied Microbiology*, **98**, 1354–80.

CDSC (1993) Shellfish-associated illness, *Communicable Disease Report Weekly*, **3**, 29–31.

CDSC (1998) Outbreaks of gastroenteritis in England and Wales associated with shellfish: 1996 and 1997, *Communicable Disease Report Weekly*, **8**, 21–4.

Chalmers JW and McMillan JH (1995) An outbreak of viral gastroenteritis associated with adequately prepared oysters, *Epidemiology and Infection*, **115**, 163–7.

Christensen BF, Lees D, Wood KH, Bjergskov T and Green J (1998) Human enteric viruses in oysters causing a large outbreak of human food borne infection in 1996/97, *Journal of Shellfish Research*, **17**, 1633–5.

Cook D (1997) Refrigeration of oyster shellstock: conditions which minimize the outgrowth of *Vibrio vulnificus*, *Journal of Food Protection*, **60**, 349–52.

Cook D and Ruple A (1992) Cold storage and mild heat treatment as processing aids to reduce the number of *Vibrio vulnificus* in raw oysters, *Journal of Food Protection*, **55**, 985–9.

Costafreda MI, Bosch A and Pinto RM (2006) Development, evaluation, and standardization of a real-time TaqMan reverse transcription–PCR Assay for quantification of hepatitis A virus in clinical and shellfish samples, *Applied and Environmental Microbiology*, **72**, 3846–55.

CSPI (2001) *Outbreak Alert. Closing the Gaps in our Federal Food-Safety Net*, Centre for Science in the Public Interest, Washington, DC.

Doré WJ, Henshilwood K and Lees DN (1998) The development of management strategies for control of virological quality in oysters, *Water Science and Technology*, **38**, 29–35.

Doré WJ and Lees DN (1995) Behaviour of *Escherichia coli* and male-specific bacteriophage in environmentally contaminated bivalve molluscs before and after depuration, *Applied and Environmental Microbiology*, **61**, 2830–4.

Doyle A, Barataud D, Thiolet JM, Le Guyager S, Kohli E and Vaillant V (2004) Norovirus foodbourne outbreaks associated with the consumption of oysters from the Etang de Thau, France, December 2002, *Eurosurveillance*, **9**, 24–6.

Drake SL, DePaola A and Jaykus LA (2007) An overview of the *Vibrio vulnificus* and *Vibrio parahaemolyticus*, Comprehensive reviews in food science and food safety, **6**, 120–44.

EC (2004) Regulation (EC) No 854/2004 of the European parliament and of the council of 29 April 2004 laying down specific rules for the organisation of official controls on products of animal origin intended for human consumption, *Official Journal of the European Union*, **L139**, 30 April, 83–127.

FDA (2001) Draft risk assessment on the public health impact of *Vibrio parahaemolyticus* in raw molluscan shellfish, US Food and Drug Administration, Center for Food Safety and Nutrition, Washington, DC.

FDA (2005) *National Shellfish Sanitation, Manual of Operations*, 2005 revision, US Food and Drug Administration, Center for Food Safety and Applied Nutrition, Washington, DC.

Gillespie IA, Adak GK, O'Brien SJ, Brett MM and Bolton FJ (2001) General outbreaks of infectious intestinal disease associated with fish and shellfish, England and Wales, 1992–1999, *Communicable Disease and Public Health*, **4**, 117–23.

Grove SF, Forsyth S, Wan JJ, Coventry Cole M, Stewart CM, Lewis T, Ross T and Lee A (2008) Inactivation of hepatitis A virus, poliovirus and a norovirus surrogate by high pressure processing, *Innovative Food Science and Emerging Technologies*, **9**(2), 206–10.

Halliday M, Kang LY, Zhou TK, Hu, MD, Pan QC, Fu TY, Huang YS and Hu SL (1991) An epidemic of hepatitis A attributable to the ingestion of raw clams in Shanghai, China, *Journal of Infectious Diseases*, **164**, 852–9.

He H, Adams R, Farkas D and Morrissey M (2002) Use of high-pressure processing for oyster shucking and shelf-life extension, *Journal of Food Science*, **67**, 640–3.

Henshilwood K, Doré WJ, Anderson A and Lees DN (2003) The development of a quantitative assay for the detection of Norwalk like virus and its application to depuration, *Proceeding of the 4th International conference on molluscan shellfish safety*, June 4–8, Santiago de Compestella, Spain.

Huss HH, Ababouch L and Gram L (2004) *Assessment and Management of Seafood Safety and Quality*, FAO Fisheries Technical Paper 444, Food and Agriculture Organization, Rome.

Hutson AM, Atmar RL and Estes MK (2004) Norovirus disease: changing epidemiology and host susceptibiluty factors, *Trends in Microbiology*, **12**, 279–87.

IPCC (2007) *Climate change 2007: the physical basis – summary for policy makers*, Intergovernmental panel on climate change, Geneva.

Jensen HH (2006) Changes in seafood consumer preference patterns and changes in risk exposure, *Marine Pollution Bulletin*, **53**, 591–8.

Jothikumar N, Lowther JA, Henshilwood K, Lees DN, Hill VR and Vinje J (2005) Rapid and sensitive detection of noroviruses by using TaqMan-based one-step reverse transcriptase-PCR assays and application to naturally contaminated shellfish samples, *Applied and Environmental Microbiology*, **71**, 1870–5.

Kageyama T, Shinohara M, Uchida K, Fukushi S, Hoshino FB, Kojima S, Takai R, Oka T, Takeda N and Katayama K (2004) Coexistence of multiple genotypes, including newly identified genotypes, in outbreaks of gastroenteritis due to norovirus in Japan, *Journal of Clinical Microbiology*, **42**, 2988–95.

Kaspar CW and Tamplin ML (1993) Effects of temperature and salinity on the survival of *Vibrio vulnificus* in seawater and shellfish, *Applied and Environmental Microbiology*, **59**, 2425–9.

Keaveney S, Guilfoyle F, Flannery J and W, Dore D (2006) Detection of human viruses in shellfish and an update on REDRISK, *6th Irish Shellfish Safety Workshop*, **23**, 69–76.

Kingsley DH, Holliman D, Calci KR, Chen H and Flick, G (2007) Inactivation of a norovirus by high-pressure processing, *Applied and Environmental Microbiology*, **73**, 581–5.

Le Guyader FS, Loisy F, Atmar RL, Hutson AM, Estes MK, Ruvoen-Clouet N, Pommepuy M and Le Pendu J (2006a) Norwalk virus-specific binding to oyster digestive tissues, *Emerging Infectious Diseases*, **12**, 931–6.

Le Guyader S, Bon F, DeMedici D, Parnaudeau S, Bertone A, Crudeli S, Doyle A, Zidane M, Suffredini E, Kohli E, Maddalo F, Monini M, Gallay A, Pommepuy M, Pothier P and Ruggeri F (2006b) Detection of multiple noroviruses associated with an international gastroenteritis outbreak linked to oyster consumption, *Journal Clinical Microbiology*, **44**, 3878–82.

Lee RJ, Rangdale RE, Croci L and Hervio-Heath D (2008) Bacterial pathogens in seafood, in T. Borresen (ed.), *Improving Seafood Products for the Consumer*, Woodhead, Cambridge, 247–91.

Lees DN (2000) Viruses and bivalve shellfish, *International Journal of Food Microbiology*, **59**, 81–116.

Mcdonnell R, Wall P, Adak GK, Evans H, Cowden J and Caul E (1995) Outbreaks of infectious intestinal disease associated with person to person spread in hospitals and restaurants, *Communicable Disease Report*, **5**, 150–2.

McLaughlin J, DePaolo A, Bopp C, Martinek K, Napolilli N, Allison C, Murray S, Thompson E, Bird M and Middaugh J (2005) Outbreak of *Vibrio parahaemolyticus* gastroenteritis associated with Alaskan oysters, *New England Journal of Medicine*, **353**, 1463–70.

Mele A, Rastelli MG, Gill ON, Di Bissceglie D, Rosmini F, Pardelli G, Valtriani, C and Patriarchi P (1990) Recurrent epidemic hepatitis A associated with consumption of raw shellfish probably controlled through public health measures, *American Journal of Epidemiology*, **130**, 540–6.

Muniain-Mujika I, Girones R, Tofiño-Quesada G, Calvo M and Lucena F (2002) Depuration dynamics of viruses in shellfish, *International Journal of Food Microbiology*, **77**, 125–33.

Murphy A and Grohmann G (1980) Epidemiological studies on oyster-associated gastroenteritis, *Food Technology Australia*, **32**, 86–8.

Nishibuchi M and Kaper JB (1995) Thermostable direct haemolysin gene of *Vibrio parahaemolyticus*: A virulence gene aquired by marine bacterium, *Infection and Immunity*, **63**, 2093–9.

Olsen SJ, MacKinnon LC, Goulding JS, Bean NH and Slutsker L (2000) Surveillance for foodborne-disease outbreaks – United States, 1993–1997, *Morbidity and Mortality Weekly Report*, **49**, 1–59.

Perret K and Kudesia G (1995) Gastroenteritis associated with oysters. *Communicable Disease Report Reviews*, **5**, 153–4.

Pommepuy M, Le Guyader F, Le Saux J, Guilfoyle F, Doré B, Kershaw S, Lees D, Lowther JA, OC M, Romalde JL, Vilarino M, Furones D and Roque A (2008) Reducing microbial risk associated with shellfish in European countries, in T. Borresen (ed.), *Improving Seafood Products for the Consumer*, Woodhead, Cambridge, 212–41.

Power UF and Collins JK (1989) Differential depuration of poliovirus, *Escherichia coli*, and a coliphage by the common mussel, *Mytilus edulis, Applied and Environmental Microbiology*, **55**, 1386–90.

Pruzzo C, Gallo G and Canesi L (2005) Persistence of vibrios in marine bivalves: the role of interactions with haemolymph components, *Environmental Microbiology*, **7**, 761–72.

Richards GP (1988) Microbial purification of shellfish a review of depuration and relaying, *Journal of Food Protection*, **51**, 218–51.

Rippey SR (1994) Infectious diseases associated with molluscan shellfish consumption, *Clinical Microbiology Reviews*, **7**, 419–25.

Roos B (1956) Hepatitis epidemic conveyed by oysters, *Svenska Lakartidningen*, **53**, 989–1003.

Tamplin ML and Capers G (1992) Persistence of *Vibrio vulnificus* in tissues of gulf coast oysters, *Crassostrea virginica*, exposed to seawater disinfected with UV light, *Applied and Environmental Microbiology*, **58**, 1506–10.

Tamplin ML, GE, R, Blake NL and Cuba T (1982) Isolation and characterisation of *Vibrio vulnificus* from two Florida estuaries, *Applied and Environmental Microbiology*, **44**, 1466–70.

10

Hygienic plant design

J. Holah, Campden BRI, UK

Abstract: The hygienic design of the food processing environment is perhaps one of the least understood of the food manufacturing disciplines, yet its significance is perhaps the most important. By designing a series of barriers surrounding manufacturing areas of increasing hygienic importance, commensurate with the risk of the food product, it is possible to limit the challenge of hazards, particularly food pathogens, into such areas and thus enhance the control of food safety and spoilage. Attention to the design of floors, drains, walls, ceilings and services within the manufacturing areas limits the harbourage of hazards and facilitates their removal via cleaning and maintenance activities.

Key words: hygienic design, segregation, low/high risk areas, air hygiene, pest control.

10.1 Introduction

The schematic diagram shown in Fig. 10.1, which is typical for all food factories, shows that the production of safe, wholesome foods stems from a thorough hazard analysis – indeed this is now a legal requirement in the EU (EC, 2004a). The diagram also shows that given specified raw materials, there are four major 'building blocks' that govern the way the factory is operated to ensure that the goal of safe, wholesome food is realised. Hygienic design dictates the design of the factory infrastructure and includes the factory site and buildings, the process lines and equipment and, until replaced by robots, the operatives! Hygienic practices maintain the integrity of the facility and include cleaning and disinfection, maintenance, personal hygiene and other good manufacturing practices (GMP). Process development enables the design of safe, validated processes whilst process control subsequently

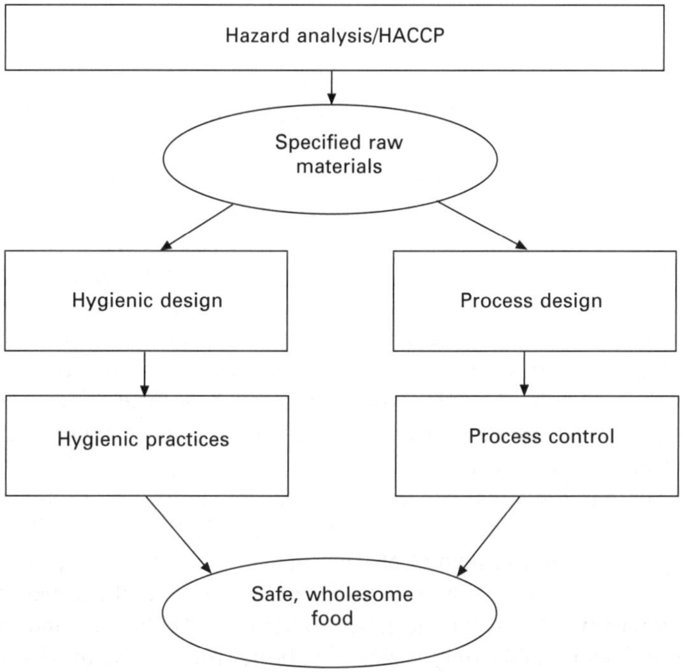

Fig. 10.1 Schematic stages required to ensure safe, wholesome food products.

ensures that each product in each batch on every day meets the process requirements.

Hazard analysis encompasses identifying the hazards that may affect the quality or safety of the food product and controlling them at all stages of the process such that product contamination is minimised. In the food industry this is commonly referred to as Hazard Analysis Critical Control Point (HACCP).

Such hazards are usually described as:

- biological, e.g. bacteria, yeasts, moulds;
- chemical, e.g. cleaning chemicals, lubricating fluids;
- physical, e.g. glass, insects, pests, metal, dust.

A hazard analysis should be undertaken at the earliest opportunity in the process of food production and, if possible, before the design and construction of the processing facility. This allows the design of the production facility to play a major role in hazard elimination or risk reduction. For example, it is possible to identify that glass is a potential hazard and one could eliminate this hazard by designing a glass-free factory. Whilst it may be impossible to design a factory that would be entirely pathogen-free, it is possible to consider how pathogen contamination could be significantly reduced. Such a hazard analysis could also be extended to cover other situations in which reducing product contamination from hazards is critical, for example the siting of a process line or the choice of existing factories.

With respect to the hygienic design building block, this chapter is concerned with the hygienic design of the food manufacturing facility. The hygienic design of the processing equipment is addressed in a separate chapter. With regard to hygienic building legislation in the EU, Regulation (EC) No 852/2004 on the hygiene of foodstuffs (EC, 2004a) details the basic requirements for food premises. Some information, at a very general level, is given on the layout, design and construction of food premises, covering, for example, the provision of cleaning and sanitary services, water supply and drainage, temperature control and the selection of floor and wall surfaces. Guidance is extended in Regulation (EC) No 853/2004 on laying down specific hygiene rules for food of animal origin to premises preparing meat, bivalve mollusc, fish, milk and egg products (EC, 2004b). Within both of these documents, however, advice is at best, concise.

Factories have always had to be compartmentalised or segregated into specific areas following good hygienic practice (GHP) principles. These were primarily due to environmental protection (i.e. protecting the product from the wind and rain, etc.), segregation of raw materials and finished product, segregation of wet and dry materials, provision of mechanical and electrical services and health and safety issues (e.g. boiler rooms, chemical stores, fire hazards, noise limitation).

Following segregation practices established in the dairy industry, which has traditionally segregated pre- and post-pasteurisation operations, the advent of ready-to-eat products has led to further segregation or 'zoning' of production areas for microbiological control reasons. A series of higher hygiene or cleaner zones have been created to help protect the product from microbiological cross-contamination events after it has been heat-treated or decontaminated. In addition, there has also been the recognition that non-microbiological hazards, particularly allergens, have to be controlled by segregating them from other product ingredients.

Finally, label declaration issues such as 'suitable for vegetarians', 'organic', 'does not contain genetically modified (GM) materials' or 'Kosher or Halal' have all caused food manufacturers to think about how raw materials are handled and processed. This is particularly true if, for example, factories are handling meat, non-organic ingredients, GM ingredients or non-religious ingredients. For example, whilst the presence of, for example, meat residues in a vegetarian product is not a safety issue, it will be an ingredients declaration issue, which could lead to poor brand perception. In many circumstances the segregation of food manufacturing sites to protect label declaration issues can have as much impact on factory design as the management of the biological, chemical and physical hazards detailed earlier.

To provide protection from these hazards food has historically been protected by a barrier system, made up of up to three barriers (Holah and Thorpe, 2000). With the advent of enhanced microbiological control in high hygiene areas, however, this has now been extended to four barriers as shown in Fig. 10.2 (Holah, 2003).

Level 1 represents the siting of the factory, the outer fence and the area up to the factory wall. This level seeks to control the degree of challenge of a hazard to the factory interior. This may be, for example, by reducing the number of pest harbourage areas, controlling pest access to waste materials, the downwind siting of effluent plants and the control of unauthorised public access.

Level 2 represents the factory wall and other processes (e.g. UV flytraps) which should separate the factory from the external environment (e.g. prevailing wind and surface water run-off). Whilst it is obvious that the factory cannot be a sealed box, the floor of the factory should ideally be at a different level to the ground outside and openings should be designed to be pest proof when not in use.

Level 3 represents the internal barriers that are used to separate manufacturing processes of different risk, e.g. pre-, and post-heat treatment. Such separation, which creates zones usually referred to as high care or risk areas, should seek to control the air, people and surfaces (e.g. the floor and drainage systems and the passage of materials and utensils across the barrier).

Level 4 represents a product enclosure zone, set within the level 3 high care/risk area. A product enclosure zone could encompass true aseptic filling or 'ultra clean' processing and packing areas such as glove boxes or the use of highly-filtered air as a barrier around the process line.

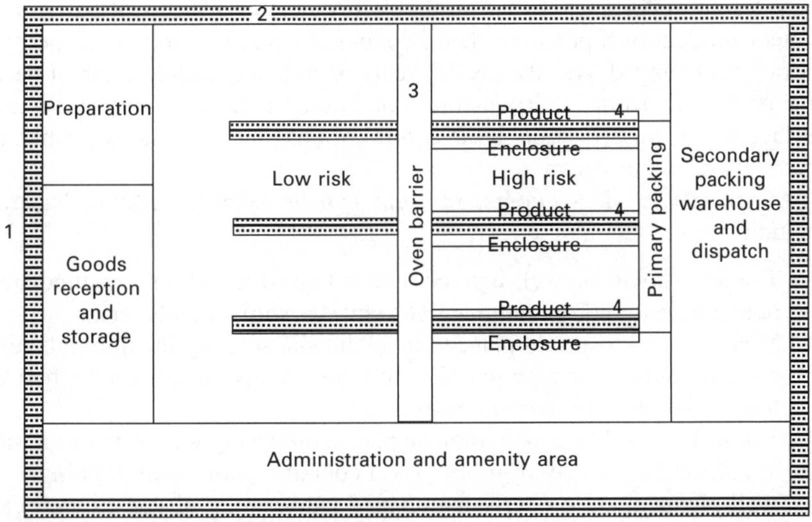

Fig. 10.2 Schematic layout of a factory site showing 'barriers' against contamination. (1) site, (2) main factory buildings, (3) walls of high risk area, (4) product enclosure within high risk.

The number of factory barriers required will be dependent on the nature of the food product, the nature of the hazards and the profile of the final consumer, and will be established from the HACCP study. Some agricultural operations, such as sorting or grading, may be undertaken in the field without any barriers. For low-risk products, e.g. fruit and vegetables, raw meat and fish, or ambient shelf-stable products, the first two barriers only are likely to be required. For high-risk products, typically short shelf-life ready-to-eat foods, the use of a third barrier is required for microbiological control. The use of third barrier level is also often applied to the manufacture of foodstuffs for 'at risk' sectors of the community, e.g. babies and the immunocompromised. The fourth barrier is necessary for aseptic products in which the elimination of external contamination is required, though some fully cooked ready-to-eat products with extended shelf-life may benefit from the additional controls this barrier affords. Whilst not absolutely necessary because of hazard control, manufacturers may choose to process food in higher hygiene zones for other reasons. This may be because of local legislation, or they believe that in the near future their product range will include higher-risk products and it makes financial sense to develop the infrastructure to produce such products at an earlier stage, or simply because they believe it will facilitate brand protection.

10.2 Barrier 1: the factory site

Attention to the design and maintenance of the site surrounding the factory provides an opportunity to set up the first (outer) of a series of barriers to protect production operations from contamination. The focus on the site barrier has changed over the last 20 years or so from consideration of pests and protection from environmental conditions, e.g. prevailing wind and surface water run-off, through the control of unwanted access by people, to the threat of bioterrorism.

At the site level, a number of steps can be taken to control hazards including:

- The site should be well-defined and/or fenced to prevent unauthorised public access and the entrance of domestic/wild animals, etc.
- Measures can be put in place to maintain site security including the use of gate houses, security patrols and maintenance schedules for barrier fencing or other protection measures.
- The factory building may often be placed on the highest point of the site to reduce the chance of ground level contamination from flooding.
- Well-planned and properly maintained landscaping of the grounds can assist in the control of rodents, insects and birds by reducing food supplies and breeding and harbourage sites. The use of two lines of rodent baits located every 15–21 m along the perimeter boundary fencing and at

the foundation walls of the factory, together with a few mouse traps near building entrances, is advocated by Imholte (1984). In addition, good landscaping of sites can reduce the amount of dust blown into the factory.

- Shapton and Shapton (1993) state that there should be a strategy of making the factory site unattractive to birds by denying food and harbourage otherwise colonies can become established and cause serious problems.
- Open waterways can attract birds, insects, vermin, etc. and should be enclosed in culverts if possible.
- Entrances that have to be lit at night should be lit from a distance with the light directed to the entrance, rather than lit from directly above. This prevents flying insects being attracted directly to the entrance.
- Some flying insects require water to support part of their life cycle, e.g. mosquitoes, and experience has shown that where flying insects can occasionally be a problem, all areas where water could collect or stand for prolonged periods of time (waterways, old buckets, tops of drums etc.) need to be removed or controlled.
- All waste containers should be lidded to prevent attraction to pests and any spillages should be cleared up as soon as possible. Waste containers should be regularly emptied.
- Processes likely to create microbial or dust aerosols, e.g. effluent treatment plants, waste disposal units or any preliminary cleaning operations, should be sited such that prevailing winds do not blow contaminants directly into manufacturing areas.
- An area of at least 3 m immediately adjacent to buildings should be kept free of vegetation and covered with a deep layer of gravel, stones, paving or roadway etc. (Katsuyama and Strachan, 1980; Troller, 1983). This practice helps weed control, assists inspection of bait boxes and traps and helps maintain control of the fabric of the factory building.
- Storage of equipment, utensils, pallets, etc. outside should be avoided wherever possible as they present opportunities for pest harbourage. Wooden pallets stacked next to buildings are also a known fire hazard.
- Siting of raw materials, process steps or finished products outside the protection of factory buildings is not desirable as this may increase the chance of product adulteration.
- Similarly, the siting outside of the factory buildings of storage facilities, e.g. silos, water tanks and packaging stores, should be avoided wherever possible. If not possible, they should be suitably locked off so that people or pests cannot gain unwanted access.
- Equipment necessary to connect transport devices to outside storage facilities (e.g. discharge tubing and fittings between tankers and silos) should also be locked away when not in use.
- Wherever possible loading and unloading operations should be undertaken under cover.

10.3 Barrier 2: the factory building

The building structure is the second and a major barrier, providing protection for raw materials, processing facilities and manufactured products from contamination or deterioration. Protection is both from the environment, including rain, wind, surface run-off, delivery and dispatch vehicles, dust, odours, microorganisms, pests and uninvited people, etc., and internally from microbiological hazards (e.g. raw material cross-contamination), chemical hazards (e.g. cleaning chemicals, lubricants) and physical hazards (e.g. from plantrooms, engineering workshops). Ideally, the factory buildings should be designed and constructed to suit the operations carried out in them and should not place constraints on the process or the equipment layout.

With respect to the external environment, whilst it is obvious that the factory cannot be a sealed box, openings to the structure must be controlled. There is also little legislation controlling the siting of food factories and what can be built around them. The responsibility, therefore, rests with the food manufacturer to ensure that any hazards (e.g. microorganisms from landfill sites or sewage works, or particulates from cement works, or smells from chemical works) are excluded from the factory interior via appropriate barriers. Detailed information on the hygienic design requirements for the construction of the external envelope and internal structure of the factory is not easily found, although Campden BRI have published a suite of design guidelines (CCFRA, 2002, 2003a,b, 2005).

New concepts of microbiological challenge control can physically start at 'barrier 2'. It can be generalised that microbiological pathogens of concern can arise from three sources: the intestines of animals (e.g. *Salmonella* spp. and pathogenic *Escherichia coli*), usually entering the factory in raw materials, the intestines and skin of operatives (e.g. *Salmonella* spp., *Staphylococcus aureus* and pathogenic *E. coli*), and the external environment (e.g. *Listeria* spp., *Bacillus* spp. and *Clostridium* spp.) entering as wind-blown dust or soil. Raw materials can be controlled via the manufacturer's quality system (specifications, inspections, etc.), operatives by the personnel hygiene policy, whilst the external environment can be controlled by the building. A typical example of a suitable outside wall structure is shown in Fig. 10.3. The diagram shows a well-sealed structure that resists pest ingress and is protected from external vehicular damage. The ground floor of the factory is also at a height above the external ground level. By preventing direct access into the factory at ground floor level, the entrance of contamination (mud, soil, foreign bodies, etc.), particularly from vehicular traffic (forklift trucks, raw material delivery, etc.), is restricted. By restricting soil ingress, the challenge of environmental pathogens is thus reduced. It should be noted, of course, that environmental contamination can also be found on some raw materials, e.g. soil on vegetables.

Key hygienic design factors include:

- Wherever possible, buildings should be single storey or with varying

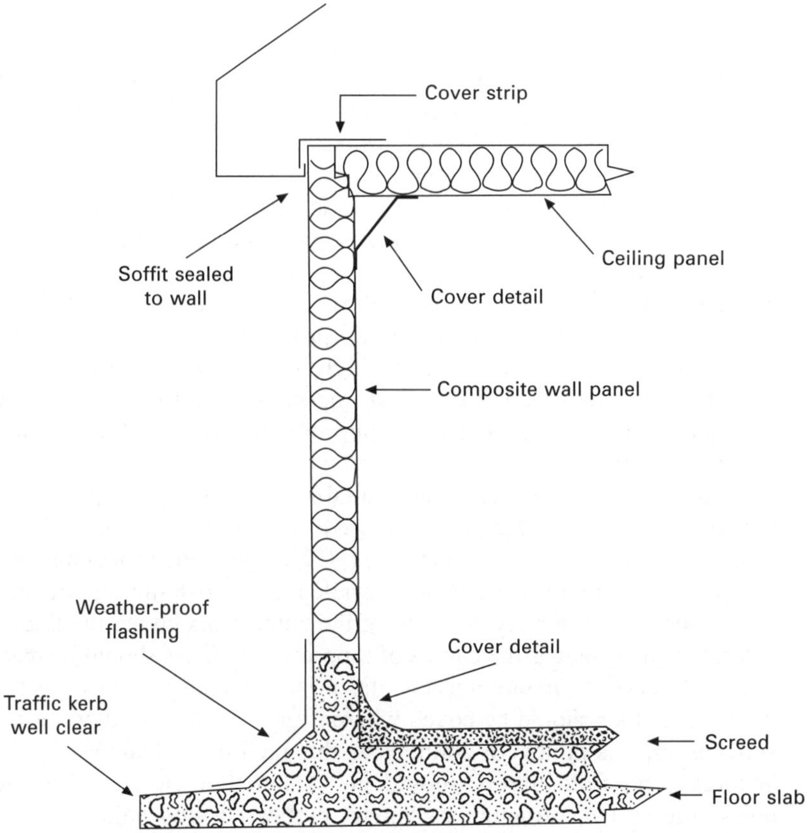

Fig. 10.3 Schematic diagram of an external wall construction preventing the ingress of soil (and associated microorganisms), foreign bodies and pests.

headroom featuring mezzanine floors to allow gravity flow of materials, where this is necessary (Imholte, 1984). This prevents any movement of wastes or leaking product moving between floors.

- Whilst ideally the process line should be straight, this is rarely possible, but there must be no backtracking and, where there are changes in the direction of process flow, there must be adequate physical barriers. Adequate chilled storage provision must be made, particularly in high-risk, to ensure product safety in the event of line stoppage, etc.
- Pipework or process lines carrying product, waste or services, etc. that present a hazard to open product should not pass directly over (above) such product lines.
- The layout should also consider that provision is made for the space necessary to undertake the process and associated quality control (QC) functions, both immediately the factory is commissioned and in the foreseeable future. In addition to process areas, provision may have to

be made for a wide range of activities including raw material storage; packaging storage; water storage; wash-up facilities; plant room; engineering workshop; cleaning stores; microbiology, chemistry and QC laboratories; test kitchens; pilot plant; changing facilities; restrooms; canteens; medical rooms; observation areas/viewing galleries; and finished goods dispatch and warehousing.

- The siting of factory openings should be designed with due consideration for prevailing environmental conditions, particularly wind direction and drainage falls.
- The floor may be considered as one of the most important parts of a building because it forms the basis of the entire processing operation. It is thus worthy of special consideration and high initial capital investment.
- The structural floor slab is the most important aspect of the floor and should be soundly constructed with appropriate waterproof membranes and movement joints.
- Both tiles or synthetic resins are suitable as a surface topping, with tiles being favoured if the predicted use of the processing area (without change) extends for five or more years. If change in the processing area is foreseen before this time an appropriate resin of 6–8 mm in thickness is appropriate. Ultimately, the skill of the contractors laying the floor is at least as important as the choice of topping and efforts should be made to get references from previous clients, or indeed to inspect previous laid floors. The floor should be coved where it meets walls or other vertical surfaces such as plinths or columns as this facilitates cleaning.
- Wherever possible, floors should be specified to be non-slip and, unless the surface structure is heavily embossed with valley to peak heights exceeding 3–4 mm (i.e. when brush bristles cannot penetrate to the bottom of the valley), this has no effect on cleanability. Flooring materials designed to be slip-resistant may have an average peak to valley height measured in millimetres, whilst the size of imperfection that could harbour microorganisms would be measured in microns (Taylor and Holah, 1996; Mettler and Carpentier, 1998; CCFRA, 2002). Such micron-sized holes are equally likely to be found on 'smooth' as 'rough' surfaces. Advice on slips and trips can be obtained from the UK Health and Safety Executive at http://www.hse.gov.uk/food/index.htm.
- Flooring and walling materials manufacturers should also be able to demonstrate that their materials have been tested for absence of water absorption and ease of cleanability so as to meet Regulation (EC) No 852/2004. Materials that absorb water also absorb microorganisms and are impossible to effectively decontaminate.
- Floors should also be effectively drained to remove water – essential to reduce microbial growth and cross-contamination. Satisfactory drainage can only be achieved if adequate falls to drainage points are provided. Falls should be in the range 1 in 50 to 1 in 80 with the slope being steeper with high water volumes or slip-resistant toppings. Falls greater than 1

in 40 may introduce operator safety hazards and also cause problems with wheeled vehicles.

- The type of drain used depends to a great extent upon the process operation involved. For operations involving a considerable amount of water and solids, channel drains are often the most suitable (Fig. 10.4) whilst gully pots are more suited to operations generating smaller volumes of water and solids. Channel gratings must be easily removable. The fall to drain should not exceed 5 metres.
- The edge of the channel rebate must be properly designed and constructed to protect it from movement damage, particularly if wheeled vehicles are in use, by the provision of sufficient anchor tangs (Fig. 10.4). The author is aware of a number of cases where the seal has been broken between the drain channel and floor structure so as to leave an uncleanable void between the two. When the channel was subsequently walked upon or traversed by wheeled vehicles, a small volume of foul liquid and *Listeria* were 'pumped' to the floor surface.
- In addition, drainage systems have been observed to act as air distribution channels, allowing contaminated air movement between rooms. This can typically occur when the drains are little used, e.g. in dry foods factories, and the water traps dry out.

Modular insulated panels are now used very widely for non-load-bearing walls. The panels are made of a core of insulating material between 50 and 200 mm thick, sandwiched between steel sheets, which are bonded to both sides of the core. Careful consideration must be given to the choice of filling which must comply with current fire retardation (and insurance) standards. They are designed to lock together and allow a silicone sealant to provide a hygienic seal between the units. The modules are best mounted on a concrete upstand or plinth or stainless steel channel filled with concrete (Fig. 10.5). The latter provides useful protection against the possibility of damage from vehicular traffic, particularly fork-lift trucks. However, it should be appreciated that this arrangement reduces the possibility of relatively easy and inexpensive changes to room layout to meet future production requirements.

Load-bearing and fire-break walls are often constructed from brick or blockwork. These walls can either be directly painted with an appropriate waterproof coating or rendered with a cement and sand screed to achieve a better surface smoothness for the coating layer. The covering of walls by other materials such as sheets of plastic or stainless steel is not recommended unless the walls are absolutely flat so as to prevent the harbourage of pests and fluids containing microorganisms between the two walling materials.

Walls and other vulnerable structures such as doors and pillars should be protected by suitable barriers. Barriers may be wall- or floor-mounted and should be designed to prevent collisions with the wall or other vulnerable structures by the specific types of transport systems, racking, totebins, etc. used within the factory.

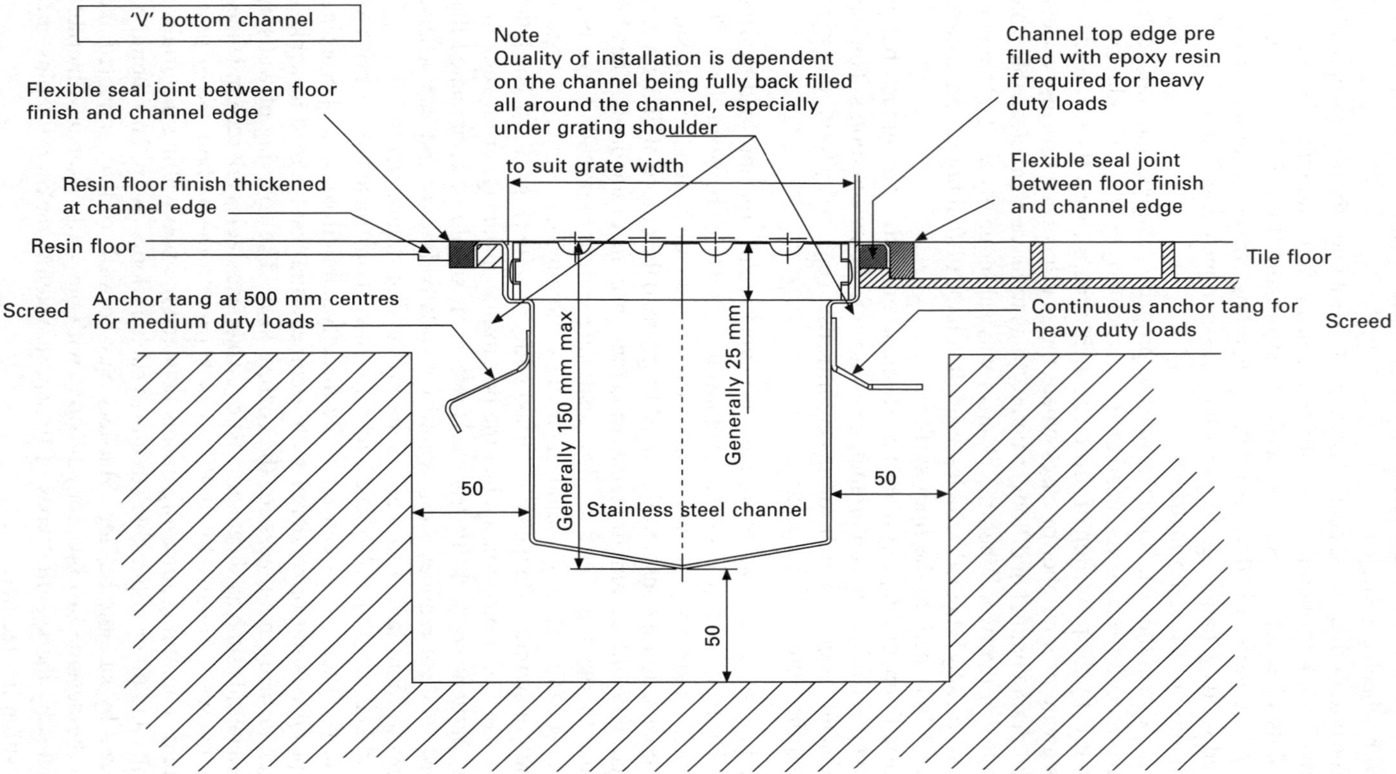

Fig. 10.4 'V' bottom stainless steel channel showing position of anchor tangs to limit drain movement.

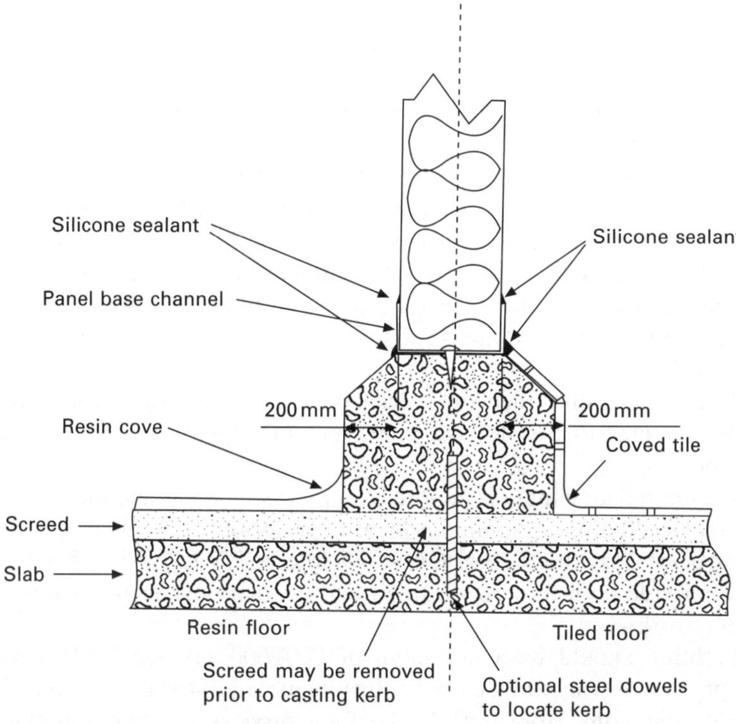

Fig. 10.5 Modular insulated panel located in U-channel and fixed to a concrete plinth.

Other hygienic design features include:

- The number and size of openings should be kept to a minimum and exterior doors should not open directly into production areas. External doors should always be shut when not in use and, if they have to be opened regularly, should be of a rapid opening and closing design.
- Doors should be constructed of metal, glass reinforced plastic (GRP) or plastic, self-closing, designed to withstand the intended use and misuse and be suitably protected from vehicular damage where applicable.
- Plastic strips/curtains are acceptable only in interior situations as they are easily affected by weather. Where necessary, internal or external porches can be provided with one door, usually the external door on an external porch, being solid and the internal door being a flyscreen door; on an internal porch it would be the opposite configuration. Air jets directed over doorways, designed to maintain temperature differentials when chiller/freezer doors are opened, may have a limited effect on controlling pest access.
- For many food manufacturers and retailers, glass is seen as the second major food hazard after pathogenic microorganisms. For this reason,

glass should be avoided as a construction material (windows, inspection mirrors, instrument and clock faces, etc.). If used, e.g. as viewing windows to allow visitor or management observation, a glass register, detailing all types of glass used in the factory and their location, should be compiled.

- Windows should be glazed with either polycarbonate or laminated. Where opening windows are specifically used for ventilation (particularly in tropical areas), these must be screened and the screens be designed to withstand misuse or attempts to remove them. Flyscreens should be constructed of stainless steel mesh and be removable for cleaning.
- The flow of air and drainage should be away from 'clean' areas towards 'dirty' ones.
- The flow of discarded outer packaging materials should not cross or run counter to the flow of either unwrapped ingredients or finished products.
- If a filtered air supply is required to processing areas and the supply will involve ducting, a minimum level of filtration of > 90 % of 5 micron particles is required, e.g. G4 or F5 filters (BS EN 779 – BSI, 2002), to provide both suitably clean air and prevent dust accumulation in the ductwork.
- Lighting should be a minimum of 500–600 lux for food inspection purposes and should ideally be flush mounted to the ceiling to facilitate bulb changing from outside the food processing area. Alternatively, lighting units should be suspended and plugged in so that in the event of a failure the entire unit can be replaced and the faulty one removed from the processing areas to a designated workshop for maintenance/ bulb replacement. Open lighting elements (i.e. those not within enclosed lighting units) should be protected by plastic sleeves.

Within the internal environment, most factories are segregated into food production areas and amenities (reception, offices, canteens, training rooms, engineering workshops, boiler houses, etc.). The prime reason for this is to clearly separate the food production processes from the other activities that the manufacturer must perform. This may be to control microbiological or foreign body hazards arising from the amenity functions, but is always undertaken to foster a 'you are now entering a food processing area' hygienic mentality in food operatives. The key area to segregate is any laboratory which is undertaking microbiological testing (particularly pathogens where positive controls are required). In some countries in the EU (e.g. Germany) it is not permitted to handle pathogens in laboratories within food factories. In other countries, however, particularly if the laboratory is accredited by an external agency, positive controls must be available in the laboratory for pathogen tests. Laboratories should have separate drainage and air supplies with no openings directly into production and should ideally be placed in a separate building.

Food production areas are typically segregated into raw material intake, raw material storage, processing, packaging and final product warehouse and dispatch. In addition, the flow of ingredients and products is such that, in ideal conditions, raw materials enter at one end of the factory (dirty end) and are dispatched at the opposite end (clean end).

The key differential between segregation barriers at this and the next level (high care/high-risk areas) is that within 'barrier 2' food operatives are freely able to move between the segregated areas without any personnel hygiene barriers (though hand washing may be required to move between some areas). Operatives cannot freely move between barriers 2 and 3.

Whilst a range of ingredients is brought together for processing, they may need to be stored separately. Storage may be temperature-orientated (ambient, chilled or frozen) or ingredient-related, and separate stores may be required for fruit and vegetable, meat, fish, dairy and dry ingredients. Other food ingredients such as allergens and non-ingredients such as packaging should also be stored separately. Segregation may also extend into the first stages of food processing, where for example the production of dry intermediate ingredients, e.g. pastry for pies, is separated from the production of the pie fillings. The degree of segregation for storage and processing of ingredients and intermediates is predominantly controlled by the exclusion of water, particularly in how they are cleaned (EHEDG, 2001), i.e.:

- **Dry cleaning** – This applies to areas where no cleaning liquids are used, only vacuum cleaners, brooms, brushes, etc. Whilst these areas are normally cleaned dry, occasionally they may be fully or partially wet cleaned, when limited amounts of water are used.
- **Wet cleaning** – This applies to areas where the entire room or zone is always cleaned wet. The contents (equipment, cable trays, ceilings, walls, etc.) are wet washed without restrictions on the amount of cleaning liquid used.

In addition to segregating dry areas from a requirement to exclude water, other areas may need to be segregated due to excessive use of water, which can lead to the formation of condensation and the generation of aerosols. Such areas include traywash and other cleaning areas.

Ideally, manufacturers who manufacture allergenic and non-allergenic products should do so on separate sites such that there is no chance of cross-contamination from different ingredients. This issue has been debated by food manufacturers in both Europe and the USA with the conclusion that it is unlikely to be economically viable to process on separate sites. Segregation of allergenic components will have to be undertaken, therefore, within the same site. As a preferred alternative to separate factories, it may be possible to segregate the whole process, from goods in through raw material storage and processing to primary packaging, on the same site. If this is not possible, segregation has to be undertaken by time, e.g. by manufacturing non-allergen containing products first and then manufacturing allergen-containing products last. Thorough cleaning and disinfection is then undertaken before the manufacture of the non-allergen containing products is then re-commenced.

If segregation by time is to be considered, a thorough HACCP study should be undertaken to consider all aspects of how the allergen is to be stored, transported, processed and packed, etc. This would include information on any dispersal of the allergen during processing (e.g. from weighing, mixing), the fate of the allergen through the process (will its allergenic attributes remain unchanged), the degree to which the allergen is removed by cleaning and the effect of any dilution of residues remaining after cleaning in the subsequent product flow. This may be complicated in that other hazards may be seen as more important. For example, some manufactures prefer to segregate their production on microbiological risk (e.g. in a sandwich operation the manufacturer produces fully cooked fillings first and then salad fillings, more likely to contain spoilage organisms, last).

10.4 Barrier 3: high risk production area

The third barrier within a factory segregates an area in which food products are further manipulated or processed following a decontamination treatment. It is, therefore, an area into which a food product is moved after its microbiological content has been reduced. Food products that are packed or further processed in these 'barrier 3' areas are all susceptible to the growth of and/or recontamination with pathogenic microorganisms during their shelf-life and they may not be further heat processed prior to consumption.

Many names have been adopted for this barrier 3 processing area including 'clean room' (or 'salle blanche' in France) following pharmaceutical terminology, 'high hygiene', 'high care' or 'high risk' area. In some sectors, particularly chilled, ready-to-eat foods, manufacturers have also adopted opposing names to describe second barrier areas such as 'low risk' or low care'.

Much of this terminology is confusing; particularly the concepts of 'low' areas which can imply to employees and other people that lower overall standards are acceptable in these areas where, for example, operations concerned with raw material reception, storage and initial preparation are undertaken. In practice, all operations concerned with food production should be carried out to the highest standard. Unsatisfactory practices in so-called low-risk areas may, indeed, put greater pressures on the 'barrier system' separating the second and third level processing areas.

The Chilled Food Association in the UK (CFA, 2006) established guidelines to describe the hygiene status of chilled foods and indicate the area status of where they should be processed after any heat treatment. Three levels were originally described; high-risk area (HRA), high-care area (HCA) and good manufacturing practice (GMP). Their definitions were:

- HRA An area to process components, **all** of which have been heat-treated to ≥ 90 °C for 10 min or ≥ 70 °C for 2 min, and in which there is a risk of contamination between heat treatment and pack sealing that may present a food safety hazard.
- HCA An area to process components, **some** of which have been heat-treated to ≥ 70 °C for 2 min, and in which there is a risk of contamination between heat treatment and pack sealing that may present a food safety hazard.
- GMP An area to process components, **none** of which have been heat-treated to ≥ 70 °C for 2 min, and in which there is a risk of contamination prior to pack sealing that may present a food safety hazard.

In practice, the definition of HCA has been extended by the author to include an area to further process components none of which has undergone a pasteurisation treatment but which have been decontaminated. For example, a high care area is appropriate for products that have undergone a wash in chlorinated water, e.g. fruit and salad products or fish after low temperature smoking and salting. In simplistic terms, products entering a HRA have undergone a 6 log pasteurisation process (≥ 70 °C for 2 min) whereas products entering a HCA may have undergone only a 1–2 log decontamination process (e.g. washing).

Much of the requirements for the design of HRA and HCA operations are the same, with the emphasis on **preventing** contamination in HRA and **minimising** contamination in HCA operations (CFA, 2006). The requirement for third barrier level high care/risk segregation for appropriate foodstuffs is now recognised by the major food retailers worldwide and is a requirement in the *BRC Global Food Standard (2005)* (www.brc.org.uk) and the *Global Food Safety Initiative Guidance Document (2004)* (www.ciesnet.com/). It must also be noted that it is the product that defines the HRA or HCA area, not the building design. For example, a high risk area could be designed for the slicing of cooked meats but, if a fresh fruit based garnish was used on the meat, the processing area would revert to a high care area because of the higher microbial risk of the fruit (it would have only received a decontamination treatment whilst the meat would have been pasteurised).

In general, high care/risk areas should be as small as possible as their maintenance and control can be very expensive. If there is more than one high care/risk area in a factory, they should be arranged together or linked as much as possible by closed corridors of the same class. This is to ensure that normal working procedures can be carried out with a minimum of different hygienic procedures applying.

Some food manufacturers design areas between the second 'low risk' and third barrier 'high risk' level zones and use these as transition areas. These are often termed 'medium care' or 'medium risk' areas. These areas are not separate areas in their own right as they are freely accessed from low risk (barrier two areas) without the need for the protective clothing and personnel hygiene barriers as required at the low/high risk area interface.

By restricting activities and access to the medium risk area from low risk, however, these areas can be kept relatively 'clean' and thus restrict the level of microbiological contamination immediately adjacent to the third level barrier. To simplify the text the term high risk is used to cover both high care and high risk, unless specified otherwise.

High risk areas are designed primarily to control microbiological risks and, in historical terms, practices have been developed for chilled, ready-to-eat food products, particularly in relation to, *Listeria monocytogenes*. For many chilled food products, *L. monocytogenes* could well be associated with the raw materials used and thus may well be found in the low risk area. After the product has been heat-processed or decontaminated, it is essential that all measures are taken to protect the product from cross-contamination from (predominantly) low risk, *L. monocytogenes* sources. Similarly, other pathogens and potential foreign body contamination that would jeopardise the wholesomeness of the finished product could also be found in low risk. A three-fold philosophy has been developed by the author to help reduce the incidence of *L. monocytogenes* in finished product and at the same time, control other microbiological contamination sources.

1. Provide as many barriers as possible related to the building structure and processing practices to prevent the entry of *Listeria* into the high risk area.
2. Prevent the growth and spread of any *Listeria* penetrating these barriers during production and minimise locations within high risk in which *Listeria* could become harboured.
3. After production, employ a suitable sanitation system to ensure that all *Listeria* are removed from high risk prior to production recommencing.

Whilst this philosophy was initially proposed for the control of food manufacturing areas in which potential pathogens (e.g. *Listeria*) could survive, grow and become adapted to a processing area, it could also be applied to the manufacturing environment of food products in which pathogens could simply survive (e.g. *Salmonella* in chocolate, cereal, tahini and milk powder plants).

The building structure, facilities and practices associated with the high risk production and assembly areas provide the third and inner barrier protecting food manufacturing operations from contamination. This barrier is built up by the use of combinations of a number of separate components or sub-barriers, to control contamination that could enter high risk from the following routes:

- structural defects;
- product entering high risk via a heat process;
- product entering high risk via a decontamination process;
- product entering high risk that has been heat-processed/decontaminated

off-site but whose outer packaging may need decontaminating on entry to high risk;
- other product transfer;
- packaging materials;
- liquid and solid waste materials;
- food operatives entering high risk;
- the air;
- utensils, which may have to be passed between low and high risk.

10.4.1 Structure

Structurally, creating a third barrier level has been described by Ashford (1986) as creating a box within a box (Fig. 10.6). In other words, the high risk area is sealed on all sides to prevent microbial ingress. Whilst this is an ideal situation, we still need openings to the box to allow access for people, ingredients and packaging and exit for finished product and wastes. Openings should be as few as possible, as small as possible (to better maintain an internal positive pressure) and should be controlled (and shut if possible) at all times. Similarly, the size of the 'box' should be as small as practically possible to reduce costs, particularly for any heating, ventilating and air conditioning (HVAC) system, and the perimeter of the box should be inspected frequently to ensure that all joints are fully sealed.

The design of the high risk food processing area must allow for the accommodation of five basic requirements, i.e.

- processed materials and possibly some ingredients;
- processing equipment;

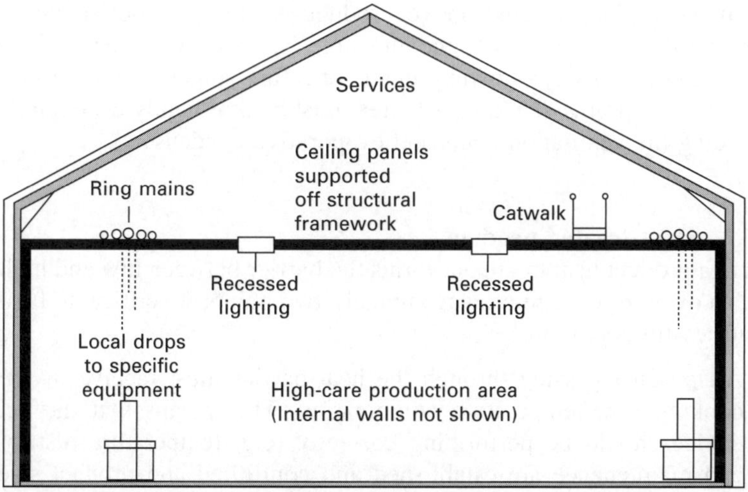

Fig. 10.6 Basic design concepts – the separation of production from services and maintenance operations.

- staff concerned with the operation of such equipment;
- packaging materials;
- finished products.

There is a philosophy which has considerable support, which states that all other requirements should be considered as secondary to these five basic requirements and, wherever possible, should be kept out of the high risk processing area. This reduces potential environmental niches for microorganisms, aids in cleaning and disinfection and thus contamination control.

These secondary requirements include:

- structural steel framework of the factory;
- service pipework for water, steam and compressed air; electrical conduits and trunking; artificial lighting units; and ventilation ducts;
- compressors, refrigeration units and pumps;
- maintenance personnel associated with any of these services;
- 'furniture' and computers, etc. associated with office facilities.

In addition to being as small as possible, the splitting down of large processing areas into smaller sub-units (e.g. a single 12-line meat slicing hall into three fully segregated sub-units of four slicing lines), can help in eliminating cross-contamination between lines. Such segregation is also now considered as a method of increasing manufacturing flexibility, for example when some lines need to be shutdown for cleaning or maintenance whilst the others need to remain in production.

If the design principles associated with the construction of high risk areas are met, problems related to food products contaminated with pathogens are rare. Pathogen contamination of foods tends to occur when the fabric of the building has been broken either accidentally (e.g. leaking waste water or rainwater pipes located above production areas) or deliberately (e.g. when the floor or floor wall junctions are disturbed during construction or refurbishment work). Following repairs or refurbishment work, particularly to the floor or drains, the high risk area must be thoroughly decontaminated (including the ventilation ducts and evaporative condensers).

10.4.2 Heat-treated product

Where a product heat treatment forms the barrier between low and high risk (e.g. an oven, fryer or microwave tunnel), two points are critical to facilitate its successful operation.

- All product passing through the heat barrier must receive its desired cooking time/temperature combination. This means that the heating device should be performing correctly (e.g. temperature distribution and maintenance are established and controlled and product size has remained constant) and that it should be impossible, or very difficult, for product to pass through the heat treatment without a cook process being initiated.

- The heating device must be designed such that as far as is possible, the device forms a solid, physical barrier between low and high risk. Where it is not physically possible to form a solid barrier, air spaces around the heating equipment should be minimised and the low/high risk floor junction should be fully sealed to the highest possible height.

The fitting of heating devices that provide heat treatment within the structure of a building presents two main difficulties. Firstly, the devices have to be designed to load product on the low risk side and unload in high risk. Secondly, the maintenance of good seals between the heating device surfaces, which cycle through expansion and contraction phases, and the barrier structure which has no thermal expansion, is problematical. Of particular concern are ovens, and the author is aware of the following issues:

- Some ovens have been designed such that they drain into high risk. This is unacceptable for the following reason. It may be possible for pathogens present on the surface of product (which is their most likely location if they have been derived from cross-contamination in low risk) to fall to the floor through the melting of the product surface layer (or exudate on overwrapped product) at a time/temperature combination that is not lethal to the pathogen. The pathogen could then remain on the floor or in the drain of the oven in such a way that it could survive the cook cycle. On draining, the pathogen would then subsequently drain into high risk. Pathogens have been found at the exit of ovens in a number of food factories and this area should always be a pathogen sampling point within the environmental sampling plan.
- Problems have occurred with leakage from sumps under the ovens into high risk. Sumps can collect debris and washing fluids from the oven operation which can facilitate the growth of *Listeria*. These areas should be routinely cleaned (from low risk).
- Where the floor of the oven is cleaned, cleaning should be undertaken in such a way that cleaning solutions do not flow from low to high risk. Ideally, cleaning should be from low risk with the high risk door closed and sealed. If cleaning solutions have to be drained into high risk, or in the case of ovens that have a raining water cooling system, a drain should be installed immediately outside the door in high risk.
- Heating devices should be designed to load product on the low risk side and unload in high risk. If oven racks of cooked product have to be transferred into high risk for unloading, these racks should be returned to low risk via the ovens, with an appropriate thermal disinfection cycle as appropriate.
- The design of small batch product blanchers or noodle cookers (i.e. small vessels with water as the cooking medium) does not often allow the equipment to be sealed into the low/high risk barrier as room has to be created around the blancher to allow product loading and unloading. Condensation is likely to form because of the open nature

of these cooking vessels, and it is important to ventilate the area to prevent microbial build-up where water condenses. Any ventilation system should be designed so that the area is ventilated from low risk; ventilation from high risk can draw into high risk large quantities of low risk air. Alternatively, the cookers can discharge at high temperature via gravity into sealable containers (e.g. tote bins) which can be pushed (and externally decontaminated) into high risk via a hatch whilst maintaining a high internal temperature.

Early installations of open cooking vessels (kettles) as barriers between low and high risk, together with (occasional) low-level retaining or bund walls to prevent water movement across the floor and barriers at waist height to prevent the movement of people, whilst innovative in their time, are now seen as hygiene hazards (Fig. 10.7a). It is virtually impossible to prevent the transfer of contamination, by people, the air and via cleaning, between low and high risk. Such barriers were then improved by fully segregating the floor junction and decanting product into high risk via open hatches (Fig. 10.7b). These are still occasionally found for products that are not suitable for pumping for quality reasons. Finally, installations had the kettles fully separated from high risk and transferred cooked product (by pumping, gravity, vacuum etc.) through into high risk via a pipe in the dividing wall. The kettles need to be positioned in low risk at a height such that the transfer into high risk is well above ground level, usually facilitated by mounting the kettles on a mezzanine (Fig. 10.7c). Pipework connections through the walls should be cleaned from high risk such that potentially contaminated low risk area cleaning fluids do not pass into high risk.

10.4.3 Product decontamination

Products that can not be heat-processed, e.g. fresh produce, should enter high care via a decontamination operation, usually involving a washing process with the wash water incorporating a biocide. Chlorinated dips, mechanically stirred washing baths or 'jacuzzi' washers are the most common methods, though alternative biocides are also used (e.g. bromine, chlorine dioxide, ozone, organic acids, peracetic acid, hydrogen peroxide).

In addition, it is now seen as increasingly important, following a suitable risk assessment, to decontaminate the outer packaging of various ingredients on entry into high risk (e.g. product cooked elsewhere and transported to be processed in the high risk area, canned foods and some overwrapped processed ingredients). Where the outer packaging is likely to be contaminated with food materials, decontamination is best done using a washing process incorporating a disinfectant (usually a quaternary ammonium compound). If the packaging is clean, the use of UV light has the advantage that it is dry and thus limits potential wet floor areas which exacerbate environmental microbial growth. For companies that also have ovens with low risk entrance

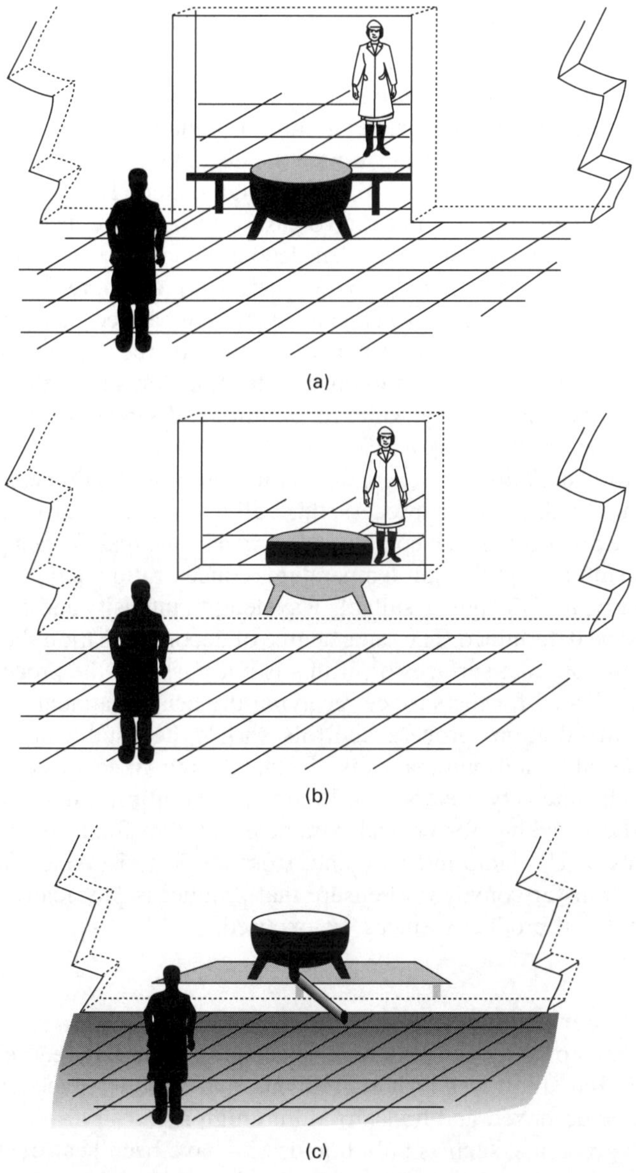

Fig. 10.7 (a) Schematic early low risk (white coated worker)/high risk divide around kettles. (b) Improved segregation in that the floor wall junction has been sealed and the opening between low and high risk has been reduced to the size of the hatch through which the product is decanted. (c), Fully segregated, mezzanine-mounted cooking equipment gravity feeding into high risk.

and high risk exit doors, it is also possible to transfer product from low to high risk via these ovens using a short steaming cycle that offers surface pasteurisation without 'cooking' the ingredients.

Decontamination systems have to be designed and installed such that they satisfy three major criteria.

1. As with heat barriers, decontamination systems need to be installed within the low/high risk barrier to minimise the free space around them. As a very minimum, the gap around the decontamination system should be smaller than the product to be decontaminated. This ensures that all ingredients in high risk must have passed through the decontamination system and thus must have been decontaminated (it is impossible to visually assess whether the outer surface of an ingredient has been disinfected, in contrast to whether an ingredient has been heat-processed). When installing washing systems, as much of the system as possible should be in low risk to minimise water transfer into high risk. Some systems incorporate an airknife for additional drying of decontaminated articles before high risk entry.
2. Prior to installation, the decontamination process should be established and validated. For a wet process, this will involve the determination of a suitable disinfectant that combines detergency and disinfectant properties and a suitable application temperature, concentration and contact time. Similarly, for UV light, a suitable wavelength, intensity and contact time should be determined. The same degree of decontamination should apply to all the product surfaces or, if this is not possible, the process should be established for the surface receiving the least treatment.
3. After installation, process controls should be established and may include calibrated, automatic disinfectant dosing, fixed-speed conveyors, UV light intensity meters, etc. In-process monitoring may include the periodic checking for critical parameters, for example blocked spray nozzles or UV lamp intensity and, from the low risk side, the loading of the transfer conveyor to ensure that product is physically separated such that all product surfaces are exposed.

10.4.4 Other product transfer

It is now poor practice to bring outer packaging materials (e.g. corrugated cardboard) into high risk. All ingredients and product packaging must, therefore, be de-boxed and transferred into high risk.

Some ingredients, such as bulk liquids that have been heat-treated or are inherently stable (e.g. oils or pasteurised dairy products), are best handled by being pumped across the low/high risk barrier directly to the point of use. Dry, stable bulk ingredients (e.g. sugar) can also be transferred into high risk via sealed conveyors.

For non-bulk quantities, it is possible to open ingredients at the low/high risk barrier and decant them through into high risk via a suitable transfer system (e.g. a simple funnel set into the wall), into a receiving container. Transfer systems should, preferably, be closeable when not in use and should be designed to be cleaned and disinfected, from the high risk side, prior to use as appropriate.

10.4.5 Packaging

Packaging materials (film reels, cartons, containers, trays, etc.) are best supplied to site 'double bagged'. This involves a cardboard outer followed by two plastic bag layers surrounding the packaging materials. The packaging is brought on site, de-boxed, and stored double-bagged in a suitable packaging store. When called for in high risk, the packaging material is brought to the low/high risk barrier, the outer plastic bag removed and the inner bag and packaging enters high risk through a suitable hatch. The second plastic bag keeps the packaging materials covered until immediately prior to use.

The hatch, as with all openings in the low/high risk barrier, should be as small as possible and should be closeable when not in use. This is to reduce airflow through the hatch and thus reduce the airflow requirements for the air-handling systems to maintain high risk positive pressure. For some packaging materials, especially heavy film reels, it may be required to use a conveyor system for moving materials through the hatch. An opening door or, preferably, double door airlock should only be used if the use of a hatch is not technically possible and suitable precautions must be taken to decontaminate the airlock after use. Airlocks should be interlocked so that only one door can be open at any one time.

10.4.6 Liquid and solid wastes

On no account should low risk liquid or solid wastes be removed from the factory via high risk, and attention is required to the procedures for removing high risk wastes. The drainage system should flow from high to low risk and, whenever possible, backflow from low risk to high risk areas should be impossible. This is best achieved by having separate low and high risk drains running to a master collection drain with an air-break between each collector and master drain. The drainage system should also be designed such that rodding points are outside high risk areas. If possible, the high risk drains should enter the collection drain at a higher point than the low risk drains, so that if flooding occurs, low risk areas may flood first.

Solid wastes that have fallen on the floor or equipment, etc. through normal production spillages should be bagged-up or placed in easily cleanable bins, on an ongoing basis commensurate with good housekeeping practices. Waste bags should leave high risk in such a way that they minimise any potential cross-contamination with processed product and should, preferably, be routed in the reverse direction to the product. For small quantities of bagged waste, existing hatches should be used e.g. the wrapped product exit hatches or the packaging materials entrance hatch, as additional hatches increase the risk of external contamination and put extra demands on the air-handling system.

For waste collected in bins, it may be necessary to decant the waste through purpose built, easily cleanable from high risk, waste chutes that deposit directly into waste skips. Waste bins should be colour-coded to differentiate them from other food containers and should only be used for waste.

10.4.7 Personnel

Within the factory building, provision must be made for adequate and suitable staff facilities and amenities for changing, washing and eating. There should be lockers for storing outdoor clothing in areas that must be separate from those for storing work clothes. Toilets must be provided and must not open directly into food processing areas, all entrances of which must be provided with handwashing facilities arranged in such a way that their ease of use is maximised.

With regard to operatives in high risk areas, personnel facilities and requirements must be provided in a way that minimises any potential contamination of high risk operations. The primary sources of potential contamination arise from the operatives themselves and from low risk operations. This necessitates further attention to protective clothing and, in particular, special arrangements and facilities for changing into high risk clothing and entering high risk. Best practice with respect to personnel hygiene is continually developing and has been reviewed by Guzewich and Ross (1999), Taylor and Holah (2000) and Taylor *et al.* (2000). In some countries, for example South Africa, best practice on the hygiene of food handlers is published by the government www.doh.gov.za/docs/factsheets/foodhandlers.pdf.

High risk factory clothing does not necessarily vary from that used in low risk in terms of style or quality, though it may have received higher standards of laundry, especially related to a higher-temperature process, sufficient to significantly reduce microbiological levels. Additional clothing may be worn in high risk, however, to further protect the food being processed from contamination arising from the operative's body (e.g. gloves, sleeves, masks, whole head coveralls, coats with hoods, boiler suits, etc.). All clothing and footwear used in the high risk area is colour-coded to distinguish it from that worn in other parts of the factory and to reduce the chance that a breach in the systems would escape early detection.

High risk footwear should be captive to high risk; i.e. it should remain within high risk, operatives changing into and out of footwear at the low/high risk boundary. This has arisen because research has shown that boot baths and boot washers are unable to adequately disinfect low risk footwear such that they cannot be worn in both low and high risk and decontaminated between the two (Taylor *et al.*, 2000). Essentially they do not remove all organic material from the treads and any pathogens within the organic material remaining are protected from any subsequent disinfectant action. In addition, boot baths and boot washers can both spread contamination via aerosols and water droplets that, in turn, can provide moisture for microbial growth on high risk floors. Bootwashers were, however, shown to be very good at removing large quantities of organic material from boots and are thus a useful tool in low risk areas to both to clean boots and to help prevent operative slip hazards.

The high risk changing room provides the only entry and exit point for personnel working in or visiting the area and is designed and built both to house the necessary activities for personnel hygiene practices and to minimise contamination from low risk. Research at Campden BRI has proposed the following hand hygiene sequence to be used on entry to high risk (Taylor and Holah, 2000). This sequence has been designed to maximise hand cleanliness, minimise hand transient microbiological levels, maximise hand dryness yet at the same time reduce excessive contact with water and chemicals that may both lead to dermatitis issues of the operatives and reduce the potential for water transfer into high risk.

1. Remove low risk or outside clothing.
2. Remove low risk/outside footwear and place in designated 'cage' type compartment.
3. Cross over the low risk/high risk dividing barrier.
4. WASH HANDS.
5. Put on in the following order:
6. high risk captive footwear;
7. hair net – put on over ears and covering *all* hair (plus beard snood if needed) and hat (if appropriate);
   Note: some RTE food manufacturers prefer to put on hairnets prior to step 3)
8. overall (completely buttoned up to neck).
9. Check dress and appearance in the mirror provided.
10. Go into the high risk production area and apply an alcohol-based sanitizer.
11. Draw and put on disposable gloves, sleeves and apron, if appropriate.

A basic layout for a changing room is shown in Fig. 10.8 and has been designed to accommodate the above hand hygiene procedure and the following requirements.

- An area at the entrance to store outside or low risk clothing. Lockers should have sloping tops.
- A barrier to divide low and high risk floors. This is a physical barrier such as a small wall (approximately 60 cm high), that allows floors to be cleaned on either side of the barrier without contamination by splashing, etc. between the two.
- Open lockers at the barrier to store low risk footwear.
- A stand on which footwear is displayed/dried, usually soles uppermost.

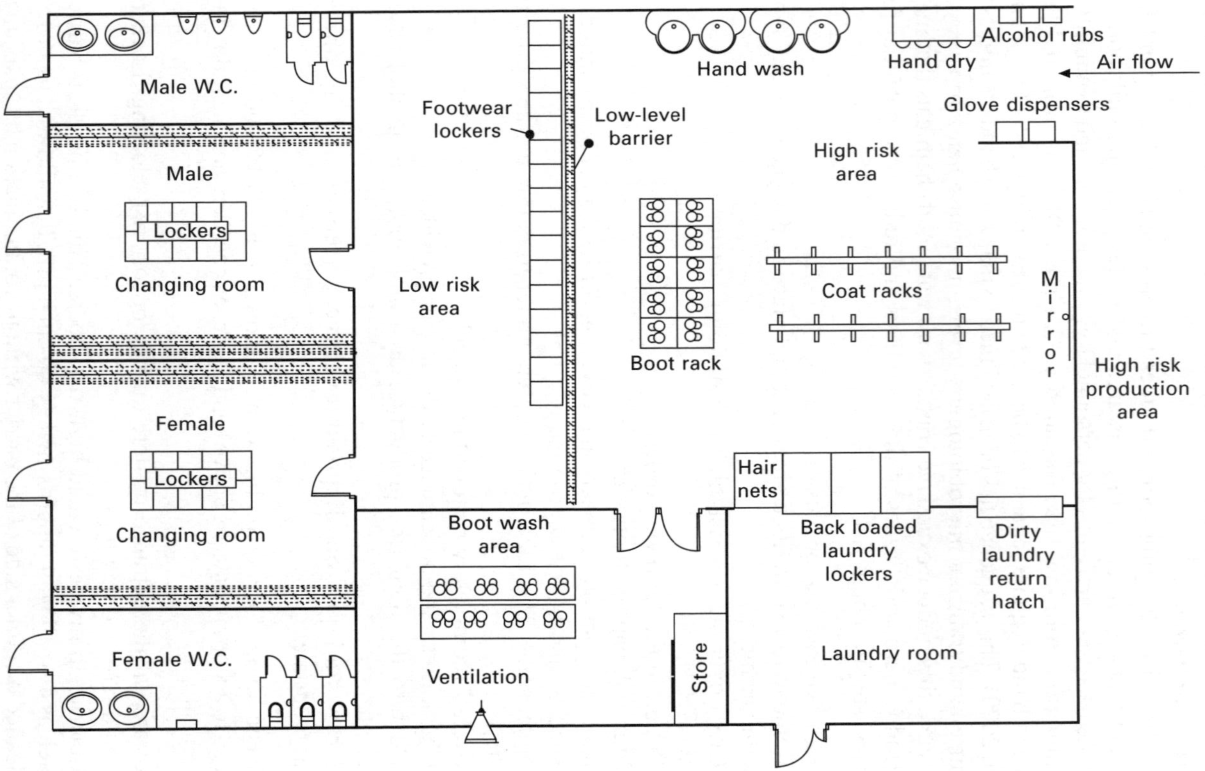

Fig. 10.8 Schematic layout for a high risk changing room.

- An area designed with suitable drainage for bootwashing operations. Research has shown (Taylor *et al.*, 2000) that manual cleaning (preferably during the cleaning shift) and industrial washing machines are satisfactory bootwashing methods.
- Hand wash basins with automatic or knee/foot operated water supplies, water supplied at a suitable temperature (that encourages hand washing) and a waste extraction system piped directly to drain. It has been shown that hand wash basins positioned at the entrance to high risk, which was the original high risk design concept to allow visual monitoring of hand wash compliance, gives rise to substantial aerosols of staphylococcal strains that can potentially contaminate the product (CCFRA, 2001).
- Suitable hand drying equipment, e.g. paper towel dispensers or hot air dryers and, for paper towels, suitable towel disposal containers.
- Access for clean factory clothing and storage of soiled clothing. For larger operations this may be via an adjoining laundry room with interconnecting hatches.
- Interlocked doors are possible such that doors only allow entrance to high risk if a key stage, e.g. hand washing, has been undertaken.
- CCT cameras as a potential monitor of hand wash compliance.
- Alcoholic hand rub dispensers immediately inside the high risk production area.

There may be the requirement to site additional hand wash basins inside the high risk area if the production process is such that frequent hand washing is necessary, and these should be clearly separated from sinks for washing utensils. As an alternative to this, Taylor *et al.* (2000) demonstrated that cleaning hands with alcoholic wipes, which can be done locally at the operative's workstation, is an effective means of hand hygiene.

10.4.8 Air

The air is an important, potential source of pathogens and the air intake into the high risk area has to be controlled. Air can enter high risk via a purpose-built air-handling system or can enter into the area from external uncontrolled sources (e.g. low risk production, packing, outside). For high risk areas, the goal of the air-handling system is to supply suitably filtered fresh air, at the correct temperature and humidity, at a slight overpressure to prevent the ingress of external air.

The cost of the air-handling systems is one of the major costs associated with the construction of a high risk area and specialist advice should always be sought before embarking on an air-handling design and construction project. Following a suitable hazard analysis, it may be concluded that the air-handling requirements for high care areas may be less stringent, especially related to filtration levels and degree of overpressure. If there are large movements of air from low risk to high care, positive pressure may be needed to reverse this flow. If airflows are more static, positive pressure in high care may not be required.

Once installed, any changes to the construction of the high risk area (e.g. the rearrangement of walls, doors or openings) should be carefully considered as they will have a major impact on the air-handling system. To aid the performance of the air-handling system, it is also important to control potential sources of aerosols, generated from personnel, production and cleaning activities, in both low and high risk.

Air quality standards for the food industry were reviewed by a Campden BRI Working Party and guidelines were produced (CCFRA, 2005). The main air flows within a high risk area are shown in Fig. 10.9, and the design of the air handling system should consider the following issues:

- degree of filtration of external air;
- overpressure;
- air flow – concerned with operational considerations and operative comfort.
- air movement;
- temperature requirements;
- local cooling and barrier control;
- humidity requirements;
- installation and maintenance.

Filtration of air is a complex matter and requires a thorough understanding of filter types and installations. The choice of filter will be dictated by the degree of external microbial and particle challenge and the degree of removal required to faciltitate the desired level of internal air cleanliness. Filter types (BS EN 779) and applications are described in detail in the Campden BRI guideline document (CCFRA, 2005). For high risk applications, a series of filters is required to provide air to the desired standard and is usually made up of a G4/F5 panel or pocket filter followed by an F9 rigid cell filter. For some high risk operations an H10 or H11 final filter may be desirable, whilst for high care operations an F7 or F8 final filter may be acceptable. The choice of filter is primarily driven by the length of time the food product is exposed to the air within high risk, as the vector of cross-contamination is via sedimentation of particles onto the food. If it is only a short time period (minutes) between the product exiting the decontamination treatment and filling into primary packaging, a lower level of filtration is required whilst if the product is exposed for longer time periods (hours) higher filtration standards may be required. Higher levels of filtration may reduce the number of particles which might sediment onto the food product.

A major risk of airborne contamination entering high risk is from low risk processing operations, especially those handling raw produce that is likely to be contaminated with pathogens. The principal role of the air-handling system is thus to provide filtered air to high risk with a positive pressure with respect to low risk. This means that wherever there is a physical break in the

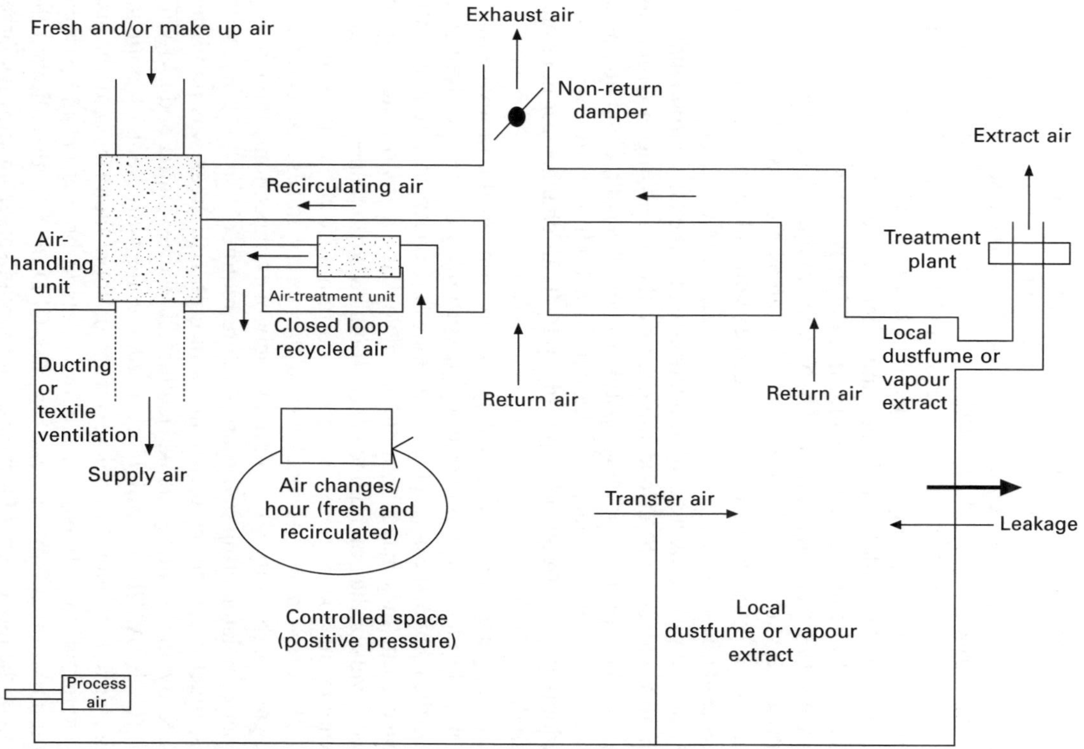

Fig. 10.9 Schematic diagram showing the airflows within a high risk or high care production area.

low/high risk barrier, e.g. a hatch, the air flow will be through the opening from high to low risk. Microbial airborne levels in low risk, depending on the product and processes being undertaken, may be quite high (Holah *et al.*, 1995), and overpressure should prevent the movement of such airborne particles, some of which may contain viable pathogenic microorganisms, entering high risk.

To be effective, the pressure differential between low and high risk should be in excess of five Pascals. The desired pressure differential will be determined by both the number and size of openings and also the temperature differentials between low and high risk. For example if the low risk area is at ambient (20 °C) and the high risk areas at 10 °C, hot air from low risk will tend to rise through any openings, e.g. hatches, whilst cold air from high risk will tend to sink through the same opening, causing two-way flow. The velocity of air through the opening from high risk may need to be 1.5 m/s or greater to ensure that one-way flow is maintained. The requirements for positive pressure in high care processing areas are less stringent, and ceiling-mounted evaporative chillers together with additional air make-up may be acceptable.

In addition to providing a positive over-pressure, the air flowrate must be sufficient to remove the heat load imposed by the processing environment (processes and people) and provide operatives with fresh air. Generally 5–25 air changes per hour are adequate, though in a high risk area with large hatches/doors that are frequently opened, up to 40 air changes per hour may be required.

Air is usually supplied to high risk by either ceiling grilles or textile ducts (socks), usually made from polyester or polypropylene to reduce shrinkage. Ceiling grilles have the advantage that they are cheap and require little maintenance, but they have limitations on velocity and flowrate without high noise levels or the potential to cause draughts. With respect to draughts, the maximum air speed close to workers to minimise discomfort through 'wind-chill' is 0.3 m s^{-1}. Air socks have the ability to distribute air, at a low draught-free velocity with minimal ductwork connections, though they require periodic laundering and spare sets are required.

A best practice guideline on air flows in high risk areas was published in the UK by the Ministry of Agriculture, Fisheries and Food (MAFF) in 2001 (CCFRA, 2001) which detailed the measurement of both air flows and airborne microbiological levels in food factories and the use of computational fluid dynamics (CFD) models to predict air and particle (including microorganism) movements. The CFD models also allowed the prediction of the movement of airborne microorganisms from known sources of microbial contamination, e.g. operatives, and visualised, for example, the influence on air flows of air intakes and air extracts, secondary ventilation systems in, for example, washroom areas, the number of hatches and doors and their degree of openings and closings. This has allowed the design of air-handling systems which provide directional air that moves particles away from the source of contamination, in a direction that does not compromise product safety.

Chilled foods manufacturers have traditionally chosen to operate their high risk areas at low temperatures, typically around 10–12 °C, both to restrict the general growth of microorganims in the environment and to prevent the growth of some (e.g. *Staphylococcus* spp.) but not all (e.g. *Listeria*) food pathogens. Chilling the area to this temperature is also beneficial in reducing the heat uptake by the product and thus maintaining the chill chain. Moreover, chilled food manufacturers have to ensure that their products meet, in the UK, the requirements of the *Food Safety (Temperature Control) Regulations 1995* (Anon., 1995) as well as those imposed by their retail customers. In the UK, *The Workplace (Health, Safety and Welfare) Regulations* (Anon., 1992), however, require that the 'temperature in all workplaces inside buildings shall be reasonable', which, in the supporting *Approved Code of Practice* (HSE, 1992), is normally taken to be at least 16 °C or at least 13 °C where much of the work involves serious physical effort.

To help solve this conflict of product and operative temperature, a document *Guidance on achieving reasonable working temperatures and conditions during production of chilled foods* (Brown, 2000) was produced which extended the information provided in HSE Food Sheet No.3 (Rev) *Workroom temperatures in places where food is handled* (HSE, 1999) (www.hse.gov.uk/pubns/fis03.pdf). The guidance document (Brown, 2000) states that employers will first need to consider alternative ways of controlling product temperatures to satisfy the *Food Safety (Temperature Control) Regulations* (Anon., 1995) rather than simply adopting lower workroom temperatures. If the alternative measures are not practical then it may be justified for hygiene reasons for workrooms to be maintained at temperatures lower than 16 °C (or 13 °C). Where such lower temperatures are adopted, employers should be able to demonstrate that they have taken appropriate measures to ensure the thermal comfort of employees.

The choice of relative air humidity is a compromise between operative comfort, product quality and environmental drying. A relative humidity of 55–65 % is very good for restricting microbial growth in the environment and increases the rate of equipment and environment drying after cleaning operations. Low humidities can, however, cause drying of the product with associated weight and quality loss, especially at higher air velocities. Higher humidities maintain product quality but may give rise to drying and condensation problems that increase the opportunity for microbial survival and growth. A compromise target humidity of 60–70 % is often recommended, which is also optimal for operative comfort.

Finally, air-handling systems should be properly installed such that they can be easily serviced and cleaned and, as part of the commissioning programme, their performance should be validated for normal use. The ability of the system to perform in other roles should also be established. These could include dumping air directly to waste during cleaning operations, to prevent air contaminated with potentially corrosive cleaning chemicals entering the air-handling unit, and recirculating ambient or heated air after cleaning operations to increase environmental drying.

10.4.9 Utensils

Wherever possible, any equipment, utensils and tools, etc. used routinely within high risk should remain in high risk, and provision for appropriate storage areas should be made. Typical examples include:

- The requirement for ingredient or product transfer containers (trays, bins, etc.) and any utensils (e.g. stirrers, spoons, ladles) should be minimised, but where these are unavoidable they should remain within high risk and be cleaned and disinfected in a separate wash room area.
- A separate wash room area should be created in which all within-production wet cleaning operations can be undertaken. The room should preferably be sited on an outside wall that facilitates air extraction and air make-up. An outside wall also allows external bulk storage of cleaning chemicals that can be directly dosed through the wall into the ring main system. The room should have its own drainage system that, in very wet operations, may include barrier drains at the entrance and exit to prevent water spread from the area. The wash area should consist of a holding area for equipment, etc. awaiting cleaning, a cleaning area for manual or automatic cleaning (e.g. traywash) as appropriate, and a holding/drying area where equipment can be stored prior to use. These areas should be as segregated as possible.
- All cleaning equipment, including hand tools (brushes, squeegees, shovels, etc.) and larger equipment (pressure washers, floor scrubbers and automats, etc.) should remain in high risk and be colour coded to differentiate between high and low risk equipment if necessary. Special provision should be made for the storage of such equipment when not in use.
- Cleaning chemicals should preferably be piped into high risk via a ring main (which should be separate from the low risk ring main). If this is not possible, cleaning chemicals should be stored in a purpose-built area.
- The most commonly used equipment service items and spares, etc., together with the necessary hand tools to undertake the service, should be stored in high risk. For certain operations, e.g. blade sharpening for meat slicers, specific engineering rooms may need to be constructed.
- Provision should be made in high risk for the storage of utensils that are used on an irregular basis but that are too large to pass through the low/high risk barrier e.g. stepladders for changing the air distribution socks.
- Written procedures should be prepared detailing how and where items that cannot be stored in high risk but are occasionally used there, or

new pieces of equipment entering high risk, will be decontaminated. If appropriate, these procedures may also need to detail the decontamination of the surrounding area in which the equipment decontamination took place.

10.5 Barrier 4: product enclosure

The fourth barrier is product enclosure and has the objective of further or totally excluding microbial contamination from product. The fourth barrier approach is essential for the production of aseptic foods, but is also being used for the production of some chilled, ready-to-eat foods, particularly sliced meat products. Product enclosure can be undertaken by physical segregation (a box within a box within a box) or by the use of highly-filtered, directional air currents.

Aseptic processing equipment is specifically built for the purpose and its description is beyond the scope of this chapter. The same microbial exclusion principals can, however, be applied to ready-to-eat products. For example, gloveboxes offer the potential to fully enclose product and work best if the product is delivered to them in a pasteurised condition, is packed within the box and involves little manual manipulation. The more complicated the product manipulation, the more ingredients to be added, the faster the production line or the shorter the product run, the less flexible gloveboxes become. Operating on a batch basis, pre-disinfected gloveboxes give the potential for a temperature-controlled environment with a modified atmosphere if required (e.g. high CO_2, or low O_2 concentrations), that can be disinfected on-line by gaseous chemicals (e.g. ozone) or UV light.

Gloveboxes may also offer some protection in the future to foodstuffs identified by risk assessments as being particularly prone to bioterrorism. Gloveboxes are only necessary, of course, if people are involved in the food production line. If robots undertook product manipulation, there would be less microbiological risk and the whole room could be temperature- and atmospherically-controlled!

Where the use of gloveboxes is impractical, partial enclosure of the product can be achieved in appropriate circumstances by the use of localised, filtered air flows. The high risk air-handling system provides control of airborne contamination external to high risk but provides only partial control of aerosols, generated from personnel, production and cleaning activities, in high risk. At best, it is possible to design an air-handling system that minimises the spread of contamination generated within high risk from directly moving over product. Localised airflows are thus designed to:

- Provide highly filtered (H11–12) air directly over or surrounding product, and its associated equipment. This could reduce the requirement to chill the whole of the high risk area to 10 °C (13 °C would be acceptable), and reduce the degree of filtration required (down to H8-9). The requirement for positive pressure in low risk is paramount, however, and the number of air changes per hour would remain unchanged.

- Provide a degree of product isolation ranging from partial enclosure in tunnels to chilled conveyor wells, where the flow of the filtered air provides a barrier that resists the penetration of aerosol particles, some of which would contain viable microorganisms.

An example of such a technology is shown in Fig. 10.10 which shows a schematic diagram of a conveyor that has chilled, filtered air directed over it, sufficient to maintain the low temperature of the product. When a microbial aerosol was generated around the operational conveyor, microbiological air sampling demonstrated a 1–2 log reduction of microorganisms within the protected zone (Burfoot *et al.*, 2001).

10.6 The design of smaller manufacturing and catering operations

The information given in this chapter has been structured towards larger food manufacturing plants producing packaged final products to be eaten directly by the consumer. However, there are many smaller food manufacturing units and catering operations that are producing a similar range of products, sometimes directly to the consumer that, for a number of reasons, could not incorporate the appropriate barrier technologies described in full. This may be due to space limitations (not allowing physically segregated areas), ownership issues (manufacturing units may be rented with strict controls over the degree of structural change that the food manufacturer can make) or costs.

The same barrier principles can, of course, be applied across the whole of the food manufacturing and service sector. For example, if a catering facility is handling products of a different risk category, e.g. raw and cooked products,

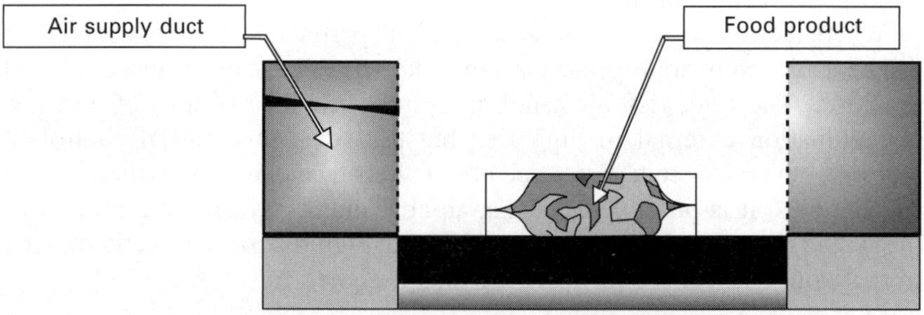

Fig. 10.10 Chilled air is supplied from air ducts on either side of a product conveyor. The chilled air retains the product temperature and its movement, spilling over the duct surfaces, and provides a barrier to microorganism penetration.

instead of a third barrier to segregate low and high risk activities, the facility can be divided into separate areas to handle raw and cooked materials. This can be further enforced by using colour-coded equipment, utensils, chopping boards, storage devices and cleaning equipment to handle raw and cooked products. In addition, it can be ensured that operatives wash their hands between handling raw and cooked products, that cooked and raw waste is handled separately and, with respect to the air, the raw and cooked operations are separated sufficiently such that any aerosols generated from the cutting, chopping, handling, etc. of the raw materials can not splash directly onto cooked products or contact surfaces. The design of other aspects of small manufacturing units and catering facilities related to goods in, de-boxing, storage, preparation, processing, packing (if appropriate) and dispatch/retail will also be similar to that of larger food factories as these are governed by product flow and manufacturing efficiency.

It is difficult to compare the microbiological quality of products prepared in large factories, that are able to implement all appropriate barriers, and smaller manufacturing units that cannot. A comprehensive review of the incidence of *L. monocytogenes* in retail and food service environments was undertaken by Lianou and Sofos (2007) which tables the incidence of the organism in a wide range of food products around the world. The authors cite a number of references which suggest that food products purchased from retail premises are more frequently contaminated with *L. monocytogenes* than from pre-packaged products. Comparisons of *L. monocytogenes* incidence between countries are very difficult as each country is likely to have slightly different products and manufacturing systems. Within countries, however, this may be more realistic. For example Holah *et al.* (2004) reported an incidence of 0.35 % and 0.25 % for *Listeria* spp. and *L. monocytogenes*, respectively, in 170 000 ready-to-eat product samples taken from four chilled food manufacturers over a three year period. In the same year, Elson *et al.* (2004) reported an incidence of 6.0 % and 2.1 % for *Listeria* spp. and *L. monocytogenes*, respectively, in 2694 product samples taken from catering and retail premises. Within catering and retail premises, Elson *et al.* (2004) also reported that more microbiologically unsatisfactory samples were found where (i), there was no low/high risk segregation and (ii), staff moved freely from low to high risk areas and did not change protective clothing. It could also be suggested that within smaller manufacturing units, the use of common product-handling utensils and cleaning equipment, the use of equipment less likely to be designed hygienically (and thus less cleanable) and the inability to readily clean surfaces to the same rigour as in larger food plants would also contribute to elevated microbiologically unsatisfactory product as compared to food factories.

With respect to the microbiological risk of the product, there is a balance of likely pathogen incidence and the length of shelf-life in which pathogens could grow to unacceptable numbers. Food products produced in smaller food manufacturing units and catering operations are likely to have a (relatively) higher incidence of pathogens though they are also likely to be consumed very quickly. Conversely, food products produced in a larger food manufacturing facility are likely to have a smaller incidence of pathogens but have a longer shelf-life in which, for pathogens that require growth, growth to unacceptable levels could occur (Lianou and Sofos, 2007). Following this basic philosophy it is safe to manufacture short shelf-life products in smaller food manufacturing units and catering operations and longer shelf-life products in appropriately hygienically designed larger facilities. The manufacture of longer shelf-life products in smaller food manufacturing units and catering operations could, however, unacceptably increase microbiological risk.

10.7 Future trends

New advances in hygiene will always drive the food industry to produce safer foods and a number of trends can be identified that may have an impact on the hygienic design of food manufacturing facilities. These include:

- **Hygiene** – Molecular identification techniques have established that specific strains of pathogens may persist in food factories for many years (Tompkin, 2002; Holah *et al.*, 2004), so much so that these are sometimes termed 'house strains'. Why they colonise plants and where they persist is currently unknown, though developments in this area will certainly influence the internal design and finishes of building materials.
- **Adulteration** – The potential to further control both accidental and deliberate adulteration of foods during manufacture will certainly be progressed. Initial guidance on this issue has been established by the WHO, though there is little direct guidance on hygienic plant design (WHO, 2002), whilst the American Meat Institute are the most progressive trade organisation in developing practical guidance (AMI, 2008). Current thinking may be to design process steps so that they can be done collectively by operatives, such that each operative observes the actions of others. Other areas of the factory such as storage would be automated, monitored by CCT cameras and not accessed by single operatives.
- **Authenticity** – As chemists become more adept at authenticity testing, smaller quantities of one product can be detected in another. For example, if a meat manufacturer produces beef sausages and then follows this down the same process line with pork sausages, how much residue of beef is acceptable in the pork sausages? If the answer to this is only 'small' quantities, we may need to consider extended cleaning times between product runs. If the answer to this is 'very small' quantities, we may need to consider separate beef and pork process lines. If the answer is 'very, very small' quantities, we may need to consider separate beef and pork production rooms to prevent, for example, the cross-contamination

from one line to the other due to cleaning aerosols or operatives moving between the lines.

- **Flexibility** – In some sectors the speed of new product design is so rapid that it demands constant reconsideration of raw materials, processes, packaging concepts and thus process lines. How can we design more modular food manufacturing areas that are both hygienic and can be rapidly reconfigured for ever changing new products?

10.8 Sources of further information and advice

There are very few academic texts describing the hygienic design of food processing facilities though there is a section on building and equipment design in the book *Handbook of Hygiene Control in the Food Industry* (Lelieveld *et al.*, 2005). The majority of information in Europe is provided by research organisations (e.g. Campden BRI) and industry focus groups (e.g. EHEDG) as noted in the text. For many food processors, information on hygienic design is either sourced in-house (for the major international companies), from their customers (particularly the major retailers) or from architects and building contractors. If the only source of information will be from the architect or building contractor, it is critical to establish whether these service providers have experience in the construction of food processing facilities. Once plans for the refurbishment or new-build have been drawn up, in addition to the often legal requirements of approval to local building and fire regulations, they can then be circulated for additional advice and comment to the local health authority, insurers and to customers. From the author's personal knowledge, it is floors and their associated drainage that are the one area in the design and building of food processing areas that has led to the most frequent problems and has also been very expensive to rectify. Appropriate design, adequate funding and the use of specialised flooring contractors are essential in this area.

10.9 References

AMI (2008) *Food Security Checklist for the Meat and Poultry Industry*, American Meat Institute, Washington, DC, available at: http://www.meatami.com/ht/a/GetDocumentAction/i/1450, accessed November 2008.

Anon (1992) *The Workplace (Health, Safety and Welfare) Regulations*, HMSO, London.

Anon. (1995) *Food Safety (Temperature Control) Regulations 1995*, HMSO, London.

Ashford, M. J. (1986) *Factory design principles in the food processing industry*. in *Preparation, Processing and Freezing in Frozen Food Production*, The Institution of Mechanical Engineers, London.

Brown, K. L. (2000) *Guidance on Achieving Reasonable Working Temperatures and Conditions During Production of Chilled Foods*, Guideline No. 26, Campden & Chorleywood Food Research Association, Chipping Campden.

BSI (2002) BSEN 779: 2002 Particulate air filters for general ventilation. Determination of the filtration performance, British Standard Institute, London.

Burfoot, D., Brown, K., Reavell, S. and Xu, Y. (2001) Improving food hygiene through localised air flows, *Proceedings International Congress on Engineering and Food, Vol. 2,* April, Puebla, Mexico, Technomic, Lancaster, PA, 1777–81.

CCFRA (2001) *Best Practice Guidelines on Air Flows in High-care and High-risk Areas*, published for the UK Ministry of Agriculture, Fisheries and Food by Campden & Chorleywood Food Research Association, Chipping Campden.

CCFRA (2002) *Guidelines For the Design and Construction of Floors for Food Production Areas*, (2nd ed), Guideline No. 40, Campden & Chorleywood Food Research Association, Chiping Campden.

CCFRA (2003a) *Guidelines For the Hygienic Design, Construction and Layout of Food Processing Factories*, Guideline No. 39, Campden & Chorleywood Food Research Association, Chipping Campden.

CCFRA (2003b) *Guidelines For the Design and Construction of Walls, Ceilings and Services for Food Production Areas*, (2nd edn), Guideline No. 41, Campden & Chorleywood Food Research Association, Chipping Campden.

CCFRA (2005) *Guidelines on Air Quality For the Food Industry*, Guideline No. 12, Campden & Chorleywood Food Research Association, Chipping Campden.

CFA (2006) *Best Practice Guidelines For the Production of Chilled Foods* (4th ed.), Chilled Food Association, Peterborough.

EC (2004a) Regulation (EC) No 852/2004 of the European Parliament and of the Council of Europe of 29 April 2004 on the hygiene of foodstuffs, *Official Journal of the European Union*, **L226**, 30 April, 3–21.

EC (2004b) Regulation (EC) No 853/2004 of the European Parliament and of the Council of 29 April 2004 laying down specific hygiene rules for the hygiene of foodstuffs, *Official Journal of the EU*, **L139**, 30 April, 55–205.

EHEDG (2001) *Hygienic Design Criteria for the Safe Processing of Dry Particulate Materials*, Guideline Document 22, European Hygienic Engineering and Design Group, Frankfurt.

Elson, R., Burgess, F., Little, C. and Mitchell, R. T. (2004) Microbiological examination of ready-to-eat cold sliced meats and pâté from catering and retail premises in the UK. *Journal of Applied Bacteriology*, **96**, 499–509.

Guzewich, J. and Ross, P. (1999) *Evaluation of risks related to microbiological contamination of ready-to-eat food by food preparation workers and the effectiveness of interventions to minimise those risks*, White Paper, Section One, Food and Drug Administration, Washington, DC.

Holah, J. T. (2003) *Guidelines for the Hygienic Design, Construction and Layout of Food Processing Factories*, Guideline No. 39, Campden & Chorleywood Food Research Association, Chipping Campden.

Holah, J. T., Bird, J and Hall, K. E. (2004) The microbial ecology of high risk, chilled food factories; evidence for persistent *Listeria* spp. and *Escherichia coli* strains, *Journal of Applied Microbiology*, **97**, 68–77.

Holah, J. T. and Hall, K. E., Holder, J., Rogers, S. J., Taylor, J. and Brown, K. L. (1995) *Airborne Microorganisms Levels in Food Processing Environments*, R&D Report No. 12, Campden & Chorleywood Food Research Association, Chipping Campden.

Holah, J. T. and Thorpe, R. H. (2000) Hygienic design considerations for chilled food plants in M. Stringer and C. Dennis (eds.) *Chilled Foods: a Comprehensive Guide*, 2nd edn, Woodhead 397–428.

HSE (1992) *Workplace Health, Safety and Welfare*, Approved Code of Practice and Guidance on Workplace (Health, Safety and Welfare) Regulations, Health and Safety Executive, London.

HSE (1999) *Workroom Temperatures in Places Where Food is Handled*, HSE Food Sheet No. 3 (revised), Health and Safety Executive, London.

Imholte, T. J. (1984) *Engineering for Food Safety and Sanitation*, Technical Institute of Food Safety, Crystal, MS.
Katsuyama, A. M. and Strachan, J. P. (eds) (1980) *Principles of Food Processing Sanitation*. The Food Processors Institute, Washington, DC.
Lianou, A. and Sofos, J. N. (2007) A review of the incidence and transmission of *Listeria monocytogenes* in ready-to-eat products in retail and food service environments, *Journal of Food Protection*, **70**, 2172–98.
Lelieveld, H. L. M., Mostert, M. A. and Holah, J. (2005) *Handbook of Hygiene Control in the Food Industry*, Woodhead, Cambridge.
Mettler, E. and Carpentier, B. (1998) Variations over time of microbial load and physicochemical properties of floor materials after cleaning in food industry premises. *Journal of Food Protection*, **61**, 57–65.
Shapton, D. A. and Shapton, N. F. (eds) (1993) *Principles and Practices for the Safe Processing of Foods*, Woodhead, Cambridge.
Taylor, J. and Holah, J. T. (1996) A comparative evaluation with respect to bacterial cleanability of a range of wall and floor surface materials used in the food industry, *Journal of Applied Bacteriology*, **81**, 262–7.
Taylor, J. H. and Holah, J. T. (2000) *Hand Hygiene in the Food Industry: a Review*, Review No. 18, Campden & Chorleywood Food Research Association, Chipping Campden.
Taylor, J. H., Holah, J. T., Walker, H. and Kaur, M. (2000) *Hand and Footwear Hygiene: An Investigation to Define Best Practice*, R&D Report No. 110, Campden & Chorleywood Food Research Association, Chipping Campden.
Tompkin, R. B. (2002) Control of *Listeria monocytogenes* in the food-processing environment, *Journal of Food Protection*, **65**, 709–25.
Troller, J. A. (1983) *Sanitation in Food Processing*, Academic Press, New York.
WHO (2002) *Terrorist Threats to Food – Guidelines for Establishing and Strengthening Prevention and Response Systems*, Food Safety Department, World Health Organisation, Geneva.

11

Hygienic equipment design

A. Hasting, Tony Hasting Consulting, UK

Abstract: The hygienic design of equipment for the food industry is a key element in ensuring the delivery of safe products to the consumer. Poor design potentially compromises the safety of the final product with serious commercial implications for the company involved. The principles of hygienic design include construction materials, equipment geometry and fabrication and equipment cleanability, while a key issue is the interaction between the product and the equipment. There are design principles that are generic in nature while major equipment items such as heat exchangers, tanks and product transfer equipment have their own specific requirements and challenges. The industry is rapidly changing in response to commercial and political pressures, and the application of hygienic design principles has to take these into account, with the trend increasingly focussing on risk assessment and its management. Such assessments need to be based on the specific application, as the choice of equipment has to be based on functional performance as well as equipment hygiene. Process lines containing unhygienic equipment may still be capable of producing acceptable product provided there is a clear understanding of the implications of such equipment and at the expense of increased costs and reduced line efficiency.

Key words: hygiene, equipment, design principles, constructional details, heat exchangers, product transfer systems, tanks, valves.

11.1 Introduction

The hygienic design of equipment for the food and allied hygienic industries is a key element in ensuring that all products produced are consistently safe from physical, chemical or microbiological contaminants. The equipment that transforms the raw materials into final product results in a direct interaction between the food materials and the equipment and, as will be described in this chapter, poor hygienic design can compromise the safety of the final product.

From an engineering perspective, equipment design while important is only one of a number of factors where an engineering understanding is essential. An emphasis on equipment design alone can increase rather than decrease the risk to the quality of the final product. These factors are:

- **Process design** – Even if the equipment meets the highest hygienic design standards, an inadequate process design will result in the product safety being compromised. For example, the intrinsic safety of product from a pasteurisation process designed with an insufficient time–temperature combination will be influenced not by the hygienic design of the equipment but by the inadequate process design.
- **Installation and layout** – Even if the individual items of equipment are acceptable from a hygienic design perspective, their integration into a complete process line can create many hygiene issues that can outweigh the benefits of using equipment designed to the most rigorous of hygiene standards.
- **Operation and control** – The control of the process plant during start-up, production, shut-down and cleaning are again critical to the safety of the final product. There can be a risk that the main focus is on the production phase and key areas such as cleaning and plant start-up are not given the attention they require. Measurement is also a key element of operation and control as, without the ability to measure process parameters reliably, it is not possible to monitor, control, validate or optimise the process.
- **Manufacturing environment** – The environment within which the process equipment is installed can have a significant impact on product quality, particularly where the product is exposed for all or part of the process.
- **Storage and distribution** – The conditions the final product is subjected to during storage and distribution, which may range from a few days to many months, are factors where engineering understanding is important but equipment design may not be a critical factor.

It is a feature of commercial operation in an increasingly competitive environment that food manufacturers often have to use equipment that is clearly of lower hygienic design quality than would be desirable, but financial pressures and time constraints do not allow for the purchase of new plant. This is particularly true for the case where products are being manufactured by a third party rather than in-house. It is therefore essential to be aware of the limitations of the equipment, evaluate the risks involved in its use and manage the overall operation to minimise such risks. It may well be possible to achieve acceptable product quality using unhygienic equipment, but compromises will have to be made with the operation of such a line, and the implications of such compromises must be addressed from the outset.

Food processing equipment is designed and built to fulfil a specific purpose such as to heat, cool or convey a product through the process line. Equipment with excellent hygienic characteristics but unable to meet its functional requirements will be of little value in a practical production environment. Upgrading existing designs to meet more rigorous hygienic requirements may well be both impractical as well as expensive, hence the importance for hygienic requirements to be taken into account as early as possible during the design stage. Waiting until the process has been fully specified is not recommended as the ability to incorporate the necessary hygienic characteristics may be compromised.

11.2 Regulatory requirements

The EU Machinery Directive 89/392/EEC (EU, 1989) together with amendments 91/368/EEC, 93/44/EEC and 93/68/EEC made it a legal obligation for machinery sold in the EU after January 1st 1995 to be safe to use, provided the manufacturer's instructions were followed. The Directive included a short section on requirements for hygiene and design, which stated that machinery intended for the preparation and processing of foods must be designed and constructed to avoid health risks. Subsequent to this Directive, a European Standard EN 1672-2, (BSI, 1997), has been adopted to further clarify the hygiene rules established in the original directive (Holah, 1998).

In the USA the 3-A Sanitary Standards are used as the source of hygienic criteria for food and dairy processing, packaging and packaging equipment. These can be applied to equipment and machinery (3-A Sanitary Standards) as well as to systems (3-A Accepted Practices) (Gilmore, 2003).

11.2.1 European Hygienic Engineering and Design Group (EHEDG)

Many groups have been involved in developing hygienic equipment design principles over a number of years. The EHEDG (www.ehedg.org) is an active, independent group, which includes amongst its members food manufacturers, equipment suppliers, research institutes and universities, thus bringing together specialists from many disciplines, from academia through to commercial operation. The objective of the group is to address specific issues relating to the hygienic manufacture of food products by providing design criteria and guidelines on all aspects of equipment, buildings and processing. The emphasis on guidelines is an important feature, which avoids the prescriptive, individual design specifications sometimes published by other international organisations. Extended summaries of many of these guidelines have been published in *Trends in Food Science and Technology* published by Elsevier, and complete versions of the documents may be obtained via the EHEDG website. Both EHEDG and 3-A are committed to collaboration in the area of hygienic design to ensure as far as possible a consistent set of standards and requirements. This aims to avoid situations where equipment manufacturers have to meet different hygiene standards in various parts of the world, which in many cases will result in additional costs for both supplier and end-user.

11.3 Hygienic design principles

The principles of hygienic equipment design can be broadly classified under three main headings:

- materials of construction;
- equipment geometry and fabrication;
- cleaning.

While the principles of hygienic design are usually accepted as being technically desirable and straightforward, they do have a number of limitations when applied in commercial practice (Hasting, 2008).

- They are difficult to apply in absolute terms.
- They are defined in isolation from the product being processed.
- They are related to specific items of equipment rather than the process itself.

The thermophysical properties of the product, in particular its rheology, will have a major impact on the interaction with the equipment geometry. A low-viscosity Newtonian fluid such as a soft drink will interact with equipment in a very different way to a viscous non-Newtonian material such as ice cream mix or tomato paste.

In many practical applications, different pieces of equipment may require different levels of hygienic design depending on the risk to the final product. In general, the closer the equipment is to the end of the line, the more likely it is that the hygiene requirements will increase. For example, the hygienic design of a mixer for raw meat need not achieve the same standards as a slicer for cooked meats (Timperley and Timperley, 1993). With this in mind the EHEDG have developed definitions associated with different hygiene levels, ranging from unhygienic to fully aseptic, Table 11.1 (EHEDG, 1993b).

It may, however, still be possible to produce microbiologically acceptable product using unhygienic equipment, but there will almost certainly be significant risk implications in such an approach as well as increased cost. Use of unhygienic equipment is likely to result in (Hasting, 2008):

- shorter run lengths between cleaning;
- longer cleaning times;

Table 11.1 Equipment definitions

Class of equipment	General definition
Aseptic equipment	Hygienic equipment that is, in addition, impermeable to microorganisms
Hygienic equipment Class 1	Equipment that can be cleaned in place and freed from relevant microorganisms without dismantling
Hygienic equipment Class 2	Equipment that is cleanable after dismantling and that can be freed from relevant microorganisms by steam sterilisation or pasteurisation after reassembly

- more aggressive cleaning regimes;
- a less consistent and robust process;
- increased product and equipment testing, thus moving from quality assurance towards quality control.

11.4 Hygienic design requirements

The European Standard EN 1672-2 (BSI, 1997) summarises hygienic design requirements under a number of generic headings.

11.4.1 Materials of construction

Food contact materials must be inert to the product as well as any chemicals used, such as detergents or biocides, under all in-use conditions. Temperature is often a key parameter as materials may need to be resistant to pressurised hot water or steam sterilisation at elevated temperatures of up 140–150 °C. They must be corrosion-resistant, non-tainting, mechanically stable, smooth and non-porous. No toxic materials should be used for food contact, and particular care must be taken when elastomers, plastics, adhesives and signal transfer liquids are used as these may contain toxic components that could be leached out into product. The suppliers of such components must provide clear evidence that the materials meet all legislative requirements.

Stainless steel

There are a wide range of stainless steels available as construction materials for food industry equipment, the choice being largely dependent on the corrosive properties of the product or other chemicals that will come into contact with the material. The most common choices are the austenitic stainless steels, AISI 304, AISI 316 and AISI 316L, which have good mechanical properties and fabrication characteristics, as well as having an attractive appearance.

All stainless steels are susceptible to corrosion in the presence of chloride-containing environments (Covert and Tuthill, 2000) and may take different forms – pitting, crevice or stress corrosion. However, a common characteristic is that they are localised and are affected by factors such as chemical environment, pH, temperature, fabrication methods, tensile stresses, oxygen concentration and surface finish.

Plastics

Plastics are increasingly widely used in a range of food applications, both in terms of equipment components as well as the final package. Plastics must conform to the following requirements.

- Approval for use with food contact: the supplier must provide evidence of the approval certification. There are particular concerns regarding the possible tainting of product due to the leaching out of potentially toxic components.
- Capability of withstanding in-use conditions during both production and cleaning, including temperatures, pressures and chemical concentrations.
- Robustness to withstand the mechanical shocks that are likely to occur during normal operation without distortion and damage.

The following plastics are considered to be easy to clean and are suitable for use in hygienic applications (Lewan, 2003):

- polypropylene
- polyvinyl chloride (unplasticised)
- acetal copolymer
- polycarbonate
- high-density polyethylene.

Polytetrafluoroethylene (PTFE) is often considered to be a potentially attractive material due to its high chemical and temperature resistance. However, care must be taken because it can be porous and thus difficult to clean. In addition, it may be insufficiently resilient to provide a permanently tight seal and it is therefore considered unsuitable for aseptic processing (Lewan, 2003).

Elastomers

Many types of rubber-based elastomer are used for seals, gaskets and joint rings due to their high elasticity and consequent ability to return to the original shape once removed from the source of the stress. The most widely used elastomers are:

- nitrile rubber
- nitrile/butyl rubber (NBR)
- ethylene propylene diene monomer (EPDM)
- silicon rubber
- fluoroelastomer (Viton®, Dupont, USA).

Of these silicon and Viton® have higher temperature capability, up to

180 °C, and chemical resistance than the others and hence are also more expensive.

Specification of such materials poses a number of challenges (Lewan, 2003).

- There are no agreed standards for compounding and hence a wide range of elastomer compounds may be used by the supplier.
- All conditions the material is subjected to must be considered including in-use process conditions and compatibility with the product. For example EPDM is not resistant to oils and fats.
- Storage conditions prior to use, e.g. exposure to UV or ozone, can affect the performance of the elastomer.
- The mechanical characteristics of the application, such as whether the elastomer is in a static or dynamic environment, can vary widely and thus impact on the selection.

Routine replacement of elastomers is a normal part of plant maintenance, the frequency being application-dependent and a function of the physical and chemical stresses imposed on the material. Symptoms of elastomer deterioration are usually seen as a hardening of the material leading to a loss of elasticity and eventually a loss of sealing function. This deterioration is frequently associated with the cleaning process with the combination of aggressive chemicals and elevated temperatures. Failure to replace elastomers in such circumstances can result in more rapid corrosion due to chemicals being able to access the area between the elastomer and the construction material.

Mechanical stresses can also create hygiene issues when the elastomer deteriorates to such an extent that the material can be abraded and physically break up. This can result in physical contamination of the product with elastomer particles as well as creating a roughened surface with cavities that is much more difficult to clean.

11.4.2 Surface finish

All food contact surfaces must be smooth, non-porous and easily cleanable (Holah and Thorpe, 1990) and free from large, randomly distributed irregularities such as pits, folds and crevices. The attachment and removal of food materials and microorganisms from surfaces is believed to be a function of the surface roughness such that the rougher the surface the longer the cleaning time. The Ra value is the most common measure of roughness, being defined as the 'average departure of the surface profile from a calculated centre line' (Verran, 2005) and is the most commonly used measure to define or specify a surface. While the Ra value gives an indication of the amplitude of the surface irregularities, it is based on a linear trace and hence gives no information on either the three-dimensional nature of the surface or the two-dimensional topography. Currently a maximum roughness of 0.8 μm has been specified for food contact surfaces both by the EHEDG and the American 3-A's organisation (Curiel *et al.*, 1993), although higher values are acceptable providing their cleanability can be demonstrated. The development of more sophisticated instruments such as atomic force microscopy (AFM) allows greater information on the surface topography to be obtained with the ability to provide Ra data on the nanometre scale together with an image of the area scanned (Verran, 2005).

11.4.3 Joints

Joints fall into two categories, permanent joints such as welds and dismountable joints such as pipe couplings. The primary function of a weld is to provide a permanent joint of the required mechanical strength whilst meeting any legislative requirements such as pressure vessel codes. Welds must be smooth and continuous, and poor-quality welding is likely to compromise the overall hygiene in an otherwise high-quality plant. Poor-quality welds can result in crevices where product can reside and be difficult to clean and, in extreme cases, may pose a foreign body hazard. Welding can be carried out manually using a TIG (tungsten inert gas) process or preferably automatically using an orbital welder. A number of key requirements for assuring the production of hygienic welds have been proposed (EHEDG, 1993a). For vessels of up to about 4 mm wall thickness, cold rolled sheet is available, which should have a surface finish well within the 0.8 μm specified. Other than being free from grease and dirt, the weld area does not generally require any special preparation. After welding, the weld must be ground flush and polished with a 150 grit abrasive to achieve the required surface finish. For thicker-walled vessels, hot-walled plate may have to be used and, since this is likely to have surface roughness, Ra, in excess of 4 μm would be unacceptable for hygienic operation. Hence, the whole vessel must be polished to the required finish after the internal weld has been ground flush with the surface.

The pipe couplings most widely used in the food industry are:

- DIN 11851;
- SMS (Swedish metric standard);
- clamp type to BS 4825-3 (BSI, 1991a);
- International Dairy Federation (IDF), often called International Sanitary Standard (ISS), (BSI, 1991b) or International Standards Organisation (ISO, 1976);
- ring joint type (RJT) and special ring joint type (SRJT) to BS 4825-5 (BSI, 1991c).

A number of practical considerations when selecting a pipe coupling were described by Sillett (2006).

- Avoid mixing metric and imperial dimension couplings as the difference in diameter will create a step in the tube wall at the joint, which may be difficult to clean

- The design of some couplings is such that a crevice is created on assembly, e.g. the RJT coupling.
- Where frequent dismantling of coupling is required, a retained gasket design can prevent the gasket falling out of the coupling and thus assists assembly.
- Some gaskets such as the IDF type are available with a metal reinforcement ring and are less prone to damage, distortion and over-tightening. Over-tightening can result in the gasket being extruded into the tube bore creating a step that cannot be cleaned as well as causing premature failure.
- Some fittings such as clamp and IDF are difficult to reassemble unless the two parts of the coupling are well aligned. Others such as RJT can withstand a significant degree of misalignment.

The general recommendation in the design of joints for pipework systems is that welding should be used where possible and pipe couplings only specified where necessary, for example where equipment components may need to be removed for inspection or maintenance.

11.4.4 Dead spaces

Dead spaces are defined as areas outside the main product flow where material can accumulate and remain for an extended period of time. This can have two important implications for process hygiene (Fig. 11.1).

1. During production, if the local environmental conditions such as temperature are favourable, microbial growth may occur within the product held up in the dead space. After a period of time the microorganisms may be able to transfer back into the main product flow, for example by diffusion, and contaminate product.

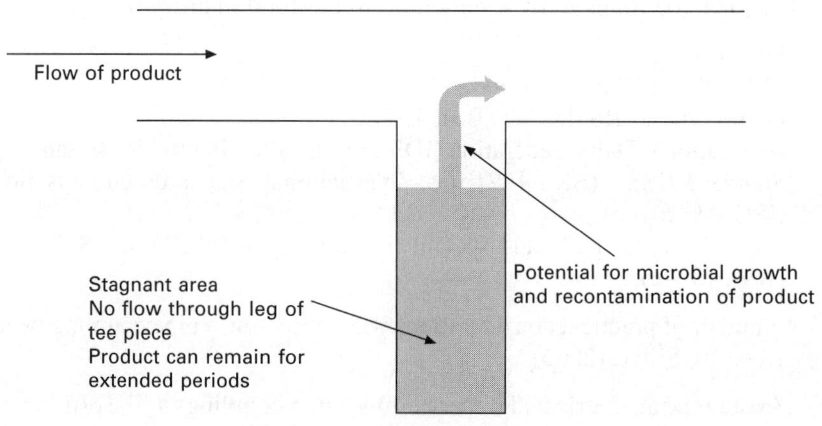

Fig. 11.1 Example of an inaccessible dead space within pipework.

2. It may also be difficult to clean and disinfect or sterilise such an area after production has finished due to the lack of fluid movement within the dead space. This would result in a potentially contaminated zone at the beginning of production, which could have a detrimental impact on product quality immediately on start-up.

The majority of such problems occur due to the physical assembly, installation and integration of individual equipment components with connecting pipework, examples being T junctions in pipework, incorrect orientation of equipment and incorrect installation of sensors into pipework.

The boundary between the bulk flow and the totally static fluid in Fig. 11.1 will not be a clearly defined interface; rather there will be a volume of fluid between the two extremes across which the fluid motion steadily reduces in intensity from the maximum at the centre of the tube to zero in the dead space. The length of this interface, L_i will be a complex function of fluid velocity and rheology as well as the geometry of the equipment. The cleaning of a range of pipework geometries was studied by Tamplin (1990). Different orientations of dead space were studied under bulk flow velocities of 0.3–1.5 m s^{-1} and the importance of fluid velocity highlighted, but it was shown that in some cases even the highest flow velocity was unable to clean the dead space. Other experimental and theoretical work has confirmed these findings (Grasshof, 1980).

It is a common misconception that a dead space will still be capable of being disinfected or sterilised using either chemical or thermal means, provided additional circulation time is used. For chemical biocides it is clearly going to be difficult to ensure that the chemical can be delivered to the whole of the dead zone as this will take place by diffusion, which is likely to take substantially longer than the circulation time of the biocide. In addition, there will also be chemical reaction occurring as the biocide reacts with the product in the dead zone, which will slow the process down even further. Heat is often considered to be more effective than chemical in penetration into a dead zone due to conduction.

However, recent studies (Asteriadou *et al.*, 2006) have shown that even with turbulent flow within a horizontal pipeline, the temperature within a dead zone orientated vertically downwards will fall rapidly with distance from the pipe centreline. This is to be expected since the natural inclination is for heat to rise and, even if heat starts to penetrate the dead space, there will be a cooling effect from the walls of the pipe in contact with the external environment. This will potentially compromise the thermal process due to a failure to reach the required minimum temperature conditions and ensure inactivation of the target microorganisms.

11.4.5 Other constructional features

There are a number of publications that provide a detailed assessment of different constructional features and the important issues regarding hygiene

(Lelieveld *et al.*, 2003, 2005) and some of the key features of concern are summarised in Table 11.2.

11.4.6 Non-product fluids

In most equipment and processes it is almost inevitable that non-product fluids will be used as:

- service fluids such as steam, water, glycol, brine;
- fluids that may be incorporated into the product deliberately, such as air, nitrogen, carbon dioxide;
- fluids that fulfil a functional purpose for the equipment, such as air to operate pneumatic equipment and lubricants for lubrication and corrosion protection.

All such fluids that may come into contact with the food product must be suitable for contact with food and thus conform to the relevant standards. For fluids that are unsuitable for food contact, risks must be managed by appropriate design and operational procedures. As an example, steam for use in direct contact with product such as in direct injection systems will have to be treated as a food ingredient and boiler feedwater chemicals will have to be acceptable for food contact.

In some applications such as the use of lubricants, the same food grade requirements must be met, but the condition of the lubricant after a period of use must be taken into account. Over time lubricants may deteriorate

Table 11.2 Key constructional features of hygiene concern

Component	Features
Fasteners	Avoid exposed screw threads, nuts, bolts and screws in food contact areas.
Drainage	Product contact surfaces should be self draining. Residual fluids can cause microbial or chemical contamination as well as increased risk of corrosion (Hasting, 2005).
Internal angles and corners	Angles and corners should be radiused, minimum 3 mm, to allow easy cleaning. Avoid sharp corners where possible.
Bearings	Mount bearings outside product area where possible to avoid contamination by lubricant or damage to bearings on contact with product. If located in product area, design must allow access of cleaning fluids to all surfaces.
Shaft seals	Must be easy to clean and to remove for replacement or maintenance.
Instrumentation	Where an instrument contains transmitting fluid, this must be acceptable for food contact. Hygiene issues frequently associated with installation rather than inherent design of sensor.

due to particle pick-up from machinery and lose performance, with adverse consequences for the machinery. If lubricants absorb moisture it is possible that microbial growth could occur leading to potential product contamination.

11.5 Hygienic design of some major equipment items

11.5.1 Heat exchangers

Heat transfer is one of the most widely used unit operations in the food industry with key processes such as pasteurisation and sterilisation reliant upon it. While heat transfer can take place in tanks and process vessels, heat exchangers are the most common way of achieving effective heating.

Heat exchangers are available in a variety of geometries, all of which are designed to achieve their many heating or cooling duties in the most cost-effective manner, whilst maintaining product quality. Their selection will invariably be strongly influenced by the characteristics of the product being handled, especially the viscosity and presence or absence of particles, as well as the capacity. Three main types of heat exchanger are commonly used each of which poses specific hygienic design challenges (Hasting, 2005).

- Plate heat exchanger:
  - standard/wide gap – fluids A and B flow through alternate flow channels;
  - dual plate with air gap – single plate unit consists of two plates with an air gap in between.
- Tubular heat exchanger (straight tubes):
  - monotube – single tube within a tube;
  - multiple tube in tube – a number of small diameter tubes within a single product shell;
  - triple tube – three concentric tubes, fluid A flows through inner tube and outer annulus, product B through inner annulus.
- Tubular heat exchanger (coiled tubes):
  - monotube and triple tube – similar principle to straight tube.
- Scraped surface heat exchanger:
  - jacketed cylinder with blades fitted on rotating shaft to scrape internal heat transfer surface;
  - jacketed cylinder with scrapers on shaft moving in a linear, reciprocating motion.

Plate heat exchangers

These are probably the most widely used types of exchanger in the food industry and have been the subject of continuous development and improvement over many years. The main features from a hygienic design perspective are the relatively narrow gap between the plates and the large number of metal-to-metal contact points in the flow channels between the plates. These contact points provide the mechanical strength necessary to withstand the required operating pressures. General hygienic design requirements do not recommend metal-to-metal contact due to concerns about retention of soil and microorganisms in and around the metal-to-metal contact area. This is a particular issue when processing products containing pulp and fibres such as fruit juices where the fibres will tend to wrap themselves around the contact points and will be difficult to remove during cleaning even if reverse flow of cleaning solution is used. Even relatively modest capacity exchangers may have many thousands of metal-to-metal contact points on the product side (Hasting, 2008). While this may be non-optimal from a hygienic design perspective, a key issue is whether the exchanger can be cleaned effectively. Practical experience gained over many years has shown that plate exchangers can be cleaned consistently to a standard compatible with the highest hygiene standards, such as aseptic processing, provided an appropriate cleaning regime is delivered.

For applications where plate heat exchangers are to be considered for processing a product containing particles, such as fruit juice with fibres, special designs termed wide-gap or free-flow are available. These have a significantly wider gap than a conventional plate exchanger (~ 6 mm compared to 2–3.5 mm) and there are few contact points within the product flow channel. This does, however, have a detrimental impact on the maximum working pressure, which may be as low as 4–5 bar compared to 20 bar for other conventional exchangers.

One of the major hygiene concerns with conventional plate exchangers is that the two fluid streams are separated by a single relatively thin metal surface. Typical metal thicknesses are 0.4–0.6 mm, significantly thinner than tubular systems, which are usually at least 1 mm thick. If the surface becomes damaged there is a risk of cross-contamination between the two fluids so, for example, if heat recovery is being used to pre-heat incoming raw product with pasteurised product, there is a risk of cross-contamination occurring between the two streams.

The susceptibility of the plates to damage is dependent on the generic method of manufacture as well as the application. Plates are manufactured by converting thin sheets of metal into a corrugated form, which both enhances heat transfer and provides the necessary mechanical strength. The pressing operation to create the final form requires imposing substantial forces on the metal to make it 'flow' into the desired shape. The finished plate is therefore a highly stressed component, particularly around the contact points. Damage to the plate can therefore occur due to, for example, corrosion or flow-induced vibration.

A number of approaches can be used to minimise this risk.

- The most common is to maintain a higher pressure on the processed product side to ensure any flow is from processed to raw, untreated product. This is good practice to incorporate into a new design but may be impractical to implement on an existing line. In addition, it does not provide total security as it has been shown that microorganisms can move against a pressure gradient.

- Another approach is to use a secondary water circuit as shown in Fig. 11.2 such that direct product-to-product contact is avoided, although this will result in a significant increase in the heat transfer area required and will entail careful maintenance of the water circuit.
- A further option is to use a modified plate configuration with an air gap as shown in Fig. 11.3.

In place of a single plate, two plates are used with a narrow air gap in between, which is connected to atmosphere. A defect in one of the plates will result in fluid passing through the plate and into the air gap, from where it can drain out to atmosphere and provide a visual indication of leakage. However, the presence of an air gap within a heat transfer system will have a

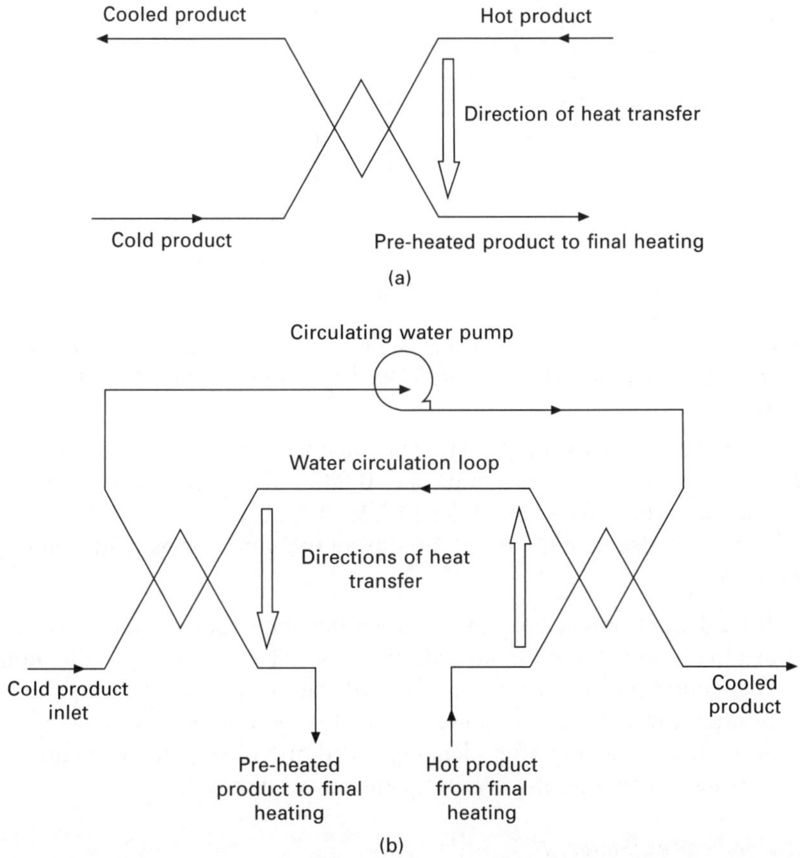

Fig. 11.2 Use of secondary water circuit for heat recovery: (a) conventional heat recovery; (b) heat recovery using a secondary water circuit.

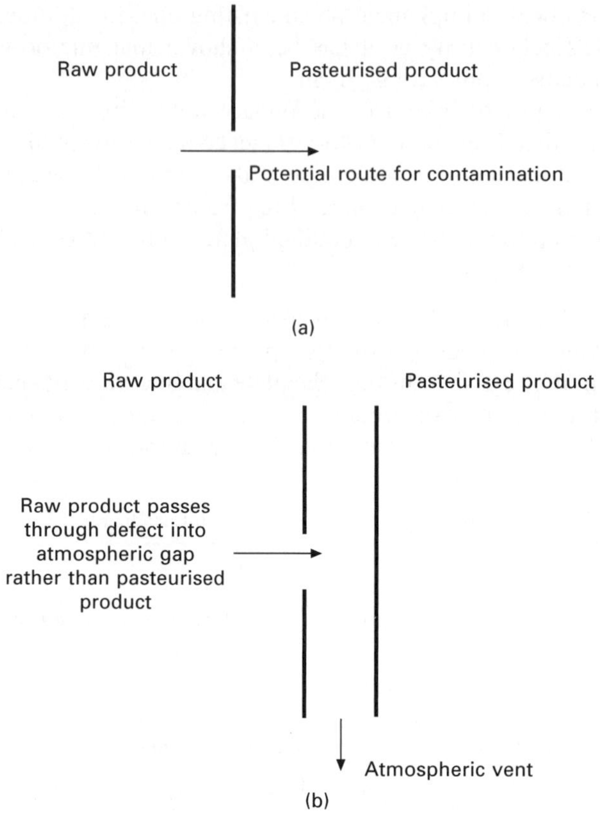

Fig. 11.3 Use of dual plate configuration for additional security: (a) conventional single plate with defect; (b) dual plate design with defect in one plate.

detrimental effect on overall heat transfer, and it has been estimated (Hasting, 2005) that even with an air gap of only 0.005 mm, the loss of heat transfer performance would be of the order of 20–30 %.

There are also a number of potential hygiene issues with such an approach.

- If fluid enters the narrow gap between the plates, surface tension effects are likely to prevent the fluid draining out of the gap by gravity alone.
- Any fluid build-up within the gap will be difficult to clean due to the limited flows of cleaning fluid that will be able to access the gap.
- Any fluid remaining after cleaning could provide a potential source for recontamination during a subsequent production run.

Tubular heat exchangers

Tubular heat exchangers can generally be classified as being of straight or coiled design, with the heat transfer surface in the form of a plain tube or an extended surface in which the tube is fabricated so as to give a corrugated geometry, claimed to enhance turbulence and hence heat transfer performance. Specific hygienic design issues for the different types of exchanger are summarised in Table 11.3.

Tubular heat exchangers are generally more robust than plate exchangers but require good mechanical design to prevent physical damage to the heat exchanger due to differential expansion between the tubes and the shell caused by large temperature differences between the hot and cold fluids. This is generally achieved by either the use of expansion bellows on the shell or a seal arrangement that allows the tubes and shell to move independently of each other. Mechanical damage or corrosion can still occur, and ensuring a higher pressure on the processed side or use of a secondary water circuit to separate raw and processed product as described above are features which are regularly incorporated.

Scraped surface heat exchangers

These heat exchangers are particularly applicable for heating and cooling duties for processing high-viscosity products or products containing particles.

Table 11.3 Hygiene issues with tubular heat exchangers

Exchanger type	Hygiene issues
Coiled tube	Some designs contain spacers to maintain the correct dimensions in the gaps between the different diameter tubes. These are potentially difficult to clean if on the product side. Coiled tubes cannot be visually inspected or dismantled; hence if severe fouling occurs resulting in blockage of the flow gap, it will be very difficult to clean the system.
Straight tube	Tubular heat exchanger modules can be linked together by pipework, for example 180 degree bends, to create a complete system. The cross-sectional area of the pipework can be greater than that of the product side of the exchanger; hence it is necessary to take this into account when specifying the CIP flow rate.
Monotube/multiple tube in tube	Straight tube designs such as single and multiple tube in tube can be visually inspected by removing the end connections from the module, which then allows access to the inside of the tubes. Many designs are also of the floating shell type such that the inner tube or tubes can be removed from the outer shell for inspection and, if necessary, manually cleaned. In order to be able to carry this out, sufficient space must be provided to allow the inner tube(s) to be removed from the shell, which may pose difficulties as such tubes are usually 6 m in length.
Triple tube units	Most complex of straight tube geometries with particular attention required in design of inlet and outlet areas including seals. Must avoid creation of stagnant, difficult to clean areas.

Conventional designs are based on a rotating approach where scrapers on a rotating shaft continually remove fluid from the heat transfer surface and mix it back into the bulk, thus allowing fresh material to reach the surface. Orientation of such units can be vertical or horizontal. A more recent design uses a linear rather than rotary motion to provide the action, with a number of scrapers attached to the reciprocating shaft.

The correct choice of materials is important for such units since, as their name implies, the blades or scrapers contact the internal surface of the cylinder and wear of both blade and cylinder can be a problem. Damage to either can result in a loss of performance as well as posing a physical contamination hazard for the final product.

Cleaning of such heat exchangers can pose problems due to their relatively large diameter, such that the cross-sectional area is often much greater than the inlet and outlet pipework. This can lead to very low linear flow velocities within the machine. Whilst the scraper system will assist in creating turbulence at the internal surface of the wall, other areas of the machine may not achieve such conditions. These include the inlet and outlet zones, the seal areas and the shaft itself, all of which are known to be potential problem areas. System design should therefore aim to ensure that sufficient cleaning flows can be delivered, and a simple way of achieving this is to use a cleaning in place (CIP) booster pump in parallel with the exchanger (Hasting, 2005).

Detection of heat exchanger defects

While the techniques described above are used to protect product from contamination, it is necessary to be aware of methods to detect such defects as early as possible. These can be based either on dismantling the exchanger or, preferably, detecting the defect *in situ*. The most widely used methods are as follows.

- **Dye penetrant** – Often used with plate heat exchangers but requires them to be dismantled. A low-viscosity coloured fluid is sprayed on one side of the plate and a developer, often in the form of a thin foam, is applied to the other. Penetration of the coloured fluid through a defect will show up as a colour change in the developer at this point. Whilst this can be carried out at the factory, it is more usual for specialist companies to carry this out off-site.
- **Pressure retention** – One side of the heat exchanger can be pressurised with liquid or gas and any reduction in pressure noted. If a gas is used it may be possible to detect its presence on the other side of the heat transfer surface, which will indicate a possible leak. With plate heat exchangers where there is a large gasket area, it is possible that false-positives could be identified due to gasket leakage rather than defective plates. The method has the advantage over the dye penetrant method in that it can be applied *in situ*.
- **Electrolytic detection** – A method has been reported by Bowling (1995) whereby an electrolyte is circulated under pressure through the product

side of the heat exchanger. On the other side water is circulated and its conductivity monitored. The product side is maintained at a higher pressure than the water side, and any defects will allow electrolyte to pass into the water stream and increase the conductivity. This method is claimed to have sensitivity similar to dye penetrant methods and can be applied *in situ* for all types of exchanger.

Whichever method is used, leak detection will always be retrospective and the concern would be how long the defect has been present. For critical application using plate heat exchangers, such as aseptic processing, it is typical for a supplier to replace the plates with a new or refurbished set as part of a preventative maintenance programme to further minimise the risk to the product.

11.5.2 Tanks

Tanks are widely used in the food industry both for storage of raw materials, intermediate or final products at different stages of the production process and as process vessels. Process vessels can be used to deliver a series of process operations such as: mixing, blending, heating, cooling, separation and fermentation operations. They are therefore likely to be more complex in design than storage tanks due to the many internal components that can be used, including: baffles, agitators, sight glasses, heating/cooling coils and pressure relief valves (Hasting, 2008). The design criteria are, however, the same for all types of tank, and any such components must conform to the appropriate hygienic design guidelines relating to materials of construction, fabrication and cleanability. This is in terms of both the individual component and the integration of the component into the overall tank.

Of the internal components, a spray device for cleaning should be included where tanks are closed and cleaning solutions can safely be sprayed on to the internal surfaces. Such devices may well be of minimal cost in comparison to the overall vessel, but they are critical design components in terms of ensuring effective cleaning and hence process hygiene. It is therefore essential that the correct cleaning requirements and spray devices are clearly specified to the tank fabricator at the design stage. Whichever type of device is selected, it is essential that it is installed such that all contact surfaces can be reached effectively. For most vessels with internal components, a single spray device will be insufficient as the presence of the component is likely to cause 'shadowing' and prevent cleaning fluid reaching all the surfaces. The spray devices themselves must also be self-cleaning and self-draining.

One important design point that is often given insufficient attention is removal of liquid from the bottom of the tank during CIP. It is always much easier to pump liquid into a vessel than it is to extract it, and failure to maintain the vessel empty during CIP may result in an increased rinsing time, inadequate cleaning of surfaces that are covered by liquid as flow velocities will be very low and may result in re-deposition of soil.

In order to maximise the drainage capability of the vessel the design should ensure:

1. provision of a sufficient slope on the base of the vessel;
2. provision of an outlet of sufficient size for the cleaning flows required;
3. selection of an appropriate extraction pump to remove liquid from the tank;
4. installation of the extraction pump as close to the vessel as practical, minimising the number of bends and avoiding changes in pipework diameter.

11.5.3 Product transfer systems

The majority of food processes consist of a number of unit operations, which are integrated to form a complete line. Each of the unit operations fulfils a functional purpose in converting the raw ingredients into the final product. In order for the complete line to function, materials have to be transferred between unit operations. For liquids or liquids containing particles, pumps are generally used for this purpose, whereas conveyors are used for solid materials. In the great majority of applications the pump or conveyor's sole purpose is to move the material through the system without any direct process purpose in terms of converting raw materials into final product. Pumps and conveyors are therefore essential for operation of a process line but are not always given the attention they deserve with regard to hygiene. There are many technical options available for both pumps and conveyors, each having their own advantages and limitations.

Pumps

Pumps can be classified into two main types (Sinnott, 2005):

1. dynamic pumps, a typical example of which is the centrifugal type;
2. positive displacement pumps, which may be of reciprocating or rotary design.

The most commonly used dynamic pump is the centrifugal type, an economic and efficient pump capable of handling large volumes of fluids, although with limited delivery pressure. It is usually the pump of choice where it is practical for the desired application. In many cases the operational requirements of the pump, such as delivery pressure and capacity, determine the type of pump used even if it is not in fact the most hygienic option available.

The general categories of pumps are shown below:

- peristaltic;
- diaphragm;
- centrifugal;
- positive pump – rotary type (gear/lobe);

- positive pump – reciprocating type (homogeniser);
- positive pump – progressive cavity type (mono pump).

Although all such pumps can be unhygienic if poorly designed, the positive type pumps are usually considered to be less hygienic and more difficult to clean than the others due their inherent design features. They are, however, widely used due to their suitability for particular applications, such as the rotary type for handling more viscous fluids, reciprocating type for generating high pressures required for homogenisation and progressive cavity type for the gentle handling of products containing fragile particles as well as viscous fluids.

Generic hygiene requirements for pumps include:

- any leakage from the pump body must be easily visible;
- shaft seals should where possible be of the mechanical type and be easily accessible for inspection, replacement and maintenance;
- passage shapes should be smooth, avoiding sharp changes in cross-section;
- bearings should be located outside the product area and be sealed or be of a self-lubricating type;
- pumps should be self-draining where possible.

Drainage is frequently an issue with many types of pump. In some cases the problem is one of installation and in others the fundamental design may limit the ability to fully drain the pump. For centrifugal and positive rotary pumps, such as the lobe or gear type, incorrect orientation may limit the ability to drain as shown in Fig. 11.4 for a centrifugal pump.

However, with the centrifugal pump drainage is not the sole issue that needs to be considered. If the fluid being pumped contains air or other gases, orientation (b), although fully drainable, may be unable to pump the gas out of the pump casing. This may result in a build-up of gas within the

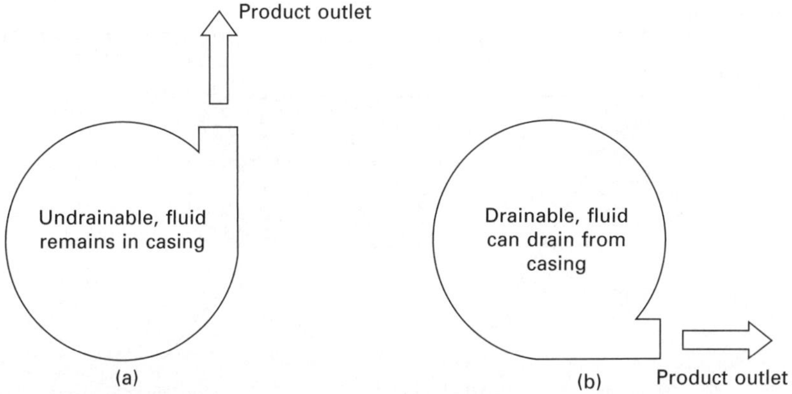

Fig. 11.4 Orientation of centrifugal pumps and the effect on fluid drainage.

casing causing cavitation of the pump, loss of performance and potential physical damage to the pump. With orientation (a), however, the vertical outlet allows any gas to be pumped out of the casing along with the fluid and minimises the risk of cavitation, albeit at the expense of the ability to drain the system. It is therefore necessary to consider the implications of residual fluid remaining in the casing after production and cleaning have been completed. The hygiene risk is one of microbial growth within the residual fluid while the unit is shutdown, depending on the length of the shutdown and the environmental temperature. This could then contaminate all the equipment downstream of the pump on start-up. It is, however, a risk that can be satisfactorily managed. The residual fluid is not in a stagnant zone and will in fact be very effectively mixed with incoming fluid when the pump is started. Where necessary the pump and downstream equipment can be disinfected or sterilised either chemically or thermally depending on the application. If it is considered essential to be able to drain the system completely, it would be straightforward to install a small drain valve at the bottom of the casing. However, in this case, care must be taken that the drain valve installed does not itself create a dead space.

Figure 11.5 shows an installation of a positive gear pump with product inlet and outlet mounted horizontally. There are similar drainage issues as with the centrifugal pump in Fig. 11.4, which could be avoided by mounting the inlet/outlet vertically. However, in this case the height of the inlet pipe is fixed due to the equipment it is connected to upstream. If the pump is then mounted vertically the pipework would have to be modified and, although the pump could then be drained, a vertical leg of pipe would be created that

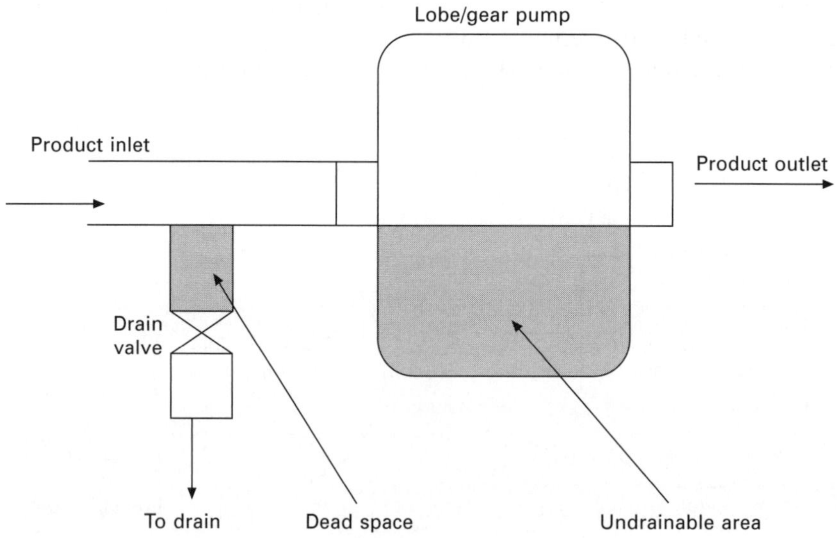

Fig. 11.5 Positive pump with hygienic design faults.

in itself may be undrainable. If the pump does need to be drained, a drain valve could be fitted on to the casing in the same way as for a centrifugal pump or the front cover loosened to allow fluid to drain out from there. Figure 11.5 also highlights a further potential unhygienic feature of such a pump installation, which is often seen in practice. There is a drain valve on the inlet line to the pump, but this will not help in draining the pump and the valve itself creates a dead zone between the valve and the product pipework. There would also be a potential safety hazard if the valve were a manual type as, if an operator opens the valve during CIP to drain or clean the line through this valve, there would be a risk of the operator being splashed with hot CIP solution.

Other types of pump are limited in ability to drain by their fundamental design, for example the progressive cavity and peristaltic pumps. Positive pumps may be either fixed or variable speed. If a variable speed type is being used, the speed and hence flow should be maximised during the CIP process to maximise cleaning efficiency.

Conveyors

Conveyors perform a similar function for open equipment as pumps do for closed plant in that they transfer materials to, from and within processing operations. Conveyors can be complex pieces of equipment, needing to be sufficiently robust for continuous operation without causing damage to fragile food materials. Depending on the product handled conveyors may be prone to heavy soiling, and hence design for cleanability is particularly important. The choice of conveyor type will be dependent on a number of factors: throughput, the nature of the solids, such as size and bulk density, length of travel and change in elevation. Hygienic design of conveyors is far from straightforward as there are conflicts between functionality, safety and easy cleaning.

Belt conveyors are the most commonly used type of equipment for the continuous transport of solids (Sinnott, 2005). They can carry a wide range of materials both horizontally or at an appreciable angle depending on the characteristics of the solid. A belt conveyor consists of an endless belt supported on rollers and passing over larger rollers at each end of the conveyor. One of the larger rollers is driven to provide the forward motion for the solids.

Screw conveyors are used for materials that are free-flowing and consist of a helical screw rotating in a U-shaped trough; they can be used horizontally or, with some loss of capacity, at an incline to lift materials. They are used to convey solids over short distances and can also be used for delivering a metered flow of solids.

When a vertical lift of solids is required the bucket elevator is widely used, where buckets are fitted to an endless chain or belt passing over a driven sprocket or roller at the top of its travel. Buckets can handle a wide range of materials including wet solids and slurries.

Some generic design considerations for conveyors are as follows.

- Use tote bins or trays in areas where debris is likely to build up, for example at the conveyor discharge, to catch debris.
- All surfaces potentially in contact with food should be accessible for cleaning.
- Avoid the use of rubber flaps to direct product flow.
- Avoid mounting drives above the conveyor where possible. Where drives are mounted above, a drip tray and drain should be fitted to protect the product from external contamination.
- Cabling should not be run below the product.
- Debris can often fall through the belt, chain or mesh of a conveyor and hence the construction of the support framework is a critical, yet often neglected, issue.
- All welds should be continuous.
- Covering panels or guards should be easily removed for cleaning and inspection.

Specific hygiene issues for different types of conveyor are summarised in Table 11.4

Table 11.4 Hygiene issues associated with conveyors

Conveyor type	Potential hygiene issues
Belt/chain/mesh	Cleaning and drying under belt difficult, particularly for large systems. Edges of belt can wear and absorb moisture unless totally sealed. Contamination build-up between belt and any side guides. For belt conveyors where there are gaps in the belt, such as the plastic modular type, product debris can accumulate in the gaps, and it is difficult to gain access for cleaning liquids. Lubricants could be a potential source of contamination and must be food grade.
Bucket	Difficult to clean both buckets and motor drives. Access for cleaning.
Vacuum	Potential contamination from air supply and lubricating oil. Difficult to clean and inspect. Product residues retained within system leading to contamination.
Screw	Difficult to clean and inspect. Potential for lubricants to pass through shaft seals into product. Drainage problems unless suitable drain holes included.
Spiral	Debris hold up within spiral. Difficult access for cleaning and inspection.

11.5.4 Valves

Valves are used for a number of different applications within a process line including:

- selection of product routing, with the valve usually either fully activated or deactivated;
- regulation of flows and pressures, where the valve opening will not be fixed but defined by the control system;
- protection, for example against overpressure, where the valve activates in response to a pressure in excess of the maximum allowable – this activation may be temporary, intermittent with partial opening of the valve.

The key areas influencing the hygiene of a valve are: the internal geometry, the inlet and outlet connections and the seal between the fluid and the external environment. The seals may be under a static load or dynamic with linear or rotary motion.

Generic requirements for valves include (EHEDG, 1994):

- ability to fully drain valve, without the need for dismantling;
- physically robust to resist wear and allow easy maintenance, since wear or distortion may affect hygiene;
- the number of seals should be minimised, positively retained and flush with adjacent surfaces;
- where unavoidable, springs in product contact should have minimum surface contact area;
- easy visual detection of internal leakage.

Aseptic applications provide the greatest challenge, particularly to dynamic seals. Dynamic seals on valve shafts in contact with product must provide an absolute barrier between the product and the environment to prevent microbial recontamination. One option is to use a flexible diaphragm or bellows type seal, but these are limited by the length of stroke they can accommodate. Where such seals cannot be used, it will be necessary to use a double seal arrangement and these have the following requirements.

- The space between the seals must be capable of being sterilised prior to production as well as maintaining asepsis during production. This is usually achieved by flushing the space between the seals with pressurised steam and monitoring the temperature. While, in principle, a chemical sterilant could equally well be used, the concern would be that failure of the seal adjacent to the product may result in chemical contamination of the product.
- The distance between the two seals must be greater than the distance moved by the shaft in order to prevent microorganisms being brought into the aseptic zone during operation. Although the shaft will move through the barrier zone, the contact time may be insufficient to inactivate any microbial contamination.

Examples of some practical issues with different types of valves are given below.

- **Diaphragm** – When used as a back pressure valve, visual detection of leakage is essential as damage to the diaphragm can result in product leaking through into the non-product side, being impossible to clean or disinfect and providing a source of potential contamination.
- **Mixproof (double seat) valve** – Vent spaces must be drainable, cleanable and avoid a pressure build-up in case of a leak from a seal. Outlet from vent line must be visible so leakage can be easily detected. In practice such valves are often part of manifolds containing many valves integrated together. It is frequently observed that the vent lines are all piped into a common line, which is led to a drain without any visual observation of whether any of the valve seats are leaking. There is also no opportunity to confirm that the vent lines are being adequately cleaned, leading to a risk that product residues remain between the two seats and can contaminate product.
- **Butterfly valves** – A circular disc of stainless steel turns through 90 degrees to seal against the elastomer and create the closed position of the valve. After a period of time depending on the number of times the valve is opened and closed, the seals can wear, break down and cause hygiene problems. Regular inspection and maintenance is essential.
- **Ball valves** – Usually considered unsuitable for CIP as the area between ball, housing and seal face is uncleanable, although improved designs have been proposed.
- **Plug valves** – Unsuitable for CIP but generally simple to dismantle for manual cleaning.
- **Pressure relief valve** – Must be installed so as to be self-draining. If the valve opens during production due to excess pressure, it is essential that the valve is opened during CIP to ensure all product residues are removed.

11.6 Conclusions

When assessing the hygienic design of a line in response to either a problem or its suitability to handle a specific product, it is essential not to focus solely on the equipment but also to consider the other engineering factors described in Section 11.1. With older factories in particular, it is likely that an assessment of equipment against the currently accepted best practice would identify many faults. However, overcoming all these faults is unlikely to be practical for a number of reasons:

- budget constraints;
- time constraints;
- problem may still remain unresolved.

A more realistic approach would be to identify, assess and prioritise hygienic design faults, agree an action plan and continually monitor the results. Ideally, the effect of resolving individual faults would be monitored to try and determine the root cause of the problems, but in reality it is unlikely that time would allow such an analysis.

Whilst it is important to learn from factory experience, the assessor should not be automatically dissuaded by suggestions that equipment identified has never caused problems. It may be true, but if the risk it poses is unacceptable, ways of managing this must be implemented.

It should not be assumed that identifying and resolving a serious design fault means that no further analysis is required. This is not the case, as a major fault may be masking a number of smaller ones and the full assessment must be completed. There is also a tendency to look for complex combinations of events as the cause of the problems and ignore the simple ones. Having an independent observer to assess the plant may also have benefits in bringing a fresh pair of eyes to the problem.

11.7 Future trends

The food industry is rapidly changing as new products come to market ever more quickly. The boundary between pharmaceutical products and foods is becoming less clear cut as food products delivering specific health benefits, such as reduced cholesterol, lowered blood pressure and improved gut health, are being successfully delivered to the market.

These trends are having a potentially significant impact on hygiene in a number of ways.

- Cost pressures from the major retailers are resulting in manufacturers reducing staff numbers, running plant for longer and reducing down-time for cleaning. This loss of experience coupled with extended run lengths and shorter cleaning increases the risk to final product quality.
- More product varieties are being produced on the same line, and this may pose different hygiene challenges.
- Manufacturers are increasingly moving from in-house to third-party manufacture to reduce costs, reducing the level of involvement and control by manufacturers over their own product.
- Manufacturers are increasingly making investments in lower-cost areas such as Eastern Europe at the expense of Western Europe.
- Equipment suppliers are increasingly using manufacturers in low labour cost economies such as China and India. This increases the risk of sub-standard machinery being purchased on the basis of capital cost, with little or no thought given to running costs, operational efficiency and potential implications for hygiene.
- Environmental issues such as sustainability place increasing emphasis

on minimising water and energy costs by reduction and re-use. More efficient use of services such as water may create additional risks due to, for example, build-up of chemicals or microorganisms and extended residence times in re-use systems, such that equipment originally fit for purpose may become compromised.

From a technical perspective, developments are likely in the areas of:

- **Surface modification** – Fouling, microbial adhesion and cleaning of surfaces have always been a cause of concern with regard to hygiene. Materials with surfaces that incorporate antimicrobial compounds are already applied within the industry, and there is also considerable activity regarding the potential to modify surfaces such that fouling rates can be reduced as well as the time taken to clean equipment. Areas of interest include the use of surface coatings (Zhao *et al.*, 2002) and surface modification by techniques such as ion implantation, sputtering or electrolytic deposition (Muller-Steinhagen and Zhao, 1997).
- **Nanoparticle technology** – May also find application within the food industry in enhanced material properties such as corrosion and abrasion resistance in addition to antimicrobial properties.
- **Improvements in measurement and monitoring** – Increasing quantities of data are available from process plant in real-time, and this can be used to demonstrate that the critical control points in the process are within the defined limits. However, further analysis of such process data can provide the opportunity to be more predictive rather than proactive in identifying potential risks to equipment and product. For example, in-line condition monitoring of equipment during production would provide a more comprehensive picture than a maintenance schedule. Improved analytical techniques may enable a more scientific understanding of the effect of issues such as surface finish to be made.
- **Risk assessment** – There will be an increasing trend for a practical risk-based approach to hygienic design, which takes into account the characteristics of the product and process being used to make a decision on how a piece of equipment or combinations of equipment can be used and define the limits on the operation. This will require a better understanding of the interaction of these factors rather than a rigid adherence to a set of rules.

11.8 Sources of further information and advice

There is a considerable and increasing amount of literature in the public domain on hygienic equipment design. There are a number of groups worldwide that continue to be actively involved in improving hygienic design guidelines. The 3-A Sanitary Standards organisation and Food and Drug Administration (FDA) in the USA and the European Hygienic Engineering

Design Group in the EU are particularly relevant. The regular publication of EHEDG guidelines provides a source of up-to-date thinking on issues of current relevance. There are also a number of recent books that provide a further source of reference material (Lelieveld *et al.*, 2003, 2005; Tamine, 2008 (see Hasting, 2008)).

11.9 References

Asteriadou K., Hasting A.P.M., Bird M.R. and Melrose J. (2006) Computational Fluid Dynamics for the prediction of temperature profiles and hygienic design in the food industry, *Food and Bioproducts Processing*, **84**(C4), 157–63.

Bowling M. (1995) *Leakage Detection*, International Patent Application WO95/16900, 22nd June 1995.

BSI (1991a) *BS 4825-3:1991 Stainless steel tubes and fittings for the food industry and other hygienic applications. Specification for clamp type couplings*, British Standards Institute, London.

BSI (1991b) *BS4825-4:1991 Stainless steel tubes and fittings for the food industry and other hygienic applications. Specification for threaded (IDF type) coupling*, British Standards Institute, London.

BSI (1991c) *BS4825-5:1991 Stainless steel tubes and fittings for the food industry and other hygienic applications. Specification for recessed ring joint type couplings*, British Standards Institute, London.

BSI (1997) BS EN 1672-2 *Food processing machinery, Basic concepts. Hygiene requirements*, British Standards Institute, London.

CCFRA (1983) *Hygienic Design of Food Processing Equipment*, Technical Manual No.7, Campden & Chorleywood Food Research Association, Chipping Campden.

Covert R.A. and Tuthill A.H. (2000) Stainless steels: an introduction to their metallurgy and corrosion resistance, *Dairy, Food and Environmental Sanitation*, **20**(7) 506–17.

Curiel G.J., Hauser G. and Peschel P. (1993) *Hygienic Equipment Design Criteria*, EHEDG Document 8, Campden and Chorleywood Food Research Association, Chipping Campden.

EHEDG (1993a) *Welding Stainless Steel to meet Hygienic Requirements*, Guideline document 9, European Hygienic Engineering and Design Group, Frankfurt.

EHEDG (1993b) *Hygienic Equipment Design Criteria*, Guideline document 8, European Hygienic Engineering and Design Group, Frankfurt.

EHEDG (1994) *Hygienic Requirements on Valves for Food Processing*, 2nd edn, Guideline document 14, European Hygienic Engineering and Design Group, Frankfurt.

EU (1989) Council Directive 89/392/EEC on the approximation of the laws of the Member States relating to machinery, *Official Journal of the European Union*, **L183**, 29 June, 9–32.

Gilmore T. (2003) Hygiene regulation in the United States, in H.L.M. Lelieveld, M.A. Mostert, J. Holah and B. White (eds), *Hygiene in Food Processing* Woodhead, Cambridge, 44–57.

Grasshof A. (1980) Untersuchungen zum Strömungsverhalten von Flüssigkeiten in zylindrischen Toträumen von Rohrleitungssystemen, *Kieler Milchwirtschaftiliche Forschungsberichte*, **32**(4) 273–98.

Hasting A.P.M. (2005) Improving the hygienic design of heating equipment, in H.L.M. Lelieveld, M.A. Mostert, J. Holah and B. White (eds) *Handbook of Hygiene Control in the Food Industry*, Woodhead, Cambridge, 212–19.

Hasting A.P.M. (2008) Designing for cleanability, in A.Y. Tamine (ed.), *Cleaning-in-Place: Dairy, Food and Beverage Operations*, Blackwell, Oxford, 81–106.

Holah J.T. and Thorpe R.H. (1990) Cleanability in relation to bacterial retention on unused and abraded domestic sink materials, *Journal of Applied Microbiology*, **69**, 599–608.

Holah J.T. (1998), Hygienic design: international issues, *Dairy, Food and Environmental Sanitation*, **18**, 212–20.

ISO (1976) *ISO 2853: 1976 Metal pipes and fittings – stainless steel screwed couplings for the food industry*, International Organisation for Standardization, Geneva

Lelieveld H.L.M., Mostert M.A. and Curiel G.J. (2003), Hygienic equipment design, in H.L.M. Lelieveld, M.A. Mostert, J. Holah and B. White (eds), *Hygiene in Food Processing* Woodhead, Cambridge, 122–65.

Lelieveld H.L.M., Mostert M.A., Holah J. and White B. (eds) (2005) *Handbook of Hygiene Control in the Food Industry*, Woodhead, Cambridge.

Lewan M. (2003) Equipment construction materials and lubricants, in H.L.M. Lelieveld, M.A. Mostert, J. Holah and B. White (eds), *Hygiene in Food Processing*, Woodhead, Cambridge, 167–78.

Muller-Steinhagen, H. and Zhao, Q. (1997) *Chemical Engineering Science*, **52**(19), 3321–32.

Sillett C.C. (2006) *IChemE Food & Drink Subject Group Newsletter*, **1**, 9–11.

Sinnott R.K. (2005) *Chemical Engineering Design*, Elsevier, Oxford, 481–2.

Tamplin T.C. (1990) Designing for cleanability, in A.J.D. Romney (ed.) *CIP:Cleaning in Place*, Society of Dairy Technology, Huntingdon, 41–103.

Timperley D.A. and Timperley A.W. (1993) *Hygienic Design of Meat Slicing Machines*, Technical Memorandum No. 679, Campden & Chorleywood Food Research Association, Chipping Campden.

Verran J. (2005) Testing surface cleanability in food processing, in H.L.M. Lelieveld, M.A. Mostert, J. Holah and B. White (eds), *Handbook of Hygiene Control in the Food Industry*, Woodhead, Cambridge, 556–72.

Zhao Q., Liu Y. and Muller-Steinhagen H. (2002) Effects of interaction energy on biofouling adhesion, in D.I. Wilson, P.J. Fryer and A.P.M. Hasting (eds), *Fouling, Cleaning and Disinfection in Food Processing*, Jesus College, University of Cambridge, 213–20.

12

Sanitation

J. Holah, Campden BRI, UK

Abstract: Sanitation is undertaken to restrict the build-up of hazards throughout the production day and significantly reduce them following production, thus returning the manufacturing areas to their inherent hygienic design. A combination of cleaning chemicals and disinfectants, temperatures, times, application methods and sequences are used to reduce soiling, including microorganisms, to a level that is not harmful to product quality and safety. Monitoring and verification of the effectiveness of the sanitation process ensures compliance with these requirements.

Key words: cleaning, disinfection, biofilms, persistence, cleaning-in-place (CIP), rapid hygiene methods.

12.1 Introduction

Contamination in food products may arise from four main vectors: the constituent raw materials, surfaces, liquids and the air. Control of the raw materials is addressed elsewhere in this book and is the only non-environmental contamination route. Food may pick up contamination as it is moved across product contact surfaces (hard surfaces) or if it is touched or comes into contact with people (food handlers) or other animals, e.g. pests (soft surfaces). Liquids include process aids (washing or cooling fluids) or residues of cleaning fluids or rinses left on surfaces. They differ from surface contact in that there is an element of uptake of the liquid into the product. The air acts as both a source of contamination, i.e. from outside the processing area, or as a transport vector, e.g. moving contamination from non-product to product contact surfaces.

Provided that the process environment and production equipment have been hygienically designed, cleaning and disinfection (referred to together as *sanitation*) are the major day-to-day controls of the hard surface vectors of food product contamination and can reduce sources of microorganisms within the processing environment. In addition, for some food manufacturing processes such as the high care/risk areas of ready-to-eat (RTE) food products, sanitation practices are the only processes that can control microbial cross-contamination from food processing areas. When undertaken correctly, sanitation programmes have been shown to be cost-effective and easy to manage and, if diligently applied, can reduce the risk of microbial or foreign body contamination. Given the intrinsic demand for high standards of hygiene in the production of foods, together with pressure from customers, consumers and legislation for ever-increasing hygiene standards, sanitation demands the same degree of attention as any other key process in the manufacture of safe and wholesome foods.

Sanitation is undertaken to:

(1) remove microorganisms or material conducive to microbial growth – this reduces the chance of contamination by pathogens and, by reducing spoilage organisms, may extend the shelf-life of some products;
(2) remove materials that could lead to foreign body contamination or could provide food or shelter for pests – this also improves the appearance and quality of product by removing food materials left on lines that may deteriorate and re-enter subsequent production runs;
(3) remove specific food products such that subsequent production runs are not cross-contaminated – such food products could include, for example, allergens, meat (for subsequent vegetarian product runs), GMO-containing ingredients, traditional products for subsequent organic or religious (e.g. Halal, Kosher) runs;
(4) extend the life of, and prevent damage to equipment and services, provide a safe and clean working environment for employees and boost morale and productivity;
(5) present a favourable image to customers and the public – on audit, the initial perception of an 'untidy' or 'dirty' processing area, and hence a 'poorly managed operation' is subsequently difficult to overcome.

12.2 Sanitation principles

Sanitation is undertaken primarily to remove all undesirable material (food residues, microorganisms, foreign bodies and cleaning chemicals) from surfaces in an economical manner, to a level at which any residues remaining are of minimal risk to the quality or safety of the product. Such undesirable material, generally referred to as 'soil', can be derived from normal production, spillages, line-breakdowns, equipment maintenance, packaging or general environmental contamination (dust and dirt). To undertake an adequate and economic sanitation programme, it is essential to characterise the nature of the soil to be removed.

Product residues are readily observed and may be characterised by their chemical composition, e.g. carbohydrate, fat or protein. It is also important to be aware of processing and/or environmental factors, however, as the same product soil may lead to a variety of cleaning problems dependent primarily on moisture levels and temperature. Generally, the higher the product soil temperature (especially if the soil has been baked) and the greater the time period before the sanitation programme is initiated (i.e. the drier the soil becomes), the more difficult the soil is to remove.

With respect to microorganisms on food production surfaces, the food industry has always recognised that they are present (either within the surface soil or attached to the surface) and that their numbers increase during production and are reduced by cleaning and disinfection. Initially it was considered that microorganisms adhered to the food product contact surfaces could be an important source of potential contamination leading to serious hygiene problems and economic losses due to food spoilage. For example, pseudomonads and many other Gram-negative organisms detected on surfaces are the spoilage organisms of concern in chilled foods. Subsequently, however, we have recognised that food contact surfaces can give rise to biofilms, can support the growth of pathogens and, more recently, that food processing environments will develop their own persistent 'house' flora, some of which can be pathogenic.

There are a number of factors that have been shown to affect attachment and biofilm formation, such as the level and type of microorganisms present, surface conditioning layer, substratum nature and roughness, temperature, pH, nutrient availability and time available. Several reviews of biofilm formation in the food industry have been published including Pontefract (1991), Holah and Kearney (1992), Mattila-Sandholm and Wirtanen (1992), Carpentier and Cerf (1993), Zottola and Sasahara (1994), Gibson *et al.* (1995), Kumar and Anand (1998), Sharma and Anand (2002) and Lindsay and von Holy (2006). In general, however, true biofilm formation (multilayered microbial communities) is usually found only on environmental surfaces (particularly drains), and progression of attached cells through microcolonies to extensive biofilm on food contact surfaces is limited by regular cleaning and disinfection. Indeed, following Hazard Analysis Critical Control Point (HACCP) principles, if the food processor believes that biofilms are a risk to the safety of the food product, appropriate steps (e.g. cleaning and disinfection) must be taken, controlled and monitored.

Following the first concerns associated with *Listeria* in chilled foods in the late 1980s, a number of authors have isolated *L. monocytogenes* from a range of food processing surfaces (Walker *et al.*, 1991; Lawrence and Gilmore, 1995; Destro *et al.*, 1996). Indeed, the assessment of the environment for *Listeria* spp. is now an essential part of the high care/risk processing area environmental sampling plan. The concept of persistence, i.e. strains of an organism becoming adapted to and surviving in an environment for considerable time periods, is more recent. Evidence for *L. monocytogenes* strain persistence within factory processing areas has been demonstrated for eight months (Rørvik *et al.*, 1995), 11 months (McLauchlin *et al.*, 1990), 14 months (Johansson *et al.*, 1999), one year (Lawrence and Gilmour, 1995), 17 months (Pourshaban *et al.*, 2000), two years (McLauchlin *et al.*, 1991), three years (Brett *et al.*, 1998; Holah *et al.*, 2004), 40 months (Unnerstad *et al.*, 1996), four years (Nesbakken *et al.*, 1996; Aase *et al.*, 2000; Fonnesbech Vogel *et al.*, 2001), seven years (Unnerstad *et al.*, 1996; Miettinen *et al.*, 1999) and 10 years (Kathariou, 2003). These authors have, between them, isolated persistent strains from the environments of factories manufacturing the following food products: cheese; mussels, shrimps and raw and smoked fish; fresh, cooked and fermented meats; pâté and pork tongue in aspic; chicken and turkey meat; pesto sauce and ready meals.

The nature of strain persistence is unknown, but it may be due to a number of factors affecting biological/physical adaptation (surface attachment, biofilm formation, attachment strength, reduced growth rate, quiescence, cleaning and disinfection resistance) to the whole variety of environmental conditions found within food factory environments (temperature range, wide pH range, fluctuating nutrient supply and moisture levels, frequency of cleaning and disinfection, etc.). Lundén *et al.* (2003a, b) have demonstrated that persistent strains may show enhanced surface adherence and increased disinfection resistance at very low disinfectant concentrations, though Holah *et al.* (2002) found no evidence of disinfection-resistance of certain persistent strains to disinfectants at their normal in-use concentration. Persistence is also known within dry food processing environments, where anecdotal evidence suggests that *Salmonella* spp. and *Enterobacter sakazakii* may survive for long time periods (years). Their persistence is thought to be related to survival mechanisms, however, such as capsule formation (Barron and Forsythe, 2007), rather than growth and adaptation mechanisms.

Within the sanitation programme, the wet cleaning phase can be divided into three sections, following the pioneering work of Jennings (1965) and interpreted by Koopal (1985), with the addition of a fourth stage to cover disinfection. These are described below:

(1) The wetting and penetration by the cleaning solution of both the soil and the equipment surface.
(2) The reaction of the cleaning solution with both the soil and the surface to facilitate: peptidisation of organic materials, dissolution of soluble organics and minerals, emulsification of fats and the dispersion and removal from the surface of solid soil components.
(3) The prevention of re-deposition of the dispersed soil back onto the cleansed surface.
(4) The wetting by the disinfection solution of residual microorganisms to facilitate reaction with cell membranes and/or penetration of the microbial

cell to produce a biocidal or biostatic action. Dependent on whether the disinfectant contains a surfactant and the disinfectant practice chosen (i.e. with or without rinsing), this may be followed by dispersion of the microorganisms from the surface.

To undertake these four stages, sanitation programmes employ a combination of four major factors as described below:

(1) mechanical or kinetic energy;
(2) chemical energy;
(3) temperature or thermal energy;
(4) time.

The combination of these four factors varies for different cleaning systems and, generally, if the use of one factor is restricted, this short-fall may be compensated for by utilising greater inputs from the others.

Mechanical or kinetic energy is used to physically remove soils and may include scraping, manual brushing and automated scrubbing (physical abrasion) and pressure jet washing (fluid abrasion). Of all four factors, physical abrasion is regarded as the most efficient in terms of energy transfer (Offiler, 1990), and the efficiency of fluid abrasion and the effect of impact pressure has been described by Anon. (1973) and Holah (1991). Mechanical energy has also been demonstrated to be the most efficient for biofilm removal (Blenkinsopp and Costerton, 1991; Wirtanen and Mattila Sandholm, 1993, 1994; Mattila-Sandholm and Wirtanen, 1992; Gibson *et al.*, 1999).

In cleaning, chemical energy is used to break down soils to render them easier to remove and to suspend them in solution to aid rinsability. At the time of writing, no cleaning chemical has been marketed with the benefit of aiding microorganism removal. In chemical disinfection, chemicals react with microorganisms remaining on surfaces after cleaning to reduce their viability. The chemical effects of cleaning and disinfection increase with temperature in a linear relationship and approximately double for every 10 C° rise. For fatty and oily soils, temperatures above their melting point are used to break down and emulsify these deposits and so aid removal. The influence of detergency in cleaning and disinfection has been described by Dunsmore (1981), Shupe *et al.* (1982), Mabesa *et al.* (1982), Anderson *et al.* (1985) and Middlemiss *et al.* (1985).

For cleaning processes using mechanical, chemical and thermal energies, generally the longer the time period employed, the more efficient the process. When extended time periods can be employed in sanitation programmes, e.g. soak-tank operations, other energy inputs can be reduced (e.g. reduced detergent concentration, lower temperature or less mechanical brushing).

Soiling of surfaces is a natural process which reduces the free energy of the system. To implement a sanitation programme, therefore, energy must be added to the soil to reduce both soil particle–soil particle and soil particle–equipment surface interactions. The mechanics and kinetics of these interactions have been discussed by a number of authors (Jennings, 1965;

Schlussler, 1975; Loncin, 1977; Corrieu, 1981; Koopal, 1985; Bergman and Tragardh, 1990), and readers are directed to these articles since they fall beyond the scope of this chapter. Research continues on the theoretical aspects of soil removal to gain a better understanding of the relative importance of both the physical and chemical removal factors and the relative influence of soil-to-soil and soil-to-surface bonding mechanisms. A useful review of these factors was undertaken by Fryer and Christian (2005) who have characterised soil removal in terms of deposit thinning (weaker soil-to-soil bonding) and deposit removal in large 'chunks' (weaker soil-to-surface bonding) for tomato and milk deposits. Suffice to say that it will be many years before soil removal mechanisms can be predicted for all soil types and that practical factory-based cleaning trials to optimise cleaning will still predominate.

In practical terms, however, it is worth looking at the principles involved in basic soil removal, as they have an influence on the management of sanitation programmes. Soil removal from surfaces decreases such that the log of the mass of soil remaining per unit area is linear with respect to cleaning time (Fig. 12.1(a)) and thus follows first-order (log linear) reaction kinetics (Jennings, 1965; Schlusser, 1975). This approximation, however, is not valid at the start and end of the cleaning process and, in practice, soil removal is initially faster and ultimately slower (dotted line in Fig. 12.1(a)) than a first-order reaction predicts. The reasons for this are unclear, though initially, non-adhered, gross soil is usually easily removed (Loncin, 1977) whilst, ultimately, soils held within surface imperfections, bound to surfaces or otherwise protected from cleaning effects, are more difficult to remove (Holah and Thorpe, 1990).

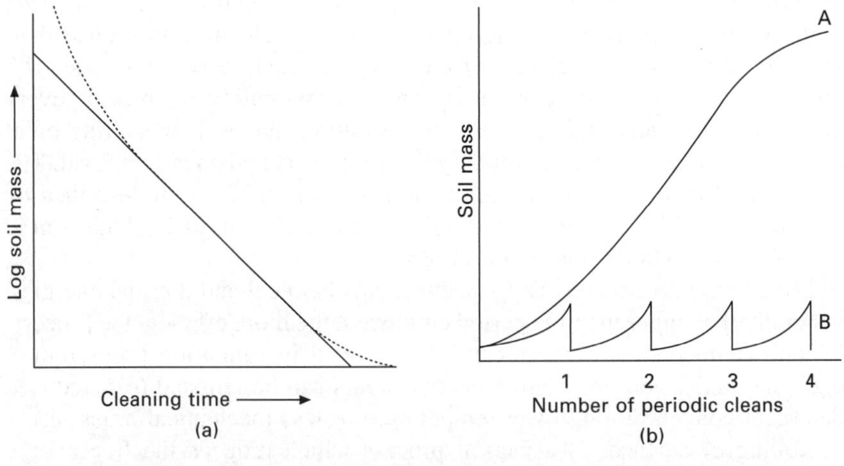

Fig. 12.1 Soil removal and accumulation. (a) Removal of soil with cleaning time. Solid line is theoretical removal, dotted line is cleaning in practice. (b) Build up of soil (and/or microorganisms): A, without periodic cleans and B, with periodic cleans. (After Dunsmore *et al.*, 1981)

Routine cleaning operations are never 100 % efficient, and over a course of multiple soiling/cleaning cycles, soil deposits (potentially including microorganisms) will be retained. As soil accumulates, cleaning efficiency will decrease and, as shown in plot A, Fig. 12.1(b), soil deposits may for a period increase exponentially. The timescale for such soil accumulation will differ between processing applications and can range from hours (e.g. heat exchangers) to typically several days or weeks, and in practice can be controlled by the application of a 'periodic' clean (Dunsmore *et al.*, 1981). Periodic cleans are employed to return the surface-bound soil accumulation to an acceptable base level (plot B, Fig. 12.1(b)) and are achieved by increasing cleaning time and/or energy input, e.g. higher temperatures, alternative chemicals (such as a periodic acid de-scale, when alkaline cleaning agents are used routinely) or manual scrubbing. A typical example of a periodic clean is the 'week-end clean down' or 'bottoming'.

12.3 Sanitation chemicals

Within the sanitation programme it has traditionally been recognised that cleaning is responsible for the removal of not only the soil but also the majority of the microorganisms present. Mrozek (1982) showed a reduction in bacterial numbers on surfaces by up to 3 log orders whilst Schmidt and Cremmling (1981) described reductions of 2–6 log orders. The results of work at Campden BRI on the assessment of well-constructed and competently undertaken sanitation programmes on food processing equipment in nine chilled food factories are shown in Table 12.1. The results suggest that both cleaning and disinfection are equally responsible for reducing the levels of adhered microorganisms and suggest a combined 3 log reduction of microorganisms. It is important, therefore, to purchase quality cleaning chemicals not only for their soil removal capabilities but also for their potential for microbial removal.

Unfortunately, no single cleaning agent is able to perform all the functions necessary to facilitate a successful cleaning programme; so a cleaning solution, or detergent, is blended from a range of typical characteristic components:

(1) water;
(2) surfactants;

Table 12.1 Arithmetic and log mean bacterial counts (per swab) on food processing equipment before and after cleaning and after disinfection

	Before cleaning	After cleaning	After disinfection
Arithmetic mean	7.41×10^7	4.48×10^5	2.35×10^3
Log mean	4.73	2.79	1.23
No. of observations	698	698	2004

(3) inorganic alkalis;
(4) inorganic and organic acids;
(5) sequestering agents.

For the majority of food processing operations it may be necessary, therefore, to employ different cleaning products, for specific operations. This requirement must be balanced by the desire to keep the range of cleaning chemicals on site to a minimum so as to reduce the risk of using the wrong product, to simplify the job of the safety officer and to allow chemical purchase to be based more on the economics of bulk quantities. The range of chemicals and their functions is well documented (Anon., 1991; Middleton and Holah, 2008) and has changed little in the last 25 years or more; only an overview of the principles is given here.

Water is the base ingredient of all 'wet' cleaning systems and must be of potable quality. Water provides the cheapest readily available transport medium for rinsing and dispersing soils, has dissolving powers to remove ionic-soluble compounds such as salts and sugars, will solubilise proteins below their coagulation point, will help emulsify fats at temperatures above their melting point and, in high-pressure cleaning, can be used as an abrasive agent. On its own, however, water is a poor 'wetting' agent and cannot dissolve non-ionic compounds.

Organic surfactants (surface-active or wetting agents) are amphipolar and are composed of a long non-polar (hydrophobic or lyophilic) chain or tail and a polar (hydrophilic or lyophobic) head. Surfactants are classified as anionic (including the traditional soaps), cationic, or non-ionic, depending on their ionic charge in solution, with anionics and non-ionics being the most common. Amphipolar molecules aid cleaning by reducing the surface tension of water and by the emulsification of fats. If a surfactant is added to a drop of water on a surface, the polar heads disrupt the water's hydrogen bonding and so reduce the surface tension of the water and allow the drop to collapse and 'wet' the surface. Increased wettability leads to enhanced penetration into soils and surface irregularities and hence aids cleaning action. Fats and oils are emulsified as the hydrophilic heads of the surfactant molecules dissolve in the water whilst the hydrophobic end dissolves in the fat. If the fat is surface-bound, the forces acting on the fat/water interface are such that the fat particle will form a sphere (to obtain the lowest surface area for its given volume) causing the fat deposit to 'roll-up' and detach itself from the surface.

Alkalis are useful cleaning agents as they are cheap, break down proteins through the action of hydroxyl ions, saponify fats and, at higher concentrations, may be bactericidal. Strong alkalis, usually sodium hydroxide (or caustic soda), exhibit a high degree of saponification and protein disruption, though they are corrosive and hazardous to operatives. Correspondingly, weak alkalis are less hazardous but also less effective. Alkaline detergents may be chlorinated to aid the removal of proteinaceous deposits, but chlorine at alkaline pH is not an effective biocide. The main disadvantages of alkalis are their potential to precipitate hard water ions, the formation of scums with soaps and their poor rinsability. It is currently not known whether alkaline cleaning agents modify allergenic protein residues sufficiently to reduce or eliminate their allergenicity.

Acids, e.g. nitric or phosphoric acid, have low detergency, although they are very useful in solubilising carbonate and mineral scales, including hard water salts, rust and proteinaceous deposits. As with alkalis, the stronger the acid the more effective it is; though, in addition, the more corrosive to plant and operatives. Acids are not used as frequently as alkalis and tend to be used for periodic cleans. During cleaning, acidic detergents must never be allowed to come into contact with chlorinated compounds because of the consequential release of toxic chlorine gas.

Sequestering agents (sequestrants or chelating agents) are employed to prevent mineral ions precipitating by forming soluble complexes with them. The calcium and magnesium ions found in hard water interfere with the action of anionic surfactants, cause precipitation and inactivation of detergent components and produce scale. Sequestrants prevent these problems by binding to the calcium and magnesium ions to form water-soluble complexes. Common sequestrants include ethylene diamine tetra acetate (EDTA), nitrilotriacetate (NTA) and polyphosphates.

A general-purpose food detergent may, therefore, contain a strong alkali to saponify fats, weaker alkali 'builders' or 'bulking' agents, oxidative agents such as chlorine to remove protein and undertake bleaching, surfactants to improve wetting, dispersion and rinsability, sequestrants to control hard water ions and solvents to stabilise the formulation. In addition, the detergent should ideally be safe, non-tainting, non-corrosive, environmentally-friendly and cheap. The choice of cleaning agent will depend on the soil to be removed and on its solubility characteristics though, essentially, alkaline detergents tend to be first choice for sugary, starchy, high-protein and fatty foods whilst acidic detergents are used for mineral and proteinaceous scales.

Because of the wide range of food soils likely to be encountered and the influence of the food manufacturing site (temperature, humidity, type of equipment, time before cleaning, etc.), there are currently no recognised laboratory methods for assessing the efficacy of cleaning compounds. Food manufacturers have to be satisfied that cleaning chemicals are working appropriately, therefore, by conducting suitable field trials. Therefore, food manufacturers should only purchase high-quality detergents because of their key role in the sanitation process.

Although the majority of the microbial contamination is removed by the cleaning phase of the sanitation programme, for some food processing sectors such as RTE foods, there are likely to be sufficient viable microorganisms remaining on the surface to warrant the application of a disinfectant. The aim of disinfection is therefore to further reduce the surface population of viable microorganisms, via removal or destruction, and/or to prevent surface microbial growth during the inter-production period. Elevated temperature is the best disinfectant as it penetrates into surfaces, is non-corrosive, is non-selective to microbial types, is easily measured and leaves no residue (Jennings, 1965). However, for open surfaces, the use of hot water or steam is uneconomic, hazardous or impossible and reliance is, therefore, placed on chemical biocides.

Whilst there are many chemicals with biocidal properties, many common disinfectants are not used in food applications because of safety or taint problems, e.g. phenolics or metal-ion-based products. In addition, other disinfectants are used to a limited extent and/or for specific purposes, e.g. peracetic acid, biguanides, formaldehyde, glutaraldehyde, organic acids, ozone, chlorine dioxide, bromine and iodine compounds. Of the acceptable chemicals, the most commonly used products for open surface disinfection are:

(1) chlorine-releasing components;
(2) quaternary ammonium compounds;
(3) quaternary ammonium/amphoteric mixtures;
(4) alcohols;

and for closed surfaces (cleaning in place, CIP)

(5) peracetic acid.

Chlorine is the cheapest disinfectant and is available as hypochlorite (or occasionally as chlorine gas) or in slow-releasing forms (e.g. chloramines, dichlorodimethylhydrantoin). Quaternary ammonium compounds (Quats or QACs) are amphipolar, cationic detergents, derived from substituted ammonium salts with a chlorine or bromine anion and amphoterics are based on the amino acid glycine, often incorporating an imidazole group.

A survey of the approved disinfectant products in Germany (DVG listed) in 1994 indicated that 36 % were QACs, 20 % were mixtures of QACs with aldehydes or biguanides and 10 % were amphoterics (Knauer-Kraetzl, 1994). In a Campden BRI survey undertaken of the UK food industry in 2000, of 117 food premises incorporating farms, food manufacturers, food hauliers and caterers 70 % used QACs, 49 % chlorine-releasing agents, 26 % alcohol, 10 % amphoterics and 7 % peracetic acid. Since the mid-1990s the synergistic combinations of QACs and amphoterics have been explored in the UK, and these compounds are now widely used.

The characteristics of the most commonly used disinfectants are compared in Table 12.2. The properties of QAC/amphoteric mixes will be similar to their parent compounds with suggested enhanced microorganism control. The table essentially confirms that QACs, amphoterics and alcohol are used predominantly for open plant cleaning because they combine sufficient antimicrobial properties yet are ‘friendly’ to operatives and the processing environment. These disinfectants are referred to as non-oxidising biocides and exert their antimicrobial action by interactions with the cell membranes, resulting in leakage of cell constituents. Chlorine and peracetic acid are

Table 12.2 Characteristics of some universal disinfectants

Property	Chlorine	QAC[a]	Amphoteric	Alcohol	Peracetic acid
Microorganism control					
Gram-positive	+ +	+ +	+ +	++	+ +
Gram-negative	+ +	+	+ +	++	+ +
Spores	+	–	–	–	+ +
Yeast	+ +	+ +	+ +	+	+ +
Developed microbial resistance	–	+	+	–	–
Inactivation by					
organic matter	+ +	+	+	++	+
Water hardness	–	+	–	–	–
Detergency properties	–	++	+	–	–
Surface activity	–	++	++	–	–
Foaming potential	–	++	++	–	–
Problems with					
taints	+/–	–	–	–	+/–
Stability	+/–	–	–	–	+/–
Corrosion	+	–	–	–	–
Safety	+	–	–	+	++
Other chemicals	–	+	–	–	–
Potential environmental impact	++	–/+	–/+	–	–
Cost	–	+ +	+ +	+++	+

a Quaternary ammonium compound
– no effect (or problem)
+ effect
+ + large effect

oxidising biocides and exert their antimicrobial action by oxidising proteins within the cells, resulting in metabolic inhibition. Chlorine has excellent antimicrobial properties though, due to its corrosiveness, it is used sparingly on open surfaces followed by thorough rinsing. Peracetic acid is perhaps the most antimicrobial chemical to microorganisms in suspension, attached to surfaces and in biofilms (Holah *et al.*, 1990) though is more of a hazard to cleaning operatives and, while the most common disinfectant used in closed plant systems, is only used under controlled conditions for open plant. The disinfectant 'active' is formulated with a range of pH regulators, buffers, surfactants (to enhance wetting), defoamers (to reduce foaming in use) and stabilisers to optimise its performance in practice.

Alcohol-based disinfectant products have traditionally been used in dry food processing operations. They are also used within the RTE food industry, particularly for mid-shift sanitation in high-risk areas. This is primarily to restrict the use of water for cleaning during production as a control measure to prevent the growth and spread of any foodborne pathogens. Ethyl alcohol (ethanol) and isopropyl alcohol (isopropanol) have bactericidal and virucidal (but not sporicidal) properties (Hugo and Russell, 1999), though they are only active in the absence of organic matter, i.e. the surfaces need to be wiped clean and then alcohol reapplied. Alcohols are most active in the 60–70 % range, and can be formulated into wipe and spray-based products. Alcohol products are used on a small, local scale because of their well recognised health and safety issues.

The efficacy of disinfectants is generally controlled by five factors: interfering substances (primarily organic matter), pH, temperature, concentration and contact time. To some extent, and particularly for the oxidative biocides, the efficiency of all disinfectants is reduced in the presence of organic matter. Organic material may react chemically with the disinfectant such that it loses its biocidal potency, or spatially such that microorganisms are protected from its effect. Other interfering substances, e.g. cleaning chemicals, may react with the disinfectant and destroy its antimicrobial properties, and it is therefore essential to remove all soil and chemical residues prior to disinfection. This is particularly important with cationic QACs such that, if used, all traces of anionic detergents need to be removed prior to disinfection.

Disinfectants should only be used within the pH range as specified by the manufacturer. Perhaps the classic example of this is chlorine, which dissociates in water to form HOCl (hypochlorous acid often termed 'free chlorine') and the OCl^- ion. From pH 3–7.5, chlorine is predominantly present as HOCl, which is a very powerful biocide, though the potential for corrosion increases with acidity. Above pH 7.5, however, the majority of the chlorine is present as the OCl^- ion which has about 100 times less biocidal action than HOCl.

In general, the higher the temperature the greater the rate of disinfection. For most food manufacturing sites operating at ambient conditions (around 20 °C) or higher this is not a problem as most disinfectants are formulated (and tested) to ensure performance at this temperature. This is not, however, the case in the chilled food industry. Taylor *et al.* (1999) examined the efficacy of 18 disinfectants at both 10 °C and 20 °C and demonstrated that for some chemicals, particularly quaternary ammonium based products, disinfection was much reduced at 10 °C. They recommended that in chilled production environments, only products specifically formulated for low-temperature activity should be used.

In practice, the relationship between microbial death and disinfectant concentration is not linear but follows a sigmoidal curve. Microbial populations are initially difficult to kill at low concentrations but, as the biocide concentration is increased, a point is reached where the majority of the population is reduced. Beyond this point the microorganisms become more difficult to kill (through resistance or physical protection) and a proportion may survive regardless of the increase in concentration. It is important, therefore, to use the disinfectant at the concentration recommended by the manufacturer. Concentrations above this recommended level may not enhance biocidal effect and will be uneconomic whilst concentrations below this level may significantly reduce biocidal action. The effect of dilution on antimicrobial activity is controlled by the concentration exponent eta (η) (Hugo and Denyer, 1987). If the disinfectant has a concentration exponent of 1, its activity will be reduced by the power of 1 on dilution. Thus for a three-fold dilution the activity will be reduced by 3^1 or 1/3. If $\eta = 4$ then a three-fold dilution would lead to a reduction in activity of 3^4 or 81 times. QACs typically have a η value of 1, whilst the value for alcohol is 10.

Sufficient contact time between the disinfectant and the microorganisms is perhaps the most important factor controlling biocidal efficiency. To be effective, disinfectants must find and concentrate at, bind to and transverse microbial cell envelopes before they reach their target site and begin to undertake the reactions which will subsequently lead to the destruction of the microorganism (Klemperer, 1982). Sufficient contact time is therefore critical to give good results, and most general-purpose disinfectants are formulated to require at least five minutes to reduce bacterial populations by 5 log orders in suspension. This has arisen for two reasons. Firstly 5 min is a reasonable approximation of the time taken for disinfectants to drain off vertical or near vertical food processing surfaces. Secondly, when undertaking disinfectant efficacy tests in the laboratory, a 5 min contact time is chosen to allow ease of test manipulation and hence timing accuracy. For particularly resistant organisms such as spores or moulds, surfaces should be repeatedly dosed to ensure extended contact times of 15–60 mins.

Ideally, disinfectants should have the widest possible spectrum of activity against microorganisms, including bacteria, fungi, spores and viruses, and this should be demonstrable by means of standard disinfectant efficacy tests. The range of currently available disinfectant test methods was reviewed by Reybrouck (1998), and these fall into two main classes, suspension tests and surface tests. Suspension tests are useful for indicating general disinfectant efficacy and for assessing environmental parameters such as temperature, contact time and interfering matter such as food residues. In reality, however, microorganisms disinfected on food contact surfaces are those that remain after cleaning and are therefore likely to be adhered to the surface. A surface test is thus more appropriate.

A number of authors have shown that bacteria attached to various surfaces are generally more resistant to biocides than are organisms in suspension (Ridgeway and Olsen, 1982; Hugo *et al.*, 1985; Le Chevalier *et al.*, 1988; Frank and Koffi, 1990; Holah *et al.*, 1990; Lee and Frank, 1991; Wright *et al.*, 1991; Dhaliwal *et al.*, 1992; Andrade *et al.*, 1998; Das *et al.*, 1998). In addition, cells growing as a biofilm have been shown to be more resistant (Frank and Koffi, 1990; Lee and Frank, 1991; Ronner and Wong, 1993). The mechanism of resistance in attached and biofilm cells is unclear but may be due to physiological differences such as growth rate, membrane orientation changes due to attachment and the formation of extracellular material which surrounds the cell. Equally, physical properties may have an effect, e.g. protection of the cells by food debris or the material surface structure or problems in biocide diffusion to the cell/material surface. To counteract such claims of enhanced surface adhered resistance, it can be argued that, in reality, surface tests do not consider the environmental stresses the organisms may encounter in the processing environment prior to disinfection (action of detergents, variations in temperature and pH and mechanical stresses) which may affect susceptibility.

Both suspension and surface tests have limitations, however, and research-based methods are being developed to investigate the effect of disinfectants against adhered microorganisms and biofilms *in situ* and in real time. Such methods have been reviewed by Holah *et al.* (1998).

The current food industry disinfectant test methods of choice in Europe for bactericidal and fungicidal action in suspension are EN 1276 (CEN, 1997) and EN 1650 (CEN, 1998), respectively, and food manufacturers should ensure that the disinfectants they use conform to these standards as appropriate. A surface test, EN 13697 (CEN, 2001), is also available for disinfectant manufactures to demonstrate efficacy for their products against surface adhered microorganisms. Because of the limitations of disinfectant efficacy tests, however, food manufacturers should always confirm the efficacy of their cleaning and disinfection programmes by 'field tests' either from evidence supplied by the chemical company or from in-house validation trials.

As well as having demonstrable biocidal properties, disinfectants must also be safe (non-toxic) and should not taint food products. Disinfectants can enter food products accidentally, e.g. from aerial transfer or poor rinsing, or deliberately, e.g. from 'no rinse status' disinfectants. The practice of rinsing or not rinsing has been under discussion for many years and has yet to be clarified. The main reason for leaving disinfectants on surfaces is to provide an alleged biocide challenge (this has not been proven) to any subsequent microbial contamination of the surface. It has been argued, however, that the low biocide concentrations remaining on the surface, especially if the biocide is a QAC, may lead to the formation of resistant surface populations (again not proven in practical situations). In Europe there is no specific legislation requiring disinfectants to be rinsed from surfaces other than the general Regulation on the hygiene of foodstuffs (852/2004/EC) which obligates food business operators to produce safe foodstuffs (EC, 2004). In other words disinfectants can be left on surfaces if the food manufacturer is happy that any residues will not affect the wholesomeness or safety of subsequently produced foodstuffs.

The larger picture concerning the use of disinfectants in the food industry and how they may affect the resistance of microorganisms, both within the food and the clinical arena, is still not clear. In the USA, the Institute of Food Technologists examined this issue in 2002 (reported by Davidson and Harrison, 2002) and 2006 (reported by Doyle, 2006). The conclusions of these reviews were that, although bacteria may be exposed to an antibiotic for an extended period of time, on the farm or in humans, bacterial exposure to food antimicrobials (e.g. disinfectants) generally occurs only once. The prevalence and mechanism of resistance among most food-use antimicrobial compounds is often unknown. When it occurs, resistance to food antimicrobials is of little practical relevance to the food industry because the antimicrobial concentrations used in food manufacturing are well above the low levels to which bacteria exhibit resistance. However, the ability of some sanitisers and disinfectants to induce multiple drug resistance pumps, which also confer antibiotic resistance, is of some concern.

In terms of the demonstration of non-toxicity, legislation will vary in each country, although in Europe this will be clarified with the implementation of Directive 98/9/EC concerning the placing of biocidal products on the market, which contains requirements for toxicological and metabolic studies. This Directive seeks to produce a list of active biocidal substances that have been assessed for both their toxicological properties and also their inherent antimicrobial properties. Following the establishment of the approved active list, formulated products (the disinfectants sold to the final user) can then only be made by incorporating an approved active and the formulated product will then itself be assessed for its toxicological and antimicrobial properties. The effects of this Directive are already being seen, as current formulated products that do not contain biocidal actives that will be supported through the approval process are being removed from the marketplace. This will undoubtedly reduce the choice of disinfectants available to the food industry in the future, though the ones that will be available will have been thoroughly assessed. In the interim, a recognised acceptable industry guideline for disinfectants is a minimum acute oral toxicity (with rats) of 2000 mg/kg bodyweight.

Approximately 30 % of food taint complaints are thought to be associated with cleaning and disinfectant chemicals and are described by sensory scientists as 'soapy', 'antiseptic' or 'disinfectant' (Holah, 1995). Campden BRI (formerly CCFRA) have developed two taint tests in which foodstuffs which have and have not been exposed to disinfectant residues are compared by a trained taste panel using the standard triangular taste test (ISO, 2004). For assessment of aerial transfer, a modification of a packaging materials odour transfer test is used (BSI, 1964) in which food products, usually of four types (high-moisture, e.g. melon, low-moisture, e.g. biscuit, high-fat, e.g. cream, high-protein, e.g. chicken) are held above a disinfectant solution or distilled water for 24 h. To assess surface transfer, a modification of a food container transfer test is used (DIN, 2004) in which food products are sandwiched between two sheets of stainless steel and left for 24 h. Disinfectants can be sprayed onto the stainless steel sheets and drained off, to simulate no-rinse status, or can be rinsed off prior to food contact. Control sheets are rinsed in distilled water only. The results of the triangular test involve both a statistical assessment of any flavour differences between the control and disinfectant treated sample and a description of any flavour changes.

12.4 Sanitation methodology

Sanitation methods used in the food industry are primarily controlled by whether the item to be cleaned is easily accessible for manual cleaning (open surfaces) or has to be cleaned via the circulation of cleaning fluids (closed surfaces). The cleaning of closed surfaces (particularly when there is no manual intervention) is usually referred to as cleaning-in-place (CIP). Open surface cleaning is further controlled by whether it can be cleaned wet (preferred) or has to be cleaned dry. Wet cleaning can then be further defined by cleaning items *in situ*, by cleaning them in specific areas using manual techniques (COP or 'cleaned-out-of-place') or by using automated devices.

All sanitation methods do, however, use a combination of the four cleaning factors as described earlier. How these factors vary for a range of cleaning techniques can perhaps be described schematically following the information detailed in Fig. 12.2. The figure details the different energy source inputs for a number of cleaning techniques and shows their ability to cope with both low and high (dotted line) levels of soiling. Some cleaning techniques may not be able to cope with high soil levels whilst for each of the ones that can, the factor(s) that can be increased to allow the cleaning of high soils are indicated.

12.4.1 Open surfaces

Cleaning and disinfection can be undertaken by hand using simple tools, e.g. brushes or cloths (manual cleaning), though as the area of open surface requiring cleaning and disinfection increases, specialist equipment becomes necessary to dispense chemicals and/or provide mechanical energy. Chemicals

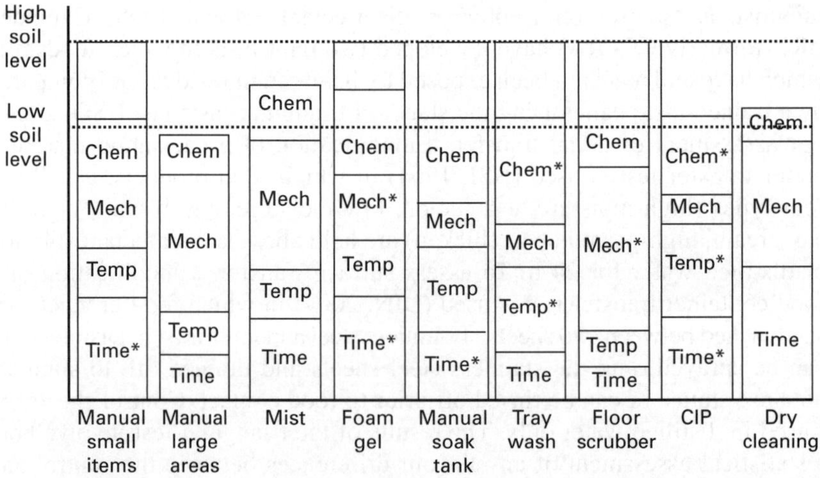

Fig. 12.2 Relative cleaning factor inputs for a range of cleaning techniques. *Factor(s) that can be increased to clean higher soil levels.

may be applied as low-pressure mists, foams or gels whilst mechanical energy is provided by high- and low-pressure water jets or water or electrically powered scrubbing brushes. These techniques have been well documented (Anon., 1991; Holah, 1991; Marriott, 2006) and, as with cleaning chemicals, have changed little since these reviews: this section considers their use in practice.

In manual cleaning of larger areas, especially above chest level, for reasons of operator safety, cleaning operatives should only use low-temperature water (< 60 °C) and low levels of chemical energy. As the surface area requiring cleaning increases, the manual techniques become uneconomic with respect to time and labour. Labour costs amount to approximately 75 % of the total sanitation programme, and for most food companies the cost of extra staff is prohibitive. Only light levels of soiling can be economically undertaken by this method.

The main difference between the mist, foam and gel techniques is in their ability to maintain a detergent/soil/surface contact time. For all three techniques, mechanical energy can be varied by the use of high- or low-pressure water rinses, though for open surface cleaning, temperature effects are minimal. Mist spraying is undertaken using small hand-pumped containers, 'knapsack' sprayers or pressure washing systems at low pressure. Misting will only 'wet' vertical smooth surfaces; therefore only small quantities can be applied and these will quickly run off to give a contact time of 5 min or less. The nature of the technique causes aerosols to form that could be an inhalation hazard, only weak chemicals can be applied, and so misting is useful only for light soiling. On cleaned surfaces, however, misting is the most commonly used method for applying disinfectants.

Foams can be generated and applied by high-pressure equipment with entrapment of air in the foam or by the addition of compressed air in low-pressure systems. Foams work on the basis of reducing the density of the cleaning agent and forming a layer of bubbles above the surface to be cleaned which then collapses and bathes the surface with fresh detergent contained in the bubble film. The critical element in foam generation is for the bubbles to collapse at the correct rate: too fast and the contact time will be minimal; too slow and the surface will not be wetted with fresh detergent.

Gels are thixotropic chemicals which are fluid at high and low concentrations but become thick and gelatinous at concentrations of approximately 5–10 %. Gels are easily applied through high- and low-pressure systems or from specific portable electric pumped units, and they physically adhere to the surface. Gels may be coloured to show coverage.

Foams and gels are more viscous than mists, are not as prone to aerosol formation, thus allowing the use of more concentrated detergents, and can remain on vertical surfaces for much longer periods (foams 10–15 min, gels 15 min to 1 h or more). Foams and gels are able to cope with higher levels of soils than misting, therefore, although in some cases rinsing of surfaces may require large volumes of water, especially with foams. Foams and gels are well liked by operatives and management as, because of the nature of the foam, a more consistent application of chemicals is possible and it is easier to identify areas that have been 'missed'.

Fogging systems have been traditionally used in the RTE food sector to create and disperse a disinfectant aerosol to reduce airborne microorganisms and to apply disinfectant to difficult to reach overhead surfaces. The efficacy of fogging has been reported (MAFF, 1998) and, providing a suitable disinfectant is used, fogging is effective at reducing airborne microbial populations by 2–3 log orders in 30–60 min. Fogging is most effective using compressed air-driven fogging nozzles producing particles in the 10–20 micron range. For surface disinfection, fogging is only effective if sufficient chemical can be deposited onto the surface. This is illustrated in Fig. 12.3 which shows the log reductions achieved on horizontal, vertical and upturned (underneath) surfaces arranged at five different heights from just below the ceiling (276 cm) to just above the floor (10 cm) within a test room. It can be seen that disinfection is greatest on surfaces closest to the floor and that disinfection is minimal on upturned surfaces close to the ceiling. Commercial fogging units are now available which electrostatically charge the disinfectant during the fogging process. The charged disinfectant droplets have been shown to be more effective at wetting overhead and vertical surfaces. To reduce inhalation risks, sufficient time (45–60 min) is required after fogging to allow the settling of disinfectant aerosols before operatives can re-enter the production area.

Cleaning chemicals are removed from surfaces by low-pressure/high-volume (LPHV) hoses operating at mains water pressure or by high-pressure/low volume-(HPLV) pressure washing systems which require a high-pressure pump. Pressure washing systems operate above mains water pressure, typically at between 25 and 100 bar through a 15° nozzle, and may be mobile units, wall-mounted units or centralised ring mains. Water jets confer high mechanical energy, can be used on a wide range of equipment and environmental surfaces, will penetrate into surface irregularities and are able to mix and apply chemicals.

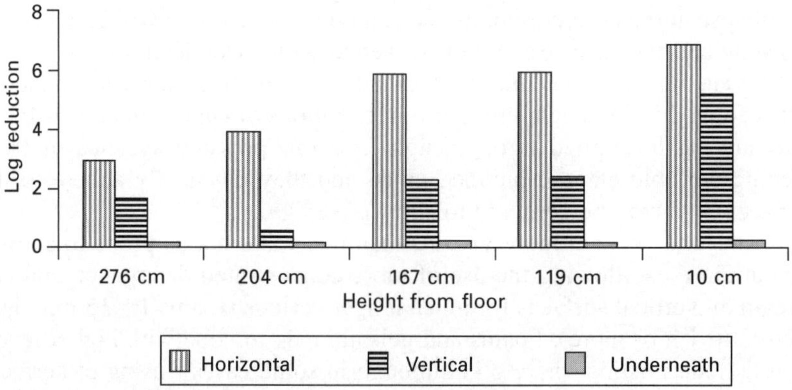

Fig. 12.3 Comparative log reductions of microorganisms adhered to surfaces and positioned at various heights and orientations.

Mechanical scrubbers include traditional floor scrubbers, scrubber/driers (automats) for floors and water-driven attachments to high-pressure systems and electrically operated small-diameter brushes that can be used on floors, walls and other surfaces. Contact time is usually limited with these techniques (though can be increased), but the combination of detergency with high mechanical input allows them to tackle most soil types. The main limitation is that food processing areas have not traditionally been designed for their use, though this can be amended in new or refurbished areas.

12.4.2 Dry cleaning

Dry cleaning techniques are used when the soil to be removed is hygroscopic or where water can react to form hard deposits that are difficult to remove. There may also be the assumption that maintaining dry conditions in the production environment will control survival or growth of microorganisms. The principal risk is that failure to control moisture can permit the growth of pathogens, e.g. *Salmonella* spp., in the processing environment, which may then become a source of food product contamination. Environments usually dry cleaned include plants producing dry ingredients, flour, chocolate, peanut butter, dry milk products and dry infant formulae, dry soup and snack mixes, and cereals and cereal products.

Dry cleaning methods include compressed air, scrapers, brushes, vacuum systems and food products and have been reviewed by Middleton *et al.* (2003). Compressed air lifts soil from surfaces by giving it kinetic energy and consequently disperses it into the surrounding environment. Another method of dry cleaning needs to be used, therefore, to remove the transferred soil. Whilst being an effective cleaning tool, the use of compressed air for cleaning has been banned in some factories due to the increased risk of dust explosions in environments where the food soils are fine, such as the flour milling industry. There are also health and safety issues relating to high noise levels if incorrect air guns are used and the risk of injury to operatives if insufficient training is given on the dangers of compressed air use.

Scrapers work on the principle of dragging the blade over a surface and lifting the soils from them by friction. Again, as this technique only lifts the soil from the surface and does not remove it, another method of dry cleaning must be used to complete the cleaning procedure. Brushes enable loosely adhered soils to be removed from surfaces and can be used to collect widespread soiling from large surface areas to one central point. They work on the principle of dragging bristles over a surface and lifting the soils from them by friction generated by the pressure between the bristle and the surface.

Vacuum cleaners use suction power to lift dry soils from the food contact surfaces. The vacuum cleaner nozzle or tool is placed approximately 1–2 cm above the surface, to ensure maximum air circulation, and the removed soil is transferred via the vacuum hose into a central collection unit. Vacuum cleaners can be fitted with HEPA (high efficiency particulate air) filters to reduce the risk of explosion and microbial contamination by preventing light dust or contaminating bacteria from becoming airborne. Vacuum cleaners may be individual units or ring main systems with a dust collection unit outside of production areas. The diameter of the vacuum hose is a significant factor in the cleaning efficiency of a ring main vacuum system, with small narrow hoses carrying heavier particles well and bigger diameter hoses suitable for finer, lighter dusts. To prevent soil build-up in pipework, and consequently prevent deposition of product back onto previously cleaned surfaces, an air velocity of 15–25 m s^{-1} may be required dependent on the nature of the soil being conveyed.

Food products can also be used as dry cleaning aids where a base ingredient is often used to flush through process lines, e.g. hot oil in fat manufacturing operations or the most bland flavoured product in ingredients manufacturing. Food products and non-food products, e.g. nut kernels, solid carbon dioxide and calcium carbonate, can also be used as abrasives and are often applied by high-pressure spray blasting.

The hygienic implications of the design and use of both wet and dry cleaning equipment should be carefully considered and should be similar to that required for food processing equipment. Sanitation equipment should be constructed out of smooth, non-porous, easily cleanable materials such as stainless steel or plastic. Mild steel or other materials subject to corrosion may be used but must be suitably painted or coated, whilst the use of wood is unacceptable. Frameworks should be constructed of tubular or box section material, closed at either end and properly jointed, e.g. welds should be ground and polished and there should be no metal-to-metal joints. Crevices and ledges where soil could collect should be avoided and exposed threads should be covered or dome nuts used. Tanks for holding cleaning chemicals or recovered liquids should be self-draining, have rounded corners and should be easily cleaned. Shrouds around brush heads or hoods and rotary scrubbing heads should be easily detachable to facilitate cleaning. Brushes should have bristles of coloured, impervious material, e.g. nylon, embedded into the head with resin so no soil trap points are apparent. Alternatively, brushes with the head and bristles moulded as one unit may be used.

Cleaning equipment is prone to contamination with *Listeria* spp. and other pathogenic microorganisms and, by the nature of its use, provides an excellent way in which contamination can be transferred from area to area. Cleaning equipment should, therefore, be specific to low and high risk areas, and after use equipment should be thoroughly cleaned and, if appropriate, disinfected and dried.

12.4.3 Cleaned-out-of-place (COP)

Designated cleaning rooms are often utilised for the cleaning and disinfection of portable and hand-held food production equipment, as by cleaning equipment within them, the movement of aerosols and the build-up of condensation in the production area can be eliminated. They also allow production areas to remain dry during production periods, which is preferable in high risk/care areas. A schematic cleaning room is illustrated in Fig. 12.4 and should ideally be positioned on the outside of the building to allow localised ventilation systems to exhaust to the external environment and facilitate the delivery of bulk chemicals. Within the wash area, the use of two or three sinks facilitates detergent cleaning, rinsing and/or disinfection of hand-held equipment and LPHV hoses can be used for the wet cleaning of larger equipment There should be drainage channels at the entrance and exit to the room and the floors should be sloped to prevent the pooling of water in the area. The layout is critical to ensure that dirty equipment does not re-contaminate cleaned equipment and is segregated with an 'In' door to the dirty area and an 'Out' door from the designated clean area. It is also good practice to provide space within the clean area for draining and air-drying equipment.

Alternatively, dismantled equipment and production utensils may undergo manual gross soil removal and then be cleaned and disinfected automatically in tray or tunnel washers. These machines use mechanical energy from the wash down jets, with high temperatures and highly alkaline or caustic-based detergents to clean the equipment. Typically a tray passing through a tunnel tray washer will be in the detergent zone for around 10–20 seconds, therefore the detergent contact time is limited. Automated washing systems have high capital cost but are able to clean a large number of utensils to a consistent quality. Cleaning rooms may also be used to isolate automated cleaning equipment, such as tray washers, from the production area and are particularly recommended for automated systems in RTE high risk/care food production areas.

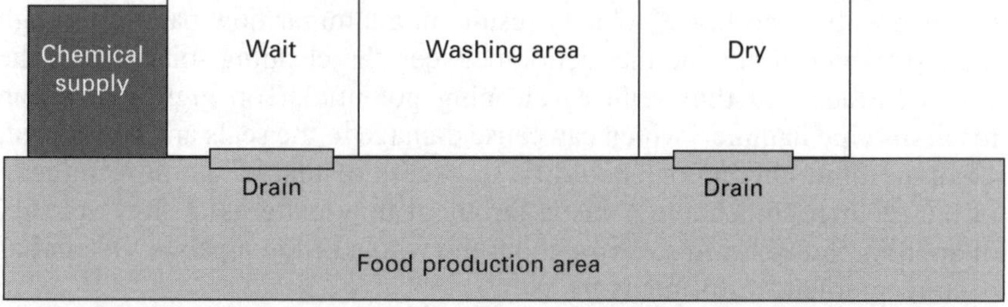

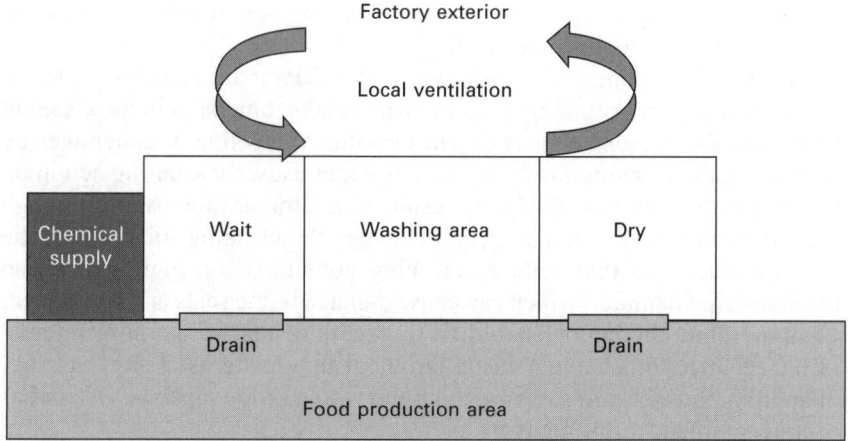

Fig. 12.4 Schematic drawing of an ideally designed cleaning room for COP operations.

12.4.5 Cleaning-in-place (CIP)

CIP is the cleaning of complete items of equipment or pipeline circuits, *in situ*, without dismantling and with little or no manual involvement. The same four cleaning factors are used as for open surface cleaning, though the relative inputs for each factor can be very different. Mechanical or kinetic energy is generally limited and is provided in pipelines by turbulent flow of cleaning solutions and in vessels by either falling films or spray impingement of cleaning solutions.

As the system to be cleaned is enclosed, there should be little if any contact between the cleaning chemicals and the operatives. The concentration of cleaning solutions and their circulation temperatures can thus exceed those used for open plant cleaning. The correct concentration of caustic or acidic detergent depends on the nature of soiling to be removed, for example, caustic detergents (e.g. sodium hydroxide) applied to vats, tanks and silos have an in-use concentration of typically 0.3–2.0 % whereas acidic detergents (e.g. phosphoric or nitric acid) on vats would be typically applied at 0.2–1.0 % acidity. When using alkaline products for the removal of organic soiling, temperatures of approximately 72–88 °C are required. When removing inorganic matter, or an acid-based detergent is being used, lower temperatures of approximately 50–60 °C are typically used. CIP cleaning enables the use of thermal disinfection, though this is only effective if all plant components are subjected to the required temperature and time. Methods of thermal disinfection include hot water, steam at temperatures between 70 and 80 °C and maintained for 15 min and hot pressurised water.

Cleaning time can be automatically controlled and can be optimised for each cleaning or disinfection stage to give reproducible sanitation programmes. Cleaning time is usually determined by the cleaning of the most difficult to clean component in the CIP circuit. The combination of longer contact times, higher temperatures and chemical strengths makes up for any deficiencies in mechanical energy as compared to open cleaning and enables CIP systems to be extremely effective and reliable.

The actual cleaning mechanism within the CIP circuit is divided into the cleaning of pipelines (and other items with total submersion in the cleaning fluids) and the cleaning of vessels. The cleaning of pipelines is undertaken by circulating the cleaning fluids at a velocity that causes a scouring action on the pipe walls. Too low a velocity results in a laminar flow pattern through the pipe which limits the interaction between the cleaning solution and the soiled surface and thus reduces cleaning potential. Too high a flow can result in 'pipe hammer' which can cause damage to the seals and equipment. Cleaning fluid circulation should be in excess of that of the flow velocity of the product and should achieve turbulent flow patterns. A flow velocity of approximately 1.5 m s^{-1} throughout the whole of the pipeline system has been recommended for many years.

The cleaning of vessels is more complex, primarily because they can differ in size from 500 litres for small cooking vessels to in excess of 100 000 litres for large brewery or dairy silos, which cannot economically be filled with cleaning fluids. A number of spraying devices are available which essentially produce a 'falling film' of cleaning fluid or an impingement jet of cleaning fluid over the whole of the vessel surface. Fixed (static) spray balls, via the use of holes drilled in specific patterns, direct cleaning fluids to cover the vessel surface. Approximately 30–50 litres of cleaning fluid per minute per metre of the vessel circumference, at 1.5–3 bar, is generated to maintain a continuous liquid film on the tank wall which falls due to gravity to provide some mechanical action. Low-pressure rotating (dynamic) sprayheads use different sizes of fan-shaped nozzles and rotate slowly inside the tank, driven by the flow of cleaning fluid. The rotating jets can generate relatively higher mechanical energy through the intermittent liquid impact on the tank wall and the spiral movement of the liquid film down the tank walls. The flow of liquid is about 50–30 % of the amount recommended for spray balls and therefore requires a lower total consumption. When using static or rotating sprayballs, the vessel must be designed to ensure adequate drainage of solution and prevent pooling of chemical and soil in the bottom of the vessel which would interrupt the falling film. This is further controlled by ensuring that the (scavenge) pump used to remove fluids from the vessels is operating at a flowrate in excess of the supply pump.

High-pressure rotating (dynamic) sprayheads are also driven by the pressure of the cleaning fluid but can also be driven mechanically. The high-pressure or mechanical drive makes the nozzles perform a geared rotation around the vertical and horizontal axis such that all points on the vessel surface are impacted over a defined time period. Once this time period has been established, cleaning must always meet or exceed this time to ensure full surface coverage. The types of spray pattern established by the different sprayheads are shown in Fig. 12.5. The choice of sprayhead is a compromise between purchase cost and maintenance (from a simple, cheap static sprayball to a complicated high-pressure sprayhead) and cleaning cost. For example, a static sprayhead may use 7.7 m^3/hour of cleaning fluid at 1 bar whilst a dynamic sprayhead uses only 1.8 m^3/hour at the same pressure. A computer-controlled high-pressure mechanically-driven sprayhead could use significantly less cleaning fluid than this.

The CIP system is configured in a number of ways, dependent on capital cost and cleaning fluid use and recovery. There are many names given to the different types of CIP sets; however, the three most common are as follows:

1. total loss;
2. partial recovery;
3. total recovery.

A total loss or a single-use system is one in which the chemicals from all stages of the cleaning process – pre-rinse, detergent application, final rinse and disinfection – are pumped around the equipment and then discharged

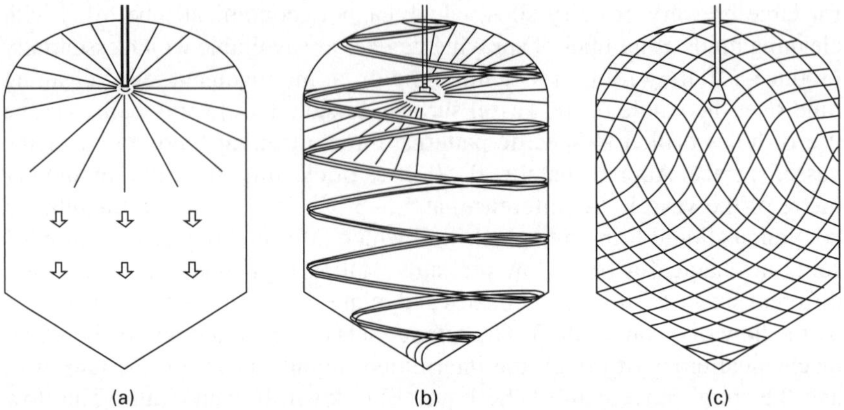

Fig. 12.5 Spray patterns inside a vessel using (a) static sprayball creating a falling film, (b) dynamic sprayhead and (c) high-pressure rotating sprayhead. (Spray pattern illustrations reproduced by kind permission of Alfa Laval)

direct to drain. The system has little capital cost and requires limited plant space though is expensive with respect to chemicals, water use and effluent disposal. The system may be used when there is a high degree of soiling or when cross-contamination must be avoided. A partial recovery system consists of a water buffer tank and a detergent recovery tank. The water buffer tank is used to supply the water for the pre-rinse. This is mains water that has been used for the final rinse and is recovered into the water buffer tank to be used for the pre-rinse stage of the next clean. The detergent tank supplies the detergent which, following the initial soil purge, is returned to this tank. The detergent volume and strength is maintained automatically by, for example, level and conductivity probes.

A total recovery system or re-use system recovers the final rinse and all cleaning chemicals. In a single-stage CIP system, tanks are used for a water buffer, alkaline detergent recovery, disinfectant recovery and final rinse recovery. A dual-stage CIP adds an acid detergent recovery tank. Total recovery systems minimise chemical, water and effluent costs but require a greater capital outlay. The selection of a type and system will depend on the size, complexity and operation of the plant as well as the type of soil and has been reviewed using breweries as an example by Manger (2006).

With CIP systems it is absolutely critical to prevent the cleaning and disinfection chemicals from contaminating the food product and, as such, the CIP circuit must always be separated from the food product. The use of single valves between the chemicals and the product is not sufficient because any leakage over the valve cannot be observed. Traditionally, the potential for contamination was minimised by bringing product and CIP lines to a flow selector plate. Here specific short lengths of pipe were manually removed and reconnected to ensure that product and cleaning fluids were separated.

Modern systems use block and bleed valves to ensure separation of chemical and product, but attention should be given to installing bleed lines correctly to ensure they drain and do not retain fluids. The valves should also be easily accessible for periodic maintenance and inspection.

Providing the CIP systems are hygienically designed, there should be no microbiological problems associated with microbial retention and harbourage in the equipment. However, it is always good practice to use separate CIP sets for raw and post-heat-treated product lines.

12.5 Sanitation procedure

Sanitation procedures are concerned with both the stage in which the sanitation programme is implemented and the sequence in which equipment and environmental surfaces are cleaned and disinfected within the processing area. Sanitation programmes are constructed to be efficient with water and chemicals, to allow selected chemicals to be used under their optimum conditions, to be safe in operation, to be easily managed and to reduce manual labour. In this way an adequate level of sanitation will be achieved, economically and with due regard to environmental friendliness. The principle stages involved in a typical sanitation programme are as follows.

12.5.1 Production periods

Production staff should be encouraged to operate good housekeeping practices (this is an aid to ensuring acceptable product quality and personnel safety), to clean their work stations prior to break periods (unless cleaning operatives are employed specifically to do this) and to leave their work stations in a hygienic condition. Food debris left in hoppers and on process lines, etc. is wasted product! Cleaning systems may spread contamination and should not be used during production periods if there is a risk of spreading microorganisms or foreign bodies onto food or food contact surfaces. For example, HPLV systems can distribute aerosols to a height of > 3 m and a distance of > 7 m whilst LPHV systems distribute aerosols to > 2 and > 3.5 m. Other techniques such as manual cleaning distribute aerosols to < 0.5 m in height and distance and are thus acceptable for use in clean-as-you-go operations. Where appropriate, care should be taken to prevent the growth and spread of pathogenic microorganisms during production periods. This is primarily undertaken by excluding or severely limiting the use of water for cleaning purposes and using manual cleaning (brushing, wiping) with alcohol as a solvent. Sound sanitation practices should be used to clean up major product spillages and re-start production, which may require the use of wet cleaning methods.

12.5.2 Preparation

As soon as possible after production, equipment for manual cleaning should be dismantled as far as is practicable or necessary to make all surfaces that microorganisms could have adhered to during production accessible to the cleaning fluids. All product and unwanted utensils/packaging/equipment should be covered or removed from the area. Dismantled equipment should be stored on racks or tables, not on the floor! Machinery should be switched off, at the machine and at the power source, and electrical and other sensitive systems protected from water/chemical ingress. Preferably, production should not be done in the area being cleaned, but in exceptional circumstances if this is not possible, other lines or areas should be screened off to prevent transfer of debris as aerosols or splashes by the sanitation process.

12.5.3 Gross soil removal

Where appropriate, all loosely adhered or gross soil should be removed by brushing, scraping, shovelling or vacuum, etc. Wherever possible, soil on floors and walls should be picked up and placed in suitable waste containers rather than washed to drains using hoses.

12.5.4 Pre-rinse

Surfaces should be rinsed with low-pressure cold water to remove loosely adhered small debris. Hot water can be used for fatty soils (approximately 60 °C), but too high a temperature (> 45 °C) may coagulate proteins.

12.5.5 Cleaning

A selection of cleaning chemicals, temperature and mechanical energy is applied to remove adhered soils.

12.5.6 Inter-rinse

Both soil detached by cleaning operations and cleaning chemical residues should be removed from surfaces by rinsing with low-pressure cold water.

12.5.7 Disinfection

Chemical disinfectants (or occasionally heat) are applied to remove and/or reduce the viability of remaining microorganisms to a level deemed to be of no significant risk. In exceptional circumstances and only when light soiling is to be removed, it may be appropriate to combine stages 5–7 by using a chemical with both cleaning and antimicrobial properties (detergent-sanitiser).

12.5.8 Post-rinse

Disinfectant residues should be removed by rinsing away with low-pressure cold water of known potable quality. Some disinfectants, however, are intended to be left on surfaces until the start of subsequent production periods and are thus so formulated to be both surface-active and of low risk, in terms of taint or toxicity, to foodstuffs.

12.5.9 Inter-production cycle conditions

A number of procedures may be undertaken, including the removal of excess water and/or equipment drying, to prevent the growth of microorganisms on production contact surfaces in the period up until the next production process. Alternatively, the processing area may be evacuated and fogged with a suitable disinfectant.

12.5.10 Periodic practices

Periodic practices increase the degree of cleaning for specific equipment or areas to return them to acceptable cleanliness levels. They include weekly acidic cleans, weekend dismantling of equipment, cleaning and disinfection of chillers and sanitation of surfaces, fixtures and fittings above 2 m.

A very similar sanitation procedure is undertaken for CIP systems and usually consists of the following operating cycles.

1. Recovery of product residues via pigging or flushing if appropriate. A 'pig', often designed from an elastic material such as silicon rubber, is sent, by means of compressed air or water as the driving force, through the pipework, to expel most of the product from the system. Alternative pigging systems include the use of ice particles (Quarini and Shire, 2007) or air flows (e.g. www.whirltech.co.uk).
2. A pre-rinse with hot or cold water to remove loosely adhered matter.
3. An alkali detergent cycle, resulting in the soil being lifted from the surface and held, suspended or dissolved in the detergent solution.
4. An intermediate rinse with cold water to remove the detergent residues.
5. If required, an acid detergent cycle, resulting in inorganic scale being dissolved.
6. An intermediate rinse with cold water to remove all traces of detergent.
7. A disinfectant stage, cycling Peroxyacetic acid (PAA) or hypochlorite around the circuit.
8. A final rinse, using clean (potable) water to rinse away the disinfectant residues.

For open surface cleaning, a sanitation sequence should be established in a processing area to ensure that the applied sanitation programme is capable of meeting its objectives and that cleaning programmes, both periodic and for areas not cleaned daily, are implemented on a routine basis. In particular, a sanitation sequence determines the order in which the product contact surfaces of equipment and environmental surfaces (walls, floors, drains, etc.) are sanitised, such that once product contact surfaces are disinfected, they should not be re-contaminated. The sequence must be performed at a 'room' level such that all environmental surfaces and equipment in the area are cleaned at the same time. It is not acceptable to clean and disinfect one line and then move on to the next and start the sequence again as this merely spreads contamination around the room. A suggested sequence is:

(1) remove gross soil from production equipment;
(2) remove gross soil from environmental surfaces;
(3) rinse down environmental surfaces (usually to a minimum of 2 m in height for walls);
(4) rinse down equipment and flush to drain;
(5) clean environment surfaces, usually in the order of drains, walls then floors;
(6) rinse environmental surfaces;
(7) clean equipment;
(8) rinse equipment;
(9) disinfect equipment and rinse if required;
(10) fog (if required).

12.6 Evaluation of sanitation effectiveness

Assessment of the effectiveness of the sanitation programme's performance is part of day-to-day hygiene testing and, as such, is linked to the factory environmental sampling plan. The control of the environmental routes of contamination is addressed via the development of a thorough hazard analysis and management strategy, typically undertaken as part of the factory HACCP study, resulting in the development of the factory environmental sampling plan. The development of environmental sampling plans has been established by a Campden BRI industrial working party and is reported in Holah (1998).

Environmental sampling is directly linked with both process development and product manufacture and as such, has three distinct phases;

- process development to determine whether a contamination route is a risk and assessing whether procedures put in place to control the risk identified are working;
- routine hygiene assessment;
- troubleshooting to identify why products (or occasionally environmental samples) may have a microbiological count that is out-of-specification or may contain pathogens.

The performance of the sanitation programme is an important aspect of due diligence and is assessed by monitoring to check sanitation process control, and verifying sanitation programme success. Monitoring is a planned sequence of observations or measurements to ensure that the control measures within the sanitation programme are operating within specification and are undertaken in a timeframe that allows sanitation programme control. Verification is the application of methods in a longer timeframe to determine compliance with the sanitation programmes specification.

Monitoring the sanitation programme is via physical, sensory and rapid chemical hygiene testing methods. Current microbiological testing procedures are not fast enough to be used for process monitoring. Physical tests are centred on the critical control measures of the sanitation programme's performance and include, for example, the correct number of cleaning operatives, the correct equipment, a sufficient cleaning window, measurement of detergent/disinfectant contact time; rinse water, detergent and disinfectant temperatures; chemical concentrations; surface coverage of applied chemicals; degree of mechanical or kinetic input; cleaning equipment maintenance and chemical stock rotation.

The primary monitoring of any cleaning programme is 'visual' cleanliness and involves the assessment of a surface as being free from food debris and other soiling by a person without any sampling aids (other than perhaps a torch). This may involve looking at the surface, feeling the surface for any signs of 'invisible' deposits such as grease and oils, and even smelling the equipment. Not all surface materials will necessarily clean to the same degree and not all surfaces will be 'visible'. Dismantling may be needed to establish that all appropriate surfaces have been visibly cleaned. As well as freedom from soiling, visual inspection should also establish freedom of any other hazards attributed to the cleaning programme. Primarily these are related to cleaning fluids and foreign bodies. Cleaning fluids may be hazards in their own right, particularly in their concentrated form or, when diluted, may become a nutrient and water source to aid residual microbial growth. Inspection should thus concentrate on the identification of any undrainable surfaces or other areas in which liquids could be contained. Foreign bodies may arise from cleaning equipment such that rough or sharp equipment surfaces may cause the entrapment of brush bristles or the disintegration of cleaning cloths/pads.

Rapid hygiene monitoring methods are defined as methods whose results are generated in a timeframe (usually regarded as within approximately 10 min) sufficiently quickly to allow process control. Current methodology allows the quantification of high numbers of microorganisms (adenosine triphosphate ATP), food soils (ATP, protein) or both (ATP). No technique is presently available which will allow the detection of specific microbial types within this timeframe.

The most popular and established rapid hygiene monitoring technique is that based on the detection of ATP by bioluminescence and is usually referred to as ATP testing. ATP is present in all living organisms, including microorganisms (microbial ATP), in a variety of foodstuffs and may also be present as free ATP (usually referred to together as non-microbial ATP). The bioluminescent detection system is based on the chemistry of the light reaction emitted from the abdomen of the North American firefly *Photinus pyralis*, in which light is produced by the reaction of luciferin and luciferase in the presence of ATP. For each molecule of ATP present, one photon of light is emitted which is then detected by a luminometer and recorded as relative light units (RLU). The reaction is very rapid and results are available within seconds of placing the sample to be quantified in the luminometer. The result, the amount of light produced, is also directly related to the level of microbial and non-microbial ATP present in the sample and is often referred to as the 'hygienic' status of the sample.

ATP testing systems were first established in the late 1980s, e.g. Bautista *et al.* (1992), Poulis *et al.* (1993), Bell *et al.* (1994), Griffiths *et al.* (1994), Hawronskyj and Holah (1997), and it is now an internationally recognised standard method. A number of companies offer ATP hygiene detection kits and they are differentiated primarily on cost, service and data-handling packages. It is possible to differentiate between the measurement of microbial and non-microbial ATP but, for the vast majority of cases, the measurement of total ATP (microbial and non-microbial) is preferred. As there is more inherent ATP in foodstuffs than in microorganisms, the measurement of total ATP is a more sensitive technique to determine remaining residues. Large quantities of ATP present on a surface after cleaning and disinfection, regardless of their source, is an indication of poor cleaning and thus contamination risk (from microorganisms or materials that may support their growth). ATP is not suitable for all food manufacturing sites, however, as some foodstuffs contain such low levels of ATP that visibly dirty surfaces can give rise to undetectable ATP readings.

Many food processors typically use the rapidity of ATP to allow monitoring of the cleaning operation such that if a surface is not cleaned to a pre-determined level it can be re-cleaned prior to production. Similarly, pieces of equipment can be certified as being cleaned prior to use in processing environments where equipment is quickly reused or when a manufacturing process has long production runs. Some processors prefer to assess the hygiene level after the completion of both the cleaning and disinfection phases whilst others monitor after the cleaning phase and only go onto the disinfection phase if the surfaces have been adequately cleaned.

Techniques have also been developed which use protein concentrations as markers of surface contamination remaining after cleaning operations. As these are dependent on chemical reactions, they are also rapid, but their applicability is perhaps less widespread as they can only be used if protein is a major part of the food product processed. As with the ATP technique, a direct correlation between the amount of protein remaining after a sanitation programme has been completed and the number of microorganisms remaining as assessed by traditional microbiological techniques, is not likely to be useful. They are cheaper in use than ATP-based systems as the end-point of the tests is a visible colour change rather than a signal which is interpreted by an instrument, e.g. light output measured by a luminometer. Protein hygiene test kits were developed later than ATP kits (e.g. Griffith *et al.*, 1997) and are also now seen as established standard methods.

Verification of the sanitation programme's performance is usually undertaken by microbiological methods, though ATP or protein levels are also used. Microbiological sampling is typically for the total number of viable microorganisms remaining after cleaning and disinfection, i.e. total viable count (TVC), both as a measurement of the ability of the sanitation programme to control all microorganisms and to maximise microbial detection. Sampling targeted at specific pathogens or spoilage organisms, which are thought to play a major role in the safety or quality of the product, is undertaken to verify the performance of the sanitation programme designed for their control. The use of TVC is applicable to both low and high risk processing areas though detection of specific pathogens should be undertaken in high risk only. Microbiological assessments have also been used to ensure compliance with external microbial standards, as a basis for cleaning operatives' bonus payments, in hygiene inspection and troubleshooting exercises, and to optimise sanitation procedures.

Traditional microbiological techniques appropriate for food factory use involve the removal or sampling of microorganisms from surfaces, and their culture using standard agar plating methods, and have been reviewed by Holah and Hall (2004). Microorganisms may be sampled via sterile cotton or alginate swabs and sponges (swabbing), after which the microorganisms are re-suspended by vortex mixing or dissolution into suitable recovery or transport media, or via water rinses for larger enclosed areas (e.g. fillers). Representative dilutions are then incubated in a range of microbial growth media, depending on which microorganisms are being selected for, and incubated for 24–18 h. Alternatively, microorganisms may be sampled directly onto self-prepared or commercial ('dip slides') agar contact plates.

The choice of sampling site will relate to risk assessment. Where there is the potential for microorganisms remaining after (poor) cleaning and disinfection to contaminate product over a period of time by direct contact, these sources would require sampling much more frequently than other sites. Other sites which may be more contaminated may present lower risks of contaminating product. For example, it is more sensible, and gives more confidence, to sample the points of the equipment that directly contact the product and that are difficult to clean than to sample non-direct contact surfaces, e.g. underneath of the equipment framework.

In relation to microorganism numbers, it is difficult to suggest what is an 'acceptable' number of microorganisms remaining on a surface after cleaning and disinfection as this is clearly dependent on the food product and process (which governs the number of microorganisms and level of soiling prior to cleaning), 'risk area' and degree of sanitation undertaken. There is also no European legislation that microbiologically describes a clean surface. A number of figures have been quoted in the past (as total viable count per square decimeter) including 100 (Favero *et al.*, 1984), 540 (Thorpe and Barker, 1987) and 1000 (Timperley and Lawson, 1980) for dairies, canneries and general manufacturing, respectively. The results in Table 12.1 show that in RTE food production, sanitation programmes should achieve levels of around 1000 microorganisms per swab, which on flat surfaces approximately equates to a square decimeter. Expressing counts arithmetically is always a problem, however, as single counts taken in areas where cleaning has been inadequate (which may be in excess of 10^8 per swab) produce an artificially high mean count, even over thousands of samples. It is better, therefore, to express counts as log to the base 10, a technique that places less emphasis on a relatively few high counts, and Table 12.1 shows that log counts of approximately 1 should be obtained.

Because of the difficulty in setting external standards, it is best to set internal standards as a measurement of what can be achieved by a given sanitation programme. A typical approach would be to assess the level of microorganisms or ATP present on a surface after a series of 10 or so sanitation programmes in which the critical control measures of the sanitation programme are carefully controlled. The mean result is the best that can be achieved when the cleaning programme is optimised and undertaken correctly. This is not likely to be repeatable on every occasion and the target may be set as the mean result plus 'a little bit for day to day error'. For example, if a mean ATP value of 100 RLU has been obtained, a target value for subsequent cleans could be set at 150 RLU. In some cases one target may not be appropriate and it may be necessary to set different targets for different areas of the process lines. For example, it may be appropriate to set an ATP target of 150 RLU for conveyor belt surfaces, which can be particularly difficult to clean, and a target of 100 RLU for all other surface material types.

The verification of freedom of allergenic residues after sanitation is now becoming an issue in parts of the food industry. A number of test kits are now available that can be used or modified for the detection of allergen residues on surfaces including the detection of peanut, hazelnut, almond, sesame, soya, egg and egg white, milk (casein, β-Lactoglobulin), crustacean, wheat gluten and sulfite. Allergen detection kits can be sourced from a number of companies including:

- Tepnel: www.tepnel.com
- R-Biopharm: www.r-biopharm.com
- Neogen: www.neogen.com
- ELISA Systems: www.elisas.com.au
- HAVen: www.hallmarkav.com

Many food manufacturers have adopted the philosophy of validating their allergen cleaning programme using the appropriate specific allergen test kit and then verifying the performance of the cleaning and disinfection programme on a day-to-day basis with an alternative hygiene monitor, e.g. ATP. For this to be an acceptable practice, the validation of the cleaning programme must be undertaken using both the allergen test kit and the hygiene monitor. In this case the validation seeks to demonstrate freedom of allergenic residues after cleaning and a positive relationship between freedom of allergens and freedom of the target for the hygiene monitor (ATP, protein, etc.). At this stage it is not possible to state that freedom of allergenic residues as demonstrated by sample results lower than the detection limit of the test kit equates to freedom of an allergenic effect in product produced from such surfaces to the final consumer. In the absence of any other information, however, allergen cleanliness as determined by results lower than the detection limit of the test kit should be the target of the sanitation programme.

As part of the assessment of sanitation programmes, it is worthwhile looking at how the programme is performing over a defined time period (weekly, monthly, quarterly, etc.) as individual sample results are only an estimate of what is happening at one specific time period. This may be to ensure that the programme remains within control, to reduce the variation within the programme or, as should be encouraged, to try and improve the programme's performance. An assessment of the performance of the programme with time, or trend analysis, can be undertaken simply, by producing a graphical representation of the results on a time basis, or it can be undertaken from a statistical perspective using statistical process control (SPC) techniques as described by Harris and Richardson (1996). Generally, graphical representation is the most widely used approach, though SPC techniques should be encouraged for more rigorous assessment of improvement in the programme's performance. As more information is obtained from routine cleaning programme verification, the targets can then be reviewed and the food manufacturer can seek to reduce them as appropriate by changing the manufacturing environment, process, equipment or cleaning programme. A review of the target standard would also be required if any of the variables controlling the target were changed; primarily the food product or process or the sanitation programme.

12.7 Future trends

Within the food industry sanitation can range from the cosmetic, to maintaining food operative health and safety through good housekeeping, to controlling food quality, to ultimately being the major pathogen reduction strategy within high risk areas of RTE manufacturers. It is at the pathogen reduction end of the scale that future improvements will be driven.

Current sanitation in RTE food manufacturing is very effective in controlling pathogens and spoilage microorganisms in the final product and thus maintaining shelf-life and product safety. It does not, however, eliminate all microorganisms within the food processing environment and the evidence of persistent strains of pathogens residing in manufacturing areas for many years bears this out. This is a potential threat as, if there was a lapse in control of the sanitation system, persistent organisms could quickly multiply in the environment and cause product contamination issues.

One goal for the future is thus to further our sanitation control beyond just the food processing surfaces to effectively control the entire processing environment. This is becoming known as 'wholeroom' disinfection. This is likely to involve additional cleaning and disinfection methods and may encompass the use of gaseous ozone or hydrogen peroxide, the use of ultraviolet light, possibly in synergistic combination with titanium dioxide in coated surfaces or the wider use of proven antimicrobial surfaces. Specific chemicals may be deployed that target the adhesion mechanisms of cells to surfaces, e.g. enzymes or biosurfactants. Alternatively, biological agents such as biosurfactants could be used to prevent the inclusion of pathogens in adhered surface communities, whilst specific pathogen phages could be employed to target persistent pathogen strains. Following the application of such wholeroom measures, it is not known how quickly persistent strains would re-colonise the environment and thus how often such measures should be applied.

For the traditional dry food industries, e.g. confectionary, cereals and oil-based products such as tahini and peanut butter, the concept of how often to dry and wet clean is being reconsidered. Historically, these types of plant have infrequently or never been wet cleaned as the wet cleaning process itself can be hazardous (water residues can allow microbial growth) if the plant and environment is not completely dried before production recommences. With the advent of genetic fingerprinting, however, authorities are able to precisely identify pathogens in food products and, if the same strain of pathogen was repeatedly found in one food manufacturer's product, derived from the same plant, over several years, questions about the management of the plant could quite rightly be asked. The balance between wet cleaning (which is recognised as being more effective than dry cleaning) and dry cleaning, particularly after a potential pathogen contamination incident, may thus need to be redefined.

12.8 Sources of further information and advice

Whilst the theory of cleaning and disinfection is global, their application in practice, particularly for open surfaces, differs somewhat. This chapter has been primarily written from a European prospective and an excellent overview of the approach in the USA is given in the book *Food Plant Sanitation* by Cramer (2006). The practical cleaning of closed surfaces, however, tends to be more universal and has recently been comprehensively reviewed and updated for the Society of Dairy Technology by Adnan (2008).

In many ways, however, successful cleaning and disinfection remains an 'art' and best practice at the plant level has to be established following a degree of trial and error. To aid in this it is essential to develop a close relationship with the chemical supplier. Good chemical suppliers are not simply purveyors of chemicals but can provide an expertise and service which can help establish, manage and review sanitation programmes to the highest level. This is particularly important in the RTE food sector.

12.9 References

Aase B, Sundheim G, Langsrud S and Rørvick L M (2000) Occurrence of and a possible mechanism for resistance to a quaternary compound in *Listeria monocytogenes*, *International Journal of Food Microbiology* **62**, 57–63.

Adnan T (2008) *Cleaning-in-place: Dairy, food and beverage operations*, 3rd edn, Society of Dairy Technology series, Wiley Blackwell, Oxford.

Anderson M E, Huff H E and Marshall R T (1985) Removal of animal fat from food grade belting as affected by pressure and temperature of sprayed water, *Journal of Food Protection*, **44**, 246–8.

Andrade N J, Bridgeman T A and Zottola E A (1998) Bacteriocidal activity of sanitizers against *Enterococcus faecium* attached to stainless steel as determined by plate count and impedance methods, *Journal of Food Protection* **661**, 833–8.

Anon (1973) The effectiveness of water blast cleaning in the food industry, *Food Technology in New Zealand*, **8**, 15 and 21.

Anon (1991) Sanitation, in D A Shapton and N F Shapton (eds), *Principles and Practices for the Safe Processing of Foods*, Woodhead, Cambridge, 117–99.

Barron J C and Forsythe S J (2007) Dry stress and survival time of *Enterobacter sakazakii* and other Enterobacteriaceae in dehydrated powdered infant formula, *Journal of Food Protection*, **70**(9), 2111–17.

Bautista D A, McIntyre L, Laleye L and Grifiths M W (1992) The application of ATP bioluminescence for the assessment of milk quality and factory hygiene, *Journal of Rapid Methods and Automation in Microbiology*, **1**, 179–93.

Bell C, Stallard P A, Brown S E and Standley J T E (1994) ATP-Bioluminescence techniques for assessing the hygienic condition of milk transport tankers, *International Dairy Journal*, **4** 629–40.

Bergman B-O and Tragardh C (1990) An approach to study and model the hydrodynamic cleaning effect, *Journal of Food Process Engineering*, **13**, 135–54.

Blenkinsopp S A and Costerton J W (1991) Understanding bacterial biofilms, *Trends in Biotechnology*, **9**, 138–43.

Brett M S Y, Short P and Mclauchlin J (1998) A small outbreak of listeriosis associated with smoked mussels, *International Journal of Food Microbiology*, **43**, 223–9.

BSI (1964) BS 3755: 1964 *Assessment of odour from packaging material used for foodstuffs*, British Standards Institute, London.

Carpentier B and Cerf O (1993) Biofilms and their consequences, with particular reference to the food industry, *Journal of Applied Bacteriology*, **75**, 499–511.

CEN (1997) EN 1276:1997 *Chemical disinfectants and antiseptics – Quantitative suspension test for the evaluation of bactericidal activity of chemical disinfectants and antiseptics used in food, industrial, domestic, and institutional areas – Test method and requirements (phase 2, step 1)*, European Committee for Standardisation, Brussels.

CEN (1998) *EN 1650:1998 Chemical disinfectants and antiseptics – Quantitative suspension test for the evaluation of fungicidal activity of chemical disinfectants and antiseptics used in food, industrial, domestic, and institutional areas – Test method and requirements (phase 2, step 1)*, European Committee for Standardisation, Brussels.

CEN (2001) EN 13697:2001 *Chemical disinfectants and antiseptics – Quantitative non-porous surface test for the evaluation of bactericidal and/or fungicidal activity of chemical disinfectants and antiseptics used in food, industrial, domestic, and institutional areas – Test method and requirements (phase 2, step 2)*, European Committee for Standardisation, Brussels.

Corrieu G (1981) State-of-the-art of cleaning surfaces, in *Proceedings of Fundamentals and Application of Surface Phenomena Associated with Fouling and Cleaning in Food Processing*, April 6–9, Lund University, Sweden, 90–114.

Cramer M M (2006) *Food Plant Sanitation*, CRC Press, Boca Raton, FL.

Das J R, Bhakoo M, Jones M V and Gilbert P (1998) Changes in biocide susceptibility of *Staphylococcus epidermidis* and *Escherichia coli* cells associated with rapid attachment to plastic surfaces, *Journal of Applied Microbiology*, **84**, 852–8.

Davidson P M and Harrison M A (2002) Resistance and adaptation of food antimicrobials, sanitizers, and other process controls, *Food Technology*, **56**(11), 69–78.

Destro M T, Leitao M F F and Farber J M (1996) Use of molecular typing methods to trace the dissemination of *Listeria monocytogenes* in a shrimp processing plant, *Applied and Environmental Microbiology*, **62**, 705–11.

Dhaliwal D S, Cordier J L and Cox L J (1992) Impedimetric evaluation of the efficacy of disinfectants against biofilms, *Letters in Applied Microbiology*, **15**, 217–21.

DIN (2004) *DIN 10955: 2004–06 Sensory Analysis – Testing of Packaging Materials and Packages for Foodstuffs*, Deutsches Institut für Normung e.V., Berlin.

Doyle M P (2006) Dealing with antimicrobial resistance, *Food Technology* **60**(8), 22–9.

Dunsmore D G (1981) Bacteriological control of food equipment surfaces by cleaning systems. 1. Detergent effects, *Journal of Food Protection*, **44**, 15–20.

Dunsmore D G, Twomey A, Whittlestone W G and Morgan H W (1981) Design and performance of systems for cleaning product contact surfaces of food equipment: A review, *Journal of Food Production*, **44**, 220–40.

EC (2004) Regulation (EC) No 852/2004 of the European Parliament and of the Council of Europe of 29 April 2004 on the hygiene of foodstuffs, *Official Journal of the European Union*, **L226**, 30 April, 3–21.

Favero M S, Gabis D A and Veseley D (1984) Environmental monitoring procedures, in M L Speck (ed.), *Compendium of Methods for the Microbiological Examination of Foods*, American Public Health Association, Washington, DC.

Fonnesbech Vogel B, Huss H H, Ojeniti B, Ahrens P and Gram L (2001) Elucidation of *Listeria monocytogenes* contamination routes in cold-smoked salmon processing plants detected by DNA-based typing methods, *Applied and Environmental Microbiology*, **67**, 2586–95.

Frank J F and Koffi R A (1990) Surface-adherent growth of *Listeria monocytogenes* is associated with increased resistance to surfactant sanitizers and heat, *Journal of Food Protection*, **53**, 550–4.

Fryer P J and Christian G K (2005) Improving the cleaning of heat exchangers, in H L M Lelieveld, M A Mostert and J T Holah (eds), *Handbook of Hygiene Control in the Food Industry*, Woodhead, Cambridge, 468–96.

Gibson H, Taylor J H, Hall K E and Holah J T (1995) *Biofilms and their Detection in the Food Industry*, R&D Report No.1, Campden & Chorleywood Food Research Association Chipping Campden.

Gibson H, Taylor J H, Hall K E and Holah J T (1999) Effectiveness of cleaning techniques used in the food industry in terms of the removal of bacterial biofilms, *Journal of Applied Microbiology* **87**, 41–8.

Griffith C J, Davidson C A, Peters A C and Fielding L M (1997) Towards a strategic cleaning assessment programme: hygiene monitoring and ATP luminometry, an options appraisal, *Food Science and Technology Today*, **11**, 15–24.

Griffiths J, Blucher A, Fleri J and Fielding L (1994) An evaluation of luminometry as a technique in food microbiology and a comparison of six commercially available luminometers, *Food Science and Technology Today*, **8**, 209–16.

Harris C S M and Richardson P S (1996) *Review of Principles, Techniques and Benefits of Statistical Process Control*, Review No. 4, Campden & Chorleywood Food Research Association, Chipping Campden.

Hawronskyj J-M and Holah J T (1997) ATP: A universal hygiene monitor, *Trends in Food Science and Technology*, **8**, 79–84.

Holah J T (1991) Food surface sanitation, in Y H Hui (ed.), *Encyclopedia of Food Science and Technology*, Wiley, New York, 1196–2001.

Holah J T (1995) Special needs for disinfectants in food-handling establishments, *Revue Scientific et Technique Office International des Épizooties*, **14**, 95–104.

Holah J T (1998) *Guidelines for Effective Environmental Microbiological Sampling*, Guideline No. 20, Campden & Chorleywood Food Research Association, Chipping Campden.

Holah J T and Hall K E (2004) *Manual of Hygiene Methods for the Food and Drink Industry*, Guideline No. 45, Campden & Chorleywood Food Research Association, Chipping Campden.

Holah J T and Kearney L R (1992) Introduction to biofilms in the food industry, in L F Melo, T R Bott, M Fletcher and B Capdeville (eds), *Biofilms – Science and Technology*, Kluwer, Dordrecht, 35–45.

Holah J T and Thorpe R H (1990) Cleanability in relation to bacterial retention on unused and abraded domestic sink materials, *Journal of Applied Bacteriology*, **69**, 599–608.

Holah J T, Higgs C, Robinson S, Worthington D and Spenceley H (1990) A conductance based surface disinfection test for food hygiene, *Letters in Applied Microbiology*, **11**, 255–9.

Holah J T, Lavaud A, Peters W and Dye K A (1998) Future techniques for disinfectant testing, *International Biodeterioration and Biodegradation*, **41**, 273–9.

Holah J T, Taylor J H, Dawson D J and Hall K E (2002) Biocide usage in the food industry and the disinfectant resistance of persistent strains of *Listeria monocytogenes* and *Escherichia coli*, *Journal of Applied Microbiology*, Symposium Supplement, **92**, 111S–120S.

Holah J T, Bird J and Hall K E (2004) The microbial ecology of high risk, chilled food factories; evidence for persistent *Listeria* spp. and *Escherichia coli* strains, *Journal of Applied Microbiology*, **97**, 68–77.

Hugo W B and Denyer S P (1987) The concentration exponent of disinfectants and preservatives (biocides), in *Preservation in the Food, Pharmaceutical and Environmental Industries*, R G Board, M C Allwoods and J G Banks (eds), Blackwell Sciences, Oxford, 281–93.

Hugo W B and Russell A D (1999) Types of antimicrobial agent, in A D Russell, W B Hugo and G A J Ayliffe (eds), *Principles and Practices of Disinfection, Preservation and Sterilization*, Blackwell Sciences, Oxford, 5–95.

Hugo W B, Pallent L J, Grant D J W, Denyer S P and Davies A (1985) Factors contributing to the survival of *Pseudomonas cepacia* in chlorhexidine, *Letters in Applied Microbiology*, **2**, 37–42.

ISO (2004) *ISO4120:2004 – Sensory analysis of food – Methodology – Triangle test*, International Organization for standardization, Geneva.

Jennings W G (1965) Theory and practice of hard-surface cleaning, *Advances in Food Research*, **14**, 325–459.

Johansson T, Rantala L, Palmu L and Honkanen-Buzalski, T (1999) Occurrence and typing of *Listeria monocytogenes* strains in retail vacuum-packed fish products and in a production plant, *International Journal of Food Microbiology* **47**, 111–19.

Kathariou S (2003) *Listeria monocytogenes* virulence and pathogenicity, a food safety perspective, *Journal of Food Protection* **65**, 1811–29.

Klemperer R (1982) Tests for disinfectants: principles and problems, in *Disinfectants: Their Assessment and Industrial Use*, Scientific Symposia Ltd, London.
Knauer-Kraetzl, B (1994) Reinigung und Desinfektion, 12, *Fleischwarenforum: Fleischhygiene, BEHR's Seminar*, Darmstadt, 14–15 April.
Koopal L K (1985) Physico-chernical aspects of hard-surface cleaning, *Netherlands Milk and Dairy Journal*, **39**, 127–54.
Kumar C G and Anand S K (1998) Significance of microbial biofilms in the food industry: a review, *International Journal of Food Microbiology*, **42**, 9–27.
Lawrence L M and Gilmore A (1995) Characterisation of *Listeria monocytogenes* isolated from poultry products and poultry-processing environment by random amplification of polymorphic DNA and multilocus enzyme electrophoresis, *Applied and Environmental Microbiology*, **61**, 2139–44.
LeChevalier M W, Cawthorn C D and Lee R G (1988) Inactivation of biofilm bacteria, *Applied and Environmental Microbiology*, **44**, 972–87.
Lee S-L and Frank J F (1991) Effect of growth temperature and media on inactivation of *Listeria monocytogenes* by chlorine, *Journal of Food Safety*, **11**, 65–71.
Lindsay D and von Holy A (2006) What food safety professionals should know about bacterial biofilms, *British Food Journal*, **108**(1), 27–37.
Loncin M (1977) Modelling in cleaning, disinfection and rinsing, in *Proceedings of Mathematical Modelling in Food Processing*, 7–9 September, Lund Institute of Technology, Lund, Sweden.
Lundén J, Autio T, Markkula A, Hellström S and Korkeala H (2003a) Adaptive and cross-adaptive responses to persistent and non-persistent *Listeria monocytogenes* strains to disinfectants, *International Journal of Food Microbiology*, **82**, 265–72.
Lundén J, Autio T, Sjöberg A-M and Korkeala H (2003b) Persistent and non-persistent *Listeria monocytogenes* contamination in meat and poultry processing plants, *Journal of Food Protection*, **66**, 2062–9.
Mabesa R C, Castillo M M, Contreras E A, Bannaad L and Bandian V (1982) Destruction and removal of microorganisms from food equipment and utensil surfaces by detergents, *The Philippine Journal of Science*, **3**, 17–22.
MAFF (1998) *A Practical Guide to the Disinfection of Food Processing Factories and Equipment Using Fogging*, Ministry of Agriculture Fisheries and Food, London.
Manger H-J (2006) CIP plants: stock cleaning or run-to-waste cleaning, *Brauwelt International*, **VI**, 380–4.
Marriot N G (2006) Sanitation equipment, in N G Marriott and R B Gravani (eds), *Principles of Food Sanitation*, 5th edn, Springer, New York, 190–212.
Mattila-Sandholm T and Wirtanen G (1992) Biofilm formation in the food industry: a review, *Food Reviews International*, **8**, 573–603.
McLauchlin J, Greenwood M H and Pini P N (1990) The occurrence of *Listeria monocytogenes* in cheese from a manufacturer associated with a case of listeriosis, *International Journal of Food Microbiology*, **10**, 255–62.
McLauchlin J, Hall S M, Velani S K and Gilbert R J (1991) Human listeriosis and paté: a possible association, *British Medical Journal*, **303**, 773–5.
Middlemiss N E, Nunes C A, Sorensen J E and Paquette G (1985) Effect of a water rinse and a detergent wash on milkfat and milk protein soils, *Journal of Food Protection*, **48**, 257–60.
Middleton K E and Holah J T (2008) *A Practical Guide to Cleaning and Disinfection of Food Factories*, Guideline No. 55, Campden & Chorleywood Food Research Association, Chipping Campden.
Middleton K E, Holah J T and Timperley A W (2003) *Guidelines for the Hygienic Design, Selection and Use of Dry Cleaning Equipment*, Campden & Chorleywood Food Research Association, Chipping Campden.
Miettinen M K, Björkroth K and Korkeala H J (1999) Characterisation of *Listeria monocytogenes* from an ice cream plant by serotyping and pulsed gel electrophoresis, *International Journal of Food Microbiology*, **46**, 187–92.

Mrozek H (1982) Development trends with disinfection in the food industry, *Deutsche-Molkerei-Zeitung*, **12**, 348–52.

Nesbakken T, Kapperud G and Caugant D A (1996) Pathways of *Listeria monocytogenes* contamination in the meat processing industry, *International Journal of Food Microbiology*, **31**, 161–71.

Offiler M T (1990) Open plant cleaning: equipment and methods in: *Proceedings of 'Hygiene for the 90's'*, November 7–8, Campden & Chorleywood Food Research Association, Chipping Campden, 55–63.

Pontefract R D (1991) Bacterial adherence: its consequences in food processing, *Canadian Institute of Food Science and Technology Journal*, **24**, 113–17.

Poulis J A, De Pijper M, Mossel D A A and Dekkers P PH A (1993) Assessment of cleaning and disinfection in the food industry with the rapid ATP-bioluminescence technique combined with tissue fluid contamination test and a conventional microbiological method, *International Journal of Food Microbiology*, **20**, 109–16.

Pourshaban M, Gianfranceschi M, Gattuso A, Menconi F and Aureli P (2000) Identification of *Listeria monocytogenes* contamination sources in two fresh sauce production plants by pulsed-field gel electrophoresis, *Food Microbiology*, **17**, 393–400.

Quarini G L and Shire G S F (2007) A review of fluid-driven pipeline pigs and their applications, Proceedings of the Institution of Mechanical Engineers, Part E: *Journal of Process Mechanical Engineering*, **221**(1), 1–10.

Reybrouck G (1998) The testing of disinfectants, *International Biodeterioration and Biodegradation*, **41**, 269–72.

Ridgeway H F and Olsen B H (1982) Chlorine resistance patterns of bacteria from two drinking water distribution systems, *Applied and Environmental Microbiology*, **44**, 972–87.

Ronner A B and Wong A C L (1993) Biofilm development and sanitizer inactivation of *Listeria monocytogenes* and *Salmonella typhimurium* on stainless steel and Buna-N rubber, *Journal of Food Protection*, **56**, 750–8.

Rørvik L M, Caugant D A and Yndestad M (1995) Contamination pattern of *Listeria monocytogenes* and other *Listeria* spp. in a salmon slaughterhouse and smoked salmon processing plant, *International Journal of Food Microbiology*, **25**, 19–27.

Schmidt U and Cremmling K (1981) Cleaning and disinfection processes. IV, Effects of cleaning and other measures on surface bacterial flora, *Fleischwirtschaft*, **61**, 1202–7.

Schlussler H J (1975) Zur Kinetic von Reinigungsvorgangen an festen Oberflachen, *Symposium über Reinigen und Desinfizieren lebensmittelverarbeitender Anlagen*, Karlsruhe.

Sharma M and Anand S K (2002) Bacterial biofilm on food contact surfaces: a review, *Journal of Food Science and Technology-Mysore*, **39**(6), 573–93.

Shupe W L, Bailey J S, Whitehead W K and Thompson J E (1982) Cleaning poultry fat from stainless steel flat plates, *Transactions of the American Society of Agricultural Engineers*, **25**, 1446–9.

Taylor J H, Rogers S J and Holah J T (1999) A comparison of the bactericidal efficacy of 18 disinfectants used in the food industry against *Escherichia coli* 0157:H7 and *Pseudomonas aeruginosa* at 10 °C and 20 °C, *Journal of Applied Microbiology*, **87**, 718–25.

Thorpe R H and Barker P M (1987) *Hygienic Design of Liquid Handling Equipment for the Food Industry*, Technical Manual No. 17, CCFRA, Chipping Campden.

Timperley D A and Lawson G B (1980) Test rigs for evaluation of hygiene in plant design, in: R Jowitt (ed.), *Hygienic Design and Operation of Food Plant*, Ellis Horwood, Chichester.

Unnerstad H, Bannerman H E, Bille J, Danielsson-Tham M-L, Waak E and Tham W (1996) Prolonged contamination of a dairy with *Listeria monocytogenes*, *Netherlands Milk and Dairy Journal*, **50**, 493–9.

Walker R L, Jensen L H, Kinde H, Alexander A V and Owens L S (1991) Environmental survey for *Listeria* species in frozen milk product plants in California, *Journal of Food Protection*, **54**, 178–82.

Wirtanen G and Mattila-Sandholm T (1993) Epifluorescence image analysis and cultivation of foodborne bacteria grown on stainless steel surfaces, *Journal of Food Protection*, **56**, 678–83.

Wirtanen G and Mattila-Sandholm T (1994) Measurement of biofilm of *Pediococcus pentosaceus* and *Pseudomonas fragi* on stainless steel surfaces, *Colloids and Surfaces B: Biointerfaces*, **2**, 33–9.

Wright J B, Ruseska I and Costerton J W (1991) Decreased biocide susceptibility of adherent *Legionella pneumophila, Journal of Applied Microbiology*, **71**, 531–8.

Zottola E A and Sasahara K C (1994) Microbial biofilms in the food processing environment – should they be a concern? *International Journal of Food Microbiology*, **23**, 125–48.

13

Safe process design and operation

M. Brown, mhb Consulting, UK

Abstract: This chapter explains the importance of product designs in setting the sensory and safety characteristics of food products and the accompanying requirements for processing and distribution. It reviews principles for designing products and processes, and then validation of designs. Important tools for managing the supply chain are described, such as good manufacturing practice, HACCP, predictive modeling and risk assessment. The chapter outlines principles for the layout and design of equipment and control systems for various types of processing. Lastly, requirements for successful supply chain management are discussed, including monitoring and verification and the key role of training.

Key words: product design, process design, microbiological safety, food manufacturing areas, food safety management tools.

13.1 Introduction: product and process design

A product design describes the anticipated product, its characteristics, positioning with respect to similar competitors' products and location within a brand portfolio. Generally, marketing and R&D will be responsible for product designs, with additional input, where necessary, from other technical specialists (e.g. microbiologists and process engineers). The product resulting from a design will always be a compromise between consumer and customer demands for safety, quality and cost on one hand, and supply chain limitations on the other (see Notermans *et al.*, 1996). The supply chain includes all the manufacturing functions from procurement of raw and packaging materials to the point of sale in the retail or food-service outlet. Translation of a sound design into specifications for the supply chain to work with represents the most important opportunity for a producer to turn a marketing concept or design into a successful and safe product. Good description, by marketing and R&D, of the key characteristics provides the basis for safe and successful products; without this guidance even the best controlled supply chain cannot succeed. With a potentially successful design, the task of the supply chain is then to consistently produce attractive and safe products, meeting consumer expectations on cost, availability, shelf-life and quality.

Each design needs to note realistic hazards and propose controls for them, if possible within the boundaries of the product characteristics noted in the original marketing brief. If the requirements (e.g. processing and preservation) for controlling the identified microbiological hazards result in a product within the typical product characteristics, then it has a chance of being successful; if the requirements are in excess (e.g. necessary heating causing colour or texture loss or reduced pH causing an excessively acid taste), then the chances of success are reduced. Within any company, a multi-disciplinary team (e.g. quality assurance (QA), R&D, buyers, production, engineering and marketing) is normally the best means of establishing working specifications. Formal techniques of hazard analysis, risk assessment and risk management may be used to guide the company towards a predictable and acceptable balance between safety, quality and cost.

Increasing globalisation of trade in foodstuffs and outsourcing of manufacturing introduces a wider, and possibly different, range of hazards whose identification and effective management requires good technical knowledge and the accurate and honest exchange of information between all the stages in the supply chain. Realistic hazards may be identified in various ways at the design stage: previous identification in similar materials, products or origins, regulatory surveillance, in-house monitoring information, knowledge of production and supply practices and any failures or consumer complaints. Unexpected hazards may arise from innovations in processing or procurement and lack of compliance with specified limits. Comprehensive identification, and hence effective control, of microbiological hazards in a variety of raw materials is sometimes difficult when a design is translated into specifications. Each specification and intake procedure may have to be tailored for ingredients from different origins (e.g. different microbiological contaminants in similar material from tropical vs. temperate climates). Globalisation also increases the range of different product storage and usage habits and attitudes by customers, retailers and consumers.

Based on the design, the degree of extra microbiological safety (e.g. pasteurising for 30, instead of the design 10 minutes) built into process specifications or work practices to meet the needs of production has to be constrained by the demands of the marketplace for particular sensory characteristics (e.g. crisp textures, mild preservation systems or freedom from preservatives), shelf-life and demands for less obvious processing. The skill of R&D and especially process engineers is to deliver the characteristics of the product design, whilst balancing these competing demands during formulation and processing. Often more severe conditions are used by manufacturing to compensate for process variability (rather than the improvement of controls) or by buyers to minimise the costs of raw materials.

The aim of the buyers and the supply chain should be to minimise the intake of pathogens and spoilage microorganisms with raw materials and make sure that process capability and any preservation systems in the product are not overwhelmed (e.g. by acid-tolerant microorganisms) and that levels of microorganisms in the product at the point of sale (and the end of shelf-life) are within the levels specified in the product design. Manufacturing risks can be minimised by using clear specifications, reliable suppliers and quality assurance procedures at intake to ensure that raw materials are always within the capability of processing (e.g. decontamination by cooking gives the required quality and recontamination afterwards is prevented) and the product usage and shelf-life (e.g. use by; see NACMCF, 2005 and http://www.fsis.usda.gov/OPHS/NACMCF/2003/rep_shelflife.htm) specified by the design can be met (e.g. ready-to-eat or cooked before eating). In principle, product designers and raw material buyers can reduce risks by avoiding materials with a substantiated history of contamination or toxicity and recognising that decontamination during processing reduces product safety risks more reliably than consumer cooking. The combination of ingredients, supply chain treatments and instructions for use on the pack should ensure that products are safe if they are used according to these instructions. How much 'cushion', especially in formulation to prevent pathogen growth, should be allowed to compensate for consumer mishandling during chilled storage is a topic often debated in companies. If the product design relies on customer cooking to free products from infectious pathogens, such as salmonellae, it is important that the design of the product, especially its dimensions, mean that cooking according to the instructions results in good quality (e.g. products are not burnt outside before they are cooked at the centre). Hence helpful, informative and validated cooking instructions must be provided, and following these instructions has to give high product quality. Buyers acting with suppliers can also contribute to safety by reducing the levels of pathogens in primary production (e.g. *Salmonella* in poultry to be sold raw) and marketing have an important role in the brand positioning (e.g. consumers are accustomed to cooking products within that brand range), the design of cooking instructions and ensuring that the product characteristics (e.g. portion size and shape) ensure good quality when cooked.

Successful designs must consider not only contamination risks, but also the impact of shelf-life and anticipated distribution and storage temperatures and times on the growth of any microorganisms in the product after manufacture (CCFRA, 2004a). Predictive modelling techniques (e.g. Pathogen Modelling Program – http://ars.usda.gov/services/docs.htm?docid=6786 or http://ars.usda.gov/Main/site_main.htm?modecode=19-35-30-00) can be used to estimate the response of microorganisms in products (defined by their intrinsic characteristics such as pH, salt level) to steady or fluctuating temperatures. Time/temperature integrators (TTI) can also be used to monitor pack temperatures (Mendoza *et al.*, 2002) and show unacceptably high surface temperatures in the distribution chain (Labuza and Fu, 1995). These indicators are not widely used for several reasons; the difficulty of matching their response kinetics to those of pathogens and also of ensuring that the temperature history they display accurately reflects conditions in the food, rather than just at the pack surface (which means product may be unnecessarily wasted). If storage conditions (e.g. chill), prevention of pack damage during distribution and preparation by consumers (e.g. cooking) are integral parts of the safety chain, risks are obviously substantially increased by mishandling or misuse. A manufacturer should decide how to account for such risks in specifications and ensure that strong messages on safety requirements are given by labelling and brand identity and in any logistics contracts.

Brackett (1992) has pointed out that chilled foods, for example, contain few, if any, preservatives to prevent growth of pathogens and high storage temperatures may allow their growth. In spite of this the incidence of food-poisoning associated with chilled foods remains very low. He also highlights related issues such as over-reliance on shelf-life as a measure of quality and the need to consider the needs of sensitive groups (such as immunocompromised consumers) in the product design. During 2006 there have been cases of botulism linked to pasteurised, refrigerated low-acid juices in the USA and Canada (see http://www.cfsan.fda.gov/~dms/juicgu15.html). It is suspected that these products may have been left unrefrigerated for an extended period, either during distribution or consumer storage allowing *Clostridium botulinum* spores to grow and produce toxin. As a result of this outbreak, FDA is now recommending that processing and product formulation include control measures to ensure that *C. botulinum* growth and toxin production will not occur should the juice, as offered for sale by the processor, be kept unrefrigerated in distribution or by consumers. Preservation could use for example acidification of the juice to a pH of 4.6 or below. Additionally the HACCP plan for these products should treat two features of the process as pre-requisites. These are the sanitation of container filling equipment to prevent post-heat process contamination of the juice with *C. botulinum* spores and secure sealing of the containers.

13.2 Principles of process design

The manufacture of safe, high-quality products relies on the use of established supply chain and preservation principles within an organised framework to target realistic microbial, chemical and physical hazards. From a microbiological safety point of view, this framework should take account of product usage, and hence the microbial hazards allowed in the product, a realistic assessment of their likely presence and levels in raw materials, the effects of heat processing and any other decontamination procedures on their numbers, the prevention of recontamination during processing and storage (including by primary packaging) and the ability of any preservation system to inhibit microbial growth. Generally the target pathogens that need to be controlled in, or eliminated from, food products include the following:

- **Infectious types** – salmonellae especially Enteritidis and Typhimurium, enterohaemorrhagic *Escherichia coli*, *Vibrio parahaemolyticus*, *Campylobacter jejuni*, *Yersinia enterocolitica* and *Listeria monocytogenes*.
- **Toxinogenic types** – *Clostridium botulinum*, *Clostridium perfringens*, *Staphylococcus aureus* and *Bacillus cereus* (also *B. subtilis* and *B. licheniformis*).

While these are the minimum range of microbial targets, designs also need to control spoilage microorganisms, especially those causing adverse quality changes. Generally, such microorganisms are more resistant than pathogens to heating, intrinsic preservation systems and inhibition by chilling. Where a decision is taken to accept limited spoilage to achieve a product with certain characteristics or shelf-life, the developer must be certain the compromise does not create an unintended safety or quality risk. For example, vacuum or modified atmosphere packaging (MAP) may be used to slow the growth of pseudomonads or moulds, but may create conditions favouring the growth of anaerobic microorganisms, such as lactic acid bacteria and clostridia.

Some products rely on minimising processing to limit quality change (e.g. salad preparation) while others rely on cooking to develop flavours and textures and eliminate microorganisms (e.g. pasteurisation and sterilisation processes). Because of the critical effect of ingredients on heating, the product design must highlight any characteristics likely to affect it (i.e. reduced heat transfer or heating rate or low initial temperature which may reduce the heat treatment provided by a specified set of processing conditions). Important factors include the size of pieces, portions or packs (e.g. thermal path length); coatings (e.g. batter or breadcrumbs) that may affect heat transfer; and frozen or chilled ingredients that may reduce temperature at the start of cooking. In the case of frozen ingredients, clumping may further slow heating.

Food may be preserved in other ways, for example by drying or reducing its water activity by the addition of humectants. Foods (such as salami, pasta and vegetable pieces) are generally dried either in continuous or batch equipment and drying conditions (e.g. time, air flow, air temperature and relative humidity) and product characteristics (e.g. initial moisture content and piece or sheet dimensions, especially thickness) must ensure that uniform drying to the required water activity (a_w) can be achieved. In many cases, the rehydration characteristics of the food and the role of packaging in preventing rehydration during storage are essential parts of the design. If humectants are used, then care has to be taken to ensure that the flavour during use is acceptable, and increasingly the use of the most popular humectant, salt (sodium chloride), presents consumer acceptability and labelling problems, as low-salt products are preferred. Where intermediate moisture foods are made combinations of humectants are used, and often the key safety boundary is effective inhibition of growth and toxin production by *Staphylococcus aureus* a_w 0.86 or slightly higher, depending on formulation (e.g. humectants, pH and storage temperature).

Acidification either by the addition of weak acids or by fermentation is used to preserve food. In meat and poultry lactic acid or lactic acid salts are often the acidulants, and the acid is a major product of fermentation giving pH values of about 5.3. With dairy products, lactic acid is also present but in combination with other acids (e.g. acetic or propionic) that give rise to distinctive flavours. Fruit and vegetable products are usually acidified with citric (which may be naturally-occurring, as in tomato products), malic and tartaric acids and have pH values around 4. The pH and undissociated acid concentration have to be fixed in the design according to the acid resistance of the target microorganisms or any other preservation factors present. The amount of acid that can be added is often limited by the sensory properties resulting from acidification; if acid taste is not part of the product character then excessive acidity will reduce consumer acceptability. If acidification relies on acid addition, then the target equilibrium pH and the uniformity of pH/acid distribution throughout the product have to be established by the design; translation of these requirements into working specifications is not an easy task and a target maximum time for reaching equilibrium (e.g. 48 hours) has to be fixed to minimise the chances of spoilage or pathogen growth. As a general rule pH should be below 4.6 within a shorter period than this to eliminate the risks of growth of *C. botulinum*. Where products contain significant amounts of fat, some preservative will partition into the fat phase (e.g. sorbate and benzoate), and this will reduce the effective level in the aqueous phase that contains the microbial contaminants. This should be taken account of by the product design.

Measures specified to control these properties and their variability, which affect the behaviour of food during processing, must be realistic and achievable with the equipment, personnel and procedures available, and operating targets may be offset (to the safe side) from the values in the design, to take account of process or ingredient variability. Specifications and work practices should identify how process conditions are to be controlled (e.g. tempering or thawing) based on the probable coldest and slowest heating ingredients. Any constraints have to be clearly translated into work instructions or software, if equipment is automatically controlled. For example, if a product requires a tight specification for one of its ingredients (e.g. particle size) to ensure a safe heat treatment, but this critical feature is variable, then the largest particle size must be used during trials to establish specifications, and this should also be fed back into the product characteristics associated with the design. The same principle must also be applied to the minimum temperatures likely to be encountered in production (e.g. chilled or frozen material is used), and these should be used for setting heating or frying conditions.

13.2.1 Heating

Product character, storage temperature and consumer usage will usually dictate heating requirements and the type of heating equipment to be used. The aim of a factory kitchen and downstream processing is often to preserve or enhance quality and make products that are ready for consumption by the customer with a minimum of further preparation. For many such products, factory cooking eliminates pathogens, and those chosen as targets will be determined by the storage temperature (e.g. ambient or chill) and any preservation system in the product or its packaging. The minimum heat treatments specified by the design must effectively eliminate or control microbial hazards, and in some cases, ensure compliance with local legislation (e.g. milk pasteurisation or sterilisation of low-acid foods). Any information necessary to achieve this should be presented in the product design and manufacturing specification, e.g. portion dimensions or raw materials quality.

Heating rates may vary in response to equipment in the factory kitchen, and it is important to verify that production-scale equipment can produce similar quality and decontamination effects to pilot-scale equipment used to prepare the concept samples used for approval of the product design. Allowances for microbiological safety and the use of larger vessels for heating should not be allowed to exert a cumulative and detrimental influence on quality by leading to over-severe heating processes. Often as vessel size increases, the amount of heating needed to give a specified heat treatment will increase considerably with a detrimental effect on quality. If heat treatments have been adapted from the design by calculation (e.g. using D and z), it may be necessary to conduct tests on the initial commercial production runs to validate that the process conditions still provide the product characteristics and heat treatment required by the original design. Thickeners such as starch or proteins will also affect product heating, particle suspension and flow characteristics and should be specified by the design as they often also have a critical effect on mouth-feel and flavour release.

QA and development staff involved in trials should understand the limitations of predictions and tests, especially when processes are scaled-up. Assessing the effects of changes in ingredients, product composition, method of product make-up, size or shape or coating of pieces must be the responsibility of a suitably experienced person in the management team, e.g. QA or development manager, who will also be responsible for making sure production understand the technical requirements for operating the process and ensuring that specifications remain up-to-date.

Microbiologically, the highest risk designs and products are those lacking any preservation system, containing materials likely to contain pathogens and intended to be consumed as ready-to-eat or after minimum heating, e.g. prepared meals, low-acid foods or salads. Hence products for microwave heating should be designed as ready-to-eat or ready-to-reheat, as some domestic cooking appliances such as microwave ovens may not be capable of uniformly achieving pasteurising temperatures. Therefore, the design principles employed in the sourcing, manufacture, distribution and sale of ready-to-eat foods, and especially salads, should be primarily designed to minimise the risks of them containing harmful concentrations of food-poisoning bacteria (e.g. *Listeria*), whose growth during ambient or chilled distribution and storage will increase risks. Ambient-stable liquid foods can be preserved either by acidification and mild heat processing (e.g. tomato-based products) or by thermal sterilisation (e.g. canned soup); both types rely on the primary packaging effectively preventing recontamination until the pack is opened (CCFRA, 2004b). With raw foods (e.g. beef and poultry) consumers may not make the link between thorough cooking and safe food and this may lead to the inadvertent consumption of pathogens if product is undercooked.

The control of spoilage microbes should always be a secondary consideration, although it may often require the application of more severe processes, formulation or conditions of hygiene than the control of safety. Sometimes the control of spoilage cannot be achieved without prejudicing sensory quality (e.g. in ambient stable foods); and there has to be a commercial decision on the acceptable balance between a controlled loss of quality and the frequency of spoilage in the marketplace. However, microbiological safety standards should never be compromised to improve sensory quality. If the required processing conditions cannot ensure safety against the background of realistic consumer usage, then the product should not reach the marketplace (see Walker and Stringer, 1990; Gould, 1992).

The manufacture of unpreserved (low-acid) ambient-stable or extended shelf-life (chill) products accentuates the microbiological hazards, from spore-forming bacteria (such as clostridia and *Bacillus* spp.), hence processing needs to eliminate them reliably (e.g. a log reduction exceeding the expected maximum contamination level). Bacterial spores can survive the mild heating processes used to destroy infectious pathogens and spoilage microorganisms. These heating processes should only be used for foods with preservation systems, including short storage times, designed to inhibit spore outgrowth (e.g. pH < 4.6) under the conditions of storage. Where this is not the case, products should be sterilised using conventional (e.g. F_0 3–8) or equivalent ultra-heat treatments (UHT), either in-pack or in-line, as part of an aseptic process (see Sastry and Cornelius, 2002). Minimum heating requirements for safe ambient stability are specified in many countries (e.g. $F_0 \geq 3$), and this time/temperature combination must be achieved in the slowest heating part of the food, for example in particulates suspended in the liquid product flow. Whereas in the case of chilled products, 90 °C > 10 min is accepted as an adequate heat treatment to destroy spores of cold-growing (≥ 3 °C) strains of *C. botulinum* and growth of the more heat-resistant spores is prevented by chill temperatures (< 12 °C).

There is still no general agreement on the risks of botulism from unpreserved chill-stored foods, but there is evidence that, in model systems inoculated with spores, growth occurs and toxin can be produced at temperatures representing commercial conditions (Notermans *et al.*, 1990). However, there is little evidence from epidemiological and survey data that these foods really constitute a realistic botulinum hazard. However, there is one documented case of botulism being caused by a product (smoked salmon) being held for longer than its shelf-life (Dressler, 2005). Spoilage of both ambient, and chill-stable foods may be caused by the survival and outgrowth of *Bacillus* or clostridial spores.

If heating of chilled foods has not been done in the primary packaging, and unwrapped, heat-treated components are used, and these should only be handled and assembled in high risk areas to prevent recontamination with spores and infectious pathogens.

Some ingredients, such as vegetables, may be incidentally heated as part of their processing (e.g. blanching to inactivate enzymes). Although blanching equipment produces temperatures of 90 °C or so, it is not intended to produce microbiologically safe products, as the equipment may not be designed to ensure minimum heating in all parts of the product flow. To minimise risks any handling or packaging equipment used after product has been blanched must be designed to be hygienic and operated to prevent recontamination. Before they are assumed to give microbiologically-effective heat treatments, any blanching operations should be validated.

13.2.2 Product grouping and process design

Chilled foods, for example, may be categorised according to their chances of carrying pathogens after processing and consumer use (Table 13.1). Some are made predominately from raw ingredients (Chilled 1a & b) and will require storage conditions to prevent growth and toxin production (e.g. by clostridia) plus cooking by the customer to eliminate any infectious pathogens (e.g. Salmonellae). Others are mixtures of raw and cooked components (Chilled 2), processed and packaged to ensure a satisfactory shelf-life. These generally contain less 'risky' materials, may not require cooking, but may from time to time contain infectious pathogens (e.g. *L. monocytogenes* or *Salmonella*). Risks from these products can be reduced by minimising numbers of pathogens on the raw materials (e.g. by careful choice of suppliers and growing areas) and preventing contamination of products during storage and processing. Shelf-life and storage temperatures should not allow pathogen numbers to increase and ensure that only 'safe' numbers are present if foods are stored for their full indicated shelf-life. For *L. monocytogenes* and chilled products this is specified in the EU Regulation 2073/2005 EC, Chapter 1: 1.1 and 1.2 (EC, 2005) where chill-stable products are categorised according to their preservation system and the risks of supporting growth of *L. monocytogenes*. For very mildly preserved products that support rapid growth of *L. monocytogenes* (e.g. it can grow to numbers exceeding 100 cfu/g during the shelf-life) *L. monocytogenes* must be undetectable in 25 g, of product at exit from the factory. For preserved products with a shelf-

Table 13.1 Representative risk classes of foods

Risk class	Typical shelf-life	Critical hazard	Relative risk	Required minimum heat treatment	Required manufacturing class[a]		
					GMP	HCA	HRA
Chilled 1a	48 h–1 week	Infectious pathogens	High	Heating (minimum 70 °C, 2 min)[b], none by consumer	*	*	
Chilled 1b	1 week	Infectious pathogens	High	Customer cook (minimum 70 °C, 2 min)	*	*	
Chilled 2	1–2 weeks	Infectious pathogens	Low	Pasteurisation by manufacturer (minimum 70 °C, 2 min)	*	*	*
Chilled 3	> 2 weeks	Infectious pathogens and spore-formers	Low	Pasteurisation by manufacturer (minimum 90 °C, 10 min)	*		*
Chilled 4	> 2 weeks	Spore-formers	Low	Pasteurisation by manufacturer (minimum 90 °C, 10 min)	*	*	
Ambient low acid	> 1 year	Spore-formers	Low	Sterilisation by manufacturer (minimum 121 °C, 3 min)	*	*	
Ambient preserved	> 1 year	Infectious pathogens and spore-formers	Medium	May or may not include pasteurisation by manufacturer (minimum 70 °C, 2 min)	*	*	*

[a] For definitions of the manufacturing requirements for each class, see CFA (2006) and Section 13.7.5.
[b] Non-thermal decontamination process, see CFA (2006).

life greater than five days (e.g. pre-packed sliced cooked meat, smoked salmon and soft cheese), the manufacturer must be able to show that with low initial levels of contamination and the designed preservation system, numbers of *L. monocytogenes* will not exceed 100/g during the product's shelf-life. In the USA, initiatives are continuing to focus on improving the safety of RTE meat and poultry products at retail deli counters. The Food Safety and Inspection Services (FSIS) continues to sample retail outlets and analyse epidemiological data and, as a result of their findings, there are recommendations for verification sampling of production facilities, hygiene training focussed on sanitation, refrigeration and safe handling of unpreserved products and the inclusion of anti-listerial agents in products.

Other chilled foods may contain only cooked or decontaminated components (Chilled 3), or may be cooked by the manufacturer within their primary packaging (Chilled 4). If manufactured under well-controlled conditions Chilled 3 and 4 foods will be free of infectious pathogens and spoilage microbes and can have a long shelf-life (up to 42 days at chilled temperatures), as they will not be subject to rapid microbial spoilage.

Foods for ambient distribution and storage, from which spores have been eliminated by severe heating (121 °C for 5–8 min) or inhibited (e.g. mild heating + acidification or low a_w, perhaps to a_w 0.6) and packaged (e.g. cans, jars or aseptic packaging) to prevent recontamination often have shelf-lives of a year or more.

13.3 Designing and validating product and process designs

Validation is a key means of ensuring that product quality and safety goals are suitable and can be met because process equipment and systems can consistently operate within established limits and tolerances. Codex Alimentarius is currently considering guidance in the area, see CAC, 200, ALINORM 08/31/13 para. 84 and Appendix III. The stages involved in validation are illustrated in Fig. 13.1.

Product design

Validation starts by examining the product design, and checks that it represents the marketing brief and contains sufficient detail for the preservation principles to be understood and approved and provisional specifications written and tested. Based on these specifications, prototype products can be made and assessed to optimise the performance of raw materials and manufacturing lines. The finished product is tested for quality and shelf-life, sometimes using accelerated storage trials (often at higher temperatures) to show it meets the requirements for functionality and safety. The results of accelerated trials must be used with caution, as they can introduce types or relative rates of quality change that are not seen under realistic conditions. Trials should be completed

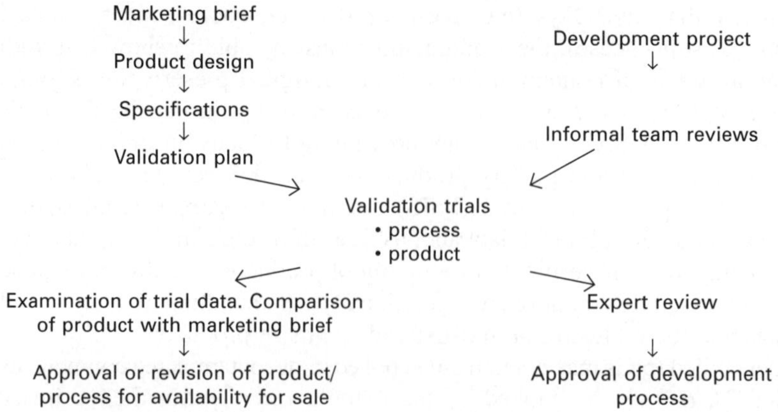

Fig. 13.1 Stages in the validation of new products and processes.

prior to commercial manufacture and distribution of the product, although retrospective validation of a product already on the market can be based on analysis of accumulated process, QA, analytical and control data.

Specifications

The quality and safety of processed foods is highly dependent on development of the initial design into a functioning supply chain. Many safety and quality problems are traceable to errors made at the design and development stages. The product design should have translated the marketing brief into technical terms that can be used to write specifications, carry out trials and choose a supply chain (e.g. process technology, material suppliers and operational procedures). The validity of specifications should be found by routine, and challenge, testing of the product during the initial development and production phase. Once a specification is demonstrated as acceptable, it is important that any changes are only made in accordance with documented change control procedures.

Specifications derived from the product design need to cover:

1. product quality characteristics (e.g. sensory quality, shelf-life and performance when used);
2. intrinsic (e.g. built-in) preservation factors such as salt (sodium chloride), pH, etc. and their target microorganisms;
3. extrinsic preservation factors such as primary packaging, heating, cooling or chill distribution;
4. any reliance on microbial interactions to provide consumer protection from pathogens or to slow spoilage.

From a microbiological perspective, specifications should highlight the processing and preservation conditions affecting realistic microbial hazards (e.g. reduction in numbers of cells of the target microorganisms).

Trials and validation of specifications, including intensive QA
Validation of the specifications (by trials, challenge testing or modelling) should be done in a planned way (see Section 13.3.1) to demonstrate that specifications:

1. have a scientifically sound basis;
2. are effective with the operational process controls available and under the consumer use conditions specified (see process validation – 13.3.2);
3. ensure production articles will be fit for their intended use and meet marketing and user requirements (see product validation – 13.3.3).

Microbiological validation can use destructive or challenge testing (e.g. by inoculation with known microbes) or predictive modelling to show if the manufacturing process or product formulation are adequate. Thermal processes may be validated by the use of time temperature integrators (TTI: see sterilisation – Serp *et al.*, 2002; pasteurisation – Tucker *et al.*, 2002). In batch heat treatment equipment these integrators may be located in containers, at the coldest point. In continuous heat processing equipment they need to move with the flow of the liquid or suspended particles. These different applications need different designs, for containers in batch equipment integrators may transmit data using wires but, where there is a flow, sensors need to be independent and microbiological integrators may be used (see CCFRA, 2002; Guiavarc'h *et al.*, 2002; de Cordt *et al.*, 2004). Critical examination of the information available after trials, etc. should be done before a new product or product design is introduced to the market or when there is a change in the manufacturing process which may affect a product's characteristics. Therefore, before production starts and the product is launched, there should be a documented and comprehensive examination of the design against the original marketing brief and safety and shelf-life requirements to record that these requirements are met and to identify any problems.

The examination should show:

1. how quality and safety are built into the product, e.g. the growth or survival of pathogens and spoilage microorganisms is within the limits imposed by product use, and if the product is used according to the instructions it is of good quality;
2. each stage of the supply chain is adequately specified, operated and controlled to maximise the probability that the finished product meets all quality and design specifications and this information is contained in a HACCP plan;
3. there are reliable and accessible data collection and analytical methods for the process parameters and critical product/material characteristics.

Any compromises proposed as a result of this examination should trigger a review of the development process.

Careful review of process operation and the performance of controls over the early days (or weeks) of operation (intensive QA) can establish a high degree of confidence that all manufactured units from successive lots will be acceptable. Successfully validating a process may reduce the dependence upon routine in-process and finished-product testing when a supply chain has been shown to work consistently well.

Review of the development process

During the development of a product there should have been informal technical reviews of progress within the development team, but for validation, a formal pre-production review of the development process needs to be done with experts from outside the development team. It should examine:

1. any future innovations or range extension plans;
2. user, supply chain and logistics requirements;
3. specifications and procedures, QA plans and procedures (including HACCP), based on documents and activities associated with the project;
4. analysis of results from each stage of the development process, challenge testing or predictive modelling, product validation trials and any validation results for new equipment.

A formal design review should allow management to confirm that all goals defined in the original brief have been achieved. Problems found at this point can be resolved more easily, save time and money, and reduce the likelihood of missing a critical issue.

13.3.1 Validation plan

The validation plan shows how validation will be carried out for a process and product, including:

1. quantities of material and process time required;
2. test parameters;
3. product characteristics and comparison with apparently similar products;
4. production equipment and controls to be used;
5. definition of acceptable test results and limits of acceptable variability.

It should produce documented evidence which provides a high degree of assurance that the process design and its operation will consistently result in a product which meets specifications and quality attributes. It should determine whether realistic hazards have been identified, and suitable process control, hygiene and monitoring measures implemented, along with safe remedial actions for use when processes go outside their control limits; and it should include a worst case study, covering upper and lower processing limits and process stoppages. Completion of the validation plan should ensure that all aspects of the product that impact on its safety and characteristics have been examined and acceptable limits and allowable variation for raw material and product acceptance have been established.

Validation should examine any relevant scientific or technical literature, previous validation studies or previous knowledge of the performance of the controls. Some of the end-product tests used as part of validation (e.g. sensory or microbiological testing) may have limited sensitivity; reliance on these tests should be minimised and there should always be clear criteria for success or failure. End-product testing cannot show all the variations that may occur during processing and may impact on safety and quality. For these circumstances, additional process data may require collection (e.g. intermediate product or heating medium temperatures).

13.3.2 Process validation

Process validation establishes documented evidence to show that a process will consistently make a product meeting its specifications and quality characteristics. Before accepting that a process has been successfully validated, it is necessary to establish that its normal operation does not adversely affect the finished product quality or safety. Validation trials can demonstrate the suitability and variability of a process by testing products and process intermediates according to a plan that specifies the tests to be done, the data to be collected and the criteria for success. Where possible data should be collected under operating conditions and be strictly related to identified hazards. Normally this is done for a specified period (e.g. weeks) of representative full-scale production and may include periods where production is increased or decreased. The purpose for which data are collected must be clear to process operators; it must reflect what happened and be collected and analysed carefully and accurately.

Before validation trials start, specifications should have been approved and the equipment must be judged to have been correctly installed and run according to agreed performance criteria. The validation plan should specify the key process variables to be monitored and documented and the number of process runs needed to provide an accurate measure of variability between successive runs. The runs should represent upper and lower processing limits and circumstances, including those within standard operating procedures, which present the greatest chance of process or product failure; such conditions are known as 'worst case' or 'most appropriate challenge'. The records from validation trials (and intensive QA) should show the suitability of materials and performance and reliability of equipment and systems. They will establish the variability of process parameters from run to run and establish whether or not the equipment and process controls are adequate to assure that product specifications are consistently met. Any parts of the process that vary enough to affect important product quality characteristics should be reviewed for improvement. Based on trial output, each stage should be defined and described sufficiently well for employees to understand what is required and to take corrective actions.

In challenging a process to assess its risks, it is important that conditions simulate those that will be encountered during actual production, including 'worst case' conditions. For example, in aseptic filling, significant process steps should be defined and challenged (e.g. the sterilisation of containers/closures, sterilisation of product, filling equipment and product contact surfaces, and the filling and sealing of containers) to show they work consistently within specification. It is dangerous to extrapolate between similar products, processes and equipment without appropriate challenges.

Process data are essential to process validation, and their value is increased if corresponding quality attributes and their variability are also measured. Where finished (or in-process) testing methods cannot adequately measure certain product attributes, process validation has to rely on comparing process (or product) data with specifications. In any such comparison, interactions between the various systems should be considered (e.g. heating and cooling stages).

13.3.3 Product validation

A product design should be validated using real products to simulate their quality and performance under real-world conditions and test 'what if'. Therefore product should be manufactured with the same equipment, methods and procedures that will be used for routine production; otherwise, the approved product may not be representative of production units and cannot be used to demonstrate that the manufacturing process complies with agreed specifications and quality attributes. Where possible, product validation should include performance testing under conditions that simulate use in each market where the product is sold. Once production units have been successfully validated, there should be a technical review before specifications are fixed. It should:

- compare the approved product specifications with the actual product;
- review the test methods used to determine compliance with the specifications;
- determine the adequacy of the change control procedures.

13.4 Modelling and product/process design

The microbiological risks and effectiveness of controls associated with particular products can be investigated either by practical trials (such as challenge testing) or by the use of predictive modelling; product approval should not be based on the performance of poorly specified manufactured products. Models can estimate interactions between supply chain times and temperatures in determining microbial growth, survival or (process) lethality (i.e. based on D and z values), and this type of modelling can indicate sensitivity to temperatures and temperature fluctuations during processing, hence improving supply chain management.

Some software packages can integrate heating and microbial inactivation during pasteurisation (e.g. destruction of *Staphylococcus aureus* during thermal processing of surimi seafood – Jaczynski and Park, 2003) or the interaction of heat with preservatives (e.g. the effects of temperature, sodium lactate and sodium diacetate on the inactivation of a serotype 4b strain of *Listeria monocytogenes* in a frankfurter slurry – Schultze *et al.*, 2006). In the UK, Food MicroModel (FMM) and in the USA, the Pathogen Modelling Program (PMP) (http://ars.usda.gov/Services/docs.htm?docid=6786) are computer-based predictive microbiology databases applicable to food products. Panisello and Quantick (1998) used FMM to make predictions on the growth of pathogens in pâté in response to variations in the pH and salt content and, specifically, the effect of lowering the pH. Zwietering and Hasting (1997) have taken this concept a stage further and developed a modelling approach to predict the effects of processing on microbial growth during food production, storage and distribution (Amézquita *et al.*, 2005a, b). Their process models were based on mass and energy balances together with simple microbial growth and death kinetics, illustrated using meat product processing lines. Such models can predict the contribution of each individual process stage to the level of microbes in a product. FMM and PMP can now be found within the ComBase portal (http://wyndmoor.arserrc.gov/combase/) which includes the Pathogen Modelling Program (http://ars.usda.gov/Services/docs.htm?docid=6786), see Baranyi and Tamplin (2004).

Zwietering *et al.* (1991, 1994a, b) have also modelled the impact of temperature, time and shifts in temperature during processing on the growth of *Lactobacillus plantarum*. Such predictive models can, in principle, be used for suggesting the conditions needed to control microbial growth or indicate the extent of the microbial 'lag' phase during processing (e.g. in pea soup during cooling – De Jong *et al.*, 2005) and distribution where temperature fluctuations may be common and could allow growth. Impe *et al.* (1992) have also built similar models describing the behaviour of bacterial populations during processing in terms of time and temperature, but have extended their models to cover inactivation at temperatures above the maximum temperature for growth. A model has been developed by Augustin *et al.* (2000) to describe the effect of temperature history and duration of pre-incubation period on the re-growth lag time of *L. monocytogenes*. Their model takes into account the influence of prolonged starvation conditions and physiological state of the cells on lag time before regrowth. Giffel *et al.* (1999) have applied mathematical models to predict the growth of spoilage and pathogenic microorganisms in food production chains and point out the potential use of such models for product development, HACCP analysis and risk assessment (Ross and McMeekin, 2003). Two examples are provided – one to predict the growth of *L. monocytogenes* and hence determine the critical control points in the production of sliced, cooked ham; the other describes the use of a model to predict growth, adherence and release of thermo-resistant streptococci during whey processing in a cheese factory. There is also a model for predicting spoilage of seafood (Dalgaard *et al.*, 2002).

Adair and Briggs (1993) have proposed the development of expert systems, based on predictive models to assess the microbiological safety of chilled food. Such systems could be used to interpret microbiological, processing (e.g. surface pasteurisation – James and Evans, 2006), formulation and usage data to predict the microbiological safety of foods. However, to be realistic, models are only as good as the inputs and at present there is both uncertainty and variability associated with the data available. Betts (1997) has also discussed the practical application of growth models to the determination of shelf-life of foods (e.g. vegetables – Carlin *et al.*, 2000) and points out the usefulness of models in speeding up product development and the importance of validating the output of models in real products (Membré *et al.*, 2005). Modelling technology can offer advantages in terms of time and cost, but is still not widely used by product developers (Pin and Baranyi, 1998). Usefulness is currently limited by variations in the microbial types present in raw materials and products, their activities and interactions, other factors altering growth or survival rates or the rate of production of metabolites causing spoilage. However, recently probabilistic models have been built (Membré *et al.*, 2005) to predict the lethality of thermal processes used to inactivate psychrotrophic and mesophilic *Bacillus cereus* spores. These take account of heat resistance, contamination frequency and concentration and difference in container heating.

Walls and Scott (1997) surveyed scientists in the food industry to determine their views on the value of predictive microbiology. They took as examples assessment of the risks of foodborne illness from a cooked meat product contaminated with *Staphylococcus aureus* and a hamburger contaminated with *Salmonella*. They found that over 80 % of the respondents had used predictive microbiology software and 36 % had developed predictive models in their companies. There was support for using models to determine critical control points in a hazard analysis plan. Based on the examples, users concluded that more data were needed for reliable microbiological risk assessments and that predictions should not be limited to single-point estimates. To take account of variability, a Monte Carlo simulation of the data distribution should be used to produce a range of risk estimates, e.g. best, average or worst.

13.5 Safety management tools: Good Manufacturing Practice (GMP), Hazard Analysis Critical Control Point system (HACCP) and risk assessment

13.5.1 Good Manufacturing Practice

Good Manufacturing Practice (GMP) or pre-requisites cover the hygiene, supply chain and control principles needed for the safe production of foods. It covers the design and operation of a plant layout, selection of equipment and controls and operational procedures for particular products. Taking a product design, Good Hygiene Practice (GHP) is the part of GMP that focuses attention on the hygiene measures needed to ensure that HACCP and pre-requisites operate effectively. In Europe, GMP has a wider meaning than in the USA where it focuses more on process hygiene. Inappropriate levels of hygiene during manufacturing will lead to microbial survival or cross-contamination or unnecessarily high costs due to rejected product. Requirement for specific levels of hygiene for particular products leads to the designation of areas for manufacturing (e.g. GMP, high care or high risk areas – CFA, 2006) or post-process areas for the production of low-acid ambient-stable foods. Choice of the level of hygiene in a manufacturing area is particularly important in chilled food manufacture, because contamination in conjunction with temperature or time abuse during storage can lead to growth of pathogenic and spoilage microorganisms within the product. GMP codes and requirements for the hygienic manufacture of foods may be formally specified (e.g. Codex Alimentarius Committee on Food Hygiene, DHSS, 1984, 1986). Codes may also be developed by food industry groups, often acting in collaboration with food inspection and control agencies or other specialist groups (Jouve *et al.*, 1998).

For any product, the GHP/GMP requirements should cover the following requirements based on the product design:

- the hygienic design and construction of food manufacturing premises;
- suitable raw and packaging materials;
- the hygienic design, construction and proper use of machinery and controls;
- cleaning and disinfection procedures (including pest control);
- general hygienic and safety practices in food processing including;
  - microbiological quality of raw materials,
  - effective and hygienic operation of each process step,
  - hygiene of personnel and their training in manufacture of safe food, including hygiene procedures,
- storage, transport and logistics;
- monitoring and record keeping.

13.5.2 HACCP

HACCP is a food safety management system that identifies hazards and controls their fate at designated points in the supply chain (critical control points, CCPs). The adequacy of the HACCP system in controlling the food safety hazards identified during the hazard analysis should be validated and that the system is being implemented effectively and is functioning as intended should be verified. HACCP can be used to ensure food safety in all scales and types of food manufacture and should provide controls for all the hazards noted in the product design and take account of product use (ILSI, 2004). The widespread introduction of HACCP, in conjunction with the recognition of the essential role of pre-requisites (e.g. steps or procedures to control the operational conditions within a food establishment essential to the production of safe food using a particular technology) has promoted a shift from end-product inspection and testing to the control during production of hazards identified by the product design and hazard analysis, especially at the CCPs. The technique is ideal for processes where many elements (e.g. formulation, processing and packaging) contribute to the chemical, physical and microbiological safety of products. Control and monitoring of compliance all along the supply chain offers the consumer better protection than testing products for pathogens. End-product testing cannot ensure safety, as at practical levels there is a very low probability of detecting product that is hazardous, because it contains pathogens. The delay to await results of microbiological testing also uses up shelf-life and leads to a requirement for greater storage capacity in the factory.

For steps in the manufacturing process that are not recognised as CCPs, the identification of pre-requisites (also known as GMP/GHP) is essential to provide assurance that suitable underpinning controls and standards are present. These cover systems and facilities (e.g. product and process design parameters, hygiene and process control) in the factory. Pre-requisites are generally technology specific and without them the CCPs in the HACCP plan cannot ensure product safety. The identification and analysis of hazards within the HACCP programme will provide information to interpret GMP/GHP requirements and indicate staff training needs for specific products or processes. There is general guidance provided by the Canadian Food Inspection Agency (CFIA, 2008), and a range of guidance for different products and processes; for example, the Microbiology and Food Safety Committee of the National Food Processors Association (NFPA, 1993) has considered HACCP systems for chilled foods produced at a central location and distributed chilled to retail establishments. Chicken salad was used as a model to propose CCPs and give practical advice on HACCP planning (i.e. development of a supply chain flow diagram, hazard identification, establishing critical limits, monitoring requirements; and verification procedures to ensure the HACCP system is working effectively). There are also US Department of Agriculture recommendations and outline HACCP flow diagrams for other processes, such as cook-in-package and cook-then-package (Snyder, 1992). Testing against microbiological criteria (regulations, standards or advisory criteria) has been widely used to determine product safety and verify the effectiveness of HACCP plans. In a HACCP plan, microbiological criteria retain their value as tools for judging the implementation and effectiveness of HACCP (verification), validating control measures and investigating problems.

13.5.3 Hazard review

Ensuring the microbiological safety and wholesomeness of food requires reliable identification of realistic hazards, their origins, levels and means of control (risk assessment) in the context of product usage. If all realistic hazards are not identified and controlled, there is an increased risk to consumers, and if too severe controls are put in place for unrealistic hazards, product quality will be reduced. Because precautions are aimed at hazards identified during the development of products, it is essential that hazards are regularly reviewed to take account of changes in the origin of raw materials or new markets or product uses.

All stages of the supply chain should be reviewed (e.g. as a flow diagram) to find out where hazards are likely to be introduced and can be controlled. The review should identify suitable controls and limits for all realistic hazards based on inputs and outputs. Controls should be based on the severity of consequences and likelihood of occurrence of the hazard, to determine the necessary measures.

Assessment of the impact of changes in a design or a process, sources of supply, and product labelling or market on the level of risk and the type of hazard is important for assuring consistent standards of food safety. The most important common changes affecting risks are development of new products and processes, new or different sources of raw materials (often from less developed countries introducing unfamiliar microbial hazards) or the targeting of new customer groups, such as children, or by demographic change. These should be regularly reviewed over the life cycle of a product or product range.

Food producers have often reviewed and assessed risks using empirical approaches but, because of the diversity of hazards, the scale of production and high consumer expectations, a more formal approach is needed, which should precede the development of the HACCP plan. Hence existing practical and empirical approaches should develop into formal systems with well-defined procedures whose overall aim is to reduce risk by:

- identifying realistic microbiological hazards and characterising them according to severity and response to processing, formulation, etc.
- examining the impact of raw material contamination, processing and use on the level of risk, and especially survival or levels of pathogens;
- communicating clearly and consistently, via the output of the study, the level of risk to the supply chain and to consumers.

From a HACCP point of view, these are known as hazard analysis and the techniques used are closely related to microbiological risk assessment (MRA) and risk management. These are described in *Microbiological Risk Assessment; an Interim Report* (ACDP, 1996) or by the Codex scheme and *Proposed Principles and Guidelines for the Conduct of Microbiological Risk Management* (CAC, 2000).

If new causal links are established between a raw material, product, foodborne illness and the presence or activity (e.g. toxigenisis) of food-poisoning microorganisms, the HACCP plans for all similar products should be formally reviewed to ensure that suitable controls are in place. When these are put together with risk communication (distribution of information on a risk and on the means of control) and used to promote sound risk management actions to eliminate or minimise risk, a risk analysis is produced (ACDP, 1996). At a more practical level these actions result in an action plan for factory management (e.g. not only changes in regulatory policy). Risk assessment has been reviewed (Jaykus, 1996) and applied to specific problems; listeriosis (Miller *et al.*, 1997), the role of indicators (Rutherford *et al.*, 1995) and links with HACCP. More details on risk assessment can be found in Chapter 4.

13.6 Process flow and equipment

From a microbiological point of view, processes should be operated to control the presence, growth and activity of target microorganisms, while producing products of good quality. Monitoring for microbiological safety and stability should concentrate on unit operations that:

- reduce risk by eliminating or reducing numbers of bacteria;
- increase risk by providing opportunities for survival, recontamination or growth.

Raw materials, product design and shelf-life/storage requirements, and even factory hygiene and layout, will determine realistic target pathogens for each process stage. At the beginning of the supply chain, agricultural produce can act as a reservoir of food-poisoning bacteria (e.g. *Salmonella*, *Campylobacter*, *E. coli* O157, *Staphylococcus aureus* and the harmful *Bacillus* and *Clostridium* spp.). Therefore, it is important that conditions during handling and processing can control this initial contamination. The extent of precautions needed (e.g. process or intake controls and monitoring of their effectiveness) will depend on the product use and needs to be proportional to the severity of the hazard, the frequency of occurrence of pathogens and the complexity and scale of the supply chain. Any interactions, for example between sourcing area and the occurrence of pathogens, should be examined by a risk assessment.

13.6.1 Process flow

The flow of materials and personnel, plant layout and equipment operation are key contributors to product safety. Translation of process designs into material flows will differ depending on the manufacturing and storage areas and equipment available and on the type of product being made. In this context, staff training and operating procedures must ensure that quality and safety aspects of the product design are consistently delivered. These should always be covered in the pre-requisites or the HACCP plan (CCPs) and may include specification of temperatures and times and rates of heating, cooling or freezing of materials.

The process design should specify heat treatments and the heat processes needed to ensure that the design requirements are always met, by taking account of process variability. As a principle, it should be assumed that unprocessed materials may contain pathogens, making the forward flow of materials essential to prevent cross-contamination and ensure that materials and products do not miss critical stages, e.g. sterilisation. Major in-factory routes for recontamination are food debris and food-handlers; therefore, layouts and procedures should be designed to minimise these risks. In some cases, inadequacies in layout may have to be compensated for by more stringent procedures, such as separation in time or personnel segregation. Food residues in equipment or processing areas during operation or after cleaning can become a major source of contamination, therefore processing equipment should minimise the amount of material 'held-up' and provide a narrow and predictable range of residence times for this material (e.g. tide marks and quantities of residual material). Airborne contaminants are a relatively minor source of contamination, but prevention of this route is an essential feature of the design of high care risk areas. Process flows should be designed for ease of operation and access for cleaning and maintenance. Procedures or corrective actions must be in place to ensure consumer safety when processes go wrong or defective material is processed.

Many types of product need to be free of pathogens when they leave the production plant, hence layouts need to be specified and the process stages operated to minimise the chances of cross- or re-contamination after decontamination and primary packaging. Wherever possible there should be physical or operational segregation of the pre- and post-cooking areas and appropriate high levels of hygiene in all the areas handling cooked product. Building and equipment hygiene, operational practices and cleaning in areas handling decontaminated products (e.g. sterilised cans) should minimise the chances of re-contamination. For example, the design and operation of areas handling blanched vegetables for salads or frozen products may need to ensure that they are not contaminated or re-contaminated after blanching. If existing equipment is unhygienic, emphasis must be placed on operational procedures, such as cleaning or restriction of periods between cleans, for ensuring safety. If a product is made from a mixture of cooked and uncooked material, then it should be handled, treated and labelled as uncooked.

13.6.2 Equipment

Many of the critical quality and safety attributes of foods rely on the technical performance and hygiene of the equipment and control systems used for cooking, cooling, cutting, shaping or packaging foods. The correct design and reliable operation of these stages, so that specifications are consistently met, is most important to product safety. There are many examples of this:

- heating equipment for particular types of material – pieces, liquid and pumpable products, etc. and controlled temperatures and residence times;
- size reduction equipment to produce pieces with a predictable size range;
- dosing equipment for acidification or for filling containers or for dosing particulates and liquids in a specified ratio;
- mixing equipment to ensure uniform distribution of components or heating or cooling;
- sealing equipment to produce seals of a known strength and integrity.

Lag periods, growth rates and heat resistance of microbial contaminants in equipment and products can be influenced by prior processing and storage conditions, including cooling rates, temperatures and levels of residual material, and should be controlled by the equipment design or operational procedures.

It is worth remembering that between cleaning and production (e.g. at weekends), chilled areas may not be chilled and, depending on temperature, *Listeria* and other pathogens may grow in food residues left on equipment. The production planning system should provide opportunities for cleaning at suitable intervals, and this is critical if allergenic materials are handled and products not labelled as 'at risk'.

13.7 Manufacturing areas

13.7.1 Raw material and packaging reception areas

Most factories will have designated areas for deliveries, which should be suitable for the type of vehicles and materials (e.g. frozen, chilled or ambient-stable) arriving and which should protect unloaded materials from changes caused by the external environment. They should allow the efficient and rapid unloading of vehicles with a minimum of temperature change and damage to packs, or packaging. Their size should promote direct removal of unloaded materials to storage areas with minimum opportunities for cross-contamination, especially if the materials handled are for direct incorporation into finished products. Segregation of receiving areas for different materials may also be governed by legislative requirements and there should be facilities for the efficient removal and disposal of secondary packaging, such as cardboard boxes. If product is delivered unpacked (e.g. vegetables), a supply of clean containers or conveyors may be required for handling, sorting and storage of delivered materials prior to use.

Areas should have facilities for the inspection, coding and maintenance of raw material batch integrity; they must be designed for effective cleaning and should not be used for storage. If materials are to be taken from them directly into hygienic or high-hygiene areas, it may be necessary to disinfect their outer packaging at a barrier station before entry. To minimise risks, these materials should be delivered to a separated area handling only low-risk ingredients and not to areas handling raw materials likely to contaminate them (e.g. raw meat).

13.7.2 Material storage areas

Food raw materials differ in their storage requirements, although all should be stored so that contamination and premature spoilage are prevented. A factory may therefore require a number of different storage areas controlled for hygiene and temperature (ambient, chilled, 0–5 °C or frozen, below –12 °C). Temperature-controlled areas should be fitted with reliable controls, monitoring and record-producing systems (to provide a record of conditions in the store) and an alarm system indicating loss of control or failure of services. A low-humidity store may be required for dry ingredients and packaging materials. Layout or racking should promote efficient stock rotation (first-in-first-out, FIFO) with batches clearly labelled with suitability, use-by dates and approval for use. Layout should allow easy access to all stored items to ensure that particular deliveries or production batches can be traced or identified.

Operation of stores and control of the means of access (such as self-closing or strip-protected doors) should ensure that the specified conditions (such as 2–4 °C in chills or –12 to –20 °C in cold stores) can be maintained during the working day. Refrigeration equipment should have sufficient capacity to maintain product temperatures when there are high outside temperatures or peak demand. All storage areas should be easily cleanable, using either wet or dry methods, as appropriate. The layout of racking and access to floors, walls and drains should allow easy cleaning, and racking should not be made of wood. Packaging stores should be laid out and operated to minimise damage to packaging and ensure requirements for machinability (e.g. temperature of plastics) and hygiene are met. Where glass containers are handled, special precautions are needed in the event of breakage.

13.7.3 Raw material preparation and cooking areas

Preparation and cooking areas receive ingredients from delivery and storage areas and convert them into ingredients by a variety of techniques, such as cutting, mixing or cooking. Batching, preparation and cooking may take place in a single area, separated from sterilisation, pasteurisation or packing areas. For pasteurised, sterilised or aseptically packed products, maintenance of 'initial temperature' for heat processing may be a key consideration for kitchen and process design and vessel (batch) sizing, to prevent uncontrolled, excessive temperature drop before sterilisation.

If long chilled shelf-life and prevention of contamination of unpacked product are important, cooking may be done in a separate area leading through suitable precautions to a higher-hygiene filling/packing area. Where the cooked product is taken to a hygienic area for cooling and packaging, it is essential that the layout, operation and access from the preparation area prevent recontamination. If space permits, cooking operations are ideally carried out in areas separated from preparation or batching areas, to minimise the chances of contamination by airborne particles, dust, aerosols or personnel or the omission of ingredients.

Where physical separation between areas cannot be achieved, cooked product should not be handled by personnel or equipment that has been in contact with uncooked material. In rooms or areas where cooking is done, the vessels or ovens may be sited to form a barrier between 'dirty' areas, i.e. those handling uncooked material, and 'clean' areas handling cooked material. Air flow should be from 'clean' to 'dirty' areas, and the supply of air to steam extraction hoods in cooking areas should ensure that air is not drawn from 'dirty' areas and condensation (water droplets) does not drop onto cooked product. If single-door ovens or autoclaves (for sterilisation) are used, there is an increased risk of cross-contamination, as it is not possible to segregate raw (pre-process) and cooked (processed) material effectively.

Pathogens may grow in warm processing areas, and particular care should be taken to ensure that *Staphylococcus aureus* cannot grow in processing areas where nutritionally-rich materials (such as egg or dairy sauces, or batters) are produced at ambient temperatures. The entry/exit areas to these heat-processing areas need to be kept clean, with loading and unloading done by designated staff, so that the opportunities for product contamination are minimised. Batches and packs pre- and post-process should be clearly identifiable to minimise the chances of unprocessed material being sold.

Short shelf-life raw products and products containing non-decontaminated components may contain infectious pathogens. If cooking is used to provide a decontaminating heat treatment (70 °C/2 min), then recontamination must be prevented; for long-life chilled products (90 °C/10 min), additional precautions (such as minimising access or build-up of dust) must be taken to prevent recontamination with bacterial spores as these may grow during the shelf-life of the product. These include a forward flow, physical separation of process stages and control of air flow away from the decontaminated product. Typical process routes for short and long shelf-life chilled products are shown in Figs 13.2–13.5.

13.7.4 Thawing of raw materials

It is usually necessary to thaw frozen ingredients before they are heated to ensure reliable and short heating times. Ingredients for pasteurisation and sterilisation should have a controlled minimum temperature and maximum particle size (clumping of frozen particles may occur) so that cold spots, which will be insufficiently heated, are not accidentally created. This should be done under conditions that minimise pathogen growth, i.e. the maximum

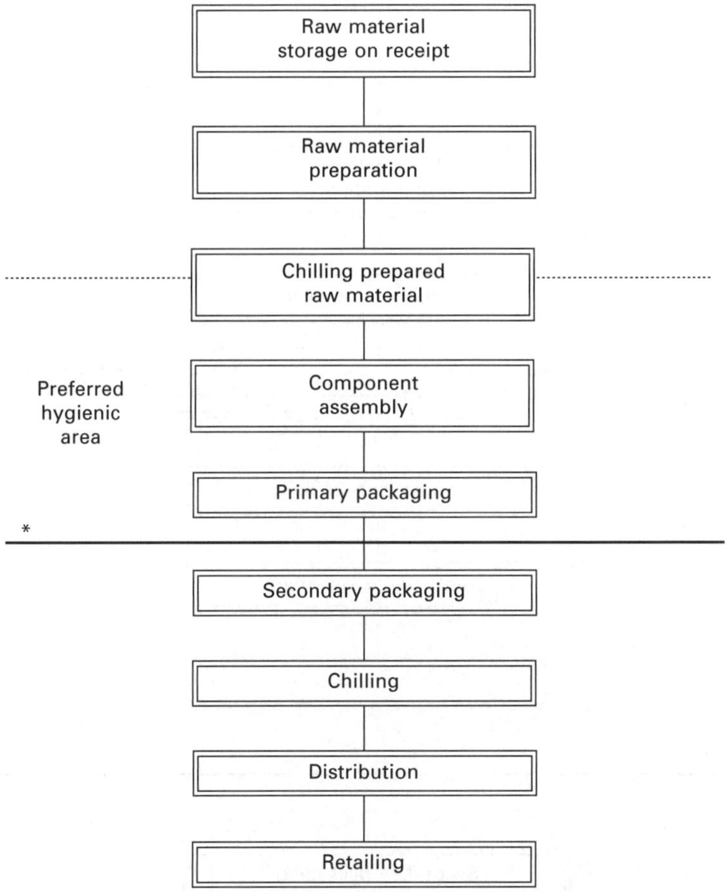

Fig. 13.2 Typical flow diagram for the production of chilled foods prepared from only raw components. *Physical and staff separation obligatory.

surface temperature of the ingredient should not be within their growth range (i.e. below 10–12 °C). If this is not possible, then thawing times should be minimised to prevent growth and quality change. Thawing at ambient temperatures can lead to the uncontrolled and unrecognised growth of pathogens. Special thawing equipment such as microwave tempering units, running-water thawing baths or air thawing units may be used for microbiologically safe thawing and their performance should be validated.

13.7.5 Hygienic areas for manufacturing foods

Manufacturing areas

The whole range of foods cannot be made in manufacturing areas that only meet basic GMP requirements. For many products, especially chilled foods,

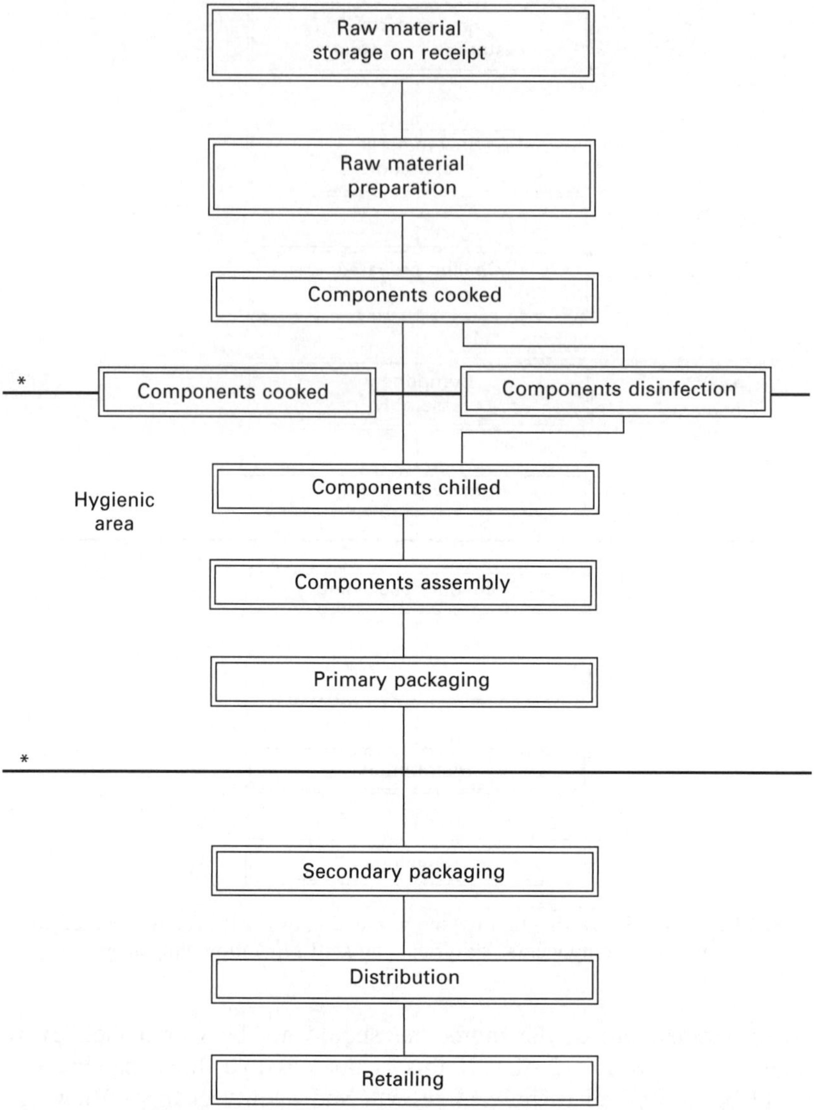

Fig. 13.3 Typical flow diagram for the production of chilled foods prepared from both cooked and raw components. *Physical and staff separation obligatory.

additional precautions and facilities are needed to ensure microbiological safety and shelf-life. High care areas (HCA) or high risk areas (HRA), separated from the low risk (or GMP) areas, should be used for handling, assembling and packing decontaminated materials for chilled distribution. Typical design features of higher-hygiene areas are segregated changing rooms, dedicated

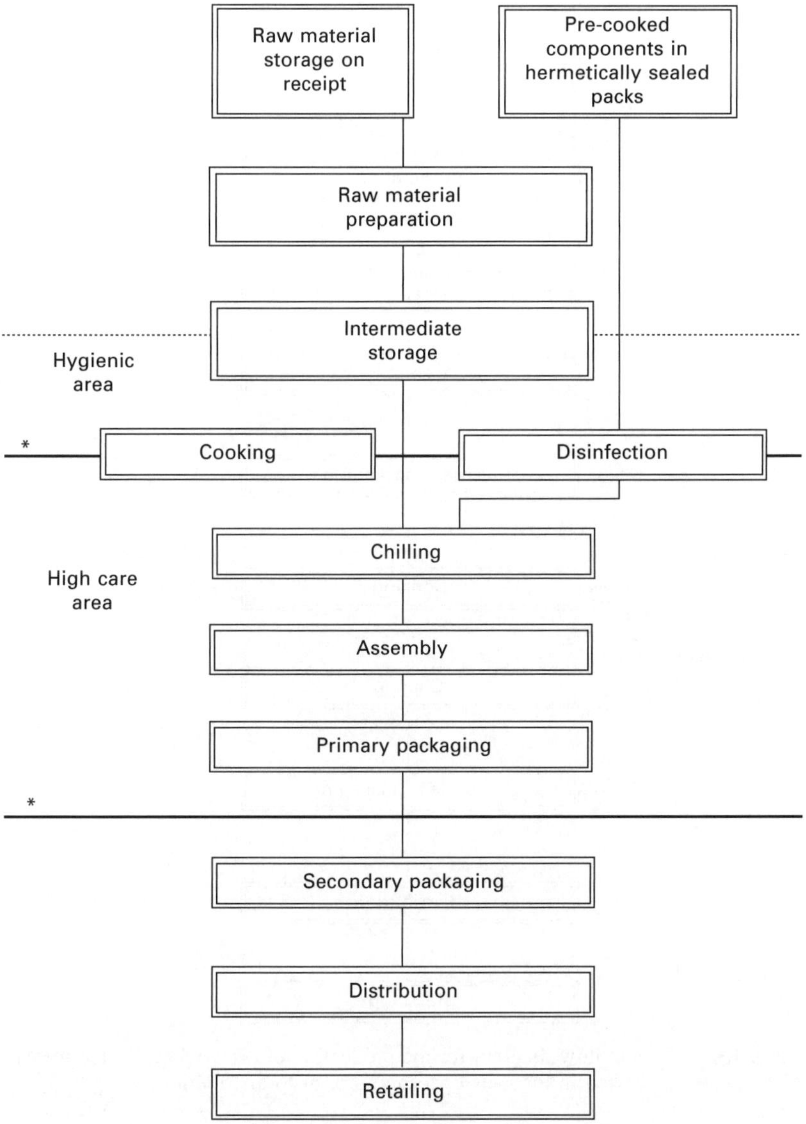

Fig. 13.4 Typical flow diagram for the production of pre-cooked chilled meals from cooked components. *Physical and staff separation obligatory.

entry routes for employees with air locks for entry and exit, dedicated chills and two-door ovens for entry of materials.

Good manufacturing practice areas
Storage areas, chills and freezers and batching areas generally are designated as GMP areas. If materials with high or medium risks of carrying microbial

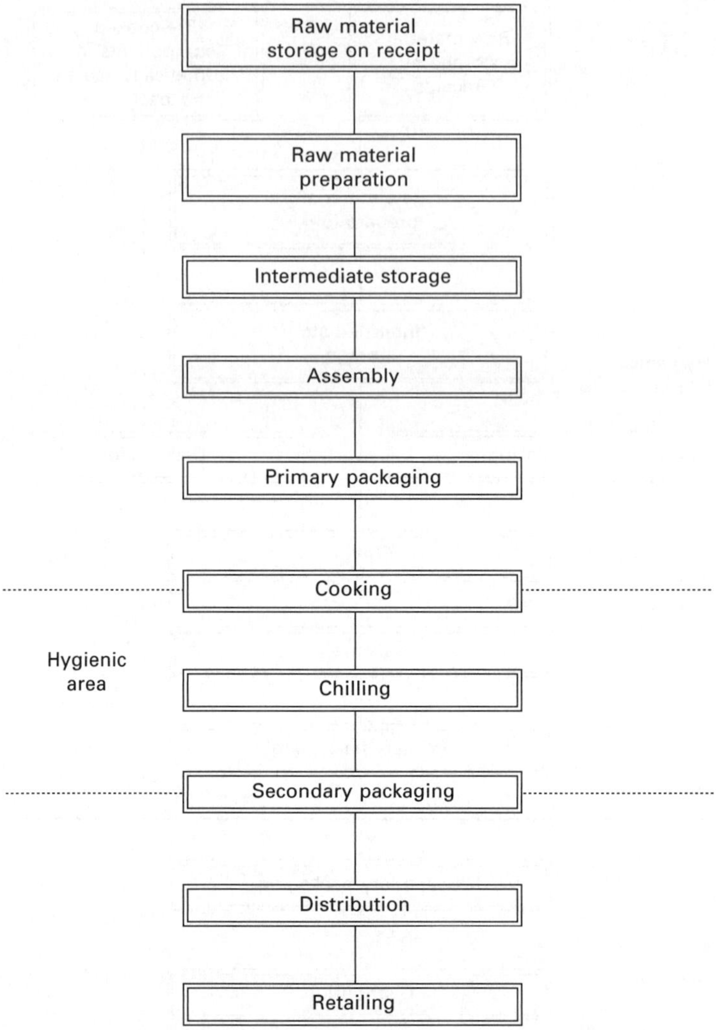

Fig. 13.5 Typical flow diagram for the production of pre-cooked chilled meals cooked in their own packaging prior to distribution.

pathogens are handled or stored in these areas, there should be effective separation from materials with a low risk. Some ambient and chilled products can be made in GMP areas (e.g. raw products intended for cooking, raw, preserved, ready-to-eat products, such as salads). As a principle, hygiene during storage and preparation should not increase product safety risks or levels of contamination. If raw materials processed in these areas are not decontaminated before being made into products, they should be low risk (e.g. salad vegetables), and the product preservation system (e.g. acid or vinegar-based) should inhibit microbial growth. If raw products are made, consumer cooking processes (giving heat treatments exceeding 70 °C for two minutes) should be specified to eliminate infectious pathogens from materials for chilled storage. Where dry products are made (e.g. dry soups or ready-meals) and the areas and equipment operate 'dry', to ensure these products can be processed, mixed and filled and do not re-hydrate, special consideration should be paid to cleaning and sanitation. These areas should be kept dry and for preference dry cleaning methods, (e.g. brushing and vacuum cleaning) should be used, but if wet cleaning is used, procedures should include a drying step. If water remains in the equipment, there is a high risk that wet or semi-dry environments that support microbial growth will be created, leading to product contamination. Depending on temperature and the nutrients provided, these areas have the potential to support the growth of pathogens (e.g. *Salmonella*) and toxin production (e.g. *Staphylococcus aureus*). Hand washing facilities should be sited at all entrances.

Higher-hygiene areas

Higher-hygiene manufacturing areas for chilled products are separated from GMP areas and can be either high care areas (HCA) or high risk areas (HRA). The latter areas share common structural features to minimise or eliminate the chances of product contamination from raw materials, food contact surfaces, air or personnel. Very reliable utilities (e.g. steam, cooling media and air supply) are essential for their safe operation, and work procedures should cover personnel and material entry and actions to be taken in the event of utility failure (e.g. temperature rise or loss of air over-pressure). Within them, precautions are specifically designed to prevent *Listeria* spp. becoming endemic (e.g. walls and floors are sealed or impermeable, to ensure cleaning is effective). Floors are sloped to assist drainage and drying, with drains always flowing away from the hygienic areas. To minimise chances of harbouring contaminants, services (steam, cooling media, compressed air and electricity) should be routed with the minimum of exposed cabling or piping. Where piping is sufficiently cold to cause condensation to form, it should be hygienically insulated with a cleanable material. Microbiological monitoring of these areas may include visual inspection (against a check list) and swabbing of key equipment or areas for *Listeria* spp., which are suitable indicators for possible presence of *Listeria monocytogenes*. If *Listeria* spp. are detected, an increased risk is indicated and the type found should be characterised and preventive measures reviewed.

Sometimes these areas may have a higher standard of hygiene than is strictly necessary to prevent pathogen contamination. This is done to limit re-contamination with spoilage microorganisms (such as yeasts, moulds and lactic acid bacteria) and to minimise spoilage risks. Only foodstuffs and packaging that have been reliably decontaminated and handled to prevent re-contamination should be admitted to these high care areas. All their food contact surfaces should be impervious to water and capable of being easily cleaned, disinfected and kept dry. After cleaning and disinfection, surfaces should have fewer than 10 microorganisms per 9 cm^2 and Enterobacteriaceae should not be detected when suitable recovery methods are used.

High care areas (HCA)

High care areas (HCA) are necessary to manufacture chilled foods that are designed as RTE and unpreserved (e.g. prepared ready meals), and they should minimise the risk of contaminating products which already contain very low levels of microorganisms. Therefore they should only handle materials that are very low risk or have been pre-decontaminated (e.g. pasteurised or chemically decontaminated, raw materials). To minimise chances of contamination, any materials or products pasteurised in-pack for use in them (e.g. for packaging or dosing) should be unloaded after cooling and stored in areas that have hygiene standards equivalent to or better than the post-sterilisation area of a cannery.

It is important to identify pathogens that can survive any decontamination procedures used (see listings by Doyle, 1998), especially if the materials potentially carry pathogens; higher-risk materials are prawns, other shellfish from warm waters and untreated herbs and spices, which may carry *Salmonella*. It must be accepted that from time to time these products and their components will contain pathogens, and hence storage conditions, hygiene in manufacturing areas, use instructions and coding practices should be designed with this in mind.

High risk areas (HRA)

High risk areas (HRA) are laid out and operated to prevent microbial (re-) contamination, especially with pathogens. They handle unpreserved products composed entirely of pasteurised ingredients, and provide facilities for their assembly in an 'open' environment. Hence they have stringent precautions against re-contamination, including limited personnel entry from segregated changing rooms. When the HRA is left after work or for a break, work clothing and footware should be removed and remain in the changing room to prevent its contamination.

13.7.6 Chillers and cooling

Chillers used for raw materials may be designed to cool materials, but are usually designed to maintain temperatures in previously cooled components. They may be an integral part of higher-hygiene areas designed to cool cooked or heated components (blast chillers – James *et al.*, 1987). Cooling of these materials should be started as soon as practicable after cooking and containers should be dimensioned to ensure it is done rapidly. Maximum cooling times should be designed to prevent the growth of surviving spore-forming bacteria. Because blast chillers may receive 'naked' product, the quality of cooling air and their environmental hygiene are critical to safety and shelf-life (see below). The design, hygiene and operating temperature of the cooling elements in their fan-driven evaporator units play a critical role in limiting contamination by aerosols. If the temperature of the heat exchanger coils is too high, water will condense on them without freezing and then be blown onto the products by the fan. Many evaporator units are designed so that condense trays do not drain and are inaccessible for cleaning; such units can harbour *Listeria* and cause contamination if the contents reach product. Similarly, if warm materials are placed in storage chillers, condensation may occur on them; therefore, product containers should be covered or chiller air temperatures or volumes managed to prevent this.

If water is used as a direct cooling medium as a spray, shower or in a bath (for products in hermetically-sealed containers), chlorination or other disinfection procedures should be used to ensure that the cooling medium does not act as a source of contamination. Stringent cleaning and disinfection systems should be used to ensure the hygiene of any circulation or recirculation systems, including heat exchangers. Packs should be dried as soon as possible after cooling and manual handling of wet packs should not be done.

13.7.7 Other environmental controls

HVAC (heating ventilation and air conditioning) systems provide refreshment air and control temperature in GMP and hygienic areas. In GMP areas air changes are only intended to provide comfort for employees. In hygienic areas, their design, operation and, especially, capacity, set temperature and rates of air replacement (> 6 changes/h) should maintain environmental temperatures under envisaged operating conditions. System capacity should provide sufficient overpressure (> 5 Pa) to ensure an outward air flow from hygienic areas during production, and prevent the inward flow of potentially contaminated air (e.g. when doors are opened). Hygiene procedures should keep air distribution ducts, filters and heat exchangers as clean and dry as possible to minimise the chances of airborne microbiological contamination. The air flow during and after equipment and area cleaning should prevent contamination and ensure rapid drying. During cleaning in one area, air should not be circulated from it to adjacent areas and it should be exhausted directly to the outside. The HVAC installation should be capable of drying the cleaned area in a reasonable time (e.g. two hours). Air refreshment rates will determine the time needed to bring the area back to its design hygiene level after the air system has been turned off.

Air can be distributed using either rigid or suspended fabric ducting, but increasingly the latter are used as they are more easily cleaned and disinfected. The air supply should be pre-filtered (to remove insects and dust) and then downstream filtered at higher filtration efficiency (e.g. semi HEPA or H11 – see CCFRA, 2005). System hygiene should be monitored using microbial and particle counts. Air exposure plates (settle plates) and battery-powered centrifugal samplers are widely used to monitor the microbiological quality of air. It is good practice to use a range of sampling points as well as different times of sampling.

Waste disposal

The efficient removal and disposal of waste from manufacturing areas is an essential part of their operation, and areas should have suitable, separate storage facilities and with distinctive containers to prevent product contamination. Any equipment or utensils used for the handling of waste should not be used for the manufacture or storage of product; but they should be maintained and cleaned to the same standards as the area in which they are used.

13.8 Processing and handling products

Based on the product design and raw materials, the design of equipment and controls should cook products and also provide predictable, reliable and adequate inactivation of microorganisms. During the manufacturing process there is a sequence of temperatures at different stages, rather than a single temperature and hence it is important to model fluctuating and different temperatures and times to predict risks of microbial survival or growth during processing. Not only do changes in temperature occur along the supply chain but, during the operating cycle of equipment, temperatures will change, especially during start-up. When heating is a key step in safety, attention has to be paid to the time taken for a pack, or vessel, to achieve its target temperature and then to cool to a safe temperature. In practice, liquid or pumpable foods will heat most rapidly because their flow properties permit the use of efficient heat continuous exchangers. For this reason, high-temperature/short-time (HTST) and UHT treatments can be used for thin liquids (e.g. soups, sauces and milk). Solid or viscous foods are usually packed and heated in containers (e.g. cans, jars or pouches). Because of their geometry (especially length of the thermal path) and the fact that heat is often transferred by conduction from the outside into the product, these containers will have relatively slow heating and cooling. The rates of temperature change at the thermal centre or cold spot will often determine the time–temperature combination used for processing. This principle is equally applicable to can or jar sterilisation for ambient stability and the design of trays for cooling chilled ready meals. For any type of heat processing it is essential that a 'safe' cool temperature is reached before the product units are palletised and put into store. The configuration of pallets dramatically slows the rate of heat exchange with the ambient air.

13.8.1 Working surfaces for manual operations

Many preparation or product assembly operations will be carried out on tables, belts or other flat surfaces. These should be hygienic and operate efficiently; for example, stainless steel is not suitable for cutting on, although it is easy to keep clean. Cutting surfaces should be made of hygienic materials such as nylon, polypropylene or occasionally Teflon®. Working surfaces should be easy to remove for cleaning or be completely cleanable without removal. It should be possible to restore the integrity of working surfaces by machining or some other process, because cut surfaces with uncleanable crevices are likely to be a potent source of contamination. The use of triclosan impregnation of surfaces has been recommended to provide a degree of antimicrobial protection. However, Cowey (1997) recommends that this is effective only as an additional line of defence, not as a replacement for hygiene procedures.

13.8.2 Cutting and slicing

To obtain consistent heating and filling characteristics it is important that particles of known and consistent dimensions are produced. Many products such as cooked meats and pâtés are prepared and cooked as blocks and consistent size is needed to ensure reliable heating. Automated slicing or cutting after cooking is an integral part of their conversion into consumer packs. Slicers for cooked products can be potent sources of contamination, because they are usually mechanically complex, providing many inaccessible sites that can harbour microorganisms living in the debris produced by slicing, which is often not completely removed by cleaning. The effectiveness of cleaning procedures is further reduced if cleaning operatives try to avoid wetting sensitive parts of machines (e.g. electronic controls, motors and sensors), which may be rendered inoperative by water penetration, especially if high-pressure cleaning is used. An integral part of the hygienic design of such machines is good access for cleaning and inspection, and effective waterproofing of controls and sensors. An insight into the routes for microbial re-contamination of product can be gained by auditing the machines for product and debris flow during operation, so that the sources of re-contamination can be identified. After production, when the equipment has been cleaned, machines should be re-examined to identify areas where product is still present. Effective cleaning schedules for them can be developed, implemented and monitored.

The difficulty of controlling the hygiene in these machines typifies the much more widespread problem of material recontamination by process equipment. Many forming and filling machines handling cooked product operate in chilled areas, both to improve their technical performance and to slow or halt the growth of microorganisms by minimising temperatures in retained product debris. Temperature auditing of machines will show that they may contain hot spots that are not effectively controlled by environmental or dedicated cooling (e.g. motors and gearboxes). In a well-designed machine, heat is conducted away without causing significant hot spots, but many machines unavoidably have localised hot spots within the growth range of pathogens (e.g. around sealing heads) in contact with food or debris, where microbial growth can occur. For example, in the area of the main cutter shaft bearing and motor of a slicer, temperatures of 25 °C+ could be found when material at below 4 °C is being processed in an area operating at 7–10 °C. During operation, these warm spaces may fill with product debris, which is retained, warmed and incubated, then may be released back into or onto product. Ideally, hot spots should be eliminated by machine design but, in practice in many existing machines, these risks can only be minimised. For safe, hygienic operation, warm areas retaining product must be identified and then controlled by cleaning, cooling or other preservation techniques, to ensure that microbial growth and product contamination are minimised. In some cases, simple engineering modifications can improve the hygiene of machines by reducing the quantities of product retained and its temperature.

13.8.3 Transport and transfer of product

From the time that a product is cooked, it will be transferred between production areas before being put into its final, primary packaging; therefore re-contamination needs to be prevented.

Containers

In simple process lines, products cooked in open vessels will be unloaded into trays for cooling or further processing. Whatever type of container is used, it should not contaminate the product, and its shape, size and loading should ensure rapid cooling or accurate dosing. Generally, stainless steel or aluminium trays are more hygienic and cool faster than plastic ones. Plastic becomes more difficult to clean as surfaces become scratched in use.

Belts

In complex production lines, transfer or conveyor belts may be used for transporting both unwrapped materials and product in intermediate wrappings from one process stage to another. Belts may also be an integral part of tunnel or spiral equipment, such as cookers, ovens and coolers. Although many types of belting material are used, they can be split into two broad groups:

- fabric or solid;
- mesh or link.

For hygienic operation, fabric conveyors should not absorb liquids and should have a smooth surface finish, so that they are easy to clean and disinfect. They are generally used only for transport under chill or ambient conditions, as they are not often heat-resistant. Solid belts for heating or cooling are usually made of (stainless) steel, which is hygienic and can be heated or cooled directly or indirectly. Belts can be cleaned in-place by in-line sprays, providing both cleaning and rinsing. A drying stage may be incorporated, using an air knife or vacuum system. The hygienic problems of conveyor belts are usually associated with wear or unhygienic design of drive axles, beds or supporting frames or poor maintenance.

Mesh or plastic link belts are used because they can carry heavier loads and form corners or curves. Metal mesh belts are widely used in ovens and chillers where circulation of hot or cold air is a part of the process, for example in spiral coolers or cookers. Belts that are regularly heated, or pasteurised, do not present a hygiene problem, as any material retained between the links will be freed of viable microbes by heating. If these belts are used in coolers, or for the transport of unwrapped product at ambient temperatures, special attention must be paid to their potential for product re-contamination via retained debris. Cleaning systems should be devised to remove the debris from between the links (for example, by high-pressure spray cleaning on the return leg of the belt). After cleaning, the belt should be disinfected or preserved, for example by chilling or drying, so that microbial growth does not occur during the processing period.

Hygienic performance becomes more difficult to achieve if routine engineering maintenance is not carried out correctly and belts become damaged or frayed. If specialised belts are incorrectly set up, or used with fragile products, this can increase the quantity of product waste generated by damage, so that even a properly designed and operated cleaning system will not keep the system hygienic. Product and pack characteristics, such as stickiness and crumbliness, should therefore be considered when transfer systems are designed, so that the generation of debris is minimised.

13.8.4 Dosing and pumping

Many food products are processed and sold in weight or volume-controlled packs, often with individual ingredients or pieces in a fixed ratio to one another, for example meat and sauce. Where ingredients are liquid or include small (*c.* 5 mm) particles suspended in a liquid, they may be dosed using filling heads or pump systems. If this type of equipment is used for dosing decontaminated materials or products where the liquid/solid ratio determines thermal processing, then its accuracy of operation and hygienic design are critical to product safety and shelf-life.

Dosing and filling systems may be operated at cold (below 8–10 °C), intermediate (20–45 °C) or hot (above 60 °C) temperatures. The most hazardous operations are those run at intermediate temperatures, within the growth range of food-poisoning bacteria and, unless the food producer is completely confident of the decontamination of the feedstock and hygiene of equipment, and is prepared for frequent cleaning/disinfection breaks, intermediate temperatures should not be used. Where hot filling is the preferred option, control of minimum temperatures and temperature drop during stoppages are critical to safety. To allow a wide window for operation, target temperatures should be set well above the growth maximum of foodborne pathogens (*c.* 55–65 °C) or above the initial temperature for heating to allow for cooling during dosing, and especially if there are stoppages and breaks in production, when the flow of product may be halted. If product is supplied to the filling heads by pipework, it is necessary to ensure that an unacceptable temperature change (leading to temperatures in the growth range) does not occur; recirculatory loops returning to heated tanks or vessels may be used to prevent this, but their hygiene must be carefully monitored. Dosing equipment is often cleaned by CIP (clean-in-place) systems, and it is essential that the pumps, valves and couplings used are suitable for this type of cleaning as well as for production. Consistent dosing plays a major role in ensuring that packs with uniform heating characteristics, headspaces and closing characteristics are produced for in-pack pasteurisation or sterilisation.

13.8.5 Packaging

Primary packaging

Most food products are sold packaged. The functions of packaging are to prevent contamination and losses and retain the product. Packing materials may be chosen for a variety of technical reasons, such as cost, appearance, machinability and heat or freeze resistance. From a microbiological safety point of view, the most important attributes of packing materials are their physical strength to provide pack integrity and exclude bacteria (especially at seams or seals) and their provision of gas and moisture barrier properties. If foods are intended for a long storage life, these properties must be maintained during the time involved. This is most important if packaging forms part of the preservation system by retaining gas mixtures. In conjunction with equipment packs need to retain their integrity and shape, especially the dimensions affecting heat penetration, during processing (e.g. during heat sterilisation at 121 °C), and thereafter exclude contaminants.

Filling equipment is important if headspace determines heating characteristics: a vacuum may be essential to ensure that a cap remains in place, and food trapped in the seal area as a result of dripping during filling can prevent a hermetic seal being formed. If residues remain in the seal area during heat sealing, then the two layers of plastic of the lid and base cannot be welded together by the sealing head; an analogous fault can occur during the double seaming of cans. Sealing problems are encountered especially when flat or unprofiled heat sealing heads are used, as these can trap material under them during sealing. Some heat seal heads are profiled or ultrasonic sealing is used to move residues out of the seal area during the sealing cycle; their effectiveness is determined by head and power delivery characteristics and the nature of the product. Careful choice of seal head may not always be an effective solution to the problem if food remains in the seal area, preventing formation of a continuous weld or causing a bridge to be formed across the seal. Problem fills are fat, water (which turns to steam on heating) and hard cellulosic fibres, such as celery.

Modified atmosphere packs

Some products designed for chilled storage rely on the gas (modified atmosphere – MA) surrounding the product forming part of the preservation system (e.g. low pO_2, high pN_2 and CO_2). This can give a considerable extension of shelf-life at chilled temperatures, for example with chilled meats, or prevent oxidative changes in long shelf-life ambient packs. In pasteurised products aerobes may be inhibited if oxygen is removed from the headspace by vacuum packing and then diffusion of air back is prevented by the barrier properties of the packaging film. Limited microbial metabolism in the pack may lead to the build-up of CO_2 and the depletion of oxygen. In such packs the potential for growth of anaerobic pathogens must always be accounted for in the design of shelf-life and distribution temperatures and times. The efficiency of flushing and replacement of oxygen with a gas mixture, control of vacuum level and the frequency with which leaking packs are produced are key operational parameters for this type of packing. Colour indicators have been advocated to indicate leaking packs (Ahvenainen *et al.*, 1997).

Vacuum or MA packs that are not gas-tight because of a faulty seal (leakers) will have a shortened shelf-life and present increased risks of contamination. This fault may be caused by incorrect choice of packaging film, or equipment faults (e.g. mis-setting of the sealing head temperature, pressure alignment or dwell time). The fault can be seen, or felt, soon after manufacture as the pack may feel soft if squeezed. If packs with an incorrect headspace volume or with weakened seals are processed in systems (such as retorts or ovens) that generate a pressure differential between the pack and its surroundings, bursting or pack weakening may result.

Other types of pack

Some short shelf-life products do not need bacteria-tight packs. Packs with a crimped seal between the lid and the base container (which may be made of aluminium or C-PET) may be used to provide a container for oven cooking or reheating by the customer. These packs carry an increased risk of product contamination, unless over-wrapped.

Any packaging materials coming into direct contact with the products must not contaminate them either chemically, physically or microbiologically. There is legislation governing the materials that may be used for food contact packaging. Toxicological clearance is linked to usage, as temperature plays a major role in determining migration of chemicals from packaging into food. Packs for microwave or oven heating carry very different risks to a pack used for the distribution of chilled meat, as they will be subjected to different conditions during use. Over-wrapping of the primary or food contact packaging and its handling within the storage and production areas should be designed to minimise the chances of contamination and pack damage.

Secondary packaging
Once the product is in its primary packaging, the pack needs to be protected from damage. To do this primary packs normally have additional, secondary packaging, which usually carries the coding needed for distribution. This packaging may be decorative and will protect it during handling or transport. For this latter function, control of the secondary pack characteristics should be part of the factory QA system.

For any product, the strength of the primary and secondary packs must be strong enough to maintain primary pack integrity and prevent damage during automated handling and transport on pallets. When chilled products are handled or marketed either in boxes or closely packed on pallets, there is only a limited opportunity for temperature change – as the surface area to volume ratio is unfavourable to rapid temperature changes, and rates of heat penetration through product and packaging are low. Therefore it is essential that product in its primary packaging is at the target temperature prior to secondary packaging and palletisation.

13.9 Control systems

13.9.1 Instrumentation and calibration

Wherever control limits and targets are in the specifications or HACCP plan, it is essential that reliable instrumentation or measurement procedures are present, correctly located and calibrated and records are kept. Their function is to ensure that the process is known to be within pre-set limits, so that the designed standards of product safety and quality are met. Sensors need to be prepared, installed and monitored, to prevent errors in readings (Sharp, 1989). Outputs can be used either for the control of process conditions (such as during sterilisation, pasteurisation, chilling or storage) or for monitoring compliance with specifications. Key outputs include temperature, time, flow and pressure measurements. For specialised pieces of equipment, such as retorts and freezers, other measurements may be important (e.g. water level, heat exchange medium or product flow or pack rotation in a retort). Sensors and their associated instruments may be in-line (e.g. oven, heat exchanger or fridge thermometers and timers or time/temperature recorders), at the side of the line (e.g. pack seam measurement, drained weight apparatus, salt or pH meters) or in the laboratory (e.g. colour measurement, gas composition or nitrogen determination). Wherever the instrument is situated, it needs to be maintained, with its sensor kept free of product debris or coatings that interfere with its signal – as this may produce errors. All sensors and equipment need regular calibration, and operatives need to have training on measurement and recording procedures.

Under quality control systems, product quality was ensured through inspection of product; acceptance or rejection was based on how well each batch met its specifications. To meet the needs of QA systems, various tools have been used to control, monitor and analyse the performance of the production process and especially critical process stages (e.g. CCPs). The information is used to predict variations and deviations that may result in rejected product, usually on a real-time basis. These tools often include statistical techniques to analyse variation around the targets, or limits, fixed in the design and specifications and hence ensure the line's ability to consistently manufacture the product as designed.

The group of production focussed techniques are known as statistical process control (SPC, see http://www.statit.com/statitcustomqc/StatitCustomQC_Overview.pdf), and similar statistical techniques can be used for sampling plans, the design of pilot plant and other R&D studies, line capability analysis and process improvement plans. Any SPC technique should distinguish between the normal variation in a process and abnormal or unacceptable variation. SPC cannot improve a poorly-designed product, but it can ensure it is made consistently as designed.

SPC techniques show whether a process operates under control and the product is manufactured as designed. Its main tool is the control chart, a graphical representation of the performance of key process parameters (e.g. flow or temperature). The control chart shows a comparison of actual performance against target over time, this generates many points forming a distribution, which can be used to detect any unusual variation in the process and indicate potential problems. Hence control charts are used to analyse process capability and for continuous improvement.

13.9.2 Supply chain monitoring and verification

The performance of manufacturing, storage and distribution operations in the supply chain should be controlled and monitored to ensure the whole process functions continually within agreed limits. For verification to be effective in protecting businesses and customers, it has to be based on product designs and supply chain operations that have been validated (see Section 13.7.3). Wherever possible the data from control systems and material testing should be recorded and used to produce information for control, quality assurance and trend analysis. Effective control along the supply chain may need relevant data to be exchanged between trading partners on a regular and frequent basis, because responsibilities for the safety and quality are often shared between suppliers (e.g. of ingredients or packaging) and producers. Even if valid process and product design principles cover realistic hazards, unsafe products can result if specifications are not met, or procedures are not carried out correctly, are not effective or have not been reviewed to take account of changed circumstances or a new hazard.

Routine auditing can support verification of the day-to-day performance of the supply chain and its HACCP plan (see Section 13.9.6). As a minimum, verification should show the producer's performance at each pre-requisite and CCP (e.g. control criteria, critical limit values, monitoring data and corrective actions) and key aspects of product quality. Operational risks highlighted during verification may include poor training and management procedures, poor hygiene, inadequate management of key process stages (e.g. heating, cooling, segregation or packaging) and the occurrence or acceptance of defective products. Process stages where safety risks are significant and those showing variability should be covered most frequently. In-house QA departments, regulatory authorities, specialist auditors and those inspecting suppliers on behalf of customers are usually responsible for this, and they should work on a continuing basis to review compliance with systems, procedures and the other outputs of the HACCP plan, or in some cases ISO 9000 series documentation.

13.9.3 Data from processes and samples

Process control data are the foundation of quality assurance and process improvement systems. Therefore its collection and analysis is critical to ensuring product safety. It should be taken at a defined frequency and kept for a period equal to at least the shelf-life plus the period of use of the product. Process control records may be generated and retained according to the framework proposed in the ISO 9000 documents on quality management. For many foods microbiological testing after manufacture does not give satisfactory assurance of safety for two reasons. Firstly, sufficient samples cannot be analysed to provide any confidence that processing and ingredient quality were under control for the duration of processing. Secondly, the time for a microbiological result may be longer than the pre-despatch time in the factory or the shelf-life, even when rapid methods are used.

In conjunction with relevant process data, the results of microbiological testing should be used for supplier monitoring, trend analysis of process control and hygiene and for 'due diligence' purposes. In some cases they may be used to supplement process data taken for verification and may provide additional criteria for product release, where process data are insufficient.

Conformance samples of products may be taken to track long-term changes in quality. This is not a problem with long shelf-life products, but with short shelf-life products, some manufacturers retain frozen samples and accept the impact of quality changes caused by freezing. Others take end-of-shelf-life samples and score their sensory attributes against fixed scales or parameters.

13.9.4 Lot tracking

Unit packs of food products and batches of ingredients and packaging should be coded to allow production lots, batches of ingredients and process periods to be identified. This is a requirement within the EU. Documentation and coding should allow any batch of finished product to be correlated with its processing period, raw materials and packaging used and with any corresponding process data and analytical/laboratory records. In practice, the better defined the lot tracking system, the better the chances of identifying and minimising the impact of a manufacturing or ingredient fault on the amount of product potentially at risk, but traceability costs money and an economic balance has to be found. Minimum traceability is usually taken as one step 'up' and one step 'down' the supply chain. Consideration should be given to allocating each batch of ingredients or process period a reference code to identify it during processing and storage. Deliveries of raw materials, any re-work and packaging should be stored so that they can be identified and identities do not become lost. If there is a fault, re-call, or good coding and tracking procedures and obvious labelling will facilitate responding to complaints and ensuring customer safety.

13.9.5 Training, operatives, supervisors and managers

All staff involved in the manufacture of food products should be adequately trained because they make an essential contribution to the control of product safety and quality. The best way of achieving this is by a period of formal or standardised training at induction, which will enable them to make their contribution to assuring the safe manufacture of high-quality products (Engel, 1998; Mortimore and Smith, 1998). After training, they should at least understand the critical aspects of hygiene for food handlers, product composition, specifications and procedures for processing, process control and records and the prevention of re-contamination. Many manufacturing operations will be done along process lines and will involve teamwork; therefore all staff must be trained in, and understand, the reasons for process controls, hygiene standards and procedures. Staff and supervisors should have defined responsibilities for ensuring that risks of making unsafe product are minimised. This is critical if products are ready-to-eat or will only receive minimal heat processing by the customer or if they are ambient-stable low-acid canned foods. The role of external HACCP consultants has been identified (White, 1998) as being especially useful to small food companies in helping them put together effective programmes for implementing HACCP and staff training. Training should also be provided for QA, laboratory managers and microbiologists to ensure that they have the required skills to advise on relevant control procedures and on data generation/challenge studies that may be used to underpin an approved product/process design or operational and cleaning procedures.

13.9.6 Process auditing

Auditing is the collection of data or information about a process or factory by a technical visit to the premises involved and/or a review of factory data. Internal or external auditors may do this, technical checklists may be used and often a scoring system or noting of non-compliances may be used for assessment of customer–supplier or departmental relationships. An integral part of an audit is to see the line operating; effective auditing does more than inspect records. It may be used to determine whether the HACCP plan and supply chain controls are correctly established, effectively implemented and are achieving their objectives. This is especially important where product safety relies on many aspects of a process being integrated and continually under control (Sperber, 1998; van Schothorst, 1998).

Any technical visit should review existing process and product-related data and findings from line visits and then make comparisons with specifications and any technical requirements of customers. Unsuitable process features that should be identified by auditing include unhygienic equipment and working practices, and equipment controls or layouts that may lead to defective products, contamination or unsuitable performance of distribution systems. Any changes in equipment performance noted should be corrected and revalidated before the process is considered 'under control' again.

Topics covered by record review should include the company structure, policy on quality and safety and the supporting systems, product validation, the capability and management of product and process development (including the safe design of products and processes), receiving, handling and processing of raw and packaging materials. Design, control, operation and maintenance of manufacturing equipment and the layout should be examined. Because of their importance, distribution and logistics should also be considered. The contribution and effectiveness of the QA department, the laboratory and any outsourced functions to the management of the operation and to training should be assessed.

A more recent development is supplier self-auditing in supplier–customer relationships with a partnership basis and high degree of trust. This starts with a clear statement of trading objectives and the resources and products involved. Next, an evidence package covering specifications and records is agreed between the supplier and customer. Because quality and process control may vary, a mechanism for managing this must also be pre-agreed, to prevent false rejection of product and enable the realistic management of any safety risks or non-compliances. Successful operation of such systems relies on the identification and allocation of responsibilities and ownership of critical technical factors by both organisations, so that the personnel best placed to make decisions can always be identified. Procedures for managing process breakdowns and other deviations must be developed and audited to ensure that risks to consumers are minimised.

Self-auditing offers lower costs than traditional audit visits and can produce information for decision making on a continuing basis; if properly handled it provides a better level of consumer protection than end-product sampling and testing. Externally accredited auditors may be placed in companies, but their ability to examine a process and derive the optimum system can be limited by access to commercially sensitive data on processing.

13.10 Conclusions

A major expansion in the sale of processed food products, such as ready-prepared foods and ingredients, continues globally to meet the constantly increasing requirements of consumers and food-service customers for convenience, variety and less severe processing and preservation. The supply chain is now supplied, and sells, on a global basis. This introduces new microbiological hazards and supplies to outlets with different degrees of control over handling plus the challenge of customers using products in different ways. Foods are processed industrially and at home using less heat; in many countries the link between cooking and food safety has been lost in the minds of consumers. Foods also contain lower levels of preservatives and less robust packaging, and these changes are not immediately compatible with an improvement in microbial stability or safety. Indeed, consumer preference has led to a lessening of the intrinsic preservation (e.g. lower salt, less acid) or stability of foods and an increase in reliance on external factors, such as in-factory decontamination and the cold chain. Hence there is greater reliance on safe product and process design and the reliable operation of all the unit operations along the supply chain. Conscientious review of the design of products and processes against new microbiological challenges and different consumer needs and expectations, with reliable monitoring and corrective action procedures is the only way to provide continuing supply of safe food and clear specifications for processing and distribution.

The assurance of food safety is an essential consumer requirement and therefore of paramount importance for the producer. It is here that modern design, processing and distribution techniques backed up by risk assessment and management techniques, such as HACCP (Mayes, 1992), properly applied, can more than compensate for the stability and safety that otherwise could be lost. Although the management of safety and prevention of new hazards is complex in detail, as indicated above, the principles of effective and safe design and processing of food products are few. They include:

- the reliable identification of hazardous microorganisms, so that design conditions will allow them to be controlled;
- design of products where safe use is obvious to consumers;
- clear identification of the criteria required for safe products, e.g. numbers or presence of pathogens or process/product conditions ensuring their numbers are below a harmful level;
- controlled and predictable raw material quality and processing leading to the reliable elimination, inactivation or inhibition of microorganisms by processes, storage conditions, product formulations and consumer use;
- control and management of process stages critical to the control of pathogens in the product and the assurance of traceability, to minimise risks in the event of a process or material problem;
- the prevention of re-contamination and cross-contamination after decontamination;

- control of the survival or persistence of any microorganisms or toxins that remain in the food;
- provision of robust packaging systems able to prevent contamination or loss of preservation conditions during distribution, in each market;
- the provision of clear consumer use instructions that are compatible with customer expectations and habits.

Safe, effective processing and distribution of any food product depends on consistent, soundly-based and robust control of these elements by the food industry. Although these principles are effective if properly carried out, their practical effectiveness can be further improved in the future (e.g. by better and wider implementation of pre-requisites and HACCP) and this may become necessary to meet consumer expectations and match their perception of safety.

13.11 Future trends

Supply chains will lengthen in the future and manufacturing sites will become larger scale. This provides opportunities for better equipment and control systems; however, it means if a mistake is made, it will be a big one. Consumer pressure for milder processing and more environmentally-friendly (i.e. less and weaker) packaging will increase risks, unless managed well by manufacturers and pressure groups. As more complex (e.g. multi-hurdle and new technology) preservation systems are used the difficulties of validating their performance will increase. With 'simple' or conventional (e.g. heat) preservation systems there are options between modelling their performance, where the interaction with food systems is known with some certainty, and doing challenge tests. With multi-factor or novel systems (e.g. natural antimicrobials based on large molecules or ultra high pressure) then key variables are much less well known and challenge testing is essential but, additionally, the impacts of microbial and food variability have to be covered to ensure the effectiveness of the chosen treatment under production conditions. There remains a big job to be done in agreeing with consumers and customers where sustainable positions are with respect to risk, milder processing and less secure packaging. As sourcing becomes increasingly global, the pressures on quality systems also increase as the range of microbiological hazards increases, with expansion in the range of growing and harvesting areas and needs for training of food handlers, especially in hygiene, also expand.

13.12 References and further reading

ACDP (1996) *Microbiological Risk Assessment: An Interim Report*, Advisory Committee on Dangerous Pathogens, The Stationery Office, London.

Adair C and Briggs P A (1993) The concept and application of expert systems in the field of microbiological safety, *Journal of Industrial Microbiology*, **12**, 263–7.

Ahvenainen R, Eilamo M and Hurme E (1997) Detection of improper sealing and quality deterioration of modified atmosphere packed pizzas by a colour indicator, *Food Control*, **8**, 177–84.

Amézquita A, Wang L and Weller C L, (2005a), Finite element modeling and experimental validation of cooling rates of large ready-to-eat meat products in small meat-processing facilities, *Transactions of the ASAE*, **48**, 287–303.

Amézquita A, Weller C L, Wang L, Thippareddi D H and Burson D E (2005b) Development of an integrated model for heat transfer and dynamic growth of *Clostridium perfringens* during the cooling of cooked boneless ham, *International Journal of Food Microbiology*, **101**, 123–44.

Augustin J C, Rosso L and Carlier V (2000) A model describing the effect of temperature history on lag time for *Listeria monocytogenes*, *International Journal of Food Microbiology*, **57**, 169–81.

Baranyi J and Tamplin M L (2004) ComBase: a common database on microbial responses to food environments, *Journal of Food Protection*, **67**, 1967–71.

Betts G (1997) Predicting microbial spoilage, *Food-Processing UK*, **66**, 23–4.

Brackett R E (1992) Microbiological safety of chilled foods: current issues, *Trends in Food Science & Technology*, **3**, 81–5.

CAC (2000) *Proposed Principles and Guidelines for the Conduct of Microbiological Risk Management*, Food and Agriculture Organization of the United Nations/World Health Organization, Rome/Geneva.

Carlin F, Guinebretiére M H, Choma C, Pasqualinin R, Bracconier A and Nguyen-The C (2000) Spore-forming bacteria in commercial cooked, pasteurised and chilled vegetable purées, *Food Microbiology*, **17**, 153–65.

CCFRA (2002) *Thermal processing: validation challenges*, CCFRA Conference Proceedings, Campden & Chorleywood Food Research Association, Chipping Campden.

CCFRA (2004a) *Evaluation of product shelf-life for chilled foods*, CCFRA Guideline No. 46, Campden & Chorleywood Food Research Association, Chipping Campden.

CCFRA (2004b) *Thermal processing: process and package innovations for convenience foods*, CCFRA Conference Proceedings, Campden & Chorleywood Food Research Association, Chipping Campden.

CCFRA (2005) *Guidelines on air quality standards for the food industry* (2nd edn) Guideline No. 12, Campden & Chorleywood Food Research Association, Chipping Campden.

CFA (2006) *Best Practice Guidelines for the Production of Chilled Foods*, Chilled Food Association, Peterborough.

CFIA (2008) *Food Safety Enhancement Program Manual*, Canadian Food Inspection Agency, Ottawa, available at: http://www.inspection.gc.ca/english/fssa/polstrat/haccp/manue/manuche.pdf, accessed December 2008.

Cowey P (1997) Anti-microbial conveyer belts, *Food-processing UK*, **66**, 10–11.

Dalgaard P, Buch P and Silberg S (2002) Seafood Spoilage Predictor – development and distribution of a product specific application software, *International Journal of Food Microbiology*, **73**, 343–9.

De Cordt S, Avila I, Hendrickx M and Tobback P (2004) DSC and protein-based time-temperature integrators: Case study of α-amylase stabilized by polyols and/or sugar, *Biotechnology and Bioengineering*, **44**(7), 859–65.

De Jong A E I, Beumer R R and Zwietering M H (2005) Modeling growth of *Clostridium perfringens* in pea soup during cooling, *Risk Analysis*, **25**, 61–73.

DHSS (1984) *Guidelines on Precooked Chilled Foods*, HMSO, London.

DHSS (1986) *Health Service Catering; Hygiene*, HMSO, London.

Doyle M P (1998) Effect of environmental and processing conditions on *Listeria monocytogenes*, *Food Technology*, **42**, 169–71.

Dressler D (2005) Botulismus durch Raucherlachsverzehr, *Der Nervenartz* **76**, 763–6.

EC (2005) Commission Regulation (EC) No 2073/2005 of 15 November 2005 on microbiological criteria for foodstuffs, *Official Journal of the European Union*, **L338**, 22 December, 1–26.

Engel D (1998) Teaching HACCP – theory and practice from the trainer's point of view, *Food Control*, **9** (2/3), 137–40.

Giffel M C T E, Jong P De and Zweitering M H (1999) Application of predictive models as a tool in the food industry, *New-Food*, **2**(38), 40–1.

Gould G W (1992) Ecosystem approaches to food preservation, *Journal of Applied Bacteriology*, **73**, Symposium Series No. 21, 585–685.

Guiavarc'h Y P, Deli V, Van Loey A M and Hendrickx M E (2002) Development of an enzymic time temperature integrator for sterilization processes based on *Bacillus licheniformis* α-amylase at reduced water content, *Journal of Food Science*, **67**(1), 285–91.

ILSI (2004) *A Simple Guide to Understanding and Applying the Hazard Analysis Critical Control Point Concept*, 3rd edn, International Life Sciences Institute, Brussels, available at: http://europe.ilsi.org/NR/rdonlyres/7FAE0B4A-53F9-417F-BF33-BF891CBFB1BC/0/ILSIHACCP3rd.pdf, accessed November 2008.

Impe J F-Van, Nicolai B M, Martens T, Baerdemaeker J-De and Vandewalle J (1992) Dynamic mathematical model to predict microbial growth and inactivation during food processing, *Applied and Environmental Microbiology*, **58**, 2901–9.

Jaczynski J and Park J W (2003) Predictive models for microbial inactivation and texture degradation in surimi seafood during thermal processing, *Journal of Food Science*, **68**(3), 1025–30.

James S J and Evans J A (2006) Predicting the reduction of microbes on the surface of foods during surface pasteurisation – the 'BUGDEATH' project, *Journal of Food Engineering*, **76**, 1–6.

James S J, Burfoot D and Bailey C (1987) The engineering aspects of ready meal production, in *Engineering Innovation in the Food Industry*, Institute of Chemical Engeers, Bath, 21–8.

Jaykus L A (1996) The application of quantitative risk assessment to microbial safety risks, *Critical Reviews in Microbiology*, **22**, 279–93.

Jouve J L, Stringer M F and Baird-Parker A C (1998) *Food Safety Management Tools*, Report ILSI Europe Risk Analysis in Microbiology Task Force, International Life Sciences Institute, Brussels.

Labuza T P and Fu B (1995) Use of time/temperature integrators, predictive microbiology, and related technologies for assessing the extent and impact of temperature abuse on meat and poultry products, *Journal of Food Safety*, **15**, 201–27.

Mayes T (1992) Simple users' guide to the hazard analysis critical control point concept for the control of food microbiological safety, *Food Control*, **3**, 14–19.

Membré J-M, Johnston M D, Bassett J, Naaktgeboren G, Blackburn C de W and Gorris L G M (2005) Application of modelling techniques in the food industry: determination of shelf-life for chilled foods, *Acta Horticulturae (ISHS)*, **674**, 407–14.

Membré J M, Amézquita A J, Bassett J, Giavedoni P, Blackburn C de W and Gorris L G M (2006) A probabilistic modeling approach in thermal inactivation: estimation of postprocess *Bacillus cereus* spore prevalence and concentration, *Journal of Food Protection*, **69**(1) 118–29.

Mendoza T F, Bruce A Welt B A, Kenneth Berger K, Otwell S, Kristonson H (2002) Verification of time temperature integrators as science based controls for safe use of vacuum packaged seafood, *Proceedings of the 6th Joint Meeting of SST and AFT Advancing Seafood Technology for Harvested and Cultured Products*, October 9–11, Orlando, FL, available at: http://sst.ifas.ufl.edu/26thAnn/file41.pdf, accessed November 2008.

Miller A J, Whiting R C and Smith J L (1997) Use of risk assessment to reduce listeriosis incidence, *Food Technology*, **51**(4), 100–3.

Mortimore S E and Smith R (1998) Standardized HACCP training: assurance for food authorities, *Food-Control*, **9**(2/3), 141–5.

NACMCF (2005) Considerations for establishing safety-based consume-by date labels for refrigerated ready-to-eat foods, National Advisory Committee on the Microbiological Safety of Foods, *Journal of Food Protection*, **68**, 1761–75.

NFPA (1993) *Guidelines for the Development, Production, Distribution and Handling of Refrigerated Foods*, National Food Processors Association, New York.

Notermans S, Dufrenne J and Lund B M (1990) Botulism risk of refrigerated, processed foods of extended durability, *Journal of Food Protection*, **53**, 1020–4.

Notermans S, Mead G C and Jouve J L (1996) Food products and consumer protection: a conceptual approach and a glossary of terms, *International Journal of Food Microbiology*, **30**, 175–85.

Panisello P J and Quantick P C (1998) Application of Food MicroModel predictive software in the development of Hazard Analysis Critical Control Point (HACCP) systems, *Food Microbiology*, **15**(4), 425–39.

Pin C and Baranyi J (1998) Predictive models as means to quantify the interaction of spoilage organisms, *International Journal of Food Microbiology*, **41**, 59–72.

Ross T and McMeekin T A (2003) Modeling microbial growth within food safety risk assessments, *Risk Analysis*, **23**, 179–97.

Rutherford N, Phillips B, Gorsuch T, Mabey M, Looker N and Boggiano R (1995) How indicators can perform for hazard and risk management in risk assessment of food premises, *Food Science and Technology Today*, **9**(1), 19–30.

Sastry S K and Cornelius B D (2002) *Aseptic Processing of Foods Containing Particulates*, Wiley, New York.

Schultze K, Linton R, Cousin M, Luchansky J B, Tamplin M L (2006) A predictive model to describe the effects of temperature, sodium lactate and sodium diacetate on the inactivation of a serotype 4b strain of *Listeria monocytogenes* in a frankfurter slurry, *Journal of Food Protection*, **69**(7), 1552–60.

Serp D, Von Stockar U and Marison I W (2002) Immobilized bacterial spores for use as bioindicators in the validation of thermal sterilization processes, *Journal of Food Protection*, **65**, 1134–41.

Sharp A K (1989) The use of thermocouples to monitor cargo temperatures in refrigerated freight containers and vehicles, *CSIRO Food Research Quarterly*, **49**, 10–18.

Snyder O P Jr (1992) HACCP – an industry food safety self-control program. XII. Food processes and controls, *Dairy, Food and Environmental Sanitation*, **12**, 820–3.

Sperber W H (1998) Auditing and verification of food safety and HACCP, *Food Control*, **9**, 157–62.

Tucker G S, Lambourne T, Adams J B and Lach A (2002) Application of a biochemical time-temperature integrator to estimate pasteurisation values in continuous food processes, *Innovative Food Science and Emerging Technologies*, **3**, 165–74.

Van Schothorst M (1998) Introduction to auditing, certification and inspection, *Food-Control*, **9**, 127–8.

Walker S J and Stringer M F (1990) Microbiology of chilled foods, in T R Gormley (ed.), *Chilled Foods: The State of the Art*, Elsevier Applied Science, London, 269–304.

Walls I and Scott V N (1997) Use of predictive microbiology in microbial food safety risk assessment, *International Journal of Food Microbiology*, **36**, 97–102.

White L (1998) The role of the HACCP consultant, *International Food Hygiene*, **9**, 29–30.

Zwietering M H and Hasting A P M (1997) Modelling the hygienic processing of foods – a global process overview, *Food and Bioproducts Processing*, **75**(C3), 159–67.

Zwietering M H, Koos J T-de, Hasenack B E, Wit J C-de and Riet K-Van T (1991) Modelling of bacterial growth as a function of temperature, *Applied and Environmental Microbiology*, **57**, 1094–101.

Zwietering M H, Cuppers H G A M, Wit J C-de and Riet K-Van T (1994a) Evaluation of data transformations and validation of a model for the effect of temperature on bacterial growth, *Applied and Environmental Microbiology*, **60**, 195–203.
Zwietering M H, Cuppers H G A M, Wit J C-de and Riet K-Van T (1994b) Modelling of bacterial growth with shifts in temperature, *Applied and Environmental Microbiology*, **60**, 204–13.

14

The effective implementation of HACCP systems in food processing

A. Mayes, Unilever Colworth, UK, and S. Mortimore, Land O'Lakes Inc, USA

Abstract: 'It is widely acknowledged that, although absolute safety in food production is unattainable, effective Hazard Analysis Critical Control Point (HACCP) implementation is the surest way of delivering safe food.' While HACCP is relatively well developed in some sectors, notably larger food processors in developed countries, it is still far from widespread in others, particularly in primary producers, small and medium-sized enterprises, most notably in the service and catering sectors and in developing countries. However, even where it has been established longest, among larger food manufacturers in Europe and the USA for example, the quality of implementation has varied. Based on the experience of those who have implemented HACCP systems in practice, this chapter outlines the key issues in HACCP implementation identified by HACCP practitioners and how they can be addressed in building a truly effective food safety regime.

Key words: Hazard Analysis Critical Control Point (HACCP) implementation.

14.1 Introduction

'It is widely acknowledged that, although absolute safety in food production is unattainable, effective Hazard Analysis Critical Control Point (HACCP) implementation is the surest way of delivering safe food' (Mayes and Mortimore, 2001). The development of HACCP from its initial conception by Pillsbury working with Natick is well documented. Initially developed as a means of ensuring the microbiological safety of foods used in the space programme, its wider potential for ensuring the control of microbiological and other food safety hazards in a broad range of foods in manufacturing, distribution, food service and retail situations was soon realised. Initially the HACCP concept developed in a somewhat piecemeal fashion, and it was not until the mid- to late-1980s that HACCP development began to accelerate in a coordinated way. The HACCP concept was endorsed by the WHO/FAO as an effective way of controlling foodborne disease in 1983 when the Joint FAO/WHO Expert Committee on Food Safety advised that HACCP should replace the traditional end-product testing approach to food safety assurance. Since then there have been many guidelines issued on the principles of HACCP implementation (e.g. ICMSF, 1988; NACMCF, 1992; CCFRA 1997; ILSI, 1997).

It was, however, only during the late 1980s and early 1990s that there was a concerted international attempt to harmonise HACCP principles and terminology. The current Codex Alimentarius HACCP document (CAC, 1997) can be seen as the first authoritative, internationally agreed document both agreeing HACCP principles and providing guidelines for their implementation in practice. This document identifies seven principles underlying the HACCP system and provides guidelines listing 12 tasks for the implementation of these principles (Tables 14.1 and 14.2). The document is widely seen as the international benchmark for effective HACCP implementation.

The seven principles, and in particular the 12 'guidelines' tasks listed in Table 14.2, outline the minimum practical steps needing to be taken to implement HACCP. Although not specifically listed in the above table, the CAC guidelines make much of the philosophy behind HACCP and the need to have the correct balance of skills and experience in the team that prepares the HACCP plan and the commitment needed to successfully implement and maintain the plan. Although there are some differences in emphasis on the guidelines for HACCP implementation in some publications (see later) the CAC document is widely seen as the international benchmark for effective HACCP implementation. While HACCP is relatively well developed in some sectors, notably larger food processors in developed countries, it is still far from widespread in others, particularly in primary producers, small

Table 14.1 Seven principles of the HACCP system

Principle 1	Conduct a hazard analysis
Principle 2	Determine the critical control points (CCPs)
Principle 3	Establish critical limit(s)
Principle 4	Establish a system to monitor control of the CCP
Principle 5	Establish the corrective action to be taken when monitoring indicates that a particular CCP is not under control
Principle 6	Establish procedures for verification to confirm that the HACCP system is working effectively
Principle 7	Establish documentation concerning all procedures and records appropriate to these principles and their application

Table 14.2 Guidelines for the application of the HACCP system

Task 1	Assemble the HACCP team
Task 2	Describe product
Task 3	Identify intended use
Task 4	Construct flow diagram
Task 5	On-site verification of flow diagram
Task 6	List all potential hazards, conduct a hazard analysis, determine control measures
Task 7	Determine CCPs
Task 8	Establish critical limits for each CCP
Task 9	Establish a monitoring system for each CCP
Task 10	Establish corrective action for deviations that may occur
Task 11	Establish verification procedures
Task 12	Establish record keeping and documentation

and medium-sized enterprises (SMEs), most notably in the service and catering sectors, and in developing countries. The problems of implementing HACCP systems in the SME sector have, for example, been identified by the WHO (WHO, 1993, 1999). Studies in the UK, for example, confirm a relative lack of investment by SMEs in HACCP (Gormley, 1999; Mortlock *et al.*, 1999; Panisello *et al.*, 1999), but research suggests that SMEs can utilise it if provided with sufficient guidance and support, which is the approach that the UK government has taken in recent years (Taylor and Kane, 2005). However, even where it has been established longest, among larger food manufacturers in Europe and the USA for example, the quality of implementation has varied (Kane, 2001).

What makes for effective HACCP implementation? Based on the experience of those who have implemented HACCP systems in practice, this chapter outlines the key issues in HACCP implementation identified by HACCP practitioners and how they can be addressed in building a truly effective food safety regime.

14.2 Hazard Analysis Critical Control Point system (HACCP) methodology and implementation

Although there is broad agreement on basic HACCP methodology, some differences in emphasis can be seen when comparing guidelines on HACCP implementation from three leading authorities:

- the Codex Alimentarius (CAC, 1997);
- Campden and Chorleywood Food Research Association (CCFRA, 1997);

- Those provided by one of the most respected texts in this field: Mortimore and Wallace's *HACCP: A Practical Approach* (1998).

These approaches are compared in Table 14.3. For the purposes of comparison, only the basic headings provided by Codex and the CCFRA are given. These are then related to the detailed step-by-step approach in *HACCP: A Practical Approach*. While the central stages covering HACCP planning and design are broadly similar across all three sets of guidelines, it is noticeable that *HACCP: A Practical Approach* lays particular emphasis on the preparatory work required, such as training in HACCP methodology and conducting a baseline audit. The importance of effective preparation is endorsed, for example, by the New Zealand government which suggests the need to agree the scope of the HACCP plan and to set appropriate food safety objectives (Lee, 2001: p. 146). As the following sections show, the experience of HACCP practitioners confirms the importance of such preparation, whether in the importance of training, defining the scope of a HACCP study (for example, what should be covered by pre-requisite programmes-PRPs), or conducting a baseline audit of a company's existing operations.

There are a number of areas that influence effective HACCP implementation:

- motivation;
- knowledge;
- resources and planning;
- pre-requisite systems;
- HACCP teams and analysis;
- implementation;
- maintenance.

These areas are discussed in the following sections.

14.3 Motivation

As with any significant new system within a company, there needs to be the requisite motivation to take on the challenge and follow it through. This is particularly important in the case of HACCP because motivation lies at the heart of any HACCP system. HACCP requires a proactive approach at all levels of an organisation. The impetus to develop a HACCP system typically comes from a number of drivers, including:

- customer pressure;
- regulatory requirements;
- desire for self-improvement.

Many SMEs in particular are inherently suspicious of HACCP for a range of reasons. They may be ignorant of food safety issues and have

Table 14.3 Approaches to HACCP implementation

Codex guidelines	Mortimore and Wallace	CCFRA manual
	Stage 1 **Preparation and planning**	
1 Assemble HACCP team	• Understanding HACCP concept • Identifying and training HACCP team • Baseline audit • Project planning (incl. improving pre-requisite systems)	2 Assemble HACCP team
	Stage 2 **HACCP studies and planning**	
	• Terms of reference	1 Define terms of reference
2/3 Describe product and intended use	• Describe product and intended use	3/4 Describe product and intended use
4/5 Construct and verify flow diagram	• Construct process flow diagram	5/6 Construct and verify flow diagram
6 Conduct hazard analysis, identify control measures	• Hazard analysis	7 Conduct a hazard analysis, identify control measures
7 Determine CCPs	• Identify CCPs	8 Identify CCPs
8 Establish critical limits	• Establish critical limits	9 Establish critical limits
9 Establish monitoring procedures	• Identify monitoring procedures	10 Establish monitoring procedures
10 Establish corrective action procedures	• Establish corrective action procedures	11 Establish corrective action procedures
	• Validate the HACCP Plan	
	Stage 3 **Implementing the HACCP plan**	
12 Establish records and documentation	• Determine implementation method	13 Establish documentation and record-keeping
11 Establish verification procedures	• Set up implementation team	12 Verification
	• Agree actions and timetable (inc. training; equipment; record keeping) • Confirm implementation complete • Verify implementation through audit	
	Stage 4 **Maintaining the HACCP system**	
	• Defined standards and regular audit • Ongoing maintenance • Data analysis and corrective action • HACCP plan re-validation • Update	14 Review the HACCP plan

The sequence of steps given by Codex and the CCFRA has been altered to fit the sequence suggested by Mortimore and Wallace for ease of comparison. The original sequences are indicated by the numbering of the steps in each case.

a low perception of the severity of hazards arising from their day-to-day operations. It has also been suggested that the early association of HACCP with larger manufacturers, who were the first to implement HACCP systems, encouraged the assumption that HACCP was a complex system only feasible for big companies with the appropriate skills and resources (Barnes, 2001: p. 211). This combination of a low perception of need and a fear of the difficulties of implementation is a potent barrier. In developing countries, consumers may not regard food safety as a major issue, reducing the commercial rationale for HACCP implementation in relation to the costs involved (Suwanrangsi, 2001: p. 190). In addition, the risk of prosecution for food safety offences may be low. In many cases, businesses face major weaknesses in infrastructure, such as poor-quality raw materials, inadequate facilities and equipment and an unreliable distribution system (Marthi, 2001: pp. 82–4). In these circumstances the demands of HACCP implementation can seem prohibitive in relation to the advantages it gives the business.

For businesses in developing countries and in the SME sector generally, the impetus for HACCP implementation may, therefore, come most strongly from external regulatory or customer pressure. A particularly important motor for change is pressure from customers, whether these be larger manufacturers supplied with semi-finished goods by a network of smaller suppliers, or retailers stocking a business's food products for sale to the consumer (Route, 2001: p. 32). As global trade has increased, businesses in developing countries seeking to export to regions such as North America or the EU have needed to develop HACCP systems to meet both import requirements and the supplier quality assurance (SQA) standards set by their retail and manufacturing customers. Such customers often have well-developed SQA schemes requiring suppliers to develop HACCP systems which are then regularly audited either by internal audit teams or accredited third-party auditors.

Whilst such external pressures have been an important way of disseminating the HACCP concept, such a pathway into HACCP can have significant disadvantages. If a business has been forced reluctantly to embrace HACCP because of pressure from outside, it is unlikely to have the level of commitment required for effective implementation. In a study of HACCP implementation in the Thai seafood export industry (Suwanrangsi, 2001: pp. 183–201), for example, businesses developing HACCP systems in reaction to the demands of their overseas customers initially relied on often conflicting customer advice and undiscriminating use of generic HACCP models. This reactive approach resulted in ill-considered, poorly understood and ineffective systems that did not reflect the particular requirements of the individual business. The real change in momentum came when businesses themselves came to see HACCP as essential for competitive success in export markets and as a basis for future growth and profitability. The best HACCP systems are developed by businesses that are self-driven, though a mandatory approach may be necessary to accelerate progress and ensure a more consistent approach and standards (Lee, 2001: pp. 142–3). Key reasons for HACCP to be embraced positively by a business include the following:

- acknowledgement of the importance of food safety and awareness of the rising trends in foodborne illness and their implications for both consumers and businesses;
- understanding the business benefits of HACCP in terms of customer reassurance, competitive advantage and process improvement;
- seeing HACCP for what it is: a 'minimal' system allowing businesses to understand and focus on the critical safety areas and use resources more effectively.

Once they have committed themselves to HACCP implementation, many businesses have realised the benefits it brings and have retained and improved their HACCP systems in preference to alternative systems. One study of Australian companies showed that, after three years, only half had retained their quality assurance (QA) systems while all had retained their HACCP systems (Sumner, 1999). SMEs in the UK have confirmed the value of HACCP as a more cost-effective alternative to the accreditation process for a quality standard such as ISO 9001 (Route, 2001: p. 33). This enthusiasm for HACCP systems derives from the real, and sometimes unexpected, advantages it can bring. Among these, those with practical experience of implementing HACCP systems have noted the following:

- increased sales because of assured product quality;
- a reduced level of auditing by customers with increased trust in a business's own systems;
- improved process control leading to reduced product recall and rework, improved productivity and lower costs;
- greater confidence in extending the range of products manufactured;
- a more proactive culture throughout the business generally with all staff more committed and more ready to suggest system improvements.

Examples of particular savings noted by individual businesses include over $200 000 a year at the US firm Cargill, and a 47 % increase in productivity in one of Hindustan Lever's ice cream plants in India owing, in part, to fewer line shutdowns and less defective product (McAloon, 2001: pp. 77–8; Marthi, 2001: pp. 91–2).

The issue of motivation remains central at all stages of HACCP implementation. As an example, at some point ownership of a HACCP system has to transfer to the line staff appointed as critical control point (CCP) monitors. The system will stand or fail on their commitment and understanding in monitoring and taking corrective action where required. Ways of motivating line staff are discussed later in the chapter. It is important that all senior managers are fully briefed on the importance of HACCP so that their cooperation and support are secured, particularly given the demands that HACCP planning and implementation will have on staff and resources (Killen, 2001: p. 122). As soon as the senior management team has recognised the value of HACCP to the organisation and decided to proceed, this commitment needs to be communicated outwards to other staff.

As a first step, the appointment of a senior person as HACCP 'champion' sends the right signal in disseminating a sense of importance and value of HACCP within the organisation. Another step is to integrate HACCP implementation into existing sets of objectives for management, setting HACCP-related goals which are monitored as part of a company's existing system for appraisal of its managers. This might even be reinforced through bonus schemes for good performance in safety management. The US firm Cargill, for example, encouraged awareness campaigns and newsletters at both corporate and plant level to emphasise the importance of food safety to all employees (McAloon, 2001: pp. 63–4). Changes such as this help to integrate HACCP into an organisation's existing management culture.

14.4 The knowledge required for HACCP

An essential consideration in using the HACCP approach is whether a company will have the knowledge and expertise needed to design, implement and maintain an effective program. Even before this, however, it is clear that a critical success factor for food safety is the operating culture within the organisation. A company with an effective food safety culture is not one that is merely able to produce a documented HACCP plan. In addition to this, it would be expected that there was evidence of strong pre-requisite programs (PRPs), including environmental monitoring, and a 24 hour 7 days a week mentality of hazard analysis. Every process, ingredient, formulation, equipment and plant layout change should be evaluated for its impact on food safety: is there for example a likelihood of a hazard occurrence that was not previously identified, and can it be controlled? Successful implementation of HACCP is most likely to occur in organisations where the culture promotes active involvement of all in delivering safe food. Ensuring access to the expertise needed for HACCP is another example of commitment to food safety and, whilst the concept of HACCP is relatively straightforward, a full understanding of HACCP methodology is more demanding. Scientific knowledge and sound judgement are needed to develop and maintain a program.

One reason that SMEs in particular give for their caution in approaching HACCP implementation is a lack of knowledge and expertise. Effective HACCP implementation depends on the following kinds of knowledge:

- understanding of HACCP methodology;
- knowledge of key hazards and control measures;
- expertise in the organisation's processes and technologies.

There is a useful role here for trade associations, research associations and consultants in providing knowledge that wouldn't normally be available to a small to medium sized organisation in particular. Being able to compare what others in a particular sector have done, or having access to sector hazard information can be of enormous value. The difficulty for many around the world is how to identify reliable sources of information and consultants from the very many that are available. This is where trade associations (in countries where they exist), can help their members by selecting and recommending references which are valid. They also frequently bring together groups of SMEs to work on generic model HACCP plans for their sector.

Experience of implementing HACCP systems in practice suggests a number of common areas of misunderstanding:

- a failure to distinguish between pre-requisite and HACCP systems;
- difficulty in separating out safety from other quality issues, and in prioritising hazards for attention;
- problems in identifying which control points are genuinely critical;
- overlooking hazards that are not biological and ones that are commonly encountered.

Such basic misunderstanding of HACCP methodology, combined with limited knowledge of relevant hazards and their control, produces inappropriate, over-complex and unmanageable HACCP systems. Such systems often tackle too many 'hazards' (for example spoilage and other quality problems which do not present a hazard to consumers, or pathogens which are not a problem for the particular sector in which the business operates) and are overloaded with control points, only a few of which are CCPs. Physical hazard analysis seems to be a difficult area too, differentiating between those contaminants that are truly significant hazards, those which could be legally classed as adulteration, and which are controlled through PRPs versus specific control measures. Chemical contaminants prove to be equally challenging, many chemicals being controlled through regulatory limits or being more typically associated with chronic illness – allergens and heavy metals being good examples of those which should be included in a HACCP plan but are frequently overlooked in countries where they are less well known as hazards. These inappropriate systems have given rise to the idea among some companies in the food industry that HACCP systems in general are complex, bureaucratic and burdensome – as opposed to being seen as an aid.

Lack of expertise in hazard analysis is still one of the most significant problems in effective HACCP implementation and not just in SMEs, though this has been the focus of recent attention. A UK study of SMEs in the ready meals sector, for example, identified that the employment of experienced technically qualified staff was the most important single factor in successful HACCP implementation (Holt, 1999). However, few companies can call on specialised microbiological expertise within the organisation, and a thorough knowledge of microbiological hazards and their methods of control can take years to acquire. Common problems resulting from lack of expertise in this area include the following:

- difficulty in prioritising risks, for example physical versus microbiological risks;
- an inability to identify which pathogens or agents present a significant risk;
- identifying which control points are genuinely critical in controlling hazards.

Ignorance of which hazards are most important and how they should be tackled produces either systems that fail to identify and control key hazards or over-complex and unwieldy systems that attempt to tackle all possible hazards. In one UK example, a technical manager from an SME initially identified over 140 'critical' control points on a process line, subsequently reduced to four CCPs after training (Taylor, 2001: p. 18). Such misconceived systems prove impossible to manage, generating confusion, overload and dilution of control.

Knowledge of its own products, processes and procedures should ideally present few problems for a business. However, even here knowledge may be diffused through various levels of an organisation. In many cases processes have evolved over a number of years and procedures have developed piecemeal with no central or formal record. Informal practices often develop, which differ significantly from those standards that do exist.

At a practical level, being unfamiliar with what is involved in implementing a HACCP system can make it very difficult to set a realistic budget for the project. HACCP is sometimes thought of as being expensive yet many are unaware of what costs are involved. This can be a barrier particularly to smaller businesses or those in developing countries. Costs are frequently incurred as a result of the need to upgrade pre-requisite hygiene programs at the start (Bas *et al.*, 2006; pp. 118–26), and also for training and motivating employees (Semos and Kontogeorgos, 2007, pp. 5–19). It could be argued that neither of these examples is directly attributable to HACCP but that HACCP is acting as the catalyst for improvement in the operating conditions at a basic level.

14.5 Initial training and preparation

There are many ways of finding out about HACCP systems and what they involve. A range of sources of information and advice which can help to get a business started is summarised in Fig. 14.1. Among these, effective training provides a first step in gaining a full understanding of HACCP methodology and can be a particularly worthwhile investment in making HACCP implementation successful. As an example, it can demonstrate the importance of distinguishing between pre-requisite and HACCP systems. Recognising the relationship between the two, and building up strong pre-requisite systems first as a foundation for HACCP implementation, has been

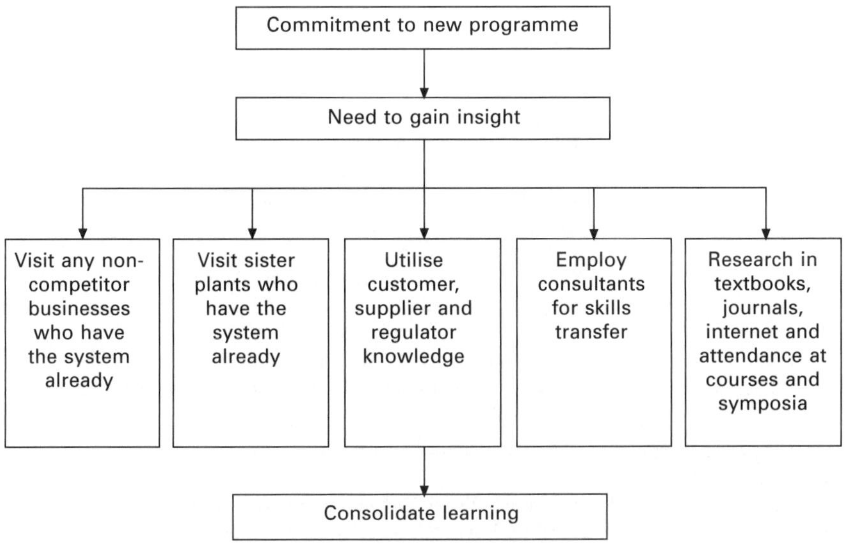

Fig. 14.1 Sources of knowledge and skills transfer.

proved to make the latter a simpler and more successful process. Key training requirements are summarised in Table 14.4. There are a number of factors determining the effectiveness of training, notably:

- The quality of training provision or advice;
- How effectively it is delivered –
  - some form of verification of learning system can be helpful as a measure of whether learning has taken place.
- Its relevance to the circumstances of the business;
- How well supported the trainees are by their managers –
  - before and after the training event.

Although, at present, there is no recognised international body that can certify training providers, there have been various national initiatives, for example in the USA, New Zealand and the UK, to establish standards against which to assess training programmes and develop accredited qualifications in HACCP and recognised training materials (Royal Institute of Public Health, 1995a, b; Meat Industry Standards Council 1999; UK Steering Group on HACCP Training Standards, 1999). More recently in the UK, the skill sector councils have taken over this role and provide a common platform for awarding bodies. These initiatives provide a benchmark against which to assure the quality of the range of training on offer. WHO has developed a standard HACCP training package that is both internationally recognised and designed to be easily translated (Motarjemi and Van Schothorst, 1999). As well as the quality of the provider, various issues impact on how effective training delivery is. Some of these are summarised in Table 14.5.

Table 14.4 Possible training objectives for different groups of staff

Group	Training objective
Senior management	1. Understand the general principles of HACCP and how they relate to the food business. 2. Demonstrate an understanding of the training and knowledge requirements for HACCP team members and the workforce as a whole. 3. Demonstrate an understanding of the links between HACCP and other quality management techniques and programmes and how a combined product management system can be developed. 4. Understand the need to plan the HACCP system and develop a practical timetable for HACCP application in the whole operation.
HACCP practitioners (including auditors of HACCP systems)	HACCP system and its management 1. Demonstrate an up-to-date general knowledge of HACCP. 2. Explain how a HACCP system supports national and international standards, trade and legislative requirements. 3. Describe the essential nature of PRPs and their relationship with HACCP. 4. Demonstrate the ability to plan an effective HACCP system. 5. Demonstrate an understanding of the practical application of HACCP principles. 6. Demonstrate the ability to design, implement and manage appropriate programmes for verification and maintenance of HACCP systems. 7. Explain the methods to be used for the effective implementation of HACCP into the process operation. 8. Demonstrate an understanding of the nature of hazards and how they are manifested in food products/operations and give relevant examples. 9. Demonstrate an understanding of the intrinsic factors governing the safety of product formulations and methods that can be used to assess safety of new products. 10. Carry out the steps to identify significant hazards relevant to the operation and determine effective control measures, i.e. assessment of risk (likelihood of occurrence and severity). 11. Demonstrate an understanding of the training and knowledge requirements for HACCP team members and the workforce as a whole. 12. Develop appropriate training programmes for CCP monitoring personnel. 13. Demonstrate an understanding of the links between HACCP and other quality management techniques and how a combined product management system can be developed. 14. Describe the method by which hazard analysis may be carried out and appropriate control measures ascertained to assess the practical problems.

Table 14.4 (Cont'd)

Group	Training objective
	15. Identify CCPs including critical limits to ensure their control. 16. Develop suitable monitoring procedures for critical points and explain the importance of corrective action procedures. 17. Validate and verify the HACCP system by the use of appropriate measures. 18. Carry out the steps to introduce and manage a fully operational HACCP system. 19. Develop appropriate training programmes for CCP monitoring personnel. 20. Perform HACCP audits using sampling, questioning, observation and assessment skills.
CCP monitors	1. Understand the general principles of HACCP and how they relate to the food handler's role. 2. Perform CCP monitoring tasks, record results and initiate appropriate actions. 3. Understanding the essential nature of PRPs and their role in food safety assurance.
General workforce	1. Understand the general principles of HACCP and how they relate to the food handler's role. 2. Understand the essential nature of PRPs and their role in assuring food safety.

Sources: UK Steering Group on HACCP Training Standards (1999); RIPHH (1995a); FSA (2000).

Research shows that training should be delivered specifically to achieve positive business results (Wick *et al.*, 2006) and that trainees should be supported as they work to implement their learning. HACCP is a preventative system, but by involving managers before and after the actual HACCP training and seeing the actual training interventions as a part of a longer learning journey can be a critical success factor in ensuring that the implementation will be truly beneficial.

There are various ways of disseminating training through an organisation. The US firm Cargill, for example, set up a Corporate Food Safety Department (CFSD) to build up a central pool of expertise in food safety, initially using external consultants to train staff in key HACCP skills (McAloon, 2001: pp. 64–6). With the help of the CFSD, each Cargill business set up its own business management team which included a CFSD member of staff. These teams were then responsible for individual plant Food Safety Committees (FSCs) which, again, included a CFSD representative and, where possible, the business management team coordinator. CFSD and the business management team coordinator assumed responsibility for training each plant FSC in HACCP methodology. In this way key skills were cascaded from external

Table 14.5 Reasons for ineffective training

Reason	Possible solution
Lack of motivation	Ensure that the trainees are briefed prior to undertaking the training. They need to know what their role is in relation to the topic and why they are being sent on the course. When they are on the course it should be made relevant to them.
Lack of understanding of the type of training needed	It is necessary to understand the task to be achieved following the training. For example, if a trainee is charged with leading a HACCP programme then a half-day 'Introduction to HACCP' course will not produce an expert who is up to job. The other aspect here is the learning approach to take. A cognitive approach combined with experiential learning is likely to be appropriate. Operatives are often not used to taking decisions or using their initiative, and a behavioural approach to training would not encourage this.
Training the wrong people or training at the wrong time	A typical mistake would be to train a HACCP team who then do not begin the project until 12 months later.
Lack of follow-up	Reinforcement after training is a well-documented requirement if the learning is to be cemented in the minds of the trainees.
Poor-quality training	This can encompass a lack of understanding on the part of the trainer, a lack of ability to transfer knowledge effectively, a poor scope and quality of training materials and inappropriate learning styles. These will all contribute to a poor outcome. There are advantages in using in-house trainers. As a result of developing in-house trainers you also create HACCP champions and, as a greater level of expertise is required to run a training course, the overall skill base within the company reaches a higher level. As there is no recognised standard for HACCP training certification, when using external trainers it can be difficult to assess the quality of trainer beforehand. Businesses have to choose between the good and the bad – and cost is not necessarily a good indicator.
Coverage – not training enough people	This can be a problem, particularly for SMEs who cannot afford to send many (if any) people on external training courses. The benefit in having in-house trainers is that more people can be trained.

consultants through the CFSD and business management teams to individual plant teams. This structure also allowed Cargill to set common standards in HACCP implementation across its individual businesses and, through the CFSD, business teams and regular liaison between plants, share knowledge between plant teams. The CFSD also built up a central resource of materials in such areas as training. A similar approach has been taken by Hindustan Lever in India (Marthi, 2001: pp. 87–90). Research is underway to aid the prediction of effective HACCP implementation based on an assessment of trainees post-training knowledge (Wallace *et al.*, 2005). This may be a helpful indicator to guide companies towards very targeted training in the future.

14.6 Building knowledge and expertise

Training is a crucial first step in building knowledge and skills, but businesses must then build on this initial foundation to utilise some of the other sources of information outlined earlier. In the case of one UK SME, the HACCP team used existing customer audits to help identify key issues and start HACCP planning, followed by visits to key customer staff for advice on particular issues. In some cases, customers supplied sample documentation and advice on the emerging HACCP plan (Route, 2001: p. 35). Such cooperation between customers and suppliers, and the readiness of retail and manufacturing customers to support their suppliers, can provide an important source of advice in such areas as:

- selecting a HACCP team and team leader;
- constructing process flow charts;
- information on microbiological and other hazards;
- procedures for classifying the severity of hazards and isolating CCPs;
- methods for auditing a HACCP plan.

However, organisations need to be careful in assessing the information and advice they get from outside sources. A number of those with direct experience of HACCP implementation highlight the limitations in advice provided by inexperienced regulatory and retail audit staff with an ill-informed approach reflecting confusion about HACCP methodology and a business's operations. Some government agencies are themselves relatively inexperienced in HACCP methodology (Mayes and Mortimore, 2001: p. 269; McEachern *et al.*, 2001: pp. 177–8). In a number of cases, for example, businesses have been advised to add additional CCPs which subsequently were accepted as not critical (Route, 2001: pp. 35–6; Taylor, 2001: pp. 27–8). In the UK, retailers have recognised that they have sometimes given conflicting advice to businesses supplying more than one retail customer (Kane, 2001: p. 44).

Using external consultants either in training or subsequently in HACCP planning has not always proved successful for SMEs in particular. Problems have included the variable quality of consultant advice, the unfamiliarity of SMEs with the key issues on which advice was needed and the consultant's unfamiliarity with the business's operations and requirements. A UK study of the use of consultants by SMEs in HACCP development showed that none of the resulting HACCP systems was adequate when independently audited (Holt, 1999).

These problems can be overcome in part by better initial training for SMEs, equipping them to respond more critically to the advice given to them. It may also be beneficial to have common agreement and a standard approach on CCPs by those providing advice, and requiring that these organisations are independently assessed.

Where cost is an issue, some governments have made financial support available to subsidise training in SMEs and in developing countries and have suggested approved providers. This initial training allows companies to do more of the groundwork before approaching a consultant, for example, using consultants to develop and help validate a HACCP plan rather than employing them right from the start. The UK government, for example, has provided a service to SMEs involving initial HACCP training followed by subsidised access to an approved consultant. Other initiatives in the UK have included setting up local HACCP resource centres to provide a range of services including the following:

- details of suitably qualified HACCP consultants and trainers;
- sources of funding (often not accessed by SMEs);
- information sources and advice;
- the opportunity for SMEs to meet to share problems and solutions;
- IT facilities with access to the Internet and HACCP software.

In developing countries, government agencies have provided similar support, particularly for their export industries (China requires a form of HACCP certification) which are both economically important and face demanding requirements from importing countries and businesses. In India, for example, the Agricultural Products Export Development Authority (APEDA) provides training, guidance and financial support for HACCP implementation. In the fisheries sector, the Central Institute for Fisheries Education (CIFE) provides training materials (in the appropriate language) and support, while the Marine Products Exports Development Agency (MPEDA) advises exporters on the range of international standards they must meet (Marthi, 2001: p. 85). Given the range of international standards, guidance in this area is particularly important for exporting countries (Suwanrangsi, 2001: pp. 192–3). In developing countries such as India there is a particular need to build up more information on microbiological and other hazards in the local environment, given that the majority of such information relates to hazards within developed countries.

As production has become more global, HACCP systems have been adopted in most countries involved in international trade in food products. However, many of the published materials, and most knowledgeable experts and longest-established HACCP systems are from developed countries such as the USA, Canada, Australia, New Zealand and those in Europe. There can be significant problems in transferring this expertise to other countries where translated materials do not exist and HACCP experts are not fluent in the local language. Practical solutions to such hurdles include the following:

- Using a bilingual expert in HACCP methodology to translate materials rather than a non-expert translator who may well produce misleading and confusing documentation.
- Where possible, using someone from the company, though it is best to avoid using staff from marketing or sales (who are most likely to have language skills), given their lack of subject expertise.
- Training a bilingual member of staff to help in teaching, which is likely to be more successful than using a third-party expert to train via an interpreter.
- Basing initial discussion around translated versions of the process flow diagrams. Local personnel will be able to relate to these much more easily.
- Allowing extra time to ensure that key issues are properly understood, particularly if discussions have to proceed via a translator. Where possible, try to check understanding independently of the translator.

There are some useful organisations who have provided training materials for use in developing countries. The Industry Council for Development / WHO/FAO is one such example where a Train the Trainers package on HACCP was developed (Motarjemi, 2006).

There can be significant cultural differences to negotiate when training in different countries. In some cultures, an overtly critical and challenging approach to problem-solving is considered rude. In one example, validation and verification procedures were initially viewed with suspicion by employees within a Chinese business since they appeared to question their competence (Route, 2001: p. 40). Such procedures have to be explained in terms that recognise these sensibilities. In some business cultures there are stronger traditions of hierarchy and paternalism with more junior staff expected, and expecting, to pass problems and decisions to their superiors. Those responsible for working with partner companies with such cultures need to be able to recognise and negotiate these differences.

In its report on SMEs and HACCP, the WHO suggests that generic plans may be a useful starting point in HACCP planning (WHO, 1999). The New Zealand and Thai governments, for example, have produced generic HACCP plans for specific sectors of the meat and seafood industries (Lee, 2001: p. 141; Suwanrangsi, 2001: p. 194). It is important, however, that businesses take a generic plan only as a starting point in developing a customised HACCP plan for the business. In Thailand, for example, government inspectors assessed the amount of reworking of generic HACCP models and used this as a measure of business commitment to and understanding of effective HACCP implementation.

In general, while governments have responded in different ways to the needs of their food industries in HACCP implementation, there has been an overall increase in the level of support provided, particularly for SMEs which often lack the resources and expertise to develop a HACCP system on their own. The New Zealand government, for example, has developed generic HACCP plans as a starting point for businesses in particular sectors, information on microbiological and other hazards to support businesses in HACCP planning, and, in collaboration with industry, HACCP competency qualifications for both industry and regulatory personnel (MAF, 1997, 1999; Meat Industry Standards Council, 1999). Similar developments have been noted in the UK and Thailand. The WHO (1996) has recommended that national governments develop national food safety strategies which identify key food safety issues and then develop a programme including the following:

- strengthening food safety legislation and enforcement;
- promoting industry food safety management systems such as HACCP;
- providing education and advice in food safety;
- developing resources such as microbiological data.

However, the effective use of the support and advice available to an organisation will always depend on the organisation taking responsibility for developing its own HACCP system and adopting a critical approach. An essential element in this process is the quality of the HACCP team and team leader, reinforcing the importance of initial training and the motivation to make HACCP implementation a central activity for the organisation.

14.7 Resources and planning

The time and material resources required for HACCP implementation depend on how well a business has been run before HACCP implementation, its size and the complexity of its products and processes. It has been estimated that a well-run business will already meet 80–90 % of HACCP requirements (McAloon, 2001: p. 68). In planning effectively, businesses often under-estimate the time and resources required in two areas in particular:

- Getting PRPs right can often be the most costly and time-consuming part of the process, particularly for businesses in developing countries. Even in the case of larger manufacturers in the USA and Europe, improvements to PRPs can take six months or more.
- The amount of training required, whether in building up basic hygiene skills for PRPs, HACCP implementation, or in maintaining the HACCP system in the long term, is often under-estimated.

Setting clear objectives and priorities is a critical first step. In the case of the US company Cargill, for example, the company undertook a risk assessment to identify where the greatest potential risks to food safety were located. Other larger manufacturers have adopted the same approach, targeting immediate food safety problems identified through a baseline audit and then concentrating on high-risk products (Rudge, 2001: pp. 101–2). Cargill focused first on training and supporting their priority plants in HACCP implementation (in meat, poultry and eggs where risks were greatest). They allowed a year to complete implementation within these plants, and two to three years for implementation within other lower-risk sectors of the business (McAloon, 2001: p. 65). This timetable compares to 18 months for implementation across the various plants at Hindustan Lever in India (Marthi, 2001: p. 87).

Cargill also decided to introduce HACCP systems within one plant at a time rather than simultaneously, despite the extra work and time involved. This more cautious approach proved effective in ensuring that HACCP plans were tailored closely to the requirements of each plant, were fully 'owned' by each plant team and benefitted from cumulative experience as the implementation programme moved from plant to plant. The same approach was used at Hindustan Lever where one factory was chosen for pilot implementation, based on risk assessment, size of operations and level of preparedness.

The approach at one of Heinz's largest facilities in Europe was to concentrate on one production line at a time because this did not overstretch their resources and gave them the benefit of cumulative experience and expertise as they moved from line to line (Killen, 2001: p. 123). Product-specific issues were then tackled within the analysis of each process step. The complete HACCP programme for all 12 production lines took three years in total, including completion of longer-term changes such as new equipment installation as part of improving PRPs. Another large European manufacturer, Kerry, adopted a different approach, looking at products individually in one of its Polish subsidiaries in hazard identification, using HACCP plans developed in sister plants, and then at common processes for these products in identifying CCPs (Rudge, 2001: pp. 105 and 107).

In the case of SMEs, it has been suggested that a developmental approach is best, concentrating in turn on the following:

- the installation of sound PRPs;
- HACCP studies to identify specific areas that need additional control, followed by development of CCP control measures and monitoring regimes;
- finally, developing appropriate systems of verification and review.

Such a process might need to be phased in over a period of up to two years (Taylor, 2001: p. 29). This approach has also been recommended for businesses in developing countries.

The examples above clearly indicate that it is possible to implement HACCP successfully using more than one approach. The approach selected must be that which is the best fit for the business circumstances. Whichever route is chosen, it is essential to plan it thoroughly with clear priorities and objectives, and to learn the lessons of experience.

Many texts on HACCP focus, not surprisingly, on the skills/knowledge/resources needed for the actual development and implementation of the HACCP plan. At this stage resource requirements are greatest and planning needs to be thorough. During the period of actual HACCP implementation, which can vary significantly (see above), HACCP has a high priority within the company/plant. However, HACCP cannot be seen as a 'one-off' application but must be seen as a long-term commitment to food safety. There is a danger that once the initial, high-priority focus on the implementation of HACCP has passed, that insufficient attention will be given to maintenance of the HACCP system to ensure it remains effective, and the initial planning must take this into account. A key element of the initial planning process must be the realisation that people move on in an organisation but systems remain. The incorporation of HACCP into the quality management system of a company will ensure that HACCP becomes embedded into the company processes (and hopefully culture as well – see Section 14.4 and 14.6). Section 14.12 details some of the practical activities required to maintain an effective HACCP plan, but planning for such maintenance activities should be done at the very early stages of HACCP plan development and implementation.

14.8 Pre-requisite programmes

The term pre-requisite programme (PRP) is defined by the WHO as 'practices and conditions needed prior to, and during, the implementation of HACCP and which are essential for food safety' (WHO, 1998). PRPs cover a range of general measures designed to promote food safety, including but not limited to:

- hygiene training, equipment and procedures for personnel;
- cleaning and disinfection procedures for plant and facilities;
- pest control;
- water and air control;
- waste management;
- equipment maintenance;
- raw materials purchasing/supplier quality assurance;
- transportation;
- product rework and recall procedures;
- labelling and traceability systems.

Although definitions differ, the concept of pre-requisite programmes does not differ significantly from what may be termed good manufacturing practice (GMP) or good hygienic practice (GHP). In effect, the concept refers to a base level of practice and procedure necessary for the production of safe food. The most successful implementation of HACCP is done within an environment of well-managed pre-requisite programmes (Codex Committee on Food

Hygiene, 1997a, b). Whilst it is in theory possible for an establishment to develop PRPs and HACCP at the same time, this is not the preferred option. In practice, however, many businesses may need to work on the PRP systems and HACCP development at the same time. If so, it can be helpful to set up separate project teams, both of which report to the HACCP team leader.

The vast majority of experienced HACCP practitioners will state that HACCP must be built on a sound base of effective PRP implementation, and this is certainly the view of the authors. The need to assess quality of PRPs within business through a baseline audit is set out by Mortimore and Wallace (1998) as well as by others with practical experience of HACCP implementation. The US Food and Drug Administration (FDA), for example, has issued detailed guidance which sets benchmarks for building effective PRPs and how HACCP controls can then be built on the foundation they provide (Taylor, 2001: p. 22).

In many cases PRPs will already be in place when establishments begin to introduce HACCP, but may just not be termed as such or standardised. They may often have been built up piecemeal in response, for example, to customer requests or changing legislative requirements. Once sound PRPs are in place most businesses realise that they need a relatively simply HACCP system with few CCPs. Although GMPs/GHPs cannot substitute for a CCP, collectively they can minimise the potential for hazards to occur, thus eliminating the need for a CCP. The implementation of effective PRPs will control 'general' or 'establishment' hazards that would otherwise have to be controlled by a CCP. This will leave the HACCP plan, and inherent CCPs, to concentrate on the control of product/process-specific hazards and thereby ensure that the HACCP plan contains only those CCPs that are essential to the safety of the product. Failure to have PRPs in place will inevitably lead to a large number of CCPs in the HACCP plan, covering both 'general/establishment' hazards and product-specific ones. This will almost certainly lead to some dilution of focus on the 'true' CCPs and make the HACCP plan unnecessarily difficult to manage. The importance of PRPs can be further illustrated by the fact that it is not uncommon for failures in product safety to be directly or indirectly attributable to lack of control over PRPs (e.g. poor personal hygiene, failure of cleaning and disinfection systems/processes, poor-quality raw materials, poor equipment maintenance, etc.).

14.9 HACCP teams

The quality of HACCP analysis depends critically on the scientific knowledge of the HACCP team, as indicated earlier, together with knowledge of the HACCP technique and detailed understanding of the products and processes under consideration. The composition of the HACCP team is very important. HACCP teams need to be multifunctional because detailed knowledge of the process and product does not usually reside in one individual. Those with experience of managing HACCP teams suggest keeping them to a maximum of five or six, to ensure a sufficient pool of expertise while preventing teams from becoming unwieldy, and keeping meetings to no more than 2–3 hours (Killen, 2001: pp. 124 and 127). It may be sensible to co-opt occasional team members to contribute at particular points where they have specific expertise. Co-opting staff may also be a way of using particularly busy members of the company who might not be free to attend all HACCP meetings. Preparation is key, for example in having line layout charts prepared in advance and in team leaders making their own preliminary analysis of key issues so as to be able to anticipate issues within meetings. In the case of SMEs, it may not be possible to set up a formal multidisciplinary team, but alternative routes to ensure that the right mix of skills and knowledge can be taken. Some SMEs enlist the help of local regulators, customers or suppliers in order to extend the resources available to them. Trade organisations can be an additional way to access knowledge and provide a forum where companies can brainstorm ideas, compare information on hazards and control measures and access validation references.

An essential element in making HACCP teams effective is involving operational staff and motivating them to contribute, particularly in putting together process flow diagrams in the early stages of HACCP planning (McAloon, 2001: p. 68). Their early involvement has a number of advantages:

- They know the relevant processes best.
- They are closest to how systems really work rather than how they may look on paper.
- They can become HACCP champions in the process area itself.
- They can help translate HACCP plans into systems and documentation their colleagues can use easily and effectively.

It is possible to secure their motivation by reinforcing their key role in the safety of products for the consumer. To be most effective, it is best to involve line staff where they have the most to contribute, particularly in describing processes and in designing CCP procedures and documentation, rather than in more unfamiliar areas such as hazard analysis.

In this respect, SMEs may be particularly well suited to HACCP teamwork. The relatively small number of employees makes it easier for the organisation to be well represented, enabling greater ownership to develop. It is also likely that many managers have worked their way up from the factory floor (and may continue to work there when production demands are high). This first-hand knowledge of operational procedures and empathy for the workforce speeds up the work of constructing a process flow diagram and CCP identification. The subsequent control strategies are more likely to be both practical and effective.

14.10 Hazard analysis

Hazard analysis has been identified as one of the most demanding tasks for a HACCP team. The approach at Cargill was to break down hazards into one of four areas:

- hazards controlled or minimised by CCPs;
- hazards controlled or minimised by GMPs;
- hazards considered to be of such low risk that no control is needed;
- hazards that can be designed out of the process.

This process significantly reduced the number of hazards that needed to be included in the HACCP study. The last option is not always considered but can be the best way of eliminating a hazard. HACCP practitioners suggest that it is often best to start with physical hazards as the easiest kind of hazard to understand and manage (Rudge, 2001: p. 111). However, in some respects this is not so easy as debates rage, for example, about what size and shape of metal constitutes a significant food safety hazard? The most common problem in hazard analysis identified by practitioners in both the larger manufacturing and SME sectors was the tendency to confuse quality (e.g. spoilage) and safety issues (Mortimore and Mayes, 2001: p. 255). This problem can be resolved by setting clear objectives at the start of the HACCP programme during the 'terms of reference' stage where clarification of two key issues should be sought:

- Is the HACCP study going to cover just product safety hazards, or also quality issues such as spoilage?
- What group of hazards will be considered (e.g. microbiological and/or physical and/or chemical hazards for safety studies)?

Even when the above questions have been answered study teams still have difficulties sometimes in selecting which specific hazards within one of the groups above should be considered during the study. As an example, which microorganisms represent a significant hazard to the product/process? It may be helpful to think in terms of groups of microorganisms with similar characteristics, e.g. vegetative pathogens, spore-forming pathogens, toxin-forming pathogens, and identify which groups contain microorganisms that represent a significant hazard. Having done so the study team could then carry out the hazard analysis using one or two specific organisms from within the relevant group. In this case, and in many others, a broad representation of skills and expertise in the study team will help in decision making. Where such skills are lacking internally, the study team can co-opt outside experts to help, or use one of the many published sources of information on significant hazards that can be found on the internet. In recent years the number of adverse reactions to allergens in foods appears to be increasing, particularly in Western culture countries. It is now considered essential to include consideration of allergens as significant hazards in food HACCP studies. Allergens may be considered as part of chemical contaminants or as a separate hazard group. A basic list of allergens recognised by the WHO as requiring labeling can be found in Codex (CAC, 1991). Directive 2003/89/EC Annex 111a contains a more detailed list (EC, 2003).

There are three important rules to follow in hazard analysis:

- Always be sure of the terms of reference for the study.
- Ensure that the team selects all the significant hazards relevant to the study.
- Start simply – do not try to cover all hazard groups at once unless the team is already experienced.

14.11 HACCP implementation

In larger plants it may be necessary to employ someone full-time on HACCP implementation: experience suggests this can halve the time required for implementation (McAloon, 2001: p. 69). However, the most critical element is securing the commitment of supervisors and operatives, not only for HACCP implementation and later maintenance but also for ensuring that PRPs are properly under control on a daily basis.

A HACCP system requires CCP monitors to be proactive and show initiative in spotting if a critical limit has been exceeded and taking the appropriate corrective action. In many cases this does not just involve learning new skills but assuming a new level of responsibility in taking important decisions for themselves. Persuading operational staff to take on this role can be difficult with a suspicion of management, a 'seen it all before' attitude and a conservatism about adopting new work practices. Such resistance may be compounded by a prescriptive management culture which restricts the independence of more junior staff. It is essential that these issues be resolved because failure by CCP monitors to understand and perform their duties is one of the biggest common weaknesses in HACCP implementation (Mortimore and Mayes, 2001: pp. 256–7). SMEs with less formal management structures and simpler channels of communication are often at an advantage in motivating line staff.

As has been noted, these problems can be resolved by active involvement of supervisors and line staff at an early stage in working in HACCP teams, writing or, at least reviewing, the HACCP plan and assisting in the design of HACCP documentation. This involvement means that the HACCP plan has the benefit of their knowledge of actual operations on the production line, and that the system takes account of their needs. Line staff are more likely to 'own' the system, to embrace the new responsibilities a HACCP system brings and to champion it among colleagues.

Effective training of line staff is also critical to their success as CCP monitors. Such training will be successful if it addresses not just what staff needs to do but why. Explaining the implications of a failure in food safety for consumers and what role employees play in preventing such failures is a proven way of gaining employee commitment. It is then important to balance the information on hazards and CCPs so that operators know enough without being swamped by unnecessary detail. Involving them in the design of documentation and procedures will help to achieve this balance. It may also be necessary to review operator knowledge of and skills in operating processes and conduct refresher training.

One of the main criticisms levelled at HACCP systems by SMEs in particular is the requirement for documentation. However, research suggests that excessive documentation is usually associated with poor design of HACCP systems (Taylor, 2000). If properly designed, a HACCP system usually has relatively few CCPs which require a basic system of record-keeping (Fig. 14.2). The US firm Cargill developed a flexible approach to document design, setting an overall standard specifying basic content as a framework for staff to develop their own documentation.

The Irish firm Kerry reviewed all its existing documentation and consolidated it into a single system. As an example, internal non-conformity and finished

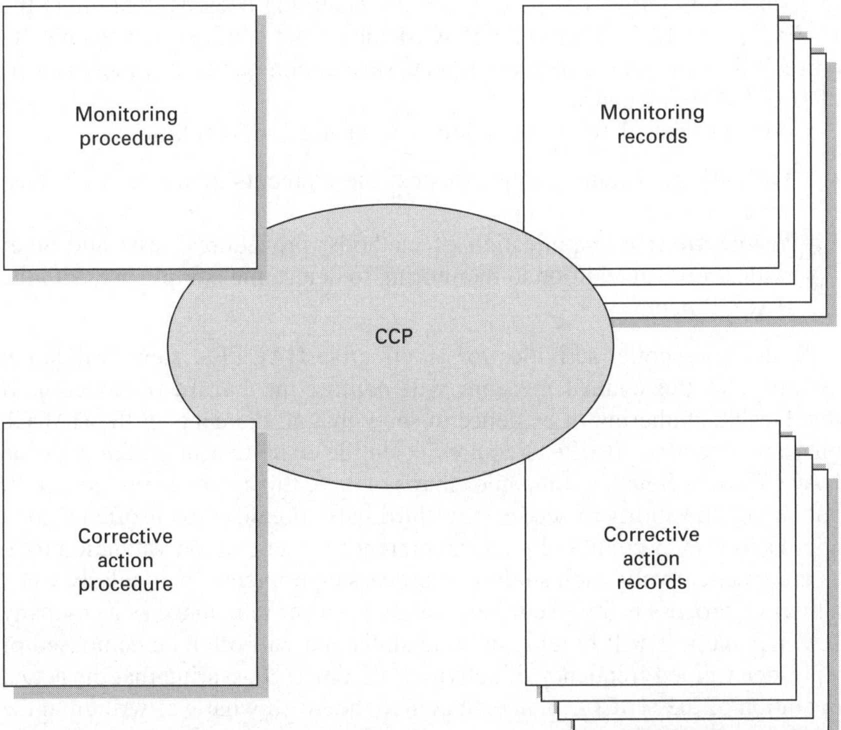

Fig. 14.2 The essential documentation required for each CCP.

product reject documents from the business's existing quality system were combined with the corrective action documentation required by the HACCP system to produce a single non-conformity report. The design of the report was made as simple as possible so that it could be readily understood and completed. The company also computerised the record-keeping system to improve its efficiency and flexibility (Rudge, 2001: p. 115).

14.12 Maintenance

Maintenance of a HACCP system will comprise a number of tasks. Verification that the system is working is a major part of this. Verification is Codex HACCP Principle 6. Confusion has arisen due to the fact that validation is included under this same principle and is therefore perceived as a subset of verification. Validation and verification occur at different times in the HACCP system development, however. Validation, whilst often referred to as part of verification, is in reality an exercise to determine that the HACCP plan is technically and scientifically sound, and should be carried out prior to HACCP plan implementation, and following receipt of new information (e.g. change in product or process design, emerging hazards) that affect the HACCP plan (ILSI, 1999). If the HACCP plan is not valid (i.e. not technically correct) then the whole plan is suspect. Verification occurs routinely as part of HACCP maintenance.

Codex (1997a) defines validation and verification as follows:

- **Validation:** obtaining evidence that the elements of the HACCP plan are effective.
- **Verification:** the application of methods, procedures, tests and other evaluations, in addition to monitoring to determine compliance with the HACCP plan.

Both are essential activities for an effective HACCP system. Validation assures that the control measure will control the hazard of concern. It involves the gathering of evidence to show that all elements of the HACCP plan are effective. It often requires scientific expertise in addition to the HACCP team. Smaller companies may not have this resource on their staff, but having the ability to access via a third-party organisation is often a good approach. Use of credible literature references is a common validation tool, as are practical validation studies on process equipment – particularly when a thermal process is involved. Verification is easier to manage as a company HACCP team. It will require auditing skills and will often be composed of a pre-determined frequency of activities all aimed at assuring that the actual operation of the HACCP plan is in compliance with what was written in the documented system.

The failure to verify that a HACCP system is working as planned is another commonly reported cause of HACCP systems failure. Effective and regular company audit procedures are the best solution. Such audits are best focused on CCP monitoring and documentation to ensure that staff genuinely understood their responsibilities and that CCPs are properly monitored (Killen, 2001: pp. 130–1; McAloon, 2001 pp. 74–5). The benefit of such regular verification activity in these companies not only was that HACCP system efficiency improved but that staff were prompted to start suggesting improvements of their own to make HACCP documentation, for example, easier to manage. HACCP auditors require a high degree of expertise and experience, and should, if possible, be drawn from those with direct experience of production operations.

The HACCP audit must be carried out at regular intervals, (typically not longer than 12 months), and cover the entire HACCP system. Many companies will select to audit part of the system on a monthly basis, reporting findings to the HACCP team. Internal audits also help companies to prepare effectively for, and make best use of, external audits, whether from government agencies, third-party certification bodies, or from customer audit staff. UK retailers and manufacturers have recognised the importance of establishing a common approach to auditing HACCP as part of an overall food safety and quality program, moving from independent auditing to the use of accredited third-party auditors and establishing minimum audit standards through, for example, the British Retail Consortium Technical Standard (Dillon and Griffith, 2001: p. 2). This common approach extends across borders through the work of the Global Food Safety Initiative (GFSI – CIES, 2007), where national standards are reviewed as being compliant with the requirements of GFSI.

The other global audit standard is ISO22000 (ISO, 2005) which is based on Codex HACCP principles. This will help to ensure a more consistent approach to assessment against the standard by the certification companies around the world.

Audits form only part of verification activities. Review of CCP records is vitally important and is usually linked to a positive release programme, i.e. review of records by a trained HACCP reviewer prior to release of product. Microbiological testing can also be useful as part of a verification programme, i.e. to confirm that the product meets its microbiological specification limits. Consumer complaint data will also provide information, particularly with regard to physical hazard control. Some companies will set up a HACCP maintenance team to manage and mutually review the activities which form the HACCP verification programme.

It is also important to have an ongoing training programme to maintain awareness of food safety and HACCP skills. Food safety needs to be made an integral part both of new employee training and regular training programmes or briefing sessions for existing staff. Regular monthly training sessions of 15–20 minutes can be effective, as can simple reward systems to recognise good food safety management. At Heinz in the UK, employees are issued with training 'passports' that record all the training they have undergone.

Employees are independently checked on their understanding after training to verify its effectiveness (Killen, 2001: pp. 131–2). Experience at large firms such as Cargill suggests the value of having back-up staff to cover the absence of key staff together with standard procedures such as checklists that can be used by a colleague when a member of staff is on holiday or sick leave (McAloon, 2001: p. 76).

Maintaining a HACCP system once implemented is where the real work begins but also where the benefits start to be fully realised.

14.13 HACCP and public health goals

It is clear that effective HACCP, built upon sound GHP and GMP, has contributed significantly to making food safety management in the supply chain more robust. However, one of the limitations of HACCP is that it is primarily a hazard-based approach rather than one based on identification and management of risk. A typical criticism is that HACCP is not outcome-orientated and that, as a result, it is difficult to measure its benefits (Orriss and Whitehead, 2000). It has been suggested that HACCP systems could be linked to Food Safety Objectives (FSOs), derived from risk assessments which set appropriate levels of protection (ALOP) (Mayes and Mortimore, 2001). The manner in which HACCP systems could possibly interact with FSOs and ALOP is described below, using microbiological food safety as an example.

The FAO and WHO have taken the lead in promoting the use of a risk-based approach to the control of public health and have adopted risk analysis as the framework around which food safety programmes should be built (Buchanan, 2004; CAC, 2004). The impact of foodborne microbiological hazards on public health can be evaluated by conducting a microbiological risk assessment (MRA) managed by a governmental authority. Increasingly MRA is being used as a framework for making decisions on microbiological safety. A number of new concepts have been introduced in the process: ALOP, FSOs, performance objectives (PO), performance criteria (PC) and microbiological criteria (MC), though the latter has been in use for some time. Whilst these concepts may appear new, they are not markedly different from some that have been used to manage food safety for many years. However, the manner in which they can be used to link the expression of a public health goal set by a governmental authority to the operational management of food processes by HACCP to meet that public health goal is relatively new and offers the possibility of a much more transparent process of food safety management than has been previously available.

14.13.1 Definitions

- **Appropriate level of protection (ALOP):** The level of protection deemed appropriate by the member (country) establishing a sanitary or phytosanitary measure to protect human, animal or plant life or health within its territory (WTO, 1995). ALOP can be regarded as an acceptable level of risk or tolerable risk. It could be a public health goal or an aim to reduce illness by a particular amount.
- **Food safety objective (FSO):** The maximum frequency and/or concentration of a hazard in a food at the time of consumption that provides or contributes to the ALOP.
- **Performance objective (PO):** The maximum frequency and/or concentration of a hazard in a food at a specified step in the food chain before the time of consumption that provides or contributes to an FSO or ALOP, as applicable.
- **Performance criterion (PC):** The effect in terms of frequency and/or concentration of a hazard that must be achieved by the application of one or more control measures to provide or contribute towards a PO or FSO (CAC, 2004).
- **Microbiological criterion (MC):** An MC defines the acceptability of a product or food lot, based on the presence, absence or number of microorganisms, including parasites, and/or the quantity of their toxins/metabolites, per unit mass, volume, area or lot (CAC, 2001).

An FSO specifies the maximum level of a hazard that can be tolerated in the consumed product and should be set by a competent authority based ideally on a governmental risk assessment and an ALOP. This will allow a consideration of exposure of the hazard on the health of the relevant population and therefore a link with a public health goal. Industry can play a key role in the governmental risk assessment leading to the establishment of an FSO. The FSO can therefore be viewed as an 'end of food chain target' which not only expresses a governmental view of what constitutes safe food but also brings in the concept of equivalence whereby the FSO specifies the food safety target but does not specify how it must be achieved. This clearly allows for a range of different food safety management operational approaches provided a common 'end of food chain target' is met.

The PO represents a target in the food chain that links the public health goal represented by the end-of-chain FSO with one further upstream in the food chain. In effect the PO can be viewed as one or many mid-chain targets that can help assure compliance with the final end-of-chain FSO target. POs need to be set after taking into account whether a microbiological hazard will increase, decrease or not change in number during subsequent processing steps prior to the FSO. Whilst FSOs are set by a competent authority, POs can be set by industry, again allowing for a range of different operational approaches.

A PC is related to a PO at any specific step in a food chain and is the effect of one or more control measures working together to meet the

PO (Gorris *et al.*, 2006). Control measures can include a wide range of activities, such as specific processes to reduce/eliminate pathogens, use of preservatives to prevent microbial growth, physical measures to prevent cross-contamination, microbial specifications, etc. A PC can be viewed as the effect of one or more control measures on the level/concentration of a hazard. In the case of microbiological hazards this can mean a reduction in the level of the hazard (e.g. heat treatment), an increase in the numbers of organisms (e.g. fermentation), or assurance of no change in numbers (e.g. chilling or freezing).

In effect a PC links the end-of-chain targets relating to public health (ALOP/FSO and by derivation PO), to defined operation control measures at specific processing steps. Industry has responsibility for establishing PCs based on available processes and control measures to deliver the required PO.

Whilst the above concepts represent the latest developments in risk-based food safety, they cannot be successfully implemented in isolation. They need to be a part of an overall food safety management system for a product/process/plant combination. HACCP provides an excellent food safety management system around which the above concepts could be implemented. The concept of PC and control measures fits readily into HACCP. It could be envisaged that a PO could be set at a number of stages in the 'farm to fork' food chain. For example it is possible to have a PO for each of the manufacturing, distribution and product preparation stages. Each PO would have PC and control measures contributing to it. Some control measures would be considered to be CCPs, which would need critical limits, corrective actions, monitoring and verification actions which could be delivered via the operational HACCP plan.

POs and FSOs represent mid-chain and end-of-chain targets that can also be part of the HACCP plan. In the case of microbiological hazards, microbiological criteria would play a role in measuring compliance with such POs and FSOs. The measurement of microbiological numbers is not a new concept, having been used for many years as a means of checking compliance with raw materials/final product specifications. The limitations of microbiological testing are well known and outside the scope of this chapter but, with appropriate sampling plan procedures, microbiological criteria can play a role in verifying the effectiveness of food safety management systems and compliance with POs and FSOs.

The HACCP food safety management system has the potential to provide an excellent operational framework in which PC, PO and FSO concepts can be expressed. The FSO concept is derived from the ALOP which is the expression of a desirable public health goal. In this way HACCP can be viewed as the operational framework allowing the measurable delivery of a public health goal. Figure 14.3 shows how ALOP and FSOs could be linked to HACCP and the delivery of safe food.

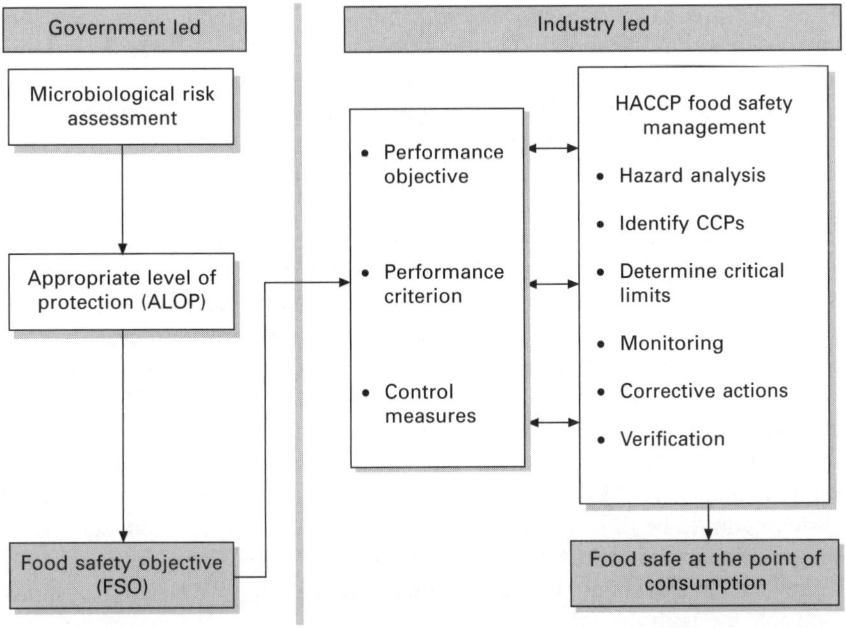

Fig. 14.3 Appropriate level of protection (ALOP) and HACCP.

14.14 Future trends

Those involved in implementing systems in practice, and with auditing them, have identified a number of common problem areas in effective implementation, notably:

- poor initial motivation and commitment;
- ignorance of HACCP methodology;
- lack of relevant technical expertise;
- a failure to define the scope of and plan effectively for HACCP implementation, notably in relationship to improvements to PRPs;
- failures in effective CCP monitoring;
- a failure to audit and maintain the HACCP system.

A lack of control over those offering HACCP training and, on a global basis, the scarcity of independent assessment for those undertaking training undoubtedly contribute to the root cause of these failures. As this chapter has shown, there are a range of practical solutions to many of these problems open to businesses, whether they are large companies or SMEs, in developed or developing countries. It is also clear that where it is well implemented, the HACCP approach can make a significant difference to food safety. In the USA, for example, where HACCP systems became mandatory for larger meat and poultry plants by 1998 and for smaller plants by 2001, contamination levels in chicken, turkey and beef fell initially by as much as

45 % in some cases (Crawford, 2000: p. 30). Where food safety incidents have occurred, it has mostly been possible to trace them to specific failures to follow correct procedures within PRPs, for example. It is also evident, even in situations where HACCP implementation has been mandated, that effective implementation rates can be disappointingly low (Losikoff, 2000). It is clear that just mandating HACCP will not be sufficient on its own to reduce the number of cases of food poisoning – effective support, training in use and monitoring of performance will be necessary.

The USA has taken the lead in the use of HACCP as a regulatory tool, but many other countries (e.g. Canada, Australia/New Zealand, EU countries and others), also have HACCP as a part of their regulatory process. Not all countries intend to use HACCP in the same way as the USA has piloted, and that is quite appropriate, since the maturity of the HACCP legislation must reflect the maturity of the HACCP process.

HACCP needs to be seen as one essential element in effective food safety management. As discussed earlier, it has been suggested that HACCP systems should be linked to FSOs based on risk assessment exercises which set specific targets or ALOPs for target groups of consumers. Although still somewhat controversial, this development may help to give HACCP systems a clearer focus in the future.

It is also increasingly clear that, from its origins in manufacturing, HACCP needs to be applied across the entire food chain from 'farm to fork'. HACCP principles are applicable to all sectors of the food supply chain, though it is important to be flexible in how they are applied. In farm and retail settings, for example, CCPs may be difficult to identify and implement, and more emphasis needs to be given to sound PRPs in controlling hazards (Johnston, 2000). In this respect, the relationship between PRPs and HACCP systems continues to need clarification (ILSI, 1998) and reinforcement.

As mentioned earlier, the development of HACCP has been led primarily by the larger food manufacturers and the application of HACCP to SMEs and even smaller micro-enterprises has always been problematical. Strong concern has been expressed about the language used, the extent of skilled knowledge and expertise needed for hazard analysis, the (some would say), burdensome record keeping required and the general lack of flexibility in Codex (1997)a. Such concerns, expressed by both regulatory and trade association representatives, have led to a number of proposals to modify the extent and rigour of HACCP application in certain businesses/business sectors.

Article 5(1) of Regulation (EC) 852/2004 (EC, 2004) on the hygiene of foodstuffs contains a requirement that food business operators put in place, implement and maintain a permanent procedure or procedures based on the HACCP principles. Recital 15 of the above Regulation specifically states that 'HACCP requirements should take account of the principles contained in the Codex Alimentarius. They should provide sufficient flexibility to be applicable in all situations, including in small businesses. In particular it is necessary to recognise that, in certain food businesses, it is not possible to identify critical control points and that, in some cases, good hygienic practices can replace the monitoring of critical control points. Similarly, the requirement of establishing "critical limits" does not imply that it is necessary to fix a numerical limit in every case. In addition, the requirement of retaining documents needs to be flexible in order to avoid undue burdens for very small businesses.' This wording clearly provides for significant flexibility in the application of HACCP.

A recent guidance document published on 16/11/2005 by the European Commission entitled 'Guidance document on the implementation of procedures based on the HACCP principles, and facilitation of the implementation of the HACCP principles in certain food businesses' (EC, 2005) has been prepared in response to requests to clarify to what extent flexibility can be applied with regard to the implementation of procedures based on the HACCP principles, particularly for small businesses.

The document provides guidance on how the use of pre-requisite requirements, guides to good practice, generic guides for the implementation of HACCP, a flexible use of HACCP principles and HACCP record keeping commensurate with the nature and size of the business can be used in a flexible way to ensure that a system based on HACCP principles can be applied in small businesses. The guidance document has taken a look at the HACCP approach, distilled from it the basic, underlying principles and provided guidance on how these principles can be applied in an extremely flexible manner to ensure safe food. Whilst this document has no legal status and has been established for information only the result of such guidance would not be the Codex HACCP system that the majority of large food manufacturers would recognise, but it is a pragmatic approach looking to stay true to the EU 852/2004 requirements of 'a procedure based on HACCP principles'.

In a similar vein there is a current EU proposal (April 2007) to amend Regulation 852/2004 such that food businesses with less than 10 employees are exempted from the whole of Article 5 (1) of the above regulation (EC, 2007). This proposal applies to businesses that sell directly to the final consumer and are considered to be micro enterprises (e.g. bakeries, butchers, grocery shops, market stalls, restaurants and bars). This proposal is part of the European Commission Strategic Review of Better Regulation in the European Union and seeks to reduce the administrative burdens on business. The proposal makes it clear that food businesses would have to comply with all other requirements of 852/2004 on the hygiene of foodstuffs. The proposal undoubtedly reflects the concern, expressed by small businesses, that HACCP is too burdensome for them. At the time of writing, the proposal is still under discussion, but there is some concern that exemption rather than application of a flexible approach could weaken control over food safety in businesses that are, despite their size, capable of causing serious food safety problems for many members of the public should a failure in food safety occur.

Another example of a modified approach to HACCP can be found in the Safer Food Better Business pack produced by the FSA in the UK (FSA,

2005). This pack, which has been produced for small catering businesses, is a very practical approach with clear guidance on safe methods (e.g. cross-contamination, cleaning, chilling, cooking and management) and a diary for record keeping. Whilst the pack is based on the principles of HACCP it contains no mention of HACCP nor terms like hazard, but instead is presented as a simple, working pack that, if followed, will help caterers produce safe food and comply with the law. It bears no resemblance to Codex HACCP but has taken the principles of HACCP and distilled them down into a series of practical guides to help caterers comply with the law.

All three examples above, although taken from the EU, probably reflect a wider view that whilst the principles of Codex HACCP are sound, the practical application of such principles needs to be tailored to meet the specific requirements of different food business sectors. This could be seen as part of the evolution of HACCP. It is better that the practical application of HACCP principles is focused appropriately for specific business sectors and thereby implemented successfully than to enforce a rigid application of the Codex HACCP across all food businesses that results in unsuccessful implementation in some businesses.

14.15 Conclusions

HACCP can be an extremely beneficial tool for food processors but only if used effectively. If it is implemented reluctantly because of regulatory or customer pressure, and in the absence of a real desire for self-improvement, then it will probably add no real value to the business and will fail to ensure safe food. Used properly it can help to focus businesses in many, if not all, of its food safety and quality control programmes, from SQA activities to finished goods process control. Understanding and applying the HACCP concept across the food industry whether Codex or a variation on the principles can undoubtedly make a substantial contribution to efficient and effective food safety assurance worldwide.

14.16 References

Barnes J (2001) *Implementation and enforcement in the United Kingdom*, in T Mayes and S Mortimore (eds), *Making the Most of HACCP*, Woodhead, Cambridge, 202–12.

Bas M, Ersun A S, Kivanc G (2006) Implementation of HACCP and prerequisite programmes in food businesses in Turkey, *Food Control*, **17**(2), 118–26.

Buchanan R L (2004) Principles of risk analysis as applied to microbial food safety concerns, *Mitteilungen aus Lebensmitteluntersuchung und Hygiene*, **95**(1), 6–12.

CAC (1991) *General Standards for the Labelling of Prepackaged Foods*, Codex Standard 1-1985 (Rev. 1-1991), Codex Alimentarius Commission, Rome.

CAC (1997) *Hazard Analysis and Critical Control Point System and Guidelines for its Application*, Alinorm 97/13A, Codex Alimentarius Commission, Rome.

CAC (2001) *Food Hygiene – Basic Texts*, 2nd edn, Joint FAO/WHO Food Standards Programme, Codex Alimentarius Commission Secretariat, Rome.

CAC (2004) *Procedural Manual*, 14th edn, Joint FAO/WHO Food Standards Progamme, Codex Alimentarius Commission Secretariat, Rome.

CCFRA (1997) *HACCP: A Practical Guide*, 2nd edn, Technical Manual 38, Campden and Chorleywood Food Research Association, Chipping Campden.

CIES (2007) *The Global Food Safety Initiative: GFSI Guidance Document*, CIES, Paris.

Codex Committee on Food Hygiene (1997a) *HACCP System and Guidelines for its Application*, Annex to CAC/RCP 1-1969, Rev 3 in Codex Alimentarius Commission Food Hygiene Basic Texts, Food and Agriculture Organisation of the United Nations/ World Health Organisation, Rome/Geneva.

Codex Committee on Food Hygiene (1997b) *Recommended International Code of Practice, General Principles of Food Hygiene*, CAC/RCP 1-1969, Rev 3 in Codex Alimentarius Commission Food Hygiene Basic Texts, Food and Agriculture Organisation of the United Nations/World Health Organisation, Rome/Geneva.

Crawford L (2000) HACCP in the United States: regulation and implementation, in M Brown (ed.), *HACCP in the Meat Industry*, Woodhead, Cambridge, 27–36.

Dillon M and Griffith C (eds) (2001) *Auditing in the Food Industry*, Woodhead Cambridge.

EC (2003) Directive 2003/89/EC of the European Parliament and the Council of 10 November 2003 amending Directive 2003/13/EC as regards indication of the ingredients present in foodstuffs, *Official Journal of the European Union*, **L308**, 15, 25 November, 15–18.

EC (2004) Regulation (EC) No 852/2004 of the European Parliament and of the Council of 29th April 2004 on the hygiene of foodstuffs, Corrigenda, *Official Journal of the European Union*, **L226/3**, 25 June, 3–21.

EC (2005) Guidance document on the implementation of procedures based on the HACCP principles, and on the facilitation of the implementation of the HACCP principles in certain food businesses, European Commission DG Health and Consumer Protection, Brussels.

EC (2007) Proposal for a Regulation of the European Parliament and of the Council amending Regulation No 11 concerning the abolition of discrimination in transport rates and conditions, in implementation of Article 79 (3) of the Treaty establishing the European Economic Community and Regulation (EC) No 852/2004 of the European Parliament and the Council on the hygiene of foodstuffs, Interinstitutional File 2007/0037 (COD), Council of the European Union, Brussels.

FSA (2000) *Syllabus for HACCP Assessment Course for Enforcement Officers*, Food Standard Agency, London.

FSA (2005) Safer Food Better Business, Foods Standards Agency, London.

Gormley R T (1999) R & D needs and opinions of European food SMEs, *Farm & Food*, **5**, 27–30.

Gorris L G M, Bassett J and Membré J-M (2006) Food safety objectives and related concepts: the roll of the food industry, in Motarjemi Y and Adams M (eds), *Emerging Foodborne Pathogens*, Woodhead, Cambridge.

Holt G (1999) Researcher investigating barriers to the implementation of GHP in SMEs, personal communication.

ICMSF (1988) *Micro-organisms in Foods 4: Applications of the Hazard Analysis and Critical Control Point (HACCP) Concept to Ensure Microbiological Safety and Quality*, Blackwell Scientific, London.

ILSI (1997) *A Simple Guide to Understanding and Applying the Hazard Analysis and Critical Control Point Concept*, 2nd edn, ILSI Europe Scientific Committee on Food Safety.

ILSI (1998) *Food Safety Management Tools*, report prepared under the responsibility of

ILSI Europe Risk Analysis in Microbiology Task Force, International Life Sciences Institute, Brussels.

ILSI (1999) *Validation and Verification of HACCP*, report prepared under the responsibility of the ILSI Europe Risk Analysis in Microbiology Task Force, International Life Sciences Institute, Brussels.

ISO (2005) ISO22000:2005 *Food Safety Management Systems – Requirements for any organisation in the food chain*, International Organisation for Standardisation, Geneva.

Johnston A M (2000) HACCP and farm production, in M Brown (ed.), *HACCP in the Meat Industry*, Woodhead, Cambridge, 37–80.

Kane M (2001) Supplier HACCP systems: a retailer's perspective, in T Mayes and S Mortimore (eds), *Making the Most of HACCP*, Woodhead, Cambridge, 43–58.

Killen D (2001) Implementing HACCP systems in Europe: Heinz, in T Mayes and S Mortimore (eds), *Making the Most of HACCP*, Woodhead, Cambridge, 119–36.

Lee J A (2001) *HACCP enforcement in New Zealand*, in T Mayes and S Mortimore (eds), *Making the Most of HACCP*, Woodhead, Cambridge, 139–64.

Losikoff M (2000) *Compliance with Food and Drug Administration's Seafood HACCP Regulations*, Presentation before the International Association for Food Protection, August, Atlanta, GA.

MAF (1997) *National Microbiological Database (Bovine and Ovine Species): Specification for Technical Procedures*, Manual 15, Ministry of Agriculture, Wellington.

MAF (1999) *A Guide to HACCP Systems in the Meat Industry*, Ministry of Agriculture and Forestry, Wellington.

Marthi B (2001) HACCP implementation: the Indian experience, in T Mayes and S Mortimore (eds), *Making the Most of HACCP*, Woodhead, Cambridge, 81–97.

Mayes T and Mortimore S (2001) The future of HACCP, in T Mayes and S Mortimore (eds), *Making the Most of HACCP*, Woodhead, Cambridge, 265–76.

McAloon T R (2001) HACCP implementation in the United States, in T Mayes and S Mortimore (eds), *Making the Most of HACCP*, Woodhead, Cambridge, 61–80.

McEachern V, Bungay A, Ippolito S B and Lee-Spiegelberg S (2001) Enforcing safety and quality: Canada, in T Mayes and S Mortimore (eds), *Making the Most of HACCP*, Woodhead, Cambridge, 165–82.

Meat Industry Standards Council (1999) *Industry Standard for a HACCP Plan, HACCP Competency Requirements and HACCP Implementation*, New Zealand Meat Industry Standards Council, Wellington.

Mortimore S and Mayes T (2001) Conclusions, in T Mayes and S Mortimore (eds), *Making the Most of HACCP*, Woodhead, Cambridge, 235–64.

Mortimore S E and Wallace C A (1998) *HACCP: A Practical Approach*, 2nd edn, Aspen, Gaithersburg, MD.

Mortlock M P, Peters A C and Griffith C J (1999) Food hygiene and the hazard analysis critical control point in the United Kingdom food industry: practices, perceptions and attitudes, *Journal of Food Protection*, **62**(7), 786–92.

Motarjemi Y (2006) ICD in perspective: putting social responsibility into practice, *Food Control*, **17**(12), 1018–22.

Motarjemi Y and van Schothorst M (1999) *HACCP Principles and Practice: Teacher's Handbook*, WHO/SDE/PHE/FOS/99.3, World Health Organization, Geneva.

NACMCF (1992) Hazard Analysis and Critical Control Point System, *International Journal of Food Microbiology*, **16**, 1–23.

Orriss G D and Whitehead A J (2000) Hazard analysis and critical control point (HACCP) as part of an overall quality assurance system in international food trade, *Food Control*, **11**, 345–51.

Panisello J P, Quantick P C and Knowles M J (1999) Towards the implementation of HACCP: results of a UK regional survey, *Food Control*, **10**, 87–98.

RIPHH (1995a) *HACCP Principles and Their Application in Food Safety (Introductory Level) Training Standard*, Royal Institute of Public Health and Hygiene, London.

RIPHH (1995b) *The First Certificate in Food Safety*, Royal Institute of Public Health and Hygiene, London.

Route N (2001) HACCP and SMEs: a case study, in T Mayes and S Mortimore (eds), *Making the Most of HACCP*, Woodhead, Cambridge, 32–42.

Rudge G (2001) *Implementing HACCP systems in Europe*: Kerry Ingredients, in T Mayes and S Mortimore (eds), *Making the Most of HACCP*, Woodhead, Cambridge, 98–118.

Semos A and Kontogeogros A (2007) HACCP Implementation in Northern Greece – food companies perception of costs and benefits, *British Food Journal*, **109**(1), 5–19.

Sumner J (1999) HACCP and quality assurance systems in the Australian seafood industry, *The 3rd International Fish Inspection and Quality Control Conference*, Oct 6–8, Halifax, Nova Scotia.

Suwanrangsi S (2001) HACCP implementation in the Thai fisheries industry, in T Mayes and S Mortimore (eds), *Making the Most of HACCP*, Woodhead, Cambridge, 183–201.

Taylor E A (2000) *Food safety in the UK catering industry*, PhD thesis, Food Policy Research Unit, University of Central Lancashire.

Taylor E (2001) HACCP and SMEs: problems and opportunities, in T Mayes and S Mortimore (eds), *Making the Most of HACCP*, Woodhead, Cambridge, 13–31.

Taylor E and Kane K (2005) Reducing the Burden of HACCP on SMEs, *Food Control*, **16**(10), 833–9.

UK Steering Group on HACCP Training Standards (1999) *HACCP Principles and Their Application in Food Safety (Advanced Level) Training Standard*, Royal Institute of Public Health and Hygiene, London.

Wallace C A, Powell S C and Hollyoak L (2005) Post Training Assessment of HACCP Knowledge: its use as a predictor of effective HACCP development, implementation and maintenance in food manufacturing, *British Food Journal*, **107**(10), 743–59, Emerald, UK.

WHO (1993) *Application of the Hazard Analysis Critical Control Point System for the Improvement of Food Safety*, WHO Supported Case Studies on Food Prepared in Homes, at Street Vending Operations, and in Cottage Industries, WHO/FNU/ FOS/993.1, World Health Organization, Geneva.

WHO (1996) *Training Aspects of the Hazard Analysis Critical Control Point System (HACCP)*, Report of a WHO Workshop on Training in HACCP, WHO/FNU/FOS/996.3, World Health Organization, Geneva.

WHO (1998) *Guidance on Regulatory Assessment of HACCP*, Report of a joint FAO/WHO consultation on the role of government agencies in assessing HACCP. WHO/FSF/FOS/98.5, World Health Organization, Geneva.

WHO (1999) *Strategies for Implementing HACCP in Small and Less Well Developed Businesses (SLWBs)*, World Health Organization, Geneva.

WTO (1995) *The WTO Agreement on the Application of Sanitary and Phytosanitary Measures (SPS Agreement)*, World Trade Organization, Geneva, available at: http://www.wto.org/english/tratop_e/sps_e/spsund_e.htm, Last accessed December 2008.

Wick C, Pollock R, Jefferson A, Flannigan R (2006) *The Six Disciplines of Breakthrough Learning*, Wiley, San Francisco, CA.

15

Good practice for food handlers and consumers

C. Griffith and E. Redmond, University of Wales Institute Cardiff, UK

Abstract: Responsibility for food safety is shared between governments, industry and food handlers (including consumers). Human error, especially in relation to cross-contamination, is critical and is reduced by using a good management system. This must be embedded within a positive food safety culture. These concepts are explored from manufacturing through food service to the home. Theoretical behavioural models can be used both to understand why food handlers do not always implement the practices they know and to improve food handler and consumer behaviour.

Key words: good hygiene practices, food handler behaviour, food safety systems, food safety culture, responsibility for food safety.

15.1 Introduction

15.1.1 Food safety, industry and the consumer

Ensuring food safety is the responsibility of all links in the food chain, including consumers, the government (via legislation, health services and the educational system) as well as the food and media industries. Their respective roles are summarised in Fig. 15.1.

The role of government includes education, health service provision, epidemiological and other research as well as legislation. The latter provides the correct framework or context in which all others will act. Around the world, for political or other reasons, governments are increasingly forming units such as the UK Food Standards Agency, which can be seen as independent bodies to manage a government's food safety role. The European Commission has formed an independent European food authority in an attempt to ensure

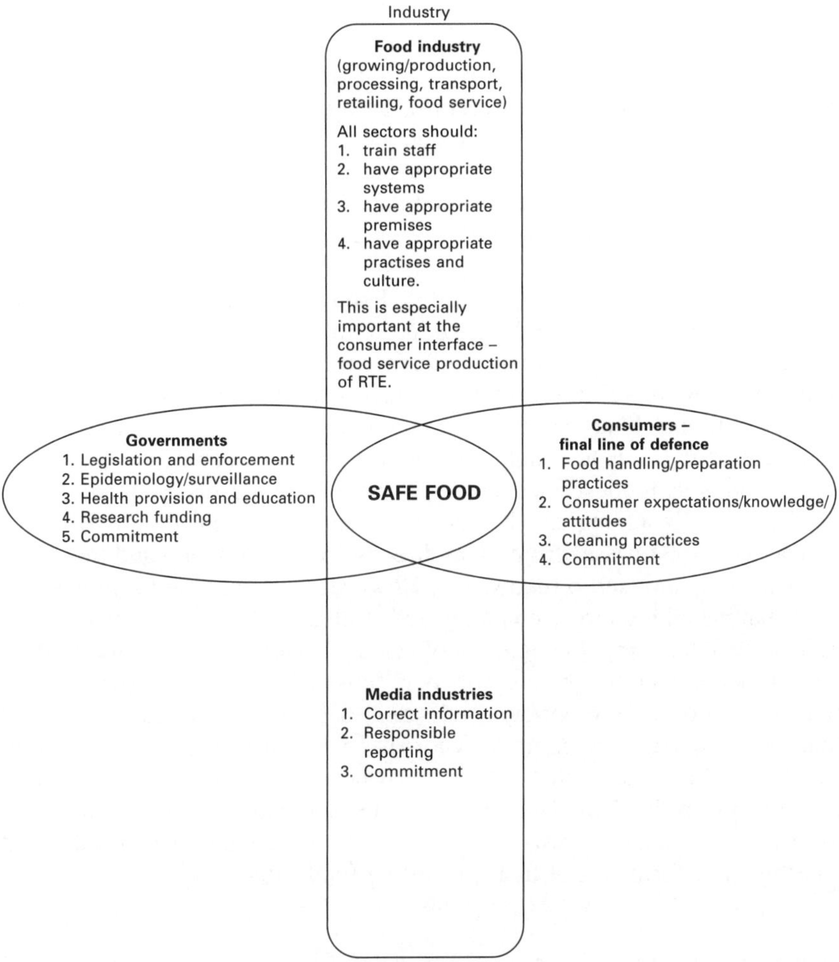

Fig. 15.1 Shared responsibility for food safety.

high food safety standards across the community.[1] The work of this authority includes enforcement and legislation as well as acting as a scientific reference point. It is perhaps inevitable given the economic importance of food and recent political and international disagreements relating to food and trade that food safety needs to be agreed at a supranational level.

Industry includes all links in the food chain from primary production through retailing to food service. Their role is to produce, market and label food of the best microbiological quality possible within moral, technical and financial constraints. Currently, nutritional and compositional labelling has a higher profile than food safety labelling although the former is highly important with respect to food allergens. Production of high-quality food requires a trained workforce and an appropriate food safety management system, which under EU law must be documented and based upon Hazard

Analysis Critical Control Point (HACCP) principles. This must then be implemented within an appropriate food safety culture.

The media industries also have an important role to play, one which perhaps they may not even recognise, let alone implement appropriately. Elements of the mass media may argue their main role is to entertain not to educate, yet they should provide adequate, accurate food safety information in an unbiased way. It has been stated that the news media are an effective food safety watchdog but are guilty of exaggerating reports on food safety concerns.[2]

Consumers also have a responsibility. Food-handling practices (including reconsitution and storage of powdered infant formula) employed by consumers in the domestic kitchen influence the risk of pathogen survival and multiplication, as well as cross-contamination to other products. They may hope or anticipate that the food reaching them should be pathogen free, but their responsibility is to handle food as if it *was* contaminated. Consumers, along with the food service industry and the producers of ready-to-eat (RTE) foods (including salads and fresh fruits) represent the final line of defence, i.e. food passes from them directly for consumption and thus their actions are paramount to food safety. However, it is possible for previously uncontaminated food to be contaminated, with a variety of pathogens, at the point of final handling. Recognition of personal responsibility for food safety during food preparation by consumers is considered to be a pre-requisite for implementation of appropriate food safety behaviours.[3] Research suggests that consumers are beginning to recognise personal responsibility for food safety, but still consider that external providers of food are also accountable to maintain levels of food safety.[4] For safe food and thus safe eating to be achieved all who have a responsibility must have a commitment to safe food and this must form part of their prevailing food safety culture.

15.1.2 Epidemiology of foodborne diseases – 'eating out and eating in'

People must eat to survive, yet they should not be come ill as a consequence. Data concerning the incidence of foodborne illness are usually incomplete but suggest that it is a major problem in many countries. A recent UK study projected one in five members of the population in England and Wales would suffer infectious intestinal disease each year of which 20–40 % were likely to be foodborne.[5,6] Other US studies indicate a similar general figure with variation depending upon pathogens involved.[7] If this were true it would indicate a significant morbidity due to food-transmitted illnesses. In broad terms the food people eat can be considered as prepared inside the home or outside the home. Whilst large outbreaks which inevitably occur outside the home capture headlines, most cases of food-borne illnesses are apparently sporadic, with food prepared in the home an important vehicle of illness.[8] With the development of better typing techniques some sporadic cases involving RTE produce, especially salad items, consumed in the home are now being identified as part of larger outbreaks.

It is difficult to determine accurately the number of cases occurring in a domestic setting; however, the probability is that the home is the single most important location for acquiring food poisoning. Table 15.1 indicates a UK ratio of general to family outbreaks of approximately 1:6. Whilst it cannot be assumed that all family outbreaks are acquired in the home, it is likely that the majority are. Furthermore approximately 12 % of all general outbreaks have their origin in a domestic setting and this seems to be more or less typical of other developed countries.[9,10,11,12] These data relate to notifications and, whilst there are likely to be more cases reported per general outbreak than per family outbreak, it is probable that the large majority of unreported cases occur in the home. Collectively, this type of data suggests that the home is likely to be a major location for the acquisition of food poisoning and highlights the need for a greater understanding of consumer food-handling behaviour and a need to educate the consumer. This need is supported by various studies of consumer knowledge, attitudes and practices.[11,12,13,14,15,16,17,18,19] Such studies indicate that consumers often do not practise appropriate food hygiene, e.g. hand washing, even if they possess the relevant knowledge. When food poisoning does occur outside the home it is most likely to be associated with a food service establishment.[20] It has been estimated that in the UK 1 in 1500 food service establishments gives rise to a notified case of food poisoning every year[21] although this is likely to be an under-estimate. The understanding of risk and the potential for causing illness is poorly understood both within food service establishments and by consumers (see Section 15.3.3).

15.1.3 Risk factors and errors associated with foodborne illnesses

Surveillance authorities around the world have attempted to identify factors contributing to outbreaks of foodborne illness. Unfortunately, different methodologies are used to collect the data with different approaches and categories used for presenting or grouping the results (see Table 15.2).[22] The rationale behind this approach is to understand and thus manage foodborne illness based upon preventing the repetition of previous malpractices. Whilst useful, this does not provide a full explanation of what may have gone

Table 15.1 The home as a location for food poisoning outbreaks

	1992	1993	1994	1995	1996
Ratio of general to family outbreaks	1:6.7	1:7.1	1:7.0	1:5.6	1:6.0
Percentage of general outbreaks of food poisoning occurring in the home	19 %	14 %	13 %	11 %	16 %

Source: Food Hygiene '98 Conference.[8]

Table 15.2 Risk factors implicated by various studies in outbreaks of food poisoning

Contributory factor	USA data %	England and Wales data %	England and Wales data %	USA data %	England and Wales data %	England and Wales data %
Inappropriate storage	21.1	38.5	24.4	23.9	45	45
Preparation of food in advance	22.6	57.1	–	9.9	–	–
Inadequate heating	15.5	15.8	23.3	20.0	50	–
Inadequate re-heating	10.6	26.4	–	8.5	–	40
Inadequate hot-holding	16.6	–	–	17.3	–	–
Cross-contamination	5.4	6.4	22.0	8.9	39	36
Infected food handlers					12	21

Source: Adapted from Coleman and Griffith,[21] with additional information from Griffith.[25]

wrong, and this is particularly true with cross-contamination. Reported as a risk factor in up to 38 % of outbreaks even this is likely to be an underestimate. For example, food handlers may remember if a burger or chicken looked undercooked but they are less likely to remember if a cloth used to wipe up after raw product was subsequently used 30 minutes later to wipe equipment/surfaces in contact with RTE foods.

Another problem with the way risk factors are reported is that they do not tell us what the underlying cause was, e.g. undercooking of food could be due to an equipment failure, food handler error, e.g. incorrect setting of cooking temperature, a cultural error, e.g. taking food out of the oven before fully cooked to serve waiting customers, or a management system error, e.g. incorrect cooking instructions. A case of food poisoning most likely represents failure of one or more types, and a generalised scheme of error categories which can, in the presence of hazards, result in illness is illustrated in Fig. 15.2.

It is suggested that human error is a factor in 97 % of general outbreaks of food poisoning.[23] However, caution is needed because, although often perceived as separate, there is frequently considerable overlap between each of the categories. It may be difficult to allocate failure, in an outbreak, to a single error category, and often a combination of two or more will be involved. For example, it has been suggested that over 50 % of human errors can in reality be related to the 'prevailing food safety culture and thus poor

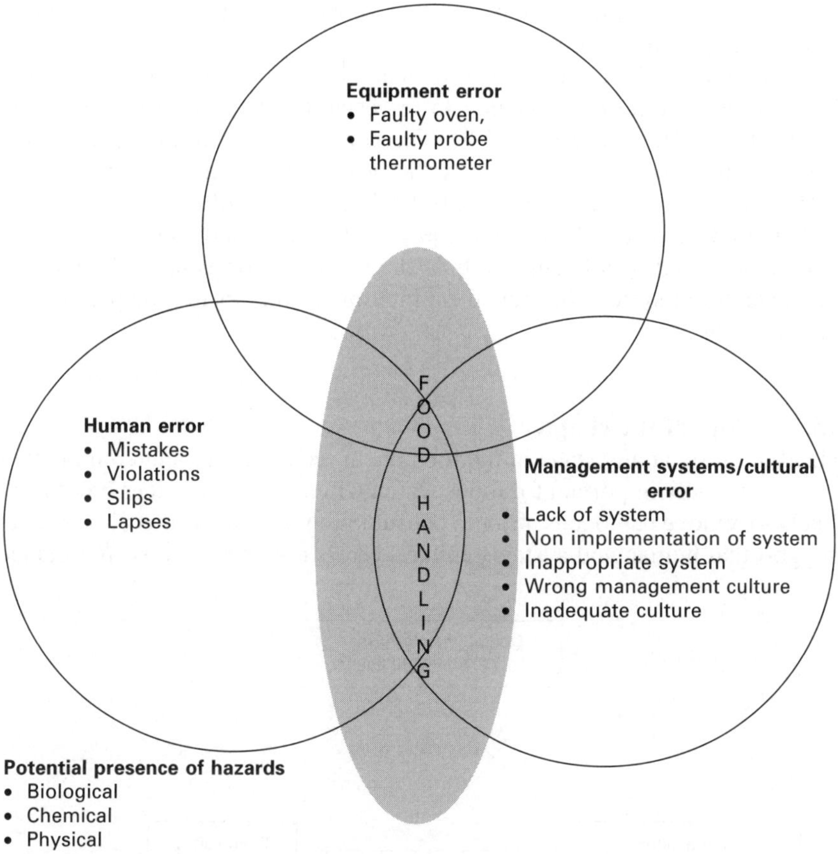

Fig. 15.2 Generalised scheme of error categories. Food handling with one or more errors, in the presence of hazards, can result in food poisoning.

leadership'. This is illustrated by an outbreak of hepatitis A in a Washington subway sandwich store. An investigation revealed that one or more of the employees was infected with the hepatitis virus at the time of the outbreak. Upon initial examination it could be recorded as human operator error with the handlers contaminating the food. However, subsequent action was taken stating that 'the store operators were negligent in their selection, training and supervision of their employees and in failing to adequately instruct and supervise them and provide them with the necessary health and safety practices to prevent contamination of food with the virus'. This suggests the error was a management/systems error. Another illustration from the UK demonstrates this difficulty and involved a chain of public houses/ restaurants. In this case poor food-handling practices were publicised by a journalist before food poisoning occurred. It involved trained staff being persuaded by a unit manager to deliberately implement poor food safety practices, using food out of shelf-life and redate-coding food, for financial reasons. The prevailing action was that economics and culture was more important than their food safety.

Superficially this could appear as human (food handlers) error but, in reality it was a managerial error leading to the establishment of an inappropriate food safety culture (see Fig. 15.3). The concept of an in appropriate food safety culture was recently used by the prosecution in an *Escherichia coli* O157 outbreak in South Wales (http://wales.gov.uk/ecoliinquiry/?lang=en). Relatively few outbreaks involve equipment failure, and using this type of generic approach would suggest that more information is needed about the human/cultural/systems/management intermix of factors in food preparation and handling.

15.1.4 Aim of the chapter

Handling food at any stage in the food chain requires care and commitment but is especially important for foods which will receive no further processing likely to remove pathogens, prior to consumption. Investigations of outbreaks suggest that human and systems/cultural errors are of particular importance

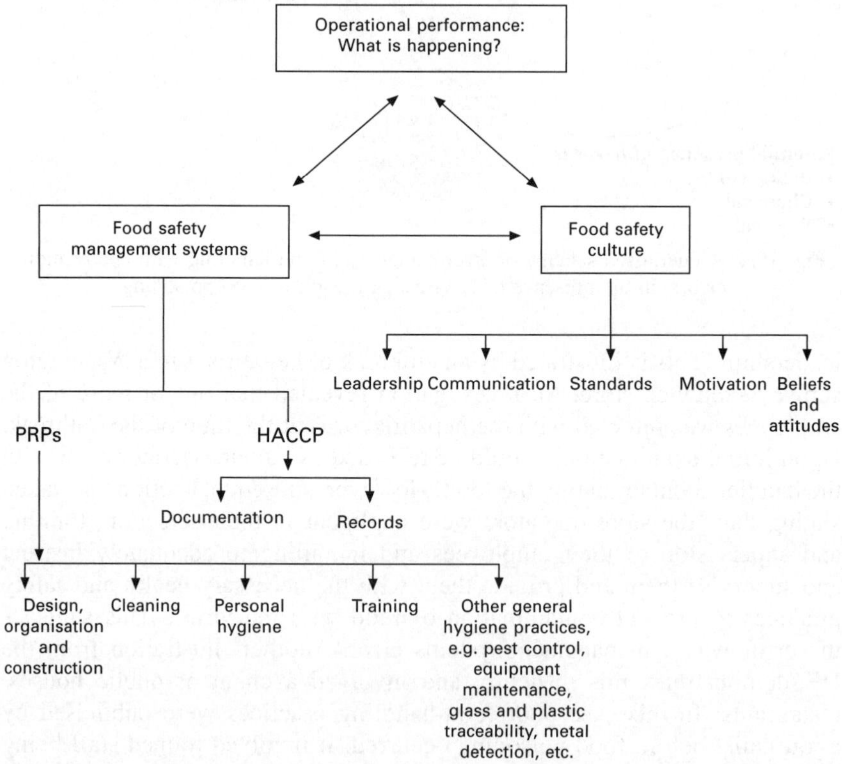

Fig. 15.3 Operational performance: what is happening?

and the aim of this chapter is to inform about factors influencing these errors, especially those close to the point of consumption.

15.2 Food-handling practices and food safety management systems

15.2.1 Food manufacturers, HACCP and GMP

Food safety legislation around the world is being updated in an attempt to minimise foodborne illness. A key feature of the newer legislation is the incorporation of a risk-based approach to management requiring the identification and control of steps which are critical to food safety as well as, if necessary, the ability to demonstrate 'due diligence'. To fulfil the latter, an appropriate food safety management system is required plus evidence that it is being effectively implemented. Failure to comply with legislation, which in the UK applies to all food businesses, can result in fines, closure or even jail.

All food businesses should have a documented food safety management system based upon Hazard Analysis Critical Control Point (HACCP) principles. This can be defined as the organisational structures, responsibilities, practices, procedures and appropriate resources to produce food safely.

Although developed for food manufacturers HACCP is a food safety management system which can be adapted for use even in the small food operations.[24] However, it cannot be used in a vacuum and should be implemented in conjunction with a range of pre-requisite programmes (PRPs). These are part of good manufacturing practice (GMP) and cover the fundamental principles, procedures, running, design and construction of an environment for the production of food of acceptable quality. The good hygiene and manufacturing processes of a company form the foundation upon which HACCP plans can be built and provide the framework within which a company can develop a positive food safety culture. Organisational culture is an elusive concept which can be defined in various ways, but most simply it can be described as 'the way we do things around here' and can help to contribute to subjective norms (see Section 15.3.2 and Fig. 15.5).

15.2.2 Food service establishments

HACCP has generally been implemented in larger food manufacturing operations in developed countries. Whilst its application to all food operations has been advocated, it is less well implemented in small or less developed businesses (SLDBs).[25] The reasons for this are summarised in Table 15.3.

Because of the difficulties in introducing HACCP into smaller, less developed food service businesses, adaptations of the HACCP approach in the form of a more 'user-friendly format' have been attempted. These

Table 15.3 Factors which could make it more difficult to implement HACCP in food service operations

Food manufacturers	Food service
Restricted range of food items	Wide range of food items
Better technical support	Less technical support (none in some cases)
Dedicated design and construction	'Ad hoc' use of premises Poorer design and construction
Production in relation to scheduling i.e. work to stock	Production in peaks and troughs to meet demand i.e. work to order
Produce only	Produce and serve
Better trained labour force	Poorer trained, high turn over, often part time labour force
Better facilities and equipment	Poorer facilities and equipment
Fewer, usually larger, businesses	Many more businesses usually much smaller
Greater business demand for HACCP, e.g. from retailers	Less business demand.

more user-friendly or simplified approaches to HACCP vary in complexity with some being less easily recognisable as HACCP was (see Table 15.4). Some make use of generic process flow diagrams, others have eliminated their use entirely. Some monitor and report critical control points in a fairly conventional way, others rely on sensory assessment with reporting only by exception.

Strategies designed to help SLDBs introduce HACCP have been reviewed[24] and include the development of sector-specific industry guides which have been produced in a number of countries. The UK version[26] assists food businesses to comply with the legal requirement to analyse and control potential food hazards, as well as providing advice on training and good catering practice (GCP). GCP is analogous to GMP and deals with the effective management of food safety and quality in catering/food service businesses.

15.2.3 Domestic kitchens – GDKP

The domestic kitchen, unless used for business, is exempt from legislation. Hence there is no requirement for HACCP, although its application has been recommended.[27] Implementation of good domestic kitchen practice (GDKP) is likely to be the consumer's approach to food safety/hygiene, is dependent upon the individual food handler (consumer), is extremely variable and is likely to be informal.

The role of the consumer in food safety, ignored for many years, is currently the subject of investigation and research.[13,14,15,16,17,18,19] In general terms studies from around the world have indicated standards of GDKP are a cause for concern with failure to implement even basic hygiene practices. Such failures

Table 15.4 Comparison of models

	SFBB[a]	Safe Catering	SafeFood[b]	CookSafe[c]	FDA[d]
1. Similarity to conventional HACCP	No	Yes	Yes	Yes	Yes
2. Flow diagram	No	No[e]	Yes	No	Yes
3. Reporting by exception	Yes	No	No	No	No
4. Emphasis	Provision of information and some management	Structured management records/forms	Comprehensive range of documents/forms/policies for small to large businesses	Management and information	Emphasis on doing not filling in
5. Information on PRPs	Integrated	Some	Extensive with key tips	Some with emphasis on house rules	Some but separate
6. Use of technical terms	No	Some with explanation	Some with explanation	Some with explanation	Extensive with additional information
7. Extent of required business input	Minimal +	Moderate ++	Moderate ++	Moderate ++	Extensive ++++
8. Specific 'Sector' packs	Yes	No	Yes	No	No
	→ Increased complexity				

[a]Safer Food Better Business.
[b]SafeFood – System developed in South Africa based on focus group evaluation of UK models.
[c]CookSafe – Model developed in Scotland.
[d]FDA – US Food and Drug Administration model.
[e]Safe Catering – Developed in NI – earlier version had a flow diagram but not in current version, although use is optional.

may be due a multitude of reasons including perceptions of under-estimated risk, optimistic bias and the illusion of control.[4] Under-estimation of personal risk to food may prevent consumers from taking appropriate steps to reduce their exposure to microbiological food-related hazards.[28] Research suggests that consumers only think about safe food preparation behaviours when they perceive a risk, yet many consumers may have mistaken ideas about which practices are effective at reducing risks.[29] Food safety failures may be due to a basic lack of awareness of all the practices that are needed to handle food safely or of how to clean properly. There is evidence that domestic kitchens are more difficult to clean than their commercial equivalents.[30] They may be subject to contamination from non-food sources, e.g. pets, and may be used for a variety of non-food handling activities, e.g. gardening, motor maintenance.[31] It is not surprising that many food pathogens have been isolated from a range of kitchen sites,[25] and the kitchen is thought to be more heavily contaminated than bathrooms and toilets.[32]

Food hygiene failure in the domestic kitchen is compounded by the fact that consumers often do not implement the food hygiene practices that they know. Although knowledge of the consequences of unsafe food-handling practices can enhance consumer motivation to implement safe behaviours, a substantial amount of research has established that provision of knowledge does not necessarily translate into practice.[33,34,35] Furthermore nearly three quarters of UK consumers have indicated they do not worry that they might not always carry out all of the food hygiene practices they know.[18] Further work is required to determine why there is such a discrepancy between knowledge and behaviour. Some evidence suggests that people find implementation of hygiene practices not personally applicable, too much trouble or too time-consuming.

Caregivers, including health professionals and mothers of young infants, have an important responsibility to prevent infant diarrhoea and infections by implementing safe-handling behaviours when preparing and storing complementary food and powdered formula milk. *Enterobacter sakazakii* (proposed *Cronobacter*) can be a problem, especially in hospitals, and has been isolated from 2–12 % of powdered infant formula milk.[36,37,38] Even low levels of contamination are considered to be a risk factor, given the potential for multiplication during the preparation and holding time prior to consumption of reconstituted formula.[38,39] The growth and survival properties of *E. sakazakii* on preparation and feeding equipment and in reconstituted formula emphasises the need to ensure adequate decontamination of feeding utensils after use and proper temperature control of reconstituted formula.

The importance of preventing foodborne infections from in-home preparation has now been realised and a number of countries have, or are planning, food safety initiatives. These should raise the general awareness and importance of GDKP (food hygiene) and should present the information within a framework based on hazard, risk and specific control measures.

15.3 Understanding food handlers' behaviour

15.3.1 Introduction

It has been suggested that poor food-handling practices contribute to 97 % of foodborne illnesses in food service establishments and the home.[23] If this is correct food-handling behaviour is the single most important factor affecting the control of food hazards and in managing risk. Although work on understanding aspects of human behaviour has been undertaken by psychologists there is very little information on why or how people implement food safety practices. Also lacking are data that link individual food handler behaviour to a company's food safety culture. A better understanding of both of these could help to facilitate correct food-handling and identify appropriate behavioural change.

The need to understand the human aspects of food-handling with the involvement of behavioural scientists, has been recognised.[40] In one recent study of food handlers' beliefs 62 % of food handlers admitted to sometimes not carrying out all food safety behaviours on every appropriate occasion with 6 % admitting that they often did not.[41] Lack of time was the most quoted reason for failure to implement relevant practices. In food service operations this could be related to unreasonable demands made on food handlers by their employers. Possible factors influencing human food safety behaviour are summarised in Fig. 15.4. However, it must be realised that food handlers, especially consumers, are not a homogeneous group, they are

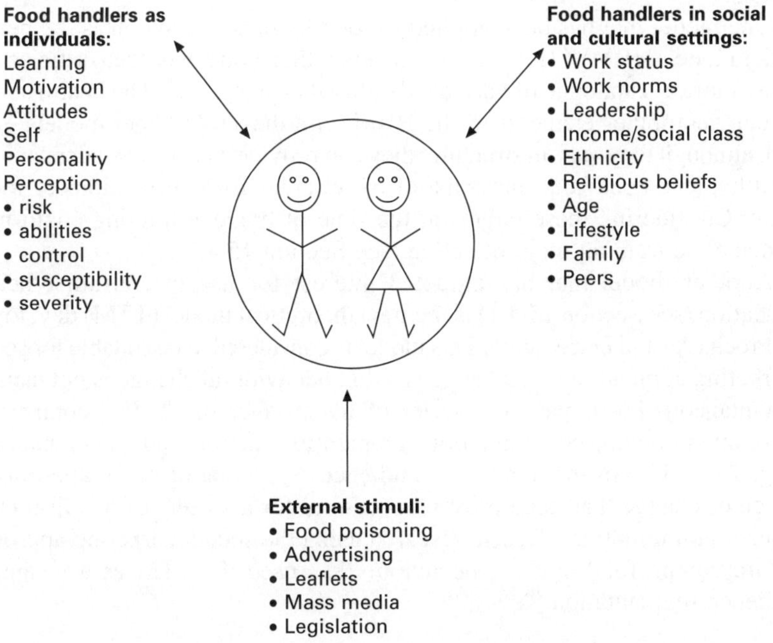

Fig. 15.4 Factors influencing food handlers' behaviour.

individuals and will not all behave in the same way or necessarily in a way food safety experts might expect.

15.3.2 Behavioural and psychological models

In attempting to interpret, study or even predict behaviour it is helpful to use existing theoretical models. The simplest of these is the knowledge, attitudes, practice (KAP) model, and this is used as the basis for much hygiene training. This relies on the provision of information (knowledge) to modify a food handler's attitudes and then behaviour. The approach takes no cogniscence of the conditions, culture and environment in which the food handler may be working.

More sophisticated and successful models try to consider many more of the factors which can influence behaviour. Social cognition models start from the assumption that an individual's behaviour is best understood in terms of his or her perceptions of their social environment. Two of the most widely used models are the theory of planned behaviour (TPB) and the theory of reasoned action (TRA) (Fig. 15.5). These theories aim to measure behavioural intentions, recognising that certain uncontrollable factors can inhibit implementation and thus prediction of actual behaviour, e.g. an individual intends to wash their hands, but arrives at the sink to find there is no soap. These models include the element of subjective norm which considers the perceived beliefs of others and the individual's desire to comply with those beliefs.

One model that has been applied to food hygiene behaviour is the health belief model (HBM)[42] (Fig. 15.6). However, this model has been criticised as being more a 'catalogue of variables' rather than a model.[43] The health action model[42] combines elements of the HBM with the TRA. Other models exist and, although they vary in structure, they can provide a basis for understanding food hygiene behaviour and a list of target areas for interventions to focus upon. Considering these targets at the time of hygiene training could help to make the training more effective (see Section 15.4.3).

Another model that has attracted interest for use in consumer health education (see Section 15.4.1) is the transtheoretical model (TTM) developed by Prochaska and Diclemente. This model is considered to be suitable for social marketing applications[44] and suggests that behavioural change is not usually instantaneous but is part of a series of five 'stages of change' comprising pre-contemplation, contemplation, preparation, action and maintenance.[45] (Fig. 15.7). Use of this model for audience segmentation to an appropriate 'stage of change' has been proven successful for a variety of health-related behavioural initiatives. Recent research using the social marketing approach for improving food hygiene behaviours has used the TTM as a means of audience segmentation.[46,47]

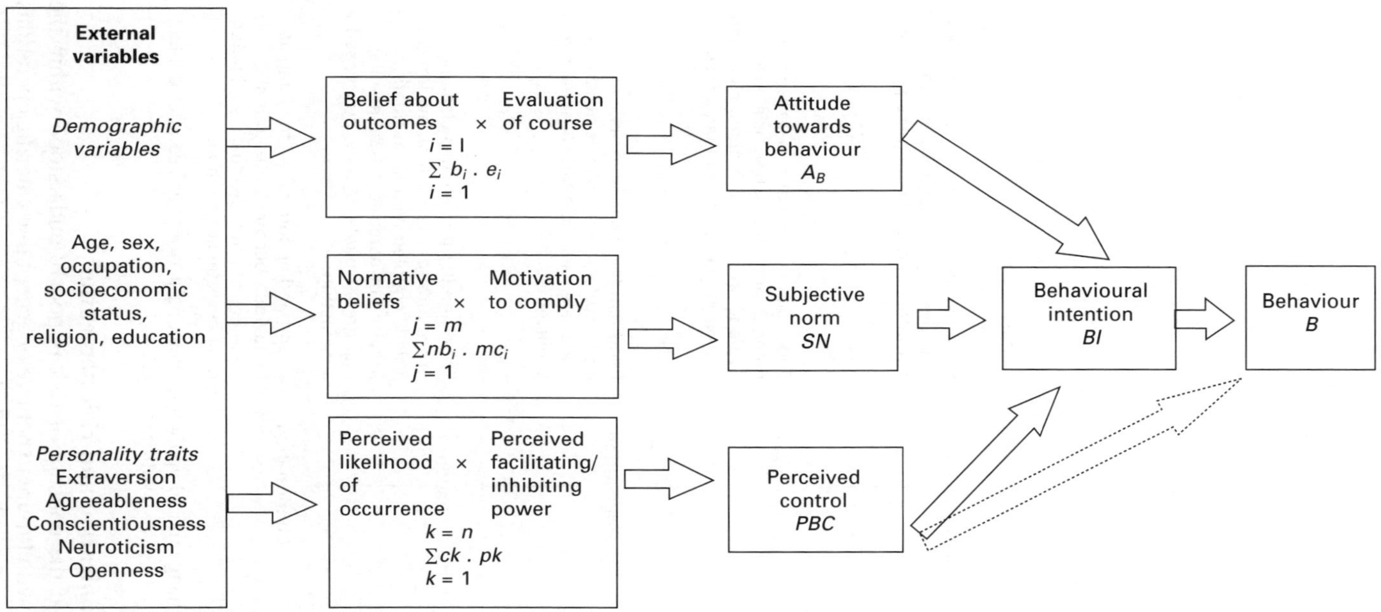

Theory of reasoned action:
This model emphasises the relationships between specific attitudes and specific behaviours, with 'intention' as the mediator between these two elements. Behaviour is a linear regression function of intention to behave and perceived control over that behaviour:

$$B = w_1\, BI + w_2\, PBC$$

Where B = behaviour, BI = behavioural intention, PBC = perceived behavioural control, w_1 w_2 = regression weights. Whilst this model has been used in food choice, there have been no publications to date of its practical application to food hygiene

Fig. 15.5 Theory of reasoned action.

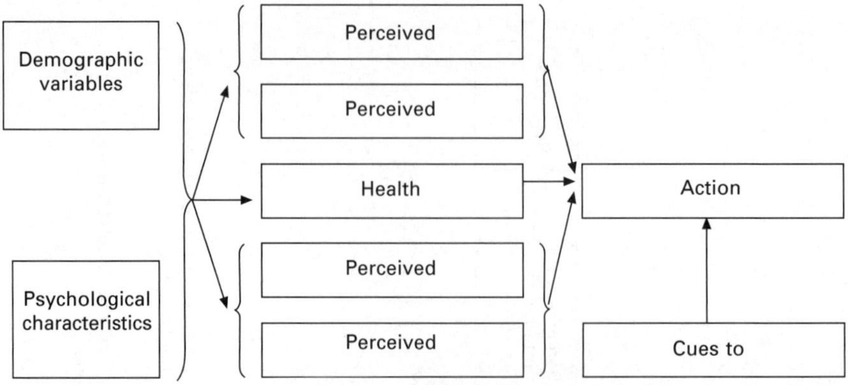

Fig. 15.6 The health belief model.

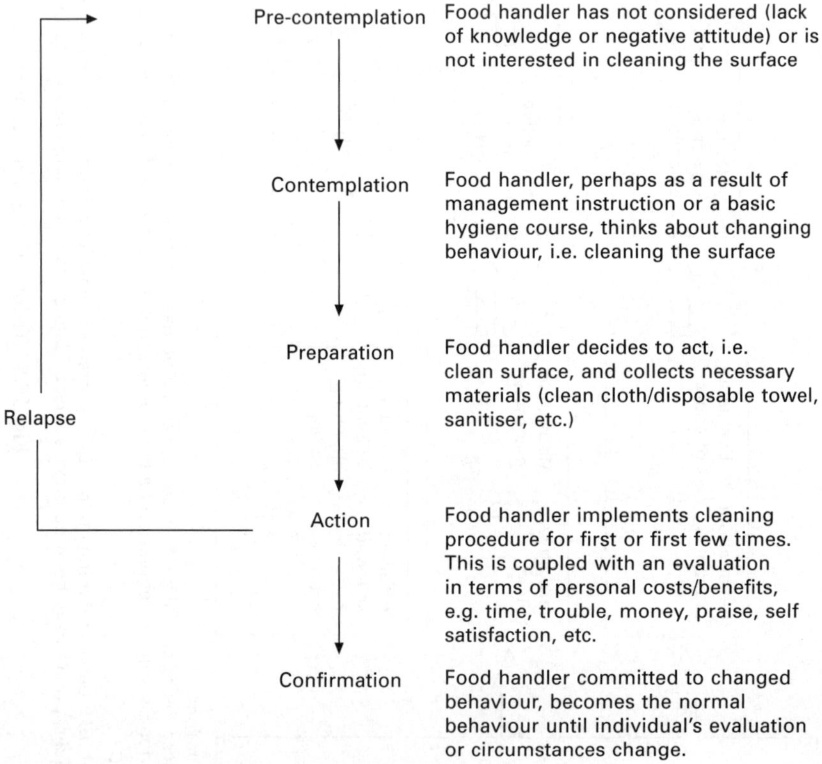

Fig. 15.7 Stages of change model applied to cleaning a work surface after contact with raw food.

15.3.3 Risk communication and risk perception

There is evidence that the concept of risk is poorly understood within the food service industry with failure to believe or accept that catering operations present a food safety risk. Behavioural change is most likely to follow self realisation of the level of risk posed, and self-assessment checklists can be useful devices for achieving this.[22]

Given that zero risk with regard to food safety is unlikely then government and the agrifood industry have a responsibility for communicating information on risk and risk management to the consumer.[17] Risk communication is an important element of the work of the Food Standards Agency (FSA) in the UK;[48] it is vital to ensure transparency and, where possible, for consumers to participate in decision making. It is important to inform and advise them about risks in a way which makes it clear that they have the ability to manage or control the risks.

This latter aspect could be of concern to specific sectors of the food industry. If there is a problem with presence of pathogens in raw foods, e.g. *Campylobacter* in poultry, then risk needs to be communicated to consumers in a way which both alerts them to the problem and informs them of how they can reduce/manage the risk. This is known as the reassurance/arousal paradox. Failure by a sector of industry to inform consumers about risk associated with a food could lead to an increase in outrage (see Table 15.5). What is clear is that risk is not just about science but is heavily influenced by psychology, and experts and the public perceive risk in very different ways.[49]

In general terms, risk communication should enable the recipients to understand the nature of the problem, help or support decision making and be unbiased. There is debate over the precise role of the government in this process – should this be restricted merely to provision of information or should it include modifying or influencing behaviour? A detailed study of risk communication is beyond the scope of this chapter and readers should be aware of useful reviews,[49,50] however, it is worthwhile considering some of the pitfalls/barriers in effective risk communication.

A common misconception is that consumers and experts view and consider risk in the same way. This applies to the levels of risk associated with food

Table 15.5 Factors relating to hazards that are likely to increase consumer outrage

Acceptance/low outrage	Rejection/high outrage
Natural	Technological
Voluntary	Compulsory
Obvious consumer benefits	No obvious consumer benefits
Enjoyable	Not enjoyable
Consumer control	No consumer control
Information readily available/open	Secrecy, information not available, restricted
Familiar	Exotic
Trustworthy sources	Untrustworthy sources
Hazard has limited damage potential (severity or dread)	Widespread damage potential, can be passed through generations
Fair	Unfair

hazards and the likely response to them, i.e. the risks considered by experts to be important may not the ones that concern the public. Experts believe all they have to do is present the public with current risk data and consumers will change their behaviour. When this does not happen a common response by experts is to believe that the public are stupid or at best ignorant. One explanation is that risk is a combination of hazard + outrage. Outrage concerns the emotive response by the public to information on hazards. It has been suggested that experts respond to hazard information and the public responds to outrage. Factors likely to increase outrage are outlined in Table 15.5.

Another problem is that whilst consumers and food businesses may accept there is a risk to the public in general, they may not realise the risk applies to them. Known as unrealistic optimism or optimistic bias, this can apply to food poisoning in the home and in food service establishments. Food handlers believe (perhaps incorrectly) that food poisoning is less likely from food prepared in their kitchens. A possible contributing explanation to this and a factor which should be considered in education strategies concerns the illusion of control. Food handlers see or hear messages about food-handling and agree this is necessary information for the public. However, what can happen is that individual food handlers fail to realise that the message applies to them, erroneously believing that their own practices and behaviour are adequately hygienic and capable of controlling food hazards.

15.3.4 Sources of food safety information

In general terms, larger food businesses have access to the latest food safety information. Whilst increasing use of the Internet may assist smaller businesses and consumers, there is evidence that they currently lack information on food safety.[20] Improved channels of communications, possibly via Environmental Health Departments or by small business/trade associations, could help small manufacturers/food services. In the UK, local authorities (LAs), especially Environmental Health departments, appear to be the most significant disseminators of consumer food hygiene information. However, for consumers to benefit from such information, LAs need to become accessible to consumers and assume a more proactive role in the promotion of home food hygiene.[18]

Consumers acquire food safety information from a variety of sources including the home, school and others, e.g. the media including television, printed materials and food safety campaigns. The role of the mass media for informing consumers about food safety risks is of critical importance, and the media have a responsibility to ensure that advice they provide is accurate and adequate.[51] Although media coverage of food safety issues in recent years has heightened consumer awareness of microbiological safety it has also increased consumer concern and confusion.[52] The media have been responsible for sensationalising so-called 'food scares' whereby consumers' initial anxieties regarding the issue are commonly amplified. For example, in response to the *Salmonella* in eggs 'scare' in 1988, consumers' risk perceptions increased not only to the same level of the actual risk, but also to a level far greater than the actual risk.[53] Consumers need to receive consistent food safety information not only to prevent panic instigated by careless communications but also to promote safe food-handling behaviours in a manner that is accurate and credible.

Channels and sources generally used in public communication efforts include a variety of formats such as television, radio, posters, leaflets, newspapers, cookery books, magazines and reminder aids. Although limited research has been conducted to evaluate the effectiveness of different intervention types, it has been reported that the potential effectiveness of different media elements does vary considerably, despite having common characteristics. Research in the UK evaluating consumer perceptions of different sources of food safety information has shown that Environmental Health Departments and the Food Standards Agency were perceived to be the most trusted and credible organisations providing food safety information. The most believable spokespersons for promotion of food safety advice were determined as Environmental Health Officers and the Chief Medical Officer and the most preferred source of food safety information identified was food packaging, followed by advice from a medical doctor.[18,54]

Schools are acknowledged as important places for developing health promotion and influencing health-related behaviours – including hygiene practices.[55] Indeed, parents may not have the correct information to give to their children and there is evidence that consumers' (parents) knowledge about food safety is inadequate. Hygiene teaching in schools has suffered, certainly in the UK, as a result of a packed school curriculum. Recently food safety material specially designed for schools has been produced and its usage could help to initiate good practice at an early age.[52] Elements of the mass media, especially TV, offer a powerful means of influencing behaviour; unfortunately they do not always project the correct message and TV cookery programmes, with highly respected or admired chefs acting as a role model, can illustrate bad practices.[51] Furthermore there are suggestions that the media present conflicting advice causing confusion in consumers.[56] In spite of this in the USA, the majority of consumers depend on the media as their primary information source. Initiatives designed to inform the consumer need to use a wide variety of strategies. Traditionally, food safety health education has overly depended upon the use of leaflets, although this is changing.[57]

15.4 Helping people to change behaviour – education and training

15.4.1 Educating the consumer

Until relatively recently, educating the consumer concerning food hygiene has been largely ignored. Publicity concerning food poisoning in the UK in the late 1980s led to the widespread distribution of standardised, knowledge based leaflets with food safety advice.[51] Although acquisition of knowledge must precede behavioural change, and knowledge of the consequences of unsafe food-handling practices can enhance consumer motivation to change behaviour, findings from other areas of health education suggest that, whilst leaflets can play a role in raising awareness and supporting communication initiatives, they generally do not bring about the behavioural change.[40] Leaflets can provide people with the information enabling them to change, if they are motivated to and want to change. Other consequences of the publicity in the UK included the realisation that more information concerning knowledge, attitudes and behaviour was required, and a food safety initiative was launched by the Food and Drink Federation (FDF)[51]. More recently, to achieve the 20 % UK FSA target for foodborne disease reduction, a national food hygiene campaign was implemented.[58] The campaign was based upon increasing awareness and understanding of 'The 4 C's' (cleanliness, cooking, chilling and cross-contamination). In the UK the FSA, FDF and the Food Safety Promotion Board (FSPB) (Ireland) are the largest providers of consumer food safety information.

In the USA a National Food Safety Initiative was set up to provide targeted information for consumers. The Fight BAC! Campaign was a product of the Partnership for Food Safety Education which is a unique public–private partnership of government and consumer groups dedicated to increasing awareness of food safety and reducing the incidence of foodborne illness. The Campaign is based on four food safety messages ('*Clean*' – wash hands and surfaces often, '*Separate*' – don't cross-contaminate, '*Chill*' – refrigerate properly and '*Cook*' – cook to proper temperatures),[59] and BAC!, a big, green 'bacterium' character, has served as the focal point to the campaign.[60] A recent addition to the Fight BAC! initiative has been the introduction of 'Thermy', a cartoon thermometer. Such a character has been used to support the Fight BAC! message of '*Cook*', based on studies that have indicated there is significant risk of foodborne illness when colour is used to judge if a food has been cooked to a safe temperature.[59] Intervention materials have not only been targeted at specific food safety behaviours but also aimed at specific groups of consumers, and they have included a wide range of media formats, some of which have been interactive.

To improve food safety at an international level The World Health Organisation has developed a food safety initiative entitled 'The Five Keys to Safer Food' to teach safe food-handling practices.[61] This initiative has been developed from the 'Ten Golden Rules for Safe Food Preparation' in the early 1990s but is simpler and generally more applicable to food handlers, especially those in high-risk groups and their caregivers. Implementation of this initiative has been undertaken by WHO regional offices, and interventions have been translated into over 40 languages.[61]

In addition to national initiatives, a number of research initiatives for collecting information concerning consumer food safety were launched, although they mostly relied upon self-report of behaviour. A few studies have focused upon actual behaviour, and these have indicated that consumers often do not know or understand hygiene advice and also that they often may not implement known hygiene practices.[16] Other behavioural and attitudinal studies have identified that consumers may believe they implement adequate safe food-handling behaviours but their perception of what is adequate has been frequently reported and observed to be inadequate.[18,62]

It has been suggested that the future of hygiene promotion should be based on the analysis of the specific needs of the target audience. A contemporary approach to structured behavioural change for health education initiatives has been the application of social marketing to a variety of public health-related disciplines[44] including consumer food safety.[46,57,59,63] Social marketing is a social change strategy that uses traditional marketing techniques from the commercial sector. The approach focuses on voluntary behavioural change to benefit the individual and society, rather than coercing consumers to adopt healthy behaviours. Use of the social marketing approach facilitates the development of a consumer-orientated strategy whereby the needs and wants of targeted consumers are actively sought and acted upon in programme planning, management and evaluation. This approach requires a careful evaluation of consumer behaviour and the identification of target risk groups using market segmentation. A precise message is identified and the target audience's response to and beliefs about the message is assessed as well as the likelihood they will change behaviour. Results from these preparatory stages are then used to plan the initiative based upon marketing principles (involving the 4 Ps or marketing mix, see Table 15.6). After implementation the initiative is evaluated, preferably based upon assessment of actual behaviour.

This type of approach is more comprehensive than an ordinary health educational approach. The latter is generally more concerned with imparting and receiving messages and with people learning facts. Social marketers argue that learning is only important if it results in the desired behavioural

Table 15.6 The 4 Ps of marketing – marketing mix

Product	What is being offered must be acceptable and as tangible, accessible and attractive as possible. The changed behaviour must not be excessively expensive, time consuming, totally impracticable, painful, etc.
Price	Decisions to act are as a result of a consideration of costs and benefits. Ways to minimise the cost as well as the benefits need to be addressed in strategies for behavioural change.
Place	The means to implement new behaviour must be readily available and convenient, e.g. availability of soap for handwashing.
Promotion	The message needs to be 'advertised' in the most appropriate way. This could include leaflets/TV/cook books/magazines or individual advice, e.g. health visitors.

change. Whilst there are differences in marketing commercial products compared to improved safety in food-handling, not least in terms of budgets available, there are potential benefits in adopting the approach. Relatively small improvements/changes in behaviour could significantly reduce the cost of foodborne disease.

15.4.2 Training food handlers

Legislation and education have been advocated as part of a dual strategy to reduce food poisoning.[40] Current UK legislation now contains a requirement for handlers to be trained commensurate with work activities. This was introduced in 1995 but, although industry considers training important, levels of training are variable, particularly with regard to type of employee and food industry sector.[64] The food service sector particularly lags behind, although there may be specific reasons for this (see Table 15.7). The legislation itself has been criticised.[65] The proprietor need not necessarily be trained and, whilst guides to implementation provide a definition of a food handler, they are not themselves legal documents.

It has been estimated that only 46 % of food handlers have been trained,[64] but even if trained there is no guarantee that food handlers will behave differently. Generally there is a lack of information on the efficacy of training, although individual studies have produced mixed results, some showing an improvement in knowledge others no improvement. In cases where knowledge improved this did not necessarily change behaviour. Many businesses do not plan to make training effective (i.e. behavioural change) but have as their prime purpose the acquisition of a certificate to show the Environmental Health Officer. Therefore it has been suggested that training should consider behavioural theory.[40] In the final analysis, whilst over four million people have been trained in the UK, given the vagueness with epidemiological data, there seems to have been no dramatic improvement in the level of food poisoning. This indicates the need for improved food safety attitudes and culture and should act as a spur to make learning more effective. In the UK the FSA claims to have reduced food poisoning, based on reported data for a restricted number of organisms, by 20 %.

Table 15.7 Factors influencing training in the food service sector

Staffing	High staff turnover Large numbers of part-time staff Low pay Low status Possible poorer educational background and language skills.
Business size	Smaller businesses can not afford time/money for training
Type of operation	Poorer facilities to practice hygiene production and service

Most studies of food hygiene learning have concentrated on formal learning as opposed to work-based learning (defined as learning at work, linked to a job requirement) although the two differ (see Table 15.8).[65] Companies should try to develop training strategies embracing both types of learning.

15.4.3 How to make training effective

Suggestions have been made to improve the quality of formal courses.[65] including changes to syllabuses, examinations and pass marks. The following should be considered in delivering or purchasing formal training:

- Employers should plan to make training effective and for employees to implement what they have learnt.
- Training should be targeted and specific and geared to the needs of audiences using work-related examples.
- Information should be delivered within a hazard/risk framework.
- The content should be simple, accurate and jargon-free, at an appropriate level and should promote understanding. It is important to cover theory and practice – what to do and why.
- Trainers should have the respect of the trainees, be considered trustworthy and reputable (and have 'street credibility').
- The trainer should examine, with trainees, likely barriers to implementation of good practice and how they can be overcome.
- Trainers should assess available facilities and equipment and ensure implementation is possible. If inadequate, management should be informed.
- The correct management culture should be in place. Trainees must work in an environment which expects implementation of all practices as the norm and establishes that inappropriate practices will not be tolerated.

Table 15.8 Formal and informal/work-based training

	Formal	Informal
Subject matter	Syllabus externally set, training courses	Content internally set, work notes/sheets/manuals.
Certification	Likely to be formally recognised	Less likely to be formally recognised
Attendance	Part-time study, often off site, day release usually in class with others	During work, one to one, peer/managerial involvement
Evaluation	Written test/knowledge based	Competency/behavioural based
Value	Valued by company	Employee valued
Relationship to organisational culture	Greater chance of being remote	Part of work culture

- A motivational framework should be provided to encourage training and implementation of learned behaviour.
- Training should address costs (including failure) as well as benefits. These may be financial, social or medical.
- Training is part of a regular programme with updating, associated with maintaining training logs and records.
- Training effectiveness should be evaluated and appropriate improvements made to training as a result.

15.5 Future trends

A substantial level of food poisoning is related to food service establishments and the home and the role of food handlers is paramount. Legislation will require food businesses to operate food safety management systems based upon good practices and increasingly in the future in conjunction with some form of specific risk management. This is less likely in the home, although elements of GDKP should be linked to risk-based advice, which should be simplified and co-ordinated at a government level.

Achieving desirable food-handling practices is likely to be of major importance but is not merely dependent upon the provision of information. Good food-handling requires understanding of the factors which influence human behaviour, and greater efforts need to be made to promote good-handling practices, including those of consumers. There have been considerable achievements with respect to food safety training in the UK and other countries, but further improvements need to be made. A greater percentage of the work force requires training, and this training needs to be made more effective. In the future more training will be needed, particularly with respect to HACCP-based approaches, however; it is paramount that business owners/managers provide the correct cultural framework and context for food safety to be practised. It is not a company's obligation merely to provide training but to allow the learning to be implemented. Ultimately failure to do so could cost a business money or even survival; however, possibly of more importance is that it may cost someone their life.

15.6 References

1. European Food Safety Authority (EFSA) – http://www.efsa.europa.eu/EFSA/efsa_locale-1178620753812_home.htm.
2. Anon. (2000) Media food safety coverage effective but accuracy concerns exist, *Food Safety Report*, 2000 **2**, 527.
3. Unklesbury N, Sneed J and Toma R (1998) College students attitudes, practices and knowledge of food safety, *Journal of Food Protection*, **61**(9), 1175–80.
4. Redmond EC and Griffith CJ (2004) Consumer perceptions of food safety risk, control and responsibility. *Appetite*, **43**, 309–19.

5. Handysides S (1999) Underascertainment of infectious intestinal disease, *Communicable Disease and Public Health*, **2**, 78–9.
6. Evans HS, Madden P, Douglas C, Adak GK, O'Brien SJ and Wall PG (1998) General outbreaks of infectious intestinal diseases in England and Wales: 1995 and 1996, *Communicable Disease Report*, **1**, 165–71.
7. Mead PS, Slutsker L, Dietz V, McCraig LF, Bresee JS, Shaprio C, Griffin PM and Tauxe RV (2000) Food-related illness and death in the United States, *Emerging Infectious Diseases,* **5**(5), 607–25.
8. Griffith CJ (1998) Food preparation risk and the consumer: food hygiene in the home, *Food Hygiene '98*, Telford.
9. Todd ECD (1992) Foodborne disease in Canada – a 10 year summary from 1975–1984, *Journal of Food Protection*, **55**, 123–32.
10. Bean NH, Goulding JS, Daniels MT, Angulo FJ (1992) Surveillance for foodborne disease outbreaks – United States, 1988–1992, *Journal of Food Protection*, **60**, 1265–86.
11. Scott E (2000) *Food safety in the home,* in JM Todd and ECD Todd (eds), *Safe Handling of Foods*, New York, Marcel Dekker, 313–34.
12. O'Brien SJ, Gillespie IA, Sivansan MA, Elson R, Hughes C and Adak GD (2007) Publication bias in foodborne outbreaks of infectious intestinal disease and its implications for evidence based policy. England and Wales, 1992–2003, *Epidemiology and Infection*, **134**, 667–74.
15. Angelillo IF, Viggiani NAM, Rizzo L and Bianco A (2000) Food handlers and foodborne diseases: knowledge, attitudes and reported behaviour in Italy, *Journal of Food Protection*, **63**, 381–5.
14. Bruhn CM, Schutz HG (1999) Consumer food safety knowledge and practices, *Journal of Food Safety*, **19**, 73–87.
15. Sammarco ML, Ripabelli G, Grasso GM (1997) Consumer attitude and awareness towards food-related hygienic hazards, *Journal of Food Safety*, **17**, 215–21.
16. Worsfold D and Griffith CJ (1997) An assessment of the standard of consumer food safety behaviour, *Journal of Food Protection*, **60**, 399–406.
17. Griffith CJ, Worsfold D and Mitchell R (1998) Food preparation, risk communication and the consumer, *Food Control*, **9**, 225–32.
18. Redmond EC, Griffith CJ, King S and Dyball M (2005) *Evaluation of consumer food safety education initiatives in the UK and determination of effective strategies for food safety risk communication (RRD–8)*, London, Food Standards Agency.
19. Mahon D, Cowan C, Henchion M and Fanning M (2006) Food handling practices of Irish beef consumers, *Journal of Food Safety*, **26**, 72–81.
20. Coleman P, Griffith CJ, Botterill D (2000) Welsh caterers: an exploratory study of attitudes towards safe food handling in the hospitality industry, *International Journal of Hospitality Management*, **19**, 145–57.
21. Coleman P and Griffith CJ (1997) Food safety legislation, risk and the caterer, *Hygiene and Nutrition in Food Service and Catering*, **1**(4), 231–44.
22. Coleman P and Griffith CJ (1998) Risk assessment: a diagnostic self assessment tool for caterers, *International Journal of Hospitality Management*, **17**, 289–301.
23. Howes MS, McEwen S, Griffiths M and Harris L (1996) Food handler certification by home study: measuring changes in knowledge and behaviour, *Dairy Food and Environmental Sanitation*, **16**, 737–44
24. WHO (1999) *Strategies for implementing HACCP in small and/or less developed businesses*, Report of a WHO consultation in collaboration with the Ministry of Health, Welfare and Sports, the Netherlands, The Hague, 16–19 June, Geneva, World Health Organization.
25. Griffith CJ (2000) *Food safety in catering establishments*, in JM Farber and ECD Todd (eds), *Safe Handling of Foods*, New York, Marcel Dekker, 235–56.
26. Food Safety and Hygiene Working Group (1997) *Industry Guide to Good Hygiene Practices: Catering Guide*, London, Chadwick House.

27. Griffith CJ and Worsfold D (1994) Application of HACCP to food preparation practices in domestic kitchens, *Food Control*, **5**, 200–4.
28. Sammarco ML and Ripabelli G (1997) Consumer attitude and awareness towards food related hygienic hazards, *Journal of Food Safety*, **17**, 215–21.
29. Levy M (1997) Consumer information center, in *Changing Strategies, Changing Behavior Conference*, Conference Proceedings, 13–14 June, Washington, DC, United States Department of Agriculture (USDA), Food and Drink Administration (FDA), Centers for Disease Control and Prevention (CDC).
30. Worsfold D and Griffith CJ (1996) An assessment of cleanliness in domestic kitchens, *Hygiene and Nutrition in Food Service and Catering*, **1**, 163–74.
31. Worsfold D and Griffith CJ (1997) Keeping it clean – a study of the domestic kitchen, *Food Science and Technology Today*, **11**, 28–35.
32. Rusin P, Orosz-Coughlin P, Gerba C (1998) Reduction of faecal coliform, and heterotrophic plate count bacteria in the household kitchen and bathroom by disinfection with hypochlorite cleaners, *Journal of Applied Microbiology*, **85**, 819–28.
33. Ackerley L (1994) Consumer awareness of food hygiene and food poisoning, *Environmental Health*, March, 70–4.
34. Curtis V, Cousens S, Mertens T, Traore T, Kanki B and Diallo I (1993) Structured observations of hygiene behaviours in Burkina Faso: validity, variability and utility, *Bulletin of the World Health Organisation*, **71**, 23–32.
35. Pinfold JV (1999) Analysis of different communication channels for promoting hygiene behaviour, *Health Education Research*, **14**(5), 629–39.
36. Nazarowec-White M and Farber JM (1997) Incidence, survival and growth of *Enterobacter sakazakii* in infant formula, *Journal of Food Protection*, **60**(3), 226–30.
37. Iversen C and Forsythe S (2004) Isolation of *Enterobacter sakazakii* and other Enterobacteriaceae from powdered infant milk formula and related products, *Food Microbiology*, **21**, 771–7.
38. EFSA (2004) Opinion of the Scientific Panel on biological hazards (BIOHAZ) related to the microbiological risks in infant formulae and follow-on formulae, European Food Safety Authority, *The EFSA Journal*, **113**, 1–35, available at: http://www.efsa.europa.eu/cs/BlobServer/Scientific_Opinion/biohaz_opinion14_ej113_microrisks_v2_en1.pdf?ssbinary=true, accessed December 2008.
39. FAO/WHO (2004) *Enterobacter sakazakii and other microorganisms in powdered infant formula: meeting report*, MRA Series 6, Rome/Geneva, Food and Agriculture Organization of the United Nations/World Health Organization, available at: http://www.who.int/foodsafety/publications/micro/es.pdf, accessed November 2008.
40. Griffith CJ, Mullan B and Price PE (1995) Food safety: implications for food, medical and behavioural scientists, *British Food Journal*, **97**, 23–8.
41. Clayton D, Griffith CJ, Peters AC and Price P (2000) Food handlers' beliefs about food safety procedures and risks, *International Association for Food Protection Conference*, Atlanta, GA.
42. Tones K, Tilford S and Robson Y (1990) *Health Education Effectiveness and Efficiency*, London, Chapman Hall.
43. Wallston BS, Wallston KA (1984) Social Psychological models of health behaviour: an examination and integration, in A Baum, S Taylor and JE Singer (eds), *Handbook of Psychology and Health*, Hillsdale, NJ, Erlbaum, 23–51.
44. Andreasen AR (1995) *Marketing Social Change*, San Fransisco, CA, Jossey-Bass.
45. Prochaska JO and Diclemente CC (1984) *The Transtheoretical Approach: Crossing Traditional Foundations of Change*, Harnewood, IL, Don Jones/Irwin.
46. Redmond EC, Griffith CJ and Peters AC (2000) Use of social marketing in the prevention of specific cross contamination actions in the domestic environment *Proceedings of the 2nd NSF International Conference on Food safety: Preventing foodborne illness through science and education*, 11–13 October, Savannah, GA.

47. Griffith CJ, Peters AC, Redmond EC and Price P (1999) *Food safety risk scores applied to consumer food preparation and the evaluation of hygiene interventions*, London, Department of Health.
48. *Foods Standards Act 1999*, London, The Stationery Office.
49. Mitchell RT (2000) *Practical Microbiological Risk Analysis*, London, Chandos.
50. Miles S, Draxton DS and Frewer LJ (1999) Public perceptions about microbiological hazards in food, *British Food Journal*, **101**, 744–62.
51. Griffith CJ, Mathias KA, and Price PE (1994) The mass media and food hygiene education, *British Food Journal*, **96**, 16–21.
52. HEA (2000) *Aliens in our food*, Food Hygiene Training Pack, London, Health Education Authority.
53. Mitchell V-W and Greatorex M (1990) Consumer percieved risk in the UK food market, *British Food Journal*, **92**(2) 16–22.
54. Redmond EC and Griffith CJ (2005) Consumer perceptions of food safety education sources: implications for effective strategy development, *British Food Journal*, **107**(7), 467–83.
55. Eves A, Bielby G, Egan B, Lumbers M, Raats M and Adams M (2006) Food hygiene knowledge and self-reported behaviours of UK school children, *British Food Journal*, **108**(9), 706–20.
56. Rowe S (2000) Media messages viewed by the public as inconsistent, consumer group says, *Food Safety Report*, **2**, 527.
57. Redmond E, Griffith CJ and Peters AC (1999) Utilising the social marketing approach to improve specific domestic food hygiene behaviours, *9th Annual Social Marketing in Public Health Conference*, Cleanwater Beach, FL, 23–26 June.
58. FSA (2002) *Microbiological Foodborne Disease Strategy July 2001: Revised post Board discussion*, London, Food Standards Agency, available at: http://www.food.gov.uk/multimedia/pdfs/fdscg-strategy-revised.pdf, accessed December 2008.
59. Food Safety Education Staff (FSES), Food Safety Inspection Service (FSIS) and United States Department of Agriculture (USDA) (2001) *Final Research Report – A Project to Apply the Theories of Social Marketing to the Challenges of Food Thermometer Education in the United States*, Washington, DC, Baldwin Group Inc.
60. Partnership for Food Safety Education (2002) Fight Bac for Education, available at: http://www.fightbac.org/, accessed November 2008.
61. WHO (2006) Prevention of foodborne disease: Five keys to safer food, INFOSAN Information Note No. 5/2006, Geneva, World Health Organization.
62. Redmond EC, Griffith CJ, Slader J and Humphrey TJ (2001) *The Evaluation and Application of Information on Consumer Hazard and Risk to Food Safety Education*, (CSA 5107), London, Food Standards Agency.
63. United States Department of Agriculture (USDA), Food Safety Inspection Service (FSIS), Food Safety Education Staff (FSES) (2003) *Food thermometer education campaign*, Washington, DC, Baldwin Group Inc, available at: http://www.fsis.usda.gov/Oa/research/boomburbs_style_guide.pdf, accessed December 2008.
64. Mortlock MP, Peters AC and Griffith CJ (2000) A national survey of food hygiene training and qualification levels in the UK food industry, *International Journal of Environmental Health*, **10**, 111–23.
65. Griffith CJ (1999) Training food handlers – new approaches, *Food Hygiene '99*, Telford.

Part II

Bacterial hazards

16

Preservation principles and new technologies

G. Gould, University of Leeds, UK

Abstract: Most food preservation techniques act by slowing down or completely inhibiting the growth of contaminating microorganisms (e.g. chilling, freezing, drying, curing, conserving, vacuum and gas packaging, acidifying, fermenting, adding preservatives). Heat remains the only substantially employed inactivation technique. Inactivation is preferable to inhibition, particularly with respect to pathogens. New and emerging techniques usefully do act mainly by inactivation (e.g. irradiation, high-hydrostatic pressure, high-voltage pulses, high-intensity light pulses and maybe high-intensity magnetic field pulses, and combinations of mild heat with ultrasound and pressure). New uses for numerous naturally-occurring antimicrobials derived from animals, plants and microorganisms are also being explored and evaluated for practical use in foods. The new and emerging techniques may be applied alone or, potentially more usefully, in synergistic combinations.

Key words: food preservation, high-hydrostatic pressure, high-voltage electric field and high-intensity light pulses, modified atmosphere packaging, natural antimicrobials, hurdle technologies.

16.1 Introduction

Foods may be contaminated by microorganisms, and multiplication may occur, at any of the stages between harvest, slaughter or manufacture and consumption of the final product by the consumer. Such contamination and multiplication may be prevented by a range of preservation techniques. Most of these act by stopping or slowing down growth, e.g. cooling (chilling, freezing), drying and freeze drying, curing with added salts, conserving with added sugars, vacuum packing, modified atmosphere packing, acidifying, fermenting and adding preservatives. A much smaller number of techniques are employed to inactivate microorganisms, predominantly pasteurisation and sterilisation by heat and, to a lesser extent, inactivation by low pH in acidified foods. Complementary techniques restrict the access of microorganisms to already treated foods, including clean and aseptic processing and packaging.

With respect to the ensurance of food safety, the inactivation of pathogenic microorganisms is more attractive than their inhibition. It is therefore interesting that a number of new and 'emerging' preservation techniques that are coming into use, are under development or are being evaluated include more that act by inactivation, e.g. high-hydrostatic pressure, irradiation, high-voltage pulses, high-intensity light pulses, combined heat, pressure and ultrasonication ('manothermosonication'), and the addition to foods of bacteriolytic enzymes (Forsyth, 2000; Gould, 2000; Devlieghere *et al.*, 2004) so these techniques are given special attention in this chapter. A further trend is towards the use of procedures that deliver products that are less heavily preserved, have higher quality and appear to the consumer to be more natural, freer from additives and nutritionally healthier than hitherto. More preservation techniques are being developed that make use of preservation factors in combinations that deliver less damage to product quality, new methods of heating that are better controlled and therefore deliver less thermal damage, cook–chill combinations that deliver extended, safe, high-quality shelf-lives, modified atmosphere packaging to retain quality longer, and use of naturally-occurring antimicrobial systems that normally operate in healthy plants and animals.

It has become clear that many preservation techniques, the inhibitory ones in particular, act by interfering with the various homeostatic mechanisms that microorganisms have evolved in order to survive extreme environmental stresses. Reaction to the stresses imposed by particular preservation techniques can lead to substantial increases in microbial resistance to those techniques and also to other, sometimes apparently unrelated ones, as well (Storz and Hengge-Aronis, 2000). Stress reactions must therefore also be taken into account in the ensurance of microbial safety of foods, and particularly when novel processes and new formulations are introduced.

16.2 Basis of food preservation, safety and the extension of shelf life

Food preservation and the ensurance of microbiological safety are based firstly on the delay or prevention of microbial growth and must therefore operate through those factors that most effectively influence the growth and survival of microorganisms. These include a number of physical factors, some predominantly chemical ones, and some that are essentially biological. These factors have been categorised in a number of ways, but the most widely quoted are those of ICMSF (1980), Mossel (1983) and Huis in't Veld (1996). These identify 'intrinsic factors', 'processing factors', 'extrinsic factors' and 'implicit factors' and, additionally, 'net effects' that take into account the fact that many of the other factors strongly influence the effects of each other, so that the overall influence of combinations of factors may not be obviously predictable, but may be derived from modern predictive modelling studies (McMeekin *et al.*, 1993; McClure *et al.*, 1994; McMeekin and Ross, 1996), and may be of greater efficacy than the perceived effects of single factors would lead one to expect.

16.3 Major food preservation and safety technologies

The major currently employed food preservation technologies are summarised in Table 16.1 in such a way as to highlight the fact that most, including the chemical preservatives (Table 16.2), act primarily by slowing down, or in some cases completely inhibiting, microbial growth. In some instances, although effective preservation results from inhibition, some degree of killing may occur and be important as well, for instance, in fermented and otherwise acidified foods such as mayonnaises and pickles. However, in contrast to the inhibitory techniques listed in Table 16.1 few of the current technologies act primarily by inactivating the target microorganisms, indeed heating remains the only technique used substantially for this purpose. Interestingly though, most of the newer, emerging or proposed physical techniques do act by direct inactivation (Table 16.3), including irradiation, high-voltage electric

Table 16.1 Major existing technologies for food preservation and safety

Technique	Examples of use
Techniques that slow or prevent growth of microorganisms	
Reduction in temperature	Chill storage, frozen storage
Reduction in water activity	Drying, curing with added salt, conserving with added sugars
Reduction in pH	Acidification (citric, acetic acids etc.), fermentation
Removal of O_2	Vacuum or modified atmosphere packaging
Modified atmosphere packaging	Replacement of air with gas mixtures
Addition of preservatives	Inorganic (e.g. sulphite, nitrite) Organic (e.g. propionate, sorbate, benzoate, parabens) Bacteriocin (e.g. nisin) Antimycotic (e.g. natamycin/pimaricin)
Control of microstructure	Water-in-oil emulsion foods
Techniques that inactivate microorganisms	
Heating	Pasteurisation, sterilisation
Techniques that restrict the access of microorganisms to foods	
Packaging	Most products
Aseptic processing and packaging	UHT products

Table 16.2 Antimicrobial food preservatives

Preservative	Examples of foods in which used
	Weak organic acids and ester preservatives
Propionate	Bread, cake, cheese, grain
Sorbate	Cheeses, syrups, cakes, dressings
Benzoate	Pickle, soft drinks, dressings
Benzoate esters (parabens)	Marinaded fish products
	Organic acid acidulants
Lactic, malic, citric, acetic	Low pH sauces, mayonnaises, dressings, salads, drinks, fruit juices and concentrates
	Inorganic acid preservatives
Sulphite	Fruit pieces, dried fruit, wine, meat sausages
Nitrite	Cured meat products
	Mineral acid acidulants
Phosphoric, hydrochloric	Drinks
	Antibiotics
Nisin	Cheese, canned foods
Natamycin (pimaricin)	Soft fruit
	Smoke
	Meat and fish

Table 16.3 New and emerging technologies for food preservation and safety

Physical processes
Gamma and electron beam irradiation
High-voltage electric gradient pulses ('electroporation')
High hydrostatic pressure
    Pressure pasteurisation
    Pressure-assisted thermal sterilisation ('PATS')
Combined ultrasonics, heat and pressure ('manothermosonication')
Intense laser and non-coherent light pulses
High magnetic field pulses

Biological additives
Animal-derived antimicrobials
    Lysozyme
    Lactoperoxidase system
    Lactoferrin, lactoferricin
Plant-derived antimicrobials
    Herb and spice extracts
Microbial products
    Bacteriocins (nisin, pediocin)
    Other bacteriocins and complex culture products
    Antimycotic (natamycin)

discharge ('electroporation'), combinations of ultrasonication with raised temperature and pressure ('manothermosonication'), as do most of the new and emerging biologically-based techniques (Table 16.3), including the use of bacteriolytic enzymes such as lysozyme.

Many of thc techniques listed in Table 16.1 are increasingly employed in combinations ('hurdle technologies'; Leistner, 1995) that allow reductions in the extreme use of any single technique. Whilst the development of many of the most used combination techniques was empirical, the basis for their efficacy has been worked out in some instances, and is leading to a more rational approach (Leistner and Gould, 2002), as indicated in some of the examples below.

16.3.1 Low pH and weak acid synergy

Low pH limits for the growth of some important pathogens are listed in Table 16.4. Whilst pH reduction alone may be employed to prevent microbial growth in foods, adjuncts are often employed in combination treatments that allow preservation with less reduction in pH than would otherwise be necessary.

Amongst the most extensively used combination treatments are those in which an antimicrobial acid is employed, and its effectiveness enhanced by lowering the pH. Indeed the majority of the useful food preservatives fall into this category. There remain few wide-spectrum food preservatives that are effective at pH values near neutrality. The acids include the inorganic

Table 16.4 pH limits for growth of food poisoning microorganisms

Microorganism	Minimum growth pH
Bacillus cereus	5.0
Campylobacter jejuni & coli	4.9
Clostridium botulinum (proteolytic)	4.6
Clostridium botulinum (non-proteolytic)	5.0
Clostridium perfringens	5.0
Escherichia coli	4.4
Listeria monocytogenes	4.3
Salmonella spp.	3.8
Salmonella typhimurium	4.0
Shigella flexneri	4.7
Shigella sonnei	4.9
Staphylococcus aureus	4.0
Vibrio cholerae	5.0
Vibrio parahaemolyticus	4.8
Vibrio vulnificus	5.0
Yersinia enterocolitica	4.2

Source: Adapted from Lund and Eklund (2000).
Note: pH limits are 'moving targets'. Reports of lower pH growth of particular strains continue to appear.

preservatives sulphite and nitrite and the weak organic acids. The latter generally increase in effectiveness in the order acetic, propionic, sorbic, benzoic, and this reflects their increasing lipophilicity. It is the lipid solubility of their mainly undissociated forms that enables them to cross the microbial plasma membrane and gain access to the cytoplasm within the cell (Booth and Kroll, 1989; Booth, 1995; Brul and Coote, 1999; Lund and Eklund, 2000). The second element important in their modes of action is the dissociation constant of the acid, because it is the undissociated form that is most lipophilic and therefore most able to permeate the membrane, and the pH value and the dissociation constant together that determine the proportion of acid that is in this form. The p*K* values of the common weak acid preservatives range from 4.2 (benzoic) to 4.87 (propionic), so that at pH values much above these antimicrobial activity is greatly reduced. At the pH value of most foods microorganisms maintain an internal pH higher than that of their surroundings. On entering the cytoplasm, the undissociated acids therefore tend to dissociate, delivering hydrogen ions, along with the particular anion. The additional hydrogen ions may be exported to maintain a suitably high internal pH, but this is energy demanding, so cell growth is restricted. If the energy demand is overcome, the pH of the cytoplasm will eventually fall to a level that is too low for growth to continue (Lund and Eklund, 2000).

In addition, careful uptake studies have shown that the preservative action of some of the acids, and of sorbic and benzoic in particular, results as well from effects additional to the membrane gradient-neutralising ones mentioned above (Eklund, 1985), and from gradient dissipation effects that are indirect (Stratford and Anslow, 1998). Lactic and citric acids do not display the same hydrophobicity as the monocarboxylic organic acid preservatives and act, at higher concentrations, differently. For example, the antibotulinal efficacy of monocarboxylic acids correlated well with their dissociation constants as expected, but citrate did not fit this pattern, implying a different mode of action (Miller *et al.*, 1993). Combinations of Na or K lactate with Na diacetate have proved to be especially practically useful in synergistically inactivating salmonellae and inhibiting the growth of *Listeria* in sliced meats (Mbandi and Shelef, 2002) and in inhibiting outgrowth from *Clostridium perfringens* spores, e.g. in processed turkey breast (Juneja and Thippareddi, 2004). The safety of some long shelf-life products such as mayonnaises is supported by the enhanced inactivation of pathogens such as salmonellae by the organic acid combinations they contain (ICMSF, 1998).

An obvious practical conclusion is that, wherever possible, microbial growth is best prevented by simultaneous reduction of pH in the presence of weak acid preservatives and, since there is then an increased energy demand placed on the cells, by anything in addition that restricts the efficient generation of cellular ATP (adenosine triphosphate). For instance, for facultative anaerobes, removal of oxygen, e.g. by vacuum or modified atmosphere packaging, is a sensible further adjunct to apply wherever possible. For example, such combinations, along with mild heat treatments, have been shown to introduce completely new approaches to the safe preservation of long-life fruits and vegetables (Alzamora *et al.*, 2000).

16.3.2 Heating with reduced pH

The synergy of low pH with mild heating is the basis of the thermal processing of high-acid foods in which the pH is below 4.5 and therefore sufficiently low to prevent growth from spores of *C. botulinum*. Reduction in pH enhances the rates of heat inactivation of spores, including those of pathogens such as *C. botulinum* (Xezones and Hutchings, 1965). Other mild heat–low-pH combinations have been proposed, though less used. For example, mild heating of spores at low pH values decreases their resistance to subsequent heating even if the subsequent heating is at a higher pH (Alderton and Snell, 1963). The mechanism seems to be based on cation exchange, whereby spores lose endogenous calcium that is an important part of their heat resistance mechanism (Marquis *et al.*, 1994). The procedure was proposed as a potentially useful method for reducing heating requirements during food sterilisation. However, difficulties in designing suitable large-scale processes were such that the procedure has never been used commercially. With the advent of modern aseptic processing and packaging techniques for pumpable liquid foods, it may be that re-examination of the potential for 'acid-assisted sterilisation' is opportune.

16.3.3 Carbon dioxide and modified atmosphere packaging

Another weak acid that is increasingly employed is carbon dioxide (Farber, 1991; Molin, 2000; Davies, 2003). The efficacy of carbon dioxide results from a number of unrelated phenomena. The elimination of oxygen that may be achieved alters the microbial flora by preventing the growth of strict aerobes and slowing the growth of facultative anaerobes through restricting the amount of energy they can obtain from substrates utilised. However, carbon dioxide has a strong specific antimicrobial action that is very microorganism specific. For instance, many oxidative Gram-negative bacteria such as species of *Pseudomonas* are sensitive to concentrations as low as 5 %, while many lactic acid bacteria and yeasts are capable of growth in the presence of 100 %, and even carbon dioxide under pressure. Carbon dioxide-enriched packaging of foods such as fresh meats therefore generally results in a shift from a rapidly growing Gram-negative to a slow-growing Gram-positive flora (Davies, 2003). Good low-temperature control is vital for effective use of modified atmosphere packaging of chill stored foods because the antimicrobial efficacy of carbon dioxide is greatly enhanced as the temperature is reduced, to some extent resulting from the increase in its solubility that accompanies reduction in temperature.

In addition to the antimicrobial effects of modified atmosphere packaging, there are often important additional effects on food colour. For instance, in the preservation of chill stored fresh meats, carbon dioxide is usually employed at concentrations of 20–30 %, with oxygen making up the remainder. While the raised carbon dioxide delays growth of Gram-negative spoilage organisms and potential pathogens, the raised level of oxygen ensures that the haem pigments remain in the bright red oxymyoglobin form that keeps the meat looking attractive to the consumer, and usually with an approximate doubling of acceptable shelf-life (Parry, 1993; Molin, 2000). Particularly for long shelf-life products, the gas permeabilities of packaging films are critical. Very long-life ambient stored products may require highly impermeable aluminium foil laminate-type packaging to prevent loss of the packaging gas or the ingress of atmospheric oxygen.

Use of CO_2 at high pressure has long been proposed as a potentially useful 'cold pasteurisation' food preservation technique, with about 60 published papers by mid 2006 (Garcia-Gonzales *et al.*, 2007). Pressures employed are generally > 20 MPa, so the technology is easily applied and controlled (Spilimbergo and Bertucco, 2003). While suggestions for the mechanisms of the observed bacteriostatic and bactericidal actions have been made, these remain to be proven (Damar and Balaban, 2006). In a recent comprehensive review, Garcia-Gonzalez *et al.* (2007) listed inactivation levels obtained for more than 40 key food spoilage and poisoning microorganisms in different foods and media. The levels ranged widely, e.g. from 9 D or more for *Listeria monocytogenes* in water (6.18 MPa, 35 °C, 2 h) to just 0.3 D in chicken meat and 2.3 D in orange juice under near-identical conditions. *Escherichia coli* was inactivated > 6.3 D (6.2 MPa, 2 h) and 8.6 D (11 MPa, 38 °C, 45 min) in broth. Lethality was reduced at low a_ws. Some of these reductions are potentially useful, so approaches to commercialisation have been made, and so far include pilot-scale evaluations and mobile demonstration equipment, and at least one commercial unit (Garcia-Gonzalez *et al.*, 2007).

16.3.4 Low water activity

Reduction of water activity (a_w) to below 0.6 prevents all microbial growth. Just above this level yeasts and moulds are most tolerant, with no bacteria of safety concern able to multiply below about 0.86, which is the limit for *Staphylococcus aureus* growing aerobically (Table 16.5). Between 0.6 and about 0.86–0.90, in shelf-stable 'intermediate moisture foods' (IMF), additional adjuncts are necessary to prevent growth of a_w-tolerant microorganisms, such as oxygen-free packaging and the addition of antimycotics such as sorbic and benzoic acids. Whilst successfully used in semi-moist pet foods, such combinations are less used in human foods because of organoleptic penalties resulting from the extensive drying or high levels of solutes needed to sufficiently reduce $a_{w,}$ unless the foods are based on traditionally low a_w ones, e.g. N_2 or vacuum-packed salamis and cheese snack products.

Table 16.5 Water activity limits for growth of food poisoning microorganisms

Microorganism	Minimum a_w for growth[a]
Aeromonas hydrophila	0.97
Bacillus cereus	0.93
Campylobacter jejuni	0.98
Clostridium botulinum (proteolytic)	0.94
Clostridium botulinum (non-proteolytic)	0.97
Clostridium perfringens	0.97
Escherichia coli	0.95
Listeria monocytogenes	0.92
Salmonella spp.	0.95
Staphylococcus aureus (aerobic)	0.86
Staphylococcus aureus (anaerobic)	0.91
Vibrio parahaemolyticus	0.95

[a]Data from salt-adjusted laboratory media.
Source: Adapted from Christian (2000).

16.3.5 Heating with reduced water activity

At pH values above 4.5 (Table 16.4) a 'botulinum cook' would normally be necessary to ensure the safety of long ambient stored foods unless some other factor was known to be operating. For example, for the safe preservation of some cured meat products mild heating is combined with reduced a_w, sometimes with a reduction in pH value, and often the presence of nitrite as well. These combination systems allow the production of a wide range of so-called SSPs (shelf-stable products; Leistner, 1995) that have long safe ambient stability, yet only require pasteurisation heat processing. Many other alternative water activity–pH–mild heat combinations have been reported, but remain little used (Leistner, 1995; Leistner and Gould, 2002).

16.3.6 Mild heating and chill storage

There has been a recent expansion in the use of the combined heating of vacuum-packaged foods with well-controlled chill storage, initially for catering, but now increasingly for retail also. These foods include 'sous vide' products (Livingston, 1985) and REPFEDS (refrigerated processed foods of extended durability; Notermans *et al.*, 1990; Mossel and Struijk, 1991). The efficacy of the procedure relies on the inactivation of vegetative flora by the mild heating, and also on the fact that spores of psychrotrophic bacteria, including those of non-proteolytic strains of *C. botulinum*, that can grow at chill temperatures are more heat sensitive than those of the mesophiles and thermophiles that cannot grow at these temperatures. In order to ensure safety, heat processes equivalent to about 90 °C for 10 min are regarded as sufficient to ensure inactivation of spores of the coldest growing pathogenic sporeformers (Notermans *et al.*, 1990; Mossel and Struijk, 1991; Lund and Peck, 1994, 2000). For lower heat treatments, strict limitation of shelf-life, efficient control of storage temperature below 3 °C, or some form of demonstrable intrinsic preservation is necessary (Anon, 1989; ACMSF, 1992; Gould, 1999).

16.4 New and emerging technologies

Of the technologies listed in Table 16.3, irradiation has the longest history of use. More than 40 countries have approved the use of ionising radiation to improve the preservation and safety of various foods (Patterson and Loaharanu, 2000). While upper dose limits still apply in most countries that allow the technique to be used, an FAO/IAEA/WHO Study Group on (the safety of) High Dose Irradiation (WHO, 1999) reconsidered in depth the potential safety issues, including the known generation in foods of low levels of 'unique radiolytic products', and concluded that food irradiated to any dose appropriate to achieve an intended technological objective is both safe to consume and nutritionally adequate. In this sense, processing by irradiation would be considered equivalently processing by heating. In the USA and some other countries where health authorities encourage use of irradiation for quarantine purposes and to improve public health, applications have advanced in recent years, but with numerous setbacks resulting from a continuing general public resistance. In EU countries progress has generally been slow (Diehl, 2002). In contrast, progress with some of the other new and emerging techniques, high-hydrostatic pressure in particular (Hendrickx and Knorr, 2001), has been steady and continuing, as indicated below.

16.4.1 High hydrostatic pressure and synergy with heat

Vegetative cells

Vegetative cells of bacteria were first shown to be inactivated by pressures above about 100 MPa (MegaPascals) more than 100 years ago by Hite (1899), while bacterial spores were shown to be much more resistant, surviving pressures above 1200 MPa (Larson *et al.*, 1918; Basset and Machebouf, 1932). Food preservation by pressure was demonstrated by shelf-life extensions of milk, fruits and vegetables (Hite *et al.*, 1914), but difficulties in applying the technology delayed commercial use for food preservation until the 1980s (Mertens, 1995), when processes were developed for the non-thermal pressure pasteurisation of a number of low-pH foods, in which survival of spores was not a problem because they were unable to outgrow under the acid conditions. Pressure-processed foods at first included jams, fruit juices, jellies, acid sauces and fruit for inclusion in yoghurts (Selman, 1992). More recently, more attention has been given to materials science and other aspects of pressure effects on foods (Knorr, 1995), and further pressure-treated foodstuffs have been marketed, including chill-stored guacamole, dairy products, fish, oysters, vacuum-packed sliced meat products and meal components, some of which have higher pH values than those marketed earlier (Palaniappan, 1996; Grant *et al.*, 2000; San Martin *et al.*, 2002; Hayman *et al.*, 2004). For example, high-pressure pasteurisation

(600 MPa, 20 °C, 3 min) of ready-to-eat meats was shown to greatly extend 4 °C shelf-life with no organoleptic penalties, whilst reducing *L. monocytogenes* numbers by > 4 log (Hayman *et al.*, 2004).

Foods sterilised by combined pressure–heat treatments can have quality advantages over traditionally heat-sterilised foods (see below) and are likely to become increasingly available in the near future. Implementation of pressure processing began slowly, due mainly to the high investment costs involved and, to some extent, to the discontinuous nature of the process, which tended to restrict use to high added value products (Devlieghere *et al.*, 2004).

The effects of pressure derive from the Le Chatelier principle, in which pressure favours any physical change or chemical reaction associated with a net volume decrease and suppresses any change or reaction involving a volume increase. In biological systems the volume decrease reactions that are most important include the denaturation of proteins, gelation, hydrophobic reactions, phase changes in lipids (and therefore in cell membranes) and increases in the ionisation of dissociable molecules due to 'electrostriction' (Heremens, 1995). The changes that are brought about therefore differ from those brought about by heat. For instance, pressure-denatured proteins differ in structure from heat-denatured ones (Hoover, 1993; Heremens, 1995). Such differing effects will probably be exploitable through the setting and validation of processes in factory situations, with potential product quality advantages. Small molecules are generally less affected than macromolecules so that low molecular weight flavour and odour compounds, etc. in foods tend to survive pressure treatment unchanged, with quality advantages in some types of products (Horie *et al.*, 1991). A structure as complex as a microorganism will clearly have many potentially pressure-sensitive sites within it, e.g. enzymes, genetic material, macromolecular assemblies such as membranes, ribosomes, etc. (Isaacs *et al.*, 1995), and pressure has been shown to induce a variety of changes in vegetative bacterial cells that may contribute to their inactivation (Hoover *et al.*, 1989). Kinetic studies have shown some examples of exponential inactivation of cells held at constant pressure (e.g. *E. coli*; Butz and Ludwig, 1986; Ludwig *et al.*, 1992), but the majority of studies have reported 'tails' on survivor curves, i.e. a decreasing death rate with the time of treatment (Metrick *et al.*, 1989; Earnshaw, 1995; Patterson *et al.*, 1995a). It has been proposed that at higher temperatures (e.g. 40 °C and above for *E. coli* at 250 MPa) inactivation is near first-order whereas at lower temperatures (e.g. 30 °C for *E. coli*) it is nearer second-order, and that temperature-dependent membrane lipid changes may account for these differences (Eze, 1990).

More generally, over the range –20 °C to +20 °C pressure is more effective in inactivating vegetative microorganisms at lower than higher temperatures, e.g. for *Staphylococcus aureus*, *Salmonella* Bareilly and *Vibrio parahaemolyticus* in buffers (Takahashi *et al.*, 1991) and for *Citrobacter freundii* in beef (Carlez *et al.*, 1992). Reduction in water activity resulting from the addition of a range of solutes led to substantial increases in pressure tolerance, e.g. of *Rhodotorula rubra* (Oxen and Knorr, 1993). Other factors that are not yet fully understood affect the pressure tolerance of microorganisms in different foods. For instance,

Salmonella Typhimurium was inactivated about 10^6-fold in pork in 10 min at 300 MPa, but only 10^2-fold in chicken baby food at 350 MPa (Metrick *et al.*, 1989). *L. monocytogenes* was more pressure-tolerant in UHT milk than in phosphate-buffered saline (Styles *et al.*, 1991), probably resulting from a protective effect of calcium (Hauben *et al.*, 1996). Strain-to-strain variability in sensitivity to pressure is greater than the variation to other inactivation techniques, such as heat (Solomon and Hoover, 2004). *L. monocytogenes* strains NCTC 11994, Scott A and an isolate from chicken were inactivated by less than ten-fold, just over 10^2-fold and about 10^5-fold, respectively, by a similar 375 MPa pressurisation for 10 min in phosphate-buffered saline (Patterson *et al.*, 1995a). Pressure-tolerance of *Campylobacter* species (*jejuni*, *coli*, *lari* and *fetus*) varied greatly and changed with growth phase, with stationary phase cultures being more resistant than log phase ones (Martinez-Rodriguez and Mackey, 2005).

Otherwise, exponential phase cells are more pressure-sensitive than stationary phase ones and Gram-positive microorganisms are generally more pressure-tolerant than Gram-negative ones (Shigahisa *et al.*, 1991). However, there are some important exceptions. For example, a strain of *E. coli* O157 H7 was found to be extremely pressure-tolerant in some foods, e.g. exposure to 800 MPa in UHT milk only brought about a 10^2-fold reduction (Patterson *et al.*, 1995b).

Spores

In contrast to vegetative cells, bacterial spores were shown in the earliest studies to be very pressure-tolerant. Pressures up to 1200 MPa failed to inactivate spores of a number of species (Larson *et al.*, 1918; Basset and Machebouf, 1932; Timson and Short, 1965). However, later it was shown that, surprisingly, under certain conditions inactivation of spores proceeded more rapidly and completely at lower than at higher pressures (Clouston and Wills, 1969; Sale *et al.*, 1970). An explanation for this was found when it was observed that inactivation of spores occurred in two stages (Clouston and Wills, 1969; Gould and Sale, 1970). First, pressure caused spores to germinate, then pressure, and/or heat if the temperature was high enough, inactivated the germinated forms. This led to the investigation of the combined use of pressure with raised temperature (Sale *et al.*, 1970) and with low-dose irradiation (Wills, 1974) to achieve a higher level of spore inactivation. The overall pattern of inactivation showed a strong pressure–heat synergism. The effect has been confirmed for a wide range of spore types, though the effectiveness of the combination varies greatly in magnitude for different spores (Murrell and Wills, 1977; Kimugasa *et al.*, 1992; Kowalski *et al.*, 1992; Seyerderholm and Knorr, 1992; Hayakawa *et al.*, 1994). For example, the survival of 40 *Bacillus* isolates subjected to pressure plus heat (600 MPa, 1 min, initial 72 °C) highlighted the great variability of spore tolerance, with kills ranging from c. 6 log reductions to no obvious inactivation (Scurrah *et al.*, 2006). The kinetics of pressure-inactivation was near exponential for spores of *Bacillus pumilus* (Clouston and Wills, 1970), but for spores of *B. coagulans*, *B. subtilis* and *C. sporogenes* concave-upward curves or long tails were reported (Sale *et al.*, 1970). Mild 'germinative' pressure treatments (e.g. 200 MPa, 45 °C,

30 min) followed by mild heating (60 °C, 10 min) to inactivate the germinated spores reduced *B. cereus* spore levels in milk > 10^6-fold (Van Opstal *et al.*, 2004). Repeated pressurisation–depressurisation cycles increased spore germination and inactivation (Furukawa *et al.*, 2003), apparently partly due to the break-up of spore clumps (Furukawa *et al.*, 2004). Pressure cycling was effective in reducing numbers of *B. cereus* spores in raw milk cheeses (Lopez *et al.*, 2003).

The fact that, although spores of some species are relatively sensitive to pressure (e.g. *B. cereus*), those of others, including some of special importance in foods such as *Geobacillus stearothermophilus* and *C. botulinum*, are very resistant (Knorr, 1995), initially prevented the use of pressure to sterilise foods (Hoover, 1993). Kills of *C. botulinum* type A strains of c. 3 log by 827 MPa at 75 °C for 15 min in buffer and crab meat emphasised the need for higher temperatures to reach levels of inactivation equivalent to those delivered by conventional thermal processing (Reddy *et al.*, 2003). Pressure resistance of spores of seven different strains of *C. botulinum* in mashed carrots varied greatly, but some exceeded the pressure tolerance of a range of *Bacillus* spores studied (Margosch *et al.*, 2004). Spores of *C. botulinum* type B, amongst the most pressure-tolerant of spores, were more rapidly inactivated by heat (90–110 °C) with increase in pressure (600–1400 MPa) in carefully controlled isothermal treatments (Margosch *et al.*, 2006). However, heating at 100–120 °C more effectively inactivated spores at ambient pressure, illustrating a protective effect of medium pressures that will be important in the development of sterilisation processes. While these phenomena must be taken into account, the difficulties in developing sterilisation processes are being overcome with the development of presses that operate at higher temperature–pressure combinations. Most advanced is the combined use of heat and pressure to deliver pressure-assisted thermal sterilisation ('PATS'). In PATS the predominant lethal effect is due to the high temperature. A key advantage is the improved product quality delivered by the process. This derives from the very high rates of adiabatic heating and cooling that accompany pressurisation and depressurisation, enabling far more precise control of heat processing overall than can be achieved during conventional processing, thus minimising thermal damage to foods. The rapid transient temperature rise accompanying pressurisation (and fall during depressurisation), to some extent dependent on food type, is about 25 °C at 800 MPa (Farr, 1990). Process parameters have now been determined for successful pressure-assisted high-temperature sterilisation of foods, using at first spores of *C. sporogenes* PA3679, then *C. botulinum* (Rovere *et al.*, 1996; Koutchma *et al.*, 2005; Reddy *et al.*, 2006; Toepfl *et al.*, 2006).

High pressure shows synergism with stresses other than heat. Raso and Barbosa-Canovas (2003) summarised effective combinations with treatments including low pH, antimicrobials (including lysozyme, nisin, pediocin AcH and CO_2) and ultrasound. High pressure apparently sensitises vegetative microorganisms to lysozyme by permeabilising the outer membrane in Gram-negative bacteria, allowing enzymatic action on the underlying peptidoglycan, e.g. in *E. coli* (Hauben *et al.*, 1996; Masschalck *et al.*, 2001, 2003), but by non-enzymatic mechanisms in Gram-positives, e.g. *Staph. aureus* (Masschalck *et al.*, 2002). Combination of relatively low (60 MPa, 'germinative pressure') with nisin usefully reduced *B. cereus* spore numbers in cheeses (Lopez-Pedemonte *et al.*, 2003). The sporicidal effects of pressure–temperature combinations were enhanced by low pH (Stewart *et al.*, 2000; Wuytack and Michiels, 2001), by nisin (Roberts and Hoover, 1996) and by argon (Fujii *et al.*, 2002).

16.4.2 High-voltage electric pulses

While the application of electric fields to heat foods has become well established (Palaniappan, 1996), e.g. through electrical resistance or 'ohmic' heating (Fryer, 1995) and through microwave heating (Mullin, 1995), the use of electric pulses to bring about the essentially non-thermal inactivation of microorganisms has only been explored and exploited more recently (Castro *et al.*, 1993). The use of the technique at lower, non-lethal, voltage gradients has become established as the basis for 'electroporation', by which genetic material can be exchanged between protoplasted cells of microorganisms, plants and animals (Neumann *et al.*, 1989). A method was patented for inactivating microorganisms with an electric field by Doevenspeck (1960). Later studies demonstrated the inactivation of bacteria and yeasts and the lysis of protoplasts and erythrocytes (Sale and Hamilton, 1967; 1968). Still later studies concentrated on varying the electrical parameters (Hulsheger and Niemann, 1980; Hulsheger *et al.*, 1981, 1983), and on the effects of environmental parameters (Mizuno and Hori, 1988; Jayaram *et al.*, 1992).

While the effects of high-voltage fields on microorganisms are not understood at the molecular level, the gross effects and the mechanisms that cause them are well established. They result from the permeablisation of the cell membrane (Hamilton and Sale, 1967) that results when the voltage gradient is high enough to overcome its intrinsic resistance. Breakdown occurs when the potential difference across the membrane exceeds about 1 V (Chernomordik *et al.*, 1987; Glaser *et al.*, 1988). Massive leakage of cell contents occurs and the cell dies (Tsong, 1991).

Pulsed electric field ('PEF') inactivation has been reported for *E. coli*, *S.* Typhimurium, *S.* Dublin, *Streptococcus thermophilus*, *Lactobacillus brevis*, *Pseudomonas fragi*, *Klebsiella pneumoniae*, *Staph. aureus*, *L. monocytogenes*, *Saccharomyces cerevisiae* and *Candida albicans* (Barbosa-Canovas *et al.*, 1995). In contrast to vegetative organisms, bacterial spores (Sale and Hamilton, 1967) and yeast ascospores (Mertens and Knorr, 1992) are resistant, even at very high voltage gradients, i.e. above 30 $kVcm^{-1}$. A number of intrinsic and extrinsic factors influence the effectiveness of the electrical treatments. Inactivation increased greatly with rise in temperature, e.g. for *E. coli* and for *Lactobacillus brevis* (Jayaram *et al.*, 1992). Low ionic strength favours inactivation. A reduction in the KCl concentration in skim milk from 0.17 to 0.03 M resulted in a fall in the surviving fraction of *E. coli* following a 55 $kV\ cm^{-1}$, 20 pulse treatment, from about 0.3 to about 0.002. Reduction in pH increased inactivation, e.g. it was doubled for *E. coli* by reducing the pH of skim milk from 6.8 to 5.7. Log phase cells were more sensitive than stationary phase ones.

Applications of electric pulse teatments to a number of liquid foods have indicated that useful 'cold pasteurisation' inactivation of vegetative bacteria and yeasts can be achieved (Molina *et al.*, 2002). For example, treatment of apple juice, at temperatures below 30 °C, with less than 10 pulses in a continuous treatment chamber brought about more than a 10^6-fold reduction in numbers of *Sacch. cerevisiae* at a voltage gradient of 35 kV cm^{-1}; 22 kV cm^{-1} caused about 10^2-fold inactivation (Qin *et al.*, 1998). Studies of inactivation rates under different conditions have generally indicated kinetics that, on a log survivor versus treatment time or versus number of pulses basis, show long tails. Near straight lines are seen when log survivors are plotted against the log of the treatment time or the log of the number of pulses (Zhang *et al.*, 1995). Some potentially useful synergies have been described. For example, electroporated cells of *E. coli*, *L. monocytogenes* and *S.* Typhimurium became much more sensitive than untreated cells to the bacteriocins nisin and pediocin (Kalchayanand *et al.*, 1994). PEF treatment (up to 25 kV cm^{-1}) followed by exposure to CO_2 at raised pressure (20 MPa, 40 °C) synergistically inactivated vegetative bacteria (*Staph. aureus*, *E. coli*) and greatly reduced numbers of *B. cereus* spores (Spilimbergo *et al.*, 2003). Whilst PEF is not being used commercially at present, it has been shown to inactivate vegetative microorganisms, and in pilot plant studies to usefully extend the shelf-life of a range of liquid foods, e.g. apple and orange juices, skim milk, beaten eggs, pea soup (Raso and Barbosa-Canovas, 2003).

Electric fields may be delivered as oscillating, bipolar, exponentially decaying or square wave pulses. Bipolar pulses were more lethal than monopolar ones because, it was presumed, rapid reversal in the direction of movement of charged molecules caused greater damage to cell membranes (Qin *et al.*, 1994). Bipolar pulses generate less electrolysis in the material being treated, which may be advantageous for organoleptic, and possibly also for toxicological and nutritional, reasons, and they are energy-efficient. It is generally most economic to raise the field strength as high as possible while reducing the duration of the pulses, without reducing pulse energy (Grahl *et al.*, 1992). On the other hand, the use of very high field strengths demands more complex and expensive engineering (Zhang *et al.*, 1994). As a result of these competing requirements, modern pulse field devices employ field strengths from about 20 up to about 70 kV cm^{-1}, with pulse durations between one and about 5 μs. Repetition rates are typically between one and up to 30 s or so at the higher voltages in order to minimise rises in temperature. Treatment chambers may operate batchwise or continuously. The earliest versions were not fully enclosed and so were limited to voltage gradients of about 25 kV cm^{-1} because this is the approximate breakdown voltage of air (Sale and Hamilton, 1967; Dunn and Pearlman, 1987). Enclosure and design improvements led to devices that could deliver 30–40 kV cm^{-1} (Grahl *et al.*, 1992; Zhang *et al.*, 1994). These were useful for laboratory studies to optimise design parameters for efficient killing of microorganisms. Continuous operation is essential for cost-effective commercial applications able to treat liquids and liquids containing particulates, and a number of such systems, mostly designed around coaxial cylindrical electrodes, have been developed (Boulart, 1983; Hoffman and Evans, 1986; Dunn and Pearlman, 1987; Sato and Kawata, 1991; Bushnell *et al.,* 1993; Sitzmann, 1995).

16.4.3 High-intensity light pulses

High-intensity laser and non-coherent light has long been known to inactivate microorganisms (Mertens and Knorr, 1992), though it was often unclear to what extent the lethal effects derived from the ultraviolet component of the radiation and, sometimes, local transient heating. The delivery of light to packaging materials, food surfaces and to transparent liquid products, in short pulses of high intensity, has been shown to be capable of inactivating vegetative and spore forms of microorganisms in these environments (Dunn *et al.*, 1988) and in the medical area, particularly dentistry (Powell and Wisenart, 1991; Cobb *et al.*, 1992; Rooney *et al.*, 1994). Commercially practicable machines for treating foods and other materials have been patented (Dunn *et al.*, 1988). These machines use broad spectrum light with pulse durations from 10^{-6} to 10^{-1} s and with energy densities from about 0.1 to about 50 J cm^{-2}. Different spectral distributions and energies are selected for different applications. For example, ultraviolet-rich light in which about 30 % of the energy is at wavelengths shorter than 300 nm is recommended for treatment of packaging materials, water or other transparent fluids. In contrast, for food surfaces, when high intensities of ultraviolet may accelerate lipid oxidation or cause colour loss, etc., the shorter wavelengths are filtered out and the killing effects are largely thermal. The advantage of delivering heat in this manner is that a large amount of thermal energy is transferred to a very thin layer of product surface very quickly while the temperature rise within the bulk of the product can be very small (Dunn *et al.*, 1988). Overall therefore, these intense light treatments are effective predominately because they deliver conventional microorganism-inactivating treatments, ultraviolet irradiation or heat, but in an unconventional and sometimes advantageous manner (Mertens and Knorr, 1992).

16.4.4 High-intensity magnetic field pulses

Exposure to oscillating magnetic fields has been reported to have a variety of effects on biological systems ranging from selective inactivation of malignant cells (Costa and Hofmann, 1987) to the inactivation of bacteria on packaging materials and in foods (Hofmann, 1985). Treatment times are very short, typically from 25 μs to a few μs, and field strengths very high, typically from 2 to about 100 Tesla at frequencies between about 5 and 500 kHz. It has been suggested that the mechanism of action could involve alteration of ion fluxes across cell membranes, but this is not really known (Pothakamury *et al.*, 1993). Efficacies of the treatments did not exceed about 10^2-fold reductions in numbers of vegetative microorganisms inoculated into milk (*Streptococcus thermophilus*), orange juice

(*Saccharomyces*), bread rolls (mould spores) and no inactivation of bacterial spores has been reported (Hofmann, 1985), so the practical potential for the technique, as it has been developed so far, appears to be limited (Mertens and Knorr, 1992; Barbosa-Canovas *et al.*, 1995).

16.4.5 Manothermosonication

The use of ultrasound to inactivate microorganisms was first reported over 80 years ago (Harvey and Loomis, 1929). The mechanism of action derives from the rapidly alternating compression and decompression zones propagating into the material being treated and the cavitation that these cause. Cavitation involves the formation and rapid collapse of small bubbles, generating shock waves with associated local but very high temperatures and pressures that can be sufficiently intense to catalyse chemical reactions and disrupt animal, plant and microbial cells (Scherba *et al.*, 1991). Generally, large cells are more susceptible than small ones. Rod-shaped bacteria are more sensitive than cocci (Alliger, 1975) and Gram-positive bacteria more sensitive than Gram-negative ones (Ahmed and Russell, 1975), while spores are so resistant as to be essentially non-disruptable (Sanz *et al.*, 1985).

A potentially useful synergy of ultrasound with heat was reported for the inactivation of bacterial spores (*B. cereus* and *B. licheniformis*; Burgos *et al.*, 1972), thermoduric streptococci (Ordonez *et al.*, 1984), *Staph. aureus* and other vegetative microorganisms (Ordonez *et al.*, 1987), and for the inactivation of enzymes (Lopez *et al.*, 1994). However, as the temperature was raised, the potentiating effect of ultrasound became less and less and (for spores) disappeared near to the boiling point of water (Garcia *et al.*, 1989). The development of manothermosonication followed the proposal that this disappearance of the synergism could be prevented if the pressure was raised slightly (e.g. by only a few tens of MPa; Sala *et al.*, 1995). The combination procedure was reported to have the effect of reducing the apparent heat resistance of microorganisms by about 5–20 °C or so, depending on the temperature, the organism and its *z*-value. *E. coli* and *Lactobacillus acidophilus* in milk and in fruit and vegetable juices were inactivated at lower temperatures in an ultrasound-assisted pilot-scale pasteurisation plant than in a conventional thermal process (Zenker *et al.*, 2003), though total energy consumption was not reduced. Ordonez *et al.*, (1987) demonstrated the combined effectiveness of ultrasonication and heat in inactivating *Staph. aureus*. It was later shown that the additional application of mild hydrostatic pressure increased lethality further (Sala *et al.*, 1995; Pagan *et al.*, 1999). A similar combination of pressure, mild heating and ultrasound reduced *S.* Enteritidis numbers by > 3 log, with the advantage over pasteurisation by heat alone of not causing protein coagulation (Huang *et al.*, 2006). While the process has not been commercialised, it is therefore claimed to offer the possibility of new sterilisation or pasteurisation processes for liquid products with reduced levels of thermal damage.

16.5 Natural antimicrobial systems

A wide range of natural antimicrobial systems has evolved in animals, plants and microorganisms (Table 16.6). A few of them are already employed for food preservation (see Table 16.3), while many more have been investigated to a limited degree, and probably have substantial future potential.

16.5.1 Animal-derived antimicrobials

The range of antimicrobial systems and substances that operate in animals is extensive (Table 16.6). It includes the immune system in higher animals and a variety of bacteriolytic and other enzymes that are components of antimicrobial systems, as well as non-enzymic proteins with antimicrobial activity. A steadily

Table 16.6 Major natural antimicrobial systems and potential for synergy

Origin	Example	
Animals – constitutive	Phagosomes	myeloperoxidase
	Serum	transferrins
	Milk	lactoperoxidase lactoferrin
	Eggs	lysozyme ovotransferrin avidin
Animals – inducible	Immune system	antibodies complement
	Frogs	magainins
	Insects	attacins cecropins
Plants – constitutive	Herbs, spices and other plants	eugenol (cloves) allicin (garlic) allyl isothyocyanate (mustard)
Plants – inducible	Injured or infected plants	low MW[a] phytoalexins high MW polyphenolics
Microorganisms	Lactic acid bacteria	nisin, pediocin, other bacteriocins
	Other microorganisms	other antibiotics (pimaricin, subtilin) bacteriophages yeast 'killer toxins' organic acids and other low MW metabolites
Potential synergists	Low pH	Organic acids
	Low a_w	Specific solutes
	Chelators	
	Low oxygen	Raised carbon dioxide
	Mild heat	Low temperature
	Raised pressure	Electroporation

[a]Molecular weight.

expanding range of small antimicrobial peptides has been discovered recently. These are usually membrane-active and lethal for a wide range of microorganisms.

Of the animal-derived antimicrobials, most commercial use has been made of hen egg white lysozyme. Lysozyme is present in many body fluids, and at levels up to about 3.5 % of dry weight in egg white, so that it is readily available, relatively inexpensive and natural. It lyses many, though not all, Gram-positive bacteria in their vegetative forms, but is inactive against Gram-negative ones because their outer membranes prevent access to the underlying peptidoglycan that is the enzyme's substrate. The major successful use of lysozyme commercially has been to lyse cells of *C. tyrobutyricum* as they outgrow from germinated spores (Carminati *et al.*, 1985), thus preventing gas formation and spoilage by 'blowing' in certain types of cheeses (Wasserfall *et al.*, 1976; Carini and Lodi, 1982). It was reported that more than 100 tonnes of lysozyme may be used annually for this purpose (Scott *et al.*, 1987). The antimicrobial spectrum of lysozyme can be broadened, e.g. by pre-treating the target microbial cells in a number of ways. For instance, pre-treatment with some chelating agents such as ethylenediaminetetraacetate (EDTA) will sensitise some Gram-negative bacteria to the action of lysozyme (Samuelson *et al.*, 1985), but this is not yet exploited commercially. Conjugation of lysozyme with dextran increased its activity against Gram-positive and Gram-negative bacteria, particularly if the temperature was raised during treatment (Nakamura *et al.*, 1990). Lysozyme and the bacteriocin nisin (see below) act synergistically under certain conditions to inhibit the growth of, and to inactivate cells of, *L. monocytogenes* (Monticello, 1989). Freeze–thaw treatments sensitised *E. coli* to lysozyme (Ray *et al.*, 1984). Gram-negative bacteria, including salmonellae, become lysozyme-sensitive following an osmotic downshift (e.g. achieved by the sudden dilution of a salt solution). This has been proposed as the basis for a dip- or spray-decontamination treatment for poultry and other animal carcases (Chatzolopou *et al.*, 1993). Bacterial spores may be sensitised to lysozyme, which then causes them to germinate and so become heat-sensitive, but only by severe treatments with reagents that break protein disulphide bonds (Gould and Hitchins, 1963), and which are unfortunately incompatible with foodstuffs.

Lactoperoxidase is an enzyme that is widely distributed, e.g. in colostrum, milk, saliva and other body fluids. It catalyses the oxidation of thiocyanate (SCN^-) by hydrogen peroxide, to form hypothiocyanite ($OSCN^-$), which is cidal or static to a wide range of vegetative bacteria (Reiter and Harnulv, 1984). The lactoperoxidase system ('LPS') has been effectively exploited to improve the keeping quality of milk e.g. in countries in which distribution infrastructure and refrigeration are poorly developed (Bjorck *et al.*, 1979; Harnulv and Kandusamy, 1982), usually by the addition of the activators of LPS (thiocyanate and H_2O_2) to the milk (FAO, 1999; Seifu *et al.*, 2005), and in animal feeds to minimise enteric infections in young farm animals (Reiter *et al.*, 1981). While ineffective against most yeasts, moulds and bacterial spores, the LPS has been shown to be bactericidal or bacteriostatic for a wide range of spoilage and food poisoning microorganisms, including *Listeria*, *Staphylococcus*, *Campylobacter*, *Salmonella* and *B. cereus* (Ekstrand, 1994) and some yeasts (Pruitt and Reiter, 1985). Suggested non-dairy applications include the eradication of *L. monocytogenes* from the surfaces of fish and meat. LPS and, less effectively, the soy bean peroxidase–thiocyanate system, inactivated *E. coli* and *Shigella* in acidic fruit and vegetable juices, particularly in apple (> 5 log kill in 24 h and 20 °C; Van Opstal *et al.*, 2006). Toothpastes are on sale that contain amyloglucosidase and glucose oxidase enzymes which, following the breakdown of food starches to maltodextrins by salivary amylase, generate hydrogen peroxide (Thomas *et al.*, 1983). This activates salivary lactoperoxidase and is claimed to promote, naturally, dental health (Tenovuo, 2002).

The iron-binding glycoproteins from eggs (ovotransferrin or conalbumin), milk (lactoferrin) and blood (serum transferrins) bind ferric iron so tightly that, in low-iron media like serum and egg white, they prevent microbial growth (Tranter and Board, 1982). The requirement for a low-iron environment limits their use in foods. They may therefore find more effective use in combination with other factors or procedures that limit the availability of iron. Tranter (1994) suggested that, like EDTA, ovotransferrin may act to enhance the bactericidal activity of lysozyme. Transferrins have been shown to increase the activity of lysozyme against *L. monocytogenes* (Johnson, 1989). Lactoferricin, a polypeptide that can be derived from lactoferrin by acid hydrolysis (Saito *et al.*, 1991) or by limited proteolysis (Tomita *et al.*, 1991), interestingly acts on microbial cells by means that do not depend on iron-binding. A potential advantage of lactoferricin in food preservation is its relative heat resistance. It has been shown to be effective in extending the shelf-life of raw milk and in reducing the microbial contamination of raw vegetables when employed as a 1 % dip (Kobayashi *et al.*, 1990). Small (usually between 20 and 40 or so amino acid residues) membrane-active peptides are common in animals. They include, for example, defensins in mammals, attacins and cecropins in insects and magainins in frogs (Board, 1995). There is considerable interest in their potential for commercial application, particularly in the pharmaceutical industry and for topical application in medicine, though not yet in foods.

16.5.2 Plant-derived antimicrobials

Wilkins and Board (1989) reported that more than about 1340 plants are known to be potential sources of antimicrobial compounds. These compounds include many low molecular weight substances (phytoalexins; Table 16.6), among which phenolic compounds predominate (e.g. caffeic, cinnamic, ferulic and gallic acids, oleuropein, thymol, eugenol). About 60 are mentioned by Nychas (1995). Beuchat (1994) listed about 60 plants that are commonly used as herbs or spices and are also known to contain low molecular weight antimicrobial substances with activity against a wide range of bacteria, yeasts and moulds. Plant-derived phenolic compounds were particularly effective as antifungal agents (Lattanzio *et al.*, 1994). Antimicrobial organic acids present in plants, of course, include many that are already employed as food preservatives or acidulants (e.g., benzoic, sorbic, acetic, citric, etc.). Many herbs and spices contain essential oils that are antimicrobial. Deans and Ritchie (1987) list over 80 plants that contain high levels of antimicrobials with potential food use (e.g. sage, rosemary, clove, coriander, garlic, onion). Some (e.g. rosemary) have been incorporated into foods for their strong antioxidant activities. So far, mainly the organic acids have been employed for food preservation, whilst the phenolic compounds, many of which are related chemically to the additives butylated hydroxyanisol (BHA) and butylated hydroxytoluene (BHT) that are employed as antioxidants, have not been exploited, although they do show antimicrobial activity as well (Kabara, 1991).

While many studies have been reported using laboratory media, too few have been undertaken using foods (Shelef, 1983). Furthermore, the levels necessary to inhibit microorganisms in foods have sometimes been found to be much higher than those determined using cultures (Farbood *et al.*, 1976). Other disadvantages associated with the use of many plant-derived antimicrobials in foods (e.g. thymol, carvacrol, citral, eugenol, europein, etc.) are the strong flavours of some of them. Use in combinations with organic acids or with other permitted antimicrobials may be more promising (Nychas, 1995; Nazer *et al.*, 2005). However, a growing number of realistic applications have recently been proposed and evaluated. For example, Burt (2004) and Holley and Patel (2005) listed over 50 studies that demonstrated the successful use of crude extracts and purified essential oils in a range of foods. They pointed out that although organoleptic problems do limit applications, they may often be minimised by the incorporation of complementary ingredients that act to soften flavour notes.

16.5.3 Microorganism-derived antimicrobials

The tetraene antimycotic natamycin (pimaricin), produced by *Streptomyces natalensis*, is used to prevent mould growth in some foods, e.g. on the surfaces of some cheeses and dry sausages (Kliss, 1960; Stark and Tan, 2003). However, the use of microbially-derived antimicrobials for food preservation is dominated by the peptide 'lantibiotic' bacteriocin nisin, produced by certain strains of *Lactococcus lactis*, and applied mainly in cheese and cheese products and in canned tomatoes and other canned vegetables (Delves-Broughton, 1990; Fowler and Gasson, 1991; Abee and Delves-Broughton, 2003) and, in some countries, soups, mayonnaise and infant foods (Montville and Winkowski, 1997). Other applications proposed have been for meat, fish, milk and milk products and alcoholic beverages (Delves-Broughton and Gasson, 1994). Poor stability in animal-derived foods will probably continue to limit applications in meat products (De Martinis *et al.*, 2002).

However, the numbers of other bacteriocins that have been discovered and that may have potential in food preservation continue to grow (Hoover, 2000). They have been categorised into three classes: (i) small, post-transcriptionally modified lantibiotics, e.g. nisin; (ii) small, heat-stable unmodified peptides, e.g.

pediocin PA-1; (iii) larger, heat-labile molecules (e.g. helveticin J). Hill (1995) listed more than 20 that have been well characterised, and many more have been reported. Many are produced by strains of starter cultures (Ray and Daeschel, 1994). While these have not been applied as additive food preservatives in the same way as nisin, the potential for such use, or by generation *in situ* in foods, e.g. from the addition of particular starter cultures, or in cultured product food ingredients such as Microgard™, derived from culture supernatants of particular *Propionibacterium* and *Lactococcus* strains, must be substantial. Microgard™ has proved to be very effective in ensuring the keepability and safety of cottage cheeses for more than two decades (Daeschel, 1989; Al-Zoreky *et al.*, 1991).

Later on, the genetic modification of bacteriocins, that is the focus for much current research, may result in derivative molecules that have usefully modified antimicrobial activity spectra, or changes in other properties that increase their potential value in food preservation and safety, if regulatory hurdles can be overcome (Fowler and Gasson, 1991; Hill, 1995). Already genetic engineering has resulted in the production of nisins with changed amino acid sequences. For example, Kuipers *et al.* (1992) changed the serine residue at the 5 position of prenisin Z to threonine, and the threonine was shown to be dehydrated to dehydroalanine during processing of the molecule. In this instance, the new nisin had lower antimicrobial activity than the wild-type molecule. Dodd *et al.* (1992) replaced the position 23 threonine of nisin A with a serine residue to produce a less antimicrobially active mature peptide. However, a nisin A variant in which the dehydroalanine was replaced by alanine at position 5 had near-normal antimicrobial activity (Delves-Broughton and Gasson, 1994).

Much recent research has concentrated on the potential for bacteriocins as antagonists for vegetative bacteria (e.g. *Listeria*). It should not be forgotten that nisin, the only widely used bacteriocin (its use is currently allowed in about 50 countries), is employed mainly as an anti-spore agent, e.g. finding substantial use to prevent the outgrowth from germinated spores of *C. tyrobutyricum* in processed cheese products (Delves-Broughton, 1990) and to prevent growth from spores of thermophiles such as *Geobacillus stearothermophilus* and *Thermoanaerobacterium thermosaccharolyticum* that may survive and grow in canned foods stored at high temperatures (Eyles and Richardson, 1988; Fowler and Gasson, 1991). However, nisin suffers from the disadvantage of being relatively ineffective against the proteolytic types of *C. botulinum* (Sommers and Taylor, 1987). Discovery of bacteriocins that are more effective against this organism, and which control the growth of a wider range of spoilage sporeformers as well, would have great potential value, because they may allow substantial reductions in the thermal processing of high water activity–low acid foods, with potential advantages in energy saving and improved product quality.

16.6 Conclusions

Whilst the preservation and safety of the vast majority of foods continue to be assured using long-standing conventional techniques, the use of the new and emerging physical techniques that inactivate microorganisms without the application of heat, or with the use of less heat than would be otherwise necessary, is attractive from the point of view of product quality. Three facts limit their usefulness at the present time. Firstly, bacterial spores remain the organisms most tolerant to all the techniques so that, with the exception of PATS and irradiation, sterilisation, as opposed to pasteurisation, is not yet possible. Secondly, the kinetics of inactivation that result from some of the techniques are different from those resulting from heating, so that careful new approaches, e.g. to product safety and risk analysis, will be needed if application of the techniques continues to be promoted. Thirdly, again with the exception of hydrostatic pressure and irradiation, the efficacy of the other techniques is impaired by product structure, and may therefore be limited to liquid products or products containing small particulates, or (for light pulses) transparent products and surfaces, etc. At the same time, combination techniques in which a new technology is only one component of a total 'hurdle' preservation system (Leistner, 1995) have already been described and, if these are further developed and proven to be effective, the opportunities for use of the new techniques are very likely to grow in the future.

It has been pointed out that 'minimal' preservation techniques may be less able than traditional ones to overcome microbial adaptation to stresses (Abee and Wouters, 1999). The major stresses to which microorganisms respond with increased resistance, and that are of importance in food processing and preservation, include cold shock, mild heating, osmotic/low a_w, low pH, and the presence of preservatives and disinfectants (Beales, 2004), and stress reactions to high pressure must now be taken into account as well (Aertsen *et al.*, 2004). In addition, stress reactions of cells in biofilms can raise the tolerance of microorganisms in processing environments. Oxidative stress occurs, but is less likely to be important, in food environments (Alter and Scherer, 2006).

With respect to the exploitation of naturally-occurring antimicrobials, it is clear that, in nature, an enormous number of very effective antimicrobial systems exists. However, it is also clear that disappointingly few have been developed for positive use in foods. To some extent, this may reflect industry's reluctance to embark on the substantial and expensive programmes of toxicological testing that can be necessary for the introduction of a new antimicrobial. To some extent, it has probably also been due to the absence of strong commercial and marketing incentives to develop their use, but this has changed in recent years in response to consumers' changing needs. Most would predict that natural systems, particularly in additive or synergistic, hurdle, combinations with other factors and techniques that we can already make use of will, as indicated at the bottom of Table 16.6, have an increasing role to play in the future, particularly if more studies are undertaken in realistic foodstuffs as well as in laboratory media.

16.7 Sources of further information and advice

Alzamora S M, Tapia M S and Lopez-Malo A (eds) (2000) *Minimally Processed Fruits and Vegetables*, Gaithersburg, MD, Aspen.

Barbosa-Canovas G V and Gould G W (eds) (2000) *Innovations in Food Processing*, Lancaster, PA, Technomic.

Davidson P M and Branen A L (eds) (1993) *Antimicrobials in foods*, New York, Marcel Dekker.

Forsythe S J (2000) *The Microbiology of Safe Food*, Oxford, Blackwell Science.

Griffiths M (ed.) (2005) *Understanding Pathogen Behaviour: Virulence, Stress Response and Resistance*, Cambridge, Woodhead.

Hendrickx M E G and Knorr D (eds) (2001) *Ultra High Pressure Treatments of Foods*, New York, Kluwer Academic/Plenum.

ICMSF (1998) *Microorganisms in Foods 6, Microbial Ecology of Food Commodities*, International Commission on Microbiological Specifications for Foods, London, Blackie Academic and Professional.

Juneja V J and Sofos J N (eds) (2002) *Control of Foodborne Microorganisms*, New York, Marcel Dekker.

Leistner L and Gould G W (2002) *Hurdle Technologies: Combination Treatments for Food Stability, Safety and Quality,* New York, Kluwer Academic.

Lelieveld H L M, Notermans S and de Haan S W H (eds) (2007) *Food Preservation by Pulsed Electric Fields: From Research to Application*, Cambridge, Woodhead.

Lund B M, Baird-Parker A C and Gould G W (eds) (2000), *The Microbiological Safety and Quality of Food*, Gaithersburg, MD, Aspen.

Ohlsson T and Bengsston N (eds) (2002) *Minimal Processing Technologies in the Food Industry*, Cambridge, Woodhead.

Russell N J and Gould G W (eds) (2003), *Food Preservatives*, 2nd edn, New York, Kluwer Academic.

Storz G and Hengge-Aronis R (eds) (2000) *Bacterial Stress Responses*, Washington, DC, ASM Press.

Zeuthen P (ed) (2003) *Food Preservation Techniques*, Cambridge, Woodhead.

16.8 References

Abee T and Delves-Broughton J (2003) Bacteriocins – nisin, in Russell N J and Gould G W (eds), *Food Preservatives*, New York, Kluwer Academic, 146–78.

Abee T and Wouters J A (1999) Microbial stress response in minimal processing, *Int J Food Microbiol*, **50**, 65–91.

ACMSF (1992) *Report on Vacuum Packaging and Associated Processes*, Advisory Committee on the Microbiological Safety of Food, London, HMSO.

Aertsen A, Van Houdt R, Vanoirbeek K and Michiels C W (2004) An SOS response induced by high pressure in *Escherichia coli*, *J Bacteriol*, **186**, 6133–41.

Ahmed F I K and Russell C (1975) Synergism between ultrasonic waves and hydrogen peroxide in the killing of microorganisms, *J Appl Bacteriol* **39**, 31–40.

Alderton G and Snell N (1963) Base exchange and heat resistance of bacterial spores, *Biochim Biophys Res Commun*, **10**, 139–43.

Alliger H (1975) Ultrasonic disruption, *Amer Lab*, **10**, 75–85.

Alter T and Scherer K (2006) Stress response of *Campylobacter* spp. and its role in food processing, *J Vet Med B*, 53, 351–7.

Alzamora S M, Tapia M S and Lopez-Malo A (2000) *Minimally Processed Fruits and Vegetables: Fundamental Aspects and Applications*, Gaithersburg, MD, Aspen.

Al-Zoreky N, Ayres J W and Sandine W E (1991). Antimicrobial action of Microgard™ against food spoilage and pathogenic microorganisms, *Dairy Sci*, 74, 748–63.

Anon (1989) *Chill and Frozen: Guidelines on Cook-chill and Cook-freeze Catering*, London, HMSO.

Barbosa-Canovas G V, Pothakamury U R and Swanson B G (1995) State of the art technologies for the sterilization of foods by non-thermal processes: physical methods, in Barbosa-Canovas G and Welti-Chanes J (eds), *Food Preservation by Moisture Control: Fundamentals and Applications*, Lancaster, PA, Technomic, 493–532.

Basset J and Machebouf M A (1932) Étude sue les effets biologiques des ultrapressions: résistance de bactéries, des diastases et de toxines aux pressions très élevées', *Compt Rend Hebd Sci Acad Sci*, **196**, 1431–42.

Beales N (2004) Adaptation of microorganisms to cold temperatures, weak acid preservatives, low pH, and osmotic stress: a review, *Comp Rev Food Sci Food Saf*, **3**, 1–20.

Beuchat L R (1994) Antimicrobial properties of spices and their essential oils, in Dillon V M and Board R G (eds), *Natural Antimicrobial Systems and Food Preservation*, Wallingford, CAB International, 167–80.

Bjorck L, Claesson Q and Schultness W (1979) The lactoperoxidase/thiocyanate/hydrogen peroxide system as a temporary preservative for raw milk in developing countries, *Milchwissenschaft*, **34**, 726–9.

Board R G (1995) Natural antimicrobials from animals, in Gould G W (ed.), *New Methods of Food Preservation*, Glasgow, Blackie Academic and Professional, 40–57.

Booth I R (1995) Regulation of cytoplasmic pH in bacteria, *Microbiol Rev*, **49**, 359–78.

Booth I R and Kroll R G (1989) The preservation of foods by low pH, in Gould G W (ed.), *Mechanisms of Action of Food Preservation Procedures*, London, Elsevier Applied Science, 119–60.

Boulart J (1983) Process for protecting a fluid and installations for realization of that process, *French Pat*, 2,513,087.

Brul S and Coote P (1999) Review: Preservative agents in foods–mode of action and microbial resistance mechanisms, *Int J Food Microbiol*, **50**, 1–17.

Burgos J, Ordonez J A and Sala F J (1972) Effect of ultrasonic waves on the heat resistance of *Bacillus cereus* and *Bacillus coagulans* spores, *Appl Microbiol*, **24**, 497–8.

Burt S (2004) Essential oils: their antibacterial properties and potential applications in foods – a review', *Int J Food Microbiol*, **94**, 223–53.

Bushnell A H, Dunn J E, Clark R W and Pearlman J S (1993) High pulsed voltage system for extending the shelf life of pumpable food products, *US Pat* 5,235,905.

Butz P and Ludwig H (1986) Pressure inactivation of microorganisms at moderate temperatures, *Physica*, 139/140B, 875–7.

Carini S and Lodi R (1982) Inhibition of germination of clostridial spores by lysozyme, *Ind Latte*, **18**, 35–48.

Carlez A, Cheftel J-C, Rosec J P, Richard N, Saldana J-L and Balny C (1992) Effects of high pressure and bacteriostatic agents on the destruction of *Citrobacter freundii* in minced beef muscle, in Balny C, Hayashi R, Heremans K and Masson P (eds), *High Pressure and Biotechnology*, Libby Eurotech Ltd, Colloque INSERM/J, 365–8.

Carminati D, Nevianti E and Muchetti G (1985) Activity of lysozyme on vegetative cells of *Clostridium tyrobutyricum*, *Latte* **10**, 194–8.

Castro A I, Barbosa-Canovas G V and Swanson B G (1993) Microbial inactivation in foods by pulsed electric fields, *J Food Proc Pres*, **17**, 47–73.

Chatzolopou A, Miles R J and Anagnostopoulos G (1993) Destruction of Gram-negative bacteria, *Int Pat Appl* WO 93/00822.

Chernomordik L V, Sukharev S I, Popov S V, Pastushenko V F, Sokirko A V, Abidor I G and Chizmadzhev Y A (1987) The electrical breakdown of cell and lipid membranes: the similarity of phenomenologies, *Biochim Biophys Acta*, **972**, 360–5.

Christian J H B (2000) Drying and reduced water activity, in Lund B M, Baird-Parker A C and Gould G W (eds), *The Microbiological Safety and Quality of Food*, Gaithersburg, MD, Aspen 146–74.

Clouston J G and Wills P A (1969) Initiation of germination and inactivation of *Bacillus pumilus* spores by hydrostatic pressure, *J Bacteriol*, **97**, 684–90.

Clouston J G and Wills P A (1970) Kinetics of germination and inactivation of *Bacillus pumilus* spores by hydrostatic pressure, *J Bacteriol*, **103**, 140–3.

Cobb C M, McCawley T K and Killoy W J (1992) A preliminary study on the effects of the Nd:YAG laser on root surfaces and subgingival microflora *in vivo*, *J Periodont*, **63**, 701–7.

Costa J L and Hofmann G A (1987) Malignancy treatment, *US Pat* 4,665,898.

Daeschel M A (1989) Antimicrobial substances from lactic acid bacteria for use as food preservatives, *Food Technol*, **43**(1), 164–7.

Damar S and Balaban M O (2006) Review of dense phase CO_2 technology: microbial and enzyme inactivation, and effects on food quality, *J Food Sci*, **71**, R1–R11.

Davies (2003) Modified atmospheres and vacuum packaging, in Russell N J and Gould G W (eds), *Food Preservatives*, New York, Kluwer Academic, 218–39.

Deans S G and Ritchie G (1987) Antibacterial properties of plant essential oils, *Int J Food Microbiol*, **5**, 165–80.

Delves-Broughton J (1990) Nisin and its uses, *Food Technol*, **44**, 100–17.

Delves-Broughton J and Gasson M J (1994), Nisin, in Dillon V M and Board R G (eds), *Natural Antimicrobial Systems and Food Preservation*, Wallingford, CAB International, 99–132.

De Martinis E C P, Alves V F and Franco B D G M (2002) Fundamentals and perspectives for the use of bacteriocins produced by lactic acid bacteria in meat products, *Food Rev Int*, **18**, 191–208.

Devlieghere F, Vermeiren L and Debevere J (2004) New preservation technologies: possibilities and limitations', *Int Dairy J*, **14**, 273–85.

Diehl J F (2002) Food irradiation-past, present and future, *Radiat Phys Chem*, **63**, 211–5.

Dodd H M, Horn N, Hao Z and Gasson M J (1992) A lactococcal expression system for engineered nisins, *Appl Environ Microbiol*, **58**, 3683–93.

Doevenspeck H (1960) *German Pat* 1,237,541.

Dunn J E and Pearlman J S (1987) Methods and apparatus for extending the shelf life of fluid food products, *US Pat* 4,695,472.

Dunn J E, Clark, R W, Asmus J F, Pearlman J S, Boyer K and Parrichaud F (1988) Method and apparatus for preservation of foodstuffs, *Int Pat* WO88/03369.

Earnshaw R G (1995) High pressure microbial inactivation kinetics, in Ledward D A, Johnston D E, Earnshaw R G and Hasting A P M (eds), *High PressureProcessing of Foods*, Nottingham, Nottingham University Press, 37–46.

Eklund T (1985) Inhibition of microbial growth at different pH levels by benzoic and propionic acids and esters of *p*-hydroxybenzoic acid, *Int J Food Microbiol*, **2**, 159–76.

Ekstrand B (1994) Lactoperoxidase and lactoferrin, in Dillon V M and Board R G (eds), *Natural Antimicrobial Systems and Food Preservation*, Wallingford, CAB International, 15–64.

Eyles M J and Richardson K C (1988) Thermophilic bacteria and food spoilage, *CSIRO Food Res Quart*, **48**, 19–24.

Eze M O (1990) Consequences of the lipid bilayer to membrane-associated reactions, *J Chem Ed*, **67**, 17–20.
FAO (1999) Manual on the use of the LP- system in milk handling and preservation, Rome, Food and Agriculture Organization of the United Nations.
Farber J M (1991) Microbiological aspects of modified atmosphere packaging: a review, *J Food Prot*, **54**, 58–70.
Farbood M I, Macneil J H and Ostovar K (1976) Effect of rosemary spice extract on growth of microorganisms in meats, *J Milk Food Technol*, **39**, 675–9.
Farr D (1990) High pressure technology in the food industry, *Trends Food Sci Technol*, **1**, 14–16.
Forsyth S J (2000) *The Microbiology of Safe Food*, Oxford, Blackwell Scientific.
Fowler G G and Gasson M J (1991) Antibiotics – nisin, in Russell N J and Gould G W (eds), *Food Preservatives*, Glasgow, Blackie Academic and Professional, 135–52.
Fryer P (1995) Electrical resistance heating of foods, in Gould G W (ed.), *New Methods of Food Preservation*, Glasgow, Blackie Academic and Professional, 205–35.
Fujii K, Ohtani A, Watanabe J, Ohgoshi H, Fujii T and Honma K (2002) High-pressure inactivation of *Bacillus cereus* spores in the presence of argon, *Int J Food Microbiol*, **72**, 239–42.
Furukawa S, Shimoda M and Hayakawa I (2003) Mechanism of the inactivation of bacterial spores by repeated pressurization treatment, *J Appl Microbiol*, **94**, 1–6.
Furukawa S, Shimoda M and Hayakawa I (2004) Effect of repeated pressure treatment on breakdown of clumps of bacterial spores, *Food Sci Technol Res*, **10**, 10–12.
Garcia M L, Burgos J, Sanz B and Ordonez J A (1989) Effect of heat and ultrasonic waves on the survival of two strains of *Bacillus subtilis*, *J Appl Bacteriol*, **67**, 619–28.
Garcia-Gonzalez L, Geeraerd A H, Spilimbergo S, Elst K, Van Ginneken L, Debevere J, Van Impe J F and Devlieghere F (2007) High pressure carbon dioxide inactivation of microorganisms in foods: the past, the present and the future, *Int J Food Microbiol*, **117**, 1–28.
Glaser R W, Leikin S L, Chernomordik L V, Pastushenko V F and Sokirko A V (1988) Reversible electrical breakdown of lipid bilayers: formation and evolution of pores, *Biochim Biophy Acta*, **940**, 275–81.
Gould G W (1999) Sous vide foods: conclusions of an ECFF botulinum working party, *Food Control*, **10**, 47–51.
Gould G W (2000) Strategies for food preservation, in Lund B M, Baird-Parker A C and Gould G W (eds), *The Microbiological Safety and Quality of Food*, Gaithersburg, MD, Aspen, 19–35.
Gould G W and Hitchins A D (1963) Sensitization of spores to lysozyme and hydrogen peroxide with agents which rupture disulphide bonds, *J Gen Microbiol*, **33**, 413–22.
Gould G W and Sale A J H (1970) Initiation of germination of bacterial spores by hydrostatic pressure, *J Gen Microbiol*, **60**, 335–46.
Grahl T, Sitzmann W and Mak, H (1992) Killing of microorganisms in fluid media by high voltage pulses, *DECHMA Biotechnol Conf Ser* 5B, 675–8.
Grant S, Patterson M and Ledward D (2000) Food processing gets freshly squeezed, *Chem Ind*, 24 Jan, 55–8.
Hamilton W A and Sale A J H. (1967) Effects of high electric fields on microorganisms 11. Mechanism of action of the lethal effect, *Biochim Biophys Acta*, **148**, 789–95.
Harnulv B G and Kandusamy C (1982) Increasing the keeping quality of raw milk by activation of the lactoperoxidase system. Results for Sri Lanka, *Milchwissenschaft*, **37**, 454–7.
Harvey E and Loomis A (1929) The destruction of luminous bacteria by high frequency sound waves, *J Bacteriol*, **17**, 373–9.
Hauben K E, Wuytack C and Michiels C W (1996) High pressure transient sensitisation of *Escherichia coli* to lysozyme and nisin by disruption of outer membrane permeability, *J Food Prott*, 59, 350–5.

Hauben K J A, Bernaerts K and Michiels C W (1998) Protective effect of calcium on inactivation of *Escherichia coli* by high hydrostatic pressure, *J Appl Microbiol*, **85**, 678–84.

Hayakawa I, Kanno T, Tomita M and Figio Y (1994) Application of high pressure for spore inactivation and protein denaturation, *J Food Sci*, **59**, 159–63.

Hayman M M, Baxter I, O'Riordan P J and Stewart C M (2004) Effects of high-pressure processing on the safety, quality, and shelf life of ready-to-eat meats, *J Food Protect*, **67**, 1709–18.

Hendrickx E G and Knorr D (eds) (2001) *Ultra High Pressure Treatments of Foods*, New York, Kluwer Academic/Plenum.

Heremans K (1995) High pressure effects on biomolecules, in Ledward D A, Johnston D E, Earnshaw R G and Hasting A P M (eds), *High Pressure Processing of Foods*, Nottingham, Nottingham University Press, 81–97.

Hill C (1995) Bacteriocins: natural antimicrobials from microorganisms, in Gould G W (ed.), *New Methods of Food Preservation*, Glasgow, Blackie Academic and Professional, 22–39.

Hite B H (1899) The effect of pressure in the preservation of milk, *Bull West Virginia Expt Station*, No. 58, 15–35.

Hite B H, Giddings N J and Weakley C W (1914) The effect of pressure on certain microorganisms encountered in the preservation of fruits and vegetables, *Bull West Virginia Expt Station*, No. 146, 3–67.

Hoffman G A (1985) Inactivation of microorganisms by an oscillating magnetic field, *US Pat* 4,524,079 and *Internat Pat* WO85/02094.

Hoffman G A and Evans E G (1986) Electronic, genetic, physical and biological aspects of electromanipulation, *IEEE Medical Biol Mag*, **5**, 6–25.

Holley R A and Patel D (2005) Improvement in shelf–life and safety of perishable foods by plant essential oils and smoke antimicrobials, *Food Microbiol*, **22**, 273–92.

Hoover D G (1993) Pressure effects on biological systems, *Food Technol*, **47**(6), 150–5.

Hoover D G (2000) Microorganisms and their products in the preservation of food, in Lund B M, Baird-Parker A C and Gould G W (eds), *The Microbiological Safety and Quality of Food*, Gaithersburg, MD, Aspen 251–76.

Hoover D G, Metrick K, Papineau A M, Farkas D F and Knorr D (1989) Biological effects of high hydrostatic pressure on food microorganisms, *Food Technol*, **43**, 99–107.

Horie Y, Kimura K, Ida M. Yosida Y and Ohki K (1991) Jam preservation by pressure pasteurization, *Nippon Nogeiki Kaisu*, **65**, 975–80.

Huang E, Mittal G S and Griffiths M W (2006) Inactivation of *Salmonella enteritidis* in liquid whole egg using combination treatments of pulsed electric field, high pressure and ultrasound, *Biosystems Eng*, **94**, 404–13.

Huis in't Velt J H J (1996) Microbiological and biochemical spoilage of foods: an overview, *Int J Food Microbiol*, **33**, 1–18.

Hulsheger H and Niemann E G (1980) Lethal effect of high voltage pulses on *E. coli* K12, *Radiat Environ Biophys*, **18**, 281–8.

Hulsheger H, Potel J and Neimann E G (1981) Killing of bacteria with electric pulses of high field strength, *Radiat Environ Biophys*, **20**, 53–61.

Hulsheger H, Potel J and Neimann E G (1983) Electric field effects on bacteria and yeast cells, *Radiat Environ Biophys*, **22**, 149–56.

ICMSF (1980) *Microbial Ecology of Foods 1, Factors Affecting Life and Death of Microorganisms*, International Commission on the Microbiological Specfications for Foods, New York, Academic Press.

ICMSF (1998) *Microorganisms in Foods 6, Microbial Ecology of Food Commodities*, Ch.11, Oil and fat-based foods, International Commission on the Microbiological Specfications for Foods, London, Blackie Academic and Professional, 390–417.

Isaacs N S, Chilton P and Mackey B (1995) Studies on the inactivation by high pressure of microorganisms, in Ledward D A, Johnston D E, Earnshaw R G and Hasting A P M (eds), *High Pressure Processing of Foods*, Nottingham, Nottingham University Press, 65–79.

Jayaram S, Castle G S P and Margaritis A (1992) Kinetics of sterilization of *Lactobacillus brevis* by the application of high voltage pulses, *Biotechnol Bioeng*, **40**, 1412–20.

Johnson E A (1989) The potential application of antimicrobial proteins in food preservation, *J Dairy Sci*, **72**, 123–4.

Juneja V K and Thippareddi H (2004) Inhibitory effects of organic acid salts on growth of *Clostridium perfringens* from spore inocula during chilling of marinated ground turkey breast *Int J Food Microbiol*, **93**, 155–63.

Kabara J (1991) Phenols and chelators, in Russell N J and Gould G W (eds), *Food Preservatives*, Glasgow, Blackie Academic and Professional, 200–14.

Kalchayanand N, Sikes T, Dunne C P and Ray B (1994) Hydrostatic pressure and electroporation have increased bactericidal efficiency in combination with bacteriocins, *Appl Environ Microbiol*, **60**, 4174–7.

Kimugasa H, Takao T, Fukumoto K and Ishihara M (1992) Changes in tea components during processing and preservation of tea extracts by hydrostatic pressure sterilization, *Nippon Nogeiku Kaichi*, **66**, 707–12.

Kliss W S (1960) Effectiveness of pimaricin as an alternative to sorbate for the inhibition of yeasts and moulds in foods, *Food Technol*, **13**, 124–32.

Knorr D (1995) Hydrostatic pressure treatment of food: microbiology, in Gould G W (ed.), *New Methods of Food Preservation*, Glasgow, Blackie Academic and Professional, 159–75.

Kobayashi S, Tomita M, Kawase K, Takase M, Miyakawa H, Yamauchi K, Saito H, Abe H and Shimamura S (1990) Lactoferrin hydrolysate for use as an antibacterial agent and as a tyrosinase inhibition agent, *European Pat Appl No* 90125035.7

Koutchma T, Guo B, Patazca E and Parisi B (2005) High pressure-high temperature sterilization: from kinetic analysis to process verification, *J Food Proc Eng*, **28**, 610–29.

Kowalski E, Ludwig H and Tausche, B (1992) Hydrostatic pressure to sterilize foods 1. Application to pepper (*Piper nigrum* L), *Deutsche Lebensmi Rundsch*, **88**, 74–5.

Kuipers O P, Rollema H S, Yap W M G, Boot H J Siezen R J and De Vos W M (1992) Engineering dehydrated amino acid residues in the antimicrobial peptide nisin, *J Biol Chem*, **267**, 2430–4.

Larson W P, Hartzel, T B and Diehl H S (1918) The effect of high pressure on bacteria, *J Infect Dis*, **22**, 271–9.

Lattanzio V, de Cicco V, Di Venere D, Lima G and Salermo M (1994) Antifungal activity of phenolics against fungi commonly encountered during storage, *Italian J Food Sci*, **6**, 23–30.

Leistner L (1995) Principles and applications of hurdle technology, in Gould G W (ed.), *New Methods of Food Preservation*, Glasgow, Blackie Academic and Professional, 1–21.

Leistner L and Gould G W (2002) *Hurdle Technologies: Combination Treatments for Food Stability, Safety and Quality,* New York, Kluwer Academic/Plenum.

Livingston G E (1985) Extended shelf life chilled prepared foods, *J Foodserv Syst*, **3**, 221–30.

Lopez P, Sala F J, Fuente J L, Condon S, Raso J and Burgos J (1994) Inactivation of peroxidase, lipoxygenase and polyphenoloxidase by manothermosonication, *J Agric Food Chem*, **42**, 552–6.

Lopez T J, Roig A X, Capellaas M, Trujillo A J, Hernandez M and Guamis B (2003) Evaluation of the importance of germinative cycles for destruction of *Bacillus cereus* spores in miniature cheeses, *High Pressure Res*, **23**, 81–5.

Lopez-Pedemonte T J, Roig-Sagues A X, Trujillo A J, Capellas M and Guamis B (2003) Inactivation of spores of *Bacillus cereus* in cheese by high hydrostatic pressure with the addition of nisin or lysozyme, *J Dairy Sci*, **86**, 3075–81.

Ludwig H, Bieler C, Hallbauer K and Scigalla W (1992) Inactivation of microorganisms by hydrostatic pressure, in Balny C, Hayashi R, Heremans K and Masson P (eds), *High Pressure Biotechnology*, Colloque INSERM/J, Libby Eurotext Ltd, 25–32.

Lund B M and Eklund T (2000) Control of pH and use of organic acids, in Lund B M, Baird-Parker A C and Gould G W (eds), *The Microbial Safety and Quality of Food*, Gaithersburg, MD, Aspen 175–99.

Lund B M and Peck M W (1994) Heat resistance and recovery of non-proteolytic *Clostridium botulinum* in relation to refrigerated processed foods with an extended shelf-life, *J Appl Bact Symp Suppl*, **76**, 115S–128S.

Lund B M and Peck M W (2000) Clostridium botulinum, in Lund B M, Baird-Parker A C and Gould G W (eds), *The Microbiological Safety and Quality of Food*, Gaithersburg, MD, Aspen, 1057–109.

Margosch D, Ehrmann M A, Ganzle M G and Vogel R F (2004) Comparison of pressure and heat resistance of *Clostridium botulinum* and other endospores in mashed carrots, *J Food Pro*, **67**, 2530–7.

Margosch D, Ehrmann M A, Buckow R, Heinz V, Vogel R F and Ganzle M G (2006) High-pressure-mediated survival of *Clostridium botulinum* and *Bacillus amyloliquifaciens* endospores at high temperatures, *Appl Environ Microbiol*, **72**, 3476–81.

Marquis R E, Sim J and Shin S Y (1994) Molecular mechanisms of resistance to heat and oxidative damage, *J Appl Microbiol Symp Suppl*, **76**, 40S–48S.

Martinez-Rodriguez A and Mackey B M (2005) Factors affecting the pressure resistance of some *Campylobacter* species, *Lett Appl Microbiol*, **41**, 321–6.

Masschalck B, Van Houdt R, Van Haver E R and Michiels C W (2001) Inactivation of Gram-negative bacteria by lysozyme, denatured lysozyme and lysozyme-derived peptides under high hydrostatic pressure, *Appl Environ Microbiol*, **67**, 339–44.

Masschalck B, Deckers D and Michiels C W (2002) Lytic and nonlytic mechanism of inactivation of Gram-positive bacteria by lysozyme under atmospheric and high hydrostatic pressure, *J Food Pro*, **65**, 1916–23.

Masschalck B, Deckers D and Michiels C W (2003) Sensitization of outer-membrane mutants of *Salmonella* Typhimurium and *Pseudomonas aeruginosa* to antimicrobial peptides under high pressure, *J Food Prot*, **66**, 1360–7.

Mbandi E and Shelef L A (2002) Enhanced antimicrobial effects of combination of lactate and diacetate on *Listeria monocytogenes* and *Salmonella* spp. in beef bologna, *Int J Food Microbiol*, **76**, 191–8.

McClure P J, Blackburn C deW, Cole M B *et al.* (1994) Modelling the growth, survival and death of microorganisms in foods: the UK Food Micromodel approach, *Int J Food Microbiol*, **23**, 265–75.

McMeekin T A and Ross T (1996) Shelf life prediction: status and future possibilities, *Int J Food Microbiol*, **28**, 65–83.

McMeekin T A, Olley J, Ross T and Ratkowsky D A (1993) *Predictive Microbiology: Theory and Applications*, Taunton, Research Studies Press.

Mertens B (1995) Hydrostatic pressure treatment of food: equipment and processing, in Gould G W (ed.), *New Methods of Food Preservation*, Glasgow, Blackie Academic and Professional, 135–8.

Mertens B and Knorr D (1992) Development of nonthermal processes for food preservation, *Food Technol*, **46**(5), 124–33.

Metrick C, Hoover D G and Farkas D F (1989) Effects of high hydrostatic pressure on heat-sensitive strains of *Salmonella*, *J Food Sci*, **54**, 1547–64.

Miller A J, Call J E and Whiting R C (1993) Comparison of organic-acid salts for *Clostridium botulinum* control in an uncured turkey product, *J Food Prot*, **56**, 958–62.

Mizuno A and Hori Y (1988) Destruction of living cells by pulsed high voltage applications, *Trans IEEE Indus Appl*, **24**, 387–95.

Molin G (2000) Modified atmospheres, in Lund B M, Baird-Parker A C and Gould G W (eds), *The Microbiological Safety and Quality of Food*, Gaithersburg, MD, Aspen 214–34.

Molina J F, Barbosa-Canovas G V, Swanson B G and Clark S (2002) Inactivation by high-intensity pulsed electric fields, in Juneja V K and Sofos J N (eds), *Control of foodborne microorganisms*, New York, Marcel Dekker, 383–97.

Monticello D J (1989) Control of microbial growth with nisin/lysozyme formulations, *European Pat Appl* No. 89123445.2.

Montville T J and Winkowski K (1997) Biologically based preservation systems and probiotic bacteria, in Doyle M P, Beuchat L R and Montville T J (eds), *Food Microbiology: Fundamentals and Frontiers*, Washington, DC ASM Press, 557–77.

Mossel D A A (1983) Essentials and perspectives of the microbial ecology of foods, in Roberts T A and Skinner F A (eds), *Food Microbiology: Advances and Prospects, Soc Appl Bacteriol Symp Ser 11,* London, Academic Press, 1–45.

Mossel D A A and Struijk C B (1991) Public health implications of refrigerated pasteurised ('sous-vide') foods, *Int J Food Microbiol*, **13**, 187–206.

Mullin J (1995) Microwave processing, in Gould G W (ed.), *New Methods of Food Preservation*, Glasgow, Blackie Academic and Professional, 112–34.

Murrell, W G and Wills P A. (1977) Initiation of *Bacillus* spore germination by hydrostatic pressure: effect of temperature, *J Bacteriol*, **129**, 1272–80.

Nakamura R, Kato A and Kobayashi K (1990) Novel bifunctional lysozyme-dextran conjugate that acts on both Gram-negative and Gram-positive bacteria, *Agric Biol Chem*, **54**, 3057–9.

Nazer A I, Kobilinsky A, Tholozan J-L and Dubois-Brisonnet F (2005) Combinations of food antimicrobials at low-levels to inhibit the growth of *Salmonella* sv. Typhimurium: a synergistic effect?, *Food Microbiol*, **22**, 391–8.

Neumann E, Sowers A E and Jordan C A (eds) (1989) *Electroporation and Electrofusion in Cell Biology*, New York, Plenum.

Notermans S, Dufrenne J and Lund B M (1990) Botulism risk of refrigerated processed foods of extended durability, *J Food Prot*, **53**, 1020–4.

Nychas G J E (1995) Natural antimicrobials from plants, in Gould G W (ed.), *New Methods of Food Preservation*, Glasgow, Blackie Academic and Professional, 58–99.

Ordonez J A, Sanz B, Hernandez P E and Lopez-Lorenzo P (1984) A note on the effect of combined ultrasonic and heat treatments on the survival of thermoduric streptococci, *J Appl Bacteriol*, **56**, 175–7.

Ordonez J A, Aguilera M A, Garcia M L and Sanz B (1987) Effects of combined ultrasonic and heat treatment (thermosonication) on the survival of a strain of *Staphylococcus aureus*, *J Dairy Res*, **54**, 61–7.

Oxen P and Knorr D (1993) Baroprotective effects of high solute concentrations against inactivation of *Rhodotorula rubra*, *Lebensmit Wiss Technol*, **26**, 220–3.

Pagan R, Manas P, Raso J and Condon S (1999) Bacterial resistance to ultrasonic waves under pressure at nonlethal (manosonication) and lethal (manothermosonication) temperatures, *Appl Environ Microbiol*, **65**, 297–300.

Palaniappan S (1996) High isostatic pressure processing of foods, in Chandarana P I (ed.), *New Processing Technologies Yearbook*, Washington, DC, National Food Processors Association, 51–66.

Parry R T (1993) *Principles and Application of Modified Atmosphere Packaging of Foods*, Glasgow, Blackie Academic and Professional.

Patterson M F and Loaharanu P (2000) Irradiation, in Lund B M, Baird-Parker A C and Gould G W (eds), *The Microbiological Safety and Quality of Food*, Gaithersburg, MD, Aspen 65–100.

Patterson M F, Quinn M, Simpson R and Gilmour A (1995a) Effects of high pressure on vegetative pathogens, in Ledward D A, Johnston D E, Earnshaw R G and Hasting A P M (eds), *High Pressure Processing of Foods*, Nottingham, Nottingham University Press, 47–63.

Patterson M F, Quinn M, Simpson R and Gilmour A (1995b) Sensitivity of vegetative pathogens to high hydrostatic pressure treatment in phosphate-buffered saline and foods, *J Food Pro*, **58**, 524–9.

Pothakamury U R, Monsalve-Gonzalea A, Barbosa-Canovas G V and Swanson B G (1993) Magnetic-field inactivation of microorganisms and generation of biological changes, *Food Technol*, **47**(12), 85–92.

Powell G L and Wisenart B (1991) Comparison of three lasers for dental instrument sterilization, *Lasers Surg Med*, **11**, 69–71.

Pruitt K M and Reiter B (1985) Biochemistry of the peroxidase system: antimicrobial effects, in Pruitt K and Tenovuo J O (eds), *The Lactoperoxidase System*, New York, Marcel Dekker, 143–65.

Qin B, Zhang Q, Barbosa-Canovas G V, Swanson B G and Pedrow PD (1994) Inactivation of microorganisms by pulsed electric fields with different voltage wave forms, *IEEE Trans Dielec Elec Insul*, 1, 1047–57.

Qin B, Barbosa-Canovas G, Swanson B and Pedrow P (1998) Inactivating microorganisms using a pulsed electric field continuous treatment system, *IEEE Trans Indust Appl*, **34**, 43–9.

Raso J and Barbosa-Canovas G V (2003) Nonthermal preservation of foods using combined processing techniques, *Crit Rev Food Sci Nutr*, **43**, 265–85.

Ray B and Daeschel M A (1994) Bacteriocins of starter cultures, in Dillon V M and Board R G (eds), *Natural Antimicrobial Systems and Food Preservation*, Wallingford, CAB International, 133–66.

Ray B, Johnson C and Wanismail B (1984) Factors influencing lysis of frozen *Escherichia coli* cells by lysozyme, *Cryo Lett*, **5**, 183–90.

Reddy N R, Solomon H M, Tetzloff R C and Rhodehamel E J (2003) Inactivation of *Clostridium botulinum* type A spores by high-pressure processing at elevated temperatures, *J Food Pro*, **66**, 1402–7.

Reddy N R, Tetzloff R C, Solomon H M and Larkin J W (2006) Inactivation of *Clostridium botulinum* nonproteolytic type B spores by high pressure processing at moderate to elevated high temperatures, *Innov Food Sci Emerg Technol*, **7**, 169–75.

Reiter B and Harnulv G (1984) Lactoperoxidase natural antimicrobial system: natural occurrence, biological functions and practical applications, *J Food Pro*, **47**, 724–32.

Reiter B, Fulford R J, Marshall V M, Yarrow N, Ducker M J and Knutsson M (1981) An evaluation of the growth promoting effect of the lactoperoxidase system in newborn calves, *Animal Production*, **32**, 297–306.

Roberts C M and Hoover D G (1996) Sensitivity of *Bacillus coagulans* spores to combinations of high hydrostatic pressure, heat, acidity and nisin, *J Appl Bacteriol*, **81**, 363–8.

Rooney J, Midda M and Leeming J (1994) A laboratory investigation of the bactericidal effect of a Nd:Yag laser, *Brit Dental J*, **176**, 61–4.

Rovere P, Maggi A, Scaramuzza N, Gola N, Miglioli L, Carpi G and Dall'aglio G (1996) High-pressure heat treatments: evaluation of the sterilizing effect of thermal damage, *Ind Conserve*, **71**, 473–83.

Saito H, Miyakawa H, Tamura Y, Shinamura Y and Tomita M (1991) Potent bactericidal activity of bovine lactoferrin hydrolysate produced by heat treatment at acidic pH, *J Dairy Sci*, **74**, 3724–30.

Sala F J, Burgos J, Condon S, Lopez P and Raso J (1995) Effect of heat and ultrasound on microorganisms and enzymes, in Gould G W (ed.), *New Methods of Food Preservation*, Glasgow, Blackie Academic and Professional, 176–204.

Sale A J H and Hamilton W A (1967) Effects of high electric fields on microorganisms I: killing of bacteria and yeasts, *Biochim Biophys Acta*, **148**, 781–8.

Sale A J H and Hamilton W A (1968) Effects of high electric fields on microorganisms II: lysis of erythrocytes and protoplasts, *Biochim Biophys Acta*, **163**, 37–45.

Sale A J H, Gould G W and Hamilton W A (1970) Inactivation of bacterial spores by hydrostatic pressure, *J Gen Microbiol*, **60**, 323–34.

Samuelson K J, Rupnow J H and Froning G W (1985) The effect of lysozyme and ethylenediaminetetraacetic acid on *Salmonella* on broiler parts, *Poultry Sci*, **64**, 1488–90.

San Martin M F, Barbosa-Canovas G V and Swanson B G (2002) Food processing by high hydrostatic pressure, *Crit Rev Food Sci Nutr*, **42**, 627–45.

Sanz B, Palacios P, Lopez P and Ordonez J A (1985) Effect of ultrasonic waves on the heat resistance of *Bacillus stearothermophilus* spores, in Dring G J, Ellar D J and Gould G W (eds), *Fundamental and Applied Aspects of Bacterial Spores*, London, Academic Press, 215–59.

Sato M and Kawata H (1991) Pasteurization method for liquid foodstuffs, *Jap Pat* 398,565.

Scherba G, Weizel R M and O'Brien J R (1991) Quantitative assessment of the germicidal efficacy of ultrasonic energy, *Appl Environ Microbiol*, **57**, 2079–84.

Scott D, Hammer F E and Szalkucki T J (1987) Bioconversions: enzyme technology, in Knorr D (ed.), *Food Biotechnology*, New York, Marcel Dekker, 413–42.

Scurrah K J, Robertson R E, Craven H M, Pearce L E and Szabo E A (2006) Inactivation of *Bacillus* spores in reconstituted skim milk by combined high pressure and heat treatment, *J Appl Microbiol*, **101**, 172–80.

Seifu E, Buys E M and Donkin E F (2005) Significance of the lactoperoxidase system in the dairy industry and its potential applications: a review, *Trends Food Sci Technol*, **16**, 137–54.

Selman J (1992) New technologies for the food industry, *Food Sci Technol Today*, **6**, 205–9.

Seyerderholm I and Knorr D (1992) Reduction of *Bacillus stearothermophilus* spores by combined high pressure and temperature treatments, *J Food Indust*, **43**(4), 17–20.

Shelef L A (1983) Antimicrobial effects of spices, *J Food Saf*, **6**, 29–44.

Shigahisa T, Ohmori T, Saito A, Tuj, S and Hayashi R (1991) Effects of high pressure on the characteristics of pork slurries and inactivation of microorganisms associated with meat and meat products, *Int J Food Microbiol*, **12**, 207–16.

Sitzmann W (1995) High voltage pulse techniques for food preservation, in Gould G W (ed.), *New Methods of Food Preservation*, Glasgow, Blackie Academic and Professional, 236–52.

Solomon E B and Hoover D G (2004) Inactivation of *Campylobacter jejuni* by high hydrostatic pressure, *Lett Appl Microbiol*, **38**, 505–9.

Sommers E B and Taylor S L (1987) Antibotulinal effectiveness of nisin in pasteurized processed cheese spreads, *J Food Prot*, **50**, 842–8.

Spilimbergo S and Bertucco A (2003) Non-thermal bacteria inactivation with dense CO_2, *Biotechnol Bioeng*, **84**, 617–38.

Spilimbergo S, Dehghani F, Bertucco A and Foster N R (2003) Inactivation of bacteria and spores by pulse electric field and high pressure CO_2 at low temperature, *Biotechnol Bioeng*, **82**, 118–25.

Stark J and Tan H S (2003) Natamycin, in Russell N J and Gould G W (eds), *Food Preservatives*, New York, Kluwer Academic/Plenum, 179–95.

Stewart C M, Dunne C P, Sikes A and Hoover D G (2000) Sensitivity of spores of *Bacillus subtilis* and *Clostridium sporogenes* PA3679 to combinations of high hydrostatic pressure and other processing parameters, *Innov Food Sci Emerg Technol*, **1**, 49–56.

Storz G and Hengge-Aronis (eds) (2000) *Bacterial stress responses*, Washington, DC, ASM Press.

Stratford M and Anslow P A (1998) Evidence that sorbic acid does not inhibit yeast as a classic 'weak acid' preservative, *Lett Appl Microbiol*, **27**, 203–6.

Styles M F, Hoover D G and Farkas D F (1991) Response of *Listeria monocytogenes* and *Vibrio parahaemolyticus* to high hydrostatic pressure, *J Food Sci*, **56**, 1404–7.

Takahashi K, Ishi, H and Ishikawa H (1991) Sterilization of microorganisms by hydrostatic pressure at low temperature, in Hayashi R (ed.), *High Pressure Science of Food*, Kyoto, San-Ei Publishing Co, 225–32.

Tenovuo J (2002) Clinical applications of antimicrobial host proteins lactoperoxidase, lysozyme and lactoferrin in xerostomia: efficacy and safety, *Oral Dis*, **8**, 23–9.

Thomas E L, Pera K A, Smith K W and Chwang A K (1983) Inhibition of *Streptococcus mutans* by the lactoperoxidase antimicrobial system, *Infect Immun*, **39**, 767–78.

Timson W J and Short A J (1965) Resistance of microorganisms to hydrostatic pressure, *Biotechnol Bioeng*, **7**, 139–59.

Toepfl S, Mathys A, Heinz V and Knorr D (2006) Review: potential of high hydrostatic pressure and pulsed electric fields for energy efficient and environmentally friendly food processing, *Food Rev Int*, **22**, 405–23.

Tomita M, Bellamy W, Takase M, Yamauchi K, Wakabayashi H and Kawase K (1991) Potent antibacterial peptides generated by pepsin digestion of bovine lactoferrin, *J Dairy Sci*, **74**, 4137–42.

Tranter H S (1994) Lysozyme, ovotransferrin and avidin, in Dillon V M and Board R G (eds), *Natural Antimicrobial Systems and Food Preservation*, Wallingford, CAB International, 65–98.

Tranter H S and Board R G (1982) Review: the antimicrobial defense of avian eggs: biological perspective and chemical basis, *J Appl Biochem*, **4**, 295–338.

Tsong T Y (1991) Minireview: electroporation of cell membranes, *Biophys J*, **60**, 297–316.

Van Opstal I, Bagamboula C F, Vanmuysen S C M, Wuytack E Y and Michiels C W (2004) Inactivation of *Bacillus cereus* spores in milk by mild pressure and heat treatments, *Int J Food Microbiol*, **92**, 227–34.

Van Opstal I, Bagamboula C F, Theys T, Vanmuysen S C M and Michiels C W (2006) Inactivation of *Escherichia coli* and *Shigella* in acidic fruit and vegetable juices by peroxidase systems, *J Appl Microbiol*, **101**, 142–50.

Wasserfall F, Voss E and Prokopek D (1976) Experiments on cheese ripening: the use of lysozyme instead of nitrite to inhibit late blowing of cheese, *Kiel Milchwirtschaft Forschung*, **28**, 3–16.

WHO (1999) High-dose irradiation: wholesomeness of food irradiated with doses above 10 kGy, *WHO Technical Report Series 890*, Geneva, World Health Organization.

Wilkins K M and Board R G (1989) Natural antimicrobial systems, in Gould G W (ed.), *Mechanisms of Action of Food Preservation Procedures*, London, Elsevier Applied Science, 285–362.

Wills P A (1974) Effects of hydrostatic pressure and ionizing radiation on bacterial spores, *Atomic Energy Australia*, **17**, 2–10.

Wuytack E Y and Michiels C W (2001) A study of the effects of high pressure and heat on *Bacillus subtilis* spores at low pH, *Int J Food Microbiol*, **64**, 333–41.

Xezones H and Hutchings I J (1965) Thermal resistance of *Clostridium botulinum* (62A) spores as affected by fundamental constituents. 1. Effect of pH, *Food Technol*, **30**, 1003–5.

Zenker M, Heinz V and Knorr D (2003) Application of ultrasound-assisted thermal processing for preservation and quality retention of liquid foods, *J Food Prot*, **66**, 1642–9.

Zhang Q, Barbosa-Canovas G V and Swanson B G (1994) Engineering aspects of pulsed electric field pasteurization, *J Food Eng*, **25**, 261–8.

Zhang Q, Qin B L, Barbosa-Canovas G V and Swanson B G (1995) Inactivation of *E. coli* for food pasteurization by high strength pulsed electric fields, *J Food Proc Pres*, **19**, 103–18.

17

Pathogenic *Escherichia coli*

C. Bell, Independent Consultant Microbiologist, UK, and
A. Kyriakides, Sainsbury's Supermarkets Ltd, UK

Abstract: Based on their extensive experience working within the food industry and supported by published reports and research work, the authors describe in sections titled with self-explanatory headings the key characteristics of pathogenic *Escherichia coli*, particularly Vero cytotoxigenic *E. coli*, that help to explain why they are important foodborne pathogens and approaches to the monitoring and control of these organisms in food production processes.

Key words: *Escherichia coli*, *E. coli* O157, Vero cytotoxigenic, entero-haemorrhagic, bacterial pathogen, foodborne illness, source, characteristics, growth, survival, control in foods, hazard analysis, risk factors, food processes, detection, meat and meat products, milk and milk products, fruit, vegetables.

17.1 Introduction

The involvement of *Escherichia coli* in human illness has been recognised virtually since its discovery in 1885. It has been associated with diarrhoea (particularly in children), haemorrhagic colitis, dysentery, bladder and kidney infections, surgical wound infection, septicaemia, haemolytic uraemic syndrome, pneumonia and meningitis; some of these conditions result in death.

Although the role of contaminated food and water in outbreaks of illness attributed to *E. coli* has been widely acknowledged for decades, it is only in more recent years that the food industry has re-focused attention on *E. coli* as a cause of significant morbidity and mortality in outbreaks of foodborne illness; Vero cytotoxin-producing (Vero cytotoxigenic) *E. coli* (VTEC), especially those causing haemorrhagic illness (entero-haemorrhagic *E. coli* – EHEC) are of particular and major concern.

Since the early 1900s, *E. coli* has been viewed by public health microbiologists as an indicator of faecal contamination in water sources and milk. The inclusion of *E. coli* in many food product specifications today also recognises its value as an indicator of the hygienic status of many food types. Although there is still much to learn concerning the epidemiology of the organism, there are strategies that can be followed and actions which can be taken by those in the business of supplying food as primary producers, processors, distributors, caterers or retailers, to minimise the incidence and level of the organism in foods, thus improving their safety. This chapter aims to give the reader an overview of *E. coli*, particularly VTEC, in respect of the hazard they present to food products and the means for controlling these organisms.

Despite the fact that a vast amount of work has been carried out on all aspects of *E. coli* since it was first described, the organism continues to provide new challenges to food safety because of the wide diversity of types within the species, ranging from harmless commensals to dangerous human pathogens. Although, in many countries including the UK and the USA, the entero-haemorrhagic *E. coli* O157:H7 is currently the most predominant foodborne VTEC, it is not the only VTEC associated with foodborne illness: *E. coli* O26, O45, O91, O103, O111 and O145 and other VTEC are causing significant morbidity in many countries, and such serogroups are increasingly being recognised as posing an equal or possibly greater threat to human health than *E. coli* O157. There is therefore a continuing need for information concerning sources, growth and survival characteristics, detection methods and effective, practical control measures that can be applied in the food industry for the control of all VTEC.

It is essential that a detailed and competent hazard analysis is carried out at an early stage in all new food product and process developments to ensure that relevant critical controls and monitoring systems can be put in place. This will help to minimise potential public health problems that could arise from the presence and outgrowth of all types of pathogenic *E. coli*, especially the Vero cytotoxigenic types. It is equally important to recognise that hazard analyses conducted for existing products and processes should be reviewed taking any 'new' pathogens such as VTEC into account so that any additional controls can be identified and implemented.

17.2 Characteristics of *Escherichia coli*

17.2.1 The organism

In 1885, Theodor Escherich described some organisms he had isolated from infant stools, one of which he named *Bacterium coli commune*. Following a great deal of work on the phenotypic characteristics of bacteria, by the 1960s, the genus *Escherichia* was described as: Gram-negative, non-sporing rods; often motile, with peritrichate flagella. It is easy to cultivate on ordinary laboratory media, aerobic and facultatively anaerobic. All species ferment glucose with the formation of acid or of acid and gas, both aerobically and anaerobically. All reduce nitrates to nitrites and are oxidase-negative, catalase-positive. Typically, they are intestinal inhabitants of humans and animals, though some species may occur in other parts of the body, on plants and in the soil and many species are pathogenic (Wilson and Miles, 1964). As bacterial classification systems have become more technically sophisticated, the conventional methods for distinguishing between strains of organisms based on biotype (Table 17.1) have been supplemented by, among others, serological techniques, phage typing and genotyping.

The DNA relatedness of *E. coli* to some other genera of the Enterobacteriaceae, particularly some notable human pathogens, has been established (Brenner, 1984) and, based on DNA homology, Jones (1988) indicates that *E. coli* and the four species of the genus *Shigella* should be considered a single species. By the mid-1940s, a serogrouping scheme was developed that allowed *E. coli* to be divided into more than 170 different serogroups based on their somatic (O) antigens (Kauffmann, 1947). In addition, over 50 flagella (H) antigens

Table 17.1 Some biotypic characteristics of *Escherichia coli*

Character	Reaction
Gram	Negative
Cell morphology	Non-sporing straight rod, 1.1–1.5 × 2.0–6.0 μm
Motility	+ by peritrichous flagellae or non motile
Aerobic growth	+
Anaerobic growth	+
Optimum growth temperature	37 °C
Catalase	+
Oxidase	–
D-Mannitol fermentation	≥ 90 % +
Lactose 37 °C and 44 °C	≥ 90 % +
D-Adonitol	≥ 90 % –
D-Glucose	acid produced
Indole 37 °C	≥ 90 % +
Indole 44 °C	≥ 90 % +
Methyl Red reaction	≥ 90 % +
Voges-Proskauer reaction	≥ 90 % –
Growth in Simmons' citrate	≥ 90 % –
Urease, Christensen's	≥ 90 % –
Phenylalanine deamination	≥ 90 % –
Lysine decarboxylase	76–89 % strains +
H_2S on TSI (triple sugar iron) medium	≥ 90 % –
Growth in KCN (potassium cyanide) medium	≥ 90 % –
Gelatin liquefaction (at 22 °C)	≥ 90 % –

+ = positive reaction; – = negative reaction.
Source: Adapted from Brenner (1984) and Ørskov (1984).

and approximately 100 capsular (K) antigens are now also recognised, and these are used to further subdivide *E. coli* into serotypes. Serogrouping and serotyping, together with other information such as biotype, phage type and enterotoxin production, now facilitate distinction between those strains able to cause infectious disease in humans and animals (Linton and Hinton, 1988). Some correlation has been established between the *E. coli* serogroup and virulence.

17.2.2 The illness caused

Depending on the virulence genes acquired, different types of pathogenicity are conferred to certain strains of *E. coli*. These strains are classified as enteropathogenic *E. coli* (EPEC), entero-toxigenic *E. coli* (ETEC), entero-invasive *E. coli* (EIEC), entero-haemorrhagic *E. coli* (EHEC) and entero-aggregative *E. coli* (EAEC or EaggEC) (Nataro and Kaper, 1998). Diffusely adherent *E. coli* (DAEC) appear to be an important cause of acute diarrhoea in young children in some parts of the world. The different virulence factors expressed by the organism, e.g. colonisation factors, ability to invade epithelial cells of the small intestine, haemolysin production and toxin production, lead to the different strains of *E. coli* being associated with a wide variety of types of disease.

Table 17.2 summarises information concerning the pathogenicity of some different serogroups together with the type of diseases and characteristics of the illnesses associated with these organisms. As can be seen, although *E. coli* is a common and harmless member of the normal commensal microflora of the distal (end or terminal) part of the intestinal tract of humans and other warm-blooded animals (constituting less than 1 % of this flora in numbers ranging up to 10^8 per gram in humans; Smith, 1961), some strains are responsible for causing severe illness, sometimes resulting in death.

Production of cytotoxins (referred to as Shiga toxins) is a common feature of VTEC which may also be referred to as Shiga toxin-producing *E. coli* (STEC). Entero-haemorrhagic *E. coli* (EHEC) are a subset of VTEC / STEC that are considered to be highly pathogenic to humans; they are, therefore, of importance in human disease (Karch *et al.*, 2005). Although there is no clear definition of EHEC, they principally include those strains that cause haemorrhagic colitis (HC), haemolytic uraemic syndrome (HUS), produce Vero cytotoxin, cause attaching-and-effacing lesions on epithelial cells and possess a 60-MDa EHEC plasmid (Levine, 1987). There is still some variation in terminology used in discussions and publications so in this chapter the term VTEC will continue to be used. It is, however, important to recognise that not all VTEC cause infections in humans.

Escherichia coli is acquired by infants within a very few days of birth. The organism is acquired predominantly from the mother by the faecal–oral route, but also from the environmental surroundings. Results from oral challenge experiments on adults suggest that levels of 10^5–10^{10} EPEC are required to

Table 17.2 Pathogenicity and characteristics of foodborne illness caused by pathogenic *E. coli*

Pathogenic type of *E. coli*	Serogroup examples[a]	Summary of *E. coli*/ host interaction	Time to onset of illness	Duration of illness	Range of symptoms
EPEC (enteropathogenic)	O18ab, O18ac, O26, O44, O55, O86, O114, O119, O125, O126, O127, O128, O142, O158	EPEC attach to intestinal mucosal cells causing cell structure alterations (attaching and effacing). EPEC cells invade the mucosal cells.	17–72 h, average 36 h	6 h to 3 d, average 24 h	Severe diarrhoea in infants which may persist for more than 14 days. Also, fever, vomiting and abdominal pain. In adults, severe watery diarrhoea with prominent amounts of mucus without blood (main symptom), nausea, vomiting, abdominal cramps, headache, fever and chills.
ETEC (enterotoxigenic)	O6, O15, O25, O27, O63, O78, O115, O148, O153, O159	ETEC adhere to the small intestinal mucosa and produce toxins that act on the mucosal cells.	8–44 h, average 26 h	3–19 d	Watery diarrhoea, low-grade fever, abdominal cramps, malaise, nausea. When severe, causes cholera-like extreme diarrhoea with rice water-like stools, leading to dehydration.
VTEC (Vero-cytotoxigenic) (Entero-haemorrhagic)	O2, O4, O5, O6, O15, O18, O22, O23, O26, O55, O75, O91, O103, O104, O105, O111, O113, O114, O117, O118, O121, O128ab, O145, O153, O157, O163, O168	EHEC attach to and efface mucosal cells and produce toxin.	3–9 d, average 4 d	2–9 d, average 4 d	Haemorrhagic colitis: sudden onset of severe crampy abdominal pain, grossly bloody diarrhoea, vomiting, no fever. Haemolytic uraemic syndrome (HUS): bloody diarrhoea, acute renal failure in children, thrombocytopaenia, acute nephropathy, seizures, coma, death. Thrombotic thrombocytopaenic purpura: similar to HUS but also

Table 17.2 Cont'd

Pathogenic type of *E. coli*	Serogroup examples[a]	Summary of *E. coli*/ host interaction	Time to onset of illness	Duration of illness	Range of symptoms
VTEC Cont'd					fever, central nervous system disorders, abdominal pain, gastrointestinal haemorrhage, blood clots in the brain, death.
EIEC (enteroinvasive)	O28ac, O29, O112ac, O121, O124, O135, O144, O152, O167, O173	Toxin lyses membranes, EIEC invade cells in the colon and spread laterally, cell to cell.	8–24 h, average 11 h	Days to weeks	Profuse diarrhoea or dysentery, chills, fever, headache, muscular pain, abdominal cramps.
Eagg EC or EAEC (enteroaggregative)	O3, O44, O51, O77, O86, O99, O111, O126	EAggEC bind in clumps (aggregates) to cells of the small intestine and produce toxins.	7–22 h	Days to weeks	Persistent diarrhoea in children. Occasionally bloody diarrhoea or secretory diarrhoea, occasionally vomiting, dehydration.
DAEC (diffusely adherent)	O1, O2, O21, O75	Fimbrial and non-fimbrial adhesins identified.	Not yet established.	Not yet established.	Childhood diarrhoea.

[a]Only certain strains within a serogroup may be associated with human illness and full characterisation, e.g. serotype, phage type, is necessary to clearly identify the causative agent.

Note: there are an increasing number of strains that do not easily fit into the main groups described above.

Source: Adapted from Bell and Kyriakides (1998); Willshaw *et al.* (2000); McClure (2005).

produce diarrhoea, 10^8–10^{10} ETEC organisms are necessary for infection and diarrhoea and 10^8 cells of EIEC are required to produce diarrhoeal symptoms in adults. However, this may vary depending on the acidity in the stomach (Doyle and Padhye, 1989; Sussman, 1997). Following an outbreak of *E. coli* O157:H7 infection in the USA (November 1992–February 1993) involving ground beef patties, a detailed investigation and microbiological testing of the meat supply indicated an infectious dose of less than 700 organisms in this outbreak (Tuttle *et al.*, 1999).

It is now accepted that the infective dose of VTEC can be very low, i.e. < 100 cells of the organism (ACMSF, 1995). The haemolytic uraemic syndrome (HUS), caused by VTEC and characterised by acute renal failure, haemolytic anaemia and thrombocytopaenia, usually occurs in young children (under five years of age). It is the major cause of acute renal failure in children in the UK and several other countries. Generally, about 5 % of cases of haemorrhagic colitis caused by VTEC progress to HUS which can prove fatal (PHLS, 2004). There are no specific treatments of the conditions caused by VTEC and each symptom is treated as it occurs in the individual (ACMSF, 1995; PHLS, 2000).

17.2.3 Growth and survival

Farm environments, particularly cattle farms, may become readily contaminated with a variety of strains of *E. coli* including *E. coli* O157 from the faeces of these animals. Once present, the organism may survive in these environments for some time. The organism may then be spread directly or indirectly from the farms to the wider environment and present a risk of contamination to other animals and growing crops as probably occurred in 2006 causing an outbreak involving spinach (CalFert, 2007).

Studies of the survival of *E. coli* O157 in different environments have been undertaken and are reported periodically. Levels of pathogens remaining viable in the soil are subject to a variety of factors, including exposure to sunlight, drying, etc. However, *E. coli* O157 and other VTEC have been shown to survive for very long periods in animal wastes and in the soil (Table 17.3). Jones (1999) has usefully summarised data on the survival of *E. coli* O157 in the environment and considered the potential health risk associated with the persistence of the organism in agriculture. The consequences of using artificially-contaminated water for spray irrigation of lettuce plants were investigated by Solomon *et al.* (2003). Intermittent irrigation with water contaminated at levels of *c.* 10^2 or *c.* 10^4 cfu/ml *E. coli* O157:H7 yielded contaminated lettuce plants at harvest.

From a study of the persistence of the organism on different types of farm surface under different conditions, Williams *et al.* (2005) reported that *E. coli* O157 inoculated into fresh cattle faeces used to contaminate pieces of farmyard surface materials, could persist for more than 28 d on moist wood and 14–28 d when desiccated on wood in cool conditions and that

Table 17.3 Results of some studies of the survival of pathogenic *E. coli* in different environments

Substrate	Organism and conditions	Survival time	Reference
Soil cores	*E. coli* O157 stored under continuous illumination	Over 120 d	Maule, 1997
Bovine faeces at 15 °C	*E. coli* O26, O111 and O157: inoculated at 10 colony-forming units (cfu)/g	Up to 8 weeks	Fukushima *et al.*, 1999
	inoculated at 10^3 or 10^5 cfu/g	Up to 18 weeks	
Bovine slurry wastes (n = 5) Ovine abattoir wastes (n = 5) Sewage wastes – treated (n = 5) Sewage wastes – untreated (n = 3) Sewage sludge wastes (n = 5) Creamery wastes (n = 4)	*E. coli* O157:H7 laboratory inoculated into replicate microcosms of each waste type then incubated at 10 °C for 64 d	Over 2 months in 21 of the 27 waste types tested	Avery *et al.*, 2005
Garden soil fertilised with cattle manure	*E. coli* O157:H7	2–3 months	Mukherjee *et al.*, 2006
Cow slurry in model of farm slurry store	Four strains of Shiga toxin-producing *E. coli* O26	At least 3 months	Fremaux *et al.*, 2007a
Cow manure heaps	Eight strains of Shiga toxin-producing *E. coli* non-O157:H7 in:		Fremaux *et al.*, 2007b
	heaps turned	Up to 42 d	
	heaps left unturned	Up to 90 d	

during brief contact, the organism could transfer to human hands from the surfaces.

Given the mounting evidence that *E. coli* O157 is not only widely disseminated in the animal farming environment but is also capable of surviving for considerable periods of time, careful consideration of the means for safe storage and effective treatment of animal wastes prior to use or discharge to other environments is required. This will help to ensure any pathogenic bacterial hazard to these environments is minimised. This includes preventing run-off to watercourses that may be used to irrigate crops. Guidance on good practice is provided by some government and industry bodies (USDA, 1998; CFA, 2007)

Table 17.4 indicates some of the key growth-limiting parameters for pathogenic *E. coli*. These and other physicochemical factors, used either singly or in combination, can be effective in controlling the survival and growth of *E. coli* during food processing and also in the finished food products

Table 17.4 Growth-limiting parameters for pathogenic *E. coli*

	Minimum	Optimum	Maximum
Temperature (°C)			
E. coli (all types)	7–8	35–40	44–46
VTEC O157:H7	6.5[a]	37	44–45
pH	4.4[b]	–	9.0
Pathogenic *E. coli*			
a_w	0.95	–	–
Pathogenic *E. coli*			
Sodium chloride[c]	Grows vigorously in 2.5 % NaCl		
Pathogenic *E. coli*	Grows slowly in 6.5 % NaCl		
	Does not grow in 8.5 % NaCl		

[a]Kauppi *et al.* (1996).
[b]*E. coli* O157 is reported to survive at pH values below 4.4 and has also been shown to grow at pH 3.6 in apple juice, pH 3.58 in hydrochloric acid, pH 3.78 in lactic acid and pH 3.96 in citric acid.
[c]Glass *et al.* (1992).
Source: adapted from ICMSF (1996).

(Bell and Kyriakides, 1998). The application of these parameters, alone or in combination in foods with other factors, are discussed in Sections 17.5, 17.6 and 17.7.

Research has been carried out to determine the survival capability of *E. coli* O157 in foods and drinks and examples of the results reported from such studies are given in Table 17.5. It is evident that the organism may survive in or on different foods under a range of conditions and, hence, addressing matters relating to the control of the presence of the organisms in the first place is of the highest priority for public health protection from foodborne VTEC infection.

There are reports indicating that heat-shocked cells of *E. coli* O157:H7 exhibit increased acid tolerance (Wang and Doyle, 1998) with 10–100 times more heat-shocked cells surviving at pH 2.5 than untreated cells; also that once habituated in an acidic environment, e.g. fruit juice, acid-adapted cells of *E. coli* O157:H7 may require a more severe heat process to reduce/destroy populations effectively than is required for non-acid-adapted cells (Sharma *et al.*, 2005). Due care to consider such enhanced survival possibilities will assist in ensuring that subsequent manufacturing processes applied will effect the required levels of pathogen reduction where the process is relied on for this purpose.

Notwithstanding such reports of survival and either natural/inherent or increased resistance to treatments caused through pre-exposure to 'habituating' or 'shocking' environmental conditions, *E. coli* O157 can be controlled by many of the common agricultural food industry practices when applied effectively, e.g. composting, chlorination, cleaning and disinfection procedures, and is readily destroyed by temperatures greater than 65 °C. Some of these are discussed in greater detail in Section 17.5.

Table 17.5 Results of some studies of the survival of pathogenic *E. coli* in different foods

Substrate	Organism and conditions	Survival time	Reference
Fermented dry sausage	Inoculation with *E. coli* O157:H7 (*c.* 10^4 cfu/g). Fermentation to pH 4.8, dried to 37 % moisture content then stored at 4 °C for 8 weeks	Approx. 2 $\log_{10}$ reduction by the end of storage over 8 weeks	Glass *et al.*, 1992
Traditional yoghurt and 'Bifido' yoghurt	Inoculation with *E. coli* O157:H7; low level (*c.* 10^3 cfu/ml); high level (*c.* 10^7 cfu/ml)	< 1 $\log_{10}$ reduction in 7 d; approx. 2 $\log_{10}$ reduction in 7 d	Massa *et al.*, 1997
Ground-beef patties	Inoculated with *E. coli* O157:H7 (four strain cocktail) at level of *c.* 10^5 cfu/g. Stored at: 2 °C for 4 weeks; –20 °C for 1 year	1.9 $\log_{10}$ cfu/g reduction; 1 – 2 $\log_{10}$ cfu/g reduction	Ansay *et al.*, 1999
Unpasteurized milk	Inoculated with *E. coli* O157:H7 (seven strains used separately) at levels *c.* 10^3 cfu/ml–*c.* 10^6 cfu/ml. Stored at 8 °C	> 9/d (some strains increased in population by approx. 2 $\log_{10}$)	Massa *et al.*, 1999
Mineral water – natural, non-carbonated	Inoculated with non-VT-producing *E. coli* O157:H7 to achieve a level *c.* 10^3/ml. Stored at 15 °C for 10 weeks	Population declined from *c.* 3.5 $\log_{10}$ cfu/ml to 0.56 $\log_{10}$ cfu/ml over 63/d. Not detectable at 70/d	Kerr *et al.*, 1999
Juice concentrates: apple, orange, pineapple, white grape	≥ 10^3 cfu/g *E. coli* O157:H7 inoculum. Storage at –23 °C	12 weeks (all juices)	Oyarzábal *et al.*, 2003

17.3 Risk factors for *Escherichia coli* O157

17.3.1 Human infection

It should be recognised that while this group of organisms can cause serious illness, it is also clear that some individuals exposed to the organism remain asymptomatic, a feature in common with a number of other enteric pathogens, e.g. *Salmonella* species. Stephan *et al.* (2000) reported the detection of the polymerase chain reaction (PCR) product for VT-encoding genes in 3.5 % of over 5500 stool samples taken from healthy employees in the Swiss meat-processing industry. Although the presence of VT-encoding genes does not necessarily indicate the presence of VTEC and that individuals were harbouring the organisms, there is an underlining concern raised about the possibility of asymptomatic human carriers in food-handling environments. Silvestro *et al.* (2004) recovered four (1.1 %) *E. coli* O157 isolates from farm workers in a survey of 350 farm workers on 276 dairy farms and 50 abattoir employees from seven different operations. None of the positive individuals were suffering intestinal symptoms or had any history of bloody diarrhoea or renal failure. Following a second sampling of the positive individuals and their household contacts, a positive isolation was obtained from one farm worker's wife.

In some countries in which surveillance of foodborne infections has become routine and considered reliably indicative of trends, e.g. some western European countries, the USA and Canada, the numbers of cases of VTEC-related illness has been shown to have steadily increased from 1982 to the late 1990s. Since then the numbers of reported cases has fluctuated but at the higher levels previously attained, e.g. ranging from approximately 600 to 1000 reported cases per year for England and Wales (Table 17.6). Each of the recorded cases represents a potential source of contamination to

Table 17.6 Trend in laboratory-confirmed cases of VTEC O157 in England and Wales

Year	Number of cases per year
1982–1984	< 10
1985	50
1986	76
1987	89
1988	49
1989	119
1990	250
1991	361
1992	470
1993	385
1994	411
1995	792
1996	660
1997	1087
1998	890
1999	1084
2000	896
2001	768
2002	595
2003	675
2004	699
2005	950
2006	1001
2007	828
2008	948

Source: Adapted from HPA (2008).

others in their vicinity, and person-to-person spread is of particular concern in relation to this organism.

In the EU, there were 4916 confirmed cases of VTEC infection in 2006 from 22 member states who reported data (EFSA, 2007a). This represented a case rate of 1.1 per 100 000 population although this varied from < 0.1–15.2 across the different member states. The serogroups recorded by The European Surveillance System (referred to as TESSy in EFSA, 2007a) as most commonly associated with infection, in descending order were, O157 (46.5 % of known serogroups), O26 (16.3 %), O126 (5.2 %), O55 (4.3 %), O127 (3.8 %), O25 (3.7 %), O103 (2.5 %), O128 (2.5 %) and O119 (2.4 %). VTEC O157 was the most common (or equal most common) serogroup in 11 of the 13 EU member states reporting data. In the USA, VTEC have been estimated to cause 73 000 illnesses per year, resulting in 2168 hospitalisations and 61 deaths (Mead *et al.*, 1999), with current rates of infection in the USA states contributing to the FoodNet surveillance programme running at three cases per 100 000 population (Gould, 2008).

It is generally agreed that there are four principal transmission routes by which VTEC are passed to humans (EFSA, 2007b):

- food, e.g. consumption of contaminated meat or meat products that are subsequently undercooked, unpasteurised or improperly processed dairy products or contaminated ready-to-eat products;
- water, e.g. drinking water contaminated with animal faeces;
- animal contact, e.g. contact on-farm with animals and inadequate hygienic precautions such as hand-washing;
- person-to-person, e.g. transmission of organism from an infected individual to others through the faecal–oral route.

17.3.2 Foods and water

Outbreaks of *E. coli* infection in developed countries (Table 17.7), and particularly those caused by entero-haemorrhagic VTEC (EHEC), continue to implicate bovine sources of contamination as a significant contributory factor. However, it is apparent that other animal species appear to be increasingly contaminated with these organisms and other routes of infection are also important, including contaminated water and ice (Jackson *et al.*, 1998; Pebody *et al.*, 1999; Anon. 1999c, Kim and Harrison, 2008), person-to-person spread and contact with contaminated animals, particularly at farms (Milne *et al.*, 1999; Møller Nielsen *et al.*, 2005). Indeed, outbreaks of *E. coli* O157 infection have also been associated with bathing beaches (Harrison and Kinra, 2004), environmental contamination in a scout camp (Howie *et al.*, 2003) and agricultural fairs (Crump *et al.*, 2003). A summary of outbreaks of pathogenic *E. coli* occurring across 10 EU member states (and two non-member states) in 2006 identified 48 outbreaks (0.8 % of the total outbreaks of foodborne disease) affecting 750 people with 13.7 % requiring hospitalization (EFSA, 2007a).

Table 17.7 Some foods associated with outbreaks of illness caused by pathogenic *E. coli*

E. coli type	Year	Country	Suspected food vehicle	Cases (deaths)	Reference[a]
EPEC O111:B4	1967	Washington D.C., USA	Water	170 (0)	
ETEC O27:H20	1983	USA, also Denmark, Netherlands, Sweden	French Brie cheese	169 (0) USA	
ETEC O6:H16 and O27:H20	1983	UK	Curried turkey mayonnaise	27 (0)	
EIEC O124	1947	UK	Canned salmon	47 (0)	
EIEC O124:B17	1971	USA	French Brie and Camembert cheese	387 (0)	
VTEC O157:H7	1982	Oregon and Michigan, USA	Hamburger patties in sandwiches	> 47 (0)	
VTEC O157	1985	Ontario, Canada	Undercooked beef patties	73 (17)	
VTEC O157:H7 phage type 49	1991	UK	Yoghurt	16 (0)	
VTEC O157:H7	1993	USA	Hamburgers	631(3)	
VTEC O111:NM	1995	South Australia	Uncooked, semi-dry fermented sausage	23 (1)	
VTEC O157	1996	Sakai City, Japan	White radish sprouts	6309 (3)	
VTEC O157 phage type 2	1996	Scotland, UK	Meat products	> 272 (20)	
VTEC O157:H7	1996	Connecticut and Illinois, USA	Lettuce	> 61 (0)	Hilborn *et al.*, 1999
VTEC O157 phage type 2	1998	UK	Unpasteurised cream	7 (0)	Anon. 1998a, b
VTEC O157:H7	1999	New York State, USA	Contaminated well water	> 1000 (2)	Charatan, 1999
VTEC O157 phage type 21/28 VT2	1999	UK	Cheese made from unpasteurised milk	3 (0)	Anon. 1999a, b
VTEC O157:H7	1999	British Columbia, Canada	Salami	143 (0)	MacDonald *et al.*, 2004

Table 17.7 (Cont'd)

E. coli type	Year	Country	Suspected food vehicle	Cases (deaths)	Reference[a]
VTEC O157 phage type 21/28 VT2	2001	UK	Cooked meats	> 13 (?)	Anon., 2001
VTEC O157	2002	USA	Ground beef	> 28	Shillam *et al.*, 2002
VTEC O157	2003–2004	Denmark	Organic milk	25	Jensen *et al.*, 2006
VTEC O157:H7	2004	Canada	Beef donair	43	Currie *et al.*, 2007
VTEC O157:H7	2005	France	Beefburgers	26	Vaillant, 2005
VTEC O157	2005	USA	Unpasteurised milk	18	Anon., 2005
VTEC O157	2005	South Wales, UK	Pre-cooked meat products	15 (1)	Simmons, 2005, Pennington, 2009
VTEC O157	2005	Washington, USA	Unpasteurised milk – cow share	18	Bhat *et al.*, 2007
VTEC O157 VT2	2005	Sweden	Lettuce	110	Hjertqvist, 2005
VTEC O103	2006	Norway	RTE cured meat sausage product	10 HUS 6 Diarrhoea 1 Asymptomatic	Schimmer, 2006
VTEC O157:H7	2006	USA	Spinach contaminated from cattle possibly via wild pigs	205 (3)	CalFert, 2007, Anon., 2006a
VTEC O26:H11	2007	Denmark	Organic fermented cured beef sausage	20	Ethelberg *et al.*, 2007
VTEC O157	2006	USA	Shredded lettuce	Approx. 80	CalFert, 2008
VTEC O157 PT2	2007	Scotland	Cold cooked meats ex supermarket	9 (1)	Stirling *et al.*, 2007
VTEC O157 & O26	2007	Belgium	Ice cream	5 (0)	De Schrijver *et al.*, 2008
VTEC O157	2007	UK	Chicken wrap	12	Syed (2008)

[a]Source: Adapted from Bell and Kyriakides (1998) unless otherwise stated.

Analysis of VTEC O157 outbreaks in England and Wales between 1992 and 2002 identified that foodborne outbreaks were more likely to occur in spring (odds ratio (OR) 2.13) and that farms (OR 27.4), the community (OR 7.52) or shops/retailers (OR 4.19) were more likely to be the setting for such outbreaks (Gillespie *et al.*, 2005). The food vehicles most likely to be associated with outbreaks were red meat/meat products (OR 2.8) and milk/milk products (OR 19.81) whilst cross-contamination was more likely to be the contributory factor (OR 2.15).

In response to outbreaks of illness occurring involving pathogenic *E. coli*, especially *E. coli* O157, associated with a food not previously or frequently implicated in such outbreaks, a considerable effort is made to investigate specific contamination sources, prevalence, survival, food treatment approaches, etc. relevant to the implicated food. Examples of such investigations include those relating to meat products (Pennington, 1996, 2009; Reilly, 2001; Simmons, 2005) and others relating to salad and other green vegetables (CalFert, 2007, 2008).

Research into related topics is also inspired by outbreaks occurring, e.g. carriage in, or by, different types of creature that may directly or indirectly facilitate contamination of food materials, e.g. houseflies, slugs. Following outbreak investigation work indicating identical strains of EHEC O157 were associated with patients and houseflies in a nursery school, Kobayashi *et al.* (1999) demonstrated that, in addition to carrying the organism on and around mouthparts, it could be cultured from fly excreta for at least 3 d. For farm environments, Alam and Zurek (2004) reported 2.9 % (52/1815) and 1.4 % (23/1625) positive results for *E. coli* O157:H7 in houseflies caught in a feed bunk and corn storage area respectively. Sproston *et al.* (2006) reported isolating *E. coli* O157 from field slugs taken from a sheep farm on which the sheep were known to have been shedding the organism and suggested this could be a means for the contamination of vegetable crops. The results of such work may prove to be supportive of other research studies and/or assist in the investigation of further outbreaks of foodborne illness.

It is clear from reports over the years as illustrated by the examples given in Table 17.7, that many different types of food material may become implicated in outbreaks of foodborne illness caused by *E. coli* O157 when circumstances combine for the hazard to be realised. It is essential that the lessons learned from investigations of outbreaks, e.g. Reilly (2001), CalFert (2007) Pennington (2009), are acted upon to prevent the severe illnesses and deaths that can and do occur from infection with VTEC.

Animals and their products

Locking *et al.* (2006) reported the results of an in-depth evaluation of VTEC infection surveillance data collected for 2004 in Scotland. Direct exposure to farm animals or other livestock and/or their faeces or indirect exposure via land or water environments contaminated with animal faeces were found to be significant sources of infection for the cases examined.

Escherichia coli O157 and other VTEC are believed to be carried asymptomatically by cattle, cows and sheep. For instance, Cerqueira *et al.* (1999) reported the results of a survey of healthy cattle in Brazil in which rectal swabs were taken from healthy animals from 10 dairy farms (n = 121), four beef farms (n = 60) and one beef slaughterhouse (n = 16); 99/121 (82 %) of the dairy cattle and 40/76 (53 %) of the beef cattle were positive for Shiga toxin gene sequences. However, the isolation of VTEC O26 from an eight-month old heifer suffering dysentery where attaching and effacing lesions were evident in the large intestine may also indicate that some strains are associated with illness in cattle (Pearson *et al.*, 1999). Seongbeom *et al.* (2006) reported an *E. coli* O157 incidence of 5.2 % and 4.7 % in faecal samples from cattle in Minnesota, USA in 2001 and 2002, respectively. This represented a farm incidence of 36.8 % in 2001 and 23.5 % in 2002. In the same study, *E. coli* O157 was isolated from cattle manure from 75 % of the 12 county fairs in 2001 and 2002, highlighting the importance of hygienic precautions at such events.

Shedding of *E. coli* O157 in the faeces of animals has been reported regularly (Lenahan *et al.*, 2007; Nastasijevic *et al.*, 2008). The organism has also been shown not only to be shed intermittently and exacerbated by a variety of factors (Garber *et al.*, 1999) including stress and feed types (Dargatz *et al.*, 1997; Kudva *et al.*, 1997) but also occasionally in very high prevalences by so-called 'super-shedding' cattle from which over 90 % of faecal pats sampled are found positive for *E. coli* O157 (Matthews *et al.*, 2006). Carriage on faecally-contaminated, dirty hides has also been reported (Brichta-Harhay *et al.*, 2007; Nastasijevic *et al.*, 2008). The complete picture in relation to carriage and excretion of *E. coli* O157 by animals is, however, still not fully understood. The results from surveys of the incidence of VTEC including *E. coli* O157 in meat and meat products clearly demonstrate the potential risk to human health presented by these materials (Table 17.8).

Raw milk is principally exposed to contamination with faecal pathogens from the faeces of the cow, which contaminate the udder and teats of the cow and pass into the milk during the milking stage. Raw drinking milk and food products made from raw milk are therefore at risk from contamination by pathogenic *E. coli*.

Microbiological surveys of raw milk frequently find a high incidence of bacterial indicators of faecal contamination, i.e. *E. coli*, and rarely find pathogenic strains. In a survey of raw milk in the UK, *E. coli* was detected (1/ml or greater) in 978 (58 %) of 1591 phosphatase-positive samples (DOH, 1998). Some 0.4 % samples had counts of 10^3cfu/ml or greater, 3 % samples were between 10^2/ml and < 10^3/ml and 21 % between 10 and < 10^2/ml. The remaining samples had counts below 10/ml. *E. coli* O157 was not detected in any 25 ml sample in this or in any of three other surveys of 1011 samples of raw milk in the Netherlands (Heuvelink *et al.*, 1998), 42 samples in the USA (Ansay and Kaspar, 1997) and a variety of raw milk, dairy and associated samples in the UK (Neaves *et al.*, 1994). *E. coli* O157

Table 17.8 Incidence of VTEC reported for some raw and processed meat products and other foods

Source of samples	Country	Occurrence (%)	Reference
Retail fresh meats and poultry	USA	6/164 (3.7) ground-beef 4/264 (1.5) pork 4/263 (1.5) poultry 4/205 (2.0) lamb all isolates *E. coli* O157:H7	Doyle and Schoeni, 1987
Retail chickens and sausages	UK	46/184 (25) pork sausages 0/71 (0) chickens All non-O157 VTEC	Smith *et al.*, 1991
Retail raw meats	Netherlands	2/770 (0.3) minced mixed beef and pork 0/1000 (0) raw minced beef 0/26 (0) minced pork 0/300 (0) poultry products Isolates confirmed as VTECs	Heuvelink *et al.*, 1996
Raw meat products	UK	3/89 (3.4) frozen beefburgers 1/50 (2.0) fresh minced beef 0/50 (0) sausages all isolates *E. coli* O157	Bolton *et al.*, 1996
Beef carcass (1600 cm^2)	Belgium Survey	67/7567 (0.89)	Chahed *et al.*, 2005
Minced beef (25 g)	1999–2003	4/2341 (0.17) all isolates *E. coli* O157	
Raw beef products	UK	36/3216 (1.1)	Chapman *et al.*, 2000
Raw lamb products		29/1020 (2.9)	
Lamb sausages		3/73 (4.1)	
Lamb burgers		18/484 (3.7)	
Mixed meat products	UK	7/857 (0.8) all isolates *E. coli* O157	Chapman *et al.*, 2000
Ground beef	USA	61/1750 (3.5) EHEC 20/1750 (1.1) *E. coli* O157	Samadpour *et al.*, 2006
Sprouts		12/200 (6.0) EHEC 3/200 (1.5) *E. coli* O157	
Mushrooms		4/100 (4.0) EHEC 0/100 (0.0) *E. coli* O157	
Raw ground beef	Canada	1/100 (1) *E. coli* O22:H8	Bohaychuk *et al.*, 2006
Raw chicken legs		0/100 (0)	
Raw pork chops		0/98 (0)	
Fermented sausage		0/100 (0)	
Roast beef		0/101 (0)	
Turkey breast		0/100 (0)	

Table 17.8 (Cont'd)

Source of samples	Country	Occurrence (%)	Reference
Beef wieners Chicken wieners		0/100 (0) 0/101 (0)	Bohaychuk *et al.*, 2006 cont'd
Beef carcasses (from four abattoirs during Jan–April 2006)	USA	256/1520 (16.8) *E. coli* O157:H7	Brichta-Harhay *et al.*, 2007
Lamb carcass	Ireland		Lenahan *et al.*, 2007
Pre-chill		6/400 (1.5)	
Post-chill		4/400 (1.0)	
Minced beef from supermarkets	Scotland		Solecki *et al.*, 2007
Urban		0/44 (0.0)	
Rural		3/101 (2.97)	
Minced lamb from supermarkets			
Urban		0/14 (0.0)	
Rural		1/32 (3.13)	

was, however, detected in three out of 1097 samples of raw milk on sale in the UK (de Louvois and Rampling, 1998). In a survey of milk from dairy farms in Trinidad, 47.4 % of bulk milk samples were contaminated with *E. coli* at levels between 8.4×10^3 and 2.0×10^5 cfu/ml (Adesiyun *et al.*, 1997). Some 27.7 % and 18.5 % of the *E. coli* strains isolated from the bulk milks were EPEC and VTEC, respectively. Caro *et al.* (2006) reported the isolation of *E. coli* O157:H7 in three samples of raw ewes' milk (n = 84) in Spain.

The incidence of VTEC found in raw cows' milk during 2006 in different European countries has been reported by EFSA (2007a). The overall incidence from the four member states and one non-member state reporting data was 0.4 % (n = 3474); this represented a reduction from the previous year (1.4 %, n = 3947). The sample size or method employed by the different reporting countries was not necessarily the same and so the results need to be interpreted carefully. The survey conducted in Germany was noteworthy in that 1.4 % (2/148) of samples of raw cows' milk intended for human consumption were positive for VTEC although these were non-O157 VTEC. The same country reported a 1 % (10/977) incidence of VTEC in raw milk intended for subsequent heat processing where the isolated serogroups included O2, O8, O21 and O22 and not O157. None of these serogroups features as significant causes of identified foodborne VTEC infections.

Fruit, vegetables and other raw materials

An increasing number of outbreaks of VTEC infection have implicated fruit as a primary vehicle, usually when pressed to make unpasteurised fruit juice

(Besser *et al.*, 1993; Mshar *et al.*, 1997). Vegetables and fresh produce have also been implicated in a variety of outbreaks throughout the world (Mermin *et al.*, 1997).

Due to the sporadic nature of contamination of vegetable crops, usually through some isolated poor practice, surveys very rarely find VTEC. No VTEC were reported in a variety of surveys across the EU in 2006 which included vegetables from Germany (n = 179), Slovenia (n = 50) and Spain (n = 51), fruits from the Netherlands (n = 816) and sprouted seeds from Slovenia (n = 30) (EFSA, 2007a).

Summary of risk factors

Historical outbreaks (Table 17.7) indicate that the factors likely to be associated with increased risk of *E. coli* infection include the following:

- raw material or product exposed to contamination from bovine origin (meat or faeces);
- product manufactured with no processing stage capable of destroying the organism, e.g. cooking or inadequately controlled or applied destruction stage;
- product exposed to post-process contamination;
- product sold as ready-to-eat;
- contact with an infected individual or animals.

In the food industry, *E. coli* is commonly included in buying specifications relating to raw materials and finished food products as an indicator of the hygienic status of the food. It is also included in some industry guidelines and in 'process hygiene criteria' as an indicator of faecal contamination in legislation (EC, 2007a). Currently, however, there is limited food-related legislation that refers to pathogenic *E. coli*, but these organisms will be included in the generic statements made in some food- and water-related legislation concerning microbiological safety requirements, e.g. 'foodstuffs should not contain micro-organisms or their toxins or metabolites in quantities that present an unacceptable risk for human health' (EC, 2005); at source and during marketing; a natural mineral water shall be free from parasites and pathogenic microorganisms (EC, 1980; EC, 2007b). Also, the European Directive on the quality of water states, among other obligations, that member states shall take the measures necessary to ensure that all 'water intended for human consumption is wholesome and clean' and that this means the water must be free from any microorganisms and parasites and from any substances which, in numbers or concentrations, constitute a potential danger to human health (EC, 1998).

Escherichia coli O157 is specifically included in some food industry buying specifications but, when present, it is likely to be present only at a low frequency in the raw material or finished product; therefore testing for the presence of pathogenic *E. coli* is likely to be of limited benefit.

17.4 Detecting *Escherichia coli*

Reliable methods for detecting and identifying pathogenic *E. coli*, especially *E. coli* O157:H7 but also other VTEC, are important in the support of properly developed and implemented HACCP systems for their control. Such methods are used appropriately in key raw material monitoring programs rather than for routine testing against buying specifications. They are also used more extensively in surveillance programs to help establish prevalence in different environments and routes of transmission to humans. Table 17.9 indicates the common conventional microbiological method used by food microbiologists to isolate and identify *E. coli* O157. Alternative selective broth and plating media are available and, as with detection and isolation methods for other bacterial pathogens, particular media and methods used should be verified as suitable for specific purposes by laboratory's personnel.

Table 17.9 Common conventional method for the detection and identification of *E. coli* O157 in foods

1:10 dilution of test sample selective enrichment broth (Modified Tryptone Soya Broth with Novobiocin)[a]
⇓
Incubate 41.5 °C for 6 h and/or 22 ± 2 h
⇓
Capture cells with immunomagnetic beads[b] after 6 h and/or 22 h incubation
⇓
⇓
Plate out onto Sorbitol MacConkey Agar containing Cefixime and Potassium Tellurite (CT-SMAC) and Sorbitol MacConkey Agar containing Cefixime and Rhamnose (CR-SMAC or other medium of choice)
⇓
Incubate 37 °C for 24 h
⇓
Inspect plates Purify suspect colonies on Nutrient Agar
⇓
Incubate 37 °C for 24 h Confirm the identity of the isolates e.g.
⇓
O157 latex bead or antisera agglutination ***if positive, then*** Biochemical profile
⇓
Incubate 37 °C for 24 h
⇓
Read reactions

[a]Stressed cells may not be detected by direct selective enrichment and may require periods of non-selective enrichment and/or longer enrichment times prior to subculture (Blackburn and McCarthy, 2000).

[b]Manufacturer's instructions must be followed.

Source: Based on ISO (2001).

A variety of technology-based methods are available for the detection and/or identification of *E. coli* O157 in foods; these may be based on antibody use such as immuno-magnetic separation (IMS), enzyme-linked immunosorbent assays (ELISA) or immuno-chromatography, nucleic acid hybridisation, polymerase chain reaction (PCR), and additional research will lead to the development of others (for a more extensive list see Baylis and Limburn, 2006).

As different phenotypes of foodborne human pathogenic *E. coli* are discovered from outbreak investigations, e.g. sorbitol-fermenting *E. coli* O157 (SF *E. coli* O157), any implications to detection methods of the new phenotypes should be considered. In this case, SF *E. coli* O157 are not readily detected using the current conventional method which relies on media containing sorbitol to distinguish the more common sorbitol non-fermenting colonies of *E. coli* O157 for further isolation and characterisation. Where clinical evidence suggests VTEC infection in a patient but suspect colonies are not observed on media containing sorbitol, clinical microbiologists have been advised to test sorbitol-positive colonies for agglutination with *E. coli* O157 antiserum; positives should be further investigated for confirmation and characterisation (Anon., 2006b).

When suspected positive results are obtained, further tests to characterise the organism may be necessary, and these can include serological tests, tests for the production of Vero cytotoxins, phage typing and pulsed field gel electrophoretic (PFGE) pattern. These latter two forms of test, however, tend to be confined to use in public health and veterinary research laboratories for identifying and tracing outbreak strains of the organism and when investigating the epidemiology of pathogenic *E. coli*. A valuable summary of methods available has been prepared by Campden and Chorleywood Food Research Association, now Campden BRI (Baylis and Mitchell, 2008). More variations on these methods are still being developed and will undoubtedly become available, but the more urgent need is for reliable methods to be developed for detecting and identifying other VTEC types of importance to food industry and public health microbiologists.

17.5 Control of pathogenic *Escherichia coli* in foods

Because of the growing variety of foods being implicated in outbreaks of illness attributed to VTEC, particularly *E. coli* O157 (Table 17.7), a continuing work programme by both researchers and food industry scientists is undertaken to answer questions that can help determine where controls may be best operated and what controls are most effective to minimise, if not eliminate, contamination of food by VTEC.

The very low infective dose of some VTEC, particularly VTEC O157:H7, underlines the importance of ensuring that the highest possible standards are maintained in agricultural practice and that food processors consistently operate well-designed and effective hygienic food production processes based on Hazard Analysis Critical Control Point (HACCP) assessments of each food process. In addition to attention to the detail of cleaning and hygiene procedures, the treatment and formulation of food products are important for controlling any residual *E. coli* and preventing their potential to cause harm to consumers.

The controls outlined here principally focus on *E. coli* O157 and other VTEC because of the severity of these hazards and their increasing incidence, but many of the controls equally apply to other pathogenic *E. coli.*

Escherichia coli O157 and other VTEC in particular do appear to be capable of surviving under adverse conditions in certain foods. Although these traits are also found in some other non-pathogenic *E. coli*, they have allowed VTEC to expose the frailties of products traditionally thought of as being safe, such as fruit juices, fermented meats and cheeses.

Control of *E. coli* and VTEC in food production is achieved through the application of effective control of raw materials, process conditions, post-process contamination and retail or catering practices. Indeed, it should be remembered that the general public also has a vital role to play in controlling the organism when handling and cooking products where the organisms may be present, e.g. raw meat and beefburgers (Table 17.8).

17.6 Raw material control

A contaminated raw material is one of the most common reasons why *E. coli* is present in a final product. Where a product receives no further process capable of eliminating the organism, these contaminants will inevitably result in the sale of potentially hazardous foods. VTEC outbreaks associated with cheese made from unpasteurised milk (Anon., 1999a), unpasteurised apple juice (Besser *et al.*, 1993) and raw, fermented meats (Cameron *et al.*, 1995) have all implicated contaminated raw ingredients as significant contributory factors. However, it should also be remembered that contaminated raw materials used for the production of cooked products have also caused outbreaks because of inadequate cooking by the manufacturer. Such products have also caused outbreaks through cross-contamination of *E. coli* O157 from the contaminated raw material to a cooked, ready-to-eat product (Ahmed, 1997). Bell and Kyriakides (1998) have extensively reviewed these outbreaks.

Effective control of the raw material to preclude or reduce levels of the organism is absolutely critical to the safety of many food industry products (ILSI, 2001). The primary raw materials implicated in foodborne outbreaks include raw milk, raw beef and raw fruit. In all cases, the principal route of contamination to the material is from exposure to animal faeces, from bovine sources in particular.

17.6.1 VTEC and animals

Animals and the products obtained from them have been clearly demonstrated to be high-risk sources of *E. coli* O157 and other pathogenic *E. coli* (see Section 17.3.2). It is unlikely to be possible to eliminate the organism from herds in the short term and, although the possibility of vaccination reported to reduce colonisation of the terminal rectum of beef cattle (Peterson *et al.*, 2007a, b) may be able to assist in reducing the hazard from an important source, it is the application of current knowledge in relation to reducing shedding and spread through effective animal husbandry practices that may be the immediate means of reducing the incidence of the organism in the faeces.

Data submitted by 12 EU countries on the incidence of VTEC in animals for 2006 showed an overall incidence for cattle of 1.8 % (742/41 983). VTEC O157 was reported at an incidence of 0.4 % (EFSA, 2007a). In one survey, carried out in Sweden, it is interesting to note that a higher incidence of VTEC O157 was found in ear samples (10.9 %) than in faecal samples (3.1 %).

Surveys in four EU member states of other ruminants such as sheep and goats generally found VTEC presence at levels equal to or higher than in cattle. EFSA (2007a) reported a 6.4 % (8/125) incidence of VTEC in goats in two member states and 1.6 % (7/428) in sheep (four member states) during 2006. Although VTEC are traditionally associated with ruminants, surveys of other animal species do occasionally find VTEC present. EFSA (2007a) reported for Germany, a 2.4 % incidence of VTEC in pigs ($n = 3308$ samples), although the majority of the serogroups isolated were unspecified; other surveys and investigations around the world have also reported carriage of *E. coli* O157:H7 in pigs, e.g. in the USA, Doane *et al.* (2007), Jay *et al.* (2007). Also for Germany, no VTEC were found in surveys of poultry ($n = 111$) and, interestingly, VTEC was also not found in surveys of dogs ($n = 576$) and cats ($n = 648$), although in Italy, two positive samples from rabbits (2.3 %, $n = 86$) were identified as VTEC non-O157.

17.6.2 Raw milk

Clearly, once introduced onto milking equipment, the organisms can then either proliferate, if the milk is not stored under effective refrigeration (< 8 °C) and/or spread to further batches through inadequate cleaning and disinfection. It should be noted that some reports indicate the growth of *E. coli* O157 in raw milk at temperatures as low as 7 °C (Heuvelink *et al.*, 1998) although only a 1.5 $\log_{10}$ increase occurred over a 144 hour period. Similar findings were reported by Kauppi *et al.* (1996) and Katić and Radenkov (1998) who demonstrated slow growth in nutrient media milk at 6.5 °C and 7 °C, respectively.

Controls that can reduce the introduction of faecal pathogens into the milk supply centre are effective milking parlour hygiene, which includes cleaning and disinfection of udders and teats, together with cleaning and disinfection of the milking equipment used for milking and the subsequent milk storage systems. This should include all transfer pipes, including portable hoses, which can often be overlooked as significant points for the build-up of contamination.

Raw milk should be subject to routine monitoring for indicators of contamination, including *E. coli* or coliform bacteria. Where the raw milk is intended to be used in the manufacture of products without effective bacterial reduction processes, such as in the manufacture of raw milk cheeses, it is important to source milk from farms known to be operating very high standards of hygiene. The farmer should be aware of the intended use of the milk, and it is common for manufacturers of such raw milk products to utilise incentive payment schemes using *E. coli* counts as a measure of the hygienic status of the milk. Consistent supply of high-quality milk is rewarded with higher payment and poor-quality milk supply is penalised. It is clear that such approaches can achieve improved hygienic quality of the raw milk supply, but it is also important to recognise that this cannot guarantee the absence of pathogens from the raw milk supply since low-level or low-frequency contamination is inevitable from time to time. Other indicators of hygiene may also be relevant for such purposes; for example, Hutchison *et al.* (2005) reported that total mesophilic aerobic colony counts were better than other bacterial indicators of on-farm hygiene they examined when assessed against routine hygiene audit scores.

17.6.3 Raw meat

Meat, and beef in particular, has been frequently implicated in outbreaks of foodborne *E. coli* O157 infection. Meat becomes contaminated through the transfer of faecal pathogens to the muscle tissue from faeces on the hide, hooves or from the intestine itself during the slaughtering operation. Minimising this contamination through the prevention of dirty animals entering the abattoir via effective farm handling, transport and lairage control is essential. The preclusion of dirty animals entering the slaughter line is part of the formal process used at inspection in UK abattoirs. Animal hide removal, evisceration and the handling of other parts of the animal carrying contaminants such as the hooves must be carefully controlled to prevent transfer from these areas to the muscle meat.

As well as effective operational procedures with well-defined process flows, subsequent thorough cleaning and disinfection of product contact surfaces, particularly when cutting into primal joints, is absolutely essential to prevent spread of any contaminants entering the plant. Like raw milk, it is not possible to preclude contamination with enteric organisms during raw meat processing and it is important to regularly monitor the hygienic status of the carcass meat. This is particularly important where the raw meat is subject to processes not capable of significantly reducing levels of enteric pathogens such as in the manufacture of raw fermented meats.

The inability to preclude such contamination has led many processors, particularly in the USA, to introduce steam pasteurisation plants, which effectively decontaminate the surfaces of the meat while maintaining the raw meat quality and appearance. These systems are capable of reducing contamination on the surface by up to 3 $\log_{10}$ units (Phebus *et al.*, 1997) and can make a significant contribution, together with effective animal husbandry and slaughter hygiene, to minimising the levels and frequency of contamination with these harmful organisms in raw meat. Surveys for *E. coli* O157 have rarely found it to be present in poultry products, and it is usually associated with ruminant animals such as cattle and sheep (Table 17.8).

Data from VTEC in food surveys in 16 EU member states and one non-member state in 2006 summarised by EFSA (2007a) continue to indicate the presence of VTEC in raw meat and meat products, albeit at low incidence. The reported prevalence in bovine meat was 0.3 % (n = 4443), which was significantly lower than was reported for the previous year (1.2 %, n = 8566). In meat from sheep the incidence was 1.8 % (n = 381), in pork 0.8 % (n = 3057) and in minced bovine meat the incidence was 1.9 % (n = 3064). No VTEC was reported detected in 862 samples of poultry meat.

17.6.4 Fruit, vegetables and other raw materials

In many cases, the source of the contamination of fruit and vegetables with pathogenic *E. coli* has been found or is believed to have been from contamination with animal faeces in the field prior to harvesting. Clearly, many fruit and vegetables are grown in soil where animal manure is used to add nutrients and conditioning to the soil. Many wild animals and birds are now known to be carriers of VTECs pathogenic to humans and these all serve to perpetuate the reservoir of these organisms in the environment. However, bovine faecal contamination is considered to be the primary source of VTEC in the environment and, if the incidence of *E. coli* O157 and other VTEC should increase in cattle and dairy herds, it follows that their incidence in animal wastes will also increase. Without effective treatment of the wastes prior to application to land, use of animal manure on soil has the potential to introduce enteric pathogens such as *E. coli* O157 to the soil where fruit and vegetables are grown.

Certain treatments of the animal waste such as composting can significantly reduce levels of contaminating pathogens owing to the high temperatures (*c.* 60 °C) achieved as a result of microbial growth and fermentation processes in the compost pile. Such approaches can make such materials safer to use. However, this is clearly dependent on effective composting practices with regular turning of the pile and breakdown of soil clumps. Much animal waste, however, is applied after only very little storage and consequently can remain potentially hazardous with respect to VTEC.

A practice often identified as contributory to outbreaks of VTEC infection attributed to unpasteurised fruit juices has been the use of 'drop fruit', i.e. fruit that has dropped to the ground and is picked up to avoid wastage. Contamination from pathogens present in or on the soil originating from the application of animal manure or via the deposition of these wastes by free-roaming animals may lead to the fruit subsequently carrying the organism into the processing plant, thereby entering the juice supply.

In a survey of agricultural practices in orchards in Virginia, USA Wright *et al.* (2000) reported 5 % of producers allowed grazing animals in the orchards, 8 % fertilised with manure and 32 % used drop apples. Proper treatment of animal wastes prior to application to soil, including long-term storage and composting, together with good agricultural practices that avoid the use of 'drop fruit' and ensuring wastes are never applied to exposed crops intended for consumption without further processing, are simple and obvious but nevertheless effective control measures that will reduce the chances of such materials being contaminated with VTEC.

Once introduced onto vegetables, *E. coli* O157 can survive for extended periods or even, grow. Beuchat (1999) reported survival of *E. coli* O157 on lettuce for 15 d even when contaminated with low initial levels (1–10 cfu/g). Indeed, it has also been shown that *E. coli* O157 can become internalised in fruit such as apples by entering through the blossom end (Buchanan *et al.*, 1999) and through scar tissue. Seeman *et al.* (2002) demonstrated the internalisation of *E. coli* in drop apples under natural conditions. Apples were gently dropped onto artificially contaminated soil and sampled after 1, 3, 8 and 10 d. Contamination occurred at highest levels on the skin of the apples, but high levels of contamination were also evident in the flesh and both the inner and outer cores. Apples showed no signs of visible damage and apples lying with the calyx facing down on the soil had a greater chance of internalisation than those where the calyx was uppermost or where the apples were on their side. Bernstein *et al.* (2007) demonstrated internalisation of *E. coli* into the aerial parts of maize via the root system and Warriner *et al.* (2003) demonstrated the same with *E. coli* during the sprouting of mung beans. Such reports add to the increasing evidence that internalization of bacterial pathogens may be a significant risk factor that should be taken into account when considering the appropriate controls for VTEC in ready-to-eat, agriculturally-derived products, and especially fruit. Whilst post-harvest washing processes may be important, there is an essential role for the consistent application of good agricultural practices in preventing contamination of the plants and fruit during growth and harvesting. Other potential sources of contamination should not be overlooked in relation to *E. coli* O157, including contaminated irrigation or processing water, other sources such as insects, rodents, and any seeds used for sprouting (Bell and Kyriakides, 1998).

Any hazard analysis of a production process should examine the raw material sources, identify whether they are likely sources of the organism through historical knowledge or knowledge of the agricultural practices involved in their production. Controls should then be introduced to minimise the occurrence of the hazard. Ideally this should involve implementing and maintaining best practices in the production of the material, such as the introduction of a destruction stage but, in the absence of such options, the use of microbiological testing, appropriately applied, can be useful in indicating general hygienic quality of the material and can allow differentiation of poor supply points from those operating better standards. For example, sprouted seed salad producers, e.g. beansprouts, should routinely monitor the microbiological quality of the seed batches for the presence of *Salmonella* and *E. coli* O157. It must be remembered, however, that such testing can never give complete assurance of a pathogen's absence.

17.7 Control in processing

The majority of food products are manufactured using some form of processing, and some processes are capable of reducing or eliminating pathogenic *E. coli*, if present. In such circumstances, it is the application of effective controls at critical stages of the manufacturing process that, if properly and consistently applied, will generate a safe finished product. Products such as cheese made from pasteurised milk, cooked meats and ready-meals are all subject to processes in which the organisms should be effectively eliminated.

Processes used to produce products such as prepared, ready-to-eat salads and raw, fermented meats together with some hard cheeses made from raw milk usually result in the reduction of contaminating pathogens, but survival may occur, particularly if initial contamination levels are high. And, of course, there are some products such as soft cheese made from raw milk or sprouted seeds such as beansprouts or alfalfa sprouts where the production process can allow growth and lead to increases in pathogen levels if present originally in the raw material.

However, even with these products it is possible to identify production methods that, if applied correctly, can reduce the risk associated with their consumption. The application of effective processes, e.g. use of washing systems, use of disinfection chemicals, incorporation of organic acids or humectants, for eliminating or at least reducing the levels of any pathogenic *E. coli* that may be present and, hence, process controls necessitates some understanding of the effect of different process stages on the growth and survival of these organisms. Once understood, it is then possible to introduce relevant systems for monitoring the processes.

17.7.1 Cooking

Heat processing is probably the most effective means of eliminating *E. coli* O157 and other pathogenic *E. coli.* from foods Cooked meats, pasteurised milk and milk products such as cheese and yoghurt, prepared ready-to-eat meals and a wide variety of other products rely on a heat process to eliminate the organism, if present, in the raw materials.

The effective application of the heat process is critical to the safety of the product and, no matter what the product, appropriate validation of the process must be conducted to ensure the consistent delivery of correct cooking times and temperatures. Cooking validation should be carried out prior to the introduction of a new product or process and must take account of all variables likely to affect the heat process. Cooking validation is intended to identify the conditions that will consistently produce a safe product. All of the following factors must be considered when developing a safe heat process:

- minimum ingoing temperature of the material;
- largest size/piece size of the product;
- cold spots in the oven/temperature distribution of the oven;
- fill load of the container, if appropriate, i.e. air spaces will affect heat transfer;
- oven load;
- minimum time setting of the process;
- minimum temperature setting of the oven.

In other words, for a cooked meat product, a process validation should establish that a cook of 70 °C for 2 min (or an equivalent process) will be achieved throughout the product when the largest size product, entering the process at its lowest initial temperature, is cooked in the slowest heating position in the oven with a full oven load and where the oven is set at the minimum designated process time and temperature for that product.

Following the identification of effective process controls it is then necessary to apply systems to monitor the efficacy of the process, and these should be built into the quality system of the manufacturing plant. Therefore, checks should be undertaken of the ingoing raw meat temperature, the size of the product and the process times and temperatures. It is usual to utilise a continuous chart recorder to monitor and record process times and temperatures. In addition, product is also temperature probed at the end of the cooking stage, and it is common to check products from the top, middle and bottom of each rack of product being cooked. Cooking temperatures applied to destroy *E. coli* O157 and VTEC vary depending on the product type. Pasteurised milk is heat-processed in the UK to comply with legislative requirements achieving 71.7 °C for 15 s. Other cooked foods such as meat are heat-processed to achieve a minimum process of 70 °C for 2 min. Studies have demonstrated that this process is sufficient to significantly reduce (> 6 $\log_{10}$ cfu/g) the levels of contaminating enteric pathogens including VTEC O157 (ACMSF, 1995). Based on studies in chicken homogenate, Betts *et al.* (1993) determined the D-value at 64 °C (D_{64}) of a strain of *E. coli* O157 to be 0.36–0.46 min with a z-value of 5.2–7.3 C°. Orta-Ramirez *et al.*

(1997) reported a D-value at 68 °C (D_{68}) of 0.12 min in ground beef (z-value 5.59C°). Stringer *et al.* (2000) collated data from different studies on the thermal destruction of *E. coli* O157:H7 and found z-values ranged between 3.9C°and 7.3C° depending on the meat type used. Such studies underline the need to carry out validation studies relating to specific products and processes to ensure product safety.

In a review of the safe cooking of burgers conducted by the ACMSF (2007), the recommended 70 °C for 2 min was still considered to reliably eliminate the organism. Modelling of published thermal destruction data indicated that the cooking process would give 95 % confidence of a 6 $\log_{10}$ reduction in *E. coli* O157. The ACMSF also recommended changing the z-value used to establish process equivalence at temperatures above and below 70 °C to 6C° (previously 7.4C°).

The development of mathematical models to predict the death of *E. coli* O157 (Murphy *et al.*, 2004) and their subsequent availability for industry use (www.ars.usda.gov/) can be helpful approaches in support of developing safe cooking processes.

17.7.2 Fermentation and drying processes

Products produced through a fermentation process such as raw, fermented meat and raw milk hard cheeses rely on the fermentation and drying process to reduce levels of contaminants present in the raw material. Although not as effective or controlled as a cooking process, these products can achieve some reduction in levels of enteric pathogens. It is essential that the conditions necessary to achieve such reductions are clearly understood by those manufacturing these products. Outbreaks occurring in these types of products have usually involved deficiencies in the fermentation or drying/maturation process, e.g. slow acid production or too short a drying period, or poor post-process hygiene conditions. As a result of some outbreaks it has also become evident that historically 'safe' processes have now been shown to be incapable of eliminating VTEC (Blue Ribbon Taskforce 1996; MacDonald *et al.*, 2004).

Soft cheese made from unpasteurised milk together with raw milk hard cheeses and raw, fermented meats undergoing only short drying/maturation periods are more vulnerable to outbreaks of *E. coli* O157 and other VTEC. This is due to the ability of the organism to survive the fermentation stage (Getty *et al.*, 2000) and, in the absence of significant drying stages, little further reduction of the organism occurs.

Challenge test studies should be undertaken, using research facilities, to identify the critical process points that require effective control to achieve a desirable reduction in *E. coli* O157 and VTEC. In general, an active fermentation process resulting in rapid acid production and associated pH reduction together with a long drying stage (several weeks) combine to achieve at least a 2–3 $\log_{10}$ reduction in *E. coli* O157 (Glass *et al.*, 1992;

Blue Ribbon Taskforce, 1996). Reduced drying times and temperature and reduced acidity development have been shown to result in less reduction of this organism in fermented meats.

Schlesser *et al.* (2006) reported on the survival of a five-strain cocktail of *E. coli* O157:H7 inoculated at three inoculum levels (10^1, 10^3 and 10^5 $\log_{10}$ cfu/ml) during the production of an unpasteurised cheddar cheese. In general, the organism increased by approximately 1–2 $\log_{10}$ cfu at all inoculum levels during the initial stages of production (draining, milling and pressing) which is consistent with both slight increases due to growth and concentration due to the production process. After storage at 7 °C, populations decreased by *c.* 1 and 2 $\log_{10}$ cfu after 60 and 120 d, respectively, and by 3 and 4 $\log_{10}$ cfu after 180 and 240 d, respectively. In the cheese made from milk inoculated at *c.* 10^5 cfu/ml, after 360 d storage the organism had reduced by approximately 5 $\log_{10}$ cfu but was still detectable and enumerable (*c.* 20 cfu/g). This was also the case in cheese made from milk inoculated at 10^3 cfu/ml (*c.* 8 cfu/ml after 360d) and even at an initial inoculum level of 10^1 cfu/ml, the organism was still detectable after 240 d storage (presence in 25 g).

Hew *et al.* (2006) used data from studies on the survival of *E. coli* O157:H7 during the production of Mexican-style chorizo with varying chemical composition to develop a death rate curve. The maximum death rate (0.3 $\log_{10}$ cfu/d) occurred at a water activity of 0.876 and decreased markedly as the water activity increased.

Riordan *et al.* (1998) studied the survival of *E. coli* O157:H7 during the manufacture of pepperoni. Batter mixtures were varied in terms of salt, sodium nitrite and dextrose, the latter to yield product with low (4.4–4.59), standard (4.69–4.86) and high (5.04–5.61) pH values. The pepperoni was fermented at 38 °C and then dried at 15 °C, both under controlled humidity conditions, until the desired pH value and water activity (< 0.80) were achieved which took approximately 7 d. In the standard formulation product (2.5 % salt, 100 ppm sodium nitrite, pH 4.69–4.86), initial levels of the organism declined by only 0.41 $\log_{10}$ cfu with a total decline after drying of 0.84 $\log_{10}$ cfu. Increased pathogen reductions occurred in product formulations with higher salt, higher sodium nitrite and lower pH value, with the death rates being greatest during the short fermentation phase. The biggest reduction was recorded in pepperoni with 3.3 % salt, 300 ppm sodium nitrite and pH 4.4–4.59 where levels decreased by 3.36 $\log_{10}$ cfu after fermentation and by a total of 4.79 $\log_{10}$ cfu by the end of the drying phase.

The processes employed for the production of these products are highly variable, and it is for this reason that it is advised that such products should be subject to process challenge tests to determine whether the process achieves a sufficient reduction of VTEC and, if not, what modifications can be introduced to achieve this. This may include lengthening the maturation stage or perhaps adding or increasing antimicrobial factors such as salt.

17.7.3 Washing processes

Many production processes rely on simple washing stages to remove extraneous dust and soil and to reduce levels of contaminating organisms including pathogens, e.g. ready-to-eat salads and vegetables and fresh, unpasteurised fruit juices. A variety of outbreaks involving these products have highlighted the vulnerability of the minimal processing employed in their production. Salad vegetables and fruit for juicing are usually washed in chlorinated water prior to further processing. Chlorine levels used in commercial practice have historically been at ≤ 200 ppm. Studies have shown this level to be capable of reducing contaminating pathogens by 1–2 $\log_{10}$ units.

Beuchat (1999) showed that *E. coli* O157 was reduced by *c.* 1–3 $\log_{10}$ units using a spray system to wash lettuce leaves with chlorinated water (200 ppm). However, rinsing in water alone achieved *c.* 1–2 $\log_{10}$ reduction and, at low levels of contamination, reductions were limited to *c.* 1 $\log_{10}$ unit with sprays of deionised water appearing as effective at reducing the organism as chlorinated water. When lettuce leaves were contaminated with *E. coli* O157 suspended in bovine faeces slurry at levels as low as 1.1 cfu/g, the organism could still be detected in 50 g samples of the product after spraying with chlorinated or deionised water.

Outbreaks of VTEC infection associated with these product groups have, however, prompted processors to investigate much higher levels of chlorination or indeed the use of other disinfection agents in an attempt to achieve greater reductions (Taormina and Beuchat, 1999). Whichever method is used, it is important to ensure that the active ingredient, most often chlorine, is present in its active form on a continuous basis. Washing systems have just as much capacity to spread contamination as they have for reducing it if the wash water is not regularly changed or the levels of active ingredient are not properly maintained. In addition, it is important to recognise that washing efficacy will also be dependent on facilitating good contact between the contaminant and the antimicrobial agent, and the surface structures of many vegetables and fruits can offer significant protection to microorganisms. Washing systems that incorporate means for agitation will clearly be helpful in reducing microorganisms on these plant material surfaces.

Washing has also been applied to decontaminate meat carcasses after slaughter. Kalchayanand *et al.* (2008) studied the effect of a variety of treatments in a spray-washing system on the reduction of *E. coli* O157:H7 artificially inoculated onto beef heads. Hot water (74 °C for 26 s) and lactic acid (2 % for 26 s) reduced counts of the organism by 1.72 and 1.52 $\log_{10}$ cfu/cm^2 relative to a standard pre-evisceration wash whereas ozonated water (*c.* 2.3 ppm) reduced counts by < 0.5 $\log_{10}$ cfu/cm^2. Bosilevac *et al.* (2006) compared the efficacy of a lactic acid wash (2 % at 42 °C) and a hot water wash (74 °C for 5.5 s) in reducing *E. coli* O157:H7 on pre-evisceration beef carcasses. In isolation, the lactic acid wash reduced the prevalence by 35 % and the hot water wash by 81 %. Applying the two treatments together did not improve the reduction (79 % reduction in prevalence).

17.7.4 Sprouting processes

Sprouted seed vegetables have been implicated in a number of outbreaks caused by *E. coli* O157 (Gutierrez, 1996; Como-Sabetti *et al.*, 1997) and other enteric bacteria such as *Salmonella* (O'Mahony *et al.*, 1990; Taormina *et al.*, 1999). These products are as vulnerable to microbial contamination as other prepared salads and fruits with one notable additional factor, germination in warm, wet conditions from seed. Because of this, any contaminating organisms, if present on the seed, are offered the opportunity for significant growth during the germination and growth of the sprouting seed. Levels of contamination on seed batches can increase by several orders of magnitude during the sprouting process (Jaquette *et al.*, 1996; Hara-Kudo *et al.*, 1997) and there is very little that can be done to prevent this from occurring if the organism is present initially on the seed. Taormina and Beuchat (1999) studied a variety of chemical treatments to reduce levels of *E. coli* O157 on alfalfa seed. Levels of 2000 ppm calcium hypochlorite, $Ca(OCl)_2$ gave a reduction of 1.84 $\log_{10}$ to $\geq$ 2.38 $\log_{10}$ cfu/g after 3 min and 1.83 log to $\geq$ 2.5 $\log_{10}$ cfu/g after 10 min. Treatment with 10 000 ppm and 20 000 ppm $Ca(OCl)_2$ gave reductions of $\geq$ 2.38 $\log_{10}$ cfu/g after 3 min and $\geq$ 2.5 $\log_{10}$ cfu/g after 10 min. These reductions were similar to those achieved with lower levels of hypochlorite. Although germination was not significantly reduced in comparison to water treatment, higher concentrations of hypochlorite reduced the rate of germination of the seed. It was also reported that *E. coli* O157, inoculated and then dried on alfalfa seeds at an initial level of 3.04 $\log_{10}$ cfu/g, were not significantly reduced when the seed was stored at 5 °C over a 20 week period and the organism could still be detected in 25 g samples of seed after 54 weeks.

Kumar *et al.* (2006) studied the inactivation of *E. coli* O157:H7 on a variety of seeds destined for sprouting using an oxychloro-based sanitiser (SOC). This was considered to have some benefits over traditional chlorine sanitisers due to its longer-term stability in use which would be more practicable for the long initial seed soaking times often employed in sprouting seed manufacture. Soaking the seed in concentrations of SOC up to 200 ppm for 24 h at 28 °C did not affect the subsequent germination rate of the seed. Mung beans contaminated with an initial loading of 10^3–10^4 cfu/g and germinated for 4 d permitted the growth to levels of 9.12 $\log_{10}$ cfu/g of sprouts. Treatment with 100–200 ppm SOC for 24 h followed by germination for 4 d eliminated *E. coli* O157:H7 (not detected in 25 g sprouts after 4 d germination). Growth of the organism was evident at a concentration of 50 ppm SOC. Different exposure times affected the survival of the organism with contamination being evident in seeds sprouted for 48 h after 4 and 8 h exposure to 200 ppm SOC but not after 19 and 24 h exposure. In addition, the ratio of mung bean to sanitiser also had an effect with ratio of 1:1 (mung bean: sanitiser wt/vol) and 1:2, resulting in survival and ratios of 1:6, 1:8 and 1:10 resulting in destruction i.e. not detected in 25 g sprouts after treatment and subsequent germination for 48 h at 28 °C. The sanitiser was not effective in eliminating *E. coli* O157:H7 for all seed types with the organism surviving in radish seeds, chick peas and broccoli seeds but not alfalfa, cress, flax or clover seeds when treated with 200 ppm SOC for 24 h.

Processors of these products usually employ extensive raw material control and testing programmes to monitor batches of seed for contamination. In addition, it is possible to test samples of spent water and germinated seed several days prior to harvesting and packing to gain an indication of contaminated production lots. Clearly, this cannot achieve total control of safety but in the absence of other suitable controls it is an important element of safety assurance for these types of product.

17.7.5 Hygiene and post-process contamination

A recurring factor often identified as a reason for outbreaks of *E. coli* O157 and VTEC infection is contamination of the product after application of the pathogen reduction process. Outbreaks implicating pasteurised milk, cheese, cooked meats and prepared salads and fruit have all indicated this area as a source of product contamination.

If a process has a recognised pathogen reduction or destruction stage, it is essential to avoid contaminating the finished product after this stage. Contamination usually arises from two sources: exposure of the finished product to contaminated raw material(s) or exposure to contamination from the environment or people.

Facilities and procedures for segregation must be in place to prevent raw materials coming into contact with finished products. In the manufacture of cooked meat, ready meals or prepared salads this is usually achieved by separating the factory into two areas, the low and high risk areas. The pathogen reduction stage is used as the division between the two areas, such as the building of an oven into a separating wall or the placing of a chlorinated wash water flume between high and low risk operations. Individuals on the low risk side handle the raw product and then, after processing, it is removed by individuals dedicated to the high risk side. The principles of segregation should be applied to all personnel moving from the low risk side of the factory to the high risk side, with appropriate coat and footwear changes and appropriate hand-washing procedures. An infectious disease policy must be in place for operatives handling ready-to-eat products, which should include notification of any instances of infectious disease and associated absence from work.

Effective cleaning and disinfection of all equipment is essential. It is particularly important to prevent the organism from building up in the equipment used for processing raw materials. High levels at these stages may result in spread of contaminants to other batches or excessive levels that may exceed the capacity of the subsequent process, e.g. cooking or fermentation, to reduce initial levels of any bacterial pathogens to those resulting in a safe finished product.

Farrell *et al.* (1998) studied the transfer of *E. coli* O157:H7 from artificially-contaminated raw meat to stainless steel plates attached to the inside of a meat grinder and their subsequent removal by various cleaning and disinfection techniques. Approximately 3–4 $\log_{10}$ cfu/cm^2 attached to the plates after grinding for 5.3 min with 4.5 kg meat initially contaminated with *c.* 6 $\log_{10}$ cfu/g. The lean to fat ratio of the meat did not affect attachment (90:10, 80:20 and 75:25 lean:fat ratios). Washing in an alkaline detergent at 60 °C for 15 min followed by treatment with a chlorine-based disinfectant (200 ppm) or peroxyacetic acid (0.2 %) for 2 min reduced the contamination levels by *c.* 2.5 $\log_{10}$/cm^2. Manual scrubbing of the plates during the washing stage reduced the recovery rate of the organism. Cleaning efficacy can be effectively monitored using indicators of hygiene such as tests for coliform bacteria or adenosine triphosphate (ATP) bioluminescence tests, which monitor the presence of residual levels of ATP from product residues and microorganisms.

Small manufacturers and butcher shops handle and process both raw and cooked meats in facilities that are not often subject to the same level of sophisticated segregation controls in place at large manufacturers. However, the principles of separation between raw and cooked meat in processing and on display together with effective cleaning and disinfection of surfaces and equipment are just as important to ensure product safety.

17.8 Final product control

Many products are sold in pre-packaged units that are not subject to further introduction of contaminants during distribution and retail display. However, other products are sold either loose or are subject to further preparation in the retail or catering outlet such as cooked meats that may be sliced. Outbreaks of *E. coli* O157 have resulted from the poor segregation of raw meats from ready-to-eat products during storage and sale.

Effective segregation of raw products from those intended to be consumed as ready-to-eat must be operated in both the retail and catering environments. This includes using separate counters for displaying raw and cooked foods or having perspex/safety glass dividers in counters to keep them apart. Safe procedures for serving, slicing or weighing raw and ready-to-eat foods must be developed. Personnel carrying out such tasks should be appropriately trained to understand the risk associated with cross-contamination from raw foods and from items coming into contact with raw foods such as equipment, surfaces and hands, and they should be provided with the appropriate facilities to operate good hygienic practice, e.g. hand-washing basins and supplies.

17.8.1 Post-process additives

Some products have additives placed on them as a final garnish including pâtés and cooked meats. Clearly, the garnish or post-process additive must be subject to appropriate treatments and controls to ensure it will not add pathogens, including VTEC, to the garnished product. It is common for many garnish materials to be subject to some form of disinfection such as the washing of fresh herbs and/or they may be subject to a testing programme or requirement for supplier certification of compliance to a specification prior to use.

17.8.2 Labelling of products

The consumer often has a significant role to play in ensuring safety with respect to many foods. It is essential that appropriate information is given on the product packaging to provide guidance concerning safe product handling and storage conditions and the preparation or cooking processes that need to be employed. Food safety tips are often present on raw meat packaging as a reminder of the need to ensure separation of raw from ready-to-eat foods during storage and preparation and also of the need for effective washing of hands, surfaces and utensils when handling raw and then ready-to-eat foods. In the USA this information is mandatory on raw meat products, but in the UK, some manufacturers and retailers choose to place this advice on products voluntarily.

Cooking instructions for raw foods such as beefburgers and other meats are important to help guide consumers as to the approximate times and temperatures necessary to achieve an appropriate cook. The generation of this guidance needs to be subject to very careful assessment owing to the extremely wide variation in cooking temperature that can result from different cooking appliances and methods. All the following variables should be carefully considered when generating cooking instructions for products:

- heating appliance (gas, electric solid fuel or microwave oven);
- pre-heating of the heat source;
- distance from the heat source;
- number of products cooked together;
- cooking time;
- temperature/power setting of the heat source;
- turning frequency of the product.

Owing to the inherent variability of cooking processes in the home it is also common for cooking instructions to include general indicators for cooking efficacy such as 'ensure it is cooked until the centre is piping hot and until the juices run clear'.

Some products, even with all of the applied process controls, are not considered to be acceptably safe and are also provided with labels of warning or, indeed, advice to avoid their consumption. An example of this includes sprouting salad vegetables such as alfalfa sprouts in the USA where, because of the high number of outbreaks of serious illness, the Food and Drug Administration issued advice to avoid the consumption of raw sprouts in order to reduce the risk of foodborne illness (FDA, 1999). Similarly, in the USA, unpasteurised fruit juice must be labelled with statements indicating that microbial risks exist with the consumption of this type of product.

17.9 Future trends

Considerable research resources continue to be expended on determining sources of pathogenic *E. coli* and the routes of infection to consumers as well as the means for control of these sources and routes to minimise or prevent foodborne illness arising through failures in controls. Reports of the isolation of VTEC O157:H7 from cloacal swabs taken from living layer hens, e.g. Dipineto *et al.* (2006), suggest further research may be required to establish the relative public health importance of different animal or avian species as sources of pathogenic *E. coli*.

Exploration of different methods and technologies to control contamination levels at each production stage from feed studies in animals, treatment of animal wastes, approaches to transport, harvesting and slaughtering processes and application of newer technologies in food production require extensive validation work to verify efficacy prior to acceptance by the relevant industry sector. Involvement of industry personnel at an early stage in such programmes of work can be beneficial to the outcome by ensuring only practical approaches are followed.

Major investigations of outbreaks, e.g. CalFert (2008), concerning a large outbreak attributed to contaminated iceberg lettuce, and outbreak and research reviews, e.g. Reilly (2001), Pennington (2009), provide clear recommendations, based on good evidence, as to those areas requiring specific attention to assist the control and prevention of foodborne infections due to pathogenic *E. coli*, particularly *E. coli* O157. It is essential that such good advice is properly disseminated to those who are in the best position to act on it and effect the necessary controls including, as applicable, producers and growers, processors and consumers.

The combined conscientious effort of all those involved in producing, processing and selling food as well as those preparing food at home is required to ensure that food is ultimately kept safe for consumption from the hazard of VTEC. Governments should facilitate a positive partnership between all these groups to ensure that relevant, clear and consistently correct information about VTEC and the means for its control is available and acted upon.

17.10 Sources of further information and advice

Duffy G (2006) Emerging pathogenic *E. coli*., in, Y Motarjemi and M Adams (eds), *Emerging Foodborne Pathogens* Woodhead, Cambridge, 253–81.

EC (2004) Regulation (EC) No 852/2004 of the European Parliament and of the Council of 29 April 2004 on the hygiene of foodstuffs, *Official Journal of the European Union*, **L139**; 20 April, 1–54

EC (2004) Regulation (EC) No 853/2004 of the European Parliament and of the Council of 29 April 2004 laying down specific hygiene rules for food of animal origin, *Official Journal of the European Union*, **L226**, 25 June, 22–82 and amendment in Commission Regulation (EC) No 1243/2007, *Official Journal of the European Union*, **L281**, 25 October, 8–11.

FSA (2008) *Review of past and current research on Verocytotoxin-producing Escherichia coli (VTEC) in relation to public health protection*, Food Standards Agency, London.

Mainil J G and Daube G (2005) A review: Verotoxigenic *Escherichia coli* from animals, humans and foods: who's who? *Journal of Applied Microbiology*, **98**, 1332–44.

Pavankumar A R and Sankaran K (2008) The need and new tools for surveillance of *Escherichia coli* pathogens, *Food Technology and Biotechnology*, **46**(2), 125–45.

Public Health Laboratory Service (2004) Preventing person-to-person spread following gastrointestinal infections: guidelines for public health physicians and environmental health officers. Prepared by a Working Group of the former PHLS Advisory Committee on Gastrointestinal Infections, *Communicable Disease and Public Health*, **7**(4), 362–84.

Stewart C S and Flint H J (eds) (1999) Escherichia coli *O157 and farm animals*, CABI, Wallingford.

Willshaw G A, Scotland S M and Rowe B (1997) Vero cytotoxin-producing *Escherichia coli*, in M Sussman (ed.), Escherichia coli: *Mechanisms of Virulence*, Cambridge University Press, Cambridge, 421–48.

WHO (2006) *Guidelines for drinking-water quality. First addendum to third edition Volume 1: Recommendations*, World Health Organization, Geneva.

Welinder-Olsson C and Kaijser B (2005) Enterohemorrhagic *Escherichia coli* (EHEC), *Scandinavian Journal of Infectious Diseases*, **37**(6 & 7), 405–16.

17.11 References

Adesiyun A A, Webb L A, Romain H *et al.* (1997) Prevalence and characteristics of strains of *Escherichia coli* isolated from milk and faeces on dairy farms in Trinidad. *Journal of Food Protection*, **60**(10), 1174–81.

ACMSF (1995) *Report on Vero cytotoxin-producing* Escherichia coli, Advisory Committee on the Microbiological Safety of Food, HMSO, London.

ACMSF (2007) *Report on the Safe Cooking of Burgers*, Advisory Committee on the Microbiological Safety of Food, Available at: http://www.food.gov.uk/multimedia/pdfs/acmsfburgers0807.pdf, accessed November 2008.

Ahmed S (1997) An outbreak of *E. coli* O157 in Central Scotland, *Scottish Centre for*

Infection and Environmental Health Weekly Report, Number 1, No. 97/13 8, Scottish Centre for Infection and Environmental Health, Glasgow.

Alam M J and Zurek L (2004) Association of *Escherichia coli* O157:H7 with houseflies on a cattle farm, *Applied and Environmental Microbiology*, **70**(12), 7578–80.

Anon. (1998a) Cases of *Escherichia coli* O157 infection associated with unpasteurised cream, *Communicable Disease Report Weekly*, **8**(43), 377.

Anon. (1998b) VTEC O157 infection and unpasteurised cream, *Communicable Disease Report Weekly*, **8**(44), 389–92.

Anon. (1999a) *Escherichia coli* O157 associated with eating unpasteurised cheese. *Communicable Disease Report Weekly*, **9**(13), 113–116.

Anon. (1999b) *Escherichia coli* O157 associated with eating unpasteurised cheese–update, *Communicable Disease Report Weekly*, **9**(15), 131–4.

Anon. (1999c) Outbreak of *Escherichia coli* O157:H7 and *Campylobacter* among attendees of the Washington County fair – New York, 1999, *Morbidity and Mortality Weekly Report*, **48**(36), 803–4.

Anon. (2001) Vero cytotoxin-producing *Escherichia coli* O157: 1999 and 2000. *Communicable Disease Report Weekly*, **11**(19), 7–8.

Anon., (2005) *E. coli* O157, unpasteurised milk – USA (Oregon, Washington), ProMed mail Archive number, 20060121.0199, www.promedmail.org.

Anon. (2006a) Ongoing multistate outbreak of *Escherichia coli* serotype O157:H7 infections associated with consumption of fresh spinach – United States, September 2006, *Morbidity and Mortality Weekly Report*, **55**(38), 1045–6.

Anon. (2006b) Sorbitol-fermenting Vero cytotoxin-producing *E. coli* O157 (VTEC O157), *Health Protection Agency Communicable Disease Report Weekly*, **16**(21), 2.

Anon. (2009) Vero cytotoxin-producing *Escherichia coli* 0157; 2007 and 2008 (corrected 20 March 2009), *Health Protection Report*, **3**(10), 10–11.

Ansay S E and Kaspar C W (1997) Survey of retail cheese, dairy processing environments and raw milk for *Escherichia coli* O157:H7, *Letters in Applied Microbiology*, **25**(2), 131–4.

Ansay S E, Darling K A, Kaspar C W *et al.* (1999) Survival of *Escherichia coli* O157:H7 in ground-beef patties during storage at 2, –2, 15 and then –2 °C, and –20 °C, *Journal of Food Protection*, **62**(11), 1243–7.

Avery L M, Killham K and Jones D L (2005) Survival of *E. coli* O157:H7 in organic wastes destined for land application, *Journal of Applied Microbiology*, **98**, 814–22.

Baylis C and Mitchell C (2008) *Catalogue of Rapid Microbiological Methods, Review No. 1*, 6th edn, Campden and Chorleywood Food Research Association, Chipping Campden.

Bell C and Kyriakides A (1998) E. coli: *A Practical Approach to the Organism and its Control in Foods*, Blackwell Science, Oxford.

Bernstein N, Sela S, Pinto R *et al.* (2007) Evidence of internalisation of *Escherichia coli* O157 into the aerial parts of maize via the root system, *Journal of Food Protection*, **70**(2), 471–5.

Besser R E, Lett S M, Weber J T *et al.* (1993) An outbreak of diarrhea and hemolytic uremic syndrome from *Escherichia coli* O157:H7 in fresh-pressed apple cider. *Journal of the American Medical Association*, **269**(17), 2217–20.

Betts G D, Lyndon G and Brooks J (1993) *Heat Resistance of Emerging Foodborne Pathogens:* Aeromonas hydrophila, Escherichia coli *O157:H7,* Plesiomonas shigelloides and Yersinia enterocolitica, Technical Memorandum No. 672, Campden and Chorleywood Food Research Association, Chipping Campden.

Beuchat L R (1999) Survival of enterohemorrhagic *Escherichia coli* O157:H7 in bovine faeces applied to lettuce and the effectiveness of chlorinated water as a disinfectant *Journal of Food Protection*, **62**(8), 845–9.

Bhat M, Denny J, MacDonald K *et al.* (2007) *Escherichia coli* O157:H7 infection associated with drinking raw milk–Washington and Oregon, November–December 2005, *Morbidity and Mortality Weekly Report*, **56**(08), 165–7.

Blackburn C de W and McCarthy J D (2000) Modifications to methods for the enumeration and detection of injured *Escherichia coli* O157:H7 in foods, *International Journal of Food Microbiology*, **55**, 285–90.

Blue Ribbon Task Force (1996) *Dry Fermented Sausage and* E. coli *O157:H7*, Research report No. 11–316, National Cattlemen's Beef Association, Chicago, IL.

Bohaychuk V M, Gensler G E, King R K *et al.* (2006) Occurrence of pathogens in raw and ready-to-eat meat and poultry products collected from the retail marketplace in Edmonton, Alberta, Canada, *Journal of Food Protection*, **69**(9), 2176–82.

Bolton F J, Crozier L and Williamson J K (1996) Isolation of *Escherichia coli* O157 from raw meat products, *Letters in Applied Microbiology*, **23**, 317–21.

Bosilevac J M, Nou X, Barkocy-Gallagher A *et al.* (2006) Treatments using hot water instead of lactic acid reduce the levels of aerobic bacteria and *Enterobacteriaceae* and reduce the prevalence of *Escherichia coli* O157:H7 on preevisceration beef carcasses. *Journal of Food Protection*, **69**(8), 1808–13.

Brenner D J (1984) Family 1. Enterobacteriaceae Rahn 1937, Nom. Fam. Cons. Opin.15, Jud. Comm. 1958, 73: Ewing, Farmer and Brenner 1980, 674; Judicial Commission 1981, 104, in N R Krieg and J G Holt (eds) *Bergey's Manual of Systematic Bacteriology Volume 1* Williams & Wilkins, Baltimore, MD 408–20.

Brichta-Harhay D M, Arthur T M, Bosilevac J M *et al.* (2007) Enumeration of *Salmonella* and *Escherichia coli* O157:H7 in ground beef, cattle carcass, hide and faecal samples using direct plating methods, *Journal of Applied Microbiology*, **103**, 1657–68.

Buchanan R L, Edelson S G, Miller R L *et al.* (1999) Contamination of apples after immersion in an aqueous environment containing *Escherichia coli* O157:H7, *Journal of Food Protection*, **62**(5), 444–50.

CalFert (2007) *Investigation of an* Escherichia coli *O157:H7 outbreak associated with Dole pre-packaged spinach. Final March 21, 2007*, prepared by California Food Emergency Response Team, California Department of Health Services, Sacramento, CA.

CalFert (2008) *Investigation of the Taco John's* Escherichia coli *O157:H7 outbreak associated with Iceberg lettuce. Final Report February 15, 2008*, prepared by The California Food Emergency Response Team, California Department of Health Services, Sacramento, CA.

Caro I, Fernández-Barata V M, Alonso-Llamazares A *et al.* (2006) Detection, occurrence and characterization of *Escherichia coli* O157:H7 in raw ewe's milk in Spain, *Journal of Food Protection*, **69**(4), 920–24.

Cameron S, Walker C, Beers M *et al.* (1995) Enterohaemorrhagic *Escherichia coli* outbreak in South Australia associated with the consumption of Mettwurst. *Communicable Disease Intelligence*, **19**(3), 70–71.

Cerqueira A M F, Guth B E C, Joaquim R M *et al.* (1999) High occurrence of Shiga toxin-producing *Escherichia coli* (STEC) in healthy cattle in Rio de Janeiro State, Brazil, *Veterinary Microbiology*, **70**, 111–21.

CFA (2007) *Microbiological guidance for produce suppliers to chilled food manufacturers*, 2nd edn, Chilled Food Association, Peterborough.

Chahed A, Ghafir Y, China B *et al.* (2005) Survey of the contamination of foodstuffs of animal origin by Shiga toxin producing *Escherichia coli* serotype O157:H7 in Belgium from 1999 to 2003, *Eurosurveillance*, **10**(1–3), available at: http://www.eurosurveillance.org/ViewArticle.aspx?ArticleId=527, accessed November 2008.

Chapman P A, Siddons C A, Cerdan Malo A T *et al.* (2000) A one year study of *Escherichia coli* O157 in raw beef and lamb products, *Epidemiology and Infection*, **124**, 207–13.

Charatan F (1999) New York outbreak of *E. coli* poisoning affects 1000 and kills two, *British Medical Journal*, **319**(2 Oct), 873.

Como-Sabetti K, Reagan S, Allaire S *et al.* (1997) Outbreaks of *Escherichia coli* O157:H7 infection associated with eating alfalfa sprouts – Michigan and Virginia, June–July 1997, *Morbidity and Mortality Weekly Report*, **46**(32), 741–4.

Crump J A, Braden C R, Dey M E *et al.* (2003) Outbreaks of *Escherichia coli* O157 infections at multiple county agricultural fairs: a hazard of mixing cattle, concession stands and children, *Epidemiology and Infection*, **131**, 1055–62.

Currie A, MacDonald J, Ellis A *et al.* (2007) Outbreak of *Escherichia coli* O157:H7 infections associated with consumption of beef donair, *Journal of Food Protection*, **70**(6), 1483–8.

Dargatz D A, Wells S J, Thomas L A *et al.* (1997) Factors associated with the presence of *Escherichia coli* O157 in feces of feedlot cattle, *Journal of Food Protection*, **60**(5), 466–70.

De Louvois J and Rampling A (1998) One fifth of samples of unpasteurised milk are contaminated with bacteria, *British Medical Journal*, **316**(21 Feb), 625.

De Schrijver K, Buvens G, Posse B *et al.* (2008) Outbreak of verocyotoxin-producing *E. coli* O157 and O26 infections associated with the consumption of ice cream produced at a farm, Belgium, *Eurosurveillance*, **13**(7), available at: http://www.eurosurveillance.org/ViewArticle.aspx?ArticleId=8041, accessed November 2008.

Dipineto L, Santaniello A, Fontanella M *et al.* (2006) Presence of Shiga toxin-producing *Escherichia coli* O157:H7 in living layer hens, *Letters in Applied Microbiology*, **43**, 293–5.

Doane C A, Pangloli P, Richards H A *et al.* (2007) Occurrence of *Escherichia coli* O157:H7 in diverse farm environments, *Journal of Food Protection*, **70**(1), 6–10.

DOH (1998) *Surveillance of the Microbiological Status of Raw Cows' Milk on Retail Sale*, Department of Health, London.

Doyle M P and Padhye V V (1989) *Escherichia coli*, in M P Doyle (ed.), *Foodborne Bacterial Pathogens*, Marcel Dekker, New York, 235–81.

Doyle M P and Schoeni J L (1987) Isolation of *Escherichia coli* O157:H7 from retail fresh meats and poultry, *Applied and Environmental Microbiology*, **53**(10), 2394–6.

EC (1980) Council Directive 80/777/EEC on the approximation of the laws of the member states relating to the exploitation and marketing of natural mineral waters, *Official Journal of the European Communities*, **L229**, 30 August, 1–10.

EC (1998) Council Directive 98/83/EC of 3 November 1998 on the quality of water intended for human consumption, *Official Journal of the European Communities*, **L330**, 5 December, 32–54.

EC (2005) Commission Regulation (EC) No 2073/2005 of 15 November 2005 on microbiological criteria for foodstuffs, *Official Journal of the European Union*, **L338**, 22 December, 1–26.

EC (2007a) Commission Regulation (EC) No 1441/2007 of 5 December 2007 amending Regulation (EC) No 2073/2005 on microbiological criteria for foodstuffs, *Official Journal of the European Union*, **L322**, 7 December, 12–29.

EC (2007b) Proposal for a Directive of the European Parliament and of the Council on the exploitation and marketing of natural mineral waters (recast), Commission of the European Communities, COM (2007) 858 final, 2007/0292 (COD), http://eur-lex.europa.eu.

EFSA (2007a) The Community Summary Report on Trends and Sources of Zoonoses, Zoonotic Agents, Antimicrobial Resistance and Foodborne Outbreaks in the European Union in 2006, European Food Safety Authority, *The EFSA Journal*, **130**, 165–78.

EFSA (2007b) Monitoring of verotoxigenic *Escherichia coli* (VTEC) and identification of human pathogenic VTEC types; Scientific Opinion of the Panel on Biological Hazards, European Food Safety Authority, *The EFSA Journal*, **579**, 1–61.

Ethelberg S, Smith B, Torpdahl M *et al.* (2007) An outbreak of Verocytotoxin-producing *Escherichia coli* O26:H11 caused by beef sausage, Denmark 2007, *Eurosurveillance*, **12**(22), available at: http://www.eurosurveillance.org/ViewArticle.aspx?ArticleId=3208, accessed November 2008.

Farrell B L, Ronner A R and Lee Wong A C (1998) Attachment of *Escherichia coli* O157:H7 in ground beef to meat grinders and survival after sanitation with chlorine and peroxyacetic acid, *Journal of Food Protection*, **61**(7), 817–22.

FDA (1999) Consumers advised of risks associated with raw sprouts, Food and Drug Administration, *HHS News*, 9 July, US Department of Health and Human Services, Washington, DC.

Fremaux B, Prigent-Combaret C, Delignette-Muller M L *et al.* (2007a) Persistence of Shiga toxin-producing *Escherichia coli* O26 in cow slurry, *Letters in Applied Microbiology*, **45**(1), 55–61.

Fremaux B, Delignette-Muller M L, Prigent-Combaret C *et al.* (2007b) Growth and survival of non-O157:H7 Shiga-toxin-producing *Escherichia coli* in cow manure, *Journal of Applied Microbiology*, **102**(1), 89–99.

Fukushima H, Hoshina K and Gomyoda M (1999) Long-term survival of Shiga toxin-producing *Escherichia coli* O26, O111, and O157 in bovine faeces, *Applied and Environmental Microbiology*, **65**(11), 5177–81.

Garber L, Wells S, Shroeder-Tucker L *et al.* (1999) Factors associated with the faecal shedding of verotoxin-producing *Escherichia coli* O157 on dairy farms. *Journal of Food Protection*, **62**(4), 307–12.

Getty K J K, Phebus R K, Marsden J L *et al.* (2000) *Escherichia coli* O157:H7 and fermented sausages: a review, *Journal of Rapid Methods and Automation in Microbiology*, **8**, 141–70.

Gillespie I A, O'Brien S J, Adak G K *et al.* (2005) Foodborne general outbreaks of Shiga toxin-producing *Escherichia coli* O157 in England and Wales 1992–2002: where are the risks? *Epidemiology and Infection*, **133**, 803–8.

Glass K A, Loeffelholz J M, Ford J P *et al.* (1992) Fate of *Escherichia coli* O157:H7 as affected by pH or sodium chloride and in fermented, dry sausage, *Applied and Environmental Microbiology*, **58**(8), 2513–6.

Gould H (2008) Shiga toxin-producing *Escherichia coli*: burden and trends, *Foodnet News*, **2**(1), 1.

Gutierrez E (1996) Is the Japanese O157:H7 *E. coli* epidemic over? *The Lancet*, **348** (16 Nov), 1371.

Hara-Kudo Y, Konuma H, Iwaka M *et al.* (1997) Potential hazard of radish sprouts as a vehicle of *Escherichia coli* O157:H7, *Journal of Food Protection*, **60**(9), 1125–7.

Harrison S and Kinra S (2004) Outbreak of *Escherichia coli* O157 associated with a busy bathing beach, *Communicable Disease and Public Health*, **7**(1), 47–50.

Heuvelink A E, Wernars K and De Boer E (1996) Occurrence of *Escherichia coli* O157 and other Verocytotoxin-producing *E. coli* in retail raw meats in the Netherlands *Journal of Food Protection*, **59**(12), 1267–72.

Heuvelink A E, Bleumink B, Van Den Biggelaar F L A M *et al.* (1998) Occurrence and survival of Verocytotoxin-producing *Escherichia coli* O157 in raw cows' milk in the Netherlands, *Journal of Food Protection*, **61**(12), 1597–601.

Hew C M, Hajmeer M N, Farver T B *et al.* (2006) Pathogen survival in Chorizos: ecological factors, *Journal of Food Protection*, **69**(5), 1087–95.

Hilborn E D, Mermin J H, Mshar P A *et al.* (1999) A multi-state outbreak of *Escherichia coli* O157:H7 infections associated with consumption of mesclun lettuce. *Archives of Internal Medicine*, **159**(15), 1758–64.

Hjertqvist M (2005) *E. coli* O157, Lettuce – Sweden (West Coast), 20050916.2738, www.promedmail.org.

Howie H, Mukerjee A, Cowden J *et al.* (2003) Investigation of an outbreak of *Escherichia coli* O157 infection caused by environmental exposure at a scout camp. *Epidemiology and Infection*, **131**, 1063–9.

HPA (2008) Vero cytotoxin-producing *E. coli* O157 strains examined by LEP reported to the Health Protection Agency Centre for Infections: Isolations from Humans, England and Wales, 1982 – 2006, Health Protection Agency, London.

Hutchison M L, Thomas D J I, Moore A *et al.* (2005) An evaluation of raw milk microorganisms as markers of on-farm hygiene practices related to milking, *Journal of Food Protection*, **68**(4), 764–72.

ICMSF (1996) *Microorganisms in Foods 5, Microbiological Specifications of Food Pathogens*, International Commission on Microbiological Specifications for Foods, Blackie Academic and Professional, London.

ILSI (2001) *Approach to the control of enterohaemorrhagic* Escherichia coli (EHEC), ILSI Europe Report Series, International Life Sciences Institute, Brussels.

ISO (2001) ISO 16654: 2001 *Microbiology of food and animal feeding stuffs – horizontal method for the detection of* Escherichia coli *O157*, International Organization for Standardization, Geneva.

Jackson S G, Goodbrand R B, Johnson R P *et al.* (1998) *Escherichia coli* O157:H7 diarrhoea associated with well water and infected cattle on an Ontario farm, *Epidemiology and Infection*, **120**, 17–20.

Jaquette C B, Beuchat L R and Mahon B E (1996) Efficacy of chlorine and heat treatment in killing *Salmonella stanley* inoculated onto Alfalfa seeds and growth and survival of the pathogen during sprouting and storage, *Applied and Environmental Microbiology*, **62**(6), 2212–5.

Jay M T, Cooley M, Carychao D *et al.* (2007) *Escherichia coli* O157:H7 in feral swine near spinach fields and cattle, Central California coast, *Emerging Infectious Diseases*, **13**(12) 1908–11.

Jensen C, Ethelberg S, Gervelmeyer A *et al.* (2006) First general outbreak of Verocytotoxin-producing *Escherichia coli* O157 in Denmark, *Eurosurveillance*, **11** (1–3), available at: http://www.eurosurveillance.org/ViewArticle.aspx?ArticleId=597, accessed November 2008.

Jones D (1988) Composition and properties of the family Enterobacteriaceae, in B M Lund, M Sussman, D Jones and M F Stringer (eds), *Enterobacteriaceae in the Environment and as Pathogens, Proceedings of a Symposium.* Nottingham 7–9 July 1987, The Society for Applied Bacteriology Symposium Series No. 17, *Journal of Applied Bacteriology* Symposium Supplement, Blackwell Scientific Publications, London, 8S.

Jones D L (1999) Potential health risks associated with the persistence of *Escherichia coli* O157 in agricultural environments, *Soil Use and Management*, **15**, 76–83.

Kalchayanand N, Arthur T M, Bosilevac J M *et al.* (2008) Evaluation of various antimicrobial interventions for the reduction of *Escherichia coli* O157:H7 on bovine heads during processing, *Journal of Food Protection*, **71**(3), 621–4.

Karch H, Tarr P I and Bielaszewska M (2005) Enterohaemorrhagic *Escherichia coli* in human medicine, *International Journal of Medical Microbiology*, **295**(6–7), 405–18.

Katić, V and Radenkov S (1998) The fate of *Escherichia coli* O157:H7 in pasteurized milk, *Acta Veterinaria (Beograd)*, **48**(5–6), 345–50.

Kauffmann F (1947) Review, the serology of the coli group, *Journal of Immunology*, **57**, 71–100.

Kauppi K L, Tatini S R, Harrell F *et al.* (1996) Influence of substrate and low temperature on growth and survival of verotoxigenic *Escherichia coli*, *Food Microbiology*, **13**, 397–405.

Kerr M, Fitzgerald M, Sheridan J J *et al.*, (1999) Survival of *Escherichia coli* O157:H7 in bottled natural mineral water, *Journal of Applied Microbiology*, **87**, 833–41.

Kim J K and Harrison M A (2008) Transfer of *Escherichia coli* O157:H7 to Romaine lettuce due to contact water from melting ice, *Journal of Food Protection*, **71**(2), 252–6.

Kobayashi M, Sasaki T, Saito N *et al.* (1999) Houseflies: not simple mechanical vectors of enterohemorrhagic *Escherichia coli* O157:H7, *American Journal of Tropical Medicine and Hygiene*, **61**(4), 625–9.

Kudva I T, Hunt C W, Williams C J *et al.* (1997) Evaluation of the dietary influences on *Escherichia coli* O157:H7 shedding in sheep, *Applied and Environmental Microbiology*, **63**(10), 3878–86.

Kumar M, Hora R, Kostrzynska M *et al.* (2006) Inactivation of *Escherichia coli* O157:H7 and *Salmonella* on mung beans, alfalfa, and other seed types destined for sprout production by using an oxychloro-based sanitiser, *Journal of Food Protection*, **69**(7), 1571–8.

Lenahan M, O'Brien S, Kinsella K *et al.* (2007) Prevalence and molecular characterization of *Escherichia coli* O157:H7 on Irish lamb carcasses, fleece and in faeces samples, *Journal of Applied Microbiology*, **103**(6), 2401–9.

Levine M M (1987) *Escherichia coli* that cause diarrhea: enterotoxigenic, enteropathogenic, enteroinvasive, enterohemorrhagic, and enteroadherent, *Journal of Infectious Diseases*, **155**(3), 377–89.

Linton A H and Hinton M H (1988) Enterobacteriaceae associated with animals in health and disease, in B M Lund, M Sussman, D Jones and M F Stringer (eds), *Enterobacteriaceae in the Environment and as Pathogens. Proceedings of a Symposium.* Nottingham, 7–9 July 1987, The Society for Applied Bacteriology Symposium Series No. 17, *Journal of Applied Bacteriology* Symposium Supplement, Blackwell Scientific Publications, London, 71S–77S.

Locking M, Allison L, Rae L *et al.* (2006) VTEC infections and livestock-related exposures in Scotland, 2004, *Eurosurveillance*, **11**(2), available at: http://www.eurosurveillance.org/ViewArticle.aspx?ArticleId=2908, accessed November 2008.

MacDonald D M, Fyfe M, Paccagnella A *et al.* (2004) *Escherichia coli* O157:H7 outbreak linked to salami, British Columbia, Canada, 1999, *Epidemiology and Infection*, **132**, 283–9.

Massa S, Altieri C, Quaranta V *et al.* (1997) Survival of *Escherichia coli* O157:H7 in yoghurt during preparation and storage at 4 °C, *Letters in Applied Microbiology*, **24**, 347–50.

Massa S, Goffredo E, Altieri C *et al.* (1999) Fate of *Escherichia coli* O157:H7 in unpasteurized milk stored at 8 °C, *Letters in Applied Microbiology*, **28**, 89–92.

Matthews L, McKendrick I J, Ternent H *et al.* (2006) Super-shedding cattle and the transmission dynamics of *Escherichia coli* O157, *Epidemiology and Infection*, **134**, 131–42.

Maule A, (1997) Survival of the Verotoxigenic strain *E. coli* O157:H7 in laboratory-scale microcosms, in D Kay and C Fricker (eds), *Coliforms and* E. coli *– Problem or Solution?* The Royal Society of Chemistry, London, 61–5.

McClure P, (2005) *Escherichia coli*: virulence, stress response and resistance, in M Griffiths (ed.), *Understanding Pathogen Behaviour – Virulence, Stress Response and Resistance*, Woodhead, Cambridge, 240–78.

Mead P S, Slutsker L, Dietz V *et al.* (1999) Food-related illness and death in the United States, *Emerging Infectious Diseases*, **5**, 607–25.

Mermin J H, Hilborn E D, Voetsch A *et al.* (1997) A multi-state outbreak of *Escherichia coli* O157:H7 infections associated with eating mesculin mix lettuce, *Abstract from VTEC '97, 3rd International Symposium and Workshop on Shiga Toxin (Verocytotoxin) Producing* Escherichia coli *Infections*, 22–26 June, Baltimore, MD.

Milne L M, Plom A, Strudley I *et al.* (1999) *Escherichia coli* O157 incident associated with a farm open to members of the general public, *Communicable Disease and Public Health*, **2**(1), 22–6.

Møller Nielsen E, Jensen C, Lau Baggesen D (2005) Evidence of transmission of verocytotoxin producing *Escherichia coli* O111 from a cattle stable to a child, *Clinical Microbiology and Infection*, **11**(9), 767–70.

Mshar P A, Dembek Z F, Cartter M L *et al.* (1997) Outbreaks of *Escherichia coli* O157:H7 infection and cryptosporidiosis associated with drinking unpasteurized apple cider – Connecticut and New York, October 1996, *Journal of the American Medical Association*, **277**(10), 781–2.

Mukherjee A, Cho S, Scheftel J *et al.* (2006) Soil survival of *Escherichia coli* O157:H7 acquired by a child from garden soil recently fertilized with cattle manure, *Journal of Applied Microbiology*, **101**, 429–36.

Murphy R Y, Beard B L, Martin E M *et al.* (2004) Predicting process lethality of

Escherichia coli O157:H7, *Salmonella* and *Listeria monocytogenes* in ground, formulated and formed beef/turkey links cooked in an air impingement oven, *Food Microbiology*, **21**, 493–9.

Nastasijevic I, Mirovic R and Buncic S (2008) Occurrence of *Escherichia coli* O157 on hides of slaughtered cattle, *Letters in Applied Microbiology*, **46**(1), 126–31.

Nataro J P and Kaper J B (1998) Diarrheagenic *Escherichia coli, Clinical Microbiology Reviews*, **11**, 142–201.

Neaves P, Deacon J and Bell C (1994) A survey of the incidence of *Escherichia coli* O157 in the UK Dairy Industry *International Dairy Journal*, **4**, 679–96.

O'Mahony M, Cowden J, Smyth) B *et al.* (1990). An outbreak of *Salmonella saint-paul* infection associated with beansprouts, *Epidemiology and Infection*, **104**, 229–35.

Ørskov F (1984) Genus 1. *Escherichia* Castellani and Chalmers 1919, 941AL, in N R Krieg, and J G Holt (eds), *Bergey's Manual of Systematic Bacteriology Volume 1*, Williams and Wilkins, Baltimore, MD, 420–23.

Orta-Ramirez A, Price J F, Hsu Y-C *et al.* (1997) Thermal inactivation of *Escherichia coli* O157:H7, *Salmonella senftenberg*, and enzymes with potential as time–temperature indicators in ground beef, *Journal of Food Protection*, **60**(5), 471–5.

Oyarzábal O A, Nogueira M C L and Gombas D E (2003) Survival of *Escherichia coli* O157:H7, *Listeria monocytogenes*, and *Salmonella* in juice concentrates, *Journal of Food Protection*, **66**(9), 1595–8.

Pearson G R, Bazeley K J, Jones J R *et al.* (1999) Attaching and effacing lesions in the large intestine of an eight-month-old heifer associated with *Escherichia coli* O26 infection in a group of animals with dysentery, *Veterinary Record*, **145**, 370–73.

Pebody R G, Furtado C, Rojas A *et al.* (1999) An international outbreak of vero cytotoxin-producing *Escherichia coli* O157 infection amongst tourists; a challenge for the European infectious disease surveillance network, *Epidemiology and Infection*, **123**, 217–23.

Pennington T H (1997) *The Pennington Group: Report on the circumstances leading to the 1996 outbreak of infection with* E. coli *O157 in Central Scotland, the implications for food safety and the lessons to be learned*, The Stationery Office Ltd, Edinburgh.

Pennington T H (2009) The Public Inquiry into the September 2005 outbreak of *E. coli* 0157 in South Wales. Available of http://www.ecoliinquiry.wales.org, accessed March 2009.

Peterson R E, Klopfenstein T J, Moxley R A *et al.* (2007a) Efficacy of dose regimen and observation of herd immunity from a vaccine against *Escherichia coli* O157:H7 for feedlot cattle, *Journal of Food Protection*, **70**(11), 2561–7.

Peterson R E, Klopfenstein T J, Moxley R A *et al.* (2007b) Effect of a vaccine product containing Type III secreted proteins on the probability of *Escherichia coli* O157:H7 fecal shedding and mucosal colonization in feedlot cattle, *Journal of Food Protection*, **70**(11), 2568–77.

Phebus R K, Nutsch A L, Schafer D E *et al.* (1997) Comparison of steam pasteurisation and other methods for reduction of pathogens on surfaces of freshly slaughtered beef, *Journal of Food Protection*, **60**(5), 476–84.

PHLS (2000) Guidelines for the control of infection with Vero cytotoxin producing *Escherichia coli* (VTEC), Subcommittee of the PHLS Advisory Committee on Gastrointestinal Infections, *Communicable Disease and Public Health*, **3**(1), 14–23.

PHLS (2004) Preventing person-to-person spread following gastrointestinal infections: guidelines for public health physicians and environmental health officers. Prepared by a Working Group of the former PHLS Advisory Committee on Gastrointestinal Infections, *Communicable Disease and Public Health*, **7**(4), 362–84.

Reilly W (2001) *Task Force on E. coli O157. Final Report June 2001*, Food Standards Agency (FSA) Scotland and Scottish Executive Health Department, Edinburgh.

Riordan D C R, Duffy C, Sheridan J J *et al.* (1998) Survival of *Escherichia coli* O157:H7 during the manufacture of pepperoni, *Journal of Food Protection*, **61**(2), 146–51.

Samadpour M, Barbour M W, Nguyen T *et al.* (2006) Incidence of Enterohaemorrhagic *Escherichia coli*, *Escherichia coli* O157, *Salmonella*, and *Listeria monocytogenes* in retail ground beef, sprouts, and mushrooms, *Journal of Food Protection*, **69**(2), 441–3.

Schimmer B (2006) Outbreak of haemolytic uraemic syndrome in Norway: update. *Eurosurveillance*, **11**(4), available at: http://www.eurosurveillance.org/ViewArticle.aspx?ArticleId=2938, accessed November 2008.

Schlesser J E, Gerdes R, Ravishankar S *et al.* (2006) Survival of a five-strain cocktail of *Escherichia coli* O157:H7 during the 60-day aging period of Cheddar cheese made from unpasteurized milk, *Journal of Food Protection*, **69**(5), 990–8.

Seeman B K, Sumner S S, Marini R *et al.* (2002) Internalization of *Escherichia coli* in apples under natural conditions, *Dairy, Food and Environmental Sanitation*, **22**(9), 667–73.

Seongbeom C, Bender J B, Diez-Gonzalez F *et al.* (2006) Prevalence and characterization of *Escherichia coli* O157 isolates from Minnesota dairy farms and county fairs, *Journal of Food Protection*, **69**(2), 252–9.

Sharma M, Adler B B, Harrison M D *et al.* (2005) Thermal tolerance of acid-adapted and unadapted *Salmonella*, *Escherichia coli* O157:H7, and *Listeria monocytogenes* in cantaloupe juice and watermelon juice, *Letters in Applied Microbiology*, **41**(6), 448–53.

Silvestro L, Caputo M, Blancato, S *et al.* (2004) Asymptomatic carriage of verocytotoxin-producing *Escherichia coli* O157 in farm workers in Northern Italy, *Epidemiology and Infection*, **132**, 915–9.

Simmons M (2005) *South Wales* E. coli *O157 outbreak – September 2005*, a Review commissioned by the Chief Medical Officer, Office of the Chief Medical Officer, Welsh Assembly Government, Cardiff.

Shillam P, Woo-Ming A, Mascola L *et al.* (2002) Multistate outbreak of *Escherichia coli* O157:H7 infections associated with eating ground beef – United States, June–July 2002, *Morbidity and Mortality Weekly Report*, **51**(29), 637–9.

Smith H W (1961) The development of the bacterial flora of the faeces of animals and man: the changes that occur during ageing, *Journal of Applied Bacteriology*, **24**(3), 235–41.

Smith H R, Cheasty T and Roberts D *et al.* (1991) Examination of retail chickens and sausages in Britain for Vero cytotoxin-producing *Escherichia coli, Applied and Environmental Microbiology*, **57**(7), 2091–3.

Solecki O, MacRae M, Ogden I *et al.* (2007) Can the high levels of human verocytotoxigenic *Escherichia coli* O157 infection in rural areas of NE Scotland be explained by consumption of contaminated meat? *Journal of Applied Microbiology*, **103**(6), 2616–21.

Solomon E B, Pang H-J and Matthews K R. (2003) Persistence of *Escherichia coli* O157:H7 on lettuce plants following spray irrigation with contaminated water, *Journal of Food Protection*, **66**(12), 2198–202.

Sproston E L, Macrae M, Ogden I D *et al.* (2006) Slugs: potential novel vectors of *Escherichia coli* O157, *Applied and Environmental Microbiology*, **72**(1), 144–9.

Stephan R, Ragettli S and Untermann F (2000) Prevalence and characteristics of verotoxin-producing *Escherichia coli* (VTEC) in stool samples from asymptomatic human carriers working in the meat processing industry in Switzerland, *Journal of Applied Microbiology*, **88**, 335–41.

Stirling A, McCartney G, Ahmed S *et al.* (2007) An outbreak of *Escherichia coli* O157 phage type 2 infection in Paisley, Scotland. Eurosurveillance. **12**(34), available at: http://www.eurosurveillance.org/ViewArticle.aspx?ArticleId=3253, accessed November 2008.

Stringer S C, George S M and Peck M W (2000) Thermal inactivation of *Escherichia coli* O157:H7 *Journal of Applied Microbiology* (Symposium Supplement) **88**, 79–89.

Sussman M (1997) *Escherichia coli* and human disease, in M Sussman (ed.), Escherichia coli: *Mechanisms of Virulence*, Cambridge University Press, Cambridge, 3–48.
Syed Q (2008) *Outbreak Report. National outbreak of vero cytotoxin-producing* Escherichia coli *O157 infection associated with lemon and coriander chicken wraps in England & Wales: June – July 2007*, Outbreak Control Committee, Health Protection Agency, London.
Taormina P J and Beuchat L R (1999) Behaviour of enterohemorrhagic *Escherichia coli* O157:H7 on alfalfa sprouts during the sprouting process as influenced by treatments with various chemicals, *Journal of Food Protection*, **62**(8), 850–6.
Taormina P J, Beuchat L R and Slutsker L (1999) Infections associated with eating seed sprouts: an international concern, *Emerging Infectious Diseases*, **5**(5), 626–34.
Tuttle J, Gomez T, Doyle M P *et al.* (1999) Lessons from a large outbreak of *Escherichia coli* O157:H7 infections: insights into the infectious dose and method of widespread contamination of hamburger patties, *Epidemiology and Infection*, **122**(2), 185–92.
USDA (1998) *Guide to minimize microbial food safety hazards for fresh fruits and vegetables*, United States Department of Agriculture, Washington, DC, available at, http://vm.cfsan.fda.gov/~dms/prodguid.html, accessed December 2008.
Vaillant V (2005) Outbreak of *E. coli* O157:H7 infections associated with a brand of beefburgers in France, *Eurosurveillance*, **10**(44), available at: http://www.eurosurveillance.org/ViewArticle.aspx?ArticleId=2825, accessed November 2008.
Wang G and Doyle M P (1998) Heat shock response enhances acid tolerance of *Escherichia coli* O157:H7, *Letters in Applied Microbiology*, **26**, 31–4.
Warriner K, Spaniolas S, Dickinson M *et al.* (2003) Internalization of bioluminescent *Escherichia coli* and *Salmonella* Montevideo in growing bean sprouts, *Journal of Applied Microbiology*, **95**, 719–27.
Williams A P, Avery L M, Killham K *et al.* (2005) Persistence of *Escherichia coli* O157 on farm surfaces under different environmental conditions, *Journal of Applied Microbiology*, **98**(5), 1075–83.
Willshaw G A, Cheasty T and Smith H R (2000) *Escherichia coli*, in B M Lund, T C Baird-Parker and W G Gould (eds), *The Microbiological Safety and Quality of Food*, Volume II, Aspen, Gaithersburg, MD, 1136–77.
Wilson G S and Miles A A (1964) *Topley and Wilson's Principles of Bacteriology and Immunity*, Volume 1, 5th edn, Arnold, London, 806–26.
Wright J R, Sumner S S, Hackney C R *et al.* (2000) A survey of Virginia apple cider producer practices, *Dairy, Food and Environmental Sanitation*, **20**, 190–5.

18

Salmonella

C. Bell, Independent Consultant Microbiologist, UK, and A. Kyriakides, Sainsbury's Supermarkets Ltd, UK

Abstract: Based on their extensive experience working within the food industry and supported by published reports and research work, the authors describe in sections titled with self-explanatory headings the key characteristics of *Salmonella* spp. that help to explain why it is an important foodborne pathogen and approaches to the monitoring and control of the organism in food production processes.

Key words: *Salmonella*, salmonellosis, bacterial pathogen, foodborne illness, source, characteristics, growth, survival, control in foods, hazard analysis, risk factors, food processes, detection, meat and meat products, milk and milk products, fruit, vegetables.

18.1 Introduction

Salmonella species have been recognised for over 100 years as the cause of illnesses ranging from mild to severe food poisoning (gastroenteritis), and even more severe typhoid (enteric fever), paratyphoid, bacteraemia, septicaemia and a variety of associated longer-term conditions (sequelae). Some of these severe conditions can incur high rates of morbidity and mortality and can occur in outbreaks involving large numbers of people, particularly in relation to typhoid outbreaks and septicaemic conditions.

From the time of probably the first laboratory confirmed outbreak of salmonellosis in 1888, when Gaertner isolated *Bacterium enteritidis* (later renamed *S. enteritidis*) from the meat of a cow and the organs of a man who consumed a large portion of the meat and who subsequently suffered fatal food poisoning (Topley and Wilson, 1929), *Salmonella* spp. have been considered some of the most important of the causal agents of foodborne illness throughout the world. The large number of outbreaks of foodborne salmonellosis that still occur in many countries are testimony to the importance of this bacterial genus in terms of morbidity and mortality.

Many foodborne incidents of salmonellosis are recognised as being sporadic in nature, occurring as apparently isolated cases. Improved methods of investigation of foodborne disease together with advancements in the collection and sharing of food poisoning statistics among developed countries has now allowed many of the sources of these organisms to be identified. This has also demonstrated quite clearly the importance of *Salmonella* spp. as some of the main causative agents of foodborne disease outbreaks, some of which have affected large numbers (up to many thousands) of individuals from a single food source.

Although a great deal is already known about *Salmonella* spp., these organisms continue to provide new challenges to food safety, particularly because of the evolution of new strains resulting from the acquisition of genes conferring characteristics such as multiple antibiotic resistance. There is, therefore, a continuing need for research and information concerning the origins and characteristics of these new strains so that practical control measures can be applied in the food industry. In addition, because of the ubiquitous nature of *Salmonella*, there is a need to further develop new production and processing methods to assist in the control of all *Salmonella* from all sources.

As for all foodborne bacterial pathogens, it is essential that a detailed and competent hazard analysis is carried out at an early stage in all food product and process developments to ensure that relevant critical controls and monitoring systems can be put in place. This will help to minimise potential public health problems that could arise from the presence and outgrowth of *Salmonella*. This chapter aims to give an overview of *Salmonella* spp. in respect of the hazard they present in food products and the means for controlling these organisms.

18.2 Characteristics of *Salmonella*

18.2.1 The organism

In 1885, an American bacteriologist D.E. Salmon characterised the hog cholera bacillus causing 'swine plague' which, at that time, was named *Bacterium suipestifer* but later renamed as the type species of the genus named after him, *Salmonella cholerae-suis*. It was not until the 1960s, however, that the name *Salmonella* became the widely accepted name for this genus of the family Enterobacteriaceae.

Salmonella spp. are facultatively anaerobic, Gram-negative, straight, small (0.7–1.5 × 2.0–5.0 μm) rods, which are usually motile with peritrichous flagella. Table 18.1 shows some key biochemical characteristics of *Salmonella*.

Table 18.1 Biochemical characteristics of *Salmonella*

Characteristic	Usual reaction
Catalase	+
Oxidase	–
Acid produced from lactose	–
Gas produced from glucose[a]	+
Indole	–
Urease produced	–
Hydrogen sulphide produced from triple-sugar iron agar	+
Citrate utilised as sole carbon source[a]	+
Methyl Red	+
Voges–Proskauer	–
Lysine decarboxylase	+
Ornithine decarboxylase	+

+ = positive reaction; – = negative reaction.
[a] an important exception is Typhi which is negative in these tests.
Source: Bell and Kyriakides (2002) adapted from Brenner (1984) and Le Minor (1984).

Strains of *Salmonella* are antigenically distinguishable by agglutination (formation of aggregates/clumping) reactions with homologous antisera, and the combination of antigens possessed by each strain, referred to as the antigenic formula, is unique to each *Salmonella* serotype. Extensive work on differentiation of the antigens on the surface of the bacterial cell where these are present, i.e. O, the somatic or outer membrane antigens; H, the flagella antigens and Vi, the capsular antigens, has led to the current recognition of around 2500 serotypes, a number that is increasing every year. Additional methods for differentiating strains of *Salmonella* serotypes are also used and phage typing (determined from the sensitivity of cells to the lytic activity of selected bacteriophages) has become a valuable method of particular importance in epidemiological studies and in tracing foodborne outbreaks of salmonellosis.

Pulsed field gel electrophoresis (PFGE) is a technique that is now used extensively in investigations of salmonellosis outbreaks to form associations between the *Salmonella* spp. isolated from the food, patient, environment or animal. PFGE involves fragmentation of the DNA of an organism using restriction enzymes that cut the DNA at specific base pairs resulting in a series of DNA fragments. These are separated using an electric current in a gel and result in sequences of like-size concentrating in regions of the gel as bands. The pattern formed by the bands, i.e. the number and distance between each band on the gel, represents a PFGE type and allows differentiation of *Salmonella* sub-types e.g. Typhimurium.

In recent years, the nomenclature of *Salmonella* has been the subject of considerable debate (Old, 1992; Old and Threlfall, 1997; Threlfall *et al.*, 1999), and most *Salmonella* serotypes are now designated to one species,

S. enterica. Six sub-species of *S. enterica* have been recognised, and one, *S. enterica* subsp. *bongori*, is now considered a distinct species, *S. bongori* (Tindall *et al*., 2005). Most of the previously named *Salmonella* species associated with foodborne illness belong to the new sub-species *enterica*, i.e. *Salmonella enterica* subsp. *enterica*. Owing to the long names now assigned to the organisms, i.e. *Salmonella enterica* subsp. *enterica* serotype Typhimurium (previously *S. typhimurium*), these are now shortened to *S*. Typhimurium or, indeed, Typhimurium. The current practical nomenclature style of serotype names will be used in this chapter together with any additional descriptive information that is supplied for some of the more predominant *Salmonella* serotypes such as provisional phage type (PT) or definitive phage type (DT), e.g. *S*. Typhimurium DT104.

A wide variety of animals and birds, both wild and domestic types, as well as amphibians and reptiles carry *Salmonella* spp. in their gastrointestinal tract and these are the primary sources of foodborne salmonellosis for humans. Some *Salmonella* serotypes are specifically host-related, for example, *S*. Gallinarum and poultry, *S*. Choleraesuis and swine and *S*. Typhi and *S*. Paratyphi A and humans. Some strains, however, are more infectious to some animals, e.g. *S*. Dublin to cattle but they may still cause infections in humans. Uzzau *et al*. (2000) advocated referring to those serotypes that are almost exclusively associated with one particular host species as host restricted, e.g. *S*. Gallinarum and *S*. Typhi; serotypes that are prevalent in one host species but that can cause disease in other host species are referred to as host-adapted, e.g. *S*. Dublin. Serotypes that cause infection in a broad range of unrelated species are referred to as unrestricted, e.g. *S*. Typhimurium and *S*. Enteritidis.

18.2.2 The illness caused

Salmonella spp. cause illness by means of infection. They multiply in the small intestine, colonising and subsequently invading the intestinal tissues, producing an enterotoxin and causing an inflammatory reaction and diarrhoea. The organisms can get into the blood stream and/or the lymphatic system and cause more severe illnesses. Table 18.2 summarises the types of illness caused by different *Salmonella* serotypes. Although most people are vulnerable to salmonellosis, certain groups are more vulnerable to infection; young children, the elderly and those with underlying chronic illness or immunocompromised individuals being particularly vulnerable to salmonellosis.

Although *Salmonella* can persist in faeces during convalescence following illness, excretion usually declines and eventually ceases. However, in a few cases, excretion can persist such that the individual becomes a chronic carrier of the organism. In such cases, the organism may be excreted only intermittently, but it makes such individuals, particularly if employed as a food handlers, a potential hazard to food safety and public health. Deaths resulting from salmonellosis (mild to moderate gastroenteritis) in outbreaks

Table 18.2 Illnesses caused by *Salmonella*

Illness	Serotypes involved	Infective dose/cause	Characteristics of the illness
Gastroenteritis	Mainly members of *S. enterica* subsp. *enterica* (the majority of serotypes are in this sub-species, e.g. Agona, Dublin, Hadar, Enteritidis, Poona, Typhi, Typhimurium, Virchow). Also, members of *S. enterica* subsp. *arizonae*	Usually high numbers (> 10 000 cells) required to cause illness but where organisms are protected, e.g. in high fat foods, low numbers (< 100 cells) may cause illness.	Incubation 6–72 h, commonly 12–36 h. Lasts 2–7 d. Symptoms: diarrhoea (dehydration if this is severe), abdominal pain, vomiting, fever, sometimes fatal. Prolonged excretion may occur.
Enteric fever	*S.* Typhi and *S.* Paratyphi	Infective dose may be < 1000 cells.	Incubation 7–28 d, average 14 d. Typhoid. High fever, malaise, nausea, abdominal pain, anorexia, delerium, constipation in the early stages, diarrhoea in the late stages. Convalescence may take up to 8 weeks. Carrier state can last several months to years.
Septicaemia or bacteraemia	Members of *S. enterica* subsp. *enterica*	Caused when *Salmonella* are present in the bloodstream.	High fever, malaise, pain in the thorax and abdomen, chills and anorexia.
Sequelae	Members of *S. enterica* subsp. *enterica*	Morbid condition that occurs as the result of a previous disease, e.g. salmonellosis	Uncommon. A variety have been identified including arthritis, osteoarthritis, appendicitis, endocarditis, pericarditis, meningitis, peritonitis and urinary tract infections.

Source: adapted from Bell and Kyriakides (2002) using information from PHLS (2004) and ICMSF (1996).

are rare although they do occur, and normally healthy individuals usually recover from salmonellosis with supportive treatment including fluid and electrolyte replacement and without the need for antibiotics.

In developing countries, infection by *Salmonella* more commonly results in severe gastroenteritis, and in some outbreaks, up to 40 % of cases may develop septicaemia and 30 % of cases may die. Multiple-antimicrobial agent resistant strains of *Salmonella* are often involved in these outbreaks (Old and Threlfall, 1997). Strains of *S.* Typhimurium DT104 commonly resistant to at least five and sometimes more antibiotics have regularly caused food-associated outbreaks of severe gastrointestinal infection (Anon., 2000; Besser *et al.*, 2000).

The increasing occurrence of antibiotic-resistant *Salmonella* in cases of human illness is now of major concern. Analysis of *S.* Typhimurium isolates ($n = 5563$) from cases of salmonellosis occurring in European Union member states in 2006 revealed 39.7 % to be resistant to four or more antimicrobials (Anon., 2007a). In contrast, multiple antimicrobial resistance in human isolates of *S.* Enteritidis ($n = 20148$) was reported to be 0.7 %. There has been much discussion about the factors contributing to the development and spread of antibiotic resistance in enteric pathogens such as *Salmonella* spp., including the subject of indiscriminate use of antibiotics in human and animal medicine, either prophylactically or therapeutically, and the mechanisms, transmission and reasons for antibiotic resistance in *Salmonella* spp. have been extensively reviewed by Alcaine *et al.* (2007).

18.2.3 Growth and survival

Once soil and environmental waters are contaminated with salmonellas, the organism can survive and these environments in turn can then play a significant part in spreading the organism, directly or indirectly, to food animals and crops. Studies of the survival in different environments of *Salmonella* spp. derived from varied sources have been reported periodically.

From a study of the prevalence of *Salmonella* spp. in urban wild brown rats, Hilton *et al.* (2002) reported 8 % (8/100) of fresh or moist faecal samples were positive for the organism which could also be recovered for up to 86 d in drying rat faeces exposed to an indoor environment. *Salmonella* spp. can be readily isolated from animal farm environments; Rodriguez *et al.* (2006) examined nearly 2500 samples over a two-year period from 18 farms in five US states including pig farms, poultry farms, dairy and beef cattle farms and reported *Salmonella* prevalence from farms of 57.3 %, 16.2 %, 17.9 % and 8.5 %, respectively. Such results suggest that a considerable amount of *Salmonella*-contaminated farm animal waste is regularly produced. Studies of the fate of salmonellas in soils that have been treated with contaminated manures or irrigation water and on food plants grown in such soils demonstrate that these organisms can survive for longer than 220 d in salmonella-contaminated soils (Islam *et al.*, 2004; You *et al.*, 2006) and on the food plants (carrots and radishes) for the full duration of their growing time in the soils (Islam *et al.*, 2004).

In a study of the rearing environment of organically produced pigs, *S.* Typhimurium was shown to persist in the soil and water in the paddock environment for several weeks, and introducing *Salmonella*-negative pigs into paddocks that were contaminated with *Salmonella* caused some infections in the pigs (Jensen *et al.*, 2006). Moore *et al.* (2003) reported that a strain of *S.* Typhimurium inoculated into aquaria survived (culture-positive samples) in the water for at least 28 d and in the sediment for over 100 days. Salmonellas have also been shown, in a number of studies, to survive for longer periods in manure/slurry, soil and water at temperatures below 10 °C, but populations decline more quickly at temperatures above 20 °C, e.g. Budzińska, 2005.

It is clear that environments that are contaminated with salmonellas from whatever source present a contamination risk to growing crops and farm animals and on into the food raw material supply chain and food manufacturing processes.

In any food production process, attention to the detail of in-process hygiene systems and procedures is important in the control of extraneous microbiological contamination. In addition, methods used in the actual manufacturing and handling processes and also the formulation of food products can be important in the control of any residual *Salmonella* in foods and in preventing their potential to cause illness in consumers. Table 18.3 indicates some of the key growth-limiting parameters for *Salmonella* of particular importance when considering matters of food safety.

Salmonella is readily destroyed by heat in foods with a high water activity, e.g. ≥ 0.98, but in foods materials with a low water activity, e.g. high fat content, much higher temperatures are needed to kill the organism (Table 18.4). The possibility of cross-protection to heat treatments has been reported whereby exposure of the organism to acidic conditions results in increased thermal tolerance (Sharma *et al.*, 2005). It is important to note, however, that in the case of acids, lethality at pasteurisation temperatures is enhanced in the presence of increasing acidity. Nevertheless, where only relatively mild heat treatments may be used, e.g. pasteurisation of fruit juices

Table 18.3 Limits for the growth of *Salmonella* under otherwise optimal conditions

Parameter	Minimum	Maximum
Temperature (°C)	5.2[a]	46.2
pH	3.8[b]	9.5
Water activity	0.94	> 0.99

[a] Most serotypes will not grow at < 7.0 °C.
[b] Most serotypes will not grow below pH 4.5.
Source: Bell and Kyriakides (2002) adapted from ICMSF (1996).

Table 18.4 *D*-values[a] for *Salmonella* in different substrates

Food substrate/conditions	Temperature (°C)	*D*-value	Reference
Aqueous sucrose solution at pH 6.9	57.2		Goepfert *et al.*, 1970
Water activity			
0.99		0.8–1.1 min	
0.90		21.1–80 min	
(six strains of *Salmonella* – not Senftenberg)			
Milk (sterile, homogenised)	68.3	0.28–10 s depending on serotype	ICMSF, 1996
Ground beef	63	0.36 min	
Milk chocolate	71	4.5–6.6 h, depending on serotype	Lee *et al.*, 1989
Liquid whole egg	60	0.55–9.5 min depending on pH and serotype	D'Aoust, 1989

[a]*D*-value: the time required to reduce the number of cells to 10% of the initial population.

or cured meat products, account may need to be taken of the possibility of induced enhanced heat tolerance to ensure pathogenic microorganisms are eliminated by the processes applied. In frozen foods or those in which the water activity is low, *Salmonella* has been shown to survive for many months or even years (Table 18.5).

Although the growth of *Salmonella* is believed to be controlled by low temperatures (refrigeration), and industry relies heavily on refrigerated storage of fresh foods to maintain their safety in relation to pathogen outgrowth, there have been some reports (summarised by D'Aoust, 1991) of growth of *Salmonella* in shell eggs at 4 °C and in minced meat and chicken parts at 2 °C. Some of these reports are, as yet, unconfirmed other than by observation of growth on microbiological media (ICMSF, 1996).

It is of concern to note that some studies have shown that *S.* Enteritidis PT4 and *S.* Typhimurium DT104 increase in biomass in milk, chicken and microbiological media at 4 °C and in matrices at low water activities (0.93–0.98) due to filament formation rather than cell division (Phillips *et al.*, 1998; ILSI, 2000; Mattick *et al.*, 2000). Kieboom *et al.* (2006) reported that filamentous cells of *S.* Enteritidis induced through exposure to low water activity conditions were more resistant to sodium hypochlorite disinfection than control cells. Such findings may have implications for the disinfection procedures used in food processing units. Further work needs to be done to understand filament formation and its wider effects on the cell's ability to survive adverse conditions and also to determine if there are any public health implications arising from this phenomenon.

Salmonellas have been shown to form biofilms on a variety of surface types commonly used in food manufacturing and catering units including metal and plastic. Stepanović *et al.* (2004) reported there was no difference

Table 18.5 Some survival data relating to different *Salmonella* serotypes

Condition	Serotype	Food	Temperature (°C)	Survival time	Reference
Freezing temperatures	Enteritidis	Poultry	–18	4 months	D'Aoust, 1989
	Cholerae-suis	Minced beef	–18	4 months	
	Typhimurium	Chow-mein	–25	9 months	
	Enteritidis and Typhimurium	Ice cream	–23	7 years	
Low water activity	Typhimurium PT10	Cheddar cheese	5	8 months	D'Aoust *et al.*, 1985
	Naturally contaminated with Typhimurium, Java, Blockley	Milk powder product samples		Up to 10 months	Ray *et al.*, 1971
	Infantis and Typhimurium	Pasta		Up to 12 months	Rayman *et al.*, 1979
	Eastbourne	Milk chocolate (a_w 0.41)		> 9 months	Tamminga *et al.*, 1976
	5-serotype culture: Agona Enteritidis Michigan Montevideo Typhimurium	Peanut butter (a_w 0.2–0.33)	5 21	up to 24 weeks up to 6 weeks	Burnett *et al.*, 2000

among the more than 100 strains of *Salmonella* they tested in the quantity of biofilm formed on plastic. Experimental work reported also indicates that *Salmonella* can form biofilms on the surface (rind) of cantaloupe melons (Annous *et al*., 2005). Microbial biofilms can afford protection to the organisms within them from disinfection treatments carried out during cleaning processes in food manufacturing units and kitchens. It is therefore important that cleaning regimes are rigorously applied to ensure that biofilms do not form and inhibit cleaning efficacy.

In routine application, temperature, pH, water activity and other physicochemical factors, used either singly or in combination, are still used effectively in the control of the survival and growth of *Salmonella* during manufacturing processes and also in the finished food products (Bell and Kyriakides, 2002) (Table 18.3).

18.3 Risk factors for *Salmonella*

Microbiological risk assessment in relation to food safety has become an 'industry' on its own now with research, publications and commercial activity all contributing to sustaining it. Common sense dictates that raw foods cannot be produced entirely free from pathogenic microorganisms. The primary environmental sources of raw food materials (meats, poultry, fish, cereals, vegetables, herbs, spices, seeds and fruits) means they will be exposed to combinations of animal and human activity, soil and water, insects, birds, air, etc. whether produced and harvested commercially or domestically. The ability of *Salmonella* spp. to survive in different environments has been discussed in Section 18.2.3.

All raw food materials must be expected, therefore, to 'carry' potentially harmful organisms at some (usually) low incidence and level, and in cases where the foodstuff is consumed raw or after only minimal 'processing' such as washing, e.g. as for many fruits and salad vegetables, a certain risk to the health of individual consumers will exist (Bell, 1997a, b). Despite this, most outbreaks of foodborne illness that have occurred were preventable. There is no complicated formula involved in generally providing a microbiologically safe food supply as the basic requirements for safe food production and supply are known and are relatively straightforward. Food primary producers and manufacturers should ensure they employ the necessary skills and apply the relevant knowledge required for producing safe foods and be able to demonstrate that critical procedures are identified and controlled. They should provide caterers and consumers with clear information about the nature of the foods offered and how to store, handle and prepare them safely for consumption. Caterers and consumers should apply common sense to food purchase and use and consistently operate good basic hygiene practices in the kitchen. There is a considerable body of information already available to draw on and apply in food production, food product development, food manufacture and supply operations to help ensure that the anticipated hazard of *Salmonella* spp. is minimised or even eliminated.

18.3.1 Human infection

The numbers of lower gastrointestinal isolates of *Salmonella* from humans, excluding *S.* Typhi and *S.* Paratyphi, recorded by the Public Health Laboratory Service for England and Wales (now the Health Protection Agency) between 1990 and 1997, ranged from 26 239 (1991) to 31 480 (1997) but fell to 23 164 in 1998 then to 16 958 in 1999 from which time, the numbers reported have remained consistently lower than approximately 16 000 per annum (12 542 cases in 2007; provisional figure 12 000) (HPA, 2008 and Anon., 2008a). These figures are recognised as being an under-estimate of the real numbers of cases occurring; for every confirmed case there are estimated to be a further 3.2 cases in the community (FSA, 2000). The more recent reduction is probably as a result of public health and industry measures to control *Salmonella*, particularly in raw foods such as eggs and poultry.

In an analysis of the 160 049 confirmed salmonellosis cases in the EU in 2006 (a drop from 173 879 in 2005), over 75 % were reported to be caused by *S.* Typhimurium and *S.* Enteritidis (Anon., 2007a). This represents a case rate of 34.6 per 100 000 population, with the rate in young children (0–4 years) of nearly 150 per 100 000, close to three times higher than the next highest rate (in 5–14 year olds) and five to seven times higher than age groups of 15 and over.

The infective dose for *Salmonella* has, for decades, been considered to be in excess of 100 000 cells, but a number of outbreaks have been recorded, particularly implicating products containing a high fat level, e.g. chocolate, cheese and salami, in which the infective dose was found to be very low, e.g. < 10–100 cells. Data from outbreaks have been used to develop more sophisticated dose–illness models which should aid in risk assessment (Bollaerts *et al.*, 2008).

The realisation that such low levels could cause illness led to a reappraisal of industry controls which has taken place over the past two decades. The development and application of Hazard Analysis Critical Control Point (HACCP) based assessments to each food process from raw materials through to expected consumer handling practices and the consequent implementation of relevant standards of agricultural practice and appropriate design and hygienic operation of all food production processes has helped to minimise the incidence and level of *Salmonella* at all stages of the food chain.

Outbreaks of salmonellosis have been caused by foods produced commercially, by private caterers, e.g. for special events, and by domestically produced foods, including hotels, homes for the elderly, hospitals, schools and individual households, and most, if not all, were preventable. Relatively simple actions such as proper attention paid to personal hygiene in the form of effective hand-washing before food preparation and immediately following a 'dirty' task are key to prevention of foodborne illness.

18.3.2 Foods

In the commercial food manufacturing industry and large catering companies, *Salmonella* is commonly included in buying specifications relating to raw materials and finished food products that may be at higher contamination risk from these organisms so that there is some evidence from specific microbiological examinations that can be used for supporting both hazard control and due diligence requirements. The organism is included in a number of microbiological food safety criteria written into legislation, e.g. for egg products, live bivalve molluscs and cooked molluscs and crustaceans, some milk-based products and minced meat and meat preparations as well as ready-to-eat sprouted seeds, pre-cut fruit and vegetables and unpasteurised fruit and vegetable juices (EC, 2005, 2007). In the UK, the Public Health Laboratory Service (PHLS) published microbiological guidelines (Gilbert *et al.*, 2000; HPA, 2009) to provide advice on the interpretation of results from certain types of microbiological examination of ready-to-eat food sampled from retail sale. Although having no statutory status, they may be used in prosecution cases by Environmental Health departments. If *Salmonella* is found to be present in 25 g of any of the product groups described, then the advice given is to consider the food 'unacceptable – potentially hazardous'. In the food manufacturing industry, confirmed detection of *Salmonella* in ready-to-eat products or those destined for vulnerable groups usually leads to complete removal of products from the distribution and retail system and, on occasion, to public recall and closure of the manufacturing line or unit pending full investigation and action (Bell and Kyriakides, 2002).

Although a wide variety of foods have been implicated in outbreaks of salmonellosis (Table 18.6), chicken and eggs, in particular, which are two of the most commonly consumed foods in many countries, are among the most common food types implicated in food poisoning outbreaks reported year after year in European countries, the USA and other industrialised countries (Evans *et al.*, 1998; Trepka *et al.*, 1999; Reporter *et al.*, 2000). *Salmonella* spp. were responsible for 53.9 % (3131) of all reported foodborne outbreaks in the EU in 2006 accounting for 22 705 cases with a hospital admission rate of 14 % and a total of 23 deaths (Anon., 2007a). Where the salmonellosis outbreak location was known (66 % of outbreaks), the majority were reported to have occurred in households (37.3 %; 1167 outbreaks) followed by catering, e.g. restaurants/cafés (6.8 %; 214 outbreaks). The foods most commonly implicated in salmonellosis outbreaks, where known (1902 individual outbreaks or 60.7 % of total outbreaks), were unsurprisingly eggs and egg products (1043 outbreaks), red meat products, e.g. sausage, ham, kebabs, etc. (210 outbreaks), dairy products (53 outbreaks), bakery products (43 outbreaks), broiler meat/products (41 outbreaks) and fish/fish products (38 outbreaks).

Table 18.6 Examples of food-associated outbreaks of illness caused by *Salmonella* Typhimurium

Salmonella type	Year	Country	Suspected food vehicle	Cases
Typhimurium PT8	1953	Sweden	Raw meat	8845
Typhimurium	1974	USA	Apple cider	> 200
Typhimurium PT204	1981	Scotland	Raw milk	654
Typhimurium PT10	1984	Canada	Cheddar cheese	> 1500
Typhimurium	1985	USA	Pasteurised milk	16 284
Typhimurium	1985	Switzerland	Vacherin Mont d'Or cheese	> 40
Typhimurium	1987	Norway and Finland	Chocolate products	361
Typhimurium DT124	1987–1988	UK	Salami sticks	101
Typhimurium DT49	1988	UK	Mayonnaise	> 76
Typhimurium DT12	1989	UK	Cooked meats	> 545
Typhimurium DT141 Enteritidis PT1A, PT4, PT24	1992	UK	Tiramisu (containing raw eggs)[a]	*c*. 98
Typhimurium	1995	Italy	Salami	83
Typhimurium 'DT12 atypical'	1997	France	Raw milk soft cheese – Morbier	113
Typhimurium DT104	1998	UK	Pasteurised milk	86
Typhimurium PT 135A	1999	Australia	Unpasteurised orange juice	> 400
Typhimurium DT104	2000	UK	Lettuce	361
Typhimurium DT104	2001	Australia, Canada, England and Wales, Sweden	Halva from Turkey	> 100
Typhimurium DT104	2004	UK	Possibly ready-to-eat foods	> 69
Typhimurium DT104A	2004	Italy	Traditional pork salami	63
Typhimurium DT12	2005	Denmark	Pork products	26

[a] Five separate outbreaks in the same year.

Sources: Adapted from Bell and Kyriakides (2002); Fisher *et al.* (2001); Horby *et al.* (2003); Luzzi *et al.* (2007); Wilson (2004); Torpdahl *et al.* (2006).

A growing number of different types of salad produce and also fruit have also been implicated in outbreaks of salmonellosis, particularly in the USA (Table 18.7) and nuts and nut products have also been increasingly associated with a number of different serotypes of *Salmonella* (Table 18.8).

It is now common to expect an increase in numbers of reports from investigations into *Salmonella* sources, prevalence, survival, food treatment approaches, etc. when outbreaks of salmonellosis have been associated with a food not previously or frequently implicated in such outbreaks, or when a specific strain of *Salmonella* becomes more frequently implicated

Table 18.7 Examples of produce-associated outbreaks of illness caused by different *Salmonella* serotypes

Salmonella serotype	Year	Country	Suspected food vehicle	Cases
Chester	1990	USA	Cantaloupe	295
Javiana	1990	USA	Fresh tomatoes	176
Poona	1991	USA	Cantaloupe	> 400
Montevideo	1993	USA	Tomatoes	100
Bovismorbificans	1994	Finland	Sprouts	210
Stanley	1995	USA and Finland	Alfalfa sprouts	242
Newport	1995	USA and Canada	Alfalfa sprouts	133
Agona	1995	UK, Canada, USA, Israel	Peanut flavoured corn snack	27 (UK) > 1000? (Israel)
Saphra	1997	USA	Cantaloupe	> 20
Poona	2000	USA	Cantaloupe	> 19
Enteritidis	2000	USA	Mung bean sprouts	45
Kottbus	2001	USA	Alfalfa sprouts	32
Newport	2005	USA (16 States)	Tomatoes	72
Braenderup	2005	USA (8 States)	Tomatoes	82
Newport	2006	USA (19 States)	Tomatoes	115
Typhimurium	2006	USA (21 States)	Tomatoes	190
Bareilly and Virchow	2006	Sweden	Mung beans	115
Weltevreden	2007	Norway, Denmark, Finland	Alfalfa sprouts	4 18 7
Stanley	2007	Sweden	Alfalfa sprouts	51
Java PT3b var 9 (Paratyphi B variant Java)	2007	UK, Sweden, Denmark, Finland, Norway, Netherlands, USA	Salad vegetables	20 172 15 1 10 2 1
Senftenberg	2007	UK, Denmark, Netherlands, USA	Fresh basil from Israel	30 (UK)
Litchfield	2008	USA, Canada	Honduran Cantaloupe melons	50 9

Source: Adapted from Bell and Kyriakides (2002); Anon. (2002); Anon. (2008b); Bidol *et al.* (2007); Denny *et al.* (2007); Emberland *et al.* (2007); de Jong *et al.* (2007); Killalea *et al.* (1996); Shohat *et al.* (1996); Pezzoli *et al.* (2007); Werner *et al.* (2007).

in outbreaks. This has occurred for example with *S.* Enteritidis and eggs, *S.* Poona and melons, salmonellas and fruit juices, and is now occurring in relation to outbreaks of salmonellosis implicating nut products (Danyluk *et al.*, 2007; Uesugi *et al.*, 2007; Eglezos *et al.*, 2008; Willford *et al.*, 2008) and seeds such as sesame seeds (Brockmann *et al.*, 2004).

It is clear from reports over the years that many types of food material,

Table 18.8 Some examples of nut and nut product-associated outbreaks of illness caused by different *Salmonella* serotypes

Salmonella serotype	Year	Country	Suspected food vehicle	Cases	Reference
Mbandaka	1996	Australia	Peanut butter	54	Ng *et al.*, 1996
Enteritidis PT 30	2000–01	Canada and USA	Raw whole almonds	157 (Canada) + 11 (USA)	Isaacs *et al.*, 2005
Stanley and Newport	2001	Australia, Canada, England and Wales, Scotland	Peanuts	97 Stanley 12 Newport	Kirk *et al.*, 2004
Stanley	2001	Australia, Canada, UK	Peanuts from China	100+	Little, 2001
Enteritidis	2003–04	USA and Canada	Raw almonds	> 29	Keady *et al.*, 2004
Enteritidis	2005–06	Sweden	Almonds	15	Müller *et al.*, 2007
Tennessee	2007	USA	Peanut butter	628	Anon, 2007b

including spices, seeds, nuts, cereal products, milk-based products, egg-based products, meat and poultry containing products, chocolate confectionery, fruit, fruit drinks, salad produce (Bell and Kyriakides, 2002 and Tables 18.6, 18.7 and 18.8), may become implicated in an outbreak of foodborne salmonellosis. National and international publications, such as the England & Wales Health Protection Agency's publication, the Health Protection Report, Eurosurveillance and the International Society for Infectious Diseases reports under ProMED-mail, relating to public health issues indicate clearly that outbreaks of salmonellosis involving many different food types and serotypes of the organism continue to occur on both a national and international scale.

The serotypes of *Salmonella* reported as the most prevalent causes of foodborne salmonellosis do change over the years. *S.* Typhimurium was the most common in the USA and UK up until the mid-1980s, then *S.* Enteritidis superceded *S.* Typhimurium because of a large rise, mainly in egg-related outbreaks, particularly involving *S.* Enteritidis PT4. Other serotypes have waxed and waned on the list of 'top ten' serotypes causing foodborne salmonellosis, and through epidemiological studies, particular food sources can be identified, e.g. in 1997, *S.* Enteritidis PT6 became the second most commonly reported phage type of *S.* Enteritidis in the UK (> 1600 cases) and was attributed to shell eggs and poultry meat. Since then, alongside the reduction in total cases of salmonellas reported, there has been a significant drop in reported cases involving *S.* Enteritidis, but this serotype still accounts for well over 50 % of the total cases reported (HPA, 2008). The most prevalent serotypes isolated from confirmed cases of salmonellosis in the EU and USA in 2006 are shown in Table 18.9.

Table 18.9 *Salmonella* serotypes in confirmed cases of salmonellosis in the EU and USA[a]

EU		USA	
Serovar	% cases	Serovar	% cases
S. Enteritidis	62.5	*S.* Typhimurium	19
S. Typhimurium	12.9	*S.* Enteritidis	19
S. Infantis	0.9	*S.* Newport	9
S. Virchow	0.7	*S.* Javiana	5
S. Newport	0.5	*S.* Montevideo	4
S. Hadar	0.5	*S.* Heidelberg	4
S. Stanley	< 0.5	*S.* I 4,5,12:i:-, 239	4
S. Derby	< 0.5		
S. Agona	< 0.5		
S. Kentucky	< 0.5		

[a]USA data from FoodNet which is based on active surveillance from 10 US states.
Source: Anon. (2007a); Vugia *et al.* (2007).

In more recent years, there has been an increase in the number of cases that were attributable to multiple antibiotic-resistant *S.* Typhimurium including DT104 involving meat products, raw milk and products made from raw milk as well as direct contact with livestock. In 2005, 41.7 % of the isolates of *S.* Typhimurium from human cases in EU member states (who reported data) were resistant to four or more antimicrobials (Anon., 2006) and this decreased slightly to 39.7 % in 2006 (Anon., 2007a).

Table 18.10 shows, from examples of a small number of outbreaks, how valuable information can be gained regarding factors that can be important for controlling the organism. Common underlying causes of foodborne outbreaks include:

- contamination of the primary food raw materials, often from direct or indirect animal faecal contamination or cross-contamination from a contaminated source;
- production processes with no stage that reduces or destroys the organism, e.g. cooking;
- product exposed to post-process contamination through poor personal or equipment hygiene practices;
- product consumed with no destruction/reduction process applied by the consumer, i.e. it is ready-to-eat.

Animals and meat

Salmonella spp. can cause severe infections in animals and, although some such strains are host-restricted, i.e. *S.* Gallinarum (poultry), many strains cause little illness in the animal which nevertheless still serves to amplify the organism until its passage to the human host, in which more severe illness can occur. A survey of rectal contents (1 g samples) of cattle and sheep presented at slaughter in Great Britain found only two out of 891 cattle and one out of 973 sheep rectal samples to be contaminated with *Salmonella* (Evans, 2000). The serotypes isolated from the cattle were *S.* Typhimurium DT193 and DT12 and from the sheep *S.* Typhimurium DT41. In contrast, a survey of caecal contents (10 g samples) from pigs taken in abattoirs in the UK found a 23 % incidence of *Salmonella* (Davies, 2000; Davies *et al.*, 2004), although clearly some of this difference must be attributable to the much larger sample size in comparison to the studies on cattle and sheep. The predominant isolates from the pig caeca were *S.* Typhimurium (11.1 %) and *S.* Derby (6.3 %) and the predominant phage types of *S.* Typhimurium were DT104 (21.9 % of isolates of Typhimurium from caeca) and DT193 (18.7 %).

Data from the EU Zoonoses report (Anon., 2007a) indicated that none of the 550 pooled faecal samples examined from 976 pig fattening herds in Sweden were positive. In the Danish surveillance targeted at pig herds with high or medium serological levels, 3.4 % ($n = 11\ 239$) of faecal samples from fattening herds were reported to be contaminated. In contrast, a Dutch survey of 100 faecal samples from pig holding units reported a 23 % incidence.

Table 18.10 Examples of foodborne outbreaks of illness caused by *Salmonella* and possible reasons for their occurrence

Outbreak	Organism	Product(s)	Possible reasons for occurrence
1973/74, USA/Canada	*S.* Eastbourne	Chocolate balls and other chocolate novelty confectionery	– Raw cocoa beans contaminated with *Salmonella* and subsequent inadequate heat processing – Inadequate segregation of roasted beans from raw beans – Heat processing of subsequent chocolate liquor not adequate to destroy *Salmonella* due to the low water activity of the liquor – Consumption by vulnerable groups
1985, England	*S.* Ealing	Spray dried milk powder used for infant formula	– Contaminated water used to test wet-down integrity of spray drier – Water and bacterial contaminant leakage into the insulation cavity of the spray drier and milk powder contamination of the insulation cavity via a hole in the inner skin – Contaminant proliferation in the insulation cavity and are re-introduction into the spray drier – Consumption of the product by vulnerable groups
1994, USA	*S.* Enteritidis	Commercially-produced ice cream	– Pasteurised ice cream premix transported in a trailer previously used for raw egg – Transport trailer inadequately cleaned/sanitised between loads – Raw material used in a product not subjected to a further pathogen destruction stage
1990 and 1993, USA	*S.* Javiana and *S.* Montevideo	Tomatoes	– Contamination of tomatoes in field from manure, irrigation water or wild animal activity and possible internalisation – Widespread cross-contamination of tomatoes in common water bath used in pack-house – Internalisation of *Salmonella* during washing due to temperature differential during washing (cold wash water and warm tomatoes) – Consumption as a ready-to-eat product

Source: Adapted from Bell and Kyriakides (2002); Corby *et al.* (2005); Hedberg *et al.* (1999).

Several countries also monitored *Salmonella* infection through lymph node sampling with incidence ranging from 0.1 % (fattening pigs in Sweden, *n* = 3153) to 58.8 % (slaughter batch in Italy, *n* = 68). The most commonly isolated serotypes were *S.* Typhimurium followed by *S.* Derby and *S.* Choleraesuis. Incidence data from cattle was limited to a few countries, but in general was very low (0.9 %, *n* = 9188, range 0–7.3 %, samples included faeces and lymph nodes). Serotypes included *S.* Typhimurium, *S.* Enteritidis and *S.* Dublin.

Although *Salmonella* is most often associated with poultry, the incidence of the organism in live birds is not well documented as most effort is directed towards assessing contamination of the flock through combined samples of litter, or of the finished product through carcass sampling. Contamination can spread quickly through a poultry house as high levels build up in the chicken intestine.

Both feed and water and external body surfaces of the bird, i.e. feathers, feet, can become contaminated with faeces, thus spreading infection. Some strains such as *S.* Enteritidis are also invasive and can spread to the reproductive tissues of the bird through which they can pass to the egg and chick. Although rather a selective study, the data from UK statutory reports of *Salmonella* isolates from chickens (layers, broilers and breeders) in 1998 show many exotic serotypes including *S.* Mbandaka (10 %), *S.* Senftenberg (8 %), *S.* Montevideo (8 %) as well as the more common *S.* Enteritidis (21 %) and *S.* Typhimurium (7 %) (MAFF, 2000). More concerted surveillance programmes to determine and monitor the prevalence of *Salmonella* at the farm level have been initiated within the EU by member states.

Data from EU member states on the incidence of *Salmonella* in farmed animals demonstrates that whilst the organism occurs infrequently in breeding flocks, it can be an occasional contaminant which, if not addressed at this stage, can result in widespread infection across the poultry industry as such flocks are used to populate laying or broiler farms (Anon., 2007a). Of parent breeding flocks for laying hen production, 2.2% (*n* = 2275) were reported to have a *Salmonella* isolation at some point during their rearing which was a reduction on previous years (6.1 % in 2005; 6.9 % in 2004). Fourteen of the 18 member states reporting did not find any positives with the incidence in the remaining four ranging from 1.5 to 5.8 %; two of these member states found *S.* Enteritidis and/or *S.* Typhimurium and the other two found other salmonellas. This resulted in a 4.8 % incidence in laying flocks (*n* = 27 108) ranging from 0 % to 31.2 %. *S.* Enteritidis and/or *S.* Typhimurium were the predominant serotypes and were identified in 15 of the 18 member states reporting a positive isolation in laying flocks.

Data for broiler flocks indicated a higher average incidence at the parent breeder level (5.2 %, *n* = 11 835, range 0–37.5 %) although the predominant serotypes were the same as for laying hen parent breeder flocks. In contrast, the incidence in broiler farms was lower (3.4 %, *n* = 72 861, range 0.2–66 %) and there was also a marked difference in the contaminating serotypes as, in general, *S.* Enteritidis and/or *S.* Typhimurium represented < 10 % of the isolates indicating that contamination at broiler level is likely to be coming from additional sources. In general, it was evident that countries that had low incidence at the breeder stage also had lower incidence at the layer or broiler stage.

Salmonella incidence in EU turkey production flocks was reported to be similar to that in chicken (5 %, n = 4646, range 0–14.7 %) although only 1.1 % was due to *S.* Enteritidis and/or *S.* Typhimurium. In contrast, the incidence in duck and geese, although from a much smaller dataset, was much higher with 44.4 % (n = 676) of duck production flocks and 10.4% (n = 1400) of geese production flocks contaminated. Like turkey, the incidence of *S.* Enteritidis and/or *S.* Typhimurium was low (4.1 % in ducks and 3.3 % in geese).

Raw poultry (including birds such as chicken, duck and turkey), beef, lamb and pork are all commonly contaminated with *Salmonella* and, not surprisingly, have all been implicated in outbreaks of salmonellosis. Although some contamination of the deep muscle of meat is possible, particularly with some invasive strains of *Salmonella*, this is the exception to the general rule that most contamination occurs on the surfaces of the raw meat/carcass. Contamination occurs through the transfer of faeces containing the pathogen to the muscle tissue during slaughter and subsequent processing. In the case of large animals, e.g. sheep and cattle, this can occur during any stage from evisceration through to hide removal and subsequently during de-boning and cutting into joints.

Milk and milk products

Milk and milk products are regularly implicated in outbreaks of salmonellosis. Contamination of raw milk with *Salmonella* usually occurs as a result of the transfer of faeces from the animal to the milk via unclean teats and udders. Such contamination can pass into the milk during milking and once present on milking parlour equipment can readily proliferate and spread if such equipment is not adequately cleaned and sanitised.

Notwithstanding the reports of growth/biomass increase at temperatures as low as 2–4 °C (D'Aoust, 1991; Phillips *et al.*, 1998; ILSI, 2000; Mattick *et al.*, 2000) growth in the milk should be limited by effective refrigeration (< 8 °C). Regular reports of cases or outbreaks of foodborne salmonellosis involving milk, however, indicate that milk production systems are not always as well-controlled as they could be and serve to underline the continuing importance of *Salmonella* contamination of milk and milk products as potential hazards presented by *Salmonella* spp. to human health.

Fruit, vegetables and other materials

A large number of outbreaks of salmonellosis have implicated fruit and vegetable produce as potential vehicles for *Salmonella*. These have focused attention on agricultural practices associated with their growth and harvesting.

Although the nature of the original contamination is not always clear in these outbreaks, it is evident that contamination of these types of primary agricultural products can occur from a number of sources. Firstly, these crops are grown in soil, and much soil has animal waste applied to it to add nutrients and provide conditioning material. If animal waste handling and application is under poor control it could lead to the deposition of microbial pathogens, present in the waste, onto the land. Then, depending on the conditions in the soil and environment and the subsequent time to harvest of the crop, such organisms may either perish, persist or even proliferate in the soil, thus offering various risks of cross-contamination to the harvested crop.

There are a large number of other possible food sources of *Salmonella*, including materials such as herbs and spices, often subject to growing, harvesting and drying in poorly controlled conditions, together with raw material seeds such as mung or alfalfa, dill, fennel, sesame and products of mixed seeds, a variety of shellfish, particularly those derived from aquaculture environments, as well as non-food cross-contamination sources such as rodents, reptiles, amphibians and wild or pet birds (Bell and Kyriakides, 2002).

18.4 Detecting *Salmonella*

It is generally understood that, if *Salmonella* is present in a food, because there are likely to be only low numbers of heterogeneously distributed cells within the batch, a large number of samples must be taken to gain a high degree of confidence in the detection of the organism (ICMSF, 1986). This is especially important when foods are destined for vulnerable groups, and it is common to find industry specifications requiring 'not detected in 250 g test portions' and several such test portions per batch being examined against this specification for materials such as milk powder destined to be used for baby food products. The buying specification for many other processed and ready-to-eat food materials and products such as herbs and spices, egg products, cooked meats and ready meals usually includes criteria to be met for *Salmonella*; a common target level applied is 'not detected in 25 g' of three or five 25 g test portions with no tolerance of acceptance, i.e. if one test portion is positive, the batch is withheld or rejected. Useful sampling plans and recommended microbiological limits for *Salmonella* have also been published in relation to some foods in international trade (ICMSF, 1986, 2007).

Reliable methods for detecting and identifying *Salmonella* are important in the support of properly developed and implemented HACCP-based systems for their control. Table 18.11 indicates the common conventional microbiological method, which is widely used by food microbiologists to isolate and identify *Salmonella*. There is also a wide range of microbiological techniques available for the detection and characterisation of *Salmonella*, and this range is increasing rapidly as the considerable investment in method/

Table 18.11 Conventional method for the isolation and identification of *Salmonella* spp.

Pre-enrichment[a]	
Inoculate pre-enrichment medium, e.g. buffered peptone water (1 part test portion + 9 parts medium)	Day 0
⇓	Incubate 37 °C/16–20 h
⇓	
Selective enrichment	
Sub-culture to 2 selective enrichment broths e.g. Muller-Kauffmann Tetrathionate broth (with novobiocin) (MKTTn) (1 + 9), Rappaport Vassiliadis Soya (RVS) broth (1+ 100)	Day 1
⇓	Incubate MKTTn 37 °C/24 h
⇓	RVS 41.5 °C/24 h
Selective plating	
Streak onto two selective agars, e.g. XLD Agar Brilliant Green Agar (modified) Hektoen Enteric Agar	Day 2
⇓	Incubate 37 °C/24 h
⇓	
Inspect plates for the presence of characteristic colonies and any primary biochemical reactions	Day 3
Confirmation of suspect colonies	
Purify suspect colonies on Nutrient Agar	
⇓	Incubate 37 °C/24 h
Serology 'O' & 'H'	Day 4
Inoculate media or test strips to obtain biochemical profile	
⇓	Incubate according to the manufacturer's
⇓	instructions, usually 37 °C/18–24 h
Read reactions	Day 5–6

[a] Alternative pre-enrichment systems may be required for some food types, e.g. chocolate and confectionery products, reconstituted skim milk powder (10 %w/v) with Brilliant Green dye (final concentration of 0.002 %w/v) is commonly used; products that may contain inhibitory substances or products that may be osmotically active, e.g. some herbs and spices (oregano, cinnamon, cloves), honey, 1 : 100 dilution in Buffered Peptone Water, e.g. 25 g + 2475 ml may be used.
Sources: Roberts and Greenwood (2003) and based on ISO (2002).

techniques development results in commercialisation of products. A valuable summary of methods together with validation information and key references has been prepared and is periodically updated by Campden BRI (Baylis and Mitchell, 2008).

There are a variety of miniaturised and automated biochemical kits/systems for distinguishing *Salmonella* from other members of the Enterobacteriaceae. However, some strains occasionally exhibit atypical biochemical reactions and some care is necessary, particularly when assessing cultures on selective plates, to ensure false negative results are minimised/avoided (Threlfall *et al.*, 1983). Because of this and also because different strains of *Salmonella* may be sensitive to different combinations of inhibitory substances and high incubation temperatures, combinations of selective enrichment broths, selective plating media and incubation conditions are used to increase confidence in results from tests for detecting *Salmonella.* It is important to ensure that the methods used for the detection of *Salmonella* in specific food types are fully validated.

Following confirmation of the identity of *Salmonella*, it is important for food surveillance purposes and the investigation of outbreaks to sub-type or 'fingerprint' *Salmonella* serotypes. There are a variety of techniques available now that can allow confident traceability of strains in factory environments. These include biotyping, serotyping (including variation in H antigens), phage typing, antibiotic resistance patterns (resistotyping), various molecular typing methods including PFGE (Old and Threlfall, 1997) and ribotyping (Jones, 2000). During an investigation of the European outbreak of illness caused by *S.* Typhimurium DT204b in 2000, a combination of phage typing, antibiotic resistance patterning, genomic DNA analysis by PFGE and calculation of plasmid DNA molecular masses allowed the isolates from the five countries involved to be confirmed as indistinguishable and therefore strongly suggesting a common source, although this was not convincingly identified (Crook *et al.*, 2003). The application of such fingerprinting tools early enough in investigations of cases and outbreaks can help to significantly reduce the number of people at risk by identifying a source and removing it from the food chain.

A wide variety of foods have been implicated in outbreaks of illness attributed to many different serotypes of *Salmonella.* Tables 18.6 and 18.7 give some indication of the wide involvement of *S.* Typhimurium in foodborne outbreaks and also the variety of *Salmonella* serotypes that have been implicated in outbreaks of illness associated with fruit, salad and vegetable products. In addition, outbreaks of salmonellosis have implicated foods and food materials as diverse as potato salad, mustard dressing, black pepper, roast cuttlefish, savoury corn snack (yeast-based flavouring), infant cereal and toasted oats cereal, peanut butter and coconut. These continuing outbreaks of *Salmonella* foodborne illness involving most food types have continued to fuel extensive work programmes by both researchers and food industry scientists to further improve methods of detection and develop new controls aiming to eliminate contamination of food by *Salmonella.*

18.5 Control of *Salmonella* in foods

As has already been noted in Section 18.3, a wide variety of food raw materials may be contaminated, albeit at low incidence and levels, by

Salmonella, and most food types have been implicated in outbreaks of salmonellosis. It is probable that detailed and structured hazard analysis followed by the implementation and maintenance of all relevant controls at the critical points identified would have prevented most of the outbreaks attributed to commercially-produced foods, and everyone involved in the primary production, processing and sale of food should adopt a hazard analysis approach to food safety considering all relevant pathogens, including *Salmonella*. Indeed, this is a requirement embodied in legislation in Europe (EC, 2004).

The International Commission on Microbiological Specifications for Foods, in a document prepared for the Codex Committee on Food Hygiene, indicated that, at present, *Salmonella* cannot be eliminated from most farms, slaughtering processes, raw milk, fruits and vegetables, although good hygienic practices and properly implemented HACCP systems may minimise or reduce levels of contamination (CAC, 1996). It is therefore prudent to expect that *Salmonella* will be present at some incidence and level in raw material foods and to ensure that adequate control measures are in place to prevent the organism from becoming a hazard to consumer health.

Whatever the *Salmonella* serotype, effective controls for controlling/eliminating the hazard of *Salmonella* in foods are the same and involve control of the following:

- raw materials;
- personal, equipment and environmental hygiene;
- manufacturing process conditions;
- post-process contamination;
- retail and catering practices;
- consumer handling.

18.6 Raw material control

Like many other enteric pathogens, contaminated raw materials are common routes by which *Salmonella* enters the food chain. Where such contaminated material is not subject to sufficient decontamination by the food processor or by the consumer, prior to consumption, or indeed, where cross-contamination occurs to ready-to-eat foodstuffs, outbreaks of food poisoning can occur. A wide range of contaminated raw materials have been implicated in outbreaks of salmonellosis, although initial contamination is often traced back to human or animal sources.

Key raw materials implicated in outbreaks include raw meat (poultry, cattle, pigs and sheep), raw milk, raw eggs, contaminated fruit and vegetables, contaminated water, seeds and nuts, and fish and shellfish. In most of these cases, it is cross-contamination of the organism from faeces (either animal or human) to the food material that presents the problem to the food processor or handler. Recognition of these sources of contamination and implementation of procedures to reduce or eliminate the organism are important components of any strategy to deal with *Salmonella*.

18.6.1 Control in animals

Control of *Salmonella* in food animals and poultry is extremely difficult and is affected by diet, health, environmental factors and exposure to the organism in the first place. Clearly, contaminated feed can be a major source of the organism to these animals and considerable effort is required to prevent such sources of contamination, particularly where feeds are derived from industrial processes such as feed mills. Use of heat-treated feed, adequately pasteurised and where the raw ingredients are properly segregated from the finished product is preferable, particularly for breeding stock animals.

The incidence of *Salmonella* in feeding stuffs in the EU demonstrates the potential for this to be a significant source of contamination during animal rearing (Table 18.12) although, in general, *S*. Enteritidis and/or *S*. Typhimurium were infrequently detected.

When grazing animals, it is clearly preferable to avoid using grazing land that may be contaminated with potential sources of pathogens such as human and animal wastes. Where these are to be applied, this should be done using properly composted manure or treated waste to prevent recycling of pathogens. Clearly, there is very little that can be done to prevent natural deposition of faeces by animals onto land, but if the animal is not infected in the first place then this source presents less of a risk.

Once introduced, *Salmonella* can spread quickly in conditions where animals are reared in close proximity, such as poultry in sheds, and every effort must be made to prevent the organism entering such sheds through control of the feed, water and environment. Pest control and control of human entry are key to a successful strategy for controlling *Salmonella*. Proper hand-washing and boot changing/disinfection are essential to keep

Table 18.12 Incidence of *Salmonella* in feeding stuffs reported by EU countries in 2006

Feedingstuff	Incidence of *Salmonella* (%)	Number of samples
Fishmeal	1.9	2 414
Meat and bone meal	2.3	12 350
Cereals	0.3	5 331
Oil seeds and products	2.5	18 449
Cattle feed	0.7	4 141
Pig feed	0.6	6 234
Poultry feed	0.8	13 819

Source: Adapted from Anon. (2007a).

the organism out of poultry houses. A strategy that has achieved successful reduction of *Salmonella* and *S.* Enteritidis in particular in recent years has been the introduction of vaccination. While this has been adopted widely for the laying flocks in the UK it has yet to gain a similar application to broiler flocks, which owing to the high cost per bird is unlikely to occur until the introduction of a live vaccine that could be given in the feed rather than the current dead vaccine injected into each bird.

Spread of the organism is also affected by animal husbandry and transportation practices where waste shed by one animal during transport to slaughter rapidly becomes spread to others owing to their close proximity in the vehicle, and in the case of poultry where cages are often stored one above another. Reducing feed prior to transport (without causing undue starvation) together with means of preventing spread of waste are important considerations in preventing extensive contamination of live animals entering the slaughterhouse.

It is also important to recognise the role of effective cleaning and decontamination procedures for sheds, lairage and transportation cages/vehicles in avoiding the spread of contamination to successive batches of animals. Schmidt *et al.* (2004) demonstrated that effective cleaning and disinfection could reduce the incidence of *Salmonella* in pig lairage pens, and on some occasions this resulted in reduced carriage in pigs, although this did not occur consistently. It is possible to reduce the incidence of pathogens such as *Salmonella* in animal herds and flocks through the application of known intervention measures, but it must be recognised that such pathogens will be present in herds and flocks, from time to time, and subsequent processes must always be operated with this in mind.

18.6.2 Raw milk

Raw milk may become contaminated with *Salmonella* from the faeces of the milking animal via teats and udders to milking equipment and thence to the milk. As described for other enteric pathogens (see Chapter 17), effective milking parlour hygiene (cleaning and disinfection of udders and teats), cleaning and sanitisation of milking equipment and subsequent milk storage systems are essential elements in preventing the spread of *Salmonella* spp. Routine microbiological monitoring of the hygienic quality of raw milk should be employed using indicator bacteria such as *E. coli* or coliforms, and incentive payment schemes should be considered where the milk is intended to be used without a bacterial destruction stage in the process, i.e. for raw milk cheeses, to encourage the adoption of high hygienic standards.

Microbiological surveys of raw milk for indicators of contamination have already been reviewed (see Chapter 17). Like many pathogens, *Salmonella* is not commonly found in surveys of raw milk owing to its relatively low incidence (usually < 1 %). Surveys of raw, bulk cow's milk in the USA and raw cow's drinking milk in the UK have found it to be present at 4.7 %

and 0.06 %, respectively (McManus and Lanier, 1987; Anon., 1998). When present, it is rarely found at enumerable levels, i.e. > 1/ml. Other studies in the UK have confirmed the low incidence; one sample out of 1674 samples (0.06 %) of raw cows' milk taken from farm and as finished products in 1995/6 was contaminated with *S.* Typhimurium (FSA, 2002a). Two samples (0.33 %) of raw cows' milk were found to be contaminated with *Salmonella* (Goldcoast and unknown serotype) in a survey conducted in the UK between 1999 and 2000 (FSA, 2003a). None of 111 samples of goats' and sheep milk taken from producers in the UK in 1997/8 were found to be contaminated with *Salmonella* spp. (FSA, 2002b).

18.6.3 Raw meat and eggs

Raw meat, including poultry meat, can be contaminated by the animal's faeces, and any bacterial pathogens these may contain, during the slaughtering and subsequent processes. Preventing dirty animals (with excessive faecal/ mud soiling of the hide) entering the abattoir is an important first stage in reducing contamination which, together with controlled procedures for hide removal, evisceration and the handling of other contaminated regions such as the hooves, can prevent transfer of pathogens to the muscle meat. Many of the same principles apply to the slaughter and processing of pigs, but the inclusion of a high-temperature singeing stage to remove hairs can reduce bacterial contamination on the pig-skin. Coupled with careful evisceration, as described for other animals, it is possible to minimise spread of intestine-associated pathogens.

In poultry processing, once a *Salmonella* contaminated flock is introduced into the slaughter and process line, it is very difficult to prevent it spreading to other birds owing to the high throughput and use of common processing equipment. As a consequence of this it is common practice for poultry processors to test all flocks prior to processing in order to determine those that are contaminated with *Salmonella* so that these can be scheduled for processing at the end of the production day. In this way, flocks that are not infected stand a chance of remaining uncontaminated during their passage through the slaughter and processing plant.

Another way to reduce spread of *Salmonella* spp. from bird to bird is to prevent defecation during transport and subsequent handling prior to slaughter. Poultry farmers therefore often withhold feed from flocks due to be de-populated in order to reduce the potential for faecal shedding during transport as this can be exacerbated by the stress incurred at this stage. Bilgili (2002) reported on the impact of feed withdrawal regimes on slaughter quality of poultry.

If present, *Salmonella* is readily spread through the scalding tank that operates at *c.* 60 °C to completely wet the feathers, facilitating effective defeathering. As the bird is yet to be eviscerated, faeces are shed into the tank and contamination is spread from bird to bird. This is further exacerbated in the defeathering equipment where rubber fingers pull the feathers out but can also spread contamination. After evisceration the bird is washed inside and outside with free-flowing clean water.

Strategies to reduce contamination on the final animal/bird carcass through carcass washing or, in the case of beef, through steam 'pasteurisation' have met with some success. Yang *et al.* (1998) demonstrated a 1.7–2.0 $\log_{10}$ cfu reduction in *S.* Typhimurium inoculated onto chicken breast, the back and in the cavity when subject to washing with four different chemicals (10 % tri-sodium phosphate (pH 12.3), 0.5 % cetyl pyridinium chloride (pH 7.6), 2 % lactic acid (pH 2.2) and 5 % sodium bisulphate (pH 1.3)). This compared to a 0.56 $\log_{10}$ cfu reduction when washing with water alone. Cutter *et al.* (2000) demonstrated nearly a 5–6 $\log_{10}$ reduction in *S.* Typhimurium when lean beef tissue was washed with 1 % cetyl pyridinium chloride, although unacceptable residual levels of the chemical remained.

A number of beef processors in the USA employ steam pasteurisation to decontaminate the external surface of the carcass after primary processing. The carcass enters a chamber where it is subject to steam that causes elevation of temperature at the beef surface. The carcass is then water-cooled and leaves the chamber for further processing. Such steam treatment has been shown to deliver a reduction approaching 3 $\log_{10}$ cfu on carcasses artificially inoculated with enteric organisms (Phebus *et al.*, 1997).

It should always be remembered that effective cleaning and disinfection of all equipment, surfaces and the environment in the slaughter and processing halls are essential for the control of enteric pathogens.

Raw meat and the associated environment should be subject to monitoring for indicators of process hygiene such as *E. coli* or coliforms/Enterobacteriaceae. With the exception of poultry and perhaps pork, it is usually not worthwhile testing for *Salmonella* in raw meat because of its infrequent occurrence. However, where such material is frozen or is to be used for processes such as fermented or dried meat production, where minimal bacterial reduction occurs, it may be useful to monitor for the organism to ensure it is not present at an unusually high frequency/level in a batch. The use of indicators such as *E. coli*/coliforms may, however, give useful information about the hygienic quality of the material, and also deliver results faster.

Salmonella is found in all types of raw meat, although the incidence varies considerably depending on the meat species and also on its processing and origin. A recent survey of 2509 carcass swab samples (0.1 m^2) taken from slaughtered pigs at abattoirs in the UK found 5.3 % (135 samples) to be contaminated with *Salmonella* (Davies, 2000). The predominant serotypes were *S.* Typhimurium (2.1 %) and *S.* Derby (1.6 %). Older surveys of pork have reported a *Salmonella* incidence ranging from 0.4 % in Sweden to over 75 % in a Dutch survey (D'Aoust, 1989). The incidence in beef is much lower than that in pork. A major study in the USA in 1992/93 found the average incidence to be 1 % with levels on positive carcasses of 0.1 MPN/ cm^2 (MPN is the most probable number; USDA, 1994). Slightly higher incidence has been reported by Sofos *et al.* (1999). Bohaychuk *et al.* (2006) found no *Salmonella* positives in a survey of 100 samples (25 g) of raw ground beef. Data from EU sampling (Anon., 2007a) of minced meat and meat preparations demonstrated an incidence of 0.5 % from 20 425 samples (25 g).

Salmonella is a frequent isolate from poultry and poultry products. Although some countries have achieved a much lower incidence, notably Sweden, where the incidence is < 1 %, the majority of the rest of the developed world has reported incidences ranging from < 10 % to 100 %. Sampling methodologies are not always easy to compare and vary from 25 g composite skin samples, i.e. neck flap, breast and thigh, to skin and meat composites to whole bird rinses. Nevertheless, it is rare to see results from surveys of raw poultry of any nature yielding an incidence of < 5 %.

A survey of raw poultry on retail sale in the UK in 2001 found an average incidence of *Salmonella* of 5.7 % (FSA, 2003b). Personal experience indicates that the incidence can vary significantly from farm to farm and is very much dependent on the farming practices employed in the rearing of the birds and subsequently on the practices employed by the broiler processor. Nevertheless, it is now common to find UK fresh poultry with incidence regularly below 10 % and some individual processors with < 5 % incidence. Chang (2000) reported a *Salmonella* incidence of 25.9 % on raw broilers on sale in Korea (25 g samples of leg, thigh and breast of chicken meat and skin) with predominant serotypes being *S.* Enteritidis, *S.* Virchow and *S.* Virginia. Bryan and Doyle (1995) summarised results from a large number of surveys of raw chicken, duck and turkey that demonstrated an incidence varying from 2 % to 100 % with the median incidence of 30 %. They also reported that the population of *Salmonella* on the carcass was typically 1–30 cells with occasionally as many as 10^4 cfu per 100 g of broiler skin. Bohaychuck *et al.* (2006) reported an incidence of 30 % in raw chicken legs ($n = 100$, 25 g sample). Extensive data in 2006 on the incidence of *Salmonella* in poultry meat is published for EU member states (Anon., 2007a) and indicated an overall incidence in fresh broiler meat (including samples from slaughterhouse, processing/cutting plant and retail) of 5.6 % ($n = 33\ 257$).

Eggs are well-recognised as a primary source of *Salmonella* spp. in outbreaks of foodborne salmonellosis involving foods made from them. Results from surveys of the incidence of *Salmonella* in eggs vary considerably depending on the sample size, time from laying, i.e. fresh eggs or stored eggs, and method of detection used. A survey carried out in Northern Ireland of 2090 packs of six eggs found an incidence of 0.43 % *Salmonella* (Wilson *et al.*, 1998) whereas a British survey between 1992 and 1993 of eggshells and contents stored at 21 °C for five weeks found an incidence of 0.2 % from a total of 7730 six-egg samples (de Louvois, 1994). Following the widespread introduction of egg layer vaccination and adherence to an industry code of practice (Lion code), the incidence of *Salmonella* in UK eggs is believed to be significantly lower than this now. Recent surveys in the UK of eggs in use in retail and catering establishments have found low incidence of *Salmonella*. Little *et al.* (2008) reported six positive isolations of *Salmonella* in 1588 samples (0.38 %) of pooled eggs (six eggs per sample) in surveys from food service outlets in the UK between 2005 and 2006. Five of these isolates were *S.* Enteritidis and one was *S.* Mbandaka. In a similar survey of eggs on sale in UK retail outlets in 2003, the calculated prevalence per box of six eggs was 0.34 % (n = 4753 samples) (FSA, 2004). Incidence in EU member states for 2006 was reported to be 0.8 % (n = 28 773, some single samples, some batch samples) (Anon., 2007a).

Although the incidence of *Salmonella* spp. in eggs may be very low, the fact that many millions of eggs are consumed each week in the world means that the risk of salmonellosis from eggs or egg-containing products can be high where these are poorly/unhygienically handled prior to consumption.

18.6.4 Fruit, vegetables and other raw materials

Outbreaks of salmonellosis associated with salad vegetables, e.g. bean sprouts and alfalfa sprouts, fruits and unpasteurised fruit juice (Tables 18.6 and 18.7), have served to focus attention on agricultural practices in these sectors. *Salmonella* can be present in very large numbers in animal wastes, particularly if derived from infected herds or flocks (Nicholson *et al.*, 2000) and can survive for periods up to 41 weeks in stored slurry depending on the slurry composition and time of year. Aeration of animal waste slurries can significantly reduce the levels of *Salmonella* (Heinonen-Tanski *et al.*, 1998). Composting of animal wastes can achieve a reduction in levels of microbial pathogens owing to the high temperatures achieved in the pile due to microbial activity. Good management of the pile through regular turning can allow a safer waste material to be produced for spreading onto agricultural land. Even in the absence of composting, storage of liquid wastes for long periods to encourage microbial activity is a sensible approach that has been advocated to reduce the potential burden of pathogens on agricultural land (Nicholson *et al.*, 2000). In a study of organic produce sampled from farms, i.e. pre-harvest, Mukherjee *et al.* (2004) reported *E. coli* contamination levels 19 times higher in products that had been grown in fields where compost or manure used was less than 12 months old in comparison to crops where it was stored for longer than 12 months.

Izumi *et al.* (2008) studied the contamination sources during the growing and packing of mandarin satsumas in a conventionally managed orchard in Japan between May and November 2005. The farm was not located near vegetable or livestock production. Samples were taken from the orchard (soil, fallen leaves, cloth mulch, agricultural water, pesticide solution, chemical fertiliser, organic fertiliser, fruit) and the packing shed (gloves, scissors, collection baskets, size sorters, fruit) at monthly frequencies, where relevant. *Salmonella* was only detected in one sample during the extensive investigation, this being the pesticide solution which was made up with agricultural water.

Several studies have shown that pesticide solutions can support the survival and growth of *Salmonella* (Guan *et al.*, 2001; Ng *et al.*, 2005) and, given the use of such solutions in direct application to exposed ready-to-eat crops, this reinforces the need for effective control of such agricultural inputs. Highest microbial counts (mesophilic bacteria) were found in fallen leaves, followed by soil, organic fertiliser and agricultural water. Microbial counts of the peel increased after fruit harvesting and sorting, although there was no evidence of contamination of the flesh.

Recent outbreaks of salmonellosis caused by contaminated peppers/tomatoes have served to highlight the potential risks associated with very common commodities (www.fda.gov) (Table 18.7). Orozco *et al.* (2008) reported on a study examining the impact of animal and environmental factors on *Salmonella* contamination of hydroponically grown tomatoes. During a one-year study, two significant natural events occurred; water run-off entered the greenhouses in one incident and, in another, wild animals gained access to several greenhouses. *Salmonella* was detected in a wide variety of samples including tomatoes, water puddles, soil, shoes and faecal samples of local wild and farm animals. Although some contamination occurred in tomatoes before the flooding and animal entry, the majority of isolations occurred during and/or after the events (1.8 % contamination before flooding, 0 % during flooding, 9.4 % after flooding; 3.3 % before animal entry, 10.4 % during animal entry, 2.5 % after entry). PFGE analysis demonstrated that many of the *Salmonella* isolates from tomatoes and wild animals were identical. In addition to these sources of contamination, workers' shoes were also considered to present a significant route of contamination of *Salmonella* into the greenhouses.

Flooding can increase the risk of enteric pathogen transfer to growing crops and, if such crops are intended for minimal subsequent processing i.e. lettuce, herbs, etc., such risks need to be assessed. Specific food safety advice has been issued to the general public in relation to food potentially affected by floodwater (FSA, 2008).

Clearly, no animal wastes should be applied to land where there are growing crops as this will inevitably lead to contaminants entering the human food chain.

Irrigation water is another possible source of potential contamination of crops. Although the source of water is much more difficult to control as it is not usually under the direct authority of the farmer, it is important that known contamination sources are not used as irrigation water for growing crops.

A variety of wild animals including birds, reptiles and rodents can carry *Salmonella* (Ward, 2000) and, although it is very difficult to control such pests in the field, it is essential to prevent their entry into subsequent storage and processing areas of fruit and vegetable produce. Another practice that may also be increasingly associated with outbreaks of salmonellosis is the use of drop fruit; also, the potential for internalisation of pathogens in the fruit may be important. These have already been described in detail (see Chapter 17).

Although outbreaks of salmonellosis continue to be reported from fresh produce, published surveys very rarely find the organism. Fröder *et al.* (2007) detected *Salmonella* spp. in four out of 133 (3 %) minimally-processed ready-to-eat salads/vegetables on sale in retail stores in Sao-Paulo, Brazil. Produce contaminated were lettuce, mixed salad (beets, carrots, lettuce, cabbage, watercress), watercress and chicories. Sagoo *et al.* (2003) did not detect *Salmonella* spp. in any 25 g sample (n = 2943) of open, prepared, ready-to-eat salad vegetables from catering and retail salad bars in the UK. Similarly, *Salmonella* was not detected in 3200 samples of ready-to-eat organic salad vegetables (Sagoo *et al.*, 2001). In a compilation of surveys across the EU comprising fruits, vegetables, salads, herbs and spices (all 25 g samples) *Salmonella* was reported in 0.14 % of 7571 samples (Anon., 2007a).

The large number of other possible food and non-food sources of *Salmonella* already noted (Section 18.3) should be clearly identified by the application of a thorough hazard analysis of any production process or new product development. In this way, suitable controls can be established to ensure the process and products are as safe as required. Additionally, some useful good agricultural practice documents are available that provide excellent general guidance on preventing microbiological contamination of fresh produce (USDA, 1998; NACMCF, 1999a, b; CFA, 2007)

18.7 Control in processing

For most processes, stages are identified where defined conditions are exerted that can reduce or eliminate enteric pathogen contamination, if present in the raw ingredients. Reliable identification and control of these stages are clearly critical to the safety of the finished product. In many processes the pathogen reduction stages are obvious, such as in cooking or washing, but in others, the reduction stage is less clear, as in raw fermented meats or raw milk cheese production. It is incumbent on the food producer to identify the stages and process conditions within those stages necessary to achieve a satisfactory reduction in pathogens to generate a safe product. This is best achieved by the application of hazard analysis-based approaches. The descriptions here are short as many of the process and finished product controls relating to *Salmonella* are very similar to those for pathogenic *E. coli*, discussed in Chapter 17.

18.7.1 Cooking

The application of high temperatures in cooking or pasteurising foods is one of the best means by which to destroy *Salmonella*. The organisms are readily destroyed by moist heat and temperatures designed to destroy pathogenic *E. coli* will equally destroy (> 6 $\log_{10}$ reduction) *Salmonella*, e.g. 70 °C for 2 min for meat and 71.7 °C for 15 seconds for milk. These heat processes generally confer a significant margin of safety in relation to most serotypes of *Salmonella*. For example, a review of heat-resistance studies by Doyle and Mazzotta (2000) concluded that it would take only 1.2 s at 71 °C to inactivate 1 $\log_{10}$ of *Salmonella* cells (*z*-value 5.3 C°). Although this estimate excludes the most heat-resistant strain, *S.* Senftenberg 775W, even the milder milk pasteurisation would be estimated to deliver > 4 $\log_{10}$ reduction in this strain (Bell and Kyriakides, 2002). Application of heat to destroy organisms in low-moisture environments, i.e. chocolate, where water activity is very low requires much longer times or higher temperatures (Table 18.4). The principles underpinning the consistent delivery of the correct temperature and subsequent segregation of cooked product to avoid recontamination after cooking have already been described in detail (see Chapter 17).

18.7.2 Fermentation and drying processes

Many products are traditional in their manufacture and have been made in the same way for centuries. Although their manufacture is often considered to be a craft rather than a science, it is possible to use modern-day knowledge to identify and control the factors responsible for the production of a safe product. Raw fermented meat and raw dry-cured meats and raw milk cheeses are good examples of this. In all cases, it is the standard of hygiene in the production of the raw ingredient that is the first important control. This is then followed by the promotion of an effective fermentation stage by lactic microflora to both prevent the proliferation of pathogens and potentially cause their reduction. The addition of salt to both meat for fermented/dry-cured, meat and to the milk for cheese adds further microbial inhibitory compounds, and the reduction in moisture achieved during the long drying periods for hard cheese and the raw dry-cured and fermented meats are also important factors in the pathogen reduction process. Reduction in *Salmonella* occurs during both the fermentation and maturation stages (Smith *et al.*, 1975) and reduced fermentation and/or drying times have been reported as reasons for previous outbreaks of salmonellosis (Pontello *et al.*, 1998). It is also for this reason that raw milk soft cheeses are often considered to represent a higher risk – they have fewer pathogen controls within the production process or intrinsic product pathogen protection characteristics than hard cheeses or fermented meat equivalent.

Hew *et al.* (2006) calculated the death rate of *Salmonella* (Enteritidis, Typhimurium, Gaminara, Newport, Montevideo) in artificially inoculated chorizo (an unfermented dried meat) at varying water activity (0.85–0.97) with fixed pH (5.0) when stored at room temperature for 7 d. The fitted death plot served to demonstrate how slowly the organism dies off in such products with the maximum calculated death rate of 0.7 $\log_{10}$ per day at a water activity of 0.873. Death rates were calculated to be negative, i.e. growth when the water activity was above *c.* 0.95, and only appreciably increased (death rate > 0.25 $\log_{10}$ per day) when water activities were below *c.* 0.93. It is also important to note that the death rate decreased (< 0.25 $\log_{10}$ per day) as the water activity dropped below *c.* 0.8. Another interesting aspect of this study was the investigation of the inhibitory effect of the various spices added to chorizo, the only spice suspensions demonstrating inhibition of growth being the fresh garlic and dried garlic powder. Black pepper, cumin and crushed pepper had no effect. This is consistent with many studies showing the antimicrobial effect of garlic. It is strongly recommended that challenge test studies be conducted in research facilities on these types of products to determine the process lethality and, together with tight control of the critical process stages, routine monitoring of the raw ingredients for pathogens and indicators of hygienic quality should be employed, i.e. *E. coli*/coliforms.

It is very unusual for surveys of these products to find contamination due to its low incidence, although it is reported occasionally. Bohaychuk *et al.* (2006) did not find any positives in a survey of fermented sausage (n = 100) in Canada. In a survey of dry and fermented meats conducted between 1997 and 1999 in the USA, Levine *et al.* (2001) reported the detection of 10 *Salmonella*-positive samples out of 698 (1.43 %). In contrast, Cabedo *et al.* (2008) reported nine out of 81 samples (11.1 %) of cured dried pork sausage to be contaminated with *Salmonella* spp. in a survey conducted in Catalonia, Spain. In the same survey none of the samples of cured dried pork loin (n = 43), cured dried ham (n = 35) or salami (n = 15) were found to be contaminated.

Little *et al.* (2007) did not detect any *Salmonella* spp. in 1819 samples of raw or thermised milk cheeses purchased from retail sale in the UK during 2004. In surveys of raw or thermised milk cheeses in EU member states during 2006, only three isolations of *Salmonella* were reported in a total of 1473 cows' milk cheese samples (1.73 %), no isolations from 954 sheep milk cheese samples and no isolations from 78 goats' milk cheese samples (Anon., 2007a).

18.7.3 Washing processes

The washing of fruit and salad vegetables in preparation processes for their sale as ready-to-eat salads or fruit juice has already been described (see Chapter 17). The same levels of disinfectants are employed in respect of controls relating to both *Salmonella* and *E. coli*. However, given the recent number of outbreaks of salmonellosis associated with unpasteurised fruit juice, some additional discussion of this sector will be made here. It has already been mentioned that fruit used for juice extraction may sometimes be picked from the ground as 'drop fruit' and, as such, is more likely to be contaminated than that picked directly from the tree. Fruits are routinely washed prior to juice extraction and, while chlorinated water is often used for the washing process, a number of studies have been made to determine the efficacy of different washing treatments.

Pao and Davis (1999) assessed the reduction in *E. coli*, an indicator of enteric contamination, inoculated onto oranges using hot water treatment (1–4 mins in water heated to 30, 60, 70 and 80 °C) and a number of chemical treatments: sodium hypochlorite (200 ppm chlorine), chlorine dioxide (100 ppm), acid anionic sanitiser (200 ppm), peroxyacetic acid (80 ppm) and tri-sodium phosphate (2 %). Chemical treatment for up to 8 min reduced *E. coli* levels by 1 $\log_{10}$ cfu to 3.1 $\log_{10}$ cfu, this varying according to the chemical and region of the inoculum (stem-scar vs non stem-scar region). In comparison, water reduced *E. coli* by 0.6–2.0 $\log_{10}$ cfu. Hot water immersion achieved a 5 $\log_{10}$ reduction in *E. coli* after treatment at 80 °C for 2 min, although it was reported to result in a discernible difference in flavour of the extracted juice. Such information can be useful in relation to a consideration of the use of such treatments in the control of *Salmonella*.

Stopforth *et al.* (2008) studied the effect of various decontamination systems (50 ppm free chlorine acidified to pH 6.5 with citric acid; acid electrolysed water; acidified sodium chlorite 1200 ppm) on the reduction of *Salmonella* artificially inoculated on leafy greens. Water-washing reduced contamination by 0.6 $\log_{10}$ cfu after 60 s and 1.1 $\log_{10}$ cfu after 90 s. Chlorine treatment and acid electrolysed water had similar effects, reducing contamination by 2.2 and 2.3 $\log_{10}$ cfu after 60 s, respectively and by 2.8 and 2.5 $\log_{10}$ cfu after 90 s, respectively. Acidified sodium chlorite had the greatest effect reducing levels by 3.6 and 3.8 $\log_{10}$ cfu after 60 and 90 s, respectively. Changing the chlorine acidulant from citric acid to sodium acid sulphate had no significant effect.

As well as the chemical disinfectant used to wash produce, the method of washing is also important. The degree of physical action employed is likely to have a significant effect on reducing microbial attachment and thus residual contaminants after washing. Large businesses often have highly sophisticated water flume/jacuzzi washing units that allow considerable agitation, but it is still common for small scale operations to have more manual dip tanks where the effect of disinfectants on microbial contaminants is likely to be limited.

The temperature of washing can also play an important role in the risk presented by certain washed produce. For example, it has been very well established that washing tomatoes in water where the temperature differential is high, i.e. low temperature water/warm temperature fruit, causes water and associated contaminants to be essentially sucked into the fruit, thereby internalising the organism and protecting it from external disinfection (Zhang *et al.*, 1995).

It is already recognised that *Salmonella* can survive for extended periods under very acidic conditions. The lag time before *S.* Typhimurium even started to decline in numbers in orange juice at pH 4.1 and pH 3.8 was

16.15 d and 7.97 d, respectively, when stored at 4 °C (Parish *et al.*, 1997). Under conditions of less harsh pH, *Salmonella* will not only survive but can grow to extremely high numbers on some fruit. Golden *et al.* (1993) demonstrated the growth potential of a mixture of *Salmonella* serotypes when inoculated onto cantaloupe, watermelon and honeydew melons. When stored at 23 °C, levels increased by over 5 log cfu within 24 h, although no growth was reported over the same time period when stored at 5 °C.

18.7.4 Sprouting processes

Sprouted seed vegetables have caused a number of salmonellosis outbreaks (O'Mahony *et al.*, 1990; Taormina *et al.*, 1999; Werner *et al.*, 2007; Emberland *et al.*, 2007). The presence of *Salmonella* in the raw material seed represents a significant hazard to this process as the conditions of germination employed for their production encourage the rapid multiplication of any contaminating pathogens. Levels of *Salmonella* contamination on seed batches have been shown to increase by over four orders of magnitude during the sprouting process (Jaquette *et al.*, 1996).

In a study of factors influencing the growth of *Salmonella* during sprouting of alfalfa seeds, Fu *et al.* (2008) compared a simulated commercial process to less sophisticated home sprouting using naturally-contaminated seeds from two salmonellosis outbreaks. Sprouts grown in glass jars (mean temperature 28.5 °C) supported increases of 3–4 $\log_{10}$ units after 48 h from initial contamination levels of 13.2–16.2 MPN per kg. In contrast, when grown in a minidrum, with irrigation every 20 min, *Salmonella* did not increase over a period of 4 d (mean temperature 20.2 °C). Varying temperature in the glass jars and drum resulted in differing levels of increase (*c.* 3 $\log_{10}$ increase in glass jars at 30 °C after 72 h but only 1 $\log_{10}$ increase at 20 °C; 1–2 $\log_{10}$ increase in the drum at 28–29 °C after 72 h and no increase at 20 °C). Irrigation frequency also played a role in allowing the organism to proliferate with counts being 2 $\log_{10}$ units higher when the sprouts were irrigated every 2 h in the minidrum (vs 20 minutes) or when irrigated every 24 h (vs 4 h) in the glass jar. This may have been due partly to the effect of lower temperature as a consequence of more frequent irrigation cooling the seeds/sprouts.

Studies on treatments to reduce *E. coli* O157 on the seed have already been reported (see Chapter 17) and they are likely to have a similar effect on *Salmonella*. Jaquette *et al.* (1996) demonstrated that treatment of contaminated alfalfa seeds with 100 ppm and 290 ppm active chlorine solution for five or 10 min achieved a reduction in contaminating *S.* Stanley, although this was less than one order of magnitude. Increasing levels to 1010 ppm resulted in approximately a 2 $\log_{10}$ reduction of the organism. Bari *et al.* (2008) studied the effect of flash heat treatment on the destruction of *Salmonella* in artificially-contaminated mung beans together with the effect on subsequent germination. Mung beans (5 g in netted bags) contaminated with 5.34 $\log_{10}$

cfu per g *S.* Enteritidis (four-strain cocktail) were immersed and shaken in hot water (80 and 90 °C) for 30, 60, 90, 120 and 150 s prior to quenching in chilled water (0 °C) for 30 s. Treatments capable of achieving a 5 $\log_{10}$ unit reduction in the organism included 80 °C for 60 s or more and 90 °C for 30 s or more. The organism was still detectable (after enrichment of the seeds) after all treatment times at 80 °C and after 30 and 60 s at 90 °C. The mildest heat treatment capable of eliminating the organism (90 °C, 90 s) resulted in 98.7 % germination of the seeds.

As for the control of *E. coli*, processors should conduct extensive testing of raw material seed batches and production samples to monitor the effectiveness of controls in place and gain greater assurance of the microbiological safety of these products in relation to *Salmonella* spp.

18.7.5 Hygiene and post-process contamination

For details of this see Chapter 17.

18.8 Final product control

Final product control of *Salmonella* spp. is similar to that for pathogenic *E. coli*, discussed in Chapter 17.

18.9 General considerations

It is clear that there are common areas that need to be addressed in relation to the control of *Salmonella* regardless of the food materials being processed:

- Process flows should be designed to eliminate any possibility of cross-contamination between raw and heat-processed products.
- Procedures for effective personal, equipment and environmental hygiene must be carefully designed, implemented, maintained and monitored for ongoing efficacy.
- Product process conditions must be designed to consistently achieve the physicochemical conditions required to control (reduce/eliminate) *Salmonella* at all stages of manufacture and the life of the finished product.

Where new technologies such as pulsed electric fields (Jeantet *et al.*, 1999; Liang *et al.*, 2002), ultrasonic energy, oscillating magnetic field pulses, high-pressure (Ritz *et al.*, 2006; Whitney *et al.*, 2008), high-intensity visible light and ultraviolet light (Kim *et al.*, 2002) and even old technologies not yet widespread in use, e.g. irradiation, are developed and exploited for use in food production processes, it is critical that they are fully assessed for safety in respect of specific pathogens including *Salmonella*.

Carefully designed challenge test studies can be useful for demonstrating effective control of *Salmonella* in relation to new and old technologies applied to many product types and have been widely used in the past to assess the survival of pathogens including *Salmonella* in foods/processes as diverse as bulk meat cooking processes, mayonnaise products and fermented dairy products. The results of such studies are valuable for helping to determine the critical controls and monitors required to effect safe processes in respect of the specific pathogen studied, e.g. *Salmonella.*

In addition to ensuring that appropriate production process practices and procedures are in place from the agricultural sector, through manufacturing and retailing processes, it is important to keep in mind that the caterer and home consumers can also play a significant part in maintaining the safety of foods. The basic rules for safe food-handling by caterers and home consumers are well publicised but, it would seem, not always well understood or applied by the consumer. The wealth of practical information available for consumers to use for playing their part in a safe food chain needs to be promulgated by a continuous education programme for consumers of all ages. Such a programme must have a long-term benefit, both physical and economic, for all. Within such a programme, the food industry must continue to supply appropriate information on the product packaging that gives clear guidance concerning safe handling and storage conditions of food and the preparation or cooking processes that need to be employed to keep the food safe to eat. Additionally, whilst food is clearly the most significant source of *Salmonella* in relation to human infections, it should be remembered that the organism can be contracted through other means such as direct animal contact; Mermin *et al.* (2004) estimated that 6 % of sporadic *Salmonella* infections in the USA were attributable to contact with reptiles or amphibians and Swanson *et al.* (2007) reported on an outbreak of salmonellosis affecting 15 people from pet rodents. Clearly, preventing salmonellosis through widespread understanding of risk factors and associated intervention measures, not least good personal practices, is key.

18.10 Future trends

The ubiquitous nature of *Salmonella* and the wide diversity of foods that are associated with outbreaks of salmonellosis already cause many hundreds of thousands (probably millions) of cases of illness every year around the world. From the evidence to date, different serotypes or phage types of *Salmonella* will continue to 'emerge' as important causes of foodborne outbreaks of salmonellosis, go into decline and then another will 'emerge'. In addition, the increasing spread of antibiotic resistance in the environmental 'pool' of *Salmonella* may lead to longer persistence of 'emerged' serotypes and an increase in numbers of types emerging. Groisman and Ochman (1997, 2000) have suggested that virulence genes were transferred into *Salmonella* from related bacterial genera, e.g. *E. coli*, or between strains of *Salmonella*. If such were needed, the prospects of further gene acquisitions/transfers resulting in a regular 'production' of new pathogens should be a clear indicator of the need to ensure that all the controls already known to be effective in the prevention of *Salmonella* contamination of foods should be actively encouraged and, where necessary, enforced by legislation.

The food industry should ensure that structured hazard analyses in respect of *Salmonella* are carried out of all food production processes from primary agriculture and aquaculture throughout the chain to the consumer. Where changes are made to product formulations to meet 'public demand', e.g. reduced salt or sugar levels, or newer food processing technologies are introduced, e.g. high-pressure treatment, ionizing radiation, their impact on the hazard analysis and food safety should be reviewed for individual products to ensure that any controls identified are implemented and maintained and also monitored for consistent efficacy. In addition, all food handlers and consumers should be provided with the necessary information and training (as applicable) in relation to food safety issues, by industry on product labelling and through structured government education programmes to ensure food is handled safely to final consumption. Such joint action should control even the, as-yet, 'un-emerged' new salmonellas and significantly reduce the numbers of cases of salmonellosis that occur.

18.11 Sources of further information and advice

EC (1998) Council Directive 98/83/EC of 3 November 1998 on the quality of water intended for human consumption, *Official Journal of the European Communities*, **L330**, 5 December, 32–54.

EC (2004) Regulation (EC) No 853/2004 of the European Parliament and of the Council of 29 April 2004 laying down specific hygiene rules for food of animal origin, *Official Journal of the European Union*, **L226**, 25 June 22–82 and amendment in Commission Regulation (EC) No 1243/2007, *Official Journal of the European Union*, **L281**, 25 October, 8–11.

EFSA (2008) A quantitative microbiological risk assessment on *Salmonella* in meat: source attribution for human salmonellosis from meat, Scientific Opinion of the Panel on Biological Hazard, European Food Safety Authority, *The EFSA Journal*, **625**, 1–32.

Heaton J C and Jones K (2008) Microbial contamination of fruit and vegetables and the behaviour of enteropathogens in the phyllosphere: a review, *Journal of Applied Microbiology*, **104**, 613–26.

ICMSF (2002) *Microorganisms in Foods 7: Microbiological testing in food safety management*, International Commission on Microbiological Specifications for Foods, Springer Science, New York.

Membré J-M, Bassett J and Gorris L G M (2007) Applying the food safety objective and related standards to thermal inactivation of *Salmonella* in poultry meat, *Journal of Food Protection*, **70**(9), 2036–44.

WHO (2006) *Guidelines for drinking-water quality. First addendum to third edition Volume 1: Recommendations*, World Health Organization, Geneva.

18.12 References

Alcaine S D, Warnick L D and Wiedmann M (2007) Antimicrobial resistance in nontyphoidal *Salmonella. Journal of Food Protection*, **70**(3), 780–90.

Annous B A, Solomon E B, Cooke P H *et al.* (2005) Biofilm formation by *Salmonella* spp. on cantaloupe melons, *Journal of Food Safety*, **25**(4), 276–87 (abstract on www.ingentaconnect.com).

Anon. (1998) *Surveillance of the Microbiological Status of Raw Cows' Milk on Retail Sale*, Microbiological Food Safety Surveillance, Department of Health, London.

Anon. (2000) National increase in *Salmonella typhimurium* DT104 – update, *Communicable Disease Report Weekly*, **10**(36), 323–6.

Anon. (2002) Outbreak of *Salmonella* serotype Kottbus infections associated with eating alfalfa sprouts – Arizona, California, Colorado and New Mexico, February–April 2001, *Morbidity and Mortality Weekly Report*, **51**(01), 7–9.

Anon. (2006) The community summary report on trends and sources of zoonoses, zoonotic agents, antimicrobial resistance and foodborne outbreaks in the European Union in 2005, *The EFSA Journal 2006*, **94**.

Anon. (2007a) The Community Summary Report on Trends and Sources of Zoonoses, Zoonotic Agents, Antimicrobial Resistance and Foodborne Outbreaks in the European Union in 2006, *The EFSA Journal*, **130**.

Anon. (2007b) Multistate outbreak of *Salmonella* serotype Tennessee infections associated with peanut butter, USA, 2006–2007, *Morbidity and Mortality Report Weekly*, **56**(21), 521–4.

Anon. (2008a) Recent trends in selected gastrointestinal infection, Health Protection Report, **2**(28), 2–3.

Anon. (2008b) *Salmonella* Litchfield, melon – North America ex Honduras: Alert, Promed Archive number, 20080323.1098, www.promedmail.org.

Bari M L, Inatsu Y, Isobe S *et al.* (2008) Hot water treatments to inactivate *Escherichia coli* O157: H7 and *Salmonella* in mung beans, *Journal of Food Protection*, **71**(4), 830–4.

Baylis C and Mitchell C (2008) *Catalogue of Rapid Microbiological Methods, Review No. 1*, 6th edn, Campden & Chorleywood Food Research Association, Chipping Campden.

Bell C (1997a) The use of risk assessment in establishing microbiological criteria for foods, *International Food Safety News*, **6**(3), 2–3.

Bell C (1997b) Correspondence: Quantitative risk assessment – musings continued, *International Food Safety News*, **6**(6), 2–4.

Bell C and Kyriakides A (2002) Salmonella*: A Practical Approach to the Organism and its Control in Foods*, Blackwell Science, Oxford.

Besser T E, Goldoft M, Pritchett L C *et al.* (2000) Multiresistant *Salmonella* Typhimurium DT 104 infections of humans and domestic animals in the Pacific Northwest of the United States, *Epidemiology and Infection*, **124**, 193–200.

Bidol S A, Daly E R, Rickert R E *et al.* (2007) Multistate outbreaks of *Salmonella* infections associated with raw tomatoes eaten in restaurants – United States, 2005–2006, *Morbidity and Mortality Report Weekly*, **56**(35), 909–11.

Bilgili S F (2002) Slaughter quality as influenced by feed withdrawal, *World's Poultry Science Journal*, **58**, 123–30.

Bohaychuk V M, Gensler G E, King R K *et al.* (2006) Occurrence of pathogens in raw and ready-to-eat meat and poultry products collected from the retail marketplace in Edmonton, Alberta, Canada, *Journal of Food Protection*, **69**(9), 2176–82.

Bollaerts K, Aerts M, Faes C *et al.* (2008) Human salmonellosis: estimation of dose-illness from outbreak data, *Risk Analysis*, **28**(2), 427–40.

Brenner D J (1984) Family 1. Enterobacteriaceae RAHN 1937, Nom. Fam. Cons. Opin. 15, Jud. Comm. 1958, 73; Ewing, Farmer, and Brenner 1980, 674; Judicial commission 1981, 104, in N R Krieg and J G Holt (eds), *Bergey's Manual of Systematic Bacteriology*, Vol. 1, Williams and Wilkins, Baltimore, MD, 411.

Brockmann S O, Piechotowski I and Kimmig P (2004) *Salmonella* in sesame seed products. *Journal of Food Protection*, **67**(1), 178–80 (abstract on www.ingentaconnect.com).

Bryan F L and Doyle M F (1995) Health risks and consequences of *Salmonella* and *Campylobacter jejuni* in raw poultry, *Journal of Food Protection*, **58**(3), 326–44.

Budzińska K (2005) Survivability of *Salmonella senftenberg* W-775 in cattle slurry under various temperature conditions, *Folia Biologica*, **53**(1), 145–50 (abstract on www.ingentaconnect.com).

Burnett S L, Gehm E R, Weissinger W R *et al.* (2000) A note: survival of *Salmonella* in peanut butter and peanut butter spread, *Journal of Applied Microbiology*, **89**(3), 472–7.

Cabedo L, Picart L, Barrot I *et al.* (2008) Prevalence of *Listeria monocytogenes* and *Salmonella* in ready-to-eat food in Catalonia, Spain, *Journal of Food Protection*, **71**(4), 855–9.

CAC (1996) *Establishment of Sampling Plans for Microbiological Safety Criteria for Foods in International Trade Including Recommendations for Control of* Listeria monocytogenes, Salmonella enteritidis, Campylobacter *and entero-haemorrhagic* E. coli, Document prepared by the International Commission on Microbiological Specifications for Foods for the Codex Committee on Food Hygiene, twenty-ninth session 21–25 October 1996, Agenda item 11, CX/FH 96/9 1–16, Codex Alimentarius Commission, Rome.

CFA (2007) *Microbiological guidance for produce suppliers to chilled food manufacturers*, 2nd edn, Chilled Food Association, Peterborough, available from: http://www.chilledfood.org/.

Chang Y H (2000) Prevalence of *Salmonella* spp. in poultry broilers and shell eggs in Korea, *Journal of Food Protection*, **63**(5), 655–8.

Corby R, Lanni V, Kistler V *et al.* (2005) Outbreaks of *Salmonella* infections associated with eating Roma tomatoes – United States and Canada 2004, *Morbidity & Mortality Weekly Report*, **54**(13), 325–8.

Crook P D, Aguilera J F, Threlfall E J *et al.* (2003) A European outbreak of *Salmonella enterica* serotype Typhimurium definitive phage type 204b in 2000, *Clinical Microbiology and Infection*, **9**(8), 839–45.

Cutter C N, Dorsa W J, Handie A *et al.* (2000) Antimicrobial activity of cetyl pyridinium chloride washes against pathogenic bacteria on beef surfaces, *Journal of Food Protection*, **63**(5), 593–600.

D'Aoust J-Y (1989) *Salmonella*, in M P Doyle (ed.), *Foodborne Bacterial Pathogens*. Marcel Dekker, New York, 327–445.

D'Aoust J-Y (1991) Psychrotrophy and foodborne *Salmonella*, *International Journal of Food Microbiology*, **13**, 207–16.

D'Aoust J-Y, Warburton D W and Sewell A M (1985) *Salmonella typhimurium* phage type 10 from cheddar cheese implicated in a major Canadian foodborne outbreak, *Journal of Food Protection*, **48**(12), 1062–6.

Danyluk M D, Jones T M, Abd S J *et al.* (2007) Prevalence and amounts of *Salmonella* found on raw California almonds, *Journal of Food Protection*, **70**(4), 820–7 (abstract on www.ingentaconnect.com).

Davies R (2000) Salmonella *monitoring and control in pigs*, presented at 'Zoonotic infections in livestock and the risk to human health', MAFF, London, 7 December.

Davies R H, Dalziel R, Gibbens J C *et al.* (2004) National survey for *Salmonella* in pigs, cattle and sheep at slaughter in Great Britain (1999–2000), *Journal of Applied Microbiology*, **96**, 750–60.

de Jong B, Åberg J and Svenungsson B (2007) Outbreak of salmonellosis in a restaurant in Stockholm, Sweden, September–October 2006, **12**(11), available at: http://www.eurosurveillance.org/ViewArticle.aspx?ArticleId=749, accessed November 2008.

de Louvois J (1994) *Salmonella* contamination of stored hens' eggs, *PHLS Microbiology Digest*, **11**(4), 203–5.

Denny J, Threlfall J, Takkinen J *et al.* (2007) Multinational *Salmonella* Paratyphi B variant Java (*Salmonella* Java) outbreak, August–December 2007, **12**(51), available at: http://www.eurosurveillance.org/ViewArticle.aspx?ArticleId=3332, accessed November 2008.

Doyle M E and Mazzaotta A S (2000) Review of studies on the thermal resistance of salmonellae, *Journal of Food Protection*, **63**(6), 779–95.

EC (2004) Regulation (EC) No 852/2004 of the European Parliament and of the Council of 29 April 2004 on the hygiene of foodstuffs, *Official Journal of the European Union*, **L139**, 30 April, 1–54.

EC (2005) Commission Regulation (EC) No 2073/2005 of 15 November 2005 on microbiological criteria for foodstuffs, *Official Journal of the European Union*, **L338**, 22 December, 1–26.

EC (2007) Commission Regulation (EC) No 1441/2007 of 5 December 2007 amending Regulation (EC) No 2073/2005 on microbiological criteria for foodstuffs, *Official Journal of the European Union*, **L322**, 7 December 12–29.

Eglezos S, Bixing H and Stuttard E D (2008) A survey of the bacteriological quality of pre-roasted peanut, almond, cashew, hazelnut, and brazil nut kernels received into three Australian nut-processing facilities over a period of 3 years, *Journal of Food Protection*, **71**(2) 402–4 (abstract on www.ingentaconnect.com).

Emberland K E, Ethelberg S, Kuusi M *et al.* (2007) Outbreak of *Salmonella* Weltevreden infections in Norway, Denmark and Finland associated with alfalfa sprouts, July–October 2007, **12**(48), available at: http://www.eurosurveillance.org/ViewArticle.aspx?ArticleId=3321, accessed December 2008.

Evans H S, Madden P, Douglas C *et al.* (1998) General outbreaks of infectious intestinal disease in England and Wales: 1995 and 1996, *Communicable Disease and Public Health*, **1**(3) 165–71.

Evans S J (2000) *Salmonella* surveillance of cattle and sheep in GB, presented at 'Zoonotic infections in livestock and the risk to human health', MAFF, London, 7 December 2000.

Fisher I, Andersson Y, de Jong B *et al.* (2001) International outbreak of *S.* Typhimurium DT104 – update from Enter-net, *Eurosurveillance Weekly*, **5**: 010809.

Fröder H, Martins C G, de Souza K L O *et al.* (2007) Minimally processed vegetable salads: Microbial quality evaluation, *Journal of Food Protection*, **70**(5), 1277–80.

FSA (2000) *A Report on the Study of Infectious Intestinal Disease in England*, Food Standards Agency, London.

FSA (2002a) *Microbiological survey of raw cows' milk on retail sale 1995/6*. Food Standards Agency, London. available at: http://www.food.gov.uk/multimedia/pdfs/raw_cows_milk.pdf, accessed November 2008.

FSA (2002b) *Microbiological survey of unpasteurised sheep and goats' drinking milk*, Food Standards Agency, London, available at: http://www.food.gov.uk/multimedia/pdfs/unpasteurisedsheep_goats.pdf, accessed November 2008.

FSA (2003a) *Report of the national study on the microbiological quality and heat processing of cows' milk*, Food Standards Agency, London, available at: http://www.food.gov.uk/multimedia/pdfs/milksurvey.pdf, accessed November, 2008.

FSA (2003b) *UK-wide survey of* Salmonella *and* Campylobacter *contamination of fresh and frozen chicken on retail sale*, Food Standards Agency, London, available at: http://www.food.gov.uk/multimedia/pdfs/campsalmsurvey.pdf, accessed November 2008.

FSA (2004) *Report of the survey of* Salmonella *contamination of United Kingdom-produced shell eggs on retail sale*, Food Standards Agency, London, available at: http://www.food.gov.uk/multimedia/pdfs/fsis5004report.pdf, accessed November 2008.

FSA (2008) *Flooding: food safety advice*, Food Standards Agency, London, available at: http://www.food.gov.uk/safereating/microbiology/flood, accessed November 2008.

Fu T-J, Reineke K F, Chirtel S *et al.* (2008) Factors influencing the growth of *Salmonella* during sprouting of naturally contaminated alfalfa seeds, *Journal of Food Protection*, **71**(5), 888–96.

Gilbert R J, De Louvois J, Donovan T *et al.* (2000) Guidelines for the microbiological quality of some ready-to-eat foods sampled at the point of sale, *Communicable Disease and Public Health*, **3**(3), 163–7.

Goepfert J M, Iskander I K and Amundson C H (1970) Relation of the heat resistance of salmonellae to the water activity of the environment, *Applied Microbiology*, **19**(3), 429–33.

Golden D A, Rhodehamel E J and Kautter D A (1993) Growth of *Salmonella* spp. in cantaloupe, watermelon, and honeydew melons, *Journal of Food Protection*, **56**(3), 194–6.

Groisman E A and Ochman H (1997) How *Salmonella* became a pathogen, *Trends in Microbiology*, **5**(9), 343–9.

Groisman E A and Ochman H (2000) The path to *Salmonella*, *ASM News*, **66**(1), 21–7.

Guan T Y, Blank G, Ismond A *et al.* (2001) Fate of foodborne bacterial pathogens in pesticide products, *Journal of the Science of Food and Agriculture*, **81**(5), 503–12.

Hedberg C W, Angulo F J, White K E *et al.* (1999) Outbreaks of salmonellosis associated with eating uncooked tomatoes: implications for public health, *Epidemiology & Infection*, **122**(3), 385–93.

Heinonen-Tanski H, Niskanen E M, Salmela P *et al.* (1998) *Salmonella* in animal slurry can be destroyed by aeration at low temperatures, *Journal of Applied Microbiology*, **85**, 277–81.

Hew C M, Hajmeer M N, Farver T B *et al.* (2006) Pathogen survival in chorizos: Ecological factors, *Journal of Food Protection*, **69**(5), 1087–95.

Hilton A C, Willis R J and Hickie S J (2002) Isolation of *Salmonella* from urban wild brown rats (*Rattus norvegicus*) in the West Midlands, UK, *International Journal of Environmental Health Research*, **12**(2), 163–8 (abstract on www.ingentaconnect.com).

Horby P W, O'Brien S J, Adak G K *et al.* (2003) A national outbreak of multi-resistant *Salmonella enterica* serovar Typhimurium definitive phage type (DT) 104 associated with consumption of lettuce, *Epidemiology and Infection*, **130**(2), 169–78.

HPA (2008) Salmonella in humans (excluding S. Typhi amp; S. Paratyphi) Faecal amp; lower gastrointestinal isolates reported to the Health Protection Agency Centre for Infections England and Wales, 1981–2006, Health Protection Agency, London, available from: www.hpa.org.uk.

HPA (2009) Guidelines for assessing the microbiological safety of ready-to-eat foods, Health Protection Agency (Draft).

ICMSF (1986) *Microorganisms in Foods 2 Sampling for Microbiological Analysis: Principles and Specific Applications*, 2nd edn, International Commission on Microbiological Specifications for Foods, University of Toronto Press, Toronto.

ICMSF (1996) *Microorganisms in Foods 5, Microbiological Specifications of Food Pathogens*, International Commission on Microbiological Specifications for Foods, Blackie Academic & Professional, London.

ICMSF (2007) *Microorganisms in Foods 7, Microbiological Testing in Food Safety Management*, International Commission on Microbiological Specifications for Foods, Springer, New York.

ILSI (2000) Salmonella *Typhimurium definitive type (DT) 104: A Multi-resistant* Salmonella, Report prepared under the auspices of the International Life Sciences Institute, European Branch: Emerging Pathogen Task Force, ILSI Europe, Brussels.

Isaacs S, Aramini J, Ciebi B (2005) An international outbreak of salmonellosis associated with raw almonds contaminated with a rare phage type of *Salmonella* Enteritidis, *Journal of Food Protection*, **68**(1), 191–8 (abstract on www.ingentaconnect.com).

Islam M, Morgan J, Doyle M P *et al.* (2004) Fate of *Salmonella enterica* serovar Typhimurium on carrots and radishes grown in fields treated with contaminated manure composts or irrigation water, *Applied and Environmental Microbiology*, **70**(4), 2497–502.

ISO (2002) ISO *6579:2002 Microbiology of Food and Animal Feeding Stuffs – Horizontal Method for the Detection of* Salmonella *spp*, International Organization for Standardization, Geneva.

Izumi H, Poubol J, Hisa K *et al.* (2008) Potential sources of microbial contamination of Satsuma mandarin fruit in Japan, from production through packing shed, *Journal of Food Protection*, **71**(3), 530–8.

Jaquette C B, Beuchat L R and Mahon B E (1996) Efficacy of chlorine and heat treatment in killing *Salmonella stanley* inoculated onto alfalfa seeds and growth and survival of the pathogen during sprouting and storage, *Applied and Environmental Microbiology*, **62**(6), 2212–15.

Jeantet R, Baron F, Nau F *et al.* (1999) High intensity pulsed electric fields applied to egg white: effect on *Salmonella* Enteritidis inactivation and protein denaturation, *Journal of Food Protection*, **62**(12), 1381–6.

Jensen A N, Dalsgaard A, Stockmarr A *et al.* (2006) Survival and transmission of *Salmonella enterica* serovar Typhimurium in an outdoor organic pig farming environment, *Applied and Environmental Microbiology*, **72**(3), 1833–42.

Jones L (2000) *Molecular Methods in Food Analysis: Principles and Examples*, Key topics in Food Science & Technology, No. 1, Campden and Chorleywood Food Research Association Group, Chipping Campden.

Keady S, Briggs G, Farrar J *et al.* (2004) Outbreak of *Salmonella* serotype Enteritidis infections associated with raw almonds – United States an Canada, 2003–2004, *Morbidity and Mortality Weekly Report*, **53** (Dispatch); 1–3.

Kieboom J, Kusumaningrum H D, Tempelaars M H *et al.* (2006) Survival, elongation, and elevated tolerance of *Salmonella enterica* serovar Enteritidis at reduced water activity, *Journal of Food Protection*, **69**(11), 2681–6 (abstract on www.ingentaconnect.com).

Killalea D, Ward L R, Roberts D *et al.* (1996) International epidemiological and microbiological study of outbreak of *Salmonella agona* infection from a ready to eat savoury snack—I: England and Wales and the United States, *British Medical Journal*, **313**, 1105–7.

Kim T, Silva J L and Chen T C (2002) Effects of UV irradiation on selected pathogens in peptone water and on stainless steel and chicken meat, *Journal of Food Protection*, **65**(7), 1142–5.

Kirk D, Little C L, Le M *et al.* (2004) An outbreak due to peanuts in their shell caused by *Salmonella enterica* serotypes Stanley and Newport – sharing molecular information to solve international outbreaks, *Epidemiology and Infection*, **132**, 571–7.

Lee B H, Kermasha S and Baker B E (1989) Thermal, ultrasonic and ultraviolet inactivation of *Salmonella* in thin films of aqueous media and chocolate, *Food Microbiology*, **6**, 143–52.

Le Minor L (1984) Genus *111. Salmonella* Lignières 1900, 389AL, in N R Krieg and J G Holt (eds), *Bergey's Manual of Systematic Bacteriology*, Vol. 1, Williams & Wilkins, Baltimore, MD, 427–58.

Levine P, Rose B, Green S *et al.* (2001) Pathogen testing of ready-to-eat meat and poultry products collected at federally inspected establishments in the United States, 1990 to 1999, *Journal of Food Protection*, **64**(8), 1188–93.

Liang Z, Mittal G and Griffiths M W (2002) Inactivation of *Salmonella* Typhimurium in orange juice containing antimicrobial agents by pulsed electric field, *Journal of Food Protection*, **65**(7), 1081–7.

Little C (2001) *Salmonella* Stanley and *Salmonella* Newport in imported peanuts – update, *Eurosurveillance Weekly*, **5**: 011025.

Little C L, Rhoades J R, Sagoo S K *et al.* (2007) European Commission co-ordinated programme for the official control of foodstuffs for 2004: Microbiological examination of cheeses made from raw or thermised milk from production establishments and retail premises in the United Kingdom, available at: http://www.food.gov.uk/multimedia/pdfs/861rawor thermisedcheeses, accessed November 2008.

Little C L, Rhoades J R, Hucklesy L *et al.* (2008) Survey of *Salmonella* contamination of raw shell eggs used in food service premises in the United Kingdom, 2005 through 2006, *Journal of Food Protection*, **71**(1), 19–26.

Luzzi I, Galetta P, Massari M *et al.* (2007) Outbreak Report: An Easter outbreak of *Salmonella* Typhimurium DT104A associated with traditional pork salami in Italy, *Eurosurveillance*, **12**(4), available at: http://www.eurosurveillance.org/ViewArticle.aspx?ArticleId=702, accessed December 2008.

MAFF (2000) *Zoonoses Report UK 1998*, Ministry of Agriculture, Fisheries and Food, London.

Mattick K L, Jørgensen F, Legan J D *et al.* (2000) Survival and filamentation of *Salmonella enterica* serovar Enteritidis PT4 and *Salmonella enterica* serovar Typhimurium DT104 at low water activity, *Applied and Environmental Microbiology*, **66**(4), 1274–9.

McManus C and Lanier J M (1987) *Salmonella*, *Campylobacter jejuni* and *Yersinia enterocolitica* in raw milk, *Journal of Food Protection*, **50**(1), 51–5.

Mermin J, Hutwagner L, Vugia D *et al.* (2004) Reptiles, amphibians and human *Salmonella* infection – a population based case-control study, *Clinical Infectious Diseases*, **38** (Suppl 3), S253–261.

Moore B C, Martinez E, Gay J M *et al.* (2003) Survival of *Salmonella enterica* in freshwater and sediments and transmission by the aquatic midge *Chironomus tentans* (Chironomidae: Diptera), *Applied and Environmental Microbiology*, **69**(8), 4556–60.

Mukherjee A, Speh D, Dyck E *et al.* (2004) Preharvest evaluation of coliforms, *Escherichia coli*, *Salmonella*, and *Escherichia coli* O157:H7 in organic and conventional produce grown by Minnesota farmers, *Journal of Food Protection*, **67**(5), 894–900.

Müller L, Hjertqvis M, Payne L *et al.* (2007) Cluster of *Salmonella* Enteritidis in Sweden 2005–2006 – suspected source: almonds, Outbreak Report, *Eurosurveillance*, **12**(6), available at: http://www.eurosurveillance.org/ViewArticle.aspx?ArticleId=718, accessed December 2008.

NACMCF (1999a) Microbiological safety evaluations and recommendations on sprouted seeds, National Advisory Committee on Microbiological Criteria for Foods, *International Journal of Food Microbiology*, **52**(3), 123–53.

NACMCF (1999b) Microbiological safety evaluations and recommendations on fresh produce, National Advisory Committee on Microbiological Criteria for Foods, *Food Control*, **10**(2), 117–43.

Ng P J, Fleet G H and Heard G M (2005) Pesticides as a source of microbial contamination of salad vegetables, *International Journal of Food Microbiology*, **101**, 237–50.

Ng S, Rouch G, Dedman R *et al.* (1996) Human salmonellosis and peanut butter. *Communicable Disease Intelligence*, **20**(14), 326.

Nicholson F A, Hutchison M L, Smith K A *et al.* (2000) *A study of farm manure applications to agricultural land and an assessment of the risks of pathogen transfer into the food chain*, Research Report, Ministry of Agriculture Fisheries and Food, London.

Old D C (1992) Nomenclature of *Salmonella*, *Journal of Medical Microbiology*, **37**, 361–3.

Old D C and Threlfall E J (1997) *Salmonella*, in A Balows and B I Duerden (eds), *Topley and Wilson's Microbiology and Microbial Infections*, 9th edn, Arnold, London, 969–97.

O'Mahony M, Cowden J, Smyth B *et al.* (1990) An outbreak of *Salmonella saint-paul* infection associated with bean sprouts, *Epidemiology and Infection*, **104**, 229–35.

Orozco L, Iturriaga M H, Tamplin M *et al.* (2008) Animal and environmental impact on the presence and distribution of *Salmonella* and *Escherichia coli* in hydroponic tomato greenhouses, *Journal of Food Protection*, **71**(4), 676–83.

Pao S and Davis C L (1999) Enhancing microbiological safety of fresh orange juice by fruit immersion in hot water and chemical sanitizers, *Journal of Food Protection*, **62**(7), 756–60.

Parish M E, Narciso J A and Freidrich L M (1997) Survival of salmonellae in orange juice, *Journal of Food Safety*, **17**, 273–81.

Pezzoli L, Elson R, Little C *et al.* (2007) International outbreak of *Salmonella* Senftenberg in 2007, *Eurosurveillance*, available at: http://www.eurosurveillance.org/ViewArticle.aspx?ArticleId=3218, accessed December 2008.

Phebus R K, Nutsch A L, Schafer D E *et al.* (1997) Comparison of steam pasteurisation and other methods for reduction of pathogens on surfaces of freshly slaughtered beef, *Journal of Food Protection*, **60**(5), 476–84.

Phillips L E, Humphrey T J and Lappin-Scott H M (1998) Chilling invokes different morphologies in two *Salmonella enteritidis* PT4 strains, *Journal of Applied Microbiology*, **84**, 820–6.

Pontello M, Sodano L, Nastasi A *et al.* (1998) A community-based outbreak of *Salmonella enterica* serotype Typhimurium associated with salami consumption in Northern Italy, *Epidemiology and Infection*, **120**, 209–14.

Public Health Laboratory Service (2004) Preventing person-to-person spread following gastrointestinal infections: guidelines for public health physicians and environmental health officers. Prepared by a Working Group of the former PHLS Advisory Committee on Gastrointestinal Infections, *Communicable Disease and Public Health*, **7**(4), 362–84.

Ray B, Jezeski J J and Busta F F (1971) Isolation of salmonellae from naturally contaminated dried milk products. II. Influence of storage time on the isolation of salmonellae, *Journal of Milk and Food Technology*, **34**(9), 423–7.

Rayman M K, D'Aoust J-Y, Aris B *et al.* (1979) Survival of microorganisms in stored pasta, *Journal of Food Protection*, **42**(4), 330–4.

Ritz M, Pilet M F, Juiau F *et al.* (2006) Inactivation of *Salmonella* Typhimurium and *Listeria monocytogenes* using high-pressure treatments: destruction or sublethal stress? *Letters in Applied Microbiology*, **42**, 357–62.

Reporter R, Mascola L, Kilman L *et al.* (2000) Outbreaks of *Salmonella* serotype Enteritidis infection associated with eating raw or undercooked shell eggs – United States, 1996–1998, *Morbidity and Mortality Weekly Report*, **49**(4), 73–9.

Roberts D and Greenwood M (2003) *Practical Food Microbiology*, 3rd edn, Blackwell, Oxford.

Rodriguez A, Pangloli P, Richards H A *et al.* (2006) Prevalence of *Salmonella* in diverse environmental farm samples, *Journal of Food Protection*, **69**(11), 2576–80 (abstract on www.ingentaconnect.com).

Sagoo S K, Little C L and Mitchell R T (2001) The microbiological examination of ready-to-eat organic vegetables from retail establishments in the United Kingdom, *Letters in Applied Microbiology*, **33**(6), 434–9.

Sagoo S K, Little C L and Mitchell R T (2003) Microbiological quality of open ready-to-eat salad vegetables: effectiveness of food hygiene training of management, *Journal of Food Protection*, **66**(9), 1581–6.

Schmidt P L, O'Connor A M, McKean J D *et al.* (2004) The association between cleaning and disinfection of lairage pens and the prevalence of *Salmonella enterica* in swine at harvest, *Journal of Food Protection*, **67**(7), 1384–8.

Sharma M, Adler B B, Harrison M D *et al.* (2005) Thermal tolerance of acid-adapted and unadapted *Salmonella*, *Escherichia coli* O157:H7, and *Listeria monocytogenes* in cantaloupe juice and watermelon juice, *Letters in Applied Microbiology*, **41**, 448–53.

Shohat T, Green M S, Merom D *et al.* (1996) International epidemiological and microbiological study of outbreak of *Salmonella agona* infection from a ready-to-eat savoury snack- II: Israel, *British Medical Journal*, **313**, 1107–9.

Smith J L, Huhtanen C N, Kissinger J C *et al.* (1975) Survival of salmonellae during pepperoni manufacture, *Applied Microbiology*, **30**(5), 759–63.

Sofos J N, Kochevar S L, Reagan J O *et al.* (1999) Incidence of *Salmonella* on beef carcasses relating to the U.S. meat and poultry inspection regulations, *Journal of Food Protection*, **62**(5), 467–73.

Stepanović S, Ćirković I, Ranin L *et al.* (2004) Biofilm formation by *Salmonella* spp. and *Listeria monocytogenes* on plastic surface, *Letters in Applied Microbiology*, **38**, 428–32.

Stopforth J D, Mai T, Kottapalli B *et al.* (2008) Effect of acidified sodium chlorite, chlorine, and acidic electrolyzed water on *Escherichia coli* O157:H7, *Salmonella*, and *Listeria monocytogenes* inoculated onto leafy greens, *Journal of Food Protection*, **71**(3), 625–8.

Swanson S J, Snider C, Braden C R *et al.* (2007) Multidrug-resistant *Salmonella enterica* serotype Typhimurium associated with pet rodents, *The New England Journal of Medicine*, **356**(1), 21–8.

Taormina P J, Beuchat L R and Slutsker L (1999) Infections associated with eating seed sprouts: an international concern, *Emerging Infectious Diseases*, **5**(5), 626–40.

Tamminga S K, Beumer R R, Kampelmacher E H *et al.* (1976) Survival of *Salmonella eastbourne* and *Salmonella typhimurium* in chocolate, *Journal of Hygiene, Cambridge*, **76**, 41–7.

Threlfall E J, Hall M L M and Rowe B (1983) Lactose-fermenting salmonellae in Britain, *FEMS Microbiology Letters*, **17**, 127–30.

Threlfall J, Ward L and Old D (1999) Changing the nomenclature of *Salmonella, Communicable Disease and Public Health*, **2**(3), 156–7.

Tindall B J, Grimont P A D, Garrity G M *et al.* (2005) Nomenclature and taxonomy of the genus *Salmonella, International Journal of Systematic and Evolutionary Microbiology*, **55**, 521–4.

Topley W W C and Wilson G S (1929) *The Principles of Bacteriology and Immunity*, Vol. II, Arnold, London, 1037–8.

Torpdahl M, Sørensen G, Ethelberg S *et al.* (2006) A regional outbreak of *S.* Typhimurium in Denmark and identification of the source using MLVA typing, *Eurosurveillance*, **11**(5), available at: http://www.eurosurveillance.org/ViewArticle.aspx?ArticleId=621, accessed November 2008.

Trepka M J, Archer J R, Altekruse S F *et al.* (1999) An increase in sporadic and outbreak-associated *Salmonella* Enteritidis infections in Wisconsin: the role of eggs, *Journal of Infectious Disease*, **180**, 1214–19.

Uesugi A R, Danyluk M D, Mandrell R E *et al.* (2007) Isolation of *Salmonella* Enteritidis phage type 30 from a single almond orchard over a 5-year period, *Journal of Food Protection*, **70**(8), 1784–9 (abstract on www.ingentaconnect.com).

USDA (1994) *Nationwide Beef Microbiological Baseline Data Collection Program: Steers and Heifers. October 1992–1993*, United States Department of Agriculture, Food Safety and Inspection Service, Washington, DC.

USDA (1998) Guide to minimize microbial food safety hazards for fresh fruits and vegetables, United States Department of Agriculture, Washington, DC, available at: http://vm.cfsan.fda.gov/~dms/prodguid.html, accessed December 2008.

Uzzau S, Brown D J, Wallis T *et al.* (2000) Review: host adapted serotypes of *Salmonella enterica, Epidemiology and Infection*, **125**, 229–55.

Vugia D, Cronquist A, Hadler J *et al.* (2007) Preliminary FoodNet data on the incidence of infection with pathogens commonly transmitted through food – 10 States, 2006, *Morbidity and Mortality Weekly Report*, **56**(14), 336–9.

Ward L (2000) *Salmonella* perils of pet reptiles, *Communicable Disease and Public Health*, **3**(1), 2–3.

Werner S, Boman K, Einemo I *et al.* (2007) Outbreak of *Salmonella* Stanley in Sweden associated with alfalfa sprouts, July – August 2007, **12**(42), available at: http://www.eurosurveillance.org/ViewArticle.aspx?ArticleId=3291, accessed November 2008.

Whitney B M, Williams R C, Eifert J *et al.* (2008) High pressures in combination with antimicrobials to reduce *Escherichia coli* O157:H7 and *Salmonella* Agona in apple juice and orange juice, *Journal of Food Protection*, **71**(4), 820–4.

Willford J, Mendonca A and Goodridge L D (2008) Water pressure effectively reduces *Salmonella enterica* serovar Enteritidis on the surface of raw almonds, *Journal of Food Protection*, **71**(4), 825–9 (abstract on www.ingentaconnect.com).

Wilson D (2004) *Salmonella* outbreak linked to food outlet in northeast England, *Eurosurveillance Weekly*, **8**(33), 2.

Wilson I G, Heaney J C N and Powell G G (1998) *Salmonella* in raw shell eggs in Northern Ireland: 1996–7, *Communicable Disease and Public Health*, **1**(3), 156–60.

Yang Z, Li Y and Slavik M (1998) Use of antibacterial spray applied with an inside–outside birdwasher to reduce bacterial contamination on pre-chilled chicken carcasses, *Journal of Food Protection*, **61**(7), 829–32.

You Y, Rankin S C, Aceto H W *et al.* (2006) Survival of *Salmonella enterica* serovar Newport in manure and manure-amended soils, *Applied and Environmental Microbiology*, **72**(9), 5777–83

Zhang R Y, Beuchat L R, Angulo F J (1995) Fate of *Salmonella* Montevideo on and in raw tomatoes as affected by temperature and treatment with chlorine, *Applied Environmental Microbiology*, **61**, 2127–31.

19

Listeria monocytogenes

C. Bell, Independent Consultant Microbiologist, UK, and
A. Kyriakides, Sainsbury's Supermarkets Ltd, UK

Abstract: Based on their extensive experience working within the food industry and supported by published reports and research work, the authors describe in sections titled with self-explanatory headings the key characteristics of *Listeria monocytogenes* that help to explain why it is an important foodborne pathogen and approaches to the monitoring and control of the organism in food production processes.

Key words: *Listeria, Listeria monocytogenes, L. monocytogenes*, bacterial pathogen, listeriosis, foodborne illness, sources, characteristics, growth, survival, control in foods hazard analysis, risk factors, food processes, detection, meat and meat products, milk and milk products, fruit, vegetables.

19.1 Introduction

Listeria monocytogenes has been recognised as a significant foodborne pathogen only since the early 1980s when outbreaks of foodborne listeriosis demonstrated the severe nature of the illness with exceptionally high levels of mortality, particularly in the most vulnerable members of the community such as unborn babies, the elderly and the immunocompromised. *L. monocytogenes* is widely distributed in the environment and occurs in almost all food raw materials from time to time. It is recognised that its presence in raw foods cannot be completely eliminated, but it is possible to reduce its incidence and levels in food products through the application of effective hygienic measures (CAC, 1996).

19.2 Characteristics of *Listeria monocytogenes*

19.2.1 The organism

Currently, six clearly distinguishable species of *Listeria* are recognised: *L. monocytogenes*, *L. innocua*, *L. welshimeri*, *L. seeligeri*, *L. ivanovii* and *L. grayi*. Proposals have also been made for two sub-species of *L. ivanovii*: *L. ivanovii* sub-species *ivanovii* and *L. ivanovii* sub-species *londoniensis* (Boerlin *et al.*, 1992). Table 19.1 indicates the common conventional tests used in the differentiation of *Listeria* spp. The most commonly occurring species in food are *L. innocua* and *L. monocytogenes* (Jay, 1996; Kozak *et al.*, 1996).

Listeria monocytogenes is the main human pathogen of the *Listeria* genus (Jones, 1990). The factors predisposing infection are not fully understood but include host immunity, level of inoculum and virulence including haemolytic activity of the specific *L. monocytogenes* strain.

The majority of large foodborne outbreaks of listeriosis have been attributed to one of two serogroups of *L. monocytogenes*, either 1/2a or, more frequently, 4b. The UK Health Protection Agency (previously Public Health Laboratory Service) defines a case of listeriosis as a 'patient with a compatible illness from whom *L. monocytogenes* was isolated from a normally sterile site (usually blood or cerebrospinal fluid, CSF)' (Anon., 1997; Gillespie *et al.*, 2005). The serogroup of the organism is believed to be important when assessing the risk to an individual consuming a food containing *L. monocytogenes* and national authorities may, when methods become available to differentiate pathogenic from non-pathogenic strains, need to consider the apparent differences in the potential of different serogroups to cause foodborne outbreaks in future risk assessments.

19.2.2 The illness caused

Individuals principally at risk from listeriosis have been reviewed by Rocourt (1996) and in order of descending risk are organ transplant patients, patients with AIDS, HIV-infected individuals, pregnant women, patients with cancer and the elderly. *L. monocytogenes* can cause a variety of infections (Table 19.2), but listeriosis most commonly takes the form of an infection of the uterus, the bloodstream or the central nervous system which in pregnant women can result in spontaneous abortion, stillbirth or birth of a severely ill baby owing to infection of the foetus. Listeriosis may also be acquired by new-born babies owing to postnatal infection from the mother or other infected babies. The mother is rarely severely affected by listeriosis as the disease appears to focus on the foetus (Rocourt, 1996). Non-pregnant adults may also suffer listeriosis, with the most vulnerable groups including the immunocompromised and the elderly because of the reduced competence of their immune systems. In such groups, listeriosis usually presents as meningitis, septicaemia or bacteraemia.

Table 19.1 Common conventional tests used to differentiate *Listeria* spp.

Species	β-haemolysis	Production of acid from:			CAMP reaction with:	
		L-rhamnose	D-xylose	D-mannitol	*Staphylococcus aureus*	*Rhodococcus equi*
L. monocytogenes	+	+ (a few strains –)	–	–	+	– (some strains +)
L. innocua	–	V	–	–	–	–
L. ivanovii	++	–	+	–	–	+
L. welshimeri	–	V	+	–	–	–
L. seeligeri	(+)	–	+	–	(+)	–
L. grayi	–	–	–	+	–	–

++ = Strong positive reaction
+ = Positive reaction
(+) = Weak positive reaction
– = Negative reaction
V = variable reaction for different isolates.
Source: Adapted from Bell and Kyriakides (2005).

Table 19.2 Types of illness caused by *Listeria monocytogenes*

Type of illness	Incubation time	Symptoms
Zoonotic infection.	1 – 2 days.	Localised skin lesions that are mild and self-resolving.
Neonatal infection: newborn babies infected from mother during birth or due to cross-infection from one neonate in the hospital to other babies.	1 – 2 days (early onset) usually from congenital infection prior to birth.	Can be severe resulting in meningitis and death.
Infection of the mother during pregnancy acquired from contaminated food. Infection is more common in the third trimester.	Varies from 1 day to several months.	Pyrexial influenza-like illnesses or asymptomatic in the mother, but serious implications for the unborn infant, including spontaneous abortion, stillbirth and meningitis.
Listeriosis acquired by non-pregnant individuals from contaminated food.	Varies from 1 day to several months.	Asymptomatic or mild illness which may progress to bacteraemia, septicaemia or central nervous system infections such as meningitis. Most common and most severe in immunocompromised or elderly people.
Listeria food poisoning caused by consumption of food containing very high levels (> 10^7/g) of *L. monocytogenes*.	< 24 h.	Vomiting and diarrhoea plus fever, sometimes progressing to bacteraemia but usually self-resolving.

Source: Adapted from Bell and Kyriakides (2005).

The symptoms of listeriosis do not usually resemble those of the more familiar types of food poisoning, but there have been several recent episodes where the presence of extremely high levels of *L. monocytogenes* has resulted in the rapid onset of symptoms of vomiting and diarrhoea with few apparent cases of classical listeriosis (Salamina *et al.*, 1996; Dalton *et al.*, 1997; Aureli *et al.*, 2000; Carrique-Mas *et al.*, 2003; Makino *et al.*, 2005). Cases of listeriosis and *Listeria* gastroenteritis have stimulated interest in the mechanisms whereby the organism effects these infection outcomes (Gahan and Hill, 2005). Where cases of listeriosis or *Listeria* gastroenteritis are diagnosed, advice given by Public Health clinicians involves treatment of individual cases and, as appropriate, removal of the contaminated foods identified as the source of infection (PHLS, 2004).

In spite of the relatively low annual incidence of disease, i.e. 1–10 cases per million in some western countries, listeriosis is a serious illness, and this is reflected by the apparent high mortality rate in many cases, with fatalities averaging approximately 30 % (Newton *et al.*, 1992). Indeed, in a review of listeriosis outbreaks in the USA between 1998 and 2002 Lynch *et al.* (2006) reported 14.8 % mortality (38 deaths out of 256 cases from 11 outbreaks).

Although the pathogenicity of individual strains of *L. monocytogenes* and hence their relevance to public health is of great concern, in a detailed review of the pathogenicity of *Listeria monocytogenes*, McLauchlin (1997) concluded 'in the interests of public safety and for considerations for public health purposes, all *L. monocytogenes*, including those recovered from food, should be regarded as potentially pathogenic'. There has been much research on the virulence and pathogenicity markers of *L. monocytogenes* over the years in an attempt to better characterise the risk associated with different strains of the organism. Studies in animal models and epidemiological evidence do suggest there are differences in virulence and pathogenicity with the occurrence of naturally-occurring avirulent and virulent strains (EFSA, 2007a). However, as there are no routine methods available to differentiate pathogenic and non-pathogenic strains of *L. monocytogenes*, it remains the case that regulators consider all *L. monocytogenes* as potentially pathogenic (EFSA, 2007a). Consequently, where *L. monocytogenes* is included in food-related legislation, guidelines or specifications, *L. monocytogenes* is not qualified as to serotype or other sub-division of the species. Indeed, the food industry always reacts to positive findings of the species, not specific serotypes of the species, and many manufacturers use the presence of *Listeria* spp. as a general indicator for the presence of *L. monocytogenes*. Industry guidelines and specifications reflect this approach.

The general approach taken to legislation in Europe and North America in the context of food safety is to indicate the clear responsibility of food business proprietors to produce and supply safe and wholesome foods using Hazard Analysis Critical Control Point (HACCP)-based approaches for the control and monitoring of food manufacturing processes, e.g. European Union Regulation (EC) No 852/2004 of the European Parliament and of the Council of 29 April 2004 on the hygiene of foodstuffs (EC, 2004). Some legislation, however, contains specific microbiological standards that include *L. monocytogenes*, and compliance is compulsory, e.g. European Union Commission Regulations on microbiological criteria for foodstuffs (EC 2005, 2007).

Listeria monocytogenes must, of course, also be considered for inclusion in generic standards that state that pathogenic organisms must be absent in a specified quantity of food. In the USA, following surveys of the incidence of *L. monocytogenes* in foods and food-related samples, the Food and Drug Administration and Food Safety and Inspection Service implemented a policy of 'zero tolerance' (negative in 25 g of food sample) for *L. monocytogenes* in cooked and ready-to-eat foods (Shank *et al.*, 1996), which was reported to have contributed to a 40 % decline in listeriosis in the USA (Tappero *et al.*, 1995). It has also resulted in *L. monocytogenes* contamination accounting for the greatest number of food product recalls in the USA between 1993 and 1998 (Wong *et al*., 2000).

19.2.3 Growth and survival

Listeria monocytogenes can survive for long periods on soil, depending on the type of soil and environmental conditions. Nicholson *et al.* (2000) reviewed studies on the presence and survival of the organism in the environment which showed it to be able to survive for 69–730 d in dry soil and 295–350 d in moist soil. Its presence in uncultivated soils (30.8 % and 44 % of the samples examined in two surveys) was more common than in cultivated soils (8.3 % and 12.2 % of the samples examined in two surveys).

Within food production processes, a variety of physicochemical factors, used either singly or in combination, can be effective in controlling the survival and growth of *L. monocytogenes* both during processing and in the finished food products. Table 19.3 indicates the growth-limiting temperature, pH and water activity for *L. monocytogenes*. It should be noted, however, that *L. monocytogenes* is known to grow at the temperatures used for refrigeration, although only slowly (Table 19.4). Hudson *et al.* (1994) reported growth of *L. monocytogenes* at temperatures as low as –1.5 °C in vacuum-packed, cooked beef, although the precision of the temperature measurement was not specified.

Although *L. monocytogenes* is not expected to proliferate in conditions outside of the minimum and maximum growth-limiting parameters, strains will and do survive for varying periods in these adverse-to-growth conditions. Studies are carried out periodically to assess the survival capability of *L.*

Table 19.3 Growth-limiting parameters for *Listeria monocytogenes*

	Minimum	Maximum
Temperature (°C)	–0.4[a]	45
pH	4.39	9.4
Water activity (a_w)	0.92	–

[a]Hudson *et al.* (1994) reported growth at temperatures of –1.5 °C.
Source: Adapted from ICMSF (1996).

Table 19.4 Growth rate and lag times for *Listeria monocytogenes* at different temperatures

Temperature (°C)	0–1	2–3	4–5	5–6	7–8	9–10	10–13
Lag time (days)	3–33	2–8	–	1–3	2	< 1.5	–
Generation time (hours)	62–131	–	13–25	–	–	–	5–9

Source: Adapted from Mossel *et al.* (1995).

monocytogenes in food types that become of particular concern through outbreak reports and surveillance, e.g. cultured dairy products, and Table 19.5 summarises a few such studies.

Listeria monocytogenes is recognised as being particularly adapted to be capable of growth at low temperatures, a feature that gives it a competitive advantage over many other microbial contaminants in food. The mechanisms by which it achieves this, however, are less well understood. Tasara and Stephan (2006) reviewed recent studies on the cold stress response of *L. monocytogenes* and the associated food safety implications. Amongst the various responses triggered by cold stress, the following key features are noteworthy:

- changes occur in cell membrane structure to maintain lipid fluidity and structural integrity;
- cells accumulate cryoprotective osmolytes and peptides to help maintain enzyme and membrane activity at low temperature;
- alterations in cell surface proteins may relate to the need to access environments offering greater potential for growth and survival for nutritional reasons when under cold stress;
- cells produce 'cold shock proteins' which may be important in protecting against oxidative stress associated with cold adaptation;
- structural or protective changes may occur in response to cold stress to adapt to the colder temperatures to maintain the structural and functional stability of the ribosome which is critical to protein synthesis.

An understanding of the mechanisms of cold stress responses at the molecular level may provide ways to intervene using common food ingredients or processes to reduce the organism's ability to adapt to low temperatures in food.

To control the growth of *L. monocytogenes* in chilled foods, it is crucial to operate and apply well-controlled chill holding and storage systems both within the production process for component or part-processed foods and for finished product storage and distribution. This and other physicochemical factors can control the survival and growth of *L. monocytogenes* during the manufacturing process and the shelf-life of the finished food products (Bell and Kyriakides, 2005).

Perhaps one additional trait of note with regards to the growth and survival characteristics of *L. monocytogenes* is its ability to grow at relatively high salt concentrations (and thus low water activity). It is capable of growing at far higher salt levels compared to other infectious foodborne pathogens. Cole et al. (1990) studied the growth of *L. monocytogenes* in nutrient media with differing salt and pH values stored at varying temperature. They reported growth at 5 °C and at pH 7.0 in a salt concentration of 8 %, although no growth occurred at 10 % salt. When the temperature was increased to 10 °C, the organism grew in 10 % salt, but levels of 12 % salt and above prevented growth. At 30 °C the organism grew in 12 % salt but not 14 %. Adjusting

Table 19.5 Results of some studies of the survival of *Listeria monocytogenes* in dairy products

Substrate	Conditions	Survival time	Reference
Yogurt	Cows' milk inoculated with *L. monocytogenes*, made into yogurt and then stored at 4 °C Product pH Day 0:0 h–6.6; 5 h–5.0 Day 2–15:–4.2 Inoculum level:		Massa *et al.* (1991)
	c. 10^3 cfu/ml	Recoverable after 2 d; not detectable after 5 d	
	c. 10^7 cfu/ml	Recoverable after 7 d; not detectable after 15 d	
Cottage cheese	Three different source batches of cottage cheese inoculated with *L. monocytogenes* at *c.* 4.4×10^4 cfu/g Stored at 4 °C for 14 d Product pH:		Hicks and Lund (1991)
	Day 0–5.06; Day 14–5.00	Little change over 14 d	
	Day 0–4.69; Day 14–4.67	*c.* 1.4 $\log_{10}$ reduction	
	Day 0–4.75; Day 14–4.60	*c.* 1.5 $\log_{10}$ reduction	
Soft cheeses:	Inoculated with acid-adapted and non acid-adapted *L. monocytogenes* at levels approx. 10^5/g Stored at 4 °C	Approximate population changes after 14 d	Cataldo *et al.* (2007)
Mozzarella	acid-adapted	1.6 $\log_{10}$ reduction	
	non acid-adapted	1.9 $\log_{10}$ reduction	
Crescenza	acid-adapted	3.0 $\log_{10}$ increase (cell morphology changes observed)	
	non acid-adapted	0.8 $\log_{10}$ reduction	
Gorgonzola	acid-adapted	1.4 $\log_{10}$ reduction	
	non acid-adapted	1.3 $\log_{10}$ reduction	
Ricotta	acid-adapted	1.5 $\log_{10}$ reduction	
	non acid-adapted	2.0 $\log_{10}$ reduction	

the pH to 5.13 resulted in marked inhibition of growth with no growth being recorded at 5 °C, even with 0 % salt and growth at 10 °C and 30 °C occurring at salt concentrations up to 6 % and 8%, respectively. *L. monocytogenes* cells have been shown to respond to the stress of high sodium chloride levels by producing elongated cells (Hazeleger *et al.*, 2006).

Listeria monocytogenes, like many other bacteria, can develop increased resistance to heat destruction, i.e. pasteurisation processes, if first subjected to sub-lethal heat processes, i.e. 44–48 °C for 20–60 min immediately before pasteurisation (Farber, 1989; Lou and Yousef, 1999). Enhanced thermotolerance is reported also to be conferred when the organism is exposed to a variety of other stresses including growing at 43 °C, cell starvation or exposure to ethanol, acid or hydrogen peroxide (Lou and Yousef, 1999; Farber and Peterkin, 2000). This heat-shock response may be an important consideration in the design of heat processes where intermediate stages may include holding the ingredients for extended periods at sub-lethal temperatures, or where heating of the product may be very slow. Carlier *et al.* (1996) reported a near doubling of the *D*-value of *L. monocytogenes* from 1.82 min to 3.48 min at 60 °C when previously exposed to temperatures of 42 °C for 1 h. Taking account of the *z*-value of the strain with increased heat tolerance (6.74 C°), the predicted reduction of the organism at 70 °C (17.5 $\log_{10}$ reduction) would still far exceed the 6 $\log_{10}$ reduction normally expected from a two minute cook.

Likewise, enhanced tolerance to acidic environments, e.g. in cheese, has been reported (Cataldo *et al.*, 2007) in strains that have been acid-adapted (Table 19.5). In manufacturing environments where acidified products are processed, organisms present in and around the environment will be exposed to acidic conditions with the associated risk they may become habituated/adapted to the conditions and exhibit enhanced tolerance within any product they may subsequently contaminate.

In any discussion of foodborne pathogens, the subject of consistent and effective cleaning/hygiene procedures continues to be a high priority. Although cleaning is, perhaps, considered to be a mundane subject that is 'easy' to resolve, evidence provided by investigations such as those of Lundén *et al.* (2002) which demonstrated the transfer of a specific strain of *L. monocytogenes* via one dicing machine to three cooked meat manufacturing plants would contradict any dismissive attitude. The machine was subject to regular cleaning, but the routine cleaning regimes employed did not remove/destroy the strain and it persisted over a period of 2–3 y on the machine from which the strain contaminated other plant and product in the manufacturing areas in which it was used. The persistence of the strain was considered to be due, at least in part, to the strong surface adherence characteristics of the strain and associated increased resistance to some disinfectants. Persistent 'in-house' strains of *L. monocytogenes* have also been reported, e.g. Mędrala *et al.* (2003) who studied strains isolated from the products of a fish processing plant in Poland, and ineffective cleaning and disinfection procedures have been considered contributory to the development of these situations.

Resistance to disinfection is an important consideration as this is a critical part of ensuring the organism is managed in the factory environment. Studies have demonstrated that biofilms of *L. monocytogenes* are more resistant to the lethal effects of chlorine than when challenged as a cell suspension or as adherent cells (Lee and Frank, 1991). Fulsom and Frank (2006) demonstrated resistance to chlorine (5 mins at 60 ppm chlorine) in nine of 13 strains of *L. monocytogenes* when exposed in solution (planktonic growth). Exposure of the same strains when grown as biofilms resulted in 11 of the 13 strains demonstrating resistance to 60 ppm chlorine for 5 min; this was not related to the cell density of the biofilm. Resistance of stains in planktonic form was not correlated with resistance when in biofilms. There was an association between genetically related strains and chlorine resistance when grown on biofilms but not when grown planktonically. It has been reported that resistance to disinfectants such as chlorine can also be induced by prior exposure to other disinfectants including chlorine itself and quaternary ammonium compounds and that this resistance could last for up to a week (Lundén *et al.*, 2003).

The potential for lubricants used in the food industry to be a means of proliferation and spread of the organism is also an important consideration. Aarnisalo *et al.* (2006) reported on the survival of *L. monocytogenes* in lubricants used in the food industry and, although the organism generally reduced in number over time in most lubricants, the results demonstrated that the organism could survive in synthetic conveyor-belt lubricant diluted in water. Additionally, the organism was shown to be transferred from stainless steel surfaces into conveyor-belt lubricants and into mineral-oil-based lubricants.

Such findings, including those associated with biofilm formation, reinforce the importance of designing cleaning and disinfection strategies to minimise adsorption and attachment of the organism to surfaces and subsequent biofilm development, and also the need to regularly review both the chemicals and the methods employed in cleaning programmes. Likewise, the report of persistent *L. monocytogenes* strains being more resistant to acid stress together with the significant variability in heat tolerance in persistent and non-persistent environmental strains (Lundén *et al.*, 2007) demonstrates the need to ensure that cleaning and disinfection regimes are designed to take account of strains with increased resistance to subsequent processing; the aim of eliminating all strains of *Listeria* species by effective cleaning and disinfection is clearly the best one in this regard.

The efficacy of a number of new technologies in eliminating *L. monocytogenes* from foods have been studied with the aim of reducing processing of food whilst delivering a safe product. The application of pulsed electric fields (PEF) to foods has been shown be effective in reducing *Listeria* species in foods (Mosqueda-Melgar *et al.*, 2007, 2008). Álvarez *et al.* (2002) studied the effect of PEF on *L. monocytogenes* under a variety of different conditions, including different growth phases, pH value, water activity and prior exposure to low temperature. They found that the organism was more resistant to PEF when in the stationary rather than exponential phase of growth and also reported that lower water activity offered increased protection to the lethal effect of PEF. Similarly, cells grown at 35 °C were more resistant to PEF than those grown at 4 °C. The application of a treatment of 25 kV/cm for 800 μs achieved a $\log_{10}$ reduction of 1.48 at pH 7.0, and as the pH was reduced the treatment delivered a greater reduction (3.86 $\log_{10}$ at pH 5.4 and 5.09 $\log_{10}$ at pH 3.8).

High-pressure treatment also has the potential to achieve 'cold-pasteurisation' of foods thereby reducing the microbial load whilst minimising the effect on the foodstuff. Vachon *et al.* (2002) applied dynamic high-pressure (DHP) treatment to *L. monocytogenes* inoculated into phosphate buffered saline (PBS) and milk. One, three or five passes through the DHP unit at 100 MPa in PBS had little effect (< 1 $\log_{10}$ cfu) on the organism. Increasing the pressure to 200 and 300 MPa increased the lethal effect of the treatment, although this was dependent on the number of passes. One pass at 200 MPa had little lethal effect, whereas three and five passes resulted in approximately a 4 and 5 $\log_{10}$ reduction, respectively. At 300 MPa, one, three and five passes resulted in approximate reductions of 6, 8 and 8 $\log_{10}$ cfu, respectively. The lethality of the DHP process was reduced when the treatment was applied to artificially inoculated raw milk. Three and five passes at 100 MPa achieved a < 1 $\log_{10}$ cfu reduction and even at 200 MPa, three and five passes only achieved < 2 and < 3 $\log_{10}$ cfu reductions, respectively. At 300 MPa, one pass achieved a reduction of < 1 $\log_{10}$ cfu, three passes resulted in an approximate reduction of 4 $\log_{10}$ cfu and even five passes only reduced the organism by 5.6 $\log_{10}$ cfu.

19.3 Risk factors for *Listeria monocytogenes*

19.3.1 Human infection

L. monocytogenes is transmitted via three main routes: contact with animals, cross-infection of new-born babies in hospital and foodborne infection. The latter two sources account for the majority of cases of listeriosis in humans. Listeriosis is a rare but serious illness with an incidence of 1–3 cases per million of the total population of the European Union (EFSA, 2007b). The incidence does vary between different EU countries and, in 2006, ranged from zero in Malta up to 10 cases per million in Denmark. Similar incidence (3.1 cases per million population) has been reported in the USA (Vugia *et al.*, 2007). Of 2449 cases of human listeriosis recorded in England and Wales between 1983 and 2000, a total of 739 (30.2 %) were associated with pregnancy (Table 19.6), although significant variations can be seen from year to year (11–48%). In addition, 2–6% of healthy individuals are reported to be asymptomatic faecal carriers of *L. monocytogenes* (Rocourt, 1996).

Table 19.6 Cases of human listeriosis in England and Wales, 1983–2006

Year	Total cases (pregnancy associated[a])
1983	111 (44)
1984	112 (35)
1985	136 (59)
1986	129 (42)
1987	238 (102)
1988	278 (114)
1989	237 (114)
1990	116 (25)
1991	128 (32)
1992	105 (25)
1993	103 (17)
1994	111 (24)
1995	87 (11)
1996	117 (17)
1997	124 (24)
1998	107 (22)
1999	105 (17)
2000	100 (13)
2001	146 (18)
2002	136 (10)
2003	237 (35)
2004	213 (20)
2005	189 (25)
2006	187 (26)
2007	230 (28)

[a] Mothers and babies are counted as one case.
Source: Adapted from HPA (2008a).

In the UK the levels of listeriosis declined following a large outbreak between 1987 and 1989 (Table 19.6) caused by pâté from a single Belgian manufacturer (McLauchlin *et al.*, 1991). The subsequent fall is believed to have been due to public health advice relating to the consumption of higher-risk food commodities including pâté and the concerted industry action at that time to control the organism (Gilbert, 1996). Similar decreases in infections to those noted in the UK were reported in the USA following improvements in regulatory/industry control and the issuing of advice to susceptible groups (Tappero *et al.*, 1995). However, in the 2000s, listeriosis has been resurgent in the UK and many other countries (EFSA, 2007b). A distinguishing feature in some countries such as Germany and England and Wales has been the upsurge in cases affecting adults over the age of 60 (Denny and McLauchlin, 2008). In England and Wales, this increase has not been associated with pregnant women, who were significantly affected during the Belgian pâté outbreak, and it has not been linked to a common food source (McLauchlin, 2007; McLauchlin and Gillespie, 2007). Another feature of this increase is the sporadic nature of the cases; sporadic cases of listeriosis are now much more common than point source outbreaks with many serotypes being involved. Additionally, the clinical presentation has also changed in that a high proportion of the cases in the over-60s in England and Wales have been diagnosed following bacteraemia, i.e. the organism recovered from the blood (McLauchlin and Gillespie, 2007) rather than as a result of central nervous system infections. This increase in cases in a specific demographic group has led to the potential that behavioural factors may be contributing to infections (ACMSF, 2008). Likewise, it has prompted specific advice to be provided to the elderly on measures to prevent listeriosis (see Table 19.10 on p. 708).

19.3.2 Foods

Listeria species are widely distributed in the environment, and the occurrence of *L. monocytogenes* in raw and processed foods has been extensively studied. Its presence in livestock and wild animals and birds has been frequently reported. In a two-year study in two areas in Ontario, Canada, Hartmann *et al.* (2007) isolated the organism from faecal samples of livestock (beef and dairy), wildlife (deer, moose, otter and raccoon) and human waste (biosolids and septic), with the isolates exhibiting overlapping but distinct population types. Some of these types were reported to be similar to isolates from surface water identified in previous studies. Bouttefroy *et al.* (1997) reported a prevalence of 33 % (37/112 samples) in droppings of urban rooks. Hellström *et al.* (2008) studied the prevalence of *L. monocytogenes* in wild bird faeces from a municipal landfill site and from urban areas in Helsinki. They reported an overall incidence of 36 %, with prevalence in landfill samples being higher than those from municipal samples. The strains were frequently similar to those detected in foods and food processing environments.

Listeria spp. are frequently present in/on raw meat of all types but appear to be less frequent in raw fish. Surveys have shown *L. monocytogenes* to be present at < 1 % to 20 % in raw meat (Jay, 1996) where it comes from faecal or environmental contamination of the meat in the abattoir and during subsequent processing. When reviewing studies of the prevalence of *L. monocytogenes* in meat and poultry products, Jay (1996) combined the reported findings from a number of countries to give an overall prevalence in the different meat products; these are shown in Table 19.7 together with the incidence of the organism reported in some other surveys. It is not common to find high levels of *L. monocytogenes*, i.e. > 100/g, on carcass meat although it may be present at enumerable levels. For example Uyttendaele *et al.* (1999) reported 12.3 % raw chicken samples to be contaminated with *L. monocytogenes* at > 1 cfu per g or cm^2. Bohaychuk *et al.* (2006) reported *L. monocytogenes* in 52 % raw ground beef (n = 100), 34 % raw chicken legs (n = 100) and 23.5 % of raw pork chops (n = 98). Raw meats and meat products are, therefore, potential sources of *Listeria* spp. in the food chain.

Ben Embarek (1994) summarised many reports relating to raw and processed seafoods in which prevalences up to 75 % in lightly preserved (cold-smoked,

Table 19.7 Incidence of *Listeria monocytogenes* reported in some raw and processed foods

Samples	Incidence (%)	Reference
Soft and semi-soft cheeses	2/374 (0.5)	Farber *et al.*, 1987
Raw chicken portions	19/32 (59)	Arumugaswamy *et al.*, 1994
Raw beef	6/12 (50)	
Fresh prawns – raw	7/16 (44)	
Leafy vegetables	5/22 (23)	
Satay, ready-to-eat	11/39 (28)	
Squids, prawns, chicken and clams, ready-to-eat	6/27 (22)	
Raw caprine milk	37/1445 (2.56)	Gaya *et al.*, 1996
Read-to-use mixed vegetable salads	21/70 (30)	García-Gimeno *et al.*, 1996
Vegetable salads	1/63 (1.6)	Lin *et al.*, 1996
Raw milk	342/9837 (3.48)	Kozak *et al.*, 1996
Soft and semi-soft cheeses	20/333 (6)	Loncarevic, 1995
Fresh and frozen pork (16 reports)	(20) n = 1033	Jay, 1996
Fresh and frozen beef and lamb (21 reports)	(16) n = 1571	
Fresh and frozen poultry (26 reports)	(17) n = 7054	
Processed meats (26 reports)	(13) n = 5089	
Ready-to-eat:		Little *et al.*, 1997
Refrigerated salads, vegetables	77/2113 (3.6)	
Crudités	2/242 (0.8)	
Live shellfish	11/120 (9.2)	Monfort *et al.*, 1998
Raw cows' drinking milk	32/1591 (2)	DoH, 1998
Cold-smoked salmon	(34–43)	Jørgensen and Huss, 1998
Cold-smoked halibut	(45–60)	
Gravad fish	(25–33)	
Heat-treated seafood	(5–12)	
Cured seafood	(4)	
Prepared salads	15/146 (10.2)	de Simón and Ferrer, 1998
Prepared meals:		
vegetables	8/217 (3.6)	
meat, poultry	16/324 (4.9)	
seafoods	2/137 (1.4)	
Preserved fish products – not heat-treated	35/335 (10.4)	Nørrung *et al.*, 1999
Preserved meat products – not heat-treated	77/328 (23.5)	
Heat-treated meat products	45/772 (5.8)	
Raw fish	33/232 (14.2)	
Raw meat	106/343 (30.9)	

Table 19.7 (Cont'd)

Samples	Incidence (%)	Reference
Prepared retail foods including meat, fish, vegetable and cheese products:		Nørrung *et al.*, 1999
1997 (*L. monocytogenes* present at >10/g)	760/7760 (9.8)	
1998 (*L. monocytogenes* present at >10/g)	670/7703 (8.7)	
Raw poultry carcasses and processed poultry products	295/772 (38.2)	Uyttendaele *et al.*, 1999
Frozen, smoked mussels – imported in Korea	3/68 (4.4)	Baek *et al.*, 2000
Ice-cream – imported	8/132 (6.1)	
Domestic foods	111/1337 (8.3)	
Raw milk	(13) n = 1300	Carlos *et al.*, 2001
Ground pork	171/340 (50.2)	Kanuganti *et al.*, 2002
Dressed lamb carcasses	3/69 (4.3)	Antoniollo *et al.*, 2003
Raw chicken	14/80 (17.5)	Soultos *et al.*, 2003
Prepared ready-to-eat salad vegetables	88/2934 (3)	Sagoo *et al.*, 2003
Cold meats	*Listeria* spp.	Elson *et al.*, 2004
Pâté	180/2874 (6.3)	
	44/1178 (3.7)	
Cold meats	*L. monocytogenes*	
Pâté	61/2874 (2.1)	
	22/1178 (1.9)	
Butter	13/3229 (0.4)	Lewis *et al.*, 2006
Fermented sausage	(4) n = 100	Bohaychuk *et al.*, 2006
Roast beef	(0) n = 101	
Turkey breast	(3) n = 100	
Beef wieners	(5) n = 100	
Chicken wieners	(3) n = 101	
Cold-smoked salmon	41/626 (6.5 %)	Beaufort *et al.*, 2007
Sampled from eight processing plants and tested at the end of storage life	Variation of 0–24 % between plants	
Minimally-processed vegetables	1/181 (0.6)	Fröder *et al.*, 2007
Ready-to-eat fish products (anchovies, herring, surimi, **bonito pies, smoked salmon**)	9/140 (6.4)	Cabedo *et al.*, 2008
Ready-to-eat meat products (cooked sausage, **frozen chicken croquettes**, **cured dried pork sausage**, dried pork loin, duck liver pâté, pork liver pâté, **cooked ham**, cured dried ham, **cooked turkey**	22/501 (4.4)	

Table 19.7 (Cont'd)

Samples	Incidence (%)	Reference
breast, salami, frankfurters, cooked restructured ham, cooked pork sausage, bologna, **pork luncheon meat**)		Cabedo *et al.*, 2008
Ready-to-eat dairy products (fresh salty cheese, Camembert, soft cheese, Suisse style cheese, Edam cheese, Manchego cheese, Roquefort, ice cream, yoghurt, **fresh cheese**, UHT cream, pasteurised whole milk)	1/462 (0.2)	
Ready-to-eat dishes & desserts (**frozen cannelloni**, frozen lasagna, potato omelette, sweet custard, custard cream, honey)	11/123 (8.9)	
Sandwiches	88/3249 (2.7 %) *Listeria* spp. 248/3249 (7.6 %)	Little *et al.*, 2008

Note: Items in bold indicate the foods where the organism was detected.

hot-smoked, marinated) fish products are noted. Other studies reinforce the potential for such commodities to be contaminated with *L. monocytogenes* at varying incidences (Nørrung *et al.*, 1999; Miettinen *et al.*, 2003; Laciar and de Centorbi, 2002; Thimothe *et al.*, 2002).

Beuchat (1996) summarised the results of studies of the prevalence of *L. monocytogenes* in raw vegetables, including bean sprouts, cabbage, cucumber, potatoes, pre-packed salads, radish, salad vegetables, tomatoes; reported prevalences ranged from 1.1 to 85.7 %. Surveys of ready-to-eat salads and vegetables often show the presence of *L. monocytogenes*, albeit at low levels. Little *et al.* (1997) reported a 3 % (77/2276) incidence of *L. monocytogenes* in ready-to-eat pre-packed salads and 0.8 % (2/247) incidence in crudités (washed, ready-to-eat vegetables for use with savoury dips or as hors d'oeuvres, e.g. carrot or celery sticks). The organism was not detected (limit of detection < 20 cfu per g) in 3200 samples of organic ready-to-eat vegetables purchased from retail establishments in the UK in 2000 (Sagoo *et al.*, 2001). A similar study of prepared, pre-packed ready-to-eat salads and vegetables (conventionally grown) in England and Wales in 2001 found 2.3 % (90/3849) contaminated with *L. monocytogenes* (Sagoo *et al.*, 2002). Sagoo *et al.* (2003) also reported on the incidence of *L. monocytogenes* in open, ready-to-eat salads sold in retail and catering establishments where 88/2934 (3 %) samples were found to be contaminated.

Listeria monocytogenes is a relatively frequent contaminant of raw milk. In a study of raw cows' milk on retail sale in England and Wales, 32 out of 1591 (2 %) of samples contained *L. monocytogenes* in 25 ml samples (DoH,

1998). None of the samples contained enumerable levels of the organism, i.e. > 100/ml and, of the positive isolates, 10 were serotype 4 and 22 were serotype 1. Higher incidence of *L. monocytogenes* has been reported in surveys of raw cows' milk sampled from farm bulk tanks. O'Donnell (1995) reported the organism in 5 % (102/2009) samples of raw milk in England and Wales between 1992 and 1993.

Recent concerns have been raised associated with the link between sandwiches and sporadic cases of listeriosis (Gillespie *et al.*, 2006). Meldrum and Smith (2007) reported on the incidence of *L. monocytogenes* in sandwiches in Wales. They found the organism present in 27 out of 950 sandwiches (2.84 %) available in hospitals, and they also found it present in two products (0.21 %) by direct enumeration (0.5 ml of a primary dilution). This compared with an incidence of 4.42 % (26/588) in retailer bought sandwiches with five samples (0.85 %) positive by direct culture. One hospital-bought sandwich had a level of *L. monocytogenes* of 1200 cfu/g.

It is clear that eliminating *L. monocytogenes* from most foods is both impractical and impossible, but it is possible to reduce and control this hazard in foods, thereby minimising the risk presented to public health. In order to ensure the safety of food products in respect of potential foodborne bacterial pathogens, growing, harvesting, handling, storage, processing and associated food supply systems must be managed by food producers and processors in such a way as to reliably control the growth of *L. monocytogenes* which must be prevented from multiplying to potentially harmful levels (> 100/g; Gilbert *et al.*, 2000). To achieve this, it is necessary to understand the conditions and factors that affect growth and survival.

Attention to the detail of cleaning and hygiene practices throughout the food supply chain is the prime essential for minimising the levels of *L. monocytogenes* contaminating plant and animal crops and food production environments. Thereafter, it is the treatment, formulation and storage conditions of the food materials themselves that will determine how any residual contaminating population of *L. monocytogenes* develops.

The European Commission has also strengthened controls in relation to *L. monocytogenes* in food in light of the recommendation of the European Commission Scientific Committee on Veterinary Measures relating to Public Health to 'reduce significantly the fraction of foods with a concentration of *L. monocytogenes* above 100 cfu/g at the point of consumption' (EC, 1999). Consequently, Regulation (EC) 2073/2005 (as amended) (EC, 2005, 2007) sets microbiological criteria for *L. monocytogenes* in ready-to-eat foods. Criteria have been set for three distinct groups of foods, based on the risk presented to target consumers, and from different foods based on their potential to support growth of the organism. Any product which does not comply with these limits must be withdrawn from the market.

- *Criterion 1.1* sets a limit of absence in 25 g throughout the product shelf-life to ready-to-eat foods intended for infants or for special medical purposes.

- *Criterion 1.2* sets limits for ready-to-eat foods that can support the growth of *L. monocytogenes* based on the ability of the manufacturer to demonstrate the potential for growth during the shelf-life of the product:
  (a) A limit of ≤ 100 cfu/g applies throughout the product shelf-life where the manufacturer can demonstrate that if the organism is present in 25 g at the beginning of the shelf-life it cannot exceed this limit by the end of shelf-life.
  (b) A limit of absence in 25 g applies at the end of the manufacturing process before the product is placed on the market where the manufacturer is unable to demonstrate compliance with the limit of ≤ 100 cfu/g throughout the shelf-life.
- *Criterion 1.3* sets a limit of ≤ 100 cfu/g throughout the product shelf-life for all other RTE foods that are *unable* to support the growth of *L. monocytogenes*. Such products are defined as those with pH ≤ 4.4 or a water activity of ≤ 0.92, products with pH ≤ 5.0 and a water activity of ≤ 0.94 and products with a shelf-life less than five days.

Criterion 1.2 is currently under review and may be revised to clarify interpretation. It is essential that food law enforcers and food business operators ensure they have the means to be kept informed of any changes in local, national or international law that may affect the operation of businesses they have jurisdiction over or responsibility for, respectively, to facilitate conformance to relevant and current legislation.

Although the USA has operated a zero tolerance policy for *L. monocytogenes* in ready-to-eat foods since the 1980s, the United States Food and Drug Administration has proposed a tolerance of up to 100 cfu/g *L. monocytogenes* for refrigerated or frozen ready-to-eat foods that cannot support the growth of *L. monocytogenes* (FDA, 2008).

Listeria monocytogenes is included in some food industry guidelines, and many food company specifications relating to individual food products, particularly chilled, ready-to-eat foods, also include specific requirements for the 'absence' (non-detection) of *L. monocytogenes* in 25 g of food sample (Bell and Kyriakides, 2005). The success or otherwise of any systems put in place to control *L. monocytogenes* is usually monitored by examining samples taken from various points in the process, e.g. incoming raw materials, food materials in process, e.g. after washing procedures, cooking procedures or slicing operations and finished products, for the presence of *Listeria* spp. and *L. monocytogenes* in particular.

Outbreaks have implicated cooked meat-based products such as pâté and rillettes (Anon., 2000a), soft ripened cheese, cooked shellfish, coleslaw and milk shake (Table 19.8). Indeed, the range of products from which the organism has been isolated (Table 19.7) is significantly broader than those so far implicated in outbreaks of foodborne disease.

The factors that appear to elevate the risk in relation to the potential for a food product to cause outbreaks of listeriosis include all of the following:

Table 19.8 Some examples of foodborne outbreaks of listeriosis

Year	Country	Cases (deaths)	Food	Outbreak serotype	Reference
1980–81	Canada	41 (18)	Coleslaw	4b	Schlech *et al.*, 1983
1983	USA	49 (14)	Pasteurised milk	4b	Fleming *et al.*, 1985
1983–87	Switzerland	122 (34)	Vacherin cheese	4b	Bille, 1990
1985	USA	142 (48)	Mexican-style soft cheese	4b	Linnan *et al.*, 1988
1987–89	UK	> 350 (> 90)	Belgian pâté	4b	McLauchlin *et al.*, 1991
1992	New Zealand	4 (2)	Smoked mussels	1/2a	Baker *et al.*, 1993
1992	France	279 (85)	Pork tongue in aspic	4b	Goulet *et al.*, 1993
1994	USA	45 (0)	Chocolate milk	1/2a	Dalton *et al.*, 1997
1995	France	20 (4)	Raw-milk soft cheese	4b	Goulet *et al.*, 1995
1997	Italy	> 1500 (0)	Corn and tuna salad	4b	Aureli *et al.*, 2000
1998–99	USA	108 (14[a])	Hot dogs and delicatessen meats	4b	Anon., 1999; Mead *et al.*, 2006
1998	USA	40 (4)	Cooked hot dogs	4b	Anon., 1998
1998–99	Finland	18 (4)	Butter	3a	Lyytikäinen *et al.*, 1999
1999–2000	France	26 (7)	Pork tongue in jelly	4b	Anon., 2000b
2000	USA	13 (8)	Mexican style soft cheese	4b	MacDonald *et al.*, 2005
2001	USA	16 (0)	Sliced cooked turkey	1/2a	Frye *et al.*, 2002
2001	Sweden	50 (0)	Cheese	?	Carrique-Mas *et al.*, 2003
2001	Japan	38 (0)	Cheese	1/2b	Makino *et al.*, 2005
2002	USA	54 (11)	Cooked turkey meat	4b	Gottlieb *et al.*, 2006
2003	England	17 (4)	Butter	4b	Gillespie *et al.*, 2006
2003	England	5 (0)	Hospital sandwiches	1/2	Dawson *et al.*, 2006
2005	Switzerland	10 (5)	Tomme cheese	1/2a	Bille *et al.*, 2006
2006	Czech Republic	78 (13)	Mature cheese	1/2b	EFSA, 2007b; Vit *et al.*, 2007
2006	Germany	6 (1)	Harz cheese	?	EFSA, 2007b

[a] Plus four miscarriages or stillbirths.
? Not reported.

- raw material or product exposed to contamination;
- product manufactured with no processing stage capable of destroying the organism, e.g. cooking;
- product with little or no preservation factors, e.g. neutral pH, low salt, high moisture;
- product exposed to post-process contamination;
- product sold with long shelf-life under chilled conditions;
- product sold as ready-to-eat.

Listeria monocytogenes is not a special organism in terms of its ability to survive adverse conditions, but the trait that has allowed it to exploit many food environments and cause outbreaks of illness is its capacity to grow at very low temperatures in foods, i.e. as low as –0.4 °C with one report of growth at –1.5 °C. Together with its capacity to colonise factory environments, particularly moist/wet environments, the organism has been able to exploit chilled, ready-to-eat food products that have minimal preservative levels and long shelf-lives.

19.4 Detecting *Listeria monocytogenes*

Reliable methods for detecting and identifying *L. monocytogenes* are important in the support of properly developed and implemented HACCP systems for its control. Table 19.9 indicates the common conventional microbiological method used by food microbiologists to isolate and identify *L. monocytogenes*. A horizontal method specifically for the enumeration of *L. monocytogenes* in foods has also been published by the International Organization for Standardization (ISO, 1998, 2004). A range of alternative microbiological techniques and methods are used for the detection and identification of *Listeria* spp. and *L. monocytogenes* from foods and food-related samples. These include antibody-based methods such as enzyme-linked immunosorbent assays (ELISA) and immuno-chromatographic assays, nucleic acid hybridisation methods and various polymerase chain reaction (PCR)-based methods; a valuable summary of many of these methods has been prepared by Campden BRI (Baylis and Mitchell, 2008).

When suspected positive results are obtained, further work to culture and characterise the isolates obtained may be necessary, and this can include serotyping, phage typing, assessing the isolates' ability to produce listeriocins and their ability to inhibit the growth of selected indicator strains. Molecular/genetically-based methods, including ribotyping, pulsed field gel electrophoresis (PFGE) and amplified restriction fragment length polymorphism, allow more detailed sub-typing of strains. Such work is usually undertaken in specialist laboratories within the public health sector, e.g. in the UK, the Health Protection Agency's Food Safety Microbiology Laboratory, and help facilitate traceability of strains in outbreak situations.

Table 19.9 Common conventional method for the detection and identification of *Listeria monocytogenes* from foods[a]

1:10 inoculation of selective enrichment broth: half Fraser Broth (others used include FDA and UVM1)
⇓
Incubate 30 °C for 24 h (FDA 48 h incubation)
⇓
Sub-culture to secondary enrichment broth: Fraser Broth (0.1:–10 ml)
(FDA medium does not require secondary enrichment stage)
⇓
Incubate 37 °C for 48 h
⇓
Streak onto selective agars: a chromogenic agar for detection of *Listeria* and a second medium, e.g. Oxford Agar, Palcam Agar
⇓
Incubate 37 °C for 24–48 h
⇓
Inspect plates
⇓
Purify suspect colonies
(Tryptone Soya Agar + 0.6 %w/v Yeast Extract)
Incubate 30 °C or 37 °C for 24 h
⇓
Confirmatory tests

- Gram reaction
- Catalase reaction
- Tumbling motility at 22–25 °C
- Carbohydrate utilisation, mannitol, rhamnose, xylose
- β-haemolysis reaction
- Christie, Atkins, Munch-Petersen (CAMP) test

⇓
Read reactions

[a] All media selected for use should be prepared and used in accordance with the manufacturer's instructions.
Source: Based on ISO 11290 – 1: 1996 Microbiology of food and animal feeding stuffs – Horizontal method for the detection and enumeration of *Listeria monocytogenes* – Part 1: Detection method and ISO 11290 – 1: 1996/Amd 1:2004 Modification of the isolation media and the haemolysis test, and inclusion of precision data. International Organization for Standardization, Geneva

A wide variety of foods has been implicated in outbreaks of illness attributed to *L. monocytogenes* (Table 19.8). This has led to an increasing amount of work by both researchers and food industry scientists to determine the specific survival characteristics of this organism, to further improve methods of detection and to develop controls aimed at minimising, if not eliminating, contamination of food by *L. monocytogenes*.

These foods continue to be implicated in outbreaks of illness attributed to *L. monocytogenes*, but the pattern of illness has been changing with sporadic cases occurring rather than outbreaks attributable to a single source and increasing numbers of adults over 60 years of age being diagnosed with listeriosis (see Section 19.3.1). It is essential that methods for the detection, isolation, identification and full characterisation of *L. monocytogenes* continue to improve to ensure that reliable information can be made available as quickly as possible, allowing effective corrective actions to be taken to protect the health of the public.

19.5 Control of *Listeria monocytogenes* in foods

Despite the heightened profile of *L. monocytogenes* brought about by increases in incidence and outbreaks of listeriosis in the 1980s, the organism is still, regrettably, all too frequently implicated in foodborne disease outbreaks and sporadic cases of listeriosis. Increased availability and consumption of ready-to-eat, extended shelf-life, chilled foods has clearly given this psychrotrophic organism greater scope to cause such problems if inadequately controlled. Outbreaks have implicated a variety of food types (Table 19.8) and, as shown by the results from surveys, some of which are summarised in Table 19.7, many more different food types harbour *L. monocytogenes* and, therefore, have the potential to become hazardous to consumers.

Control of *Listeria* spp. in foods is dependent on four key factors:

- preventing contamination of raw materials or growth in raw materials, if present;
- destroying or reducing it, if present in the raw materials;
- preventing recontamination from the factory environment, equipment or personnel after applying a reduction or destruction stage;
- preventing or minimising its growth, if present, during the shelf-life of the final product.

19.6 Raw material control

Listeria monocytogenes is ubiquitous and can be found in a wide variety of environments, including soil, water and vegetation. As a consequence, it will be present in most raw materials used in the food industry which are not subject to a process that will kill the organism. Raw vegetables, meat and milk will be contaminated with the organism and, at best, controls at these stages will help to reduce the levels of contamination entering the process but cannot guarantee absence from most of these materials.

Clearly, the presence of *L. monocytogenes* represents the greatest hazard to those products whose raw materials are not subject to a subsequent process stage capable of reducing or eliminating the organism. Such products include cheeses made from raw milk, cold-smoked fish and raw fermented meats. However, high levels of the organism in raw materials intended for products that only receive mild processing, such as prepared salads and coleslaw, where the materials are only washed, can also represent a significant risk. Evidence of this was seen in the large outbreak associated with coleslaw (sliced cabbage and carrots, without dressing) that occurred in Canada in 1981 (Schlech *et al.*, 1983).

19.6.1 Raw milk

Listeria spp. most commonly gain access to the milk from the cow's udder during milking. *L. monocytogenes* can be found in the faeces of animals and, indeed, can cause serious infection in ruminant animals, especially goats and sheep. Animals are most likely to become colonised through consumption of the organism in grass or feed and, in particular, silage. As with enteric contaminants, *Listeria* spp. can gain entry to the milk from faecal contamination of the udder. Hygienic milking practices involving udder and teat cleaning and disinfection can help reduce contamination of the milk. However, equally important in the context of *Listeria* spp. is the cleaning and disinfection regime applied to the milking and milk storage equipment.

Once introduced into the milking parlour or equipment, *L. monocytogenes* can readily colonise these moist environments. The capacity to grow at low temperatures also means that refrigerated storage, while useful in slowing the growth of the organism (Table 19.4), cannot prevent it from increasing in number completely. It is therefore essential to ensure that cleaning and disinfection are carried out properly. This must include the milking parlour environment and all the equipment, and must also take account of all transport vehicles and transfer pipes used in the movement of milk from the farm to the processing factory. Failure to do this will inevitably lead to increased incidence and levels of the organism in the raw milk itself. For the production of raw milk soft cheese, where the organism can grow during its production and storage, this could be the difference between a safe and unsafe product.

It is usual to subject farm milk used in the production of raw milk cheeses to routine monitoring for *Listeria* spp. and indicators of dairy parlour hygiene such as *Escherichia coli*. The latter is often used in incentive payment schemes where consistent high-quality milk is rewarded with higher payment and poor-quality milk supplies are excluded from production of raw milk products.

19.6.2 Raw meat and fish

Periodically, reported surveys of the prevalence of *Listeria* spp. and *L. monocytogenes* in foods indicate that *Listeria* spp. are frequently present in/on raw meat of all types but are less prevalent in raw fish (see Section 19.3.2). Raw meats and meat products must be regarded as potential sources of *Listeria* spp. in manufacturing and catering units and the home and may therefore contribute a cross-contamination risk to food processing environments and other foods processed in the same environment. Although controls in the abattoir and meat processing rooms are not targeted at controlling *Listeria* spp. specifically, the controls in place to minimise contamination by faecal pathogens will also aid in the control of *Listeria* spp.

Raw fish used for manufacture of products such as smoked salmon or trout may occasionally be contaminated with *Listeria* species at an incidence of < 5 % and with only low levels, i.e. < 100/g (Bell and Kyriakides, 2005). Contamination may occur on the fish farm, where most salmon used for such products are grown or, more commonly, arise and spread during subsequent gutting and processing.

Control of hygiene during the slaughtering and preparation stages is critical for minimising the level of contamination entering a cold-smoked fish processing unit. Results from studies of fish processing plants have revealed significant differences in the occurrence of *L. monocytogenes*. Autio *et al.* (1999) reported only one sample (2 %) of raw trout entering the processing unit to be contaminated with *L. monocytogenes*, and none of the 49 samples of filleted fish, skinned fish or pooled skin were found to harbour the organism. However, Eklund *et al.* (1995) found the organism to be a common contaminant of raw, eviscerated salmon supplied to salmon processors, with the organism being isolated from 4/19 samples of slime, 30/46 skins, 8/17 heads and 1/15 belly cavity and belly flap trimmings. Hansen *et al.* (2006) assessed the prevalence of *L. monocytogenes* in fish processing environments and reported an increasing incidence as the degree of human activity increased, i.e. 2 % in seawater fish farms, 10 % in freshwater fish farms, 16 % in fish slaughterhouses and 68 % in a fish smokehouse. The strains found inside processing plants were more homogenous, i.e. similar types, than those found externally, indicating that control of the organism inside the processing plant is critical to reducing the risk from these products. Purchasers of raw meat for products such as fermented meat or raw fish for cold-smoking where no cooking process takes place would normally monitor the material for incidence and levels of *Listeria* spp. in order to ensure high levels are not being routinely supplied on raw materials, thereby entering the process.

19.6.3 Fruit, vegetables and other raw materials

Listeria spp. will be present at low levels on many raw fruits and vegetables from time to time. They may be contaminated from soil and water sources, or indeed from animal wastes applied to land. The presence of *L. monocytogenes* is less of a concern in most fruits as the pH usually precludes its growth (Table 19.3) and, as relatively higher levels (> 10^2/g) are normally required to cause infection and illness, any very low levels that may be present on fruit are not considered to represent a significant risk. However, its presence on salad materials, vegetables and fruit of low acidity intended for minimal further processing such as prepared salads is of more concern.

An outbreak of listeriosis has occurred because of extensive contamination of cabbages thought to be due to the use, on the growing field, of raw and composted sheep manure from animals with known recent incidence of listeriosis. Prazak *et al.* (2002) demonstrated the different points at which *L. monocytogenes* could contaminate cabbage during production and storage. Over a seven-month period the organism was isolated from 20 samples of cabbage (seven from farms and 13 from packing sheds), three water samples (two from field furrows on farms and one from wash water used for cleaning/cooling produce in a packing shed) and three packing shed surfaces (two conveyor belts and one cooler surface). Some of the isolates, i.e. from conveyor belts, shared identical DNA patterns with cabbage isolates.

In addition, many raw material vegetables are stored for long periods in cold stores to facilitate continued supply through winter months. Clearly, the use of animal wastes on land for crops intended to be consumed with little or no further processing, like many salads, needs to be carefully controlled. Animal wastes should be properly composted to ensure sufficient heat is generated to achieve a reduction in contaminating microorganisms before application to land. Alternatively, artificial fertilisers should be employed.

Salad materials and vegetables used for direct consumption, if stored, should be stored under conditions that preclude or minimise the growth of *Listeria* spp. either at low temperatures or for short periods. Most vegetables will not support the growth of *Listeria* spp. if they are intact and if moisture is not available; dry storage conditions should be maintained and the produce protected from physical damage. If good conditions are not maintained, such produce can succumb to spoilage through rots or mould growth and, in many situations, this serves as a useful 'natural' control of psychrotrophic pathogens insofar as damaged or rotten produce is rejected and thus prevents potentially hazardous food materials from entering the food chain. However, the use of technologies involving modified atmosphere storage to extend shelf-life by inhibition of 'normal' spoilage microflora must take account of the opportunity this provides for the growth of other contaminants such as *Listeria* spp. It is essential that process innovations to improve quality are carefully considered to avoid solving a spoilage problem at the expense of safety.

19.7 Control in processing

Control of raw material quality is an important element of many, if not all, processes. However, many products are manufactured employing a stage that, if carried out properly and consistently, will deliver a significant reduction or achieve the destruction of *Listeria* spp. if present. Key to the safety of such products is an understanding of the process and the conditions required to achieve the desired reduction or destruction of the organism. Process validation studies allow a processor to understand where to apply the controls necessary to ensure the critical process is conducted correctly, thereby delivering the required reduction or destruction of *L. monocytogenes* on every occasion. The proper application and operation of a HACCP-based system will naturally incorporate such approaches. It is also important to revisit the safety assessments of different foods in light of new research. For example, outbreaks of listeriosis in butter have highlighted the potential for the organism to both survive for long periods and even grow in butter and spreads depending on the formulation and storage conditions of the product (Bell and Kyriakides, 2005).

19.7.1 Cooking

Cooking is often used as a primary process in the manufacture of a variety of products, including cooked meats, ready meals, dairy desserts and this process will also destroy *Listeria* spp. The controlling factors important in ensuring the consistent delivery of correct cooking processes have already been described in this book for pathogenic *E. coli* and include all of the following:

- minimum ingoing temperature of the material;
- largest size/piece size of the product;
- cold spots in the oven/temperature distribution of the oven;
- fill load of the container, if appropriate, i.e. air-spaces will affect heat transfer;
- oven load;
- minimum time setting for the load;
- minimum temperature setting of the oven.

Process validation should be used to identify, for implementation, the process controls and checks that will ensure the process is delivered consistently on each occasion, such as the monitoring of the time and temperature of the cook using thermographs (see Chapter 17).

Listeria monocytogenes is readily destroyed by pasteurisation temperatures applied to meat, which in the UK are specified as 70 °C for two minutes (DoH, 1992). This process is capable of delivering at least a 6 $\log_{10}$ reduction of the organism in a food. Studies with cooked chicken indicated a *D*-value for *L. monocytogenes* of 0.133 minutes at 70 °C (Murphy *et al.*, 1999). Pasteurised milk is heat-processed in the UK to comply with legislative requirements and achieves 71.7 °C for 15 seconds, significantly less than the process applied to meat but nevertheless still sufficient to achieve a 3–4 $\log_{10}$ reduction in *L. monocytogenes* (using the data of Murphy *et al.* and based on *z*-values of 5–7.5 C°). Concerns have been expressed relating to the potential protective effect afforded to *L. monocytogenes* by its capacity to invade somatic cells. However, the levels of contamination of *L. monocytogenes* in raw milk and the frequency with which this phenomenon occurs do not appear to have resulted in *Listeria* spp. being found as common contaminants of pasteurised milk. It is therefore not believed to be a significant safety consideration.

19.7.2 Fermentation processes

Products produced using a fermentation process, such as raw, fermented meat and raw milk hard cheeses, rely on the fermentation and drying process to reduce levels of microbial contaminants present in the raw material. It is apparent from the frequency with which *L. monocytogenes* occurs in raw, fermented meats (MAFF, 1997; Bohaychuk *et al.*, 2006) that the organism can survive the traditional salami process. Gianfranceschi *et al.* (2006) studied the prevalence of *L. monocytogenes* in uncooked Italian dry sausage (salami), taking 1020 samples over a one-year period from 19 manufacturing plants. Contamination rates found ranged from 1.6 % to 58.3 % between producers with an overall average of 22.7 %. As reported in most other surveys of this type of product, all positives were at low levels (< 10 cfu per g). These findings reinforce the fact that whilst the organism is a frequent contaminant of the products, its growth is restricted.

Studies on the survival of *L. monocytogenes* in raw milk, mould-ripened, soft cheese, e.g. camembert, indicate that growth is much greater during the ripening stage where levels can increase by 4–5 $\log_{10}$ units (Ryser and Marth, 1987). The growth coincides with the elevation in pH that occurs owing to the growth of mould on the surface of the cheese. The risk of *L. monocytogenes* growth during ripening of mould-ripened soft cheeses relates just as much to pasteurised varieties as it does to raw milk varieties if the organism is introduced as a post-pasteurisation contaminant. This clearly can occur with both varieties during the ripening stages.

Hard cheeses such as cheddar usually have a sufficiently low pH (*c.* 5.0) that, together with the low moisture and high aqueous salt content, serve to restrict the growth of *L. monocytogenes* during the ripening stages. These cheeses are therefore considered to represent less of a risk in relation to listeriosis. *L. monocytogenes* represents greatest risk in soft ripened cheeses and, indeed, other soft non-ripened cheese where the pH is not low, e.g. mozzarella, mascarpone.

Control is exerted by effective pasteurisation, if employed, and by the avoidance of post-process contamination, particularly in the ripening and slicing stages. The safety of these products can only be ensured through the adoption of effective controls outlined above and, where necessary, by the restriction of shelf-life. Wherever possible, the safety of these types of products should be validated using challenge test studies, using research facilities, to establish the measures necessary to preclude growth of the organism.

19.7.3 Washing processes

For many products including salads and vegetables only a simple washing stage is used in their preparation. Increasing quantities of these are sold as ready-to-eat products where washing by the consumer is not deemed or advised as necessary. Outbreaks of listeriosis implicating these products are rare, and it is probably the short shelf-life assigned to these products that is the principal reason for their apparent safety. Nevertheless, the washing stage for these products is essential to reduce levels of contaminating *Listeria* spp.

Prepared salad materials are washed in chlorinated water with levels normally ranging from 50–200 ppm chlorine. Evidence from the incidence and levels of *Listeria* in washed produce indicates that these levels of chlorine are capable of achieving approximately a 1 $\log_{10}$ reduction in the organism. Zhang and Farber (1996) reported a 1.2 $\log_{10}$ reduction and 1.6 $\log_{10}$ reduction after one minute at 4 °C and 22 °C, respectively, in washed shredded lettuce, inoculated with *L. monocytogenes* and then washed in sodium hypochlorite (200 ppm chlorine, pH 9.24–9.31). Reductions were similar with cabbage and were not significantly greater when exposure to chlorine was extended to 10 minutes. At 100 ppm chlorine concentrations, reductions were reduced to $< 1 \log_{10}$ after one minute. Recent studies with chlorinated water (100 ppm), peracetic acid (0.05 %) and a commercial citric acid-based produce wash (0.25 %) demonstrated reductions of 1.0, 1.7 and 0.8 $\log_{10}$ cfu per g, respectively (Hellström *et al.*, 2006).

Whatever the method used, it is important to ensure that the active ingredient, most often chlorine, is present in its active form on a continuous basis. Washing systems have just as much capacity to spread contamination as they have for reducing them if the wash water is not regularly changed or the levels of active ingredient are not properly maintained. It is therefore common to employ systems that continuously dose and/or continuously monitor the active concentration of the antimicrobial compound used.

19.7.4 Smoking process

Outbreaks of listeriosis have implicated smoked fish and shellfish. Two methods of smoking are usually employed: a cold smoking process at 18–28 °C and a hot smoking process at > 68 °C. The capacity to achieve reductions in contaminating *L. monocytogenes* clearly differs between the two, with the latter process essentially representing a pasteurisation stage. Cold smoking takes place for *c.* 18 hours and in theory represents a significant opportunity for growth of contaminating *Listeria* species. The organism commonly contaminates the fish during slaughtering and subsequent processing in the gutting, filleting and salting stages. Reducing contamination through the operation of good hygienic practices at these stages is critical to the safety of this production process.

Smoking processes involve the burning of damp wood chips or wood shavings. Although there are reports of the antimicrobial effects of smoke on *Listeria* spp., it is evident that the nature of the smoke and the site of contamination are important in the potential for the survival and growth of *Listeria* spp. Eklund *et al.* (1995) showed the organism either did not grow or decreased in number if it was inoculated on the outside of the fish. However, levels increased by several orders of magnitude during the smoking process when inoculated into the flesh.

19.7.5 Hygiene and post-process contamination

There is probably no bacterial pathogen that exploits the food processing environment better than *L. monocytogenes*. Investigation of a large number of outbreaks throughout the world indicates that contamination of the product from the environment is a significant contributory factor. The organism survives well in cold, wet environments and can exploit deficiencies in cleaning and hygiene practices.

Product contamination by *Listeria* spp. occurs from either the environment or from direct product contact surfaces, including utensils and equipment. The organisms are transferred from the environment to the product or via product contact surfaces from aerosols or from poor personnel handling practices. To date, it has not been common for listeriosis outbreak investigations to have implicated products that have been directly cross-contaminated from raw materials, although, clearly, this is an important potential route that should not be overlooked. It is, however, much more common for the organism to colonise the post-process environment and equipment and be a direct contaminant from such sources, e.g. slicers.

The main means for the effective control of *Listeria* spp., and *L. monocytogenes* in particular, is to reduce or eliminate it from the post-processing environment. The procedures necessary to achieve this are usually applied in the manufacture of those foods where its presence in the finished product represents a significant hazard. These include ready meals, sliced cooked meats and sliced soft ripened cheeses. For these products, the raw material processing and finished product areas are segregated within the factory so that personnel or food materials cannot pass between the two areas without the application of procedures designed to prevent contaminants passing from one area to the other. The two areas, often referred to as 'low risk' and 'high risk', are separated by dividing walls and the pathogen reduction or destruction stage is constructed within the division, e.g. double entry cookers (see Chapter 17) or chlorinated water wash flumes. The food materials progress through cooking or washing from the low risk to the high risk side.

To further minimise entry of bacterial pathogens, a positive air pressure may be maintained on the high risk side of the factory, thereby ensuring air flows from the high to the low risk side of the factory. Such flows are also in place for drainage. Staff entering the high risk side must change clothes in a segregated changing area, removing shoes and coats and putting on overalls/coats, shoes and hair covering, all dedicated to the high risk side of the factory. Detailed guidance on such controls can be found in appropriate texts (CFA, 2006).

Effective cleaning and disinfection of the environment and equipment are absolutely essential to prevent the organism from building up and presenting a contamination risk to the product. All equipment such as slicers, mixing bowls, pipework and conveyor belts should be routinely dismantled or parts loosened for effective cleaning and disinfection as the organism readily colonises such equipment. 'Dead' spots, such as redundant sample valves or poorly designed pipe ends, are often found in factories where the organism has persisted. Such areas should be designed out of the manufacturing unit.

In addition to obvious product contact surfaces, it should be remembered that the organism may be found in 'reservoirs' throughout the factory on a wide variety of non-food contact surfaces. As these are often cleaned inadequately, the organisms can readily grow to high levels on residual product debris and then spread to product or product contact surfaces on the hands of personnel or by aerosols. Table ledges, the underside of tables, door handles, extraction hoods, overhead pipes and many other areas should be routinely and properly cleaned and disinfected.

The organism is frequently found on floors, walls, in cracks and in drains. Although it is difficult to eliminate it completely from these sources, effective maintenance of floors and walls ensuring they are properly sealed and routinely cleaned and disinfected can contain the hazard. Cleaning practices themselves can also spread the organism, as in the use of power hoses to 'chase' food debris into drains, or if the cleaning utensils themselves are not cleaned or disinfected properly as is often the case with handles of brushes or spray guns, wheels and brush heads. Many of these factors have been reviewed in more detail in other texts (Bell and Kyriakides, 2005).

Recently, risk assessors have started to explore risk assessment methodologies to help quantify the risk in different microbiological cross-contamination scenarios and thereby identify the most effective strategies to reduce risk in food production/handling. Pérez-Rodríguez *et al.* (2006) calculated the risk of high levels of *L. monocytogenes* being present on cooked ham slices at consumption when using different handling exposures (bare hands, washed hands and gloved hands) in its preparation and found the highest risk when using gloved hands without washing.

In addition to cleaning and disinfection, it is also important to employ routine monitoring of cleaning efficacy using both indicators of effective cleaning and tests for the organism itself. Indicators such as coliform bacteria or adenosine triphosphate (ATP) bioluminescence tests, which monitor the presence of residual levels of product debris and microorganisms, can be useful, but effectively targeted tests for *Listeria* species can yield significant information about the organism and its control in the factory environment. Advice on sampling plans for such purposes is given by Bell and Kyriakides (2005).

19.8 Final product control

Finished products that are packed and sold as single units are protected from bacterial contaminants that may otherwise enter during distribution and retail display. It is common, however, for many items, such as cooked meats, cheese, salads, etc., to be sold to retailers for slicing or open display on the delicatessen or food service counter. The potential for storage, slicing, preparation and display equipment and associated utensils to harbour *Listeria* spp. and subsequently contaminate products is high indeed. Contamination by *L. monocytogenes* on delicatessen counters and subsequent spread to other products was a contributory factor in a large outbreak of listeriosis in France; 279 people were affected, 63 died and 22 suffered abortions (Goulet *et al.*, 1993). A survey of cooked meats, sliced in retail stores in the North-West region of the UK found *L. monocytogenes* contamination in 7.3 % (82/1127) of products tested on the day of purchase; five had levels of 100 cfu/g or more (HPA, 2008b). Storage of the 82 contaminated samples at 6 °C for 48 h resulted in a further 26 giving counts of *L. monocytogenes* ≥100 cfu/g. Such reports clearly indicate the potential risk presented by sliced cooked meats when their production is not controlled effectively.

Cleaning and disinfection of utensils and surfaces in the retail environment are as important to product safety as the same disciplines in a manufacturing operation.

19.8.1 Shelf-life

Although it is possible to exclude *Listeria* spp. from products cooked in their containers or even hot-filled products, the organisms will be occasional contaminants of many other prepared foods. It may be possible to limit the frequency and level of contamination, but it is not always possible to exclude contamination completely.

Listeria monocytogenes will be an occasional contaminant of chill stored products such as cooked sliced meats, ready meals, prepared salads and ripened cheeses. Clearly, storage at low chill temperatures will reduce the rate of growth of the organism but, as it can grow at low temperatures, the ultimate safety of the product rests on ensuring high levels of the organism are not allowed to develop in the product.

The addition of inhibitory factors such as salt or acid to the product can help reduce the growth of the organism but, as most of these products have fairly neutral pH and only low salt levels, the only practical means of ensuring growth limitation is to restrict the product shelf-life. In neutral pH, high water activity, chilled products where *L. monocytogenes* may be a contaminant, e.g. ready meals, the organism can increase in numbers by several orders of magnitude over 10–12 d and shelf-lives should be set accordingly. Safe shelf-lives can be estimated using predictive models for the growth of *L. monocytogenes* in foods under varying physicochemical conditions of the product (ComBase – common database for predictive microbiology accessed via www.combase.cc). Although such models can assist in growth prediction, in many cases they may need to be supplemented with challenge tests in the actual product, under laboratory conditions, to determine the real growth potential of the organism in the food in question.

19.8.2 Post-process additives

A number of ready-to-eat products are garnished or have ingredients added after the main processing stage and prior to final packing. This may include the addition of cheese to ready meals or herbs to cooked pâté. It is normal for ingredients entering the high risk side of a factory to be dipped directly, or contained in their packaging, into a sanitiser such as chlorine to ensure they are not introducing contaminants into the high risk side of the factory. The ingredients themselves should have been subject to some form of processing capable of reducing bacterial contamination loads including heating or washing, and it is usual for such materials also to be subject to some form of testing programme on receipt or have a relevant supplier certificate of compliance against a specification prior to use.

19.8.3 Consumer advice

It is important to recognise that the consumer often plays an important role in maintaining the safety of products after their purchase from the retail store. In relation to *L. monocytogenes*, probably the most critical factor is effective temperature control. Products requiring refrigeration should be clearly marked 'Keep Refrigerated' and, as the shelf-life of the product is often also critical to product safety, a clear durability indicator should also be printed on the pack. This is usually in the form of a 'use-by' date, which indicates when a product should be consumed by.

Some sectors of the population are more vulnerable to infection by *L. monocytogenes* than others, namely pregnant women, the young, the elderly and the immunocompromised. It is common practice to advise vulnerable groups of the need to avoid certain foods where the organism may be an occasional contaminant, e.g. soft cheeses, pâté. It is normal to do this using consumer advice leaflets or through media advertising and general awareness campaigns. Such advice may also be found on the packaging of high-risk foods such as mould-ripened soft cheese made from unpasteurised milk. In such circumstances it is important that products are clearly labelled with suitable advice for the vulnerable groups to avoid eating such cheese.

The changing demographics associated with listeriosis, which now appears to be increasingly affecting the older age groups, may necessitate the provision of better advice on food safety to such individuals as getting older is probably not something that such groups would consider necessitates a different approach to food behaviour (Cates *et al.*, 2006). Indeed, changing food behaviour in older age groups is intuitively more difficult to achieve. Advice regarding correct food safety behaviour in older age groups is not likely to differ from other vulnerable groups, e.g. temperature control of chilled foods and adhering to durability indicators, e.g. 'use by' dates (Table 19.10). However, the particular difficulties that may lead to poor adherence to this advice are important, e.g. problems may occur due to difficulty in reading the labels due to declining eyesight. Recent advice regarding elderly

Table 19.10 Official advice to vulnerable groups on foods to avoid

Country	Target group	Advice	Website
Australia and New Zealand	Pregnant women; elderly (> 65–70 years); people with immune systems weakened by disease or illness; anyone taking medication that suppresses the immune system	• Cold meats: unpackaged ready-to-eat from delicatessen counters, sandwich bars, etc, packaged, sliced ready-to-eat. • Cold cooked chicken: purchased (whole, portions, or diced) ready-to-eat. • Pâté: refrigerated pâté or meat spreads. • Salads (fruit and vegetables): pre-prepared or pre-packaged salads (e.g. from salad bars, smorgasbords, etc.). • Chilled seafood: raw (e.g. oysters, sashimi or sushi), smoked ready-to-eat; or ready-to-eat peeled prawns (cooked), e.g. in prawn cocktails, sandwich fillings and prawn salads. • Cheese: soft, semi-soft and surface-ripened (pre-packaged and delicatessen), e.g. Brie, Camembert, ricotta, feta and blue. • Ice cream: soft-serve. • Other dairy products: unpasteurised dairy products (e.g. raw goats' milk).	http://www.foodstandards.gov.au/_srcfiles/Listeria.pdf
Germany	People with a weakened immune system, older individuals and pregnant women	• Do not eat any foods of animal origin raw. • Refrain from eating smoked or marinated fish products, particularly vacuum-packed smoked salmon and gravadlax. • Do not eat any raw milk soft cheese and always remove the cheese rind. • Always prepare green salads themselves and do not use any cut packaged salads. • Use foods, in particular vacuum-packed foods, as soon as possible after purchase and well in advance of the 'best before' date.	http://www.bfr.bund.de/cd/10966
UK	Pregnant women, unborn and newborn babies and people with	• All types of pâté (including vegetable). • Mould-ripened soft cheeses of the Brie, Camembert and	

Table 19.10 (Contd.)

Country	Target group	Advice	Website
	reduced immunity, particularly those over 60 (these could include people who've had transplants, are taking drugs that weaken the immune system or with cancers affecting the immune system, such as leukaemia or lymphoma).	blue veined types. • Any raw meat, shellfish or ready-meals should be well cooked until they are piping hot throughout. • Chilled foods should be kept out of the fridge for the shortest time possible and you shouldn't use food after its 'use by' date. Additional advice is given to pregnant women, the elderly and people with weakened immune systems during the lambing season involving avoiding direct or indirect (clothing, boots, etc.) contact with new-born lambs and any materials associated with these (HPA, 2008c).	http://www.eatwell.gov.uk/healthissues/food poisoning/abugslife/
USA	Pregnant women; older adults; people with cancer, AIDS, and other diseases that weaken the immune systems.	• Soft cheeses: Mexican-style soft cheeses (including queso blanco, queso fresco, queso de hoja, queso de crema and asadero), feta, Brie and Camembert, blue cheeses and Roquefort. Cheeses made from raw milk. • Refrigerated pâtés or meat spreads. • Raw (unpasteurised) milk or foods that contain unpasteurised milk. • Hot dogs, luncheon meats and other ready-to-eat foods – *unless they're reheated until steaming hot.* • Meats and seafood – *unless it's in a cooked dish.*	http://www.fda.gov/

Source: adapted from ACMSF (2009)

groups and listeriosis has been issued by the UK Food Standards Agency. This advice, together with advice provided in some other countries to vulnerable groups is summarised in Table 19.10.

Where cooking is recommended as an important control for the organism as in the case of ready meals for vulnerable groups, it is essential that the cooking guidance printed on the pack is clear and correct. Factors to take into consideration when establishing such guidance have already been described in Chapter 17.

19.9 Future trends

Much has been learnt since the 1980s about the factors necessary for the control of this organism in foods and food production environments. Listeriosis had generally been in decline in many countries, and certainly in the UK. However, it is apparent from recent increases in sporadic cases that the disease is resurgent with particular focus on the elderly. This may require further emphasis on educating and informing these vulnerable groups in a similar way that such information helped to reduce cases in pregnant women in the 1980s. However, it remains the case that reducing the prevalence of the organism in foods, and particularly reducing the presence of high levels in foods through effective application of HACCP principles, is central to preventing listeriosis, and the main challenge remains the consistent implementation of the known control strategies. It is this approach that will make human listeriosis a disease of the past.

19.10 Sources of further information and advice

CAC (2007) Guidelines on the application of general principles of food hygiene to the control of *Listeria monocytogenes* in ready-to-eat foods. CAC/GL 61-2007, Codex Alimentarius Commission, Rome, available at: www.codexalimentarius.net/download/standards/10740/cxg_061e.pdf, accessed November 2008.

Farber J M and Peterkin P I (2000) *Listeria monocytogenes*, in B M Lund, T C Baird-Parker and G W Gould (eds), *The Microbiological Safety and Quality of Food*, Vol. II, Aspen, Gaithersburg, MD, 1178–1232.

Griffiths M (ed.) (2005) *Understanding Pathogen Behaviour – Virulence, Stress Response and Resistance*, CRC, New York and Woodhead, Cambridge.

ICMSF (1996) *Microorganisms in Foods 5, Microbiological Specifications of Food Pathogens*, Chapter 8, International Commission on Microbiological Specifications for Foods, Blackie Academic and Professional, London.

Ryser E T and Marth E H (2007) *Listeria, Listeriosis, and Food Safety*, CRC, Boca Raton, FL.

Stringer M and Dennis C (eds) (2000) *Chilled Foods – a Comprehensive Guide*, 2nd edn, CRC Press, New York and Woodhead, Cambridge.

US Department of Health and Human Sciences and US Department of Agriculture (2003) *Quantitative assessment of relative risk to public health from foodborne* Listeria monocytogenes *among selected categories of ready-to-eat foods*, FDA/Center for Food Safety and Applied Nutrition, USDA/Food Safety and Inspection Service, Centers for Disease Control and Prevention, September 2003, available at: www.foodsafety.gov/~dms/lmr2-toc.html, accessed November 2008.

19.11 References

Aarnisalo K, Raaska L and Wirtanen G (2006) Survival and growth of *Listeria monocytogenes* in lubricants used in the food industry, *Food Control*, **18**(9), 1019–25.

ACMSF (2009) *Increased incidence of listeriosis in the UK*, Advisory Committee on the Microbiological Safety of Food, London.

Álvarez A, Pagán R, Raso J *et al.* (2002) Environmental factors influencing the inactivation of *Listeria monocytogenes* by pulsed electric fields, *Letters in Applied Microbiology*, **35**, 489–93.

Anon. (1997) Listeriosis in England and Wales: 1983 to 1996, *Communicable Disease Report*, **7**(11), 95.

Anon. (1998) Multi-state outbreak of Listeriosis – United States, 1998, *Morbidity and Mortality Weekly Report*, **47**(50), 1085–6.

Anon. (1999) Update: multistate outbreak of listeriosis – United States, 1998–1999, *Morbidity and Mortality Weekly Report*, **47**(51 & 52), 1117–18.

Anon. (2000a) Outbreak of listeriosis linked to the consumption of rillettes in France. *Eurosurveillance Weekly*, **3**, 19 January 2–3.

Anon. (2000b) Outbreak of *Listeria monocytogenes* serovar 4b infection in France, *Communicable Disease Report*, **10**(9), 81–4.

Antoniollo P C, Bandiera F da S, Jantzen M M (2003) Prevalence of *Listeria* spp. in feces and carcasses at a lamb packing plant in Brazil, *Journal of Food Protection*, **66**(2), 328–30.

Arumugaswamy R K, Ali G R R and Hamid S N BA (1994) Prevalence of *Listeria monocytogenes* in foods in Malaysia, *International Journal of Food Microbiology*, **23**, 117–21.

Aureli P, Fiorucci G C, Caroli D *et al.* (2000) An outbreak of febrile gastroenteritis associated with corn contaminated by *Listeria monocytogenes*, *New England Journal of Medicine*, **342**(17), 1236–41.

Autio T, Hielm S, Miettinen M *et al.* (1999) Sources of *Listeria monocytogenes* contamination in a cold smoked rainbow trout processing plant detected by pulsed-field gel electrophoresis typing, *Applied and Environmental Microbiology*, **65**, 150–5.

Baek S-Y, Lim S-Y, Lee D-H *et al.* (2000) Incidence and characterization of *Listeria monocytogenes* from domestic and imported foods in Korea, *Journal of Food Protection*, **63**(2), 186–9.

Baker M, Brett M, Short P *et al.* (1993) Listeriosis and mussels, *Communicable Disease New Zealand*, **93**(1), 13–14.

Baylis C and Mitchell C (2008) *Catalogue of Rapid Microbiological Methods, Review No. 1*, 6th edn, Campden & Chorleywood Food Research Association, Chipping Campden.

Beaufort A, Rudelle S, Gnanou-Besse N *et al.* (2007) Prevalence and growth of *Listeria monocytogenes* in naturally contaminated cold-smoked salmon, *Letters in Applied Microbiology*, **44**, 406–11.

Bell C and Kyriakides A (2005) Listeria: *A Practical Approach to the Organism and its Control in Foods*, Blackwell, Oxford.

Ben Embarek P K (1994) Presence, detection and growth of *Listeria monocytogenes* in seafoods: a review, *International Journal of Food Microbiology*, **23**, 17–34.

Beuchat L R (1996) *Listeria monocytogenes*: incidence on vegetables, *Food Control*, **7** (4/5), 223–8.

Bille J (1990) Epidemiology of human listeriosis in Europe, with special reference to the Swiss outbreak, in A J Miller, J L Smith and G A Somkuti (eds), *Foodborne Listeriosis*, Elsevier, Amsterdam, 71–4.

Bille J, Blanc D S, Schmid H *et al.* (2006) Outbreak of human listeriosis associated with Tomme cheese in NorthWest Switzerland, 2005, *Eurosurveillance*, **11**(6), 1st June 2006.

Boerlin P, Rocourt J, Grimont F *et al.* (1992) *Listeria ivanovii* subsp. *londoniensis* subsp. nov, *International Journal of Systematic Bacteriology*, **42**(1), 69–73.

Bohaychuk V M, Gensler G E, King R K *et al.* (2006) Occurrence of pathogens in raw and ready-to-eat meat and poultry products collected from the retail marketplace in Edmonton, Alberta, Canada, *Journal of Food Protection*, **69**(9), 2176–82.

Bouttefroy A, Lemaître J P and Rousset A (1997) Prevalence of *Listeria* sp. in droppings from urban rooks (*Corvus frugilegus*), *Journal of Applied Microbiology*, **82**, 641–7.

Cabedo L, Picart L, Barrot I *et al.* (2008) Prevalence of *Listeria monocytogenes* and *Salmonella* in ready-to-eat food in Catalonia, Spain, *Journal of Food Protection*, **71**(4), 855–9.

CAC (1996) *Establishment of Sampling Plans for Microbiological Safety Criteria for Foods in International Trade Including Recommendations for Control of Listeria monocytogenes, Salmonella enteritidis, Campylobacter and entero-haemorrhagic E. coli*, Document prepared by the International Commission on Microbiological Specifications for Foods for the Codex Committee on Food Hygiene, twenty-ninth session, 21–25 October 1996, Agenda item 11, CX/FH 96/9 1–16, Codex Alimentarius Commission, Rome.

Carlier V, Augustin J C and Rozier J (1996) Heat resistance of *Listeria monocytogenes* (phagovar 2389/2425/3274/2671/47/108/340): D- and z-values in ham, *Journal of Food Protection*, **59**(6), 588–91.

Carlos V -S, Oscar R -S and Irma Q -R E (2001) Occurrence of *Listeria* species in raw milk in farms on the outskirts of Mexico City, *Food Microbiology*, **18**(2), 177–81.

Carrique-Mas J J, Hokeberg I, Andersson, Y *et al.* (2003) Febrile gastroenteritis after eating on-farm manufactured cheese – an outbreak of listeriosis? *Epidemiology and Infection*, **130**(1), 79–86.

Cataldo G, Conte M P, Chiarini F *et al.* (2007) Acid adaptation and survival of *Listeria monocytogenes* in Italian-style soft cheeses, *Journal of Applied Microbiology*, **103**, 185–93.

Cates S C, Kosa K M, Teneyck T *et al.* (2006) Older adults' knowledge, attitudes, and practices regarding listeriosis prevention, *Food Protection Trends*, **26**(11), 774–85.

CFA (2006) *Best Practice Guidelines for the Production of Chilled Foods*, 4th Edn, Chilled Food Association, Peterborough.

Cole M B, Jones M V and Holyoak C (1990) The effect of pH, salt concentration and temperature on the survival and growth of *Listeria monocytogenes*, *Journal of Applied Bacteriology*, **69**, 63–72.

Dalton C B, Austin C C, Sobel J *et al.* (1997) An outbreak of gastro-enteritis and fever due to *Listeria monocytogenes* in milk, *The New England Journal of Medicine*, **336**(2), 100–5.

Dawson S J, Evans M R W, Willby D *et al.* (2006) Listeriosis outbreak associated with

sandwiches consumption from a hospital retail shop, United Kingdom, *Eurosurveillance*, **11**(6), 1st June 2006.
De Simón M and Ferrer M D (1998) Initial numbers, serovars and phagovars of *Listeria monocytogenes* isolated in prepared foods in the city of Barcelona (Spain), *International Journal of Food Microbiology*, **44**, 141–4.
Denny J and McLauchlin J (2008) Human *Listeria monocytogenes* infections in Europe – an opportunity for improved European surveillance, *Eurosurveillance*, **13**(13), 27 March 2008.
DoH (1992) *Safer Cooked Meat Production Guidelines. A 10 Point Plan*, Department of Health, London.
DoH (1998) *Surveillance of the Microbiological Status of Raw Cows' Milk on Retail Sale*, Microbiological Food Safety Surveillance, Department of Health, London.
EC (1999) *Opinion of the Scientific Committee on the Veterinary Measures Relating to Public Health on* Listeria monocytogenes. EC Health and Consumer Protection Directorate, European Commission, Brussels.
EC (2004) Regulation (EC) No 852/2004 of the European Parliament and of the Council of 29 April 2004 on the hygiene of foodstuffs, *Official Journal of the European Union*, **L139**, 30 April, 1–54.
EC (2005) Commission Regulation (EC) No 2073/2005 of 15 November 2005 on microbiological criteria for foodstuffs, *Official Journal of the European Union*, **L338**, 22 December, 1–26.
EC (2007) Commission Regulation (EC) No 1441/2007 of 5 December 2007 amending Regulation (EC) No 2073/2005 on microbiological criteria for foodstuffs, *Official Journal of the European Union*, **L322**, 7 December, 12–29.
EFSA (2007a) Scientific Opinion of the Panel on Biological Hazards on a request from the European Commission on Request for updating the former SCVPH opinion on *Listeria monocytogenes* risk related to ready-to-eat foods and scientific advice on different levels of *Listeria monocytogenes* in ready-to-eat foods and the related risk for human illness, European Food Safety Authority, *The EFSA Journal*, **599**, 1–42.
EFSA (2007b) The Community Summary Report on Trends and Sources of Zoonoses, Zoonotic Agents, Antimicrobial Resistance and Foodborne Outbreaks in the European Union in 2006, European Food Safety Authority, *The EFSA Journal*, **130**, 135–51.
Eklund M W, Poysky F T, Paranjpye R N *et al.* (1995) Incidence and sources of *Listeria monocytogenes* in cold–smoked fishery products and processing plants, *Journal of Food Protection*, **58**(5), 502–8.
Elson R, Burgess F, Little C L *et al.* (2004) Microbiological examination of ready-to-eat cold sliced meats and pâté from catering and retail premises in the UK, *Journal of Applied Microbiology*, **96**, 499–509.
Farber J M (1989) Thermal resistance of *Listeria monocytogenes* in foods, *International Journal of Food Microbiology*, **8**, 285–91.
Farber J M and Peterkin P I (2000) *Listeria monocytogenes*, in B M Lund, T C Baird-Parker and G W Gould (eds), *The Microbiological Safety and Quality of Food*, Vol. II, Aspen, Gaithersburg, MD, 1178–232.
Farber J M, Johnston M A, Purvis U *et al.* (1987) Surveillance of soft and semi-soft cheeses for the presence of *Listeria* spp, *International Journal of Food Microbiology*, **5**, 157–63.
FDA (2007) Guidance for Industry: *Guide to Minimize Microbial Food Safety Hazards of Fresh-cut Fruits and Vegetables*, Draft Final Guidance Contains Non-binding Recommendations, US Food and Drug Administration, Center for Food Safety and Applied Nutrition, Washington, DC, available at: www.cfsan.fda.gov/~dms/prodgui3.html, accessed November 2008.
FDA (2008) Guidance for Industry: *Control of Listeria monocytogenes in Refrigerated or Frozen Ready-To-Eat Foods*, Draft Guidance Contains Nonbinding Recommendations, US Food and Drug Administration, Center for Food Safety and Applied Nutrition,

Washington, DC, available at: http://www.cfsan.fda.gov/~dms/lmrtegui.html#measure, accessed December 2008.

Fleming D W, Cochi S L, Macdonald K L *et al.* (1985) Pasteurized milk as a vehicle of infection in an outbreak of listeriosis, *New England Journal of Medicine*, **312**(7), 404–7.

Fröder H, Martins C G, de Souza K L O *et al.* (2007) Minimally processed vegetable salads: microbial quality evaluation, *Journal of Food Protection*, **70**(5), 1277–80.

Frye D M, Zweig R, Sturgeon J *et al.* (2002) An outbreak of febrile gastroenteritis associated with delicatessen meat contaminated with *Listeria monocytogenes*, *Clinical Infectious Diseases*, **35**(8), 943–9.

Fulsom J P and Frank F F (2006) Chlorine resistance of *Listeria monocytogenes* biofilms and relationship to subtype, cell density, and planktonic cell chlorine resistance, *Journal of Food Protection*, **69**(6), 1292–6.

Gahan C G M and Hill C (2005) A Review: Gastrointestinal phase of *Listeria monocytogenes* infection, *Journal of Applied Microbiology*, **98**, 1345–53.

García-Gimeno R M, Zurera-Cosano G and Amaro-López M (1996) Incidence, survival and growth of *Listeria monocytogenes* in ready-to-use mixed vegetable salads in Spain, *Journal of Food Safety*, **16**, 75–86.

Gaya P, Saralegui C, Medina M *et al.* (1996) Occurrence of *Listeria monocytogenes* and other *Listeria* spp. in raw caprine milk, *Journal of Dairy Science*, **79**, 1936–41.

Gianfranceschi M, Gattuso A, Fiore A *et al.* (2006) Survival of *Listeria monocytogenes* in uncooked Italian dry sausage, *Journal of Food Protection*, **69**(7), 1533–8.

Gilbert R J (1996) Zero tolerance for *Listeria monocytogenes* in foods – is it necessary or realistic? *Food Australia*, **48**(4), 169–70.

Gilbert R J, De Louvois J, Donovan T *et al.* (2000) Guidelines for the microbiological quality of some ready-to-eat foods sampled at the point of sale, *Communicable Disease and Public Health*, **3**(3), 163–7.

Gillespie I, McLauchlin J, Adak R *et al.* (2005) Changing pattern of human listeriosis in England and Wales, 1993–2004, *Report for the ACMSF. September 2005*, available at: http://www.food.gov.uk/multimedia/pdfs/acm753.pdf, accessed November 2008.

Gillespie I A, McLauchlin J, Grant K A *et al.* (2006) Changing pattern of human listeriosis, England and Wales, 2001–2004, *Emerging Infectious Disease*, **12**(9), 1361–6.

Gottlieb S L, Newbern C, Griffin P M *et al.* (2006) Multistate outbreak of listeriosis linked to turkey deli meat and subsequent changes in US regulatory policy, *Clinical Infectious Diseases*, **42**, 29–36.

Goulet V, Lepoutre A, Rocourt J *et al.* (1993) Épidémie de listériose en France – Bilan final et résultats de l'enquête épidémiologique, *Bulletin Épidémiologie Hebdomaire*, **4**, 13–14.

Goulet V, Jacquet C, Vaillant V *et al.* (1995) Listeriosis from consumption of raw milk cheese, *The Lancet*, **345**, 1581–2.

Hansen C H, Vogel B F and Gram L (2006) Prevalence and survival of *Listeria monocytogenes* in Danish aquatic and fish-processing environments, *Journal of Food Protection*, **69**(9), 2113–22.

Hartmann A, Pagotto F, Tyler K *et al.* (2007) Characteristics and frequency of detection of fecal *Listeria monocytogenes* shed by livestock, wildlife, and humans, *Canadian Journal of Microbiology*, **53**(10), 1158–67.

Hazeleger W C, Dalvoorde M and Beumar R R (2006) Fluorescence microscopy of NaCl-stressed, elongated *Salmonella* and *Listeria* cells reveals the presence of septa in filaments, *International Journal of Food Microbiology*, **112**(3), 288–90.

Hellström S, Kervinen R, Lyly M *et al.* (2006) Efficacy of disinfectants to reduce *Listeria monocytogenes* on precut iceberg lettuce, *Journal of Food Protection*, **69**(7), 1565–70.

Hellström S, Kiviniemi K, Autio T *et al.* (2008) *Listeria monocytogenes* is common in wild birds in Helsinki region and genotypes are frequently similar with those found along the food chain, *Journal of Applied Microbiology*, **104**(3), 883–8.

Hicks S J and Lund B M (1991) The survival of *Listeria monocytogenes* in cottage cheese, *Journal of Applied Bacteriology*, **70**, 308–14.

HPA (2008a) *Case reports 1983–2006 Listeria monocytogenes. Human cases in residents of England and Wales reported to the Health Protection Agency Centre for Infections 1983–2006*, Health Protection Agency, London, available from: http://www.hpa.org.uk/.

HPA (2008b) *Reminder on safe storage of freshly cooked sliced meats*, Press release on 23 May 2008, Health Protection Agency, London available from: http://www.hpa.org.uk.

HPA (2008c) Advice to pregnant women during the lambing season, *Health Protection Report*, **2**(4), January.

Hudson J A, Mott S J and Penney N (1994) Growth of *Listeria monocytogenes*, *Aeromonas hydrophila*, and *Yersinia enterocolitica* on vacuum and saturated carbon dioxide controlled atmosphere-packaged sliced roast beef, *Journal of Food Protection*, **57**(3), 204–8.

ICMSF (1996) *Microorganisms in Foods 5, Microbiological Specifications of Food Pathogens*, International Commission on Microbiological Specifications for Foods, Blackie Academic and Professional, London.

ISO (1998) *ISO 11290-2:1998 Microbiology of Food and Animal Feeding Stuffs – Horizontal Method for the Detection and enumeration of Listeria monocytogenes*, Part 2: Enumeration method, International Organization for Standardization, Geneva.

ISO (2004) *ISO 11290-2:1998/Amd 1:2004 Modification of the enumeration medium*, International Organization for Standardization, Geneva.

Jay J M (1996) Prevalence of *Listeria* spp. in meat and poultry products, *Food Control*, **7**(4/5), 209–14.

Jones D (1990) Foodborne illness – foodborne listeriosis, *The Lancet*, **336** (Nov. 10), 1171–4.

Jørgensen L V and Huss H H (1998) Prevalence and growth of *Listeria monocytogenes* in naturally contaminated seafood, *International Journal of Food Microbiology*, **42**, 127–31.

Kanuganti S R, Wesley I V, Reddy P G *et al.* (2002) Detection of *Listeria monocytogenes* in pigs and pork, *Journal of Food Protection*, **65**(9), 1470–4.

Kozak J, Balmer T, Byrne R *et al.* (1996) Prevalence of *Listeria monocytogenes* in foods: incidence in dairy products, *Food Control*, **7**(4/5), 215–21.

Laciar A L and de Centorbi O N P (2002) *Listeria* species in seafood: isolation and characterization of *Listeria* spp. from seafood in San Luis, Argentina, *Food Microbiology*, **19**, 645–51.

Lee, S-H and Frank, J F (1991) Inactivation of surface-adherent *Listeria monocytogenes* by hypochlorite and heat, *Journal of Food Protection*, **60**, 43–7.

Lewis H C, Little C L, Elson R *et al.* (2006) Prevalence of *Listeria monocytogenes* and other *Listeria* species in butter from United Kingdom production, retail, and catering premises, *Journal of Food Protection*, **69**(7), 1518–26.

Lin C-M, Fernando S Y and Wei C-I (1996) Occurrence of *Listeria monocytogenes*, *Salmonella* spp., *Escherichia coli* and *E. coli* O157:H7 in vegetable salads, *Food Control*, **7**(3), 135–40.

Linnan M J, Mascola L, Xiao Dong Lou *et al.* (1988) Epidemic listeriosis associated with Mexican-style cheese, *The New England Journal of Medicine*, **319**(13), 823–8.

Little C L, Monsey H A, Nichols G L *et al.* (1997) The microbiological quality of refrigerated salads and crudités, *PHLS Microbiology Digest*, **14**(3), 142–6.

Little C L, Barrett N J, Grant K *et al.* (2008) Microbiological safety of sandwiches from hospitals and health care establishments in the United Kingdom with a focus on *Listeria monocytogenes* and other *Listeria* species, *Journal of Food Protection*, **71**(2), 309–18.

Loncarevic S, Danielsson-Tham M-L and Tham W (1995) Occurrence of *Listeria monocytogenes* in soft and semi-soft cheeses in retail outlets in Sweden, *International Journal of Food Microbiology*, **26**, 245–50.

Lou Y and Yousef A E (1999) Characteristics of *Listeria monocytogenes* important to food producers, in E T Ryser and E H Marth (eds), *Listeria, Listeriosis, and Food Safety*, Marcel Dekker, New York, 131–224.

Lundén J M, Autio T J and Korkeala H J (2002) Transfer of persistent *Listeria monocytogenes* contamination between food-processing plants associated with a dicing machine, *Journal of Food Protection*, **65**(7), 1129–33.

Lundén J T, Autio A, Markkula S *et al.* (2003) Adaptive and cross-adaptive responses of persistent and non persistent *Listeria monocytogenes* strains to disinfectants, *International Journal of Food Microbiology*, **82**, 265–72.

Lundén J, Tolvanen R and Korkeala H (2007) Acid and heat tolerance of persistent and nonpersistent *Listeria monocytogenes* food plant strains, *Letters in Applied Microbiology*, **46**, 276–80.

Lynch M, Painter J, Woodruff R *et al.* (2006) Surveillance for foodborne-disease-outbreaks – United States, 1998–2002, *Morbidity & Mortality Weekly Report*, **55**(SS10), 1–34.

Lyytikäinen O, Ruutu P, Mikkola J *et al.* (1999) An outbreak of listeriosis due to *Listeria monocytogenes* serotype 3a from butter in Finland, *Eurosurveillance Weekly*, **11**(11 March), 1–2.

MacDonald P D, Whitwam R E, Boggs J D *et al.* (2005) Outbreak of listeriosis among Mexican immigrants as a result of consumption of illicitly produced Mexican-style cheese, *Clinical Infectious Disease*, **40**(5), 677–82.

MAFF (1997) *Report on the National Study of Ready-to-eat Meats and Meat Products*, Part 4, Ministry of Agriculture Fisheries and Food, London.

Makino S I, Kawamoto K, Takeshi K *et al.* (2005) An outbreak of foodborne listeriosis due to cheese in Japan during 2001, *International Journal of Food Microbiology*, **104**(2), 189–96.

Massa S, Trovatelli L D and Canganella F (1991) Survival of *Listeria monocytogenes* in yogurt during storage at 4 °C, *Letters in Applied Microbiology*, **13**, 112–14.

McLauchlin J (1997) The pathogenicity of *Listeria monocytogenes*: a public health perspective, *Reviews in Medical Microbiology*, **8**(1), 1–14.

McLauchlin J (2007) Update on listeriosis in the United Kingdom, June 2007, *Report for the ACMSF*, June 2007, available at: http://www.food.gov.uk/multimedia/pdfs/847listeria, accessed November 2008.

McLauchlin J Hall S M, Velani S K *et al.* (1991) Human listeriosis and pâté: a possible association, *British Medical Journal*, **303**, 773–5.

McLauchlin J and Gillespie I (2007) Update on listeriosis in England and Wales, December 2007, *Report for the ACMSF, December 2007*, available at: http://www.food.gov.uk/multimedia/pdfs/committee/879listeria.pdf, accessed November 2008.

Mead P S, Dunne E F, Graves L *et al.* (2006) Nationwide outbreak of listeriosis due to contaminated meat, *Epidemiology and Infection*, **134**(4), 744–51, Abstract.

Mędrala D, Dąbrowski W, Czekajło-Kołodziej U *et al.* (2003) Persistence of *Listeria monocytogenes* strains isolated from products in a Polish fish-processing plant over a 1-year period, *Food Microbiology*, **20**, 715–24.

Meldrum R J and Smith R M M (2007) Occurrence of *Listeria monocytogenes* in sandwiches available to hospital patients in Wales, United Kingdom, *Journal of Food Protection*, **70**(8), 1958–60.

Miettinen H, Arvola A, Luoma T *et al.* (2003) Prevalence of *Listeria monocytogenes* in, and microbiological and sensory quality of, rainbow trout, whitefish, and vendace roes from the Finnish retail markets, *Journal of Food Protection*, **66**(10), 1832–9.

Monfort P, Minet J, Rocourt J *et al.* (1998) Incidence of *Listeria* spp. in Breton live shellfish, *Letters in Applied Microbiology*, **26**, 205–8.

Mosqueda-Melgar J, Raybaudi-Massilia, Martin-Belloso O (2007) Influence of treatment time and pulse frequency on *Salmonella* Enteritidis, *Escherichia coli*, and *Listeria monocytogenes* populations inoculated in melon and watermelon juices treated by pulsed electric fields, *International Journal of Food Microbiology*, **117**(2), 192–200.

Mosqueda-Melgar J, Raybaudi-Massilia, Martin-Belloso O (2008) Combination of high-intensity pulsed electric fields with natural antimicrobials to inactivate pathogenic microorganisms and extend the shelf-life of melon and watermelon juices, *Food Microbiology*, **25**(3), 479–91.

Mossel D A A, Corry J E L, Struijk C B and Baird R M (1995) *Essentials of the Microbiology of Foods. A Textbook for Advanced Studies*, Wiley, New York.

Murphy R Y, Marks B P, Johnson E R *et al.* (1999) Inactivation of *Salmonella* and *Listeria* in ground chicken breast meat during thermal processing, *Journal of Food Protection*, **62**(9), 980–5.

Newton L, Hall S M, Pelerin M *et al.* (1992) Listeriosis surveillance: 1991, *Communicable Disease Report*, **2**, Review No. 12, R142–R144.

Nicholson F A, Hutchison M L, Smith K A *et al.* (2000) *A study on farm manure application to agricultural land and an assessment of the risks of pathogen transfer into the food chain*, Research Report, Ministry of Agriculture Fisheries and Food, UK.

Nørrung B, Andersen J K and Schlundt J (1999) Incidence and control of *Listeria monocytogenes* in foods in Denmark, *International Journal of Food Microbiology*, **53**, 195–203.

O'Donnell E T (1995) The incidence of *Salmonella* and *Listeria* in raw milk from farm bulk tanks in England and Wales, *Journal of the Society of Dairy Technology*, **48**, 25–9.

Pérez-Rodríguez F, Todd E C D, Valero A *et al.* (2006) Linking quantitative exposure assessment and risk management using food safety objective concept; An example with *Listeria monocytogenes* in different cross contamination scenarios, *Journal of Food Protection*, **69**(10), 2384–94.

PHLS (2004) Preventing person-to-person spread following gastrointestinal infections: guidelines for public health physicians and environmental health officers, Prepared by a Working Group of the former Public Health Laboratory Service Advisory Committee on Gastrointestinal Infections, *Communicable Disease and Public Health*, **7**(4), 362–84.

Prazak A-M, Murano E A, Mercado I *et al.* (2002) Prevalence of *Listeria monocytogenes* during production and post-harvest processing of cabbage, *Journal of Food Protection*, **65**(11), 1728–34.

Rocourt J (1996) Risk factors for listeriosis, *Food Control*, **7**(4/5), 195–202.

Ryser E T and Marth E H (1987) Fate of *Listeria monocytogenes* during the manufacture and ripening of Camembert cheese, *Journal of Food Protection*, **50**(5), 372–8.

Sagoo S K, Little C L and Mitchell R T (2001) The microbiological examination of ready-to-eat organic vegetables from retail establishments in the United Kingdom, *Letters in Applied Microbiology*, **33**, 434–9.

Sagoo S K, Little C L, Ward L (2002) Microbiological examination of bagged prepared ready-to-eat salad vegetables from retail establishments, *LACORS/PHLS Coordinated Food Liaison Group Studies*, available from: www.lacors.gov.uk.

Sagoo S K, Little C L and Mitchell R T (2003) Microbiological quality of open ready-to-eat salad vegetables: effectiveness of food hygiene training of management, *Journal of Food Protection*, **66**(9), 1581–6.

Salamina G, Dalle Donne E, Niccolini A *et al.* (1996) A foodborne outbreak of gastroenteritis involving *Listeria monocytogenes, Epidemiology and Infection*, **117**, 429–36.

Schlech W F, Lavigne P M, Bortolussi R A *et al.* (1983) Epidemic listeriosis – evidence for transmission by food, *The New England Journal of Medicine*, **308**(4), 203–6.

Shank F R, Elliot E L, Wachsmuth I K *et al.* (1996) US position on *Listeria monocytogenes* in foods, *Food Control*, **7**(4/5), 229–34.

Soultos N, Koidis P and Madden R H (2003) Presence of *Listeria* and *Salmonella* spp. in retail chicken in Northern Ireland, *Letters in Applied Microbiology*, **37**, 421–3.

Tappero J N, Schuchat A, Deaver K A *et al.* (1995) Reduction in the incidence of human listeriosis in the United States. Effectiveness of prevention efforts, *Journal of the American Medical Association*, **273**, 1118–22.

Tasara T and Stephan R (2006) Cold stress tolerance of *Listeria monocytogenes*: A review of molecular adaptive mechanisms and food safety implications, *Journal of Food Protection*, **69**(6), 1473–84.

Thimonthe J, Walker J, Suvanich V *et al.* (2002) Detection of *Listeria* in crawfish processing plants and in raw, whole crawfish and processed crawfish (*Procambarus* spp.), *Journal of Food Protection*, **65**(11), 1735–9.

Uyttendaele M, De Troy P and Debevere J (1999) Incidence of *Salmonella*, *Campylobacter jejuni*, *Campylobacter coli*, and *Listeria monocytogenes* in poultry carcasses and different types of poultry products for sale on the Belgian retail market, *Journal of Food Protection*, **62**(7), 735–40.

Vachon J F, Kheadr E E, Giasson J *et al.* (2002) Inactivation of foodborne pathogens in milk using dynamic high pressure, *Journal of Food Protection*, **65**(2), 345–52.

Vít M, Olejník R, Dlhý J *et al.* (2007) Outbreak of listeriosis in the Czech Republic, late 2006 – preliminary report, *Eurosurveillance*, **12**(6), 08 February 2007.

Vugia D, Cronquist A, Hadler J *et al.* (2007) Preliminary FoodNet data on the incidence of infection with pathogens transmitted commonly through food – 10 states, 2006, *Morbidity & Mortality Weekly Report*, **56**(14), 336–9.

Wong S, Street D, Delgado S I *et al.* (2000) Recalls of food and cosmetics due to microbial contamination reported to the US Food and Drug Administration, *Journal of Food Protection*, **63**, 1113–16.

Zhang S and Farber J M (1996) The effects of various disinfectants against *Listeria monocytogenes* on fresh-cut vegetables, *Food Microbiology*, **13**, 311–21.

20

Campylobacter and *Arcobacter*

C. de W. Blackburn and P. J. McClure, Unilever, UK

Abstract: Campylobacteriosis continues to be one of the leading causes of foodborne illness throughout industrialised countries, while arcobacters are much less well recognised as foodborne pathogens. This chapter covers the general characteristics of *Campylobacter* and *Arcobacter* species, the nature of infections they cause and their growth and survival characteristics. Risk factors associated with infection are addressed, together with methods for their detection and procedures and strategies for their control throughout the food chain.

Key words: campylobacteriosis, viable but non-culturable, *Campylobacter jejuni*, thermophilic *Campylobacter*, aerotolerant *Campylobacter*, *Arcobacter butzleri*.

20.1 Introduction

Campylobacteriosis continues to be one of the leading causes of foodborne illness throughout industrialised countries, causing mild to severe symptoms. In Europe, the total number of reported cases of campylobacteriosis was about 194 000 in 2005, resulting in an overall incidence of approximately 39.5 cases per 100 000 population, although this figure was much higher for some countries in Europe, e.g. 296 per 100 000 population in the Czech Republic (EC, 2008). The major agent causing disease in humans is *Campylobacter jejuni*, but illness is also caused by *C. coli*, *C. lari* and *C. upsaliensis* and, with *C. jejuni*, this group is referred to as thermophilic *Campylobacter* species. These four *Campylobacter* species are zoonotic pathogens naturally present in the gastrointestinal tract of both domestic and wild animals. Arcobacters are much less well-recognised as foodborne pathogens and they are also associated with products of animal origin. These are commonly referred to as 'aerotolerant campylobacters'. The species most often associated with enteritis is *A. butzleri*, although *A. skirrowii* and *A. cryaerophilus* are also sometimes associated with human foodborne disease.

Campylobacter species were first described by Theodore Escherich in 1886 and isolated from the stools of patients with diarrhoea in 1972 but were not commonly recognised as a cause of human illness until the late 1970s, after the development of selective media by Butzler and co-workers in Belgium and Skirrow in the UK. However, an association between microaerophilic *Vibrio*-like bacteria and diarrhoea in cattle and pigs resulted in the naming of *Vibrio jejuni* in the early 1930s and *Vibrio coli* in the late 1940s, respectively. Similar organisms were found in the blood of patients with diarrhoea as long ago as 1948.

A group of aerotolerant campylobacters was identified in 1977, from the aborted foetuses of cattle, pigs and sheep in the UK (Ellis *et al.*, 1977). The organisms were also found in the milk of cows with mastitis and in blood and faeces from humans with diarrhoea, and *Campylobacter cryaerophila* was the name proposed for this group. Following this, the name *Campylobacter butzleri* was proposed for the group of strains most frequently isolated from human and non-human primates with diarrhoea. A number of detailed DNA:rRNA hybridisation studies followed, and this resulted in a revision of the taxonomy of these organisms. *Arcobacter* was the name proposed for a new genus consisting of *C. cryoaerophila* and *C. butzleri*, amongst others. This genus is closely related to the genus *Campylobacter* and, together with *Bacteroides ureolyticus*, make up the family Campylobacteraceae.

Campylobacter jejuni subsp. *jejuni* (*C. jejuni*) is responsible for 80–90 % of campylobacteriosis (Vandamme, 2000). Despite the huge number of *C. jejuni* cases currently being reported, the organism does not generally cause the same degree of concern as some other foodborne pathogens, since it rarely causes death and is not commonly associated with newsworthy outbreaks of food poisoning. It is, however, among the most common causes of sporadic bacterial foodborne illness. *Campylobacter jejuni* is associated with warm-blooded animals, but, unlike salmonellae and *Escherichia coli*, does not survive well outside of the host. *Campylobacter jejuni* is susceptible to environmental conditions and does not survive well in food and is, therefore, fortunately relatively easy to control. *Campylobacter coli* is responsible for about 7 % of human campylobacteriosis cases but, in some areas (e.g. Central African Republic and Zagreb), this number can be as high as 35–40 %. *Campylobacter upsaliensis* and *C. lari* have been isolated from patients suffering diarrhoea, and are responsible for about 1 % of human cases of illness associated with campylobacters.

Food-associated illness usually results from eating foods that are re-contaminated after cooking or eating foods of animal origin that are raw or inadequately cooked. The organisms are part of the normal intestinal flora of a wide variety of wild and domestic animals, and have a high level of association with poultry (Ketley, 1997). The virulence of the organisms, as suggested by the relatively low infectious dose of a few hundred cells, and its widespread prevalence in animals are important features that explain why this relatively sensitive organism is a leading cause of gastroenteritis in humans. *Campylobacter jejuni* tends to predominate in cattle, broiler chickens and turkeys while *C. coli* is more commonly found in pigs.

20.2 General characteristics of *Campylobacter* and *Arcobacter* species

20.2.1 Taxonomy

Véron and Chatelain (1973) published a comprehensive study on the taxonomy of '*Vibrio*-like' organisms, proposing, amongst others, the names *Campylobacter jejuni* and *Campylobacter coli*, and, after much confusion, this was officially accepted and included in the 'Approved Lists of Bacterial Names' (Skerman *et al.*, 1980). Following this, Vandamme *et al.* (1991) carried out an extensive polyphasic study of the entire *Campylobacter* complex using DNA-rRNA hybridisations, and this study still forms the basis of the current taxonomic structure. On (2001) described 16 species of *Campylobacter* and six further sub-species. An additional species has been described since that time, named *C. insulaenigrae*. Campylobacters are Gram-negative, microaerophilic, non-sporing, small vibroid (spiral-shaped) cells that have rapid, darting reciprocating motility. They reduce nitrate and nitrite (apart from *C. jejuni* subsp. *doylei* and *C. fennelliae*), are unable to oxidise or ferment carbohydrates and obtain energy from amino acids, or tricarboxylic acid cycle intermediates. The thermophilic species are characterised by their ability to grow best between 37 and 42 °C and their inability to grow at 25 °C and, in that sense, they are probably better described as thermotolerant rather thermophilic (often refers to growth at 55 °C). Strains of the two genera *Campylobacter* and *Arcobacter* have similar morphology and metabolism, and share several other genotypic and phenotypic features. Arcobacters are able to grow at 15 °C, and this enables distinction between *Campylobacter* and *Arcobacter* species.

Isolates of *Campylobacter* are extremely diverse, compared to some other enteropathogens. There are more than 60 different heat-stable serotypes, more than 100 heat-labile serotypes, differences in adherence properties, invasive properties, toxin production, serum resistance, colonisation potential, aerotolerance and temperature tolerance. This diversity may be partly explained by the natural competency of campylobacters to take up DNA. The high levels of multiple-strain colonisation and high frequency of incidence in mammals and birds mean there is substantial opportunity for exchange of genetic material. Genetic diversity, as evidenced by the different genotypes observed – through pulsed field gel electrophoresis (PFGE), randomly amplified polymorphic DNA (RAPD), ribotyping, amplified fragment length polymorphism (AFLP, etc.) – is indicative of genomic plasticity – the order of genes on the chromosome is not conserved between isolates of the same species. Sub-types recognised by one phenotypic or genotypic method often do not correlate with other techniques.

20.2.2 Virulence factors

Since most disease is associated with *C. jejuni*, more is known about the virulence factors of this species. That said, there is still a degree of confusion over the number and specificity of various toxins, due partly to the use of different assays and also to the status of culture filtrates, with some preparations being crude and others being relatively pure. The virulence factors currently identified include heat-labile enterotoxin (CJT), cytotoxins including cytolethal distending toxin (CLDT) and cytotoxin designated CT, enterocyte-binding factors such as *Campylobacter* adhesion factor, flagella, outer membrane protein that binds fibronectin and lipopolysaccharide (LPS) required for invasion of intestinal embryonic cells, capsule formation and a virulence plasmid that allows invasion of epithelial cells and also codes for a type IV secretion system. These are described in more detail by Levin (2007). The mechanism of invasion is still not completely understood, but the flagella filament is thought to act as a type III secretion system through secretion of Cia proteins. Interactions with epithelial cells and role of other virulence factors in *Campylobacter* pathogenesis are reviewed by Poly and Guerry (2008). The invasion of intestinal epithelial cells and presence of cytotoxins are consistent with the bloody inflammatory disease caused by campylobacters.

20.3 Nature of *Campylobacter* and *Arcobacter* infections

Campylobacter jejuni and *C. coli* are clinically indistinguishable, and most laboratories do not attempt to distinguish between the two organisms. Although *C. jejuni* is the predominant cause of campylobacteriosis, there are studies reporting 7 % of isolates are *C. coli*, in the UK and 18.6 % in Germany (Sopwith *et al.*, 2003; Gürtler *et al.*, 2005).

Campylobacteriosis in man is usually characterised by an acute, self-limiting enterocolitis, lasting up to a week, and clinical illness is often preceded by a prodrome of fever, headache, myalgia and malaise. Symptoms of disease often include abdominal pain, fever and cramps (frequently severe) and diarrhoea, which may be inflammatory, with slimy/bloody stools, or non-inflammatory, with watery stools and absence of blood. These clinical symptoms are often indistinguishable from gastrointestinal infections caused by *Shigella*, *Salmonella* and *Yersinia* species. The incubation period is usually 24–72 h after ingestion, but may extend to 7 d (Skirrow, 1994). Although not all those exposed to the organism will develop symptoms, human volunteer studies indicate that the infectious dose can be low, of the order of a few hundred cells. Occurrence of waterborne outbreaks and person-to-person spread also imply that the infective dose can be low.

In many regions, where antibiotics are used in the empirical treatment of *Campylobacter* infections, e.g. in the elderly, prolonged cases of enteritis, septicaemia or other extraintestinal infections, some antibiotics such as fluoroquinolones are of limited use, due to antibiotic resistance. Erythromycin is still the drug of choice. There is some evidence of protective immunity after infection from volunteer studies, and this may explain the higher incidence of disease in very young children. The efficacy of artificial immunogens is questionable since there is a wide variety of virulence between different phenotypes and an absence in increase of specific antibody. A small proportion (5–10 %) of affected individuals suffer relapses, possibly caused by an incomplete immune response since immunocompromised hosts often have severe, extraintestinal and prolonged illness.

Reactive arthritis and bacteraemia are rare complications, and infection with *C. jejuni* is also associated with Guillain-Barré syndrome (GBS), an autoimmune peripheral neuropathy causing limb weakness that is sometimes fatal. GBS has become the most common cause of acute, flaccid paralysis since the eradication of polio in most parts of the world, and *Campylobacter* infection is now known as the single most identifiable antecedent infection associated with the development of GBS. Schiellerup *et al.* (2003) considered 1339 cases of campylobacteriosis and reported 171 (19.9 %) of these involved joint pain. In another study in Finland, Hannu *et al.* (2002) reported 7 % of those suffering illness with reactive arthritis. It is thought that *C. jejuni* lipopolysaccharides mimic nerve gangliosides, causing an autoimmune response. Engberg (2002) concluded that 30–40 % of GBS cases are triggered by *Campylobacter* enteritis. This condition is thought to be associated with particular serotypes (e.g. O:19, O:4, O:5, O:41, O:2 and O:1) capable of producing structures (epitopes) that mimic ganglioside motor neurons. There are several pathological forms of GBS including demyelinating (acute inflammatory demyelinating polyneuropathy) and axonal (acute motor axonal neuropathy) forms. Host factors also play an important role in development of disease. Mortality rates of GBS have been reduced to 2–3 % in the developed world but remain higher in much of the developing world. Incidence of bacteraemia and systemic infection in immunocompromised individuals indicates that the immune system is important for confining the disease associated with campylobacters to the intestine.

Prior to 1991, *Arcobacter butzleri* and *A. cryaerophilus* were known as aerotolerant *Campylobacter*. These organisms have been associated with abortions and enteritis in animals and enteritis in man. Although both species are known to cause disease in man (i.e. clinical isolates have been reported in individuals with diarrhoea), most human isolates come from the species *A. butzleri*. There is very little known about the epidemiology, pathogenesis and real clinical significance of arcobacters, but it is thought that consumption of contaminated food may play a role in transmission of this group of organisms to man. Musmanno *et al.* (1997) tested a number of strains from river water for toxic effects and showed that a number of these induced cytotoxic effects on cell cultures, but these were not invasive. Although arcobacters are rarely associated with outbreaks of foodborne illness, they have been isolated from domestic animals, poultry, ground pork, beef and water (Lehner *et al.*, 2005). Symptoms of *Arcobacter* infections are similar to those caused by campylobacters, and include diarrhoea associated with abdominal pain, nausea, vomiting and fever (Vandenberg *et al.*, 2004). There are documented cases of extraintestinal disease, e.g. bacteraemia. Since primary screening procedures for campylobacters use high temperatures (42 °C) for recovery, *Arcobacter* spp. would be overlooked.

20.4 Growth and survival characteristics of *Campylobacter* and *Arcobacter*

Survival of *C. jejuni* outside the gut is poor, and the organism is sensitive to drying, freezing and low pH (pH ≤ 4.7). Survival kinetics tend to follow a rapid decline in numbers followed by a slower rate of inactivation. *Campylobacter jejuni* fails to grow in the presence of 2 % NaCl. Studies investigating the survival of campylobacters show that the potential increases with decreasing temperature, with survival lasting a few hours at 37 °C and several days at 4 °C. Campylobacters can survive several weeks in groundwater, and survival is enhanced in the presence of other organisms and in biofilms (Buswell *et al.*, 1998). The survival of *Campylobacter* in the environment through washing and disinfection, and even through repeated washings, disinfection and production cycles (in poultry slaughterhouses) may be due to *Campylobacter* remaining in biofilm layers (Johnsen *et al.*, 2007). *Campylobacter* has been shown to survive for more than one week in biofilms and an increased resistance to disinfection agents has been reported for *Campylobacter* present in biofilms (Buswell *et al.*, 1998; Trachoo *et al.*, 2002; Johnsen *et al.*, 2007). Snelling *et al.* (2005, 2006) reported that campylobacters are commonly found in communities in biofilms. Campylobacters lack the stress response mechanisms, such as the global stress response factor RpoS and other factors such as the oxidative stress response factors SoxRS, the cold-shock response protein CspA, RpoH or Lrp that are found in many other bacteria. Even though the organisms are not able to grow below 30 °C, probably due to absence of CspA and other cold-shock proteins, metabolic activity has been measured at 15 °C and at even lower temperatures (Kelly *et al.*, 2003).

Survival in HCl solutions at pH values below 3.0 is poor, whilst at pH values of 3.6 and above survival is unaffected. Protection against gastric acid may be afforded by ingestion with buffered foods, e.g. milk, or with water, which is rapidly washed through. In survival studies investigating acid stress, Kelly *et al.* (2001) reported mid-exponential cells to be more tolerant of low pH than stationary phase cells, which is not consistent with the behaviour of other bacteria that are generally more resistant in stationary phase. Murphy *et al.* (2003) provided an explanation for this, linking this response to production of a phase-specific extracellular component secreted during growth, contributing to both heat and low pH stresses. In foods, survival at low pH in plain marinade is short-lived, with a > 5 log reduction occurring after 48 h (Björkroth, 2005). In the presence of meat, however, survival can be extended, with organisms surviving for at least 9 d. The author attributed this to the buffering effect of meat.

Campylobacters are particularly susceptible to drying. For example, below an a_w of 0.97, cells die quickly, particularly at higher temperatures. Kusumaningrum *et al.* (2003) did not detect any surviving campylobacters on stainless steel surfaces that had been stored at room temperature for 4 h, starting with an inoculum as high as 10^7 cfu/100 cm^2. Modified atmospheres appear to have little effect on survival.

Chlorine (e.g. as sodium hypochlorite) is effective for inactivation of campylobacters in water and biofilms, and other disinfectants, such as benzalkonium chloride and peracetic acid, are effective at commonly used concentrations (Trachoo and Frank, 2002; Avrain, *et al.* 2003). However, chlorinated water is not very effective for campylobacters attached to poultry carcasses. Northcutt *et al.* (2005) reported that chlorine at 50 ppm and temperatures up to 54.4 °C had no effect on numbers of campylobacters on chicken carcasses.

The decimal reduction time (or *D*-value) for campylobacters is *c.* 1–6.6 min at 55 °C and the *z*-value is about 5–6.3 °C (Park *et al.*, 1991; Hong *et al.*, 2006). This is dependent on the heating medium. For example, in milk, *D*-values at 48 and 55 °C are 7–13 min and 0.7–1.0 min, respectively (Moore and Madden, 2000) and at 49, 53 and 57 °C, *D*-values in ground chicken meat were 20.5, 4.9 and 0.8 min, respectively (Blankenship and Craven, 1982). In a study using differential scanning calorimetry to determine the mechanism of heat inactivation, Hong *et al.* (2006) concluded that cell death in *C. jejuni* and *C. coli* coincided with the most thermally labile regions of the ribosome. Cells in exponential phase are more resistant than cells in stationary phase (Kelly *et al.*, 2001) and, at temperatures above 55 °C, the kinetics of inactivation are reported to be non-log linear, with tailing effects observed.

Alternative processing technologies are generally effective in destroying campylobacters. A *D*-value of 0.19 kGy has been demonstrated for *C. jejuni* during irradiation of vacuum-packaged ground pork (Collins *et al.*, 1996a), and Lewis *et al.* (2002) reported that electron-beam irradiation doses of 1.0 and 1.8 kGy destroyed campylobacters on poultry meat. Ultraviolet irradiation requires a dose of 1.8 mWs/cm^2 to effect a 3-log kill (Butler *et al.*, 1987). Solomon and Hoover (2004) studied the effects of high hydrostatic pressure on *C. jejuni* in chicken meat and demonstrated that pressures of 300–325 MPa could reduce numbers by 2–3 logs and 400 MPa inactivated the organism completely. A different response between exponential and stationary-phase cells has been reported for pressure treatment compared to low pH and heating effects, by Martinez-Rodriguez and Mackey (2005), with early stationary-phase cells being more resistant.

Campylobacter spp. exhibit a relatively high susceptibility to the effects of freezing (Park, 2002) and this has been attributed to the absence of cold-shock proteins (CSPs). A recent study by Georgsson *et al.* (2006) has confirmed results from previous studies, reporting that *Campylobacter* levels in broilers were decreased by between 0.7 and 2.9 $\log_{10}$ units after freezing and 31 d storage. Numbers of faecal coliforms, often used as indicator organisms, in the same samples remained largely unchanged. The significant die-off of *Campylobacter* is consistent with previous studies (Humphrey and Cruikshank, 1985; Stead and Park, 2000; Moorhead and Dykes, 2002) which reported approximately 2–3 $\log_{10}$ decreases in numbers after freeze–thaw stress, although, again, there is strain-to-strain variability reported with some strains showing higher sensitivity to freezing (Chan *et al.*, 2001). According to Stern *et al.* (2003) one of the major factors responsible for the reduction in cases of poultry-borne campylobacteriosis in Iceland was the introduction of programme for freezing carcasses from *Campylobacter*-positive flocks. Since then Iceland has mandated a policy for the continued use of this procedure. Other countries in Europe, such as Denmark and Norway, have also introduced controls for *Campylobacter* spp. that include freezing of poultry carcasses. Ritz *et al.* (2007) studied the survival of *Campylobacter* on different surfaces of frozen chicken meat, but difficulties in modelling the data prevented development of an acceptable predictive model. At refrigeration temperatures, numbers of campylobacters show a slow decrease with increased storage time. Bhaduri and Cottrell (2004) reported decrease of 0.3–0.81 logs in chicken (skin and ground meat) after 3–7 d storage.

Campylobacter jejuni is regarded by some to be capable of forming so-called viable but non-culturable (VBNC) cells. Although these cells are metabolically active and show signs of respiratory activity, they are not able to resuscitate by recovery through conventional culturing techniques. The VBNC state has been proposed as a survival strategy or as a moribund condition where cells become progressively debilitated, until they finally 'die'. Induction of the VBNC state in *C. jejuni* comes about through exposure to sub-lethal adverse environmental conditions, such as prolonged exposure to water or freeze–thaw injury, and recovery is effected by passage of the organism through a susceptible host (Rollins and Colwell, 1986; Jones *et al.*, 1991; Saha *et al.*, 1991). This probably reflects the inability of some culturing methods to provide suitable conditions for the resuscitation of 'injured' cells. The VBNC campylobacters sometimes form a coccoid shape, although non-coccoid VBNCs have also been described (Federighi *et al.*, 1998). There have been examples of outbreaks of campylobacteriosis where cases were observed after attempts to culture the organism from the identified source were no longer positive (Palmer *et al.*, 1983). Treatment of the water supply on one farm also resulted in the disappearance of a particular serotype that had colonised most of the chickens on the farm, although no campylobacters had been isolated from the chicken-shed water supply (Pearson *et al.*, 1988). The significance of the VBNC is still unclear and reversion of coccoid cells is not easily initiated, requiring very specific conditions. The existence of this state may exert an important influence on considerations in the epidemiology of human and animal campylobacteriosis, and should not be ignored.

Arcobacters are, generally speaking, much less well characterised than campylobacters. *Arcobacter* spp. are so named because of their bow-shape and are motile by a single unsheathed polar flagellum, exhibiting darting or corkscrew movement, whereas *Campylobacter* spp. may have a single flagellum at one or both ends. The growth and survival characteristics of *Arcobacter* spp. are summarised in Table 20.1. Their growth range is between 15 and 37 °C, with the upper limit dependent on growth conditions (Mansfield and Forsythe, 2000). *Arcobacter* spp. do not generally show the same thermotolerance as *Campylobacter* spp. and arcobacters may grow aerobically at 30 °C. The pH range for growth is between 5.5 and 9.5, and the limited amount of information available on inactivation by irradiation suggests *Arcobacter* spp. are slightly more resistant than *Campylobacter* spp. *Arcobacter cryaerophilus* is divided into two subgroups, 1A and 1B, based on DNA-DNA hybridisation, cellular protein composition (using sodium dodecyl sulfate–polyacrylamide gel electrophoresis – SDS-PAGE) and fatty acid analysis. However, there are no phenotypic analyses to differentiate

Table 20.1 Growth and survival characteristics of *Arcobacter* spp.

Characteristic		Values	Reference
Heat resistance	$D_{60\,°C}$ (PBS)[a]	0.07–0.12 min	D'Sa and Harrison, 2005
	$D_{55\,°C}$ (PBS)	0.38–0.76 min	D'Sa and Harrison, 2005
	$D_{50\,°C}$ (PBS)	5.12–5.81 min	D'Sa and Harrison, 2005
	z	5.6–6.3 C°	D'Sa and Harrison, 2005
	$D_{55\,°C}$ (ground pork)	2.81 min	D'Sa and Harrison, 2005
	$D_{50\,°C}$ (ground pork)	18.5 min	D'Sa and Harrison, 2005
Growth limits	pH (lower)	4.9	Park, 2002
	Temperature (lower)	10–15	D'Sa and Harrison, 2005; Hilton *et al.*, 2001
	Temperature (upper)	37	Hilton *et al.*, 2001
	NaCl	3.5–4 %	Atabay *et al.*, 1998; D'Sa and Harrison, 2005
Generation times	Laboratory medium, 30 °C	34 min	Hilton *et al.*, 2001

[a]Phosphate-buffered Saline.

these two groups. Several treatment regimes have been tested to reduce numbers of *Arcobacter* spp. in foods of animal origin. Cervenka *et al.* (2004) reported on the lethal effects of > 0.2 % acetic and citric acids after 4–5 h incubation. Freezing reduces viable numbers by 2 logs after 24 h but numbers remain constant after this time (Hilton *et al.*, 2001). More recent results of a freezing study by D'Sa and Harrison (2005) showed a 0–1.5 log decrease in numbers after storage at –20 °C for six months in a laboratory medium. There are relatively few studies describing effects of heating on arcobacters, and those that are available indicate similar heat resistance characteristics to *Campylobacter* spp., with reported *D*-values at 50 °C of 1.9–18.5 min and 0.75–2.81 min at 55 °C, depending on strains used and heating medium (Hilton *et al.*, 2001; D'Sa and Harrison, 2005). Arcobacter *butzleri* is closely related to *A. skirrowii*, and serotypes 1 and 5 are currently recognised as the main pathogenic strains (Lior and Woodward, 1991).

Investigation of the use of irradiation treatment for vacuum-packaged ground pork demonstrated that *A. buzleri* had a *D*-value (0.27 kGy) that was 1.4 times higher than that of *C. jejuni* (0.19 kGy) (Collins *et al.*, 1996a). Detection of *Arcobacter* spp. in piggery effluent and effluent-irrigated soil in Queensland was reported by Chinivasagam *et al.* (2007), suggesting that there was good survival in the effluent pond environment. The survival and inactivation of *Arcobacter* spp. has been recently reviewed by Cervenka (2007) and this article provides more detail on effects of other factors such as acid washes that have been applied to carcasses.

20.5 Risk factors for *Campylobacter*

Human infectious campylobacters have been found to be environmentally ubiquitous, being readily isolated from numerous wild and domesticated mammals in addition to wild avian species, poultry, fresh water and marine aquatic environments (Levin, 2007). *Campylobacter* is part of the normal intestinal flora of a wide variety of wild and domestic animals, particularly avian species including poultry, where the intestine is colonised resulting in healthy animals as carriers (Shane, 1992; Levin, 2007). *Campylobacter jejuni* has become recognised worldwide as a leading cause of diarrhoeal disease and foodborne gastroenteritis, but the epidemiology of this common infection is not well understood. Most cases are sporadic, as opposed to being associated with large outbreaks, and there is a high frequency in which the vehicle is not identified (Neal and Slack, 1997). Where identified, poultry, raw milk, contaminated drinking water, contact with pet animals and travel abroad have been recognised as the most common vehicles and/or risk factors associated with *Campylobacter* infection in humans (Schorr *et al.*, 1994; Neal and Slack, 1997; Solomon and Hoover, 1999; Rautelin and Hanninen, 2000).

Wildlife has long been considered an infectious reservoir for campylobacters because of their close association with surface waters (Levin, 2007). *Campylobacter* spp. have been frequently isolated from waterfowl and other wild birds (Abulreesh *et al.*, 2006). *Campylobacter jejuni* has been shown to be more prevalent in the faeces of waterfowl than *C. coli* or *C. lari*. *Campylobacter* spp. have been isolated from faecally-contaminated aquatic environments, with the diversity of *Campylobacter* populations dependent on the sources of contamination (Abulreesh *et al.*, 2006).

A variety of infection vehicles for transmitting *Campylobacter* enteritis in man have been identified, but there is general agreement that contaminated poultry meat is the most important (Humphrey *et al.*, 2007) and serotypes associated with poultry are also frequently associated with illness in humans (Shane, 1992). Other avian species, including geese, have been shown to be potential reservoirs for human and animal campylobacteriosis (Aydin *et al.*, 2001). Although consumption of poultry and poultry products has been frequently quoted as an important risk factor for *Campylobacter* infection (Schorr *et al.*, 1994; Solomon and Hoover, 1999; Rautelin and Hanninen, 2000; Humphrey *et al.*, 2007), it has been hypothesised that in a rural setting contact with livestock is of greater importance as compared with the poultry-related factors identified in urban studies (Garrett *et al.*, 2007).

Other foods implicated as risk factors for *Campylobacter* infection include: barbecued/grilled meat; undercooked meat; raw milk; bird-pecked milk; bottled mineral water; untreated water; salad vegetables and grapes (Humphrey *et al.*, 2007). In addition to risks from food and water consumption, the following factors also have an increased risk of *Campylobacter* infection: contact with either domestic pets or farm animals, including occupational exposure; problems with the household sewage system; and underlying medical conditions like diabetes or reduced gastric acidity due to the use of proton pump inhibitors (Humphrey *et al.*, 2007).

Numerous surveys of retail raw poultry and raw poultry products have been made in different countries and *Campylobacter* contamination rates have ranged from 3.7 % to 92.6 % (Fernandez and Pison, 1996; Ono and Yamamoto, 1999; Osano and Arimi, 1999; Jørgensen *et al.*, 2002; Scherer *et al.*, 2006; Sallam, 2007). In one study 241 whole raw chickens purchased from retail outlets in England during 1998–2000 were examined with *Campylobacter* isolated from 83 % of the chicken samples, 56 % of entire packaging samples (representing contamination present on both the inside and outside of the packaging) and 6 % of the outer packaging (Jørgensen *et al.*, 2002). Although there has been a lack of association of the organism with shell eggs (Zanetti *et al.*, 1996), some surveys have shown a low prevalence (up to 1.1 %) of the organism contaminating egg shells or their contents, if the shells are cracked or flawed (Jones and Musgrove, 2007). *Campylobacter* was detected at a frequency of 0.5 % on commercially cool water-washed shell eggs, with all 384 samples negative for *Campylobacter* in the egg contents (Jones *et al.*, 2006).

Where species identifications have been made, *C. jejuni*, *C. coli* and *C. laridis* have all been isolated, but the predominant species has varied (Fernandez and Pison, 1996; Osano and Arimi, 1999; Sallam, 2007). Antimicrobial resistance amongst *Campylobacter* strains has been shown to be common, with 73.3 % of isolates resistant to three or more antimicrobials (Sallam, 2007). Where quantitative methods have been used, levels of *Campylobacter* of 10^2–10^4/g of poultry (McClure, 2000) and 10 to > 230 campylobacters/100 ml chicken liver exudate have been reported (Fernandez and Pison, 1996). Enumeration of *Campylobacter* spp. in raw retail chicken leg samples yielded a medium of 4 log cfu/leg surface in skin samples (standard deviation of 0.6) and a medium of 4.3 log cfu/leg surface in rinse samples (Scherer *et al.*, 2006).

Due to the high level of *Campylobacter* on raw poultry, many sporadic cases of campylobacter enteritis are probably caused by cross-contamination in the domestic environment. It has been shown that during preparation of chicken for cooking, *Campylobacter* became widely disseminated to hand and food contact surfaces (Cogan *et al.*, 1999). Consequently, cross-contamination is one of the most important infection risk factors, and control can be particularly difficult in the household environment (Humphrey *et al.*, 2007). A mechanistic model for bacterial cross-contamination during food preparation in the domestic kitchen applied to *Campylobacter*-contaminated chicken breast demonstrated that cross-contamination can contribute significantly to the risk of *Campylobacter* infection and that cleaning frequency of kitchen cleaning utensils and thoroughness of rinsing of raw food items after preparation has more impact on cross-contamination than previously emphasised (Mylius *et al.*, 2007).

Although *Campylobacter* is a relatively heat-sensitive vegetative bacterium, undercooking is another potential risk factor, and a combination of inadequate cooking time and the use of large chicken pieces was the likely cause of a *Campylobacter* outbreak in a restaurant specialising in stir-fried food (Evans *et al.*, 1998). Barbecues appear to present special hazards for infection because, in addition to the potential for undercooking, they permit easy transfer of bacteria from raw meats to hands and other foods (Butzler and Oosterom, 1991).

A model has been developed to examine how chicken prepared in (German) homes exposes the consumer to *Campylobacter* spp. and the level of resulting illness (Brynestad *et al.*, 2008). Resulting simulations predicted no cases of illness for frozen chicken products, based on the very low levels of viable bacteria left on the chicken surface after freezing, and more than 90 % of campylobacteriosis cases were caused by fresh chicken legs. Undercooking was found in the simulations to be responsible for around 3 % of the cases of illness, but the greatest impact for consumer exposure came from cross-contamination events caused by hygienic failures in private kitchens.

The high incidence of *Campylobacter* in retail poultry and poultry products is a direct reflection of the prevalence of the organism in live chickens and turkeys and the ease with which the organism can be spread during poultry processing. A quantitative microbiological risk assessment model describing the transmission of *Campylobacter* through the broiler meat production chain and at home has been developed, which demonstrated that flocks with high concentrations of *Campylobacter* in the faeces that leak from the carcasses during industrial processing seem to have a dominant impact on the human incidence (Nauta *et al.*, 2007). The organism is morphologically and physiologically well adapted to its ecological niche in the mucus-filled crypts of villi of the gastrointestinal tract of poultry, but establishment and maintenance of colonisation involve a complex interaction between bacterium and host that has yet to be fully elucidated (Mead, 2002). Slaughtering of *Campylobacter*-positive broilers has been shown to result in extensive contamination of the slaughterhouse, including the air which in itself constituted a risk of human infection (Johnsen *et al.*, 2007). In this study the high proportion of *Campylobacter*-positive sites at the beginning of the slaughter line was attributed to the extremely humid or dust-filled air in the arrival, defeathering and evisceration areas. The decline in *Campylobacter* presence in the later stages was explained by a combination of factors: a less humid atmosphere; dilution of contamination due to continuous irrigation of carcasses; removal of feather and intestines; and daily disinfection. In a study investigating the influence of relative humidity (RH) on transmission of *C. jejuni* in broiler chickens, a significant delay in colonisation was observed in birds raised under low (approximately 30 %) compared with high (*c*. 80 %) RH conditions (Line, 2006).

In surveys of broiler farms, *Campylobacter* spp. have been detected in 32–57 % of flocks (Kapperud *et al.*, 1993; van de Giessen *et al.*, 1996; Hald *et al.*, 2000) with the proportion of colonised flocks varying geographically and seasonally with a peak in the autumn (Kapperud *et al.*, 1993). It has been estimated that when carrying the organism in their intestinal tract poultry typically has populations of 10^4–10^7 *Campylobacter*/g of intestinal content with *C. jejuni* as the most prevalent species (Nielsen *et al.*, 1997; Hald *et al.*, 2000). Once *Campylobacter* colonise the poultry intestinal tract, the subsequent shedding of the organism can result in environmental contamination, with a subsequent increased risk for the rest of the flock. In a study to determine the daily shedding pattern of *C. jejuni* in 24 artificially infected broiler chickens it was found that *C. jejuni* failed to colonise 16.6 % of birds, whereas 12.5 % of birds were observed to be chronic shedders (Achen *et al.*, 1998). Throughout the sampling period from days one to 43, a cyclic pattern of shedding was observed in individual birds with *C. jejuni* excreted on an average of 25 out of 43 days. Enumeration of *C. jejuni* in the crop, jejunum and caecum on day 43 revealed that the caecum was the major colonisation site (Achen *et al.*, 1998). In a separate study, the lack of colonisation of one-day-old chicks by VBNC *C. jejuni* has been proposed as casting serious doubts on the significance of the VNBC state in environmental transmission of *C. jejuni* (Medema *et al.*, 1992).

Several studies have been made in order to identify the risk factors associated with the occurrence of, and increase in the incidence of, *Campylobacter* in broiler flocks. These risk factors have included: lack of a hygiene barrier when entering the broiler house; presence of other farm animals in the vicinity of the broiler house on farms with a missing hygiene barrier, dividing the flock into batches for staggered slaughter; the speed of catching and slaughtering of the flock; and the use of undisinfected drinking water (Kapperud *et al.*, 1993; van de Giessen *et al.*, 1996; Hald *et al.*, 2000, 2001). In poultry processing plants significant contamination of *C. jejuni* has been demonstrated in chicken carcasses, processing equipment and workers' hands, and this contamination increased during the defeathering and evisceration processes (Nielsen *et al.*, 1997; Ono and Yamamoto, 1999).

A study to identify risk factors for *Campylobacter* spp. colonisation in French free-range broiler flocks demonstrated that risk increased: in spring/summer and autumn compared with winter; on total freedom rearing farms compared with farms with a fenced run; when the first disinfection of the poultry house was performed by the farmer instead of a hygiene specialist; when rodent control was carried out by a contractor and not by the farmer; when the farmer came into the house twice a day as opposed to three or more (Huneau-Salaun *et al.*, 2007)

Prevalence of *Campylobacter*-positive turkey carcasses slaughtered in Quebec, Canada, was 36.9 %, as compared with 31.2 % for *Salmonella* (Arsenault *et al.*, 2007). The proportion of positive carcasses was significantly higher in lots: with *Campylobacter*-positive caecal cultures, undergoing > 2 h of transit to slaughterhouse, visible carcass contamination, crating for > 8 h before slaughtering; and no antimicrobials used during rearing.

Campylobacter spp. have been shown to be widespread in the environment, with a wide variation of genotypes, and a higher frequency of isolation on rainy days when compared with sunny days (Hansson *et al.*, 2007b). However, there were no differences in the presence of *Campylobacter* spp. outside broiler houses between farms that often deliver *Campylobacter*-positive slaughter batches and those that rarely deliver positive batches.

In Sweden, there has been a *Campylobacter* programme with the objective of achieving a 0–2 % *Campylobacter*-positive prevalence of batches at slaughter (Hansson *et al.*, 2007a). Although this was not achieved, the annual incidence of *Campylobacter*-positive broiler chicken batches at slaughter was progressively reduced from 20 % (in 2002) to 13 % (in 2005) during the period of the Swedish *Campylobacter* programme (2001–2005), possibly because of producers' increased knowledge of the importance of hygiene barriers. During all five years, a seasonal peak of incidence was observed in the summertime.

Carcass contamination has been shown to be related to the within-flock prevalence of *Campylobacter* colonisation with higher numbers of campylobacters per carcass found originating from fully-colonised flocks (average of 5.3 $\log_{10}$ cfu; range 1.3 to > 8.0 $\log_{10}$ cfu) compared to those from low-prevalence flocks (average of 2.3 $\log_{10}$ cfu; range 1.1–4.1 $\log_{10}$ cfu) (Allen *et al.*, 2007). Contamination in the slaughterhouse from previously processed flocks was shown to be significant, especially on carcasses from low-prevalence flocks, and forced air cooling of carcasses reduced contamination levels (Allen *et al.*, 2007).

In a study of Icelandic broiler flocks it was found that approximately 15 % of the flocks were positive for *Campylobacter* spp., with most (95 %) of the infected flocks being raised during the months of April–September (Barrios *et al.*, 2006). The odds of a flock being positive for *Campylobacter* increased with age and flock size, and vertical ventilation systems were strongly associated with positive flocks.

In addition to the high prevalence, the increasing resistance to antimicrobial drugs of *Campylobacter* isolates obtained from poultry and humans has been documented, with a recent study demonstrating multi-resistance in 38.8 % of poultry isolates, and nearly 50 % resistant to quinolones which, along with macrolides and tetracyclines, are used in humans to treat campylobacteriosis (Neubauer and Hess, 2006a).

In contrast to poultry, there appears to be comparatively low incidence of *Campylobacter* in meat products with detection rates of up to 2 % (Loewenherz-Luning *et al.*, 1996; Ono and Yamamoto, 1999; Osano and Arimi, 1999). In a study involving 3959 raw meat samples in the UK during 2003–2005, *Campylobacter* contamination was more prevalent (7.2 %) than *Salmonella* (2.4 %), with *Campylobacter* contamination rates, in decreasing order: other meats such as mutton and rabbit (19.8 %); lamb (12.6 %); pork (6.3 %); and beef (4.9 %) (Little *et al.*, 2008). In this study, *C. jejuni* predominated in all meat types, but *C. coli* isolates were more likely to exhibit antimicrobial drug resistance. *Campylobacter* isolation rates of 47 % and 46 % have been found in cattle and swine, respectively, in slaughterhouses (Nielsen *et al.*, 1997). *Campylobacter jejuni* was the most prevalent species associated with cattle while *C. coli* comprised the majority of isolates from swine. In one study *Campylobacter* spp. were isolated from 70–100 % of the pigs on a pig farm (Harvey *et al.*, 1999) with further evidence that *C. coli* predominates over *C. jejuni*. Concentrations of *C. coli* and *C. jejuni* ranged from 10^3–10^7 cfu/g of caecal content.

In cattle, where colonisation of dairy herds has been associated with drinking unchlorinated water, young animals are more often colonised than older animals, and feedlot cattle are more likely to be carriers than grazing animals (McClure, 2000). It has been proposed that *C. jejuni* is a cause of winter dysentery in calves and older cattle, and experimentally infected calves have shown some clinical signs of disease such as diarrhoea and sporadic dysentery. *Campylobacter jejuni* is a known cause of bovine mastitis, and the organisms associated with this condition have been shown to cause gastroenteritis in persons consuming unpasteurised milk from affected animals. Drinking pasteurised milk from bottles with tops damaged by birds has been shown to be a risk factor (Neal and Slack, 1997). However,

Campylobacter have not been associated with dairy products such as raw milk cheeses (Butzler and Oosterom, 1991; Federighi *et al.*, 1999).

Healthy swine, cattle and sheep have been shown to be important reservoirs of thermophilic campylobacters of different species and high genetic diversity (Oporto *et al.*, 2007). A study in the Basque Country showed herd positive rates of 67.1 % for dairy cattle, 58.9 % for beef cattle, 55.0 % for sheep and 52.9 % for pigs, which were within the contamination rates found in other European countries. The proportion of animals shedding *Campylobacter* in the different herds varied widely. Particularly remarkable was the difference found between dairy and beef cattle (66.7 % vs 5.4 %), and different husbandry systems was proposed as a reason to account for these differences. Dairy cattle are reared indoors compared with beef cattle which in the Basque Country spend most of the year grazing outdoors. Although pasture grazing increases exposure to multiple sources of contamination and hampers cleaning operations, indoor housing favours the chance of reinfection from faecal material.

A study to assess whether *C. coli* isolated from different sources (human, broilers pigs and cattle) in Denmark constitute separate populations demonstrated greater genetic diversity compared with studies from other countries (Litrup *et al.*, 2007). In particular, the pig isolates seemed to comprise a population different to the others and only consisted of a small fraction of the *C. coli* causing human disease. It has been concluded that, although meat from pigs and ruminants is considered to present a relatively low risk to consumers, undercooked offal from these food animals is likely to present a considerable risk (EFSA, 2005).

Shellfish, such as oysters, mussels and clams, can be contaminated by *Campylobacter* spp. and it has been presumed that they originate from faeces from gulls feeding in the growing or relaying waters (Butzler and Oosterom, 1991; Teunis *et al.*, 1997; Federighi *et al.*, 1999). The usual steaming process of mussels was found to completely inactivate *Campylobacter* spp. so that the risks associated with shellfish are restricted to consumption of the raw commodity (Teunis *et al.*, 1997).

Contaminated drinking water is one of the vehicles for the transmission of *Campylobacter* infection, and the possible role of VBNC forms of the organism in the large number of waterborne gastroenteritis outbreaks from which a disease agent cannot be isolated remains to be fully clarified (Thomas *et al.*, 1999). Differences in the survival of *C. jejuni*, *C. coli* and *C. lari* in water together with their differing incidence in the sources of water contamination (e.g. birds, sewage effluent) are reflected in the isolation rates from surface waters (Korhonen and Martikainen, 1991; Obiri-Danso *et al.*, 2001). An observed association between higher rainfall and the incidence of *Campylobacter* infections in humans has been proposed as being caused by the washing-out of *Campylobacter*-contaminated faeces from the soil into the drinking water sources for humans and animals (Sandberg *et al.*, 2006).

Contact with wild animals can increase exposure to campylobacters. *Campylobacter jejuni*, but not *C. coli*, have been isolated from 16.8 % of stray cats in southern Italy, and this was highlighted as a potential source of *C. jejuni* infection given that the cats cohabit with humans in places like parks and public gardens (Gargiulo *et al.*, 2008).

It is generally accepted that *Campylobacter* does not survive as well as other pathogens such as *Salmonella*, although it has been shown to survive in foods for longer at lower temperatures (Curtis *et al.*, 1995). *Campylobacter* is probably very vulnerable to factors such as high temperature and dry environments and also to the presence of oxygen in atmospheric conditions. Therefore, it is assumed that the organism does not persist in products like pelleted food, meals, egg powder and spices, which are often contaminated with *Salmonella* (Butzler and Oosterom, 1991).

20.6 Risk factors for *Arcobacter*

There is very little known about the epidemiology, pathogenesis and real clinical significance of arcobacters, but it is thought that consumption of contaminated food may play a role in transmission of this group of organisms to man. Similar to *Campylobacter*, *Arcobacter* spp. have been isolated from poultry, cattle, swine, water and domestic animals (Mansfield and Forsythe, 2000; McClure, 2000). *Arcobacter* spp. has been frequently isolated from products of animal origin with the highest prevalence in chicken (15.1–100 %), followed by pork (0.5–51.1 %) and beef (1.5–34 %) and other kinds of meat (Cervenka, 2007). Although meat is the main source of arcobacters, other studies have found the organisms in raw milk (46 %) and mussels, but not in eggs (Cervenka, 2007). It has been highlighted that the lack of a standard isolation method for *Arcobacter* spp. detection means that the true occurrence of this organism in foods could be underestimated (Cervenka, 2007). However, to date arcobacters have not been associated with outbreaks of foodborne illness, although person-to-person transmission has been reported (Mansfield and Forsythe, 2000).

The first description of *Arcobacter* spp. was from pigs in the late 1970s (Phillips, 2001), and pork products are routinely found to be contaminated with the organism (Wesley, 1996). In contrast, the first documented report of *Arcobacter* spp. in clinically healthy dairy cattle occurred more recently (Wesley *et al.*, 2000). In this study, 71 % of dairy operations and 14.3 % of individual dairy cattle faecal samples were positive for *Arcobacter*, compared with 80.6 % of farm operations and 37.7 % of individual dairy cattle faecal samples positive for *C. jejuni* and figures of 19.4 % and 1.8 %, respectively, for *C. coli*.

Arcobacter spp. have been isolated from piggery effluent ponds and effluent-treated soils (immediately after effluent irrigation) at levels of

6.5×10^5–1.1×10^8 MPN 100 ml^{-1} and 9.5×10^2–2.8×10^4 MPN 100 g^{-1}, respectively (Chinivasagam *et al.*, 2007). Three species were identified: *A. butzleri*, *A. cryaerophilus* and *A. cibarius*, with the latter species being associated with pigs for the first time (Chinivasagam *et al.*, 2007). In surveys of pork processing plants the analysis of ground pork samples has led to *Arcobacter* spp.-positive rates of 5–90 % (Collins *et al.*, 1996b).

Arocobacter spp., including *A. butzleri*, *A. cryaerophilus* and *A. skirrowii* (Atabay *et al.*, 1998), have been isolated from poultry with recovery rates of 0–97 %, with variations in isolation rate dependent on the environmental conditions of the originating flock, although eggs do not seem to be infected (Phillips, 2001). For example, *Arcobacter* spp. were detected in 53 % of retail-purchased chicken samples (Gonzalez *et al.*, 2000) and 77 % of mechanically-separated turkey samples of which 74 % were positive for *A. butzleri* (Manke *et al.*, 1998). Despite its common isolation from poultry carcasses, *A. butzleri* is infrequently isolated from caecal samples suggesting that arcobacters are probably not normal inhabitants of the poultry intestine and that contamination may be post-slaughter (Atabay and Corry, 1997; Phillips, 2001).

The prevalence of *Arcobacter* in various food, animal and water sources in Turkey has been studied (Aydin *et al.*, 2007). *Arcobacter butzleri* was the predominant species isolated (92 out of 99), with five *A. skirrowii* and two *A. cryaerophilus* isolates. *Arcobacter* spp. were isolated from chicken carcasses (68 %), turkey carcasses (4 %), minced beef (37 %), and rectal swabs (6.9 %) and gall bladders of cattle (8 %). No arcobacters were obtained from the rectal swabs of sheep and dogs, cloacal swabs of broilers and layers and water samples. The 68 % of positive chicken carcasses samples was within the range (23–95 %) from previous studies, the differences being attributed to several factors such as hygiene conditions during the processing and sensitivity of the isolation methods used (Aydin *et al.*, 2007).

The prevalence and diversity of different *Arcobacter* spp. in various poultry species in Denmark have been investigated (Atabay *et al.*, 2006). Very high prevalence of *Arcobacter* spp. was detected in different duck flocks (75 %), whereas a lower rate of recovery was found in the intestinal samples from broiler chickens (4.3 %) and turkey flocks (11 %). However, all chicken carcasses examined were found contaminated with *Arcobacter* spp. suggesting post-slaughter contamination. *Arcobacter* isolates were identified as *A. butzleri* (62 %), *A. cryaerophilus* (28 %) and *A. skirrowii* (10 %). It was concluded that poultry are naturally colonised by various species of *Arcobacter* (Atabay *et al.*, 2006).

Water probably has a significant role in the transmission of *Arcobacter* spp. both to animals and to humans (Phillips, 2001), and it has been estimated that 63 % of *A. butzleri* infection in humans is from the consumption of, or contact with, potentially contaminated water (Mansfield and Forsythe, 2000). Due to its sensitivity to chlorine, infection is probably due to improper chlorination procedures or post-treatment contamination (Phillips, 2001) and hence *Arcobacter* spp. may be more common in developing nations with inadequate water supplies (Mansfield and Forsythe, 2000). Arcobacters have been isolated from drinking water reservoirs and treatment plants, canal water, river water, raw sewage and disinfected effluent (Mansfield and Forsythe, 2000), and it has been suggested that the land application of anaerobically-digested sludge may cause high risk of infection (Phillips, 2001).

20.7 Methods for *Campylobacter*

Campylobacter is a notoriously difficult organism to culture and maintain in the laboratory (Solomon and Hoover, 1999), and as a result there have been many methods developed, and method modifications proposed, for its detection in foods. As with the development of methods for other pathogens, media originally used for the isolation of the organism from faeces were used. Subsequent modifications have been required to enable the detection of low numbers of sub-lethally injured cells in the presence of higher numbers of competitor organisms and this has led to methods based on liquid enrichment prior to selective agar plating with colony identification. The history of the development of selective media for isolation of campylobacters, including the rationale for the choice of selective agents, has been described by Corry *et al.* (1995). Most of the media include ingredients intended to protect campylobacters from the toxic effect of oxygen derivatives. Most commonly used are lysed or defibrinated blood; charcoal; a combination of ferrous sulphate, sodium metabisulphite and sodium pyruvate; and haemin or haematin.

A number of approaches have been taken to avoid the inhibitory effects of the toxic components in the media on sub-lethally injured cells including a preliminary period of incubation at reduced temperature and a delay in the addition of antibiotics. A 4 h pre-enrichment time at 37 °C was found to be optimal for allowing the recovery of low levels of *C. jejuni* while preventing competitive inhibition due to the outgrowth of the accompanying flora (Uyttendaele and Debevere, 1996). A delay of 4–8 h before adding antibiotics to broth was found to significantly increase the *Campylobacter* isolation rate from naturally-contaminated river water compared with direct culture in selective broth. However, with chicken samples, significantly better results were obtained with selective broth as the primary medium (Mason *et al.*, 1999). These findings probably reflect the varying degrees of sub-lethal injury of the *Campylobacter* cells in the different environments.

A number of enrichment broths for *Campylobacter* have been developed. In a recent study three of these broths, Bolton broth (BB), Campylobacter Enrichment broth (CEB) and Preston broth (PB), were compared for the isolation of *Campylobacter* from foods (Baylis *et al.*, 2000). Both BB and CEB were better than PB for the isolation of *Campylobacter* from naturally-contaminated foods, although BB yielded more confirmed *Campylobacter* growth than CEB. The use of selective enrichment broths supplemented with the enzyme Oxyrase®, a membrane-bound enzyme derived from *E. coli*, has been used and shown to be effective in order to provide an oxygen-reduced atmosphere for optimal *Campylobacter* recovery (Abeyta *et al.*, 1997). In contrast, a blood-free enrichment broth (BFEB) has been investigated under aerobic conditions and compared favourably with the US Food and Drug Administration's Bacteriological Analytical Manual (BAM) method for the recovery of *C. jejuni* from inoculated foods. The BFEB method had the advantage of not requiring the use of blood, Oxyrase® or special equipment (Tran, 1998).

Several agars have also been developed for the isolation of *Campylobacter*. Three of the most commonly used agars are Butzler agar, charcoal cefoperazone desoxycholate agar (CCDA) and Preston agar, and in a comparison for the detection of campylobacters in manually shelled egg samples and raw meat samples no substantial difference was seen (Zanetti *et al.*, 1996), although CCDA was preferred. In this study the variable that had the most influence was the incubation temperature with a higher number of strains isolated at 42 °C than at 37 °C. Three isolation media (Karmali, Butzler and Skirrow agar) were compared for the detection of *Campylobacter* in 1500 samples of oysters, ready-to-eat vegetables, poultry products and raw milk cheeses following enrichment in Preston or Park and Sanders broths (Federighi *et al.*, 1999). Park and Sanders enrichment and isolation on Karmali agar appeared to be the most efficient combination.

Cefoperazone amphotericin teicoplanin (CAT) agar was developed from modified charcoal cefoperazone deoxycholate agar (mCCDA) by modification of the selective antibiotics in order to permit growth of strains of *C. upsaliensis* (Corry and Atabay, 1997). CAT agar supported the growth of a wider variety of *Campylobacter* species than mCCDA, which was attributed to the level of cefoperazone in mCCDA being inhibitory to some *Campylobacter* strains.

There has been an International Organisation for Standardisation (ISO) standard method for the detection of *Campylobacter* in food and animal feeding stuffs (ISO, 1995). This method involves enrichment in either Preston broth or Park and Sanders broth. If Preston broth is used, incubation is at 42 °C in a microaerophilic atmosphere for 18 h. If Park and Sanders broth is used the initial suspension is incubated in a microaerophilic atmosphere at 32 °C for 4 h, then antibiotic solution is added and incubated at 37 °C for 2 h followed by transfer to 42 °C for 40–42 h. The enrichment broth cultures are streaked out onto Karmali agar and a second selective agar from the following: modified Butzler agar, Skirrow agar, CCDA and Preston agar. Agar plates are incubated at 42 °C in a microaerophilic atmosphere for up to 5 d prior to a series of confirmatory tests on characteristic colonies. Since then, this horizontal ISO standard method for the detection of *Campylobacter* spp. has been updated (ISO, 2006a). The method now comprises enrichment in

Bolton broth in a microaerophilic atmosphere for 4–6 h at 37 °C and then 40–48 h at 41.5 °C, followed by plating on mCCDA and a second medium of own choice and confirmatory tests. The need for quantitative assessment of *Campylobacter* in risk assessment and monitoring the effect of control measures has been met with the development of an ISO standard method using a colony count technique with mCCDA (ISO, 2006b).

Methods for the enumeration of *Campylobacter* have been developed. A direct plating method (Karmali agar, biochemical confirmation) and a MPN technique (Preston broth enrichment, Karmali agar, biochemical confirmation) were developed for the quantification of *Campylobacter* spp. in raw retail chicken legs (Scherer *et al.*, 2006). The direct plating method was considered superior to the MPN technique because it was more rapid, less laborious and more reproducible for certain sample types, but the MPN technique had the advantage of a lower detection level.

A comparison of fluorescence *in situ* hybridisation (FISH) with traditional culture methods has shown that *C. jejuni* was detectable for only one day by culture methods after spiking of water in a biofilm reactor, whereas bacteria were found after 1–3 weeks using FISH (Lehtola *et al.*, 2006). The authors concluded that culture methods probably seriously under-estimate the real number in water and in biofilms.

In recent years, numerous rapid methods have been developed for the detection of *Campylobacter*, some of which have been evaluated for application with foods. However, very few of these methods have been commercialised, which probably reflects the fact that the food industry is not doing a lot of testing for *Campylobacter* and the market for rapid method test kits is therefore small. The reason for the lack of testing by the food industry is probably due to a number of factors including: the organism does not grow in food under most normal storage conditions; it does not survive well and is relatively easily controlled in processed foods; it is prevalent in raw foods where the ultimate critical control is in the hands of the consumer; and the organism is fastidious, and its detection and maintenance in the laboratory are not easy.

Latex agglutination tests for *Campylobacter* have been available for several years (Wilma *et al.*, 1992). These tests enable convenient and easy visualisation of serological agglutination of *Campylobacter* and they are intended for confirmation of presumptive isolates. The concentration of cells needed for agglutination ranged from 10^6 to 10^8 cfu/ml and similar sensitivities were seen with the non-culturable coccoid forms of *Campylobacter*. In an attempt to improve the speed of the *Campylobacter* detection method, latex tests have been applied to enrichment broth cultures (Wilma *et al.*, 1992). The Microscreen® Campylobacter test has been used after incubation in CCD broth (42 °C, 8 h), filtration (0.45 μm) and incubation in blood-free modified CCD broth (42 °C, 16–40 h) for the detection of *Campylobacter* in fresh and frozen raw meat (Baggerman and Koster, 1992). The Microscreen® Campylobacter latex kit has also been used for the testing of water samples following a physical enrichment (filtration and centrifugation) rather than a cultural enrichment (Baggerman and Koster, 1992).

The importance of the choice of recovery method for maximising the efficacy of recovery of *Campylobacter* from different meat matrices has been highlighted (Fosse *et al.*, 2006). For example, for pork skin samples mechanical pummelling was more effective than swabbing, and peptone water and glucose serum were more effective than demineralised water, whereas no such differences were seen with skinless chine samples.

Significantly higher numbers of injured campylobacters from chicken carcass rinse samples have been recovered using a biphasic enumeration method compared with conventional direct plating techniques (Line and Siragusa, 2006). The biphasic method comprised 96-well microtiter plates consisting of an agar layer containing selective antibiotics (Campy-Line agar at double strength apart from agar component) covered by an equal volume of liquid sample, which were inoculated and incubated microaerophilically for up to 48 h at 42 °C. The greater efficacy of the method for injured campylobacters was attributed to the slow release of selective antibiotics into the system.

To counter the complex, labour-intensive and time-consuming nature of culture procedures for the isolation of *Campylobacter*, a simple procedure for isolation of campylobacters after enrichment in Rosef's enrichment broth was developed using a hydrophobic grid membrane filter on semisolid medium (Valdivieso-Garcia *et al.*, 2007). The improved isolation of campylobacters found using this method was attributed to the utilisation of differential motility, which negated the need for antibiotics in the semisolid medium.

A commercial kit, the API® Campy, is available for the differentiation of *Campylobacter* spp., although identification of species within the family Campylobacteraceae using standard biochemical tests can be problematical because of the variability and atypical reactions of some strains (Phillips, 2001). The heat-stable serotyping system (the 'Penner' Scheme) has been used for *Campylobacter*. However, in a study in Denmark it was found that this system had limitations in that 16 % of the strains were untypable (Nielsen and Nielsen, 1999).

Numerous rapid methods have been developed for *Campylobacter* based on the polymerase chain reaction (PCR) technique (Nogva *et al.*, 2000; O'Sullivan *et al.*, 2000). A magnetic immuno-PCR assay (MIPA) was developed for the detection of *C. jejuni* in milk and chicken products (Docherty *et al.*, 1996). Target bacteria were captured from the food sample by magnetic particles coated with a specific antibody and the bound bacteria then lysed and subjected to PCR. The MIPA could detect 420 cfu/g of chicken after 18 h enrichment, 42 cfu/g after 24 h and 4.2 cfu/g after 36 h. For artificially-contaminated milk, 63 cfu/ml could be detected after 18 h and 6.3 cfu/ml after 36 h. Immunocapture has also been combined with PCR for the detection of *Campylobacter* in foods, and this enabled detection of the pathogen without an enrichment step and took about 8 h to perform (Waller and Ogata, 2000). The assays were quantitative and the limit of detection was one cell/ml, and this was not affected when tested in spiked milk samples and chicken skin washes.

Many food samples and enrichment media, e.g. the presence of charcoal and iron (Thunberg *et al.*, 2000), are inhibitory to the PCR, thereby lowering its detection capacity. Sample preparation methods using buoyant density centrifugation (Uyttendaele *et al.*, 1999; Wang *et al.*, 1999) and heat treatment at 96 °C (Uyttendaele *et al.*, 1999) have been used to overcome this inhibition. While Oxyrase™ has been shown to significantly enhance the growth of *C. jejuni*, it appeared not to improve the PCR detection of *C. jejuni* in naturally-contaminated chickens (Wang *et al.*, 1999).

Enrichment in selective broth will always be a compromise between inhibition of competitive flora and the recovery and growth of the target microorganism (Krause *et al.*, 2006). Differences in reported performances of Preston and Bolton broth have been attributed to the matrices and the background flora of the samples, and it has been concluded that the two enrichment broths can be used equally for routine testing (Krause *et al.*, 2006). A rapid (< 4 h) real-time TaqMan® PCR method, used directly on samples without any preceding enrichment, was successfully validated by NordVal (the Nordic system for validation of alternative microbiological methods) at an abattoir by correctly identifying *Campylobacter*-positive flocks (10 out of 20) and has been implemented for use in separated-slaughter practice by the leading poultry producers in Denmark as part of a risk management programme, and for the certified production of *Campylobacter*-free chicken (Krause *et al.*, 2006).

A quantitative real-time PCR method for enumerating *C. jejuni* in chicken rinse without a culturing step has been developed (Ronner and Lindmark, 2007). The procedure involved: filtration of chicken rinse; centrifugation; and DNA extraction. A close correlation with direct plating was obtained for spiked chicken rinse ($r = 0.99$), but there was a much lower correlation ($r = 0.76$) when retail samples were analysed, attributed to the greater variation in the proportion of dead and/or viable but not culturable *Campylobacter* types.

A real-time PCR method has been developed capable of detecting and quantifying *C. jejuni* from chicken rinses without an enrichment step, the whole assay being completed in 90 min with a detection limit of 1 cfu (Debretsion *et al.*, 2007). A real-time NASBA assay for the detection of *C. jejuni* mRNA was developed with a detection limit of 10^2 cells per Nucleic acid sequence-based amplification (NASBA) reaction, but as the assay was able to detect dead cells it was concluded that it cannot be used to demonstrate viability (Cools *et al.*, 2006).

Microarray technology has been utilised for campylobacters. An electronic oligonucleotide microarray technique has been developed for detection and differentiation of *C. jejuni*, *C. coli* and *C. lari* using mRNA of the 60 kDa heat-shock protein as a viability marker so that dead cells are not detected (Zhang *et al.*, 2006). The technique was able to detect as few as two viable

Campylobacter cells within 8 h from RNA extraction to microarray analysis. A multiplex PCR method was developed for the detection and differentiation of *Campylobacter*, *Arcobacter* and *Helicobacter* in poultry and poultry products in a single-step procedure (Neubauer and Hess, 2006b). A DNA oligonucleotide array for the simultaneous detection of *Arcobacter* and *Campylobacter* in retail chicken samples was developed (Quinoñes *et al.*, 2007). Probes, which showed a high level of specificity, were selected that target housekeeping and virulence-associated genes in *A. butzleri*, *C. jejuni* and *C. coli*. The DNA microarray had a detection sensitivity threshold of approximately 10 000 cells and was successfully applied to package liquid from whole chicken carcasses and enrichment broths.

For epidemiological studies, molecular typing techniques have been preferred (Rautelin and Hanninen, 2000; Phillips, 2001). Pulsed-field gel electrophoresis was used for the first time in a *Campylobacter* outbreak in the USA, where a cafeteria worker was the likely source, and was critical in determining that community cases were not linked (Olsen *et al.*, 2001). The use of PCR with a RAPD protocol has also proved a useful tool for the epidemiological analysis of *Campylobacter* (Hilton *et al.*, 1997).

20.8 Methods for *Arcobacter*

Due to only fairly recent interest in *Arcobacter* there are only a limited number of detection methods and a standardised method does not yet exist (Johnson and Murano, 1999b; Cervenka, 2007). Although *Arcobacter* strains are capable of aerobic growth, the optimum growth condition for primary isolation is microaerophilic (3–10 % oxygen). Several different methods using both aerobic and microaerophilic conditions have been proposed based on media for *Campylobacter* and *Leptospira* (Mansfield and Forsythe, 2000; Phillips, 2001).

A modified cefsulodin–irgasan–novobiocin (CIN) medium has been developed for the recovery of *Arcobacter* spp. from meats and used with two other media: brain heart infusion agar supplemented with 10 % bovine blood and cephalothin, vancomycin and amphotericin B; and brain heart infusion agar supplemented with 10 % bovine blood and no antibiotics (Collins *et al.*, 1996b).

CAT agar was shown to support a wider range of arcobacters than mCCDA, and the sub-optimal growth on mCCDA was considered to be due to the synergistic interaction between deoxycholate and cefoperazone (Corry and Atabay, 1997). A pre-enrichment stage in either CAT broth or *Arcobacter* enrichment broth together with a filter method onto mCCDA or CAT agar has been suggested for optimum isolation form chicken carcasses (Phillips, 2001). Another combination of pre-enrichment (in an *Arcobacter* selective broth) and plating onto a semi-solid *Arcobacter* selective medium has been used and, in this case, an isolation temperature of 24 °C was employed and piperacillin was added to prevent the outgrowth of *Pseudomonas* spp. from raw meats (Phillips, 2001).

The productivity of an arcobacter enrichment medium (AM) (Oxoid) was compared with two campylobacter enrichment media (Preston broth, Oxoid; LabM broth) (Atabay and Corry, 1998). Twenty strains of *Arcobacter* and *Campylobacter* were tested for growth. None of the *Campylobacter* spp. grew in AM and the medium supported good growth of all strains of *Arcobacter*.

An agar medium (JM agar) containing a basal nutrient mix along with 0.05 % thioglycolic acid, 0.05 % sodium pyruvate and 5 % sheep's blood (pH 6.9 +/– 0.2) was found to be the most effective formulation for the growth of *A. butzleri*, *A. cryaerophilus* and *A. nitrofigilis* (Johnson and Murano, 1999a). In addition to superior growth characteristics, a deep red colour around the colonies was also observed with this formulation. The use of an aerobic pre-enrichment in JM broth prior to plating on JM agar was subsequently found to be more effective at isolating arcobacters from broiler chicken samples than two existing methods (Johnson and Murano, 1999b).

Being an organism of recent interest, no standard, widely accepted methodologies for the isolation and enumeration of levels of *Arcobacter* exist (Chinivasagam *et al.*, 2007). Several studies have compared various media formulations and enrichment procedures, but there is a need for suitable, optimal recovery media and conditions that can detect the levels of *Arcobacter* spp. in a range of different sources such as faeces, carcasses and the environment (Chinivasagam *et al.*, 2007). A MPN method was developed for enumeration of *Arcobacter* spp. in piggery effluent (Chinivasagam *et al.*, 2007). The MPN approach was developed by adopting an existing method for the selective isolation of *Arcobacter* spp. reported by Johnson and Murano (1999a, b) and enabled isolation of the faster growing *A. butzleri*, as well as the slower growing *A. cryaerophilus* and *A. cibarius* after 48 h at 30 °C (Chinivasagam *et al.*, 2007).

Methods that have been used, or proposed, for the identification and characterisation of *Arcobacter* isolates have included: biochemical profiling; antimicrobial resistance testing; serotyping based on heat-labile antigens; SDS-PAGE of whole cell proteins; fatty acid profiles; PCR-based assays; and ribotyping (Harrass *et al.*, 1998; Mansfield and Forsythe, 2000; Phillips, 2001). It has been suggested that some of these methods might represent valuable tools for epidemiological analyses of *Arcobacter* isolates. However, reliable commercially-available tests are lacking. For example, the API Campy, although effective for the differentiation of *Campylobacter* spp., does not allow similar identification of *Arcobacter* at species level (Phillips, 2001).

Rapid method development for *Arcobacter* has centred on PCR-based methods. A multiplex PCR assay to identify *Arcobacter* isolates, and to distinguish *A. butzleri* from other arcobacters, has been developed (Harmon and Wesley, 1997). Upon PCR amplification all the *Arcobacter* isolates yielded a 1233 bp product, whereas the *A. butzleri* exhibited an additional

686 bp product. It was claimed that the assay was specific, rapid and easy to interpret. A multiplex PCR method has also been developed to specifically detect both *C. jejuni* and *A. butzleri* in the same reaction tube (Winters and Slavik, 2000). The organisms were differentiated by 159 bp and 1223 bp products, respectively, and the method has been evaluated in a range of spiked foods. Real-time PCR (TaqMan™) assays developed for the specific detection of *A. butzleri* and *A. cryaerophilus* from DNA samples were claimed to provide improvements in sensitivity (< 10 cfu per PCR reaction) and species representation over other published *Arcobacter* PCR assays (Brightwell *et al.*, 2007). A multiplex PCR method, applied to enrichment broths, was compared with traditional culture methods for the detection and identification of *Arcobacter* spp. in chicken (livers and carcasses) and wastewater (Gonzalez *et al.*, 2007). The *Arcobacter* detection rate for PCR amplification was higher (91 % for chicken and 100 % for wastewater) than for culture isolation, which highlighted the inadequacy of the latter.

20.9 Control procedures for *Campylobacter*

Although *Campylobacter* has been demonstrated to be susceptible to a wide variety of antimicrobial treatments, food processing methods and environmental stresses (Solomon and Hoover, 1999; McClure, 2000), it still continues to cause an increasing level of human foodborne disease. Although campylobacters have long been regarded as very sensitive to the extra-intestinal environment and thus easier to kill than either salmonellas or *E. coli*, it would be wrong to regard them as highly resistant, and it is becoming clear that they are more robust than previously thought (Humphrey *et al.*, 2007). It is obvious that there is no single approach to controlling this organism and a number of preventative measures are needed throughout the farm-to-table continuum in order to reduce the incidence of campylobacteriosis in humans.

As poultry meat products appear to be a major source of campylobacteriosis, through cross-contamination to ready-to-eat foods, through direct hand-to-mouth transfer during food preparation and to a lesser extent from the consumption of undercooked poultry meat (EFSA, 2005), the rest of this section focuses on controls and risk management options that can reduce this risk. The following control options have been recommended by EFSA (2005) to reduce the risk of *Campylobacter* infection via some of the other important transmission routes:

- Milkborne campylobacteriosis can be controlled by proper pasteurisation or by applying alternative measures that eliminate *Campylobacter*. Alternatively, consumers of untreated raw milk should be aware of the associated risk.
- The risk associated with fresh produce consumed raw requires implementation of good agricultural practice (GAP) and good hygienic

practice (GHP) throughout the food chain, and avoiding the use of untreated faecally-contaminated water for irrigation and washing is essential.
- The application of GHP and the control of water quality before harvesting and during depuration are important in reducing the risk associated with bivalve molluscs, including oysters.

To reduce the risk associated with water supply systems requires a reliable safety assurance system to exclude *Campylobacter* from drinking water (EFSA, 2005). Disinfection of drinking water at the water works, improvement of water pipelines and control of pressure drops have been recommended as interventions to reduce the risk of *Campylobacter* infection from drinking water (Sandberg *et al.*, 2006). As with the consumption of untreated raw milk, consumers of water from uncontrolled sources should be aware of the associated risk (EFSA, 2005).

20.9.1 Control procedures for farms

Various modelling and risk assessment approaches have demonstrated that strategies to reduce the *Campylobacter* load on chicken may result in a reduction in the incidence of human illness. EFSA (2005) concluded that reducing the proportion of *Campylobacter*-infected poultry flocks and/or reducing the numbers of *Campylobacter* in live poultry and on poultry carcasses will lower the risk to consumers considerably.

A model developed to examine how chicken prepared in (German) homes exposes the consumer to *Campylobacter* spp. and the level of resulting illness was used to simulate the impact of different risk mitigation strategies (Brynestad *et al.*, 2008). It was shown that reducing the *Campylobacter* load on chicken may result in a greater reduction in the incidence of human illness than reducing prevalence of contaminated products. This highlights the need for quantitative data of the surface level of *Campylobacter* on poultry as well as prevalence data.

A mathematical model developed to evaluate human health risks from foodborne pathogens associated with changes in animal illness was applied to a hypothetical example of *Campylobacter* from chicken, in general, the model predicted that very minor perturbations in microbial loads on meat products could have relatively large impacts on human health and, consequently, small improvements in food animal health might result in significant reductions in human illness (Singer *et al.*, 2007).

Due to the prevalence of *Campylobacter* in poultry, control measures to reduce infection and spread of the organism on broiler farms would reduce the risk of transmission to humans further down the food chain. Control measures that have been shown to be effective have included: strict hygienic routines when the farm workers enter the rearing room (washing hands, the use of separate boots for each broiler house and the use of footbath disinfection), disinfection of drinking water, and depopulation of broiler houses as quickly as possible and in one batch only (van de Giessen *et al.*, 1996 Hald *et al.*, 2000, 2001; EFSA, 2005).

A systematic review of data extracted from research papers has enabled evidence-based ranking of the major contributing factors and sources of *Campylobacter* occurrence in broilers in the UK (Adkin *et al.*, 2006). The top three sources and contributing factors to *Campylobacter* occurrence were (i) movement on farm during depopulation, (ii) cross-broiler house transfer and (iii) transfer by staff; with vertical transmission not a major source of *Campylobacter* transmission. The main factor contributing to decreasing the risk was use of a hygiene barrier (Adkin *et al.*, 2006).

With poultry that are housed, control is possible by using strict hygiene measures to prevent horizontal transmission from the outside environment (Humphrey *et al.*, 2007). It has been shown that the widespread prevalence of *Campylobacter* spp. in the environment outside broiler houses is not correlated to the extent to which the farms deliver *Campylobacter*-positive slaughter batches, which suggests that physical barriers play an important role in preventing transmission (Hansson *et al.*, 2007b). Control measures for birds reared in free-range systems present even greater challenges (Humphrey *et al.*, 2007).

While successful, strict hygiene control measures are not absolutely effective and other measures are required to reduce flock colonisation to the lowest possible level, e.g. vaccination, bacteriophages and improving flock health and welfare (Humphrey *et al.*, 2007).

Studies have indicated that diet formulations excluding animal proteins and fat may help towards reducing colonisation of *C. jejuni* in the caeca of poultry, and it has been suggested that combining this strategy with other methods, including the use of probiotics or prebiotics is worth studying (Hariharan *et al.*, 2004). A bacteriocin-based treatment (secreted by *Paenibacillus polymyxa*, purified and microencapsulated in polyvinylpyrroloidone and incorporated into chicken feed) was shown to reduce both intestinal levels and frequency of chicken colonisation by *C. jejuni* (Stern *et al.*, 2005).

Attempts to prevent or reduce *Campylobacter* colonisation by establishing a competitive or antagonistic microflora in chicks have not always been successful (Mead, 2002). Some protection has been obtained with bacteria such as mucin-utilising coliforms, but the protective potential of these organisms appears to be limited (Mead, 2002). Chicks dosed with anaerobic preparations of caecal mucus from *Campylobacter*-free adult hens were shown to be partly protected against *C. jejuni* (Mead *et al.*, 1996). Vaccination and drug therapy have also been proposed (White *et al.*, 1997) although the use of antibiotics has been a factor in the rapid emergence of antibiotic-resistant *Campylobacter* strains all over the world (Allos, 2001). This trend has further emphasised the need for appropriate and safe use of antibiotics in animal production (Pedersen *et al.*, 1999).

Acidification treatments for broiler litter are commercially available. Two such treatments, aluminium sulphate and sodium bisulphate, were tested to determine their effect on *Campylobacter* and *Salmonella* levels associated with commercial broilers during a six-week grow out period (Line and Bailey, 2006). Both treatments, at the application rates investigated, caused a slight delay in the onset of *Campylobacter* colonisation in broiler chicks (not seen with *Salmonella*), but this was not statistically significant and *Campylobacter* populations (and *Salmonella* incidence) associated with unprocessed, whole-carcass rinse samples analysed at the end of production (week 5 and 6) were unaffected by the treatments.

Bacteriophage therapy is one possible means by which colonisation of broiler chickens by campylobacters could be controlled (Carrillo *et al.*, 2005). Phage treatment of *C. jejuni*-colonised birds resulted in *Campylobacter* counts falling between 0.5 and 5 $\log_{10}$ cfu/g (dependent on the phage combination, dose applied and time elapsed after administration) of caecal contents compared to untreated controls over a five-day period post administration. *Campylobacter* resistant to bacteriophage infection were recovered at a frequency of < 4 %, but these resistant types were compromised in their ability to colonise experimental chickens and rapidly reverted to a phage-sensitive phenotype *in vivo*. The selection of appropriate phage and their dose optimisation are key elements for the success of phage therapy to reduce campylobacters in broiler chickens (Carrillo *et al.*, 2005). It has been reported that in chickens bacteriophage resistance is infrequent because the mutants that escape bacteriophage are not proficient in poultry colonisation but readily revert back to colonisation-proficient phage-sensitive types (Scott *et al.*, 2007).

20.9.2 Control procedures for slaughterhouses

Certified *Campylobacter*-free poultry products have been produced in Denmark since 2002, the first example of fresh (unprocessed and non-frozen) chickens labelled '*Camplyobacter*-free' (Krause *et al.*, 2006). This success occurred partly through the use of a 4-h gel-based PCR testing scheme on faecal swabs, enabling the implementation of a separated-slaughter practice as part of a risk management programme. Measures in the slaughterhouse can reduce contamination in the final product, and the most promising control strategies are: keep colonised and non-colonised flocks separate during slaughter ('scheduled processing') (Wagenaar *et al.*, 2006; Havelaar *et al.*, 2007); limiting faecal leakage during processing, followed by decontamination of the contaminated flock (Havelaar *et al.*, 2007).

A quantitative microbiological risk assessment model describing the transmission of *Campylobacter* through the broiler meat production chain was developed with the ultimate objective to serve as a tool to assess the effects of interventions to reduce campylobacteriosis in the Netherlands (Nauta *et al.*, 2007). The results show that, on average, there is a continuous decrease in the level of *Campylobacter* on the carcasses and the meat, which is the consequence of the absence of growth and the repeated inactivation and removal of *Campylobacter* during processing, as a consequence of submersion, washing, defeathering, skinning, storage, etc. Contamination of the carcass exterior by faeces leaking from the carcass (predominantly during defeathering) may give a temporary increase in the level of *Campylobacter*, and cross-contamination during processing and food handling leads to a dispersal of campylobacters and so the prevalence of contaminated units (birds, carcasses, fillets) my increase, whereas the variability between units decreases (Nauta *et al.*, 2007).

When birds are sent for slaughter, there is a high risk that *Campylobacter* present will be transmitted from carrier birds to the carcasses being processed. This underlines the need for HACCP principles to be applied to processing plants, to minimise product contamination (Mead, 2000). Subsequent critical control points and good manufacturing practices that have been identified and implemented have included: temperature controls (washer and product); chemical interventions, 'water replacements', counter-flow technology in the scalder and chiller, 'equipment maintenance', chlorinated-water sprays for equipment and working surfaces; increase in chlorine concentrations in process water; and removal of unnecessary carcass contact surfaces (Mead *et al.*, 1995; White *et al.*, 1997). To reduce the risk of infection of workers and the levels of subsequent contamination of *Campylobacter*-negative flocks, implementation of preventative measures, such as washing the equipment after slaughtering of *Campylobacter*-positive flocks or logistic slaughter when practically achievable, has been recommended (Johnsen *et al.*, 2007).

20.9.3 Control procedures for manufacture, retail and domestic kitchen

Campylobacter is relatively heat-sensitive and so commercial heat processes within a HACCP framework, together with pre-requisite programmes to address any cross-contamination risks, should guarantee control of the pathogen. However, it should be borne in mind that campylobacters appear to attach well to biological material and this can have a marked effect on heat tolerance under commercial conditions (Humphrey *et al.*, 2007).

Decontamination of poultry carcasses using steam or hot water in combination with rapid cooling, chilling or freezing of carcass surfaces has been evaluated (James *et al.*, 2007). The optimum treatment for reducing *Campylobacter* numbers without extensive degradation of carcass appearance was treatment with water at 80 °C for 20 s followed by crust freezing, which reduced the numbers of artificially-inoculated *C. jejuni* by about 2.9 $\log_{10}$ cfu cm^{-2}.

The efficacy of different chemical decontamination methods (spraying with, or dipping in, sodium chlorite, lactic acid and citric acid) for *C. jejuni* on poultry carcasses was studied (Ellebroek *et al.*, 2007). Dipping was found to be more effective than spraying, with the highest log reductions of 1.51,

1.24 and 1.41 achieved with lactic acid (15 %, 30 s), citric acid (15 %, 30 s) and sodium chlorite (0.4 %, 30 s). Citric acid appeared to be more effective than lactic acid and sodium chlorite for spray-washing.

It has been suggested that freezing offers a short- to long-term solution to the safe storage of chicken meat with respect to the survival of *Campylobacter* (Ritz *et al.*, 2007). While it does not entirely eliminate potentially infectious *Campylobacter*, the initial reduction achieved in its population using this method may have a substantial impact on reducing the incidence of campylobacteriosis (Ritz *et al.*, 2007). The implementation of the freezing of *Campylobacter*-infected carcasses for three weeks as a risk-reducing measure in Norway has been credited for the reduction in *Campylobacter* counts on carcases that enter the market (Sandberg *et al.*, 2006).

The feasibility of using irradiation to render meat products (such as vacuum-packaged ground pork) safe from *C. jejuni* and *A. butzleri* has been demonstrated, with a 1.5 kGy treatment yielding 7 and 5 log reductions, respectively (Collins *et al.*, 1996). *Campylobacter* have been shown to be more susceptible to radiation than *Samonella* and *Listeria monocytogenes*, with *D*-values varying between 0.12 and 0.25 kGy (Humphrey *et al.*, 2007).

Wine has been shown to inactivate *C. jejuni*, and results have suggested that the antimicrobial fractions (ethanol and certain organic acids) act synergistically, and exposure of contaminated foods to wine, as in marinade conditions, significantly reduces the number of viable *C. jejuni* (Carneiro *et al.*, 2008). Interestingly, results from a stomach model containing food matrix and synthetic gastric juice suggest that ingestion of wine during a meal may greatly diminish the quantity of *C. jejuni* persisting further in the alimentary tract, thereby decreasing the risk of infection (Carneiro *et al.*, 2008).

Food-handling at retail outlets and by the consumer often provides the last chance for control of this organism in the farm-to-table continuum. Prevention of cross-contamination and adequate cooking are probably the most important control measures in this environment. Following a prescribed cleaning procedure using hypochlorite in addition to cleaning with detergent and hot water has been shown to reduce the level of contamination of *Campylobacter* in the domestic kitchen (Cogan *et al.*, 1999). It is the effective promotion of good hygienic practices in such food preparation areas that could bring about the greatest benefits in terms of improved food safety, but this still presents an enormous challenge. Quantitative risk assessment of *Campylobacter* spp. in poultry-based meat preparations has shown that eating raw chicken meat products can give rise to exposures that are 10^{10} times higher than when the product is heated, indicating that campaigns are important to inform consumers about the necessity of an appropriate heat treatment of these types of food products (Uyttendaele *et al.*, 2006).

The presence of *Campylobacter* in raw meats reinforces the importance of adequate cooking of meat and good hygiene to avoid cross-contamination

(Little *et al.*, 2008). As meal preparation with heavily-contaminated food like raw chicken can lead to widespread dissemination of campylobacters it is important that cleaning is applied to a wide area around the food production site (Humphrey *et al.*, 2007). Control can be particularly difficult in the household environment and there is a need to identify the best ways to advise consumers on this subject (Humphrey *et al.*, 2007).

Due to the many sources of *Campylobacter* and the difficulty of controlling all possible transmission routes, the importance of maintaining a satisfactory overall level of hygiene, especially for children and people with compromised immunological status, has been highlighted (Sandberg *et al.*, 2006). For example, the isolation of *Campylobacter* from stray cats highlights the importance of following good hygienic practices and reducing contact with stray animals (Gargiulo *et al.*, 2008).

20.10 Control procedures for *Arcobacter*

The risk of transmission to humans of *Arcobacter* via properly cooked foods and chlorinated water is negligible as it is for *Campylobacter* (Wesley, 1996). As with all foodborne pathogens, effective implementation of HACCP is important in the control of *Arcobacter* (Phillips, 2001).

Although *Arcobacter* and *Campylobacter* are closely related, this does not necessarily mean that all treatments are similarly effective (Phillips, 2001). For example, the fact that, unlike *Campylobacter*, arcobacters are probably not normal inhabitants of the poultry intestine (Atabay and Corry, 1997; Phillips, 2001) means that control measures should be focused on preventing contamination of, and proliferation in, the broiler environment and post-slaughter. Although *A. butzleri* is more resistant than *C. jejuni* to irradiation treatment, doses of irradiation currently allowed for pork in USA (0.3–1.0 kGy) provide an effective method of reducing, if not completely eliminating, *A. butzleri* from pork (Phillips, 2001). However, treatment with sprays of organic acids has been shown to be effective at reducing *Salmonella* and *Campylobacter* spp. on pork carcasses (Epling *et al.*, 1993), but there is evidence that this type of treatment may not be as effective against *A. butzleri* (Phillips, 1999). However, some treatments involving organic acids alone or in combination with nisin and/or low temperature have shown some effectiveness and application potential for controlling arcobacters on meat surfaces (Cervenka, 2007).

Arcobacters also appear to be resistant to antimicrobial agents typically used in the treatment of diarrhoeal illness caused by *Campylobacter* spp., e.g. erythromycin, other macrolide antibiotics, tetracycline and choramphenicol and clindamycin (Mansfield and Forsythe, 2000).

20.11 Future trends

The reservoirs of *Campylobacter* spp. in the environment are extensive and include wild and domestic birds and animals, and fresh water and marine aquatic environments. Despite this, there are examples of production facilities that rear infection-free stock, through use of strict hygiene and control procedures and practices, and through sourcing pathogen-free materials. In some countries, there has been sufficient motivation and resources, from government and the animal/poultry production industry, to realise *Campylobacter*-free production at the farm level, although this can be difficult to maintain for long periods of time. The public health impact of providing *Campylobacter*-free animal/poultry produce is unclear, however. In many regions where this produce is consumed, the incidence of campylobacteriosis remains high, and there are likely to be other sources of infection that will still cause significant morbidity. Despite this, it is clear that human exposure to *Campylobacter* via retail chicken is important. Gormley *et al.* (2008) concluded that changes in the population structure of campylobacters in this reservoir need to be taken into account in investigating human infection.

Public awareness of some of the hazards associated with foods has been increasing over the past 20 years. However, since large outbreaks of campylobacteriosis are relatively uncommon or go unrecognised, *Campylobacter* rarely hits the headlines in the popular press, and public awareness of this pathogen is probably still relatively small. With better methods, particularly genotypic, for the characterisation of microorganisms (e.g. genetic 'fingerprinting') and common standardised approaches being used and synchronised in the investigation of foodborne diseases, there is likely to be an increasing number of links made between 'sporadic' cases of illness. The low biochemical activity of *Campylobacter* spp. and frequent variability made characterisation/identification on the basis of phenotype very difficult. Despite the advances in molecular techniques, this difficulty has hampered epidemiological investigations and our understanding of the incidence and behaviour of *Campylobacter* spp. in the environment and animals. The development in molecular methods has changed bacterial diagnostics and taxonomy. At present the methods available have different pros and cons, and a combination of methods is needed to secure identification of all *Campylobacter* spp. These methods will continue to be developed and refined to enable more reliable characterisation/identification to be carried out in non-specialist laboratories.

With *Arcobacter* spp., there is currently little direct evidence of this group causing foodborne illness. However, since the organisms are known to cause disease in animals (including humans) and have been associated with meat and poultry products, there seems little doubt that they should be regarded as potential foodborne pathogens. Methods for characterisation/identification, as for *Campylobacter* spp., are being improved in terms of specificity and sensitivity. The trend toward use of molecular techniques has also enabled identification and direct detection of virulence genes, in isolates from different environments. This ability of linking pathogenicity markers (virulence factors) to isolates was not possible with the more classical characterisation techniques.

The impact of molecular approaches is also important for understanding pathogenesis. Since genotypic methods are becoming reliable and easily employed, genotypic analysis of isolates from different reservoirs should indicate whether animals (including humans) and birds share the same bacterial populations or whether there are sub-populations only pathogenic to susceptible groups or hosts. The genomic sequence of *C. jejuni* is now available and, if homology with genes known to be responsible for virulence traits (e.g. invasiveness or toxin production) is identified, a better understanding of pathogenesis can be expected.

Another trend that deserves further elaboration is the change in surveillance strategies. Existing surveillance systems in many countries are limited and will only detect outbreaks of established and easily recognised pathogens. Both national and international surveillance strategies are now being enhanced by employing common, standardised sub-typing techniques and also by more active surveillance of populations. For example, in the USA, FoodNet has been designed as a platform to collect/collate information from different sources (e.g. case-control studies) so that true disease incidence can be determined. Other initiatives, such as CAMPYNET and PulseNet, should also help to address the problems associated with methodology and communication between laboratories. These efforts will result in identification of diffuse outbreaks that may have little impact on a local basis but, when taken together with other related cases, may have a much bigger impact. Immediate responses to these outbreaks will result in product recalls and longer-term activities will focus on prevention of future similar outbreaks. Environmental sampling and end-product testing is another trend that is increasing in some countries, and detection of campylobacters will be facilitated through development of new approaches and techniques.

The antibiotic resistance profile of isolates associated with foodborne illness is another feature that will be determined in surveillance activities. The development of antibiotic resistance in campylobacters has been attributed, at least in part, to use of antibiotics in agriculture. This has led to the banning of particular types of antibiotics for some uses in agriculture, and more restrictive controls are likely to implemented both in veterinary and human medicine in the future.

With campylobacters, as for many other enteric pathogens, there are a number of measures that may be used for control. If used independently, these measures are often not sufficient to reduce the risk of infection to an acceptable level, and combinations of measures are necessary. For many national and international regulatory bodies, the emphasis is increasingly toward an integrated approach to food safety. For campylobacters, this means looking at reservoirs, where future control measures may include selective sourcing of 'clean' raw materials and feed, changing hygienic practices on the farm, modification of the slaughter process and consideration of alternative processing technologies, such as irradiation. Many of these changes and improvements come with a price, and successful implementation will be dependent on an acceptance by the consumer and producers that these changes are a real benefit to them.

It is likely that risk assessment approaches will increasingly be used to evaluate the effect of interventions and help make risk management decisions to control *Campylobacter* and reduce human campylobacteriosis. However, the usefulness of current models is limited by the uncertainty of the model outputs due to the gaps in the knowledge and data used for their construction. To lower the uncertainty in the model results, future data collection should be focused more on quantitative data; in the case of broiler meat, particularly the prevalence and concentration data for campylobacters in chicken faeces and on birds entering the processing plant and the effect of subsequent processing steps on these data (Nauta *et al.*, 2007).

20.12 Sources of further information

Alter T and Scherer K (2006) Stress response of *Campylobacter* spp. and its role in food processing, *Journal of Veterinary Medicine*, **B 53**, 351–7.

Brown M (ed.), (2000) *HACCP in the Meat Industry*, Woodhead, Cambridge.

Humphrey T, O'Brien S and Madsen M, (2007) Campylobacters as zoonotic pathogens: a food production perspective, *International Journal of Food Microbiology*, **117**, 237–57.

ICMSF (1996) *Microorganisms in Foods 5, Microbiological Specifications of Food Pathogens*, International Commission on the Microbiological Safety of Foods, Blackie Academic and Professional, London.

Wassenar T M and Blaser M J (1999) Pathophysiology of *Campylobacter jejuni* infections in humans, *Microbes and Infection*, **1**, 1023–33.

20.13 References

Abeyta C, Trost P A, Bark D H, Hunt J M, Kaysner C A, Tenge B J and Wekell M M (1997) The use of bacterial membrane fractions for the detection of *Campylobacter* species in shellfish, *Journal of Rapid Methods and Automation in Microbiology*, **5**, 223–47.

Abulreesh H H, Paget T A and Goulder R (2006) *Campylobacter* in waterfowl and aquatic environments: incidence and methods of detection, *Environmental Science and Technology*, **40**, 7122–31.

Achen M, Morishita T Y and Ley E C (1998) Shedding and colonisation of *Campylobacter jejuni* in broilers from day-of-hatch to slaughter age, *Avian Diseases*, **42**, 732–7.

Adkin A, Hartnett E, Jordan L, Newell D and Davison H (2006) Use of a systematic review to assist the development of *Campylobacter* control strategies in broilers, *Journal of Applied Microbiology*, **100**, 306–15.

Allen V M, Bull S A, Corry J E L, Domingue G, Jorgensen F, Frost J A, Whyte R, Gonzalez A, Elviss N and Humphrey T J (2007) *Campylobacter* spp. contamination of chicken carcasses during processing in relation to flock colonisation, *International Journal of Food Microbiology*, **113**, 54–61.

Allos B M (2001) *Campylobacter jejuni* infections: update on emerging issues and trends, *Clinical Infectious Diseases*, **32**, 1201–16.

Arsenault J, Letellier A, Quessy S, Morin J-P and Boulianne M (2007) Prevalence and risk factors for *Salmonella* and *Campylobacter* spp. carcass contamination in turkeys slaughtered in Quebec, Canada, *Journal of Food Protection*, **70**, 1350–59.

Atabay H I and Corry J E L (1997) The prevalence of campylobacters and arcobacters in broiler chickens, *Journal of Applied Microbiology*, **83**, 619–26.

Atabay H I and Corry J E L (1998) Evaluation of a new arcobacter enrichment medium and comparison with two media developed for enrichment of *Campylobacter* spp., *International Journal of Food Microbiology*, **41**, 53–8.

Atabay H I, Corry J E L and On S L W (1998) Diversity and prevalence of *Arcobacter* spp. in broiler chickens, *Journal of Applied Microbiology*, **84**, 1007–16.

Atabay H I, Wainø M and Madsen M (2006) Detection and diversity of various *Arcobacter* species in Danish poultry, *International Journal of Food Microbiology*, **109**, 139–45.

Avrain L, Allain C, Vernozy-Rozand C and Kempf I (2003) Disinfectant susceptibility testing of avian and swine *Campylobacter* isolates by a filtration method, *Veterinary Microiology*, **96**, 35–40.

Aydin F, Atabay H I and Akan M (2001) The isolation and characterization of *Campylobacter jejuni* subsp. *jejuni* from domestic geese (*Anser anser*), *Journal of Applied Microbiology*, **90**, 637–42.

Aydin F, Gümüşsoy K S, Atabay H I, Iça, T and Abay S (2007) Prevalence and distribution of *Arcobacter* species in various sources in Turkey and molecular analysis of isolated strains by ERIC-PCR, *Journal of Applied Microbiology*, **103**, 27–35.

Baggerman W I and Koster T (1992) A comparison of enrichment and membrane filtration methods for the isolation of *Campylobacter* from fresh and frozen foods, *Food Microbiology*, **9**, 87–94.

Barrios P R, Reiersen J, Lowman R, Bisaillon J R, Michel P, Fridriksdottir V, Gunnarsson E, Stern N, Berke O, McEwen S and Martin W (2006) Risk factors for *Campylobacter* spp. colonization in broiler flocks in Iceland, *Preventive Veterinary Medicine*, **74**, 264–78.

Baylis C L, MacPhee S, Martin K W, Humphrey T J and Betts R P (2000) Comparison of three enrichment media for the isolation of *Campylobacter* spp. from foods, *Journal of Applied Microbiology*, **89**, 884–91.

Bhaduri S and Cottrell B (2004) Survival of cold-stressed *Campylobacter jejuni* on ground chicken and chicken skin during frozen storage, *Applied and Environmental Microbiology*, **70**, 7103–9.

Björkroth J (2005) Microbiological ecology of marinated meat products, *Meat Science*, **70**, 477–80.

Blankenship L C and Craven S E (1982) *Campylobacter jejuni* survival in chicken meat as a function of temperature, *Applied and Environmental Microbiology*, **44**, 88–92.

Brightwell G, Mowat E, Clemens R, Boerema J, Pulford D J and On S L (2007) Development of a multiplex and real time PCR assay for the specific detection of *Arcobacter butzleri* and *Arcobacter cryaerophilus*, *Journal of Microbiological Methods*, **68**, 318–25.

Brynestad S, Braute L, Luber P and Bartelt E (2008) Quantitative microbiological risk assessment of campylobacteriosis cases in the German population due to consumption of chicken prepared in homes, *International Journal of Risk Assessment and Management*, **8**, 194–213.

Buswell C M, Herlihy Y M, Lawrence L M, McGuiggan J T M, Marsh P D, Keevil C W and Leach S A (1998) Extended survival and persistence of *Campylobacter* spp. in water and aquatic biofilms and their detection by immunofluorescent-antibody and -rRNA staining, *Applied and Environmental Microbiology*, **64**, 733–41.

Butler R C, Lund V and Carlson D A (1987) Susceptibility of *Campylobacter jejuni* and *Yersinia enterocolitica* to UV radiation, *Applied and Environmental Microbiology*, **53**, 375–8.

Butzler J P and Oosterom J (1991) *Campylobacter* – Pathogenicity and significance in foods, *International Journal of Food Microbiology*, **12**, 1–8.

Carneiro A, Couto J A, Mena C, Queiroz J and Hogg T (2008) Activity of wine against *Campylobacter jejuni*, *Food Control*, **19**, 800–5.

Carrillo C L, Atterbury R J, El-Shibiny A, Connerton P L, Dillon E, Scott A and Connerton I F (2005) Bacteriophage therapy to reduce *Campylobacter jejuni* colonization of broiler chickens, *Applied and Environmental Microbiology*, **71**, 6554–63.

Cervenka L (2007) Survival and inactivation of *Arcobacter* spp., a current status and future prospect, *Critical Reviews in Microbiology*, **33**, 101–18.

Cervenka L, Malikova Z, Zachova I and Vytrasova J (2004) The effect of acetic acid, citric acid and trisodium citrate in combination with different levels of water activity on the growth of *Arcobacter butzleri* in culture, *Folia Microbiology*, **49**, 8–12.

Chan K F, Tran H L, Kanenaka R Y and Kathariou S (2001) Survival of clinical and poultry-derived isolates of *Campylobacter jejuni* at a low temperature (4 °C), *Applied and Environmental Microbiology*, **67**, 4186–91.

Chinivasagam H N, Corney B G, Wright L L, Diallo I S and Blackall P J (2007) Detection of *Arcobacter* spp. in piggery effluent and effluent-irrigated soils in southeast Queensland, *Journal of Applied Microbiology*, **103**, 418–26.

Cogan T A, Bloomfield S F and Humphrey T J (1999) The effectiveness of hygiene procedures for prevention of cross-contamination from chicken carcasses in the domestic kitchen, *Letters in Applied Microbiology*, **29**, 354–8.

Collins C I, Murano E A and Wesley I V (1996a) Survival of *Arcobacter butzleri* and *Campylobacter jejuni* after irradiation treatment in vacuum-packaged ground pork, *Journal of Food Protection*, **59**, 1164–6.

Collins C I, Wesley I V and Murano E A (1996b) Detection of *Arcobacter* spp. in ground pork by modified plating methods, *Journal of Food Protection*, **59**, 448–52.

Cools I, Uyttendaele M, D'Haese E, Nelis H J and Debevere J (2006) Development of a real-time NASBA assay for the detection of *Campylobacter jejuni* cells, *Journal of Microbiological Methods*, **66**, 313–20.

Corry J E L and Atabay H I (1997) Comparison of the productivity of cefoperazone amphotericin teicoplanin (CAT) agar and modified charcoal cefoperazone deoxycholate (mCCD) agar for various strains of *Campylobacter*, *Arcobacter* and *Helicobacter pullorum*, *International Journal of Food Microbiology*, **38**, 201–19.

Corry J E L, Post D E, Colin P and Laisney M J (1995) Culture media for the isolation of campylobacters, *International Journal of Food Microbiology*, **26**, 43–76.

Curtis L M, Patrick M and Blackburn C de W (1995) Survival of *Campylobacter jejuni* in foods and comparison with a predictive model, *Letters in Applied Microbiology*, **21**, 194–7.

Debretsion A, Habtemariam T, Wilson S, Nganwa D and Yehualaeshet T (2007) Real-time PCR assay for rapid detection and quantification of *Campylobacter jejuni* on chicken rinses from poultry processing plant, *Molecular and Cellular Probes*, **21**, 177–81.

Docherty L, Adams M R, Patel P, McFadden J (1996) The magnetic immuno-polymerase chain reaction assay for the detection of *Campylobacter* in milk and poultry, *Letters in Applied Microbiology*, **22**, 288–92.

D'Sa, E M and Harrison M A (2005) Effect of pH, NaCl content, and temperature on growth and survival of *Arcobacter* spp, *Journal of Food Protection*, **68**, 18–25.

EC (2008) *Campylobacteriosis*, DG Health and Consumers, European Commission,

Brussels, available at: http://ec.europa.eu/health/ph_information/dissemination/echi/docs/campylobacteriosis_en.pdf.

EFSA (2005) Opinion on the Scientific Panel on Biological Hazards on *Campylobacter* in animals and foodstuffs, European Food Safety Authority, *The EFSA Journal*, **173**, 1–10.

Ellebroek L, Lienau J A, Alter T and Schlichting D (2007) Effectiveness of different chemical decontamination methods on the *Campylobacter* load of poultry carcasses, *Fleischwirtschaft*, **87**, 224–7.

Ellis W A, Neill S D, O'Brien J J and Alam M (1977) Isolation of *Spirillium*/*Vibrio*-like organisms from bovine fetuses, *Veterinary Record*, **199**, 451–2.

Engberg J H, (2002) Guillaine Barre syndrome and *Campylobacter*, *Ugeskr Laeger*, **164**, 5905–8.

Epling L K, Carpenter J A and Blankenship L C (1993) Prevalence of *Campylobacter* spp. and *Salmonella* spp. on pork carcasses and the reduction effected by spraying with lactic acid, *Journal of Food Protection*, **56**, 536–7.

Evans M R, Lane W, Frost J A and Nylen G (1998) A campylobacter outbreak associated with stir-fried food, *Epidemiology and Infection*, **121**, 275–9.

Federighi M, Tholozan J L, Cappelier J M, Tissier J P and Jouve J L (1998) Evidence of non-coccoid viable but non-culturable *Campylobacter jejuni* cells in microcosm water by direct viable count, CTC-DAPI double staining, and scanning electron microscopy, *Food Microbiology*, **15**, 539–50.

Federighi M, Magras C, Pilet M F, Woodward D, Johnson W, Jugiau F and Jouve J L (1999) Incidence of thermotolerant *Campylobacter* in foods assessed by NF ISO 10272 standard: results of a two-year study, *Food Microbiology*, **16**, 195–204.

Fernandez H and Pison V (1996) Isolation of thermotolerant species of *Campylobacter* from commercial chicken livers, *International Journal of Food Microbiology*, **29**, 75–80.

Fosse J, Laroche M, Rossero A, Federighi M, Seegers H and Magras C (2006) Recovery methods for detection and quantification of *Campylobacter* depend on meat matrices and bacteriological of PCR tools, *Journal of Food Protection*, **69**, 2100–106.

Gargiulo A, Rinaldi L, D'Angelo L, Dipineto L, Borrelli L, Fioretti A and Menna L F (2008) Survey of *Campylobacter jejuni* in stray cats in southern Italy, *Letters in Applied Microbiology*, **46**, 267–70.

Garrett N, Devane M L, Hudson J A, Nicol C, Ball A, Klena J D, Scholes P, Baker M G, Gilpin B J and Savill M G (2007) Statistical comparison of *Campylobacter jejuni* subtypes from human cases and environmental sources, *Journal of Applied Microbiology*, **103**, 2113–21.

Georgsson F, Þorkelsson Á E, Geirsdóttir M, Reiersen J and Stern N (2006) The influence of freezing and duration of storage on *Campylobacter* and indicator bacteria in broiler carcasses, *Food Microbiology*, **23**, 677–83.

Gonzalez I, Garcia T, Antolin A, Hernandez P E and Martin R (2000) Development of a combined PCR-culture technique for the rapid detection of *Arcobacter* spp. in chicken meat, *Letters in Applied Microbiology*, **30**, 207–12.

Gonzalez A, Botella S, Montes R M, Moreno Y and Ferrus M A (2007) Direct detection and identification of *Arcobacter* species by multiplex PCR in chicken and wastewater samples from Spain, *Journal of Food Protection*, **70**, 341–7.

Gormley F J, MacRae M, Forbes K J, Ogden I D, Dallas J F and Strachan N J C (2008) Has retail chicken played a role in the decline of human campylobacteriosis?, *Applied and Environmental Microbiology*, **74**, 383–90.

Gürtler M, Alter T, Kasimir S, Fehlhaber K (2005) The importance of *Campylobacter coli* in human campylobacteriosis: prevalence and genetic characterization, *Epidemiology and Infection*, **133**, 1081–7.

Hald B, Wedderkopp A and Madsen M (2000) Thermophilic *Campylobacter* spp. in Danish broiler production: a cross-sectional survey and a retrospective analysis of risk factors for occurrence in broiler flocks, *Avian Pathology*, **29**, 123–31.

Hald B, Rattenborg E and Madsen M (2001) Role of batch depletion of broiler houses on the occurrence of *Campylobacter* spp. in chicken flocks, *Letters in Applied Microbiology*, **32**, 253–6.

Hannu T, Mattila L, Rautelin H, Pelkonen P, Lahdenne P, Siitonen A and Leirisalo-Repo M (2002) *Campylobacter*-triggered reactive arthritis: a population-based study, *Rheumatology*, **41**, 312–18.

Hansson I, Plym Forshell L, Gustafsson P, Boqvist S, Lindblad J, Olsson Engvall E, Andersson Y and Vågsholm I (2007a) Summary of the Swedish *Campylobacter* program in broilers, 2001 through 2005, *Journal of Food Protection*, **70**, 2008–14.

Hansson I, Vågsholm I, Svensson L and Olsson Engvall E (2007b) Correlations between *Campylobacter* spp. prevalence in the environment and broiler flocks, *Journal of Applied Microbiology*, **103**, 640–49.

Hariharan H, Murphy G A and Kempf I (2004) *Campylobacter jejuni*: public health hazards and potential control methods in poultry: a review, *Veterinarni Medicina*, **49**, 441–6.

Harmon K M and Wesley I V (1997) Multiplex PCR for the identification of *Arcobacter* and differentiation of *Arcobacter butzleri* from other arcobacters, *Veterinary Microbiology*, **58**, 215–27.

Harrass B, Schwarz S and Wenzel S (1998) Identification and characterization of *Arcobacter* isolates from broilers by biochemical tests, antimicrobial resistance patterns and plasmid analysis, *Journal of Veterinary Medicine Series B – Infectious Diseases and Veterinary Public Health*, **45**, 87–94.

Harvey R B, Young C R, Ziprin R L, Hume M E, Genovese K J, Anderson R C, Droleskey R E, Stanker L H and Nisbet D J (1999) Prevalence of *Campylobacter* spp. isolated from the intestinal tract of pigs raised in an integrated swine production system, *Journal of the American Veterinary Medical Association*, **215**, 1601–4.

Havelaar A H, Mangen M-J J, de Koeijer A A, Bogaardt M-J, Evers E G, Jacobs-Reitsma W F, van Pelt W, Wagenaar J A, de Wit G A, van der Zee H and Nauta M J (2007) Effectiveness and efficiency of controlling *Campylobacter* on broiler chicken meat, *Risk Analysis*, **27**, 831–44.

Hilton A C, Mortiboy D, Banks J G and Penn C W (1997) RAPD analysis of environmental, food and clinical isolates of *Campylobacter* spp., *FEMS Immunology and Medical Microbiology*, **18**, 119–24.

Hilton C L, Mackey B M, Hargreaves A J and Forsythe S J (2001) The recovery of *Arcobacter butzleri* NCTC 12481 from various temperature treatments, *Journal of Applied Microbiology*, **91**, 929–32.

Hong N T T, Corry J E L and Miles C A (2006) Heat resistance and mechanism of heat inactivation in thermophilic campylobacters, *Applied and Environmental Microbiology*, **72**, 908–13.

Humphrey T J and Cruikshank J G (1985) Antibiotic and deoxycholate resistance in *Campylobacter jejuni* following freezing or heating, *Journal of Applied Bacteriology*, **59**, 65–71.

Humphrey T, O'Brien S and Madsen M (2007) Campylobacters as zoonotic pathogens: A food production perspective, *International Journal of Food Microbiology*, **117**, 237–57.

Huneau-Salaun A, Denis M, Balaine L and Salvat G (2007) Risk factors for *Campylobacter* spp. colonization in French free-range broiler-chicken flocks at the end of the indoor rearing period, *Preventive Veterinary Medicine*, **80**, 34–48.

ISO (1995) *ISO 10272: (1995)E Microbiology of food and animal feeding stuffs – Horizontal method for detection of thermotolerant Campylobacter*, International Organization for Standardization, Geneva.

ISO (2006a) *ISO 10272-1:2006 Microbiology of food and animal feeding stuffs – Horizontal method for detection and enumeration of Campylobacter* spp. *– Part 1: Detection method*, International Organization for Standardization, Geneva.

ISO (2006b) *ISO/TS 10272-2:2006 Microbiology of food and animal feeding stuffs – Horizontal method for detection and enumeration of Campylobacter spp. – Part 2: Colony-count technique*, International Organization for Standardization, Geneva.

James C, James S J, Hannay N, Purnell G, Barbedo-Pinto C, Yaman H, Araujo M, Gonzalez M L, Calvo J, Howell M and Corry J E L (2007) Decontamination of poultry carcasses using steam or hot water in combination with rapid cooling, chilling or freezing of carcass surfaces, *International Journal of Food Microbiology*, **114**, 195–203.

Johnsen G, Kruse H and Hofshagen M (2007) Genotyping of thermotolerant *Campylobacter* from poultry slaughterhouse by amplified fragment length polymorphism, *Journal of Applied Microbiology*, **103**, 271–9.

Johnson L G and Murano E A (1999a) Development of a new medium for the isolation of *Arcobacter* spp., *Journal of Food Protection*, **62**, 456–62.

Johnson L G and Murano E A (1999b) Comparison of three protocols for the isolation of *Arcobacter* from poultry, *Journal of Food Protection*, **62**, 610–4.

Jones D R and Musgrove M T (2007) Pathogen prevalence and microbial levels associated with restricted shell eggs, *Journal of Food Protection*, **70**, 2004–7.

Jones D M, Sutcliffe, E M and Curry A (1991) Recovery of viable but non-culturable *Campylobacter jejuni*, *Journal of General Microbiology*, **137**, 2477–82.

Jones D R, Musgrove M T, Caudill A B and Curtis P A (2006) Frequency of *Salmonella*, *Campylobacter*, *Listeria* and Enterobacteriaceae detection in commercially cool water-washed shell eggs, *Journal of Food Safety*, **26**, 264–74.

Jørgensen F, Bailey R, Williams S, Henderson P, Wareing D R A, Bolton F J, Frost J A, Ward L and Humphrey T J (2002) Prevalence and numbers of *Salmonella* and *Campylobacter* spp. on raw, whole chickens in relation to sampling methods, *International Journal of Food Microbiology*, **76**, 151–64.

Kapperud G, Skjerve E, Vik L, Hauge K, Lysaker A, Aalmen I, Ostroff S M and Potter M (1993) Epidemiologic investigation of risk factors for *Campylobacter* colonisation in Norwegian broiler flocks, *Epidemiology and Infection*, **111**, 245–55.

Kelly A F, Park S F, Bovill R and Mackey B M (2001) Survival of *Campylobacter jejuni* during stationary phase: evidence for the absence of a phenotypic stationary-phase response, *Applied and Environmental Microbiology*, **67**, 2248–54.

Kelly A F, Martinez-Rodriguez A, Bovill R and Mackey B M (2003) Description of a 'phoenix' phenomenon in the growth of *Campylobacter jejuni* at temperatures close to the minimum for growth, *Applied and Environmental Microbiology*, **69**, 4975–8.

Ketley J M (1997) Pathogenesis of enteric infection by *Campylobacter*, *Microbiology*, **143**, 5–21.

Korhonen L K and Martikainen P J (1991) Comparison of the survival of *Campylobacter jejuni* and *Campylobacter coli* in culturable form in surface water, *Canadian Journal of Microbiology*, **37**, 530–33.

Krause M, Josefsen M H, Lund M, Jacobsen N R, Brorsen L, Moos M, Stockmarr A and Hoorfar J (2006) Comparative, collaborative, and on-site validation of a TaqMan PCR method as a tool for certified production of fresh, *Campylobacter*-free chickens, *Applied and Environmental Microbiology*, **72**, 5463–8.

Kusumaningrum H D, Riboldi G, Hazeleger W C and Beumer R R (2003) Survival of foodborne pathogens on stainless steel surfaces and cross-contamination to foods, *International Journal of Food Microbiology*, **85**, 227–36.

Lehner A, Tasara T and Stephan R (2005) Relevant aspects of *Arcobacter* spp. as potential foodborne pathogen, *International Journal of Food Microbiology*, **102**, 127–35.

Lehtola M J, Pitkanen T, Miebach L and Miettinen I T (2006) Survival of *Campylobacter jejuni* in potable water biofilms: a comparative study with different detection methods, *Water Science and Technology*, **54**, 57–61.

Levin R E (2007) *Campylobacter jejuni*: a review of its characteristics, pathogenicity, ecology, distribution, subspecies characterisation and molecular methods of detection, *Food Biotechnology*, **21**, 271–347.

Lewis S J, Velasquez A, Cuppett S L, McKee S R (2002) Effect of electron beam irradiation on poultry meat safety and quality, *Poultry Science*, **81**, 896–903.

Line J E (2006) Influence of relative humidity on transmission of *Campylobacter jejuni* in broiler chickens, *Poultry Science*, **85**, 1145–50.

Line J E and Bailey J S (2006) Effect of on-farm litter acidification treatments on *Campylobacter* and *Salmonella* populations in commercial broiler houses in northeast Georgia, *Poultry Science*, **85**, 1529–34.

Line J E and Siragusa G R (2006) Biphasic microtiter method for campylobacter recovery and enumeration, *Journal of Rapid Methods and Automation in Microbiology*, **14**, 182–8.

Lior H and Woodward D (1991) A serotyping scheme for *Campylobacter butzleri*, *Microbial Ecology and Health Diseases*, **4**, S93.

Litrup E, Torpdahl M and Nielsen E M (2007) Multilocus sequence typing performed on *Campylobacter coli* isolates from humans, broilers, pigs and cattle originating in Denmark, *Journal of Applied Microbiology*, **103**, 210–8.

Little C L, Richardson J F, Owen R J, de Pinna E and Threlfall E J (2008) *Campylobacter* and *Salmonella* in raw meats in the United Kingdom: prevalence, characterization and antimicrobial resistance pattern, 2003–2005, *Food Microbiology*, **25**, 538–43.

Loewenherz-Luning K, Heitmann M and Hildebrandt G, (1996) Survey about the occurrence of *Campylobacter jejuni* in food of animal origin, *Fleischwirtshaft*, **76**, 958–61.

Manke T R, Wesley I V, Dickson J S and Harmon K M (1998) Prevalence of genetic variability of *Arcobacter* species in mechanically separated turkey, *Journal of Food Protection*, **61**, 1623–4.

Mansfield L P and Forsythe S J (2000) *Arcobacter butzleri*, *A. skirrowii* and *A. cryaerophilus* – potential emerging human pathogens, *Reviews in Medical Microbiology*, **11**, 161–70.

Martinez-Rodriguez A and Mackey B M (2005) Physiological changes in *Campylobacter jejuni* on entry into stationary phase, *International Journal of Food Microbiology*, **101**, 1–8.

Mason M J, Humphrey T J and Martin K W (1999) Isolation of sublethally injured campylobacters from poultry and water sources, *British Journal of Biomedical Science*, **56**, 2–5.

McClure P J (2000) Microbiological hazard identification in the meat industry, in M Brown (ed.), *HACCP in the Meat Industry*, Woodhead, Cambridge, 157–76.

Mead G C (2000) HACCP in primary processing: poultry, in M Brown (ed.), *HACCP in the Meat Industry*, Woodhead, Cambridge, 123–53.

Mead G C (2002) Factors affecting intestinal colonisation of poultry by campylobacter and role of microflora in control, *Worlds Poultry Science Journal*, **58**, 169–78.

Mead G C, Hudson W R and Hinton M H (1995) Effect of changes in processing to improve hygiene control on contamination of poultry carcasses with campylobacter, *Epidemiology and Infection*, **115**, 495–500.

Mead G C, Scott M J, Humphrey T J and McAlpine K (1996) Observations on the control of *Campylobacter jejuni* infection of poultry by 'competitive exclusion', *Avian Pathology*, **25**, 69–79.

Medema G J, Schets F M, Vandegiessen A W and Havelaar A H (1992) Lack of colonization of 1-day-old chicks by viable, non-culturable *Campylobacter jejuni*, *Journal of Applied Bacteriology*, **72**, 512–16.

Moore J E and Madden R H (2000) The effect of thermal stress on *Campylobacter coli*, *Journal of Applied Microbiology*, **89**, 892–9.

Moorhead S M and Dykes G A (2002) Survival of *Campylobacter jejuni* on beef trimmings during freezing and frozen storage, *Letters in Applied Microbiology*, **34**, 72–6.

Murphy C, Carroll C and Jordan K N (2003) Identification of a novel stress response mechanism in *Campylobacter jejuni*, *Journal of Applied Microbiology*, **95**, 704–8.

Musmanno R A, Russi M, Lior H and Figura N (1997) *In vitro* virulence factors of

Arcobacter butzleri strains isolated from superficial water samples, *Microbiologica*, **20**, 63–8.

Mylius S D, Nauta M J and Havelaar A H (2007) Cross-contamination during food preparation: a mechanistic model applied to chicken-borne *Campylobacter*, *Risk Analysis*, **27**, 803–13.

Nauta M J, Jacobs-Reitsma W F and Havelaar A H (2007) A risk assessment model for *Campylobacter* in broiler meat, *Risk Analysis*, **27**, 845–61.

Neal K R and Slack R C B (1997) Diabetes mellitus, anti-secretory drugs and other risk factors for *Campylobacter* gastro-enteritis in adults: a case-control study, *Epidemiology and Infection*, **119**, 307–11.

Neubauer C and Hess M (2006a) Breakpoint-method to determine the antibiotic resistance of *Campylobacter* isolates from Austrian broiler flocks, *Wiener Tierarztliche Monatsschrift*, **93**, 4–9.

Neubauer C and Hess M (2006b) Detection and identification of food-borne pathogens of the genera *Campylobacter*, *Arcobater* and *Helicobacter* by multiplex PCR in poultry and poultry products, *Journal of Veterinary Medicine Series B – Infectious Diseases and Veterinary Public Health*, **53**, 376–81.

Nielsen E M and Nielsen N L (1999) Serotypes and typability of *Campylobacter jejuni* and *Campylobacter coli* isolated from poultry products, *International Journal of Food Microbiology*, **46**, 199–205.

Nielsen E M, Engberg J and Madsen M (1997) Distribution of serotypes of *Campylobacter jejuni* and *C. coli* from Danish patients, poultry, cattle and swine, *FEMS Immunology and Medical Microbiology*, **19**, 47–56.

Nogva H K, Bergh A, Holck A and Rudi K (2000) Application of the 5'-nuclease PCR assay in evaluation and development of methods for quantitative detection of *Campylobacter jejuni*, *Applied and Environmental Microbiology*, **66**, 4029–36.

Northcutt J K, Smith D P, Musgrove M T and Ingram K D (2005) Microbiological impact of spray washing broiler carcasses using different chlorine concentrations and water temperatures, *Poultry Science*, **84**, 1648–52.

Obiri-Danso K, Paul N and Jones K (2001) The effects of UVB and temperature on the survival of natural populations of pure cultures of *Campylobacter jejuni*, *Camp. coli*, *Camp. lari* and urease-positive thermophilic campylobacters (UPTC) in surface waters, *Journal of Applied Microbiology*, **90**, 256–67.

Olsen S J, Hansen G R, Bartlett L, Fitzgerald C, Sonder A, Manjrekar R, Riggs T, Kim J, Flahart R, Pezzino G and Swerdlow D L (2001) An outbreak of *Campylobacter jejuni* infections associated with food handler contamination: the use of pulsed-field gel electrophoresis, *Journal of Infectious Diseases*, **183**, 164–7.

On S L W (2001) Taxonomy of *Campylobacter*, *Arcobacter*, *Helicobacter* and related bacteria: current status, future prospects and immediate concerns, *Journal of Applied Microbiology*, **90**, 1S–15S.

Ono K and Yamamoto K (1999) Contamination of meat with *Campylobacter jejuni* in Saitama, Japan, *International Journal of Food Microbiology*, **47**, 211–19.

Oporto B, Esteban J I, Aduriz G, Juste R A and Hurtado A (2007) Prevalence and strain diversity of thermophilic campylobacters in cattle, sheep and swine farms, *Journal of Applied Microbiology*, **103**, 977–84.

Osano O and Arimi S M (1999) Retail poultry and beef as sources of *Campylobacter jejuni*, *East African Medical Journal*, **76**, 141–3.

O'Sullivan N A, Fallon R, Carroll C, Smith T and Maher M (2000) Detection and differentiation of *Campylobacter jejuni* and *Campylobacter coli* in broiler chicken samples using a PCR/DNA probe membrane based colorimetric detection assay, *Molecular and Cellular Probes*, **14**, 7–16.

Palmer S R, Gully P R, White J M, Pearson A D, Suckling W G, Jones D M, Rawes J C L and Penner J L (1983) Water-borne outbreak of *Campylobacter* gastroenteritis, *Lancet*, **i**, 287–90.

Park S (2002) The physiology of *Campylobacter* species and its relevance to their role as foodborne pathogens, *International Journal of Food Microbiology*, **74**, 177-88.

Park R W A, Griffiths P L and Moreno G S (1991) Sources and survival of campylobacters – relevance to enteritis and the food-industry, *Journal of Applied Bacteriology*, **70**, S97–S106.

Pearson A D, Colwell R R, Rollins D, Watkin-Jones M, Healing T, Greenwood M, Shahamat M, Jump E, Hood M and Jones D M (1988) Transmission of *C. jejuni* on a poultry farm, in B Kaijser and E Falsen (eds), *Campylobacter* IV, University of Goteberg, Goteberg, 281–4.

Pedersen K B, Aarestrup F M, Jensen N E, Bager F, Jensen L B, Jorsal S E, Nielsen T K, Hansen H C, Meyling A and Wegener H C (1999) The need for a veterinary antibiotic policy, *The Veterinary Record*, **10 July**, 50–53.

Phillips C A (1999) The effect of citric acid, lactic acid, sodium citrate and sodium lactate, alone and in combination with nisin, on the growth of *Arcobacter butzleri*, *Letters in Applied Microbiology*, **29**, 424–8.

Phillips C A (2001) Arcobacters as emerging human foodborne pathogens, *Food Control*, **12**, 1–6.

Poly F and Guerry P (2008) Pathogenesis of *Campylobacter*, *Current Opinion in Gastroenterology*, **24**, 27–31.

Quinoñes B, Parker C T, Janda J M, Miller W G and Mandrell R E (2007) Detection and genotyping of *Arcobacter* and *Campylobacter* isolates from retail chicken samples by use of DNA oligonucleotide arrays, *Applied and Environmental Microbiology*, **73**, 3645–55.

Rautelin H and Hanninen M L (2000) Campylobacters: the most common bacterial enteropathogens in the Nordic countries, *Annals of Medicine*, **32**, 440–45.

Ritz M, Nauta M J, Teunis P F M, van Leusden F, Federighi M and Havelaar A H (2007) Modelling of *Campylobacter* survival in frozen chicken meat, *Journal of Applied Microbiology*, **103**, 594–600.

Rollins D M and Colwell R R (1986) Viable but non-culturable stage of *Campylobacter jejuni* and its role in survival in the natural aquatic environment, *Applied and Environmental Microbiology*, **52**, 531–8.

Ronner A C and Lindmark H (2007) Quantitative detection of *Campylobacter jejuni* on fresh chicken carcasses by real-time PCR, *Journal of Food Protection*, **70**, 1373–8.

Sandberg M, Nygard K, Meldal H, Valle P S, Kruse H and Skjerve E (2006) Incidence trend and risk factors for campylobacter infections in humans in Norway, *BMC Public Health*, **6**, Art. No. 179.

Saha S K, Saha S and Sanyal S C (1991) Recovery of injured *Campylobacter jejuni* cells after animal passage, *Applied and Environmental Microbiology*, **57**, 3388–9.

Sallam K I (2007) Prevalence of *Campylobacter* in chicken and chicken by-products retailed in Sapporo area, Hokkaido, Japan, *Food Control*, **18**, 1113–20.

Scherer K, Bartelt E, Sommerfeld C and Hildebrandt G (2006) Comparison of different sampling techniques and enumeration methods for the isolation and quantification of *Campylobacter* spp. in raw retail chicken legs, *International Journal of Food Microbiology*, **108**, 115–19.

Schiellerup P, Locht H and Krogfelt K A (2003) Comparison of the clinical manifestations between patients with reactive joint symptoms and gastroenteritis only in a cohort of *Campylobacter* spp. infected, Abstracts of the 12th International Workshop on *Campylobacter Heliobacter* and Related Organisms. Aarhus, Denmark, 6–10 September 2003 *International Journal of Medical Microbiology*, **293**, 35–6.

Schorr D, Schmid H, Rieder H L, Baumgartner A, Vorkauf H and Burnens A (1994) Risk factors for *Campylobacter* enteritidis in Switzerland, *Zentralblatt fur Hygiene und Umweltmedizin*, **196**, 327–37.

Scott A E, Timms A R, Connerton P L, Carrillo C L, Radzum K A and Connerton I F (2007) Genome dynamics of *Campylobacter jejuni* in response to bacteriophage predation, *PLoS Pathogens*, **3**, 1142–51.

Shane S M (1992) The significance of *Campylobacter jejuni* infection in poultry: a review, *Avian Pathology*, **21**, 189–213.

Singer R S, Cox L A, Dickson J S, Hurd H S, Phillips I and Miller G Y (2007) Modelling the relationship between animal health and foodborne illness, *Preventative Veterinary Medicine*, **79**, 186–203.

Skerman V B D, McGowan V and Sneath P H A (1980) Approved lists of bacterial names, *International Journal of Systematic Bacteriology*, **30**, 225–420.

Skirrow M B (1994) Review: diseases due to *Campylobacter*, *Helicobacter* and related bacteria, *Journal of Comparative Pathology*, **111**, 113–49.

Snelling W J, McKenna J P, Lecky J M and Dooley J S G (2005) Survival of *Campylobacter jejuni* in waterborne protozoa, *Applied and Environmental Microbiology*, **71**, 5560–71.

Snelling W J, Moore J E, McKenna J P, Lecky J M and Dooley J S G (2006) Bacterial-protozoa interactions: an update on the role these phenomena play towards human illness, *Microbes and Infection*, **8** , 578–87.

Solomon E B and Hoover D G (1999) *Campylobacter jejuni*: A bacterial paradox, *Journal of Food Safety*, **19**, 121–36.

Solomon E B and Hoover D G (2004) Inactivation of *Campylobacter jejuni* by high hydrostatic pressure, *Letters in Applied Microbiology*, **38**, 505–9.

Sopwith W, Ashton M, Frost J A, Tocque K, O'Brien S, Regan M, Syed Q (2003) Enhanced surveillance of *Campylobacter* infection in the North West of England 1997–1999, *Journal of Infection*, **46**, 35-45.

Stead D and Park S F (2000) Roles of Fe superoxide dismutase and catalase in resistance of *Campylobacter coli* to freeze–thaw stress, *Applied and Environmental Microbiology*, **66**, 110–12.

Stern N J, Hiett K L, Alfredsson G A, Kristensson K G, Reierson J, Hardardottir H, Briem H, Gunnarsson E, Georgsson F, Lowman R, Berndtson E, Lammerding A M, Paoli G M and Musgrove M T (2003) *Campylobacter* spp. in Icelandic poultry operations and human disease, *Epidemiology and Infection*, **130**, 20–32.

Stern N J, Svetoch E A, Eruslanov B V, Kovalev Y N, Volodina L I, Perelygin V V, Mitsevich E V, Mitsevich I P and Levchuk V P (2005) *Paenibacillus polymyxa* purified bacteriocin to control *Campylobacter jejuni* in chickens, *Journal of Food Protection*, **68**, 1450–53.

Teunis P, Havelaar A, Vliegenthart J and Roessink C (1997) Risk assessment of *Campylobacter* species in shellfish: identifying the unknown, *Water Science and Technology*, **35**, 29–34.

Thomas C, Gibson H, Hill D J and Mabey M (1999) *Campylobacter* epidemiology: an aquatic perspective, *Journal of Applied Microbiology*, **85** (Supplement), 168S–177S.

Thunberg R L, Tran T T and Walderhaug M O (2000) Detection of thermophilic *Campylobacter* spp. in blood-free enriched samples of inoculated foods by the polymerase chain reaction, *Journal of Food Protection*, **63**, 299–303.

Trachoo N and Frank J F (2002) Effectiveness of chemical sanitizers against *Campylobacter jejuni*-containing biofilms, *Journal of Food Protection*, **65**, 1117–21.

Trachoo N, Frank J F and Stern N J (2002) Survival of *Campylobacter* in biofilms isolated from chicken houses, *Journal of Food Protection*, **65**, 1110–16.

Tran T T (1998) A blood-free enrichment medium for growing *Campylobacter* spp. under aerobic conditions, *Letters in Applied Microbiology*, **26**, 145–8.

Uyttendaele M and Debevere J (1996) Evaluation of Preston medium for detection of *Campylobacter jejuni in vitro* and in artificially and naturally contaminated poultry products, *Food Microbiology*, **13**, 115–22.

Uyttendaele M, Debevere J and Lindqvist R (1999) Evaluation of buoyant density centrifugation as a sample preparation method for NASBA-ELGA detection of *Campylobacter jejuni* in foods, *Food Microbiology*, **16**, 575–82.

Uyttendaele M, Baert K, Ghafir Y, Daube G, De Zutter L, Herman L, Dierick K, Pierard

D, Dubois J J, Horion B and Debevere J (2006) Quantitative risk assessment of *Campylobacter* spp. in poultry based meat preparations as one of the factors to support the development of risk-based microbiological criteria in Belgium, *International Journal of Food Microbiology*, **111**, 149–63.

Valdivieso-Garcia A, Harris K, Riche E, Campbell S, Jarvie A, Popa M, Deckert A, Reid-Smith R and Rahn K (2007) Novel *Campylobacter* isolation method using hydrophobic grid membrane filter and semisolid medium, *Journal of Food Protection*, **70**, 355–62.

van de Giessen A W, Bloemberg B P M, Ritmeester W S and Tilburg J J H C (1996) Epidemiological study on risk factors and risk reducing measures for campylobacter infections in Dutch broiler flocks, *Epidemiology and Infection*, **117**, 245–50.

Vandamme P (2000) Taxonomy of the family *Campylobacteraceae*, in I Nachamkin and M J Blaser (eds), *Campylobacter*, ASM, Washington, DC, 3–27.

Vandamme P, Falsen E, Rossau R, Hoste B, Segers P, Tytgat R. and DeLey J (1991) Revision of *Campylobacter*, *Helicobacter*, and *Wolinella* taxonomy: emendation of generic descriptions and proposal of *Arcobacter* gen. nov., *International Journal of Systematic Bacteriology*, **41**, 88–103.

Vandenberg O, Dediste A, Houf K, Ibekwem S, Souayah H, Cadranel S, Douat N and Zissis G (2004) *Arcobacter* species in humans, *Emerging Infectious Diseases*, **10**, 1863–7.

Véron M and Chatelain R (1973) Taxonomic study of the genus *Campylobacter* Sebald and Véron and designation of the neotype strain for the type species *Campylobacter fetus* Sebald and Véron, *International Journal of Systematic Bacteriology*, **23**, 122–34.

Wagenaar J A, Mevius D J and Havelaar A H (2006) *Campylobacter* in primary animal production and control strategies to reduce the burden of human campylobacteriosis, *Revue Scientifique et Technique -de 'Office International des Epizooties*, **25** 581–94.

Waller D F and Ogata S A (2000) Quantitative immunocapture PCR assay for detection of *Campylobacter jejuni* in foods, *Applied and Environmental Microbiology*, **66**, 4115–18.

Wang H, Farber J M, Malik N and Sanders G (1999) Improved PCR detection of *Campylobacter jejuni* from chicken rinses by a simple sample preparation procedure, *International Journal of Food Microbiology*, **52**, 39–45.

Wesley I V (1996) *Helicobacter* and *Arcobacter* species: risks for foods and beverages, *Journal of Food Protection*, **59**, 1127–32.

Wesley I V, Wells S J, Harmon K M, Green A, Schroeder-Tucker L, Glover M and Siddique I (2000) Fecal shedding of *Campylobacter* and *Arcobacter* spp. in dairy cattle, *Applied and Environmental Microbiology*, **66**, 1994–2000.

White P L, Baker A R and James W O (1997) Strategies to control *Salmonella* and *Campylobacter* in raw poultry products, *Revue Scientifique et Technique de L'Office International des Epizooties*, **16**, 525–41.

Wilma C, Hazeleger, R R, Beumer F and Rombouts F M (1992) The use of latex agglutination tests for determining *Campylobacter* species, *Letters in Applied Microbiology*, **14**, 181–4.

Winters D K and Slavik M F (2000) Multiplex PCR detection of *Campylobacter jejuni* and *Arcobacter butzleri* in food products, *Molecular and Cellular Probes*, **14**, 95–9.

Zanetti F, Varoli O, Stampi S and DeLuca G (1996) Prevalence of thermophilic *Campylobacter* and *Arcobacter butzleri* in food of animal origin, *International Journal of Food Microbiology*, **33**, 315–21.

Zhang H, Gong Z, Pui O, Liu Y and Li X-F (2006) An electronic DNA microarray technique for detection and differentiation of viable *Campylobacter* species, *Analyst*, **131**, 907–15.

21

Yersinia, *Shigella*, *Vibrio*, *Aeromonas*, *Plesiomonas*, *Cronobacter*, *Enterobacter*, *Klebsiella* and *Citrobacter*

S. Forsythe, Nottingham Trent University, UK, J. Sutherland and A. Varnam (deceased), University of North London, UK

Abstract: Although bacterial gastroenteritis is a leading cause of illness and deaths worldwide, no infectious agent is recovered from 30 % of cases. Therefore it is probable that either there are still a number of new enteric pathogens yet to be discovered or some already recognised organisms have yet to be linked to gastroenteritis. The bacteria discussed in this chapter are diverse in terms of nature, physiology, ecology and diseases caused. They include *Vibrio cholerae*, *V. parahaemolyticus*, *Yersinia enterocolitica*, *Y. pseudotuberculosis*, *Cronobacter* spp. *(Enterobacter sakazakii)* and other *Enterobacteriaceae* such as the *Klebsiella*, *Citrobacter* and *Enterobacter*. Although they are not as well understood as *Salmonella*, *Campylobacter* and verotoxigenic *E. coli* they constitute a significant threat to human health.

Key words: emergent pathogens, *Vibrio*, *Shigella*, *Yersinia*, *Cronobacter*.

21.1 Introduction

Bacterial gastroenteritis is a leading cause of illness and deaths worldwide. It is notable that no infectious agent can be recovered using current technologies in over 30 % of infections which involve diarrhoea. Therefore it is reasonable to conjecture that there are still a number of new enteric pathogens yet to be discovered or already recognised organisms which are yet to be linked to gastroenteritis. The bacteria discussed in this chapter are principally from the bacterial Family *Enterobacteriaceae*, as well as *Aeromonadaceae* and *Pleisiomonadaceae*. They are diverse in terms of nature, physiology, ecology and diseases caused. Although these particular organisms receive relatively little attention in comparison with *Salmonella*, *Campylobacter* and verocytotoxigenic *Escherichia coli*, their importance should not be under-estimated. The under-reporting of foodborne disease is recognised as a significant problem in all countries, especially where symptoms are 'mild' and self-limiting. *Vibrio cholerae* is a well-known cause of severe, even life-threatening gastroenteritis through contaminated water, but there are other *Vibrio* species such as *V. parahaemolyticus* which is associated with shellfish that causes illness. Similarly, *Y. enterocolitica* and *Y. pseudotuberculosis* are less pathogenic, yet still infectious, than the better known *Y. pestis*. These have caused illness through the ingestion of contaminated food. The microbial flora of the human intestinal tract contains a considerable number of *Enterobacteriaceae* which are opportunistic pathogens. Our intestines become colonised soon after birth from the mother, environment, and feed. The newborn baby is immunocompromised, and is more susceptible to infections in the early months of development. *Cronobacter* (*Enterobacter sakazakii*) has caused much concern as the causative agent of, albeit rare, neonatal necrotising enterocolitis and meningitis in newborn babies through the ingestion of reconstituted infant formula. Other *Enterobacteriaceae* spp such as the *Klebsiella*, *Citrobacter* and *Enterobacter* include species which are opportunistic pathogens. They may occur at low levels as part of the normal intestinal flora. However, there is always a susceptible portion of the human population, and ingestion of these organisms may result in illness. Because they will primarily cause sporadic illnesses that are not as easily identifiable as outbreaks, their true incidence of illness remains unrecognised and under-reported.

21.2 Characteristics of the genus *Yersinia*

21.2.1 Introduction

The genus *Yersinia* comprises Gram-negative, rod-shaped bacteria, which are capable of both oxidative and fermentative metabolism of carbohydrates. The genus is a member of the Family *Enterobacteriaceae*, although it has a number of distinctive characteristics, including small colony size and, in many circumstances, coccoid morphology of cells. Members of the *Yersinia* genus are also considered as 'low-temperature' pathogens and, unlike most other members of the *Enterobacteriaceae*, are able to grow at 4 °C. Optimal growth temperature of *Y. enterocolitica* is 28–30 °C with a doubling time ~34 min, compared with 5 h at 7 °C. Incubation temperature has a profound effect on growth characteristics between 28–30 °C and 37 °C (Sheridan *et al.*, 1998). Biochemical characteristics and suppression of expression of cellular components are involved. Both *Y. enterocolitica* and *Y. pseudotuberculosis* can grow between pH 4 and 10, the optimal being around pH 7.6. Both organisms are readily killed by pasteurisation. *Yersinia enterocolitica* can tolerate NaCl up to 5 % w/v.

The genus comprises environmental organisms, which may be opportunistic pathogens, and three specific pathogens, *Y. enterocolitica*, the cause of foodborne yersiniosis, *Y. pseudotuberculosis*, which may be foodborne under some circumstances, and *Y. pestis*, the causative organism of plague. *Yersinia pestis* is a clone of *Y. enterocolitica* that separated between 1500 and 20 000 years ago. The three species share some virulence determinants, analogs of which can be found in enteropathogenic and enterohaemorrhagic *E. coli*, *Shigella* and *Salmonella*. In *Y. enterocolitica*, pathogenicity is largely restricted to certain bio-serotypes, others being considered non-pathogenic. Biovar 4 is the most prevalent worldwide. There are approximately 60 serogroups of *Y. enterocolitica* which coincide to some degree with the biovar divisions. Most pathogenic strains from humans are in five serogroups: O:3, O:8, O:9 and O:5,57. The other pathogenic groups are O:1,2a,3; O:2a: 2b,3; O:4,32; O:13a,13b; O:18; O:20; and O:21. Serogroup O:3 strains dominate in the USA and Europe, almost all of which are biovar 4. Because strains become rough on repeated sub-culturing (particularly at 37 °C), isolates should be maintained at 22–25 °C before serotyping. Although biotype 1A has been considered non-pathogenic, this is primarily due to lack of response in a mouse model, and needs to be reconsidered due to a number of infections associated with this biotype (Grant *et al*., 1999; Falcão *et al*., 2006).

National surveillance suggests that the importance of *Y. enterocolitica* varies geographically. It is of high prevalence in Belgium and the third most common bacterial infection in Sweden. FoodNet (USA) reports that between 1996 and 1999, the annual incidence of *Y. enterocolitica* was 0.9/100 000 population. In the UK, 42 cases of foodborne yersiniosis were reported to the Public Health Laboratory Service in 2000. Numbers of reported cases have fallen from a peak of 726 in 1989. In England and Wales, the sentinent study of infectious intestinal diseases found the incidence of *Yersinia* recovery was 6.8 and 0.58/1000 person-years for those in the community and patients, respectively.

21.2.2 Nature of foodborne *Yersinia* infections

Yersinia enterocolitica

Infection by *Y. enterocolitica* can involve many symptoms, which vary according to host and bacterium-associated factors (Table 21.1). The most common are gastrointestinal, appearing within 2–3 d and characterised by abdominal pain and diarrhoea. The temperature is usually raised and vomiting may be a secondary symptom. Severity of symptoms can vary considerably and pain can be very severe, resembling that of appendicitis. This results from acute terminal ileitis and inflammation of the mesenteric lymph nodes. In adolescents, pain may be confined to the right quadrant of the trunk (right fossa iliaca), leading to possible confusion with appendicitis. *Yersinia enterocolitica* is also associated with extraintestinal symptoms, which may not be preceded by gastroenteritis. Systemic spread may occur, especially with

Table 21.1 Symptoms of *Yersinia enterocolitica* infections in relation to host- and bacterial related factors

Symptoms	Predisposing factors Host-related	Bacterium-related
Gastroenteritis: abdominal pain, diarrhoea	All	All
Erythema nodosum	Adults	Serotype O:3
Rheumatoid arthritis	Adults	Serotype O:3
Sjøgren's syndrome	Females: middle-aged	All
Autoimmune disease	Adults	Serotype O:3
Septicaemia	Elderly, immunocompromised	Serotype O:8
Osteomyelitis	All	Serotype O:8
Pharyngitis	All	Serotypes O:3, 13a, 13b

Table 21.2 Temperature-regulated phenotypes of *Yersinia enterocolitica*

Phenotype	Expressed at 25 °C	 37 °C
Motility	+	–
Voges–Proskauer reaction	+	–
O-Nitrophenyl galactoside (ONPG) hydrolysis	+	–
Ornithine decarboxylase	+	–
HeLa cells		
Adherence	+	±
Invasion	+	±
Serum resistance	–	+

the invasive serotype O:8, predominant in North America. Symptoms include septicaemia, most common amongst the elderly and immunocompromised, soft tissue infections, conjunctivitis and exudative pharyngitis. There may also be long-term sequelae, especially arthritis.

Yersinia enterocolitica is an infectious pathogen, virulence of which is mediated partially by a 48MDa plasmid, pVYe. This plasmid, which is similar in size and function to the virulence plasmids of other pathogenic yersinias, mediates a number of temperature-regulated phenotypes associated with pathogenicity (Table 21.2). Initial stages of infection, adherence and penetration, however, appear to be primarily under chromosomal control. Spread of most serotypes beyond the epithelial cells is limited, except in susceptible persons, and Peyer's patches are the most heavily colonised tissues. A systemic infection can result from *Y. enterocolitica* draining into the mesenteric lymph nodes (Bottone, 1997). Serotype O:8 is of greater invasive capability than others, possibly because of its lower requirement for iron, *Yersinia* spp. being unusual *Enterobacteriaceae* in not producing siderophores. *Yersinia enterocolitica* produces a heat-stable enterotoxin, Y-ST, which has a similar biological activity to *E. coli* STA toxin. The role of the toxin in pathogenicity, however, is uncertain.

Yersinia pseudotuberculosis

Yersinia pseudotuberculosis infections are characterised by fever with abdominal pain caused by mesenteric lymphadenitis which is often indistinguishable from acute appendicitis. The organism is found in water, the environment and various wild and domesticated animals. However, unlike *Y. enterocolitica*, *Y. pseudotuberculosis* sources and vehicles of transmission are poorly documented. In a few outbreaks, transmission by direct contact with infected animals, or by drinking contaminated water, vegetable juice or pasteurised milk have been suspected sources of infection, but no definitive specific vehicle was identified. Finland has reported nationwide outbreaks associated with iceberg lettuce and grated carrots (Nuorti *et al.*, 2004; Jalava *et al.*, 2006). These are the first reports in which a specific food has been implicated through epidemiological investigations to the source of contamination. The outbreaks were recognised promptly because of ongoing laboratory-based surveillance. Forty-seven cases of *Y. pseudotuberculosis* serotype O:3 occurred over the period 15th October to 6th November 1998. Following an epidemiological investigation, these were strongly associated with iceberg lettuces served through four lunch cafeterias. The bacterial contamination had possibly been through the use of irrigation water which was intermittently contaminated with animal faeces at a farm before distribution. In 2006, Jalava *et al.* reported an outbreak of gastroenteritis and erythema nodosum due to *Y. pseudotuberculosis* serotype O:1 that occurred in May 2003. One hundred and eleven cases were identified, and indistinguishable pulsetypes were isolated from carrot residues at the production farm.

21.2.3 Risk factors

As with many other gastrointestinal diseases, age is an important factor in determining risk of *Y. enterocolitica* infection. Most infections occur in children less than five years of age, although susceptibility to *Yersinia* infection remains high until about 14 years of age. There is a general susceptibility to *Y. enterocolitica* among the immunocompromised, and the most important predisposing conditions involve cirrhosis, or other liver disorders, and iron overload. Liver disorders and iron overload also predispose to more serious systemic infections. Hospitalised patients, in general, are at enhanced risk of *Yersinia* infection. Bio-serotypes not usually considered pathogenic have been involved on some occasions. Hospitalised patients, especially children, may also be at risk from *Y. frederickensii* and other species not usually considered pathogenic.

Most cases of *Yersinia* infections are sporadic. The organism has been isolated from a wide range of foods, but has been associated only with outbreaks involving milk, pork, tofu and water. The carriage rate of *Y. enterocolitica* is high in pigs, the tonsil and surrounding tissue usually being involved, although the incidence can be particularly high when diarrhoeal symptoms are present in the animals. Carriage of *Y. enterocolitica* in pigs usually involves bio-serotype O:3/4, most commonly involved in human infections, although typing by pulsed-field gel electrophoresis suggests strain differences (Asplund *et al.*, 1998). Contamination of meat inevitably occurs during butchery, although a direct causal link between pork and yersiniosis has been established in only a small number of cases. A strong epidemiological link has been established between consumption of raw pork mince and yersiniosis in Belgium, one of the few countries where raw pork is widely eaten (Tauxe *et al.*, 1988).

The highly virulent serotypes O:8 and O:21 are not associated with pigs and it is unlikely that pork is a vehicle of infection. It has been suggested that wild rodents form a reservoir for these serotypes and that transmission to humans largely involves fleas, although contamination of water is also a possibility. Fleaborne transmission implies poor living conditions as a risk factor for *Y. enterocolitica* infections, although there is likely to be occupational risk for workers in agriculture, sewer maintenance and pest control.

21.2.4 Detection methods

It is not usual to examine for *Y. enterocolitica* on a routine basis. Where examination is made, cultural methods are most commonly used. Most media will not differentiate between species of *Yersinia*, and identification is required as part of the detection and enumeration procedure. The situation is further complicated by the fact that only certain bio-serotypes are recognised pathogens and that further testing may be required. The level to which putative isolates of *Y. enterocolitica* are identified depends on circumstances and the purpose of the work. In surveys, for example, identification to species level is usually sufficient and, in cooked foods, the presence of other *Yersinia* species may well be considered cause for concern.

In common with any specimen for the isolation of enteric pathogens, if they are not cultured within 2–4 h of collection then the sample should be refrigerated or placed in a transport medium. Detection procedures, based on cultural methodology, have been developed specifically for *Y. enterocolitica*. Enrichment is necessary and may involve either incubation at low temperatures (4–15 °C) in non-selective broth (cold enrichment), or selective enrichment at 25 °C. Selective enrichment is now preferred, bile–oxalate–sorbitol (BOS) usually being the medium of choice. Alkali treatments may be applied in conjunction with enrichment where the number of competing microorganisms is high, but its value in many circumstances is doubtful (Varnam and Evans, 1996). Cephaloridin–irgasan–novobiocin medium (CIN) is widely used and is generally effective, although serotype O:3 biotype 3b (an unusual biotype) may be inhibited. This medium contains cefsulodin and irgasan (a salicylate compound) to inhibit the Gram-negative organisms other than *Yersinia*, and novobiocin to inhibit Gram-positive organisms. Differentiation is achieved by the inclusion of D-mannitol. *Yersinia enterocolitica* colonies (all biotypes) appear convex red with a bulls-eye. Incubation at 25–29 °C is preferred, as *Y. enterocolitica* may be outgrown by mixed flora at higher incubation temperatures. CIN plates should be read after 24 and 48 h. If necessary, a second medium should be used to recover this serotype, a modification of *Salmonella–Shigella* agar being recommended. A weakness with CIN agar, however, is that *Yersinia* colonies are not well distinguished from other *Enterobacteriaceae* such as *Citrobacter*, *Serratia*, *Enterobacter* and *Aeromonas*. A chromogenic agar that distinguishes virulent *Y. enterocolitica* biotypes 1B, 2-5 from non-virulent biotype 1A and other *Enterobacteriaceae* has been developed (Weagant, 2008). The medium contains cellobiose as a fermentable sugar, neutral red dye and a chromogen 5-bromo-4-choloro-3-indolyl-β-D-glucopyranoside (X-β-Glu) and inducer methyl-D-glucoside. If the chromogen is cleaved by β–glucosidase it liberates 5-bromo-4-chloro-3-indolol which dimerises in the presence of oxygen to form a coloured precipitate (bromo-chloro-indigo) on the colony. Crystal violet, cefsulodin, vancomycin and Irgasan DP300 are included as selective agents. After incubation at 30 °C for 24 and 48 h, potentially virulent *Y. enterocolitica* (β-glucosidase negative) form convex, red with a bulls-eye colonies due to cellobiose fermentation, whereas *Y. enterocolitica* biotype 1A and other related *Yersinia* species form colonies that are purple/blue due to the additional β–glucosidase hydrolysis of X-β-Glc.

Confirmation of identity of *Y. enterocolitica* may be made using commercial identification kits (incubated at 30 °C). Relatively simple cultural methods also exist for determination of virulence, although interpretation can be difficult (Varnam and Evans, 1996). Rapid genetic methods for detection of *Y. enterocolitica* have been developed. A number of workers have developed polymerase chain reaction (PCR) methods, the most effective selecting primers directed both at chromosomal genes and the plasmid-borne virulence, *virF*, gene. Advantages have been claimed in terms of selectivity and specificity (Thisted-Lambertz *et al.*, 1996), but use in the food industry is limited and commercial kits are not currently available. Several detection methods identify *Y. enterocolitica* as '*Y. enterocolitica* group', which does not differentiate the species from the generally non-pathogenic species *Y. kristensenii*, *Y. frederiksenii* and *Y. intermedia*, whereas, certain phenotyping kits (API20E and RapID onE) identify *Y. enterocolitica* as a separate species, but cannot distinguish between pathogenic, and non-pathogenic strains. One unusual trait of *Y. enterocolitica* is that it degrades pectin, and Yersinia pectin agar can be used to isolate the organism from stool samples. *Yersinia pestis* and *Y. pseudotuberculosis* can be distinguished from *Y. enterocolitica* by negative reactions of ornithine decarboxylase and sucrose fermentation. However, distinguishing *Y. enterocolitica* from *Y. bercovieri* and *Y. mollaretii* is more problematic since the later two used to be biotypes of *Y. enterocolitica*.

21.2.5 Control procedures

Control procedures for foodborne *Y. enterocolitica* are similar to those for other zoonotic pathogens; reduction of levels of contamination in the pre-process food chain, destruction by thermal processing and adequate domestic cooking and prevention of recontamination. *Yersinia enterocolitica*, however, is generally considered to be much less important a cause of human morbidity than *Campylobacter* or *Salmonella* and relatively little effort has been put into specific precautions against the organism.

Reduction of contamination in the pre-process food chain

A high incidence of *Y. enterocolitica* carriage in pigs has been associated with particular agricultural practices, especially 'buying in' of pigs to the farm. This system is dictated by economic circumstances and is likely to continue. Some control of carriage rates can be exerted by good hygiene, and it is common practice to quarantine incoming pigs before entry into a herd. Carriage of *Y. enterocolitica*, however, is long term and, usually, asymptomatic and the scope for control of infection at this stage is limited. Control by vaccination is a possibility, but introduction seems unlikely at present. Incidence of carriage can be modulated by a number of factors, including feeding of growth-promoting antimicrobials (Asplund *et al.*, 1998).

Contamination of meat with zoonotic pathogens at a high level is often associated with poor slaughterhouse practice, and improvements in technology may reduce carriage (Andersen, 1988). Concern over *E. coli* O157:H7 has enhanced efforts to reduce carcass contamination by decontamination procedures. Many decontamination treatments have been developed, 'steam-vacuuming' being considered particularly effective (Corry *et al.*, 1995). Although some decontamination procedures are effective in reducing the incidence of pathogens, the possibility of enhanced growth if recontamination occurs is a general concern (Jay, 1996). Work with *Y. enterocolitica* on pork suggests little difference in growth rate on decontaminated meat (Nissen *et al.*, 2001), although longer shelf-life, resulting from reduction in the spoilage microflora could, potentially, permit a significant increase in numbers.

Although porcine carriage of *Y. enterocolitica* may involve colonisation of the gastrointestinal tract, the tonsils are more commonly involved. Control by improving standards of butchery has been proposed, with particular emphasis being placed on preventing contamination of pork mince. Although obviously desirable, the extent to which this can be achieved in industrial-scale butchery is questionable.

Elimination by processing

Yersinia enterocolitica is not unusually heat resistant and will be eliminated by adequate thermal processing, including high-temperature short-time pasteurisation of milk and generally applied processes for cooking of meat. Although there has been relatively little work on alternative processing methods in relation to the bacterium, it is unlikely that response will be significantly different from that of other members of the *Enterobacteriaceae*. Equally, generic measures taken to prevent recontamination after processing are likely to be effective against *Y. enterocolitica*. Contamination with *Y. enterocolitica*, however, is potentially serious owing to its ability to grow at 4 °C. Vacuum-packing offers no protection, and sensitivity to preservatives is similar to that of other *Enterobacteriaceae*, although sodium lactate, in combination with mild heat processing, has been found to be effective in control of *Y. enterocolitica* in *sous vide* foods (McMahon *et al.*, 1999). Carbon dioxide is also effective in limiting growth of *Y. enterocolitica* in modified atmosphere packed fish (Davies and Slade, 1995). Growth models for the survival and growth of *Y. enterocolitica* in foods according to temperature, pH, [NaCl], [sodium nitrite] and [lactic acid] have been published (Sutherland and Bayliss, 1994; Bhaduri *et al.*, 1995).

With respect to risk of contamination from food handlers, human carriage is known, although there is disagreement over the extent and significance. Figures suggesting a higher carriage rate than non-typhoid salmonellas have been quoted, but it is not clear if this is true, or convalescent, carriage. Precautions against contamination by human carriers are the same as those devised for *Salmonella* and include exclusion from work during acute illness, but return to be permitted after recovery, providing stools are well formed and personal hygiene good. Convalescent persons should not, however, be allowed to handle foods to be consumed without cooking until three consecutive stool samples have tested negative for *Y. enterocolitica*.

21.2.6 Future trends

The relative lack of information concerning causal relationships between yersioniosis and specific foods means that predicting future trends is difficult. It is probable that the number of cases will follow general trends, although these are likely to be biased by consumption of pork. Consumption of raw meat has gained some popularity in the USA as a supposed means of ensuring general well-being and also as 'therapy' for conditions including AIDS. Protagonists of raw meat consumption can only be described as irresponsible in the extreme and the practice can only lead to increased morbidity due to zoonotic pathogens, including *Y. enterocolitica*. There is also evidence of culinary fashion involving consumption of raw, or lightly cooked, meat in marinades. *Yersinia enterocolitica* is relatively sensitive to low pH, which reduces risk. In pork-eating countries, however the lack of knowledge of parasitic infections associated with the meat means that there is considerable resistance to consumption of meat which is less than fully cooked. In addition it is plausible that biotype 1A will be recognised as a foodborne infectious agent despite the lack of a mouse model (Howard *et al.*, 2006).

21.2.7 Further information

Cornelis, G. R. (1992) Yersiniae, finely tuned pathogens, in C. Hormaeche, C. W. Penn and C. J. Smythe (eds) *Molecular Biology of Bacterial Infection: Current Status and Future Perspectives*, Cambridge University Press, Cambridge, 231–65.

Robins-Browne, R. M. (2007) *Yersinia enterocolitica*, in M. P. Doyle, L. R. Beuchat and T. J. Montville (eds) *Food Microbiology, Fundamentals and Frontiers*, ASM Press, Washington, DC, 215–45.

21.3 Characteristics of the genus *Shigella*

21.3.1 Introduction

The genus *Shigella* is a member of the *Enterobacteriaceae* and conforms to the general characteristics of that group. The *Shigella* genus contains four main sub-groups differentiated by a combination of biochemical and serological characteristics, A (*Sh. dysenteriae*), B (*Sh. flexneri*), C (*Sh. boydii*) and D (*Sh. sonnei*). The genus is non-motile and is relatively inert in tests for carbohydrate fermentation and other biochemical properties. *Shigella* has been thought of as a biochemically inactive variant of *E. coli* and shares some common antigens. In fact, with the exception of *S. boydii* type 13, the *Shigella* species can be regarded as virulent classes of *E. coli* that diverged between 35 000 and 270 000 years ago.

Shigella is highly host-adapted and usually infects only humans and some other primates, although a few infections of dogs have been reported. Humans are usually the main reservoir of infection and person-to-person transmission is the most common route of infection. The organism is also transmitted by contaminated water and food. There is a common conception that water is the more important vehicle and that food is of relatively little significance. Public health data suggest, however, that food is more important than water (ICMSF, 1996). The infectious dose of these bacteria is low, ranging from $1–10^4$ cells (Lampel and Maurelli, 2007).

Shigellosis is common in countries where hygiene standards are low. The infection is less common elsewhere, although there may be importation by travellers. In the UK, 966 cases were reported in 2000, falling from a peak of 18 069 in 1992, whereas the US CDC using data from 1982 to1997 estimated there were approximately 450 000 shigellosis cases, making it the third most common cause of bacterial illness. The World Health Organization has calculated that there are in the order of 165 million shigella cases per year (Kotloff *et al.*, 1999). The incidence is greater in developing countries than industrialised countries; 163 million and 1.5 million cases, respectively. Additionally, in developing countries the mortality rate is 1.1 million deaths per year, of which the majority are in children under five years of age.

21.3.2 Nature of *Shigella* infections

Classic *Shigella* (bacillary) dysentery is characterised by frequent passing of liquid stools, which contain blood, mucus and inflammatory cells. The incubation period is usually 12–50 h; onset is rapid and accompanied by fever and severe abdominal pain. Symptoms usually last 3–4 d, but can persist for 14 d, or longer. In healthy adults, death is rare, but *Shigella* dysentery is a major cause of death among infants in countries where hygiene is poor. Bacteraemia is an unusual symptom, being most common amongst persons aged less than 16 years, but also occurring in compromised persons, the death rate approaching 50 %.

Infections with *Sh. dysenteriae* almost always develop full and severe symptoms of dysentery. Similar symptoms, although often less severe, can also be associated with *Sh. boydii* and *Sh. flexneri*. Most adult infections by these species, however, and virtually all by *Sh. sonnei* do not progress beyond relatively mild, non-bloody diarrhoea. Symptoms may differ in young children and be of greater severity, possibly involving extraintestinal symptoms, including convulsions, headaches and delirium.

Shigella is primarily an invasive pathogen, the primary site of invasion being the colon. Cells multiply in the lumen and penetrate the colonic epithelium by receptor-mediated endocytosis. Lesions form in the gastric mucosa, covered by a pseudomembrane composed of polymorphonuclear leukocytes, bacteria and cell debris in a fibrin network. Entry into the lamina propria probably results in formation of ulcerative lesions and the appearance of bloody, mucosal stools. The mechanism of ulceration is cell death caused by accumulation of metabolic products and release of endotoxin.

All shigella contain a ~220 kb plasmid, called the virulence or invasion (pINV) plasmid which carries a 32 kb pathogenicity island. This is composed of 38 genes on an operon *ipa-mxi-spa*. The *ipa* genes encode a series of effector proteins that are delivered into the host cell through a type III secretory system. Other virulence factors are encoded on the bacterial chromosome, such as aerobactin, and genomic islands for multiple antibiotic resistance. Many shigella produce two iron-regulated enterotoxins which have roles in the symptoms of hemorrhagic colitis and haemolytic ureamic syndrome. Invasive strains of *Sh. dysenteriae* Type 1 produce high levels of a heat-sensitive cytotoxin, commonly designated Shiga toxin. The toxin has a molecular weight of *c.* 70 kDa and consists of two polypeptide sub-units, A and B, which are combined in the ratio 1:5.7. The B sub-unit mediates binding of the toxin to surface receptors on target cells. The A sub-unit enters cells by endocytic transport, binding to 60S ribosomes and inhibiting protein and DNA synthesis, leading to cell death (O'Brien and Holmes, 1987). There is evidence for enterotoxic and neurotoxic activity as well as cytotoxic, but the mechanism is not fully understood. Although only invasive strains of *Sh. dysenteriae* 1 produce large quantities of Shiga toxin, low levels of a Shiga-like toxin are produced by some strains of *Sh. flexneri* and *Sh. sonnei*.

Full expression of virulence by *Shigella* requires at least three genetic determinants, but the large plasmid, present in all virulent shigellas and enteroinvasive strains of *E. coli*, determines invasive ability. A second small (6 kDa) plasmid is also carried by virulent shigellas and codes for O antigen production, but not invasion (Watanabe and Timmis, 1984), while plasmid genes are also involved in early blockage of respiration in infected cells (Scotland, 1988). Chromosomal genes are involved in stability of the small plasmid and Shiga toxin production is probably chromosomally mediated (Pal *et al.*, 1989).

21.3.3 Risk factors

Individual susceptibility to *Shigella* infection varies according to age and the presence of predisposing conditions. Infection is rare, however, in neonates and very young children owing to lack of receptor sites on colon epithelial cells.

Shigella is not an environmental organism, and the overwhelming source of contamination of foods is humans. The organism is readily destroyed by processing and foods involved are invariably consumed without cooking, or processing, after contamination. Any food may, in principle, be contaminated and a wide range has been implicated as vehicles of shigellosis (Table 21.3). Recent outbreaks have included fresh parsley and layered bean dip (CDC, 1999, 2000). In the former outbreak, investigations revealed that the water used for chilling the parsley was unchlorinated and susceptible to bacterial contamination. A number of people became ill with *Sh. sonnei* after consuming (uncooked) chopped parsley. In the bean dip outbreak, the product was in five layers, one of which was cheese which was contaminated with *Sh. sonnei.*

In practice, foods that receive significant handling during preparation are of greatest risk where contamination is direct from handlers. However, the bacterium can survive for extended periods on foods, and contamination of salads, etc. can occur before harvest from soil, water, etc., containing infected faecal material. In either case, risk is greatest where hygiene standards are poor, and the incidence of *Shigella* infection is high among the general population. In developed countries, shigellosis can be associated with localised failure of hygiene, leading to sporadic outbreaks in schools, military establishments, summer camps, religious communities, etc. A very large outbreak in the USA involved more than 1300 culture-confirmed cases of *Sh. sonnei* infection

Table 21.3 Foods implicated as vehicles for *Shigella* infection

Salads, various	Spaghetti
Tuna salads	Shrimp cocktail
Lettuce	Potato salad
Milk	Mashed potato
Soft cheese	Chocolate pudding
Cooked rice	Stewed apples

during 1986–87, most of the affected being tradition-observant Jews. In developed countries, shigellosis is also a disease of poverty and in the USA and Australia, the urban poor, migrant workers and native peoples are at a continuing high risk (Varnam and Evans, 1996).

21.3.4 Detection methods

It is not general practice to examine foods for *Shigella* in non-outbreak situations. Examination for the organism on a regular basis is unlikely to be of any benefit in safety assurance and is a diversion of resources from management of critical control points.

Isolation is difficult owing to relatively poor growth on commonly-used media, especially when other members of the *Enterobacteriaceae* are present. A number of media have, however, been reformulated to improve recovery of *Shigella* spp., although use is primarily with clinical samples. Direct plating of food is unlikely to be successful, but little attention has been given to enrichment media. Direct plating of stools is preferable to the use of enrichment broths. Other than Hajna's Gram-negative broth, enrichment broths are generally designed for the isolation of salmonella, and do not enable the growth of shigellae and similar organisms. Selenite broth and Gram-negative (GN) broth have been used, in combination with selective plating. *Salmonella–Shigella* agar, deoxycholate–citrate agar, xylose lysine–deoxycholate agar, various MacConkey agars and eosin–methylene blue agar have all been used. Increased interest in foodborne shigellosis in the 1980s led to attempts to develop new media. The most promising approach has been supplementation of various media with novobiocin. Despite this, the scope for further improvement of traditional, cultural methods appears limited.

There is notable variation in phenotypes between *Shigella* species and types. *Shigella sonnei* is the most common species and can be easily distinguished from other shigella as it is o-nitrophenyl-β-D-galactopyronosidase (ONPG)- and ornithine decarboxylase-positive. Most shigellae ferment mannitol, except *Sh. flexneri* types 4 and 6, *Sh. boydii* type 14, and *Sh. dysenteriae* which are mannitol-negative.

Both serological and genetic approaches have been used in attempts to develop alternative methods for detection of *Shigella* in foods. The major concern has been detection of *Shigella* spp. in stool samples and, while possibly useful, none is fully satisfactory with foods. Some effort has been made with the fluorescent antibody technique, but non-specific reactions limit use. A more promising method, the Bactigen® slide agglutination test, separately detects *Salmonella* and *Shigella*, but has not been fully validated for foods. A further approach is use of enzyme-linked immunosorbent assay (ELISA) to detect the virulence marker antigen of virulent shigellas and enteroinvasive *E. coli*, or Shiga toxin itself, but use in foods has been limited. DNA hybridisation methods using invasion-essential gene segments as probes are successful with virulent strains, but problems occur due to loss of the plasmid and selective deletion of invasion associated genes during cultivation and storage of isolates. A preferred approach is detection of the *ipaH* gene sequence which is present on both plasmid and chromosome (Venkatesan *et al.*, 1989). Invasive strains of *Shigella* may also be detected in foods using PCR-based assays (Vantarakis *et al.*, 2000; Lampel, 2005). None is available commercially.

21.3.5 Control procedures

The epidemiology of foodborne *Shigella* infections is such that control procedures must be directed against contamination from human sources. One of the main features of foodborne *Shigella* outbreaks is that contamination occurrs via a food handler rather than the production or processing plant. Shigellosis in communities is difficult to control, owing to a high incidence of person-to-person infection, multiple exposure to the bacterium and significant secondary spread. Recognition that a problem exists is often considered the first step in control (ICMSF, 1996), accompanied by hygiene education and measures such as supervised hand washing by children.

With respect to shigellosis associated with contamination of particular foods, the critical control points are prevention of contamination of salads and other crops eaten without cooking and control of contamination during handling, especially after heating or other lethal processing has been applied. In contrast to previous opinion that shigellas are delicate and have only limited survival capability in foods, the organism can survive refrigerated and frozen storage on a variety of foods. Illness may therefore occur at a time remote from the point of contamination. At the level of primary agriculture, risk has been associated with use of nightsoil as manure, or irrigation with polluted water. Leafy salad vegetables, such as lettuce, are most commonly involved, and washing is unreliable even where extensive use is made of disinfectants.

Contamination during handling almost invariably involves direct human contact, although there has been one, apparently authentic, case of *Sh. flexneri* infection caused by a monkey touching a child's ice cream (Rothwell, 1990). Control involves minimising hands-on procedures, exclusion of potential excretors of *Shigella* from handling food and ensuring good standards of personal hygiene. Several outbreaks of shigellosis attributable to contamination by food handlers have been described by Smith (1987). These included an outbreak of *Sh. boydii* infection, which affected 176 persons following a pasta meal. The situation can be complicated by atypical symptoms making shigellosis difficult to recognise. The greatest risk is from convalescent excretors and return to work must be carefully monitored, three consecutive negative stool samples usually being required.

21.3.6 Future trends

In countries where *Shigella* infection is endemic, only improvement in standards of hygiene and provision of clean water supplies will alleviate the problem. It is possible that climate change leading to reduction in water supplies will exacerbate problems and that areas of endemic infection will increase. This is likely to be a general pattern involving other enteric pathogens, such as *Vibrio* spp., as well as *Shigella*. There are implications for areas outside those that may be directly affected by climate change. These include a higher incidence of imported shigellosis amongst returning travellers and contacts. There may also be a general food-associated increase, if hygiene standards fall in countries supplying salad crops and fruit.

21.3.7 Further information

Dupont, H. I. (1995) *Shigella* species (bacillary dysentery), in G. L. Mandel, J. E. Bennett and R. Dolin (eds) *Principles and Practices of Infectious Diseases*, Churchill Livingstone, Edinburgh, 381–94.

Lampel, K. A. and Maurelli, A.T. (2007) *Shigella* species, in M. P. Doyle, L. R. Beuchat and T. J. Montville (eds) *Food Microbiology, Fundamentals and Frontiers*, ASM Press, Washington, DC, 247–61.

21.4 Characteristics of the genus *Vibrio*

21.4.1 Introduction

The genus *Vibrio* comprises Gram-negative, oxidase-positive, facultatively anaerobic rods. Differentiation from *Aeromonas* may be made on the basis of the sensitivity of most vibrios to the vibriostat O/129. Cells are often curved, or comma-shaped, and motile by a characteristic sheathed, polar flagellum. *Vibrio* species are frequently isolated from estuarine waters and are therefore associated with a great variety of fish and seafoods. There are 20 recognised species of *Vibrio*, of which 12 are pathogenic to humans; *V. cholerae*, *V. parahaemolyticus* and *V. vulnificus* being of primary concern. *Vibrio cholerae* is the causative agent of cholera, and is one of the few foodborne diseases which has both epidemic and pandemic potential. Seafood is the most common vehicle of foodborne infection, although water is historically associated with *V. cholerae* infection. *Vibrio vulnificus* is also an important cause of wound infections.

Enteropathogenic *Vibrio* species are generally found in greater numbers in warm waters and may show a marked seasonal variation in occurrence, which correlates with temperature (Varnam and Evans, 2000). The genus is among those able to enter the putative viable non-recoverable (VNR) state (Oliver *et al*., 1995), which can complicate the epidemiology of infections.

Foodborne infections with *Vibrio* spp. are most common in Asia. Infections are rare in Europe and usually associated with imported foods, or travellers returning from affected areas. Not all US states report *Vibrio* infections, but 30–40 cases of *V. parahaemolyticus* infection occur each year in Gulf Coast states, such as Texas and Alabama. *Vibrio vulnificus* infection is less common, *c.* 300 cases being reported between 1988 and 1995. Carriage of the organism by infected humans is important in the disease transmission, as water can become contaminated by faecal matter. Ingestion of water containing *V. cholerae* or of foods that are washed with contaminated water, but not disinfected, can lead to widespread transmission of cholera.

21.4.2 *Vibrio cholerae, Vibrio parahaemolyticus, Vibrio vulnificus*: infection and epidemiology

The three main enteropathogenic species are distinguishable on the basis of phenotypic properties, have a distinct ecology within the aquatic environment and cause illness of different types and severity.

Vibrio cholerae

The cause of serious human disease since antiquity, *V. cholerae* is the causative agent of Asiatic cholera. The current cholera pandemic is the eighth and the disease is present in many parts of the world. Cholera is classically a waterborne disease, but food is also an important vehicle, especially in countries with a sea-coast. Classic Asiatic cholera is caused only by strains of *V. cholerae* serogroups O1 and O139, which also produce cholera toxin. Serogroup O1 has two biotypes, classic and El Tor, and three serotypes, Inaba, Ogawa and Hikojima. The incubation period of cholera varies from 6 h to as long as 3 d. Although initial symptoms may be mild, the infection rapidly progresses to copious diarrhoea ('rice water stools'), with rapid loss of body fluids and mineral salts, especially potassium. Fluid loss is up to 1 l/h, resulting in dehydration, hypertension and salt imbalance. Where treatment is unavailable, death may be rapid, but recovery in 1–6 d is normal where rehydration and salt replacement are possible.

Other illnesses are caused by *V. cholerae* other than O1 and O139, and by strains of these serotypes which lack the ability to produce cholera toxin. Symptoms are usually milder, but severity varies and can approach that of cholera. Gastroenteritis produced by *V. cholerae* other than O1 and O139 is characterised by diarrhoea, which is bloody in *c.* 25 % of cases, abdominal cramps and fever *(c.* 70 %). Nausea and vomiting accompany diarrhoea in *c.* 20 % of cases. Illness caused by non-toxigenic strains of O1 and O139 usually involves diarrhoea only. Mild diarrhoea is also caused by a separate species within the *V. cholerae* group, *V. mimicus*.

Vibrio cholerae adheres to the surface of the small intestine, little being known of the mechanism of adhesion, although there is an important role for cell-associated haemagglutinins. Cells grow, producing cholera toxin, a true enterotoxin, which is an oligomeric protein (MW 84000), composed of two sub-units, A1 (MW 21000) and A2 (MW 7000), which is linked by covalent bonds to five B sub-units (MW 10 000). The B sub-units bind to a receptor ganglioside in mucosal cell membrane, while the A1 sub-unit enters the cells, irreversibly activating adenylate cyclase, leading to increased intracellular levels of cyclic AMP, and a biochemical reaction cascade, resulting in hypersecretion of water and salts. The role of the A2 sub-unit is probably in mediating entry into the cell.

Other toxins are produced by *V. cholerae*, although their significance is uncertain. These include heat-stable (ST) and heat-labile (LT) enterotoxins, a Shiga-like cytotoxin, a haemolysin similar to that of *V. parahaemolyticus* and an endotoxin. This is produced in significantly greater quantities by non-toxigenic strains. At least some of the pathogenic mechanisms of *V. cholerae* are shared by *V. mimicus*.

Vibrio parahaemolyticus

Vibrio parahaemolyticus causes a predominantly diarrhoeal syndrome. Typical symptoms are diarrhoea, abdominal cramps and nausea. Vomiting occurs in *c.* 50 % of cases and there may be a headache. Mild fever and chills occur in *c.* 25 % of cases. Onset is usually after 4–24 h, although longer and shorter periods have been reported. Symptoms usually abate within 2–3 d and death is very rare. A second, more severe, dysenteric form has been described (Oliver and Kaper, 2007), which is characterised by bloody, or mucoid, stools. Human pathogenicity of *V. parahaemolyticus* usually correlates with presence of direct, thermostable haemolysin activity (Kanagawa reaction). Kanagawa-negative strains can be a cause of diarrhoea, however, and are important outside tropical countries.

Vibrio parahaemolyticus differs from *V. cholerae* in being invasive and in not producing a classic enterotoxin. Much attention has been paid to the role of the direct, thermostable haemolysin (Vp-TDH), but relatively little is known of other aspects of pathogenicity. Both flagella and cell-associated haemagglutinins are involved in adhesion to the epithelial cells before entry and penetration to the lamina propria. The direct, thermostable haemolysin promotes *in vitro* vascular permeability and has enterotoxin properties, but its precise role in induction of diarrhoea is unknown. Other toxins may be produced, including a Shiga-like cytotoxin, which may be involved in the less common dysenteric form of infection. Pathogenicity in these strains is associated with production of a second haemolysin, Vp-TDH related haemolysin.

Vibrio vulnificus

Foodborne *V. vulnificus* infections are almost invariably associated with underlying medical conditions (see below). Although similar to *V. parahaemolyticus*, *V. vulnificus* differs from this, and other enteropathogenic vibrios, in having a septicaemic rather than an enteric human pathology. In healthy persons, infection may involve no more than mild and self-limiting diarrhoea, but most cases involve primary septicaemia with fever, chills and, in many cases, nausea. Hypotension occurs in *c.* 33 % of cases and *c*. 66 % of patients develop skin lesions on extremities and trunk. The normal incubation period of *V. vulnificus* infection is 16–48 h, septicaemia progressing rapidly and being difficult to treat. Mortality is 40–60 %.

Invasion occurs rapidly after adhesion and is accompanied by extensive tissue damage. The organism has a number of virulence factors which protect against host defences. These include cell-associated factors protective against normal serum and complement-induced lysis and a surface component (capsule) which is antiphagocytic and anticomplementary. *Vibrio vulnificus* also produces active siderophores, which facilitate iron acquisition and which are associated with enhanced virulence when blood iron levels are high.

A number of products may be responsible for tissue damage. At least two haemolysins, which possess cytotoxic activity, are produced but, while contributing to virulence, these are not essential determinants. Specific proteinases, collagenase and phospholipases are also produced, which affect vascular permeability and cause tissue destruction.

21.4.3 Risk factors

Vibrio cholerae

Cholera is a disease of poverty, and the malnourished and individuals at extremes of age are at particularly high risk. Genetic factors are also involved and persons of blood group O are more highly susceptible than those of other groups. Low stomach acidity, due to over-medication with antacids, or to clinical factors, also increases susceptibility to *V. cholerae* infection. People with blood group O are more susceptible to severe cholera. However, they are less likely to become infected with *V. cholerae* O1 (Harris *et al.*, 2005). This protection, however, does not extend to *V. cholerae* O139. The mechanism for this protection is unknown, but may be related to the interference by A and B histo-blood group glycoconjugates with the binding of cholera toxin to ganglioside receptor on the intestinal cells.

In countries where cholera is endemic, contaminated water, used for drinking or food preparation, is the most common vehicle of infection. *Vibrio cholerae* O1 can become established in the environment, which was probably the origin in an outbreak involving crabmeat in Louisiana (Blake *et al.*, 1980). Most outbreaks, however, have been attributed to direct sewage pollution. Many have involved molluscs and, to a lesser extent, other shellfish and crustaceans, but outbreaks have been caused by other foods, including lettuce. An outbreak of cholera involving bottled mineral water in Portugal was attributed to contamination of the source with sewage-contaminated river water, which had percolated through sub-surface strata. Despite these examples, shellfish can be a high-risk food, especially when consumed raw.

Vibrio parahaemolyticus

Little is known of individual variations in susceptibility to *V. parahaemolyticus*, with the exception of a higher risk of infection when stomach acidity is low. Isolation of *V. parahaemolyticus* is made only occasionally from foods other than those of marine origin. Isolation can be made at any time in warm climates, but is normally a problem only in summer in temperate countries. In Europe and the USA, there is an enhanced risk of infection associated with consumption of raw molluscs (oysters, clams). In warmer areas, such as India, Japan and Africa, a similar association exists with raw fin fish. *Vibrio parahaemolyticus* infection is also associated with cooked seafood, especially crab and shrimp. This is not a consequence of inherent risk, but of poor hygiene, including the practice of cooling cooked fish with untreated sea water (Varnam and Evans, 1996).

Vibrio vulnificus

There is a strong association between invasive *V. vulnificus* infections and underlying medical conditions. Underlying predisposing conditions include chronic cirrhosis, hepatitis, thalassemia major and haemachromatosis, and there is often a history of alcohol abuse. Less commonly, *V. vulnificus* infections occur where there are underlying malignancies, gastric disease, including inflammatory bowel disease and achlorhydria, steroid dependency and immunodeficiency. Males are markedly more susceptible than females and account for over 80 % of infections (Oliver, 1989).

Vibrio vulnificus infections are almost uniquely associated with consumption of raw oysters. The incidence of infection follows a seasonal distribution and numbers of the bacterium may be high in oysters when water temperatures are > 21 °C. Isolation, either from water or fish, is rare when water temperature is below 10–15 °C. Geographical factors are also involved, however, and while, during summer months, *V. vulnificus* is readily isolated in significant numbers in California, for example, the incidence is low in the Spanish Mediterranean sea, probably because of the high salinity value (Arias *et al.*, 1999).

Other Vibrio species

A number of other *Vibrio* species causing human illness have been reported periodically. Their reservoirs are normally marine and estuarine water, marine molluscs, crustaceans and sediments. These are generally related to the consumption of contaminated seafood, but little information is known regarding their modes of infection. In general they are less common and less severe than *V. cholerae*, *V. vulnificus* and *V. parahaemolyticus*. Although *V. alginolyticus* is possibly the most frequently isolated *Vibrio* species from seafoods, it has rarely been linked to foodborne illness. Most cases are immunocompromised individuals, and alcohol abuse may be a predisposing factor. *Vibrio fluvialis* and *V. furnissi* were initially considered as different biogroups of the same species. *Vibrio fluvialis* cases have symptoms of vomiting and diarrhoea, frequently with bloody stools. Outbreaks have been linked to the consumption of raw oysters, and shrimps. A few outbreaks due to *V. furnissii* have been reported, but the cause has seldom been identified. An outbreak in 1969 was linked to shrimp and crab salad, or the cocktail sauce which was additionally served. Otherwise outbreaks have rarely been reported, let alone investigated. *Vibrio hollisae* infections have been linked to the consumption of raw seafood: oysters, clams, crabs and shrimp. Symptoms include severe abdominal cramping, vomiting, fever and watery diarrhoea. This organism is unable to grow on MacConkey agar and could be missed in routine laboratory analysis. *Vibrio mimicus* has been associated with the consumption of raw oysters, leading to diarrhoea, nausea, vomiting and abdominal cramps in most cases.

21.4.4 Detection methods

Laboratory testing for enteropathogenic *Vibrio* plays an important part in control of foodborne infection. In many situations determination of numbers in the water of the growing beds is of greatest importance. Under most conditions, numbers are small ($< 10^3$/ml), although significantly higher numbers can be present if water temperatures are very high.

Conventional cultural methods are well-established for enteropathogenic vibrios. Enrichment is usually necessary, and the relatively non-selective alkaline peptone water is widely used. This medium is not always effective, especially in isolation of *V. vulnificus*, if numbers of competing bacteria are high and supplementation with polymixin B can improve performance (O'Neill *et al.*, 1992). Several, more selective, enrichment media have been developed, including the widely used glucose–salt–teepol broth (Beuchat, 1977), although the value of these is not universally accepted.

Thiosulphate–citrate–bile salts–sucrose (TCBS) agar is the most widely used selective plating medium for the common enteropathogenic species. *Vibrio cholerae* (sucrose fermenting) may be distinguished from *V. parahaemolyticus* and *V. vulnificus* (sucrose non-fermenting), but other bacteria may be able to grow. Confirmatory testing is, therefore, required. Commercial identification kits primarily intended for the *Enterobacteriaceae* (e.g. API 20E) have been used with success, although Dalsgaard *et al.* (1996) found the API 20E to be unsatisfactory in identification of *V. vulnificus* from the environment.

A particular problem with *V. parahaemolyticus* is that most isolates from water and foods *(c.* 98 %) are non-pathogenic. Where specific information is required, the Kanagawa test may be required, involving determination of β-haemolysis on Wagatsuma agar (Oliver and Kaper, 2007).

Rapid genetic methods, mostly based on PCR and colony hybridisation techniques, have been developed for each of the main enteropathogenic species of *Vibrio*, including toxigenic *V. cholerae* O1 (Shangkuan *et al.*, 1996). None of these has been applied on a large scale and none is commercially available.

21.4.5 Control procedures

A number of risk assessments have been undertaken concerning *Vibrio* species and seafood. The documents by FAO/WHO and CFSAN (FDA, 2005) are extensive and only general summaries are given here. Of particular interest are two by FAO/WHO, numbered 8 and 9 in their MRA series. Issue 8 is entitled '*V. vulnificus* in raw oysters'. However, it is composed of five separate risk assessments, three of which concern *V. parahaemolyticus*. These are:

- *V. parahaemolyticus* in raw oysters harvested and consumed in Japan, New Zealand, Australia, Canada and the United States of America;
- *V. parahaemolyticus* in finfish consumed raw;
- *V. parahaemolyticus* in bloody clams harvested and consumed in Thailand;
- *V. vulnificus* in raw oysters harvested and consumed in the United States of America;
- choleragenic *V. cholerae* O1 and O139 in warm-water shrimp in international trade.

The second FAO/WHO risk assessment (MRA 9, FAO/WHO, 2005a) is entitled 'Choleragenic *Vibrio cholerae* O1 and O139 in warm-water shrimp in international trade', and therefore links with MRA 8 (FAO/WHO, 2005b).

Vibrio cholerae

In most circumstances, *V. cholerae* is not present in the food production and processing environment and no specific measures are required. In general, low temperature retards *Vibrio* multiplication, and thorough heating of shellfish will kill any *Vibrio* present. As a general precaution applying to all enteropathogenic vibrios, however, harvesting of shellfish from inshore waters should cease when water temperatures exceed 25 °C, unless full processing is applied.

Where cholera is endemic, it is necessary to break the cycle of infection by protecting food from sewage contamination. This means that foods that are to be eaten without full processing should be harvested only where absence of contamination can be guaranteed. In the case of cooked foods, precautions against recontamination must be taken. Cooked crabmeat, which is picked by hand, is a particular risk, and a secondary pasteurisation to 80 °C internally is recommended. The bacterium is also highly radiation-sensitive and irradiation may be preferred where heat-induced changes are undesirable, or for secondary pasteurisation.

In areas of endemic cholera, risk of contamination is low, but constant. Regular bacteriological testing of waters of shellfish beds and less frequent checks on shellfish offer significant, but not absolute, protection. Bacteriological testing is also required for water used in food processing and for bottled water. Bacteriological testing should be used in conjunction with general precautions, including control of shellfish growing beds, application of Hazard Analysis Critical Control Point (HACCP) in food processing operations and education of food handlers. It is likely to be necessary to apply regulatory control to prevent unsafe food entering the food supply system. Control of unofficial food supply can be difficult where shortages exist.

Vibrio parahaemolyticus

Control of *V. parahaemolyticus* at source can have some effect on incidence of infection, and harvesting from obviously polluted beds should be prevented. Depuration or relaying in clean water is used for reducing levels of enteric pathogens in oysters and other shellfish. Properly controlled, this procedure removes loosely attached sewage-derived bacteria, but is of less efficiency with *V. parahaemolyticus*, which is a normal inhabitant of the intestinal tract of some shellfish. Under most circumstances, however, *V. parahaemolyticus* is unlikely to be present in sufficient numbers to cause illness, unless growth in the food occurs. Refrigeration to below 5 °C offers a significant level of control, and such temperatures should be maintained at all stages of handling of shellfish, such as oysters and clams, and dishes, such as sushi, which are eaten raw. *Vibrio parahaemolyticus* is readily destroyed by heating, death occurring at temperatures as low as 47 °C.

Recontamination must be prevented by good practice. Growth on cooked fish is very rapid at temperatures above 20 °C and efficient cooling to below 5 °C is necessary if the product is not to be consumed within 2 hours. The US FDA (2005) risk assessment listed the risk mitigation strategies as (1) cooling oysters immediately after harvest and being kept refrigerated, during which time the viability slowly decreased, (2) mild heat treatment (5 min, 50 °C) giving a > 4.5 log decrease in viability and almost eliminating the likelihood of illness, and (3) quick-freezing and frozen storage giving a 1–2 log decrease in viability and hence reducing likelihood of illness.

Yamamoto *et al.* (2008) in their risk assessment of *V. parahaemolyticus* in bloody clams estimated a mean number of times a person would become ill of 5.6×10^{-4}/person/year. Increased illness was particularly likely for the portion of people who did not properly boil the clams. Phillips *et al.* (2007) combined the US FDA risk assessment with sea surface temperatures to predict the numbers of *V. parahaemolyticus* in oysters and therefore assist in risk management decisions.

Vibrio vulnificus

Infection is a significant problem only in the restricted circumstances of raw oyster consumption by susceptible persons. A high level of complete control could, therefore, be achieved by ensuring that such individuals do not eat raw oysters. This requires extensive and continuing public education, including warnings on packaging, and is unlikely to be entirely successful. Risk can be reduced by suspending harvesting when water temperatures are above 25 °C, rapid chilling to less than 15 °C and maintaining strict temperature control during storage at below 5 °C. Further measures, including fishing only at cool times of day, harvesting from deeper, colder waters and protecting catch from sunlight have been described by Brenton *et al.* (2001). *Vibrio vulnificus* is heat-sensitive, but cooking is not acceptable to those who insist on consuming oysters raw. Post-harvest irradiation at 1.0–1.5 kGy, which would kill *V. vulnificus*, but not the oyster, has been proposed, but is unlikely to be applied.

21.4.6 Future trends

Within the enteropathogenic vibrios as a whole, an increase in the incidence of infection may be predicted as a consequence of climate change. Higher water temperatures are likely to mean an increase in numbers of vibrios, especially *V. parahaemolyticus* and *V. vulnificus*, and a higher rate of infection in areas where a significant problem already exists. At the same time, warming of seas in temperate regions is likely to introduce the problem to areas where non-imported infections are rare. Problems are likely to be compounded by increasing salinity in some waters and by a general shortage of clean water for washing, etc. In the specific case of *V. cholerae*, there is a well-recognised relationship between drought and epidemic cholera (Fleurat, 1986). It is, therefore, possible that the disease will spread to, and become established in, regions where it is currently rare. These include southern European countries, such as Spain and Portugal.

The strong association of illness caused by *V. parahaemolyticus* and *V. vulnificus* with consumption of raw shellfish means that dietary fashion may directly affect the incidence of infection with these organisms. A factor in the rise of *V. vulnificus* infections in California and elsewhere in the USA was the increased popularity of the 'raw bar' (Blake, 1983). Shellfish bars are becoming increasingly popular in the UK, although the risk of *Vibrio* infection is low, owing to the very low incidence in fish caught in UK waters.

21.4.7 Further information

Elliott, E. L., Kaysner, C. A., Jackson, I. and Tamplin M. L. (1995) *Vibrio cholerae*, *Vibrio parahaemolyticus*, *Vibrio vulnificus* and other *Vibrio* spp., in *Bacteriological Analytical Manual*, 8th edn, Association of Official Analytical Chemists, Washington, DC, 9.01–9.27.

Oliver, J. D. and Kaper, J. B. (2007) *Vibrio* species, in M. P. Doyle, L. R. Beuchat and T. J. Montville (eds) *Food Microbiology, Fundamentals and Frontiers*, ASM Press, Washington, DC, 263–300.

21.5 Characteristics of the genera *Aeromonas* and *Plesiomonas*

21.5.1 Introduction

The genus *Aeromonas* comprises Gram-negative, oxidase-positive, facultatively anaerobic rods. These are insensitive to the vibriostat O/129, have no requirement for Na^+ and further differ from vibrios in producing gas during sugar fermentation (most strains). There is considerable phenotypic variation among *Aeromonas*. Gas production is variable and may be temperature-dependent, while there can also be marked differences in cellular morphology.

Aeromonas hydrophilia is present in all freshwater environments and in brackish water. It has frequently been found in fish and shellfish and in market samples of red meats (beef, pork, lamb) and poultry. The organism is able to grow slowly at 0 °C. It is presumed that, given the ubiquity of the organisms, not all strains are pathogenic. Two distinct groups of *Aeromonas* exist, the psychrophilic group represented by *A. salmonicida*, a fish pathogen, rarely isolated in a free-living state, and the mesophilic (more correctly psychrotrophic) genera, which are usually motile. The genus has undergone taxonomic reassessment in recent years and now comprises ten phenotypic species, of which eight have been associated with human enteric disease (Table 21.4). The *Aeromonas* species now form a separate Family Aeromonadaceae.

The taxonomy of *Aeromonas* is complex and has undergone many changes. The genus has been refined by DNA–DNA hybridisation and now comprises of 14–17 genospecies (DNA hybridisation groups: HGs). Eight HGs have been associated with human illness. Some species of *Aeromonas* are pathogens of fish and reptiles, while others are associated with human disease (Altwegg *et al.*, 1991; Deodhar *et al.*, 1991). Although epidemiological links are relatively easily established, these have not been confirmed with feeding trials, and direct causal links exist in only a few cases. Despite a probable involvement in foodborne disease, the status of *Aeromonas* remains as a putative enteropathogen.

In contrast to the taxonomy of *Aeromonas*, the *Plesiomonas* genus comprises a single species *Pl. shigelloides*, which is now classified in the Family Plesiomonadaceae. *Plesiomonas* shares characteristics with both *Vibrio* and *Aeromonas*, although it contains the *Enterobacteriaceae* common antigen. Like *Aeromonas*, it has been regarded as a putative enteropathogen, although a direct causal relationship has been established following outbreaks of illness associated with water and oysters (Wadström and Ljungh, 1991).

The primary reservoirs for *Pl. shigellosis* are aquatic ecosystems such as rivers and lakes. It can also be recovered from marine environments but at a lower frequency. Dogs and cats are also reservoirs of *Pl. shigelloides*, with a range of colonisation rates, 2–10 %. Farm animals carrying *Pl. shigelloides* include cattle, pigs and sheep. There have been a number of outbreaks in which *Pl. shigelloides* has been implicated as the causative agent, but not proven. Seafoods have been associated as the route of infection in all outbreaks.

Table 21.4 Phenotypic species of *Aeromonas*

A. caviae	*A. eucrenophilia*[a]
A. hydrophila	*A. jandei*
A. media	*A. salmonicida*[a]
A. schubertii	*A. sobria*
A. trota	*A. veronii*

[a] Fish pathogen.

The putative nature of *Aeromonas* as a pathogen means that data concerning incidence are not readily available. In the UK, 261 isolations were made from faecal samples during 2000, of which 144 were identified with *A. hydrophila* and 75 with unidentified species. In the same year 31 isolations of *Plesiomonas* from faeces were made.

21.5.2 *Aeromonas* and *Plesiomonas* as human pathogens

There is controversy as to whether *A. hydrophila* is a cause of human gastroenteritis. The uncertainty is because volunteer human feeding studies (10^{11} cell dose) have failed to demonstrate any associated human illness. However, its presence in the stools of individuals with diarrhoea, in the absence of other known enteric pathogens, suggests that it has some role in disease and has been included in this survey of foodborne pathogens. *Aeromonas* has been associated with both gastroenteritis and extraintestinal disease, usually in immunocompromised hosts. Traditionally, *A. hydrophila* has been most commonly involved, although descriptions of outbreaks have not always distinguished between this species as well as *A. caviae*, *A. sobria* and *A. veronii*. Most human pathogenic strains are now recognised as grouping in three genomic species, *A. hydrophila* HG 1, *A. caviae* HG 4 and *A. veronii* biovar sobria HG 8. As the role as an enteropathogen is not definitely established, descriptions of symptoms must be treated with care. Strains vary in pathogenicity, Kirov *et al.* (1994) postulating that this results from differences in the degree to which virulence factors can be expressed. Gastroenteritis associated with *Aeromonas* varies in severity from mild diarrhoea to a life-threatening, cholera-like illness (ICMSF, 1996). Two syndromes have been described. The first, which accounts for *c.* 75 % of cases is characterised by watery diarrhoea with mild, or absent, fever, possibly accompanied by abdominal pain and nausea. The second form is dysentery-like and characterised by bloody, mucoid stools. The incubation period is not known. Mild infections are self-limiting and recovery is complete within 1–7 d. In severe cases, symptoms are prolonged and may persist for a month or more, often requiring antibiotic therapy. The infectious dose of this organism is unknown. *Aeromonas hydrophila* may spread throughout the body in the bloodstream and cause a general infection in persons with impaired immune systems. *Aeromonas hydrophila* produces a cytotonic enterotoxin(s), haemolysins, acetyl transferase and a phospholipase.

Aeromonas is associated with a range of extraintestinal disease, of which septicaemia and meningitis are most common. Other symptoms include endocarditis, eye infections and osteomyelitis. Portal of entry may be the intestinal tract, and mortality can exceed 60 %.

Aeromonas is able to colonise the intestine and invade epithelial cells. Significant systemic spread, however, usually occurs only in the immunocompromised. The bacterium produces a number of putative toxins, although their relationship with illness in man is not fully resolved. Both a cytotonic enterotoxin, with genetic relatedness to cholera toxin, and a cytotoxic enterotoxin, with β-haemolysin activity, have been described (Varnam and Evans, 1996). The relative importance of the two enterotoxins is disputed, but it has been suggested that the toxins of *Aeromonas* act by facilitating colonisation of the intestine and/or invasion rather than directly inducing enteritis (Todd *et al.*, 1989).

Symptoms of *Pl. shigelloides* infection appear within 24–48 h and are predominantly diarrhoeal (94 %). Diarrhoea is accompanied by abdominal pain (74 %), nausea (72 %), chills (49 %), fever (37 %), headache (34 %) and vomiting (33 %). Recovery is usually within 1–9 d, but may be prolonged. Three types of diarrhoeal symptoms have been described, secretory, shigella-like and cholera-like. Secretory diarrhoea is most common and can vary greatly in severity. The shigella-like type is also potentially severe, with symptoms of prolonged duration, while the cholera-type illness is very rare.

Plesiomonas shigelloides is also a relatively uncommon cause of extraintestinal infection, usually where predisposing conditions are present. The intestine is the portal of entry in at least some cases and there may be prodromal enteritis. Meningitis is most common and has a mortality rate of *c.* 80 %.

Although *Pl. shigelloides* has a number of putative virulence factors, including ability to adhere to epithelial cells, invasive ability (possibly only in certain strains) and toxin production, direct relationships with pathogenicity have been difficult to demonstrate. Two distinct enterotoxins are produced and there is also evidence for endotoxin, protease, elastase and haemolysin production.

21.5.3 Risk factors

Greatest risk of *Aeromonas* infection involves persons with predisposing conditions, especially leukaemia and a deficient immune system. Persons with liver disease are also at higher risk. Males are generally more susceptible and account for *c.* 80 % of cases involving persons with underlying cancer. Among healthy persons, young children are at greatest risk. Most cases involve children aged six months to two years and there is a markedly lower incidence beyond five years. It has been suggested that, owing to reduced intestinal pH, breast-fed children are at lower risk than formula-fed with respect to *A. caviae* infection.

Aeromonas can be isolated from a wide range of foods and is capable of growth during storage at low temperature. Isolation has been made from plant and animal foods, and some strains are probably part of the normal spoilage microflora of meat, poultry and fish. Neyts *et al.* (2000), for example, recovered *Aeromonas* from 26 % of vegetable samples, 70 % of meat and poultry and 72 % of fish and shrimps at levels of 10^2–10^5 cfu/g. Despite variation between surveys, these results are probably typical. The significance in foods, in relation to human disease, is not known and many isolates may be non-toxigenic. The potentially pathogenic genospecies HG4 (*A. caviae* complex) was, however, isolated by Neyts *et al.* (2000), especially from vegetables. *Aeromonas* has been isolated from piped water supplies, as well as bored well water and bottled mineral water (Slade *et al.*, 1986; Krovacek *et al.*, 1989). The bacterium can be present in high numbers in river water and there is a risk in any water abstracted from such sources for human use (Schubert, 1991), although survival rates have been reported to be highest in mineral water (Brandi *et al.*, 1999). *Aeromonas* does not have enhanced resistance to chlorine and its presence in chlorinated water close to point of treatment is indicative of high organic loading, resulting in inadequate chlorination (Schubert, 1991).

There has been considerable discussion concerning the relative importance of water and food in *Aeromonas* infections, but the two sources are probably interrelated. The bacterium does appear to have an association with water, and it is probable that risk of exposure is greatest through consumption of contaminated water, or food processed with contaminated water.

Plesiomonas shigelloides potentially infects all ages but, while there is disagreement over the age group at highest risk, it is generally considered that the very young, the elderly and the infirm are most susceptible. Susceptibility to infection is strongly influenced by a number of underlying conditions. Malignancies, especially leukaemia, are of importance, while low gastric acidity is a general risk factor, with a number of possible causes. These include malnutrition, gastric surgery, malignancy and, possibly, over-use of antacid preparations. The majority of *Pl. shigelloides* septicaemia cases do not involve the central nervous system (CNS). However a number of CNS disease cases in neonates have occurred with a high mortality rate (63 %).

Plesiomonas shigelloides is widely distributed in the wild and is a common contaminant of natural, non-saline waters, especially in warmer months. Although the bacterium is also associated with a number of animals, risk of exposure is most likely to involve drinking contaminated water, or consumption of raw foods, such as oysters which have had contact with contaminated water.

21.5.4 Detection methods

Several authors have stated the need for monitoring for *Aeromonas* as a parameter of water quality. Although the organism can readily be isolated from foods, it is not generally examined for on a routine basis. Detection and enumeration are based on conventional cultural methods. Media were originally devised for use with water, but are suitable for foods, starch–ampicillin medium being considered most effective (ICMSF, 1996). Ampicillin–bile salts–inositol–xylose agar (*Aeromonas* medium), however, is more effective for recovery of *Aeromonas* from water when other bacteria are in high numbers (Villari *et al.*, 1999) and may also be superior with unprocessed foods, especially if *A. sobria* is present (Pin *et al.*, 1994). Enrichment before plating is not normally used, but tryptone–soy broth plus ampicillin is effective when other organisms are present in large numbers. Biochemical confirmation of identity is likely to be required and commercial kits are suitable, but may differ in performance (Ogden *et al.*, 1994). Rapid methods have not been developed.

Foods are not usually examined for *Pl. shigelloides* and, while selective media exist, these have been developed for clinical and environmental samples. Media designed for *Salmonella* are still widely used, but are over-selective and not recommended. Two media have been specifically devised for *Pl. shigelloides*: *Plesiomonas* medium and inositol–brilliant green–bile salts medium (IBB). The organism is selected due to its tolerance of brilliant green and bile salts, and distinguished from *Aeromonas* by the fermentation of *m*-inositol resulting in pinkish colonies. Use of both media is likely to be most satisfactory, but IBB is recommended for single use. If enrichment is considered necessary, alkaline peptone water or tetrathionate broth have been used, although neither is considered particularly effective (Varnam and Evans, 1996). Rapid methods have not been developed.

21.5.5 Control procedures

Basic control procedures for both *Aeromonas* spp. and *Pl. shigelloides* are those used for all Gram-negative pathogens: prevention of contamination, thermal or other processing for safety and prevention of recontamination. In the case of salads and vegetables eaten raw, water used for washing should be chlorinated or otherwise disinfected and care should be taken to ensure that water distribution systems are not colonised by *Aeromonas*. Studies in a remote area of Italy, however, showed that while *Aeromonas* was frequently present in water sources, there was no evidence that it grew in the water distribution system after chlorination (Legnani *et al.*, 1998). Presence of *Aeromonas* is independent of anthropic pollution, and these authors, and others, consider that monitoring of the organism should be used as a further measure of water quality.

Water is also a potential source of contamination of processed foods and similar precautions must be taken to prevent contamination of distribution systems. Water and soil may be considered as part of an ecological continuum, and laboratory survival experiments suggest that soil may play an important role in the epidemiology of human *Aeromonas* infection (Brandi *et al.*, 1996).

Human carriage of *Aeromonas* spp. or *Pl. shigelloides* is not a significant hazard, and special precautions are unnecessary in food handling. *Aeromonas*, but probably not *Plesiomonas*, is capable of growth at 4 °C or below, and growth cannot be totally prevented by refrigeration. There is doubt, however, that strains capable of growth at low temperatures are enteropathogenic. Little is known of the effect of preservatives, although *Aeromonas* is sensitive to NaCl and pH below 6.0 and has been shown to be inhibited by CO_2 in modified atmosphere packaging of fish (Davies and Slade, 1995).

Risk of *Aeromonas* and *Pl. shigelloides* infections is significantly greater among the immunocompromised and may be reduced by avoidance of suspect foods. This is difficult, however, in the absence of information concerning the relationship between infection and specific foods.

21.5.6 Future trends

Uncertainties over the role of *Aeromonas* and *Plesiomonas* mean that future trends are difficult to predict. There do not, however, appear to be any factors specific to these organisms. Numbers in the environment may increase with higher overall temperatures, while some strains of *Aeromonas* are able to grow at low temperatures, and there is a possibility of a higher incidence of infection associated with further dependence on refrigeration for controlling microbial growth on foods.

21.5.7 Further information

Kirov, M. S. (2007) *Aeromonas* and *Plesiomonas* species, in M. P. Doyle, L. R. Beuchat and T. J. Montville (eds) *Food Microbiology, Fundamentals and Frontiers*, ASM Press, Washington, DC, 301–27.

21.6 Characteristics of the genera *Cronobacter*

21.6.1 Introduction

The newly designated genus *Cronobacter* is composed of Gram-negative, facultative anaerobic rods, which are members of the *Enterobacteriaceae* Family (Iversen *et al.*, 2008). The genus is composed of five species, plus a possible sixth species. The new species are described in Table 21.5. They were formerly known as '*Enterobacter sakazakii*,' and this name was used in many publications before mid-2008. Unfortunately, it is uncertain which *Cronobacter* species is referred to in many publications, if the strain was

Table 21.5 *Cronobacter* species, including former *Enterobacter sakazakii*

Cronobacter species	Previous 16S cluster group[a]	Biotype(s)[b]
Cr. sakazakii	1	1,2,3,4,7,8,11,13
Cr. malonaticus	1	5,9,14
Cr. turicensis	2	16
Cr. muytjensii	3	15
Cr. dublinensis	4	6,10,12
Genomospecies 1	4	(16)

[a] Iversen *et al.* (2004).
[b] Iversen *et al.* (2006).
Source: Iversen *et al.* (2008).

identified using phenotyping rather than genotyping. However, the majority of isolated strains are usually *Cr. sakazakii*. *Cronobacter* species differentiation is primarily based on DNA sequence analysis, supported by biochemical differentiation, and not vice versa. The organisms grow well on non-selective media at 37 °C, and approximately 80 % of strains produce a yellow pigment at temperatures < 30 °C. The organism probably colonises plant material. This would account for its yellow carotenoid-based pigmentation to tolerate sunlight-generated oxygen radicals, and mucoid nature enhancing attachment to plant surfaces. Adult carriage of the organism has been reported.

21.6.2 *Cronobacter* as human pathogens

Cronobacter sakazakii and related species has come to prominence due to the association of '*E. sakazakii*' with infant infections, primarily associated with infant formula (Forsythe, 2005). In fact '*E. sakazakii*' does cause infections across all age groups, although these are principally immunocompromised individuals. Neonates, particularly those of low birth weight, are the major identified group at risk, with a high mortality rate. A number of outbreaks of *Cr. sakazakii* have been reported in neonatal intensive care units (Van Acker *et al*., 2001; Himelright *et al*., 2002; Coignard *et al*., 2006; Caubilla-Barron *et al*., 2007). However, the number of cases directly linked to intrinsically contaminated powdered infant formula (PIF) are few. Additionally, feeds may be contaminated during preparation, and administration.

The *Cronobacter* are opportunistic pathogens, and few virulence factors have been identified. Certain strains can invade human intestinal cell, replicate in macrophages and invade the blood–brain barrier (Townsend *et al*., 2008). Based on virulence studies of different pulsetypes from an outbreak, it is proposed that certain serotypes or biotypes are particularly virulent (Townsend *et al*., 2008). Of particular interest are *Cr. sakazakii*, *Cr. malonaticus* and *Cr. turicensis* which have been isolated from neonatal infections.

21.6.3 Risk factors

PIF is not a sterile product, and can contain a range of bacteria. As reported by Caubilla-Barron and Forsythe (2007), *Cr. sakazakii* can persist in PIF for up to two years. Nevertheless, there are no published reports of *Cronobacter* in PIF exceeding one cell/g PIF. Since the organism is ubiquitous, and found in many other foods, it is reasonable to assume that the neonates are primarily at risk due to their under-developed immune status and lack of competing intestinal flora. Reconstituted PIF can be contaminated by mixing utensils, and maybe by personnel. Infant formula is nutritious and therefore the control of bacterial growth must be limited by refrigeration and limited time before discarding to restrict bacterial multiplication. The FAO/WHO (2004, 2006) has proposed that PIF is reconstituted at 70 °C, and that it is used immediately (within 3 h). Keeping the time between reconstitution and feeding will reduce the risk of infection, as it limits bacterial multiplication. A number of outbreaks have been reported, and a common feature has been a lack of adequate hygienic preparation and temperature control of the reconstituted infant formula.

Following two FAO/WHO meetings (2004, 2006), a risk model has been constructed for '*E. sakazakii*' in reconstituted infant formula. The model allows the user to compare the level of risk between different levels of contamination and reconstitution practices. The model is available online at http://www.mramodels.org/ESAK/default.aspx.

21.6.4 Detection methods

A number of methods for the recovery of *Cronobacter* from PIF have been developed. Since the organism is normally present at such low numbers (< 1 cfu/g), a large volume of material is tested. The CCFH (2008) requirement is 30 × 10 g quantities. Therefore presence/absence testing is applied rather than direct enumeration. Due to the stressed state of the cells, initial step involves resuscitation with buffered peptone water, followed by enrichment in a selective broth (i.e. EE), and plating on a chromogenic agar. Commonly the expression of α-glucosidase is the basis of most chromogenic agars (Iversen and Forsythe, 2007). The current US FDA (FDA, 2002) and ISO (2006) methods use phenotyping as their end test, and are applicable for *Cronobacter* genus detection rather than any specific species. Commercial companies producing phenotyping kits are currently updating their databases in the light of the new taxonomy; the former '*E. sakazakii*' Preceptrol strain (ATCC 51329) is *Cr. muytjensii*, not *Cr. sakazakii*. DNA-based PCR probes have also been designed (Fanning and Forsythe, 2007). However, due to changes in taxonomy, these are now specific for the *Cronobacter* genus, rather than being species specific.

21.6.5 Control procedures

PIF is not manufactured as a sterile product, but must meet certain microbiological standards. Recently, CCFH (2008) revised the codes of hygienic practice for infant formula, including microbiological criteria as described above. As well as end-product analysis, environmental samples are taken in the production facility, and ingredients (especially starches, plant-derived material) are screened. PIF can be prepared by either dry-blending or wet-blending of the ingredients. Spray-drying does not act as a sterilisation step, as the organism can survive the drying process; nevertheless, there will be a considerable reduction in viability and the surviving cells may be severely damaged. In addition, production facilities and processes are designed to control enteric pathogens, especially *Salmonella*. These control measures are also applicable to the control of *Cronobacter*. For more detail on microbiological control in the production of PIF see Cordier (2007).

In addition to control of intrinsic contamination, control must be applied during preparation and handling. By keeping reconstituted feeds refrigerated, bacterial multiplication is reduced. The FAO/WHO (2004, 2006) equate reduced bacterial multiplication with reduced risk of infection. The additional step of reconstituting with hot water (i.e. 70 °C) should further reduce the total number of vegetative bacteria within the *Enterobacteriaceae* and therefore provide additional protection. Care must be shown, however, to ensure the feed is adequately cooled to prevent scalding of the baby.

21.6.6 Future trends

Cronobacter is primarily regarded as a 'one product-population problem', i.e. PIF and low-birth weight neonates. It is evident that since *Cronobacter* was recognised as being transmitted through contaminated feed, the incidence of the organism in PIF on the market has considerably reduced. Outbreaks are rarely reported and, as hygiene measures are improved, *Cronobacter* may have served as a wake-up call on neonatal care and, combined with improved general hygiene this will lead to reduced infection from other *Enterobacteriaceae* as well. It is plausible that *Cronobacter* may cause infections in the elderly and other immunocompromised individuals through the consumption of contaminated foods. Of particular interest are the range of powdered nutritional formulas that are consumed as dietary supplements by the elderly, and sports people. Although the frequency and level of contamination may be low, the lack of adequate cleaning of utensils and temperature abuse may result in bacterial multiplication and an increased risk of infection, just as occurs in infant infections.

21.6.7 Further information

FAO/WHO (2004) Enterobacter sakazakii *and other microorganisms in powdered infant formula: meeting report*, MRA Series 6, Rome/Geneva, Food and Agriculture Organization of the United Nations/World Health Organization, available at: http://www.who.int/foodsafety/publications/micro/es.pdf, accessed November 2008.

FAO/WHO (2006) Enterobacter sakazakii and Salmonella *in powdered infant formula: Meeting report*, MRA Series 10, Rome/Geneva, Food and Agriculture Organization of the United Nations/World Health Organization, available at: http://www.who.int/foodsafety/publications/micro/mra10.pdf, accessed November 2008.

Farber, J.M. and Forsythe, S.J. (eds.) (2007) *Emerging issues in food safety.* Enterobacter sakazakii, ASM Press, Washington, DC.

21.7 *Enterobacter, Klebsiella, Pantoea* and *Citrobacter* species

21.7.1 Introduction

There are a number of remaining genera in the *Enterobacteriaceae* which are considered opportunistic pathogens. These include *Enterobacter*, *Klebsiella*, *Pantoea* and *Citrobacter*, and these can cause a range of infections: gastrointestinal, septicaemia, chest infections (*K. pneumoniae*) and even meningitis (*Cit. koseri*). Since they may also be part of the normal (adult) human intestinal tract, their likelihood of causing foodborne illness is reduced. However, it is plausible that they cause infections in neonates and infants since the intestinal flora will not have been established and the immune system will be poorly developed (Townsend and Forsythe, 2007). In 2004, the FAO/WHO considered organisms reported to occur in powdered infant formula, and causative agents of neonatal infections. Since epidemiological evidence was lacking, *Pantoea agglomerans*, *Esch. vulneris* (both formerly known as *Ent. agglomerans*), *Ent. cloacae*, *K. pneumoniae*, *K. oxytoca*, *Hafnia alvei*, *Cit. freundii*, *Cit. koseri* and *Serratia* were designated as 'Causality plausible, but not yet demonstrated'. This list was revised in 2006 to include *E. coli* and *Acinetobacter* spp.

21.7.2 Risk factors, detection methods and control procedures

As given above, this section primarily considers the possible role of a number of genera in the *Enterobacteriaceae* in neonatal infections through contaminated infant formula. These organisms are not normally regarded as severe pathogens, and may be part of the normal intestinal flora in adulthood. However, the consumption by infants of exceptionally large numbers of these organisms, combined with the absence of a competing intestinal flora or immunity, could lead to infections. Caubilla-Barron and Forsythe (2007) reported the persistence of desiccated *Pantoea*, *K. oxytoca* and *E. vulneris* in infant formula for up to two years, whereas *Citrobacter* species and *Ent. cloacae* were no longer recoverable after six months, and *Salmonella* Enteritidis, *K. pneumoniae* and *E. coli* could not be recovered after 15 months. With respect to control of these organisms in the finished product, the CCFH (2008) revised guidelines recommend microbiological criteria for Enterobacteriaceae of $n = 10$, $c = 2$, $m = 0/10$ g for PIF. In addition, there is a three-class plan of $n = 5$, $c = 2$, $m = 500$/g and $M = 5000$/g for mesophilic aerobic bacteria as hygiene indicators. Unlike a number of named bacterial pathogens, the detection of *Enterobacteriaceae* is well-established, and is already routinely used in production facilities. These organisms are non-spore formers, and are not noted for any particular resistance to heat treatment. Therefore they will be susceptible to heat treatments during production and reconstitution with water at 70 °C. As is the case for *Cr. sakazakii* and *Salmonella* control, hygienic preparation of the infant feed is important in reducing the risk of infection.

21.7.3 Future trends

There is an increasing incidence of reported *Enterobacter*, *Citrobacter* and *Klebsiella* infections in the general public, including enhanced antibiotic resistance. This will increase attention on this group of organisms. The current *Enterobacter* genus is polyphyletic, and it is highly likely that it contains hitherto unrecognised species and genera. The reclassification of *Enterobacter* is highly probable following the increased use of DNA sequence analysis of clinical isolates, as opposed to phenotyping. The latter approach can be flawed when the associated database was based on misidentified organisms.

It is plausible that in the future at least some of these organisms may be formally recognised as foodborne pathogens of neonates and infants. However, it is also foreseeable that control measures for *Cr. sakazakii* and *Salmonella* will be directly applicable, and the risk minimised through control during production and control of multiplication prior to ingestion.

21.7.4 Further information

FAO/WHO (2004) Enterobacter sakazakii *and other microorganisms in powdered infant formula: meeting report*, MRA Series 6, Rome/Geneva, Food and Agriculture Organization of the United Nations/World Health Organization, available at: http://www.who.int/foodsafety/publications/micro/es.pdf, accessed November 2008.

FAO/WHO (2006) Enterobacter sakazakii *and* Salmonella *in powdered infant formula: Meeting report*, MRA Series 10, Rome/Geneva, Food and Agriculture Organization of the United Nations/World Health Organization, available at: http://www.who.int/foodsafety/publications/micro/mra10.pdf, accessed November 2008.

21.8 Acknowledgement

The editors would like to acknowledge that this chapter, written by Stephen Forsythe, was based on a chapter from the first edition of this book written by Jane Sutherland and the late Alan Varnum.

21.9 References

Altwegg M, Martinetti Luchini G, Lüthy-hoffenstein J and Rohrbach M (1991) *Aeromonas*-associated gastroenteritis after consumption of contaminated shrimp, *European Journal of Clinical Microbiology and Infectious Diseases*, **10**, 44–5.

Andersen J K (1988) Contamination of freshly slaughtered pig carcasses with human pathogenic *Yersinia enterocolitica, International Journal of Food Microbiology*, **7**, 193–202.

Arias C R, Maclan M C, Aznar R *et al.* (1999) Low incidence of *Vibrio vulnificus* among

Vibrio isolates from sea water and shellfish of the western Mediterranean coast, *Journal of Applied Microbiology*, **86**, 125–34.

Asplund K, Hakkinen M, Okkonen T *et al.* (1998) Effects of growth-promoting antimicrobials on inhibition of *Yersinia enterocolitica* O:3 by porcine ileal microflora, *Journal of Applied Microbiology*, **85**, 164–70.

Beuchat L R (1977) Evaluation of enrichment broths for enumerating *Vibrio parahaemolyticus* in chilled and frozen crabmeat, *Journal of Food Protection*, **40**, 592–5.

Bhaduri S, Buchanan R L and Phillips J G (1995) Expanded surface response model for predicting the effects of temperature, pH, sodium chloride contents and sodium nitrite concentrations on the growth rate of *Yersinia enterocolitica*, *Journal of Applied Bacteriology*, **9**, 163–70.

Blake P A (1983) Vibrios on the half shell. What the walrus and the carpenter didn't know, *Annals of Internal Medicine*, **99**, 558–9.

Blake P A, Allegra D T, Snyder J G *et al.* (1980) Cholera: a possible endemic focus in the United States, *New England Journal of Medicine*, **302**, 305–9.

Bottone E J (1997) *Yersinia enterocolitica*: the charisma continues, *Clinical Microbiology Reviews*, **10**, 257–76.

Brandi G, Sisti M, Schiavano G F *et al.* (1996) Survival of *Aeromonas hydrophila*, *Aeromonas caviae* and *Aeromonas sobria* in soil, *Journal of Applied Bacteriology*, **81**, 439–44.

Brandi G, Sisti M, Giardini F *et al.* (1999) Survival ability of cytotoxic strains of motile *Aeromonas* spp. in different types of water, *Letters in Applied Microbiology*, **29**, 211–15.

Brenton C E, Flick G J, Pierson M D *et al.* (2001) Microbiological quality and safety of quahog clams, *Mercanaria mercenaria*, during refrigeration and at elevated storage temperatures, *Journal of Food Protection*, **64**, 341–7.

Caubilla-Barron J and Forsythe S J (2007) Dry stress and survival time of *Enterobacter sakazakii* and other *Enterobacteriaceae*, *Journal of Food Protection* **70**, 2111–7.

Caubilla-Barron J, Hurrell E, Townsend S *et al.* (2007) Genotypic and phenotypic analysis of *Enterobacter sakazakii* strains from an outbreak resulting in fatalities in a neonatal intensive care unit in France, *Journal of Clinical Microbiology*, **45**, 3979–85.

CCFH (2008) *Code of Hygienic Practice for Powdered Formulae for Infants and Young Children*, Thirty-first Session of the Codex Alimentarius Commission in Geneva, Switzerland, 30 June–5 July 2008, available at: http://www.codexalimentarius.net/download/report/686/al31_13e.pdf, accessed November 2008.

CDC (1999) Outbreaks of *Shigella sonnei* infection associated with eating fresh parsley – United States and Canada, July–August, 1998, *Morbidity and Mortality Weekly Report*, **48**, 285–9.

CDC (2000) Outbreaks of *Shigella sonnei* infections associated with eating a nationally distributed dip – California, Oregon, and Washington, January 2000, *Morbidity and Mortality Weekly Report*, **49**, 60–1.

Coignard B, Vaillant V, Vincent J-P *et al.* (2006) Infections severes á *Enterobacter sakazakii* chez des nouveau-nés ayant consomme une preparation en poudre pour nourrissons, France, octobre-decembre 2004, *Bulletin Epidemiology Hebdomadaire* **2–3**, 10–13.

Cordier J-L (2007) Production of powdered infant formulae and microbiological control measures, in J M Farber and S J Forsythe (eds), *Emerging issues in food safety* Enterobacter sakazakii, ASM Press, Washington, DC, 145–86.

Corry J E L, James C, James S J and Hinton M (1995) *Salmonella*, *Campylobacter* and *Escherichia coli* O157:H7 decontamination techniques for the future, *International Journal of Food Microbiology*, **28**, 187–96.

Dalsgaard A, Dalsgaard I, Hoi L and Larsen J L (1996) Comparison of a commercial biochemical kit and an oligonucleotide probe for identification of environmental isolates of *Vibrio vulnificus, Letters in Applied Microbiology*, **22**, 184–8.

Davies A P and Slade A (1995) Fate of *Aeromonas* and *Yersinia* on modified atmosphere-packaged (MAP) cod and trout, *Letters in Applied Microbiology*, **21**, 354–9.

Deodhar L P, Saraswathi K and Varudkar A (1991) *Aeromonas* spp. and their association with human diarrhoeal disease, *Journal of Clinical Microbiology*, **29**, 853–6.

Falcão J P, Falcão D P, Pitondo-Silva A, Malaspina A C and Brocchi M (2006) Molecular typing and virulence markers of *Yersinia enterocolitica* strains from human, animal and food origins isolated between 1968 and 2000 in Brazil, *Journal of Medical Microbiology*, **55**, 1539–48.

Fanning S and Forsythe S J (2007) Isolation and identification of *Enterobacter sakazakii*, in J Farber and S J Forsythe (eds), *Emerging issues in food safety. Enterobacter sakazakii*, ASM Washington, DC, 27–59.

FAO/WHO (2004) Enterobacter sakazakii *and other microorganisms in powdered infant formula: meeting report*, MRA Series 6, Rome/Geneva, Food and Agriculture Organization of the United Nations/World Health Organization, available at: http://www.who.int/foodsafety/publications/micro/es.pdf, accessed November 2008.

FAO/WHO (2005a) *Risk assessment of* Vibrio vulnificus *in raw oysters*, MRA Series 8, Rome/Geneva, Food and Agriculture Organization of the United Nations/World Health Organization, available at: http://www.who.int/foodsafety/publications/micro/mra8.pdf, accessed November 2008.

FAO/WHO (2005b) *Risk assessment of choleragenic Vibrio cholerae O1 and O139 in warm-water shrimp in international trade*, MRA Series 9, Rome/Geneva, Food and Agriculture Organization of the United Nations/World Health Organization, available at: http://www.who.int/foodsafety/publications/micro/mra9.pdf, accessed November 2008.

FAO/WHO (2006) Enterobacter sakazakii *and* Salmonella *in powdered infant formula: Meeting report*, MRA Series 10, Rome/Geneva, Food and Agriculture Organization of the United Nations/World Health Organization, available at: http://www.who.int/foodsafety/publications/micro/mra10.pdf, accessed November 2008.

FDA (2002) *Isolation and enumeration of* Enterobacter sakazakii *from dehydrated powdered infant formula*, US Food and Drug Administration, Center for Food Safety and Applied Nutrition, Washington DC, available at: http://www.cfsan.fda.gov/~comm./mmesakaz.html, accessed November 2008.

FDA (2005) *Quantitative risk assessment on the public health impact of pathogenic* Vibrio parahaemolyticus *in raw oysters*, Center for Food Safety and Applied Nutrition, US Food and Drug Administration, US Department of Health and Human Services, Washington, DC, available at: http://www.cfsan.fda.gov/~dms/vpra-toc.html, accessed November 2008.

Fleurat A (1986) Indigenous response to drought in sub-Saharan Africa, *Disasters*. **10**, 224–9.

Forsythe S (2005) *Enterobacter sakazakii* and other bacteria in powdered infant milk formula, *Mother and Child Nutrition*, **1**, 44–50.

Grant T, Bennett-Wood V and Robins-Browne R M (1999) Characterization of the interaction between *Yersinia enterocolitica* Biotype 1A and phagocytes and epithelial cells *in vitro*, *Infection and Immunity*, **67**, 4367–75.

Harris J B, Khan A I, LaRocque R C *et al.* (2005) Blood group, immunity, and risk of infection with *Vibrio cholerae* in an area of endemicity, *Infection and Immunity*, **73**, 7422–7.

Himelright I, Harris E, Lorch V and Anderson M (2002) *Enterobacter sakazakii* infections associated with the use of powdered infant formula–Tennessee, 2001, *Journal of the American Medical Association*, **287**, 2204–5.

Howard S L, Gaunt M W, Hinds J, Witney A A, Stabler R and Wren B W (2006) Application of comparative phylogenomics to study the evolution of *Yersinia enterocolitica* and to identify genetic differences relating to pathogenicity, *Journal of Bacteriology*, **188**, 3645–53.

ICMSF (1996) *Micro-organisms in Foods 5, Microbiological Specification of Food Pathogens*, International Commission on Microbiological Specifications for Foods, Blackie, London.

ISO (2006) ISO/TS 22964:2006 Milk and milk products – Detection of *Enterobacter sakazakii*, International Organization for Standardization, Geneva.

Iversen C and Forsythe S (2007) Comparison of media for the isolation of *Enterobacter sakazakii*, *Applied and Environmental Microbiology*, **73**, 48–52.

Iversen C, Waddington M, On S L W and Forsythe S (2004) Identification and phylogeny of *Enterobacter sakazakii* relative to *Enterobacter* and *Citrobacter*, *Journal of Clinical Microbiology*, **42**, 5368–70.

Iversen C, Waddington M, Farmer III J and Forsythe S (2006) The biochemical differentiation of *Enterobacter sakazakii* genotypes, *BMC Microbiology*, **6**, 94.

Iversen C, Mullane N, McCardell B D *et al.* (2008) *Cronobacter* gen. nov., a new genus to accommodate the biogroups of *Enterobacter sakazakii*, and proposal of *Cronobacter sakazakii* gen. nov., comb. nov., *Cronobacter malonaticus* sp. nov., *Cronobacter turicensis* sp. nov., *Cronobacter muytjensii* sp. nov., *Cronobacter dublinensis* sp. nov., *Cronobacter* genomospecies 1, and of three subspecies, *Cronobacter dublinensis* subsp. *dublinensis* subsp. nov., *Cronobacter dublinensis* subsp. *lausannensis* subsp. nov. and *Cronobacter dublinensis* subsp. *lactaridi* subsp. nov., *International Journal of Systematic and Evolutionary Microbiology*, **58**, 1442–7.

Jay J M (1996) Microorganisms in fresh ground meats: the relative safety of products with low versus high numbers, *Meat Science*, **43**, 559–66.

Jalava K, Hakkinen M, Valkonen M, Nakari U-M, Palo T, Hallanvuo S, Ollgren J, Siitonen A and Nuorti J P (2006) An outbreak of gastrointestinal illness and erythema nodosum from grated carrots contaminated with *Yersinia pseudotuberculosis*, *Journal of Infectious Disease*, **194**, 1209–16.

Krovacek K, Peterz M, Faris A and Mansson I (1989) Enterotoxigenicity and drug sensitivity of *Aeromonas hydrophila* isolated from well water in Sweden: a case study, *International Journal of Food Microbiology*, **8**, 149–54.

Kirov S M, Hudson J A, Hayward L J and Mott S J (1994) Distribution of *Aeromonas hydrophila* hybridization groups and their virulence properties in Australian clinical and environmental isolates, *Letters in Applied Microbiology*, **18**, 71–3.

Kotloff K L, Winickoff J P, Ivanoff B, Clemens J D, Swerdlow D L, Sansonetti P J, Adak G K and Levine M M (1999) Global burden of *Shigella* infections: implications for vaccine development and implementation of control strategies, *Bulletin of the World Health Organization*, **77**, 651–66.

Lampel K A (2005) *Shigella* species, in P M Fratamico, A K Bhunia and J L Smith (eds), *Foodborne Pathogens. Microbiology and Molecular Biology*, Caister Academic Press, Norwich, 341–56.

Lampel K A and Maurelli A T (2001) *Shigella* species, in M P Doyle, L R Beuchat and T J Montville (eds), *Food Microbiology, Fundamentals and Frontiers*, ASM Press, Washington, DC, 247–61.

Legnani P, Leoni E, Soppelsa F and Burigo R (1998) The occurrence of *Aeromonas* species in drinking water supplies of an area of the Dolomite mountains, Italy, *Journal of Applied Microbiology*, **85**, 271–6.

McMahon C M M, Doherty A M, Sheridan J J *et al.* (1999) Synergistic effect of heat and sodium lactate on the thermal tolerance of *Yersinia enterocolitica* in minced beef, *Letters in Applied Microbiology*, **28**, 340–4.

Neyts K, Huys G, Uyttendale M *et al.* (2000) Incidence and identification of mesophilic *Aeromonas* from retail foods, *Letters in Applied Microbiology*, **31**, 359–63.

Nissen H, Maugesten T and Lea P (2001) Survival and growth of *Escherichia coli* O157:H7, *Yersinia enterocolitica* and *Salmonella enteritidis* on decontaminated and untreated meat, *Meat Science* **57**, 291–8.

Nuorti J P, Niskanen T, Hallanvuo S, Mikkola J, Kela E, Hatakka M, Fredriksson-Ahomaa

M, Lyytikäinen O, Siitonen A, Korkeala H and Ruutu P (2004) A widespread outbreak of *Yersinia pseudotuberculosis* O:3 infection from iceberg lettuce, *Journal of Infectious Disease*, **189**, 766–74.

O'Brien A D and Holmes K R (1987) Shiga and Shiga-like toxins, *Microbiological Reviews*, **51**, 206–20.

Ogden I D, Millar I G, Watt A J and Wood L (1994) A comparison of three identification kits for the confirmation of *Aeromonas* spp., *Letters in Applied Microbiology*, **18**, 97–9.

Oliver J D (1989) *Vibrio vulnificus*, in M P Doyle (ed.), *Foodborne Bacterial Pathogens*, Marcel Dekker, New York, 569–600.

Oliver J D and Kaper J B (2007) *Vibrio* species, in M P Doyle, L R Beuchat and T J Montville (eds), *Food Microbiology, Fundamentals and Frontiers,* ASM Press, Washington, DC, 263–300.

Oliver J D, Hite E, McDougald P *et al.* (1995) Entry into, and resuscitation from, the viable but non culturable state by *Vibrio vulnificus* in an estuarine environment, *Applied and Environmental Microbiology*, **61**, 2624–30.

O'Neill K R, Jones S H and Grimes D J (1992) Seasonal incidence of *Vibrio vulnificus* in the Great Bay estuary of New Hampshire and Maine, *Applied and Environmental Microbiology*, **58**, 737–9.

Pal T, Formal S B and Hale T (1989) Characterization of virulence marker antigen of *Shigella* sp. and enteroinvasive *E. coli, Journal of Clinical Microbiology*, **27**, 561–3.

Pin C, Marin M L, Garcia M L *et al.* (1994) Comparison of different media for the isolation and enumeration of *Aeromonas* spp. in foods, *Letters in Applied Microbiology*, **18**, 190–92.

Phillips A M B, DePaola A, Bowers J, Ladner S and Grimes D J (2007) An evaluation of the use of remotely sensed parameters for prediction of incidence and risk associated with *Vibrio parahaemolyticus* in Gulf Coast oysters (*Crassostrea virginica*), *Journal of Food Protection*, **70**, 879–84.

Rothwell J (1990) Microbiology of ice cream and related products, in R K Robinson (ed.), *Dairy Microbiology, Vol. 2, The Microbiology of Milk Products*, Applied Science Publishers, London, 1–39.

Schubert R H W (1991) Aeromonads and their significance as potential pathogens in water, in B Austin (ed.), *Pathogens in the Environment*, Society for Applied Bacteriology, Symposium Series 20, Journal of Applied Bacteriology, **65** (Suppl.), 131–7.

Scotland S M (1988) Toxins, in B M Lund, M Sussman, D Jones and M F Stringer (eds), *Enterobacteriaceae in the Environment and as Pathogens*, Society for Applied Bacteriology, Symposium series 17, *Journal of Applied Bacteriology* **65** (Suppl.), 117–19.

Shangkuan Y H, Show Y S and Wang T M (1996) Multiplex polymerase chain reaction to detect toxigenic *Vibrio cholerae* O1 and to biotype *Vibrio cholerae* O1, *Journal of Applied Bacteriology*, **79**, 264–73.

Sheridan J J, Logue C M, McDowell D A *et al.* (1998) A study of growth characteristics of *Yersinia enterocolitica* serotype O:3 in pure and meat culture systems, *Journal of Applied Microbiology*, **85**, 293–301.

Slade P J, Falah M A and Al-Ghady M R (1986) Isolation of *Aeromonas hydrophila* from bottled water and domestic supplies in Saudi Arabia, *Journal of Food Protection*, **49**, 471–6.

Smith J L (1987) *Shigella* as a foodborne pathogen, *Journal of Food Protection*, **50**, 788–801.

Sutherland J P and Bayliss A J (1994) Predictive modeling of growth of *Yersinia enterocolitica*: the effects of temperature, pH and sodium chloride, *International Journal of Food Microbiology*, **21**, 197–215.

Tauxe R V, Holmberg S D, Dolin A *et al.* (1988) Epidemic cholera in Mali: high mortality and multiple routes of infection in a famine area, *Epidemiology and Infection*, **100**, 279–88.

Thisted-Lambertz S T, Ballagi-Pordany A, Nilsson A *et al.* (1996) A comparison between a PCR method and a conventional culture method for detecting pathogenic *Yersinia enterocolitica* in food, *Journal of Applied Microbiology*, **81**, 303–8.

Todd L S, Hardy J C, Stringer M F and Bartholomey B A (1989) Toxin production by strains of *Aeromonas hydrophila* grown in laboratory media and prawn puree, *International Journal of Food Microbiology*, **9**, 145–56.

Townsend S and Forsythe S J (2007) The neonatal intestinal microbial flora, immunity, and infections, in J Farber and S J Forsythe (eds), *Emerging Issues in Food Safety*, Enterobacter sakazakii, ASM Press, Washington, DC, 61–100.

Townsend S, Hurrell E and Forsythe S J (2008) Virulence studies of *Enterobacter sakazakii* isolates associated with a neonatal intensive care unit outbreak, *BMC Microbiology*, **8**, 64.

Van Acker J, De Smet F, Muyldermans G, Bougatef A, Naessens A and Lauwers S (2001) Outbreak of necrotizing enterocolitis associated with *Enterobacter sakazakii* in powdered milk formula, *Journal of Clinical Microbiology*, **39**, 293–7.

Vantarakis A, Komninou G, Venieri D and Papapetropoulou M (2000) Development of a multiplex PCR detection of *Salmonella* and *Shigella* in mussels, *Letters in Applied Microbiology*, **31**, 105–9.

Varnam A H and Evans M G (1996) *Foodborne Pathogens: an Illustrated Text*, Manson Publishing, London.

Varnam A H and Evans M G (2000) *Environmental Microbiology*, American Society for Microbiology, Washington, DC, Manson Publishing, London.

Venkatesan M, Buyesse J M and Kopecko D J (1989) Use of *Shigella flexneri ipaC* and *ipaH* gene sequences for the general identification of *Shigella* spp. and entero-invasive *Escherichia coli, Journal of Clinical Microbiology*, **27**, 2687–91.

Villari P, Pucino A, Santagata N and Torre I (1999) A comparison of different culture media for the membrane filter quantification of *Aeromonas* spp. in water, *Letters in Applied Microbiology*, **29**, 253–8.

Wadström T and Ljungh A (1991) *Aeromonas* and *Plesiomonas* as food and waterborne pathogens, *International Journal of Food Microbiology*, **12**, 202–12.

Watanabe H and Timmis K N (1984) A small plasmid in *Shigella dysenteriae* I specifies one or more functions essential for O-antigen production and bacterial virulence, *Infection and Immunity*, **46**, 55–63.

Weagant S D (2008) A new chromogenic agar medium for detection of potentially virulent *Yersinia enterocolitica, Journal of Microbiological Methods*, **72**, 185–90.

Yamamoto A, Iwahori J, Vuddhakul V *et al.* (2008) Quantitative modeling for risk assessment of *Vibrio parahaemolyticus* in bloody clams in southern Thailand, *International Journal of Food Microbiology*, **124**, 70–8.

22

Staphylococcus aureus and other pathogenic Gram-positive cocci

M. Adams, University of Surrey, UK

Abstract: The role of Gram-positive cocci as agents of foodborne illness is described along with methods for their control. The chapter focuses primarily on staphylococcal food poisoning caused by *Staphylococcus aureus*, but the enterococci and pathogenic streptococci that can be transmitted by food are also discussed.

Key words: *Staphylococcus aureus*, enterococci, *Streptococcus*.

22.1 Introduction

This chapter discusses a number of bacteria which share a Gram-positive coccoid morphology and an association with foodborne illness. It focuses primarily on *Staphylococcus aureus* since it is the most significant cause of foodborne illness among these organisms.

Other organisms discussed here are the enterococci which are commonly found in the gastrointestinal tract of humans and animals and in a variety of foods. These are associated with a variety of infections, and have been implicated in outbreaks of foodborne illness in the past. Some of the evidence relating to this is discussed here. Finally, some other pathogenic streptococci which are occasionally spread by food are considered.

22.2 *Staphylococcus aureus* and other enterotoxigenic staphylococci

22.2.1 The organism and its characteristics

In recent years *Staph. aureus* has attracted much attention and notoriety as a cause of human infections, and it was in this role that it was first described by Ogston in 1882. He coined the name staphylococcus to describe the microscopic appearance of pyogenic bacteria which resembled bunches of grapes. The first recognised description of its association with foodborne illness came very shortly afterwards when Vaughan identified coccoid bacteria as responsible for illness caused by cheese and, later, ice cream in Michigan (Vaughan, 1884; Anon., 1886). He also isolated and crystallised a putative toxin which he named tyrotoxicon. Subsequent reports in the early 20th century associated the organism with illness from meat products in Belgium and mastitic milk in the Philippines, and in 1930 Dack, investigating an outbreak caused by a cream filled cake, showed that staphylococcal food poisoning was caused by a filterable enterotoxin (Dack *et al.*, 1930).

Methicillin-resistant *Staph. aureus* (MRSA) was first reported in 1961 and has since become an important cause of primary infections and bacteraemia. Initially these were primarily hospital-acquired infections but, more recently, the incidence of community-acquired infections has been increasing. A number of studies have reported on the incidence of MRSA on animals and foods. A survey in the USA conducted in 2001–2003 examined 1913 samples from cattle, pigs and chickens. Of these, 421 contained *Staph. aureus* and 15 (12 from dairy cows and three from chickens) were positive for the *mecA* gene (a gene that encodes a penicillin binding protein with low affinity for β-lactams) (Lee, 2003). In Italy 1634 food samples yielded 160 *Staph. aureus* isolates of which six were *mecA*-positive (four from cow's milk and two from cheeses) (Normanno *et al.*, 2007). A similar level of isolation (0.45 %) was reported from 444 raw chicken meat samples examined in Japan although, more recently, a higher incidence (2.5 %) has been reported from a relatively small survey of beef and pork products in the Netherlands (n = 79) (Kitai *et al.*, 2005; van Loo *et al.*, 2007).

An outbreak of MRSA infection initiated by food was reported in the Netherlands affecting 21 patients in a hospital haematology unit of whom five died (Kluytmans *et al.*, 1995). MRSA was detected in food and in the throat of a health care worker who prepared food for the patients, although airborne contamination played an important role in the outbreak. Since staphylococcal food poisoning is an intoxication, the antibiotic resistance of *Staph. aureus* is of no direct relevance, although its occurrence could increase with spread of the organism to the community. An outbreak of foodborne illness caused by MRSA has been recorded in the USA where the same organism was isolated from food handlers at the store, samples of pork and coleslaw purchased, as well as the three affected patrons (Jones *et al.*, 2002).

In terms of official statistics, foodborne illness caused by *Staph. aureus* would appear to be of minor importance compared with that caused by pathogens such as *Salmonella*, *Campylobacter* and Verocytotoxin-producing *E. coli*. In the USA between 1998 and 2002 there was an average of 20 *Staph. aureus* food poisoning outbreaks per year (Lynch *et al.*, 2006). The equivalent figure for England and Wales over the same period was 2.4 outbreaks per year, affecting a total number of only 120 people over the whole period. More recent data suggest an even lower incidence with less than one outbreak being recorded each year (HPA, 2008). However, it is likely that these data seriously under-report the true incidence of staphylococcal food poisoning. The short duration of the relatively mild symptoms means that many of those affected will not seek medical advice, and official statistics are confined to recording general outbreaks and do not include sporadic cases. In the USA it has been estimated that the official statistics under-report by a factor of 38 and Adak *et al.* estimated an incidence of 9196 cases per year in England and Wales based on 1996–2000 data (Mead *et al.*, 1999; Adak *et al.*, 2005). In France, it is recognised as the second most common cause of foodborne illness after *Salmonella* and was the confirmed cause of 86 out of 1787 outbreaks reported in 2001–2003 and the suspected cause of a further 173 (Kérouanton *et al.*, 2007).

Staphylococus aureus is a non-motile, catalase, positive, facultatively anaerobic Gram-positive coccus typically about 1 μm in diameter. Cells divide in more than one plane to form irregular clumps (*staphyle* – Greek, bunch of grapes). Colonies often have a golden yellow pigmentation on many media, particularly on prolonged incubation, giving rise to the epithet aureus (*aureus* – Latin, golden). A mesophile, it grows optimally around 37 °C and over a range of temperature from 7 °C to 48 °C. Growth is favoured around neutral pH (6–7) but will occur between pH 4 and pH 10, albeit much more slowly. It grows optimally at high water activity but is the most a_w/salt tolerant foodborne bacterial pathogen, being capable of growth in up to 20 % sodium chloride (a_w 0.86) or down to an a_w of 0.83 (Adams and Moss, 2008). It generally survives well under adverse conditions and is resistant to drying and freezing but does not have exceptional heat resistance with *D*-values of the order of seconds to a few minutes at 70 °C depending on the growth phase of the cells and the heating medium (Table 22.1).

22.2.2 The nature of staphylococcal enterointoxications

Illness caused by *Staph. aureus* is a foodborne intoxication where the organism has to grow in a food material in order to produce sufficient enterotoxin to cause illness. The incubation period is short, typically 2–4 h, and nausea, vomiting, stomach cramps and retching predominate although diarrhoea does also occur. Recovery is usually complete within about 24 h. Although there are more than 40 species of *Staphylococcus*, outbreaks of staphylococcal food poisoning are almost always caused by *Staph. aureus*. Some other species

Table 22.1 Heat resistance of *Staphylococcus aureus*

Temperature (°C)	*D*-value (min)				Reference
	54.5	55	60	70	
Phosphate buffer	–	3.0	–	–	Halpin-Dohnalek and Marth, 1989
	4.9 (pH 4.5)	–	1.0 (pH 4.5)		Stiles and Witter, 1965
	8.4 (pH 5.5)	–	1.4 (pH 5.5)		
	11.9 (pH 6.5)	–	2.5 (pH 6.5)		
Brain heart infusion	–	17.8	–	–	Rosenberg and Sinell, 1989
Milk	–	3.0	0.9	0.1	Firstenberg-Eden *et al.*, 1977

of both coagulase-positive and coagulase-negative staphylococci have been shown to be enterotoxigenic or to carry enterotoxin genes (Fukuda *et al.*, 1984; Aliu and Bergdoll, 1988; Marín *et al.*, 1992; Becker *et al.*, 2001), but there is very little evidence of them causing outbreaks. *Staphylococcus intermedius* caused one outbreak associated with a butter spread which affected more than 265 people in the USA in 1991 (Khambaty *et al.*, 1994), and an earlier outbreak caused by a coagulase-negative *Staphylococcus* has been reported (Breckinridge and Bergdoll, 1971).

The staphylococcal enterotoxins (SEs) are single-chain polypeptides (194–245 amino acid residues) with molecular weights in the range 24 800–28 500 (Thomas *et al.*, 2007). The SEs share significant amino acid homology and a common two-domain structure. Most have a cystine loop structure which is thought to be associated with emetic activity.

One consequence of their structure is their resistance to digestive proteases such as trypsin, chymotrypisn and pepsin which allows them to persist in the gastrointestinal tract after ingestion. They also possess a marked heat resistance which can be an important factor in some outbreaks. It was demonstrated quite early on that SEs in a culture filtrate survived boiling for 30 min (Jordan *et al.*, 1931), and a *D*-value for SEB in milk at 100 °C of about an hour has been recorded (Read and Bradshaw, 1966). Heat sensitivity depends on the nature of the heating medium. SEs have been shown to be much less heat-stable at acidic pH values (Humber *et al.*, 1975) and, while 11 min at 121 °C was reported to remove activity in one report (Denny *et al.*, 1966), even 28 min at 121 °C did not eliminate all activity from canned mushrooms (Anderson *et al.*, 1996). The intensity of heat treatment required to detoxify a food will also depend on the level of toxin present; toxin levels of 5, 20 and 60 μg/ml in beef bouillon required 27, 37 and 40 min heating at 121.1 °C, respectively, to render them non-detectable (Denny *et al.*, 1971) (Table 22.2).

The results from heat resistance studies should be treated with some caution as they also reflect the method used to detect toxin presence/activity.

Table 22.2 Thermal stability of *Staphylococcus aureus* enterotoxin

	Temperature (°C)	*D*-value (min.)	Time (min) to last positive serological reaction [initial concentration]	Percentage loss after 20 min heating
Phosphate buffer	80	–	–	23
	100	–	–	70
	120	–	–	100
Milk	104	46.2	90 [1 μg/ml]	–
	121	9.4	20 [1 μg/ml]	–
Beef bouillon	104	–	100 [5 μg/ml]	–
	121	–	27 [5 μg/ml]	–

Source: Adapted from ICMSF (1996); Read and Bradshaw (1966)

Immunological detection, while technically more attractive, may detect epitopes not associated with activity, and heat-inactivated samples of SEA and SED were shown to retain biological activity in an *in vivo* assay when they were undetectable immunologically (Bennett, 1992).

Currently about 20 antigenic variants of SE are recognised and these have been designated SEA, SEB, through to SEV including three SECs. There is no SEF; the toxin originally designated SEF does not have emetic activity and has now been recognised as the toxin responsible for toxic shock syndrome. The emetic activity of some other SEs has not been investigated and in some it is reported to be negative (SEL). SEA is the toxin most frequently associated with foodborne outbreaks, although SEB, SEC and SED are also fairly common.

SEs are thought to act by binding to specific receptors in the abdominal viscera stimulating the vomiting centre in the brain stem via the vagus and sympathetic nerves. SEs also have the property of being super antigens with the ability to bind to and stimulate a much higher proportion of T cells than normal antigens by interacting with them in a non-specific manner (Balaban and Rasooly, 2000). The relevance of this to their enterotoxic activity is not clear.

Production of some, but not all, of the SEs is controlled by the *agr* operon, a quorum sensing system that leads to pleiotropic changes in gene expression when the cell concentration reaches a threshold level (Novick, 2003). Even in situations where SE production is not under this control, constitutive expression of SE would require a certain cell density for there to be sufficient toxin present to cause illness. This level is reported at around 10^5–10^6 cfu/ml or g. Thus some growth of the organism in the food is usually necessary for staphylococcal food poisoning to occur.

22.2.3 Risk factors

There are two basic preconditions for staphylococcal food poisoning to occur

1. *Enterotoxigenic staphylococci must be present or introduced into the food, at some stage in its production, processing or storage.*

Staphylococcus aureus is a common organism on the skin, the skin glands and mucous membranes of warm-blooded animals. It can get into foods from food animals themselves, but contamination from humans is the most common source in food poisoning outbreaks. Human carriage is primarily in the nose and throat, but the organism is also isolated in small numbers from healthy skin. Higher numbers can be isolated from damaged skin, sores and infected lesions such as boils. It is present as a persistent coloniser in 20–30 % of the population and occurs intermittently in about 60 %. Only about 20 % of people almost never carry *Staph. aureus* (Kluytmans *et al.*, 1997). Studies on the enterotoxigenicity of strains from foods and human carriers indicate that a variable but usually high proportion of strains are enterotoxigenic (Marín *et al.*, 1992; Lawrynowicz-Paciorek *et al.*, 2007; Morandi *et al.*, 2007).

2. *The organism must have the opportunity to grow to levels sufficient to produce enough enterotoxin to cause illness.*

Volunteer feeding studies indicated that the dose of toxin required to elicit emetic symptoms is around 10–20 μg for an adult. Sensitivity can vary widely between individuals, and it is acknowledged that in practice the toxic dose can be considerably lower, less than 1 μg in many cases. In an outbreak in the USA, the chocolate milk implicated contained 0.09–0.18 μg per carton and the attack rate in those consuming one carton was 31.5 % (Evenson *et al.*, 1988). More recently, in an extensive outbreak in Japan involving 13 420 cases, the low-fat milk implicated contained $\leq$ 0.38 ng/ml of SEA and the total per capita intake was estimated at 0.02–0.1 μg (Asao *et al.*, 2003).

It is generally accepted that levels of 10^5–10^6 *Staph. aureus*/g of food are required to produce sufficient toxin to cause illness. Staphylococcal growth does not invariably lead to toxin production which occurs over a narrower range of environmental conditions (ICMSF, 1996; Le Loir *et al.*, 2003). For example, toxin production does not occur at salt levels above 12 % (Notermans and Heuvelman, 1983). Table 22.3 presents the reported ranges of some key environmental factors for growth and toxin production. These data do, however, represent worst-case scenarios since the limits they describe were determined when all other conditions were optimal. Growth of *Staph. aureus* is also very susceptible to competition from other organisms that might be present in a mixed microflora or in a food fermentation process.

Meat, poultry and their products are most commonly implicated in staphylococcal enterointoxications. In a study of cases in the UK over a 21-year period between 1969 and 1990 they were the vehicle in 75 % of incidents. Poultry and poultry products were responsible for 39 % of the meat-related cases and ham was responsible for 25 % (Wienecke *et al.*, 1993). Since the very earliest recorded outbreaks, dairy products have been a common vehicle. *Staphylococcus aureus* is a common cause of mastitis in cows, goats and

Table 22.3 Limits for growth and enterotoxin production by *Staphylococcus aureus* under otherwise optimal conditions

Factor	Growth		Toxin production	
	Optimum	Range	Optimum	Range
Temperature (°C)	37	7–48	40–45	10–48
pH	6–7	4–10	7–8	4.5–9.6†
Water activity	0.98	0.83–> 0.99*	0.98	0.87–> 0.99‡

† Aerobic (anaerobic lower pH limit is 5.0).
* Aerobic (anaerobic 0.90–> 0.99).
‡ Aerobic (anaerobic 0.92–> 0.99).
Source: Adapted from ICMSF (1996).

sheep and can also be introduced into milk by a contaminated food handler. In the study dairy products were responsible for 8 % of incidents (Wienecke *et al.*, 1993).

Since *Staph. aureus* is generally regarded as being a poor competitor it is most likely to pose problems in situations where it has been given a competitive advantage. For example:

- post-process contamination of heat-processed foods where the food is otherwise supportive of growth, but competitors have been eliminated by the heat process, e.g. sliced cooked meats, cream cakes, canned foods;
- fermented foods such as cheese or salami where failure of the starter culture allows staphylococci to grow;
- salted or partially dried foods where the reduced a_w gives the staphylococci a growth advantage.

22.2.4 Control measures

Prevention of contamination

Staphylococcus aureus is likely to be a natural contaminant of raw foods of animal origin, although this can be minimised through good hygienic practices. Its presence in well-cooked foods, however, indicates cross-contamination from raw product or, more often, contamination from a food handler. It is not feasible to screen food handlers and exclude carriers of *Staph. aureus* from work except in the case of those with obviously infected skin lesions. More effective measures would be to minimise direct handling of foods and to enforce strictly injunctions to wash hands thoroughly and to wear disposable gloves when handling foods.

Staphylococci can adhere to surfaces and participate in biofilm formation (Gotz, 2002). Inadequate cleaning of work surfaces and equipment can lead to the development of biofilms in food processing equipment that could act as a source of chronic contamination, a problem that could be compounded by the greater resistance to biocides exhibited by adhered organisms (Bolton *et al.*, 1988; Chmielewski and Frank, 2003). *Staphylococcus aureus* is sensitive to disinfecting agents such as hypochlorite, although MRSA displays greater resistance to quaternary ammonium compounds than methicillin-sensitive strains, and some markedly resistant strains have been isolated (Al-Masaudi *et al.*, 1988; Akimitsu *et al.*, 1999). In a comparison of hypochlorite and peracetic acid disinfection of stainless steel surfaces contaminated with *Staph. aureus* peracetic acid was much less effective, achieving a little over a 50 % reduction in adhered numbers at a concentration of 250 mg l^{-1} compared to a reduction of more than 96 % with 100 mg l^{-1} hypochlorite (Rossoni and Gaylarde, 2000). Staphylococcal food poisoning is particularly associated with restaurant and catering operations. The sources of contamination and their control in such settings have been reviewed by Soriano *et al.* (2002).

Elimination/inactivation

Heat processing is the most common microbial inactivation treatment in food production. The unexceptional heat resistance of *Staph. aureus* has already been referred to in Section 22.2.1 and quantitative estimates in terms of *D*-values presented in Table 22.1. Assuming a *z*-value of between 10 C° and 5 C°, the data of Firstenberg-Eden *et al.* (1977) indicate that a high temperature/short time (HTST) process (72 °C for 15 s) applied to milk will reduce viable *Staph. aureus* by 4 to > 6 log cycles.

The sensitivity of *Staph. aureus* to alternative inactivation technologies such as irradiation, high hydrostatic pressures and pulsed electric fields has also been investigated. Irradiation at doses of 2–7 kGy can effectively eliminate *Staph. aureus* from foods where, depending on the irradiation conditions and the substrate, D_{10} values range from 0.26 to 0.57 kGy (Farkas, 1998). Gram-positive bacteria are generally more resistant to high hydrostatic pressure than Gram-negatives, but a treatment of 600 MPa at 20 °C for 15 min reduced *Staph. aureus* numbers by 2 log cycles in UHT milk and by 3 log cycles in poultry meat (Patterson *et al.*, 1995). Comparable results were reported for a model ham system by Tassou and co-workers who noted that the lethality of a process was increased at higher temperatures (Tassou *et al.*, 2008). *Staphylococcus aureus* strains have been found to show more uniform sensitivity to pulsed electric fields than to heating at 58 °C (Cebrian *et al.*, 2007). Synergies between high pressure or pulsed electric field treatments and bacteriocins such as nisin and Pediocin AcH have also been noted (Patterson, 2005; Sobrino-Lopez and Martin-Belloso, 2006).

In the case of staphylococcal food poisoning, heat processes such as cooking are not the important control measure they are with most other foodborne pathogens. Heating can reduce risk by eliminating or reducing numbers of staphylococci before they have multiplied sufficiently to pose a direct hazard, but the heat resistance of any toxin they have produced will mean that it will survive all but the most extreme of processes. It is therefore possible to have toxic food containing few or no viable staphylococci.

The EC Regulation on Microbiological Criteria for Foodstuffs recognises this by having Process Hygiene Criteria for coagulase-positive staphylococci in milk powder and some cheeses based on microbial counts, while the Food Safety Criteria specify testing for staphylococcal toxin rather than the organisms. The criteria also specify that in cases where coagulase-positive staphylococci exceed 10^5 cfu/g then the batch has to be tested for staphylococcal enterotoxins (EC, 2004, 2005). A survey of retail cheeses in the UK using these criteria found that their microbiological quality with respect to *Staph. aureus* was generally very good, though the incidence of cheeses deemed borderline or unsatisfactory was much higher in those made from unpasteurised or thermised milk (26 out of 1795 samples) compared with pasteurised milk (three out of 2618 samples) (Little *et al.*, 2008).

Inhibition of growth

The limits for growth and toxin production were presented in Table 22.3. These data represent extreme cases since they are generally determined when all other conditions are optimal. The effects of combinations of sub-optimal factors can be determined using predictive software such as ComBase (http://wyndmoor.arserrc.gov/combase/). The single most effective and widely applied measure that controls *Staph. aureus* is chill storage – growth will not occur below 6–7 °C and the minimum temperature for toxin production is slightly higher and is likely to be higher still when other growth factors are sub-optimal. The decline in the number of outbreaks reported over recent years may reflect the more widespread use of refrigeration in food processing, distribution and storage. The large Japanese outbreak described previously was thought to be a consequence of a power failure that cut off refrigeration allowing temperatures to rise and growth of *Staph. aureus* to occur in the milk before it was dried.

A number of natural antimicrobial compounds such as plant essential oils exhibit activity against *Staph. aureus* and other organisms (Nychas and Skandamis, 2003). Where these occur in food ingredients they may well exert an inhibitory effect on microbial growth in a product, although they are not generally employed deliberately for this purpose. Despite a plethora of reports on the activity of plant antimicrobials relatively little is known about their mode of action. Polyphenols associated with tea have been shown to interact with the cell envelope of *Staph. aureus* to inhibit growth, and the activity of the olive metabolite oleuropein against *Staph. aureus* has been shown to be mediated in part by hydrogen peroxide production (Zanichelli *et al.*, 2005; Stapleton *et al.*, 2007).

The bacteriocin food preservative nisin is most active against bacterial spores, although at higher concentrations (approximately 10-fold) it is inhibitory to vegetative Gram-positive organisms such as *Staph. aureus*. Practical applications of nisin are largely confined to the inhibition of spores, and most interest in its use against vegetative pathogens has focused on *Listeria monocytogenes* rather than *Staph. aureus*. A competitive microflora is known to inhibit growth of *Staph. aureus*, and starter culture failure in the production of cheese or salami could lead to its multiplication to dangerous levels. Production of nisin *in situ* by a *Lactococcus lactis* starter culture was found to add nothing to the inhibition of *Staph. aureus* achieved as a result of lactic acid production (Yusof *et al.*, 1993).

22.2.5 Detection of the organism and enterotoxins

Because of the nature of staphylococcal intoxications, there are two approaches to determining the risk posed by particular foods. Enumeration of *Staph. aureus* is an indirect approach where the presence of low numbers may highlight a concern over contamination and high numbers (> 10^5 cfu/g) can indicate the possibility that sufficient toxin has been produced to cause illness. Detection of the toxin itself is of course the most direct method of determining risk, but the techniques involved are more complex and expensive.

Baird-Parker (BP) agar is the most widely used selective and diagnostic plating medium for the detection and enumeration of *Staph. aureus*. This contains tellurite, glycine and lithium chloride as selective agents, pyruvate to assist the recovery of injured cells and egg yolk and tellurite as the diagnostic system (Baird and Lee, 1995). Presumptive *Staph. aureus* appear as jet black colonies, due to the reduction of tellurite, surrounded by a zone of clearing produced by proteolysis of lipovitellin and sometimes an inner cloudy zone caused by precipitation of fatty acid salts. Some strains, particularly those from cattle which might therefore occur in dairy products, do not produce the typical egg yolk reaction. In such situations media based on BP agar but containing plasma or fibrinogen instead of egg yolk can be used. *Staphylococcus aureus* produce colonies with the normal appearance associated with BP agar but are surrounded by a zone of plasma coagulation and precipitation.

More recently, a number of chromogenic media have been developed for the isolation and identification *Staph. aureus* (Merlino *et al.*, 2000; Hedin and Fang, 2005; Cherkaoui *et al.*, 2007). These were designed principally for clinical application and are not widely used in food analysis.

Staphylococcus aureus is confirmed by its ability to produce coagulase. It is the only human *Staphylococcus* to produce coagulase, but animal strains such as *Staph. hyicus* and *Staph. intermedius* are also coagulase-positive. *Staphylococcus intermedius* does not reduce tellurite and therefore does not produce the characteristic jet black colonies on BP agar. Traditionally, the coagulase test is performed using the tube coagulase test where the use of citrated plasma should be avoided due to the ability of some bacteria such as enterococci to utilise citrate and produce false-positives. Nowadays, there are numerous kits where latex particles or red blood cells have been sensitised with fibrinogen or plasma to detect bound coagulase (clumping factor) and, in the case of latex particles, immunoglobulin to detect protein A on the *Staph. aureus* cell wall. Coagulase-positive *Staph. hyicus* can be differentiated from *Staph. aureus* using biochemical test kits such as the API® Staph.

The balance between the selectivity of a medium and its recovery of a target pathogen is a perennial problem in food microbiology. Sub-lethal injury caused by stresses such as acidification, heating, freezing or drying can affect recovery as it often results in a target organism becoming sensitive to selective agents normally tolerated. In BP agar this is addressed by the inclusion of pyruvate, as described above. Anaerobic incubation can also improve the recovery of *Staph. aureus* cells injured by heat treatment (Ugborogho and Ingham, 1994). Other approaches have employed non-selective solid media, which allow recovery of injured cells, with an overlay of selective media. In the case of *Staph. aureus*, use of a more selective medium than BP agar with this overlay technique appeared to offer some advantages in situations where samples are heavily contaminated with competing bacteria (Sandel *et al.*, 2003).

Molecular techniques based on the polymerase chain reaction (PCR) have been developed using specific primers directed against a variety of sequences in the *Staph aureus* genome including genes for thermostable nuclease and enterotoxins (Wilson *et al.*, 1991; Brakstad *et al.*, 1992; Mantynen *et al.*, 1997; Becker *et al.*, 1998; Cremonesi *et al.*, 2007; Riyaz-Ul-Hassan *et al.*, 2008). There are also kits available for detection based on nucleic acid hybridisation, such as AccuPROBE® which is based on detection of specific ribosomal RNA sequences.

Enterotoxins can be detected based on immunological methods following extraction from the food matrix, and there are numerous kits available using enzyme linked immunoassay or latex agglutination (Su and Wong, 1997). All procedures involve some form of work-up to extract the toxin, and the immunoassays themselves are occasionally subject to interference from food components.

22.3 Other Gram-positive cocci

22.3.1 Enterococci

Enterococcus faecalis and *Enterococcus faecium*, formerly classified as members of the Group D or faecal streptococci, are common inhabitants of the human and animal gut. They were first associated with food poisoning in an outbreak caused by cheese in the USA in the 1920s (Linden *et al.*, 1926 cited in Deibel and Silliker, 1963) and occasional reports implicating them appeared up to the 1950s. The reported symptoms were mild, vague and variable and volunteer feeding studies gave conflicting results. Reviewing the published literature and some original experimental data relating to a total of 23 different enterococcal strains fed to human volunteers, Deibel and Silliker concluded that 'the association of enterococci and food poisoning is questionable' (Deibel and Silliker, 1963).

Enterococci are relatively resistant organisms and occur in a wide range of environmental niches such as soil, water, plants and the gastrointestinal tract of warm-blooded animals. They also occur quite commonly in high numbers (up to 10^7cfu/g) in many foods, particularly fermented products such as Mediterranean cheeses and fermented meats, and they are used as probiotics in humans and animals (Giraffa, 2002; Foulquié Moreno *et al.*, 2006). Their presence in the gut of humans and animals has led to a role as indicators of faecal contamination in water (Clesceri *et al.*, 1998).

They are also important opportunistic pathogens and a major cause of nosocomial infections such as bacteraemia and endocarditis. They are recognised as possessing a number of virulence factors, although the association between these and clinical illness is not always well established. Studies do, however, suggest a lower incidence of virulence factors in food isolates. There is also considerable concern over the resistance of enterococci to a wide variety of antibiotics and the role that antibiotic usage in animal husbandry may have played in this process (Franz and Holzapfel, 2006).

This dual role of the enterococci has led the Scientific Committee of the European Food Safety Authority (EFSA) to be cautionary in its approach to the enterococci when considering the safety of a range of organisms used in food and feed production. While it was content to recommend 'Qualified Presumption of Safety (QPS)'* status for a number of species of other non-spore-forming Gram-positive genera such as *Lactobacillus*, *Lactococcus*, *Leuconostoc*, *Pediococcus* and *Bifidobacterium*, it concluded that 'Due to safety concerns and the lack of information on safety, no members of the genus *Enterococcus* can be proposed for QPS status' (EFSA, 2007).

22.3.2 Group A Streptococci

Beta haemolytic Group A streptococci (*Streptococcus pyogenes*) are associated with pharyngitis ('strep throat'), known as scarlet fever when accompanied by a red rash and fever, and sometimes rheumatic fever. Although it is frequently spread by aerosols and fomites, milk was a common vehicle prior to the advent of pasteurisation, and foodborne outbreaks are still occasionally reported. Foodborne outbreaks have an incubation period ranging from 12 h to several days and are characterised by a clustering of cases and a high attack rate. They are generally associated with institutional settings such as prisons, dormitories and receptions (Martin *et al.*, 1985; Levy *et al.*, 2003; Sarvghad *et al.*, 2005). Eggs (egg mayonnaise, egg salad, curried egg rolls) are a commonly implicated vehicle, but other foods have been reported (e.g. chicken salad, ice cream, steamed lobster). The food is normally contaminated by aerosols or more directly by an infected food handler, although the possibility of deliberate contamination has sometimes been alluded to (Kaluski *et al.*, 2006). Temperature abuse which will allow

* QPS is similar in concept and purpose to the Generally Recognised as Safe (GRAS) definition used in the USA modified to take account of different regulatory practices in Europe.

low levels of contamination to multiply in the food can also be a factor. Control can be exercised by minimising the handling of food, excluding workers suffering from streptococcal infections and by ensuring adequate temperature control.

22.3.3 *Streptococcus equi* subspecies *zooepidemicus*

Streptococcus equi subsp. *zooepidemicus* is a Group C beta-haemolytic streptococcus. It can be found on the mucus membranes of healthy horses and cattle but it does also cause infections such as mastitis in sheep, goats, cows and other animals. Human infections are rare and are generally related to contact with animals such as horses. Symptoms include pharyngitis, septicaemia, pneumonia, septic arthritis, endocarditis and meningitis, and post-infective glomerulonephritis also occurs. Foodborne outbreaks have been associated with unpasteurised or inadequately pasteurised milk or cheese (CDC, 1983; Francis *et al.*, 1993; Balter *et al.*, 2000; Kuusi *et al.*, 2006).

22.4 Future trends

We have a good understanding of staphylococcal food poisoning and how it can be prevented by the correct handling and storage of foods. The widespread accessibility and use of chill storage in both the commercial and domestic situation is undoubtedly an important factor in its decreasing incidence. However, when lapses occur as a result of human and/or mechanical failure, so too do outbreaks. To prevent these requires the constant reinforcement of basic food hygiene messages and rigorous application of appropriate Hazard Analysis Critical Control Point (HACCP) schemes. To paraphrase a well known quotation, 'the price of control of foodborne illness is eternal vigilance'.

22.5 Further reading

Baird-Parker T.C. (2000) *Staphylococcus aureus*, in B M Lund, T C Baird-Parker and GW Gould (eds), *The Microbiological Safety and Quality of Food*, Vol. 2, Aspen, Gaithersburg, MD 1317–35.

22.6 References

Adak GK, Meakins SM, Yip H, Lopman BA and O'brien SJ (2005) Disease risks from foods, England and Wales, 1996–2000, *Emerg Infect Dis*, **11**, 365–72.

Adams MR and Moss MO (2008) *Food Microbiology*, The Royal Society of Chemistry, Cambridge.

Akimitsu N, Hamamoto H, Inoue R, Shoji M, Akamine A, Takemori K, Hamasaki N and Sekimizu K (1999) Increase in resistance of methicillin-resistant *Staphylococcus aureus* to β-lactams caused by mutations conferring resistance to benzalkonium chloride, a disinfectant widely used in hospitals, *Antimicrob Agents Chemother*, **43**, 3042–3.

Aliu B and Bergdoll MS (1988) Characterization of staphylococci from patients with Toxic Shock Syndrome, *J Clin Microbiol*, **26**, 2427–8.

Al-Masaudi S, Day MJ and Russell AD (1988) Sensitivity of methicillin-resistant *Staphylococcus aureus* strains to some antibiotics, antiseptics and disinfectants, *J Appl Bacteriol*, **65**, 329–37.

Anderson JE, Beelman RR and Doores S (1996) Persistence of serological and biological activities of staphylococcal enterotoxin A in canned mushrooms, *J Food Prot*, **59**, 1292–9.

Anon. (1886) A new poison in ice cream, *The New York Times* June 26.

Asao T, Kumeda Y, Kawai T, Shibata T, Oda H, Haruki K, Nakazawa H and Kozaki S (2003) An extensive outbreak of staphylococcal food poisoning due to low-fat milk in Japan: estimation of enterotoxin A in the incriminated milk and powdered skim milk, *Epidemiol Infect*, **130**, 33–40.

Baird RM and Lee WH (1995) Media used in the detection and enumeration of *Staphylococcus aureus*, in JEL Corry, GDW Curtis and RM Baird (eds), *Culture Media for Food Microbiology*, Elsevier, Amsterdam, 77–87.

Balaban N and Rasooly A (2000) Staphylococcal enterotoxins, *Int J Food Microbiol*, **61**, 1–10.

Balter S, Benin A, Pinto SW, Teixeira LM, Alvim GG, Luna E, Jackson D, Laclaire L, Elliott J, Facklam R and Schuchat A (2000) Epidemic nephritis in Nova Serrana, Brazil, *Lancet*, **355**, 1776–80.

Becker K, Roth R and Peters G (1998) Rapid and specific detection of toxigenic *Staphylococcus aureus*: use of two multiplex PCR enzyme immunoassays for amplification and hybridization of staphylococcal enterotoxin genes, exfoliative toxin genes, and toxic shock syndrome toxin 1 gene, *J Clin Microbiol*, **36**, 2548–53.

Becker K, Keller B, Von Eiff C, Bruck M, Lubritz G, Etienne J and Peters G (2001) Enterotoxigenic potential of *Staphylococcus intermedius*, *Appl Environ Microbiol*, **67**, 5551–7.

Bennett RW (1992) The biomolecular temperament of staphylococcal enterotoxin in thermally processed foods, *J Assoc Off Agric Chem*, **75**, 6–12.

Bolton KJ, Dodd CER, Mead GC and Waites WM (1988) Chlorine resistance of strains of *Staphylococcus aureus* isolated from poultry processing plants, *Appl Microbiol*, **6**, 31–4.

Brakstad OG, Aasbakk K and Maeland JA (1992) Detection of *Staphylococcus aureus* by polymerase chain reaction amplification of the *nuc* gene, *J Clin Microbiol*, **30**, 1654–60.

Breckinridge JC and Bergdoll MS (1971) Outbreak of foodborne gastroenteritis due to a coagulase negative enterotoxin producing staphylococcus, *New England J Medicine*, **248**, 541–3.

CDC (1983) Group C streptococcal infections associated with eating homemade cheese – New Mexico, *MMWR*, **32**(39), 510, 515–6.

Cebrian G, Sagarzazu N, Pagan R, Condon S and Manas P (2007) Heat and pulsed electric field resistance of pigmented and non-pigmented enterotoxigenic strains of *Staphylococcus aureus* in exponential and stationary phase of growth, *Int J Food Microbiol*, **118**, 304–11.

Cherkaoui A, Renzi G, Francois P and Schrenzel J (2007) Comparison of four chromogenic media for culture-based screening of meticillin-resistant, *Staphylococcus aureus*, *J Med Microbiol*, **56**, 500–3.

Chmielewski RAN and Frank JF (2003) Biofilm formation and control in food processing facilities, *Comp Rev Food Sci Technol*, **2**, 22–32.

Clesceri LS, Greenberg AE and Eaton AD (eds) (1998) *Standard Methods for the Examination of Water and Wastewater*, 20th edn, 9–32 and 9–75, American Public Health Association, Washington, DC.

Cremonesi P, Perez G, Pisoni G, Moroni P, Morandi S, Luzzana M, Brasca M and Castiglioni B (2007) Detection of enterotoxigenic *Staphylococcus aureus* isolates in raw milk cheese, *Lett Appl Microbiol*, **45**, 586–91.

Dack GM, Cary WE, Woolpert O and Wiggins HJ (1930) An outbreak of food poisoning proved to be due to a yellow haemolytic *Staphylococcus*, *J Prev Med*, **4**, 167–75.

Deibel RH and Silliker JH (1963) Food-poisoning potential of the enterococci, *J Bacteriol*, **85**, 827–32.

Denny CB, Tan PL and Bohrer CW (1966) Heat inactivation of staphylococcal enterotoxin A, *J Food Sci*, **31**, 762–7.

Denny CB, Humber JY and Bohrer CW (1971) Effect of toxin concentration on the heat inactivation of staphylococcal enterotoxin A in beef bouillon and phosphate buffer, *Appl Microbiol*, **21**, 1064–6.

EC (2004) Commission Recommendation of 19 December 2003 concerning a co-ordinated programme for the official control of foodstuffs for 2004, *Official Journal of the European Union*, **L6**, 10 January, 29–37.

EC (2005) Commission Recommendation of 1 March 2005 concerning a co-ordinated programme for the official control of foodstuffs for 2005, *Official Journal of the European Union*, **L59**, 5 March, 27–39.

EFSA (2007) Scientific report on the assessment of Gram-positive, non-sporulating bacteria, European Food Safety Authority, *The EFSA Journal*, **587**, 1–16.

Evenson ML, Hinds MW, Bernstein RS and Bergdoll MS (1988) Estimation of human dose of staphylococcal enterotoxin A from a large outbreak of staphylococcal food poisoning involving chocolate milk, *Int J Food Microbiol*, **7**, 311–6.

Farkas J (1998) Irradiation as a method for decontaminating food: a review, *Int J Food Microbiol*, **44**, 189–204.

Firstenberg-Eden B, Rosen B and Mannheim CH (1977) Death and injury of *Staphylococcus aureus* during thermal treatment of milk, *Canadian J Microbiol*, **23**, 1034–7.

Foulquié Moreno MR, Sarantinopoulos P, Tsakalidou E and De Vuyst L (2006) The role and application of enterococci in food and health, *Int J Food Microbiol*, **106**, 1–24.

Francis AJ, Nimmo GR, Efstratiou A, Galanis V and Nuttall N (1993) Investigation of milk-borne *Streptococcus zooepidemicus* infection associated with glomerulonephritis in Australia, *J Infect*, **27**, 317–23.

Franz CMAP and Holzapfel WH (2006) Enterococci, in Y Motarjemi and M Adams (eds), *Emerging Foodborne Pathogens*, Woodhead, Cambridge, 557–613.

Fukuda S, Tokuno H, Ogawa H, Sasaki M, Kishimoto T, Kawano J, Shimizu A and Kimura S (1984) Enterotoxigenicity of *Staphylococcus intermedius* strains isolated from dogs, *Zentralbl Bakteriol, Mikrobiol Hyg (A)*, **258**, 360–7.

Giraffa G (2002) Enterococci from foods, *FEMS Microbiol Rev*, **26**, 163–71.

Gotz F (2002) *Staphylococcus* and biofilms, *Mol Microbiol*, **43**, 1367–78.

Halpin-Dohnalek MI and Marth EH (1989) *Staphylococcus aureus*: production of extracellular compounds and behavior in foods – a review, *J Food Prot*, **52**(4), 267–82.

Hedin G and Fang H (2005) Evaluation of two new chromogenic media, CHROMagar MRSA and *S. aureus* ID, for identifying *Staphylococcus aureus* and screening methicillin-resistant *S aureus*, *J Clin Microbiol*, **43**, 4242–4.

HPA (2008) *Staphylococcus aureus*, Health Protection Agency, London, available at: www.hpa.org.uk/infections/topics_az/staphylo/food/data_ew.htm, accessed November 2008.

Humber JY, Denny CB and Bohrer CW (1975) Influence of pH on the heat inactivation of staphylococcal enterotoxin A as determined by monkey feeding and serological assay, *Appl Microbiol*, **30**, 755–8.

ICMSF (1996) *Micro-organisms in Foods Vol. 5 Microbiological Specifications of Food Pathogens*, International Commission on Microbiological Specification for Foods, Blackie Academic and Professional, London.

Jones TF, Kellum ME, Porter S, Bell M and Schaffner W (2002) An outbreak of community-acquired foodborne illness caused by methicillin-resistant *Staphylococcus aureus*, *Emerg Infect Dis*, **8**, 82–4.

Jordan EO, Dack GM and Woolper O (1931) The effect of heat, storage, and chlorination on the toxicity of staphylococcal filtrates, *J Prevent Med*, **5**, 383–6.

Kaluski DN, Barak E, Kaufman Z, Valinsky L, Marva E, Korenman Z, Gorodnitzki Z, Yishal R, Koltai D, Leventhal A, Levine S, Havkin O and Green MS (2006) A large food-borne outbreak of Group A streptococcal pharyngitis in an industrial plant: potential for deliberate contamination, *Isr Med Assoc J*, **8**, 618–21.

Kérouanton A, Hennekinne JA, Letertre C, Petit L, Chesneau O, Brisabois A and De Buyser ML (2007) Characterization of *Staphylococcus aureus* strains associated with food poisoning outbreaks in France, *Int J Food Microbiol*, **115**, 369–75.

Khambaty FM, Bennett RW and Shah DB (1994) Application of pulsed-field gel electrophoresis to the epidemiological characterization of *Staphylococcus intermedius* implicated in a food-related outbreak, *Epidemiol Infect*, **113**, 75–81.

Kitai S, Shimizu A, Kawano J, Sato E, Nakano C, Uji T and Kitigawa H (2005) Characterization of methicillin resistant *Staphylococcus aureus* isolated from raw chicken meat in Japan, *J Vet Med Sci*, **67**, 107–10.

Kluytmans J, Van Leeuwen W, Goessens W, Hollis R, Messer S, Herwaldt L, Bruining H, Heck M, Rost J and Van Leeuwen N (1995) Food initiated outbreak of methicillin-resistant *Staphylococcus aureus* analyzed by phenol- and genotyping, *J Clin Microbiol*, **33**, 1121–8.

Kluytmans J, Van Belkum A and Verbrugh H (1997) Nasal carriage of *Staphylococcus aureus*: epidemiology, underlying mechanisms, and associated risks, *Clin Microbiol Rev*, **10**, 505–20.

Kuusi M, Lahti E, Virolanian A, Hatakka M, Vuento R, Rantala L, Vuopio-Varkila J, Seuna E, Karppelin M, Hakkinen M, Takkinen J, Gindonis V, Siponen K and Huotari K (2006) An outbreak of *Streptococcus equi* subspecies *zooepidemicus* associated with consumption of fresh goat cheese, *BMC Infect Dis*, **6**, 36–43.

Lawrynowicz-Paciorek M, Kochman M, Piekarska K, Grochowska A and Windyga B (2007) The distribution of enterotoxin and enterotoxin-like genes in *Staphylococcus aureus* strains isolated from nasal carriers and food samples, *Int J Food Microbiol*, **117**, 319–23.

LEE JH (2003) Methicillin (Oxacillin)-resistant *Staphylococcus aureus* strains isolated from major food animals and their potential transmission to humans, *Appl Environ Microbiol*, **69**, 6489–94.

Le Loir Y, Baron F and Gautier M (2003) *Staphylococcus aureus* and food poisoning, *Genet Mol Res*, **2**, 63–76.

Levy M, Johnson CG and Karaa E (2003) Tonsillopharyngitis caused by foodborne Group A Streptococcus: a prison-based outbreak, *Clin Infect Dis*, **36**, 175–82.

Little CL, Rhoades JR, Sagoo SK, Harris J, Greenwood M, Mithani V, Grant K and McLauchlin J (2008) Microbiological quality of retail cheeses made from raw, thermized or pasteurized milk in the UK, *Food Microbiol*, **25**, 304–12.

Lynch M, Painter J, Woodruff R and Braden C (2006) Surveillance for foodborne disease outbreaks United States 1998–2002, *MMWR*, **55**(10), 1–34.

Mantynen V, Niemala S, Kaijalainen S, Pirhonen T and Lindstrom K (1997) MPN-PCR-quantification method for staphylococcal enterotoxin cl gene from fresh cheese, *Int J Food Microbiol*, **36**, 135–43.

Marín ME, De La Rosa MC and Cornejo I (1992) Enterotoxigenicity of *Staphylococcus* strains isolated from Spanish dry-cured hams, *Appl Environ Microbiol*, **58**, 1067–9.

Martin TA, Hoff GL, Gibson V and Biery RM (1985) Foodborne streptococcal pharyngitis Kansas City, Missouri, *Am J Epidemiol*, **122**, 706–9.

Mead PS, Slutsker L, Dietz V, McCaig LF, Bresee JS, Shapiro C, Griffin PM and Tauxe RV (1999) Food-related illness and death in the United States, *Emerg Infect Dis*, **5**, 607–25.

Merlino J, Leroi M, Bradbury R, Veal D and Harbour C (2000) New chromogenic identification and detection of *Staphylococcus aureus* and methicillin-resistant *S. aureus*, *J Clin Microbiol*, **38**, 2378–80.

Morandi S, Brasca M, Lodi R, Cremonesi P and Castiglioni B (2007) Detection of classical enterotoxins and identification of enterotoxin genes in *Staphylococcus aureus* from milk and dairy products, *Vet Microbiol*, **124**, 66–72.

Normanno G, Corrente M, La Salandra G, Dambrosio A, Quaglia NC, Parisi A, Greco G, Bellacicco AL, Virgilio S and Celano GV (2007) Methicillin-resistant *Staphylococcus aureus* (MRSA) in foods of animal origin product in Italy, *Int J Food Microbiol*, **117**, 219–22.

Notermans S and Heuvelman CJ (1983) Combined effects of water activity, pH and suboptimal temperature on growth and enterotoxin production, *J Food Sci*, **48**, 1832–5.

Novick RP (2003) Autoinduction and signal transduction in the regulation of staphylococcal virulence, *Mole Microbiol*, **48**, 1429–49.

Nychas G-JE and Skandamis PN (2003) Antimicrobials from herbs and spices, in S. Roller (ed.), *Natural Antimicrobials for the Minimal Processing of Foods*, Woodhead Cambridge, 176–200.

Patterson MF (2005) Microbiology of pressure-treated foods, *J Appl Microbiol*, **98**, 1400–9.

Patterson MF, Quinn M, Simpson R and Gilmour A (1995) Sensitivity of vegetative pathogens to high hydrostatic pressure in phosphate-buffered saline and foods, *J Food Prot*, **58**, 524–9.

Read RB and Bradshaw JG (1966) Staphylococcal enterotoxin B thermal inactivation in milk, *J Dairy Sci*, **49**, 202–3.

Riyaz-Ul-Hassan S, Verma V and Qazi GN (2008) Evaluation of three different molecular markers for the detection of *Staphylococcus aureus* by polymerase chain reaction, *Food Microbiol*, **25**, 452–9.

Rosenberg U and Sinell HJ (1989) The effect of microwave heating on some food spoilage or pathogenic bacteria, *Zentralbl Hyg Umweltmed*, **188**, 271–83.

Rossoni EMM and Gaylarde CC (2000) Comparison of sodium hypochlorite and peracetic acid as santising agents for stainless steel food processing surfaces using epifluorescence microscopy, *Int J Food Microbiol*, **61**, 81–5.

Sandel MK, Wu Y-FG and McKillip JL (2003) Detection and recovery of sublethally-injured enterotoxigenic *Staphylococcus aureus*, *J Appl Microbiol*, **94**, 90–4.

Sarvghad MR, Naderi HR, Naderi-Nassab M, Majdzadeh R, Javanian M, Faramarzi H and Fatehmanish P (2005) An outbreak of food-borne group A *Streptococcus* (GAS) tonsillopharyngitis among residents of a dormitory, *Scand J Infect Dis*, **37**, 647–50.

Sobrino-Lopez A and Martin-Belloso O (2006) Enhancing inactivation of *Staphylococcus aureus* in skim milk by combining high-intensity pulsed electric fields and nisin, *J Food Prot*, **69**, 345–53.

Soriano JM, Font G, Molto JC and Manes J (2002) Enterotoxigenic staphylococci and their toxins in restaurant foods, *Trends Food Sci Technol*, **13**, 60–7.

Stapleton PD, Shah S, Ehlert K, Hara Y and Taylor PW (2007) The β-lactam resistance modifier (-)-epicatechin gallate alters the architecture of the cell wall of *Staphylococcus aureus*, *Microbiology*, **153**, 2093–103.

Stiles ME and Witter LD (1965) Thermal inactivation, heat injury and recovery of *Staphylococcus aureus*, *J Dairy Sci*, **48**, 677–81.

Su H YC and Wong AC (1997) Current perspectives on detection of staphylococcal enterotoxins, *J Food Prot*, **60**, 195–202.

Tassou CC, Panagou EZ, Samaras FJ, Galiatsatou P and Mallidis CG (2008) Temperature-assisted high hydrostatic pressure inactivation of *Staphylococcus aureus* in a ham model system: evaluation in selective and non-selective medium, *J Appl Microbiol*, **104**, 1764–73.

Thomas D, Chou S, Dauwalder O and Lina G (2007) Diversity in Staphylococcus enterotoxins, *Chem Immunol Allergy*, **93**, 24–41.

Ugborogho TO and Ingham SC (1994) Increased D-values of *Staphylococcus aureus* resulting from anaerobic heating and enumeration of survivors, *Food Microbiol*, **11**, 275–80.

Van Loo IHM, Diederen BMW, Savelkoul PHM, Woudenberg JHC, Roosendaal R, Van Belkum A, Lemmens-Den Toom N, Verhulst C, Van Keulen PHJ and Kluytmans JAJW (2007) Methicillin-resistant *Staphylococcus aureus* in meat products, the Netherlands, *Emerg Infect Dis*, **13**, 1753–5.

Vaughan VC (1884) Poisonous or sick cheese, *Public Health*, **10**, 241–5.

Wienecke AA, Roberts D and Gilbert RJ (1993) Staphylococcal food poisoning in the United Kingdom 1969–90, *Epidemiol Infect*, **110**, 519–31.

Wilson IG, Cooper JE and Gilmour A (1991) Detection of enterotoxigenic *Staphylococcus aureus* in dried skim milk powder; use of polymerase chain reaction for amplification and detection of staphylococcal enterotoxin genes *entB* and *entC1* and the thermonuclease gene *nuc*, *Appl Environ Microbiol*, **57**, 1793–8.

Yusof RM, Morgan JB and Adams MR (1993) Bacteriological safety of a fermented weaning food containing L-lactate and nisin, *J Food Prot*, **56**, 414–7.

Zanichelli D, Baker TA, Clifford MN and Adams MR (2005) Inhibition of *Staphylococcus aureus* by oleuropein is mediated by hydrogen peroxide, *J Food Prot*, **68**, 1492–6.

23

Pathogenic *Clostridium* species

P. A. Gibbs, Leatherhead Food International, UK and Universidade Católica Portuguesa, Portugal

Abstract: This chapter considers those species of the genus *Clostridium* which are responsible for serious foodborne diseases of man and animals, e.g. botulism and enterotoxæmias including food poisoning. The disease syndromes, occurrence, growth, survival and detection methods for the neurotoxic clostridia (*Clostridium botulinum*) are reviewed. The efficacy of the various methods for controlling the growth and toxin production by the botulinal species in foods, are examined. The characteristics of the gastrointestinal infective species *Cl. perfringens* and *Cl. difficile*, disease symptoms, sporulation, toxins, foods involved and detection and control methods, are reviewed. Opinions on the likely future developments in the occurrence and food vehicles of case and outbreaks of these foodborne pathogens are also expressed.

Keywords: botulism, neurotoxin, perfringens food poisoning, *Cl. difficile*.

23.1 Introduction

The spore-forming bacteria belong to the prokaryotic Family *Bacillaceae*, a heterogeneous group of rod-shaped, Gram-positive bacteria. Bacterial spores are essentially 'resting stages' in the life cycle of the bacteria, their formation generally resulting from a depletion of some essential nutrient in the growth milieu, or the accumulation of a signalling compound ('quorum sensing') to a critical level (see Miller and Bassler, 2001; Novak and Fratamico, 2004).[1,2] Species of the Family are widely distributed in nature, mainly as a result of the resistance and longevity of their spores, and readily contaminate foods. Spores are dormant structures, with a central 'core', where the genetic element(s) of the cell resides, together with the enzymes necessary for the germination and initial outgrowth of the germinated spore. The core is highly protected from the external environment by a thick protein coat and a cortex. The protein coat is highly resistant to chemical and radiation damage, whereas it is thought that the cortex provides a very low a_W surrounding for the core, resulting in a high resistance to heat. Spores are very resistant to many environments lethal to vegetative cells, e.g. high temperatures, drying, UV and other ionising irradiation, high pressure, disinfectants and chemical sterilants, etc. These properties result in difficulties for the safe processing of foods, especially canning.

For rapid germination, clostridial spores usually require some form of shock treatment (activation), usually applied as a sub-lethal heat shock (*c.* 75–80 °C for 10 min), although high pressures or low pH values, can also activate spores. A supply of specific germinants, generally an organic acid (and/or CO_2) and one or more aminoacids (e.g. lactic acid, alanine, and phenylalanine for *Cl. bifermentans*), is necessary for germination. During germination, clostridial spores lose their intrinsic properties of heat and other resistances, including to oxygen (or high Eh), becoming almost as sensitive as vegetative cells. Thus a sequence of heating, cooling and reheating ('tyndallisation') has been suggested as a means of processing foods for safety. However, such treatments have been demonstrated not to be successful in practice, as not all spores germinate, even after several sequential treatments (Gould, 2006).[3] Also, the time and temperature between heat treatments needs careful control if the surviving spores are not to germinate, grow and produce a further crop of spores before subsequent heating. However, as a means of reducing the numbers of heat-resistant spores, or increasing the lag time for spore germination, especially at low temperatures, tyndallisation may still have advantages. Outgrowth of the germinated spores usually requires more nutrients than germination, since the spore contains few reserves and needs to synthesise most of the enzyme systems necessary for vegetative growth.

Clostridial species show a wide variety of metabolic types, from strictly saccharolytic (e.g. the butyric species) to strictly proteolytic – using only amino acids for energy production (e.g. *Cl. tetani*). Species of the genus *Clostridium* are responsible for several serious diseases of man and animals, e.g. botulism, tetanus, several different types of 'black disease' of cattle, enterotoxæmias including food poisoning, gas gangrene, etc. Many of these species produce powerful extra-cellular enzymes responsible for degradation of carbohydrates, proteins and fats, including haemolysins that in some species are used for strain identification (e.g. *Cl. perfringens*), and may also produce large quantities of gas (restricting blood flow as in gas gangrene), that contribute to the efficacy of infective processes.

23.2 Neurotoxic *Clostridium* species; *Clostridium botulinum*, general characteristics

Botulism as a neuroparalytic intoxication was recognised at least one century ago as often being derived from eating sausage (Latin, *botulus*). The producing organism is a strictly anaerobic, Gram-positive, spore-forming bacillus and is widely distributed in soils, freshwater and marine sediments (see tables in Dodds and Austin, 1997).[4] The toxin type E strain is particularly prevalent in marine and freshwater sediments where there is considerable land drainage from rivers and little water movement, e.g. the Baltic and coasts of Scandinavia, some areas of the Great Lakes in North America (see e.g. Merivirta *et al.*, 2006).[5] However, other authors regard Type E as a true aquatic species, as it grows well in decomposing carcases of fish, shellfish and invertebrates, and outbreaks of botulism in fish and water birds are generally correlated with warmer temperatures, low redox potential (low oxygen levels) in and close to the bottom sediment, and low levels of mixing of the bottom sediment and of the lower water levels (see e.g. Bott *et al.*,1968; Huss, 1980; Hielm *et al.*, 1988; Hyytiä-Trees 1999).[6,7,8,9] The recently described ability of the organism to degrade chitin, may enhance growth in aquatic insect larvae (Sebaihia *et al.*, 2007).[10]

The frequency of occurrence of *Cl. botulinum* in soils and sediment samples varied from 3–6 % (UK; South Africa; Thailand; Rome, Italy) to *c.* 50 % (USA; Switzerland; China; Japan; New Zealand) and *c.* 100 % (Scandinavian coastal sediment; Netherlands soil; Green Bay sediment, USA). The MPN/kg in soil and sediment samples varied from 1–3 (South Africa; UK; Thailand), *c.* 50–100/kg (Brazil; Iran Caspian sea sediment; New Zealand sediment) to > 500/kg (Green Bay sediment, USA; Scandinavian coast sediment; Netherlands soil; Sinkiang, China soil [*c.* 25 000/kg]). The psychrotrophic, non-proteolytic types (B, E) were more prevalent in colder soils and sediments (e.g. Scandinavian, Alaskan samples), whereas proteolytic, mesophilic strains (A, B) were more prevalent in samples from warmer sites. Type F strains were isolated relatively infrequently. The frequency, MPN/kg and toxigenic types isolated from food samples, especially fish, tended to reflect the frequency and occurrence in the local environment (Dodds and Austin, 1997),[4] and agricultural or other human uses of soil (non-virgin soils) appear to increase the frequency and occurrence of *Cl. botulinum* compared with virgin soils (see Luquez *et al.*, 2005 for references to occurrence in several countries).[11]

Although all strains of clostridia capable of producing the neurotoxin were originally called *Cl. botulinum*, it is now recognised that this species is heterogeneous and certain strains of other clostridial species may be capable of producing the toxin, e.g. *Cl. butyricum*, *Cl. barati*. There are four groups of strains of *Cl. botulinum* identified on the basis of their physiological characteristics and the immunological properties of the toxin produced (see Table 23.1), plus two other neurotoxigenic species, *Cl. butyricum* and

Table 23.1 Characteristics of groups of *Clostridium botulinum* strains[a]

Characteristic	Group			
	I	II	III	IV[b]
Neurotoxin type	A, B, F	B, E, F	C, D	G
Strain origin	Soil	Aquatic environments	Soil, water, dead animals	ND
Typical food vehicles of cases	Meats, vegetables, canned foods, honey for infant botulism	Minimally processed fish, meats and packaged foods	Animal intoxications from forage, dead invertebrates or other animals	ND
P/NP[c]	P	NP	NP	ND
Min. growth temp.	10 °C	3.3 °C	15 °C	ND
Opt. growth temp.	35–40 °C	20–30 °C	40 °C	37 °C
Min. pH for growth	4.6	5.0	ND	ND
Min. a_W for growth [%NaCl w/v][d]	0.94 [10]	0.97 [5]	ND	ND
$D_{100\,°C}$ of spores (min)	25	< 0.1	0.1–0.9	0.8–1.1
$D_{121\,°C}$ of spores (min)	0.1–0.2 ['Bot cook' F_o3]	< 0.001	ND	ND

[a] Adapted from Dodds and Austin (1997)[4].
[b] Now recognised as *Cl. argentinense*.
[c] P, proteolytic, NP non-proteolytic.
[d] % NaCl-on-water.
ND, not determined, or unknown.

Cl. baratii (produces Type F toxin), which form two further groups, and *Cl. argentinense* (produces Type G toxin: see Poulain *et al.*, 2006; Bohnert *et al.*, 2006; Alouf and Popoff, 2006).[12,13,14] For food microbiologists only the groups I and II strains of *Cl. botulinum* are of major concern, producing toxin types A, B, E and, rarely, type F. A particularly useful reference on laboratory diagnosis of *Cl. botulinum* is by Lindström and Korkeala (2006).[15] Genomic research has confirmed the phenotypic classification of *Cl. botulinum* as a heterogeneous collection of organisms into four groups, all producing botulinal neurotoxins, and the species identity of strains not *Cl. botulinum* but producing botulinal neurotoxins, e.g. strains of *Cl. butyricum* and *Cl. baratii* (Hutson *et al.*, 1993; East *et al.*, 1996; Sebaihia *et al.*, 2007).[16,17,10]

Ingestion of botulinal toxins results in muscle paralysis by blockade of the release of acetylcholine from the nerve cell across the neuromuscular synapse. Occasionally in infants, generally of < 12 months of age, botulinal spores can germinate and colonise the intestine, releasing toxin towards the end of the growth phase and on lysis of the producing cells, resulting in the so-called 'floppy baby' syndrome. Infant botulism is often related to ingestion of honey that has high levels of botulinum spores (*c.* 10^{3-4}/kg in honey samples associated with botulism, but *c.* 1–10 spores/kg in other samples of honey), and is mainly caused by strains of Group I (toxin types A and B), although rarely other toxin types and other sources of botulinal spores have been recorded, including types C, E and F (e.g. Midura and Arnon 1976; Arnon *et al.*, 1979; Midura, 1996; Fenicia *et al.*, 2002; Barash *et al.*, 2005; Brett *et al.*, 2005; Johnson *et al.*, 2005; Keet *et al.*, 2005).[18,19,20,21,22,23,24,25] This syndrome is also not unknown in adults with major changes in gastrointestinal flora, e.g. as a result of surgery or broad spectrum antibiotic treatment.

The initial symptoms of ingestion of toxic food can occur within 12–36 h but may be delayed up to 10 d or longer; the earlier the symptoms occur, the more serious is the intoxication. The early symptoms are generally nausea and/or vomiting, closely followed by neurological symptoms such as double vision and fixed dilated pupils, speech impediment, difficulty in swallowing, etc. These symptoms then develop into a more general lack of muscular coordination and weakness, fatigue and respiratory distress. The intoxication has often been misdiagnosed as Guillain-Barré syndrome but, whereas this begins in the peripheral systems and ascends towards the head, botulism affects the cranial nerves first. The toxic dose for humans is unknown, but estimated to be of the order of 10–30 ng, but for mice the dose is less than 0.1 ng/kg, among the most toxic of natural poisons.

Botulinal toxins are secreted by the bacterial cells as a multimeric complex containing several other components (e.g. haemagglutinins, non-toxin, non-haemagglutinin), that is stable in acid pH conditions and may serve to protect the neurotoxin in the stomach. The complex dissociates spontaneously at physiological pH values. The neurotoxins consist of a heavy chain and a light chain joined by a disulphide bridge. The heavy chain is responsible for binding the complex to the exterior of the synaptic nerve cell, and the light chain is then internalised. The light chain is a zinc metalloprotease that specifically cleaves proteins forming the synaptic vesicle docking and fusion complex responsible for transporting acetylcholine across the cell membrane into the neuromuscular junction, thus paralysing that muscle.

Treatment of the intoxication is by rapid administration of antitoxin although, once the toxin has entered the nerve cells at neuromuscular junctions, it is unavailable for neutralisation by the antitoxin. Other medical treatment is restricted to supportive interventions to alleviate the symptoms, particularly by artificial ventilation. In infant botulism the use of antitoxin is contra-indicated and supportive ventilation is needed in only a minority of cases (*c.* 25 %). Since the availability of antitoxin, the case fatality rate has fallen considerably. In China before antitoxin was available, *c.* 50 % of all cases were fatal but, more recently, with antitoxin treatment this has fallen to *c.* 8 % (Dodds and Austin 1997).[4]

23.2.1 Incidence of botulism

Because botulism is a serious illness, the statistics of occurrence probably more accurately reflect the true incidence of this form of food poisoning than others, at least in well-developed countries. Misdiagnosis is still a problem in some outbreaks as many medical practitioners will not have seen botulism (St Louis *et al.*, 1988; McLauchlin *et al.*, 2006).[26,27]

Commercial products generally have a good record with respect to botulism since the canning industry adopted safe heat processes and good manufacturing practices. There have been some specific exceptions, however, that arose from not understanding the ubiquitous nature and characteristics of the organism and changing preservation mechanisms without due consideration. Thus a well-researched outbreak in the UK involved substitution of high levels of sugar by an intense sweetener in the preparation of a hazelnut flavouring for yoghurt without consideration of the resulting high a_W and altering the pasteurisation process to a full 12-D canning process. Similarly, inclusion of fresh herbs and spices in oils, such as garlic in oil, is risky as shown by cases in the USA and Canada (see Dodds and Austin, 1997).[4] Most cases of botulism arise from home-prepared foods such as home-canned vegetables (e.g. in USA), improperly cured meats and salted or lightly cooked fish (e.g. in Scandinavia, USA, Egypt) or improper time and temperature of holding after cooking (see Dodds and Austin, 1997).[4]

In Northern colder countries, and generally food from marine or freshwater sources, the main causative group is Group II (psychrotrophic, non-proteolytic) strains, toxins type E and occasionally B. However, from meat and vegetables (generally previously cooked or cured), the most common type of toxin detected in outbreaks is B and occasionally A (Group I strains). Cases of infant botulism, resulting from colonisation of the gastrointestinal tract in infants < 12 months of age, have now been reported from all inhabited continents except Africa. In the USA, the most extensively studied country, with four exceptions, almost all hospitalised cases, were caused either by *Cl. botulinum* type A or type B (or, in one case, both), and the distribution of cases by toxin type appeared to parallel the distribution of toxin types in US soils (see Dodds and Austin, 1997).[4] Although the only food vehicle definitively identified is ingestion of honey contaminated with botulinal spores, other sources of the organism have been suspected, e.g. soils, household dust and sharing of a cot. Honey from California, although also from Utah, Pennsylvania and Hawaii, is most commonly contaminated; toxin types A and B are involved. Levels of spores in honey samples associated with botulism cases are *c.* $10^{3–4}$/kg, but much lower or undetectable in other honeys (see Nevas, 2006).[28]

The statistics of outbreaks, cases and deaths from botulism in different countries and the foods involved indicate that in European countries one to five outbreaks per year occur on average. Germany tends to record about 20 outbreaks per year as does Portugal, and Poland recorded over 400 outbreaks per year over the three-year period between 1984 and 1987; in general, these were the result of improper home-curing of meat products or fish. The number of cases per outbreak is generally low, averaging one to five cases although some outbreaks involved more than ten persons, and the most common food type involved is meat.

As consumers demand ever decreasing processing of foods for freshness of appearance, texture and taste, and lower levels of preservatives – lower salt, nitrite, sugar, etc. – combined with requirements for longer shelf-life and ready-to-eat products, there are distinct possibilities for an increase in botulism cases. Food processors would be well-advised to increase their vigilance regarding the potential for the presence, survival and growth of *Cl. botulinum* in re-formulated products. The use of models for predicting the survival or growth of *Cl. botulinum* is to be highly recommended (e.g. Growth Predictor, IFR, Norwich, UK; USDA Pathogen Modelling Program).

23.2.2 Growth/survival characteristics

Most foods contain sufficient nutrients for the growth of *Cl. botulinum*. The limiting physicochemical parameters for growth under otherwise ideal conditions for the two main groups are summarised in Table 23.1 The solute used for lowering of a_W is salt (NaCl), but different solutes have differing effects on the growth of cells at any given a_W and should be investigated carefully before committing to changes in formulations of a food. Combinations of less than limiting conditions can be efficacious in preventing growth. A good example is that of pasteurised cured meats, in which salt (NaCl), nitrite, pH below neutral, perhaps with added polyphosphates and ascorbates, mild heat treatment and cool storage temperatures can be combined to produce a safe stable product. If other organisms can grow in a product initially safe against *Cl. botulinum*, conditions may be sufficiently altered to permit toxin production; examples are pickled vegetables in which equilibration of internal pH has been too slow and/or yeast or moulds have grown, raising the local pH to permissive values (e.g. Montville, 1982; Odlaug and Pflug, 1979).[29,30] New combinations of preservation conditions should be confirmed as safe by using either predictive models or challenge studies.

Reliance on a single preservative characteristic, particularly low temperature, is unwise as consumers or retailers can abuse these conditions resulting in growth and toxin production. Two examples are that of the outbreak of botulism from mascarpone cheese in which the intrinsic properties (pH, a_W and absence of lactic acid bacteria) are not growth limiting and storage at 4 °C is the only preservative factor (Aureli *et al.*, 1996; Franciosa *et al.*, 1999),[31,32] and also that of pasteurised chilled carrot juice that was stored incorrectly either whilst for sale or in the patients' homes and is now recommended to be either fully heat treated or have the pH adjusted to < 4.6 (see US FDA Enforcement Report, 2008).[33] Under certain special conditions, the absolute lower limits for pH and of temperature have been decreased marginally, but growth of strains of botulinum is very slow. Incorporation of oxygen in packs of foods will not prevent the growth of the obligately anaerobic clostridia, as fresh foods generally have active enzyme systems and active microbial populations that reduce the oxidation/reduction potential (Eh) to values permitting clostridial growth which may be as high as +150 mV (Kim and Foegeding, 1993),[34] and cooked foods have a low Eh intrinsically.

Inactivation

The lower heat resistance of Group II botulinum spores enables the use of comparatively mild heat treatments for products to be stored under refrigeration (< 8 °C). However, initial research showed that lysozyme and certain other similar enzymes present in foods stimulate the germination of heat-damaged Group II spores, and can persist through mild heat treatments (Peck *et al.*, 1993, 1995).[35,36] On this basis it was recommended that some currently used heat processes may need re-evaluation (Gould, 1999),[37] although there are no known cases of botulism resulting from this potential cause. However, later work (Stringer *et al.*, 1999)[38] could not confirm in vegetables naturally containing lysozyme that heat-damaged spores of Group II were able to germinate and grow. Nevertheless, for long-life chilled foods the recommended heat treatment of 90 °C for 10 min gives the desired safety level and is retained.

In order to eliminate the most heat-resistant spores of *Cl. botulinum*, the canning industry adopted the concept of a 12 log reduction in spores as the benchmark of thermal processes for foods with pH values > 4.5. This reduction is achieved by the 'Botulinum Cook', based on a $D_{121\,°C}$-value of 0.2 min (i.e. 1 log reduction is achieved by 0.2 min at 121 °C). The addition of a further safety margin (12 log reduction at 121 °C = 2.4 min) results in a required thermal process of 3 min at 121 °C (or the equivalent) at the slowest heating point in a product. Adopting a *z*-value (increase or decrease in temperature required to change the *D*-value by 1 log) of 10 C°, allows the calculation of any equivalent process at alternative temperatures and integration of effective heat processes.

However, there are other factors present in foods that may affect heat resistance of spores, e.g. sporulation medium, pH, a_W, fat content, etc. Esty and Meyer (1922)[39] showed that *Cl. botulinum* spores had maximum heat resistance at pH values *c.* 6.3 but that resistance decreased markedly at pH values below 5 and above 9. Lowered a_W values increased spore heat resistance, whether by addition of NaCl (Esty and Meyer, 1922)[39] or sucrose (Sugiyama, 1951).[40] Sugiyama (1952)[41] also showed that spores grown in media containing fatty acids were more heat resistant, while spores coated in oil (as in certain canned fish products) were also more heat resistant than those suspended in water (Lang, 1935).[42]

More modern physical processes of inactivation of spores have been investigated, such as high pressure, radiation, both ultraviolet (UV) and ionising radiations (gamma or electron beam), pulsed electrical fields, and also chemical sterilants (e.g. chlorine or hypochlorite, hydrogen peroxide or peroxyacetic acid). Anaerobe spores in general are susceptible to oxidising agents, and chlorine derivatives and peroxides are efficient lethal agents for botulinum spores (Kim and Foegeding, 1993)[34] and are used in the canning industry and fish processing. Spores are sensitive to moderate pressures, by virtue of germination of the spores and subsequent rapid killing of the germinated spore (Clouston and Willis, 1969; Gould and Sole, 1970)[43,44], whereas spores are much more resistant to high pressures (> 1200 MPa; Sole *et al.*, 1970).[45] Spores of *Cl. botulinum* are highly resistant to ionising radiations; *D*-values vary from 0.1–4.5 kGy, and it is not realistic to use this physical method of inactivation routinely in foods, although it can been used in special circumstances. Pulsed electric fields have been investigated for their effects on spores, but the equipment is difficult to handle and does not provide sufficiently reliable results for adoption in the food industry at the present state of development (see www.cfsan.fda.gov/~comm/ift-pef.html (accessed April 2009)).

23.2.3 Detection of *Cl. botulinum*

An excellent review of most of the methods of detecting the organism or toxins, is that by Lindström and Korkeala (2006).[15] There are no specific selective-indicator media available for detecting this organism. The classical method of detection, and relied upon by many laboratories, is based on the neuroparalytic action of the botulinal toxins by intra-peritoneal injection of food extracts or cultures into mice, inactivation of the toxins by boiling for 10 min, and neutralisation of the action by specific antibodies. Various alternative methods of detection of botulinal toxins based on antibody-antigen methods, e.g. ELISA, have been published, but these methods generally are not as sensitive as the mouse test, and the antibodies are often not highly specific for the toxic moiety, reacting with one or other of the components of the multimeric complex secreted by or liberated from the lysed cells. However, there are some modern approaches to detection of toxins that may be worthy of further development for routine testing, e.g. mass spectrometry (van Baar, *et al.*, 2002; Baar *et al.*, 2005)[46,47] or combining the specific endopeptidase activity of the neurotoxins and mass spectrometry (Boyer *et al.*, 2005),[48] or toxin proteomics (Kalb *et al.*, 2005).[49]

The action of the botulinal toxins on the synapse vesicle docking proteins is a result of their very specific proteinase activity, and a highly sensitive assay based upon this activity has been developed using recombinant or synthetic peptides (Schiavo *et al.*, 1993; Binz *et al.*, 1994; Ekong *et al.*, 1997; Shone *et al.*, 1995; Wictome *et al.*, 1999),[50,51,52,53,54] that may eventually replace *in vivo* tests. However, although a prototype kit has been tested successfully by several laboratories, it has not been made available commercially. The mouse test is still the preferred and definitive method for detection of *Cl. botulinum* toxins.

Detection of the organism and toxin genes by polymerase chain reaction (PCR), has been developed and applied to various environments (e.g. Aranda *et al.*, 1997; Nevas, 2006),[55,28] but there is always the possibility of detection of dead cells, although this may be overcome by a short incubation period to increase the number of the target genes. The various types of *Cl. botulinum* may be also differentiated by ribotyping (Hielm *et al.*, 1999)[56] or pulsed field gel electrophoresis (PFGE) of macrorestriction profiles (Lin and Johnson, 1995; Hielm *et al.*, 1998).[57,58]

23.2.4 Control measures and possible future trends

Botulinal spores must be recognised as ubiquitous, and food preservation processes therefore must be adequate to eliminate these spores or inhibit germination and outgrowth, and subsequent toxin production. Foods with a pH value < 4.5 have been shown to be intrinsically safe from botulinal growth, but some studies in foods with high protein levels (*c.* 3 % soy or casein) and pH adjusted with weak organic acids, have shown that toxin production can occur at pH values slightly lower than 4.5 (Smelt *et al.*, 1982),[59] although this is generally regarded as a consequence of higher microenvironment pH values. Nevertheless, foods with a pH < 4.5 do not require a full 'Botulinum Cook', i.e. 121 °C for 3 min, or the equivalent based on a *z*-value of 10 C°.

For control of the mesophilic strains in salted meats or fish, 10 % w/v NaCl on the water phase is required; control of the psychrotrophic strains is achieved with only 3.5 % w/v salt-on-water. However, together with nitrite in meats (*c.* 150 ppm), moderately low pH and polyphosphates (*c.* 0.5 % w/w) and a pasteurisation heat process, 3.5 % salt provides an adequate control of all strains of *Cl. botulinum*, although large cans of pasteurised cured meats must be stored below 8 °C. Salting, drying and smoking of fish, particularly cold smoking (< 30 °C), must be carried out with great care and with a limited shelf-life at chill temperatures, if the risk of botulism is to be avoided.

Storage temperatures of < 10 °C will control the growth of mesophilic (Group I) strains of *Cl. botulinum*, but temperatures of < 3 °C are needed to prevent growth of the psychrotrophic (Group II) strains, although temperature control alone should not be relied upon for limiting botulinal growth. There are also differences in the controlling levels of a_w for Group I and II strains of *Cl. botulinum* (Table 23.1), but this should be confirmed by trials for solutes other than NaCl.

23.3 Other food poisoning clostridia; *Clostridium perfringens*, general characteristics

Clostridium perfringens, previously known as *Cl. welchii*, is an anaerobic, Gram-positive bacterium widely found in soils at levels of 10^3–10^4 cfu g^{-1} and the intestinal tract of vertebrates, typically at levels of 10^4–10^6 cfu g^{-1} in humans, which may explain the prevalence of perfringens foodborne toxico-infections. Generally, *c.* 50 % of fresh or frozen meat samples are positive for the organism, although only a few strains from foods and the environment appear to be capable of eliciting food poisoning (see Section 23.3.2; Saito, 1990).[60]

Clostridium perfringens is responsible for three clinical syndromes, including gas gangrene, enteritis necroticans ('pig-bel') and food poisoning. At least 12 different toxins are produced by *Cl. perfringens*, including four lethal toxins, hemolysins, proteases, neuraminidase and enterotoxin. The ability of individual strains to produce the four lethal toxins forms the basis for a classification scheme that divides the species into five types, A to E. Enterotoxin is produced only by Type A strains, and is the primary mediator of *Cl. perfringens* food poisoning. The enterotoxin is produced in the gastrointestinal tract during sporulation and can be detected, by antibody-based tests, in stool samples of affected patients.

Cells of *Cl. perfringens* are short rod-shaped bacilli, typically with blunt or square ends, and non-motile. Spores of food poisoning strains, although formed, are not commonly found in laboratory media, and special techniques or additives must be used to encourage sporulation. If using the differential reinforced clostridial medium (DRCM) technique for isolation, omission of the heat treatment of the first (black) culture and direct streaking on a differential indicator medium (e.g. lactose–egg yolk–milk agar, LEYM) will result in many more positive isolations of this organism (Gibbs, 1973).[61] The organism, like all clostridia, is a strict anaerobe, but is rather more aerotolerant for short periods (e.g. 15–30 min exposure of colonies on plates on the laboratory bench will still yield viable cells), than many other clostridial species. The organism, although not a thermophile, has quite a high optimum temperature for growth, 43–45 °C, and will continue to grow up to *c*. 50 °C. Growth slows significantly below 15 °C and ceases at 6 °C. *Clostridium perfringens* is not particularly tolerant of low a_W or low pH values (Table 23.2). Under optimum conditions for growth (see Table 23.2) the organism exhibits probably the fastest growth rate of all foodborne bacteria, doubling in number every 8–10 min.

The symptoms of *Cl. perfringens* type A food poisoning occur within 8–22 h, usually 12–18 h, after eating contaminated foods. The infective dose of vegetative cells is typically a total ingestion of 10^6–10^8 cells of a food poisoning strain. The symptoms consist of severe abdominal pain, flatulence and diarrhoea; nausea, vomiting and fever are rare. Recovery is generally complete within a further 24 h and the attack is relatively mild; in a few cases in elderly persons rapid dehydration can occur, but death is rare. There is no specific treatment necessary other than to replace lost fluid and salts, as the disease is self-limiting.

Clostridium perfringens enterotoxin (CPE) is a single 3.5-kDa polypeptide of 319 amino acids with an isoelectric point of 4.3, and the amino acid structure appears to be highly conserved between the CPEs produced by several strains (Cornillot *et al.*, 1995).[62] Since the biological activity of CPE increases 2–3 fold when it is digested *in vitro*, it is possible that intestinal enzymes such as trypsin and chymotrypsin have a role in activating CPE once the toxin is released in the lumen of the intestine. CPE is released upon sporulation of the organism in the small intestine, and synthesis is controlled by the same sigma factors controlling sporulation. Its proposed mode of action involves (i) activation, (ii) binding to cellular receptors, and (iii) creating perforations in intestinal epithelial cells through a series of reactions, thus increasing permeability of the membrane to small macromolecules (< 200 Daltons in size), decreasing the internal a_W and increasing water gain until the cell lyses releasing fluids into the gastrointestinal lumen resulting in diarrhoea.

Table 23.2 Growth characteristics of *Clostridium perfringens*

Min. growth temp.	*c*. 12 °C
Opt. growth temp.	43–45 °C
Max. growth temp.	50 °C
Min. pH for growth	5
Min. a_w for growth [NaCl][a]	0.95 [8 % w/v]
$D_{95\,°C}$ of spores heat-sensitive	1–3 min
$D_{95\,°C}$ of spores heat-resistant	18–64 min
Eh range for growth[b]	–125 – +350 mV

[a] Different a_w limits may apply with other solutes.
[b] Growth will not occur in the presence of air, e.g. on surfaces of agar media, although colonies are aerotolerant for a few minutes.

23.3.1 Incidence of *Clostridium perfringens* food poisoning

As the disease is of short duration and the symptoms are relatively mild, there is significant under-reporting of cases. Only when there are large numbers of persons involved simultaneously, such as in a large catering operation, is there likely to be public health involvement and investigation. Thus Todd (1989)[63] estimated that reporting in Canada and the USA is of the order of only 0.15 % of actual cases. There is a downward trend in the number of outbreaks and cases in both the UK and North America, by *c.* 50 % in the UK over the past ten years and a rather greater fall in the USA. This fall may be related to better awareness of the importance of prompt chilling and refrigeration of cooked foods and also perhaps to a decline in mass catering in factories. Although type A food poisoning can occur at any time of the year following temperature abuse, cases are slightly more common in summer months, probably as a result of warmer temperatures and difficulties in keeping cooked foods cool.

The main vehicles of *Cl. perfringens* type A food poisoning are cooked meats, particularly beef but also including poultry. In the USA, Mexican foods that include cooked meats are emerging as a vehicle for *Cl. perfringens* food poisoning. The major factor contributing to this type of food poisoning is temperature abuse after cooking, but incomplete cooking can stimulate the rapid germination of *Cl. perfringens* spores on cooling.

23.3.2 Growth/survival characteristics of *Clostridium perfringens*

The organism, although not a thermophile, has quite a high optimum temperature for growth, 43–45 °C, and will continue to grow up to *c.* 50 °C. Growth is generally accepted to cease below 12–15 °C (Labbe, 1981, 1989).[64,65] Vegetative cells of *Cl. perfringens* are not particularly tolerant of low temperatures, refrigeration or freezing resulting in death. Spores are more resistant to low temperatures and may survive in chilled or frozen cooked foods, only to germinate and begin rapid growth on reheating the food. Cultures to be stored should not be chilled but stored at ambient temperatures (spores are only poorly formed in general laboratory media), or freeze-dried.

Clostridium perfringens is not particularly tolerant of low a_W or low pH values (Table 23.2). For complete inhibition of growth, 6–8 %w/v salt is required, and up to 400 ppm of nitrite. Thus commercially cured products of acceptable preservative levels (i.e. containing 3.5 % NaCl on water phase, 100–150 ppm nitrite) may not completely inhibit the growth of this organism. However, the observation that commercially cured meats are relatively rarely involved in type A food poisoning indicates that curing salts are effective in significantly reducing growth of *Cl. perfringens* (Juneja and Majka, 1995);[66] combined with chilled storage, now commonly practised for cured meats, this severely limits the potential for *Cl. perfringens* growth.

As indicated previously, *Cl. perfringens* is very tolerant of Eh conditions that would inhibit many other clostridia (Table 23.2). Under optimum conditions for growth (see Table 23.2), the organism exhibits probably the fastest growth rate of all foodborne bacteria, doubling in number every 8–10 min.

23.3.3 Detection of *Clostridium perfringens*

Early methods relied on the rapid formation of a 'stormy clot' reaction in milk, but this is not necessarily definitive and is easily misread. Agar or liquid media containing sulphite and iron salts are commonly used to detect clostridia by formation of black precipitates (iron sulphides), but are not sufficiently specific for *Cl. perfringens*. Liquid iron–sulphite containing media (e.g. DRCM; Freame and Fitzpatrick, 1972),[67] may be used in most probable number (MPN) techniques. Since the organism does not readily form spores in laboratory media, heating presumptive positive (black) cultures in liquid media and sub-culturing into fresh medium and examining for growth, is rarely successful (Gibbs, 1973).[61] However, streaking of the presumptive positive (black) culture on to a diagnostic medium often yields characteristic colonies of *Cl. perfringens* (Gibbs, 1973).[61] Solid sulphite–iron media suffer from the problem of excessive gas production from fermentable carbohydrates, separation of the agar from the plate surface and overall blackening. Also, large amounts of acid can mask the production of iron sulphide.

Selective agar media are now generally employed for the isolation and enumeration of *Cl. perfringens*, one of the most commonly used being egg-yolk free tryptose-sulphite-cycloserine agar incubated anaerobically at 37 °C (see Hauschild *et al.*, 1979).[68] Confirmation of the identity of the organism is by relatively few physiological tests; reduction of nitrate to nitrite, fermentation of lactose, lack of motility and indole production, gelatin digestion (but not of casein) being some of the more important tests. A very useful diagnostic agar medium for the common clostridia in foods was devised by combining lactose fermentation, egg yolk reaction (Nagler; lecithinase activity), lipase and proteolysis of milk (LEYM; Willis and Hobbs, 1959).[69]

A comparison of four routine methods for the detection of *Cl. perfringens* was made by Eisgruber *et al.* (2000).[70] These authors found that gas production from lactose and sulphite reduction was less reliable for identification than the reverse CAMP test or acid phosphatase test. Alternatively, Schalch *et al.*, (1999)[71] used molecular methods for analysing isolates of *Cl. perfringens* below the species level; ribotyping and PFGE patterns were recommended for epidemiological investigations of food poisoning outbreaks.

However, detection of *Cl. perfringens* type A does not necessarily indicate that a foodpoisoning hazard exists. Only a small proportion of strains are capable of producing CPE, and Saito's (1990)[60] survey of a range of foods and faecal samples from food handlers using very sensitive CPE serological assays indicated that only shellfish and faecal samples yielded CPE-positive cultures. Although no CPE-positive cultures were detected in poultry, pigs or cattle in Saito's (1990)[60] survey, in a later survey using *cpe*-gene technology, positive cells were recovered from a variety of animals including food animals (Kokai-Kun *et al.*, 1994).[72] However, it also appears that the gene controlling CPE production can be either plasmid- or chromosomally-borne, giving rise to speculation that there may be two distinct populations of food poisoning strains of the organism (Cornillot *et al.*, 1995).[62] There is also the possibility that CPE-negative strains may acquire a plasmid-borne *cpe*-gene under certain conditions.

For most practical purposes in a food microbiology laboratory, the demonstration of large numbers in an incriminated food or in faecal samples from patients with typical symptoms, preferably backed up by CPE-positive results, is sufficient evidence for *Cl. perfringens* type A food poisoning. The commercial kits for CPE detection are generally only used on faecal samples since CPE is produced only during sporulation, and special culturing methods must be used in the laboratory to encourage sporulation, even of food-poisoning strains. However, tracing the definitive food source of the organism requires the use of specific agglutinating antisera, of which 75 were described by Stringer *et al.* (1982).[73] In general, these are only available in public health or reference laboratories, although ribotyping and PFGE may be more easily and more effectively used by laboratories with the appropriate equipment.

23.3.4 Control measures and possible future trends

As *Cl. perfringens* spores are resistant to normal cooking temperatures and cells resistant to curing salts, their presence in a raw food item cannot be eliminated by cooking, and other control measures must be relied upon to limit their growth. Alternative preservative combinations of food-grade additives were investigated by Sikes and Ehioba (1999).[74] These authors found that a combination of sucrose laurate, ethylene diamine tetra acetate (EDTA) and the antioxidant butylated hydroxyl anisole (all generally recognised as safe – GRAS), at > 100 ppm, inhibited *Cl. perfringens* growth in an optimal medium for > 8 d. However, these inhibitors have not been tried in foods.

The best control measure is prompt cooling of cooked foods, particularly meat dishes, to 15 °C, followed by chilling to < 8 °C. Comparing four different methods of food service cooling of cooked turkey roasts, Olds *et al.* (2006)[75] found that total times of > 12 h taken to cool from 74 °C to 5 °C of turkey roasts permitted growth of *Cl. perfringens* by > 1.5 log. These results re-emphasised the conclusions and advice in the original and definitive paper on *Cl. perfringens* food poisoning by Hobbs *et al.* (1953)[76]: 'Outbreaks of this kind should be prevented by cooking meat immediately before consumption, or if this is impossible, by cooling the meat rapidly and keeping it refrigerated until it is required for use'. It is generally recommended that joints of meat should not exceed *c.* 2.5 kg in weight if they are to be cooled effectively after cooking.

Amézquita *et al.* (2005)[77] developed a dynamic integrated model for the cooling of meats and growth of *Cl. perfringens*, and the 'Perfringens Predictor' model is available (free download) on the Institute of Food Research, Norwich, UK website (http://www.ifr.ac.uk/safety/growthpredictor/). The latter is a valuable resource for food processors and service personnel involved in cooking and cooling of meats, and should be used as part of the Hazard Analysis Critical Control Point (HACCP) system.

The possible future trend of food poisoning by type A *Cl. perfringens* seems likely to be downward if the pattern of the past five or more years continues. However, the increasing availability of ready-to-eat meat dishes, e.g. ethnic foods, and the potential for temperature abuse, particularly in the summer months, may reverse this trend. It also seems unlikely that the rate of reporting the symptoms will increase since they are so mild. As the population of the elderly increases, there may be opportunities for more outbreaks in residential homes, where preparation of food in bulk may occur for reasons of economy. Investigations of such outbreaks will, in future, probably concentrate more on identification of CPE-producing strains and their origin, rather than just *Cl. perfringens*. Kits are already available for detecting CPE in faecal samples, and may become available for detecting the *cpe* gene, as induction of sporulation and production of CPE in laboratory media is difficult to achieve routinely.

23.4 Other food-poisoning clostridia; *Clostridium difficile*, general characteristics

Clostridium difficile infection is the most important cause of hospital-acquired diarrhoea, but is not generally regarded as a foodborne infection. A good general introduction and discussions of recent developments is given in the Society of General Microbiology Symposium (Poxton, 2005).[78] Early investigators noted that the bacterium produced a potent toxin, but the role of *Cl. difficile* in antibiotic-associated diarrhoea and pseudomembranous colitis was not elucidated until the 1970s. *Clostridium difficile* infection is usually spread on the hands of healthcare staff and other people who come into contact with infected patients or with environmental surfaces (e.g. floors, bedpans, toilets) contaminated with the bacteria or its spores. It seems possible that food could be a vehicle of infection if contaminated by a handler, but there is little or no evidence for this as a common occurrence. Spores, produced when *Cl. difficile* encounters unfavourable conditions, such as being outside the body, are very hardy and can survive on clothes and environmental surfaces for long periods. Few of the commonly-used disinfectants are effective against spores; chlorine (as solutions) and formaldehyde are effective treatments but corrosive and unpleasant to use in quantity (Mayfield *et al.*, 2000).[79]

23.4.1 Incidence

Clostridium difficile is naturally present in the gut of up to 3 % of healthy adults and 66 % of infants. However, *Cl. difficile* rarely causes problems in children or healthy adults, as it is kept in check by the normal bacterial population of the intestine. When certain antibiotics disturb the balance of bacteria in the gut, *Cl. difficile* can multiply rapidly and produce toxins which cause illness. Mortality rates in more seriously ill patients, is 10–30 %. In a prospective study of patients in Swedish hospitals, Wiström *et al.* (2001)[80] reported a case rate of *c.* 2–7 %, with antibiotic-treated patients associated with an increased risk of developing the infection. Residents of homes for the elderly are also at risk of becoming infected.

Clostridium difficile infection ranges from mild to severe diarrhoea to, more unusually, severe inflammation of the bowel (known as pseudomembranous colitis). People who have been treated with broad spectrum antibiotics (those that affect a wide range of bacteria), people with serious underlying illnesses and the elderly are at greatest risk – over 80 % of *Cl. difficile* infections reported are in people aged over 65 years. Pathogenic strains of *Cl. difficile* produce two distinct toxins. Toxin A is an enterotoxin, and toxin B is a cytotoxin; both are high-molecular weight proteins capable of binding to specific receptors on the intestinal mucosal cells. Receptor-bound toxins gain intracellular entry where they catalyse a specific alteration of Rho proteins, small glutamyl transpeptidase (GTP)-binding proteins that assist in actin polymerisation, cytoskeletal architecture and cell movement. Both toxin A and toxin B appear to play a role in the pathogenesis of *Cl. difficile* colitis in humans.

Recommendations for treatment of severely ill patients are *per oral* administration of metronidazole, or vancomycin in patients sensitive to the former drug. However, vancomycin can give rise to vancomycin-resistant strains of enterococci that may cause other gastrointestinal tract complications.

23.4.2 Growth/survival characteristics of *Clostridium difficile*

Clostridium difficile is a Gram-positive spore-forming anaerobe, with an optimum temperature for growth of 35–40 °C; it requires five amino acids (Leu, Ile, Pro, Trp and Val) for energy metabolism and addition of glycine increases growth significantly. *Clostridium difficile* undergoes amino acid fermentation in order to create adenosine triphosphate (ATP) as a source of energy and can also utilise sugars. It expresses two S-layer proteins, one of which is highly conserved among strains and one which shows sequence diversity. Both proteins are derived from a single gene product and both are associated with amidase activity. During *Cl. difficile*'s exponential phase of growth (when sugars are available in the medium, i.e. glucose) expression of toxins A and B is repressed but shows a significant increase once the bacteria enter the stationary phase. By prioritising growth over toxin production the bacteria can grow and colonise as much as possible before starting to produce toxins, which also indicates that the genes for toxin A and B undergo some type of catabolite repression. Considerable research on the genome of *Cl. difficile* has been done to understand the mechanisms of infection and pathogenesis. The genome consists of a single chromosome and a single plasmid, which has been sequenced. The genome includes 11 % as mobile genetic elements that provide antibiotic resistance, virulence and host interaction via surface proteins. Like many other bacterial spores, those of *Cl. difficile* are resistant to many inimical environments and can survive for at least two years, although they are susceptible to oxidising agents, e.g. hypochlorite, used as disinfectants.

23.4.3 Detection of *Clostridium difficile*

Following presumptive clinical diagnosis of infection, culture of stool samples on the selective and differential media cyclosenine cefoxitin fructose agar (CCFA) (George *et al.*, 1979)[81] or *Clostridium difficile* moxalactam norfloxacin (CDMN) (Aspinall and Hutchinson, 1992),[82] and detection of toxins in faecal samples, are the recommended methods of detection and confirmation of the infective organism. Immunodetection systems (e.g. ImmunoCard *C. difficile* test) are less sensitive and/or reliable (Staneck *et al.*, 1996)[83] than culture followed by cytotoxin tests.

In an outbreak situation, it has been recommended that sub-species typing is done as there may be more than one type present in the outbreak. PFGE (Alonso *et al.*, 2005),[84] and ribotyping (Arroyo *et al.*, 2005)[85] have been suggested as the most useful techniques.

23.4.4 Control measures and possible future trends

The continued use of general disinfectants, e.g. quaternary ammonium compounds, does little to reduce the rates of infection in hospitals, as the spores are highly resistant to these disinfectants, but switching to use of chlorine (as hypochlorite) reduced the rate of infection markedly (Mayfield *et al.*, 2000[79]). In general, clostridia are more sensitive to strongly oxidising chemicals used as disinfectants.

The current trend is for *Cl. difficile* infections in hospitals and retirement or nursing homes to increase in both number and severity as the population ages and broad spectrum antibiotics are prescribed. There are also indications of more pathogenic – toxigenic – strains emerging (http://www.medscape.com/viewarticle/546729). Decreasing use of broad spectrum antibiotics, particularly those for gastrointestinal tract infections, will limit this cause of *Cl. difficile* infections, while simultaneous prophylactic administration of probiotic bacteria resistant to the antibiotic prior to or together with the antibiotic, is likely to minimise gastrointestinal tract infections by *Cl. difficile*.

23.5 Future trends

The trend towards preparation of foods with long shelf-lives, for consumption at or away from home, together with 'grazing eating' and the potential for improper storage of meal items, seems likely to increase the numbers of cases and outbreaks of clostridial food poisoning. This has been noted in the USA already with the introduction of Mexican-style wrapped meat-containing foods. With the current rapidly increasing costs of energy, for cooking and refrigeration, there may also be a tendency to undercook foods and restrict the rates of cooling and low temperatures of storage. With respect to the neurotoxigenic species, it seems probable that there will be an increase in the non-botulinum species acquiring and expressing one or more of the neurotoxin genes, as has been noted for some butyric species of clostridia. There has already been work to identify more effective methods for controlling *Cl. difficile* with appropriate disinfectants, and if this knowledge, together with information on infected sites, is transferred into actual practice, problems of large numbers of cases in hospitals and nursing and residential homes may steadily decline.

23.6 Sources of further information and advice

23.6.1 *Clostridium botulinum*

Several sources of advice are available to the food industry. Leatherhead Food International UK, has operated a Botulinum Laboratory for many years and offers Member Companies advice and challenge-testing facilities for assessing the botulinal safety of food products. Campden and Chorleywood Food Research Association (CCFRA, UK) also has similar facilities for work with clostridia. The Central Science Laboratory of DEFRA/FSA, York, and the Institute for Food Research, Norwich, also have botulinal laboratory facilities in the UK.

The Health Protection Agency Laboratory, Colindale, London (HPA), has facilities for examining foods for botulinal toxin and *Cl. perfringens*, but generally is active only in outbreak situations. Much of the recent research work on alternative methods for detecting botulinal toxins, has been in conjunction with the Centre for Applied Microbiological Research (CAMR), part of the Microbiological Research Authority within the Department of Health.

Most of the recent research on heat resistance of the spores of psychrotrophic (Group II) strains of *Cl. botulinum* and lysozyme activation of germination of heat damaged spores, has been done at the Institute for Food Research, Norwich, together with determination of the limitation of growth by combinations of low temperature and salt.

In the USA, the Centers for Disease Control, Atlanta, has extensive facilities and expertise in all aspects of food poisoning, including the clostridial species.

In France, the Institut Pasteur is a private non-profit foundation dedicated to the prevention and treatment of diseases through biological research, education and public health activities.

Several predictive models have been developed for assessing the probability of toxin production by *Cl. botulinum*, or rate of growth, survival or death. Most of the dynamic (growth) models have been combined into one freely available from the Institute of Food Research, Norwich UK (http://www.combase.cc/toolbox.html). It is strongly recommended that food manufacturers consult these models if they are contemplating altering the preservative characteristics of foods, including storage regimes.

Useful texts on botulism and the organism include the following:

Dodds K L (1993) *Clostridium Botulinum*, in Y H Hui, J R Gorham, K D Murrell and D O Cliver (eds), *Foodborne Disease Handbook*, Marcel Dekker, New York, 97–131.

Doyle M P (1989) *Foodborne Bacterial Pathogens*, Marcel Dekker, New York, A H W Hauschild, 111–89.

Doyle M P, Beuchat L R and Montville T J (eds) (1997) *Food Microbiology, Fundamentals and Frontiers*, Washington DC, ASM Press, K L Dodds and J W Austin, 288–304.

Hauschild A H W and Dodds K L (1993) *Clostridium botulinum*: *Ecology and Control in Foods*, New York, Marcel Dekker.

23.6.2 *Clostridium perfringens*

Useful texts on *Cl. perfringens* include the following:

Labbe R (1989) *Clostridium perfringens*, in M P Doyle (Ed.), *Foodborne Bacterial Pathogens*, Marcel Dekker New York, 192–227.

Wrigley D M (1994) *Clostridium perfringens*, in Y H Hui, J R Gorham, K D Murrell and D O Cliver (eds), *Foodborne Diseases Handbook, Vol. 1. Diseases Caused by Bacteria*, Marcel Dekker New York, 133–67.

23.7 References

1. Miller M B and Bassler B L (2001) Quorum sensing in bacteria, *Ann Rev Microbiol*, **55**, 165–99.
2. Novak J S and Fratamico P M (2004) Evaluation of ascorbic acid as a quorum-sensing analogue to control growth, sporulation and enterotoxin production in *Clostridium perfringens*, *J Food Sci*, **69**, FMS72–FMS78 .
3. Gould G W (2006) History of science – spores, *J Appl Microbiol*, **101**(3), 507–13.
4. Dodds K L and Austin J W (1997) *Clostridium botulinum* in M P Doyle L R Beuchat and T J Montville (eds), *Food Microbiology: Fundamentals and Frontiers*, Washington, DC, ASM Press, 288–304.
5. Merivirta L O, Lindström M, Björkroth K J and Korkeala H J (2006) The prevalence of *Clostridium botulinum* in European river lamprey (*Lampetra fluviatilis*) in Finland, *Int J Food Microbiol*, **109**(3), 234–7.

6. Bott T L, Johnson J, Foster E M and Sugiyama H (1968) Possible origin of the high incidence of *Clostridium botulinum* in an inland bay (Green Bay of Lake Michigan) *J. Bacteriol*, **95**, 1542–7.
7. Huss H H (1980) Distribution of *Clostridium botulinum*, *Appl Environ Microbiol*, **39**, 764–9.
8. Hielm S, Hyytiä E, Andersin A.-B and Korkeala H (1998) A high prevalence of *Clostridium botulinum* type E in Finnish freshwater and Baltic Sea sediment samples, *J Appl Microbiol*, **84**, 133–7.
9. Hyytiä-Trees E (1999) *Prevalence, molecular epidemiology and growth of* Clostridium botulinum *type E in fish and fishery products*, PhD thesis, University of Helsinki, ISBN 952-45-8684-0 (PDF version).
10. Sebaihia M, Peck M W, Minton N G, Thomson N R *et al.* (2007) Genome sequence of a proteolytic (Group I) *Clostridium botulinum* strain Hall A and comparative analysis of the clostridial genomes, *Genome Res*, **17**, 1082–92.
11. Luquez C, Bianco M I, de Jong L I T, Sagua M D, Arenas G N, Ciccarelli A S, Rafael A and Fernández R A (2005) Distribution of botulinum toxin-producing clostridia in soils of Argentina, *Appl Environ Microbiol*, **71**(7), 4137–9.
12. Poulain B, Stiles B G, Popoff M R and Molgo J (2006) Attack of the nervous system by clostridial toxins, in J E Alouf and M R Popoff (eds) *The Comprehensive Sourcebook of Bacterial Protein Toxins*, 3rd edn, Academic Press, London, 348–89.
13. Bohnert B, Deinhardt K, Salinas S and Schiavo G (2006) Uptake and transport of clostridial neurotoxins, in J E Alouf and M R Popoff (eds), *The Comprehensive Sourcebook of Bacterial Protein Toxins*, 3rd edn, Academic Press, London, 390–408.
14. Alouf J E and Popoff M R (eds) (2006) *The Comprehensive Sourcebook of Bacterial Protein Toxins*, 3rd edn, Academic Press, London.
15. Lindström M and Korkeala H (2006) Laboratory diagnostics of botulism, *Clin Microbiol Revs*, **19**(2), 298–314.
16. Hutson R A, Thompson D E, Lawson P A, Schocken-Itturino E R, Bottger P C and Collins M D (1993) Genetic interrelationships of proteolytic *Clostridium botulinum* type A, B, and F and other members of the *Clostridium botulinum* complex as revealed by 16S rRNA gene sequences, *Ant van Leeuwen*, **64**, 278–83.
17. East A K, Bhandari M, Stacey J M, Campbell K D and Collins M D (1996) Organization and phylogenetic interrelationships of genes encoding components of the botulinum toxin complex in proteolytic *Clostridium botulinum* types A, B, and F: evidence of chimeric sequences in the gene encoding the nontoxic nonhamagglutinin component, *Int J Syst Bacteriol*, **46**(4), 1105–12.
18. Midura T F and Arnon S S (1976) Infant botulism: identification of *Clostridium botulinum* and its toxin in faeces, *Lancet*, **ii**, 934–6.
19. Arnon S S, Midura T F, Damus K, Thompson B, Wood R M and Chin J (1979) Honey and other environmental risk factors for infant botulism, *J Pediatr*, **94**, 331–6.
20. Midura T F (1996) Update: infant botulism, *Clin Microbiol Rev*, **9**, 119–25.
21. Fenicia L, Da Dalt L, Anniballi F, Zanconato S and Aureli P (2002) A case of infant botulism due to neurotoxigenic *Clostridium butyricum* type E associated with *Clostridium difficile* colitis, *Eur J Clin Microbiol Infect Dis*, **21**, 736–8.
22. Barash J R, Tang T W H and Arnon S S (2005) First case of infant botulism caused by *Clostridium baratii* type F in California, *J Clin Microbiol*, **43**, 4280–2.
23. Brett M M, McLaughlin J, Harris A, O'Brien S, Black N, Forsyth R J, Roberts D and Bolton F J (2005) A case of infant botulism with a possible link to infant formula milk powder: evidence for the presence of more than one strain of *Clostridium botulinum* in clinical specimens and food, *J Med Microbiol*, **54**, 769–76.
24. Johnson E A, Tepp W H, Bradshaw M, Gilbert R J, Cook P E and McIntosh D G (2005) Characterization of *Clostridium botulinum* strains associated with an infant botulism case in the United Kingdom, *J Clin Microbiol*, **43**, 2602–7.

25. Keet C A, Fox C K, Margeta M, Marco E, Shane A L, Dearmond S J, Strober J B and Miller S P (2005) Infant botulism, type F, presenting at 54 hours of life, *Pediatr Neurol*, **32**, 193–6.
26. St Louis M E, Peck S H S, Morgan G B, Blatherwick J, Banerjee S, Kettyls G D M, Black W A, Milling M E, Hauschild A H W, Tauxe R V and Blake P A (1988) Botulism from chopped garlic: delayed recognition of a major outbreak, *Ann Intern Med*, **108**, 363–8.
27. McLauchlin J, Grant K A and Little C J (2006) Food-borne botulism in the United Kingdom, *J Public Health*, **28**(4), 337–42.
28. Nevas M (2006) *Clostridium botulinum in honey production with respect to infant botulism*, PhD thesis, University of Helsinki.
29. Montville T J (1982) Metabiotic effect of *Bacillus licheniformis* on *Clostridium botulinum*: implications for home-canned tomatoes, *Appl Environ Microbiol*, **44**, 334–8.
30. Odlaug T E and Pflug I J (1979) *Clostridium botulinum* growth and toxin production in tomato juice containing *Apergillus gracilis*, *Appl Env Microbiol*, **37**, 496–504.
31. Aureli P, Franciosa G and Pourshaban M (1996) Foodborne botulism in Italy, *Lancet*, **348**, 1594.
32. Franciosa G, Pourshaban M, Gianfranceschi M, Gattuso A, Fenicia L, Ferrini A M, Mannoni V, de Luca G and Aureli P (1999) *Clostridium botulinum* and toxin spores in mascarpone cheese and other milk products, *J Food Prot*, **62**(8), 867–71.
33. US FDA Enforcement Report 08-02 (Jan 9, 2008) Recalls and Field Corrections: Class 1, Food and Drug Administration, Washington, DC, available at: www.fda.gov/bbs/topics/ENFORCE/2008/ENF01038.html, accessed November 2008.
34. Kim J and Foegeding P M (1993) Principles of control, in A H W Hauschild and K L Dodds (eds), Clostridium botulinum: *Ecology and Control in Foods*, Marcel Dekker, New York, 121–76.
35. Peck M W, Fairbairn D A and Lund B M (1993) Heat-resistance of spores of non-proteolytic *Clostridium botulinum* estimated on medium containing lysozyme, *Lett Appl Microbiol*, **16**, 126–31.
36. Peck M W and Fernandez P S (1995) Effect of lysozyme concentration, heating at 90 °C, and then incubation at chilled temperatures on growth from spores of non-proteolytic *Clostridium botulinum*, *Lett Appl Microbiol*, **21**, 50–54.
37. Gould G W (1999) Sous vide foods: conclusions of an ECFF botulinum working party, *Food Control*, **10**, 47–51.
38. Stringer S C, Haque N and Peck MW (1999) Growth from spores of nonproteolytic *Clostridium botulinum* in heat-treated vegetable juice, *Appl Environ Microbiol*, **65**(5), 2136–42.
39. Esty J R and Meyer K F (1922) The heat resistance of the spores of *Bacillus botulinus* and allied anaerobes, *J Infect Dis*, **31**, 650–63.
40. Sugiyama H (1951) Studies on factors affecting the heat resistance of spores of *Clostridium botulinum*, *J Bact*, **62**, 81–96.
41. Sugiyama H (1952) Effect of fatty acids on the heat resistance of *Clostridium botulinum* spores, *Bact Rev*, **16**, 125–6.
42. Lang O W (1935) Thermal processes for canned marine products, *Univ Calif Publ Health*, **2**, 1–182.
43. Clouston J G and Willis P A (1969) Initiation of germination and inactivation of *Bacillus pumilis* spores by hydrostatic pressure, *J Bact*, **97**, 684–90.
44. Gould G W and Sole A J H (1970) Initiation of germination of bacterial spores by hydrostatic pressure, *J Gen Microbiol*, **60**, 335–46.
45. Sole A J H, Gould G W and Hamilton W A (1970) Inactivation of bacterial spores by hydrostatic pressure, *J Gen Microbiol*, **60**, 323–34.
46. van Baar B L M, Hulst A G, de Jong A L and Wils E R J (2002) Characterisation of botulinal toxins type A and B, by matrix-assisted laser desorption ionisation and electrospray mass spectrometry, *J Chromatog*, **970** (1–2), 95–115.

47. Baar J R, Moura H, Boyer A E, Woolfit A R, Kalb S R, Pavlopoulos A, McWilliams L G, Schmidt J G, Martinez R A and Ashley D L (2005) Botulinum neurotoxin detection and differentiation by mass spectrometry, *Emerg Inf Dis*, **11**(10), 1578–83.
48. Boyer A E, Moura H, Woolfit A R, Kalb S R, McWilliams L G, Pavlopoulos A, Schmidt J G, Ashley D L and Baar J R (2005) From the mouse to the mass spectrometer: detection and differentiation of the endoproteinase activities of botulinal neurotoxins A-G by mass spectrometry, *Anal Chem*, **77**(13), 3916–24.
49. Kalb S R, Goodnough M C, Malizio C J, Pirkle J L and Baar J R (2005) Detection of botulinum neurotoxin A in a spiked milk sample with subtype identification through toxin proteomics, *Anal Chem*, **77**(19), 6140–46.
50. Schiavo G, Santucci A, Dasgupta B R, Mehta P P, Jontes J, Benfenati F, Wilson M C and Montecucco C (1993) Botulinum neurotoxins serotypes A and E cleave SNAP-25 at distinct COOH-terminal peptide bonds, *FEBS Lett*, **335**, 99–103.
51. Binz T, Blasi J, Yamasaki S, Baumeister A, Link E, Sudhof T C, Jahn R and Niemann H (1994) Proteolysis of SNAP-25 by types E and A botulinal neurotoxins, *J Biol Chem*, **269**, 1617–20.
52. Ekong T A N, Feavers I M and Sesardic D (1997) Recombinant SNAP-25 is an effective substrate for *Clostridium botulinum* type A toxin endopeptidase activity *in vitro*, *Microbiol*, **143**, 3337–47.
53. Shone C C, Hallis B, James B A F and Quinn C P (1995) Toxin assay, PCT Patent Application WO 95/33850.
54. Wictome M, Newton K, Jameson K, Hallis B, Dunnigan P, Mackay E, Clarke S, Taylor R, Gaze J, Foster K and Shone C (1999) Development of an *in vitro* bioassay for *Clostridium botulinum* Type B neurotoxin in foods that is more sensitive than the mouse bioassay, *Appl Environ Microbiol*, **65**(9), 3787–92.
55. Aranda E, Rodriguez M M, Asensio M A and Córdoba J J (1997) Detection of *Clostridium botulinum* types A,B,E and F in foods by PCR and DNA probe, *Lett Appl Microbiol*, **25**, 186–90.
56. Hielm S, Björkroth J, Hyytiä E and Korkeala H (1999) Ribotyping as an identification tool for *Clostridium botulinum* strains causing human botulism, *Int J Food Microbiol*, **47**(1–2), 121–31.
57. Lin W J and Johnson E A (1995) Genome analysis of *Clostridium botulinum* Type A by pulsed-field electrophoresis, *Appl Environ Microbiol*, **61**(12), 4441–7.
58. Hielm S, Björkroth J, Hyytiä E and Korkeala H (1998) Genomic analysis of *Clostridium botulinum* group II by pulsed-field electrophoresis, *Appl Environ Microbiol*, **64**, 703–8.
59. Smelt J P P, Raatjes J G M, Crowther J C and Verrips C T (1982) Growth and toxin formation by *Clostridium botulinum* at low pH values, *J Appl Bact*, **52**, 75–82.
60. Saito M (1990) Production of enterotoxin by *Clostridium perfringens* derived from humans, animals, foods and the natural environment in Japan, *J Food Prot*, **53**, 115–18.
61. Gibbs P A (1973) The detection of *Clostridium welchii* in the Differential Reinforced Clostridial Medium Technique, *J Appl Bact*, **36**, 23–33.
62. Cornillot E, Saint-Joannis B, Daube G, Katayama S, Granum P E, Canard B and Cole S T (1995) The enterotoxin gene (*cpe*) of *Clostridium perfringens* can be chromosomal or plasmid-borne, *Mol Microbiol*, **15**, 639–47.
63. Todd E C D (1989) Cost of acute bacterial foodborne disease in Canada and the United States, *Int J Food Microbiol*, **52**, 313–26.
64. Labbe R (1981) Enterotoxin formation by *Clostridium perfringens* type A in a defined medium, *Appl Environ Microbiol*, **41**, 315–17.
65. Labbe R (1989) *Clostridium perfringens*, in M P Doyle (ed.), *Foodborne Bacterial Pathogens*, Marcel Dekker, New York, 192–227.
66. Juneja V K and Majka W M (1995) Outgrowth of *Clostridium perfringens* spores in cook-in-bag beef products, *J Food Safety*, **15**, 21–34.

67. Freame G and Fitzpatrick BWF (1972) The use of Differential Reinforced Clostridial Medium for the isolation of clostridia from food, in D A Shapton and R G Board (eds), *Isolation of Anaerobes*, Academic Press, London, 48–55.
68. Hauschild A, Desmarchelier P, Gilbert R, Harmon S and Vahlefeld R (1979) ICMSF methods studies. XII. Comparative study for the enumeration of *Clostridium perfringens* from feces, *Can J Microbiol*, **25**(9), 953–63.
69. Willis A T and Hobbs G (1959) Some new media for the isolation and identification of clostridia, *J Path Bact*, **77**, 511–21.
70. Eisgruber H, Schalch B, Sperner B and Stolle A (2000) Comparison of four routine methods for the confirmation of *Clostridium perfringens* in food, *Int J Food Microbiol*, **57**, 135–40.
71. Schalch B, Sperner B, Eisgruber H and Stolle A (1999) Molecular methods for the analysis of *Clostridium perfringens* relevant to food hygiene, *FEMS Immunol Med Microbiol*, **24**(3), 281–6.
72. Kokai-Kun J F, Songer J G, Czeczulin J R, Chen F and McClane B A (1994) Comparison of immunoblots and gene detection assays for identification of potentially enterotoxigenic isolates of *Clostridium perfringens*, *J Clin Microbiol*, **32**, 2533–9.
73. Stringer M F, Watson G N and Gilbert R J (1982) *Clostridium perfringens* Type A: Serological typing and methods for the detection of enterotoxin, in J E L Corry, D Roberts and F A Skinner (eds), *Isolation and Identification Methods for Food Poisoning Organisms*, Academic Press, London, 111–35.
74. Sikes A and Ehioba R (1999) Feasibility of using food-grade additives to control the growth of *Clostridium perfringens*, *Int J Food Microbiol*, **46**, 179–85.
75. Olds D A, Mendonca A F, Sneed J and Bisha B (2006) Influence of four retail service cooling methods on the behaviour of *Clostridium perfringens* ATCC 10388 in turkey roasts following heating to an internal temperature of 74 degrees C, *J Food Prot*, **69**(1), 112–7.
76. Hobbs B C, Smith M E, Oakley C L, Warrack G H and Cruickshank J C (1953) *Clostridium welchii* food poisoning, *J Hyg*, **51**, 75–101.
77. Amézquita A, Weller C L, Wang l, Thippareddi H and Burson D E (2005) Development of an integrated model for heat transfer and dynamic growth of *Clostridium perfringens* during the cooling of cooked boneless ham, *Int J Food Microbiol*, **101**(2) 123–44.
78. Poxton I R (2005) *Clostridium difficile*, *J Med Microbiol*, **54**, 97–100.
79. Mayfield J L, Leet T, Miller J and Mundy L M (2000) Environmental control to reduce transmission of *Clostridium difficile*, *Clin Infect Dis*, **31**, 995–1000.
80. Wiström J, Norrby S R, Myhre E B, Eriksson S, Granström G, Largergren L, Englund G, Nord C E and Svenungsson B (2001) Frequency of antibiotic-associated diarrhoea in 2462 antibiotic-treated hospitalized patients: a prospective study, *J Antimicrob Chemother*, **47**(1), 43–50.
81. George W L, Sutter V L, Citron D and Finegold S M (1979) Selective and differential medium for isolation of *Clostridium difficile*, *Clin Microbiol*, **9**(2), 214–9.
82. Aspinall S T and Hutchinson D N (1992) New selective medium for isolating *Clostridium difficile* from faeces, *J Clin Pathol*, **45**, 812–4.
83. Staneck J L, Weckbach L S, Allen S D, Siders J A, Gilligan P H, Coppitt G, Kraft J E and Willis D H (1996) Multicenter evaluation of four methods for *Clostridium difficile* detection: ImmunoCard *C. diff*, cytotoxin assay, culture, and latex agglutination, *J Clin Microbiol*, **34**, 2718–21.
84. Alonso R, Martín A, Peláez T, Marín M, Rodríguez-Creixéms M and Bouza E (2005) An improved protocol for pulsed-field gel electrophoresis typing of *Clostridium difficile*, *J Med Microbiol*, **54**, 155–7.
85. Arroyo L G, Kruth S A, Willey B M, Staempfli H R, Low D E and Weese J S (2005) PCR ribotyping of *Clostridium difficile* isolates originating from human and animal sources. *J Med Microbiol*, **54**, 163–6.

24

Pathogenic *Bacillus* species

C. de W. Blackburn and P. J. McClure, Unilever, UK

Abstract: *Bacillus* spp. are ubiquitous spore-forming bacteria that typically occur as saprophytes in soil and water, but some species, *B. cereus* in particular, can cause food poisoning. This chapter covers the general characteristics of pathogenic *Bacillus* species, the nature of *Bacillus* food poisoning and their growth and survival characteristics. Risk factors associated with *Bacillus* food poisoning are addressed, together with methods for their detection and procedures and strategies for their control throughout the food chain.

Key words: *Bacillus cereus*, emetic toxin, diarrhoeal toxin, psychrotrophic *B. cereus*, mesophilic *B. cereus*.

24.1 Introduction

The spore-forming bacteria belong to the prokaryotic family Bacillaceae, a heterogeneous group of rod-shaped, Gram-positive bacteria. Bacterial spores are essentially 'resting stages' in the life cycle of the bacteria, their formation generally resulting from a depletion of some essential nutrient in their environment. Spores are dormant structures, with a central 'core', where the genetic elements of the cell reside, together with the enzymes necessary for the germination and initial outgrowth of the germinated spore. The core is highly protected from the external environment by a thick protein coat and a cortex. The protein coat is highly resistant to chemical and radiation damage, whereas it is thought that the cortex provides a very low a_w surrounding for the core, resulting in a high resistance to heat. It is hypothesised that the dormancy of bacterial spores is due to at least the central protoplast region being maintained in a low moisture content glassy state (Ablett *et al.*, 1999). Spores are very resistant to many environments lethal to vegetative cells, e.g. high temperatures, drying, ultra violet (UV) radiation and other ionising radiations, disinfectants and chemical sterilants. These properties make for difficulties in the safe processing of foods, especially canning.

Bacillus spp. are ubiquitous and typically occur as saprophytes in soil and water, but some species, such as *B. cereus*, *B. anthracis* and to a lesser extent *B. subtilis*, are pathogenic in humans and other mammals. There is evidence that some strains of other species including *B. lichenifiormis*, *B. thuringiensis*, *B. pumilus* and *Brevibacillus* (formerly *Bacillus*) *brevis* can cause foodborne illness (e.g. gastrointestinal symptoms that are self-limiting) typical of either *B. cereus* or *B. subtilis*. Nevertheless, *B. subtilis*, *B. licheniformis*, *B. pumilus* and *Br. Brevis* are commonly regarded as spoilage organisms. The genus *Bacillus* contains other pathogens of man, animals and insects, such as the greater wax moth *Galleria mellonella* (Ali *et al.*, 1973). Some strains of *B. thuringiensis* are used in insecticide preparations. Many species of the genus produce powerful extra-cellular enzymes responsible for degradation of carbohydrates, proteins and fats and, as a consequence, some species are used in 'biological detergents'. Some species, such as *B. subtilis*, *B. amyloliquefaciens*, *B. licheniformis*, *B. alkalophilus* and *B. lentus*, have properties that are utilised by fermentation industries (Priest, 1990). Some strains are also used as probiotics. Members of the genus *Bacillus* are aerobic or facultative, and can only sporulate under aerobic conditions.

24.2 General characteristics of pathogenic *Bacillus* species

24.2.1 Taxonomy

The genus *Bacillus* was divided into three broad groups by Drobniewski (1993), depending on the morphology of the spore and sporangium and into five groups by Ash *et al.* (1991), depending on comparative analysis of small-subunit-ribosomal RNA sequences. Since then, at least eight different phylogenetic groups have been identified (Shida *et al.*, 1996). Garrity *et al.* (2003) provided an update of the grouping of aerobic endospore-forming bacteria within the order Bacillales, and recent taxonomic changes have been recently reviewed by Jay (2003) and Fritze (2004). The taxonomy of the genus *Bacillus* is determined by comparison of the sequences of 16SrRNA/DNA (Stackebrandt and Swidersky, 2002) but, where this method is unsatisfactory, classical methods such as morphology, biochemistry, physiology and chemotaxonomic traits are still used. The two groups of importance in the context of foodborne illness are the *Bacillus cereus* group, and the *Bacillus subtilis* group, and both of these are in Group 1 of the genus *Bacillus*.

Studies referring to the *B. cereus* group (*B. cereus sensu lato* group) generally recognise four valid species, namely *B. cereus*, *B. anthracis*, *B. thuringiensis* and *B. mycoides*. The members of this group are easily distinguished from other aerobic endospore-forming bacteria by their inability to form acid from mannitol and their production of lecithinase, but difficult to distinguish from each other. Cells of this group are wider than 1 μm, sporangia are not swollen and spores are ellipsoidal and centrally or paracentrally placed. Members of this group are generally mesophilic and neutrophilic, although psychrotolerant strains do occur. Vegetative cells are motile, Gram-positive, facultatively anaerobic, rod-shaped, sporulating readily only in the presence of oxygen, and are widely distributed in the environment. The resistance of the spores to adverse environmental conditions, and the ability to produce a range of food-degrading enzymes, e.g. protease, amylase, lecithinase, enable *B. cereus* to survive and grow well in many different conditions. It is typically responsible for the defect in temperature-abused pasteurised milk called 'bitty cream' that results from degradation of lecithin, the cream-stabilising emulsifier in milk, during growth of the organism from spores surviving pasteurisation.

The four species above are very closely related showing 99 % sequence similarity in the primary structures of 16S rRNA (Ash *et al.*, 1991), similarity in the 23S rRNA gene sequences and similarity in the 16S-23S spacer regions. This spacer region is regarded as a good marker for comparing closely related and distantly related organisms because the selective pressure influencing this area of the genome is different from that for the rDNA structural genes. However, the genomic structure and organisation between different strains of *B. cereus* and *B. thuringiensis* suggests that there is as much intraspecies diversity as there is interspecies diversity. The distinguishing characteristics are the two virulence plasmids carried by *B. anthracis*, crystalline parasporal inclusions (plasmid-encoded) and delta-endotoxin production in *B. thuringiensis*, and rhizoid growth on blood agar by *B. cereus* var. *mycoides*. DNA–DNA hybridisation has also been used for species differentiation, and 70 % homology has been suggested as the limit for separation of species. In studies that have looked at the DNA–DNA relatedness of *Bacillus* spp., results are contradictory with some showing large heterogeneity between strains and others showing close relatedness. Hybridisation data appear to vary unsystematically and are not correlated with 16S rDNA analysis and RAPD analysis (Lechner *et al.*, 1998). Fatty acid profiles have also been used to differentiate between groups.

The similarities between *B. anthracis* and some non-anthrax *Bacillus* species have prompted many calls for *B. anthracis*, *B. thuringiensis* and *B. cereus* to be regarded as varieties of a single species. There is also significant cross-agglutination between the spore antigens of *B. cereus*, *B. anthracis* and *B. thuringiensis* and the flagellar antigens of *B. cereus* and *B. thuringiensis*. However, the group comprises both medically and economically important species and proposals for changes in the nomenclature are therefore opposed. Recently, there have been some changes to this group, with the naming of *B. weihenstephanensis* and *B. pseudomycoides* as new species. *Bacillus pseudomycoides* is genetically distinct from *B. mycoides* but is separable on the basis of fatty acid composition. The newly named *B. weihenstephanensis* is supposed to represent 'psychrotolerant' or 'psychrotrophic', but not mesophilic, *B. cereus* strains. However, Stenfors and Granum (2001) concluded that there are psychrotrophic *B. cereus* strains that cannot be classified as *B. weihenstephanensis* and that intermediate forms exist between the two species.

The *B. subtilis* group comprises *B. subtilis*, *B. licheniformis*, *B. pumilus*, *B. firmus* and *B. lentus*. More recently, *B. subtilis* has been split, with new species being distinguished such as *B. amyloliquefaciens*, *B. atrophaeus*, *B. mojavensis* and *B. vallismortis*. Members of the group are typically referred to as the small-cell sub-group because cells are less than 1 μm wide. The sporangia are not swollen and spores are ellipsoidal. Members of this group are mesophilic and neutrophilic but often tolerant of higher pH values. Some members of the group, such as *B. pumilus* and *B. licheniformis*, can be distinguished by phenotypic tests, such as starch hydolysis and hippurate reaction (negative and positive respectively for *B. pumilus*) or propionate reaction (*B. licheniformis* is positive and facultatively anaerobic and grows up to 55 °C). For other members, phenotypic discrimination is not as straightforward.

Strains of species from this group are widely used in traditional and industrial fermentation processes and in agriculture due to their enzyme production, antibiotic and probiotic characteristics and ability to produce surfactants and degrade xenobiotics. However, there are some species that have been associated with cases of foodborne disease, although these appear to be less significant in number compared to *B. cereus*.

24.2.2 Virulence factors

Only in the early 1950s was *B. cereus* recognised as a food poisoning bacterium (Hauge, 1955), although the earliest report is thought to be that of Lubenau (1906), affecting 300 hospitalised individuals and hospital staff, and Plazikowski (1947) reported that 117 out of 367 foodborne outbreaks in Sweden were most likely due to aerobic sporeformers belonging to the subtilis-mesentericus group. It is now known that a restricted number of strains of this organism are responsible for two types of foodborne illness, emetic and diarrhoeal. The emetic form is associated with toxin produced in food. Emetic intoxication following ingestion of a preformed cyclic (dodeca-) depsipeptide, cereulide, which consists of a ring structure of three repeats of four amino and/or oxy acids. Chemically, it is closely related to the ionophore linomycin, and it is recognised as a powerful mitochondrial toxin. It is also reported to have an immunomodulatory effect, by toxic activity, towards natural killer cells (Paananen *et al.*, 2002). Cereulide is sometimes referred to as the 'sperm toxin' since one of the commonly used tests for presence of this toxin is the boar spermatozoa motility test. Production of the emetic toxin is known to depend on the composition of the contaminated food, with low concentrations of amino acids promoting toxin production (Agata *et al.*, 1999). In a study describing a method for detection of emetic toxin (Mikami *et al.*, 1994), 310 fresh/fermented food *B. cereus* isolates (using NaCl glycine Kim Goepfert agar) were assayed for their ability to produce cereulide and only 16 strains were shown to be positive. Many of these were isolated from fermented bean paste. All cereulide-positive strains were serotype H1, which is frequently detected in emetic food poisoning. None of these strains was shown to produce enterotoxin. Many of the strains linked with outbreaks of emetic-type illness are 'overproducers' of the emetic toxin (Agata *et al.*, 1996). Jaaskelainen *et al.* (2003) reported the toxic dose in humans to be less than or equal to 8 μg per kg body weight.

In a separate study, comparing environmental isolates but also including strains associated with foodborne illness outbreaks, Pirttijävi *et al.* (1999) also looked at ability to produce toxins, hydrolyse starch and lecithin and ability to grow at refrigeration temperatures. Results from this study agreed with the previous study – none of the strains shown to produce emetic toxin produced hemolysin BL (HBL) toxin, none was able to hydrolyse starch and all were serotype H1 – these characteristics appear to be typical of those isolates associated with emetic-type syndrome (Agata *et al.*, 1996). In addition, some were shown capable of producing non-haemolytic enterotoxin (see below), none was capable of growth at 6 °C and only two were capable of growth at 10 °C. Ribotyping of the isolates also grouped all the emetic-toxin producing strains together with very similar ribopatterns. These ribotypes were rarely found amongst other *B. cereus* group isolates. In a study using phenotypic and genotypic analysis of ninety *B. cereus* isolates, Ehling-Schulz *et al.* (2005) reported evidence for a clonal population structure of cereulide-producing emetic *B. cereus* amongst a weakly clonal background population structure of the species, suggesting these may have arisen from recent acquisition of specific virulence factors such as the cereulide synthetase gene.

In studies looking at the prevalence of emetic toxin-producing strains, Altayar and Sutherland (2006) reported that these were relatively rare, with only three out of 177 *B. cereus* isolates from soils, animal faeces, raw and processed vegetables showing this ability. Ueda and Kubawara (1993), however, showed that 44 % of isolates from paddy field soil, the processing environment and rice product were capable of emetic toxin production.

When a range of *Bacillus* species were tested for their ability to produce toxin(s) (Beattie and Williams, 1999), using two commercially-available kits and a cytotoxicity assay, a number of species were positive, indicating that they produce similar protein toxins to those produced by *B. cereus*. In the cytotoxicity assay, which included heated samples (100 °C for 10 min), samples from some strains (*Br. brevis*, *B. cereus*, *B. circulans*, *B. lentus*, *B. licheniformis*, *B. mycoides*, *B. subtilis* and *B. thuringiensis*) were also shown to be toxic, indicating presence of heat resistant toxin(s). Taylor *et al.* (2005) reported heat-stable toxin production in strains of *B. firmus*, *B. megaterium*, *B. simplex* and *B. licheniformis*. Partial purification of the toxins produced by non-*cereus* strains showed they had similar physical characteristics to cereulide. Thorsen *et al.* (2006) reported cereulide production in two psychrotolerant strains of *B. weihenstephanensis*, but the significance of this finding is not clear since production at lower temperatures was not considered sufficient to cause foodborne illness. From *et al.* (2007) recently reported production of pumilacidin, which closely resembles other cyclic peptides produced by *Bacillus* spp., in a strain of *B. pumilus* associated with a case of foodborne illness, although vomiting was not reported. The main symptoms reported were back and neck pain, headache, dizziness, chills, followed by stomach cramps and diarrhoea in the days following acute symptoms.

The diarrhoeal disease is caused by one or more toxins produced in the gastrointestinal tract. *Bacillus cereus* can produce at least three different enterotoxins (Arnesen *et al.*, 2008); two of these are tripartite toxins and both are associated with foodborne outbreaks. They cause diarrhoea by disrupting the integrity of the plasma membrane of epithelial cells in the small intestine. The most studied enterotoxin is HBL, a three-component haemolysin consisting of two lytic components (L_1 and L_2) and a binding component (B). This enterotoxin is thought to be the primary virulence factor in *B. cereus* diarrhoea and has haemolytic, dermonecrotic and vascular permeability activities. The *hbl* operon includes three known proteins encoded by *hblA*, *hblC* and *hblD*. The second enterotoxin, named NHE, is also made up of three components and is non-haemolytic. Another enterotoxin is coded by the *bceT* gene and exhibits Vero cell cytotoxicity and also causes vascular permeability but is not haemolytic. The role of this enterotoxin in foodborne disease, if it has any at all, is unclear (Choma and Granum, 2002). Another potential enterotoxin, EntFM, is described by Asano *et al.* (1997) but the role of this in foodborne disease is also unclear.

In the study of food/environmental isolates referred to above (Mikami *et al.*, 1994), 228 out of 310 isolates were shown to produce enterotoxin (using the CRET-RPLA kit). In the study by Pirttijävi *et al.* (1999), 30 out of 56 industrial/environmental strains were positive for HBL and 24 of 44 produced the non-haemolytic enterotoxin. Enterotoxin-positive isolates were associated with a number of different ribogroups, and 78 % of strains capable of growth at 6 °C were HBL producers.

Determination of pathogenicity through detection of genes encoding for enterotoxins, presence of the complete HBL complex (through immunological assays), cytotoxic and haemolytic activity is, however, not straightforward. In a survey (Prüß *et al.*, 1999) of different species of the *B. cereus* sub-group looking for presence of HBL, the enterotoxin was found in all species of the sub-group, suggesting that the HBL is broadly distributed among the group. However, the same study showed that the majority of *B. weihenstephanensis* strains tested expressed little or no HBL complex and were only slightly cytotoxic, even though they possessed the *hblA* gene. The study by Prüß *et al.* (1999) also showed that only half of the *B. cereus* strains tested possessed the *hblA* gene. The amount of enterotoxin produced by *B. cereus* strains can vary by more than 100-fold, and only the high enterotoxin-producing strains seem to cause food poisoning (Granum *et al.*, 1993).

In a study looking at environmental (including food) *B. cereus* and *B. thuringiensis* isolates (Mikami *et al.*, 1995), enterotoxin was found to be produced by 47 out of 90 *B. thuringiensis* strains tested. In Sweden, 23 out of 40 strains of *B. thuringiensis* isolated from soil were shown to produce *B. cereus*-diarrhoeal-type enterotoxin (Abdel-Hameed and Landen, 1994). More recently, Hendriksen *et al.* (2006) reported presence of multiple genes coding for virulence factors in all isolates representing *B. cereus*, *B. weihenstephanensis*, *B. weihenstephanensis*-like *B. mycoides* and mesophilic *B. mycoides* in sandy loam in Denmark. Enterotoxigenic *B. thuringiensis* strains have also been detected in commercial preparations of microbial pesticides (Damgaard, 1995). Damgaard *et al.* (1996) found by Vero cell assay that *B. thuringiensis* isolated from food was enterotoxigenic. Shinagawa (1990) investigated a number of *B. cereus* and *B. thuringiensis* isolates with an immunological assay and concluded that 43 % of the *B. thuringiensis* isolates had the same level of enterotoxins as enterotoxigenic *B. cereus*. Gaviria Rivera *et al.* (2000) found genes/operons encoding the three enterotoxins in 63 out of 74 strains (representing 24 serovars) of *B. thuringiensis* examined. Culture supernatents from 73 of these showed toxicity against Vero cells, generally with equivalent toxicity to that seen in strains of *B. cereus* isolated from incidents of food poisoning. In a study of 86 *B. cereus* strains isolated from milk, in't Veld *et al.* (2001) detected all three genes for the HBL enterotoxin in 55 % of these.

There are several haemolytic factors (other than the haemolytic enterotoxins mentioned above) expressed in *B. cereus* group bacteria, including cereolysin, sphingomyelinase, cereolysin AB and cerolysin-like haemolysin (also known as thuringiolysin). The role of these in foodborne disease is not clear, but they are thought to play a role in infection. Studies looking for haemolytic activity in the *B. cereus* group (e.g. Prüß *et al.*, 1999) show that most isolates tested will show some haemolytic activity. Haemolysin and cereolysin genes are found in non-*cereus Bacillus* species, such as *B. thuringiensis*.

A number of toxicity tests have been used to assess the pathogenicity of *Bacillus* spp. The boar spermatozoa motility test was originally used for the detection of *B. cereus* emetic toxin (Andersson *et al.*, 1998a) and has been used to assess the 'toxicity' of *B. licheniformis* isolates implicated in cases of foodborne illness (Salkinoja-Salonen *et al.*, 1999). In this study, inhibition of *Corynebacterium renale* was also used to assess 'toxicity'. This paper claims to be the first to describe a toxin produced by *B. licheniformis* associated with human illness. The toxic effects associated with growth of *B. licheniformis* were different to those previously observed for *B. cereus* emetic toxin, interfering with energy metabolism of sperm motility in a different way. It is stated that the sperm-toxic agent is similar in many physicochemical properties to cereulide, i.e. both are resistant to heat, acid and alkali and soluble in methanol. It is worth noting that only eight out of

210 (4 %) illness-associated isolates tested were positive in either toxicity test. Further work (Mikkola *et al.*, 2000) on the newly recognised toxin, produced by strains isolated from an infant formula (in which more than one *Bacillus* spp. was present) linked to fatal illness, has identified the toxin as lichenysin A. This is a cyclic heptalipopeptide, known as a biosurfactant and antibacterial agent, that can be produced aerobically or anaerobically. It is a heat-stable lipopeptide lactone and is structurally related to surfactin and other biologically active surfactants. There is little real evidence that this newly reported toxic agent had or has any role in causing foodborne illnesses.

Other virulence factors that have been suggested as having a role in disease and that are present in *Bacillus* species include phospholipase C, proteases and Beta-lactamases. Antibiotic resistance is an important consideration for the application and use of products containing viable microorganisms, such as drain cleaning products, where *Bacillus* species are a common ingredient. Surface proteins (S-layer) in spores of *B. cereus* are thought to have a key role in the adhesion process (Andersson *et al.*, 1998b) and these are also present in other *Bacillus* species. Sporulation and production of enterotoxin at the site of colonisation are supposed to cause more serious symptoms in diarrhoea-type disease.

A study by Lund *et al.* (2000) identified a cytotoxin (CytK) produced by a strain of *B. cereus* that caused a severe food poisoning outbreak killing three people in France, in March 1998. The outbreak occurred in a home for elderly people and involved 44 people, causing diarrhoea, where six patients developed bloody diarrhoea and three died. The organism was isolated from a vegetable purée at a level of 3×10^5 per gram. The protein was highly cytotoxic, necrotic and haemolytic, and the strain did not produce any other known *B. cereus* enterotoxins. The structure (amino acid sequence) of the protein showed similarities to a family of β-barrel channel-forming toxins that includes *Clostridium perfringens* β-toxin and *B. cereus* haemolysin II. Through further genetic characterisation, this strain together with two others have been identified as a separate cluster and are genetically sufficiently distant from other *B. cereus* to warrant novel species status, proposed *B. cytotoxicus* or *B. cytoxicus* according to Lapidus *et al.* (2008). These strains are capable of growth at temperatures 6–8 °C higher than other mesophilic strains (Auger *et al.*, 2008).

24.3 Nature of *Bacillus* food poisoning

The incidence of either form of this disease is seriously under-reported as the symptoms in most cases are relatively mild and of short duration. Only when an outbreak involves several cases and arises from a single source is the disease likely to be reported and investigated. The diagnosis of *B. cereus* foodborne illness is usually confirmed by isolation of $\geq 10^5$ organisms per gram of implicated food. However, once diagnosed, the disease is not reportable in many countries. The emetic syndrome is most commonly associated with consumption of rice and some other farinaceous foods in which the spores have survived cooking, germinated and allowed sufficient time at moderate temperatures for growth and toxin production to occur. The diarrhoeagenic syndrome is more often associated with meat-containing dishes such as stews and soups, although vegetables, sauces and milk products may also be vehicles.

The diarrhoeal form of illness causes abdominal pain and profuse diarrhoea, but rarely nausea and vomiting, *c.* 8–16 h after ingestion of vegetative cells or spores, and closely resembles the symptoms caused by *Clostridium perfringens* enterotoxin. The effective dose is 10^5–10^7 total cells to infect the small intestine for the diarrhoeagenic syndrome (see Table 24.1). The emetic syndrome causes nausea and vomiting and occasionally diarrhoea, *c.* 1–5 h after consumption of a contaminated meal, and closely resembles symptoms of food poisoning caused by *Staphylococcus aureus*. The effective dose for the emetic syndrome is 10^5–10^8 cells/g of food where the toxin is pre-formed in the food. For either form of illness, symptoms resolve after 6–24 h, but some patients may be sufficiently severely affected to be hospitalised, with mortality being very rare. Recovery is usually complete in 24–48 h, and no specific treatment other than replacement of fluid and salts loss is indicated. The other species of the genus *Bacillus* that have occasionally been shown to cause food poisoning of either type are described later in this chapter.

Differences in the reported incidences of *B. cereus* seem to reflect differences in eating habits but also the frequencies of other food-poisoning incidents in various countries. Thus in Japan, North America and Europe (in general), the reported outbreaks are between 1 % and 22 % of all outbreaks. However,

Table 24.1 Characteristics of *B. cereus* food poisoning syndromes

Characteristic	Emetic syndrome	Diarrhoeal syndrome
'Infective' dose	10^5–10^8/g food	10^5–10^7 total cells consumed
Production site of toxin	In food	In small intestine
Symptoms	Nausea and vomiting; possibly also diarrhoea	Abdominal pain, diarrhoea; occasionally nausea
Incubation time	1–5 h	8–16 h
Duration of symptoms	6–24 h	12–24 h
Toxin characteristics	Small peptide; heat-resistant; acid-stable; hydrophobic; protease stable; non-immunogenic	At least two protein enterotoxins recognised; heat and protease sensitive; immunogenic
Most common food vehicles	Reheated rice, farinaceous products	Meat products, stews, soups, sauces, puddings, milk products

the frequency of the emetic and diarrhoeal syndromes differ; in Japan the emetic syndrome is ten times more frequently reported than the diarrhoeal syndrome, whereas in Europe and North America the latter is more frequently reported. In the UK the most common source of the emetic syndrome used to be fried or reheated rice dishes, until the importance of rapid cooling and chill storage in prevention was realised and advice provided (Kramer and Gilbert, 1989). Hughes *et al.* (2007) reported a total of 43 outbreaks between 1992 and 2003, with still almost half of these associated with rice. In the USA, Mead *et al.* (1999) estimated the total number of cases per year to be 27 360, assuming this to be 38 times the number of reported cases by extrapolation. In Norway, however, whereas the country is almost free of *Salmonella* and *Campylobacter* food poisoning, *B. cereus* is the most commonly reported food poisoning syndrome. During 1991–2000, *B. cereus* was reported to be the leading cause of bacterial foodborne outbreaks in Taiwan, responsible for 113 out of 171 outbreaks of illness (Chang and Chen, 2003). Kotiranta *et al.* (2000) report that the incidence of *B. cereus*-related foodborne disease is increasing in industrialised countries.

In the literature the value quoted for the minimum infective dose for the diarrhoeal illness caused by *B. cereus* is generally $> 10^5$ cells/g (Notermans *et al.*, 1997; Granum and Baird-Parker, 2000; Schoeni and Wong, 2005). The reference frequently cited as the basis of this number is Kramer and Gilbert (1989). In this study, the authors state that the levels of *B. cereus* recovered from foods implicated in outbreaks of the diarrhoeal-type illness have almost always been within the range 5×10^5–9.5×10^8 cfu/g. Kramer and Gilbert (1989) quote reviews by Goepfert *et al.* (1972) and Gilbert (1979).

Goepfert *et al.* (1972) and Gilbert (1979) reviewed a number of diarrhoeal-type *B. cereus* outbreaks. The levels recovered from implicated foods were on the whole $> 10^5$ cfu/g, and commonly much higher, e.g. $> 10^8$ cfu/g. However, in a selected number of outbreaks a lower infective dose was stated; details of these outbreaks are covered below.

- Todd *et al.* (1974) reported a count in an implicated chicken dish of 6.0×10^4 cfu/g. The dying-off of viable cells from initially higher numbers is stated by the authors as one explanation for the low count from the implicated food.
- Nikodémusz *et al.* (1962) summarised 35 outbreaks attributed to *B. cereus*. These 35 outbreaks affected about 800 people. Foods often incriminated were sausage, vegetable dishes, cream pastries and meat or vegetable soups. The level of *B. cereus* contamination ranged from 3.6×10^4–9.5×10^8 cfu/g. Only one outbreak reported levels of $< 10^5$ cfu/g of food (a tripe soup with 3.6×10^4 cfu/g). Of the other 34 outbreaks, five were associated with levels of 10^5 cfu/g, and an equal number of outbreaks were associated with $10^6/10^7/10^8$ cfu/g. Importantly, the authors concluded that the level that could be harmful for consumers was identified as 10^5 viable organisms per gram of food and that this was in agreement with previous reports on *B. cereus*-related illness.

- Giannella and Brasile (1979) reported an outbreak involving 28 patients on two wards of a chronic disease hospital. The average age of the patients was 72.7 years (range 50–83 years), and the patients were mostly bedridden because of cerebrovascular disease. Turkey loaf served for an evening meal was incriminated as the source of the outbreak. A *B. cereus* count of 1.2×10^3 cfu/g was found in samples of the implicated turkey loaf. The underlying disease state of the patients may have made them more susceptible to *B. cereus* infection, and may explain the potentially low infective dose in this outbreak.
- Ghelardi *et al.* (2002) investigated two simultaneous outbreaks. In these outbreaks 173 people presented with illness after attending one of two banquets (95 people from one banquet and 78 from the other banquet). Cakes prepared by the same confectioner were served at each banquet and were implicated. The food samples tested ($n = 4$) were found to contain $> 10^2$ cfu/g of *B. cereus*. As the level is quoted as a 'greater than' value, the actual level present is unknown.
- Gaulin *et al.* (2002) reported an outbreak of *B. cereus* implicating a part-time caterer. In the outbreak 70 % (25 of 36) of attendees at a banquet had gastroenteritis. A cohort study showed that potato salad served at the party was significantly associated with illness. No food prepared for that particular meal was available for testing because all items had been eaten. Samples of the mayonnaise used to prepare the potato salad were taken for analysis from the premises of the implicated caterer. *Bacillus cereus* was isolated at a level of 10^3 cfu/g from the mayonnaise. The outbreak investigation concluded that bacterial multiplication was facilitated at several points in the interval between the preparation of the meal and consumption by the guests. Poor temperature control by an inexperienced caterer was stated as a key factor in this outbreak. It is possible that the levels of *B. cereus* increased in the salad as it was subjected to time/temperature abuse and the pH of the mixed salad was probably higher than the pH of the mayonnaise alone. Therefore levels of *B. cereus* in the implicated potato salad may have been significantly greater than the 10^3 cfu/g recovered from the mayonnaise.
- Penman *et al.* (1996) reported an outbreak of acute gastrointestinal illness among 142 persons who had attended a hotel-catered wedding reception banquet. A profuse monoculture of *B. cereus* (> 100 cfu/g) was grown from one sample of quiche from a single leftover plate of food. As the level is quoted as a 'greater than' value, the actual level present is unknown.
- In two outbreaks in Norway in which stew (meat, potatoes and vegetables) was implicated, the infective dose was stated to be approximately 10^4–10^5 total cells (Granum and Baird-Parker, 2000). Given that the cases would probably have consumed of the order of 10^2 g of the stew then the concentration of *B. cereus* in food implicated in these outbreaks would be 10^2–10^3 cfu/g. Further investigation showed that the actual numbers

were closer to 10^4 cfu g^{-1} food and that the under-estimate was due to the bacilli being present as aggregated spores (Arnesen *et al.*, 2008).

In a double-blind study of healthy adults ingesting pasteurised milk, Langeveld *et al.* (1996) concluded that there was no evidence that *B. cereus* concentrations less than 10^5/ml would cause intoxication. Examples of different foods linked to outbreaks of foodborne illness caused by *B. cereus* are shown in Table 24.2. More extensive lists of foodborne outbreaks are provided by Goepfert *et al.* (1972), Gilbert (1979) and Johnson (1984).

24.4 Growth and survival characteristics

Bacillus cereus produces a wide range of extra-cellular enzymes capable of hydrolysing proteins, fats and starch, as well as a range of other carbohydrates. The organism can therefore utilise a wide range of food materials to support growth, but starch-containing foods seem to be optimal for growth. Even though the diarrhoeal syndrome is mainly associated with meat-containing dishes, many of these will have been thickened with starch or contain spices that are often heavily contaminated with *B. cereus* spores.

Bacillus cereus is a mesophilic organism with an optimum temperature for growth of 30–35 °C, although some strains are psychrotrophic, growing at temperatures down to *c.* 5 °C (EFSA, 2005). These latter strains are mostly associated with milk and dairy products, and no doubt some that were previously identified as *B. cereus* are actually *B. weihenstephanensis.* The optimum growth temperature for psychrotrophic strains is between 30 and 37 °C. The upper temperature limit for germination of spores (i.e. loss of the specific spore characteristics of heat resistance, refractility under phase contrast microscopy, etc.) is rather higher than for vegetative growth of cells. Thus germination will occur during cooling of boiled rice (at 55–60 °C) with rapid generation of vegetative cells on further cooling. The maximum temperature for growth is between 37 and 42 °C. Some strains are even capable of growth as high as 50 °C. Whilst *B. cereus* is not generally regarded as tolerant of low pH and a_w values, it will grow down to a_w values of *c.* 0.93. Kramer and Gilbert (1989) stated that because fried rice readily supports the growth of *B. cereus*; the lower a_w value (0.912) of a range of a_w values reported for fried rice (0.912–0.961) could be reasonably regarded as a working minimum limit. We are not aware of growth being reported below a a_w of 0.93. When other conditions are considered, growth may be more limited, e.g. at 5 % NaCl, growth only occurs above 21 °C and the minimum pH allowing growth increases to 5.5 (see EFSA, 2005). Modified atmosphere packaging conditions using increased CO_2 levels slow down but do not completely prevent growth. Rusul and Yaacob (1995) reported growth of 42 % of strains tested at pH 4.5 within 7 days at 37 °C.

The effects of heat on spores of *B. cereus* are strain-dependent, with various

Table 24.2 Foods linked to outbreaks of illness caused by *B. cereus* and other *Bacillus* spp.

Food	Country	Number of people affected (deaths)	Organism	Type of illness	Reference
Chicken and rice meal	UK	300	*B. cereus*	Diarrhoeal	Ripabelli *et al.*, 2000
Quiche	US	79	*B. cereus*	Diarrhoea and vomiting	Penman *et al.*, 1996
Cakes at 2 banquets	Italy	96 and 76	*B. cereus*	Diarrhoeal	Ghelardi *et al.*, 2002
Vegetable puree	France	44 (3)	*B. cereus*	Bloody diarrhoea	Lund *et al.*, 2000
Potato salad with mayonnaise at a banquet	Canada	25	*B. cereus*	Diarrhoeal	Gaulin *et al.*, 2002
Stew	Norway	17	*B. cereus*	Diarrhoeal	Granum *et al.*, 1995
Salad sprouts	US	4	*B. cereus*	Vomiting and diarrhoea	Portnoy *et al.*, 1976
Green bean salad	Canada	> 300	*B. cereus*	Diarrhoeal	Schmitt *et al.*, 1976
Orange juice	France	43	*B. cereus*	Vomiting and diarrhoea	Talarmin *et al.*, 1993
Pasteurised milk	Japan	201	*B. cereus*	Vomiting and diarrhoea	Shinagawa, 1993
Barbeque pork	US	139	*B. cereus*	Diarrhoeal	Luby *et al.*, 1993
Reconstituted milk drink	Croatia	12	*B subtilis/ B. licheniformis*	Nausea, vomiting	Pavic *et al.*, 2005
Rice	Norway	3	*B. pumilus*	Stomach cramps, diarrhoea	From *et al.*, 2007
Pasta salad	Belgium	3 (1)	*B. cereus*	Vomiting and respiratory distress	Dierick *et al.*, 2005

reports showing a broad range of *D*-values at different temperatures. Whilst it is generally the case that more psychrotolerant strains are more susceptible to heat, there are exceptions to this rule, with some psychrotrophic strains showing relatively high heat resistance. Heat resistance characteristics are also shown in Table 24.3. The reported $D_{130\,°C}$ of around 0.3 min is extremely unusual and indicates that commercial sterilisation can, where very resistant strains are involved, allow survival of these strains. Normally speaking, however, commercial sterilisation temperatures that target *Clostridium botulinum* will also control *B. cereus*. As with other spores and vegetative cells, a decrease in pH will accelerate inactivation compared to higher pH values. For example, reduction in pH to 4.7 results in a three-fold decrease in *D*-values (Fernández *et al.*, 2002) and heating at pH 4.0 results in a five-fold decrease in *D*-values (Mazas *et al.*, 1998).

Non-thermal processing technologies are generally not very effective in destroying bacterial spores compared to vegetative cells. For example, pressures over 1000 MPa are not sufficient to inactivate spores at 25 °C (Sale *et al.*, 1970). To inactivate bacterial spores, high pressure must be used in combination with moderate heat (60–80 °C), and this can be enhanced by other factors such as nisin, low pH or oscillatory compression procedures.

Table 24.3 Growth and survival characteristics of *B. cereus*

Characteristic		Values
Spore heat resistance	$D_{90\,°C}$	2–100[d] min
	$D_{100\,°C}$	1.2–27[e] min
	$D_{130\,°C}$	0.3[f] min
	z	6–9 C°
Germination temperatures	limits	5–50 °C
	optimum	30–35 °C
Growth limits	pH	4.3–9.3
	temperature[a]	5/15–35/50 °C
	a_w, broth	0.93
Generation times	broth culture 35 °C	*c.* 20 min
	boiled rice 30 °C	*c.* 30 min
	broth culture 19.5 °C[b]	1.6 h
	broth culture 14.2 °C[b]	2.9 h
	broth culture 9.6 °C[b]	4 h
	broth culture 6.5 °C[b]	6.7 h
Lag times	broth culture 30 °C[c]	1.96 h
	broth culture 5 °C[c]	149 h

[a] 5/35 °C limits for psychrotrophic strains; 15/50 °C for mesophilic strains. Spore germination may occur up to about 60 °C
[b] Choma *et al.*, 2000.
[c] Valero *et al.*, 2000.
[d] Dufrenne *et al.*, 1994.
[e] Rajkowski and Mikolajcik, 1987.
[f] Bradshaw *et al.*, 1975.

Recently, Margosch *et al.* (2004) reported on the strong variability of resistance to pressure and temperature within *Bacillus* spp., including *B. subtilis*, *B. licheniformis* and *B. amyloliquefaciens*. The authors concluded that at pressures between 600 and 800 MPa combined with temperature > 60 °C, dipicolinic acid (DPA) is released and the DPA-free spores are then inactivated by the moderate heat. Strains of *B. amyloliquefaciens* were shown to form highly pressure-resistant spores. Scurrah *et al.* (2006) included six strains of *B. cereus* in a study looking at effects of temperature and high pressure on different *Bacillus* spp., also concluding the response was variable both within and between species. They reported 3–5 $\log_{10}$ reduction in numbers of *B. cereus* with a combined pressure–heat treatment of 600 MPa and initial temperature of 72 °C for 1 min. A review of high pressure effects on bacterial spores is provided by Zhang and Mittal (2008). Pulsed electric fields do not inactivate bacterial spores, and irradiation doses required to destroy *B. cereus* are of the order of 1.6 kGy, for a 10-fold reduction.

Bacterial spores are sensitive to some commercial sanitisers used in food processing environments. In a study comparing effects of sanitisers using the actives NaOCl, HCl, H_2O_2, acetic acid, quaternary ammonium compounds, ammonium hydroxide, citric acid, isopropanol, NaOH or pine oil, Black *et al.* (2008) concluded that hypochlorite-containing preparations were most effective against *B. cereus* spores, and HCl- or H_2O_2-containing products also significantly reduced numbers of spores although at a slower rate. Most NaOCl-containing sanitisers reduced spores to the level of sensitivity in 10 min at 25 °C. Work has also been carried out looking at the effects of sanitisers in cooked rice and apples (Kreske *et al.*, 2006; Lee *et al.*, 2006), although this type of application would require careful consideration of toxicological effects. The main growth/survival characteristics of the organism are summarised in Table 24.3. In spore form, *B. cereus* survives for very long periods of time (e.g. many years) in adverse conditions.

The finding that a psychrotrophic isolate of *B. cereus* can produce emetic toxin (claimed to be the first such observation) has led to the suggestion that refrigerated foods could be the vehicle of emetic *B. cereus* food poisoning (Altayar and Sutherland, 2006). However, this study only demonstrated the ability to produce emetic toxin by a strain that could grow at 7 °C and not emetic toxin production at 7 °C. Thorsen *et al.* (2006) reported toxin production at chill but not at clinically significant levels. There are no reports of foods that have been stored properly at refrigeration temperatures causing emetic illness. Generally speaking, emetic toxin-producing strains are not capable of growth at refrigeration temperatures. Formation of emetic toxin is dependent on a number of factors including background microflora, aeration, type of food, temperature, pH and the strains involved (Rajkovic *et al.*, 2006b). The same study reported that a total *B. cereus* count of more than 10^6 cfu/ml did not necessarily lead to emetic toxin formation and also concluded that aeration had a negative effect on cereulide production. The high incidence of emetic-producing *B. cereus* in rice paddy fields has led to suggestions that rice is 'selective area' for growth of the emetic strain of *B. cereus* (Altayar and Sutherland, 2006). Although some researchers have reported on production of diarrhoeal toxin in foods, this is thought to be of little relevance since the enterotoxin is destroyed by low pH (e.g. pH 3.0) and would not survive passage through the stomach.

24.5 Other *Bacillus* species

24.5.1 *Bacillus subtilis*

There have been a number of cases of foodborne illness associated with *B. subtilis*, implicating a wide variety of foods from which the organism was frequently recovered in substantial numbers (i.e. > 10^6). In these incidents, summarised by Kramer and Gilbert (1989), *B. subtilis* was sometimes isolated in pure culture, and other aerobic and anaerobic bacteria, fungi and yeasts were often not detected. These incidents occurred in Canada, Australia and New Zealand. During the 1980s, the PHLS in the UK periodically received non-*cereus Bacillus* isolates associated with food poisoning. In 49 of these episodes, involving > 175 cases, *B. subtilis* was identified as the most probable cause of illness. Although complete epidemiological information confirming *B. subtilis* as the causative agent was missing in most episodes, Kramer and Gilbert concluded that the data, when taken together, 'strongly suggests a causal relationship'.

Illness in these episodes was characterised by: short onset period (median 2.5 h), and in almost one-third of episodes onset times were 60 min or less (range 10 min to 14 h); vomiting being the predominant symptom in most cases (80 %), followed by diarrhoea (49 %) and some individuals (27 %) with abdominal cramps/pain; approximately 10 % patients reporting headaches, flushing sensations or sweating as additional symptoms; large numbers (up to 10^7 cfu/g) of the organism in the vomitus and acute-phase faecal specimens; 10^5–10^9 cfu/g were isolated from 30 foods tested. Duration of illness was 1.5–8 h. No other recognised bacterial foodborne pathogens were isolated in these episodes. There have been relatively few reports of illness attributed to *B. subtilis* in recent years. The data collated by Kramer and Gilbert are often referred to as 'circumstantial' evidence of foodborne illness.

24.5.2 *Bacillus licheniformis*

An early large outbreak of illness initially attributed to *B. subtilis* (Tong *et al.*, 1962) is now thought to have been caused by *B. licheniformis* (Kramer and Gilbert, 1989). Gilbert and Kramer (1989) also cite eight other published reports of illness associated with *B. licheniformis*, where known pathogens were not found. The same authors describe 24 episodes (> 218 cases) of illness, in the UK, occurring between 1975 and 1986 attributed to *B. licheniformis*. Illness was characterised by: onset of symptoms between 2–14 h (median

8 h); diarrhoea was the predominant symptom (92 %), with vomiting also a common feature (54 %), and abdominal cramps/pain (46 %) (no reports of sweating/flushing, headaches or nausea); large numbers (> 10^6 cfu/g) were found in the 10 foods tested and also in acute-phase faecal specimens (no vomitus was tested). Duration of illness was 6–24 h.

The speed of onset of symptoms with some of the cases described above suggests ingestion of preformed toxin, but the pattern of symptoms (diarrhoea vs vomiting) is different to that observed with *B. subtilis*. The duration of illness is also different, with illness resolving more quickly with *B. subtilis*-related cases.

More recently, a study (Salkinoja-Salonen *et al.*, 1999) reported on toxin-producing isolates of *B. licheniformis* obtained from foods involved in food-poisoning incidents. 210 isolates involved in food poisoning or suspected food poisoning were looked at. One of these incidents involved an infant feed formula (responsible for one fatal case), and isolates came from a hospital patient with acute-phase food poisoning and from food poisoning cases connected with curry rice and fast foods. Detailed descriptions of the illnesses observed in these episodes is not available and, in some cases, more than one *Bacillus* spp. was present in the associated food. In the case of infant food-related illness, the three strains isolated were shown to produce lichenysin A and this was toxic to boar spermatozoa (Mikkola *et al.*, 2000). For the episodes where more detail is available (data from UK), onset of symptoms was 5–12 h, diarrhoea and vomiting were observed (not always together) in 67 % of the incidents, nausea was reported in 56 % of incidents, abdominal pain and stomach cramps were observed (never together) in 34 % of the incidents and oral burning was reported on one occasion (11 %). Where numbers in the implicated food were reported, there was at least 10^5 cfu/g food. Niemenen *et al.* (2007) recently tested a number of strains of *Bacillus* species that were associated with mastitis in cows and found strains of *B. licheniformis* and *B. pumilus* capable of producing toxins. Ribotyping of these strains showed then to be very similar to strains associated with foodborne disease.

24.5.3 *Bacillus pumilus*

There are a small number of cases of foodborne illness attributed to *B. pumilus*. Kramer and Gilbert (1989) cite five episodes of illness where the symptoms included vomiting (time of onset varying between 15 min and 11 h), diarrhoea (from 2–3 h to 11 h), dizziness (15 min) and 'acute gastroenteritis'. In all episodes, large numbers of *B. pumilus* were found in the foods implicated and established bacterial pathogens were not found. From *et al.* (2007) investigated a strain of *B. pumilus* that was linked to an outbreak of illness affecting three individuals who had eaten in a Chinese restaurant, where the same organism was isolated from cooked reheated rice. Symptoms reported indicated ingestion of pre-formed toxin. The strain was shown to produce a complex of peptides known as pumilacidins. These are known to be toxic towards boar spermatozoa and exert potentially destructive biological effects. They have also been implicated in causing sore eyes, sore throats, severe sinusitis and lung infections (Peltola *et al.*, 2001).

24.5.4 *Brevibacillus* (formerly *Bacillus*) *brevis*

There are a few reports of incidents where *Br. brevis* was isolated in large numbers, either alone or with *B. cereus*, from suspected foods. In one case the organism was isolated from the patient's faeces. Incubation times were from 1–9.5 h, with predominant symptoms of vomiting and diarrhoea.

24.5.5 *Bacillus thuringiensis*

Bacillus thuringiensis was recently reported to be associated with a gastrointestinal disease outbreak (Jackson *et al.*, 1995) in a chronic care institution in Canada, involving 18 individuals. Predominant symptoms in this outbreak were nausea, vomiting and watery diarrhoea. Significant isolations/identifications were made in five cases. In three of the five cases, no other enteric pathogens were detected, and the same phage type was isolated from one patient with persistent illness. Although *B. thuringiensis* was isolated from spice (onion powder) submitted by the institution, this was a different phage type to that associated with stool specimens from individuals affected. Therefore, the epidemiological evidence for this 'outbreak' being caused by a food is weak. However, the isolates were shown to have cytoxic effects characteristic of *B. cereus*.

In a study by te Giffel *et al.* (1997), two strains originally classified as *B. cereus* and associated with food poisoning were subsequently identified as *B. thuringiensis*, on the basis of sequences of the variable regions of 16S rRNA. Using classical biochemical tests, all strains initially examined (65 in all) were identified as *B. cereus*. After further examination of 20 of these strains, six were found to be *B. thuringiensis*, two of which were originally associated with cases of foodborne illness. It has been suggested that there may be other examples of foodborne illness mistakenly attributed to *B. cereus*, where isolates may have been misclassified.

Pivovarov *et al.* (1977) reported that ingestion of foods contaminated with a commercial preparation of *B. thuringiensis* (Btg, for use in agriculture) at concentrations between 10^5 and 10^9 cells/g caused nausea, vomiting, diarrhoea and tenesmus, colic-like pains in the abdomen, and fever in three of the four volunteers studied. According to Ray (1990), the toxicity of the Btg strain may have been due to β-exotoxin.

24.5.6 *Bacillus weihenstephanensis*

In the study identifying *B. weihenstephanensis* as a new species of the *B. cereus* sub-group (Lechner *et al.*, 1998), it is proposed that this new species comprises 'psychrotolerant' but not mesophilic *B. cereus* strains. The isolates tested were taken from milk samples from three different South German dairies. None of the strains tested had been associated with outbreaks of foodborne illness. To date, there is no direct evidence that strains belonging to the newly named *B. weihenstephanensis* have been the cause of foodborne disease. No strains previously identified as 'psychrotolerant' or psychrotrophic *B. cereus* and implicated in outbreaks of foodborne disease have so far been identified as *B. weihenstephanensis*. There are some studies reporting on the pathogenic potential of strains from this species. Stenfors *et al.* (2002) concluded that the majority of strains that they tested were not cytotoxic. Svennson *et al.* (2007) did not find any producers of high levels of toxins in 156 strains of *B. weihenstephanensis* isolated from the dairy production chain.

24.6 Risk factors

Bacillus cereus is a ubiquitous bacterium in the environment and it can be present in a wide range of different foodstuffs (EFSA, 2005). The organism has been isolated from soil (especially rice paddy soil), fresh water and sediments, dust, cereal crops and cereal derivatives, vegetables, green salads, spices, dried foods, milk and dairy products, meat and animal hair (Kramer and Gilbert, 1989; Lund, 1990; ICMSF, 1996; Gilbert and Humphrey, 1998; Batt, 2000; Dahl, 2000; Granum, 2001).

Foodborne outbreaks caused by *B. cereus* have been associated with many types of foodstuffs including food of both animal and plant origin (EFSA, 2005). Diarrhoeal *B. cereus* food poisoning is commonly associated with a wide range of proteinaceous foods such as milk and dairy products, meats, vegetable dishes, sauces, soups and desserts (Batt, 2000). In contrast, emetic *B. cereus* food poisoning is commonly associated with farinaceous type foods such as rice, pasta, noodles and pastry (Batt, 2000; Beattie and Williams, 2000). Production of the emetic toxin is known to depend on the composition of the contaminated food, with the inability to hydrolyse starch is a common characteristic of isolates associated with the emetic-type syndrome (Agata *et al.*, 1996) and low concentrations of amino acids promoting toxin production (Agata *et al.*, 1999). High emetic toxin production has been demonstrated in rice and other starchy foods such as noodles, spaghetti and potatoes, while production was very low in eggs, milk, fish, meat and their products (Agata *et al.*, 2002). Overall, the evidence suggests that the risk from emetic type illness is greater from farinaceous foods than from proteinaceous foods.

For both types of illness, the food involved has usually been heat-treated, with *B. cereus* spores surviving, and subsequent storage conditions that permit the spores to germinate and grow. Practices of cooking foods in bulk and cooling at room temperature are more likely to lead to conditions that could lead to such food poisoning. *Bacillus cereus* food poisoning does not appear to exhibit a strong seasonal trend in the rate of disease, but there is a potentially greater risk of outbreaks in the summer months when the warmer temperatures make it more difficult to keep cooked foods cool.

Soil is the primary reservoir of *B. cereus* and can contain between 10^3–10^5 *B. cereus* spores/g with psychrotrophic strains more frequent in soil from cold regions (EFSA, 2005). Soil from parks, woodland and fields has been found to be a common source of *B. cereus* with large proportions of samples positive (75–80 % and 25–100 % on plating at 30 °C and 7 °C, respectively), with mean counts in the range 1.0–4.4 $\log_{10}$ cfu/g (3.0–3.6 and 1.0–4.4 $\log_{10}$ cfu/g on plating at 30 °C and 7 °C, respectively) (Altayar and Sutherland, 2005). Animal faeces have also been shown to be a common source of *B. cereus* with detection rates of 50–70 % in cow and horse faeces (59–70 % and 50–57 % positive on plating at 30 °C and 7 °C, respectively), with mean counts ranging from 1.8–2.9 $\log_{10}$ cfu/g (2.2–2.7 and 1.8–2.9 $\log_{10}$ cfu/g on plating at 30 °C and 7 °C, respectively) (Altayar and Sutherland, 2006). Consequently, *Bacillus cereus* is frequently isolated from growing plants and therefore it is easily spread to foods, particularly those of plant origin, and through cross-contamination can spread to other foods (Granum, 2001).

The role of soil as a reservoir for *B. cereus* contamination of foods has been highlighted in a study in which soil was found to be the initial source of contamination for zucchinis (Guinebretiére and Nguyen-The, 2003). Spores from soil were highly diverse and persisted throughout a pasteurised zucchini purée processing line and adapted to storage conditions of the processed food. In contrast, texturing agents such as milk proteins and starch did not contribute to the contamination of the product when properly stored at cold temperatures.

In a study to evaluate the frequency of occurrence and composition of *Bacillus* species naturally present in urban aerosols in US cities, the detection frequency was highly variable among cities, across seasons of the year and across locations within a city (Merrill *et al.*, 2006). It was stated that results suggested that the presence of *Bacillus* species (differentiated into three groups which included *B. cereus*) in the air may not depend strongly on seasonal cycles tied to plant growth or senescence and decay, but may be influenced by wind movement, aerosolisation of soil particles, local weather conditions including sunlight intensity, humidity and rain events.

There have been numerous surveys of foods for the incidence of *B. cereus* and some of these studies are summarised below.

- Samples of 124 different foods purchased from local markets in India were examined for the presence of *B. cereus* (Kamat *et al.*, 1989). The incidence of *B. cereus* was 28.5 % in rice and rice products (at 2×10^1–6×10^2 cfu/g) (100 % of boiled rice at 1.5×10^3–4×10^4 cfu/g), 40 % of fish (at 5×10^2–9×10^4 cfu/g), 80 % of chicken and meat products (at 1.2×10^3–3.2×10^4 cfu/g), 30 % of spices (at 5×10^1–3×10^4 cfu/g),

87 % of ice creams (at 2×10^3–1.5×10^5 cfu/g), 100 % of milk and milk products (at 3.5×10^3–5×10^5 cfu/g), and 0 % of oils.

- The incidence of *B. cereus* and *B. subtilis* in 229 samples of various food products (including milk, yeast, flour, pasta products, Chinese meals, cocoa, chocolate, bakery products, meat products, herbs and spices) in the Netherlands was determined (te Giffel *et al.*, 1996). *Bacillus cereus* was present in 109 (48 %) samples (at 10^2–10^6 cfu/g or ml) and *B. subtilis* was present in 18 of 72 (25 %) samples (at 10^2–10^6 cfu/g or ml).
- Samples (320) of vegetarian food purchased from local Taiwanese markets were found to contain vegetative *B. cereus* at an incidence of 3.4 % at levels up to 10^3 cfu/g and *B. cereus* spores at an incidence of 20.9 % up to 10^4 cfu/g (Fang *et al.*, 1999). *Bacillus cereus* was found in vegetarian food products containing soybean (5.1 %), cereal (9.2 %) and konjac (2.4 %), but not in products where the major component was flour or vegetable.
- Ready-to-eat food products (48 901) sampled during 2000–2003 from the Danish retail market yielded 0.5 % with counts of '*B. cereus*-like bacteria' above 1×10^4 cfu/g, a level that was deemed to be unacceptable (Rosenquist *et al.*, 2005). Most (98.7 %) of the samples contained satisfactory levels (< 1000 cfu/g), with 0.7 % having an unsatisfactory level (1000–1×10^4 cfu/g). The high counts were most frequently found in starchy, cooked products, but also in fresh cucumbers and tomatoes.
- Sampling a wide range of food commodities in the Netherlands it was found that the incidence of psychrotrophic *B. cereus* strains was low (9.6 %) compared with mesophilic (82.6 %) and intermediate (4.8 %) strains, but in products that require low-temperature storage the incidence is significantly higher (e.g. 27.5 % in milk and milk products) (Wijnands *et al.*, 2006).

The common occurrence of *B. cereus* in rice presents a significant safety risk associated with the consumption of rice and rice products, which is borne out by several studies. *Bacillus cereus* was found to be present in all 61 samples of raw rice tested, with lower levels in husked and white rice (18–102 MPN/g) than in unhusked rice (1.2–3.5×10^3 cfu/g) (Sarrías *et al.*, 2002). In a study of 4162 samples of cooked rice from restaurants and take-away premises in the UK, 15 (1 %) were deemed unacceptable with *B. cereus* counts greater than or equal to 10^5 cfu/g (Nichols *et al.*, 1999). *Bacillus cereus* is a significant hazard associated with Kimbab, which is rolled cooked rice and other foodstuffs in a dried layer, and an exposure assessment was made of this product being sold in Korea (Bahk *et al.*, 2007). The estimated contamination levels ranged from a minimum (5th percentile) of –3.63 log cfu/g to a maximum (95th percentile) of 7.31 log cfu/g, with median (50th percentile) and mean values of 1.39 log cfu/g and 1.57 log cfu/g, respectively. However, the paucity of data to conduct a full exposure assessment with associated variability and uncertainty, coupled with a validation step, was highlighted.

As *B. cereus* is common in soil and growing plants it is frequently isolated from foods of plant origin. A number of studies have shown that *B. cereus* is present in around 30 % of raw vegetables (Roberts *et al.*, 1982; Choma *et al.*, 2000) and levels of *B. cereus* in foods of plant origin such as seeds, legumes, cereals and flour are typically >10^2 cfu/g (Harmon *et al.*, 1987; Kramer and Gilbert, 1989). In a review of the literature to determine the levels of *B. cereus* present in vegetables, spices and foods containing vegetables, Carlin *et al.* (2000) found that of 1007 samples containing *B. cereus* about 69 % (697/1007) contained < 10^2 cfu/g; 30 % (304/1007) contained 10^2–10^5 cfu/g; and only 0.3 % (27/1007) contained > 10^5 cfu/g.

Valero *et al.* (2002) carried out work to detect, enumerate and characterise strains of *B. cereus* in a wide range of fresh vegetables commonly used to manufacture refrigerated minimally-processed foods. *Bacillus cereus* was isolated from 29 out of 56 (52 %) fresh vegetable samples analysed, ranging from 0 % for garlic and 14 % for onion, to 90 % for tomato and 100 % for pepper samples, with levels up to 10^3 cfu/g, but not exceeding 10^4 cfu/g. *Bacillus cereus* was detected in 20 % of unstored vegetable purées pasteurised in their final package at less than 10 cfu/g (Choma *et al.*, 2000). *Bacillus cereus* was detected in 25 % of samples of broccoli, carrot, courgette and potato (at 0.75–0.8 $\log_{10}$ cfu/g) and 0 % of potato samples (< 0.8 $\log_{10}$ cfu/g).

Bacillus cereus has been commonly isolated from dried potatoes at a prevalence of 10–40 % (10 cfu/g limit of detection) at a level ranging from < 10 to 4 × 10^3 cfu/g (King *et al.*, 2007). It was highlighted that they are likely to be present as spores that are able to survive drying of the raw vegetable and may represent a significant inoculum in the reconstituted (rehydrated) product where conditions favour germination of, and outgrowth from, spores. *Bacillus cereus* has been found in unwashed potato, carrot and lettuce with 80–88 % and 36–92 % samples positive by plating at 30 °C and 7 °C, respectively, with associated mean counts of 2.2–3.7 and 0.9–3.5 $\log_{10}$ cfu/g (Altayar and Sutherland, 2005). Further production of potatoes reduced isolation rates and counts at 37 °C and 7 °C: washed potatoes had an incidence of 32 % (1.1 $\log_{10}$ cfu/g) and 0 % and potato powder had an incidence of 20 % (0.7 $\log_{10}$ cfu/g) and 0 %. One of 148 *B. cereus* isolated at 7 °C was positive for emetic toxin, and it was suggested that it is possible for psychrotrophic strains to grow in refrigerated foods and cause emesis.

The association of *B. cereus* with milk and milk products stems from its presence in soil and grass and subsequent transfer to the udders of cows and hence into raw milk (Granum, 2001). Rates of contamination of raw milk with *B. cereus* range from 9–38 % (Ahmed *et al.*, 1983; Rangasamy *et al.*, 1993; te Giffel *et al.*, 1996; Vyletělová *et al.*, 2001), with levels ranging from < 10 cfu/ml to 10^4 cfu/ml (Stewart, 1975; Ahmed *et al.*, 1983; Rangasamy *et al.*, 1993; te Giffel *et al.*, 1996). Rates of contamination of pasteurised milk with *B. cereus* range from 8–100 % (Ahmed *et al.*, 1983; van Netten *et al.*, 1990; Rangasamy *et al.*, 1993; Notermans *et al.*, 1997; te Giffel *et al.*, 1997), with levels generally low at 0.9 spores per 100 ml to 28 cfu/ml (Rangasamy *et al.*, 1993; te Giffel *et al.*, 1996, 1997; Notermans *et al.*, 1997), with a maximum of 1000 cfu/ml (Ahmed *et al.*, 1983), although much higher levels are sometime reported (Schoder *et al.*, 2007). In a study in which pasteurised fluid milk samples were collected from three commercial dairy plants, predominantly Gram-positive rods were isolated, with 31 % of isolates identified as *Bacillus*, comprising 12 % *B. cereus*, 12 % *B. licheniformis*, 4 % *B. pumilus* and 4 % *B. lentus* (Fromm and Boor, 2004). The presence of psychrotrophic strains in milk and cream, depending on the storage conditions, has the potential to lead to high levels of *B. cereus* (Schoder *et al.*, 2007). It is worth noting that, despite the worldwide occurrence of this bacterium in milk, sometimes at high levels (> 10^6/ml), few incidents of food poisoning caused by *B. cereus* from milk have been reported (Christiansson *et al.*, 1989b; Schoder *et al.*, 2007). *Bacillus cereus* was, however, incriminated in a large food poisoning outbreak attributed to pasteurised milk (van Netten *et al.*, 1990). In fermented milk and cheese products vegetative *B. cereus* cells are rapidly eliminated during the production process (Schoder *et al.*, 2007).

Bacillus cereus is also isolated from milk powder and infant formula foods, but usually at low numbers (Schoder *et al.*, 2007). *Bacillus cereus* contamination of non-fat dry milk and infant formula, as determined by most probable number, in the range of < 0.04–2.1 cells/g has been reported (Harmon and Kautter, 1991). However, the storage under ambient temperatures fosters growth of *B. cereus*, thus making this group of foods vulnerable for contamination at high numbers at the post-processing level (Schoder *et al.*, 2007).

In a study comparing the incidence of *B. cereus* in dairy products, a total of 293 samples purchased from local markets in Taiwan were analysed and found to contain *B. cereus* at the following rates: fermented milks, 17 %; ice creams, 52 %; soft ice creams, 35 %; pasteurised milks and milk products, 2 %; milk powders, 29 %; with levels ranging from 5–800 cfu/ml or cfu/g, and average levels of 15–280 cfu/ml or cfu/g (Wong *et al.*, 1988).

Following a survey of 100 *B. cereus* strains in which none of the emetic-toxin producing strains were able to grow below 10 °C (i.e. 7 °C), it was concluded that this poor ability to grow at low temperatures shows that emetic toxin-producing strains of *B. cereus* will pose a low risk in refrigerated foods; however, due to the high spore heat resistance, they produce a special risk in heat-processed foods or pre-heated foods that are kept warm (Carlin *et al.*, 2006).

The extent to which *B. cereus* may pose a safety risk for Refrigerated Processed Foods of Extended Duration (REPFED) depends on many factors including: spore prevalence and concentration in raw materials, heat treatment, heat resistance of spores, product formulation and supply-chain storage temperatures and times (Membré *et al.*, 2008). Consequently, designing a product that is safe is a potentially complex problem. In particular, the effect of the heat treatment and product formulation on the subsequent lag time of surviving spores, both individually and as a population, can have a significant impact on the safe shelf-life.

Due to its ubiquitous nature and its ability to form spores that allow it to survive pasteurisation and cooking treatments, *B. cereus* will be present in many heat-processed and cooked foods. The heat treatment will leave a *B. cereus* spore population in the absence of competitive microorganisms due to a greatly reduced vegetative cell flora and hence the opportunity to germinate and grow given enough time at temperatures within the growth range of the organism (ICMSF, 1996). Consequently, strict control of time and temperature of product storage is often critical to reduce the risk of *B. cereus* food poisoning.

Although the food processing environment is not considered to be as significant in terms of product contamination with *B. cereus* compared with the presence of the organism in ingredients and raw materials, several studies have highlighted the risk that this source poses in food manufacture.

- Epidemiological typing of *B. cereus* isolated from milk samples from two different dairies yielded plant-specific distribution of the organism suggesting that the strains may originate from common reservoirs within dairy plants or within certain farms supplying raw milk (Schraft *et al.*, 1996).
- Raw milk collected from silo tanks at eight different dairy plants yielded on average a higher level of *B. cereus* spores and percentage of psychrotrophic isolates in summer (median 198 spores/l; 48 % psychrotrophic) than in winter (median 86 spores/ml; 35 % psychrotrophic) (Svensson *et al.*, 2004). Some strains appeared to be widely spread amongst dairies and certain strains may be selected for in the silo tank environment. It was proposed that the presence of an in-house flora of mesophilic strains indicates that the cleaning system of the silo tanks may not be satisfactory, due to their higher heat resistance.
- In the dairy industry, *B. cereus* biofilms inside storage tanks and associated piping can create problems for product sterility and longevity and, using a proteomic approach, it has been shown that expression of peptides may play an important role in the maintenance of the biofilm and/or represent metabolic changes triggered when the bacterium switches from a planktonic to a sessile lifestyle (Rasko *et al.*, 2005).
- The main approach to control of *B. cereus* in REPFEDs is prevention of spore germination and subsequent multiplication, and this can be achieved by limiting shelf-life and refrigeration below 7 °C, preferably at 4 °C (Rajkovic *et al.*, 2006a). However, in a study of the manufacture of a vacuum-packed potato purée the importance of preventing an 'in house' colonisation and contamination was identified and the importance of planned and thoroughly performed cleaning and disinfection of processing equipment was highlighted (Rajkovic *et al.*, 2006a).

- *Bacillus cereus* strains have been isolated from alkaline wash solutions (pH 11–13) used for CIP in South African diary factories and they have been reported to be capable of growth up to a pH of 10.5 (Lindsay *et al.*, 2000, 2002). It was concluded that this ability to grow at alkaline pH might be a general trait of *B. cereus* species rather than a specific phenomenon of isolates from alkaline ecosystems.

Spores of *B. cereus* have also been found in paper mill industries and in packaging materials, and this could represent an additional route of contamination in foods (EFSA, 2005).

24.7 Methods of detection and enumeration

Almost all techniques for the detection of *B. cereus* are based on plating media formulated to demonstrate one or more of the physiological properties of the growing cells. Thus many indicator agar media incorporate egg yolk to demonstrate lecithinase activity (a white precipitate around colonies), or haemolytic activity (horse, rabbit, sheep or human red cells may be used). Various egg yolk-containing agar media have been described, but the most commonly used is that described by Holbrook and Anderson (1980) (polymyxin–egg yolk–mannitol–bromothymol blue agar [PEMBA]) or a derivative of the formulation (e.g. polymyxin–egg yolk–mannitol–bromothymol purple agar [PEMPA]; Szabo *et al.*, 1984). These contain a low level of peptone to stimulate sporulation, pyruvate to reduce the size of colonies, polymixin as a selective agent, egg yolk to demonstrate lecithinase production, mannitol and a pH indicator (*B. cereus* does not ferment mannitol unlike many other *Bacillus* spp.). PEMPA is claimed to result in typical colonial appearance in 24 h rather than 48 h for PEMBA, with formation of spores that may be observed by phase contrast microscopy. If determination of a spore count is required, then a pasteurisation treatment (e.g. 70 °C for 10 min) may be applied to samples before plating. However, spores may germinate once hydrated from dry foods, giving a low spore count.

The occurrence of high numbers of *Bacillus* spp. from a suspect food should always be regarded with suspicion, whether or not the typical characteristics of *B. cereus* are present. The species of the genus *Bacillus* are often difficult to differentiate as the metabolic and physiological properties of the genus seem to be a continuum of small differences. *Bacillus subtilis* can be differentiated from *B. cereus* and *B. licheniformis* by its inability to grow anaerobically, although even the smallest amount of oxygen will permit limited growth. *Bacillus cereus* is the only one of these three species that produces lecithinase, but does not ferment mannitol, xylose or arabinose. Thus *B. cereus* can readily be distinguished from the other two species on PEMBA or PEMPA. Currently, PEMBA and mannitol-egg yolk-polymixin (MYP) agar are recommended for the identification and enumeration of presumptive *B. cereus* colonies by food authorities like ISO, LFGB (German Food and Feed Code and FDA (Fricker *et al.*, 2008).

In order to improve the sensitivity of standard agar plate-based enumeration, a filtration method was developed for detection of *B. cereus* spores in milk (Christiansson *et al.*, 1989a). The method included heat treatment at 72 °C, followed by treatment with trypson and Triton-X100® (DOW Chemical Co., USA) at 55 °C and membrane filtration with transfer of the membrane filter to the surface of an agar medium. Blood agar with 10 ppm polymyxin was found to be superior to MYP and PEMBA for detection, and spore recovery was typically 85 ± 13 %.

Following the trend applied to the detection and enumeration of other pathogenic bacteria, chromogenic agars have been developed for *B. cereus*. A chromogenic, selective and differential plating medium that detects the bacterial phoshatidyl inositol-specific phospholipase C (PI-PLC) enzyme in *B. cereus* and *B. thuringiensis*, BCM® *B. cereus/B. thuringiensis* chromogenic agar plating medium (Biosynth AG, Switzerland), was compared with MYP for the isolation and enumeration of *B. cereus* from foods (Peng *et al.*, 2001). No significant difference was found between the two media, but colony enumeration and isolation was easier on BCM® because of higher selectivity and formation of discrete, non-coalescing colonies. This agar (subsequently renamed as BCM® *B. cereus* group plating medium) was compared with another chromogenic plating medium: Chromogenic Bacillus Cereus agar (CBC, Oxoid; and subsequently renamed as Brilliance™ Bacillus cereus Agar) (Fricker *et al.*, 2008). Both agars were concluded to represent a good alternative to the conventional standard media (PEMBA and MYP), particularly if laboratory staff are not highly trained in identification of *B. cereus*, which can lead to misidentification on conventional plating media and hence under-estimation of *B. cereus*.

Immunoassays have been developed to detect the *B. cereus* diarrhoeal enterotoxin. A colony immunoblot assay was developed which allows identification of enterotoxin-producing colonies from blood agar plates (Moravek *et al.*, 2006). There are two commercially available immunoassays for the *B. cereus* diarrhoeal toxin: the *B. cereus* Enterotoxin Reverse Passive Latex Agglutination toxin detection kit (BCET-RPLA; Oxoid) and the 3M™ TECRA™ *Bacillus* Diarrhoeal Enterotoxin Visual Immunoassay (VIA™). The BCET-RPLA is based on antibody-coated polystyrene latex particles, while the (VIA™) is an ELISA format. Results demonstrated that these two assays did not detect the same antigen (Day *et al.*, 1994). It was shown subsequently that the BCET-RPLA kit detects in particular the L_2 component (encoded by *hblC*) of the Hbl enterotoxin complex and the VIA™ detection kit detects specifically the NheA, a 41 kDA component of the Nhe enterotoxin complex (Ouoba *et al.*, 2008). Commercial kits for detection of the emetic toxin (cereulide) are not yet available, but now that the structure of cereulide is known, it is at least feasible that a kit will be available in the future (Granum, 2001).

Molecular approaches have been employed in the development of methods for the detection of *B. cereus*. *Bacillus cereus* group bacteria share a significant degree of genetic similarity and therefore, in an attempt to differentiate and identify them from food samples in a single tube, a multiplex PCR method using the *gyrB* and *groEL* genes as diagnostic markers was developed (Park *et al.*, 2007). The assay yielded a 400 bp amplicon for the *groEL* gene from all the *B. cereus* group bacteria and a 475 bp amplicon for *B. cereus*, with other differentiating amplicons for *B. anthracis*, *B. thuringiensis* and *B. mycoides*. Several molecular methods targeting virulence traits have also been developed. A DNA probe for the detection of pathogenic *B. cereus* strains was developed based on a 21 kb virulence plasmid responsible for the production of non-haemolytic toxin (Subramanian *et al.*, 2006). PCR-based assays have been developed for the detection of *B. cereus* strains able to produce emetic toxin (cereulide). Nakano *et al.* (2004) developed a method based on a sequence-characterised amplified region (SCAR) derived from a random amplified polymorphic DNA (RAPD) fragment, and the assay was able to detect *B. cereus* inoculated at 10–70 cfu/g in food following a 7 h enrichment culture step. Fricker *et al.* (2007) developed a 5' nuclease (TaqMan™) real-time PCR assay that provided a rapid and sensitive method for the specific detection of emetic *B. cereus* in food. Yang *et al.* (2005) developed a multiplex PCR assay for the detection of toxigenic strains of the species in the *B. cereus* group, based on amplification of the enterotoxin and emetic-specific sequences. When the assay was applied to 162 food poisoning and food-related strains, all the *B. cereus* strains were found to carry at least one toxin gene while more than 70 % of *B. mycoides* strains carried no known toxin genes, had significantly different toxin profiles and toxin genes to the *B. cereus* strains and hence concluded that they are less prone to cause food poisoning. The authors also concluded the importance of detecting the toxin genes together with the detection of the species in the *B. cereus* group.

24.8 Control procedures

In order to determine whether control procedures are required and whether their application will be effective to manage the risk, safe levels of *B. cereus* need to be defined, and this has been the subject of much conjecture. An expert opinion from the UK Public Health Laboratory Service (PHLS, 1992) stated that, for *B. cereus* in ready-to-eat foods, a level of < 1000/g is satisfactory; 10^3– < 10^4 cfu/g is acceptable; 10^4– < 10^5 cfu/g is unsatisfactory and > 10^5 cfu/g is categorised as unacceptable or potentially hazardous. It has been stated that food processors should ensure that numbers of *B. cereus* between 10^3 and 10^5/g are not reached (or exceeded) at the stage of consumption under anticipated conditions of storage and handling, and that this should also apply for rehydrated foods reconstituted with hot water before consumption (EFSA, 2005).

The likely occurrence of *B. cereus* in many raw materials means that this hazard needs to be identified and, if appropriate, controlled in the product design. Control usually relies heavily on a combination of extrinsic (temperature) and intrinsic factors (pH, a_W and preservatives), but it may be appropriate to limit the level of *B. cereus* and/or *Bacillus* species in ingredients and raw materials via supplier assurance and appropriate microbiological specifications. Although the elimination of *B. cereus* in finished products is not necessarily required, satisfactory levels of the organisms must be achieved. The ability of some strains of *B. cereus* to grow at refrigeration temperatures means that storage times must be carefully defined. Due to the difficulty in species identification and the evidence of non-*cereus Bacillus* species as pathogens, the presence of large numbers of any *Bacillus* in any food (or ingredient) should be considered as potentially hazardous.

Vegetative cells of *B. cereus* are readily destroyed by pasteurisation or equivalent heat treatments. However, inactivation of *B. cereus* spores can only be guaranteed by a heat process equivalent to that used for low-acid canned foods ($F_0 = 3$), and this control measure cannot be used with most foods without dramatically altering their quality (EFSA, 2005). It should be noted that this high heat resistance reported for some *B. cereus* strains under certain conditions might even cast some doubt on the effectiveness of an $F_0 = 3$ process for completely eradicating high levels of *B. cereus* in all situations. It should also be noted that sporulation temperature can have a strong influence on the heat resistance with a 10-fold increase in *D*-values with a sporulation temperature increase from 32–52 °C. Under dry heat conditions *B. cereus* spores are likely to increase, for example *B. subtilis* *D*-values at 160 °C of 0.1–3.5 min have been reported (Brown, 2000).

Intrinsic factors may decrease apparent heat resistance, and models have been produced to determine the effects of factors such as pH on the thermal inactivation of *B. cereus* spores. For example, at 95 °C, a reduction in pH from 6.9 to 4.1 led to a decrease in *D*-value of 5.7 times (Gaillard *et al.*, 1998). The dichotomy of heat resistance between vegetative cells and spores is also reflected in toxin stability with the diarrhoeal toxin destroyed by relatively mild heat treatments (e.g. 56 °C for 30 min; ICMSF, 1996) while the emetic toxin can withstand sterilisation.

Novel processes have the potential to serve as alternatives to sterilisation heat treatments to inactivate *B. cereus*. Spores of *B. cereus* show higher gamma radiation resistance (D_{10} 2.5–4 kGy) than the vegetative cells (D_{10} 0.3–0.9 kGy) (Kamat *et al.*, 1989). Two samples of fresh shrimp irradiated at 2.5 kGy were both found to contain *B. cereus* at (at 2×10^2–5×10^2 cfu/g). 12D values of 43.1, 49.0 and 47.9 kGy were obtained for *B. cereus* spores in irradiated mozzarella cheese, ice cream and frozen yoghurt, respectively (Hashisaka *et al.*, 1990). This confirmed the general lower radiation resistance of vegetative cells compared to spores and the fact that *B. cereus* spores are not unusually resistant to irradiation compared with other spore-forming bacteria (Hashisaka *et al.*, 1990; ICMSF, 1996). UV treatments (30 W lamp for 30 s) have been shown to give a 2–3 $\log_{10}$ reduction of *B. cereus* spores on paper and plastic surfaces (ICMSF, 1996) indicating the potential application for packaging decontamination.

The inherent bacterial spore resistance to pressure limits the potential use of high hydrostatic pressure processing, but there is potential for its use in combination with elevated temperature. In addition, spores can be inactivated by alternating pressure cycles which initiate spore germination under moderate pressure (e.g. 30 min at 80 MPa) and then killing the more pressure-sensitive germinated spores by high pressure (e.g. 30 min at 350 MPa).

Depending on the amount of heat applied, pasteurisation heat treatments have the potential to activate *B. cereus* spores, thereby decreasing lag phase and potential shelf-life, or to impart injury, thereby doing the opposite. The effect of heat treatment conditions (90, 95 and 100 °C) on the resistance and recovery of *B. cereus* spores has been studied (Laurent *et al.*, 1999). The lag time increased with the duration of the thermal stress up to an optimum, but subsequently decreased with increasing stress duration to a threshold. The important conclusion that a longer heat treatment does not necessarily produce an extension of the shelf-life of the product has important implications for the food industry in the control of *B. cereus*. This biphasic response is likely to be due to the interplay between spore injury and activation, which does not manifest itself in a linear relationship.

After a heat treatment of 90 °C for 10 min, decreasing pH values and increasing levels of sodium chloride decreased growth rate and increased the lag phase of *B. cereus* (Martínez *et al.*, 2007). For example, the combination of pH $\leq$ 4.5, NaCl concentration $\geq$ 1.0 % and temperatures of $\leq$ 12 °C was sufficient to inhibit heat-injured (90 °C, 10 min) *B. cereus* for at least 50 d.

To control the risk from *B. cereus* requires storage conditions (time/temperature) that will prevent surviving spores from germinating and growing to hazardous levels. This requires cooked foods to be eaten soon after cooking, or kept hot (above 63 °C) or cooled rapidly and kept below 7–8 °C (ideally below 4 °C), preferably for a short time, e.g. a few days, depending on the product (EFSA, 2005). Temperatures below 10 °C greatly slow multiplication and temperatures below 4 °C prevent it. For manufactured foods this requires a Hazard Analysis Critical Control Point (HACCP)-based approach in order to define the storage conditions and shelf-life of the product depending on its formulation, or to formulate the product sufficiently to meet the desired shelf-life. Key factors affecting growth, and that can be used singly or in combination to control the risk of *B. cereus*, include pH, a_W and preservatives. Some studies demonstrating potential use of these factors to control *B. cereus* growth are summarised below.

- Although growth at pH 4.0 has been reported (Raevuori and Genigeorgis, 1975), the lower pH growth limit has more frequently been shown or stated to be 4.35–4.9. In the presence of organic acids the lower pH growth limit appears to be around 4.8 (Prokopova, 1970; Valero *et al.*, 2000). There is evidence that the minimum pH for spore germination is lower than that for subsequent vegetative cell growth (Wong and Chen, 1988).
- Del Torre *et al.* (2001) found that lactic acid was more inhibitory than citric acid for growth of a spore cocktail of three *B. cereus* strains in gnocchi at pH 5.0. Unlike citric acid, lactic acid prevented growth at 8 °C, but at 12 °C and 20 °C growth was seen at pH 5.0 with both acidulants.
- Valero *et al.* (2003) demonstrated that as the pH of carrot purée (natural pH 5.1; adjustments made with HCl or NaOH) increased (from 5.1 to 5.5 in 0.1 increments), so the temperature (5, 8, 12 and 16 °C) required to inhibit growth of *B cereus* (two-strain spore cocktail) decreased. No growth (after 60 d) was seen at pH 5.0; at pH 5.1 and 5.2 growth was only seen at 16 °C; at pH 5.3 growth was seen at 12 °C and above; and at pH 5.4 and 5.5 growth at 8 °C and above was observed.
- Control of the growth of *B. cereus* by using nisin has been demonstrated in potato purée and mashed potatoes (King *et al.*, 2007). The use of sorbic acid in combination with lactic or citric acids at pH 5.0 in samples of gnocchi stored at chilled temperatures was seen to inhibit the outgrowth of artificially-inoculated *B. cereus* spores.
- The effect of sodium chloride, benzoic acid and nisin on the growth of *B. cereus* has been studied (Roy *et al.*, 2007). All the isolates tested were inhibited by the acids in their permissible range, and the authors concluded that these acids can be used to control the pathogen. In this study nisin had a low level of inhibitory activity (≥175 $\mu g\ ml^{-1}$) and considering that multiple exposures of a pathogen to nisin would greatly increase the probability of generating stable resistant mutants, coupling nisin with several common food preservation strategies greatly reduces the frequency at which resistance arises, and nisin is usually used in a multiple-barrier inhibitory system.
- Sorbic acid (0.26 %) at pH 5.5 (equivalent to 0.05 % undissociated sorbic acid) and 0.39 % potassium sorbate at pH 6.6 (equivalent to 0.005 % undissociated sorbic acid) have been shown to be inhibitory to *B. cereus* (ICMSF, 1996).
- For bread making the addition of 0.2 % calcium propionate delayed germination and subsequent growth sufficiently to render risk of food poisoning due to *B. cereus* in bread negligible (Kaur, 1986).
- Nisin has been shown to inhibit the growth of *B. cereus*, being more effective at lower pH and temperatures. Outgrowth of psychrotrophic strains of *B. cereus* spores was inhibited by nisin at 5 and 50 µg/ml in broth (pH 5.5–6.1) at 8 and 15 °C, respectively. Nisin was less effective

against vegetative cells compared with spores (Jaquette and Beuchat, 1998).

- At concentrations as low as 1 μg/ml, nisin was lethal to *B. cereus*, the effect being more pronounced in brain heart infusion (BHI) broth than in beef gravy. A concentration of 5 and 50 μg/ml inhibited growth in gravy inoculated with vegetative cells and stored at 8 or 15 °C, respectively, for 14 d (Beuchat *et al.*, 1997). The inhibitory effect of nisin (1 μg/ml) was greater on vegetative cells than on spores inoculated into beef gravy and was more pronounced at 8 °C than at 15 °C.
- Glucose, and to a lesser extent lactose and fructose, when present in high concentrations (50–200 g/l) can reduce or eliminate diarrhoeagenic toxin production (Sutherland and Limond, 1993). Levels of starch of 10–50 g/l enhanced toxin production.
- The minimum inhibitory concentrations of a variety of preservatives against 48 strains of *B. cereus* isolated from six different kinds of legume-based Indian fermented foods were determined (Roy *et al.*, 2007). It was concluded that as all the strains were inhibited by permissible concentrations (0.5–2.0 mg ml^{-1}) of benzoic and sorbic acids, these acids can be used effectively in controlling the pathogen. However, nisin had only a low level of inhibitory activity (≥ 175 μg ml^{-1}).

Predictive growth models can help in the design and validation of a safe product and process, and some of the studies that have led to growth models for *B. cereus* are summarised below.

- The model for *B. cereus* in Growth Predictor was developed using vegetative cells and has temperature, pH (acidulant, HCl), a_W/NaCl concentration and CO_2 as controlling factors (Sutherland *et al.*, 1996). Growth models for *B. licheniformis* and *B. subtilis* with temperature, pH and a_W/NaCl concentration are also available.
- The two growth models for vegetative *B. cereus* available via the Pathogen Modelling Program have temperature, pH (acidulant, HCl), NaCl concentration and sodium nitrite concentration as controlling factors, with one produced under aerobic conditions (Benedict *et al.*, 1993), and the other under anaerobic conditions (unpublished USDA ARS Eastern Regional Research Centre data).
- The growth of toxigenic, psychrotrophic *B. cereus* spores as affected by temperature, pH and water activity was modelled and results showed that all strains were unable to grow within 42 d at 4 °C, pH 5 or a_W (glycerol as humectant) ≤ 0.95 (Chorin *et al.*, 1997).
- In addition to pH influencing the thermal inactivation of *B. cereus* spores, the type of organic acid has been shown to have an effect with a protective effect of the dissociated acid form having been demonstrated (Leguerinel and Mafart, 2001).
- Modelling studies on *B. cereus* have mainly concentrated on temperature, pH, water activity, sodium chloride, sodium nitrite and CO_2 concentrations,

but a model has been developed that incorporates the effects of sodium lactate (in combination with temperature, pH and sodium chloride) on the growth kinetics of *B. cereus* vegetative cells (Ölmez and Aran, 2005).

To demonstrate that the risk of *B. cereus* in food is effectively being managed using a HACCP-based approach requires verification, and it is in this context that microbiological criteria can play a valuable role. In addition, successful HACCP implementation requires that effective pre-requisite programmes are in place. *Bacillus cereus* spores may be present as general contaminants in the manufacturing and food preparation environment, and hence validated cleaning and disinfection procedures need to be implemented. *Bacillus cereus* strains have no unusual resistance to disinfectants compared with other spore-forming bacteria (ICMSF, 1996), but spores of *B. cereus* are more hydrophobic than other *Bacillus* spp. spores, therefore making them potentially more difficult to remove from equipment during cleaning (Granum, 1997). The ability of *B. cereus* to survive and even grow at high pH values (Lindsay *et al.*, 2000, 2002), means that the potential for contamination of cleaning fluids and CIP lines should be considered.

The food and beverage industry is beginning to apply risk assessment approaches in order to help better manage microbial pathogens (Membré *et al.*, 2005, 2006; Syposs *et al.*, 2005). Risk assessment approaches have particular value when a risk management decision is required related to a critical and complex food safety issue where there may be a high degree of uncertainty and variability in the relevant information and data. For example, risk assessment outputs have enabled decisions to be made about heat process optimisation (Membré *et al.*, 2006) and shelf-life determination (Membré *et al.*, 2005) with respect to *B. cereus*. Risk assessment approaches are not without disadvantages in that they can be time-consuming and require a great deal of detailed knowledge, as well as considerable skill, to implement.

A probabilistic model of heat-treated *B. cereus* spore lag time has been developed in order to help in assessing safe shelf-life under realistic heat-treatment profiles and supply-chain conditions, i.e. when the model inputs might not be single-point estimates but a distribution of values with their associated probabilities (Membré *et al.*, 2008). It was shown that the sensitivity analysis technique could help in interpreting output of a probabilistic model, i.e. in understanding the consequence of variability and uncertainty on the model output; this should help in choosing between different management options. For example, it was found that the impact of variability has several practical consequences. Firstly, the difference of lag time associated with the retailer refrigerator temperature in the US and European markets, indicates that the product heat treatment and shelf-life setting should be done for the two markets separately. This result has practical consequences in terms of logistics, e.g. product shelf-life may have to be adapted to the market to maintain the same level of safety/quality during the product's distribution.

Risk assessment approaches have also revealed the data gaps that need to be filled in order to better control *B. cereus*. As part of a quantitative risk assessment, Bayesian belief methods have been used to model the uncertainty and variability of the number of *B. cereus* spores that can be found in packets of minimally-processed vegetable purée, and it was revealed that spore contamination of added ingredients contributed most uncertainty to the assessment (Malakar *et al.*, 2004). A model that covers the retail and consumer part of the food chain for a cooked chilled vegetable product has been constructed to aid in the exposure assessment of psychrotrophic and mesophilic *B. cereus* as part of quantitative risk assessment (Nauta *et al.*, 2003). As well as enabling an evaluation of food product safety and a comparison of risk management options, important gaps were identified, including the lack of sporulation and germination models and data, and validated non-isothermal growth models. As the knowledge of the dose–response relationship for *B. cereus* is limited it is not possible to conduct a full microbial risk assessment, but an exposure assessment from retail store to the point of consumption of Kimbab (rolled cooked rice and other foodstuffs in dried layer, representative of ready-to-eat food in Korea) enabled estimates of levels of contamination ranges at different points in the food chain, but also highlighted the data gaps that would require filling to provide a more realistic and accurate exposure assessment in the future (Bahk *et al.*, 2007).

24.9 Future trends

The methods that are now available to test for various toxins produced by *Bacillus* spp. are now being used to identify potentially pathogenic species in different environments and, in many cases, studies raise concern about the frequent occurrence of these species in raw materials and foods. Despite these studies and other studies that show clear links between *Bacillus* spp. and foodborne disease, these organisms are not considered as strict pathogens. Indeed, in some applications, these species are included in products that will result in exposure of humans and other animals to very high numbers. *Bacillus* spp. are currently used in a number of applications including crop-protection, fat/grease degradation and probiotics. A number of different species, including those that have sometimes been reported to be associated with foodborne illness such as *B. licheniformis*, *B. subtilis*, *B. cereus* and *B. pumilus*, are currently used in products marketed as prophylactics and health foods and dietary supplements (Hong *et al.*, 2005). The large numbers of organisms (up to 10^9 cfu/g) contained in these preparations means that some strains can be used without apparently causing harm even though other strains of the same species clearly have the potential to cause disease. There is an increasing trend to use spores of *Bacillus* species in this manner, because of their heat-resistance properties and ability to survive transit across the stomach barrier, but their safety must be thoroughly investigated and should not be taken for granted, with each strain being assessed for safety on an individual basis. These investigations would need to take account of potential toxin-producing capability, as described by Ouoba *et al.* (2008), and also aspects such as antimicrobial resistance characteristics, as described in the Qualified Presumption of Safety approach developed by EFSA (2007).

In the past few years, the taxonomy of the genus *Bacillus* has undergone dramatic changes brought about by molecular techniques and also by identification of new species, such as *B. weihenstephanensis*. The clinical significance, in terms of foodborne disease, of this new species is still not clear. The ecological diversity of the different species in the *B. cereus* group has been recently reviewed by Guinbretiére *et al.* (2008), discussing adaptation to novel environments by modification of temperature tolerance limits. In this study, the authors concluded that the emergence of cold-adapted populations corresponds to colonisation of new or different environments and that this adaptive capability guarantees development of new evolutionary lines and long-term maintenance of the *B. cereus sensu lato* group. Currently, the public health risk posed by various species is restricted to the current phylogenetic groupings, and Guinbretiére *et al.* (2008) have proposed an alternative approach based on seven major phylogenetic groups.

Proteomic approaches may lead to increased understanding of *B. cereus* mechanisms which in turn may lead to improved controls. For example, greater understanding of the mechanisms involved in the switching between planktonic and sessile states may lead to better control of biofilm formation (Rasko *et al.*, 2005). Although potentially complex and time-consuming, targeted risk assessment-based approaches may provide further opportunities to improve product quality by reducing thermal processing, removing preservatives and exploiting heat-treatment injury effects to extend the lag time and hence shelf-life of products while still controlling *B. cereus*.

It seems that foodborne disease caused by *Bacillus* species will continue to be under-reported because of the relatively mild illness caused and also because many public health authorities and regulatory authorities do not consider these species to be 'frank' pathogens. However, there is continued interest in some research groups in better defining their current and potential role in foodborne disease, and this should lead to improved understanding of the various groups that harbour pathogenicity markers. For the time being, a number of *Bacillus* species should be regarded as having the potential to cause foodborne illness and must be considered in HACCP studies that will identify appropriate control procedures. Consumer interest in precooked, chilled food products with long shelf-lives may lead to products well suited for *B. cereus* survival and growth, and such foods could increase the prominence of *B. cereus* as a foodborne pathogen (Granum, 2001).

24.10 Sources of further information and advice

Growth Predictor has models, developed in laboratory media, for the growth of: *B. cereus* (vegetative cells, mesophilic and psychrotrophic strains) with variables of storage temperatures, pH value, sodium chloride content (a_w) and CO_2 concentration; *B. subtilis* (vegetative cells) with variables of storage temperatures, pH value and sodium chloride content (a_w); and *B. licheniformis* (vegetative cells) with variables of storage temperatures, pH value and sodium chloride content (a_w).

Pathogen Modelling Programme has two models, developed in laboratory media, for the growth of vegetative cells of *B. cereus*, one under aerobic conditions, one under anaerobic conditions, both with variables of storage temperature, pH value, sodium chloride concentration and sodium nitrite concentration.

Useful texts on *B. cereus* include:

Bouwer-Hertzberger S A and Mossel D A A (1982) Quantitative isolation and identification of *Bacillus cereus*, in J E L Corry, D Roberts and F A Skinner (eds), *Isolation and Identification Methods for Food Poisoning Organisms*, Academic Press, London, 255–9.

Granum P E (2001) *Bacillus cereus*, in M P Doyle, L R Beuchat and T J Montville (eds), *Food Microbiology: Fundamentals and Frontiers*, 2nd edn, ASM, Washington DC, 373–81.

Kramer J M, Turnbull P C B, Munshi G and Gilbert R J (1982) Identification and characterization of *Bacillus cereus* and other *Bacillus* species associated with foods and food poisoning, in J E L Corry, D Roberts and F A Skinner (eds), *Isolation and Identification Methods for Food Poisoning Organisms*, Academic Press, London, 261–86.

Schultz F J and Smith J L (1994) *Bacillus*: recent advances in *Bacillus cereus* food poisoning research, in Y H Hui, J R Gorham, K D Murrell and D O Cliver (eds), *Foodborne Disease Handbook, Volume 1: Diseases Caused by Bacteria*, Marcel Dekker, New York, 29–62.

24.11 Acknowledgements

The authors would like to thank Dr Paul Gibbs for contributing the chapter 'Characteristics of spore-forming bacteria' in the first edition of this book, some of which has been incorporated into this chapter on 'Pathogenic *Bacillus* species'.

24.12 References

Abdel-Hameed A and Landen R (1994) Studies on *Bacillus thuringiensis* strains isolated from Swedish soils: insect toxicity and production of *B. cereus*-diarrhoeal-type toxin, *World Journal of Microbiology and Biotechnology*, **10**, 406–9.

Ablett S, Darke A H, Lillford P and Martin D R (1999) Glass formation and dormancy in bacterial spores, *International Journal of Food Science and Technology*, **34**, 59–69.

Agata N, Ohta M and Mori M (1996) Production of an emetic toxin, cereulide, is associated with a specific class of *Bacillus cereus*, *Current Microbiology*, **33**, 67–9.

Agata N, Ohta M, Mori M and Shibayama K (1999) Growth conditions of and emetic toxin production by *Bacillus cereus* in a defined medium with amino acids, *Microbiology and Immunology*, **43**, 15–18.

Agata N, Ohta M and Yokoyama K (2002) Production of *Bacillus cereus* emetic toxin (cereulide) in various foods, *International Journal of Food Microbiology*, **73**, 23–7.

Ahmed A A-H, Moustafa M K and Marth E H (1983) Incidence of *Bacillus cereus* in milk and some milk products, *Journal of Food Protection*, **46**, 126–8.

Ali A U D D, Abdellatif M A, Bakry N M and El Sawaf S (1973) Studies on biological control of the greater wax moth *Galleria mellonella*. I. Susceptibility of wax moth larvae and adult honeybee workers to *Bacillus thuringiensis*, *Journal of Apicultural Research*, **12**, 125–30.

Altayar M and Sutherland A D (2006) *Bacillus cereus* is common in the environment but emetic toxin producing isolates are rare, *Journal of Applied Microbiology*, **100**, 7–14.

Andersson M A, Mikkola R, Helin J, Andersson M C and Salkinoja-Salonen M (1998a) A novel sensitive bioassay for detection of *Bacillus cereus* emetic toxin and related depsipeptide ionophores, *Applied and Environmental Microbiology*, **164**, 4185–90.

Andersson A, Granum P E and Rönner U (1998b) The adhesion of *Bacillus cereus* spores to epithelial cells might be an additional virulence mechanism, *International Journal of Food Microbiology*, **39**, 93–9.

Arnesen L P S, Fagerlund A and Granum P E (2008) From soil to gut: *Bacillus cereus* and its food poisoning toxins, *FEMS Microbiology Reviews*, **32**, 579–606.

Asano S-I, Nukumizu Y, Bando H, Iizuka T and Yamamoto T (1997) Cloning of novel enterotoxin genes from *Bacillus cereus* and *Bacillus thuringiensis*, *Applied and Environmental Microbiology*, **63**, 1054–7.

Ash C, Farrow J A E, Dorsch M, Stackebrandt E, Collins M D (1991) Comparative-analysis of *Bacillus anthracis*, *Bacillus cereus*, and related species on the basis of reverse-transcriptase sequencing of 16S ribosomal-RNA, *International Journal of Systematic Bacteriology*, **41**, 343–6.

Auger S, Galleron N, Bidnenko E, Ehrlich S D, Lapidus A and Sorokin A (2008) The genetically remote pathogenic strain NVH391-98 of the *Bacillus cereus* group is representative of a cluster of thermophilic strains, *Applied and Environmental Microbiology*, **74**, 1276–80.

Bahk G-J, Todd E C D, Hong C-H, Oh D-H and Ha S-D (2007) Exposure assessment for *Bacillus cereus* in ready-to-eat Kimbab selling at stores, *Food Control*, **18**, 682–8.

Batt C A (2000) *Bacillus cereus*, in R K Robinson, C A Batt and P D Patel (eds), *Encyclopaedia of Food Microbiology*, Vol. 1, Academic Press, London, 119–24.

Beattie S H and Williams A G (1999) Detection of toxigenic strains of *Bacillus cereus* and other *Bacillus* spp. with an improved cytotoxicity assay, *Letters in Applied Microbiology*, **28**, 221–5.

Beattie S H and Williams A G (2000) Detection of toxins, in R K Robinson, C A Batt and P D Patel (eds), *Encyclopaedia of Food Microbiology*, Vol. 1, Academic Press, London, 141–58.

Benedict R C, Partridge T, Wells D and Buchanan R L (1993) *Bacillus cereus*: aerobic growth kinetics, *Journal of Food Protection*, **56**, 211–14.

Beuchat L R, Clavero M R S and Jaquette C B (1997) Effects of nisin, and temperature on survival, growth, and enterotoxin production characteristics of psychrotrophic *Bacillus cereus* in beef gravy, *Applied and Environmental Microbiology*, **63**, 1953–8.

Black D G, Taylor T M, Kerr H J, Padhi S, Montville T J and Davidson P M (2008) Decontamination of fluid milk containing *Bacillus* spores using commercial household products, *Journal of Food Protection*, **71**, 473–8.

Bradshaw J G, Peeler J T and Twedt R M (1975) Heat resistance of ileal loop reactive

Bacillus cereus strains isolated from commercially canned food, *Applied Microbiology*, **30**, 943–5.

Brown K L (2000) Control of bacterial spores, *British Medical Bulletin*, **56**, 158–71.

Carlin C, Guinebretiére M H, Choma C, Pasqualini R, Braconnier A and Nguyen-The C (2000) Spore-forming bacteria in commercial cooked pasteurized and chilled vegetable purées, *Food Microbiology*, **17**, 153–65.

Carlin F, Fricker M, Pielaat A, Heisterkamp S, Shaheen R, Salonen M S, Svensson B, Nguyen-the C and Ehling-Schulz M (2006) Emetic toxin-producing strains of *Bacillus cereus* show distinct characteristics with the *Bacillus cereus* group, *International Journal of Food Microbiology*, **109**, 132–8.

Chang J M and Chen T H (2003) Bacterial foodborne outbreaks in central Taiwan, 1991-2000, *Journal of Food and Drug Analysis*, **11**, 53–9.

Choma C and Granum P E (2002) The enterotoxin T (BcET) from *Bacillus cereus* can probably not contribute to food poisoning, *FEMS Microbiology Letters*, **217**, 115–19.

Choma C, Guinebretiére M H, Carlin F, Schmitt P, Velge P, Granum P E and Nguyen-The C (2000) Prevalence, characterisation and growth of *Bacillus cereus* in commercial cooked chilled foods containing vegetables, *Journal of Applied Microbiology*, **88**, 617–25.

Chorin E, Thuault D, Cléret J-J and Bourgeois C-M (1997) Modelling *Bacillus cereus* growth, *International Journal of Food Microbiology*, **38**, 229–34.

Christiansson A, Ekelund K and Ogura H (1989a), Membrane filtration method for enumeration and isolation of spores of *Bacillus cereus* from milk, *International Dairy Journal*, **7**, 743–8.

Christiansson A, Naidu A S, Nilsson I, Wadstrom T and Pettersson H E (1989b) Toxin production by *Bacillus cereus* dairy isolates in milk at low temperatures, *Applied and Environmental Microbiology*, **55**, 2595–600.

Dahl M K (2000) *Bacillus* – Introduction, in R K Robinson, C A Batt and P D Patel (eds), *Encyclopaedia of Food Microbiology*, Vol. 1, Academic Press, London, 113–19.

Damgaard P H (1995) Diarrhoeal enterotoxin production by strains of *Bacillus thuringiensis* isolated from commercial *Bacillus thuringiensis*-based insecticides, *FEMS Immunological and Medical Microbiology*, **12**, 245–50.

Damgaard P H, Larsen H D, Hansen B M, Bresciani J and Jorgensen K (1996) Enterotoxin-producing strains of *Bacillus thuringiensis* isolated from food, *Letters in Applied Microbiology*, **23**, 146–50.

Day T L, Tatani S R, Notermens S and Bennett R W (1994) A comparison of ELSA and RPLA for detection of *Bacillus cereus* diarrhoeal enterotoxin, *Journal of Applied Bacteriology*, **77**, 9–13.

Del Torre M, Corte M D and Stecchini M L (2001) Prevalence and behaviour of *Bacillus cereus* in a REPFED of Italian origin, *International Journal of Food Microbiology*, **63**, 199–207.

Dierick K, van Collie E, Swiecicka I, Meyfroidt G, Devlieger H, Meulemans A, Hoedemaekers G, Fourie L, Heyndrickx M and Mahillon J (2005) Fatal family outbreak of *Bacillus cereus*-associated food poisoning, *Journal of Clinical Microbiology*, **43**, 4277–9.

Dufrenne J, Soentoro P, Tatini S, Day T and Notermans S (1994) Characteristics of *Bacillus cereus* related to safe food production, *International Journal of Food Microbiology*, **23**, 99–109.

Drobniewski F A (1993) *Bacillus cereus* and related species, *Clinical Microbiology Reviews*, **6**, 324–38.

EFSA (2005) Opinion of the Scientific Panel on Biological Hazards on *Bacillus cereus* and other *Bacillus* spp. in Foodstuffs, European Food Safety Authority, *The EFSA Journal*, **175**, 1–48.

EFSA (2007) Introduction of a Qualified Presumption of Safety (QPS) approach for assessment of selected microorganisms referred to EFSA – Opinion of the Scientific Committee, European Food Safety Authority, *The EFSA Journal*, **587**, 1–16, available at: http://www.efsa.europa.eu/cs/BlobServer/Scientific_Opinion/sc_op_ej587_qps_en.pdf?ssbinary=true, accessed November 2008.

Ehling-Schulz M, Svensson B, Guinebretiére M H, Lindback T, Andersson M, Schulz A, Fricker M, Christiansson A, Granum P E, Martlbauer E, Nguyen-The C, Salkinoja-Salonen M and Scherer S (2005) Emetic toxin formation of *Bacillus cereus* is restricted to a single evolutionary lineage of closely related strains, *Microbiology*, **151**, 183–97.

Fang T J, Chen C-Y and Kuo W-Y (1999) Microbiological quality and incidence of *Staphylococcus aureus* and *Bacillus cereus* in vegetarian food products, *Food Microbiology*, **16**, 385–91.

Fernández A, Collado J, Cunha L M, Ocio M J and Martínez A (2002) Empirical model building based on Weibull distribution to describe the joint effects of pH and temperature on the thermal resistance of *Bacillus cereus* in vegetable substrate, *International Journal of Food Microbiology*, **77**, 147–53.

Fricker M, Messelhausser U, Busch U, Scherer S and Ehling-Schulz M (2007) Diagnostic real-time PCR assays for the detection of emetic *Bacillus cereus* strains in foods and recent food-borne outbreaks, *Applied and Environmental Microbiology*, **73**, 1892–8.

Fricker M, Reissbrodt R and Ehling-Schulz M (2008) Evaluation of standard and new chromogenic selective plating media for isolation and identification of *Bacillus cereus*, *International Journal of Food Microbiology*, **121**, 27–34.

Fritze D (2004) Taxonomy of the genus *Bacillus* and related genera: the aerobic endospore-forming bacteria, *Phytopathology*, **94**, 1245–8.

From C, Hormazabal V and Granum P E (2007) Food poisoning associated with pumilacidin-producing *Bacillus pumilus* in rice, *International Journal of Food Microbiology*, **115**, 319–24.

Fromm H I and Boor K J (2004) Characterization of pasteurised fluid milk shelf-life attributes, *Journal of Food Science*, **69**, 207–14.

Gaillard S, Leguerinel I and Mafart P (1998) Modelling combined effects of temperature and pH on the heat resistance of spores of *Bacillus cereus*, *Food Microbiology*, **15**, 625–30.

Garrity G M, Bell J A and Lilburn T G (2003) Taxonomic outline of the prokaryotes, in *Bergey's Manual of Systematic Bacteriology*, Springer-Verlag, New York.

Gaulin C, Viger Y B and Fillion L (2002) An outbreak of *Bacillus cereus* implicating a part-time banquet caterer, *Canadian Journal of Public Health*, **93**, 353–5.

Gaviria Rivera A M, Granum P E and Priest F G (2000) Common occurrence of enterotoxin genes and enterotoxicity in *Bacillus thuringiensis*, *FEMS Microbiology Letters*, **190**, 151–5.

Ghelardi E, Celandroni F, Salvetti S, Barsotti C, Baggiani A and Senesi S (2002) Identification and characterization of toxigenic *Bacillus cereus* isolates responsible for two food-poisoning outbreaks, *FEMS Microbiology Letters*, **208**, 129–34.

Gianella R A and Brasile L (1979) A hospital food-borne outbreak of diarrhea caused by *Bacillus cereus*: clinical, epidemiologic, and microbiologic studies, *Journal of Infectious Diseases*, **139**, 366–70.

Gilbert R J (1979) *Bacillus cereus* gastroenteritis, in H Reimann and F L Bryan (eds), *Foodborne Infections and Intoxications*, 2nd edn, Academic Press, New York, 495–518.

Gilbert R J and Humphrey T J (1998) Foodborne bacterial gastroenteritis, in L Collier, A Balows and M Sussman (eds), *Topley and Wilson's Microbiology and Microbial Infections*, 9th edn, Vol. 3, 540–65.

Goepfert J M, Spira W M and Kim H U (1972) *Bacillus cereus*: food poisoning organism, a review, *Journal of Milk and Food Technology*, **35**, 213–27.

Granum P E (1997) *Bacillus cereus*, in *Foodborne Pathogenic Bacteria*, M P Doyle, L R Beuchat and T J Montville (eds), ASM Press, Washington, DC, 327–36.

Granum P E (2001) *Bacillus cereus*, in M P Doyle, L R Beuchat and T J Montville (eds), *Food Microbiology: Fundamentals and Frontiers*, 2nd edn, ASM Press, Washington, DC, 373–81.

Granum P E and Baird-Parker T C (2000) *Bacillus* species, in B M Lund, T C Baird-Parker and G W Gould (eds), *The Microbiological Safety and Quality of Food*, vol. 2, Aspen, Gaithersburg, MD, 1029–39.

Granum P E, Brynestad S, O'Sullivan K and Nissen H (1993) Enterotoxin from *Bacillus-cereus* – production and biochemical-characterization, *Netherlands Milk and Dairy Journal*, **47**, 63–70.

Granum P E, Naestvold A and Gundersby K N (1995) An outbreak of *Bacillus cereus* food poisoning in the junior Norwegian ski championships, *Norsk Veterinaertidsskrift*, **107**, 945–8.

Guinebretiére M-H and Nguyen-The C (2003) Sources of *Bacillus cereus* contamination in a pasteurized zucchini purée processing line, differentiated by two PCR-based methods, *FEMS Microbiology Ecology*, **43**, 207–15.

Guinebretiére M-H, Thompson F L, Sorokin A, Normand P, Dawyndt P, Ehling-Schulz M, Svensson B, Sanchis V, Nguyen-The C, Heydrickx M and de Vos P (2008) Ecological diversification in the *Bacillus cereus* group, *Environmental Microbiology*, **10**, 851–65.

Harmon S M and Kautter D A (1991) Incidence and growth potential of *Bacillus cereus* in ready-to-serve foods, *Journal of Food Protection*, **54**, 372–4.

Harmon S M, Kautter D A and Solomon H M (1987) *Bacillus cereus* contamination of seeds and vegetable sprouts grown in a home sprouting kit, *Journal of Food Protection*, **50**, 62–5.

Hashisaka A E, Matches J R, Batters Y, Hungate F P and Dong F M (1990) Effects of gamma irradiation at –78 °C on microbial populations in dairy products, *Journal of Food Science*, **55** 1284–9.

Hauge S (1955) Food poisoning caused by aerobic spore forming bacilli, *Journal of Applied Bacteriology*, **18**, 591–5.

Hendriksen N B, Hansen B M and Johansen J E (2006) Occurrence and pathogenic potential of *Bacillus cereus* group bacteria in a sandy loam, *Antonie Van Leeuwenhoek International Journal of General and Molecular Microbiology*, **89**, 239–49.

Holbrook R and Anderson J M (1980) An improved selective diagnostic medium for the isolation and enumeration of *Bacillus cereus* in foods, *Canadian Journal of Microbiology*, **26**, 753–9.

Hong H A, Duc L H and Cutting S M (2005) The use of bacterial spore formers as probiotics, *FEMS Microbiology Reviews*, **29**, 813–35.

Hughes C, Gillespie I A, O'Brien S J and The Breakdowns in Food Safety Group (2007) Foodborne transmission of infectious intestinal disease in England and Wales, 1992–2003, *Food Control*, **18**, 766–72.

ICMSF (1996) *Bacillus cereus*, in *Microorganisms in Foods 5, Microbiological Specifications of Food Pathogens*, International Commission for the Microbiological Safety of Foods, Blackie Academic and Professional, London, 20–35.

in't Veld P H, Ritmeester W S, Delfgou-van Asch E H M, Dufrenne J B, Wernars K, Smit E and van Leusden F M (2001) Detection of genes encoding for enterotoxins and determination of the production of enterotoxins by HBL blood plates and immunoassays of psychrotrophic strains of *Bacillus cereus* isolated from pasteurised milk, *International Journal of Food Microbiology*, **64**, 63–70.

Jaaskelainen E L, Teplova V, Andersson M A, Andersson L C, Tammela P, Andersson M C, Pirhonen T I, Saris N E L, Vuorela P and Salkinoja-Salonen M S (2003) *In vitro* assay for human toxicity of cereulide, the emetic mitochondrial toxin produced by food poisoning *Bacillus cereus*, *Toxicology in Vitro*, **17**, 737–44.

Jackson S G, Goodbrand R B, Ahmed R and Kasatiya S (1995) *Bacillus cereus* and *Bacillus thuringiensis* isolated in a gastroenteritis outbreak investigation, *Letters in Applied Microbiology*, **21**, 103–5.

Jaquette C B and Beuchat L R (1998) Combined effects of pH, nisin, and temperature on growth and survival of psychrotrophic *Bacillus cereus*, *Journal of Food Protection*, **61**, 563–70.

Jay J M (2003) A review of recent taxonomic changes in seven genera of bacteria commonly found in foods, *Journal of Food Protection*, **66**, 1304–9.

Johnson K M (1984) *Bacillus cereus* foodborne illness – an update, *Journal of Food Protection*, **47**, 145–53.

Kamat A S, Nerkar D P and Nair P M (1989) *Bacillus cereus* in some Indian foods, incidence and antibiotic, heat and radiation resistance, *Journal of Food Safety*, **10**, 31–41.

Kaur P (1986) Survival and growth of *Bacillus cereus* in bread, *Journal of Applied Bacteriology*, **60**, 513–16.

King N J, Whyte R and Hudson J A (2007) Presence and significance of *Bacillus cereus* in dehydrated potato products, *Journal of Food Protection*, **70**, 514–20.

Kotiranta A, Lounatmaa K and Haapasalo M (2000) Epidemiology and pathogenesis of *Bacillus cereus* infections, *Microbes and Infection*, **2**, 189–98.

Kramer J M and Gilbert R J (1989) *Bacillus cereus* and other *Bacillus* species, in M P Doyle (ed.) *Foodborne Bacterial Pathogens*, Marcel Dekker, New York, 21–70.

Kreske A, Ryu J-H and Beuchat L R (2006) Evaluation of chlorine, chlorine dioxide, and peroxyacetic acid acid-based sanitizer for effectiveness in killing *Bacillus cereus* and *Bacillus thuringiensis* spores in suspensions, on the surface of stainless steel, and on apples, *Journal of Food Protection*, **69**, 1892–1903.

Langeveld L P M, van Spronsen W A, van Berenstein C H, Notermans S H W (1996) Consumption by healthy adults of pasteurised milk with a high concentration of *Bacillus cereus*: a double blind study, *Journal of Food Protection*, **59**, 723–6.

Lapidus A, Goltsman E, Auger S, Galleron N, Segurens B, Dossat C, Land M L, Broussolle V, Brillard J, Guinebretiere M H, Sanchis V, Nguen-the C, Lereclus D, Richardson P, Wincker P, Weissenbach J, Ehrlich S D and Sorokin A (2008) Extending the *Bacillus cereus* group genomics to putative food-borne pathogens of different toxicity, *Chemico-Biological Interactions*, **171**, 236–49.

Laurent Y, Arino S and Rosso L (1999) A quantitative approach for studying the effect of heat treatment conditions on resistance and recovery of *Bacillus cereus* spores, *International Journal of Food Microbiology*, **48**, 149–57.

Lechner S, Mayr R, Francis K P, Pruß B M, Kaplan T, Wiessner-Gunkel E, Stewart G S and Scherer S (1998) *Bacillus weihenstephanensis* sp. nov. is a new psychrotolerant species of the *Bacillus cereus* group, *International Journal of Systematic Bacteriology*, **48**, 1373–82.

Lee M J, Bae D H, Lee D H, Jang K H, Oh D H and Ha S D (2006) Reduction of *Bacillus cereus* in cooked rice treated with sanitizers and disinfectants, *Journal of Microbiology and Biotechnology*, **16**, 639–42.

Leguerinel I and Mafart P (2001) Modelling the influence of pH and organic acid types on thermal inactivation of *Bacillus cereus* spores, *International Journal of Food Microbiology*, **63**, 29–34.

Lindsay D, Brözel V S, Mostert J F and von Holy A (2000) Physiology of dairy-associated *Bacillus* spp. over a wide pH range, *International Journal of Food Microbiology*, **54**, 49–62.

Lindsay D, Oosthuizen M C, Brözel V S and von Holy A (2002) Adaptation of a neutrophilic dairy-associated *Bacillus cereus* isolate to alkaline pH, *Journal of Applied Microbiology*, **92**, 81–9.

Lubenau C (1906) *Bacillus peptonificans* als erreger einer gastroentyeritis-epidemie, *Zentralblatt für Bakteriologie*, **40**, 433–7.

Luby S, Jones J, Dowds H, Kramer J and Horan J (1993) A large outbreak of gastroenteritis caused by diarrhoeal toxin-producing *Bacillus cereus*, *Journal of Infectious Disease*, **167**, 1452–5.

Lund B M (1990) Foodborne disease due to *Bacillus* and *Clostridium* species, *The Lancet*, **20**, 982–6.

Lund T, de Buyser M-L and Granum P E (2000) A new cytotoxin from *Bacillus cereus* that may cause necrotic enteritis, *Molecular Microbiology*, **38**, 254–61.

Malakar P K, Barker G C and Peck M W (2004) Modeling the prevalence of *Bacillus cereus* spores during the production of a cooked chilled vegetable product, *Journal of Food Protection*, **67**, 939–46.

Margosch D, Gänzle M G, Ehrmann M A and Vogel R F (2004) Pressure inactivation of *Bacillus* endospores, *Applied and Environmental Microbiology*, **70**, 7321–8.

Martínez S, Borrajo R, Franco I and Carballo J (2007) Effect of environmental parameters on growth kinetics of *Bacillus cereus* (ATCC 7004) after mild heat treatment, *International Journal of Food Microbiology*, **117**, 223–7.

Mazas M, Lopez M, Gonzalez I, Bernardo A and Martin R (1998) Effects of the heating medium pH on heat resistance of *Bacillus cereus* spores, *Journal of Food Safety*, **18**, 25–36.

Mead P S, Slutsker L, Dietz V, McCaig L F, Bresee J S, Shapiro C, Griffin P M and Tauxe R V (1999) Food-related illness and death in the United States, *Emerging Infectious Diseases*, **5**, 607–25.

Membré J-M, Johnston M D, Bassett J, Naaktgeboren G, Blackburn C de W and Gorris L G M (2005) Application of modelling techniques in the food industry: Determination of shelf-life for chilled foods, in Hertog M L A T M and Nicolaï B M (eds), *III International Symposium on Applications of Modelling as an Innovative Technology in the Agri-Food Chain; MODEL-IT*, Leuven, Belgium, *ISHS Acta Horticulture*, **674**, 407–14.

Membré J-M, Amézquita A, Bassett J, Giavedoni P, Blackburn C de W and Gorris L G M (2006) A probabilistic modeling approach in thermal inactivation: estimation of post-process *Bacillus cereus* spore prevalence and concentration, *Journal of Food Protection*, **69**, 118–29.

Membré J-M, Kan-King-Yu D and Blackburn C de W (2008) Use of sensitivity analysis to aid interpretation of a probabilistic *Bacillus cereus* spore lag time model applied to heat-treated chilled foods (REPFEDs), *International Journal of Food Microbiology*, in press.

Merrill L, Dunbar J, Richardson J and Kuske C R (2006) Composition of *Bacillus* species in aerosols from 11 U.S. cities, *Journal of Forensic Science*, **51**, 559–65.

Mikami H, Horikawa T, Murakami T, Matsumoto T, Yamakawa A, Murayama S, Katagiri S, Shinagawa K and Suzuki M (1994) An improved method for detecting cytostatic toxin (emetic toxin) of *Bacillus cereus* and its application to food samples, *FEMS Microbiology Letters*, **119**, 53–8.

Mikami T, Horikawa T, Murakami T, Sato N, Ono Y, Matsumoto T, Yamakawa A, Murayama S, Katagiri S and Suzuki M (1995) Examination of toxin production from environmental *Bacillus cereus* and *Bacillus thuringiensis*, *Yakugaku Zasshi Journal of the Pharmaceutical Society of Japan*, **115**, 742–8.

Mikkola R, Kolari M, Andersson M A, Helin J and Salkinoja-Salonen M S (2000) Toxic lactonic lipopeptide from food poisoning isolates of *Bacillus licheniformis*, *European Journal of Biochemistry*, **267**, 4068–74.

Moravek M, Wegscheider M, Schulz A, Dietrich R, Bürk C and Märtlbauer E (2006) Colony immunoblot assay for the detection of hemolysin BL enterotoxin producing *Bacillus cereus*, *FEMS Microbiology Letters*, **238**, 107–13.

Nakano S, Maeshima H, Matsumura A, Ohno K, Ueda S, Kuwabara Y and Yamada T (2004) A PCR assay based on a sequence-characterized amplified region marker for detection of emetic *Bacillus cereus*, *Journal of Food Protection*, **67**, 1694–701.

Nauta M J, Litman S, Barker G C and Carlin F (2003) A retail and consumer phase model for exposure assessment of *Bacillus cereus*, *International Journal of Food Microbiology*, **83**, 205–18.

Nichols G L, Little C L, Mithani V and de Louvois J (1999) The microbiological quality of cooked rice from restaurants and take-away premises in the United Kingdom, *Journal of Food Protection*, **62**, 877–82.

Nieminen T, Rintaluoma N, Andersson M, Taimisto A-M, Ali-Vehmas T, Seppa¨la¨ A, Priha O and Salkinoja-Salonen M (2007) Toxinogenic *Bacillus pumilus* and *Bacillus licheniformis* from mastitic milk, *Veterinary Microbiology*, **124**, 329–39.

Nikodémusz I, Bodnar S, Bojan M, Kiss M, Kiss P, Laczko M, Molnar E and Papay D (1962) Aerobe sporenbildner als lebensmittelvergifter, *Zentralblatt Bakteriologie I Originale*, **184**, 462–70.

Notermans S, Dufrenne J, Teunis P, Beumer R, Giffel M T and Weem P P (1997) A risk assessment study of *Bacillus cereus* present in pasteurized milk, *Food Microbiology*, **14**, 143–51.

Ölmez H K and Aran N (2005) Modeling the growth kinetics of *Bacillus cereus* as a function of temperature, pH, sodium lactate and sodium chloride concentrations, *International Journal of Food Microbiology*, **98**, 135–43.

Ouoba L I, Thorsen L and Varnum A H (2008) Enterotoxin and emetic toxin production by *Bacillus cereus* and other species of *Bacillus* isolated from Soumbala and Bikalga, African alkaline fermented food condiments, *International Journal of Food Microbiology*, **124**, 224–30.

Paananen A, Mikkola R, Sareneva T, Matikainen S, Hess M, Andersson M, Julkunen I, Salkinoja-Salonen M S and Timonen T (2002) Inhibition of human natural killer cell activity by cereulide, an emetic toxin from *Bacillus cereus*, *Clinical and Experimental Immunology*, **129**, 420–8.

Park S H, Kim H J, Kim J H, Kim T W and Kim H Y (2007) Simultaneous detection and identification of *Bacillus cereus* group bacteria using multiplex PCR, *Journal of Microbiology and Biotechnology*, **17**, 1177–82.

Pavic S, Brett M, Petric I, Lastre D, Smolanovic M, Atkinson M, Kovacic A, Cetinic E and Ropac D (2005) An outbreak of food poisoning in a kindergarten caused by milk powder containing toxigenic *Bacillus subtilis* and *Bacillus licheniformis*, *Archiv für Lebensmittelhygiene*, **56**, 1–24.

Peltola J, Andersson M A, Haahtela T, Mussalo-Rauhamaa H, Rainey F A, Kroppenstedt R M, Samsom R A and Salkinoja-Salonen M S (2001) Toxic-metabolite-producing bacteria and fungus in an indoor environment, *Applied and Environmental Microbiology*, **67**, 3269–74.

Peng H, Ford V, Frampton E W, Restaino L, Shelef L A and Spitz H (2001) Isolation and enumeration of *Bacillus cereus* from foods on a novel chromogenic plating medium, *Food Microbiology*, **18**, 231–8.

Penman A D, Webb R M, Woernle C H and Currier M M (1996) Failure of routine restaurant inspections: restaurant-related foodborne outbreaks in Alabama, 1992, and Mississippi, 1993, *Journal of Environmental Health*, **58**, 23–5.

PHLS (1992) Microbiological guidelines for some ready-to-eat foods sampled at the point of sale: an expert opinion from the Public Health Laboratory Service, *PHLS Microbiology Digest*, **13**, 41–3.

Pirttijärvi T S, Andersson M A, Scoging A C and Salkinjoa-Salonen M S (1999) Evaluation of methods for recognising strains of the *Bacillus cereus* group with food poisoning potential among industrial and environmental contaminants, *Systematic and Applied Microbiology*, **22**, 133–44.

Pivovarov Yu P, Ivashina S P and Padalkin V P (1977) Hygienic assessment of investigation of insecticide preparation, *Gig I Sanit.*, **8**, 45–8 (in Russian).

Plazikowski U (1947) Further investigations regarding the cause of food poisoning, *Congress of International Microbiology*, Copenhagen, **4**, 510–2.

Portnoy B L, Goepfert J M and Harmon S M (1976) An outbreak of *Bacillus cereus* food poisoning resulting from contaminated vegetable sprouts, *American Journal of Epidemiology*, **103**, 589–94.

Priest F G (1990) Products and applications, in C R Harwood (ed.), *Bacillus*, Plenum, New York, 293–320.

Prokopova L L (1970) Multiplication and toxigenicity of *Bacillus cereus* contained in food products stored under different thermal conditions, *Voprosy Pitaniia*, **29**, 56–61 (in Russian, English summary).

Prüß B M, Dietrich R, Nibler B, Märtlbauer E and Scherer S (1999) The hemolytic enterotoxin HBL is broadly distributed among species of the *Bacillus cereus* group, *Applied and Environmental Microbiology*, **65**, 5436–42.

Raevuori M and Genigeorgis C (1975) Effect of pH and sodium chloride on growth of *Bacillus cereus* in laboratory media and certain foods, *Applied Microbiology*, **29**, 68–73.

Rajkovic A, Uyttendaele M, Courtens T, Heyndrickx M and Debevere J (2006a) Prevalence and characterisation of *Bacillus cereus* in vacuum packed potato puree, *International Journal of Food Science and Technology*, **41**, 878–84.

Rajkovic A, Uyttendaele M, Ombregt S-A, Jaaskelainen E, Salknoja-Salonen M and Debevere J (2006b) Influence of type of food on the kinetics and overall production of *Bacillus cereus* emetic toxin, *Journal of Food Protection*, **69**, 847–52.

Rajkowski K T and Mikolajcik E M (1987) Characteristics of selected strains of *Bacillus cereus*, *Journal of Food Protection*, **50**, 199–205.

Rangasamy P N, Iyer M and Roginski H (1993) Isolation and characterization of *Bacillus cereus* in milk and dairy products manufactured in Victoria, *Australian Journal of Dairy Technology*, **48**, 93–5.

Rasko D A, Altherr M R, Han C S and Ravel J (2005) Genomics of the *Bacillus cereus* group of organisms, *FEMS Microbiology Reviews*, **29**, 303–29.

Ray D E (1990) Pesticides derived from plants and other organisms, in W J Hayes and E R Laws (eds), *Handbook of Pesticide Toxicology*, Academic Press, New York, London, 585–636.

Ripabelli G, McLauchlin J, Mithani V and Threlfall E J (2000) Epidemiological typing of *Bacillus cereus* by amplified fragment length polymorphism, *Letters in Applied Microbiology*, **30**, 358–63.

Roberts D, Watson G N and Gilbert R J (1982) Contamination of food plants and plant products with bacteria of public health significance, in M E Rhodes-Roberts and F A Skinner (eds), *Bacteria and Plants*, Academic Press, London, 169–95.

Rosenquist H, Smidt L, Andersen S R, Jensen G B and Wilcks A (2005) Occurrence and significance of *Bacillus cereus* and *Bacillus thuringiensis* in ready-to-eat food, *FEMS Microbiology Letters*, **250**, 129–36.

Roy A, Moktan B and Sarkar P K (2007) Characteristics of *Bacillus cereus* isolates from legume-based Indian fermented foods, *Food Control*, **18**, 1555–64.

Rusul G and Yaacob N H (1995) Prevalence of *Bacillus cereus* in selected foods and detection of enterotoxin using TECRA-VIA and BCET-RPLA, *International Journal of Food Microbiology*, **25**, 131–9.

Sale A J H, Gould G W and Hamilton W A (1970) Inactivation of bacterial spores by high hydrostatic pressure, *Journal of General Microbiology*, **60**, 323–34.

Salkinoja-Salonen M S, Vuorio R, Andersson M A, Kämpfer P, Andersson M C, Honkanen-Buzalski T and Scoging A C (1999) Toxigenic strains of *Bacillus licheniformis* related to food poisoning, *Applied and Environmental Microbiology*, **65**, 4637–45.

Sarrías J A, Valero M and Salmerón M C (2002) Enumeration, isolation and characterization of *Bacillus cereus* strains from Spanish raw rice, *Food Microbiology*, **19**, 589–95.

Schraft H, Steele M, McNab B, Odumeru J and Griffiths M W (1996) Epidemiological typing of *Bacillus* spp. isolated from food, *Applied and Environmental Microbiology*, **62**, 4229–32.

Schmitt N, Bowmer E J and Willoughby B A (1976) Food poisoning outbreak attributed to *Bacillus cereus*, *Canadian Journal of Public Health*, **67**, 418–22.

Schoder D, Zangerl R, Manafi M, Wagner M, Lindner G and Foissy H (2007) *Bacillus cereus* – a foodborne pathogen of different risk potential for the dairy industry, *Wiener Tierarztliche Monatsschrift*, **94**, 25–33.

Schoeni J L and Wong A C (2005) *Bacillus cereus* food poisoning and its toxins, *Journal of Food Protection*, **68**, 636–48.

Scurrah K J, Robertson R E, Craven H M, Pearce L E and Szabo E A (2006) Inactivation of *Bacillus* spores in reconstituted skim milk by combined high pressure and heat treatment, *Journal of Applied Microbiology*, **101**, 172–80.

Shida O, Takagi H, Kadowaki K and Komagata K (1996) Proposal for two new genera, *Brevibacillus* gen nov and *Aneurinibacillus* gen nov., *International Journal of Systematic Bacteriology*, **46**, 939–46.

Shinagawa K (1990) Purification and characterisation of *Bacillus cereus* enterotoxin and its application to diagnosis, in A E Pohland, V R Dowell Jr and J L Richard (eds), *Microbial Toxins in Foods and Feeds: Cellular and Molecular Modes of Action*, Plenum, New York, 181–93.

Shinagawa K (1993) Serology and characterisation of toxigenic *Bacillus cereus*, *Netherlands Milk and Dairy Journal*, **47**, 89–103.

Stackebrandt E and Swidersky J (2002) From phylogeny to systematics, in R Berkeley, M Heyndrickx, N Logan and P Devos (eds), *Applications and Systematics of Bacillus and Relatives*, Blackwell, Oxford, 8–22.

Stenfors L P and Granum P E (2001) Psychrotolerant species from the *Bacillus cereus* group are not necessarily *Bacillus weihenstephanensis*, *FEMS Microbiology Letters*, **197**, 223–8.

Stenfors L P, Mayr R, Scherer S and Granum P E (2002) Pathogenic potential of fifty *Bacillus weihenstephanensis* strains, *FEMS Microbiology Letters*, **215**, 47–51.

Stewart D B (1975) Factors influencing the incidence of *Bacillus cereus* spores in milk, *Journal of the Society of Dairy Technology*, **28**, 80–85.

Subramanian S B, Kamat A S, Ussuf K K and Tyagi R D (2006) Virulent gene based DNA probe for the detection of pathogenic *Bacillus cereus* strains found in food, *Process Biochemistry*, **41**, 783–8.

Sutherland A D and Limond A M (1993) Influence of pH and sugars on the growth and production of diarrhoeagenic toxin by *Bacillus cereus*, *Journal of Dairy Research*, **60**, 575–80.

Sutherland J P, Aherne A and Beaumont A L (1996) Preparation and validation of a growth model for *Bacillus cereus*: the effects of temperature, pH, sodium chloride, and carbon dioxide, *International Journal of Food Microbiology*, **30**, 359–72.

Svensson B, Ekelund K, Ogura H and Christiansson A (2004) Characterisation of *Bacillus cereus* isolated from milk silo tanks at eight different dairy plants, *International Dairy Journal*, **14**, 17–27.

Svensson B, Monthán A, Guinebretiére M-H, Nguyen-Thé C and Christiansson A (2007) Toxin production potential and the detection of toxin genes among strains of the *Bacillus cereus* group isolated along the dairy production chain, *International Dairy Journal*, **17**, 1201–8.

Syposs Z, Reichart O and Mészáros L (2005) Microbiological risk assessment in the beverage industry, *Food Control*, **16**, 515–21.

Szabo R A, Todd E C D and Rayman H K (1984) Twenty four hour isolation and confirmation of *Bacillus cereus* in foods, *Journal of Food Protection* **47**, 856–60.

Talarmin A, Nicand E, Doucet M, Fermanian C, Baylac P and Buisson Y (1993) Toxi-infection alimentaire collective à *Bacillus cereus*, *Bulletin Epidémiologique Hebdomadaire*, **33**, 154–5.

Taylor J M W, Sutherland A D, Aidoo K and Logan N A (2005) Heat-stable toxin production by strains of *Bacillus cereus*, *Bacillus firmus*, *Bacillus megaterium*, *Bacillus simplex* and *Bacillus licheniformis*, *FEMS Microbiology Letters*, **242**, 313–17.

te Giffel M C, Beumer R R, Leijendekkers S and Rombouts F M (1996) Incidence of *Bacillus cereus* and *Bacillus subtilis* in foods in the Netherlands, *Food Microbiology*, **13**, 53–8.

te Giffel M C, Beumer R R, Klijn N, Wagendorp A and Rombouts F M (1997) Discrimination between *Bacillus cereus* and *Bacillus thuringiensis* using specific DNA probes based on variable regions of 16S rRNA, *FEMS Microbiology Letters*, **146**, 47–51.

Thorsen L, Hansen B M, Nielsen K F, Hendriksen N B, Phipps R K and Budde B B (2006) Characterization of emetic *Bacillus weihenstephanensis*, a new cereulide-producing bacterium, *Applied and Environmental Microbiology*, **72**, 5118–21.

Todd E, Park C, Clecner B, Fabricius A, Edwards D and Ewan P (1974) Two outbreaks of *Bacillus cereus* food poisoning in Canada, *Canadian Journal of Public Health*, **65**, 109–113.

Tong J L, Engle H M, Cullyford J S, Shimp D J and Love C E (1962) Investigation of an outbreak of food poisoning traced to turkey meat, *American Journal of Public Health*, **52**, 976–90.

Ueda S and Kuwabara Y (1993) An ecological study of *Bacillus cereus* in rice crop processing, *Journal of Antibacterial and Antifungal Agents*, **21**, 499–502.

Valero M, Leontidis S, Fernández P S, Martínez A and Salmerón M C (2000) Growth of *Bacillus cereus* in natural and acidified carrot substrates over the temperature range 5–30 °C, *Food Microbiology*, **17**, 605–12.

Valero M, Hernández-Herrero L A, Fernández P S and Salmerón M C (2002) Characterization of *Bacillus cereus* isolates from fresh vegetables and refrigerated minimally processed foods by biochemical and physiological tests, *Food Microbiology*, **19**, 491–9.

Valero M, Fernández P S and Salmerón M C (2003) Influence of pH and temperature on growth of *Bacillus cereus* in vegetable substrates, *International Journal of Food Microbiology*, **82**, 71–9.

van Netten P, van de Moosdijk A, van Hoensel P, Mossel D A A and Perales I (1990) Psychrotrophic strains of *Bacillus cereus* producing Enterotoxin, *Journal of Applied Bacteriology*, **69**, 73–9.

Vyletlova M, Hanus O, Pacova Z, Roubal P and Kopunecz P (2001) Frequency of *Bacillus* bacteria in raw cow's milk and its relation to other hygienic parameters, *Czech Journal of Animal Science*, **46**, 260–67.

Wijnands L M, Dufrenne J B, Rombouts F M, in't Veld P H and van Leusden F M (2006) Prevalence of potentially pathogenic *Bacillus cereus* in food commodities in The Netherlands, *Journal of Food Protection*, **69**, 2587–94.

Wong H-C and Chen Y-L (1988) Effects of lactic acid bacteria and organic acids on growth and germination of *Bacillus cereus*, *Applied and Environmental Microbiology*, **54**, 2179–84.

Wong H-C, Chang M-H and Fan J-Y (1988) Incidence and characterisation of *Bacillus cereus* isolates contaminating dairy products, *Applied and Environmental Microbiology*, **54**, 699–702.

Yang I C, Shih D Y C, Huang T P, Huang Y P, Wang J Y and Pan T M (2005) Establishment of a novel multiplex PCR assay and detection of toxigenic strains of the species in the *Bacillus cereus* group, *Journal of Food Protection*, **68**, 2123–30.

Zhang H and Mittal G S (2008) Effects of high-pressure processing (HPP) on bacterial spores: an overview, *Food Reviews International*, **24**, 330–51.

Part III

Other agents of foodborne disease

25

Hepatitis viruses and emerging viruses

K. Mattison, S. Bidawid and J. Farber, Health Canada, Canada

Abstract: This chapter describes viruses that are enterically transmitted and cause systemic disease. These are recognized as important food and waterborne pathogens. The chapter first summarizes the general characteristics of the viruses, then describes their typical epidemiological patterns. The chapter then discusses methods to detect and to inactivate the viruses, with an emphasis on strategies that can be implemented for food safety.

Key words: foodborne viruses, hepatitis A virus, hepatitis E virus, inactivation of viruses, emerging viruses.

25.1 Introduction

Viruses that are recognized as causes of foodborne and waterborne illness can be separated into four general groups: the viruses that cause gastroenteritis (Chapter 32), the viruses that cause hepatitis, the enteroviruses and emerging viruses.

The enterically transmitted hepatitis viruses (hepatitis A and hepatitis E) replicate and cause disease in the liver, while the enteroviruses (polio, echo and coxsackie) replicate in the small intestine but migrate to and cause illness in other organs. These two groups of viruses are transmitted by the faecal–oral route, either directly from person-to-person or indirectly when food or water is contaminated with faecal material from infected individuals. There are also some emerging viruses that are not usually transmitted via the faecal–oral route, but recent reports indicate they may infect via the gastrointestinal tract and have the potential to emerge as a food safety concern (influenza A, coronavirus, tickborne encephalitis). See Table 25.1 for a summary of the viruses discussed in this chapter.

Table 25.1 Summary of enteric viruses and their main characteristics as discussed in this chapter

Virus	Family	Target organs	Illness	Severity
Hepatitis A virus	Picornaviridae	Liver	Hepatitis	Variable, severity increases with age
Hepatitis E virus	Hepeviridae	Liver	Hepatitis	Mild, except in pregnant women
Enteroviruses (Poliovirus) (Echovirus) (Coxsackie virus)	Picornaviridae	Central nervous system, heart, lungs, skin	Poliomyelitis, meningitis, encephalitis, myocarditis, pericarditis, pleurodynia, hand foot and mouth disease	Variable, most infections are mild
Avian Influenza virus (High pathogenicity)	Orthomyxoviridae	Intestine, lungs	Gastroenteritis, respiratory disease	Variable, certain strains have high fatality rates in humans
Coronavirus	Coronaviridae	Intestine, lungs	Gastroenteritis, respiratory disease	Variable, one strain has had high fatality rates in humans
Tick-borne encephalitis virus	Flaviviridae	Central nervous system	Encephalitis	Severe, infections are rare

These organisms are described in the following text, with an emphasis on their potential to cause foodborne illness. Methods to detect and inactivate the major viral pathogens are discussed, with indications for control measures that can be used to limit the spread of viral disease through the food supply. Each virus has particular challenges and considerations, but general themes emerge that can be used to mitigate current and future food safety concerns.

25.2 Description of the organisms

25.2.1 Hepatitis A virus

The hepatitis A virus is a member of the Picornaviridae family, in the genus Hepatovirus (Fauquet *et al.*, 2005). Particles are non-enveloped, icosahedral capsids of approximately 30 nm in diameter (Feinstone *et al.*, 1973). They enclose a 7.5 kb single-stranded, polyadenylated RNA genome that codes for the viral structural and non-structural proteins (Cohen *et al.*, 1987).

The hepatitis A virus infects epithelial cells of the small intestine and hepatocytes in primates (Balayan, 1992). Naturally-occurring disease is seen only in the human population and causes viral hepatitis characterized by fever, jaundice, light coloured stools, dark coloured urine, abdominal pain and occasional diarrhea in older children and adults (Nainan *et al.*, 2006). Infection is generally asymptomatic or anicteric in children under six years of age (Hadler *et al.*, 1980). The infection is acute and its resolution provides lifelong immune protection against future infections, as there is only one serotype of hepatitis A (Jacobsen and Koopman, 2004; Cristina and Costa-Mattioli, 2007).

In developing regions of the world where hygienic standards (i.e. clean water, sewage systems, availability of adequate hygiene facilities and proper hygienic practices) may be below acceptable standards, children are commonly exposed to the hepatitis A virus at an early age. Since most infections in children are asymptomatic, the rates of disease are low and outbreaks are uncommon in these areas (Shapiro and Margolis, 1993). In contrast, in the developed countries, with low hepatitis A endemicity, hepatitis A infections and outbreaks are a serious public health concern (AAPCID, 2007).

The hepatitis A virus is spread via the faecal–oral route, typically by person-to-person contact or by ingestion of contaminated food or water (Brundage and Fitzpatrick, 2006). The virus has a particularly high rate of person-to-person transmission, and it is common for foodborne outbreaks of hepatitis A to be amplified when primary cases transfer the infection to members of their households (Fiore, 2004). The long incubation period between infection and symptomatic disease (15–50 d) also contributes to difficulties in accurately identifying foodborne sources of hepatitis A virus infection (Fiore, 2004). However, a large number of outbreaks have been characterized, and contamination can occur at every stage in food production, from cultivation through final preparation by food handlers.

25.2.2 Hepatitis E virus

The hepatitis E virus (HEV) is a member of the genus Hepevirus in the Hepeviridae family, which shares features with the Caliciviridae, Togaviridae and Picornaviridae families (Fauquet *et al.*, 2005). The hepatitis E particle is a non-enveloped icosahedron of approximately 30 nm in diameter (Arankalle *et al.*, 1988). The 7.5 kb genome is a single-stranded RNA molecule that codes for three open reading frames (Tam *et al.*, 1991).

A single known serotype of this virus infects humans and animal species; the virus has been isolated from a range of hosts including primates, domestic swine, wild deer, wild boar and birds (Bradley *et al.*, 1987; Meng *et al.*, 1997; Haqshenas *et al.*, 2001; Matsuda *et al.*, 2003; Tei *et al.*, 2003). Anti-HEV antibodies have been detected in an even wider range of host species, including food source animals such as cattle, sheep and goats (Meng, 2000; Lu *et al.*, 2006; Shukla *et al.*, 2007).

The hepatitis E virus causes similar clinical symptoms to those associated with hepatitis A infection. Recent analysis of case studies indicates that there are two clinical forms of hepatitis E infection: a classical type associated with large waterborne outbreaks in the developing world (Acharya and Panda, 2006) and an emerging disease associated with endemically-acquired cases in the developed world (Teo, 2007). This latter form has been detected only sporadically and may be transmitted through contaminated food products.

In classical hepatitis E infections, the cases occur predominantly in young adults (Teo, 2006). The disease presentation is generally more severe than for hepatitis A. Childhood infections are generally symptomatic (Hyams *et al.*, 1992; Tsatsralt-Od *et al.*, 2007). Hyperbilirubinaemia is more pronounced, protracted cholestasis is more common, and a larger proportion of hepatitis E patients present with jaundice (Chau *et al.*, 2006). High (10 % or more) rates of fulminant hepatitis and fatality have been reported for classical hepatitis E infections in pregnant women, particularly in the third trimester (Rab *et al.*, 1997; Boccia *et al.*, 2006).

By contrast, emerging hepatitis E infections in the developed world are more frequently associated with advanced age (Dalton *et al.*, 2007; Teo, 2007). There is no evidence that the water supply is contaminated in the developed world, and this has led to suggestions that such cases may be transmitted through food, although definitive evidence is lacking in almost all cases (Dalton *et al.*, 2007). In these cases, underlying liver disease and excessive alcohol intake are risk factors for fulminant hepatitis, and there is no reported association with pregnancy (Peron *et al.*, 2007).

25.2.3 Poliovirus

Polioviruses are classified in the Enterovirus genus of the Picornaviridae family (Fauquet *et al.*, 2005). They too form 30 nm particles containing a single-stranded RNA genome (Schwerdt and Fogh, 1957; Young, 1973). The genome is 7.5 kb in length and codes for a single, long open reading frame that is post-translationally processed to yield all of the viral proteins (Kitamura *et al.*, 1981).

Infection with poliovirus occurs following the ingestion of contaminated food or water. The virus replicates in the human intestinal tract resulting in asymptomatic infection or minor malaise in over 90 % of cases (Melnick, 1996). Mild illness characterized by fever, headache, nausea and sore throat can occur in 4–8 % of infections. If the virus spreads to infect the nervous system, a further 1–2 % of cases experience stiffness in the back and neck, as well as mild muscle weakness. The major illness, paralytic poliomyelitis, occurs in approximately 1 % of cases and consists of meningitis plus persisting weakness of one or more muscle groups. The amount of damage to neurons is highly variable (Melnick, 1996). In endemic countries, the virus affects predominantly young children, with most cases of poliomyelitis occurring in those below five years of age (Singh *et al.*, 1996).

Poliovirus was the first virus shown to be transmitted through food, and a number of outbreaks were associated with raw milk prior to the introduction of routine pasteurization (Dingman, 1916; Sullivan and Read, 1968). Polio can be prevented using a live attenuated vaccine administered orally or by an injectable killed vaccine (Melnick, 1996). As a result of the poliovirus eradication campaign led by the World Health Organization (WHO), infection with this virus has been eradicated or significantly reduced in many regions of the world and, as such, it is of questionable importance for routine food safety considerations (Lahariya, 2007). However, due to its historic importance and the early development of an attenuated vaccine strain, there have been many studies on the spread and inactivation of poliovirus in food products.

25.2.4 Avian Influenza virus

The Influenza virus A genus is classified in the family Orthomyxoviridae (Fauquet *et al.*, 2005). This family is characterized by pleiomorphic, enveloped virions with a segmented single-stranded RNA genome (Fauquet *et al.*, 2005). The genome is 14 kb of negative-sense RNA in eight segments, and its complement codes for structural and non-structural proteins (Fauquet *et al.*, 2005). They are classified into sub-types on the basis of the two envelope glycoproteins, the hemagglutinase (H type) and the neuraminidase (N type) (Knossow and Skehel, 2006). There are 16 known H types and 9 known N types, which can theoretically be mixed and matched in all possible combinations (Knossow and Skehel, 2006).

All of the known sub-types of Influenza A viruses (82 H/N combinations out of a possible 144) have been found in wild birds (Van Reeth, 2007). Viruses containing combinations of the H1, H2, H3, N1 and N2 types are considered to be established in the human population; i.e. there are circulating strains of these viruses that predominantly infect humans (Hay *et al.*, 2001). In addition, viruses of the H5, H7 and H9 sub-types have been found to cause sporadic human infections (Peiris *et al.*, 1999; Shortridge *et al.*, 2000;

Koopmans *et al.*, 2004). The subtype of recent worldwide concern is the highly pathogenic H5N1 virus (Abdel-Ghafar *et al.*, 2008). This particular avian influenza virus sub-type has been detected in poultry from over 50 countries on three continents (Van Reeth, 2007). It has infected humans in Vietnam, Thailand, Indonesia, Cambodia, China, Turkey, Iraq, Azerbaijan, Egypt and Djibouti (de Jong and Hien, 2006). This virus replicates to abnormally high levels in the upper respiratory tract, causing an intense inflammatory response and an extremely high case fatality rate of over 60 % (de Jong *et al.*, 2006). The H5N1 avian influenza virus currently has a limited ability to spread from person to person. However, there is concern that this virus could acquire the ability to spread effectively in humans, leading to a worldwide pandemic (Rajagopal and Treanor, 2007). The Spanish influenza pandemic, which killed approximately 50 million people worldwide is thought to have arisen from an avian H1N1 strain (Rajagopal and Treanor, 2007).

To date, H5N1 avian influenza infections of humans have nearly all been linked to close contact between the afflicted individuals and infected poultry (de Jong and Hien, 2006; Van Reeth, 2007; Abdel-Ghafar *et al.*, 2008). However, the virus can be isolated from all parts of infected poultry including the blood, bones and meat (Lu *et al.*, 2003b; Swayne, 2006a). As a result, the consumption of raw or undercooked poultry products represents a potential source of infection (Swayne, 2006b).

25.2.5 Coronavirus

The members of the Coronaviridae family are pleiomorphic, enveloped, single-stranded RNA viruses (Fauquet *et al.*, 2005). The positive-sense, 30 kb genome codes directly for viral proteins (Fauquet *et al.*, 2005). These viruses typically cause mild respiratory disease; however, a particularly virulent strain known as the Sudden Acute Respiratory Syndrome Coronavirus (SARS-CoV) emerged in 2003 and caused over 8000 cases, with a nearly 10 % case fatality rate (Wang and Chang, 2004). This virus caused systematic infections as well as respiratory illness, and was identified in the digestive tract, as well as in faeces and sewage (Chan-Yeung and Xu, 2003; Zhang, 2003; Wang *et al.*, 2005). This raises the possibility that the virus may have had the potential to spread through the faecal–oral route and food products. Although these viruses are able to persist for days in some buffered media, they are sensitive to heating, UV light and disinfection (Duan *et al.*, 2003; Rabenau *et al.*, 2005a, b). No cases of foodborne spread were documented during this outbreak, although one cluster of cases in a housing complex was linked to an index patient suffering from diahrroea (McKinney *et al.*, 2006).

25.2.6 Other enteroviruses

The echoviruses and coxsackieviruses are enteroviruses of the family Picornaviridae and share many features with the polioviruses. They have 30 nm, non-enveloped particles enclosing a positive-sense single-stranded RNA genome (Fauquet *et al.*, 2005). These viruses are common and infections are mostly asymptomatic. They can occasionally spread outside of the gastrointestinal tract to cause aseptic meningitis, myocarditis or encephalitis. Infection of newborns with these viruses can lead to serious or fatal infection (Abzug, 2004). These viruses are transmitted via the faecal–oral route, but are not frequently associated with foodborne illness. A recent report identified milk from an infected mother as the source of coxsackievirus B infection (Chang *et al.*, 2006).

25.2.7 Tickborne Encephalitis virus

The tickborne encephalitis virus is an enveloped, single-stranded RNA virus of the family Flaviviridae (Fauquet *et al.*, 2005). It is endemic to Europe and Russia, causing 10 000–12 000 cases of encephalitis every year, with a case fatality rate of 0.5 % (Gunther and Haglund, 2005). Survivors can develop long-term neurological sequelae at rates of up to 40 % (Dumpis *et al.*, 1999). The majority of cases are transmitted by tick bites, but some have been associated with the consumption of raw milk from infected cattle or goats (Sixl *et al.*, 1989; WHO, 1994; Dumpis *et al.*, 1999; Kerbo *et al.*, 2005).

25.2.8 Emerging viruses of gastroenteritis

Additional emerging viruses of interest cause symptoms of gastroenteritis. These include the parvoviruses, toroviruses and picobirnaviruses. See Chapter 32 for a detailed description of the causative agents of viral gastroenteritis.

25.3 Risk factors

25.3.1 Hepatitis A virus

There is an inverse correlation between hepatitis A disease rates and viral endemicity (Shapiro and Margolis, 1993). The Centres for Disease Control and Prevention in the USA has classified regions according to high (Central and South America, Africa, South Asia, Greenland), intermediate (Russia and Eastern Europe) and low (North America, Western Europe, Scandinavia, Australia) seroprevalence rates for hepatitis A (CDC, 2007). This is important because many of the cases and outbreaks of hepatitis A in developed countries can be traced to travel in endemic countries (Steffen, 2005). There may also be a higher risk for food produced in the highly endemic countries to be contaminated with the hepatitis A virus (Fiore, 2004).

Disease rates in low endemicity countries are difficult to generalize. They are characterized by cyclical peaks and troughs, and they are changing in response to recent vaccination campaigns (Pham *et al.*, 2005; Wasley *et al.*,

2005). Undiagnosed children may be an important source of the hepatitis A virus in communities, as a high percentage of hepatitis A infections have no known source (Staes *et al.*, 2000; Nainan *et al.*, 2005). There has been a safe and effective vaccine available against hepatitis A since the mid-1990s; however, this virus continues to cause outbreaks, with contaminated food being one important source of infection (Fiore *et al.*, 2006; Craig *et al.*, 2007).

Foodborne outbreaks of hepatitis A have been associated with many different food types and different settings (see Table 25.2 for a summary of some outbreaks published over the past 10 years). Viruses are inert particles when present outside their host. The spread of hepatitis A infection through contaminated food, water or fomites is therefore dependent on the persistence of the virus after the introduction of contaminated human waste.

Documented outbreaks of hepatitis A due to shellfish have been caused by faecal contamination of shellfish growing waters (Conaty *et al.*, 2000; Bosch *et al.*, 2001; Bialek *et al.*, 2007; Pontrelli *et al.*, 2007; Shieh *et al.*, 2007). The hepatitis A virus survives for up to one month in dried faecal matter (McCaustland *et al.*, 1982). In mixtures of human and animal waste designed for waste treatment prior to disposal, it takes 7 d to reduce viral titre by 1 $\log_{10}$ at 37 °C (Deng and Cliver, 1995). Once introduced into the water environment, the hepatitis A virus also remains infectious for months, associating with marine sediment (Arnal *et al.*, 1998; Bosch, 1998). The hepatitis A virus has been shown experimentally to accumulate in mussels and oysters, and depuration is not effective at eliminating the virus from the shellfish tissue (Franco *et al.*, 1990; Enriquez *et al.*, 1992; Abad *et al.*, 1997b; De Medici *et al.*, 2001; Kingsley and Richards, 2003). Extensive depuration for more than three weeks was required to completely eliminate detectable hepatitis A virus from oyster tissue (Kingsley and Richards, 2003).

Fresh or frozen produce is another important source of foodborne hepatitis A outbreaks (Hutin *et al.*, 1999; Dentinger *et al.*, 2001; Calder *et al.*, 2003; CDC, 2003b; Amon *et al.*, 2005; Wheeler *et al.*, 2005; Frank *et al.*, 2007). In these reports, the human fecal contamination might have been introduced during produce growth (Amon *et al.*, 2005), harvest (Calder *et al.*, 2003), processing (Frank *et al.*, 2007) or distribution. Consequently, the point of viral contamination can be difficult to identify (Hutin *et al.*, 1999; Dentinger *et al.*, 2001; CDC, 2003b; Howitz *et al.*, 2005; Wheeler *et al.*, 2005; Tekeuchi *et al.*, 2006). Once fresh produce is contaminated, the hepatitis A virus adsorbs to its surface and persists for days (Croci *et al.*, 2002; Stine *et al.*, 2005). Freezing allows the virus to survive for months to years (Niu *et al.*, 1992). In addition, washing the contaminated produce does not usually substantially reduce the level of contamination (Croci *et al.*, 2002).

The remaining source of reported hepatitis A outbreaks are infected food handlers (CDC, 2003a; Prato *et al.*, 2006; Schenkel *et al.*, 2006; Hasegawa *et al.*, 2007). Hepatitis A virus has been shown to persist on experimentally-contaminated hands for more than 4 h (Mbithi *et al.*, 1992). The virus was

Table 25.2 Selected foodborne outbreaks of hepatitis A virus

Year	# of cases	Country	Food type	Contamination	Reference
1997	467	Australia	Oysters	Growth water	Conaty *et al.*, 2000
1997	256	USA	Frozen strawberries	Unidentified	Hutin *et al.*, 1999
1998	43	USA	Green onions	Prior to restaurant delivery (unknown)	Dentinger *et al.*, 2001
2001	183	Spain	Clams	Growth water	Bosch *et al.*, 2001
2001	46	USA	Sandwiches	Food handler	CDC, 2003a
2002	26	Italy	Various	Food handler	Prato *et al.*, 2006
2002	43	New Zealand	Raw blueberries	Harvest	Calder *et al.*, 2003
2003	601	USA	Green onions	Prior to restaurant delivery (unknown)	CDC, 2003b; Wheeler *et al.*, 2005
2003	422	USA	Green onions	Farm	Amon *et al.*, 2005
2004	64	Germany	Baked goods	Food handler	Schenkel *et al.*, 2006
2004	1180	India	Water	Sewage contamination	Arankalle *et al.*, 2006
2004	884	Italy	Raw shellfish	Growth water	Pontrelli *et al.*, 2007
2004	351	Egypt	Orange juice	Processing plant	Frank *et al.*, 2007
2005	4	Denmark	Ice cream	Unknown	Howitz *et al.*, 2005
2005	39	USA	Oysters	Growth water	Bialek *et al.*, 2007; Shieh *et al.*, 2007
2006	9	Japan	Sushi bar (undetermined)	Unknown	Tekeuchi *et al.*, 2006
2006	15	Japan	Restaurant (undetermined)	Food handler	Hasegawa *et al.*, 2007

readily transferred between inoculated fingers, to foods or stainless steel surfaces and from stainless steel surfaces (Mbithi *et al.*, 1992; Bidawid *et al.*, 2000a). The virus has been shown to attach to many of the surfaces common in food preparation settings, i.e. stainless steel, copper, polythene and polyvinyl chloride (Kukavica-Ibrulj *et al.*, 2004) and to survive for more than 60 d on aluminium, china, latex and paper surfaces (Abad *et al.*, 1994).

The most effective prevention of hepatitis A virus transmission is to prevent faecal contamination of foods and food preparation surfaces (see Section 25.5.1). In addition, there are some physical and chemical procedures known to effectively eliminate the hepatitis A virus from contaminated foods and surfaces (Tables 25.3 and 25.4)

The hepatitis A virus is resistant to acid treatment, both in buffered solutions and in food products (Scholz *et al.*, 1989; Hewitt and Greening, 2004). Heating is an effective mode of viral inactivation, but higher temperatures than those generally used to reduce bacterial counts are required (Parry and Mortimer, 1984). Protection of the virus is also conferred by food matrices, and longer times and/or higher temperatures are required to completely inactivate the virus (Croci *et al.*, 1999; Bidawid *et al.*, 2000d; Deboosere *et al.*, 2004). For example, increasing the fat content (Bidawid *et al.*, 2000d) or the sucrose concentration (Deboosere *et al.*, 2004) of food preparations has been shown to increase the heat resistance of the virus. This may be due to a general effect of decreasing the water activity, but this possibility has not been systematically addressed. The result is that many traditional heat-inactivation or cooking protocols, such as pasteurization, steaming or baking, are insufficient to ensure the safety of hepatitis A-contaminated foods (Bidawid *et al.*, 2000d; Croci *et al.*, 2005; Hewitt and Greening, 2006).

Heating to 85 °C is not possible for fresh fruits and vegetables, and even for milk and shellfish the procedures required may render foods unpalatable. Surface decontamination may be possible using targeted applications. Alternative physical inactivation methods show some promise for the inactivation of hepatitis A in food products. The virus is sensitive to UV light, but the light must access viral particles to inactivate them, which makes the technique difficult to apply to most food products (Nuanualsuwan *et al.*, 2002). High hydrostatic pressure is a newer technique that has the potential to inactivate the hepatitis A virus in food products while maintaining the organoleptic properties of raw products, particularly shellfish. The technique shows promise for the inactivation of hepatitis A virus in buffer, oysters, strawberries and green onions, although high pressure may reduce the palatability of fresh fruits and vegetables (Kingsley *et al.*, 2002, 2005; Calci *et al.*, 2005).

Chemical methods are frequently used to eliminate virus dried on hands or on surfaces. The hepatitis A virus is more resistant to chemical disinfection than many enteric bacteria and enveloped viruses. Many commercially-available disinfectants are not effective when used on hepatitis A virus-contaminated

Table 25.3 Physical inactivation of enteric viruses discussed in this chapter

Virus	Matrix	Treatment	Time	Log reduction	Conclusion	Reference
Hepatitis A virus	Buffer	pH 1.0 (RT[a])	5 h	5	Acid stable	Scholz *et al.*, 1989
	Mussels	pH 3.75 (4 °C)	4 wk	None	Acid stable	Hewitt and Greening, 2004
	Buffer (pH 7.4)	75 °C	30 s	5	Inactivated by high heat	Parry and Mortimer, 1984
	Milk	75 °C	7 min	5	Protected by matrix	Bidawid *et al.*, 2000d
	Mussels	80 °C	15 min	5	Protected by matrix	Croci *et al.*, 1999
	Strawberry mash	85 °C	10 min	5	Protected by matrix	Deboosere *et al.*, 2004
	Buffer (pH 7.4)	36.5 mW/cm^2 UV light (RT)	1 s	1	UV sensitive	Nuanualsuwan *et al.*, 2002
	Strawberries	3 kGy gamma irradiation (RT)	N/A	1	Higher dose than regulators allow	Bidawid *et al.*, 2000b
	Buffer (pH 7.4)	450 MPa (RT)	5 min	7	Pressure-sensitive	Kingsley *et al.*, 2002
	Oysters	400 MPa (9 °C)	1 min	3	Potential for use in matrix	Calci *et al.*, 2005
	Strawberry puree	375 MPa (RT)	5 min	4	Potential for use in matrix	Kingsley *et al.*, 2005
	Green onions	375 MPa (RT)	5 min	5	Potential for use in matrix	Kingsley *et al.*, 2005
Hepatitis E virus	Buffer (pH 7.4)	56 °C	60 min	2	Moderately heat-resistant	Emerson *et al.*, 2005
	Liver	56 °C	60 min	Incomplete[b]	Moderately heat-resistant	Feagins *et al.*, 2008
	Liver	71 °C	5 min	Complete[b]	Moderately heat-resistant	Zafrullah *et al.*, 2004
Poliovirus	Buffer	pH 1.0 (RT)	1 min	4	Unstable	Scholz *et al.*, 1989
	Buffer	pH 3.0 (RT)	30 min	0	Moderately acid stable	Eubanks and Farrah, 1981
	Buffer (pH 7.4)	50 °C	2 min	4	Heat-sensitive	McGregor and Mayor, 1971
	Septage	55 °C	15 min	5	Heat-sensitive	Stramer and Cliver, 1984
	Milk	72 °C	15 s	<4	Resistant to some pasteurization	Strazynski *et al.*, 2002
	Yoghurt	72 °C	15 s	<4	Resistant to some pasteurization	Strazynski *et al.*, 2002
	Buffer (pH 7.4)	24 mW/cm^2 UV light (RT)	1 s	0.5–1	UV sensitive	Ma *et al.*, 1994; Nuanualsuwan *et al.*, 2002

Table 25.3 (Cont'd)

Virus	Matrix	Treatment	Time	Log reduction	Conclusion	Reference
Avian Influenza virus	Buffer	pH 2.0 (RT)	5 min	5	Acid labile	Lu *et al.*, 2003a
	Buffer	pH 5.0 (RT)	15 min	0	Stable	Lu *et al.*, 2003a
	Buffer (pH 7.4)	63 °C	2 min	5	Heat-sensitive	Isbarn *et al.*, 2007
	Whole egg	63 °C	2 min	5	Heat-sensitive	Swayne and Beck, 2004
	Dried egg white	63 °C	1 d	5	Protected by matrix	Swayne and Beck, 2004
	Chicken meat	70 °C	2 s	5	Heat-sensitive	Thomas and Swayne, 2007
	Buffer (pH 7.4)	500 MPa (15 °C)	15 s	5	Pressure-sensitive	Isbarn *et al.*, 2007
SARS coronavirus	Buffer (pH 7.4)	56 °C	30 min	6	Heat-sensitive	Duan *et al.*, 2003; Rabenau *et al.*, 2005a
	Buffer (pH 7.4) + 20 % fetal calf serum	56 °C	30 min	2	Protein protects virus	Rabenau *et al.*, 2005a
	Buffer (pH 7.4) + 20 % fetal calf serum	60 °C	30 min	6	Heat-sensitive	Rabenau *et al.*, 2005a
	Buffer (pH 7.4)	90 $\mu W/cm^2$ UV light (RT)	60 min	6	UV-sensitive	Duan *et al.*, 2003

[a] Room temperature.
[b] Initial titre of virus stock not available.

Table 25.4 Chemical disinfection/inactivation of enteric viruses discussed in this chapter

Virus	Active compound	Contact time	Log reduction	Conclusion	Reference
Hepatitis A virus	Quaternary ammonium	5 min	< 2	Not effective	Mbithi *et al.*, 1990 (0.1–2.7 %); Jean *et al.*, 2003 (10 %)
	Glutaraldehyde	1 min	> 4	Effective	Mbithi *et al.*, 1990 (2 %)
	Sodium hydroxide	1 min	1	Not effective	Terpstra *et al.*, 2007 (0.1 N)
	Ethanol	1–5 min	< 1	Not effective	Abad *et al.*, 1997a (70 %); Mbithi *et al.*, 1990 (70 %); Bidawid *et al.*, 2000a (75 %), van Engelenburg *et al.*, 2002 (80 %)
	Iodide	1–5 min	< 1	Not effective	Mbithi *et al.*, 1990 (0.07 %, 75 ppm); Jean *et al.*, 2003 (2 %)
	Phenol	1 min	1–2	Not effective	Mbithi *et al.*, 1990 (0.1 %), Abad *et al.*, 1997a (1.41 %)
	Sodium hypochlorite	1 min	< 1–5	5000 ppm free chlorine and full 1 minute contact time are essential	Grabow *et al.*, 1983 (0.4 mg/l); Abad *et al.*, 1997a (0.125 %); Jean *et al.*, 2003 (12 %); Terpstra *et al.*, 2007 (0.1 %, 1000 ppm); Mbithi *et al.*, 1990 (5000 ppm)
	Sodium chlorite	1 min	< 1–3	Time and concentration dependent, not as effective as sodium hypochlorite	Mbithi *et al.*, 1990 (0.23 %); Abad *et al.*, 1997a (30 %)
	Hypochlorous acid	1 min	< 1	Not effective	Bigliardi and Sansebastiano, 2006 (1.6 mg/l)
Hepatitis E virus	Sodium hypochlorite	30 min	Incomplete	0.6 ppm free chlorine residual in drinking water is not effective	Guthmann *et al.*, 2006 (0.6 mg/l)

Table 25.4 (Cont'd)

Virus	Active compound	Contact time	Log reduction	Conclusion	Reference
Poliovirus	Quaternary ammonium	30 s		Effective	Weber *et al.*, 1999 (0.06 %)
	Glutaraldehyde	3 h	5	Effective	Stramer and Cliver, 1984 (1 mg/ml)
	Ethanol	1–2 min	1–4	Moderately effective	Kramer *et al.*, 2006 (70 %); Mbithi *et al.*, 1993 (62 %); Abad *et al.*, 1997a (70 %)
	Phenol	1 min	<1	Not effective	Weber *et al.*, 1999 (0.06 %); Abad *et al.*, 1997a (1.41 %)
	Sodium hypochlorite	1 min	4	Effective	Weber *et al.*, 1999 (5000 ppm); Ma *et al.*, 1994 (0.5 mg/l); Lukasik *et al.*, 2003 (300 ppm); Abad *et al.*, 1997a (0.12 %)
	Sodium chlorite	1 min	< 1	Not effective	Abad *et al.*, 1997a (30 %)
	Hypochlorous acid	1 min	< 1	Not effective	Bigliardi and Sansebastiano, 2006 (1.6 mg/l)
Avian Influenza virus	Quaternary ammonium	1 min	5	Effective	Suarez *et al.*, 2003 (0.06 %)
	Formaldehyde	1 min	5	Effective	King, 1991 (0.04 %)
	Ethanol	15 min	5	Effective	Lu *et al.*, 2003a (70 %)
	Phenol	1 min	5	Effective	Suarez *et al.*, 2003 (0.06 %)
	Sodium hypochlorite	1 min	5	Effective	Suarez *et al.*, 2003 (0.4 %)
SARS coronavirus	Ethanol	15 s	4	Effective	Rabenau *et al.*, 2005a (2 %); Rabenau *et al.*, 2005b (2 %)
	Benzalkonium chloride	30 min	4	Effective	Rabenau *et al.*, 2005b (0.5 %);
	Glutaraldehyde	15 min	4	Effective	Rabenau *et al.*, 2005a (0.5 %); Rabenau *et al.*, 2005b (0.5 %)

surfaces (see Table 25.4) (Mbithi *et al.*, 1990; Abad *et al.*, 1997a; Bidawid *et al.*, 2000a; van Engelenburg *et al.*, 2002; Jean *et al.*, 2003; Bigliardi and Sansebastiano, 2006; Terpstra *et al.*, 2007). Notably, the active ethanol component of many commercial hand disinfectants has a very limited ability to reduce infectious hepatitis A virus titre from contaminated hands or surfaces (Mbithi *et al.*, 1990; Bidawid *et al.*, 2000a). The most effective disinfectants are 2 % glutaraldehyde and 10 % sodium hypochlorite (5000 ppm free chlorine) (Grabow *et al.*, 1983; Mbithi *et al.*, 1990; Abad *et al.*, 1997a; Jean *et al.*, 2003). Care must always be taken to follow appropriate time/concentration combinations for effective disinfection.

25.3.2 Hepatitis E virus

The hepatitis E virus is considered to be an endemic human pathogen in Central America, North Africa and South Asia (Panda *et al.*, 2007). It causes epidemic outbreaks of acute, enterically transmitted hepatitis in these countries, usually associated with faecally-contaminated water (Rab *et al.*, 1997; Guthmann *et al.*, 2006; Panda *et al.*, 2007). Such outbreaks have been particularly associated with the Indian subcontinent, although this may reflect a reporting bias rather than the true level of incidence (Panda *et al.*, 2007).

During these classical outbreaks in endemic countries, person-to-person transmission does not contribute greatly to the number of cases (Hla *et al.*, 1985; Somani *et al.*, 2003). The viral genome has been detected in raw and treated sewage, and outbreaks have been associated with chlorinated water supplies, indicating that basic treatment designed to eliminate faecal coliforms may not be sufficient to inactivate the hepatitis E virus (Jothikumar *et al.*, 1993; Guthmann *et al.*, 2006). Sporadic cases of hepatitis E infection also occur in endemic countries and are likely contribute to the spread of the virus in the population (Nanda *et al.*, 1994).

In Western Europe, North America, Japan and Australia, the hepatitis E virus is an emerging concern (Teo, 2006). Hepatitis E infections in these countries have historically been associated with travel to endemic regions (Zaaijer *et al.*, 1993; Skidmore and Sherratt, 1996). Recently, however, there have been reports of locally-acquired cases in developed countries, but the source of these infections remains uncertain (Mansuy *et al.*, 2004; Ijaz *et al.*, 2005; Waar *et al.*, 2005; Reuter *et al.*, 2006; Dalton *et al.*, 2007; Perez-Gracia *et al.*, 2007; Peron *et al.*, 2007). In two cases, the disease has been linked to the consumption of raw or undercooked meat from animals naturally infected with hepatitis E (Matsuda *et al.*, 2003; Tei *et al.*, 2003). This has led to the hypothesis that the hepatitis E virus is an emerging foodborne zoonosis in developed nations (Teo, 2006).

The hepatitis E virus is known to infect a wide range of animal species (Goens and Perdue, 2004; Vasickova *et al.*, 2007). Most of the studies and evidence for zoonotic transmission to humans have focused on strains isolated from swine (Goens and Perdue, 2004). Studies have shown that these strains are very closely related to the human viruses and that they productively infect primates (Meng *et al.*, 1997, 1998). Phylogenetic analysis indicates that swine viruses are more closely related to their human counterparts within a country than to other swine isolates around the world (Meng *et al.*, 1997; Hsieh *et al.*, 1999; van der Poel *et al.*, 2001; Pei and Yoo, 2002; Nishizawa *et al.*, 2003; Banks *et al.*, 2004). Zoonotic transmission of this virus to the human population has been postulated to occur through the consumption of raw or lightly cooked meat from a naturally infected animal (Teo, 2006; Vasickova *et al.*, 2007). The evidence for this is strong in a few cases in Japan, but weak elsewhere (Matsuda *et al.*, 2003; Tei *et al.*, 2003). However, infectious hepatitis E virus and/or viral RNA has been isolated from commercial pig livers in the USA, the Netherlands and Japan, indicating that this is a possible transmission route (Yazaki *et al.*, 2003; Bouwknegt *et al.*, 2007; Feagins *et al.*, 2007).

Testing the physical and chemical stability of the hepatitis E virus is challenging due to the lack of a cell culture system to propagate the virus. As such, the pH stability of this virus has yet to be formally demonstrated. The capsid protein acquires an increased heat stability at low pH, and the virus is known to infect via the gastrointestinal tract, implying some resistance to acidic conditions (Zafrullah *et al.*, 2004). The virus resists inactivation by heating at 56 °C, but heating to 71 °C completely inactivated the virus in naturally-contaminated pig livers (Emerson *et al.*, 2005; Feagins *et al.*, 2008). Other physical and chemical methods of inactivation have not been tested, although an outbreak linked to chlorinated drinking water indicates that free chlorine residuals known to reduce fecal coliforms are not sufficient to inactivate the hepatitis E virus (Guthmann *et al.*, 2006).

25.3.3 Poliovirus

The poliovirus is no longer a prevalent enteric pathogen around the world. The widespread use of the oral polio vaccine, which confers intestinal immunity, has led to the eradication of the virus from three of the six regions of the world as defined by the WHO (Sabin, 1991; Lahariya, 2007). These three regions are: the Americas, representing 35 countries, polio-free since 1994; the Western Pacific, representing 37 countries and territories, polio-free since 2000; and the European, representing 51 countries, polio-free since 2002 (CDC, 1994, 2001, 2002). The disease is still endemic in Afganistan, India, Nigeria and Pakistan (Lahariya, 2007). In recent years, there have also been other countries within the WHO regions of Africa, South-East Asia and the Eastern Mediterranean that have experienced outbreaks of poliomyelitis caused by both wild and vaccine-derived strains (Melnick, 1996; Chumakov *et al.*, 2007). Due to the prevalence of asymptomatic disease, the absence of poliomyelitis does not always correlate with the absence of the poliovirus, and intensive vaccination campaigns are still warranted (Chumakov *et al.*,

2007). Individuals in polio-free countries still routinely follow a course of vaccination with the killed vaccine (Wood *et al.*, 2000, 2006).

Similarly to the hepatitis A virus, the poliovirus can be introduced at any stage in the farm-to-fork continuum. The virus has a low infectious dose (Plotkin *et al.*, 1959) and is frequently isolated from sewage during outbreaks and in endemic countries (Arya and Agarwal, 2007). Once again, shellfish, fresh produce and food handlers are the most significant sources of foodborne virus transmission.

Poliovirus accumulates readily in shellfish and concentrates in the digestive tract (Di Girolamo *et al.*, 1975). Depuration of poliovirus is more effective than for hepatitis A, and most of the virus can be eliminated in free-flow depuration system after 5 d (Di Girolamo *et al.*, 1975; Franco *et al.*, 1990; Enriquez *et al.*, 1992).

The stability and accumulation of poliovirus is of significant concern in agricultural systems. The poliovirus is inactivated more readily than the hepatitis A virus by heating and storage treatments used to prepare manure for spreading on lands (Stramer and Cliver, 1984; Deng and Cliver, 1992). However, once the environment is contaminated, the poliovirus survives for weeks to months in groundwater (Yates *et al.*, 1985; Gordon and Toze, 2003) and in soil (Yeager and O'Brien, 1979; Hurst *et al.*, 1980). This virus has also been shown to persist for weeks to months on vegetables irrigated by spraying or flooding with contaminated waters (Tierney *et al.*, 1977). It has also been demonstrated to survive for weeks to months on fresh and frozen produce, and simple washing does not appear to effectively eliminate poliovirus from food surfaces (Kurdziel *et al.*, 2001; Lukasik *et al.*, 2003).

Poliovirus is moderately acid stable, resistant to incubation at pH 3.0 for 30 min, but sensitive to pH 1.0 (Eubanks and Farrah, 1981; Siegl *et al.*, 1984; Scholz *et al.*, 1989). Heating is extremely effective for the inactivation of poliovirus in buffered solutions, although food matrices may provide some protection and require longer heating times (see Table 25.3) (McGregor and Mayor, 1971; Milo, 1971; Stramer and Cliver, 1984; Strazynski *et al.*, 2002). This may be due to protection provided by protein, fat, lowered a_w or a combination of these parameters. Both UV light and ozone are effective in elimination of the poliovirus, and have been considered for the treatment of wastewater to be used in irrigation (Ma *et al.*, 1994; Nuanualsuwan *et al.*, 2002; Lazarova and Savoys, 2004; Tanner *et al.*, 2004).

Food handlers with inadequate hygiene are frequently implicated in the transmission of viruses via the fecal–oral route. The virus can be transferred from contaminated hands to surfaces, where it resists drying and can persist for days to weeks (Mbithi *et al.*, 1993; Abad *et al.*, 2001). In a food service environment, there are a number of chemicals available to disinfect hands and surfaces potentially contaminated with the poliovirus (Table 25.4). Reagents that implement quaternary ammoniums, glutaraldehyde or sodium hypochlorite as the active ingredient are effective against poliovirus on surfaces (Stramer and Cliver, 1984; Ma *et al.*, 1994; Abad *et al.*, 1997a; Weber *et al.*, 1999;

Lukasik *et al.*, 2003). The poliovirus is removed by soap and water hand-washing for 5 min, but ethanol-based hand disinfectants (60–70 % ethanol) are only moderately effective against the virus (Schurmann and Eggers, 1985; Mbithi *et al.*, 1993; Kramer *et al.*, 2006).

25.3.4 Avian Influenza virus

The WHO coordinates a Global Influenza Surveillance Network, with 119 National Influenza Centres in 90 countries (WHO, 2007). These centres monitor the incidence of influenza, identify circulating strains and constantly update the risk for the emergence of a pandemic strain (Stohr, 2003). The highly pathogenic H5N1 influenza virus of recent public health significance has caused a limited number of human infections worldwide, but has been circulating in poultry since 2003 with no signs of abating (ECDC, 2007). These natural infections continue to pose a potential risk to humans who interact with infected flocks or who consume raw or undercooked poultry products (Swayne, 2006b).

All influenza viruses are predominantly spread via aerosolized droplets that enter the respiratory tract of a susceptible host. Experimental models have shown that this transmission is most efficient at low temperatures and low relative humidity, in correlation with the winter seasonal peaks in influenza infections (Lowen *et al.*, 2007). The virus can persist for days when exposed to levels of solar radiation predicted for wintertime in temperate regions (Sagripanti and Lytle, 2007). The H5 strains have been shown to persist for weeks or months in water (Brown *et al.*, 2007). Although the influenza virus is enveloped, it resists drying to some extent and exhibits a 1 log reduction in 24 h on stainless steel surfaces (Noyce *et al.*, 2007).

There is also a risk that the avian influenza viruses will spread through contaminated food products. Natural infections in chickens and ducks have been demonstrated to produce contaminated meat, blood and bone (Lu *et al.*, 2003b; Swayne, 2006a; Thomas and Swayne, 2007). The efficacy of physical and chemical inactivation methods has been shown to vary with influenza virus type, and not all methods have been tested with the relevant H5N1 strain (De Benedictis *et al.*, 2007). Tables 25.3 and 25.4 summarize the information available.

The influenza viruses have a highly variable response to acid (Scholtissek, 1985a, b). The viruses are sensitive to heating, in buffer and in poultry products (Swayne and Beck, 2004; Swayne, 2006a, Thomas and Swayne, 2007). Although the highly pathogenic H5 strains have an increased heat resistance compared to low pathogenic avian influenza strains, they are inactivated at 70 °C, a temperature to which poultry is usually cooked (Swayne and Beck, 2004; Swayne, 2006a; Thomas and Swayne, 2007). High-pressure treatment at 500 MPa has also been shown to be an effective means of inactivating the viruses (Isbarn *et al.*, 2007).

Influenza viruses are enveloped, which confers susceptibility to many types of disinfectants (De Benedictis *et al.*, 2007). Some of the agents specifically tested using avian influenza strains are listed in Table 25.4 (King, 1991; Lu *et al.*, 2003a; De Benedictis *et al.*, 2007; Suarez *et al.*, 2003).

25.4 Detection methods

25.4.1 Overview

Classical virus detection methods involve the inoculation of cell cultures to amplify and detect infectious virus particles. These methods are rarely applicable to the isolation of foodborne viruses. Many of the important viruses do not grow in cell culture and the food extracts that need to be tested may be toxic to the cells. For example, although a model system exists for the study of inactivation kinetics, wild-type isolates of the hepatitis A virus grow very slowly, if at all, in culture and accumulate many mutations during this process (Cromeans *et al.*, 1987; Konduru and Kaplan, 2006). A recently developed hepatitis E culture system takes 60 d to amplify the virus to high levels (Tanaka *et al.*, 2007). Alternative methods that rely on detection of the viral particle, such as electron microscopy and enzyme-linked immunosorbent assays, typically have detection limits on the order of 10^5 particles per gram of sample, while the infectious dose for foodborne viral infections has been estimated to be as few as 10–100 particles (Fiore, 2004; Koopmans and Duizer, 2004).

The advent of molecular methods for the detection of viral genomes has provided the increased detection sensitivity required to allow for a more accurate assessment of the viral contamination of food products (Sanchez *et al.*, 2007). Although these methods do not distinguish between infectious and non-infectious particles, their advantages are so great that they are now used for the detection of all viruses in many food virology laboratories (Jothikumar *et al.*, 2006; Sanchez *et al.*, 2007). Molecular techniques also have the advantage that they provide information on the genotype of the strains involved in outbreaks (Nainan *et al.*, 2006). This information can be useful in establishing an epidemiological link between cases of foodborne illness (Hutin *et al.*, 1999). Microarrays are also being developed that could provide genotyping without the need for sequencing amplicons (Pagotto *et al.*, 2008). The procedures outlined in the following sections describe viral extraction methods that are specific to different food types. In most cases, the extracted virus can be subsequently detected by a conventional, immunological or molecular method, as desired.

25.4.2 Detection of viruses in shellfish

Methods to extract viruses from shellfish have been extensively studied. Most methods begin by homogenizing the shellfish tissue prior to viral extraction (Sanchez *et al.*, 2007). Changing pH can be used to concentrate enteroviruses from oysters (Sobsey *et al.*, 1975). At low pH, virus adsorbs to shellfish tissue and can be concentrated by centrifugation. At high pH, the virus is eluted from the tissue. Ultrafiltration or ultracentrifugation are then able to concentrate the viral particles in solution. The main disadvantage of this type of system is that many contaminants remain in solution with the extracted virus and may interfere with downstream detection by either cell culture or molecular methods (Speirs *et al.*, 1987). Additional concentration steps using organic flocculation or polyethylene glycol precipitation can improve the detection efficiency (Traore *et al.*, 1998). Extremely sensitive detection is obtained by lysing the viral particles with a phenol-guanidinium chloride reagent and purifying the genome using magnetic poly(dT) beads (Kingsley and Richards, 2001). This method has been standardized and published in the Health Canada Compendium of Analytical Methods for the microbiological analysis of potentially contaminated shellfish (Trottier *et al.*, 2006).

25.4.3 Detection of viruses in other foods

The development of procedures to isolate viruses from non-shellfish food samples has been more recent. Methods for viral isolation from fruits and vegetables employ washes or elution from the food surface, because homogenization releases many inhibitors of molecular reactions (Sanchez *et al.*, 2007). Methods have been developed mainly for leafy greens, green onions and berries using a variety of concentration procedures (Bidawid *et al.*, 2000c; Shan *et al.*, 2005; Guevremont *et al.*, 2006; Rzezutka *et al.*, 2006; Butot *et al.*, 2007; Papafragkou *et al.*, 2008). Polyethylene glycol precipitation (Guevremont *et al.*, 2006), immunomagnetic concentration (Bidawid *et al.*, 2000c; Shan *et al.*, 2005), charge-based concentration (Bidawid *et al.*, 2000c; Papafragkou *et al.*, 2008), ultracentrifugation (Rzezutka *et al.*, 2006) and ultrafiltration (Butot *et al.*, 2007) have all been reported to be useful for the isolation of virus particles from fruits and vegetables. For example, positively charged beads circulated and captured using the Pathatrix™ machinery yield detection limits below one plaque forming unit of the hepatitis A virus in some artificially-inoculated samples (Papafragkou *et al.*, 2008). In most cases, the choice of the method used is based on the reagents available to the testing laboratory, and the use of internal controls is not consistent. This is a concern because of the potential release of inhibitory compounds and should be addressed in future studies (Sanchez *et al.*, 2007).

25.4.4 Detection of viruses in drinking water

The detection of viruses in drinking water has traditionally involved concentration from large volumes of water using charged filters (Hill *et al.*, 1976). These methods used conventional cell culture methods to detect virus, and were thus hampered by a lack of sensitivity as well as the inability to detect non-culturable viruses. The advent of molecular methods has reduced the time and cost required for detecting viruses from water samples (Pillai, 1997). These new methods have successfully been used to detect viral genomes or genome fragments in many types of raw and treated water samples (Kittigul *et al.*, 2000; Albinana-Gimenez *et al.*, 2006). It has, however, been demonstrated that viral genome fragments (non-viable) are detected after wastewater treatment protocols that eliminate infectious virus (Simonet and Gantzer, 2006). This raises concerns that the presence of genome fragments from enterically infecting viruses may not accurately predict the level of infectious virus in water samples.

25.5 Control issues

25.5.1 Hepatitis A virus

The sources of hepatitis A in the food supply are highly variable. The long incubation period before infection is clinically apparent and the high rate of person-to-person transmission during hepatitis A outbreaks makes point sources difficult to identify (Nainan *et al.*, 2005; Fiore *et al.*, 2006). The high proportion of asymptomatic infections in children under the age of five years generates another source of uncertainty in epidemiological investigations (Staes *et al.*, 2000). The accurate identification of disease transmission routes is a significant barrier to the implementation of effective control strategies to prevent hepatitis A virus transmission. The use of molecular detection and genotyping methods is one way to increase the odds of identifying linked cases of hepatitis A infection (see Section 25.4.1).

Due to the high degree of uncertainty in hepatitis A virus transmission, vaccination is one potential strategy for the control of infection (AAPCID, 2007). There is a safe and effective vaccine, and it is currently recommended for use in travellers to and residents of areas of high endemicity (AAPCID, 2007; CDC, 2007). Universal vaccination would be expensive, and has an unattractively high cost/benefit ratio when the healthy adult population of the developed world is the intended target (Anonychuk *et al.*, 2008). The use of sanitary measures to eliminate fecal contamination of foodstuffs should effectively limit foodborne hepatitis A outbreaks.

Pre-harvest control strategies are attractive because, when properly implemented, they minimize the need for downstream interventions. The most effective approach to prevent shellfish contamination is to prevent human sewage from entering shellfish growing waters. This sounds straightforward, but its enforcement can be difficult, particularly in remote areas with both commercial and recreational boat traffic. Imposing monetary penalties for waste dumping, mandating the use of waste containers that cannot easily be dumped overboard, and developing education outreach programs are three strategies with the potential to limit hepatitis A contamination of shellfish (Papafragkou *et al.*, 2006). If waters are contaminated, depuration can be used to reduce the levels of hepatitis A virus in shellfish prior to harvest, but viruses in shellfish tissue are purged more slowly than bacteria, and the hepatitis A virus in particular is not as readily depurated as other viruses (Richards, 2001; Chironna *et al.*, 2002). Unfortunately, current routine testing procedures do not look for viral contaminants in growing waters, and it has been repeatedly shown that the traditional bacterial indicators are not indicative of hepatitis A virus contamination (Croci *et al.*, 2000; Muniain-Mujika *et al.*, 2003; Pusch *et al.*, 2005; Phanuwan *et al.*, 2006; Villar *et al.*, 2007).

For the pre-harvest control of produce contamination, it is important to prevent human waste contamination of irrigation water. Although treatment regimens are available to allow the reuse of wastewater for irrigation, they are not necessarily effective against the hepatitis A virus (Gantzer *et al.*, 1998; Skraber *et al.*, 2007). Further research is necessary into the effectiveness of various water treatment protocols to determine if they are appropriate for the control of hepatitis A virus contamination. Produce is also sensitive to contamination introduced by human handling during harvest. Control procedures at this stage should include provision of toilet and hand-washing facilities, education on hygienic practices, reporting of active illnesses and provision of childcare so that young children, a prominent source of asymptomatic hepatitis A infections, are not present in the fields (Fiore, 2004; Koopmans and Duizer, 2004).

After harvest, the physical and chemical decontamination methods discussed in detail in Section 25.3.1 can be used to eliminate the hepatitis A virus from contaminated foods and/or processing areas. These are more stringent procedures than those necessary to reduce most bacterial contamination, and must be implemented properly in order to be effective. Cooking will inactivate the hepatitis A virus, but the entire product must reach 85 °C to ensure viral reduction (Parry and Mortimer, 1984). This is not suitable for fresh produce, and new technology must be developed to inactivate the virus in these products. Categories of food matrices must be individually tested to develop protocols that adequately reduce hepatitis A titre (Croci *et al.*, 1999; Bidawid *et al.*, 2000d; Deboosere *et al.*, 2004). The use of UV light and high hydrostatic pressure are promising, but their effectiveness must be further investigated before they will be useful for routine decontamination procedures (Nuanualsuwan *et al.*, 2002; Kingsley *et al.*, 2006). Gamma irradiation is somewhat effective at reducing hepatitis A titre on lettuce and strawberries, but the dose for a 1 log reduction is approximately 3 kGy (Bidawid *et al.*, 2000b), while current regulations in the USA only allow doses up to 1 kGy for fresh foods (CFSAN, 2007).

Food handlers are another source of hepatitis A virus infections (Fiore, 2004; Greig *et al.*, 2007; Todd *et al.*, 2007a, b). Contamination may be introduced during the final preparation stages for ready-to-eat foods (Greig *et al.*, 2007; Todd *et al.*, 2007a, b). The hepatitis A virus is excreted for up to two weeks prior to the development of symptoms (Fiore, 2004). It is therefore important to stress proper hygiene and hand-washing practices for all food service workers. Proper hand-washing with soap and water has been shown to be more effective at removing hepatitis A virus from hands than ethanol-based hand rubs (Mbithi *et al.*, 1993; Bidawid *et al.*, 2000a). Educational programs for food service workers must be designed with care to ensure the correct message is communicated. For example, gloved hands are frequently viewed as safer for food handling than bare skin, but care must still be taken to avoid cross-contamination of foods or surfaces. Preliminary data from our laboratory indicates that contaminated gloves spread virus very effectively (Bidawid *et al.*, 2007). Effective surface decontamination can also be used to interrupt hepatitis A virus transmission in food service settings, but as described in Section 25.3.1, not all commercial disinfectants are effective against the hepatitis A virus (Mbithi *et al.*, 1990; Abad *et al.*, 1997a; Bidawid *et al.*, 2000a; van Engelenburg *et al.*, 2002; Jean *et al.*, 2003; Bigliardi and Sansebastiano, 2006; Terpstra *et al.*, 2007). For effective disinfectants, such as sodium hypochlorite, concentration and contact time must be followed precisely to effectively inactivate the virus on a contaminated surface (Grabow *et al.*, 1983; Mbithi *et al.*, 1990; Abad *et al.*, 1997a; Jean *et al.*, 2003).

25.5.2 Hepatitis E virus

The control of hepatitis E infections in developing countries can be achieved by improving the availability of clean drinking water. This is linked to the availability of adequate hygienic facilities and improved hygienic practices. There is no vaccine available against hepatitis E, and the administration of pooled immunoglobulin from endemic areas does not appear to be protective (Khuroo and Dar, 1992; Panda *et al.*, 2007). Since it has been shown that the virus can survive the levels of chlorination currently recommended by the WHO, research into the physical and chemical inactivation of hepatitis E is urgently required to provide protocols that ensure the disinfection of contaminated water supplies (Guthmann *et al.*, 2006). Experimental studies in pig livers and epidemiological evidence indicate that boiling water is sufficient to inactivate the virus (Velazquez *et al.*, 1990; Feagins *et al.*, 2008).

As an emerging foodborne zoonotic agent, there is little information available about control measures that will prevent the spread of hepatitis E infection. The infection is endemic in many swine populations that have been examined, but does not cause overt disease. As a result, there is no incentive for control measures to improve animal health (Goens and Perdue, 2004). If the link between infected animals and transmission to humans can be established outside of Japan, this might provide a rationale for the development of animal-specific prevention strategies. At this time, however, the most effective control measure against infection is thorough cooking of meats and organ meats prior to consumption (Feagins *et al.*, 2008). Because of the low rates of person-to-person transmission in documented outbreaks, transmission via food handlers and ready-to-eat foods is not expected to be a major source of infection (Hla *et al.*, 1985; Somani *et al.*, 2003).

25.5.3 Poliovirus

Control measures against poliovirus involve vaccination programs and the global eradication initiative (Arya and Agarwal, 2007; Chumakov *et al.*, 2007). The virus does not have a non-human host, and it cannot circulate if the human population has a high level of mucosal immunity to infection (Melnick, 1996). Unfortunately, in addition to the four remaining endemic countries, 21 countries have experienced a resurgence or importation of poliomyelitis in recent years (Lahariya, 2007). This is due to the high prevalence of asymptomatic infections, as well as to vaccine-derived strains causing disease in communities (Chumakov *et al.*, 2007; Lahariya, 2007).

Fortunately, some relatively straightforward measures can be implemented to ensure that the poliovirus, if circulating, does not enter the food supply. Pasteurization of milk products has been shown historically to disrupt poliovirus transmission (Sattar *et al.*, 2001). Similar time/temperature combinations (72 °C, 30 s) can be used to inactivate the virus in other potentially contaminated liquids (Strazynski *et al.*, 2002). Depuration of shellfish greatly reduces the risk of poliovirus transmission by this route, and proper hand-washing by food handlers interrupts the chain of transmission during final preparation of foods (Schurmann and Eggers, 1985; Franco *et al.*, 1990; Enriquez *et al.*, 1992). It should be noted that for all of these measures, the conditions required to eliminate polio are less stringent than those required for the inactivation of hepatitis A (Table 25.3).

The presence of the poliovirus in wastewater remains the most important means of transmission in endemic countries, and effective methods exist to remove polio from sewage (Pavlov, 2006; Arraj *et al.*, 2005; Belguith *et al.*, 2007; Dedepsidis *et al.*, 2007). New methods under development, such as ozone and UV light, are able to decontaminate poliovirus-contaminated wastewater (Lazarova and Savoys, 2004; Tanner *et al.*, 2004). The required focus on sewage treatment and clean water in the developing world is reminiscent of control measures to prevent the spread of many bacterial illnesses (Berry *et al.*, 2006). The integration of clean water programs aimed at reducing the burden of bacterial and viral illness can only serve to increase the likelihood that these programs will see some successes in limiting the spread of enteric disease.

25.5.4 Avian Influenza virus

Current strategies to mitigate the human health risks associated with the H5N1 avian influenza virus are mainly focused on preventing the disease from spreading in the animal population (Rajagopal and Treanor, 2007). The elimination of H5N1 infections in birds would of course remove the risk to humans. Unfortunately, H5N1 infections in birds have become endemic in many countries (ECDC, 2007, Rajagopal and Treanor, 2007). Therefore, accurate monitoring and understanding the circulation of the virus in wild and domestic birds is critical to the success of control programs (Olsen *et al.*, 2006). All H5 or H7 type avian influenza infections are notifiable to the World Organization for Animal Health (OIE, 2007). Early detection of the H5N1 infection in a local poultry population is a key step in preventing the disease from becoming widespread (Sims, 2007).

A vaccine is available against the H5 and H7 avian influenza virus sub-types, and its use in the poultry population is one way to control the spread of emerging highly pathogenic viruses (Capua and Marangon, 2007). It is not possible to vaccinate all birds in areas where these viruses are endemic, and additional measures must be in place to prevent the spread of disease (Guan *et al.*, 2007; Sims, 2007). A combination of surveillance, vaccination, culling of infected birds and segregation of wild and domestic poultry is recommended to control the spread of emerging epidemic avian influenza strains (Guan *et al.*, 2007).

Control measures to prevent transmission of avian influenza through the food supply are more straightforward. All of the strains and sub-types of influenza are more susceptible to heating and to chemical disinfection than the other foodborne viruses described in this chapter (see Tables 25.3 and 25.4). Thorough cooking of poultry products and basic disinfection of food preparation surfaces is sufficient to prevent foodborne transmission of the influenza virus (De Benedictis *et al.*, 2007).

25.6 Future trends

Two of the four viruses discussed in detail in this chapter are emerging zoonoses (hepatitis E and avian influenza). It is important that public health programs continue to monitor these diseases in both animals and humans in order to develop accurate risk assessment and prevention planning (Merianos, 2007). The other two viruses (hepatitis A and polio) cause human illnesses that are vaccine-preventable. Calls from experts to continue and expand vaccine coverage with the goal of reducing the disease burden from these viruses is likely to continue (AAPCID, 2007; Chumakov *et al.*, 2007).

In addition to public health measures, the food production and processing industries can take action to reduce the spread of viruses through food. A recurring point in the above discussion is the remarkable resistance of viruses to decontamination procedures. Heating and disinfection protocols for viruses and food preparation surfaces are being defined in the literature, but they must also be recognized and implemented along the food production continuum. In addition, these viruses typically have a very low infectious dose (10–100 particles) and they can be excreted at high levels (10^6–10^{11} particles per gram of faeces). Since contamination of food products is typically a secondary event, a 5 log reduction in infectious particles has been considered effective for control. The recent development of sensitive and semi-quantitative detection methods will help to identify the critical control points along the food production and preparation continuum where virus contamination can be reduced. Unfortunately, many of these methods detect viral genomes instead of infectious virus particles. The regulation of viruses in foods will require a more detailed understanding of the correlation between the presence of viral nucleic acid fragments and a human health risk.

25.7 Sources of further information and advice

A book dedicated to food virology:

Goyal S M and Doyle M P (2006) *Viruses in foods*, New York, Springer.

Comprehensive reviews of the four viruses discussed in detail in this chapter:

Abdel-Ghafar A N, Chotpitayasunondh T, Gao Z, Hayden F G, Nguyen D H, De Jong M D, Naghdaliyev A, Peiris J S, Shindo N, Soeroso S and Uyeki T M (2008) Update on avian influenza A (H5N1) virus infection in humans, *N Engl J Med*, **358**, 261–73.

Melnick J L (1996) Current status of poliovirus infections, *Clin Microbiol Rev*, **9**, 293–300.

Nainan O V, Armstrong G L, Han X H, Williams I, Bell B P and Margolis H S (2005) Hepatitis A molecular epidemiology in the United States, 1996–1997: sources of infection and implications of vaccination policy, *J Infect Dis*, **191**, 957–63.

Vasickova P, Psikal I, Kralik P, Widen F, Hubalek Z and Pavlik I (2007) Hepatitis E virus: a review, *Vet Medicina*, **52**, 365–84.

25.8 References

AAPCID (2007) Hepatitis A vaccine recommendations, *Pediatrics*, **120**, 189–99.

Abad F X, Pinto R M and Bosch A (1994) Survival of enteric viruses on environmental fomites, *Appl Environ Microbiol*, **60**, 3704–10.

Abad F X, Pinto R M and Bosch A (1997a) Disinfection of human enteric viruses on fomites, *FEMS Microbiol Lett*, **156**, 107–11.

Abad F X, Pinto R M, Gajardo R and Bosch A (1997b) Viruses in mussels: Public health implications and depuration, *J Food Prot*, **60**, 677–81.

Abad F X, Villena C, Guix S, Caballero S, Pinto R M and Bosch A (2001) Potential role of fomites in the vehicular transmission of human astroviruses, *Appl Environ Microbiol*, **67**, 3904–7.

Abdel-Ghafar A N, Chotpitayasunondh T, Gao Z, Hayden F G, Nguyen D H, De Jong M D, Naghdaliyev A, Peiris J S, Shindo N, Soeroso S and Uyeki T M (2008) Update on avian influenza A (H5N1) virus infection in humans, *N Engl J Med*, **358**, 261–73.

Abzug M J (2004) Presentation, diagnosis, and management of enterovirus infections in neonates, *Paediatr Drugs*, **6**, 1–10.

Acharya S K and Panda S K (2006) Hepatitis E virus: epidemiology, diagnosis, pathology and prevention, *Trop Gastroenterol*, **27**, 63–8.

Albinana-Gimenez N, Clemente-Casares P, Bofill-Mas S, Hundesa A, Ribas F and Girones R (2006) Distribution of human polyomaviruses, adenoviruses, and hepatitis E virus in the environment and in a drinking-water treatment plant, *Environ Sci Technol*, **40**, 7416–22.

Amon J J, Devasia R, Xia G, Nainan O V, Hall S, Lawson B, Wolthuis J S, Macdonald P D, Shepard C W, Williams I T, Armstrong G L, Gabel J A, Erwin P, Sheeler L, Kuhnert W, Patel P, Vaughan G, Weltman A, Craig A S, Bell B P and Fiore A (2005) Molecular epidemiology of foodborne hepatitis A outbreaks in the United States, 2003, *J Infect Dis*, **192**, 1323–30.

Anonychuk A M, Tricco A C, Bauch C T, Pham B, Gilca V, Duval B, John-Baptiste A, Woo G and Krahn M (2008) Cost-effectiveness analyses of hepatitis A vaccine: a systematic review to explore the effect of methodological quality on the economic attractiveness of vaccination strategies, *Pharmacoeconomics*, **26**, 17–32.

Arankalle V A, Ticehurst J, Sreenivasan M A, Kapikian A Z, Popper H, Pavri K M and Purcell R H (1988) Aetiological association of a virus-like particle with enterically transmitted non-A, non-B hepatitis, *Lancet*, **1**, 550–54.

Arankalle V A, Sarada Devi K L, Lole K S, Shenoy K T, Verma V and Haneephabi M (2006) Molecular characterization of hepatitis A virus from a large outbreak from Kerala, India, *Indian J Med Res*, **123**, 760–69.

Arnal C, Crance J M, Gantzer C, Schwartzbrod L, Deloince R and Billaudel S (1998) Persistence of infectious hepatitis A virus and its genome in artificial seawater, *Zentralbl Hyg Umweltmed*, **201**, 279–84.

Arraj A, Bohatier J, Laveran H and Traore O (2005) Comparison of bacteriophage and enteric virus removal in pilot scale activated sludge plants, *J Appl Microbiol*, **98**, 516–24.

Arya S C and Agarwal N (2007) Poliomyelitis: concerns for polio-free countries, *Vaccine*, **25**, 3245–6.

Balayan M S (1992) Natural hosts of hepatitis A virus, *Vaccine*, **10** Suppl 1, S27–31.

Banks M, Heath G S, Grierson S S, King D P, Gresham A, Girones R, Widen F and Harrison T J (2004) Evidence for the presence of hepatitis E virus in pigs in the United Kingdom, *Vet Rec*, **154**, 223–7.

Belguith K, Hassen A, Bouslama L, Khira S and Aouni M (2007) Enterovirus circulation in wastewater and behavior of some serotypes during sewage treatment in Monastir, Tunisia, *J Environ Health*, **69**, 52–6.

Berry D, Xi C and Raskin L (2006) Microbial ecology of drinking water distribution systems, *Curr Opin Biotechnol*, **17**, 297–302.

Bialek S R, George P A, Xia G L, Glatzer M B, Motes M L, Veazey J E, Hammond R M, Jones T, Shieh Y C, Wamnes J, Vaughan G, Khudyakov Y and Fiore A E (2007) Use of molecular epidemiology to confirm a multistate outbreak of hepatitis A caused by consumption of oysters, *Clin Infect Dis*, **44**, 838–40.

Bidawid S, Farber J M and Sattar S A (2000a) Contamination of foods by food handlers: experiments on hepatitis A virus transfer to food and its interruption, *Appl Environ Microbiol*, **66**, 2759–63.

Bidawid S, Farber J M and Sattar S A (2000b) Inactivation of hepatitis A virus (HAV) in fruits and vegetables by gamma irradiation, *Int J Food Microbiol*, **57**, 91–7.

Bidawid S, Farber J M and Sattar S A (2000c) Rapid concentration and detection of hepatitis A virus from lettuce and strawberries, *J Virol Methods*, **88**, 175–85.

Bidawid S, Farber J M, Sattar S A and Hayward S (2000d) Heat inactivation of hepatitis A virus in dairy foods, *J Food Prot*, **63**, 522–8.

Bidawid S, Sattar S A, Tetro J A, Pagotto F and Mattison K (2007) The potential for pathogen cross-contamination of foods with gloved hands: experiments with feline calicivirus as a surrogate for human enteric viruses, *International Association for Food Protection 94th Annual Meeting*, Lake Buena Vista, FL, July 8–11.

Bigliardi L and Sansebastiano G (2006) Study on inactivation kinetics of hepatitis A virus and enteroviruses with peracetic acid and chlorine. New ICC/PCR method to assess disinfection effectiveness, *J Prev Med Hyg*, **47**, 56–63.

Boccia D, Guthmann J P, Klovstad H, Hamid N, Tatay M, Ciglenecki I, Nizou J Y, Nicand E and Guerin P J (2006) High mortality associated with an outbreak of hepatitis E among displaced persons in Darfur, Sudan, *Clin Infect Dis*, **42**, 1679–84.

Bosch A (1998) Human enteric viruses in the water environment: a minireview, *Int Microbiol*, **1**, 191–6.

Bosch A, Sanchez G, Le Guyader F, Vanaclocha H, Haugarreau L and Pinto R M (2001) Human enteric viruses in Coquina clams associated with a large hepatitis A outbreak, *Water Sci Technol*, **43**, 61–5.

Bouwknegt M, Lodder-Verschoor F, Van Der Poel W H, Rutjes S A and De Roda Husman A M (2007) Hepatitis E virus RNA in commercial porcine livers in The Netherlands, *J Food Prot*, **70**, 2889–95.

Bradley D W, Krawczynski K, Cook E H Jr, McCaustland K A, Humphrey C D, Spelbring J E, Myint H and Maynard J E (1987) Enterically transmitted non-A, non-B hepatitis: serial passage of disease in cynomolgus macaques and tamarins and recovery of disease-associated 27- to 34-nm viruslike particles, *Proc Natl Acad Sci USA*, **84**, 6277–81.

Brown J D, Swayne D E, Cooper R J, Burns R E and Stallknecht D E (2007) Persistence of H5 and H7 avian influenza viruses in water, *Avian Dis*, **51**, 285–9.

Brundage S C and Fitzpatrick A N (2006) Hepatitis A, *Am Fam Physician*, **73**, 2162–8.

Butot S, Putallaz T and Sanchez G (2007) Procedure for rapid concentration and detection of enteric viruses from berries and vegetables, *Appl Environ Microbiol*, **73**, 186–92.

Calci K R, Meade G K, Tezloff R C and Kingsley D H (2005) High-pressure inactivation of hepatitis A virus within oysters, *Appl Environ Microbiol*, **71**, 339–43.

Calder L, Simmons G, Thornley C, Taylor P, Pritchard K, Greening G and Bishop J (2003) An outbreak of hepatitis A associated with consumption of raw blueberries, *Epidemiol Infect*, **131**, 745–51.

Capua I and Marangon S (2007) Control and prevention of avian influenza in an evolving scenario, *Vaccine*, **25**, 5645–52.

CDC (1994) Certification of poliomyelitis eradication – the Americas, 1994, *MMWR Morb Mortal Wkly Rep*, **43**, 720–32.

CDC (2001) Certification of poliomyelitis eradication – Western Pacific Region, October 2000, *MMWR Morb Mortal Wkly Rep*, **50**, 1–3.

CDC (2002) Certification of poliomyelitis eradication – European Region, June 2002, *MMWR Morb Mortal Wkly Rep*, **51**, 572–4.

CDC (2003a) Foodborne transmission of hepatitis A – Massachusetts, 2001, *MMWR Morb Mortal Wkly Rep*, **52**, 565–7.

CDC (2003b) Hepatitis A outbreak associated with green onions at a restaurant – Monaca, Pennsylvania, 2003, *MMWR Morb Mortal Wkly Rep*, **52**, 1155–7.

CDC (2007) *Health Information for International Travel 2008*, Centers for Disease Control and Prevention, Atlanta, GA, US Department of Health and Human Services, Public Health Service.

CFSAN (2007) *Foods Permitted to be Irradiated Under FDA's Regulations*, US Food and Drug Administration, Washington, DC.

Chan–Yeung M and Xu R H (2003) SARS: epidemiology, *Respirology*, **8** Suppl, S9–14.

Chang M L, Tsao K C, Huang C C, Yen M H, Huang C G and Lin, T Y (2006) Coxsackievirus B3 in human milk, *Pediatr Infect Dis J*, **25**, 955–7.

Chau T N, Lai S T, Tse C, Ng T K, Leung V K, Lim W and Ng M H (2006) Epidemiology and clinical features of sporadic hepatitis E as compared with hepatitis A, *Am J Gastroenterol*, **101**, 292–6.

Chironna M, Germinario C, De Medici D, Fiore A, Di Pasquale S, Quarto M and Barbuti S (2002) Detection of hepatitis A virus in mussels from different sources marketed in Puglia region (South Italy), *Int J Food Microbiol*, **75**, 11–18.

Chumakov K, Ehrenfeld E, Wimmer E and Agol V I (2007) Vaccination against polio should not be stopped, *Nat Rev Microbiol*, **5**, 952–8.

Cohen J I, Ticehurst J R, Purcell R H, Buckler-White A and Baroudy B M (1987) Complete nucleotide sequence of wild-type hepatitis A virus: comparison with different strains of hepatitis A virus and other picornaviruses, *J Virol*, **61**, 50–59.

Conaty S, Bird P, Bell G, Kraa E, Grohmann G and McAnulty J M (2000) Hepatitis A in New South Wales, Australia from consumption of oysters: the first reported outbreak, *Epidemiol Infect*, **124**, 121–30.

Craig A S, Watson B, Zink T K, Davis J P, Yu C and Schaffner W (2007) Hepatitis A outbreak activity in the United States: responding to a vaccine-preventable disease, *Am J Med Sci*, **334**, 180–83.

Cristina J and Costa-Mattioli M (2007) Genetic variability and molecular evolution of hepatitis A virus, *Virus Res*, **127**, 151–7.

Croci L, Ciccozzi M, De Medici D, Di Pasquale S, Fiore A, Mele A and Toti L (1999) Inactivation of hepatitis A virus in heat-treated mussels, *J Appl Microbiol*, **87**, 884–8.

Croci L, De Medici D, Scalfaro C, Fiore A, Divizia M, Donia D, Cosentino A M, Moretti P and Costantini G (2000) Determination of enteroviruses, hepatitis A virus, bacteriophages and *Escherichia coli* in Adriatic Sea mussels, *J Appl Microbiol*, **88**, 293–8.

Croci L, De Medici D, Scalfaro C, Fiore A and Toti L (2002) The survival of hepatitis A virus in fresh produce, *Int J Food Microbiol*, **73**, 29–34.

Croci L, De Medici D, Di Pasquale S and Toti L (2005) Resistance of hepatitis A virus in mussels subjected to different domestic cookings, *Int J Food Microbiol*, **105**, 139–44.

Cromeans T, Sobsey M D and Fields H A (1987) Development of a plaque assay for a cytopathic, rapidly replicating isolate of hepatitis A virus, *J Med Virol*, **22**, 45–56.

Dalton H R, Thurairajah P H, Fellows H J, Hussaini H S, Mitchell J, Bendall R, Banks M, Ijaz S, Teo C G and Levine D F (2007) Autochthonous hepatitis E in southwest England, *J Viral Hepat*, **14**, 304–9.

De Benedictis P, Beato M S and Capua I (2007) Inactivation of avian influenza viruses by chemical agents and physical conditions: a review, *Zoonoses Public Health*, **54**, 51–68.

De Jong M D and Hien T T (2006) Avian influenza A (H5N1), *J Clin Virol*, **35**, 2–13.

De Jong M D, Simmons C P, Thanh T T, Hien V M, Smith G J, Chau T N, Hoang D M, Chau N V, Khanh T H, Dong V C, Qui P T, Cam B V, Ha Do Q, Guan Y, Peiris J S, Chinh N T, Hien T T and Farrar J (2006) Fatal outcome of human influenza A (H5N1) is associated with high viral load and hypercytokinemia, *Nat Med*, **12**, 1203–7.

De Medici D, Ciccozzi M, Fiore A, Di Pasquale S, Parlato A, Ricci-Bitti P and Croci L (2001) Closed-circuit system for the depuration of mussels experimentally contaminated with hepatitis A virus, *J Food Prot*, **64**, 877–80.

Deboosere N, Legeay O, Caudrelier Y and Lange M (2004) Modelling effect of physical and chemical parameters on heat inactivation kinetics of hepatitis A virus in a fruit model system, *Int J Food Microbiol*, **93**, 73–85.

Dedepsidis E, Kyriakopoulou Z, Pliaka V, Kottaridi C, Bolanaki E, Levidiotou-Stefanou S, Komiotis D and Markoulatos P (2007) Retrospective characterization of a vaccine-derived poliovirus type 1 isolate from sewage in Greece, *Appl Environ Microbiol*, **73**, 6697–704.

Deng M Y and Cliver D O (1992) Inactivation of poliovirus type 1 in mixed human and swine wastes and by bacteria from swine manure, *Appl Environ Microbiol*, **58**, 2016–21.

Deng M Y and Cliver D O (1995) Persistence of inoculated hepatitis A virus in mixed human and animal wastes, *Appl Environ Microbiol*, **61**, 87–91.

Dentinger C M, Bower W A, Nainan O V, Cotter S M, Myers G, Dubusky L M, Fowler S, Salehi E D and Bell B P (2001) An outbreak of hepatitis A associated with green onions, *J Infect Dis*, **183**, 1273–6.

Di Girolamo R, Liston J and Matches J (1975) Uptake and elimination of poliovirus by West Coast oysters, *Appl Microbiol*, **29**, 260–64.

Dingman C (1916) Report of a possibly milk-borne epidemic of infantile paralysis, *New York State J Med*, **16**, 589.

Duan S M, Zhao X S, Wen R F, Huang J J, Pi G H, Zhang S X, Han J, Bi S L, Ruan L and Dong X P (2003) Stability of SARS coronavirus in human specimens and environment and its sensitivity to heating and UV irradiation, *Biomed Environ Sci*, **16**, 246–55.

Dumpis U, Crook D and Oksi J (1999) Tick-borne encephalitis, *Clin Infect Dis*, **28**, 882–90.

ECDC (2007) Avian influenza update: recent outbreaks of H5N1 in poultry worldwide, *Euro Surveill*, **12**, E070125 2.

Emerson S U, Arankalle V A and Purcell R H (2005) Thermal stability of hepatitis E virus, *J Infect Dis*, **192**, 930–33.

Enriquez R, Frosner G G, Hochstein-Mintzel V, Riedemann S and Reinhardt G (1992) Accumulation and persistence of hepatitis A virus in mussels, *J Med Virol*, **37**, 174–9.

Eubanks R D and Farrah S R (1981) Inactivation of poliovirus and other enteroviruses by solutions of sodium fluoride at low pH, *Appl Environ Microbiol*, **41**, 686–9.

Fauquet C M, Mayo M A, Maniloff J, Desselberger U and Ball L A (2005) *Virus Taxonomy: Eigth Report of the International Committee on the Taxonomy of Viruses*, San Diego, CA, Elsevier.

Feagins A R, Opriessnig T, Guenette D K, Halbur P G and Meng X J (2007) Detection and characterization of infectious hepatitis E virus from commercial pig livers sold in local grocery stores in the USA, *J Gen Virol*, **88**, 912–17.

Feagins A R, Opriessnig T, Guenette D K, Halbur P G and Meng X J (2008) Inactivation of infectious hepatitis E virus present in commercial pig livers sold in local grocery stores in the United States, *Int J Food Microbiol*, **123**, 32–7.

Feinstone S M, Kapikian A Z and Purceli R H (1973) Hepatitis A: detection by immune electron microscopy of a viruslike antigen associated with acute illness, *Science*, **182**, 1026–8.

Fiore A E (2004) Hepatitis A transmitted by food, *Clin Infect Dis*, **38**, 705–15.

Fiore A E, Wasley A and Bell B P (2006) Prevention of hepatitis A through active or passive immunization: recommendations of the Advisory Committee on Immunization Practices (ACIP), *MMWR Recomm Rep*, **55**, 1–23.

Franco E, Toti L, Gabrieli R, Croci L, De Medici D and Pana A (1990) Depuration of *Mytilus galloprovincialis* experimentally contaminated with hepatitis A virus, *Int J Food Microbiol*, **11**, 321–7.

Frank C, Walter J, Muehlen M, Jansen A, Van Treeck U, Hauri A M, Zoellner I, Rakha M, Hoehne M, Hamouda O, Schreier E and Stark K (2007) Major outbreak of hepatitis A associated with orange juice among tourists, Egypt, 2004, *Emerg Infect Dis*, **13**, 156–8.

Gantzer C, Maul A, Audic J M and Schwartzbrod L (1998) Detection of infectious enteroviruses, enterovirus genomes, somatic coliphages, and *Bacteroides fragilis* phages in treated wastewater, *Appl Environ Microbiol*, **64**, 4307–12.

Goens S D and Perdue M L (2004) Hepatitis E viruses in humans and animals, *Anim Health Res Rev*, **5**, 145–56.

Gordon C and Toze S (2003) Influence of groundwater characteristics on the survival of enteric viruses, *J Appl Microbiol*, **95**, 536–44.

Grabow W O, Gauss-Muller V, Prozesky O W and Deinhardt F (1983) Inactivation of hepatitis A virus and indicator organisms in water by free chlorine residuals, *Appl Environ Microbiol*, **46**, 619–24.

Greig J D, Todd E C, Bartleson C A and Michaels B S (2007) Outbreaks where food workers have been implicated in the spread of foodborne disease. Part 1. Description of the problem, methods, and agents involved, *J Food Prot*, **70**, 1752–61.

Guan Y, Chen H, Li K S, Riley S, Leung G M, Webster R, Peiris M and Yuen K Y

(2007) A model to control the epidemic of H5N1 influenza at the source, *BMC Infect Dis*, **7**, 132.

Guevremont E, Brassard J, Houde A, Simard C and Trottier Y L (2006) Development of an extraction and concentration procedure and comparison of RT-PCR primer systems for the detection of hepatitis A virus and norovirus GII in green onions, *J Virol Methods*, **134**, 130–35.

Gunther G and Haglund M (2005) Tick-borne encephalopathies: epidemiology, diagnosis, treatment and prevention, *CNS Drugs*, **19**, 1009–32.

Guthmann J P, Klovstad H, Boccia D, Hamid N, Pinoges L, Nizou J Y, Tatay M, Diaz F, Moren A, Grais R F, Ciglenecki I, Nicand E and Guerin P J (2006) A large outbreak of hepatitis E among a displaced population in Darfur, Sudan, 2004: the role of water treatment methods, *Clin Infect Dis*, **42**, 1685–91.

Hadler S C, Webster H M, Erben J J, Swanson J E and Maynard J E (1980) Hepatitis A in day-care centers. A community-wide assessment, *N Engl J Med*, **302**, 1222–7.

Haqshenas G, Shivaprasad H L, Woolcock P R, Read D H and Meng X J (2001) Genetic identification and characterization of a novel virus related to human hepatitis E virus from chickens with hepatitis-splenomegaly syndrome in the United States, *J Gen Virol*, **82**, 2449–62.

Hasegawa Y, Matsumoto F, Tanaka C, Ouchi Y and Hayashi K (2007) Outbreak of hepatitis A virus infection caused by food served in a restaurant, *Jpn J Infect Dis*, **60**, 150–51.

Hay A J, Gregory V, Douglas A R and Lin Y P (2001) The evolution of human influenza viruses, *Philos Trans R Soc Lond B Biol Sci*, **356**, 1861–70.

Hewitt J and Greening G E (2004) Survival and persistence of norovirus, hepatitis A virus, and feline calicivirus in marinated mussels, *J Food Prot*, **67**, 1743–50.

Hewitt J and Greening G E (2006) Effect of heat treatment on hepatitis A virus and norovirus in New Zealand Greenshell mussels (*Perna canaliculus*) by quantitative real-time reverse transcription PCR and cell culture, *J Food Prot*, **69**, 2217–23.

Hill W F Jr, Jakubowski W, Akin E W and Clarke N A (1976) Detection of virus in water: sensitivity of the tentative standard method for drinking water, *Appl Environ Microbiol*, **31**, 254–61.

Hla M, Myint Myint S, Tun K, Thein-Maung M and Khin Maung T (1985) A clinical and epidemiological study of an epidemic of non-A non-B hepatitis in Rangoon, *Am J Trop Med Hyg*, **34**, 1183–9.

Howitz M, Mazick A and Molbak K (2005) Hepatitis A outbreak in a group of Danish tourists returning from Turkey, October 2005, *Euro Surveill*, **10**, E051201.2.

Hsieh S Y, Meng X J, Wu Y H, Liu S T, Tam A W, Lin D Y and Liaw Y F (1999) Identity of a novel swine hepatitis E virus in Taiwan forming a monophyletic group with Taiwan isolates of human hepatitis E virus, *J Clin Microbiol*, **37**, 3828–34.

Hurst C J, Gerba C P and Cech I (1980) Effects of environmental variables and soil characteristics on virus survival in soil, *Appl Environ Microbiol*, **40**, 1067–79.

Hutin Y J, Pool V, Cramer E H, Nainan O V, Weth J, Williams I T, Goldstein S T, Gensheimer K F, Bell B P, Shapiro C N, Alter M J and Margolis H S (1999) A multistate, foodborne outbreak of hepatitis A. National Hepatitis A Investigation Team, *N Engl J Med*, **340**, 595–602.

Hyams K C, McCarthy M C, Kaur M, Purdy M A, Bradley D W, Mansour M M, Gray S, Watts D M and Carl M (1992) Acute sporadic hepatitis E in children living in Cairo, Egypt, *J Med Virol*, **37**, 274–7.

Ijaz S, Arnold E, Banks M, Bendall R P, Cramp M E, Cunningham R, Dalton H R, Harrison T J, Hill S F, Macfarlane L, Meigh R E, Shafi S, Sheppard M J, Smithson J, Wilson M P and Teo C G (2005) Non-travel-associated hepatitis E in England and Wales: demographic, clinical, and molecular epidemiological characteristics, *J Infect Dis*, **192**, 1166–72.

Isbarn S, Buckow R, Himmelreich A, Lehmacher A and Heinz V (2007) Inactivation of avian influenza virus by heat and high hydrostatic pressure, *J Food Prot*, **70**, 667–73.

Jacobsen K H and Koopman J S (2004) Declining hepatitis A seroprevalence: a global review and analysis, *Epidemiol Infect*, **132**, 1005–22.

Jean J, Vachon J F, Moroni O, Darveau A, Kukavica-Ibrulj I and Fliss I (2003) Effectiveness of commercial disinfectants for inactivating hepatitis A virus on agri-food surfaces, *J Food Prot*, **66**, 115–19.

Jothikumar N, Aparna K, Kamatchiammal S, Paulmurugan R, Saravanadevi S and Khanna P (1993) Detection of hepatitis E virus in raw and treated wastewater with the polymerase chain reaction, *Appl Environ Microbiol*, **59**, 2558–62.

Jothikumar N, Cromeans T L, Robertson B H, Meng X J and Hill V R (2006) A broadly reactive one-step real-time RT-PCR assay for rapid and sensitive detection of hepatitis E virus, *J Virol Methods*, **131**, 65–71.

Kerbo N, Donchenko I, Kutsar K and Vasilenko V (2005) Tickborne encephalitis outbreak in Estonia linked to raw goat milk, May–June 2005, *Euro Surveill*, **10**, E050623 2.

Khuroo M S and Dar M Y (1992) Hepatitis E: evidence for person-to-person transmission and inability of low dose immune serum globulin from an Indian source to prevent it, *Indian J Gastroenterol*, **11**, 113–16.

King D J (1991) Evaluation of different methods of inactivation of Newcastle disease virus and avian influenza virus in egg fluids and serum, *Avian Dis*, **35**, 505–14.

Kingsley D H and Richards G P (2001) Rapid and efficient extraction method for reverse transcription-PCR detection of hepatitis A and Norwalk-like viruses in shellfish, *Appl Environ Microbiol*, **67**, 4152–7.

Kingsley D H and Richards G P (2003) Persistence of hepatitis A virus in oysters, *J Food Prot*, **66**, 331–4.

Kingsley D H, Hoover D G, Papafragkou E and Richards G P (2002) Inactivation of hepatitis A virus and a calicivirus by high hydrostatic pressure, *J Food Prot*, **65**, 1605–9.

Kingsley D H, Guan D and Hoover D G (2005) Pressure inactivation of hepatitis A virus in strawberry puree and sliced green onions, *J Food Prot*, **68**, 1748–51.

Kingsley D H, Guan D, Hoover D G and Chen H (2006) Inactivation of hepatitis A virus by high-pressure processing: the role of temperature and pressure oscillation, *J Food Prot*, **69**, 2454–9.

Kitamura N, Semler B L, Rothberg P G, Larsen G R, Adler C J, Dorner A J, Emini E A, Hanecak R, Lee J J, Van Der Werf S, Anderson C W and Wimmer E (1981) Primary structure, gene organization and polypeptide expression of poliovirus RNA, *Nature*, **291**, 547–53.

Kittigul L, Raengsakulrach B, Siritantikorn S, Kanyok R, Utrarachkij F, Diraphat P, Thirawuth V, Siripanichgon K, Pungchitton S, Chitpirom K, Chaichantanakit N and Vathanophas K (2000) Detection of poliovirus, hepatitis A virus and rotavirus from sewage and water samples, *Southeast Asian J Trop Med Public Health*, **31**, 41–6.

Knossow M and Skehel J J (2006) Variation and infectivity neutralization in influenza, *Immunology*, **119**, 1–7.

Konduru K and Kaplan G G (2006) Stable growth of wild-type hepatitis A virus in cell culture, *J Virol*, **80**, 1352–60.

Koopmans M and Duizer E (2004) Foodborne viruses: an emerging problem, *Int J Food Microbiol*, **90**, 23–41.

Koopmans M, Wilbrink B, Conyn M, Natrop G, Van Der Na H, Vennema H, Meijer A, Van Steenbergen J, Fouchier R, Osterhaus A and Bosman A (2004) Transmission of H7N7 avian influenza A virus to human beings during a large outbreak in commercial poultry farms in the Netherlands, *Lancet*, **363**, 587–93.

Kramer A, Galabov A S, Sattar S A, Dohner L, Pivert A, Payan C, Wolff M H, Yilmaz A and Steinmann J (2006) Virucidal activity of a new hand disinfectant with reduced ethanol content: comparison with other alcohol-based formulations, *J Hosp Infect*, **62**, 98–106.

Kukavica-Ibrulj I, Darveau A, Jean J and Fliss I (2004) Hepatitis A virus attachment to agri-food surfaces using immunological, virological and thermodynamic assays, *J Appl Microbiol*, **97**, 923–34.

Kurdziel A S, Wilkinson N, Langton S and Cook N (2001) Survival of poliovirus on soft fruit and salad vegetables, *J Food Prot*, **64**, 706–9.

Lahariya C (2007) Global eradication of polio: the case for 'finishing the job', *Bull World Health Organ*, **85**, 487–92.

Lazarova V and Savoys P (2004) Technical and sanitary aspects of wastewater disinfection by UV irradiation for landscape irrigation, *Water Sci Technol*, **50**, 203–9.

Lowen A C, Mubareka S, Steel J and Palese P (2007) Influenza virus transmission is dependent on relative humidity and temperature, *PLoS Pathog*, **3**, 1470–76.

Lu H, Castro A E, Pennick K, Liu J, Yang Q, Dunn P, Weinstock D and Henzler D (2003a) Survival of avian influenza virus H7N2 in SPF chickens and their environments, *Avian Dis*, **47**, 1015–21.

Lu X H, Cho D, Hall H, Rowe T, Mo I P, Sung H W, Kim W J, Kang C, Cox N, Klimov A and Katz J M (2003b) Pathogenesis of and immunity to a new influenza A (H5N1) virus isolated from duck meat, *Avian Dis*, **47**, 1135–40.

Lu L, Li C and Hagedorn C H (2006) Phylogenetic analysis of global hepatitis E virus sequences: genetic diversity, subtypes and zoonosis, *Rev Med Virol*, **16**, 5–36.

Lukasik J, Bradley M L, Scott T M, Dea M, Koo A, Hsu W Y, Bartz J A and Farrah S R (2003) Reduction of poliovirus 1, bacteriophages, *Salmonella montevideo*, and *Escherichia coli* O157:H7 on strawberries by physical and disinfectant washes, *J Food Prot*, **66**, 188–93.

Ma J F, Straub T M, Pepper I L and Gerba C P (1994) Cell culture and PCR determination of poliovirus inactivation by disinfectants. *Appl Environ Microbiol*, **60**, 4203–6.

Mansuy J M, Peron J M, Abravanel F, Poirson H, Dubois M, Miedouge M, Vischi F, Alric L, Vinel J P and Izopet J (2004) Hepatitis E in the south west of France in individuals who have never visited an endemic area, *J Med Virol*, **74**, 419–24.

Matsuda H, Okada K, Takahashi K and Mishiro S (2003) Severe hepatitis E virus infection after ingestion of uncooked liver from a wild boar, *J Infect Dis*, **188**, 944.

Mbithi J N, Springthorpe V S and Sattar S A (1990) Chemical disinfection of hepatitis A virus on environmental surfaces, *Appl Environ Microbiol*, **56**, 3601–4.

Mbithi J N, Springthorpe V S, Boulet J R and Sattar S A (1992) Survival of hepatitis A virus on human hands and its transfer on contact with animate and inanimate surfaces, *J Clin Microbiol*, **30**, 757–63.

Mbithi J N, Springthorpe V S and Sattar S A (1993) Comparative in vivo efficiencies of hand-washing agents against hepatitis A virus (HM-175) and poliovirus type 1 (Sabin), *Appl Environ Microbiol*, **59**, 3463–9.

McCaustland K A, Bond W W, Bradley D W, Ebert J W and Maynard J E (1982) Survival of hepatitis A virus in feces after drying and storage for 1 month, *J Clin Microbiol*, **16**, 957–8.

McGregor S and Mayor H D (1971) Internal components released from rhinovirus and poliovirus by heat, *J Gen Virol*, **10**, 203–7.

McKinney K R, Gong Y Y and Lewis T G (2006) Environmental transmission of SARS at Amoy Gardens, *J Environ Health*, **68**, 26–30; quiz 51–2.

Melnick J L (1996) Current status of poliovirus infections, *Clin Microbiol Rev*, **9**, 293–300.

Meng X J (2000) Novel strains of hepatitis E virus identified from humans and other animal species: is hepatitis E a zoonosis? *J Hepatol*, **33**, 842–5.

Meng X J, Purcell R H, Halbur P G, Lehman J R, Webb D M, Tsareva T S, Haynes J S, Thacker B J and Emerson S U (1997) A novel virus in swine is closely related to the human hepatitis E virus, *Proc Natl Acad Sci USA*, **94**, 9860–5.

Meng X J, Halbur P G, Shapiro M S, Govindarajan S, Bruna J D, Mushahwar I K, Purcell R H and Emerson S U (1998) Genetic and experimental evidence for cross-species infection by swine hepatitis E virus, *J Virol*, **72**, 9714–21.

Merianos A (2007) Surveillance and response to disease emergence, *Curr Top Microbiol Immunol*, **315**, 477–509.

Milo G E (1971) Thermal inactivation of poliovirus in the presence of selective organic molecules (cholesterol, lecithin, collagen, and beta-carotene), *Appl Microbiol*, **21**, 198–202.

Muniain-Mujika I, Calvo M, Lucena F and Girones R (2003) Comparative analysis of viral pathogens and potential indicators in shellfish, *Int J Food Microbiol*, **83**, 75–85.

Nainan O V, Armstrong G L, Han X H, Williams I, Bell B P and Margolis H S (2005) Hepatitis A molecular epidemiology in the United States, 1996–1997: sources of infection and implications of vaccination policy, *J Infect Dis*, **191**, 957–63.

Nainan O V, Xia G, Vaughan G and Margolis H S (2006) Diagnosis of hepatitis A virus infection: a molecular approach, *Clin Microbiol Rev*, **19**, 63–79.

Nanda S K, Yalcinkaya K, Panigrahi A K, Acharya S K, Jameel S and Panda S K (1994) Etiological role of hepatitis E virus in sporadic fulminant hepatitis, *J Med Virol*, **42**, 133–7.

Nishizawa T, Takahashi M, Mizuo H, Miyajima H, Gotanda Y and Okamoto H (2003) Characterization of Japanese swine and human hepatitis E virus isolates of genotype IV with 99 % identity over the entire genome, *J Gen Virol*, **84**, 1245–51.

Niu M T, Polish L B, Robertson B H, Khanna B K, Woodruff B A, Shapiro C N, Miller M A, Smith J D, Gedrose J K, Alter M J and Margolis H S (1992) Multistate outbreak of hepatitis A associated with frozen strawberries, *J Infect Dis*, **166**, 518–24.

Noyce J O, Michels H and Keevil C W (2007) Inactivation of influenza A virus on copper versus stainless steel surfaces, *Appl Environ Microbiol*, **73**, 2748–50.

Nuanualsuwan S, Mariam T, Himathongkham S and Cliver D O (2002) Ultraviolet inactivation of feline calicivirus, human enteric viruses and coliphages, *Photochem Photobiol*, **76**, 406–10.

OIE (2007) Avian Influenza, *Terrestrial Animal Health Code*, 16th edn, Paris, World Organization for Animal Health.

Olsen B, Munster V J, Wallensten A, Waldenstrom J, Osterhaus A D and Fouchier R A (2006) Global patterns of influenza A virus in wild birds, *Science*, **312**, 384–8.

Pagotto F, Mattison K, Corneau N and Bidawid S (2008) Development of a DNA microarray for the simultaneous detection and genotyping of noroviruses, *J Food Prot*, **71**, 1434–41.

Panda S K, Thakral D and Rehman S (2007) Hepatitis E virus, *Rev Med Virol*, **17**, 151–80.

Papafragkou E, D'Souza D H and Jaykus L A (2006) Foodborne viruses: prevention and control, in Goyal S M and Doyle M P (eds), *Viruses in foods*, New York, Springer, 289–330.

Papafragkou E, Plante M, Mattison K, Bidawid S, Karthikeyan K, Farber J M and Jaykus L A (2008) Rapid and sensitive detection of hepatitis A virus in representative food matrices, *J Virol Methods*, **147**, 177–87.

Parry J V and Mortimer P P (1984) The heat sensitivity of hepatitis A virus determined by a simple tissue culture method, *J Med Virol*, **14**, 277–83.

Pavlov D N (2006) Poliovirus vaccine strains in sewage and river water in South Africa, *Can J Microbiol*, **52**, 717–23.

Pei Y and Yoo D (2002) Genetic characterization and sequence heterogeneity of a Canadian isolate of swine hepatitis E virus, *J Clin Microbiol*, **40**, 4021–9.

Peiris M, Yuen K Y, Leung C W, Chan K H, Ip P L, Lai R W, Orr W K and Shortridge K F (1999) Human infection with influenza H9N2, *Lancet*, **354**, 916–17.

Perez-Gracia M T, Mateos M L, Galiana C, Fernandez-Barredo S, Garcia A, Gomez M T and Moreira V (2007) Autochthonous hepatitis E infection in a slaughterhouse worker, *Am J Trop Med Hyg*, **77**, 893–6.

Peron J M, Bureau C, Poirson H, Mansuy J M, Alric L, Selves J, Dupuis E, Izopet J and Vinel J P (2007) Fulminant liver failure from acute autochthonous hepatitis E in

France: description of seven patients with acute hepatitis E and encephalopathy, *J Viral Hepat*, **14**, 298–303.

Pham B, Duval B, De Serres G, Gilca V, Tricco A C, Ochnio J and Scheifele D W (2005) Seroprevalence of hepatitis A infection in a low endemicity country: a systematic review, *BMC Infect Dis*, **5**, 56.

Phanuwan C, Takizawa S, Oguma K, Katayama H, Yunika A and Ohgaki S (2006) Monitoring of human enteric viruses and coliform bacteria in waters after urban flood in Jakarta, Indonesia, *Water Sci Technol*, **54**, 203–10.

Pillai S D (1997) Rapid molecular detection of microbial pathogens: breakthroughs and challenges, *Arch Virol Suppl*, **13**, 67–82.

Plotkin S A, Koprowski H and Stokes J (1959) Clinical trials in infants of orally administered attenuated poliomyelitis viruses, *Pediatrics*, **23**, 1041–62.

Pontrelli G, Boccia D, Di Renzi M, Massari M, Giugliano F, Celentano L P, Taffon S, Genovese D, Di Pasquale S, Scalise F, Rapicetta M, Croci L and Salmaso S (2007) Epidemiological and virological characterization of a large community-wide outbreak of hepatitis A in southern Italy, *Epidemiol Infect*, **136**, 1027–34.

Prato R, Lopalco P L, Chironna M, Germinario C and Quarto M (2006) An outbreak of hepatitis A in Southern Italy: the case for vaccinating food handlers, *Epidemiol Infect*, **134**, 799–802.

Pusch D, Oh D Y, Wolf S, Dumke R, Schroter-Bobsin U, Hohne M, Roske I and Schreier E (2005) Detection of enteric viruses and bacterial indicators in German environmental waters, *Arch Virol*, **150**, 929–47.

Rab M A, Bile M K, Mubarik M M, Asghar H, Sami Z, Siddiqi S, Dil A S, Barzgar M A, Chaudhry M A and Burney M I (1997) Water-borne hepatitis E virus epidemic in Islamabad, Pakistan: a common source outbreak traced to the malfunction of a modern water treatment plant, *Am J Trop Med Hyg*, **57**, 151–7.

Rabenau H F, Cinatl J, Morgenstern B, Bauer G, Preiser W and Doerr H W (2005a) Stability and inactivation of SARS coronavirus, *Med Microbiol Immunol (Berl)*, **194**, 1–6.

Rabenau H F, Kampf G, Cinatl J and Doerr H W (2005b) Efficacy of various disinfectants against SARS coronavirus, *J Hosp Infect*, **61**, 107–11.

Rajagopal S and Treanor J (2007) Pandemic (avian) influenza, *Semin Respir Crit Care Med*, **28**, 159–70.

Reuter G, Fodor D, Katai A and Szucs G (2006) Identification of a novel variant of human hepatitis E virus in Hungary, *J Clin Virol*, **36**, 100–102.

Richards G P (2001) Enteric virus contamination of foods through industrial practices: a primer on intervention strategies, *J Ind Microbiol Biotechnol*, **27**, 117–25.

Rzezutka A, D'Agostino M and Cook N (2006) An ultracentrifugation-based approach to the detection of hepatitis A virus in soft fruits, *Int J Food Microbiol*, **108**, 315–20.

Sabin A B (1991) Perspectives on rapid elimination and ultimate global eradication of paralytic poliomyelitis caused by polioviruses, *Eur J Epidemiol*, **7**, 95–120.

Sagripanti J L and Lytle C D (2007) Inactivation of influenza virus by solar radiation, *Photochem Photobiol*, **83**, 1278–82.

Sanchez G, Bosch A and Pinto R M (2007) Hepatitis A virus detection in food: current and future prospects, *Lett Appl Microbiol*, **45**, 1–5.

Sattar S A and Tetro J A (2001) Other foodborne viruses, in Hui Y H, Sattar S A, Murrell K D, Nip W-K and Stanfield P S (eds), *Foodborne Disease Handbook: Viruses, Parasites, Pathogens and HAACP*, New York, Marcel Dekker, 127–36.

Schenkel K, Bremer V, Grabe C, van Treeck U, Schreier E, Hohne M, Ammon A and Alpers K (2006) Outbreak of hepatitis A in two federal states of Germany: bakery products as vehicle of infection, *Epidemiol Infect*, **134**, 1292–8.

Scholtissek C (1985a) Stability of infectious influenza A viruses at low pH and at elevated temperature, *Vaccine*, **3**, 215–18.

Scholtissek C (1985b) Stability of infectious influenza A viruses to treatment at low pH and heating, *Arch Virol*, **85**, 1–11.

Scholz E, Heinricy U and Flehmig B (1989) Acid stability of hepatitis A virus, *J Gen Virol*, **70**(pt 9), 2481–5.

Schurmann W and Eggers H J (1985) An experimental study on the epidemiology of enteroviruses: water and soap washing of poliovirus 1 – contaminated hands, its effectiveness and kinetics, *Med Microbiol Immunol*, **174**, 221–36.

Schwerdt C E and Fogh J (1957) The ratio of physical particles per infectious unit observed for poliomyelitis viruses, *Virology*, **4**, 41–52.

Shan X C, Wolffs P and Griffiths M W (2005) Rapid and quantitative detection of hepatitis A virus from green onion and strawberry rinses by use of real-time reverse transcription-PCR, *Appl Environ Microbiol*, **71**, 5624–6.

Shapiro C N and Margolis H S (1993) Worldwide epidemiology of hepatitis A virus infection, *J Hepatol*, **18** Suppl 2, S11–4.

Shieh Y C, Khudyakov Y E, Xia G, Ganova-Raeva L M, Khambaty F M, Woods J W, Veazey J E, Motes M L, Glatzer M B, Bialek S R and Fiore A E (2007) Molecular confirmation of oysters as the vector for hepatitis A in a 2005 multistate outbreak, *J Food Prot*, **70**, 145–50.

Shortridge K F, Gao P, Guan Y, Ito T, Kawaoka Y, Markwell D, Takada A and Webster R G (2000) Interspecies transmission of influenza viruses: H5N1 virus and a Hong Kong SAR perspective, *Vet Microbiol*, **74**, 141–7.

Shukla P, Chauhan U K, Naik S, Anderson D and Aggarwal R (2007) Hepatitis E virus infection among animals in northern India: an unlikely source of human disease. *J Viral Hepat*, **14**, 310–17.

Siegl G, Weitz M and Kronauer G (1984) Stability of hepatitis A virus, *Intervirology*, **22**, 218–26.

Simonet J and Gantzer C (2006) Degradation of the poliovirus 1 genome by chlorine dioxide *J Appl Microbiol*, **100**, 862–70.

Sims L D (2007) Lessons learned from Asian H5N1 outbreak control, *Avian Dis*, **51**, 174–81.

Singh J, Sharma R S and Verghese T (1996) Epidemiological considerations on age distribution of paralytic poliomyelitis, *J Trop Pediatr*, **42**, 237–41.

Sixl W, Stunzner D, Withalm H and Kock M (1989) Rare transmission mode of FSME (tick-borne encephalitis) by goat's milk, *Geogr Med Suppl*, **2**, 11–4.

Skidmore S J and Sherratt L M (1996) Hepatitis E infection in the UK, *J Viral Hepat*, **3**, 103–5.

Skraber S, Helmi K, Willame R, Ferreol M, Gantzer C, Hoffmann L and Cauchie H M (2007) Occurrence and persistence of bacterial and viral faecal indicators in wastewater biofilms, *Water Sci Technol*, **55**, 377–85.

Sobsey M D, Wallis C and Melnick J L (1975) Development of a simple method for concentrating enteroviruses from oysters, *Appl Microbiol*, **29**, 21–6.

Somani S K, Aggarwal R, Naik S R, Srivastava S and Naik S (2003) A serological study of intrafamilial spread from patients with sporadic hepatitis E virus infection, *J Viral Hepat*, **10**, 446–9.

Speirs J I, Pontefract R D and Harwig J (1987) Methods for recovering poliovirus and rotavirus from oysters, *Appl Environ Microbiol*, **53**, 2666–70.

Staes C J, Schlenker T L, Risk I, Cannon K G, Harris H, Pavia A T, Shapiro C N and Bell B P (2000) Sources of infection among persons with acute hepatitis A and no identified risk factors during a sustained community-wide outbreak, *Pediatrics*, **106**, E54.

Steffen R (2005) Changing travel-related global epidemiology of hepatitis A, *Am J Med*, **118** Suppl 10A, 46S–49S.

Stine S W, Song I, Choi C Y and Gerba C P (2005) Effect of relative humidity on preharvest survival of bacterial and viral pathogens on the surface of cantaloupe, lettuce, and bell peppers, *J Food Prot*, **68**, 1352–8.

Stohr K (2003) Overview of the WHO Global Influenza Programme, *Dev Biol (Basel)*, **115**, 3–8.

Stramer S L and Cliver D O (1984) Septage treatments to reduce the numbers of bacteria and polioviruses, *Appl Environ Microbiol*, **48**, 566–72.
Strazynski M, Kramer J and Becker B (2002) Thermal inactivation of poliovirus type 1 in water, milk and yoghurt, *Int J Food Microbiol*, **74**, 73–8.
Suarez D L, Spackman E, Senne D A, Bulaga L, Welsch A C and Froberg K (2003) The effect of various disinfectants on detection of avian influenza virus by real time RT-PCR, *Avian Dis*, **47**, 1091–5.
Sullivan R and Read R B Jr (1968) Method for recovery of viruses from milk and milk products, *J Dairy Sci*, **51**, 1748–51.
Swayne D E (2006a) Microassay for measuring thermal inactivation of H5N1 high pathogenicity avian influenza virus in naturally infected chicken meat, *Int J Food Microbiol*, **108**, 268–71.
Swayne D E (2006b) Occupational and consumer risks from avian influenza viruses, *Dev Biol (Basel)*, **124**, 85–90.
Swayne D E and Beck J R (2004) Heat inactivation of avian influenza and Newcastle disease viruses in egg products, *Avian Pathol*, **33**, 512–18.
Tam A W, Smith M M, Guerra M E, Huang C C, Bradley D W, Fry K E and Reyes G R (1991) Hepatitis E virus (HEV): molecular cloning and sequencing of the full-length viral genome, *Virology*, **185**, 120–31.
Tanaka T, Takahashi M, Kusano E and Okamoto H (2007) Development and evaluation of an efficient cell-culture system for Hepatitis E virus, *J Gen Virol*, **88**, 903–11.
Tanner B D, Kuwahara S, Gerba C P and Reynolds K A (2004) Evaluation of electrochemically generated ozone for the disinfection of water and wastewater, *Water Sci Technol*, **50**, 19–25.
Tei S, Kitajima N, Takahashi K and Mishiro S (2003) Zoonotic transmission of hepatitis E virus from deer to human beings, *Lancet*, **362**, 371–3.
Tekeuchi Y, Kobayashi G, Matui Y, Miyajima Y, Tanahashi S, Honma M, Takahashi M, Eguchi H and Tanaka M (2006) Outbreak of food-borne infection with hepatitis A virus, *Jpn J Infect Dis*, **59**, 346.
Teo C G (2006) Hepatitis E indigenous to economically developed countries: to what extent a zoonosis? *Curr Opin Infect Dis*, **19**, 460–6.
Teo C G (2007) The two clinico-epidemiological forms of hepatitis E, *J Viral Hepat*, **14**, 295–7.
Terpstra F G, Van Den Blink A E, Bos L M, Boots A G, Brinkhuis F H, Gijsen E, Van Remmerden Y, Schuitemaker H and Van't Wout A B (2007) Resistance of surface-dried virus to common disinfection procedures, *J Hosp Infect*, **66**, 332–8.
Thomas C and Swayne D E (2007) Thermal inactivation of H5N1 high pathogenicity avian influenza virus in naturally infected chicken meat, *J Food Prot*, **70**, 674–80.
Tierney J T, Sullivan R and Larkin E P (1977) Persistence of poliovirus 1 in soil and on vegetables grown in soil previously flooded with inoculated sewage sludge or effluent, *Appl Environ Microbiol*, **33**, 109–13.
Todd E C, Greig J D, Bartleson C A and Michaels B S (2007a) Outbreaks where food workers have been implicated in the spread of foodborne disease. Part 2. Description of outbreaks by size, severity, and settings, *J Food Prot*, **70**, 1975–93.
Todd E C, Greig J D, Bartleson C A and Michaels B S (2007b) Outbreaks where food workers have been implicated in the spread of foodborne disease. Part 3. Factors contributing to outbreaks and description of outbreak categories, *J Food Prot*, **70**, 2199–217.
Traore O, Arnal C, Mignotte B, Maul A, Laveran H, Billaudel S and Schwartzbrod L (1998) Reverse transcriptase PCR detection of astrovirus, hepatitis A virus, and poliovirus in experimentally contaminated mussels: comparison of several extraction and concentration methods, *Appl Environ Microbiol*, **64**, 3118–22.
Trottier Y L, Houde A, Buenaventura E, Muller P, Brassard J, Christensen K and Liu J (2006) *Concentration of Norovirus Genogroups I and II from Contaminated Oysters*

and their Detection Using the Reverse-Transcription Polymerase Chain Reaction, Ottawa, Health Canada, available at: http://www.hc-sc.gc.ca/fn-an/res-rech/analy-meth/microbio/volume5/opflp_01_e.html, accessed November 2008.

Tsatsralt-Od B, Takahashi M, Endo K, Agiimaa D, Buyankhuu O and Okamoto H (2007) Comparison of hepatitis A and E virus infections among healthy children in Mongolia: evidence for infection with a subgenotype IA HAV in children, *J Med Virol*, **79**, 18–25.

Van Der Poel W H, Verschoor F, Van Der Heide R, Herrera M I, Vivo A, Kooreman M and De Roda Husman A M (2001) Hepatitis E virus sequences in swine related to sequences in humans, The Netherlands, *Emerg Infect Dis*, **7**, 970–6.

Van Engelenburg F A, Terpstra F G, Schuitemaker H and Moorer, W R (2002) The virucidal spectrum of a high concentration alcohol mixture, *J Hosp Infect*, **51**, 121–5.

Van Reeth K (2007) Avian and swine influenza viruses: our current understanding of the zoonotic risk, *Vet Res*, **38**, 243–60.

Vasickova P, Psikal I, Kralik P, Widen F, Hubalek Z and Pavlik I (2007) Hepatitis E virus: a review, *Vet Medicina*, **52**, 365–84.

Velazquez O, Stetler H C, Avila C, Ornelas G, Alvarez C, Hadler S C, Bradley D W and Sepulveda J (1990) Epidemic transmission of enterically transmitted non-A, non-B hepatitis in Mexico, 1986–1987, *JAMA*, **263**, 3281–5.

Villar L M, De Paula V S, Diniz-Mendes L, Guimaraes F R, Ferreira F F, Shubo T C, Miagostovich M P, Lampe E and Gaspar A M (2007) Molecular detection of hepatitis A virus in urban sewage in Rio de Janeiro, Brazil, *Lett Appl Microbiol*, **45**, 168–73.

Waar K, Herremans M M, Vennema H, Koopmans M P and Benne C A (2005) Hepatitis E is a cause of unexplained hepatitis in The Netherlands, *J Clin Virol*, **33**, 145–9.

Wang J T and Chang S C (2004) Severe acute respiratory syndrome. *Curr Opin Infect Dis*, **17**, 143–8.

Wang X W, Li J S, Guo T K, Zhen B, Kong Q X, Yi B, Li Z, Song N, Jin M, Wu X M, Xiao W J, Zhu X M, Gu C Q, Yin J, Wei W, Yao W, Liu C, Li J F, Ou G R, Wang M N, Fang T Y, Wang G J, Qiu Y H, Wu H H, Chao F H and Li J W (2005) Excretion and detection of SARS coronavirus and its nucleic acid from digestive system, *World J Gastroenterol*, **11**, 4390–95.

Wasley A, Samandari T and Bell B P (2005) Incidence of hepatitis A in the United States in the era of vaccination, *JAMA*, **294**, 194–201.

Weber D J, Barbee S L, Sobsey M D and Rutala W A (1999) The effect of blood on the antiviral activity of sodium hypochlorite, a phenolic, and a quaternary ammonium compound, *Infect Control Hosp Epidemiol*, **20**, 821–7.

Wheeler C, Vogt T M, Armstrong G L, Vaughan G, Weltman A, Nainan O V, Dato V, Xia G, Waller K, Amon J, Lee T M, Highbaugh-Battle A, Hembree C, Evenson S, Ruta M A, Williams I T, Fiore A E and Bell B P (2005) An outbreak of hepatitis A associated with green onions, *N Engl J Med*, **353**, 890–97.

WHO (1994) Outbreak of tick-borne encephalitis, presumably milk-borne, *Wkly Epidemiol Rec*, **69**, 140–41.

WHO (2007) Meeting of National Influenza Centres – Western Pacific and South-East Asia regions, *Wkly Epidemiol Rec*, **82**, 389–95.

Wood D J, Sutter R W and Dowdle W R (2000) Stopping poliovirus vaccination after eradication: issues and challenges, *Bull World Health Organ*, **78**, 347–57.

Wood N, McIntyre P and Wong M (2006) Vaccination for the paediatrician, *J Paediatr Child Health*, **42**, 665–73.

Yates M V, Gerba C P and Kelley L M (1985) Virus persistence in groundwater, *Appl Environ Microbiol*, **49**, 778–81.

Yazaki Y, Mizuo H, Takahashi M, Nishizawa T, Sasaki N, Gotanda Y and Okamoto H (2003) Sporadic acute or fulminant hepatitis E in Hokkaido, Japan, may be food-borne, as suggested by the presence of hepatitis E virus in pig liver as food, *J Gen Virol*, **84**, 2351–7.

Yeager J G and O'Brien R T (1979) Enterovirus inactivation in soil, *Appl Environ Microbiol*, **38**, 694–701.

Young N A (1973) Polioviruses, coxsackieviruses, and echoviruses comparison of the genomes by RNA hybridization, *J Virol*, **11**, 832–9.

Zaaijer H L, Kok M, Lelie P N, Timmerman R J, Chau K and Van Der Pal H J (1993) Hepatitis E in The Netherlands: imported and endemic, *Lancet*, **341**, 826.

Zafrullah M, Khursheed Z, Yadav S, Sahgal D, Jameel S and Ahmad F (2004) Acidic pH enhances structure and structural stability of the capsid protein of hepatitis E virus, *Biochem Biophys Res Commun*, **313**, 67–73.

Zhang J Z (2003) Severe acute respiratory syndrome and its lesions in digestive system, *World J Gastroenterol*, **9**, 1135–8.

26

Parasites: *Cryptosporidium*, *Giardia*, *Cyclospora*, *Entamoeba histolytica*, *Toxoplasma gondii* and pathogenic free-living amoebae (*Acanthamoeba* spp. and *Naegleria fowleri*) as foodborne pathogens

H. Smith, Stobhill Hospital, UK, and R. Evans, Raigmore Hospital, UK

Abstract: This chapter describes the protozoan parasites *Giardia*, *Cryptosporidium*, *Cyclospora*, *Entamoeba histolytica*, *Toxoplasma gondii* and pathogenic free-living amoebae which may be implicated as significant contaminants of food. It relates the complex life cycles of these organisms to the human host and discusses the incidence, transmissibility and clinical consequences of infection. The methods of detection, control of infection and the current regulations for public health are described. Future trends and developments are identified.

Key words: protozoan parasites, coccidia, giardiasis, cryptosporidiosis, cyclosporiasis, toxoplasmosis, primary amoebic meningoencephalitis (PAM), granulomatous amoebic encephalitis (GAE), foodborne outbreaks, detection, control of infection, regulations, public health.

26.1 Introduction

Giardia, *Cryptosporidium*, *Cyclospora*, *Entamoeba histolytica*, *Toxoplasma gondii* and pathogenic free-living amoebae (FLA) are protozoan parasites which parasitise both human and non-human hosts. Increasing evidence, over the last 30 years, has implicated protozoan parasites as significant contaminants of food. Protozoan parasite life cycles can be complex and consist of reproductive stages, which infect human and non-human hosts, and transmissive stages, with the exception of FLA, which are excreted by infected definitive hosts. Definitive (final) hosts are those in which the parasite's life cycle can be completed.

Giardia, *Cryptosporidium*, *Cyclospora* and *E. histolytica* complete their life cycles in the human host, but *Toxoplasma gondii* can only complete its life cycle in the feline host. As the name implies, the life cycle of FLA is completed in the environment (e.g. water, silt, wet soil).

The life cycles of *Giardia*, *Cryptosporidium*, *Cyclospora* and *E. histolytica* consist of reproductive stages, which infect the intestine, and transmissive stages (cysts of *Giardia* and *E. histolytica* and oocysts of *Cryptosporidium*, *Cyclospora* and *T. gondii* {(oo)cysts}) which are excreted in the faeces of infected definitive hosts. In contrast to *Giardia*, *Cryptosporidium*, *Cyclospora* and *E. histolytica*, *T. gondii* requires two separate hosts (definitive and intermediate) to complete its life cycle, humans being intermediate hosts. Of great importance is that (oo)cysts are environmentally robust, being capable of prolonged survival in moist dark environments. Whereas cysts of *Giardia* and oocysts of *Cryptosporidium* are infectious to susceptible hosts immediately following excretion, oocysts of *Cyclospora* and *T. gondii* are not infectious when excreted and require a period of maturation in the environment before they become infective to other hosts. The trophozoites and cysts of FLA are natural contaminants of our environment.

26.2 Description of the organisms

26.2.1 Protozoan parasites and their taxonomy

Host species, site of development and morphometry (size and shape) are the historical criteria used to classify parasites. Latterly, molecular methods for determining species and genotypes have also been described and, where possible, should be used in conjunction with more conventional biological parameters. Current classification is based upon a variety of parameters including host preference and cross transmissibility, morphological differences, sites of infection, etc., and molecular taxonomic methods. Appropriate molecular methods have provided insights into the taxonomy and epidemiology of these important protozoal diseases of humans (Cacciò *et al.*, 2005; Smith *et al.*, 2006, 2007). Previously unrecognised differences in disease, symptomatology, zoonotic potential, risk factors and environmental contamination have been identified using molecular tools appropriate for species, genotype and sub-genotype analysis (Cacciò *et al.*, 2005; Smith *et al.*, 2006, 2007).

26.2.2 Life cycles

Giardia

Within the genus *Giardia* are species that are host-specific and species that have zoonotic potential. Based on trophozoite morphology, there are six

Giardia species (*G. muris* infects rodents, *G. agilis* infects amphibians, *G. psittaci* and *G. ardeae* infect birds, *G. ardeae* infects great blue herons, *G. microti* infects the prairie vole and *G. duodenalis* infects mammals). The parasites which infect humans belong to the species '*duodenalis*', but historically have also been named *G. intestinalis* and *G. lamblia*, which has led to a degree of confusion. As humans can be infected by parasites from non-human hosts, *Giardia* parasites that infect humans are now referred to as *G. duodenalis*.

Because of the lack of morphological differences, an informal categorisation based on genetic differences between genetic variants of *G. duodenalis* found in mammals has been used to classify this species (Thompson and Monis, 2004). Using a variety of genetic loci, including small sub-unit (SSU) rDNA, surface protein genes, glutamate dehydrogenase, triosephosphate isomerase, beta-giardin and other catabolic enzyme genes (Homan *et al.*, 1992; Meloni *et al.*, 1995; Hopkins *et al.*, 1997; Monis *et al.*, 1999; Thompson, 2000; Homan and Mank 2001; Lalle *et al.*, 2005), at least seven genetic groups or 'assemblages' of *G. duodenalis* have been recognised (Table 26.1). Some of these assemblages are so genetically distinct and/or have a limited or very specific host range, that it is possible they may be distinct (cryptic) species (Table 26.1) (Thompson and Monis, 2004). *Giardia duodenalis* assemblages A and B infect humans and most mammalian orders including Artiodactyla, Rodentia, Primates, Carnivora, Pinnipedia and Hyracoidea (Appelbee *et al.*, 2005). The prevalences of *G. duodenalis* assemblages A and B that infect humans vary considerably from country to country (Cacciò *et al.*, 2005).

Following ingestion, exposure to the acidity of the stomach and the alkalinity of the jejunum induce the quadrinucleate trophozoite within the cyst to excyst. The quadrinucleate trophozoite undergoes cytokinesis, whereby the cytoplasm of the trophozoite, but not the nuclei, undergoes division to produce two pyriform (pear-shaped) *G. duodenalis* trophozoites. Binucleate trophozoites attach onto the apical surfaces of enterocytes (the epithelial cells

Table 26.1 Genotypes (assemblages) of *Giardia duodenalis* found in mammals

Genotype	Host range
Assemblage A	humans, slow loris, livestock, deer, cats, dogs, beavers, muskrats, voles, guinea pigs, ferret
Assemblage B	humans, siamung, slow loris, livestock, chinchillas, dogs, beavers, muskrats, voles, rats, marmoset
Assemblage C/D	dogs, coyotes
Assemblage E	alpaca, cattle, goats, pigs, sheep
Assemblage F	cats
Assemblage G	rats

lining the intestinal tract) and divide by binary fission. Detachment from enterocytes, together with exposure to increased concentration of bile salts and elevated pH during passage through the lumen of the small intestine, cause trophozoites to encyst into ovoid cysts which are excreted in faeces. Mature cysts are quadrinucleate. The life cycle of *Giardia* is presented in Fig. 26.1.

Cryptosporidium
Originally described by Tyzzer (1910, 1912), *Cryptosporidium* has emerged as an important pathogen of human beings in the last 25 years. Although more than 20 'species' of this coccidian parasite have been described on the basis of the animal hosts from which they were isolated, host specificity as a criterion for speciation appears to be ill-founded as some 'species' lack such specificity (Xiao *et al.*, 2004). Oocysts can vary in shape and size (Table 26.2), but these often overlap and, for the majority of species, (oo)cyst morphometry cannot be used to ascribe *Cryptosporidium* species.

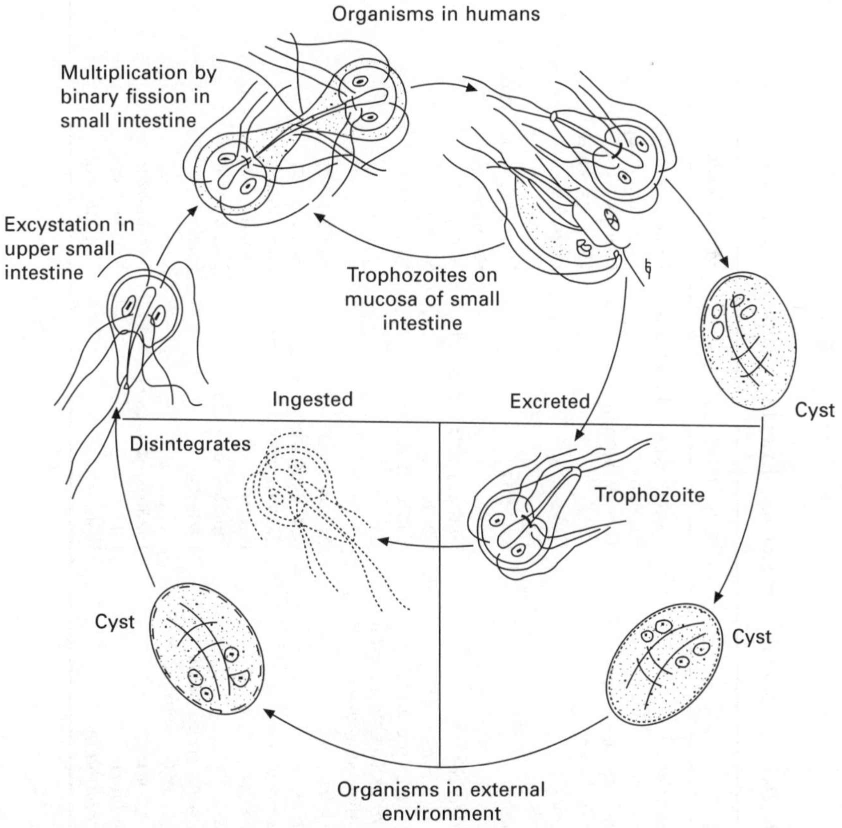

Fig. 26.1 Life cycle of *Giardia duodenalis.*

Table 26.2 Some biological features of valid *Cryptosporidium* species

Species	Major hosts	Minor hosts	Site of infection	Dimensions of oocysts (μm)
C. parvum	Cattle, livestock, humans	Deer, mice, pigs	Small intestine	4.5 × 5.5
C. hominis	Humans	Dugong, sheep	Small intestine	4.5 × 5.5
C. meleagridis	Turkey, humans	Parrots	Small intestine	4.5–4.0 × 4.6–5.2
C. felis	Cats	Humans, cattle	Small intestine	4.5 × 5.0
C. canis	Dogs	Humans	Small intestine	4.95 × 4.71
C. suis	Pigs	Humans	Small and large intestine	4.9–4.4 × 4.0–4.3
C. muris	Rodents	Humans, rock hyrax, mountain goat	Stomach	5.6 × 7.4
C. andersoni	Cattle, Bactrian camel	Sheep	Abomasum	5.5 × 7.4
C. bovis	Cattle	Sheep	Small intestine	4.7–5.3 × 4.2–4.8
C. ryanae (previously known as *Cryptosporidium* deer-like genotype)	Cattle, Bos taurus	Not known	Not known	2.94–4.41 × 2.94–3.68
C. wrairi	Guinea pigs	Not known	Small intestine	4.9–5.0 × 4.8–5.6
C. fayeri	Red kangaroo	Not known	Small intestine	4.5–5.1 × 3.8–5.0
C. macropodum	Eastern grey kangaroo	Not known	Not known	4.5–6.0 × 5.0–6.0
C. baileyi	Poultry	Quails, ostriches, ducks	Bursa	4.6 × 6.2
C. galli	Finches, chicken	Not known	Proventriculus	8.25 × 6.3
C. varanii (previously known as *C. saurophilum*)	Lizards	Snakes	Stomach and small intestine	4.2–5.2 × 4.4–5.6
C. serpentis	Lizards, snakes	Not known	Stomach	5.6–6.6 × 4.8–5.6
C. molnari	Fish	Not known	Stomach (and intestine)	4.7 × 4.5
C. scophthalmi	Fish	Not known	Intestine (and stomach)	3.7–5.0 × 3.0–4.7

Cryptosporidium infects a wide range of vertebrate hosts including mammals, rodents, birds, reptiles and fish. There are 19 'valid' *Cryptosporidium* species (Table 26.2) and more than 40 genotypes, which differ significantly in their molecular signatures but, as yet, have not been ascribed species status. Eight described *Cryptosporidium* species (*C. hominis*, *C. parvum*, *C. meleagridis*, *C. felis*, *C. canis*, *C. suis*, *C. muris* and *C. andersoni*) and five undescribed species of *Cryptosporidium* (cervine, monkey, skunk, rabbit and chipmunk genotypes) infect immunocompetent and immunocompromised humans (Xiao *et al.*, 2004; Cacciò *et al.*, 2005; Feltus *et al.*, 2006; Nichols *et al.*, 2006a), but *C. hominis* and C. *parvum* are the most commonly detected (Cacciò *et al.*, 2005). *Cryptosporidium parvum* is the major zoonotic species and causes neonatal diarrhoea in livestock, with consequent economic loss. Furthermore, *C. parvum* cryptosporidiosis in livestock is a major contributor to environmental contamination with oocysts. Molecular investigations into the species and genotypes isolated from wildlife indicate that they also harbour their own host-adapted species, which may not be infectious to humans. Increasing the breadth of molecular studies is necessary, since they help define the zoonotic potential of oocysts found as contaminants of water and food.

Cryptosporidium parvum is the best studied species and its life cycle is complex (Fig. 26.2), comprising asexual, sexual and transmissive stages in a single host (monoxenous). The spherical oocyst measures 4.5–5.5 μm in diameter and contains four naked (not within a sporocyst) crescentic sporozoites (Table 26.2; Fig. 26.2). Fayer (2008) provides a good account of the general biology of *Cryptosporidium.*

Cyclospora

Cyclospora organisms have been implicated in human enteritis since 1977, but *C. cayetanensis*, which infects humans, was only identified as a coccidian parasite in 1992. Prior to 1992, its classification remained in doubt, being referred to, among others as 'cyanobacterium-like bodies' and 'coccidia-like bodies'. *Cyclospora cayetanensis* (Ortega *et al.*, 1993) is closely related to the genus *Eimeria* (Relman *et al.*, 1996). Eleven species of *Cyclospora* have been described from moles, rodents, insectivores, snakes and humans. Three new species of *Cyclospora* isolated from monkeys and baboons from western Ethiopia have been proposed: *C. cercopitheci* from green monkeys, *C. colobi* from colobus monkeys and *C. papionis* from baboons (Eberhard *et al.*, 1999). *Cyclospora* oocysts are spherical, measuring 8–10 μm in diameter, and are excreted unsporulated.

The life cycle of *Cyclospora* is not fully understood, but involves both sexual and asexual stages of development in a single host (Fig. 26.3). As for *Giardia* and *Cryptosporidium*, exposure to the acidity of the stomach and the alkalinity of the jejunum causes the sporozoites contained within sporocysts to excyst. Two types of meronts and sexual stages were observed in the jejunal enterocytes of biopsy material from oocyst excreting humans (Ortega *et al.*, 1997a). Under laboratory conditions, 40 % of oocysts exposed to

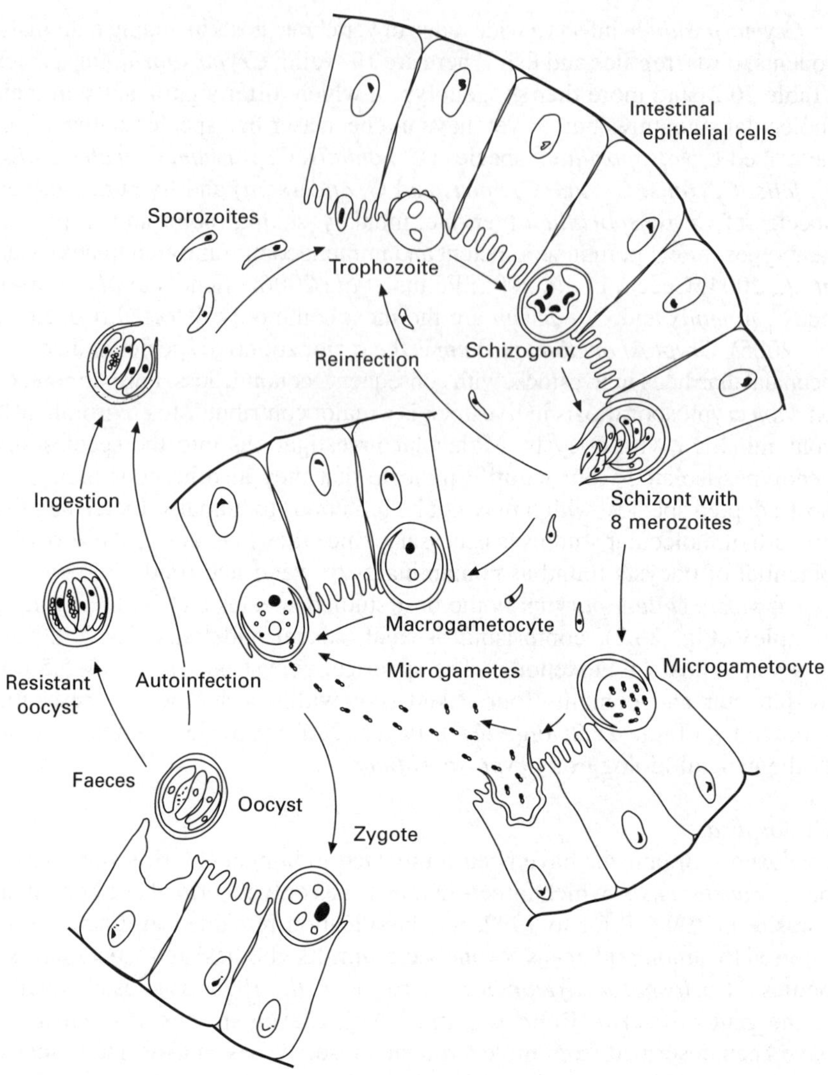

Fig. 26.2 Life cycle of *Cryptosporidium parvum*.

temperatures of 25–30 °C sporulated after 1–2 weeks, each oocyst containing two sporocysts with two sporozoites within each sporocyst (Ortega *et al.*, 1993; Smith *et al.*, 1997).

Entamoeba histolytica

Of the six *Entamoeba* species that infect humans, only *E. histolytica* is known to cause disease. *Entamoeba histolytica* infects predominantly humans and other primates, although other mammals such as dogs and cats can become infected but usually do not shed cysts and hence contribute

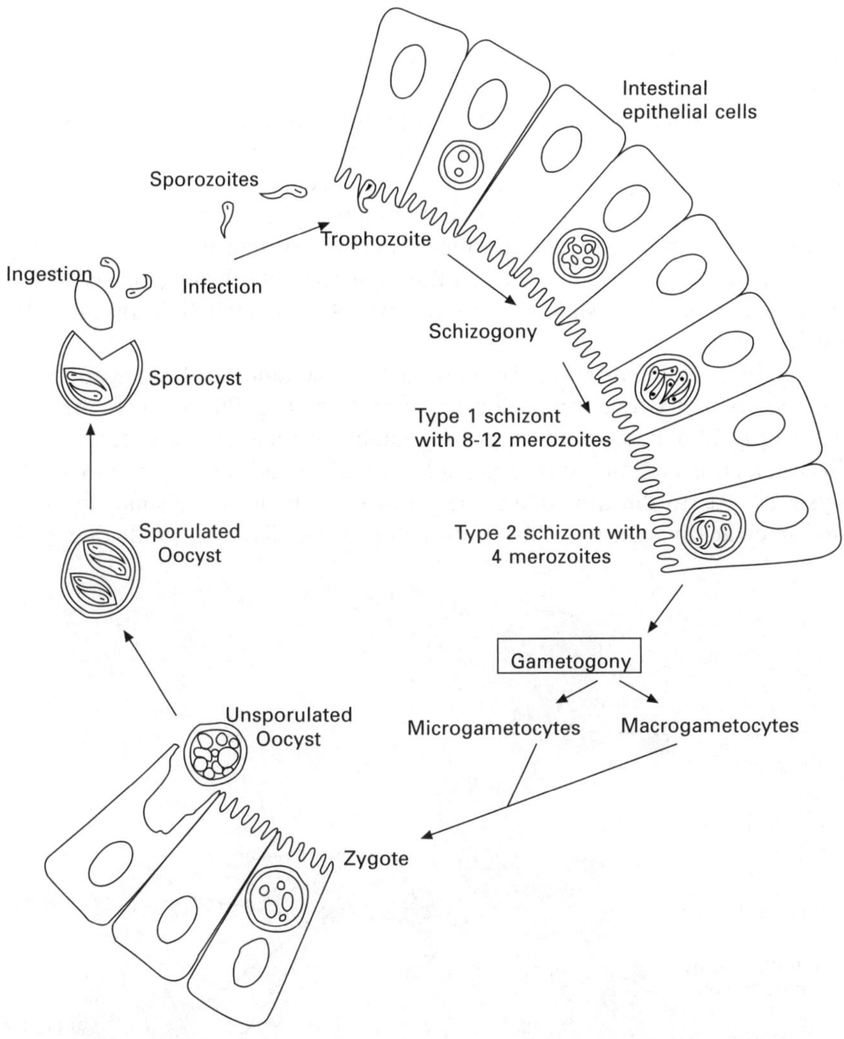

Fig. 26.3 Life cycle of *Cyclospora cayatanensis*.

minimally to transmission. Molecular analyses revealed that the parasites that were previously regarded as *E. histolytica*, based on cyst morphology and morphometry, are a species complex, consisting of two species *E. histolytica* and *E. dispar*, with morphologically identical cysts (Clark and Diamond, 1991; Diamond and Clark, 1993). These were previously known as 'pathogenic *E. histolytica*' and 'non-pathogenic *E. histolytica*' respectively. *Entamoeba histolytica* and *E. dispar* differ genetically and in their capacity to cause disease. *Entamoeba histolytica* is an invasive pathogen exhibiting varying degrees of virulence, while *E. dispar* is non-invasive (Diamond and Clark, 1993; Clark and Diamond, 1994).

Clearly, methods that can distinguish between these two *Entamoeba* species are important for accurate diagnosis, for preventing unnecessary or inappropriate chemotherapy and for generating effective baseline data of the global risk from *E. histolytica*. In older publications, *E. histolytica* was cited as infecting one tenth of the world population (~500 million people) but, while the relative prevalence of these two species is not yet fully known, it is clear that *E. dispar* is the more common in most regions of the world. *Entamoeba histolytica* is the third leading cause of morbidity and mortality due to parasitic disease in humans (after malaria and schistosomiasis) and is estimated to be responsible for between 50 000 and 100 000 deaths annually (WHO, 1997).

The life cycle consists of the reproductive and amoeboid trophozoite and the cyst which is quadrinucleate when mature (Fig. 26.4). Cysts measure 10–15 μm in diameter and typically contain four nuclei. Cysts can survive in water, soils and on foods, especially in moist conditions. When ingested, cysts excyst (to the trophozoite stage) in the lumen of the small intestine and nuclear division is followed by cytoplasmic division, producing eight

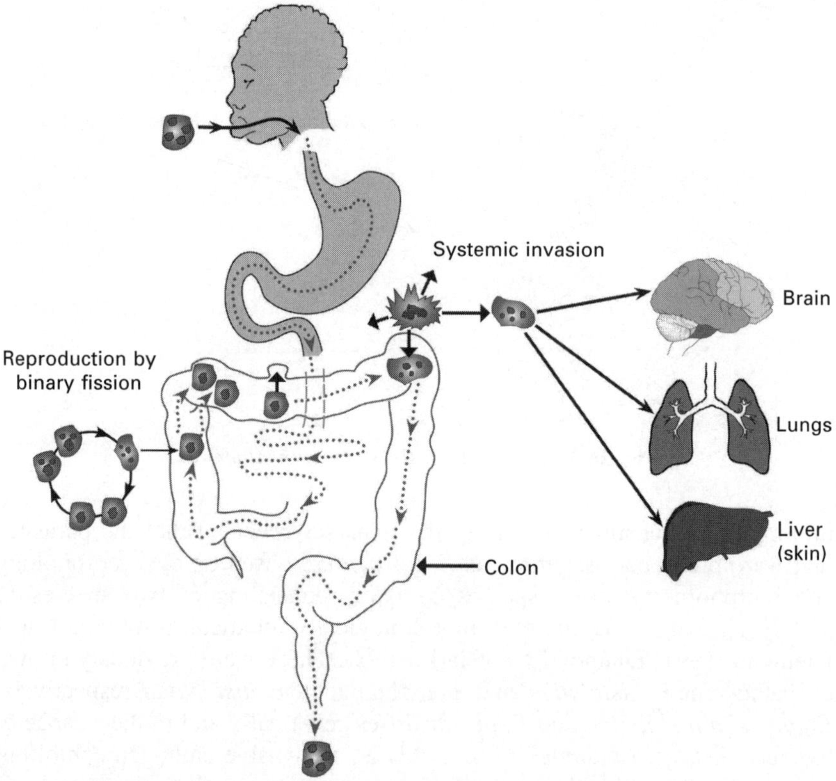

Fig. 26.4 Life cycle of *Entamoeba histolytica*.

trophozoites. Trophozoites measure 10–50 μm in diameter and contain a single nucleus with a central karyosome, and reside in the lumen of the caecum and large intestine, where they adhere to the colonic mucus and epithelial layers lining the caecum and large intestine. In asymptomatic infections, trophozoites encyst in the lumen of the colon, and cysts are excreted in the faeces. Trophozoites can invade the colonic epithelium, causing amoebic colitis. Amoebic dysentery usually occurs gradually. If trophozoites penetrate the colonic epithelium, *E. histolytica* can disseminate *via* the bloodstream (haematogenous spread) and can establish persistent extraintestinal infections, most commonly as abscesses (amoebic liver abscesses) in the liver, but spread to other organs, such as the brain and lungs, causing further pathological manifestations.

Toxoplasma gondii

Toxoplasma gondii is a coccidian parasite of man and animals with a worldwide distribution (Dubey and Beattie, 1988). Although toxoplasmosis has been associated with food and food products, the extent of its morbidity and mortality is probably greater than has been previously recognised. *Toxoplasma gondii* is now considered one of the most significant foodborne parasite pathogens in the USA (Morris, 1996; Mead *et al.*, 1999) and Europe (Vaillant *et al.*, 2005).

The genus *Toxoplasma* consists of only one species, *T. gondii* and, based on multilocus restriction fragment length polymorphism (RFLP) techniques, there are three genotypes (genotypes I, II and III) which correlate with different patterns of human disease (Howe and Sibley, 1995). The majority of human toxoplasmosis cases belong to genotype II. Unlike *Giardia*, *Cryptosporidium*, *Cyclospora* and *E. histolytica*, *T. gondii* requires two separate hosts (definitive and intermediate) to complete its life cycle. The life cycle consists of asexual and sexual cycles, with the sexual cycle, resulting in the formation and excretion of oocysts, occurring only in the intestinal epithelial cells of the definitive host, namely domestic cats and other felids (Fig. 26.5). The intermediate hosts consist of a wide range of birds and mammals, including man. There are three infective stages of the parasite: the oocyst (the environmentally robust transmissive stage), the tachyzoite and the tissue cyst (Fig. 26.5). Felines acquire the infection from eating infected tissues (e.g. musculature and brain infected with asexual stages) of mammals and birds or, less commonly, from ingesting sporulated oocysts. Oocyst excretion occurs for 7–20 d and an infected cat can excrete ten million oocysts in 50 g of faeces (Dubey and Frenkel, 1972). As with *Cyclospora*, unsporulated oocysts are excreted in feline faeces, and an external maturation period of 1–5 days is required before they become infective, depending on aeration, temperature and humidity (Jackson and Hutchison, 1989). Sporulated oocysts contain two sporocysts, with each sporocyst containing four sporozoites. Oocysts can remain infectious for greater than one year in the environment (Yilmaz and Hopkin, 1972, Frenkel *et al.*, 1975).

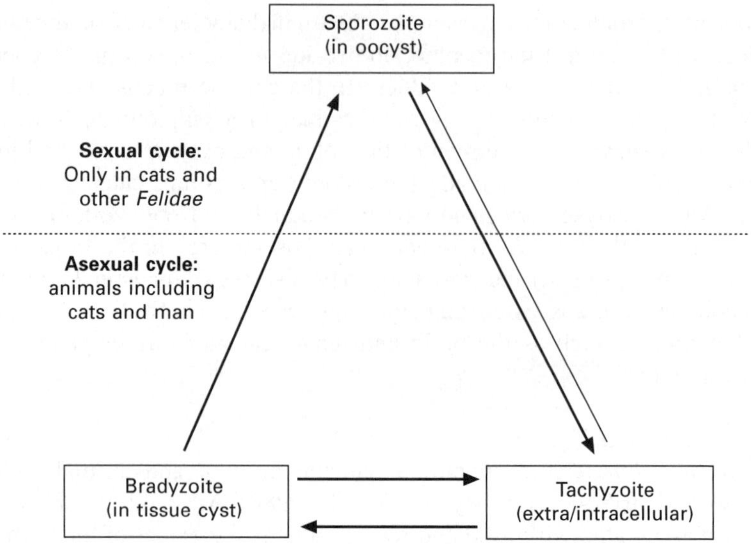

Fig. 26.5 Life cycle of *Toxoplasma gondii*.

The tachyzoite (from the Greek tachys = rapid and zōon = animal) is the rapidly dividing form and divides by endodyogeny within cells of the infected host. Dissemination throughout the body of the intermediate host occurs either intracellularly, as an intracellular parasite of macrophages and lymphocytes, or extracellularly, *via* the bloodstream. Tachyzoites transform to bradyzoites (from the Greek brady = slow), the slowly dividing form, and encyst in host tissue as tissue cysts. Many thousands of bradyzoites can occur in a tissue cyst (Garnham *et al.*, 1962). Tissue cysts can be found in host tissues within days of infection but are most numerous weeks later (Dubey and Frenkel, 1976; Derouin and Garin, 1991), and can persist for the life of the host.

Free-living amoebae (FLA)
The FLA that infect humans belong primarily to the genera *Acanthamoeba* and *Naegleria*. FLA can be either free-living or parasitic (amphizoic; Page, 1974). Following the demonstration that *Acanthamoeba* species caused disease in an animal model in the 1950s (Culbertson *et al.*, 1958), human cases of amoebic encephalitis, some of which were caused by *Naegleria* spp., were reported thereafter (Marciano-Cabral and Cabral, 2003). More recently two other genera of FLA have been associated with human disease, namely, *Balamuthia mandrillaris* and *Sappinia diploidea* (Visvesvara *et al.*, 1993; Gelman *et al.*, 2001); however, in this chapter, we focus on *Acanthamoeba* and *Naegleria* as most is known about them. The general principles identified for *Acanthamoeba* and *Naegleria* as foodborne pathogens are likely to apply to *B. mandrillaris* and *S. diploidea*.

Based on phenotypic and morphologic characteristics, the genus *Acanthamoeba* consists of at least 17 species, which can be divided further into 15 different lineages following 18S rDNA sequencing (Hewett *et al.*, 2003). The species most associated with human disease are *A. polyphaga*, *A. castellani*, *A. culbertsoni*, *A. hatcherii* and *A. healyii* (Schuster and Visvesvara, 2004). The trophozoite measures 15–45 μm and exhibits fine thorn-like protrusions (acanthapodia) and the genus derives its name from this feature (the Greek word acanth = thorn) (Marciano-Cabral and Cabral, 2003). Sexual reproduction does not seem to occur and there are no intermediate hosts in the life cycle. Cysts have a double wall, the outermost being wrinkled.

Of the genus *Naegleria*, only *N. fowleri* causes human disease, although at least five other *Naegleria* species inhabit the environment (Marshall *et al.*, 1997). The life cycle of FLA consists of an amoeboid trophic or feeding stage, a motile, pyriform, biflagellate stage and an environmentally robust cyst stage. All stages can cause infection (Schuster and Visvesvara, 2004). The motile, pyriform, biflagellate stage occurs when environmental factors such as temperature, food sources, pH or organic levels are optimal, and is present in the surface layers of water. It can differentiate into the trophozoite stage within a 24 hour period (Parija and Jaykeerthee, 1999). Cysts develop when growth conditions become poor, for example, lack of food, desiccation or toxin production (Marciano-Cabral and Cabral, 2003). Cysts, which can survive these harsher conditions, differentiate to the amoeboid stage when conditions become more favourable.

Acanthamoeba and *Naegleria* species are cosmopolitan inhabitants of wet soil, silt and water and feed on bacteria (Page, 1988). They have a worldwide distribution, but are reported more frequently from warmer climates. In temperate, developed countries, FLA prefer fresh water with a higher than average ambient temperature, such as public swimming pools and thermal discharges from factories and power stations. *Acanthamoeba* spp. are the most commonly found amoebae in the environment (Page, 1988) and can be found in soils from Antarctica to deserts; gardens and vegetables; fresh, brackish and sea water; sewage; swimming pools; aquaria; contact lens equipment; medicinal pools; dental treatment units; dialysis machines; heating, ventilating and air conditioning systems; mammalian cell cultures; human nostrils and throats; human and animal brain, skin and lung tissues (Schuster and Visvesvara, 2004). Soil is believed to be the primary habitat for *N. fowleri* (Parija and Jaykeerthee, 1999), but it can be found in bodies of warm water including ponds, lakes, coastal waters, hot springs, thermal discharges of power plants, heated swimming pools, hydrotherapy and medicinal pools, aquaria, sewage and irrigation ditches. *Naegleria* can tolerate temperatures of 40–45 °C. *Naegleria fowleri* has a worldwide distribution (Cabanes, 2001) and the majority of identified cases occur in countries with tropical and sub-tropical climates (Parija and Jaykeerthee, 1999).

26.2.3 Symptoms caused in humans

Giardiasis

Giardiasis is self-limiting in most people. The short-lived acute phase is characterised by flatulence with sometimes sulphurous belching and abdominal distension with cramps. Diarrhoea is initially frequent and watery, but later becomes bulky, sometimes frothy, greasy and offensive. Stools may float on water. Blood and mucus are usually absent and pus cells are not a feature on microscopy. In chronic giardiasis, malaise, weight loss and other features of malabsorption may become prominent. Stools are usually pale or yellow and are frequent and of small volume and, occasionally, episodes of constipation intervene with nausea and diarrhoea precipitated by the ingestion of food. Malabsorption of vitamins A and B_{12} and D-xylose can occur. Disaccharidase deficiencies (most commonly lactase) are frequently detected in chronic cases. In young children, 'failure to thrive' is frequently due to giardiasis, and all infants being investigated for causes of malabsorption should have a diagnosis of giardiasis excluded (Smith *et al.*, 1995a; Girdwood and Smith, 1999a).

Cyst excretion can approach 10^7 g^{-1} faeces (Danciger and Lopez, 1975). The pre-patent period (time from infection to the initial detection of parasites in stools) is on average 9.1 d (Rendtorff, 1979). The incubation period is usually 1–2 weeks. As the pre-patent period can exceed the incubation period, initially a patient can have symptoms in the absence of cysts in the faeces.

Cryptosporidiosis

Cryptosporidium hominis, *C. parvum*, *C. meleagridis*, *C. felis*, *C. canis*, *C. suis*, *C. muris* and *C. andersoni* and *Cryptosporidium* cervine, monkey, skunk, rabbit and chipmunk genotypes can infect immunocompetent and immunocompromised humans (Cacciò *et al.*, 2005; Feltus *et al.*, 2006; Nichols *et al.*, 2006a). In immunocompetent patients *Cryptosporidium* is a common cause of acute self-limiting gastroenteritis, symptoms commencing on average 3–14 d post-infection. Symptoms include a 'flu-like illness, diarrhoea, malaise, abdominal pain, anorexia, nausea, flatulence, malabsorption, vomiting, mild fever and weight loss (Fayer and Ungar, 1986). From two to more than 20 bowel motions a day have been noted, with stools being described as watery, light-coloured, malodorous and containing mucus (Casemore, 1987). Often severe, cramping (colicky) abdominal pain is experienced by about two-thirds of patients, and vomiting, anorexia, abdominal distension, flatulence and significant weight loss occur in less than 50 % of patients. Gastrointestinal symptoms usually last about 7–14 d, unusually five or six weeks, while persistent weakness, lethargy, mild abdominal pain and bowel looseness may persist for a month (Casemore, 1987). In young malnourished children, symptoms may be severe enough to cause dehydration, malabsorption and even death. Histopathology of infected intestinal tissue reveals loss of villus height, villus oedema and an inflammatory reaction. Mechanisms of severe diarrhoea are primarily consequences of malabsorption, possibly due to a reduction of lactase activity. The ratio of symptomatic to asymptomatic cases is not known.

Illness and oocyst excretion patterns may vary due to factors such as immune status, infective dose, host age and possible variations in the virulence of the organism; however, oocyst shedding can be intermittent and can continue for up to 50 days after the cessation of symptoms (mean: seven days). In humans, the pre-patent period is between seven and 28 days. The mean incubation period (time from infection to the manifestation of symptoms) is 7.2 days (range 1–12) with a mean duration of illness of 12.2 days (range 2–26) (Jokipii and Jokipii, 1986). As the pre-patent period can exceed the incubation period, initially a patient can have symptoms in the absence of oocysts in the faeces. Oocyst excretion by either human or non-human hosts can be up to 10^7 g^{-1} during the acute phase of infection. Infected calves and lambs excrete up to 10^9 oocysts daily for up to 14 days (Blewett, 1989).

In patients with Acquired Immune Deficiency Syndrome (AIDS), other acquired abnormalities of T lymphocytes, congenital hypogammaglobulinaemia, severe combined immunodeficiency syndrome, those receiving immunosuppressive drugs and those with severe malnutrition, symptoms include very frequent episodes of watery diarrhoea (between six and 25 bowel motions daily, passing between one and 20 litres of stool daily). Associated symptoms include cramping, upper abdominal pain, often associated with meals, profound weight loss, weakness, malaise, anorexia and low-grade fever (Whiteside *et al.*, 1984). Infection can involve the pharynx, oesophagus, stomach, duodenum, jejunum, ileum, appendix, colon, rectum, gall bladder, bile duct, pancreatic duct and the bronchial tree (Soave and Armstrong, 1986; Cook, 1987). Except in those individuals in whom suppression of the immune system can be relieved by discontinuing immunosuppressive therapies, symptoms can persist unabated until the patient dies (Soave and Armstrong, 1986). Cryptosporidiosis in the immunocompromised can be a common and life-threatening condition in developing countries, causing profuse intractable diarrhoea with severe dehydration, malabsorption and wasting. Highly active antiretroviral therapy (HAART) for AIDS patients can reduce the severity of the clinical consequences of cryptosporidiosis (Smith and Corcoran, 2004). Oocyst excretion can continue for two to three weeks after the disappearance of symptoms (Soave and Armstrong, 1986).

Cyclosporiasis

Symptoms of cyclosporiasis include a flu-like illness, diarrhoea with weight loss, low-grade fever, fatigue, anorexia, nausea, vomiting, dyspepsia, abdominal pain and bloating (Ortega *et al.*, 1993; Huang *et al.*, 1995; Fleming *et al.*, 1998). The incubation period is between two and 11 days (Soave, 1996) with moderate numbers of unsporulated oocysts being excreted for up to 60 days or more. In immunocompetent individuals the symptoms are self-limiting and oocyst excretion is associated with clinical illness, whereas in immunocompromised individuals diarrhoea may be prolonged.

The self-limiting watery diarrhoea can be explosive, but leukocytes and erythrocytes are usually absent. Often, diarrhoea can last longer than six weeks in immunocompetent individuals. The diarrhoeal syndrome may be characterised by remittent periods of constipation or normal bowel movements (Ortega *et al.*, 1993). Malabsorption with abnormal D-xylose levels has also been reported (Connor *et al.*, 1993).

Amoebiasis

Amoebiasis, caused by *E. histolytica*, is self-limiting in most people. Its distribution is global, but amoebiasis is most common where living conditions are crowded and/or there is poor hygiene and sanitation. Amoebiasis is common in parts of Africa, central and south America, India and southeast Asia. *Entamoeba histolytica* has also been diagnosed frequently among homosexual men.

Infections can persist for years and may be asymptomatic. Most infections are limited to the colon, but other tissues may be invaded. In the majority of cases, trophozoites remain in the intestinal tract but, in some cases, trophozoites invade to extraintestinal sites, most commonly the liver. Symptoms of intestinal amoebiasis can be chronic, ranging from vague gastrointestinal distress, mild diarrhoea to fulminant dysentery (blood and mucus in the stool). Complications include severe ulceration of colonic mucosal surfaces (less than 16 % of cases) and, rarely, intestinal obstruction due to trophozoite masses (amoebomas). The onset of symptoms can be highly variable. Fatalities are infrequent. Parasite- and host-derived factors influence the presence/absence/intensity of symptoms and, as for other protozoan pathogens, individuals with a damaged or undeveloped immunity may suffer more severe forms of the disease.

Symptoms occur normally two to four weeks following infection in approximately 10 % of those infected, but may take months or years to manifest. In intestinal amoebiasis, symptoms are mild and include diarrhoea, stomach pain and cramps and fever. Infection with invasive trophozoites can lead to amoebic dysentery, resulting from trophozoites invading the intestinal wall, and the development of amoebic abscesses in extraintestinal tissues, due to the dissemination of intestinal trophozoites primarily *via* the bloodstream. Amoebic dysentery is a severe form of amoebiasis, and symptoms include dysentery, severe stomach pains, loss of appetite and weight, fever and chills. The most commonly affected organ of extraintestinal spread is the liver, and symptoms of hepatic amoebiasis include fever, weakness, abdominal swelling, upper-right quadrant pain, nausea, jaundice, cough and loss of appetite and weight, which can persist from a few days to several weeks.

Toxoplasmosis

The majority of *Toxoplasma* infections in the immunocompetent host are generally asymptomatic and infection is often self-limiting. However, a review of outbreaks of toxoplasmosis indicated that only 25 % of cases were asymptomatic, the majority of patients demonstrating non-specific clinical symptoms, namely pyrexia, lymphadenopathy, malaise and myalgia (Ho-Yen, 1992). The incubation period is between four and 21 d, but this can vary greatly depending on various parasite- and host-derived factors at the time of infection, including the route of infection, the infectious dose, the stage of the parasite, the virulence of the strain and host factors.

A serious consequence of *Toxoplasma* infection occurs in the pregnant woman (Thulliez, 2001). An acute primary infection can result in infection of the foetus, the severity of symptoms in the foetus being related to its gestational age at infection (Chatterton, 1992). Infection during the first trimester can result in foetal death *in utero* and spontaneous abortion or delivery of a live child with central nervous system (CNS) symptoms, particularly hydrocephalus, meningoencephalitis, intracranial calcifications and chorioretinitis (Thulliez, 2001). Generalised symptoms of hepatosplenomegaly, jaundice, rash, anaemia and thrombocytopenia may also be present. Infection during late pregnancy is often sub-clinical with symptoms often presenting months or years after birth, usually as ocular lesions. Infection in the mother, prior to conception, provides immunity and congenital infection is unlikely. Rarely, there have been cases where the mother is immunocompromised and experiences either a reactivated *Toxoplasma* infection or an infection within days of conception (Thulliez, 2001).

Ocular toxoplasmosis is generally considered to be a result of congenital infection; however, evidence from toxoplasmosis outbreaks worldwide indicates that ocular symptoms may present after an acute, primary infection (Silveira *et al.*, 1988; Glasner *et al.*, 1992; Choi *et al.*, 1997). Congenital toxoplasmosis is the cause of blind registration in 0.1–0.2 % of patients in the UK (Dutton, 2001).

Encephalitis is the most frequent sign of toxoplasmosis in patients with AIDS (Mariuz and Steigbigel, 2001). Occasionally other organ systems are also affected, but it is rare to have only extra-cerebellar involvement. Multiple brain lesions of varying sizes are observed, particularly in the cerebral hemispheres. Clinical manifestations of toxoplasmic encephalitis are dependent on the number, site and size of the lesions (Mariuz and Steigbigel, 2001). The onset of illness is usually sub-acute with obvious symptoms developing within four to six weeks. Focal neurological defects with general cerebral dysfunction are most frequently observed (Luft and Remington, 1992). A diffuse form of toxoplasmic encephalitis has been uniquely reported in AIDS patients, which is usually fatal (Gray *et al.*, 1989). Sites of extra cerebral toxoplasmosis include the lung (Pomeroy and Felice, 1992), eyes (Weiss *et al.*, 1986), heart, gut, liver and kidney (Mariuz and Steigbigel, 2001).

Severe toxoplasmosis is not restricted to AIDS patients. A growing number of immunosuppressed patients, who are HIV-negative, may also suffer from toxoplasmosis. These patient groups include organ and bone marrow transplant recipients, those with malignancies, those on anticancer chemotherapy, those receiving corticosteroid treatment and those with connective tissue/collagen vascular disease (Wreghitt and Joynson, 2001). *Toxoplasma* is most commonly associated with heart and heart-lung transplants, primary infections being acquired from the donor organ(s). Reactivated infection is rare and was usually reported in patients undergoing aggressive immunosuppression. However, in bone marrow transplants the infection is usually due to reactivation in *T. gondii* antibody-positive patients (Derouin *et al.*, 1992). In symptomatic patients receiving solid organ transplants, fever is always present and can have one or more of a variety of symptoms, including pneumonia, encephalitis and myocarditis (Wreghitt *et al.*, 1989). *Toxoplasma* infection in other immunosuppressed patients has been reviewed extensively elsewhere (Israelski and Remington, 1993).

Free-living amoebae

Serological studies indicate that the human population is frequently exposed to FLA (Cursons *et al.*, 1980a; Marciano-Cabral *et al.*, 1987; Cerva, 1989), yet human disease is infrequent. The number of cases of disease caused by FLA worldwide is estimated to be very small: *Acanthamoeba* encephalitis ~ 200 cases, *Acanthamoeba* keratitis > 3000 cases, *Naegleria* meningoencephalitis ~ 200 cases (Schuster and Visvesvara, 2004). Infection with *Acanthamoeba* species may cause granulomatous amoebic encephalitis (GAE), usually with no meningeal involvement (Marshall *et al.*, 1997; Martinez and Visvesvara, 1997; Marciano-Cabral and Cabral, 2003). Entry into the host from contaminated soil is via breaks in the skin or by cysts inhaled within dust or airborne soil particles. Organisms spread haematogenously to the central nervous system (Martinez *et al.*, 1975). Infection is chronic and insidious, the incubation period being weeks to years (Schuster and Visvesvara, 2004). Disease presents as a sub-acute granulomatous meningoencephalitis with signs of an abscess. Abscesses in locations outside the brain and granulomatous skin lesions containing amoebae also occur.

GAE is frequently associated with HIV infection. However, as for many opportunistic pathogens, the incidence of GAE has decreased with the introduction of more effective therapies for HIV infection. Other immunocompromised patients such as those undergoing organ transplants or receiving anti-inflammatory steroids remain at risk (Steinberg *et al.*, 2002). Nasopharyngeal and cutaneous infections are observed when the organisms remain localised to the site of entry, but often dissemination occurs.

Acanthamoeba can also infect the eye, causing a characteristic corneal keratitis. Corneal keratitis caused by *Acanthamoeba* species is much commoner than GAE and is closely associated with the use of contact lenses. The source of amoebae is often in contaminated potable water used to rinse contact lenses. Trophozoites infect corneal lesions and trigger an ulcerative keratitis which may develop into panophthalmitis. Infection is normally uniocular. Amoebae are often scarce in corneal smears. Treatment is available, but reactivated infection can occur.

Amoebae can act as reservoirs for human infection by ingesting viable bacterial pathogens (Marciano-Cabral and Cabral, 2003). *Salmonella typhimurium*, *Yersinia enterocolitica*, *Shigella sonnei*, *Legionella gormanii* and *Campylobacter jejuni* can grow within FLA (King *et al.*, 1988), and once ingested can survive chlorination levels that would kill their free living counterparts.

Naegleria fowleri causes primary amoebic meningoencephalitis (PAM) and is a consequence of contaminated water penetrating the nasal cavities whilst bathing in warm fresh water. The amoeba enters the host through the nasopharyngeal mucosa and travels along the olfactory nerves to enter the brain through the cribriform plate (Martinez *et al.*, 1973; Bottone, 1993). The infection leads to the production of a toxin, which liquefies brain tissue, and the immune response leads to the symptoms of PAM (Marshall *et al.*, 1997). The incubation time can range between two and 15 d but, unlike acanthamoebiasis, often onset is rapid (1–2 d) and death usually occurs within 7–10 d. Fever, headache, vomiting and smell and taste disturbances occur in fulminant meningoencephalitis. A history of bathing in surfacewater during the preceding two weeks is significant. Unlike *E. histolytica* infections, the spherical cysts may be found in infected tissues.

26.2.4 Infectious dose and treatment

Pathogenic protozoan parasites generally have a low infectious dose for human beings.

Giardia

The infectious dose to human beings is between 25 and 100 cysts for *G. intestinalis* (= duodenalis) (Rendtorff, 1954, 1979), although a volunteer study demonstrated that a human-source isolate can vary in its ability to colonise other humans (Nash *et al.*, 1987), suggesting that certain isolates may be less infectious to some humans than others.

Cryptosporidium

Cryptosporidium parvum

The infectivity of different *C. parvum* isolates can vary in healthy human adult volunteers. These volunteer studies indicate that the infectious dose is 1042 oocysts for the UCP (UCP = Ungar *C. parvum*; bovine, received from Dr Beth Ungar in 1989, originally from Dr R. Fayer at the United States Department of Agriculture and passaged in calves by ImmuCell Corp., Maine, USA) isolate, between 30 and 132 oocysts for the Iowa (bovine, originally isolated by Dr H Moon, University of Iowa, Ames, Iowa from a calf and passaged in calves at the Sterling Parasitology Laboratory, University of Arizona, Tucson, Arizona, USA) isolate of *C. parvum* (DuPont *et al.*, 1995), and nine oocysts for the TAMU (Texas A & M University;

equine, isolated from a human exposed to an infected foal and passaged in calves) *C. parvum* isolate (Okhuysen *et al.*, 1999). Thus, isolates differ in their ID_{50}, in their attack rate and in the duration of diarrhoea they induce (Okhuysen *et al.*, 1999).

The same volunteers that were infected with the Iowa isolate (infectious dose between 30 and 132 oocysts) were each given a challenge dose of 500 oocysts one year later. A decline in the number of subjects shedding oocysts occurred together with reduced clinical severity, although the rate of diarrhoea was comparable. The serum antibody response did not correlate with the presence or absence of infection. A 14-fold increase in ID_{50} occurred in volunteers with pre-existing anti-*C. parvum* serum IgG (Chappell *et al.*, 1999).

Cryptosporidium hominis
Of 21 healthy adult volunteers experimentally infected with ten to 500 *C. hominis* (TU502; an isolate of human origin, which can also infect pigs and calves, and is maintained at Tufts University School of Veterinary Medicine, Massachusetts, USA) oocysts, 16 (76.2 %) exhibited evidence of infection. Using clinical and microbiologic definitions of infection, respectively, the ID_{50} was estimated to be ten to 83 oocysts. Diarrhoea occurred in 40 % of volunteers receiving ten oocysts, which increased to 75 % in those receiving 500 oocysts. In contrast to *C. parvum*, *C. hominis* elicited a serum IgG response in volunteers receiving > 30 oocysts, but volunteers with no evidence of infection had indeterminate or negative IgG responses (Chappell *et al.*, 2006)

Cyclospora
An infectious dose of between ten and 100 oocysts has been suggested for *C. cayetanensis* (Adams *et al.*, 1999), but a human volunteer trial failed to demonstrate gastroenteritis or oocyst shedding in seven *Cyclospora*-seronegative adults infected with between 200 and 49 000 oocysts over a 16-week period (Alfano-Sobsey *et al.*, 2004).

Entamoeba
Experimental infections of humans were produced consistently with 2000–4000 cysts (Feacham *et al.*, 1983), but the experiments were performed before the realisation that *E. histolytica* was a species complex, consisting of *E. histolytica* and *E. dispar*, both with morphologically identical cysts. The ID_{50} for *E. histolytica* is not known.

Toxoplasma
One infectious *T. gondii* oocyst, ingested orally, can cause infection in mice and pigs (Dubey *et al.*, 1996) but more than 100 oocysts are required to establish infection in the cat (Dubey, 1996). In contrast, one bradyzoite can infect a cat, while 100 bradyzoites may not be infective for mice by

the oral route (Dubey, 2001). It is not known how these figures relate to human infection.

Free-living amoebae
There are insufficient data to determine the infectious dose for FLA contracted from foods. Cabanes (2001) assessed the risk of PAM for swimmers potentially exposed to *N. fowleri*, based upon murine experimental curve data that were adjusted to make them correspond to the available human data. This provides the risk of PAM in humans as a function of the *N. fowleri* concentration in the body of water. Based upon this quantitative risk assessment, the risk is 8.5×10^{-8} at a concentration of 10 *N. fowleri* amoebae per litre.

26.2.5 Chemotherapy

Effective chemotherapy is available for giardiasis (nitroimidazole compounds, quinacrine, furazolidone, albendazole and mebendazole), cyclosporiasis (trimethoprim-sulfamethoxazole, excluding those who are intolerant to sulpha drugs) and amoebiasis (metronidazole and tinidazole for invasive stages and paromomycin and diloxanide furoate for luminal stages). Luminal, asymptomatic *E. histolytica* infections should be treated to combat colonisation, transmission and the delayed development of intestinal symptoms. Nitazoxanide is effective for cryptosporidiosis in immunocompetent and, probably, immunocompromised patients, although this will require alteration in the duration of treatment and/or the dosing regimen (Smith and Corcoran, 2004). Treatment for toxoplasmosis in the majority of immunocompetent, non-pregnant individuals is not normally considered, but in those individuals where the consequences of disease are serious, chemotherapy (pyremethamine, sulphonamides [usually sulphadiazine], spiramycin and clindamicin) is essential (Joss, 1992).

Various drugs have been used to treat disseminated *Acanthamoeba* infections including amphotericin B, azithromycin, fluconazole, flucytosine, pentamidine isethionate and sulpha drugs (Schuster and Visvesvara, 2004). Combination therapy is more effective than single therapy, possibly because many drugs are amoebostatic and not amoebicidal and because it is effective for both lifecycle stages of the organism (Marciano-Cabral and Cabral, 2003). There are few reports of successful GAE and PAM treatment. Treatment of *Acanthamoeba* keratitis can prove difficult as the trophozoites can encyst. Propamidine isethionate eye drops (Brolene®), topical neomycin, polyhexamethylene biguanide eye drops (Lavasept®) and/or topical chlorhexidine (Hibitane®) have been used, as have combinations of the above. The antiseptic Brolene® is moderately toxic to corneal epithelium. Combinations of topical steroids and anti-amoebic interventions have also been suggested. Amphotericin B is the drug of choice for PAM, but an early diagnosis is essential and few cases survive.

26.3 Current levels of incidence

Protozoan parasites can contaminate food either as a surface contaminant or within the food matrix. Given the low infectious doses of protozoan pathogens, both surface and food matrix contamination with small numbers of viable parasites, in produce that receives minimal washing/treatment prior to ingestion, poses a threat to public health. Increases in demand, global sourcing, chilling and rapid transport of foods enhance both the likelihood of contamination and the survival of the transmissive stages of these parasites. Food normally becomes a potential source of human infection by contamination, during production, collection, transport and preparation or during processing. The number of contaminating organisms will vary dependent upon the route or vehicle of contamination, therefore the sensitivity of the methods developed will have to address the detection of the smallest numbers of contaminants, practicable (Cook *et al.*, 2006a,b, 2007; Smith *et al.*, 2006, 2007; Smith and Cook, 2009).

It is often difficult to associate an outbreak with a particular food item and, furthermore, if the foodborne route is suspected, to identify how the food implicated became contaminated. Because of these difficulties, the acquisition of protozoan pathogens *via* the foodborne route is almost certainly under-detected. Given these current limitations, it not surprising to realise that documented protozoan foodborne outbreaks are rare, although some foods can be important vehicles of transmission, especially in situations of poor hygiene and endemnicity of infection (e.g. Slifco *et al.*, 2000; Nichols and Smith, 2002).

Currently, foodborne giardiasis and cryptosporidiosis are of significance because of the low infectious doses, the robustness and disinfection insensitivity of their transmissive stages and the development of specific methods for their detection on/in foods (Slifco *et al.*, 2000; Cook *et al.*, 2006a,b, 2007; Smith, 2008; Smith and Cook, 2009). Contamination of fresh produce, especially fruit, vegetables, salads and other foods consumed raw or lightly cooked, with viable (oo)cysts has been responsible for several outbreaks of giardiasis, cryptosporidiosis and cyclosporiasis (Tables 26.3, 26.4 and 26.5). Other food types known to have been contaminated or epidemiologically associated with outbreaks of giardiasis, cryptosporidiosis and cyclosporiasis include Christmas pudding, home-canned salmon, chicken salad, sandwiches, fruit salad, ice, raw sliced vegetables, unpasteurised milk, cold pressed (non-alcoholic) apple cider, raspberries, noodle salad, basil pesto pasta salad and mesclun lettuce (Tables 26.3, 26.4 and 26.5).

Meatborne parasitic diseases and meatborne protozoan diseases, particularly toxoplasmosis, remain an important cause of illness and economic loss, globally. While efforts to control this zoonosis continue, overall progress is slow. It is uncertain to what extent the oocyst or tissue cyst is implicated in the transmission of *Toxoplasma* as neither stage can be excluded as a possible source of infection in many of the outbreaks reported. Oocyst

Table 26.3 Some documented foodborne outbreaks of giardiasis

Number of persons affected	Suspected foodstuff	Probable/possible source of infection	Reference
3	Christmas pudding	Rodent faeces	Conroy, 1960
29	Home-canned salmon	Food handler	Osterholm *et al.*, 1981
13	Noodle salad	Food handler	Petersen *et al.*, 1988
88	Sandwiches	–	White *et al.*, 1989
10	Fruit salad	Food handler	Porter *et al.*, 1990
–	Tripe soup	Infected sheep	Karabiber and Aktas, 1991
27	Ice	Food handler	Quick *et al.*, 1992
26	Raw sliced vegetables	Food handler	Mintz *et al.*, 1993

transmission appears to be essential: studies of island communities have indicated that the presence of cats is essential, indicated by the lack of *Toxoplasma* infection in the intermediate host (Munday, 1972; Wallace *et al.*, 1972). Nevertheless, transmission of *Toxoplasma* infection by tissue cysts in raw or undercooked meats is significant, as suggested by the large body of evidence from descriptive and analytical epidemiological studies (Hall *et al.*, 2001). Several outbreaks have implicated meat as the source of infection (Table 26.6). In addition to eating infected meat bought over the counter, eating inadequately cooked game (e.g. bear, boar), during or after hunting, fishing and shooting expeditions, also contributes to the increased reporting of protozoan meatborne infections. Suspected foodstuffs associated with foodborne toxoplasmosis include rare lamb, raw goats' milk, raw venison, water, kibbi (a raw meat), raw pig's liver and water/ice cream (Table 26.6). Contaminated drinking water has been responsible for outbreaks of giardiasis, cryptosporidiosis, cyclosporiasis, amoebiasis and toxoplasmosis (Karanis *et al.*, 2007), and the food industry should remain aware of the possibility that (oo)cyst-contaminated drinking water can be incorporated into foods which then receive minimal heating or treatment.

Infectious (oo)cysts can be transmitted to a susceptible host *via* any faecally-contaminated matrix, including water, aerosol, food and transport hosts. Food products can become contaminated with (oo)cysts in a variety of ways, and it is likely that more than one route may be involved in transmission, particularly in endemic areas.

Person-to-person transmission

Person-to-person transmission has been documented, particularly for foods which are intended to be consumed raw, or for those that are handled after being cooked. Direct contamination, by symptomatic or asymptomatic (oo)cyst excretors, during food preparation or following food handler contact with (oo)cyst excretors is a frequently reported route for foodborne giardiasis

Table 26.4 Some documented foodborne outbreaks of cryptosporidiosis

Number of persons affected	Suspected foodstuff	Probable/possible source of infection	*Cryptosporidium* species/genotype	Reference
160	Cold pressed (non-alcoholic) apple cider	Contamination of fallen apples from infected calf	NA	Millard *et al.*, 1994
25	Cold pressed (non-alcoholic) apple cider	? Contaminated water used to wash apples	NA	Anon., 1997a
15	Chicken salad	Food handler	NA	Anon., 1996
48 school children	School cow's milk	Failure of pasteurisation plant	NA	Gelletlie *et al.*, 1997
54	Not identified	Common food ingredient	*C. hominis*	Anon., 1998a
152	Eating in one of two university campus cafeterias	Food handler	*C. hominis*	Quiroz *et al.*, 2000
8 children	Cow's milk	Drinking commercial unpasteurised milk (5/10 milk samples tested positive for *Cryptosporidium* antigen by ELISA)	NA	Harper *et al.*, 2002
12	Ozonated apple cider from local orchard	Drinking ozonated apple cider	*C. parvum* subtypes IIaA15G2R1 and IIaA17G2R1	Blackburn *et al.*, 2006

NA = not available.

Table 26.5 Some documented foodborne outbreaks of cyclosporiasis

Number of persons affected	Suspected foodstuff	Probable/possible source of infection	Reference
1465	Guatemalan raspberries	? Aerosolisation of oocysts during application of insecticides or fungicides	Herwaldt *et al.*, 1997
1450	Guatemalan raspberries	? Aerosolisation of oocysts during application of insecticides or fungicides	Anon., 1998b
48	Basil pesto pasta salad	Unknown	Anon., 1997b
Unknown	Mesclun lettuce	Unknown	Anon., 1997c

Table 26.6 Some documented foodborne outbreaks of toxoplasmosis

Number of persons affected	Suspected foodstuff	Probable/possible source of infection	Reference
6	Rare lamb	Tissue cyst	Masur *et al.*, 1978
10	Raw goat's milk	Tachyzoite	Sacks *et al.*, 1982
3	Raw venison	Tissue cyst	Sacks *et al.*, 1983
39	Water	Oocyst from jungle cat faeces	Benenson *et al.*, 1982
5	Kibbi, raw meat	Tissue cyst	De Silva *et al.*, 1988
8	Raw liver (pig)	Tissue cyst/trophozoite	Choi *et al.*, 1997
40	? Unknown	? Oocysts/tissue cysts	Doganci *et al.*, 2006
294	Water/ice cream	Oocysts	De Moura *et al.*, 2006

and cryptosporidiosis (Tables 26.3 and 26.4), and are due to poor personal hygiene standards of that food handler. Amoebiasis also has the potential to be transmitted by this route. The hygienic practice of washing hands before preparing food can minimise (oo)cyst contamination and transmission. Guidelines exist for foodhandlers suffering diarrhoea, or those with recent symptoms. A documented foodborne outbreak of cryptosporidiosis, involving 88 cases, originated from a foodhandler who continued to work in spite of having gastroenteritis (Quiroz *et al.*, 2000). Washing uncooked fruit and vegetables before consumption is also recommended, however, one study indicates that washing may not be sufficient to remove all *C. parvum* oocysts seeded onto lettuce surfaces (Ortega *et al.*, 1997b). As humans are the hosts of *C. cayetanensis*, person to-person-transmission is the major route, although transmission is indirect, as oocysts require a period of environmental maturation before they become infectious. There is no recorded outbreak of amoebiasis, since the re-description of the *E. histolytica* species complex, although food handlers are suspected of causing numerous scattered infections.

There is no evidence that anthroponotic transmission plays a role in the transmission of *Toxoplasma*. Washing uncooked fruit and vegetables that may be contaminated with *Toxoplasma* oocysts is a recognised preventive measure, but here transmission is from the environment to the person. Nevertheless the principles of good hygienic practice are essential. There are no reported cases of proven anthroponotic transmission of FLA (Schuster and Visvesvara, 2004).

Animal-to-person (zoonotic) transmission

There are no recorded outbreaks of zoonotic foodborne transmission of *Cyclospora*. There is no evidence that infections by FLA are zoonotic (Schuster and Visvesvara, 2004). Direct contact of food with bovine faeces was the suggested cause of the largest foodborne outbreak of cryptosporidiosis, which occurred in Maine, USA. In this outbreak, apples collected from an orchard in which a *Cryptosporidium*-infected calf grazed were made into non-alcoholic cider (Millard *et al.*, 1994) (Table 26.4). A further outbreak involving eight children was associated with drinking commercially obtained unpasteurised cow's milk, where five of ten milk samples tested positive for *Cryptosporidium* antigen (Harper *et al.*, 2002) (Table 26.4). There is no recorded outbreak of amoebiasis, since the re-description of the *E. histolytica* species complex.

Zoonotic transmission is the major foodborne route for *Toxoplasma*, either following the consumption of tissue cysts in meat products or oocysts in drinking water, and all three infective stages of *T. gondii* have been implicated in outbreaks. Outbreaks of oocyst contamination of drinking water supplies, or for use in food production, have been recorded (Dubey, 2004; de Moura *et al.*, 2006). Oocysts can remain viable in seawater for months (Lindsay *et al.*, 2003), and filters feeding bivalves can concentrate oocysts (Lindsay *et al.*, 2004). Seroprevalence studies in farm animals have shown a wide range of prevalence of *Toxoplasma* infection: sheep (10–37 %), pigs (1–69 %) and cattle (3–25 %) (Hall *et al.*, 2001). Horsemeat, game and chicken also harbour the organism. Studies indicating prevalence in retail meat and meat products are limited, although the prevalence of *Toxoplasma* in beef is less than that for pork and lamb. It has been suggested that beef products can be contaminated, deliberately or otherwise, with these other meats thus increasing the risk of infection (Dubey, 1988). Infection of humans by tachyzoites in unpasteurised goat milk has been indicated in milk-related outbreaks (Skinner *et al.*, 1990).

26.3.1 Foodborne giardiasis

Foodborne transmission was suggested in 1920s (Musgrave, 1922; Lyon and Swalm, 1925) when water, vegetables and other foods were found to be contaminated with cysts. Since then, cysts have been detected on vegetables including lettuce (Mastrandrea and Micarelli, 1968; Barnard and Jackson, 1980) and soft fruit (e.g. strawberries, Barnard and Jackson, 1980; Kasprzak *et al.*, 1981). One report identifies the possibility of offal (tripe) being intrinsically infected (Karabiber and Aktas, 1991). The seven documented outbreaks presented in Table 26.3 occurred from 1977 onwards.

26.3.2 Foodborne cryptosporidiosis

Nine outbreaks of foodborne transmission have been documented, seven of which occurred in the USA (Table 26.4). Two occurred following the consumption of non-alcoholic, pressed, apple cider, in 1993 and 1996, affecting a total of 185 individuals. In the first outbreak, apples were collected from an orchard in which an infected calf grazed. Some apples had fallen onto the ground (windfalls) and had probably been contaminated with infectious oocysts then (Millard *et al.*, 1994). The source of oocysts in the second outbreak is less clear as windfalls were not used and waterborne, as well as other, routes of contamination were suggested (Anon., 1997a). A foodborne outbreak affecting 15 individuals occurred in 1995 with chicken salad, contaminated by a food handler, being the probable vehicle of transmission (Anon., 1996). In 1997, an outbreak was documented in Spokane, Washington. Amongst 62 attendees of a banquet dinner, 54 (87 %) became ill. Eight of ten stool specimens obtained from ill banquet attendees were positive for *Cryptosporidium*. Epidemiological investigation suggested that foodborne transmission occurred through a contaminated ingredient in multiple menu items (Anon., 1998a). All *Cryptosporidium* faecal samples analysed from this outbreak contained *C. hominis* oocysts (previously known as C. *parvum* genotype 1) (Quiroz *et al.*, 2000).

During September and October 1998, a cryptosporidiosis outbreak affecting 152 individuals occurred on a university campus in Washington, DC, USA. A case control study with 88 case patients and 67 control subjects revealed that eating in one of the two cafeterias was associated with illness. One food-handler, positive for *Cryptosporidium*, had prepared raw produce on the 20–22nd of September. All samples analysed by molecular typing (25 cases, including the food-handler) were *C. hominis* (previously known as C. *parvum* genotype 1) (Quiroz *et al.*, 2000). An outbreak involving eight children was associated with drinking commercially obtained unpasteurised cow's milk, where five of ten milk samples tested positive for *Cryptosporidium* antigen (Harper *et al.*, 2002) (Table 26.4). This outbreak was suggested to be due to poor udder hygiene.

In October 2003, 12 individuals with laboratory confirmed cryptosporidiosis were reported to a northeast Ohio health department. Eleven of the 12 had drunk a locally produced, ozonated apple cider in the two weeks before illness (Blackburn *et al.*, 2006) (Table 26.4). No *Cryptosporidium* oocysts were detected in the water supply and the employees reportedly used a 'few' dropped apples for cider production. Ozonation was used to treat the cider during production, which was stored in refrigerated tanks. Most of the cider was ozonated a second time and then sold in plastic jugs on-site and at nearby grocery stores. The remaining cider was sold through a tap in the orchard's store, but this cider was not re-ozonated. Six individuals drank once-ozonated cider and six drank twice-ozonated cider, suggesting that even repeated ozonation was insufficient to kill *Cryptosporidium* oocysts (Blackburn *et al.*, 2006) (Table 26.4).

Molecular typing identified *C. parvum* as the infectious agent in the faeces of 11 of 12 individuals and the cervine genotype of *Cryptosporidium* in one individual. Of the eleven *C. parvum* cases, sub-typing, by sequencing the glycoprotein-60 (*GP60*) gene, identified two common bovine *C. parvum* sub-types, IIaA15G2R1 (six cases) and IIaA17G2R1 (five cases). The remainder of the cider drunk by a laboratory positive case (IIaA17G2R1) was also positive by polymerase chain reaction (PCR) for *C. parvum* sub-type IIaA17G2R1. The authors stated that the two *C. parvum* sub-types found in this outbreak are likely to represent a common contamination source as both are common in cattle, and multiple sub-types are commonly found on farms (Blackburn *et al.*, 2006) (Table 26.4).

This outbreak illustrates the importance of molecular typing and sub-genotyping for determining the *Cryptosporidium* species responsible for the outbreak and the source of the infective *C. parvum* oocysts. Molecular methods are required to determine species, genotypes and sub-genotypes and, as these source and disease tracking tools are only recent developments, there is a lack of public health understanding of the significance of the foodborne route of transmission, currently.

Either the inherent inability of ozone to inactivate all *Cryptosporidium* oocysts in apple cider or its improper use in this outbreak can account for the ozonisation failure. This outbreak highlighted failures of ozonation in this setting as well as the need to collect further evidence of its usefulness for apple cider disinfection, before its use can be accepted. Because of the paucity of evidence supporting ozonation for apple cider disinfection or for killing *Cryptosporidium* in this product, and its apparent failure in this outbreak, the FDA issued an addendum to its Hazard Analysis Critical Control Point (HACCP) rule, which advises that juice makers should not use ozone in their manufacturing process unless they can prove a 5 log pathogen reduction through ozonation (http://www.cfsan.fda.gov/~dms/juicgu12.html).

26.3.3 Foodborne cyclosporiasis

The first report of foodborne transmission of *Cyclospora* may have occurred in 1995, when an airline pilot presented with diarrhoeal illness after eating food prepared in a kitchen in Haiti that was then brought on board the aeroplane (Connor and Shlim, 1995). The majority of documented outbreaks have been recorded from north America, but one was recorded in Germany (Döller *et al.*, 2002) (Table 26.5). Many documented foodborne outbreaks of cyclosporiasis are associated with soft-skinned fruit and leafy vegetables, such as berries and lettuce, that receive no, or minimal heat treatment. To date, no frozen, processed or peeled fruit and vegetables have been implicated in disease transmission, although oocysts and/or *Cyclospora* DNA have been detected in produce that was later frozen (see below).

In 1996, outbreaks of cyclosporiasis affecting more than 1465 individuals occurred in the USA and Canada, with imported raspberries being implicated (Herwaldt *et al.*, 1997). Most infections were in immunocompetent adults, with only 40 childhood cases, 50 % being clustered cases and the remainder sporadic cases. The 55 clusters occurred between early May and mid-June following events at which Guatemalan berries, primarily raspberries, were served (Herwaldt *et al.*, 1997). At that time, no method existed for detecting *Cyclospora* oocysts on foods and their presence on the foods implicated could not be confirmed. In 1997, outbreaks in the USA were also associated with imported raspberries and, later that year, with contaminated basil and lettuce (Anon., 1997b, c; Table 26.5). In 1998, clusters of cases, again associated with fresh berries from Guatemala, were recorded in Ontario, Canada (Anon., 1998b). Most cases occurred during spring and summer. Herwaldt *et al.* (1997) identifies likely vehicles of transmission as well as the difficulty in tracing outbreak sources in the absence of supportive parasitological tools.

Various fresh fruit, lettuce and herbs have been implicated in north American outbreaks. These include non-Guatemalan raspberries, or fruit other than raspberries or blackberries; multiple combinations of many fruits including raspberries and blackberries of undetermined sources; fresh Guatemalan blackberries, frozen Chilean raspberries, and fresh US strawberries; non-Guatemalan raspberries (both foreign and domestic sources); imported blackberries (possibly Guatemalan?), strawberries and blueberries (Herwaldt *et al.*, 1997). Fresh mesclun lettuce (also known as spring mix or baby greens) and basil have also been implicated in outbreaks (Herwaldt, 1997).

The Missouri (1999) outbreak was the first US outbreak where *Cyclospora* was found, by both microscopy and PCR, in frozen leftovers of an epidemiologically implicated food item (Lopez *et al.*, 2001) (Table 26.5). A wedding cake, containing a cream filling that included raspberries, was the food item most strongly associated with illness in 68.4 % of 79 interviewed wedding party members and guests (Ho *et al.*, 2002) (Table 26.5). Leftover frozen cake was positive for *Cyclospora* DNA, and sequencing of the amplicons confirmed that the DNA was *C. cayetanensis*. Again, this outbreak illustrates the importance of applying molecular methods for determining the presence of *Cyclospora* DNA in the sample species responsible for this outbreak. As for *Cryptosporidium*, their recent development and use should augment our public health understanding of the significance of the foodborne route of transmission.

The year 2000 was the fifth year that Spring outbreaks of cyclosporiasis, definitely or probably associated with Guatemalan raspberries, occurred in North America. Thai basil, imported via the USA, was implicated in the May 2001 outbreak in British Columbia, Canada (Huang *et al.*, 2005).

The first documented European foodborne outbreak of cyclosporiasis occurred in December 2000 (Döller *et al.*, 2002) (Table 26.5). Illness was reported in 34 persons who attended luncheons at a German restaurant, and the overall attack rate was 85 %. Two salad side dishes prepared from lettuce imported from southern Europe and dressed with fresh green leafy herbs (dill, chives, parsley, green onions) were the only foods associated with significant disease risk. All 25 persons who ate a salad consisting of butterhead lettuce, mixed lettuce (lollo rosso, lollo bianco, oak leaf and romaine lettuce), red cabbage, white cabbage, carrots, cucumbers and celery became ill, and the vehicle of transmission was deduced to be one or more of the lettuce varieties. The salad products were obtained from different European countries. The butterhead lettuce was grown in southern France and the mixed lettuce, dill, parsley and green onions in southern Italy. The chives were grown in Germany. The most probable routes of contamination included the use of human waste as fertiliser; faecally-contaminated water used to irrigate crops, prepare pesticides, to refresh/clean produce at their origin; and inappropriate sanitary facilities for seasonal field workers (Döller *et al.*, 2002).

In February 2004, clusters of cases of cyclosporiasis affecting ~95 cases and associated with events held in Texas and Illinois were identified. Fresh produce, including basil and mesclun lettuce/spring mix salad products, were served at these events. During May and June, 2004, outbreaks and sporadic cases of cyclosporiasis occurring in four US states and one Canadian province were also reported (http://www.dsf.health.state.pa.us/health/cwp/view.asp?A=171&Q=239018). The Pennsylvania, USA outbreak which occurred between June and July 2004 was reported from a single residential facility, and attendance at five special events was linked to the onset of symptoms, with 96 cases attending at least one of the events. The onset of illness was 1–14 d after consuming food or beverages at one or more of these events. Forty cases were laboratory confirmed (Pennsylvanian health officials and Centers for Disease Control – CDC) and 56 cases were assessed as being probable for cyclosporiasis. The source of the infection was traced to a single container of Guatemalan snow peas used in the preparation of a pasta salad and was the first documented outbreak linked to the consumption of raw snow peas (Anon., 2004).

26.3.4 Foodborne amoebiasis

There has been no recorded outbreak of foodborne amoebiasis, since the re-description of the *E. histolytica* species complex.

26.3.5 Foodborne toxoplasmosis

Various reports in the 1950s indicated that meat was a source of *Toxoplasma* infection (Jacobs *et al.*, 1960a) and, since then, various reports of outbreaks associated with the consumption of raw or undercooked meats have been recorded (Table 26.6). A European multicentre case-controlled study (Cook *et al.*, 2000) identified that eating undercooked lamb, beef or game, contact with soil and travel outside Europe, the USA and Canada were the risk factors most strongly predictive of acute *Toxoplasma* infection in pregnant women.

In a small Korean outbreak of ocular toxoplasmosis, infected raw boar viscera and meat was the source implicated (Choi *et al.*, 1997). The boar had been freshly caught and eaten at a feast, and it was believed that the raw viscera and meat would give special health benefits. Interestingly, at feasts in previous years, the same three men had eaten boar meat that had been partially frozen. Freezing is a well-documented procedure for rendering tissue cysts non-viable (Table 26.7).

Contaminated drinking water was the suspected source of a large outbreak of toxoplasmosis in Brazil (de Moura *et al.*, 2006). The authors concluded that the outbreak was due to the consumption of water or ice cream made from water from a specific reservoir, which was vulnerable to faecal contamination from cats. Kittens were born and found living on top of the reservoir container prior to the outbreak. As the levels of chlorination in the water were insufficient to destroy *Toxoplasma* oocysts, this circumstantial evidence led the authors to conclude that the outbreak was most likely to be caused by oocyst-contaminated water.

Tachyzoites are vulnerable to the acidity in the stomach; therefore foodborne transmission following tachyzoite ingestion is uncommon. Nevertheless, a family outbreak of toxoplasmosis indicated that the source was tachyzoites in unpasteurised goat's milk (Skinner *et al.*, 1990). Although the family had various risk factors for toxoplasmosis, only the children were infected and only they drank unheated, unpasteurised goat's milk. The adults drank the milk in their tea, but the temperature of the tea would have killed any tachyzoites present. It was suggested that the tachyzoites entered the body via the buccal or throat epithelium. *Toxoplasma* was isolated from a milk sample.

26.3.6 Foodborne pathogenic free-living amoebae

FLA are not thought to be transmitted by the oral route (Schuster and Visvesvara, 2004). Transmission of FLA via the nasal epithelium is most likely to occur when humans are in contact with contaminated water (e.g. recreational water sports). It may be possible for amoebae to infect via the buccal mucosa, but the risk of infection from the direct contamination of food and food products must be considered to be very low. *Naegleria fowleri* has been identified as a concern in supermarkets in the USA. Supermarket displays release misted, cool water onto vegetables and fruit, and this was considered to be a potential source of infection. However, as *N. fowleri* are thermophilic, cool water would not be optimal for survival, but other *Naegleria* and *Acanthamoeba* species may occur (Schuster and Visvesvara, 2004).

Table 26.7 Some conditions for inactivating *Cryptosporidium*, *Giardia* (oo)cysts, *Toxoplasma gondii* oocysts, tissue cysts and tachyzoites and FLA cysts

Parasite	Physical inactivation	Chemical inactivation
C. parvum	• 71.7 °C for 15 s (pasteurisation) for bovine oocysts in either water or milk completely abrogates infectivity to neo-natal mice (Harp *et al.*, 1996). • 64.2 °C for 2 min or 72.4 °C for 1 min for bovine oocysts: no infection established in BALB/c mice (Fayer, 1994). • Air-drying (18–20 °C; cervine–ovine oocysts, isolate MD) for 2 h: 97 % death; for > 2 h: 100 % death (Robertson *et al.*, 1992). • Bovine oocysts (AUCP-1 isolate) at –70 °C for 1 h; –20 °C up to 168 h; –15 °C for 168 h: no developmental stages observed in BALB/c mice; –20 °C up to 5h; –15 °C up to 24 h and at –10 °C for 168 h: developmental stages observed in mice (Fayer and Nerad, 1996). • Snap-freezing (MD isolate) in liquid nitrogen: 100 % death (Robertson *et al.*, 1992). • Infectious to mice after storage at 4 °C (isolate GCH1) for 39 weeks or after 20 weeks at 20 °C (Widmer *et al.*, 1999). • Storage at –20 °C for 775 h, 168 h and 24 h reduces viability to 1.8 %, 7.9 % and 33 %, respectively (isolate MD) (Robertson *et al.*, 1992). • Viability of *C. parvum* oocysts unaffected by storage at 4 °C for 12 weeks in still natural mineral waters, but at 20 °C, ~ 30 % remained viable after 12 weeks (Nichols *et al.*, 2004).	• Free chlorine: ineffective. Up to 16 000 mg l^{-1} for 24 h at 5 or 20 °C pH 6, 7 and 8 or 8000 mg l^{-1} for 24 h at 5 °C pH 6 and 7 or at 20 °C pH 8 required for total inactivation of human or bovine oocyst isolates (Smith *et al.*, 1989) • Chlorine dioxide: 78 mg l^{-1} min^{-1} required for 90 % inactivation (Korich *et al.*, 1990). • Ozone: at 20 and 25 °C up to 3–8 mg l^{-1} min^{-1} required for up to 4 log_{10} inactivation. Ozone requirements increase with decrease in temperature (Smith *et al.*, 1995b). • UV irradiation: a dose of 8748 mW s cm^{-2} during two 5 min periods, produces >2 log_{10} reduction in oocyst viability (Campbell *et al.*, 1995). • 4 % iodophore; 5 % cresylic acid; 3 % NaOCl; 5 % benzylkonium chloride and 0.02M NaOH for 18 h: viable oocysts observed. 5 % ammonia or 10 % formol saline for 18 h: completely destroy oocyst viability (Campbell *et al.*, 1982)

Giardia intestinalis; G. muris	• Freezing (–13 °C) for 14 days and thawing (<1 % viable cysts) (Bingham *et al.*, 1979). • *G. intestinalis* cysts at 8 °C for 77 days (< 5 % excystation) (Bingham *et al.*, 1979). • *G. muris* cysts 3 months in cold raw water sources (De Regnier *et al.*, 1989).	• Free chlorine: up to 142 mg l^{-1} min^{-1} required for 99 % *G. intestinalis* inactivation (Hoff, 1986). At 5 °C, pH 8, 2 mg mL^{-1} for 30 min produces <30 % *G. intestinalis* cyst inactivation. At 25 °C, pH 8, 1.5 mg ml^{-1} for 10 min produces > 99 % cyst inactivation (Jarroll *et al.*, 1981). • Chlorine dioxide: 11.2 mg l^{-1} min^{-1} required for 99 % inactivation (Hoff, 1986). • Ozone: up to 2.57 mg l^{-1} min^{-1} required for up to 4 log_{10} *G. intestinalis* inactivation (Finch *et al.*, 1993).
Toxoplasma gondii oocysts	• Exposure to heat greater than 55 °C for 30 min (Dubey *et al.*, 1970). • Exposure to heat in water at 40 °C for 28 days (Dubey, 1998). • Exposure to heat in water at 45 °C for 3 days (Dubey, 1998).	• Exposure to 5 % ammonium hydroxide for 30 min (Dubey *et al.*, 1970). • 63 % sulphuric acid/ 7 % dichromate for 24 h (Dubey *et al.*, 1970). • Undiluted liquid household ammonia (5.5 %) for 3 h (Frenkel and Dubey, 1972). • 2 % tincture of iodine for 3 h or 7 % tincture of iodine for 10 min (Frenkel and Dubey, 1972). • Drying at 19 % humidity for 11 days (Frenkel and Dubey, 1972). • Lomasept for 3 h (Ito *et al.*, 1975). • 5 % paracetic acid for 48 h (Ito *et al.*, 1975).
Toxoplasma gondii tissue cysts	• Heating meat (pork) to an internal temperature of 67 °C for 3 min (Dubey *et al.*, 1990). • Cooling meat to –13 °C (Dubey, 1974; Kotula *et al.*, 1991). • Exposure to 0.5 Krad of γ irradiation (Dubey and Thayer, 1994).	
Toxoplasma gondii tachyzoites	• Sensitive to extremes of temperature (Dubey *et al.*, 1970).	• Destruction by gastric juices (Jacobs *et al.*, 1960b).

Table 26.7 (Cont'd)

Parasite	Physical inactivation	Chemical inactivation
Acanthamoeba	• Exposure to temperature of greater than 40 °C (Schuster and Visvesvara, 2004).	• *A. culbertsoni* trophozoites killed at 1.25 mg/l total available chlorine in axenic conditions (Cursons *et al.*, 1980b). • 10^3 *Acanthamoeba* sp. cysts **not** sterilized in 40 mg/l chlorine (De Jonckheere and van der Voorde, 1976). • *Acanthamoeba* sp. inactivated with 3 mg ClO_2 L^{-1} and 7 mg O_3 L^{-1} (Cursons *et al.*, 1980b; Dawson and Brown, 1987; Beltrán and Jiménez, 2008). • 20 % isopropyl alcohol and 10 % formalin kill *A. castellanii* cysts (Aksozek *et al.*, 2002). • H_2O_2 (as 3 %, AOSept®, contact lens disinfectant) kill *A. castellanii* cysts after 4 h exposure, but a 3 % laboratory grade H_2O_2 from different vendor did not (Aksozek *et al.*, 2002). • Polyhexamethylene biguanide (0.02 %), clotrimazole (0.1 %), or propamidine isethionate (Brolene®) kill *A. castellanii* cysts *in vitro* (Aksozek *et al.*, 2002). • *A. castellanii* cysts resistant to 250 K rads of γ irradiation (Aksozek *et al.*, 2002). • UV dose > 800 mW·s/cm^2 required for *A. castellanii* trophozoites (Aksozek *et al.*, 2002; Beltrán and Jiménez, 2008). • UV dose of approx. 60 mW·s/cm^2 for 2–3 log inactivation of *A. culbertsoni* (ATCC 30171; 1.6×10^4 trophozoites) (Maya *et al.*, 2003), and >173 mW·s/cm^2 required for total (up to 5 log) inactivation (Maya *et al.*, 2003).

Naegleria	• Exposure to temperatures > 45 °C (Schuster and Visvesvara, 2004).	• *N. fowleri* trophozoites killed in 0.79 mg/l total available chlorine in axenic conditions after 30 min (Cursons *et al.*, 1980b). • 10^3 *Naegleria* sp. cysts sterilized by 2 mg/l chlorine in 30 min (De Jonckheere and van der Voorde, 1976). • Free chlorine Ct (contact × time) value for *N. fowleri* trophozoites was 6 for a 99 % reduction at pH 7.5 (Gerba *et al.*, 2008). • Free chlorine Ct value for *N. fowleri* cysts was 31 for a 99 % reduction at pH 7.5 (Gerba *et al.*, 2008). • UV dose of 63 mWs/cm^2 and 12.6 mWs/cm^2 attains 2 log inactivation of *N. fowleri* cysts and trophozoites, respectively (Sarkar, 2008).

Naegleria fowleri has been detected in domestic water supplies, recreational waters and thermally-polluted run off from industrial zones, and in the USA, *N. fowleri* has been detected in surfacewaters (Wellings *et al.*, 1977; Lee *et al.*, 2002; Marciano-Cabral *et al.*, 2003). A recent study of unchlorinated well water systems used as drinking water sources in Arizona, USA, prompted by the deaths of two children due to *N. fowleri* in the Phoenix metropolitan area in 2002, identified *N. fowleri* DNA in 8 % (n = 143) of the wells and 16 % (n = 188) of the samples tested (Gerba *et al.*, 2008). The food industry should be remain aware of the potential for *N. fowleri* contamination of unchlorinated groundwater of drinking water quality which may be used in the production of foods that receive minimal heat or disinfection treatment prior to consumption, particularly in warmer climates.

26.4 Conditions of growth

Giardia, *Cryptosporidium*, *Cyclospora*, *E. histolytica* and *T. gondii* are obligate parasites and require a host to reproduce. The current *in vitro* methods for amplifying and encysting *G. duodenalis* trophozoites and for amplifying the asexual life cycle of *C. parvum* and *C. hominis* are primarily for research use. The amphizoic amoebae, *Acanthamoeba* and *Naegleria* have no requirement for a host to reproduce. *Acanthamoeba* species that cause systemic disease and *N. fowleri* are thermophilic organisms. Both can be grown in *in vitro* culture with *N. fowleri* requiring the presence of bacteria, usually *E. coli*, to grow. *Naegleria fowleri* (Lee strain, ATCC 30894) can be maintained axenically in tissue culture flasks (Corning, Lowell, MA, USA) in liquid Oxoid modified Nelson's medium at 37 °C (Marciano-Cabral *et al.*, 2003). Stock cultures can be propagated by refreshing the medium every five days. Samples should be examined regularly for the presence, and number, of trophozoites. FLA can harbour intracellular pathogens such as *Legionella* spp., *Burkholderia picketti* and *Cryptococcus neoformans*. *Naegleria* is most easily recognised by its motility in a fresh preparation. The trophozoite measures 10–35 μm and has smooth hemispherical pseudopodia. (Oo)cysts can be transmitted *via* water, food or transport hosts (e.g. seagulls, waterfowl, flies, bivalves); however, excluding FLA, these parasites cannot replicate in these matrices.

26.4.1 Infectivity and viability

In vitro methods, particularly for assessing the viability of *Cryptosporidium* and *Giardia* (oo)cysts (reviewed by Smith, 1998), have been developed as surrogates for *in vivo* models since the latter are expensive, and they require closely regulated government (e.g. UK Home Office) licenses and specialised animal housing facilities.

Giardia

Assessment of *G. intestinalis* cyst infectivity can be undertaken *in vivo* in neonatal mice or adult gerbils (Faubert and Belosevic, 1990), while *Giardia* cyst viability (reviewed by Smith, 1998: O'Grady and Smith 2002) can be undertaken *in vitro* by (i) excystation (Bhatia and Warhurst, 1981; Smith and Smith, 1989), (ii) fluorogenic vital dyes (Schupp and Erlandsen, 1987a,b; Schupp *et al.*, 1988; Smith and Smith 1989; Sauch *et al.*, 1991; Taghi-Kilani *et al.*, 1996; Smith, 1998), (iii) propidium iodide vital dye staining and morphological assessment of cysts observed under Nomarski DIC optics (Smith, 1996) or (iv) reverse transcriptase (RT)-PCR to amplify a sequence of the mRNA of *Giardia* heat shock protein 70 (*hsp 70*) (Abbaszadegan *et al.*, 1997; Kaucner and Stinear, 1998).

Cryptosporidium

Determination of *Cryptosporidium* oocyst viability (reviewed by Smith, 1998; O'Grady and Smith, 2002) can be undertaken *in vitro* by (i) excystation (Blewett, 1989, Robertson *et al.*, 1993), (ii) fluorogenic vital dyes (Campbell *et al.*, 1992; Belosevic *et al.*, 1997), (iii) *in vitro* infectivity (Upton *et al.*, 1994; Rochelle *et al.*, 1997; Slifco *et al.*, 1997; Arrowood, 2002, 2008), (iv) *in vitro* excystation followed by PCR (Wagner-Wiening & Kimmig, 1995; Deng *et al.*, 1997), (v) RT-PCR of *C. parvum hsp 70* (Stinear *et al.*, 1996; Kaucner and Stinear, 1998), (vi) fluorescence *in situ* hybridisation (Vesey *et al.*, 1998; Smith *et al.*, 2004; Graczyk *et al.*, 2006) or (vii) electrorotation (Goater and Pethig, 1998).

Cryptosporidium parvum (as well as *C. muris* and *C. meleagridis*) oocysts can establish infection in neonatal or immunosuppressed mice (reviewed in O'Grady and Smith, 2001). An *in vitro* cell culture infectivity method, developed for *C. parvum* oocysts (Upton *et al.*, 1994; Slifco *et al.*, 1997) has been used to determine the infectivity of oocysts subjected to (i) drug treatment, (ii) disinfectants, (iii) treatments in food processing, as well as pathophysiological changes in cell permeability due to *C. parvum* infection and the secretion of interleukin-8. *Cryptosporidium hominis* oocysts do not infect neonatal mice nor do they initiate infection as readily as *C. parvum* isolates in currently established cell culture systems, however, successful propagation of *C. hominis* oocysts was demonstrated in gnotobiotic piglets (Widmer *et al.*, 2000).

A variety of methods have been developed that support the growth of asexual stages of *C. parvum in vitro* (Upton *et al.*, 1994; Slifco *et al.*, 1997; Arrowood, 2002, 2008), yet the complete development of the life cycle and the production of oocysts in culture remains elusive. *In vitro* infectivity (cell culture) has been used primarily as a research tool, although there is some evidence for its use routinely in detecting infectious oocysts in some drinking water laboratories (Rochelle *et al.*, 1997; Di Giovanni *et al.*, 1999). Limited parasite growth reduces the sensitivity of the *in vitro* infectivity assay, and detection requires dedicated methods (immunofluorescence microscopy, immunoperoxidase microscopy, PCR, RT-PCR) to detect the intracellular, asexual forms of the parasite. However, within the limitations of the small number of *C. parvum* isolates tested, *in vitro* infectivity was equivalent to a CD-1 strain mouse infectivity assay for measuring the infectivity of untreated *C. parvum* oocysts, although both assays exhibited high levels of variability (Rochelle *et al.*, 2002).

Of the methods tested, *in vitro* infectivity in HCT-8 cells detected by immunofluorescence microscopy, followed by detection using RT-PCR, displayed the closest agreement with the CD-1 neonatal mouse infectivity assay. The authors recommended that disinfection studies should be limited to discerning relatively large differences between test and control samples (Rochelle *et al.*, 2002). The authors confirmed that *C. hominis* could infect HCT-8 cells in the same study.

Cyclospora

No animal model or *in vitro* viability assay has been identified for *C. cayetanensis*. Observation of *in vitro* sporulation is the only criterion available for assessing oocyst viability. *Cyclospora* oocysts sporulate maximally at 22 °C and 30 °C (Ortega *et al.*, 1993; Smith *et al.*, 1997); however, storage at either 4 °C or 37 °C for 14 days retards sporulation. Up to 12 % of human- and baboon-associated oocysts previously stored at 4 °C for 1–2 months sporulate when stored for 6–7 d at 30 °C (Smith *et al.*, 1997).

Entamoeba

As for the *Giardia, Cryptosporidium* and *Cyclospora* (oo)cysts, a major drawback is our inability to cultivate *E. histolytica* cysts *in vitro*. This results in a lack of adequate infection models. Animal models of intestinal amoebiasis attempt to reproduce invasive disease in laboratory animals, including cats, dogs, monkeys, rabbits and rats, but none reproduces the typical colonic lesions observed in human intestinal amoebiasis. More recently, gerbil and mouse models have been developed to mimic chronic non-healing lesions, characterised by inflammation and epithelial ulceration. Some mouse strains (e.g. BALB/c and C57BL/6) exhibit innate resistance to experimental intestinal amoebiasis, while other strains (e.g. C3H/Hej and CBA/J) are susceptible to disease and develop colitis (Ivory *et al.*, 2007). These methods are time-consuming and expensive, and *in vitro* methods that determine cyst viability or infectivity are required. Eosin exclusion has been used to determine cyst viability but, as for *G. duodenalis* cysts, its limitations make it an inconsistent reporter of viability (O'Grady and Smith, 2001). Further work investigating the usefulness of (i) fluorogenic vital dyes, (ii) *in vitro* infectivity, (iii) RT-PCR of *E. histolytica* heat shock protein (and other) genes and (iv) fluorescence *in situ* hybridisation should be performed.

Toxoplasma

The infectivity of the three different stages of *T. gondii* can be assessed by *in vivo* culture, normally in mice (Dubey, 2004). Infectivity using the *in vivo* bioassay can be determined by observing tissue cysts in the tissues of infected mice. Alternatively, mice can be bled and an antibody response to *Toxoplasma* antigens can be detected serologically, although the disadvantage of serology is that parasite viability is not determined and further passage of material may be required to ascertain parasite viability. *In vivo* methods are time-consuming, expensive and complex to maintain. *In vitro* methods, using cell culture lines such as HeLa cells, are also available, less expensive and are not reliant on government licenses and specialised animal housing facilities (Evans *et al.*, 1999).

Sensitive PCR methods have been devised and used to detect *Toxoplasma* DNA in animals and meat products (Wastling *et al.*, 1993; Warnekulasuriya *et al.*, 1998) using two gene loci: the B1 gene locus and the P30 gene locus. A nested PCR based at the B1 gene locus was more sensitive than one at the P30 gene. The B1 gene has approximately 40 copies in the *Toxoplasma* genome (Wastling *et al.*, 1993). PCR methods based on amplifying *Toxoplasma* DNA are not capable of determining viability, therefore, *in vivo* culture methods are often used in parallel (Warnekulasuriya *et al.*, 1998). In one Brazilian study (de Oliveira Mendonça *et al.*, 2004), 33 of 70 samples of fresh sausages collected from butchers were *Toxoplasma* PCR positive. Another Brazilian study, using a mouse bioassay to detect *Toxoplasma* infection, demonstrated seroconversion in 12 mice and *T. gondii* tachyzoites in a further mouse fed 149 samples of factory-produced, fresh pork sausage meat (8.7 % positivity), indicating that fresh pork sausage could be important in the transmission of toxoplasmosis (Dias *et al.*, 2005). As food processing can render tissue cysts non-viable (Table 26.7), it is difficult to determine the liklelihood of transmission of infection in the first study, as it is unknown whether the DNA detected was extracted from viable organisms.

Various serological methods have been used to determine seroprevalence studies in animals (Evans, 1992). The 'gold standard' for detecting anti-*Toxoplasma* antibody is the Sabin-Feldman dye test which uses living tachyzoites to detect antibody from all immunoglobulin classes. The dye test is unreliable in cattle as false-positive results are caused by bovine serum proteins (Dubey *et al.*, 1990). Less expensive and simpler to use serological tests (e.g. latex agglutination, enzyme-linked immunoassay (ELISA), indirect fluorescent test) are used to determine antibody positivity in numerous laboratories.

Free-living amoebae

There are two primary methods of *in vitro* culture for *Acanthamoeba* and *Naegleria* spp.: monoxenic culture in which Enterobacteriaceae are added to non-nutrient agar and axenic culture in which enriched nutrient broth medium is used with a bacterial food source (Ma *et al.*, 1990; Martinez and Visvesvara, 1991; Marciano-Cabral and Cabral, 2003). Tissue culture is also available. Identification of *Acanthamoeba* spp. to the genus level is standardised with morphological variations of trophozoites and cysts sufficient to identify these amoebae. However, species identification is more subjective, and a variety of techniques have been developed to improve this area of uncertainty. Molecular techniques, such as RFLP, sequencing and riboprinting, have been developed (Marciano-Cabral and Cabral, 2003).

Once isolated, *N. fowleri* has to be distinguished from other, non-pathogenic *Naegleria* spp. present in the environment. Monoclonal antibodies have been developed to all three stages of the amoeba that differentiate *N. fowleri* from non-pathogenic species (Sparagano *et al.*, 1993). RFLP and isoenzyme analysis methods require large numbers of axenically cultured organisms and, therefore, are time-consuming, while PCR methods reduce the time required to determine species identity (Kilvington and Beeching, 1995). These methods use primary cultures as their starting material, and the direct detection of amoebae from environmental samples by PCR is likely to be limited by the nucleic acid extraction method(s) used.

26.4.2 Resistance to physical and chemical treatments

(Oo)cysts are resistant to a variety of environmental pressures and chemical disinfectants normally used in water treatment (Table 26.7). *Giardia* and *Cryptosporidium* (oo)cysts survive better at lower than at higher temperatures. Smith and Grimason (2003) identify the effectiveness of disinfectants used commonly in the water industry. While less is known of *C. cayetanensis* oocyst survival in the environment, current evidence indicates that sporulation is delayed (presumably increasing survival) at lower temperatures (Smith *et al.*, 1997). The details of experimental procedures and data analyses for both *Giardia* and *Cryptosporidium* have critical effects on both the interpretation and comparability of results, often making comparisons between studies difficult. In addition, for *Giardia*, the more resistant *Giardia muris* cysts have often been used as a surrogate for those of the human parasite, *Giardia duodenalis*. There have been no comprehensive studies on E. *histolytica* cyst resistance to physical and chemical treatments, applicable to the food industry, since the re-description of the *E. histolytica* species complex.

The three infectious *Toxoplasma* stages have differing abilities to survive different physical and chemical treatments. The tachyzoite form is most sensitive to temperature extremes and does not readily survive outside the host (Evans, 1992). Tissue cysts can remain viable for 68 days at 4 °C but freezing meat to –13 °C renders tissue cysts non-viable (Dubey, 1974; Kotula *et al.*, 1991). Prolonged exposure to salt, as in the salting of meat, can also destroy tissue cysts. Oocysts are quite resistant and can survive in moist, warm environments for >1 year (Yilmaz and Hopkins 1972; Frenkel *et al.*, 1975). A number of *in vitro* experiments have been carried out to determine the effect of different physical and chemical treatments, and these are presented in Table 26.7.

Some environmental factors that influence *T. gondii* oocyst survival have been identified (Jackson and Hutchinson, 1989) and, together with known resistances to physical and chemical treatments, risk assessments for foodstuffs have been developed (e.g. http://www.nzfsa.govt.nz/science/risk-profiles/toxoplasma-gondii-in-red-meat.pdf; etc.). As published data for all these protozoan pathogens are numerous and varied, we summarise some of the more important trends and findings in Table 26.7.

Few data are available on inactivation studies for FLA. *Acanthamoeba* spp. cysts can be inactivated with 3 mg ClO_2 l^{-1} and 7 mg O_3 l^{-1}, (Cursons *et al.*, 1980b; Dawson and Brown, 1987; Beltrán and Jiménez, 2008). Aksozek *et al.* (2002) studied the resistance of *A. castellanii* cysts to physical, chemical and radiological conditions and found that solutions containing 20 % isopropyl alcohol, 10 % formalin, 3 % hydrogen peroxide, 0.02 % polyhexamethylene biguanide, 0.1 % clotrimazole or propamidine isethionate (Brolene) all killed *A. castellani* cysts *in vitro* (Table 26.7). However, *A. castellani* cysts were resistant to both 250 K rads of γ irradiation and 800 mW·s/cm^2 of UV irradiation (Aksozek *et al.*, 2002), whereas *A. culbertsoni* trophozoites (1.6 × 10^4) required > 173 mW·s/cm^2 for up to 5 log inactivation (Maya *et al.*, 2003). Some studies use much higher *Acanthamoeba* spp. densities (10^4 cysts ml^{-1}) than might be present in water or wastewater; therefore, the actual numbers in water or wastewater might be inactivated with lower UV doses (Beltrán and Jiménez, 2008).

Existing data indicate that both *N. fowleri* trophozoites and cysts are also fairly resistant to chlorine (de Jonckheere and Van de Voorde, 1976; Chang, 1978; Cursons *et al.*, 1980b; Cassells *et al.*, 1995). Gerba *et al.* (2008) calculated that the Ct of free chlorine for *N. fowleri* trophozoites was six for a 99 % reduction at pH 7.5, and that for *N. fowleri* cysts was 31 for a 99 % reduction at pH 7.5 (calculated using the Efficiency Factor Hom model, John *et al.*, 2005). Thus, the Ct values for *N. fowleri* cysts appear comparable to published values for *Giardia* cysts, but less than those for *Cryptosporidium* oocysts. A UV dose of 63 mW.s/cm^2 and 12.6 mW.s/cm^2 attains 2 log inactivation of *N. fowleri* cysts and trophozoites, respectively (Sarkar, 2008). The UV dose for a 2 log *N. fowleri* cyst inactivation is less than that of *Acanthamoeba* spp. cysts but much greater than that of *Cryptosporidium* spp. oocysts (Table 26.7).

26.5 Methods of detection and enumeration

Methods developed for detecting (oo)cysts as surface contaminants on foods are modifications of those used for water (e.g. Anon., 1994, 1999a, b; EPA, 1998; Smith, 1995, 1998; Smith and Hayes, 1996) and must be capable of detecting small numbers of (oo)cysts (1–100). Methods for detecting (oo) cysts on foods can be subdivided as follows: (i) desorption of organisms from the matrix, (ii) concentration and (iii) identification.

Desorption can be accomplished by mechanical agitation, stomaching, pulsifying and sonication of leafy vegetables or fruit suspended in a liquid that encourages desorption of (oo)cysts from the food. Detergents including Tween® 20, Tween® 80, sodium lauryl sulphate and Laureth 12 have been used to encourage desorption. Lowering or elevating pH affects surface charge and can also increase desorption. Depending on the turbidity and pH of the eluate, (oo)cysts eluted into non-turbid, neutral pH eluates can be concentrated by filtration through a 1 μm flat bed cellulose acetate membrane (Girdwood and Smith, 1999b). (Oo)cysts eluted into turbid eluates can be concentrated by immunomagnetisable separation (IMS) (Campbell and Smith, 1997; Deng *et al.*, 1997, 2000; Table 26.8). Commercial IMS kits are available for concentrating *Giardia* and *Cryptosporidium* (oo)cysts (Smith, 2008). The use of IMS has increased (oo)cyst recoveries, but matrix effects can influence recoveries. (Oo)cysts are identified by epifluorescence and Nomarski differential interference contrast microscopy and enumerated using specific criteria (Table 26.9). Standardised methods for detecting and enumerating (oo)cysts in water concentrates are available (http://www.epa.gov/microbes/1623de05.pdf; http://www.dwi.gov.uk/regs/crypto/pdf/sop%20part%202.pdf), and these form the basis for detecting and enumerating (oo)cysts in food concentrates. A standardised method for detecting *Cyclospora* on foods, in the absence of an IMS stage, is available http://www.cfsan.fda.gov/~mow/cyclmet.html.

The methods used for detecting *Giardia* and *Cryptosporidium* (oo)cysts in water have been used to detect *Toxoplasma* oocysts (Dumètre and Dardé, 2003). Direct microscopic examination of *Toxoplasma* oocysts is unhelpful because they are morphologically identical to four closely related coccidia, namely *Hammondia hammondi*, *H. heydorni*, *Neospora caninum* and *Besnoitia* species (Dubey, 2004). *Toxoplasma gondii* (as well as *H. hammondi*, *H. heydorni*, *N. caninum*, *B. darlingi* and *Sarcocystis neurona*) oocysts and sporocysts (and other coccidia) autofluorescence a pale blue colour when illuminated with an ultraviolet light source and viewed with the correct UV excitation and emission filter set of an epifluorescence microscope (Lindquist *et al.*, 2003) (see Table 26.9 for *Cyclospora* spp. oocysts). Bioassays are the 'gold standard' for detecting viable *Toxoplasma* parasites and, as sensitivity is an important consideration, Dubey (2004) demonstrated that the use of bioassays of soil samples fed to pigs and chickens may be more sensitive than the use of *in vitro* methods.

As *E. histolytica* and *E. dispar* cysts are morphologically identical, differentiating between them relies on DNA analyses following PCR. Isolation and detection methods for FLA, particularly from the environment, are well established (Page, 1988; Anon., 1989; Marshall *et al.*, 1997), although they are not particularly sensitive. *Acanthamoeba* can be differentiated at the genus level by morphology, but further identification is reliant on molecular

Table 26.8 Some reported recovery rates of *Cryptosporidium*, *Giardia* and *Cyclospora* (oo)cysts from foods

Food matrices	Extraction and concentration methods	ID methods	Recovery rates	References
200 g seeded with *Cryptosporidium* spp. oocysts (1/g) • cabbage and lettuce leaves	FDA method: sonication in 1 % SDS and 0.1 % Tween® 80, centrifugation	IF[a]	• 1 %	Bier, 1990
• Milk • Orange juice • White wine	Seed 10–1000 *C. parvum* oocysts in 70–200 mL; filtration	IF on filters	• 4–9.5 % • < 10 % • > 40 %	Bankes, 1995 (Experimental recoveries)
• Cilantro leaves • Cilantro roots • Lettuce	Rinse according to Speck (1984); centrifugation	*Cryptosporidium* oocysts by Koster stain	• 5.0 % (4 samples) • 8.7 % (7 samples) • 2.5 % (2 samples)	Monge and Chinchilla, 1996 (Costa Rica)
Raw milk	Bacto®-Trypsin and Triton® X-100 treatment of seeded samples (20 ml); centrifugation	PCR[b] and probe hybridisation	10 seeded *C. parvum* oocysts	Laberge *et al.*, 1996 (Experimental recoveries)
Seeded lettuce leaves • *Cyclospora* (50) • *C. parvum* (100)	Rinse in tap water; centrifugation	Acid-fast stain, IF and direct wet mount	• *C. parvum* 25–36 % • *Cyclospora* 13–15 %	Ortega *et al.*, 1997a (Experimental recoveries)
(i) 110 vegetables and herbs from 13 markets; (ii) 62 vegetables and herbs from 15 markets	Rinse in tap water; centrifugation	Acid-fast stain, IF and direct wet mount	• *Cryptosporidium* • (i) 14.5 %; (ii) 19.35 % • *Cyclospora* • (i) 1.8 %; (ii) 1.6 %	Ortega *et al.*, 1997b (Peru)

Table 26.8 (Cont'd)

Food matrices	Extraction and concentration methods	ID methods	Recovery rates	References
Clams (*Corbicula fluminea*)	Seeded with ~1.9×10^5 *C. parvum* oocysts per clam	IF	• 97.5 % prevalence when gills and gi tract examined; 86.7 % in haemolymph • 41.7 % of oocysts found in faeces at bottom of aquarium	Graczyk *et al.*, 1998
Homogenised milk	100 ml samples; centrifugation followed by IMS[c]	Direct PCR	10 *C. parvum* oocysts	Deng and Cliver, 1999 (Experimental recoveries)
Apple juice	• Sucrose gradient, IMS • Flotation or flotation and IMS • Flotation	• IF • direct PCR • acid-fast stain	• 10–30 *C. parvum* oocysts • 30–100 oocysts • 3000–10 000 oocysts	Deng and Cliver, 2000 (Experimental recoveries)
• Four leafy vegetables and strawberries • Bean sprouts	Rotation and sonication in elution buffer; centrifugation and IMS	IF	• *Giardia* 67 %; *C. parvum* 42 % • *Giardia* 4–42 %; *C. parvum* 22–35 %	Robertson and Gjerde, 2000 (Experimental recoveries)
Lettuce leaves	100 oocysts *C. parvum* seeded, elution buffer (pH 5.5) Stomaching Centrifugation and IMS	IF	59.0 ± 12.0 % (n = 30)	Cook *et al.*, 2006a (Experimental recoveries)
Raspberries	100 oocysts *C. parvum* seeded, elution buffer (pH 5.5)	IF	41.0 ± 13.0 % (n = 30)	Cook *et al.*, 2006a (Experimental recoveries)

	• Rolling Centrifugation and IMS			
Lettuce leaves	100 cysts *G. duodenalis* seeded, Elution buffer (pH 5.5) Stomaching Centrifugation and IMS	IF	46.0 ± 19.0 % (n = 30)	Cook *et al.*, 2007 (Experimental recoveries)

[a] direct immunofluorescence assay.
[b] polymerase chain reaction.
[c] immunomagnetic separation.

Table 26.9 Characteristic features of *G. duodenalis* cysts and *C. parvum* and *C. cayetanensis* oocysts by epifluorescence microscopy and Nomarski differential interference contrast (DIC) microscopy

- *G. duodenalis* cysts and *C. parvum* oocysts: appearance under the FITC filters of an epifluorescence microscope:
  The putative organism must conform to the following fluorescent criteria:
  Uniform apple green fluorescence, often with an increased intensity of fluorescence on the outer perimeter of an object of the appropriate size and shape (see below).
- *C. cayetanensis* oocysts: appearance under the UV filters of an epifluorescence microscope:
  The putative organism must conform to the following fluorescent criteria:
  Uniform sky blue autofluorescence on the outer perimeter of an object of the appropriate size and shape (see below).

Appearance under Nomarski differential interference contrast (DIC) microscopy:

Giardia duodenalis cysts	*Cryptosporidium parvum* oocysts	*Cyclospora cayetanensis* oocysts
• ellipsoid to oval, smooth walled, colourless and refractile • 8–12 × 7–10 μm (length × width).	• spherical or slightly ovoid, smooth, thick walled, colourless and refractile • 4.5–5.5 μm	• spherical, smooth, thin walled, colourless and refractile • 8–10 μm
• Mature cysts contain four nuclei displaced to one pole of the organism. • Axostyle (flagellar axonemes) lying diagonally across the long axis of the cyst. • Two 'claw-hammer'-shaped bodies lying transversely in the mid-portion of the organism.	• Sporulated oocysts contain four nuclei. • Four elongated, naked (i.e. not within a sporocyst(s)) sporozoites and a cytoplasmic residual body within the oocyst.	• Unsporulated oocysts contain developing sporocysts (which can look like a raspberry) • Sporulated oocysts contain two ovoid sporocysts, each containing two sporozoites.

techniques. However, it is possible that all strains are pathogenic and further differentiation is not necessary (Szénási *et al.*, 1998). Differentiation of *N. fowleri* from other *Naegleria* species is essential. Unfortunately morphology and biological characteristics are variable and other methods are required (Marshall *et al.*, 1997). Monoclonal antibodies have a distinct advantage for identifying individual or small numbers of *N. fowleri* (Visvesvara *et al.*, 1987). PCR methods have been developed that can detect *N. fowleri* from the environment (Kilvington and Beeching, 1995). There is a need for quicker, simpler and more specific methods to identify and differentiate FLA.

PCR can be used to determine the presence of parasite DNA in water concentrates (reviewed in Smith, 1998; Girdwood and Smith, 1999a, b; Smith *et al.*, 2006, 2007; Table 26.8), either following oocyst detection and enumeration by microscopy (Nichols *et al.*, 2003, 2006b; Ruecker *et al.*, 2005, 2007) or without prior identification and enumeration by microscopy (Xiao *et al.*, 2000, 2001; Jiang *et al.*, 2005; Smith and Nichols, 2009). The conservative approach would be to identify oocysts containing sporozoites by microscopy followed by the demonstration of parasite DNA in eluates of samples positive by microscopy (Smith, 2008; Smith and Nichols, 2009). The sensitivity of PCR can be better than microscopy (< 1 oocyst equivalent of DNA), but matrix effects and other PCR interferents can reduce sensitivity drastically. In the author's (HVS) laboratory, the method developed by Nichols *et al.* (2003, 2006b) is consistently the most sensitive, partly because DNA extraction has been maximised (Nichols and Smith, 2004), PCR interferents have been minimised, there are 20 copies of the targeted locus in each intact oocyst and the amplicon length is small (~500 bp) (Smith, 2008).

Published PCR primers are available for detecting the genus and species of all the protozoan parasites discussed in this chapter, however, there are insufficient studies performed on standardised DNA extraction methods, sets of primers and PCR procedures to determine their sensitivity, particularly in food matrices. For *Toxoplasma*, PCR amplification is sensitive and capable of detecting one tachyzoite in some samples (Joss *et al.*, 1993), but there are limitations. The DNA extraction method used can limit the volume of sample that can be analysed, and PCR inhibitors may be a component part of tissue samples that are co-extracted with parasite DNA. A range of 4–10 % of meat products has been shown to be infected (Hall *et al.*, 2001). One of 67 (1.5 %) cured meat products from a retail outlet contained amplifiable *T. gondii* DNA (Warnekulasuriya *et al.*, 1998). Importantly, PCR of *Toxoplasma* DNA does not determine viability.

26.5.1 Occurrence in/on foods

The foods most at risk of being contaminated with human infectious protozoan parasites are those intended to be consumed raw or lightly cooked. These include raw salads, fresh fruits, shellfish, bottled waters, unpasteurised cold drinks and potable quality water used as a food ingredient or ice (Nichols and Smith 2007, 2009). Changes in consumer consumption patterns, influenced by increased international travel, globalisation of the food supply, changes towards a 'healthier' diet and lifestyle, identify that options such as fresh fruit, salad vegetables and seafood, all of which can be at risk of protozoan parasite contamination, are now more popular. As the ID_{50} for humans can be small, the risk of infection can be significant, even when low numbers of infectious protozoan parasites are present.

Numerous studies have investigated the presence of *Giardia* cysts on/in foods (reviewed in Nichols and Smith, 2002, 2009). *G. duodenalis* cysts have been detected on produce in several countries (Amahmid *et al.*, 1999; Robertson and Gjerde 2000, 2001; Robertson *et al.*, 2002; Cook *et al.*, 2007; Smith and Paget, 2007), and contaminated irrigation water, especially, appears to constitute a major route of contamination of fresh produce (Thurston-Enriquez *et al.*, 2002; Chaidez *et al.*, 2005; Sutthikornchai *et al.*, 2005). *Giardia duodenalis* cysts (Assemblage A) were detected in clam (*Macoma balthica* and *M. mitchelli*) tissues. *Macoma* spp. clams, which have no economic value, burrow in mud or sandy-mud substrata and preferentially feed on the surface sediment layer. *Macoma* spp. clams can serve as biologic indicators of sediment contamination with *Giardia* sp. cysts of public health importance (Graczyk *et al.*, 1999).

Numerous studies have investigated the presence of *Cryptosporidium* spp. oocysts on/in foods. *Cryptosporidium* oocysts have been detected on fresh produce including herbs, leafy salads, fruit and in shellfish, mineral waters, unpasteurised milk and non-alcoholic apple cider (Nichols and Smith, 2007). *Cryptosporidium* spp. oocysts have been detected on vegetables (e.g. coriander leaves and roots, lettuce, tomato, cucumber, carrot and fennel) and in shellfish (e.g. oysters, mussels, clams and cockles) and unpasteurised apple cider and mineral water (Nichols and Smith, 2007). For *C. parvum* oocyst contamination, Nichols and Smith (2007) classified foods into (i) high-risk produce (e.g. vegetables and fruit; shellfish; unpasteurised cold drinks) and (ii) low-to medium-risk produce (e.g. meat and meat products).

Cyclospora (and *Cryptosporidium*) oocysts were detected on vegetables for sale in market places in Peru (Ortega *et al.*, 1997b), in tap water and on lettuce heads in Egypt (Abou el Naga, 1999) and on green leafy vegetables in Nepal (Sherchand *et al.*, 1999). Data on the occurrence of *Giardia*, *Cryptosporidium* and *Cyclospora* (oo)cysts on /in foods are presented in Table 26.10.

Our knowledge of the level of (oo)cyst contamination of foodstuffs has been limited due to the lack of reproducible, sensitive detection methods (reviewed in Smith *et al.*, 2007; see Table 26.8). A sensitive, reproducible, optimised and validated method is now available for detecting *Cryptosporidium* oocysts on lettuce and raspberries and *Giardia* cysts on lettuce (Cook *et al.*, 2006a, b, 2007; Smith and Cook, 2009), although the matrices of differing salad vegetables influence recovery (Cook *et al.*, 2007; Smith and Cook, 2009). No such methods are available for detecting *E. histolytica* cysts, *T. gondii* oocysts or cysts of FLA on foods, although the methods developed by Cook *et al.* (2006 a, b, 2007) could prove suitable and should be investigated on the basis that both the release of *Cryptosporidium* oocysts from the food matrix and their concentration by IMS were maximised following a critical assessment of (i) the pH of the elutant and (ii) the microenvironment required to obtain maximum interaction between antibody paratope and oocyst epitope for efficient recovery by IMS (Smith and Cook, 2009).

Table 26.10 Some published levels of *Cryptosporidium* and *Giardia* (oo)cyst contamination of foods

Food type	*Cryptosporidium*/*Giardia* (oo)cyst contamination levels and methods of detection	Food source/Country	Reference
Vegetable/ herb/fruit	*Cryptosporidium* oocysts recovered from cilantro (coriander) leaves (n = 7; 5 %); cilantro roots (n = 7; 8.7 %); lettuce (n = 2; 2.5 %); radish, tomato, cucumber and carrot (1.2 %) after washing and bright field microscopy.	Open market/Costa Rica	Monge and Chinchilla, 1995
	Cryptosporidium oocysts recovered from 16 of 110 (14.5 %) of vegetables (8 types, 3 collections) after washing, centrifugation and IF[a].	Open market/Peru	Ortega *et al.*, 1997b
	Cryptosporidium oocysts recovered from 19 of 74 (4 %) of vegetables (5 lettuce, 14 mung bean sprouts) after washing, IMS[b] and IF. Average ~3 oocysts per 100 g produce.	Commercial vegetable distributors/Norway	Robertson and Gjerde, 2000; Robertson *et al.*, 2002
	Cryptosporidium oocysts recovered from fennel and lettuce samples following irrigation with filtered wastewater (?caused by filtration system failure in pilot treatment plant). Detected by IF.	Experimental crop fields/Italy	Lonigro *et al.*, 2006
	Giardia cysts recovered from lettuce heads (64 samples) collected from 3 of 4 markets studied. Cysts detected by bright field microscopy.	Vegetables from markets/Rome, Italy	Mastrandrea and Micarelli, 1968
	Giardia cysts recovered from: • coriander (44.4 %; 2.5×10^2 cysts kg^{-1}) • carrots (33.3 %; 1.5×10^2 cysts kg^{-1}) • mint (50 %; 96 cysts kg^{-1}) • radish (83.3 %; 9.1 cysts kg^{-1}) Cysts detected by bright field microscopy. Crops irrigated with treated water (sedimentation and 16 days retention) were free from contamination.	Crops irrigated with untreated raw wastewater/ Marrakech, Morocco	Amahmid *et al.*, 1999
	Giardia cysts recovered from 10 of 475 (4.75 %) vegetable/ fruit samples (2 dill, 2 lettuce, 3 mung bean sprouts,	Commercial vegetable distributors/Norway	Robertson and Gjerde, 2000;

Table 26.10 (Cont'd)

Food type	*Cryptosporidium*/*Giardia* (oo)cyst contamination levels and methods of detection	Food source/Country	Reference
	1 radish sprouts, and 2 strawberries) after washing, IMS and IF. Average ~ 3 cysts per 100 g produce.		Robertson *et al.*, 2002
	Giardia cysts recovered from lettuce by membrane filtration, sedimentation and bright field microscopy (Faust method). Nine of 10 schools had at least one sample (n = 90) contaminated as follows: • before washing (9.5 %; n = 30) • after washing with tap water (21.4 %; n = 30) • school tap water (no cysts detected; n = 30)	Vegetables consumed at schools/São Paulo, Brazil	da Silva Coelho *et al.*, 2001
	One *Giardia* cyst recovered from 1 of 18 types of 'ready to eat foods' including leafy salads, carrots, mange touts, herbs, chillies (sample sizes varied from 14 to 150 g or package contents) after washing, IMS and IF. The *Giardia* cyst-positive sample was an organic watercress, spinach and rocket salad (50 g portion).	Salad vegetables from local retail outlets/UK	Cook *et al.*, 2007
Shellfish	30 oysters from 43 sites examined on 3 occasions for oocysts in haemolymph and gill washings. 7 of 7 sites were positive by IF and/or PCR–RFLP[c]. *C. parvum* was major contaminant but *C. hominis* also detected: • 3 of 16 sites yielded oocysts infectious to neonatal mice	Commercial oyster harvesting sites/Chesapeake Bay, USA	Fayer *et al.*, 1999
	Infectious *C. parvum* (PCR–RFLP) oocysts recovered from mussels (*Mytilus galloprovincialis*) harvested near river mouth where cattle grazed on its banks (> 10^3 oocysts per shellfish by IF).	Shellfish producing bay sites/Spain	Gomez-Bautista *et al.*, 2000
	Cryptosporidium oocysts recovered from 21 of 38 (55.2 %) samples of 17 clam, 15 mussel and 6 oyster samples (12 molluscs per sample) by IF.	Galicia, Spain/Italy/ England	Freire-Santos *et al.*, 2000

C. parvum DNA (PCR–RFLP) detected in cockles (*Cerastoderma edule*) collected from a region of intense sedimentation of river material.	Shellfish producing bay site/Spain	Gomez-Bautista *et al.*, 2000
C. *hominis* DNA (PCR–RFLP) detected in 2 of 16 mussels (*Mytilus edulis*).	Commercial shellfish beds/Northern Ireland	Lowery *et al.*, 2001
Cryptosporidium oocysts recovered from 18 clam, 22 mussel and 9 oyster samples (56 % by IF and 44 % by ABC–PCR[d]). 20 mussels and 18 cockle samples (11 % detection).	Spain	Gómez-Couso *et al.*, 2003a
C. parvum (22 of 26), *C. hominis* (1 of 26), mixed *C. parvum* and *C. hominis* (3 of 26).	UK	
Cryptosporidium oocysts recovered from oysters (54 %), mussels (32.7 %), clams (29.4 %) and cockles (20.8 %) and detected by IF. Viable oocysts recovered from 60 % clams and cockles, 47.8 % oysters and 51.4 % mussels: • 34 % positives (69 of 203) and 49 % viable oocysts from Spain • 14 % positive (14 of 38) and 71.4 % viable oocysts from European Union	Spain and EU countries	Gómez-Couso *et al.*, 2003b
C. parvum DNA (PCR-sequencing) detected in 2 of 16 samples collected from 2 sites. 480 clams (*Chamelea gallina*) collected from 3 river mouth sites sampled with 30 clams pooled for each sample.	Abruzzo, Adriatic sea/Italy	Traversa *et al.*, 2004
C. parvum, *C. felis*, *C. andersoni* and 2 novel *Cryptosporidium* spp. DNA identified (PCR-sequencing) from 156 batches of 30 mussels (*Mytilus* spp.) laced as sentinels in rivers. On average 2 oocysts (range 1–7) detected per pool of 30 sentinel clams (*Corbicula fluminea*). *C. parvum* DNA identified by PCR-sequencing.	California/USA	Miller *et al.*, 2005a,b
C. parvum DNA (PCR–RFLP) detected in 31.1 % of 222 samples of 6–8 mussel (*Mytilus galloprovincialis*) pools.	Spain	Gómez-Couso *et al.*, 2006
Cryptosporidium oocysts recovered from 9 of 133 (6.7 %) non-commercial and 6 of 46 (13.0 %) commercial oysters (*Crassostrea gigas*) and detected by IF. Viable oocysts found in surface waters that entered the harvesting area.	Oosterschelde oyster harvesting area/Netherlands	Schets *et al.*, 2007

Table 26.10 (Cont'd)

Food type	*Cryptosporidium*/*Giardia* (oo)cyst contamination levels and methods of detection	Food source/Country	Reference
	C. parvum detected in all samples from cultured edible mussels (*Mytilus edulis*) collected from 3 sites and during all seasons by IF and PCR-sequencing. Oocysts from one location caused infection in one neonatal mouse.	Normandy/France	Li *et al.*, 2006
	G. duodenalis Assemblage B detected in 1 of 9 oyster samples by multiplexed nested ABC–PCR[d] (Spain). None detected in 38 UK samples (20 mussels and 18 cockles).	Spain and UK	Gómez-Couso *et al.*, 2004
	Giardia spp. cysts recovered from sentinel clams from all 3 riverine regions studied and detected by IMS and IF: • 3 cysts per clam pool (30 clams per pool; range: 1–26 cysts)	Sentinel clams in rivers in California, USA.	Miller *et al.*, 2005a
	Giardia spp. cysts recovered from 9 of 133 (6.7 %) non-commercial and 6 of 46 (13.0 %) commercial oysters (*Crassostrea gigas*) and detected by IF. Viable cysts found in surface waters that entered the harvesting area.	Oosterschelde oyster harvesting area, Netherlands	Schets *et al.*, 2007
Unpasteurised apple cider	*C. parvum* oocysts recovered from washed, but not, unwashed apples. Oocysts detected by IF were recovered from: • water from 2/5 producers • washed apples from 1/5 producers • fresh cider from 3/5 producers • finished cider from 1/5 producers	Five cider producers/ Canada	Garcia *et al.*, 2006
Mineral water	*Cryptosporidium* oocysts recovered from 2 of 13 brands of still, bottled water on 3 occasions at densities of 0.2 to 0.5 oocysts l^{-1}. Oocysts concentrated from 20 l samples by filtration and centrifugation and detected by IF.	Supermarkets in Campinas, Brazil	Franco and Cantusio Neto, 2002

[a]Immunofluorescence assay.
[b]Immunomagnetic separation.
[c]Polymerase chain reaction followed by restriction fragment length polymorphism analysis of the amplicon.
[d]A multiplex nested PCR assay to detect *Cryptosporidium* spp. and *Giardia* assemblages A and B.

26.6 Future trends

There are no national or international guidelines for determining (oo)cyst contamination in/on foodstuffs, but further development, validation and harmonisation of methods should encourage the formulation of international standards. Currently, the only standardised, validated methods for detecting *G. duodenalis* cysts and *Cryptosporidium* spp. oocysts on foods are those of Cook *et al.* (2006a, b, 2007). A standardised method for detecting *Cyclospora* on foods, in the absence of an IMS stage, is available http://www.cfsan.fda.gov/~mow/cyclmet.html. There are no such methods for detecting *E. histolytica* cysts, *T. gondii* oocysts or cysts of FLA on foods. Methods to detect *E. histolytica* cysts, *T. gondii* oocysts or cysts of FLA on foods should be developed in order to determine their involvement in the foodborne transmission of protozoan parasite diseases.

Modifications to existing methods described will undoubtedly occur, in order to optimise recovery from specific matrices. Here, the most important 'front end of method' stage for maximising the detection of protozoan parasite transmissive stages is their efficient extraction from the food matrix in question and the need to understand parasite surface–food surface interactions so that suitable physicochemical extraction procedures can be developed (Smith and Cook, 2009). (Oo)cyst-food matrix interactions should be investigated further so that (oo)cysts can be maximally released from the surfaces of foodstuffs and from within other high-risk products such as shellfish. Both method availability and outbreaks of foodborne disease drive investigations into foodborne occurrence. The availability of IMS kits for *Giardia* and *Cryptosporidium* (oo)cysts in water (as well as outbreaks of foodborne cryptosporidiosis) are major drivers of *Giardia* and *Cryptosporidium* occurrence studies, whereas the numerous outbreaks of cyclosporiasis, in the absence of effective methods, particularly IMS, has driven method development in *C. cayetanensis* occurrence studies. At present, many publications report on seeding experiments, using a small range of purified protozoan parasite isolates (Table 26.8), although *Giardia*, *Cryptosporidium* (Table 26.10) and *Cyclospora* (oo) cysts have also been detected on vegetables for sale in market places.

Until analyte separation and concentration methods improve further, IMS is likely to remain an important component of methods for extracting protozoan parasites from food matrices. While current IMS methods impart obvious benefits to existing methods, they also have limitations, being antibody-based. Cook *et al.* (2006a) and Smith and Cook (2009) recognised that the extractants they used influenced the performance of commercially-available IMS kits for concentrating *Giardia* and *Cryptosporidium* (oo)cysts from food matrices. They found that when lettuce and raspberries were extracted with extractants that were too acidic, the low pH of the eluate prevented the pH of the final extract from reaching neutrality, causing sub-optimal binding of *Cryptosporidium* oocysts to bead-bound antibody if a pH of 7, where paratope–epitope interactions are maximised for IMS, was not reached (Smith and Cook, 2009).

Little work has been undertaken to determine the reactivity that the monoclonal antibodies used to capture *Giardia* and *Cryptosporidium* (oo)cysts from water concentrates using current IMS kits possess with respect to (i) all *Giardia* and *Cryptosporidium* species/genotypes/assemblages, (ii) all human infectious *Giardia* and *Cryptosporidium* species/genotypes/assemblages or (iii) whether they recognise all known isolates in each *Giardia* and *Cryptosporidium* species/genotype/assemblage. Further work is required to determine the reactivity of monoclonal antibodies with respect to (i), (ii) and (iii), above, so that IMS kits can be quality assured to bind to the major species/genotypes/assemblages of protozoan parasites that cause human disease. The lack of monoclonal antibodies reactive against exposed antigens on *Cyclospora* sp. oocysts has precipitated the development of an alternative method. In the absence of oocyst-reactive, fluorescence-labelled antibodies, *Cyclospora* oocysts can be visualised and enumerated under the UV filter of a fluorescence microscope (Table 26.9; http://www.cfsan.fda.gov/~mow/cyclmet.html).

Cook *et al.* (2007) used an internal control (reporter) system consisting of a known number of Texas Red™-stained *Cryptosporidium* and *Giardia* (oo)cysts (ColorSeed™) to determine the efficiency of their method in recovering *Cryptosporidium* oocysts from different seeded vegetable matrices. Their (oo)cyst recoveries from 20 separate vegetable matrices were 36.2 ± 19.7 % ($n = 20$), and oocyst recoveries differed considerably for individual matrices. They reported that the use of internal control reporter system offered much needed quality assurance and that the differences in recoveries, when using a variety of fresh produce, are probably due to the variability of the non-covalent interactions between oocyst surfaces and surfaces of the various fresh produce tested, and that 1M glycine (pH 5.5) may not optimise oocyst release from all salad vegetables. Because of this inherent variation, Cook *et al.* (2007) recommended that a reporter system should be incorporated into each set of analyses, particularly when a variety of matrices are analysed. ColorSeed™ is a commercial product consisting of a *Cryptosporidium* and *Giardia* (oo)cyst suspension containing 100 inactivated *Cryptosporidium* and *Giardia* (oo)cysts, permanently labelled with the fluorogen Texas Red (http://www.btfbio.com/product.php?nav=ColorSeed). ColorSeed™ is an internal control that can be used when analysing foods (and water). ColorSeed™ are Texas Red-stained, gamma-irradiated (oo)cysts which will fluoresce under the green filter block (535 nm excitation, > 590 nm emission) of a fluorescence microscope. The percentage of (oo)cysts that can be recovered from a sample is determined by comparing the number of Texas Red-stained (oo)cysts recovered from each sample, compared to the number of (oo)cysts contaminating the same sample. All (oo)cysts recovered that stain with the fluorescein isothiocyanate-labelled anti-*Cryptosporidium* monoclonal antibody (FITC-*C*-mAb) are enumerated. Only the Texas Red-stained reporter (oo)cysts can be visualised and enumerated under the green filter of the epifluorescence microscope. By subtracting the number of Texas Red (red)-stained (oo)cysts from the total number of fluorescein isothiocyanate FITC (green)-stained oocysts, the number of (oo)cysts naturally contaminating the product can be calculated.

As morphology cannot be used to determine human infectious species/genotypes of protozoan parasites in a contaminated sample, PCR-RFLP and sequencing methods become essential for understanding the epidemiology of infection and disease and for source and disease tracking. Importantly, empty (oo)cysts, which do not contain parasite DNA, but are an indicator of contamination, will not be detected using molecular methods. A variety of methods have been published for waterborne protozoan parasites (see above), but there are few, if any, comparisons for protozoan parasite contaminants of foods. Here, standardisation of methods, from DNA extraction to RFLP and sequence analysis, is a fundamental requirement. As the abundance of transmissive stages of protozoan parasites is expected to be low, nested PCR methods are required to obtain sufficient sensitivity for detection. For nested PCR-based assays developed to determine *Cryptosporidium* species/genotype in water samples, a major limitation of is the DNA extraction method used (Nichols and Smith, 2004), and this is expected to be the case for the other protozoan parasites discussed in this chapter. Maximising DNA extraction from (oo)cysts is fundamental to maximising PCR outcomes (Smith and Nichols, 2008). For the majority of protozoan parasites discussed in this chapter, PCR primers were developed primarily for clinical diagnosis, where a much larger abundance of target DNA is generally present in a matrix containing fewer non-target DNA templates that are likely to cross-react. Environmental (food and water) samples are the exact opposites of clinical samples, in that the abundance of target DNA is generally far smaller and is present in a matrix containing more non-target DNA templates that are likely to cross-react (Smith and Nichols, 2008). Thus the judicious choice of locus (loci) and nested primer sets is a pre-requisite for environmental samples. In addition, the information available regarding matrix effects and PCR interferents in food matrices will prove useful when developing PCR-based methods for protozoan parasites (Smith and Nichols, 2008).

PCR detection platforms are expected to change, with greater focus being placed on real-time PCR, microarrays and rapid testing. Rapid testing could prove important in some critical areas as current nested PCR-based methods are not sufficiently rapid or standardised to be applied to all levels of HACCP systems. Better methods, established by international collaboration between expert laboratories, and round robin testing, are required to address these issues.

Infectivity and viability testing of protozoan parasites have received some attention (Section 26.4.1), but they must be developed further if they are to prove useful for HACCP in the food industry. Currently, there are some data available that can support HACCP (e.g. Table 26.7), but the development and use of further infectivity and/or viability testing parameters to address food industry requirements is required.

26.6.1 Control issues

Analyses of foodborne outbreaks of giardiasis, cryptosporidiosis and cyclosporiasis identify two major sources of contamination: surface contamination and food handlers. Surface contamination of produce can be a frequent occurrence, and foods that receive minimal treatment pose the greatest risks. Surface contamination of such produce at source can be direct or indirect (Table 26.11), and effective methods for the isolation and identification of these parasites from foods should be developed and validated.

(Oo)cyst-contaminated water can be a further source of contamination. Mains potable water or treated borehole-derived water used by the food industry for its manufacturing and ancillary processes and all water used for direct food contact and food contact surfaces must be at least of potable quality and should be free of pathogens. Water uses in the food industry include: direct incorporation into foods as an ingredient, washing of food containers (e.g. cans prior to passing in to high-risk processing areas), washing raw vegetables, raw fruits, animal carcasses, etc. Water used for cleaning, which has the potential to become contaminated with pathogens washed from produce, is increasingly being reconditioned, especially for the preparation and processing of food as long as the microbiological safety and quality of each food can be assessed.

Standardised methods for detecting *Giardia* and/or *Cryptosporidium* and FLA in mains potable water are available (Anon., 1989, 1994, 1998b, 1999a, b, EPA, 2001a,b, DWI 2005), but not for *E. histolytica* or *T. gondii*, and should be applicable to treated borehole-derived water. Where undertaken, results of routine water testing for *Cryptosporidium* (and possibly *Giardia*) will be held by water companies. In England and Wales, *Cryptosporidium* non-compliances are reported to the Drinking Water Inspectorate (in Scotland, the Drinking Water Quality Regulator for Scotland, and in northern Ireland,

Table 26.11 Possible sources of food contamination

- Use of cyst and oocyst contaminated faeces (night soil), farmyard manure and slurry as fertiliser for crop cultivation.
- Pasturing infected livestock near crops.
- Defaecation of infected feral hosts onto crops.
- Direct contamination of foods following contact with cyst and oocyst contaminated faeces transmitted by coprophagous transport hosts (e.g. birds and insects).
- Use of contaminated wastewater for irrigation.
- Aerosolisation of contaminated water used for insecticide and fungicide sprays and mists.
- Aerosols from slurry spraying and muck spreading.
- Poor personal hygiene of food handlers.
- Washing 'salad' vegetables, or those consumed raw, in contaminated water.
- Use of contaminated water for making ice and frozen / chilled foods.
- Use of contaminated water for making products which receive minimum heat or preservative treatment.

the Drinking Water Inspectorate for Northern Ireland). The presence of *Giardia*, *Cryptosporidium* and FLA in reconditioned water can be tested using the above methods, although less data are available regarding their recovery efficiencies.

The lack of food sampling and surveillance programmes denotes that foodborne protozoan parasites cases are most likely confirmed only in outbreak situations or during investigations of clusters of disease, and that seldom will there be the assignation of infection source(s) for sporadic cases. While some data on the survival of (oo)cysts in food industry processes and product are available, most are generic or extrapolatory, with few pertaining to actual processes undertaken by specific manufacturers at full scale. Because of the robustness of these parasites, issues regarding the survival of (oo)cysts in treatment and production processes, on surfaces, in disinfectants used in the food industry and in/on product must be addressed, otherwise effective HACCP strategies cannot be implemented.

As poor personal hygiene is a major contributor to protozoan parasite contamination incidents, guidelines currently in use for individuals working in the preparation of food in restaurants and industry with infectious, diarrhoeal disease should be extended to incorporate giardiasis, cryptosporidiosis and cyclosporiasis. Measures include obligatory, paid abstinence from work during gastrointestinal illness and strict procedures on hand-washing and the use of clean facilities in the work place. Further evidence of foodborne amoebiasis, and the potential for contamination of foods by food handlers, is required before amoebiasis can be added to this list. There is no similar requirement for toxoplasmosis, PAM or GAE.

Globalisation of food production and new food trends can contribute to increasing opportunities for protozoan parasite contamination. Sourcing ingredients from various countries or regions, which are then incorporated into a final food product, can complicate the tracing of a particular contaminated constituent. Knowledge of parasite biology and survival as well as local agricultural practices can assist risk assessment. Effective risk assessment requires confidence in recovery and identification methods and survival data. For *Giardia*, *Cryptosporidium*, *Cyclospora* and *Toxoplasma*, the contamination of food, its surveillance and control are measures that are still being developed, however, for *E. histolytica* and FLA, there has been little development in this area.

More focused quality issues have followed the adoption of HACCP-based control systems. This approach requires substantial information on the significance of transmission routes and (oo)cyst contamination of, and survival in, matrices commonly encountered by the food industry. While such criteria are being sought for *Giardia*, *Cryptosporidium* and, to a lesser extent, *Toxoplasma*, many remain unknown for newer emerging foodborne parasites such as *Cyclospora*, FLA and the microsporidia.

26.6.2 Regulations and guidance

Public health

While not reportable in England and Wales, giardiasis and cryptosporidiosis were made laboratory reportable diseases in Scotland in 1989. Cyclosporiasis is not reportable in the UK. Following numerous outbreaks in the eastern USA, in 1996 and 1997, and Canada, the Centers for Disease Control and Prevention established cyclosporiasis as a reportable disease. *Cyclosporia cayetanensis* is also included as an emerging pathogen in the Food Safety initiative which focuses on monitoring outbreaks, research into the selected pathogens and enforcement of regulations.

Water

The numerous waterborne outbreaks of cryptosporidiosis and giardiasis that have been documented (Karanis *et al.*, 2007) have provided a better understanding of the significance and the control of *Cryptosporidium* and *Giardia* (oo)cysts in water in many developed countries. The European Union 'drinking water' directive requires that 'water intended for human consumption should not contain pathogenic organisms' and 'nor should such water contain: parasites, algas, other organisms such as animalcules'. In recognising the impracticality of the current zero standard, the proposed revision to this directive made it a general requirement 'that water intended for human consumption does not contain pathogenic micro-organisms and parasites in numbers which constitute a potential danger to health'. No numerical standard for *Giardia* or *Cryptosporidium* (or other human-infectious protozoan parasites) is proposed.

In England and Wales, *Cryptosporidium* was regulated in drinking water from 1999 until December 2007, when it was deregulated. The regulation set a treatment standard at water treatment sites determined to be of significant risk following risk assessment. Daily continuous sampling of at least 1000 l over at least 22 h period from each point at which water leaves the water treatment works is required, and the goal is to achieve less than an average density of one oocyst per 10 l of water (UK Statutory Instruments 1999 No. 1524, Anon., 1999b).

The US Environmental Protection Agency has issued several rules to address the control of *Cryptosporidium* and *Giardia* (but not other human-infectious protozoan parasites). One of the goals of the Surface Water Treatment Rule (SWTR), formulated to address the control of viruses, *Giardia* and *Legionella*, was to minimise waterborne disease transmission to levels below an annual risk of 10^{-4}. In order to accomplish this goal, treatment, through filtration, and disinfection requirements were set to reduce *Giardia* cysts and viruses by 99.9 % and 99.99 %, respectively. The SWTR also lowered the acceptable limit for turbidity in finished drinking water from a monthly average of 1.0 nephelometric turbidity unit (NTU) to a level not to exceed 0.5 NTU in 95 % of four-hour measurements. The Enhanced Surface Water Treatment Rule (ESWTR) include regulation of *Cryptosporidium* but, in order to implement this rule, a national database on the occurrence of oocysts in surface and treated waters was required to be collected under the Information Collection Rule (ICR). Water authorities serving more than 10 000 individuals began an 18-month monitoring programme for *Cryptosporidium* under the ICR, which was issued in 1996 and is now completed.

New rules to complement existing legislation to control *Cryptosporidium* in surfacewaters (lakes, rivers, reservoirs, ground water aquifers) have been approved. This is termed the Long Term 2 Enhanced Surface Water Treatment Rule (LT2 rule) (see Table 26.12, Sources of further advice). Water facilities will be required to monitor their water sources for two years to determine treatment requirements. Small facilities (serving < 10 000 people) will sample and test for *E. coli* initially and for *Cryptosporidium* only if their *E. coli* results exceed specified concentration levels. Larger facilities (serving > 10 000 people) will monitor for *Cryptosporidium*, and filtered water systems will be classified according to the expected levels of contamination and placed in one of four categories. Most facilities will not require further treatment, but some will require additional 90–99.7 % (1–2.5 $\log_{10}$) reduction in *Cryptosporidium* levels. Unfiltered waters must provide at least 99 or 99.9 % (2–3 $\log_{10}$) inactivation of *Cryptosporidium*. Facilities that store treated water in open reservoirs must either cover the reservoir or treat the reservoir discharge to inactivate 2 $\log_{10}$ *Cryptosporidium* oocysts. A further rule addresses the control of disinfectant by-products and requires that the facilities review their microbial treatment process (whenever using any disinfectant other than UV light or ozone) before making a change in their disinfection practices under the Stage 2 Disinfection Byproducts Rule, which EPA is finalising along with the LT2ESWTR.

In Sydney, Australia, the response to increased numbers of oocysts found in the water supply in 1998, albeit without increased infection levels in the population served, prompted the water industry to opt for the voluntary introduction of externally audited risk management based on HACCP (Ferguson *et al.*, 2006).

The presence of *N. fowleri* in drinking or bathing water is currently not regulated in the USA. The deaths of two children in Arizona due to *N. fowleri*, linked to unchlorinated well water used for drinking water, precipitated a need to develop guidance for the disinfection of *N. fowleri* to reduce the risk of transmission via drinking water and swimming pools, which was accomplished by developing Ct (chlorine concentration x time) values/dose requirements for the inactivation of *N. fowleri* in drinking water. This information will be used to develop a guidance document for Arizona water utilities that use groundwater. Based upon quantitative risk assessment, the French health authorities have set a maximum level of 100 *N. fowleri* amoebae per litre, not to be exceeded in watercourses where human exposure is possible (Cabanes *et al.*, 2001).

Table 26.12 Source of further advice: websites and URLs

Organisation/name	Uniform Resource Locator (URL)
FoodNet surveillance	http://www.cdc.gov/foodnet/
FoodLink	http://www.foodlink.org.uk/
UK Food Surveillance System	http://www.food.gov.uk/enforcement/foodsampling/fss/
UK Food Standards Agency	http://www.food.gov.uk/
CDC	http://www.cdc.gov/
US Food and Drug Administration Agency	http://www.fda.gov
US Department of Agriculture, Food Safety and Inspection Service; Foodborne illness and disease – Parasites and foodborne illness	http://www.fsis.usda.gov/factsheets/parasites_and_foodborne_illness/index.asp
National Shellfish Sanitation Program (USA)	http://seafood.ucdavis.edu/guidelines/usguidelin.htm
Istituto Superiore di Sanità, Community Reference Laboratory for Parasites	http://www.iss.it/crlp/food/index.php?lang=2&tipo=10
Risk communication for *Cryptosporidium*	http://es.epa.gov/ncer/fellow/progress/99/wufe00.html
Risk communication analysis for *Cryptosporidium*	http://www.scotland.gov.uk/Publications/2003/01/16129
Guide to Minimize Microbial Food Safety Hazards for Fresh Fruits and Vegetables In Brief – US Food and Drug Administration, Center for Food Safety and Applied Nutrition	http://vm.cfsan.fda.gov/~dms/prodglan.html
Long term 2 enhanced water treatment rule (LT2), US Environmental Protection Agency	http://www.epa.gov/safewater/disinfection/lt2/compliance.html#quickguides
Joint FAO/WHO Expert Consultation on the Application of Risk Communication to Food Standards and Safety Matters, held from 2–6 February 1998, Italian Ministry of Health, Rome	http://www.foodrisk.org/risk_communication.cfm
Health People 2000: Status Report Food Safety Objectives – FDA and ISC for Diseases Control and Surveillance	http://www.foodsafety.gov/~dms/hp 2k.html
Cryptosporidium oocyst detection in water	http://www.epa.gov/waterscience/methods/cryptsum.html http://www.dwi.gov.uk/regs/crypto/legalindex.htm
Giardia cyst detection in water	http://www.epa.gov/waterscience/methods/cryptsum.html
UK National Reference Method for the Examination of Faeces for *Cryptosporidium*	http://www.defra.gov.uk/animalh/diseases/vetsurveillance/pdf/nrm-002crypto.pdf

Food

The UK Food Safety Act (1990) requires that food, not only for resale but throughout the food chain, must not have been rendered injurious to health; be unfit; or be so contaminated – whether by extraneous matter or otherwise – that it would be unreasonable for it to be eaten. A set of horizontal and vertical regulations (which are provisions in food law) covering both foods in general and specific foodstuffs also ensure that food has not been rendered injurious to health. Other than being potential microbiological contaminants, none of the protozoan parasites in this chapter is identified in these regulations. There is very little evidence to indicate the extent of the survival of *T. gondii* in food industry processes. This was realised when in December 2006, the Food Standards Agency sought the view of the Advisory Committee on the Microbiology Safety of Food with regards to toxoplasmosis and food. In particular, members were invited to identify investigations that may be necessary to produce robust data on UK prevalence and foodborne transmission and assess whether there is a need to review current food safety advice.

The US Food and Drug Administration is responsible for enforcing the regulations detailed in the Federal Food, Drug and Cosmetic Act. Regulations do not address these protozoan parasites specifically, but contaminated products are covered by sections of the Act depending on whether the foods are domestic (either produced within the USA or already imported and in the domestic market [section 402(a)(l)],) or imported (at the port of entry [section 801(a)(1)]). Analysis of regulatory samples by FDA follows the procedures contained within the *Bacteriological Analytical Manual* (BAM) (2001 http://www.911emg.com/pages/library/FDA%20Bacteriological%20 Analysis.pdf). FDA strongly favours HACCP systems over 'end product' testing, being a better suited practice for controlling microorganisms in food. This and Good Manufacturing Practices (GMPs) throughout the food industry should help ensure safe food products.

Guidelines for safe shellfish food include harvesting from approved areas. FDA has also issued HACCP guidelines for seafood production and commercialisation (FDA, 1995). FDA introduced the HAACCP Juice Rule that requires wholesale fruit juice and cider producers to implement HACCP programme capable of 5 log reduction of pertinent pathogens in 2001 (FDA, 2001). Manufactures will have to draw from scientific expertise to implement control systems based on research carried out for each pathogen.

Essential guidance for the safe production of fruit and vegetables using good hygienic practice (GHP) is indicated in the Codex Alimentarius document (CAC, 1997) on 'General Principles of Food Hygiene'. More specific approaches to 'Ready-to-eat Fresh Pre-cut fruits and vegetables' and on 'Sprout Production' have been produced (CAC, 2001) and should be used in combination with HACCP (EC, 2002). The application of HACCP for fruit and vegetables depends on the identification of robust critical control points for the destruction of pathogens on/in products.

Due to the paucity of evidence supporting ozonation for apple cider disinfection or for killing *Cryptosporidium* in this product, and its apparent failure in a documented outbreak (Blackburn *et al.*, 2006), the FDA issued an addendum to its HACCP rule, which advises that juice makers should not use ozone in their manufacturing process unless they can prove a 5 log pathogen reduction through ozonation (US Food and Drug Administration/ CFSAN, Guidance for industry: recommendations to processors of apple juice or cider on the use of ozone for pathogen reduction purposes, 2004 Aug. [cited 2005 May 24], available at: http://www.cfsan.fda.gov/~dms/ juicgu12.html). Blackburn *et al.* (2006) stated that, to their knowledge, no studies had established this reduction at the time of publication.

Agricultural practices and wastes

Apart from regulations governing agricultural wastes, current regulations in the UK and USA do not require risk-based standards or guidance based on protection from microbial contaminants such as human-infectious protozoan parasites. GMPs are also suggested for farms and agricultural wastes. The EC Agri-Environment Regulation, under the Common Agricultural Policy, promotes schemes which encourage farmers to undertake positive measures to conserve and enhance the rural environment in Europe.

In the USA, guidance on GMPs and good manufacturing practices for fruits and vegetables is available on FDA's web page (http://vww.fda.gov). The guide consists of recommendations to growers, packers, transporters and distributors of produce to minimise the risks of foodborne diseases. Its purpose is the prevention of microbial contamination, by applying basic principles to the use of water and organic fertilisers, employee hygiene, field and facility sanitation and transportation. Advice is given on the establishment of a system for accountability to monitor personnel and procedures from the producer to the distributor.

26.7 Sources of further information and advice

- Table 26.12 gives sources of further advice together with websites and URLs.
- For research organisations offering advice, key texts to refer to:

Published reviews: Smith J L (1993); Jaykus (1997); Tauxe R V (1997); Millar *et al.* (2002), Doyle (2003); Dawson (2005); Weintraub (2006); Ortega (2006); Marshall *et al.* (1997).

26.8 Glossary

Anthroponotic – transmission of an infectious agent to a human host from another infected human.

Antibody – protein of the γ globulin type, secreted by plasma cells, which functions as an effector molecule of the host immune response by binding to its respective antigen.

Antigen – a chemical substance capable of eliciting an immune response.

Binary fission – process of multiplication by simple cell division observed in *Giardia* trophozoites.

Cysts – environmentally resistant transmissive stage of *Giardia* that give rise to infection following ingestion by a susceptible host.

Excystation – biological process of liberating sporozoites from *Cryptosporidium* and *Cyclospora* oocysts or trophozoites from *Giardia* cysts. Excystation is triggered by physiological criteria, including elevated temperature (incubation at 37 °C) and exposure to the acidic environment of the stomach, followed by the alkaline environment of the small intestine. Sporozoites and trophozoites are released in the intestinal tract of the host.

Macrogametocyte – in *Cryptosporidium*, macrogametocytes and microgametocytes develop from type II merozoites. Macrogametocytes are sessile, within the host cell, and are fertilized by microgametocytes.

Merogony – asexual multiplication (cell division) process resulting in the production of merozoites which, for *Cryptosporidium*, occurs in the epithelial cells lining the intestine (enterocytes).

Merozoites – crescentic shaped organisms resulting from merogony in the *Cryptosporidium* life cycle. Merozoites develop within the meront as type I or type II merozoites. When mature, merozoites are liberated into the lumen of the intestine and infect adjacent epithelial cells. Two types of meronts have also been observed in the *Cyclospora* life cycle.

Microgametocytes – gametes are the sexual stages of development of *Cryptosporidium* and probably for *Cyclospora*. Microgametocytes develop from type II merozoites and, when mature, release microgametes which invade macrogametocytes to fertilise macrogametes. Fertilised macrogametes develop into zygotes.

Microvilli – minute, finger-like extensions of the luminal membrane of epithelial cells lining the intestines.

Monoclonal antibody – laboratory derived antibody of a single paratope, isolated from a single immortalised antibody-producing cell and produced *in vitro*.

Oocyst – environmentally resistant transmissive stage of *Cryptosporidium*, *Cyclospora* and *Toxoplasma* which gives rise to infection following ingestion by a susceptible host.

Sporozoites – invasive form of the parasites *Cryptosporidium* and *Cyclospora*, crescentic in shape and contained within the oocyst. In *Cyclospora* two sporozoites are contained within each sporocyst, and two sporocysts are contained within each oocyst. In *Cryptosporidium* the oocysts are 'naked' i.e. not contained within a sporocyst. There are four *Cryptosporidium* sporozoites within each oocyst.

Sporulation – development of zygote into sporozoites which, with respect to *Cryptosporidium*, occurs during the maturation of the oocyst within the host and, with respect to *Cyclospora*, occurs in the external environment after oocyst excretion by the host.

Trophozoites – vegetative forms of *Giardia*. Trophozoites are transmitted to susceptible hosts with environmentally resistant cysts. The pear-shaped trophozoites excyst in the small intestine and attach onto the surfaces of the enterocytes lining the intestinal tract.

Zoonotic – transmission of an infectious agent to a human host by non-human hosts.

Zygote – the zygote is the product of fusion between a micro- and macrogamete. Zygotes develop into environmentally resistant oocysts in *Cryptosporidium* and *Cyclospora* (although in *Cryptosporidium* some researchers have noted the development of thin walled, auto-infective oocysts).

26.9 References

Abbaszadegan M, Huber M S, Gerba C P and Pepper I (1997) Detection of viable *Giardia* cysts by amplification of heat shock-induced mRNA, *Appl Environ Microbiol*, **63**(1), 324–8.

Abou el Naga I (1999) Studies on a newly emerging protozoal pathogen: *Cyclospora cayetanensis*, *J Egypt Soc Parasitol*, **29**, 575–86.

Adams A, Jinneman K and Ortega Y (1999) *Cyclospora*, in R Robinson, C Batt and P Patel (eds), *Encyclopaedia of Food Microbiology*, Academic Press, London and New York, 502–13.

Aksozek A, McClellan K, Howard K, Niederkorn J and Alizadeh H (2002) Resistance of *Acanthamoeba castellanii* cysts to physical, chemical and radiological conditions, *J Parasitol*, **88**(3), 621–3.

Alfano-Sobsey E M, Eberhard M L, Seed J R, Weber D J, Won K Y, Nace E K and Moe C L (2004) Human challenge pilot study with *Cyclospora cayetanensis*, *Emerg Infect Dis*, **10**, 726–8.

Amahmid O, Asmama S and Bouhoum K (1999) The effect of waste water reuse in irrigation on the contamination level of food crops by *Giardia* cysts and *Ascaris* eggs, *Int J Food Microbiol*, **49**, 19–26.

Anon. (1989) *Isolation and identification of Giardia cysts, Cryptosporidium oocysts and free living pathogenic amoebae in water etc.*, HMSO, London.

Anon. (1994) Proposed ICR protozoan method for detecting *Giardia* cysts and *Cryptosporidium* oocysts in water by a fluorescent antibody procedure, *Federal Register* **59**(28), February 10th, 6416–29.

Anon. (1996) Foodborne outbreak of diarrhoea illness associated with *Cryptosporidium parvum* – Minnesota, 1995, *MMWR*, **45**(36), 783–4.

Anon. (1997a) Outbreaks of *Escherichia coli* O157:H7 infection and cryptosporidiosis associated with drinking unpasteurized apple cider-Connecticut and New York, October 1996, *MMWR*, **46**(1), 4–8.

Anon. (1997b) Update, outbreaks of cyclosporiasis – United States and Canada, 1997, *MMWR*, **46**, 521–3.

Anon. (1997c) Outbreak of cyclosporiasis – Northern Virginia–Washington, D.C.–Baltimore, Maryland, Metropolitan Area, 1997, *MMWR*, **46**, 689–91.

Anon. (1998a) Foodborne outbreak of cryptosporidiosis – Spokane, Washington, 1997, *MMWR*, **47**(27), 565–7.

Anon. (1998b) Outbreak of cyclosporiasis – Ontario, Canada, May 1998, *MMWR*, **47**, 806–9.

Anon. (1999a) Isolation and identification of *Cryptosporidium* oocysts and *Giardia* cysts in waters 1999. Methods for the examination of waters and associated materials, HMSO, London.

Anon. (1999b) UK Statutory Instruments 1999 No. 1524. The Water Supply (Water Quality) (Amendment) Regulations 1999, The Stationery Office, London.

Anon. (2004) Outbreak of cyclosporiasis associated with snow peas–Pennsylvania, 2004, *MMWR*, **53**, 876–8.

Appelbee A J, Thompson R C A and Olson M E (2005) *Giardia* and *Cryptosporidium* in mammalian wildlife – current status and future needs, *Trends Parasitol*, **21**, 370–76.

Arrowood M J (2002) *In vitro* cultivation of *Cryptosporidium* species, *Clin Microbiol Rev*, **15**, 390–400.

Arrowood M J (2008) *In vitro* cultivation, in R Fayer and L Xiao (eds), *Cryptosporidiosis of Man and Animals*, CRC and IWA, Boca Raton, FL, 499–526.

Bankes P (1995) The detection of *Cryptosporidium* oocysts in milk and beverages, in W B Betts, D Casemore, C R Fricker, H V Smith and J Watkins (eds), *Protozoan Parasites in Water*, The Royal Society of Chemistry, Cambridge, Part 3, 152–3.

Barnard R J and Jackson G J (1980) *Giardia lamblia*. The transfer of human infections by foods, in S L Erlandsen and E A Meyer (eds), *Giardia and Giardiasis. Biology, Pathogenesis and Epidemiology*, Plenum, New York, 365–76.

Belosevic M, Guy R A, Taghi-Kilani R, Neumann N F, Gyürék L L, Liyanage R J, Millard P J and Finch G R (1997) Nucleic acid stains as indicators of *Cryptosporidium parvum* oocyst viability, *Int J Parasitol*, **27**, 787–98.

Beltrán N A and Jiménez B E (2008) Faecal coliforms, faecal enterococci, *Salmonella typhi* and *Acanthamoeba* spp. UV inactivation in three different biological effluents, *Water SA*, **34**(2), 261–9, available at: http://www.wrc.org.za/downloads/watersa/2008/April/2183.pdf, accessed November 2008.

Benenson M W, Takafuji E T, Lemon S M, Greenup R L and Sualzer A J (1982) Oocyst transmitted toxoplasmosis associated with infection of contaminated water, *N Engl J Med*, **307**, 666–9.

Bhatia V N and Warhurst D C (1981) Hatching and subsequent cultivation of cysts of *Giardia intestinalis* in Diamond's medium, *J Trop Med Hyg*, **84**, 45.

Bier J W (1990) Isolation of parasites on fruits and vegetables, *South East Asian J Trop Med Public Health*, **22**, Supplement, 144–5.

Bingham A K, Jarroll E L Jr and Meyer E A (1979) *Giardia* spp.: physical factors of excystation *in vitro*, and excystation vs. eosin as determinants of viability, *Exp Parasitol*, **47**, 284–91.

Blackburn B G, Mazurek J M, Hlavsa M, Park J, Tillapaw M, Parrish M, Salehi E, Franks W, Koch E, Smith F, Xiao L, Arrowood M, Hill V, da Silva A, Johnston S and Jones J L (2006) Cryptosporidiosis associated with ozonated apple cider, *Emerg Infect Dis*, **12**, 684–6.

Blewett D A (1989) Disinfection and oocysts, in KW Angus and DA Blewett (eds), *Cryptosporidiosis*, Proceedings of the First International Workshop, The Animal Disease Research Association, Edinburgh, 107–16.

Bottone E J (1993) Free-living amebas of the genera *Acanthamoeba* and *Naegleria*: an overview and basic microbiologic correlates, *Mt Sinai J Med*, **60**, 260–70.

Cabanes P A, Wallet F, Pringuez E and Pernin P (2001) Assessing the risk of primary amoebic meningoencephalitis from swimming in the presence of environmental *Naegleria fowleri*, *Appl Environ Microbiol*, **11**, 2927–31.

CAC (1997) *Recommended International Code of Practice, General Principles of Food Hygiene. Supplement to volume 1B*, Joint FAO/WHO Food Standards Programme, Food and Agriculture Organization of the United Nations, Rome.

CAC (2001) *Draft Report on the 34th session of the Codex Committee on Food Hygiene*, Bangkok, 8–13 October, Alinorm 03/13, *Joint FAO/WHO Food Standards Programme*, Food and Agriculture Organization of the United Nations, Rome.

Cacciò S M, Thompson R C A, McLauchlin J and Smith H V (2005) Unravelling *Cryptosporidium* and *Giardia* epidemiology, *Trends Parasitol*, **21**, 430–37.

Campbell A and Smith H (1997) Immunomagnetic separation of *Cryptosporidium* oocysts from water samples: round robin comparison of techniques, *Wat Sci Tech*, **35**, 397–402.

Campbell L, Tzipori S, Hutchison G and Angus K W (1982) Effect of disinfectant on survival of *Cryptosporidium* oocysts, *Vet Rec*, **111**, 414–15.

Campbell A, Robertson L J and Smith H V (1992) Viability of *Cryptosporidium parvum* oocysts: correlation of *in vitro* excystation with inclusion or exclusion of fluorogenic vital dyes, *Appl Environ Microbiol*, **58**, 3488–93.

Campbell A T, Robertson L J, Snowball M R and Smith H V (1995) Inactivation of oocysts of *Cryptosporidium parvum* by ultraviolet irradiation, *Wat Res*, **29**(11), 2583–6.

Casemore D P (1987) Cryptosporidiosis, *Public Health Laboratory Service Microbiology Digest*, **4**, 1–5.

Cassells J M, Yahya M T, Gerba C P and Rose J B (1995) Efficacy of a combined system of copper and silver and free chlorine for inactivation of *Naegleria fowleri* amoebas in water, *Water Sci Technol*, **31**, 119–22.

Cerva L (1989) *Acanthamoeba culbertsoni* and *Naegleria fowleri*: occurrence of antibodies in man, *J Hyg Epidemiol Microbiol Immunol*, **33**, 99–103.

Chaidez C, Soto M, Gortares P and Mena K (2005) Occurrence of *Cryptosporidium* and *Giardia* in irrigation water and its impact on the fresh produce industry, *Int J Environ Health Res*, **15**, 339–45.

Chang S L (1978) Resistance of pathogenic *Naegleria* to some common physical and chemical agents, *Appl Environ Microbiol*, **35**, 368–75.

Chappell C L, Okhuysen P C, Sterling C R, Wang C, Jakubowski W and Dupont H L (1999) Infectivity of *Cryptosporidium parvum* in healthy adults with pre-existing anti-*C. parvum* serum immunoglobulin G, *Am J Trop Med Hyg*, **60**, 157–64.

Chappell C L, Okhuysen P C, Langer-Curry R, Widmer G, Akiyoshi D E, Tanriverdi S and Tzipori S (2006) *Cryptosporidium hominis:* experimental challenge of healthy adults, *Am J Trop Med Hyg*, **75**, 851–7.

Chatterton JMW (1992) Pregnancy, in D O Ho-Yen and A W L Joss (eds), *Human Toxoplasmosis*, Oxford University Press, Oxford, 144–83.

Choi W-Y, Nam H-W, Kwak N-H, Huh W, Kim Y-R, Kang M-W, Cho S-Y and Dubey J P (1997) Foodborne outbreaks of human toxoplasmosis, *J Inf Dis*, **175**, 1280–82.

Clark C G and Diamond L S (1991) Ribosomal RNA genes of 'pathogenic' and 'nonpathogenic' *Entamoeba histolytica* are distinct, *Mol Biochem Parasitol*, **49**, 297–302.

Clark C G and Diamond L S (1994) Pathogenicity, virulence and *Entamoeba histolytica* [Commentary], *Parasitol Today*, **10**, 46–7.

Connor B A and Shlim D R (1995) Foodborne transmission of *Cyclospora*, *Lancet*, **346**, 1634.

Connor B A, Shlim D R, Scholes J V, Rayburn J L, Reidy J and Rajah R (1993) Pathologic changes in the small bowel in nine patients with diarrhea associated with a coccidia-like body, *Ann Intern Med*, **119**, 377–82.

Conroy D A (1960) A note on the occurrence of *Giardia* sp. in a Christmas pudding, *Rev Iber Parasitol*, **20**, 567–71.

Cook G C (1987) Opportunistic parasitic infections associated with the Acquired Immune Deficiency Syndrome (AIDS): parasitology, clinical presentation, diagnosis and management, *Quart J Med*, **65**, 967–84.

Cook A J C, Gilbert R E, Buffolano W *et al.* (on behalf of the European Research Network on Congenital Toxoplasmosis) (2000) Sources of toxoplasma infection in pregnant women: European multicentre case-controlled study, *Br Med J*, **321**, 142–7.

Cook N, Paton C A, Wilkinson N, Nichols R A B, Barker K and Smith H V (2006a) Towards standard methods for the detection of *Cryptosporidium parvum* on lettuce and raspberries. Part 1: Development and optimization of methods, *Int J Food Microbiol*, **109**, 215–21.

Cook N, Paton C A, Wilkinson N, Nichols R A B, Barker K and Smith H V (2006b) Towards standard methods for the detection of *Cryptosporidium parvum* on lettuce and raspberries. Part 2: Validation, *Int J Food Microbiol*, **109**, 222–8.

Cook N, Nichols R A B, Wilkinson N, Paton C A, Barker K and Smith H V (2007) Development of a method for detection of *Giardia duodenalis* cysts on lettuce and for simultaneous analysis of salad products for the presence of *Giardia* cysts and *Cryptosporidium* oocysts, *Appl Environ Microbiol*, **73**, 7388–91.

Culbertson C G, Smith J W and Minner H (1958) *Acanthamoebae*: observations on animal pathogenicity, *Science*, **127**, 1506.

Cursons R T M, Brown T J, Keys E A, Monarty K M and Till D (1980a) Immunity to pathogenic free-living amoebae: role of humoral antibody, *Infect Immun*, **29**, 401–7.

Cursons R T M, Brown T J and Keys E A (1980b) Effect of disinfectants on pathogenic free-living amoebae: in axenic conditions, *Appl Environ Microbiol*, **40**, 62–6.

da Silva Coelho LMP, de Oliveira S M, de Sá Adami Milman M H, Karasawa K A and de Paiva Santos R (2001) Detecção de formas transmissíveis de enteroparasitas na água e nas hortaliças consumidas em comunidades escolares de Sorocaba, São Paulo, Brasil, *Rev Soc Brasil Med Trop*, **34**, 479–82.

Danciger M and Lopez M (1975) Numbers of *Giardia* in the feces of infected children, *Am J Trop Med Hyg*, **24**, 232–42.

Dawson D (2005) Foodborne protozoan parasites, *Int J Food Micro*, **103**(2), 207–27.

Dawson M and Brown T (1987) The effect of chlorine and chlorine dioxide on pathogenic free-living amoebae (PFLA) in simulated natural conditions: the presence of bacteria and organic matter, *N Z J Mar Fresh*, **21**, 117–23.

De Jonckheere J and Van de Voorde H (1976) Differences in destruction of cysts of pathogenic and nonpathogenic *Naegleria* and *Acanthamoeba* by chlorine, *Appl Environ Microbiol*, **31**, 294–7.

De Moura L, Bahia-Oliveira L M G, Wada M Y, Jones J L, Tuboi S H, Carmo E H, Ramalho W M, Camargo N J, Trevisan R, Graça R M T, da Silva A J, Moura I, Dubey J P and Garrett D O (2006) Waterborne toxoplasmosis, Brazil, from field to gene, *Emerg Inf Dis*, **12**, 326–9.

de Oliveira Mendonça A, Domingues P F, da Silva A V, Pezerico S B and Langoni H (2004) Detection of *Toxoplasma gondii* in swine sausages, *Parasitol Latinoam*, **59**, 42–5.

De Regnier D P, Cole L, Schupp D G and Erlandsen S L (1989) Viability of *Giardia* cysts suspended in lake, river and tap water, *Appl Environ Microbiol*, **55**, 1223–9.

De Silva L, Mulcahy D L and Ramananda Kamaath K (1988) A family outbreak of toxoplasmosis: a serendipitous finding, *J Inf*, **8**, 163–7.

Deng M Q and Cliver D O (1999) *Cryptosporidium parvum* studies with dairy products, *Int J Food Microbiol*, **46**, 113–21.

Deng M Q and Cliver D O (2000) Comparative detection of *Cryptosporidium parvum* oocysts from apple juice, *Int J Food Microbiol*, **54**(3), 155–62.

Deng M Q, Cliver D O and Mariam T W (1997) Immunomagnetic capture PCR to detect viable *Cryptosporidium parvum* oocysts from environmental samples, *Appl Environ Microbiol*, **63**, 3134–8.

Deng M Q, Lam K M and Cliver D O (2000) Immunomagnetic separation of *Cryptosporidium parvum* oocysts using MACS MicroBeads and high gradient separation columns, *J Microbiol Meth*, **40**, 11–17.

Derouin F and Garin Y J F (1991) *Toxoplasma gondii*: blood and tissue kinetics during acute and chronic infections in mice, *Exp Parasitol*, **73**, 460–68.

Derouin F, Devergie A, Auber P, Gluckman E, Beauvais B, Garin Y J and Lariviere M (1992) Toxoplasmosis in bone marrow-transplant recipients: report of seven cases and review, *Clin Infect Dis*, **15**, 267–70.

Di Giovanni G D, Hashemi F H, Shaw N J, Abrams F A, LeChevallier M W and Abbaszadegan M (1999) Detection of infectious *Cryptosporidium parvum* oocysts in surface and filter backwash water samples by immunomagnetic separation and integrated cell culture-PCR, *Appl Environ Microbiol*, **65**, 3427–32.

Diamond L S and Clark C G (1993) A redescription of *Entamoeba histolytica* Schaudinn, 1903 (emended Walker, 1911) separating it from *Entamoeba dispar* Brumpt, 1925, *J Eukaryot Microbiol*, **40**, 340–44.

Dias R A F, Navarro I T, Ruffolo B B, Bugni F M, de Castro M V and Freire R L (2005) *Toxoplasma gondii* in fresh pork sausages and seroprevalence in butchers from factories in Londrina, Paraná state, Brazil, *Rev Inst Med Trop S Paulo*, **47**, 185–9.

Doganci L, Tanyukel M, Araz E R, Besirbellioglu B A, Erdem U, Ozogug C A, Yucel N and Ciftaioglu A (2006) A probable outbreak of toxoplasmosis among boarding school students in Turkey, *Clin Microbiol Inf*, **12**, 672–4.

Döller P C, Dietrich K, Filipp N, Brockmann S, Dreweck C, Vonthein R, Wagner-Wiening C and Wiedenmann A (2002) Cyclosporiasis outbreak in Germany associated with the consumption of salad, *Emerg Infect Dis*, **8**, 992–4.

Doyle E (2003) Foodborne parasites: a review of the scientific literature, available at: http://www.wisc.edu/fri/briefs/parasites.pdf, accessed November 2008.

Dubey J P (1974) Effect of freezing on the infectivity of *Toxoplasma* cysts to cats, *J Am Vet Med Assoc*, **165**, 534–6.

Dubey J P (1988) Long-term persistence of *Toxoplasma gondii* in tissues of pigs inoculated with *T. gondii* oocysts and effect of freezing on viability of tissue cysts in pork, *Am J Vet Res*, **49**, 910–13.

Dubey J P (1996) Infectivity and pathogenicity of *Toxoplasma gondii* oocysts for cats, *J Parasitol*, **82**, 957–60.

Dubey J P (1998) *Toxoplasma gondii* oocyst survival under defined temperatures, *J Parasitol*, **84**, 862–5.

Dubey J P (2001) Oocyst shedding by cats fed isolated bradyzoites and comparison of infectivity of bradyzoites of the VEG strain *Toxoplasma gondii* to cats and mice, *J Parasitol*, **87**, 215–19.

Dubey J P (2004) Toxoplasmosis – a waterborne zoonosis, *Vet Parasitol*, **126**, 57–72.

Dubey J P and Beattie C P (1988) *Toxoplasmosis of Animals and Man*, CRC, Boca Raton, FL.

Dubey J P and Frenkel J K (1972) Cyst-induced toxoplasmosis in cats, *J Protozool*, **19**, 155–77.

Dubey J P and Frenkel J K (1976) Feline toxoplasmosis from acutely infected mice and the development of Toxoplasma cysts, *J Protozool*, **23**, 537–46.

Dubey J P and Thayer D W (1994) Killing of different strains of *Toxoplasma gondii* tissue cysts by irradiation under defined conditions, *J Parasitol*, **80**, 764–7.

Dubey J P, Miller N L and Frenkel J K (1970) Characterization of the new fecal form of *Toxoplasma gondii*, *J Parasitol*, **56**, 447–56.

Dubey J P, Kotula A W, Sharar A, Andrews C D and Lindsay D S (1990) Effect of high temperature on infectivity of *Toxoplasma gondii* tissue cysts in pork, *J Parasitol*, **76**, 201–4.

Dubey J P, Lunney J K, Shen S K, Kwok O C H, Ashford D A and Thulliez P (1996) Infectivity of low numbers of *Toxoplasma gondii* oocysts to pigs, *J Parasitol*, **82**, 438–43.

Dumètre A and Dardé M L (2003) How to detect *Toxoplasma gondii* oocysts in environmental samples? *FEMS Microbiol Rev*, **27**, 651–61.

DuPont H L, Chappell C L, Sterling C R, Okhuysen P C, Rose J B and Jakubowski W (1995) The infectivity of *Cryptosporidium parvum* in health volunteers, *N Engl J*

Med, **332**(13), 855–9.

Dutton G N (2001) Ocular infection, in D H M Joynson and T G Wreghitt (eds), *Toxoplasmosis a comprehensive clinical guide*, Cambridge University Press, Cambridge, 277–95.

DWI (2005) *Cryptosporidium*, Legal Requirements and Standard Operating Protocols (SOPs), Drinking Water Inspectorate, London, available at: http://www.dwi.gov.uk/regs/crypto/, accessed December 2008.

Eberhard M L, da Silva A J, Lilley B G and Pieniazek N J (1999) Morphologic and molecular characterization of new *Cyclospora* species from Ethiopian monkeys: *C. cercopitheci* sp.n., *C. colobi* sp.n., and *C. papionis* sp.n., *Emerg Infec Dis*, **5**(5), 651–8.

EC (2002) *Risk Profile on the Microbiological Contamination of Fruit and Vegetables Eaten Raw*, Report of the Scientific Committee on Food, SCF/CS/FMH/SURF/Final 29 April, European Commission DG Health and Consumer Protection, Brussels.

EPA (1998) Consumer Confidence Reports Final Rule, *Federal Register*, **63**(160), August 19, 44511–36, available at: http://permanent.access.gpo.gov/lps21800/www.epa.gov/safewater/ccr/ccr-frn.pdf, accessed November 2008.

EPA (2001a) *Method 1622: Cryptosporidium in water by filtration/IMS/FA*, EPA 821-R01-026, US Environment Protection Agency, Washington, DC.

EPA (2001b) *Method 1623: Giardia* and *Cryptosporidium in water by filtration/IMS/FA*, EPA 821-R01-026, *US Environmental Protection Agency, Washington*, DC.

Evans R (1992) Life cycle, in D O Ho-Yen and A W L Joss (eds), *Human Toxoplasmosis*, Oxford University Press, Oxford, 26–55.

Evans R, Chatterton J M W, Ashburn D, Joss A W L and Ho-Yen D O (1999) Cell-culture system for continuous production of *Toxoplasma gondii* tachyzoites, *Eur J Clin Microbiol Inf Dis*, **18**, 879–84.

Faubert G M and Belosevic M (1990) Animal models for *Giardia duodenalis* type organisms, in E A Meyer (ed.), *Giardiasis*, Elsevier Science Publishers Biomedical Division, Amsterdam, 77–90.

Fayer R (1994) Effect of high temperature on infectivity of *Cryptosporidium parvum* oocysts in water, *Appl Environ Microbiol*, **60**(8), 2732–5.

Fayer R (2008) General biology of *Cryptosporidium*, in R Fayer and L Xiao (eds), *Cryptosporidiosis of Man and Animals*, CRC and IWA, Boca Raton, FL, 1–42.

Fayer R and Nerad T (1996) Effects of low temperatures on viability of *Cryptosporidium parvum* oocysts, *Appl Environ Microbiol*, **62**(4), 1431–3.

Fayer R and Ungar B L P (1986) *Cryptosporidium* and cryptosporidiosis, *Microbiol Rev*, **50**, 458–83.

Fayer R, Lewis E J, Trout J M, Graczyk T K, Jenkins M C, Higgins J, Xiao L and Lal A A (1999) *Cryptosporidium parvum* in oysters from commercial harvesting sites in the Chesapeake Bay, *Emerg Infect Dis*, **5**(5), 706–10.

FDA (1995) Procedures for the Safe and Sanitary Processing and Importing of Fish and Fishery Products; Final Rule, Food and Drug Administration, USA, *Federal Register*, 60 FR 65095, December 18, 65197–202, available at: http://www.cfsan.fda.gov/~lrd/searule3.html.

FDA (2001) Hazard analysis and critical control point (HACCP); procedures for the safe and sanitary processing and importing of juice, Final rule, 21 CFR Part 120, *Federal Register*, **66**(13), 6137–202, available at: http://www.cfsan.fda.gov/~lrd/fr01119a.html, accessed November 2008.

Feacham R G, Bradley D J, Garelick H and Mara D D (1983) *Entamoeba histolytica* and amebiasis, in R G Feacham, D J Bradley, H Garelick and D D Mara (eds), *Sanitation and Disease: Health Aspects of Excreta and Wastewater Management*, World Bank Studies in Water Supply and Sanitation, Wiley, Chichester, 337–48.

Feltus D C, Giddings C W, Schneck B L, Monson T, Warshauer D and McEvoy J M (2006) Evidence supporting zoonotic transmission of *Cryptosporidium* spp. in Wisconsin, *J Clin Microbiol*, **44**, 4303–8.

Ferguson C, Deere D, Sinclair M, Chalmers R M, Elwin K, Hadfield S, Xiao L, Ryan U, Gasser R, El-Osta Y A and Stevens M (2006) Meeting Report: Application of genotyping methods to assess risks from *Cryptosporidium* in watersheds, *Environ Health Perspec*, **114**(3), 430–34.

Finch G R, Black E K, Labatiuk C W, Gyürék L and Belosevic M (1993) Comparison of *G. lamblia* and *G. muris* cysts inactivation by ozone, *Appl Environ Microbiol*, **59**, 3674–80.

Fleming C A, Carron D, Gunn J E and Barry M A (1998) A foodborne outbreak of *Cyclospora cayetanensis* at a wedding, *Arch Intern Med*, **158**, 1121–5.

Franco R M B and Cantusio Neto R (2002) Occurrence of cryptosporidial oocysts and *Giardia* cysts in bottled mineral water commercialized in the city of Campinas, State of Sao Paulo, Brazil, *Mem Inst Oswaldo Cruz*, **97**(2), 205–7.

Freire-Santos F, Oteiza-López M, Vergara-Castibianco C A, Ares-Mazás E, Álvarez-Suárez E and García-Martín O (2000) Detection of *Cryptosporidium* oocysts in bivalve molluscs destined for human consumption, *J Parasitol*, **86**(4), 853–4.

Frenkel J K and Dubey J P (1972) Toxoplasmosis and its prevention in cats and man, *J Infect Dis*, **126**, 664–73.

Frenkel J K, Ruiz A and Chinchilla M (1975) Soil survival of toxoplasma oocysts in Kansas and Costa Rica, *Am J Trop Med Hyg*, **24**, 439–43.

Garcia L, Henderson J, Fabri M and Oke M (2006) Potential sources of microbial contamination in unpasteurised apple cider, *J Food Prot*, **69**(1), 137–44.

Garnham P C C, Baker J R and Bird R G (1962) Fine structure of cystic form of *Toxoplasma gondii*, *Br Med J*, **1**, 83–4.

Gelletlie R, Stuart J, Soltanpoor N, Armstrong R and Nichols G (1997) Cryptosporidiosis associated with school milk, *Lancet*, **4**; 350(9083), 1005–6.

Gelman B B, Rauf S J, Nader R, Popov V, Borkowski J, Chaljub G, Nauta H W and Visvesvara G S (2001) Amoebic encephalitis due to *Sappinia diploidea*, *JAMA*, **285**, 2450–51.

Gerba C P, Blair B L, Sarkar P, Bright K R, Maclean R C and Marciano-Cabral F (2008) Occurrence and control of *Naegleria fowleri* in drinking water wells, in M G Ortega-Pierres, S Cacciò, R Fayer, T Mank, H Smith and R C A Thompson (eds), *Giardia and Cryptosporidium: from Molecules to Disease*, CAB International, Wallingford, in press.

Girdwood R W A and Smith H V (1999a) *Giardia*, in R Robinson, C Batt and P Patel (eds), *Encyclopaedia of Food Microbiology*, Vol. 1, Academic Press, London and New York, 946–54.

Girdwood R W A and Smith H V (1999b) *Cryptosporidium*, in R Robinson, C Batt and P Patel (eds), *Encyclopaedia of Food Microbiology*, Vol. 1, Academic Press, London and New York, 487–97.

Glasner P D, Silveira C, Kruszon-Moran D, Martins M C, Burnier M, Silveira S, Camargo M E, Nussenblatt R B, Kaslow R A and Belford R (1992) An unusually high prevalence of ocular toxoplasmosis in southern Brazil, *Am J Ophthalmol*, **114**, 136–44.

Goater A D and Pethig R (1998) Electrorotation and dielectrophoresis, *Parasitology*, **117**, S117–90.

Gomez-Bautista M, Ortega-Mora L M, Tabares E, Lopez-Rodas V and Costas E (2000) Detection of infectious *Cryptosporidium parvum* oocysts in mussels (*Mytilus galloprovincialis*) and cockles (*Cerastoderma edule*), *Appl Environ Microbiol*, **66**(5), 1866–70.

Gómez-Couso H, Freire-Santos F, Amar C F L, Grant K A, Willianson K, Ares-Mazás M E and McLauchlin J (2003a) Detection of *Cryptosporidium* and *Giardia* in molluscan shellfish by multiplexed nested-PCR, *Int J Food Microbiol*, **91**(3), 279–88.

Gómez-Couso H, Freire-Santos F, Martínez-Urtaza J, García-Martín O and Ares-Mazás M E (2003b) Contamination of bivalve molluscs by *Cryptosporidium* oocysts: the need for new quality control standards, *Int J Food Microbiol*, **87**, 97–105.

Gómez-Couso H, Freire-Santos F, Amar C F L, Grant K A, Williamson K, Ares-Mazás M E and McLauchlin J (2004) Detection of *Cryptosporidium* and *Giardia* in molluscan shellfish by multiplexed nested-PCR, *Int J Food Microbiol*, **91**(3) 279–88.

Gómez-Couso H, Mendez-Hermida F and Ares-Mazás E (2006) Levels of detection of *Cryptosporidium* oocysts in mussels (*Mytilus galloprovincialis*) by IFA and PCR methods, *Vet Parasitol*, **141**(1–2), 60–65.

Graczyk T K, Fayer R, Cranfield M C and Conn D B (1998) Recovery of waterborne *Cryptosporidium parvum* oocysts by freshwater benthic clams (*Corbicula fluminea*), *Appl Environ Microbiol*, **64**(2), 427–30.

Graczyk T K, Thompson R C A, Fayer R, Adams P, Morgan U M and Lewis E J (1999) *Giardia duodenalis* cysts of genotype A recovered from clams in the Chesapeake Bay subestuary, Rhode River, *Am J Trop Med Hyg*, **61**, 526–9.

Graczyk T K, Girouard A S, Tamang L, Nappier S P and Schwab K J (2006) Recovery, bioaccumulation, and inactivation of human waterborne pathogens by the Chesapeake Bay nonnative oyster, *Crassostrea ariakensis*, *Appl Environ Microbiol*, **72**(5), 3390–95.

Gray F, Gherardi E L, Wingate E, Wingate J, Fenelon G, Gaston A, Sobel A and Poirier J (1989) Diffuse encephalitic cerebral toxoplasmosis in AIDS. Report of four cases, *J Neurol*, **236**, 273–7.

Hall S, Ryan M and Buxton D (2001) The epidemiology of toxoplasma infection, in D H M Joynson and T G Wreghitt (eds), *Toxoplasmosis: A Comprehensive Clinical Guide*, Cambridge University Press, Cambridge, 58–124.

Harp J A, Fayer R, Pesch B A and Jackson G J (1996) Effect of pasteurization on infectivity of *Cryptosporidium parvum* oocysts in water and milk, *Appl Environ Microbiol*, **62**(8), 2866–8.

Harper C M, Cowell N A, Adams B C, Langley A J and Wohlsen T D (2002) Outbreak of *Cryptosporidium* linked to drinking unpasteurised milk, *Commun Dis Intell*, **26**, 449–500.

Herwaldt D L, Ackers M-L and the *Cyclospora* working group (1997) An outbreak in 1996 of cyclosporiasis associated with imported raspberries, *New Eng J Med*, **336**, 1548–58.

Hewett M K, Robinson B S, Monis P T and Saint C P (2003) Identification of a new *Acanthamoeba* 18S rRNA gene sequence type, corresponding to the species *Acanthamoeba jacobsi*, *Acta Protozool*, **42**, 325–9.

Ho A Y, Lopez A S, Eberhart M G, Levenson R, Finkel B S, da Silva A J, Roberts J M, Orlandi P A, Johnson C C and Herwaldt B L (2002) Outbreak of cyclosporiasis associated with imported raspberries, Philadelphia, Pennsylvania, 2000, *Emerg Infect Dis*, **8**, 783–8.

Hoff J C (1986) *Inactivation of Microbial Agents by Chemical Disinfectants*, EPA-600/2-86-067, US Environmental Protection Agency, Cincinnati, OH.

Homan W L and Mank T G (2001) Human giardiasis: genotype linked differences in clinical symptomatology, *Int J Parasitol*, **31**, 822–6.

Homan W L, van Enckevort F H, Limper L, van Eys G J, Schoone G J, Kasprzak W and Majewska A C (1992) Comparison of *Giardia* isolates from different laboratories by isoenzyme analysis and recombinant DNA probes, *Parasitol Res*, **78**, 316–23.

Hopkins R M, Meloni B P, Groth D M, Wetherall J D, Reynoldson J A and Thompson R C A (1997) Ribosomal RNA sequencing reveals differences between the genotypes of *Giardia* isolates recovered from humans and dogs living in the same locality, *J Parasitol*, **83**, 44–51.

Howe D K and Sibley L D (1995) *Toxoplasma gondii* comprises three cloned lineages: correlation of parasite genotype with human disease, *J Inf Dis*, **172**, 1561–6.

Ho-Yen D O (1992) Clinical Features, in D O Ho-Yen and A W L Joss (eds), *Human Toxoplasmosis*, Oxford University Press, Oxford, 56–78.

Huang P, Weber J T, Sosin D M, Griffin P M, Long E G, Murphy J J, Kocka F, Peters C and Kallick C (1995) The first reported outbreak of diarrheal illness associated with *Cyclospora* in the United States, *Ann Intern Med*, **123**, 409–14.

Huang L M, Fyfe M, Ong C, Harb J, Champagne S, Dixon B and Isaac-Renton J (2005) Outbreak of cyclosporiasis in British Columbia associated with imported Thai basil, *Epidemiol Infect*, **133**, 23–7.

Israelski D M and Remington J S (1993) Toxoplasmosis in the non-AIDS immunocompromised host, in J S Remington and M N Swartz (eds), *Current Clinical topics in Infectious Diseases*, Blackwell Scientific Publications, Oxford, 322–56.

Ito S, Tsunoda K, Shimada K, Taki T and Matsui T (1975) Disinfectant effects of several chemicals against *Toxoplasma* oocysts, *Jpn J Vet Sci*, **37**, 229–34.

Ivory C, Kammanadiminti S and Chadee K (2007) Innate resistance to *Entamoeba histolytica* in murine models, *Trends Parasitol*, **23**, 46–8.

Jackson M H and Hutchison W M (1989) The prevalence and source of *Toxoplasma* infection in the environment, *Adv Parasitol*, **28**, 55–105.

Jacobs L, Remington J S and Melton M L (1960a) A survey of meat samples from swine, cattle and sheep for the presence of encysted *Toxoplasma*, *J Parasitol*, **46**, 23–48.

Jacobs L, Remington J S and Melton M L (1960b) The resistance of the encysted form of *Toxoplasma gondii*, *J Parasitol*, **46**, 11–21.

Jarroll E, Hoff J C and Meyer E A (1981) Effect of chlorine on *Giardia lamblia* cyst viability, *Appl Environ Microbiol*, **41**, 483–7.

Jaykus L A (1997) Epidemiology and detection as options for control of viral and parasitic foodborne disease, *Emerg Infec Dis*, **3**(4), 529–39.

Jiang J, Alderisio K A and Xiao L (2005) Distribution of *Cryptosporidium* genotypes in storm event water samples from three watersheds in New York, *Appl Environ Microbiol*, **71**, 4446–54.

John D E, Haas C N, Nwachuku N and Gerba C P (2005) Chlorine and ozone disinfection of *Encephalitozoon intestinalis* spores, *Water Res*, **39**, 2369–75.

Jokipii L and Jokipii A M M (1986) Timing of symptoms and oocyst excretion in human cryptosporidiosis, *N Engl J of Med*, **315**(26), 1643–7.

Joss A W L (1992) Treatment, in D O Ho-Yen and A W L Joss (eds), *Human Toxoplasmosis*, Oxford University Press, Oxford, 119–43.

Joss A W L, Chatterton J M W, Evans R and Ho-Yen D O (1993) *Toxoplasma* polymerase chain reaction on experimental blood samples, *J Med Microbiol*, **38**, 38–43.

Karabiber N and Aktas F (1991) Foodborne giardiasis, *Lancet*, **377**, 376–7.

Karanis P, Kourenti K and Smith H V (2007) Waterborne transmission of protozoan parasites: a worldwide review of outbreaks and lessons learnt, *J Water Health*, **5**, 1–38.

Kasprzak W, Rauhut W and Mazur T (1981) Transmission of *Giardia* cysts. Fruit and vegetables as the source of infection, *Wiad Parazytol*, **27**(4–5), 565–71.

Kaucner C and Stinear T (1998) Sensitive and rapid detection of viable *Giardia* cysts and *Cryptosporidium parvum* oocysts in large-volume water samples with wound fiberglass cartridge filters and reverse transcription-PCR, *Appl Environ Microbiol*, **64**(5), 1743–9.

Kilvington S and Beeching J (1995) Development of a PCR for identification of *Naegleria fowleri* from the environment, *Appl Env Microbiol*, **61**(10), 3764–7.

King C H, Shotts E B, Wooley R E and Porter K G (1988) Survival of coliforms and bacterial pathogens within protozoa during chlorination, *Appl Env Microbiol*, **54**(12), 3023–33.

Korich D G, Mead J R, Madore M S, Sinclair N A and Sterling C R (1990) Effects of ozone, chlorine dioxide, chlorine, and monochloramine on *Cryptosporidium parvum* oocysts viability, *Appl Environ Microbiol*, **56**, 1423–8.

Kotula A W, Dubey J P, Sharar A K, Andrew C D, Shen S K and Lindsay D S (1991) Effect of freezing on infectivity of *Toxoplasma gondii* tissue cysts in pork, *J Food Prot*, **54**, 687–90.

Laberge I, Ibrahim A, Barta J R and Griffiths M W (1996) Detection of *Cryptosporidium parvum* in raw milk by PCR and oligonucleotide probe hybridisation, *Appl Environ Microbiol*, **62**(9), 3259–64.

Lalle M, Pozio E, Capelli G, Bruschi F, Crotti D and Cacciò S M (2005) Genetic heterogeneity at the beta-giardin locus among human and animal isolates of *Giardia duodenalis* and identification of potentially zoonotic subgenotypes, *Int J Parasitol*, **35**, 207–13.

Lee S H, Levy D A, Craun G F, Beach M J and Calderon R L (2002) Surveillance for waterborne-disease outbreaks–United States–1999–2000, *MMWR Surveillance Summaries*, **51**, 1–47.

Li X, Guyot K, Dei-Cas E, Mallard J P, Ballet J J and Brasseur P (2006) *Cryptosporidium* oocysts in mussels (*Mytilus edulis*) from Normandy (France), *Int J Food Microbiol*, **1**(108), 321–5.

Lindquist H D A, Bennet J W, Hester J D, Ware M W, Dubey J P and Everson W P (2003) Autofluorescence of *Toxoplasma gondii* and related coccidian oocysts, *J Parasitol*, **89**, 865–7.

Lindsay D S, Collins M V, Mitchell S M, Cole R, Flick G, Wetch C N, Lindquist A and Dubey J P (2003) Sporulation and survival of *Toxoplasma gondii* oocysts in sea water, *J Eukaryot Microbiol*, **50**, S687–S688.

Lindsay D S, Collins M V, Mitchell S M, Wetech C N, Rosypal A C, Flick G J, Lindquist A and Dubey J P (2004) Survival of *Toxoplasma gondii* oocysts in eastern oysters (*Crassostrea virginica*), *J Parasitol*, **90**, 1053–6.

Lonigro A, Pollice A, Spinelli R, Berrilli F, Di Cave D, D'Orazi C, Cavalo P and Brandonisio O (2006) *Giardia* cysts and *Cryptosporidium* oocysts in membrane-filtered municipal wastewater used for irrigation, *Appl Environ Microbiol*, **72**, 7916–18.

Lopez A S, Dodson D R, Arrowood M J, Orlandi Jr P A, da Silva A J, Bier W J, Hanauer S D, Kuster R L, Oltman S, Baldwin M S, Won K Y, Nace E M, Eberhard M L and Herwaldt, B L (2001) Outbreak of cyclosporiasis associated with basil in Missouri in 1999, *Clin Infect Dis*, **32**, 1010–1017.

Lowery C J, Nugent P, Moore J E, Millar B C, Xiru X and Dooley J S G (2001) PCR-IMS detection and molecular typing of *Cryptosporidium parvum* recovered from a recreational river source and an associated mussel (*Mytilus edulis*) bed in Northern Ireland, *Epidemiol Infect*, **127**, 545–53.

Luft B J and Remington J S (1992) Toxoplasmic encephalitis in AIDS, *Clin Inf Dis*, **15**, 211–22.

Lyon B B V and Swalm W A (1925) Giardiasis, its frequency, recognition and certain clinical factors, *Am J Med Sci*, **170**, 348–64.

Ma P, Visvesvara G S, Martinez A J, Theodore F H, Caggett P and Sawyer T K (1990) *Naegleria* and *Acanthamoeba* infections: Review, *Rev Inf Dis*, **12**, 490–513.

Marciano-Cabral F and Cabral G (2003) *Acanthamoeba* spp. as agents of disease in humans, *Clin Microbiol Rev*, **16**, 273–307.

Marciano-Cabral F, Cline M L and Bradley S G (1987) Specificity of antibodies from human sera for *Naegleria* species, *J Clin Microbiol*, **25**, 692–7.

Marciano-Cabral F, MacLean R, Mensah A and LaPat-Polasko L (2003) Identification of *Naegleria fowleri* in domestic water sources by nested PCR, *Appl Environ Microbiol*, **69**, 5864–9.

Mariuz P and Steigbigel R T (2001) *Toxoplasma* infection in HIV-infected patients, in D H M Joynson and T G Wreghitt (eds), *Toxoplasmosis: a Comprehensive Clinical Guide*, Cambridge University Press, Cambridge, 147–77.

Marshall M M, Naumovitz D, Ortega Y and Sterling C R (1997) Waterborne protozoan pathogens, *Clin Microbiol Rev*, **10**, 67–85.

Martinez A J and Visvesvara G S (1991) Laboratory diagnosis of pathogenic, free-living amoebas: *Naegleria*, *Acanthamoeba* and *Leptomyxid*, *Clin Lab Med*, **11**, 861–72.

Martinez A J and Visvesvara G S (1997) Free-living, amphizoic and opportunistic amoebas, *Brain Pathol*, **7**, 583–98.

Martinez A J, Nelson E C, Duma R J and Moretta F L (1973) Experimental *Naegleria* meningoencephalitis in mice. Penetration of the olfactory mucosal epithelium by

Naegleria and pathological changes produced. A light and electron microscopy study, *Lab Invest*, **29**, 121–33.

Martinez A J, Markowitz S M and Duma R J (1975) Experimental acanthamoebic pneumonitis and encephalitis in mice, *J Infect Dis*, **131**, 692–9.

Mastrandrea G and Micarelli A (1968) Search for parasites in vegetables from the local markets in the city of Rome, *Arch Ital Sci Med Trop Parasitol*, **49**, 55–9.

Masur H, Jones T C, Lempert J A and Cherubini T D (1978) Outbreak of toxoplasmosis on a family and documentation of acquired retinochoroiditis, *JAMA*, **64**, 396–402.

Maya C, Beltrán N, Jiménez B and Bonilla P (2003) Evaluation of the UV disinfection process in bacteria and amphizoic amoebae inactivation, *Water Sci Technol: Water Supply*, **3**(4), 285–91.

Mead P S, Slukster L, Dietz V, McCaig L F, Bresce J S, Shapiro C, Griffin P M and Tauxe R V (1999) Food-related illness and death in the United States, *Emerg Inf Dis*, **5**(5), 607–25.

Meloni B P, Lymbery A J and Thompson R C A (1995) Genetic characterization of isolates of *Giardia duodenalis* by enzyme electrophoresis: implications for reproductive biology, population structure, taxonomy, and epidemiology, *J Parasitol*, **81**, 368–83.

Millar B C, Finn M, Xiao L, Lowery C J, Dooley J S G and Moore J E (2002) *Cryptosporidium* in foodstuffs – an emerging aetiological route of human foodborne illness, *Trends Food Sci Technol*, **13**(5), 168–87.

Millard P S, Gensheimer K F, Addiss D G, Sosin D M, Beckett G A, Houck-Jankoski A and Hudson A (1994) An outbreak of cryptosporidiosis from fresh-pressed apple cider, *JAMA*, **271**, 1592–6.

Miller W A, Atwill E R, Gardner I A, Miller M A, Fritz H M, Hedrick R P, Melli A C, Barnes N M and Paricia A C (2005a) Clams (*Corbicula fluminea*) as bioindicators of fecal contamination with *Cryptosporidium* and *Giardia* spp. in freshwater ecosystems in California, *Int J Parasitol*, **35**(6), 673–84.

Miller W A, Miller M A, Gardner I A, Atwill E R, Harris M, Ames J, Jessup D, Melli A, Paradies D, Worcester K, Olin P, Barnes N and Conrad P A (2005b) New genotypes and factors associated with *Cryptosporidium* detection in mussels (*Mytilus* spp.) along the California coast, *Int J Parasitol*, **35**(10), 1103–13.

Mintz E D, Hudson-Wragg M, Mshar P, Cartr M L and Hadler J L (1993) Foodborne giardiasis in a corporate office setting, *J Infect Dis*, **167**, 250–53.

Monge R and Chinchilla M (1996) Presence of *Cryptosporidium* oocysts in fresh vegetables, *J Food Protec*, **59**(2), 202–3.

Monis P T, Andrews R H, Mayrhofer G and Ey P L (1999) Molecular systematics of the parasitic protozoan *Giardia intestinalis*, *Mol Biol Evol*, **16**, 1135–44.

Morris J G (1996) Current trends in human diseases with food of animal origin, *J Am Vet Med Assoc*, **209**(12), 2045–7.

Munday B L (1972) Serological evidence of toxoplasma infection in isolated groups of sheep, *Res Vet Sci*, **13**, 100–102.

Musgrave W E (1922) Flagellate infestations and infections, *JAMA*, **79**, 2219–20.

Nash T E, Herrington D A, Losonsky G A and M M Levine (1987) Experimental human infections with *Giardia lamblia*, *J Infect Dis*, **156**, 974–84.

Nichols G L, Chalmers R M, Sopwith W, Regan M, Hunter C A, Grenfell P, Harrison F and Lane C (2006a) *Cryptosporidiosis: A report on the surveillance and epidemiology of Cryptosporidium infection in England and Wales*, Drinking Water Directorate Contract Number DWI 70/2/201, UK Drinking Water Inspectorate, London.

Nichols R A B and Smith H V (2002) *Cryptosporidium*, *Giardia* and *Cyclospora* as foodborne pathogens, in C Blackburn and P McClure (eds), *Foodborne Pathogens: Hazards, Risk and Control*, Part III, Non-bacterial and emerging foodborne pathogens, Woodhead, Cambridge, 453–78.

Nichols R A B and H V Smith H V (2004) Optimisation of DNA extraction and molecular detection of *Cryptosporidium parvum* oocysts in natural mineral water sources, *J Food Prot*, **67**, 524–32.

Nichols R A B and Smith H V (2007) *Cryptosporidium parvum* as a foodborne pathogen, in *Animal Health and Production Compendium*, CAB International, Wallingford. http://www.cabicompendium.org/ahpc/datasheet.asp?ccode=CRYPTV

Nichols R A B and Smith H V (2009) *Giardia duodenalis* as a foodborne pathogen, in *Animal Health and Production Compendium*, CAB International, Wallingford, http://www.cabicompendium.org/ahpc, In press.

Nichols R A, Campbell B M and Smith H V (2003) Identification of *Cryptosporidium* spp. oocysts in United Kingdom noncarbonated natural mineral waters and drinking waters by using a modified nested PCR-restriction fragment length polymorphism assay, *Appl Environ Microbiol*, **69**, 4183–9.

Nichols R A, Paton C A and Smith H V (2004) Survival of *Cryptosporidium parvum* oocysts after prolonged exposure to still natural mineral waters, *J Food Protec*, **67**, 517–23.

Nichols R A, Campbell B M and Smith H V (2006b) Molecular fingerprinting of *Cryptosporidium* oocysts isolated during water monitoring, *Appl Environ Microbiol*, **72**, 5428–35.

O'Grady J E and Smith H V (2002) Methods for determining the viability and infectivity of *Cryptosporidium* oocysts and *Giardia* cysts. (Chapter) in: *Detection methods for algae, protozoa and helminths*, (eds, Ziglio G and Palumbo F), John Wiley and Sons, Chichester, UK, 193–220.

Okhuysen P C, Chappell C L, Crabb J H, Sterling C R, and DuPont H L (1999) Virulence of three distinct *Cryptosporidium parvum* isolates for healthy adults, *J Infect Dis*, **180**(4), 1275–81.

Ortega Y (ed.) (2006), *Foodborne Parasites*, Food Microbiology and Food Safety Series, Springer-Verlag, New York.

Ortega Y R, Sterling C R, Gilman R H, Cama V A and Diaz F (1993) *Cyclospora* species: a new protozoan pathogen of humans, *N Engl J Med*, **328**, 1308–12.

Ortega Y R, Nagle R, Gilman R H, Watanabe J, Miyagui J, Quispe H, Kanaguusuku P, Roxas C and Sterling C R (1997a) Pathologic and clinical findings in patients with cyclosporiasis and a description of intracellular parasite life-cycle stages, *J Infect Dis*, **176**, 1584–9.

Ortega Y R, Roxas C R, Gilman R H, Miller N J, Cabrera L, Taquiri C and Sterling C R (1997b) Isolation of *Cryptosporidium parvum* and *Cyclospora cayetanensis* from vegetables collected in markets of an endemic region in Peru, *Am J Trop Med Hyg*, **57**(6), 683–6.

Osterholm M T, Forfang J C, Ristinen T L, Dean A G, Washburn J W, Godes J R, Rude R A and McCullough J G (1981) An outbreak of foodborne giardiasis, *N Engl J Med*, **304**, 24–8.

Page F C (1974) A further study of taxonomic criteria for *Limax* amoebae, with descriptions of new species and a key to genera, *Arch Protistenkd*, **116**, S149–84.

Page F C (1988) *A New Key to Freshwater and Soil Gymnamoebae*, Freshwater Biological Association, Ambleside.

Parija S C and Jaykeerthee S R (1999) *Naegleria fowleri*: a free living amoeba of emerging medical importance, *J Commun Dis*, **31**, 153–9.

Petersen L R, Cartter M L and Hadler J L (1988) A food-borne outbreak of *Giardia lamblia*, *J Infect Dis*, **157**, 846–8.

Pomeroy C and Felice G A (1992) Pulmonary toxoplasmosis, A review, *Clin Infect Dis*, **14**, 863–70.

Porter J D H, Gaffney C, Heymann D and Parkin W (1990) Food-borne outbreak of *Giardia lamblia*, *Am J Public Health*, **80**, 1259–60.

Quick R, Paugh K, Addiss D, Kobayashi J and Baron R (1992) Restaurant associated outbreak of giardiasis, *J Infect Dis*, **166**, 673–6.

Quiroz E S, Bern C, MacArthur J R, Xiao L H, Fletcher M, Arrowood M J, Shay D K,

Levy M E, Glass R I and Lal A (2000) An outbreak of cryptosporidiosis linked to a foodhandler, *J Infect Dis*, **181**(2), 695–700.

Relman D A, Schmidt T A, Gajadhar A, Sogin M, Cross J, Yoder K, Sethabutr O and Echeverria P (1996) Molecular phylogenetic analysis of *Cyclospora*, the human intestinal pathogen, suggests that it is closely related to *Eimeria* species, *J Infect Dis*, **173**, 440–45.

Rendtorff R C (1954) The experimental transmission of human intestinal protozoan parasites. II. *Giardia lamblia* cysts given in capsules, *Am J Hyg*, **59**, 209–20.

Rendtorff R C (1979) The experimental transmission of *Giardia lamblia* among volunteer subjects, in W Jakubowski and J C Hoff (eds), *Waterborne Transmission of Giardiasis*, US Environmental Protection Agency, Office of Research and Development, Environmental Research Centre, Cincinnati, OH, 64–81.

Robertson L J and Gjerde B (2000) Isolation and enumeration of *Giardia* cysts, *Cryptosporidium* oocysts, and *Ascaris* eggs from fruits and vegetables, *J Food Prot*, **63**(6), 775–8.

Robertson L J and Gjerde B (2001) Occurrence of parasites on fruits and vegetables in Norway, *J Food Prot*, **64**, 1793–8.

Robertson L J, Campbell A T and Smith H V (1992) Survival of *Cryptosporidium parvum* oocysts under various environmental pressures, *Appl Environ Microbiol*, **58**(11), 3494–500.

Robertson L J, Campbell A T and Smith H V (1993) *In vitro* excystation of *Cryptosporidium parvum*, *Parasitology*, **106**, 13–19.

Robertson L J, Johannessen G S, Gjerde B K and Loncarevic S (2002) Microbiological analysis of seed sprouts in Norway, *Int J Food Microbiol*, **75**, 11–126.

Rochelle P A, Ferguson D M, Handojo T J, De Leon R, Stewart M H and Wolfe R L (1997) An assay combining cell culture with reverse transcriptase PCR to detect and determine the infectivity of waterborne *Cryptosporidium parvum*, *Appl Environ Microbiol*, **63**, 2029–37.

Rochelle P A, Marshall M M, Mead J R, Johnson AM, Korich D G, Rosen J S and De Leon R (2002) Comparison of *in vitro* cell culture and a mouse assay for measuring infectivity of *Cryptosporidium parvum*, *Appl Environ Microbiol*, **68**, 3809–17.

Ruecker N J, Bounsombath N, Wallis P, Ong C S, Isaac-Renton J L and Neumann N F (2005) Molecular forensic profiling of *Cryptosporidium* species and genotypes in raw water, *Appl Environ Microbiol*, **71**, 8991–4.

Ruecker N J, Braithwaite S L, Topp E, Edge T, Lapen D R, Wilkes G, Robertson W, Medeiros D, Sensen C W and Neumann N F (2007) Tracking host sources of *Cryptosporidium* spp. in raw water for improved health risk assessment, *Appl Environ Microbiol*, **73**, 3945–57.

Sacks J J, Roberto R R and Brooks N F (1982) Toxoplasmosis infection associated with raw goat's milk, *JAMA*, **248**, 1728–32.

Sacks J J, Delgado D G, Lobel H O and Parker R L (1983) Toxoplasmosis infection associated with eating under-cooked venison, *Am J Epidemiol*, **118**, 832–8.

Sarkar P (2008) *Occurrence and Inactivation of Emerging Pathogens in the Environment*, PhD thesis, Department of Microbiology and Immunology, Graduate College, University of Arizona, Copyright© Payal Sarkar, 2008.

Sauch J F, Flannigan D, Galvin M L, Berman D and Jakubowski W (1991) Propidium iodide as an indicator of *Giardia* cyst viability, *Appl Environ Microbiol*, **57**, 3243–7.

Schets F M, Van den Berg H H J L, Engels G B, Lodder W J and de Roda Husman A M (2007) *Cryptosporidium* and *Giardia* in commercial and non-commercial oysters (*Crassostrea gigas*) and water from the Oosterschelde, the Netherlands, *Int J Food Microbiol*, **113**(2), 189–94.

Schupp D E and Erlandsen S L (1987a) A new method to determine *Giardia* cyst viability, correlation between fluorescein diacetate/propidium iodide staining and animal infectivity, *Appl Environ Microbiol*, **53**, 704–7.

Schupp D E and Erlandsen S L (1987b) Determination of *Giardia muris* cyst viability by differential interference contrast, phase or bright field microscopy, *J Parasitol*, **73**, 723–9.

Schupp D E, Januschka M M and Erlandsen S L (1988) Assessing *Giardia* cyst viability with fluorogenic dyes, comparisons to animal infectivity and cyst morphology by light and electron microscopy, in P M Wallis and B R Hammond (eds), *Advances in Giardia Research*, University of Calgary Press, Calgary, 265–9.

Schuster F L and Visvesvara G S (2004) Amebae and ciliated protozoa as causal agents of waterborne zoonotic disease, *Vet Parasitol*, **126**, 91–120.

Sherchand J B, Cross J H, Jimba M, Sherchand S and Shrestha M P (1999) Study of *Cyclospora cayetanensis* in health care facilities, sewage water and green leafy vegetables in Nepal, *Southeast Asian J Trop Med Public Health*, **30**, 58–63.

Silveira C, Belfort R, Burnier M and Nussenblat R (1988) Acquired toxoplasmic infection as the cause of toxoplasmic retinochoroiditis in families, *Am J Ophthalmol*, **106**, 362–4.

Skinner L J, Timperley A C, Wightman D, Chatterton J M W and Ho-Yen D O (1990) Simultaneous diagnosis of toxoplasmosis in goats and goatowner's family, *Scand J Inf Dis*, **22**, 359–61.

Slifco T M, Friedman D E, Rose J B and Jakubowski W (1997) Unique cultural methods used to detect viable *Cryptosporidium* oocysts in environmental samples, *Water Sci Technol*, **35**(11–12), 363–8.

Slifco T M, Smith H V and Rose J B (2001) Emerging parasite zoonoses associated with food and water, *Int J Parasitol*, **30**, 1379–93.

Smith J L (1993) *Cryptosporidium* and *Giardia* as agents of foodborne disease, *J Food Prot*, **56**, (5), 451–61.

Smith H V (1995) Emerging technologies for the detection of protozoan parasites in water, in W B Betts, D Casemore, C Fricker, H V Smith and J Watkins (eds), *Protozoan Parasites and Water*, The Royal Society of Chemistry, Cambridge, 108–14.

Smith H V (1996) Detection of *Cryptosporidium* and *Giardia* in water, in R W Pickup and J R Saunders (eds), *Molecular Approaches to Environmental Microbiology*, Ellis-Horwood, Hemel Hempstead, 195–225.

Smith H V (1998) Detection of parasites in the environment, *Parasitology*, **117**, 113–41.

Smith H V (2008) Diagnostics in R Fayer and L Xiao (ed.), *Cryptosporidiosis of Man and Animals*, CRC and IWA, Boca Raton, FL, 173–208.

Smith H V and Cook N (2009) Lessons learnt in development and application of detection methods for zoonotic foodborne protozoa on lettuce and fresh fruit, in A Mortimer, P Colonna, D Lineback, W Spiess, K Buckle and G V Barbosa-Cánovas (eds), *Global Issues in Food Science and Technology*, Elsevier, in press.

Smith H V and Corcoran G D (2004) New drugs and treatment for cryptosporidiosis, *Curr Op Infect Dis*, **17**, 557–64.

Smith H V and Grimason A M (2003) *Giardia* and *Cryptosporidium* in water and wastewater, in D Mara and N Horan (eds), *The Handbook of Water and Wastewater Microbiology*, Elsevier Science, Oxford, 619–781.

Smith H V and Hayes C R (1996) The status of UK methods for the detection of *Cryptosporidium* sp. oocysts and *Giardia* sp. cysts in water concentrates and their relevance to water management, *Water Sci Technol*, **35**, 369–76.

Smith H V and Nichols R A B (2008) Preparation of parasitic specimens for direct molecular applications, in D Liu (ed.), *Handbook of Nucleic Acid Purification*, Taylor & Francis CRC Press, London, in press.

Smith H V and Paget T (2007) *Giardia*, in S Simjee (ed.), *Foodborne Diseases*, Humana Press, Totowa, NJ, 303–36.

Smith A L and H V Smith (1989) A comparison of fluorescein diacetate and propidium iodide staining and *in vitro* excystation for determining *Giardia intestinalis* cyst viability, *Parasitology*, **99**, 329–31.

Smith H V, Smith A L, Girdwood R W A and Carrington E G (1989) *The effect of free chlorine on the viability of Cryptosporidium spp. oocysts*, WRC Report PRU 2023-M, Water Research Council, Marlow.

Smith H V, Robertson L J, Campbell A T and Girdwood R W A (1995a) *Giardia* and giardiasis, what's in a name?, *Microbiol Europe*, **3**, 22–9.

Smith H V, Robertson L J and Ongerth J E (1995b) Cryptosporidiosis and giardiasis, the impact of waterborne transmission, *J Water Supp Res Tech – Aqua*, **44**, 258–74.

Smith H V, Paton C A, Mtambo M M A and Girdwood R W A (1997) Sporulation of *Cyclospora* sp. oocysts, *Appl Environ Microbiol*, **63**, 1631–2.

Smith J J, Gunasekera T S, Barardi C R, Veal D and Vesey G (2004) Determination of *Cryptosporidium parvum* oocysts viability by fluorescence in situ hybridization using a ribosomal RNA-directed probe, *J Appl Microbiol*, **96**(2), 409–17.

Smith H V, Cacciò S M, Tait A, McLauchlin J and Thompson R C A (2006) Tools for investigating the abiotic transmission of *Cryptosporidium* and *Giardia* infections in humans, *Trends Parasitol*, **22**, 160–66.

Smith H V, Cacciò S M, Cook N, Nichols R A B and Tait A (2007) *Cryptosporidium* and *Giardia* as foodborne zoonoses, *Vet Parasitol*, **149**, 29–40.

Soave R (1996) *Cyclospora*: an overview, *Clin Infect Dis*, **23**, 429–37.

Soave R and Armstrong D (1986) *Cryptosporidium* and cryptosporidiosis, *Rev Infect Dis*, **8**, 1012–23.

Sparagano O, Drouet E, Brebant R, Manet E, Denoyel G A and Pernin P (1993) Use of monoclonal antibodies to distinguish pathogenic *Naegleria fowleri* (cysts, trophozoites, or flagellate forms) from other *Naegleria* species, *J Clin Microbiol*, **31**(10), 2758–63.

Steinberg J P, Galindo R L, Kraus E S and Ghanem K G (2002) Disseminated acanthamoebiasis in a renal transplant patient with osteomyelitis and cutaneous lesions: case report and literature review, *Clin Infect Dis*, **35**, e43–e49.

Stinear T, Matusan A, Hines K and M Sandery (1996) Detection of a single viable *Cryptosporidium parvum* oocyst in environmental water concentrates by reverse transcription-PCR, *Appl Environ Microbiol*, **62**, 3385–90.

Sutthikornchai C, Jantanavivat C, Thorungrunkiat S, Harnroongraj T and Sukantha Y (2005) Protozoal contamination of water used in the Thai food industry, *Southeast Asian J Trop Health*, **36**, 41–5.

Szénási Z, Endo T, Yagita K and Nagy E (1998) Isolation, identification and increasing importance of free-living amoebae causing human disease, *J Med Microbiol*, **47**, 5–16.

Taghi-Kilani R, Gyürék L L, Millard P J, Finch G R and Belosevic M (1996) Nucleic acid stains as indicators of *Giardia muris* viability following cyst inactivation, *Int J Parasitol*, **26**(6), 637–46.

Tauxe R V (1997) Emerging foodborne diseases: an evolving public health challenge, *Emerg Infec Dis*, **3**(4), 425–34.

Thompson R C A (2000) Giardiasis as a re-emerging infectious disease and its zoonotic potential, *Int J Parasitol*, **30**, 1259–67.

Thompson R C and Monis P T (2004) Variation in *Giardia*: implications for taxonomy and epidemiology, *Adv Parasitol*, **58**, 69–137.

Thulliez P (2001) Maternal and foetal infection, in D H M Joynson and T G Wreghitt (eds), *Toxoplasmosis a Comprehensive Clinical Guide*, Cambridge University Press, Cambridge, 193–213.

Thurston-Enriquez J A, Watt P, Dowd S E, Enriquez R, Pepper J L and Gerba C P (2002) Detection of parasites and microsporidia in irrigation waters used for crop production, *J Food Prot*, **65**, 378–82.

Traversa D, Giangaspero A, Molini U, Iorio R, Paoletti B, Otranto D and Giansante C (2004) Genotyping of *Cryptosporidium* isolates from *Chamelea gallina* clams in Italy, *Appl Environ Microbiol*, **70**(7), 4367–70.

Tyzzer E E (1910) An extracellular coccidium *Cryptosporidium muris* of the gastric glands of the common mouse, *J Med Res*, **23**, 487–509.

Tyzzer E E (1912) *Cryptosporidium parvum* (sp. nov.) a coccidian found in the small intestine of the common mouse, *Arch Protistenkd*, **26**, 94–412.

Upton S J, Tilley M and Brillhart D B (1994) Comparative development of *Cryptosporidium parvum* (Apicomplexa) in 11 continuous host cell lines, *FEMS Microbiol Letters*, **118**, 233–6.

Vaillant V, de Vask H, Baron E, Ancelle T, Delmon M-C, Dufour B, Pouillot R, Le Strat Y, Wanbreck P and Desenclos J C (2005) Food-borne infections in France, *Foodborne Path Dis*, **2**, 221–32.

Vesey G, Ashbolt N, Fricker E J, Deere D, Williams K L, Veal D A and Dorsch M (1998) The use of a ribosomal RNA targeted oligonucleotide probe for fluorescent labelling of viable *Cryptosporidium parvum* oocysts, *J Appl Microbiol*, **85**, 429–40.

Visvesvara G S, Peralta M J, Brandt F H, Wilson M, Aloisio C and Franko E (1987) Production of monoclonal antibodies to *Naegleria fowleri*, agent of primary amoebic meningoencephalitis, *J Clin Microbiol*, **25**, 1629–34.

Visvesvara G S, Schuster F L and Martinez A J (1993) *Balamuthia mandrillaris*, n.g., n.sp., agent of meningoencephalitis in humans and other animals, *J Eukaryot Microbiol*, **40**, 504–14.

Wagner-Wiening C and Kimmig P (1995) Detection of viable *Cryptosporidium parvum* oocysts by PCR, *Appl Environ Microbiol*, **61**, 4514–16.

Wallace G D, Marshall L and Marshall M (1972) Cats, rats and toxoplasmosis on a small Pacific island, *Am J Trop Med Hyg*, **25**, 48–53.

Warnekulasuriya M R, Johnson J D and Holliman R E (1998) Detection of *Toxoplasma gondii* in cured meats, *Int J Food Microbiol*, **45**, 211–15.

Wastling J M, Nicoll S and Buxton D (1993) Comparison of two gene amplification methods for the detection of *Toxoplasma gondii* in experimentally infected sheep, *J Med Microbiol*, **38**, 360–65.

Weintraub J M (2006) Improving *Cryptosporidium* testing methods: a public health perspective, *J Water Health*, **4** Suppl 1, 23–6.

Weiss A, Margo C E , Ledford D K, Lockey R F and Brinser J H (1986) Toxoplasmic retinochoroiditis as an initial manifestation of the acquired immune deficiency syndrome, *Am J Ophthalmol*, **101**, 248–9.

Wellings F M, Amuso P T, Chang S L and Lewis A L (1977) Isolation and identification of pathogenic *Naegleria* from Florida lakes, *Appl Environ Microbiol*, **34**, 661–7.

White K E, Hedberg C W, Edmonson L M, Jones D B W, Osterholme M T and MacDonald K L (1989) An outbreak of giardiasis in a nursing home with evidence for multiple modes of transmission, *J Infect Dis*, **160**, 298–304.

Whiteside M E, Barkin J S, May R G, Weiss S D, Fischl M A and MacLeod C L (1984) Enteric coccidiosis among patients with the Acquired Immunodeficiency Syndrome, *Am J Trop Med Hyg*, **33**, 1065–72.

Widmer G, Orbacz E A and Tzipori S (1999) β-Tubulin mRNA as a marker of *Cryptosporidium parvum* oocyst viability, *Appl Environ Microbiol*, **65**(4), 1584–8.

Widmer G, Akiyoshi D, Buckholt M A, Feng X, Rich S M, Deary K M, Bowman C A, Xu P, Wang Y, Wang X, Buck G A and Tzipori S (2000) Animal propagation and genomic survey of a genotype 1 isolate of *Cryptosporidium parvum*, *Mol Biochem Parasitol*, **108**, 187–97.

World Health Organization (1997) Amoebiasis, *Wkly Epidemiol Rec*, **72**, 97–9.

Wreghitt T G and Joynson D H M (2001) *Toxoplasma* infection in immunosuppressed (HIV-negative) patients, in D H M Joynson and T G Wreghitt (eds), *Toxoplasmosis a Comprehensive Clinical Guide*, Cambridge University Press, Cambridge, 178–92.

Wreghitt T G, Hakim M, Gray J J, Balfour A H, Stovin P G, Stewart S, Scott J, English T A H and Wallwork J (1989) Toxoplasmosis in heart and heart and lung transplant recipients, *J Clin Pathol*, **42**, 194–9.

Xiao L, Alderisio K, Limor J, Royer M and Lal A A (2000) Identification of species and sources of *Cryptosporidium* oocysts in storm waters with a small-subunit rRNA-based diagnostic and genotyping tool, *Appl Environ Microbiol*, **66**, 5492–8.
Xiao L, Singh A, Limor J, Graczyk T K, Gradus S and Lal A (2001) Molecular characterization of *Cryptosporidium* oocysts in samples of raw surface water and wastewater, *Appl Environ Microbiol*, **67**, 1097–101.
Xiao L, Fayer R, Ryan U and Upton S J (2004) *Cryptosporidium* taxonomy: recent advances and implications for public health, *Clin Microbiol Rev*, **17**, 72–97.
Yilmaz S M and Hopkin S H (1972) Effects of different conditions on the duration of infectivity of *Toxoplasma gondii* oocysts, *J Parasitol*, **58**, 938–9.

27

Foodborne helminth infections

K. D. Murrell, University of Copenhagen, Denmark, and D. W. T. Crompton, University of Glasgow, UK

Abstract: Many helminth infections of humans are recognised as zoonoses and the route of infection invariably involves one of the pathogen's natural hosts having become a human food item. Some of these zoonotic infections are deadly while others cause chronic ill health. This chapter presents a selection of foodborne species of trematode, cestode and nematode parasites, their distribution and abundance, and their public health and economic importance. The helminths include those transmitted by infected meat, fish, invertebrates and plants. Finally, a selection of approaches to the prevention and control of foodborne helminth infections is presented.

Key words: helminths, zoonosis, foodborne, control, prevention.

27.1 Introduction

Humans suffer from zoonoses or zoonotic diseases; some may be deadly (e.g. trichinellosis or cerebral cysticercosis) while others are chronic and either cause only mild illness (e.g. intestinal trematodiasis) or they may lead to serious complications (e.g. liver flukes and cholangiocarcinoma). A zoonosis is a disease caused by a pathogen that becomes established in a human host when human activities encroach into the pathogen's natural life cycle (WHO, 2006). Many helminth infections of humans are recognised as zoonoses, and the route of infection invariably involves one of the pathogen's natural hosts having become a human food item (Coombs and Crompton, 1991). In this chapter we are concerned with a selection of foodborne species of trematode, cestode and nematode (Table 27.1), their distribution and abundance, and their public health and economic importance. We also

Table 27.1 Estimates of numbers of human infections with a selection of foodborne helminths

Helminth species	Numbers (millions)[a,b]	References
TREMATODA		
Clonorchis sinensis	7	WHO, (1995)
Fasciola gigantica]	2.4–17	Rim *et al.*, 1994, Mas-Coma
F. hepatica]		*et al.* (2007)
Fasciolopsis buski	0.2	WHO, (1995)
Opisthorchis viverrini	10.5[c]	Watanapa and Watanapa, 2002
Paragonimus westermani	20	Toscano *et al.*, 1995
Small intestinal flukes[d]	4.4 (Republic of Korea)	Chai and Lee, 2002
CESTODA		
Taenia saginata	77	Keymer, 1982
	10	Craig *et al.*, 1996
T. solium	50	Eddi *et al.*, 2003
Diphyllobothrium spp.[e]	9	Von Bonsdorff, 1977
NEMATODA		
Anisakis simplex[f]	0.33	Ishakura *et al.*, 1998
Trichinella spiralis[g]	1.5	Dupouy-Camet, 2000

[a]Problems encountered in making estimates of numbers of infections are discussed by Crompton (1999).
[b]Maps of the distribution of these and other infectiuons have been published by Stürchler (1988). See also Table 27.2.
[c]Includes 1.5 million infections with *O. felineus*.
[d]A collective title for species of flukes belonging to the Heterophyidae, Echinostomatidae, Neodiplostomidae, Plagiorchiidae and Gymnophallidae (Chai and Lee, 2002; Coombs and Crompton, 1991). Secure identification is extremely difficult.
[e]At least four species are zoonotic.
[f]Representative member of a complex of species.
[g]At least six species of the genus *Trichinella* are zoonotic.
Source: Adapted from, Murrell and Crompton (2006), Table 9.1.

outline a selection of approaches for the prevention and control of foodborne helminth infections.

Such 'dangerous' food may be found wherever poor people have (i) a dependency on subsistence food production, (ii) cultural methods of food preparation, (iii) traditional attitudes to food and meals, sometimes with a gender bias, (iv) inadequate knowledge of infection and hygiene, (v) poor standards of animal husbandry, (vi) insufficient resources for the safe management and disposal of human and animal waste products and (vii) membership of societies where resources for public health measures are overstretched. Foodborne helminth infections invariably form part of daily life for millions of poor people living in developing countries.

Despite high standards of public health required by law in affluent countries, people there should not be complacent about the risk of foodborne helminth infections. Although far fewer people are at risk when compared with the citizens of developing countries, the impact can be serious for those who succumb. Infection occurs mainly through (i) lapses in food hygiene during food production and provision, and (ii) the relatively recent interest that has developed in experiencing foods from other cultures as evidenced by the bewildering range of restaurants in London, New York and other large cities. Globalisation, that ever-expanding interconnectedness between societies, has the potential to increase opportunities for the spread of foodborne helminth infections. In developed countries, the seasonality of foods has largely disappeared; modern air transport means that great diversity of food items is always available for legitimate importation into countries with strong economies. Globalisation has provided much greater mobility and travel opportunities for people, thereby spreading food habits and traditions around the world. Globalisation has also created porous borders leading to the illegal importation of such foods as bushmeat that have not been subjected to any form of quality control. There is now evidence (WHO, 1995, 2004, 2006; Dupouy-Camet and Murrell, 2007) to show that foodborne helminth infections represent a real and an emerging global health problem that merits serious consideration in the list of public health priorities. There is the likelihood that the rise in spending power of people in middle income countries will drive a demand for fish and meat and so increase opportunities for the spread of foodborne helminth infections unless preventive measures are in place. The intensification of agriculture in developing countries, due in part to expanding export opportunities, has had an impact, particularly in aquaculture.

27.2 Main features of foodborne helminth infections

The helminth infections described here represent challenges faced when attempting to implement control measures, although all share the common epidemiological trait of being transmitted to humans in inadequately prepared food. The World Health Organization has now incorporated clonorchiasis, opisthorchiasis and taeniosis/cysticercosis, which includes neurocysticercosis, into the group of so-called neglected tropical diseases (NTDs). This decision stresses the importance attached to foodborne helminth infections. Major biological and epidemiological features of foodborne helminths transmitted through dietary contact with plants, fish or meat are summarised in Table 27.2 and in Chapters 5, 7 and 13 of Crompton and Savioli (2007).

27.2.1 Plantborne helminths

Fasciola spp. (Trematoda)

The flukes *Fasciola hepatica* and *F. gigantica* are the important species causing liver disease in humans (and livestock) (Table 27.2). *Fasciola hepatica* is

Table 27.2 Features of the distribution, biology, transmission and public health risks of foodborne helminth infections

Zoonoses and helminth species responsible	Major geographic distribution	Biology and transmission features	Risk factors
Trematodiasis (flukes)			
PLANTBORNE			
Liver flukes[a] *Fasciola hepatica* *F. gigantica*	*F. hepatica*: Europe, Africa, North and South America, Northern Asia, Oceania. *F. gigantica*: Africa, SE Asia, Japan, China, Korea, Pacific region, Middle East, USA.	Adult worms in ruminants (especially cattle, sheep, goats), pigs and in humans. Intermediate host: aquatic snails, in which asexual reproduction occurs, producing swimming cercariae that attach to aquatic vegetation and encyst (metacercaria). Transmitted passively to definitive host by ingestion of vegetation, or water containing detached metacercaria.	Animal and human consumption of raw aquatic plants (e.g. watercress) and vegetables. Fertilisation of farm ponds or crops with water contaminated with human and livestock waste a major risk factor.
FISHBORNE			
Clonorchis sinensis	Asia.	Adult worms of three species in fish-eating mammals, including humans, dogs, cats, pigs and rats.	Infection of definitive host through ingestion of raw or insufficiently cooked, pickled or smoked infected fish.
Opisthorchis viverrini	Thailand, Laos, Viet Nam, Cambodia.	First intermediate host: snails, in which parasite reproduces	Eating pickled or raw fish at parties or restaurants involving alcohol is especially
O. felineus	Central and Eastern Europe, Turkey, eastern Siberia.	asexually, and emerges as a swimming cercaria stage, seeking fresh water fish, especially carp, which are important in aquaculture. Cercariae invade muscles, viscera, gills, fins and scales	risky for adults. Allowing domestic animals to eat raw fish increases risk of establishing reservoir hosts. Use of human and animal waste for pond fertilisation is an important risk for fish. Failure to control snails in aquaculture

Table 27.2 (Cont'd)

Zoonoses and helminth species responsible	Major geographic distribution	Biology and transmission features	Risk factors
		and encyst (metacercariae).	systems also an important risk factor.
Intestinal flukes			
PLANTBORNE			
Fasciolopsis buski	Central and East Asia.	Adults: Major reservoir host are pigs, secondarily humans. Aquatic snails are first intermediate host, in which asexual reproduction occurs, producing swimming cercariae which seek out aquatic plants on which to encyst (metcercariae). Some cysts may float on water and remain infective. Transmitted to definitive host when plants or water are consumed.	Major risk factor for humans is consumption of infected plants or contaminated water. Important aquatic plants are caltrop, water chestnuts, watercress, water hyacinth, water morning glory and lotus, cultivated in ponds exposed to fecal waste. Pigs likewise become infected when fed aquatic plants and/ or contaminated water.
CRUSTACEANBORNE			
Lung flukes[b,c] *Paragonimus* spp.	East and Central Asia, Central and Western Africa and Central and South America.	Adult worms in wild animals, felids, dogs, pigs and humans. Snails are first intermediate host, in which asexual reproduction occurs to produce ceracariae.	Chief risk is consumption of infected raw or insufficiently cooked crabs by definitive hosts (humans and reservoir hosts). Wild boar meat also implicated in transmission to humans.
		Second intermediate hosts include freshwater crabs and crayfish (many species) which eat infected snails. In the crustacean, the cercariae encyst (metacercariae), mainly on the gills.	Especially risky is the consumption of wine- or soy-soaked crabs and dishes that use crab products such as crab juice, crab jam and crab curd.

Table 27.2 (Cont'd)

Zoonoses and helminth species responsible	Major geographic distribution	Biology and transmission features	Risk factors
		Some mammals (e.g. wild boars) can serve as paratenic hosts.	
FISHBORNE			
This group includes many members of the Heterophyidae and Echinostomatidae families[d], e.g. *Heterophyes* *Metagonimus* *Echinochasmus*	Zoonotic species are distributed worldwide, especially Asia, Central Asia (India), Middle East.	Adult worms in fish-eating birds and humans and reservoir hosts such as dogs, cats, pigs, rats, various wild animals. Asexual multiplication occurs in snails which yields swimming cercariae that seek out intermediate hosts (usually fish). Cercariae penetrates gills, viscera, fins, scales and muscles to encyst (metacercariae). At least 45 genera of fresh and brackish water fish are susceptible, many of which are important in aquaculture.	For humans and reservoir hosts, consumption of raw, insufficiently cooked, pickled or smoked fish is most important risk. Parties accompanied by alcohol may be responsible for frequent gender difference in prevalence in humans. Use of human and animal waste for fish pond fertilisation, a high risk for fish infection.
Cestodiasis (tapeworms)			
FISHBORNE			
Diphyllobothrium spp.[e] (broad tapeworm); several species have been identified as zoonotic.	North and South America, and Eurasia.	Adult worms in intestines of wild fish-eating mammals and birds, especially dogs, bears, fur seals and sea lions, and humans. Copepods are the major first intermediate host for the first development stage, the procercoid. Ingestion by a fish releases the procercoid, which	Consumption of raw or insufficiently cooked, smoked, dried or pickled fish the major risk factor. Wild animal reservoirs ensure presence of these helminths in endemic areas and intrusion of humans into aquatic habitats increases exposure. Insufficient handling of human waste in wilderness areas is a

Table 27.2 (Cont'd)

Zoonoses and helminth species responsible	Major geographic distribution	Biology and transmission features	Risk factors
		invades the tissue and develops to the second stage (plerocercoid). Both marine and freshwater fish are important, especially pike, salmon, trout, ruff, white fish, and perch. When uncooked fish is eaten by mammal or bird host (depending upon species of *Diphyllobothrium*), the plerocercoid develops to adult stage in the intestine.	significant risk factor for both fish, animals and humans. Importation and stocking of fish may be significant factor in increasing spread.
MEATBORNE			
Taeniosis/ cysticercosis, neurocysticercosis *Taenia solium* (pork) *T. saginata* (beef)	Worldwide, especially in Asia, Africa, Latin America.	Adult tapeworms infect only the human intestine. Intermediate host: *T. solium* larvae (cysticercus) normally in pig muscle and viscera. However, if humans accidentally ingest eggs, may also serve as intermediate host with cysticerci in muscle, viscera and brain (neurocysticercosis). *T. saginata* larvae (cysticercus) only in cattle muscle.	Consumption of raw or insufficiently cooked pork or beef produces adult worm infection (taeniosis). Indiscriminate human defecation, access of pigs to latrines and use of sewage effluent for irrigation of crops and pastures are critical risk factors. Presence of people with intestinal adult tapeworm a major risk for transmission of eggs to other people and causing cysticercosis.
Nematodiasis (roundworms)			
FISHBORNE			
Anisakiasis *Anisakis simplex* (herring worm) and *Pseudoterranova decipiens* (cod worm)	Worldwide but especially important in Northern Europe, Japan, Korea, North America, and Pacific Islands.	Complex life cycles involving marine mammals (definitive hosts) such as dolphins, porpoises, and whales	Most important risk factor, as with flukes and broad tapeworm, is the consumption of raw, insufficiently cooked, salted, pickled

Table 27.2 (Cont'd)

Zoonoses and helminth species responsible	Major geographic distribution	Biology and transmission features	Risk factors
		(*Anisakis*) or seals, sea lions and walrus (*Pseudoterranova*). First intermediate hosts are marine crustaceans, in which early parasite larval development to the 3rd stage occurs. When eaten by a fish or squid (paratenic host) the larvae penetrate the intestine and invade the tissues. The parasites complete their development to adults in the intestines of marine mammals when infected fish or squids are eaten. However, in humans that eat raw intermediate hosts (fish, squids) the larvae may remain in the intestine (asymptomatic) or encapsulate in the stomach wall; sometimes penetrate the stomach wall and cause tissue pathology. Major fish host species are, herring, cod, mackerel, salmon, tuna, whiting, haddock, smelt and plaice.	or smoked fish or squid. Examples are traditional celebration and wedding dishes such as raw herring, lomi lomi, marinated salmon, sushi, sashimi, ceviche salad, sunomono are important risks.
MEATBORNE			
Trichinellosis			
Trichinella spp.[f]	*T. spiralis* is the most common zoonotic species. Distribution is cosmopolitan,	Life cycle for all is unusual in that all stages occur in one host. Most species infect a wide range	The major risk factor is ingestion of raw, insufficiently cooked, smoked or cured meat (pork, horse, many

Table 27.2 (Cont'd)

Zoonoses and helminth species responsible	Major geographic distribution	Biology and transmission features	Risk factors
	reported from nearly all areas where European domestic pigs have been imported. Other species are primarily sylvatic have more restricted geographical restrictions:[g] in North America, *T. spiralis*, *T. murrelli*, *T. nativa*, and *T. pseudospiralis* are endemic. In Eurasia, *T. spiralis*, *T. britovi*, *T. nativa* and *T. pseudospiralis* are widespread. In Africa, *T. britovi*, *T. nelsoni* and *T. zimbabwensis* (zoonotic?) are endemic. In Papua, *T. papuae* is found in pigs and reptiles (e.g. crocodiles), and humans.	of mammalian species (*T. pseudospiralis* can also infect birds). Adults in the intestine produce larvae that invade the circulatory system and filter out in striated muscle where they become intracellular (most but not all develop a collagenous capsule surrounding them). If muscle larvae ('trichinae') are eaten by a suitable host, larvae complete the cycle by developing to the adult stage in the small intestine.	game species). Outbreaks in humans are often associated with social events or seasonal celebrations (i.e. Christmas) when special ethnic dishes are prepared such as sausages. Large European outbreaks have occurred among ethnic groups eating raw or lightly cooked infected horse meat. Hunters often at risk from infection with sylvatic species when consuming game, especially wild boar and bear. Domestic pig and horse infections with *T. spiralis* caused by feeding uncooked meat products or animal waste, and to poor husbandry practices (rat infestation, exposure to sylvatic reservoir hosts).

[a]Liver infection sites; bile ducts, gall bladder, pancreatic duct.
[b]The systematics of *Paragonimus* is complex and controversial and has been recently revised (Blair *et al.*, 1999; WHO, 1995; Miyazaki, 1991).
[c]Lung: parasite locates usually in lung tissue and pleurocavity, sometimes in extrapulmonary sites such as the central nervous system.
[d]Intestinal flukes of the family Heterophyidae, commonly known as 'minute flukes', comprised many species, but the most important zoonotic ones are: *Heterophyes*, *Metagonimus* and *Haplorchis*. All have similar life cycles.
[e]This genus is comprised of at least 13 species, but the most important as a zoonotic risk are *D. latum* (Northern Hemisphere), *D. dendriticum* (sub-Arctic), *D. nihonkaiensis* (Japan) and *D. pacificum* (Pacific coast of South America) (Dick, 2007).
[f]Of the eight described species of *Trichinella*, only one (*T. zimbabwensis*) has not been yet reported from humans. From experimental studies domestic pigs are not very susceptible to species other than *T. spiralis*.
[g]See Murrell *et al.*, (2000) and Pozio and Murrell (2006) for review of taxonomy and geographic distribution.
Sources: WHO (1995); Mas-Coma and Bargues (1997); Miyazaki (1991); Murrell (2002); MacLean *et al.*, (1999); Cross (2001); Pawlowski and Murrell (2001). Adapted from Murrell and Crompton (2006), Table 9.2.

widely distributed over Europe, south-east Africa, the Americas, Oceania and Japan, and *F. gigantica* is endemic in Africa, Asia and Australia, although both species co-exist in some areas, notably south-east Asia (Miyazaki, 1991). Recent molecular studies have demonstrated that in some Asian areas (e.g. Japan) the two species hybridise (Mas-Coma *et al.*, 2007). Human fascioliasis has been reported from over 51 countries on five continents, affecting 2.4–17 million people (Mas-Coma *et al.*, 2007); WHO (1995) estimates that over 180 million people worldwide are at risk. In Africa, prevalence estimates for people in the Nile Delta are 18–19 % and in Bolivia, prevalence in communities residing in the Altiplano ranges from 70–90 % (Mas-Coma *et al.*, 2007). In Peru, there are estimates that as many as 750 000 people are infected (Kumar, 1999). Because detection of eggs by stool examination is difficult in people, it is not clear whether these data may be an underestimate (WHO, 1995). Livestock infections (cattle, goats and sheep) are common and the cause of serious economic loss (Mas-Coma and Bargues, 1997). Although human cases attributed to *F. gigantica* are fewer, infections may be common but unrecognised, especially in Asia, either because of a milder form of liver disease, or low availability of public health services (Mas-Coma and Bargues, 1997; De *et al.*, 2003).

The morphology and life cycle of both species of fasciolids are similar (Fig. 27.1). The morphological similarities in eggs make specific diagnosis from stool specimens problematical. The definitive hosts are mammals and, in addition to ruminants, dogs, cats, pigs and rodents may be infected as well as humans (Miyazaki, 1991). The intermediate host is a freshwater snail. When the eggs are deposited into water, they develop and release a ciliated miracidium which swims about until it encounters a suitable species of snail. Then it invades the snail's tissues and develops asexually to the redial stage, which in turn releases many free-swimming cercariae. The cercariae attach to the underside of water plants and encyst to form the infective metacercarial stage. When the raw contaminated plant is eaten by a suitable definitive host the metacercaria excysts in the intestine to release a minute fluke which penetrates the intestinal wall and migrates through the abdominal cavity. After migration through the liver parenchyma the growing fluke reaches the biliary system where it matures to an egg-laying adult. During migration the immature fluke may cause severe damage to the liver parenchyma, resulting in bleeding and scar formation (Miyazaki, 1991). When located in the bile duct, its mechanical and toxic effects may cause enlargement of the duct, thickening of the duct wall and degenerative lesions in the liver tissue. Chronic fascioliasis is generally considered asymptomatic or, when symptoms are reported, they are milder in form than that experienced during the acute phase. Anaemia is a prominent characteristic of chronic fascioliasis in heavy infections (WHO, 1995).

Fascioliasis in humans results from eating raw aquatic plants such as watercress (*Nasturtium officinale*) and other aquatic plants (e.g. parsley). Although eating water plants is a prime source of infection, metacercariae

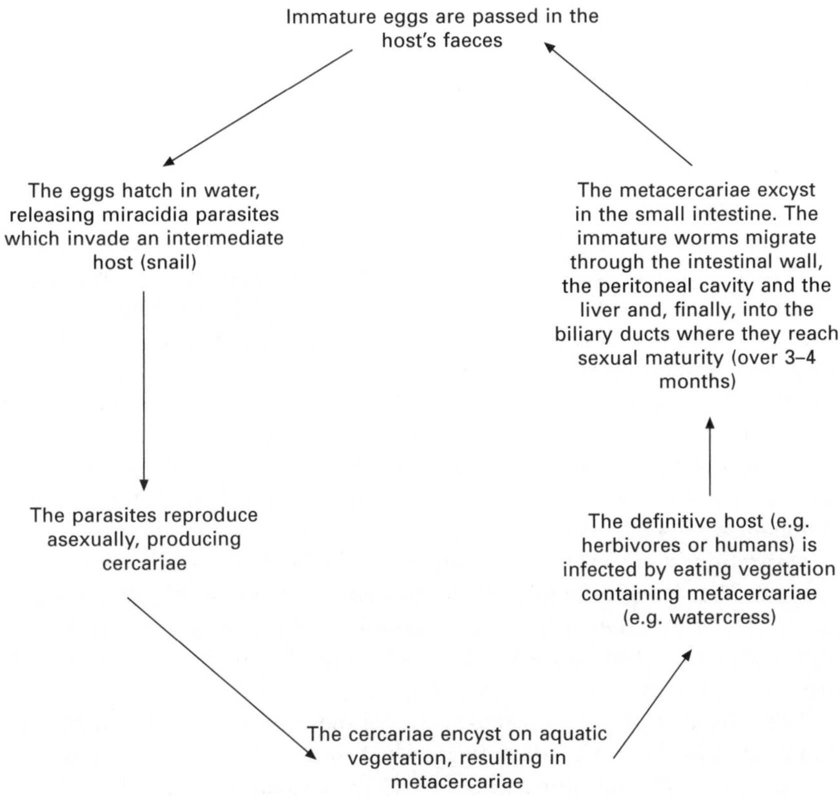

Fig. 27.1 Life cycle of *Fasciola hepatica* (the sheep liver fluke). (Adapted from Fig. 9.2 Murrell and Crompton, 2006)

will float in water if dislodged from plants, therefore water used for drinking and washing vegetables and utensils becomes a source of infection (Mas-Coma and Bargues, 1997). Metacercarial contamination of cutting boards and other kitchen utensils may also put humans at risk of infection. Some evidence suggests that people may acquire the immature worms directly by eating raw intestine and liver of cattle (Miyazaki, 1991). Infection often shows familial clustering within communities and, although generally there is little bias towards sex or age (Chen and Mott, 1990), there are instances in which children exhibit significantly higher prevalence, and may be the major source of egg contamination in the environment (Esteban *et al.*, 1997; Mas-Coma and Bargues, 1997).

The main reservoir hosts in the epidemiology of fascioliasis are sheep and goats; cattle are also important, particularly for *F. gigantica*, but eventual acquisition of immunity to infection may render them less important in sustaining endemicity (Mas-Coma and Bargues, 1997). The sources of infection for livestock include contaminated fodder such as rice, straw, hay and aquatic plants on pasture (Miyazaki, 1991). As with the fishborne helminths, the deliberate fertilisation of farm ponds with human and animal faecal waste creates another major risk factor.

27.2.2 Fishborne helminths

The liver flukes (Trematoda)

The major liver flukes, *Clonorchis sinensis*, *Opisthorchis viverrini* and *O. felineus* are closely related and share many biological traits (Table 27.2). *Clonorchis sinensis*, the Chinese liver fluke, is widely distributed in East Asia. The number of currently infected people in Korea is estimated at about 1.5 million (Chai *et al.*, 2005b). In a nationwide survey in China, the prevalence of *C. sinensis* was 0.4 % among almost 1.5 million people examined and, based on this, the number of infected people in China may be about 6 million (Xu *et al.*, 1995). In Viet Nam, clonorchiasis has been endemic mainly in the north, especially along the Red River Delta including Haiphong and Hanoi (Rim, 1986; De *et al.*, 2003).

Opisthorchis viverrini is prevalent in Thailand, Laos, Cambodia and southern Viet Nam, affecting 10 million people or more (Yossepowitch *et al.*, 2004). *Opisthorchis felineus* is a natural parasite of dogs, cats, foxes and pigs in eastern and south-eastern Europe and the Asiatic parts of Russia, but there are few prevalence data available.

Typically this trio of trematodes live in the biliary tract of humans and domestic animals and produce eggs which pass out through the common bile duct into the intestines (Fig. 27.2). The metacercaria, which is the infective stage for humans and other mammals, can be found in more than 100 species of fresh water and brackish water fish belonging to 13 families, especially the Cyprinidae which are important in aquaculture (WHO, 1995). The mode of transmission to the definitive host is through consumption of raw, undercooked or improperly pickled or poorly smoked infected fish. In the definitive host, the metacercariae excyst in the duodenum/migrate to the common bile duct and then to the extrahepatic and intrahepatic bile ducts. The metacercariae grow to the adult stage in about four weeks after infection (Rim, 1986). The hepatic lesions and clinical manifestations in infected people are similar for all the liver fluke infections. Complications such as pyogenic cholangitis, biliary calculi, cholecystitis, liver cirrhosis, pancreatitis and cholangiocarcinoma are often associated with infection (Sripa, 2003). High incidences of cholangiocarcinoma, based on both necropsy and liver biopsy data, have been reported for *O. viverrini* in north-eastern Thailand, where overall cholangiocarcinoma is estimated to be 129 per 100 000 in males and 89 per 100 000 for females, compared to 1–2 per 100 000 in western countries (Vatanasapt *et al.*, 1990). The severity of the pathology is associated with both intensity and duration of infection and the location of the lesions (Rim, 1986).

The major risk for acquiring liver flukes in endemic areas is related to

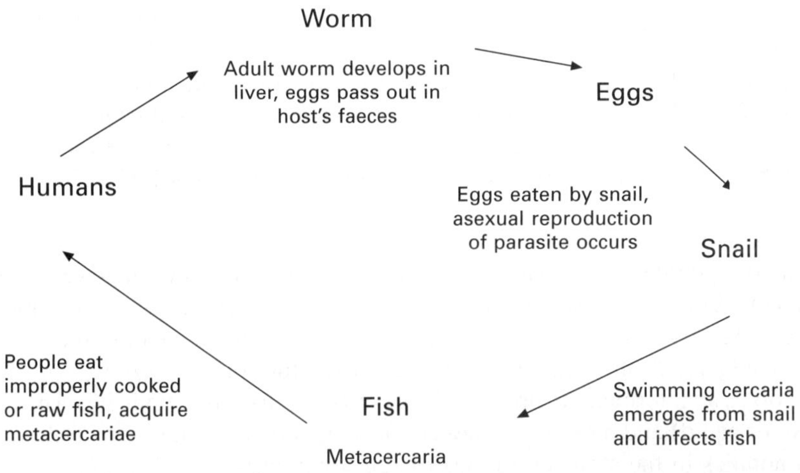

Fig. 27.2 Life cycle of *Clonorchis sinensis* and fishborne liver flukes.

the human custom of eating raw fish. The morning congee (rice gruel) with slices of raw freshwater fish (southern China and Hong Kong) or slices of raw freshwater fish with red pepper sauce (Korea) are examples of major dietary sources of *C. sinensis* infection. In north-eastern Thailand and Laos, it is well established that '*Koi pla*', the most popular raw fish dish, is an important food source of infection with *O. viverrini*. The major risk factor for infection of fish is the exposure of water bodies containing susceptible species of snail to faeces from infected humans and reservoir hosts. In many countries, because of the cost benefits to aquaculture, human and animal faeces are utilised as pond fertiliser.

There are characteristic patterns of age and sex prevalence among residents of endemic areas. The infection rates are generally higher in men than in women, and higher in adults than in children (Rim, 1982; Sithithaworn and Haswell-Elkins, 2003). This probably reflects the cultural behaviour pattern of men, who more often gather together for dinners of raw or pickled fish (usually accompanied by alcohol) than to any biological gender differences. It should be noted, however, that this pattern is changing leading to increasing cases among women. Chronic exposure to liver flukes apparently does not induce immunity to reinfection, and worm accumulations can be very large.

The role of reservoir hosts, especially cats, dogs and pigs, in maintaining liver fluke transmission is not clear, and conflicting evidence has been reported (reviewed in Chai *et al.*, 2005b). In some areas, infection may be high among people and low among domestic animals and *vice versa* (China), but in other endemic areas reservoir hosts may also have comparable infection rates (WHO, 1995; Mas-Coma and Bargues, 1997; De *et al.*, 2003). This issue is of importance because the role of reservoir hosts may have a decisive bearing on the outcome of mass drug control programmes targeted to human communities (WHO, 2004). The complex set of factors that may expose people to the risk of infection by fishborne trematodes is set out in Table 27.2. People must be protected from the many ways in which infective trematode metacercariae (sizes range around 1 mm) may enter the food chain. Further information is to be found in Crompton and Savioli (2007).

Tissue nematodes

Anisakis simplex and *Pseudoterranova decipiens* are representative species of nematode from fish that may become established as tissue- or organ-invading larvae in humans who have eaten contaminated raw, pickled or smoked fish (Table 27.2). The complex life history of *A. simplex* (Fig. 27.3) involves an intermediate host (euphasid crustacean), a paratenic host (marine fish or squid) and a definitive host (marine mammal). Adult nematodes live in nodules in the stomach linings of seals and porpoises (1, 2 and 3 in the figure). Eggs are passed into the sea (4) and the embryos develop to release second-stage larvae (5) which change into third-stage larvae after being eaten by crustacean intermediate hosts (6). When infected crustaceans are eaten by fish, the third-stage larvae *encyst* in the fish tissues without further development. Muller (2002) states that 164 species of marine fish can serve as paratenic hosts for *Anisakis*. In the natural course of events, predatory marine mammals acquire *Anisakis* by catching infected fish (8, 9 and 10) and

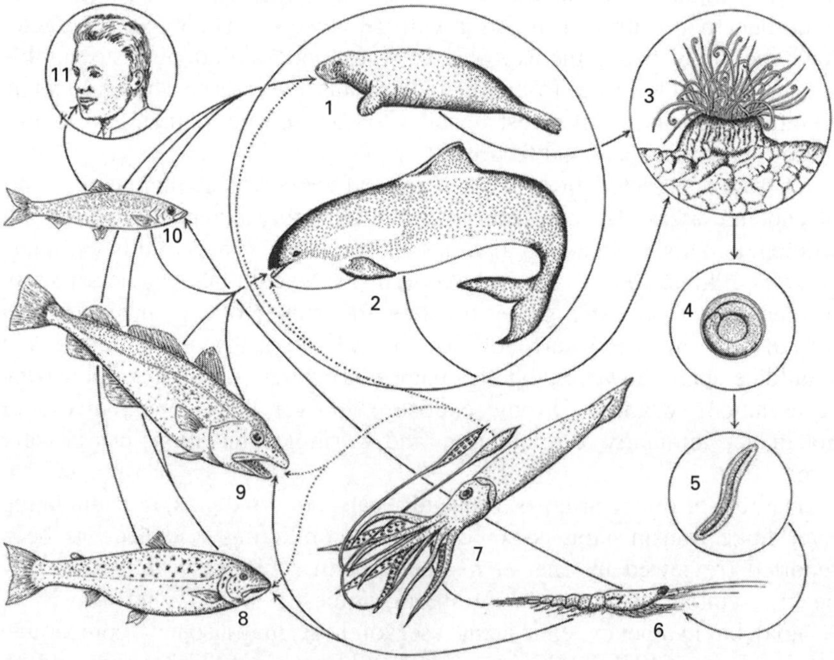

Fig. 27.3 Life cycle of *Anisakis simplex*. (From Fig. 6.3) Williams and Jones, 1994, with permission from Taylor & Francis Group LLC.

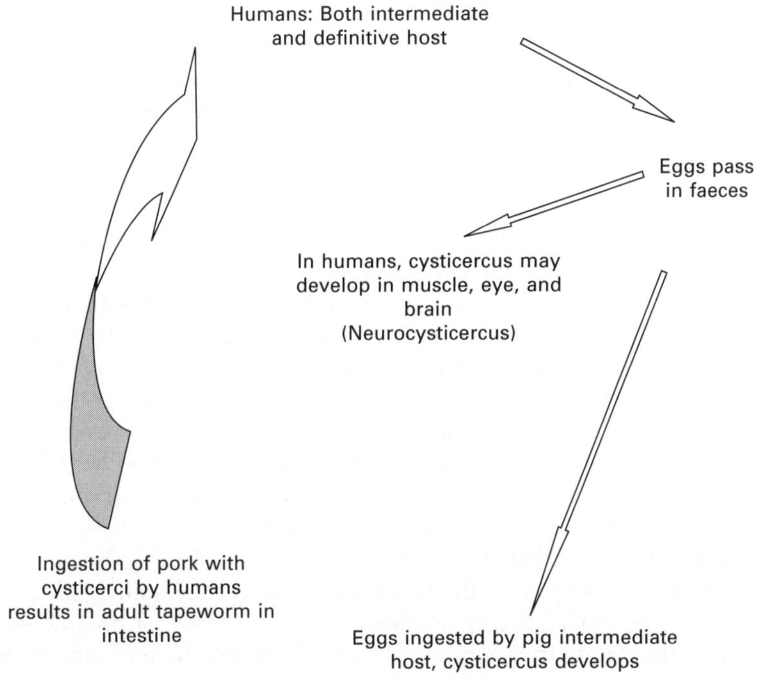

Fig. 27.4 Life cycle of *Taenia solium*.

the helminth's life cycle is completed (Williams and Jones, 1994). The life cycle of *Pseudoterranova decipiens* (Table 27.2) is similar. When humans eat infected fish harbouring live third-stage larvae, the larvae initially become active but do not attain adult development. Then they congregate mainly in the gastrointestinal mucosa, die and so induce the formation of abscesses (Gutierrez, 2006).

Anisakiasis is more prevalent in Japan than elsewhere because raw marine fish is a popular dietary item, but anisakiasis and pseudoterranoviasis remain as a health risk wherever marine fish is eaten unless properly cooked or deep frozen first. Although the gastrointestinal tract is the common site of the lesion, the larvae may invade other organs. There is a marked eosinophilia, oedema and inflammation, abdominal pain, nausea and vomiting, diarrhoea, constipation and blood in the stools may be seen. Diagnosis is difficult without access to a modern hospital and surgical procedures may be required to remove the worms. There is no medical treatment (Gutierrez, 2006).

27.2.3 Meatborne helminths

Taeniosis/Cysticercosis (Cestoda)

The terms cysticercosis and taeniosis refer to foodborne zoonotic infections with larval and adult tapeworms, respectively. The important feature of this zoonosis is that the larvae are meatborne (generally beef or pork) and the adult stage is an obligate parasite of the human intestine (Fig. 27.4) (Murrell, 2005). *Taenia saginata* ('beef tapeworm'), *T. saginata asiatica* ('Taiwan Taenia') and *T. solium* ('pork tapeworm') are the most important causes of taeniosis in humans (Flisser, 2006). When humans ingest meat with a cysticercus, the cyst evaginates and begins to grow a large series of proglottids, each of which can produce large numbers of eggs that reach the environment in the faeces. Cysticercosis is a tissue infection with the larval stage (cysticercus or metacestode) acquired when animals ingest the eggs, most commonly in pigs and cattle. *Taenia saginata* occurs only in beef, *T. asiatica* in pig organs and *T. solium* in pork. Humans, however, are unique in that they may also serve as an intermediate host for *T. solium*, in which the cysticercus stage may develop. When one of these animal species, or humans in the case of *T. solium*, ingests an egg, an onchosphere is released which penetrates the intestinal tissue and enters the bloodstream where it circulates until filtering out into striated and cardiac muscle, the eye or the central nervous system (CNS) and develops into an encysted cysticercus.

There is consensus that from the standpoint of disease in humans (and pigs) and the maintenance of the tapeworm's life cycle, adult *T. solium* are of primary importance. The epidemiology of *T. solium* is of broad concern because of the helminth's public health significance. *Taenia solium* has a cosmopolitan distribution and is highly endemic in Latin America, Africa and Asia where poverty, poor sanitation and intimate contact between humans and livestock are commonplace. It has been estimated that 2.5 million persons worldwide are infected with adult *T. solium* (carriers) and that 20 million are infected with cysticerci (Bern *et al.*, 1999). Consequently, human carriers (egg shedders) should be a major target for initial control efforts (see Section 27.4).

Infection with *T. solium* is of greatest concern where cysticerci invade the CNS thereby giving rise to neurocysticercosis. One estimate cites 50 000 human deaths each year from neurocysticercosis (Mafojanee *et al.*, 2003). This zoonosis is now receiving greater attention in sub-Saharan Africa because of the growing recognition of the importance of neurocysticercosis as a cause of epilepsy. The syndrome has become the subject of a global public health campaign ('out of the shadows') in less developed areas where neurocysticercosis is considered to be the leading cause of epilepsy (Diop *et al.*, 2003). The 56th World Health Assembly has issued a report on the Control of Neurocysticercosis, stating that cysticercosis of the CNS is the most important neurological disease of parasitic origin in humans. The rapid expansion of smallholder pig production in Africa and elsewhere has led to a significant increase in cysticercosis in pigs and humans, presenting governments with an important challenge as they seek to increase livestock production and rural incomes.

The other two species, *T. saginata* and *T. asiatica*, cause only taeniosis (adult tapeworm stage) in humans and the clinical problems are mainly minor, other than being a nuisance. Human carriers are also, however, responsible for infections in cattle (bovine cysticercosis) which can have an economic impact because of meat condemnations at slaughter house inspections.

Important risk factors in the epidemiology of *T. solium* cysticercosis in pigs include:

- extensive or free-range pig rearing in households lacking latrines and outdoor human defecation near or in pig rearing areas – this includes connecting pig pens to human latrines ('pigsty privies') and deliberate use of human faeces as pig feed;
- use of sewage effluent, sludge or 'night soil' to irrigate and/or fertilise pig pastures and food crops.

The risk factors important to *T. solium* cysticercosis in humans include:

- presence of an infected person(s) in a household – clustering of cases is commonly observed; history of passing proglottids by a member of a household or a member of the community in frequent contact with the household;
- household or community food handlers and child care givers (carriers) are potentially very high risk factors;
- cultural preferences for eating raw or improperly cooked pork.

Trichinellosis

Humans and other mammals become infected with any of seven species of the nematode *Trichinella* (Table 27.2) through the ingestion of larval stages present in muscle (Pozio and Murrell, 2006; Crompton and Savioli, 2007; Dupouy-Camet and Murrell, 2007). The larvae (newborn larvae) are released by adult worms living in close association with the definitive host's small intestine mucosa. The larvae invade the host's tissues, but can only develop to the infective stage in striated muscle cells. *Trichinella* species such as *T. spiralis*, *T. nativa*, *T. britovi*, *T. murrelli* and *T. nelsoni* stimulate the development of a surrounding collagen capsule (called a nurse cell), in which they may remain infective for the next host for months to years, before dying and becoming calcified. Two zoonotic species (the so-called non-encapsulated forms), *T. pseudospiralis* and *T. papuae*, are distinct in that they remain un-encapsulated in the muscle cell. Details of the biology, epidemiology and distribution of *Trichinella* are to be found in Pozio and Murrell (2006).

Two patterns are recognised in the life cycle of *Trichinella spiralis*; the synanthropic cycle which commonly brings humans and domestic animals (e.g. pigs, horses, dogs, cats) into contact with the helminth and a sylvatic cycle involving wild animal reservoirs (Fig. 27.5). The sylvatic species of *Trichinella* infect a wide variety of wild mammals, including walruses, seals, bush pigs, wart hogs, bears, dogs, badgers and cougars. Although *T. spiralis* can infect a wide variety of domestic and wild animals, it is the only species which can infect significantly pigs and rats, and, therefore, it is the major cause of human trichinellosis worldwide; the sylvatic species are most commonly transmitted to humans through consumption of wild game, although *T. pseudospiralis* has been documented as the cause of human outbreaks from feral pigs in Thailand and Russia. The remarkably broad host range that characterises *Trichinella*, especially *T. spiralis*, means that humans will always be at risk of infection in places where (i) statutory requirements for meat inspection are lacking, (ii) undercooked meat from whatever source is eaten and (iii) standards of animal husbandry and food hygiene are inadequate.

Trichinellosis was a major public health problem in Europe until meat inspection was introduced into slaughterhouses, thereby protecting the public from contaminated pork and pork products. Trichinellosis is not confined, however, to developing countries. Outbreaks have occurred in France and Italy since 1975 through the consumption of infected horse meat (Pozio and Zarlenga, 2005). Horses, being herbivores, might not have been expected to become infected (Fig. 27.5), but deliberately supplementing their feed with meat products would explain how transmission began (Murrell *et al.*, 2004).

Trichinellosis has an enteral (intestinal) phase and a parenteral (tissue) phase (Capo and Despommier, 1996; Bruschi and Murrell, 2002). Adult worms in the intestine may initiate diarrhoea, nausea, vomiting, fever and abdominal pain; the extent and severity depends on the number of worms in association with the intestinal mucosa. Symptoms and morbidity during the tissue phase also depend on the number of migrating larvae and the number that establish in muscle and other tissues including the brain. Muscle pain (myalgia), facial aedema, headache, skin rash, convulsions, weight loss,

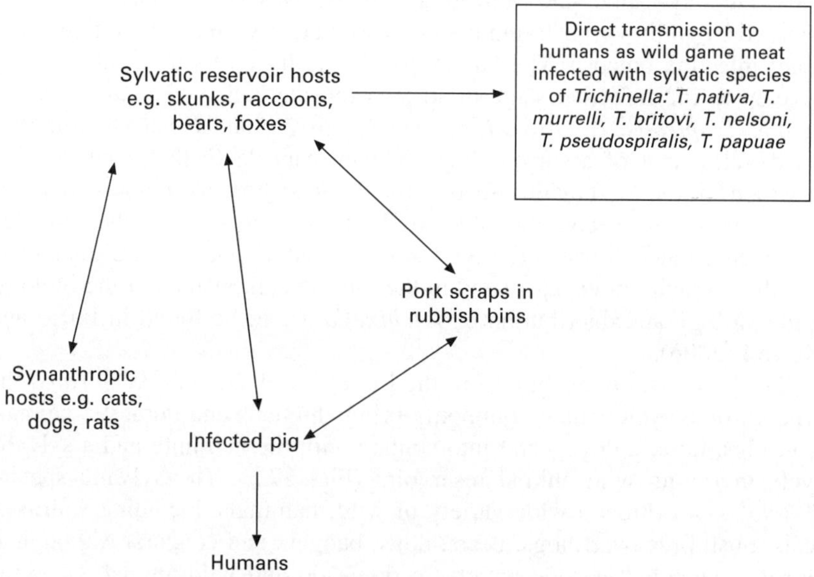

Fig. 27.5 Life cycle of *Trichinella spiralis*.

meningitis, encephalitis, vertigo are some of the complications attributed to *Trichinella* larvae during the early parenteral phase (Crompton and Savioli, 2007; Dupouy-Camet and Bruschi, 2007). Death may occur in very heavily infected individuals. Takahashi *et al.*, (2000) reported on over 500 outbreaks of trichinellosis in China affecting 20 000 people of whom over 200 died. Pigs abound in many rural villages in China.

27.3 Helminth detection and diagnosis

Different approaches to the detection and identification of foodborne helminths are required to meet the needs of the different groups concerned with prevention, control and treatment. In choosing appropriate technologies, clinicians, public health officials, commercial food producers and researchers are advised to consider the following points:

1. Information about locally prevailing foodborne helminth infections maintained by Ministries of Health, at hospitals or recorded in published biomedical literature will be useful.
2. Similar information should be sought by those responsible for the importation of food products from countries where the infections are suspected of being endemic.
3. The long-standing experience of local health professionals is a valuable resource for detection of foodborne helminth infections.
4. Detailed knowledge of the life histories, transmission routes and host ranges of the infections should be reviewed.
5. Familiarity with the accuracy and reliability of diagnostic techniques should be obtained and estimates of their cost-effectiveness in relation to available resources should be made.
6. If practical, application of more than one method of detection, such as comparing the results of stool examinations with the interpretation of clinical signs and symptoms, is desirable.
7. Databases concerning the detection of foodborne helminth infections should be established, maintained and placed in the public domain.

Basic information about the detection of a selection of major foodborne helminth infections is given in Table 27.3. In most cases, the collection and microscopic examination of stool samples remains the most realistic method of detection, particularly for public health programmes. For tissue infections such as anisakiasis, cysticercosis and trichinellosis, however, these methods are not relevant. The examination of stool specimens for eggs is not free from problems and pitfalls. Most importantly, failure to find eggs in a stool sample does not mean that the subject is free from infection and should not be assumed to be a negative result. If the infection is not patent (reproductive development of the helminth is not yet at the point when eggs are released), eggs will not be found. In patent infections consisting of very few worms,

Table 27.3 Detection and diagnosis of major foodborne helminth infections in humans

Zoonotic parasites	Detection and diagnosis methods
TREMATODA (flukes)	
Liver flukes	
PLANTBORNE *Fasciola hepatica* *F. gigantica*	Examination of stool samples for eggs. Difficult to distinguish eggs of *F. gigantica* from *F. hepatica* although PCR[a] can differentiate species. ELISA[b] useful. Clinical diagnosis by imaging methods much improved.
Lung flukes	
CRUSTACEANBORNE *Paragonimus* spp.	Detection of eggs in sputum and/or in stool samples. History, radiology, and serology important for diagnosis; various ELISA tests and immunoblotting valuable. Metacercariae in crustacean hosts can be isolated and identified morphologically. Haemoptysis may indicate tuberculosis; diagnostic confusion is problematic (Toscano *et al.*, 1995).
FISHBORNE *Clonorchis sinensis* *Opisthorchis viverrini* *O. felineus*	Detection of eggs in stool samples or in biliary drainage. Difficult to speciate morphologically eggs of these liver flukes, and to differentiate them from small intestinal flukes (Heterophyidae). PCR can differentiate *C. sinensis* eggs from those of *Opisthorchis* spp. Serology, especially ELISA, is useful provided cross reactivity with other species is considered. Metacercariae of liver flukes in fish can be isolated by pepsin digestion and differentiated by either morphology or PCR.
Intestinal flukes	
PLANTBORNE *Fasciolopsis buski*	Detection of eggs in stool samples. Eggs of *F. buski* may be confused with those of *Fasciola* spp.
FISHBORNE Heterophyidae; Echinostomatidae (many species)	Detection of eggs in stool samples using concentration techniques effective. Heterophyid eggs difficult to speciate and to differentiate from *Clonorchis/Opisthorchis*. Metacercariae easily isolated from fish and can be identified morphologically, and to species in some cases.
CESTODA (tapeworms)	
FISHBORNE	Coprological exams for characteristic eggs and proglottids. PCR useful for differentiating adults as adjunct to morphological methods. Inspection of fish for larval stage (plerocercoid) effective, but speciation morphologically difficult; some advances in developing molecular techniques.

Table 27.3 (Cont'd)

Zoonotic parasites	Detection and diagnosis methods
MEATBORNE	
Adult *Taenia* spp. infections (taeniosis)	Detection of eggs in stool samples indicates presence of *Taenia* but does not identify species. Coproantigen assays highly sensitive for taeniosis, but not specific for *Taenia* species. Multiplex PCR can differentiate parasite DNA in stools. If proglottids present in stools they can be speciated by morphological features.
Cysticercosis (*T. solium*)	Neurocysticercosis requires MRI[c] or CT[d] scans to confirm. ELISA or Western Blots highly effective in detecting multiple cysticerci infections, but single cyst infections in CNS difficult. Organoleptic meat inspection of pigs and cattle for infection has low sensitivity; serological methods promising, but only used for epidemiological studies at present.
NEMATODA (roundworms)	
FISHBORNE	
Anisakis simplex *Pseudoterranova decipiens*	Parasitological diagnosis is difficult because disease mimics other gastric infections. History, clinical exam and detection of worms by endoscopy or in biopsy samples necessary for definitive diagnosis. Serology problematical because of cross-reactivity. Antigen-specific IgE assay promising. Inspection of fish standardised in many countries, and relies either on observation by candling of fish fillet or recovery of larvae by digestion.
MEATBORNE	
Trichinella spiralis; sylvatic species	Enteral phase: Difficult to diagnose, mimics many other conditions (flu, bacterial enteritis, etc.). History, clinical signs and symptoms important. Muscle phase: Hypereosinophilia, facial edema characteristic. Serology with ELISA, IFA[e] very sensitive and specific. Muscle biopsies usually unnecessary except to isolate larvae for identification. Multiplex PCR effective in identification of all stages of the parasite. Meat inspection by digestion of animal muscle to recover larvae.

[a]Polymerase chain reaction.
[b]Enzyme-linked immunoabsorbent assay.
[c]Magnetic resonance imaging.
[d]Computerised tomography.
[e]Indirect fluorescence antibody (test).

the numbers of eggs may be too few to be observed unless concentration methods have been used in the processing of samples before microscopic examination. The reliability of any diagnostic technique depends mainly on the skill and dedication of technical staff; eggs in stool preparations may be missed when tiredness sets in. In large-scale programmes, stool samples may be wrongly identified or even substituted with samples from people outside the programme in an attempt to satisfy those in charge. A helpful critique of problems related to helminth detection, including methods relevant to tissue helminths, is given in Mas-Coma and Bargues (1997) and recent Guidelines issued by WHO and FAO (Murrell, 2005; Dupouy-Camet and Murrell, 2007).

The WHO-recommended method for the examination of stools is the Kato Katz technique (WHO, 1994; Montresor *et al.*, 2002). This procedure has been tested worldwide and has the advantage of facilitating egg counts (expressed as epg, eggs per gram) (Chai, 2007). Egg counts provide an indirect measure of the intensity of infection; generally the higher the epg, the greater the number of worms present. There is now wide agreement that estimates of intensity provide information about the likelihood and severity of morbidity and the rate of transmission (see Anderson and May, 1992). The Kato Katz technique readily lends itself to quality control whereby an independent observer can examine a selection of randomly chosen slides to see if the results tally with those of the first observer.

Three other points should be noted regarding the detection and identification of foodborne helminth infections. First, reliance on clinical signs and symptoms alone may be misleading. Paragonimiasis has often been misdiagnosed as tuberculosis (Toscano *et al.*, 1995). Secondly, finding eggs in stool samples may not allow identification. The assemblage of species of foodborne intestinal trematodes is extremely difficult to differentiate on the basis of egg morphology (Chai *et al.*, 2005a). Thirdly, detection becomes much more difficult and time consuming when the investigation of intermediate and reservoir hosts is necessary (see De *et al.*, 2003).

Diagnosis of liver fluke infections can be made by the recovery of eggs from faeces. However, the eggs must be differentiated from those of other trematodes, especially liver and intestinal flukes, a task which requires considerable training and experience, and even then the general lack of specific diagnostic tools such as molecular probes presents a challenge (Lee *et al.*, 1984; Ditrich *et al.*, 1992). Serological tests such as ELISA using excretory–secretory antigens are helpful in some cases (Sithithaworn *et al.*, 2007; Thompson *et al.*, 2007). Significant advances have been made in utilising refined parasite antigens, especially proteases, for immunological tests (see Maruyama and Nawa, 2007 for review). Serological tests for tissue helminths such as *Trichinella*, *Paragonimus* and *T. solium* cysticercosis and *Fasciola* spp. are, in comparison with liver and intestinal fluke infections, quite reliable (Dorny *et al.*, 2005; Mas-Coma *et al.*, 2005; Blair *et al.*, 2007; Dupouy-Camet and Bruschi, 2007).

Molecular detection methods, particularly DNA-based approaches, are also now becoming more commonplace (Table 27.3) (Wongratanacheewin *et al.*, 2003). Progress in this area is encouraging, for example a polymerase chain reaction (PCR) based technique to detect snail and fish infections with *O. viverrini* has been developed (Maleewong *et al.*, 2003). PCR tests for parasite DNA in faecal samples are being reported and evaluated for liver, lung and intestinal helminths such as flukes and tapeworms (reviewed by Pawlowski *et al.*, 2005; Traub *et al.*, 2005; Thompson *et al.*, 2007). Molecular methods have become especially important in the detection and identification/typing of helminth species among such genera as *Taenia*, *Diphyllobothrium*, *Trichinella*, *Fasciola*, *Paragonimus* and *Anisakis* (Traub *et al.*, 2005; Thompson *et al.*, 2007). Molecular methods will, undoubtedly, play a critical role in the future in improving the ability to detect infective stages of zoonotic helminths in food inspection activities. There is a particular requirement for methods with applicability to inspection needs for high-throughput food processing such as farmed fish at risk of infection with zoonotic flukes and nematodes (Table 27.3).

27.4 General approaches to prevention and control

Prevention and control are challenging tasks, and years of dedicated effort will be needed if foodborne helminth infections are to be eliminated and the ensuing public health benefits sustained. There are seven principles that in theory can underpin successful prevention and control.

1. Government commitment and political will, including assigning foodborne infections high priority for public health attention.
2. Advocacy to encourage donors and agencies outwith endemic countries to support prevention and control campaigns.
3. Health education and awareness programmes to improve individual and community food hygiene.
4. Health education and training programmes for people concerned with all aspects of food production including animal husbandry and aquaculture at both domestic and commercial levels.
5. Access for people in endemic areas to free diagnosis and anthelminthic treatment.
6. Provision and development of a legally enforceable food inspectorate dealing with all aspects of food production on a commercial basis.
7. Strengthening of veterinary services to increase efforts to control infections in domesticated animals.

This ideal approach raises many problems. First, since foodborne helminth infections are mainly prevalent in low-income countries, prevention and control measures must compete for funds against many other needs including HIV/AIDS, malaria, TB and so on. Secondly, prevention and control measures must not threaten the food security of poor people. Thirdly, prevention and control measures must be sensitive to the cultural traditions and values of the people in need of health care.

27.4.1 Specific control approaches for animals and fish

Several preventive measures recommended for food preparation in the home and restaurant are set out in Table 27.4. Generally, the thermal death points (cooking, freezing) sufficient to kill nematode helminths (e.g. *Trichinella*) are adequate for inactivating trematodes and cestodes such as *Clonorchis* and *Taenia*. The essential measure is the avoidance of eating raw or inadequately cooked, pickled or cured animal and plant food, particularly in regions where these helminths are endemic. Details on safe food preparation methods are normally available from public health agencies.

Although food animals collected from the wild (e.g. crabs, game, marine fish) are not subject to the preventive measures that can be applied in aquaculture and livestock farming, the detection and inactivation of helminths in infected fish and game animals is possible (and required in many countries, e.g. for *Anisakis/Pseudoterranova* and *Diphyllobothrium* in marine fish).

Specific measures have been developed and adopted in many countries (e.g. European Union and USA) for detection and decontamination of fish (both wild and farmed), either imported or consumed locally. Among the methods allowed for parasite decontamination are retorting, cooking, pasteurisation, freezing, brining/pickling and physical removal. These procedures are detailed in government manuals and can be seen online, for example, the US Food and Drug Administration Center for Food Safety & Applied Nutrition's Bacteriological Analytical Manual *Online* (Chapters 5 and 19, decontamination and detection, respectively) (http://www.cfsan.fda.gov/~ebam/bam-toc.html). The UK has promulgated similar regulations in The Food Safety (Fishery Products and Live Shellfish) (Hygiene) Regulations, 1998 (http://www.Legislation.hmso.gov.uk/si/si1998/80994-jj.htm).

In many countries, in which zoonotic helminth infection of food animals such as cattle and pigs may be detected through inspection at slaughter, procedures have been approved to allow infected meat to be rendered safe for human consumption. For example, beef lightly contaminated with *T. saginata* cysticerci can be treated by freezing, boiling or pickling. Similarly, pork contaminated with *T. solium* cysticerci can be heated to 80 °C (Murrell, 2005). Pork with *T. spiralis* larvae (trichinae) can be rendered safe for consumption and export by freezing, following strict protocols (detailed in Dupouy-Camet and Murrell, 2007, Chapter 3).

27.4.2 Irradiation methods

Although the amount of research carried out on irradiation for the control of foodborne helminth parasites is limited, compared to that for microbial

Table 27.4 Approaches to prevention and control of major foodborne helminthiasis

Transmitted by	Prevention/control of infection of food sources[a,b]	Prevention of human infection
AQUATIC PLANTS		
Fasciola, Fasciolopsis	Safe disposal of human and animal faeces. Safe preparation of domestic animal plant feed. Control of snail vectors.	Prepare plant foods using methods to eliminate/kill infective larvae. Boil or purify drinking water. Treat infected human and domestic animals associated with the household to break life cycle of parasite.
FISH		
Liver and intestinal trematodes Tapeworms (*Diphyllobothrium*) Nematodes (*Anisakis, Pseudoterranova*)	In aquaculture, control of snail vectors and prevention of contamination of farm ponds with human and animal waste; faecal waste may contain eggs of flukes. Drug treatment of infected people and domestic animals associated with fish farm recommended. Inspection of both farmed and wild caught fish. Application of appropriate processing treatments in the freezing, pickling and smoking/curing of fish.	Home preparation: Cook fish to temperature sufficient to kill parasites, normally 70 °C for 15 min. Freeze fish (eg –25 °C for 7 d). Pickle in 20 % brine for at least 10 d. For *Anisakis* and *Pseudoterranova*, prompt evisceration of marine fish after capture is critical.
CRUSTACEANS		
Lung fluke (*Paragonimus*)	Prevent contamination of wild water with human and domestic animal faecal waste.	Thorough cooking of fresh-water crabs and wild boar, pork or chicken meat.
MEAT		
Meatborne tapeworms *Taenia solium*	Prevent exposure of pigs to human faeces. Effective meat inspection for cysticerci.	Cooking of pork to 60 °C, or freezing to –25 °C for 10 d. Treatment of infected households to stop dissemination of eggs to pigs and to other humans.[c]
Taenia saginata	Prevention of cattle infections through exposure to human faeces. Effective meat inspection for cysticerci.	Cook beef to 60 °C or freeze to –25 °C for 10 d.
Nematode *Trichinella spiralis*[d]	Prevent pig access to uncooked meat of any type. If outdoor rearing, prevent	Cook pork, game and horse meat to 70 °C, or freeze at –23 °C for 10 d.

Table 27.4 (Cont'd)

Transmitted by	Prevention/control of infection of food sources[a,b]	Prevention of human infection
	exposure to domestic and wild animals, especially rats. Inspect pork, game and horse meat for presence of muscle larvae.	Cure or smoke meat according to government regulations. Discard household waste carefully.

[a]For plantborne parasites (e.g. *Fasciola*), plant may function as an 'intermediate host' in transmitting the larval parasite to the definitive host.
[b]These measures are not effective for parasitic zoonoses in which a wild animal reservoir may be involved (e.g. fish-eating birds, marine mammals, sylvatic mammals). Human consumption of wild caught fish, crustacean, mammals, etc. should assume a high risk infection is present and cook, freeze and cure the flesh according to public health recommendations.
[c]Humans, like pigs, can serve as an intermediate host for *T. solium*; this is not the case for *T. saginata* in which humans are only a definitive host (intestinal adult tapeworm).
[d]Although *T. spiralis* is still the leading cause of human trichinellosis worldwide, several other *Trichinella* species are responsible for infections through improperly cooked/frozen/cured game and horse meat (see Fig. 27.5).

organisms, enough data has been generated to clearly show that dose levels below 1kGy are sufficient to kill meat and fishborne parasites such as *Trichinella*, *Taenia* and *Opisthorchis* (Farkas, 1987; Loaharanu and Murrell, 1994). The data obtained by Brake *et al.*, (1985) led to approvals by the US FDA and the US Department of Agriculture for use of irradiation to destroy *T. spiralis* in pork intended for human consumption and export (FDA, 1985). To date, this method for the control of trichinellosis has not been adapted commercially, primarily because of consumer resistance. Preliminary work has been reported for other parasites, particularly *Anisakis*, but the data are not sufficient to determine the efficacy of irradiation as a potential control technology.

27.4.3 Pre-harvest control approaches

Preventive measures for zoonotic helminths at the animal production end of the food chain have been enhanced by the greatly expanded efforts to develop measures for foodborne microbial infections in livestock. The application of Hazard Analysis Critical Control Point (HACCP) procedures for the latter objective has had an influence on the design of parasite infection prevention, particularly the safe disposal of human and animal waste which will reduce exposure of intermediate host animals to infection (see for example, Figures 27.1, 27.2 and 27.3) which can transmit the parasite to humans. Guidelines for the surveillance, prevention and control of the meatborne helminths such as *Taenia* and *Trichinella* have recently been issued by WHO, FAO and OIE (Murrell, 2005; Dupouy-Camet and Murrell, 2007), and these should be consulted for more detailed information on prevention at the production level. These Guidelines also describe the meat inspection procedures, which many countries mandate.

Public health control for foodborne helminths such as the liver and intestinal flukes is increasingly utilising mass application of anthelminthic drugs in human populations where infection is highly prevalent. Targeted drug administration (based on diagnosis of infection in community surveys) is being evaluated for helminths such as *T. solium* in areas such as Latin America and Africa, where emphasis on health education, prevention of pig infection and changing food behaviours (consumption of inadequately prepared meat) have not been entirely successful (see description of these programmes in Murrell, 2005).

27.4.4 Anthelminthic drugs for treatment of humans infected with foodborne cestodes and trematodes

Effective anthelminthic drugs are available for the treatment of humans diagnosed to be suffering from foodborne trematodiasis or cestodiasis (Table 27.5). Guidance for the treatment of trichinellosis using albendazole, mebendazole or pyrantel is given by Couper and Mehta (2004) and Murrell

Table 27.5 WHO-recommended anthelminthic drugs for the human treatment of adult foodborne trematodes and cestode infections based on single oral doses[a,b]

Infections	Drug	Dose
Cestodiasis		
Taenia saginata and *T. solium*[c]	Niclosamide	Adult and child over 6 years, 2 g as a single dose after a light breakfast, followed by a purgative after 2 h. Child of 2–6 years, 1 g; child under 2 years, 500 mg
	Praziquantel	Adult and child over 4 years, 5–10 mg/kg as a single dose
Foodborne trematodiasis		
Clonorchis sinensis	Praziquantel	Adult and child over 4 years, 40 mg/kg
Fasciolopsis buski	Praziquantel	Adult and child over 4 years, 25 mg/kg
Heterophyes heterophyes[d]	Praziquantel	Adult and child over 4 years, 25 mg/kg
Opisthorchis felineus	Praziquantel	Adult and child over 4 years, 40 mg/kg
Opisthorchis viverrini	Praziquantel	Adult and child over 4 years, 40 mg/kg
Paragonimus westermani	Praziquantel	Adult and child over 4 years, 40 mg/kg
Fasciola gigantica	Triclabendazole	Adult and child over 4 years, 10 mg/kg
Fasciola hepatica	Triclabendazole	Adult and child over 4 years, 10 mg/kg

[a]Advice about adverse reactions, contraindications and precautions set out in the *WHO Model Formulary* (Couper and Mehta, 2004) should be followed.
[b]Anthelminthic drugs must be of good quality from a reputable manufacturer. Information supplied by the manufacturer must be respected.
[c]For *T. saginata*, half the dose of Niclosamide may be taken after breakfast and the remainder an hour later followed by a purgative 2 h after that.
[d]Representing species of minute intestinal flukes.
Source: From data in Couper and Mehta (2004) and adapted from Table 14.1 (Crompton and Savioli, 2007) with permission from Taylor and Francis Group LLC.

(2005). The guidance concerns treatment for trichinellosis in adults and children but, since the children's age is not specified, the manufacturer's instructions must be followed unless a physician takes responsibility for the treatment. These anthelminthic drugs are effective against adult *Trichinella* and so help to reduce the release of larval stages. Prednisolone (40–60 mg daily) may be needed to alleviate allergic and inflammatory symptoms (Couper and Mehta, 2004). Anisakiasis is best dealt with by preventive measures; anthelminthic treatment is considered to be rarely necessary (Couper and Mehta, 2004). If control interventions are to include treatment of domesticated animals for foodborne helminth infections, information about available anthelminthic drugs may be found in *Plumb's Veterinary Drug Handbook: Desk*, 6th edn (Plumb, 2008).

27.5 Future trends

More than a decade ago, the International Task Force for Disease Eradication (ITFDE) declared *T. solium* to be 'a potentially eradicable parasite' (Pawlowski *et al.*, 2005). More recently, ITFDE called for a large demonstration project on effective control to be carried out; such a 'proof of concept' would probably be the greatest single stimulus to further action against this potentially eradicable disease. There are, as yet, however, no truly successful intervention programmes instituted anywhere at a national level to achieve this goal. Several pilot programmes have been initiated, mostly in Latin America and Africa (Pawlowski *et al.*, 2005). For example, the Bill and Melinda Gates Foundation has provided funding to a consortium of researchers and public health agencies, including the Centers for Disease Control and Prevention, the Peruvian Ministry of Health, several Peruvian universities and John Hopkins University, that has the goal of eliminating cysticercosis from a major disease-endemic area in Peru. The approach is diverse and includes detection and treatment of infected people, detection and control of infected pigs, trials of a pig vaccine and health education. Another example is the Cysticercosis Working Group in Eastern and Southern Africa (CWGESA) Long-term Control Program. This multinational task force is composed of representatives of Ministries of Health, agriculture, environment and education, plus scientists from relevant academic and research organisations. The emphasis of this programme is the prevention and control of *T. solium* cysticercosis in both pigs and humans, and includes surveillance through regional centres using reporting systems and rapid epidemiological assessment and evaluation of cost-effective, sustainable and integrated methods. Attention to needed legislative action is also a priority, along with training and capacity building at the regional centres.

Until then, both the WHO and the ITFDE continue to urge medical and veterinary sectors to co-operate in determining national prevalences, economic impact studies and establishing surveillance reporting programmes. These efforts should adopt up-to-date diagnostic tools and clinical management procedures, employ anthelminthic drugs for the treatment of humans and pigs, set high standards of meat hygiene and greater veterinary control over pig husbandry and slaughter practices and increase health education on the risks and prevention of cysticercosis. Constraints on resources and expertise must be overcome to implement control programmes in communities where disease is judged to range from significant to severe. The declaration that *T. solium* might be eradicable is particularly important because it applies to neurocysticercosis (WHO, 2006).

While the health and economic impact of neurocysticercosis is well-documented, this is not the case for some of the other helminths discussed in this chapter. Impact evaluations of the liver, lung and intestinal flukes on the health and economies of the countries with high endemicity are needed if these zoonoses are to be assigned their appropriate national priority for prevention and control resources. Obstacles to acquiring the data for such evaluations include poor diagnostic and surveillance methodologies, limited research on clinical importance and lack of economic studies. An example of the importance of economic assessments is the significance of metacercarial infections in farmed fish. In south-east Asia, rapidly becoming a major fish production region, small-scale aquaculture is important to the need to alleviate rural poverty and to help meet the national demand for food. Large-scale fish production is also promoted for its contribution to exports. Because of rising concerns over food safety and quality on the part of both national consumers and importing countries, the impact of these infections on the aquaculture industry has received little of the attention which it urgently needs. If the impact is shown to be significant, it is more likely that resources needed for required research on developing the means to prevent fluke infections in aquaculture enterprises will be made available.

27.6 Further reading/additional sources

For additional information on the general topic of foodborne parasites the reader is directed to the following:

Food-Borne Parasitic Zoonoses (2007) D Murrell and B Fried (eds) New York, Springer.

A Guide to Human Helminths (1991) Coombs I and Crompton DWT, London and Philadelphia, Taylor & Francis.

Handbook of Helminthiasis for Public Health (2007) Crompton DWT and Savioli L, Boca Raton, FL, Taylor & Francis.

Guidelines for the Surveillance, Prevention and Control of Trichinellosis (2007) J Dupouy-Camet and KD Murrell (eds) FAO/WHO/OIE, Paris, World Organization for Animal Health.

Liver, lung and intestinal fluke infections (2006). MacLean J D, Cross J,

Mahanty S, in Guerrant R L, Walker D H, and Weller P F (eds), *Tropical Infectious Diseases*, 2nd edn, Philadelphia, PA, Churchill Livingstone, 1349–69.

27.7 References

Anderson R M and May R M (1992) *Infectious Diseases of Humans Dynamics and Control*, Oxford, Oxford University Press.

Bern C, Garcia H H, Evans C, Gonzalez A E, Verastegui M, Tsang V and Gilman R H (1999) Magnitude of the disease burden from neurocysticercosis in a developing country, *Clin Infect Dis*, **29**(5), 1203–9.

Blair D, Xu Zhi Blao and Agatsuma T (1999) Paragonimiasis and the genus *Paragonimus*, *Adv Parasitol*, **42**, 113–222.

Blair D, Agatsuma T and Wanf W (2007), Paragonimiasis, in K D Murrell and B Fried (eds), *Food-borne Parasitic Zoonoses*, New York, Springer, 117–50.

Brake R J, Murrell K D, Ray E E, Thomas J D, Muggenberg B A and Sivinsky J S (1985) Destruction of *Trichinella spiralis* by low-dose irradiation of infected pork, *J Food Prot*, **7**, 127–43.

Bruschi F and Murrell K D (2002) New aspects of human trichinellosis: the impact of new *Trichinella* species, *Postgrad Med J*, **78**, 15–22.

Capo V and Despommier D D (1996) Clinical aspects of infection with *Trichinella* spp., *Clin Microbiol Rev*, **9**, 47–54.

Chai J Y (2007) Intestinal flukes, in *Food-borne Parasitic Zoonoses*, Murrell, K D and Fried B (eds), New York, Springer, 53–115.

Chai J Y and Lee S-H (2002) Food-borne intestinal trematode infections in the Republic of Korea, *Parasitol Int*, **51**, 129–54.

Chai J Y, Murrell K D and Lymberry A (2005a) Fishborne parasitic zoonoses: status and issues, *Int J Parasitol*, **35** (Supplement 11–12), 1233–54.

Chai J Y, Park J H, Han E T, Guk S M, Shin E H, Lin A, Kim J L, Sohn W M, Yong T S, Eom K S, Hwang E H, Phommasack B, Insisiengmai B and Rim H J (2005b) Mixed infections with *Opisthorchis viverrini* and intestinal flukes in residents of Vientiane Municipality and Saravane Province in Laos, *J Helminthol*, **79**, 283–9.

Chen M G and Mott K E (1990) Progress in assessment of morbidity due to *Fasciola hepatica* infection: a review of recent literature, *Trop Dis Bull*, **87**, R1–R38.

Coombs I and Crompton D W T (1991) *A Guide to Human Helminths*, London and Philadelphia, Taylor & Francis.

Couper M R and Mehta D K (2004) *WHO Model Formulary 2004*, Geneva, World Health Organization.

Craig P S, Rogan M T and Allan J C (1996) Detection, screening and community epidemiology of taeniid cestode zoonoses: cystic echinococcosis, alveolar echinococcosis and neurocysticercosis, *Adv Parasitol*, **38**, 169–250.

Crompton D W T (1999) How much human helminthiasis is there in the world? *J Parasitol*, **85**, 397–403.

Crompton D W T and Savioli L (2007) *Handbook of Helminthiasis for Public Health*, Boca Raton, FL, Taylor & Francis.

Cross J (2001) Fish and invertebrate-borne helminths, in Y H Hui, S A Sattar, K D Murrell and P S Stanfield (eds), *Foodborne Diseases Handbook*, New York, Marcel Dekker, 249–88.

De N V, Murrell K D, Cong L D, Cam P D, Chau L V, Toan N D and Dalsgaard A (2003) The food-borne trematode zoonoses of Vietnam, *Southeast Asian J Trop Med Public Health*, **34** (Supplement), 12–34.

Dick T A (2007) Diphylobothriasis: The *Diphyllobothrium latum* human infection conundrum and reconciliation with a world-wide zoonosis, in K D Murrell and B Fried (eds), *Food-borne Parasitic Zoonoses*, New York, Springer, 151–84.

Diop A G, de Boer H M, Mandlhate C, Prilipko L and Meinardi H (2003) The global campaign against epilepsy in Africa, *Acta Trop*, **87**, 149–60.

Ditrich O, Giboda M, Scholtz T and Beer S A (1992) Comparative morphology of eggs of the Haplorchiinae (Trematodea: Heterophyidae) and some other medically important heterophyid and opisthorchiid flukes, *Folia Parasitol*, **39**, 123–32.

Dorny P, Brandt J and Geerts S (2005) Detection and diagnosis, in K D Murrell (ed.), *WHO/FAO/OIE Guidelines for the Surveillance, Prevention and Control of Taeniosis/ Cysticercosis*, Paris, World Organization for Animal Health, 45–55.

Dupouy-Camet J (2000) Trichinellosis: a worldwide zoonosis, *Vet Parasitol*, **93**, 191–200.

Dupouy-Camet J and Bruschi F (2007) Management and diagnosis of human trichinellosis, in J Dupouy-Camet and K D Murrell (eds), *FAO/WHO/OIE Guidelines for the Management, Prevention and Control of Trichinellosis*, Paris, World Organization for Animal Health, 37–56.

Dupouy-Camet J and Murrell K D (eds) (2007) *FAO/WHO/OIE Guidelines for the Surveillance, Prevention and Control of Trichinellosis*, Paris, World Organization for Animal Health.

Eddi C, Nari A and Amanfu W (2003) *Taenia solium* cysticercosi/taenisosis: potential linkage with FAO activities: FAO support possibilities, *Acta Trop*, **87**, 145–8.

Esteban J G, Flores A, Angles R, Strauss W, Aguirre C and Mas-Coma S (1997) Presence of very high prevelence and intensity of infection with *Fasciola hepatica* among Aymara children from Northern Bolivian Altiplano, *Acta Trop*, **66**, 1–14.

Farkas J (1987) Decontamination, including parasite control, of dried, chilled and frozen foods by irradiation, *Acta Alimentaria*, **16**, 351–84.

FDA (1985) Irradiation in the production, processing and handling of food, *Federal Register*, **50** (140) (21 CFR Part 179), 658–9.

Flisser A (2006) Where are the tapeworms?, *Parasitol Int*, **55** (Supplement), S117–20.

Gutierrez Y (2006) Other tissue nematode infections, in R L Guerrant, D H Walker and P F Weller (eds), *Tropical Infectious Diseases* vol. 2, 2nd edn, Philadelphia, PA, Elsevier, 1231–47.

Ishakura H, Takahashi S, Yagu K, Nakamura K, Kon S, Matsura A and Kikuchi K (1998) Epidemiology: Global aspects of anisakidosis, in I Tadi, S Kojima and M Tsuji (eds), *Proceedings, of ICOPA IX*, Bologna, Monduzi Editore Sp A, 379–82.

Keymer A (1982) Tapeworm infections, in R M Anderson (ed.), *Population Dynamics of Infectious Diseases*, London, Chapman and Hall, 109–38.

Kumar V (1999) *Trematode Infections and Diseases of Man and Animals*, Boston, MA, Kluwer.

Lee S H, Hwang S W, Chai J Y and Seo B S (1984) Comparative morphology of eggs of heterophyids and *Clonorchis sinensis* causing human infections in Korea, *Korean J Parasitol*, **22**, 171–80.

Loaharanu P and Murrell D M (1994) A role for irradiation in the control of foodborne parasites, *Trends in Food Science & Technology*, **5**, 190–95.

MacLean J D, Cross J and Mahanty S (2006) Liver, lung and intestinal fluke infections, in R L Guerrant, D H Walker and P F Weller (eds), *Tropical Infectious Diseases*, Philadelphia, PA, Churchill Livingstone, 1349–69.

Mafojanee N A, Appleton C C, Krecek R C, Michael L M and Willingham A L (2003) The current status of neurocysticercosis in Eastern and Southern Africa, *Acta Trop*, **87**, 25–33.

Maleewong W, Intapan P M, Wongkham C, Wongsaroj T, Kowsuwan T, Pumidonming W, Pongsaskulchoti P and Kitikoon V (2003) Detection of *Opisthorchis viverrini* in experimentally infected bithynid snails and cyprinoid fishes by a PCR-based method, *Parasitology*, **126**, 63–7.

Maruyama H and Nawa Y (2007) Immunology of infection, in K D Murrell and B Fried (eds), *Food-Borne Parasitic Zoonoses*, Springer, New York, 337–81.

Mas-Coma S and Bargues M D (1997) Human liver flukes: a review, *Res Rev Parasitol*, **57**, 145–218.

Mas-Coma S, Bargues M D and Valero M A (2005) Fascioliasis and other plant-borne trematode zoonoses, *Int J Parasitol*, **35** (Supplement 11–12), 1255–78.

Mas-Coma S, Bargues M D and Valero M A (2007) Plant-borne trematode zoonoses: Facioliasis and Fasciolopsiasis, in K D Murrell and B Fried (eds), *Food-Borne Parasitic Zoonoses*, Springer, New York, 293–334.

Miyazaki I (1991) *Helminthic Zoonoses*, Tokyo, International Medical Foundation of Japan.

Montresor A, Crompton D W T, Gyorkos T W and Savioli L (2002) *Helminth Control in School-age Children*, Geneva, World Health Organization.

Muller R (2002) *Worms and Human Disease*, 2nd edn, Wallingford, CAB International.

Murrell K D (2002) Fishborne zoonotic parasites: epidemiology, detection and elimination, in H A Bremner (ed.), *Safety and Quality Issues in Fish Processing*, Cambridge, Woodhead, 114–41.

Murrell K D (2005) *WHO/FAO/OIE Guidelines for the Surveillance, Prevention and Control of Taenisosis/Cysticercosis*, Paris, World Organization for Animal Health.

Murrell K D and Crompton D W T (2006) Foodborne trematodes and helminths, in *Emerging Foodborne Pathogens*, Y Motarjemi and M Adams (eds), Cambridge, Woodhead, 222–52.

Murrell K D, Lichtenfels R J, Zarlenga D S and Pozio E (2000) The systematics of *Trichinella* with a key to species, *Vet Parasitol*, **93**, 293–307.

Murrell K D, Djordjevic K, Cuperlovic K, Sofronic L, Savic M, Djordjevic M and Damjanovic S (2004) Epidemiology of *Trichinella* infection in the horse: the risk from animal feeding practices, *Veterinary Parasitology*, **123**, 223–33.

Pawlowski Z S and Murrell K D (2001) Taeniasis and cysticercosis, in Y H Hui, S A Sattar, K D Murrell, W-K Nip and P S Stanfield (eds), *Foodborne Diseases Handbook*, New York, Marcel Dekker, 217–27.

Pawlowski Z S, Allan J and Sarti E (2005) Control of *Taenia solium* Taeniasis/Cysticercosis: from research towards implementation, *Int J Parasitol*, **35**, 1221–32.

Pozio E and Murrell K D (2006) Systematics and epidemiology of *Trichinella*, *Adv Parasitol*, **63**, 367–439.

Pozio E and Zarlenga D S (2005) Recent advances on the taxonomy, systematics and epidemiology of *Trichinella*, *Int J Parasitol*, **35**, 1191–204.

Plumb D (2008) *Plumbs Veterinary Drug Handbook: Desk*, 6th edn, Oxford, Blackwell Publishing.

Rim H J (1982) Clonorchiasis, in J H Steele (ed.), *Handbook Series in Zoonoses*, Section C: Parasitic Zoonoses, Vol III (*Trematode Zoonoses*), Boca Raton, FL, CRC Press, 17–32.

Rim H J (1986) The current pathobiology and chemotherapy of clonorchiasis, *Korean J Parasitol*, **24** (Supplement), Monograph Series No 3, 1–141.

Rim H J, Farag H F, Sornmani S and Cross J H (1994) Foodborne trematodes: ignored or emerging, *Parasitol Today*, **10**, 207–9.

Sithithaworn P and Haswell-Elkins M (2003) Epidemiology of opisthorchiasis, *Acta Trop*, **88**, 187–94.

Sithithaworn P, Yongvanit P, Tesana S and Pairojkul C (2007) Liver flukes, in *Foodborne Parasitic Zoonoses*, K D Murrell and B Fried (eds) New York, Springer, 3–52.

Sripa B (2003) Pathobiology of opisthorchiasis: an update, *Acta Trop*, **88**, 209–20.

Stürchler D (1988) *Endemic Areas of Tropical Infections*, 2nd edn, Toronto, Hans Huber.

Takahashi Y, Mingyuan L and Waikagul J (2000) Epidemiology of trichinellosis in Asia and the Pacific Rim, *Vet Parasitol*, **93**, 227–39.

Thompson R C A, Traub R J and Parameswaran N (2007) Molecular epidemiology of food-borne parasitic zoonoses, in *Foodborne Parasitic Zoonoses*, K D Murrell and B Fried (eds), New York, Springer, 383–415.

Toscano C, Yu Sen Hai, Nunn P and Mott K E (1995) Paragonimiasis and tuberculosis – diagnostic confusion: a review of the literature, *Trop Dis Bull*, **92**, R1–R26.

Traub R J, Monis P T and Robertson J D (2005) Molecular epidemiology: A multidisciplinary approach to understanding parasitic zoonoses, *Int J Parasitol*, **35**, 1295–307.

Vatanasapt V, Tangvoraphonkchai V, Titapant V, Pipitgool V, Viriyapap D and Sriamporn S (1990) A high incidence of liver cancer in Khon Kaen Province, Thailand, *Southeast Asian J Trop Med Public Health*, **21**, 489–94.

Von Bonsdorff B (1977) *Diphyllobothriasis in Man*, London and New York, Academic Press.

Watanapa P and Watanapa W B (2002) Liver-fluke associated cholangiocarcinoma, *Br J Surg*, **89**, 962–70.

WHO (1994) *Bench Aids for the Diagnosis of Intestinal Parasites*, Geneva, World Health Organization.

WHO (1995) *Control of Foodborne Trematode Infections*, WHO Technical Report Series 849, Geneva, World Health Organization.

WHO (2004) *Report of joint WHO/FAO workshop on food-borne trematode infections in Asia, Ha Noi, Vietnam, 26–28 November, 2002*, Geneva, World Health Organization.

WHO (2006) *The control of neglected zoonotic diseases. A route to poverty alleviation*, Report of a joint WHO/DFID-AHP meeting with the participation of FAO and OIE, Geneva, World Health Organization (WHO/SDE/FOS/2006.1).

Williams H and Jones A (1994) *Parasitic Worms of Fish*, London, Taylor & Francis.

Wongratanacheewin S, Sermswan R W and Sirisinha S (2003) Immunology and molecular biology of *Opisthorchis viverrini* infection, *Acta Trop*, **88**, 196–207.

Xu L Q, Jiang Z, Yu S H, Xu S, Huang D, Yang S, Zhao G, Gan Y, Kang Q and Yu D (1995) Nationwide survey of the distribution of human parasites in China, *Chinese Journal of Parasitology and Parasitic Diseases*, **13**, 1–7 (in Chinese).

Yossepowitch O, Gotesman T, Assous M, Marva E, Zimlichman R and Dan M (2004) Opisthorchiasis from imported raw fish, *Emerg Infec Dis*, 10, 2122–6.

27.8 Acknowledgements

The authors would like to thank Dr D P J Daumerie, Dr D Engels, Dr S Petkevicius and Professor H H Williams for sending us information. We are most grateful to Mr E Peters and Mrs P A Peters of Kirriemuir Secretarial for much help with the preparation and submission of the manuscript.

28

Toxigenic fungi

M. Moss, University of Surrey, UK

Abstract: This chapter discusses the interactions between fungi, food and the raw materials used in their production, concentrating on those rendering the food toxic for human consumption. It deals with those mycotoxins considered still to have a major influence on human health and to have had an impact on food-related legislation and world trade in food commodities. The chapter concentrates especially on the toxic metabolites of the mould genera *Aspergillus*, *Penicillium* and *Fusarium* and deals specifically with the occurrence, significance and control of aflatoxins, ochratoxins, patulin, trichothecenes and fumonisins.

Key words: mycotoxins in foods, toxic species of *Aspergillus*, toxic species of *Penicillium*, toxic species of *Fusarium*.

28.1 Introduction

The terrestrial filamentous fungi include species that produce macroscopic fruiting bodies, such as mushrooms and toadstools, as well as many species that produce microscopic sporing structures and are generally referred to as moulds. Some of the macrofungi, such as the cultivated mushroom *Agaricus bisporus*, are valuable foods, but some, such as the deathcap *Amanita phalloides*, produce virulent toxins including amatoxin, and the consumption of a single fruit body may lead to death through liver and kidney failure. Other toxic macrofungi include *Cortinarius orellanus*, producing orellanin and causing an irreversible kidney damage, and *Inocybe patouillardii*, producing muscarine and causing vomiting and diarrhoea. Cases of poisoning from macrofungi usually arise through them being mistaken for edible wild fungi and such events are isolated. This chapter will deal with filamentous microfungi generally referred to as moulds.

Fruits and vegetables have a high water activity and many fruits have a low pH making them ideal substrates for mould growth. Fresh plant products are living tissues and plants have evolved many barriers, both physical and chemical, to inhibit invasion by microorganisms. Nevertheless, mould spoilage of fruits and vegetables is a common phenomenon. However, in general, mould spoilage itself does not lead to a health hazard because the mouldy products are usually rejected, and the main problem is one of economic loss. It is usually only when toxic fungal metabolites are produced that there is the possibility of a health hazard. However, in some cases, such as *Penicillium italicum* and *P. digitatum* on citrus fruits, vast numbers of small dry spores may be produced on the surface of the fruit and a few people may be allergic to such mould spores (Rippon, 1982). There has been a long tradition of using moulds in the production of foods, such as the mould-ripened cheeses and salami, tempeh, miso and soya sauce, but there are also species that can produce toxic metabolites, and when these contaminate foods, or animal feeds, they are referred to as mycotoxins (Moss, 1998). These can certainly be considered as foodborne hazards and will be dealt with in this chapter. Their importance is highlighted by the number of recent books, journals and reports dealing with mycotoxins in food and the food safety issues arising from such contamination (see Section 28.15).

Many toxic mould metabolites are known, but only a few are consistently associated with foodborne illness to the extent that individual countries may set maximum tolerated limits for their presence in foods (Boutrif and Canet, 1998). Mycotoxins are produced primarily in foods of plant origin, the most important of which are cereals (especially maize), legumes (especially groundnuts), pulses, oilseeds and treenuts, and they may be formed following contamination and growth after or during processing and not just from raw material contamination. However, they can also pass through the food chain into foods of animal origin such as meat and milk. It has been estimated that as much as 25 % of the world's food crops may be contaminated with mycotoxins (Boutrif and Canet, 1998), and such materials may be either destroyed, diverted to non-food uses, or be consumed in those countries that do not have legislation regulating their use as food or animal feed. It is possible to obtain information on mycotoxin-contaminated produce using the Rapid Alert System for Food and Feed (RASFF) the purpose of which is 'to provide control authorities with an effective tool for exchange of information on measures taken to ensure food safety' (http://ec.europa.eu/food/food/rapidalert/index-en.htm).

Most mycotoxins are acutely poisonous, but they are not as virulent as some of the toxins of cyanobacteria, dinoflagellates or food poisoning bacteria such as *Clostridium botulinum* (see Table 28.1). The justification for concern about their presence in food is not their acute toxicity but their

Table 28.1 Acute toxicities of some mycotoxins and other microbial toxins

Compound	Producing microorganism[a]	LD50 ($mgkg^{-1}$)
Aflatoxin B_1	*Aspergillus flavus*	0.5 (dog)
Sterigmatocystin	*Aspergillus versicolor*	166 (rat)
Patulin	*Penicillium expansum*	35 (mouse)
Ochratoxin A	*Penicillium verrucosum*	20 (rat)
Citrinin	*Penicillium citrinum*	67 (rat)
T-2 toxin	*Fusarium sporotrichioides*	5 (mouse)
Deoxynivalenol	*Fusarium graminearum*	70 (mouse)
Fumonisin B_1	*Fusarium verticillioides*	1–4 (horse)[b]
Aeruginosin	*Microcystis aeruginosa*	0.05
Saxitoxin	*Alexandrium tamarense*	0.03
Botulinum toxin	*Clostridium botulinum*	10^{-6}

[a]Many of these toxins are produced by several species.
[b]1–4 mg/kg body mass/day over 29–33 days will induce equine leucoencephalomalacia in horses (Kellerman *et al.*, 1990).

chronic toxicity which may include carcinogenicity, chronic organ damage and immunosuppressive activity.

28.2 Aflatoxins: occurrence and significance

The aflatoxins are produced by a small number of species of *Aspergillus* which currently includes *A. flavus*, *A. parasiticus*, *A. nomius* and a species originally isolated from Ivory Coast soil, *A. ochraceoroseus* (Bartoli and Maggi, 1978; Klich *et al.*, 2000). A strain of what was initially thought to be *A. tamarii* has also been shown to produce aflatoxin (Klich *et al.*, 2000), but it is more like a brown-spored form of *A. flavus*, is certainly very closely related, and is now referred to as *A. pseudotamarii* (Ito *et al.*, 2001). It should be emphasised that not all strains within a species are toxigenic. On a worldwide basis the majority of strains of *A. parasiticus* are aflatoxigenic, but only about 35 % of strains of *A. flavus* produce aflatoxins. Indeed some strains of *A. flavus* have been used to produce koji and the more widely used species, *A. oryzae*, is probably a domesticated form of this species (Cruikshank and Pitt, 1990). At the level of acute toxicity the aflatoxins cause liver damage and, in animals such as cattle, the farmer will initially see reduced feed consumption, depressed milk production, weight loss and eventually death due to severe liver damage. They were discovered following an outbreak of acute toxicosis affecting turkey poults and game birds in which large numbers of birds died (Sargeant *et al.*, 1961).

The aflatoxins are a family of compounds the most toxic of which is aflatoxin B_1 (Fig. 28.1). They are produced by a complex biosynthetic pathway starting as a decaketide and involving intermediates such as the

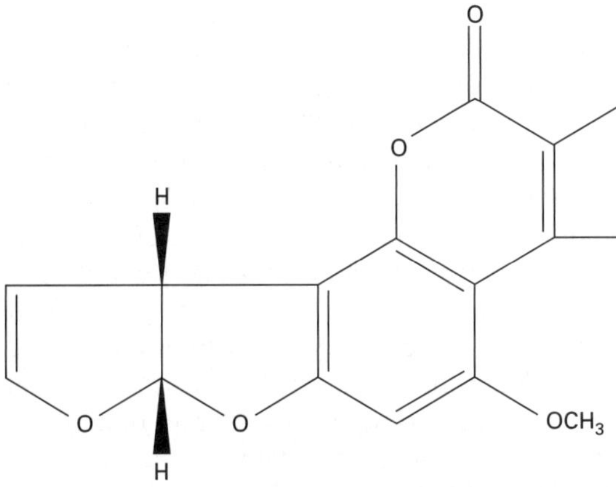

Fig. 28.1 Aflatoxin B_1.

anthraquinones averufin and versicolorin, and the xanthone derivative sterigmatocystin, the last two of which share the bis-furan ring system of the aflatoxins. In humans aflatoxin B_1 is both acutely toxic and probably carcinogenic. It was responsible for a major outbreak of toxicosis in India, initial reports of which found that 397 people were ill of whom 106 died, following consumption of aflatoxin-contaminated maize (Krishnamachari *et al.*, 1975). This outbreak affected people and dogs in several villages, and the most overt symptom was jaundice preceded by vomiting and anorexia. A more recent outbreak of aflatoxin poisoning occurred in Eastern Kenya during January–June 2004 (Lewis *et al.*, 2005). It resulted in 317 cases and 125 deaths, and the presence of aflatoxin in home grown, poorly stored maize, combined with the prevalence of hepatitis B, were clearly implicated. There have also been reports of aflatoxin and ochratoxin contamination of paprika imported into Europe during 2004 (Connolly, 2004), and in 2005 dog food sold in 22 US states was found to be contaminated with aflatoxin following the death of 23 dogs and severe liver illness in a further 18 (Stenske *et al.*, 2006). The dog is especially sensitive to the acute toxicity of aflatoxin.

A fascinating aspect of the toxicology of aflatoxin B_1 is the variation in response among different species and even between the sexes within a species. Indeed, for some animal species, such as the rat, it is among the most carcinogenic compounds known and yet in others, such as the guinea pig, it may not be carcinogenic. Although it is difficult to disentangle the interactions of aflatoxin with other carcinogenic agents, such as hepatitis B virus, it is clearly prudent to assume that aflatoxin B_1 is a human carcinogen and a review of the risk assessment leading to this assumption is presented by Castegnaro and McGregor (1998) and Moss (2002). The reason for the diversity of response to aflatoxin B_1 is because it has to be metabolised first and the balance of metabolism varies from one species to another. This aspect of the toxicology is reviewed by Moss (1998).

The moulds producing aflatoxins are ubiquitous in their ocurrence, but they are normally only active and competitive in the warmer parts of the world, the optimum temperature for growth being 35–37 °C, and the optimum temperature for aflatoxin production 30 °C. The commodities most usually contaminated are maize and groundnuts, but a wide range of other plant products may also contain aflatoxin B_1. These include Brazil nuts, almonds, figs and spices for human consumption as well as cottonseed meal and copra meal for animal feeds. Levels can range from a few μg/kg to several mg/kg (Pittet, 1998). The moulds producing aflatoxins grow particularly well on appropriate substrates when these are stored at high moisture content and elevated temperatures. Under these conditions, levels of contamination can be high enough to cause acute toxicity as occurred in India in 1974 when it was estimated that as much as 2–6 mg of aflatoxin could be consumed daily over a period of several weeks (Krishnamachari *et al.*, 1975), although the maize may not be so obviously mouldy as to be rejected.

However, these same species can infect crops such as maize and groundnuts in the field, before harvest, by establishing an endophytic relationship with the plant. When such an infected plant is stressed by, for example, a period of drought, aflatoxin may be produced which will then be present in the crop at harvest even though it may look perfectly sound. Under these conditions only relatively low concentrations are formed, but they may be higher than the maximum tolerated levels set by individual countries. Incidents of field contamination of crops by aflatoxin B_1 have been well-documented for the southern states of the USA (Hagler, 1990). In 1980, for example, 65.7 % of samples of maize from the harvest in North Carolina contained more than 20 μg/kg of aflatoxin with an estimated cost to producers and handlers of $30 816 000 (Nichols, 1983). These losses arose because 20 μg/kg was the maximum tolerated level in commodities for human consumption in the USA at that time. Attempts to harmonise national legislation at the international level have so far failed, and maximum permitted levels for aflatoxin B_1 in foods may be as low as 2 μg/kg (European Union) or as high as 30 μg/kg (India).

Although aflatoxin M_1, which is secreted in milk after consumption and metabolism of aflatoxin B_1, is less toxic than its precursor, most countries set much more stringent levels (0.05 μg/kg in the EU) because milk may be consumed by very young children who may be at high risk because of an intrinsically higher sensitivity and light body weight. Mothers exposed to aflatoxin B_1 in their diet also secrete aflatoxin M_1 in their milk exposing babies to both poor-quality milk because of mother's poor diet and aflatoxin M_1.

28.3 Aflatoxins: control measures

Because aflatoxins can be produced in the field, in stored commodities, in food products following contamination by mould spores (e.g. from packaging) and outgrowth, or as a result of passage through the food chain to contaminate foods that had not been mouldy, their control requires an integrated management programme (Lopez-Garcia *et al.*, 1999). Such a programme must include the control of formation in the field by preventing insect damage, alleviating drought stress and ensuring that potential inoculum does not build up in plant residues after harvest. Following harvest, it is essential that crops be dried to reduce their water activity and stored in cool, dry conditions. Details of the control of mycotoxin contamination during processing and in storage have been provided by Scudamore (2004) and Shapira (2004) who also provide comprehensive bibliographies.

For example, the cleansing of whole grain to remove dust and damaged broken grains may itself remove some mycotoxin-contaminated material. When mould and aflatoxin contamination occur they are usually very heterogeneous, and the recognition and removal of hot spots of contamination can be effective. In the case of relatively large particulate commodities such as groundnuts and pistachio nuts it may be possible to physically remove highly-contaminated nuts. In a particular batch of pistachio nuts it has been shown that 90 % of the aflatoxin was present in 4.6 % of the product and, if this could be removed, it would be possible to reduce the average aflatoxin contamination of this commodity from 1.2 to 0.12 μg/kg for product going to the market for human consumption (Schatzki, 1995; Schatzki and Pan, 1996). A similar study of aflatoxin contamination in a consignment of figs also demonstrated a very skewed distribution (Sharman *et al.*, 1991). There have been many studies of the possible inactivation of aflatoxins by physical, chemical or biological processes, but only the use of ammonia at either high pressure/high temperature or atmospheric pressure/ambient temperature has proved efficacious (Lopez-Garcia *et al.*, 1999), and such treatment is only of use for animal feeds and not for human foods.

28.4 Ochratoxin A: occurrence and significance

Ochratoxin A (Fig. 28.2), which is an L-phenylalanyl derivative of a pentaketide derivative, is produced by *P. verrucosum* in temperate climates and by a number of species of *Aspergillus* in the warmer parts of the world, the most important of which are *A. ochraceus*, *A. westerdjikiae* (Frisvad *et al.*, 2004) and, more recently, *A. carbonarius*, and other members of the black-spored *A. niger* group have been shown to be responsible for the presence of ochratoxin A in grapes and wine (Battilani and Pietri, 2002). In temperate climates it is most commonly associated with barley and other cereals but, in tropical and sub-tropical regions, a much wider range of foodstuffs may be contaminated, including coffee, cocoa, vine fruits and spices (Pittet, 1998). The importance of ochratoxigenic fungi in the production of coffee, cocoa and wine was a topic of major importance at a recent meeting of the International Commission on Food Mycology at their 2007 Workshop

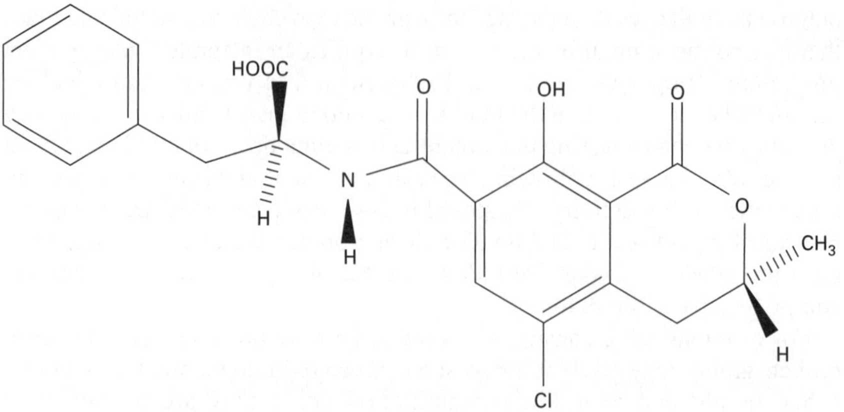

Fig. 28.2 Ochratoxin A.

(www.foodmycology.org). During the last few years levels of contamination in foods for human consumption sampled in Europe range from as little as 0.01–7.0 μg/kg in wine to 0.05–121 μg/kg in rye (Pittet, 1998).

Ochratoxin is primarily a nephrotoxin and has been associated with Balkan endemic nephropathy, although a case for maintaining an open mind on this issue was made by Mantle (2002). This condition in turn is linked to the occurrence of urinary tract tumours, and current regulations in the EU set maximum tolerated levels in human foods based on the assumption that ochratoxin A should be considered carcinogenic to humans. In 2002 the European Commision set maximum levels of 5 μg/kg in raw cereals, 3 μg/kg in cereal products and 10 μg/kg in dried vine fruits. In 2004 these were extended to include roasted coffee (5 μg/kg) and soluble coffee (10 μg/kg). A few other countries already had maximum tolerated levels in place which vary from 5–50 μg/kg (Boutrif and Canet, 1998). Ochratoxin A has a relatively long residence time in the animal body and so can be found in the flesh of animals, such as pigs, after slaughter for human consumption. There is not yet any evidence that ochratoxin is produced in plants in the field, but this may reflect our current lack of knowledge of the detailed ecology of the moulds known to produce ochratoxins. The possibility of uptake from the soil has been suggested and supported by studies with radiolabelled ochratoxin A (Mantle, 2000). A review of recent developments and significance of ochratoxin A in food was organised by the International Life Sciences Institute in 2005 and has been published (Hazel and Walker, 2005).

28.5 Ochratoxin A: control measures

Ochratoxin is a relatively stable metabolite as demonstrated by the survival of significant amounts in roasted coffee and in wine after fermentation. Its control must thus depend on the prevention of mould growth at every stage in the production of foods for human consumption. In the case of cereals, it is clear that care in harvesting and subsequent drying and adequate storage is required. However, some commodities such as coffee involve complex processes, some of which are inevitably at a high water activity, and there is a need to understand the relationship between the ecology of ochratoxin-producing fungi and the stages in these processes. Frank (1999) has demonstrated the importance of the drying process in the production of ochratoxin-free coffee beans and has attempted to apply the Hazard Analysis Critical Control Point (HACCP) approach to the study of coffee production. The results from detailed studies of the formation, prevention and fate of ochratoxin A in cereals, grapes and wine and coffee beans have made it possible to develop Codes of Practice for prevention of ochratoxin in these commodities based on Good Agricultural Practice (GAP) and Good Manufacturing Practice (GMP) (Scudamore, 2004).

28.6 Patulin: occurrence and significance

Patulin (Fig. 28.3) is one of the smallest of the mycotoxin molecules, being derived from a tetraketide precursor, and is produced by a number of species of *Penicillium*, *Aspergillus* and *Byssochlamys* but, in the context of human foods, the most important is *P. expansum*. This species is responsible for a soft rot of apples and a number of other fruits. It would not normally be a problem in fresh fruits because the rot is usually obvious and the fruit discarded. There are a few cultivars of apple, such as the Bramley, with an open core structure in which infection and mould growth can occur without any external evidence of a problem. Again, in the case of fresh fruit, the moulding would be apparent on cutting up the apple and it would most probably be discarded. The major source of patulin in the human diet is from fresh apple juice (and other fruit juices) because a perfectly palatable juice can be expressed from fruit in which some fruits are mouldy. Levels of contamination of fruit juices (mainly apple juices) over the last few years

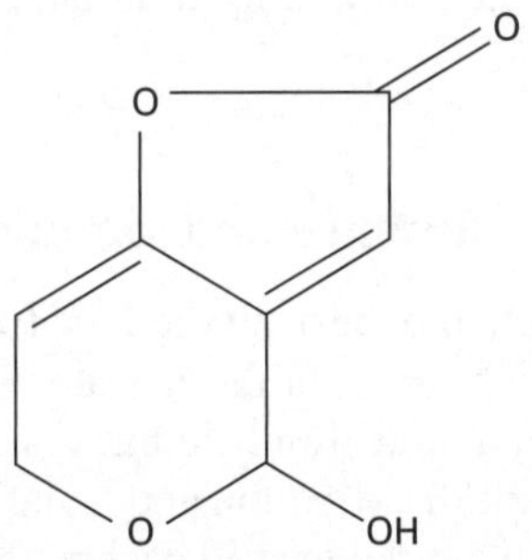

Fig. 28.3 Patulin.

have ranged from 5–1130 µg/l and the incidence of positive samples has ranged from 21 to 100 % (Pittet, 1998).

In the past the UK had an advisory level of 50 µg/l, and those few countries that had set maximum tolerated levels also used this figure although, in the Czech Republic, the level was reduced to 30 µg/l for children and 20 µg/l for infants (Boutrif and Canet, 1998). In February 2004 the EC set the following limits; for all fruit juices 50 µg/kg; for solid apple products such as purées 25 µg/kg; and for apple juice and solid apple products for infants 10 µg/kg. Although patulin has a significant acute toxicity at concentrations ranging from 9–55 mg/kg body weight, the main symptoms of which are agitation, convulsions, dyspnoea, pulmonary congestion, oedema and ulceration (Beretta *et al.*, 2000), early suggestions that it is carcinogenic have not been confirmed for oral administration. It is excreted very rapidly from the body (Dailey *et al.*, 1977) but clearly, with a vehicle such as fresh apple juice which is increasingly popular and may be consumed on a daily basis, it is sensible to ensure that exposure is as low as practically possible. Speijers (2004) concluded after a detailed review of patulin that 'In general, if the producers adhere to the tolerance limits set and maintain good manufacturing practice (GMP) there is no appreciable health risk from patulin'.

28.7 Patulin: control measures

Patulin is not particularly stable in an aqueous environment except at reduced pH when it will even survive elevated temperatures (Lovett and Peeler, 1973), hence its occurrence in pasteurised apple juice. Fresh apple juice is the commodity of most concern, and the removal of mouldy apples, or even the overtly mouldy part of individual apples, is an effective control measure. As indicated above, there is the possibility of mould infection and patulin contamination in some apple cultivars that do not appear to be overtly mouldy, and further study is required to understand the circumstances in which they become infected. Patulin is destroyed by the active fermentation of apple juice to cider by *Saccharomyces cerevisiae* (Moss and Long, 2002). It can also be removed with activated charcoal and by treatment with sulphur dioxide.

28.8 Fumonisins: occurrence and significance

The fumonisins, of which the best studied is fumonisin B_1 (Fig. 28.4), are polar compounds which are soluble in water and aqueous solutions of such solvents as methanol and acetonitrile but not in the non-polar solvents traditionally used for the initial extraction and isolation of mould metabolites. They have a long-chain polyketide-derived backbone with amino and hydroxyl groups esterified with tricarballic acid giving the molecule four carboxyl

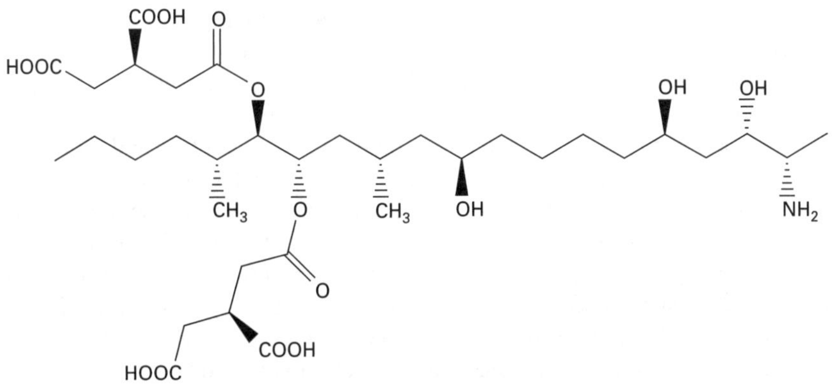

Fig. 28.4 Fumonisin B_1.

groups and adding to its polar properties. They are produced by *Fusarium verticillioides* (previously known as *F. moniliforme*) and a few closely related species of *Fusarium*, the most important of which is *F. proliferatum*, as well as *Alternaria alternata* f. sp. *lycopersici*. By far the most important substrate is maize in which it can be found in concentrations ranging from 6 μg/kg to > 37 mg/kg with incidences from 20–100 %.

The fumonisins, which are the subject of a major monograph (Jackson *et al.*, 1996) are clearly widespread in maize and maize products and may occur at relatively high concentrations, and yet it is unclear what their role may be in human health. Fumonisin B_1 is certainly responsible for a number of serious animal illnesses, such as encephalomalacia in horses (Kellerman *et al.*, 1990), pulmonary oedema in pigs and hepatocarcinoma in rats. The presence of *F. verticillioides* in maize had been linked to human oesophageal carcinoma as long ago as 1981 (Marasas *et al.*, 1981) and it is possible that, of the several toxic metabolites produced by this species, the fumonisins are implicated in this particular form of human cancer. Based on the toxicological evidence, the International Agency for Research on Cancer (IARC) classified *F. verticillioides* toxins as potentially carcinogenic to humans. In 2001 the US Food and Drug Administration set guidance levels for fumonisins in corn based foods which range from 2000–4000 μg/kg. The EC established detailed regulations for *Fusarium* toxins in 2005, most of which were implemented in 2006, but those for fumonisins in maize and maize products have only been implemented since July 2007. For unprocessed maize these are 4000 μg/kg and maize products range from 800–1000 μg/kg with the exception of those in processed baby foods which have been set at 200 μg/kg.

28.9 Fumonisins: control measures

Fusarium verticillioides can be a seedborne endophyte in maize and its elimination is difficult. When maize is growing well there may be no overt symptoms, but warm dry weather early in the growing season, followed by wet weather during the development of the cob, has been associated with ear rot disease of maize which in turn is associated with high levels of fumonisin.

Insect damage is also associated with high levels of infection and fumonisin contamination, and the breeding of cultivars resistant to such damage is a possible control strategy. Although fumonisin levels can increase during poor storage, it is clearly important to control disease and infection in the field. The possibility of using biological control with biocompetitive bacteria and non-toxigenic strains of *F. verticillioides* is being actively investigated. This range of possible control strategies has been reviewed by Riley and Norred (1999) and more recently by Jackson and Jablonski (2004).

28.10 Deoxynivalenol: occurrence and significance

Deoxynivalenol (DON), a member of a large family of mevalonate-derived sesquiterpene mould metabolites known as the trichothecenes, is recognised as an increasingly important mycotoxin in crops such as wheat when unfavourable weather conditions can lead to increased levels of *Fusarium* infection. It is produced by a number of species of *Fusarium*, but the most important are *F. graminearum*, *F. culmorum* and *F. pseudograminearum* which has only recently been segregated from the *F. graminearum* complex. Concern about the widespread presence of DON is reflected in the report of an ILSI workshop on trichothecenes with a focus on DON (Page, 2004).

Known also as vomitoxin, DON (Fig. 28.5) is much less acutely toxic than some other trichothecenes such as T-2 toxin, but it is much more common in cereals grown throughout the world. The presence of DON in wheat and barley is linked with *Fusarium* head blight in quite a complex way; the ecology and physiology of this association have been reviewed by Miller (2002) and the influence of agricultural practices on it has been discussed by Edwards (2004). The phytotoxic properties of DON are certainly important

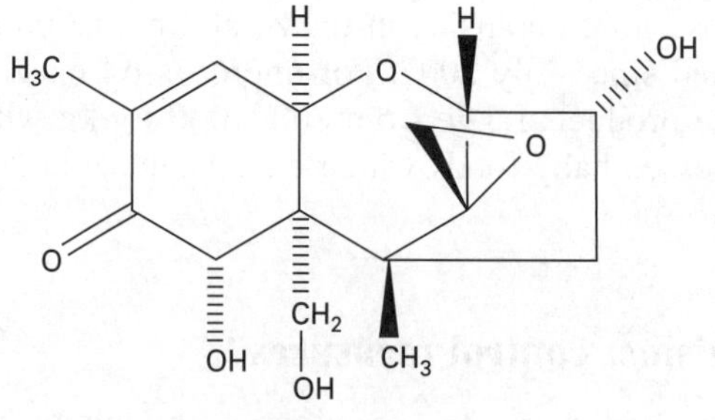

Fig. 28.5 Deoxynivalenol.

in the development of the mould during the initial infection of the living plant and the continuing invasion of plant tissue.

Despite its low acute toxicity DON is biologically active, especially immunosuppressive, at very low concentrations, and Pieters *et al.* (2002) derives a provisional Total Daily Intake (TDI) of 1.1 μg/kg body weight. A range of symptoms have been reported in humans following ingestion of DON including abdominal pain, dizziness, headache, throat irritation, nausea, vomiting and diarrhoea.

Regulations for DON were introduced by the EC in 2005 and implemented in July 2006 (with the exception of those for maize which were implemented in July 2007). They range from 1250–1750 μg/kg for unprocessed cereals, 750 μg/kg for cereal flour, maize flour, grits, meal and pasta, 500 μg/kg for bread, bakery products and breakfast cereals and 200 μg/kg for processed cereal-based foods for infants.

28.11 Deoxynivalenol: control measures

Aldred and Magan (2004) have described the key components, including increased knowledge of the ecology of the *Fusarium* species implicated, breeding cereals for resistance, more effective fungicides at the pre-harvest stage and effective drying, storage and preservation at the post-harvest stage, which need to inform a HACCP approach to reducing DON accumulation in cereal grain in Europe. The possibility of cultivars of cereals resistant to *Fusarium* head blight combined with the use of fungicides is considered a possible strategy. There are a number of field management strategies which could also contribute to the control of infection of the growing crop, such as deep ploughing to remove residual fungal material in debris from a previous crop, as well as crop rotation.

It is inevitable that grain at harvest will be contaminated with a wide range of microorganisms including those which could cause spoilage and toxin production. The interaction between these organisms and others such as insects is complex, but rapid moisture removal and storage at a moisture content equivalent to a water activity below 0.7 (i.e. < 14.5 % moisture by weight) would be ideal.

28.12 Other mycotoxins

Although the aflatoxins, ochratoxins, patulin, the fumonisins and deoxynivalenol are presently perceived to be the most important mycotoxins in the context of human health, there are many more that undoubtedly affect farm animals, have been implicated in human health in the past, and may become important in the future. The trichothecenes are produced by several species of

Fusarium in cereals such as barley, wheat and maize. As a group they are all immunosuppressive, and this may be a matter of concern following long-term exposure at low concentrations. The acute toxicity varies considerably from one member of the family to another. T-2 toxin is very toxic and was implicated in the serious outbreak of alimentary toxic aleukia in Russia at the beginning of the last century, but it is relatively uncommon. Another *Fusarium* toxin is zearalenone, often produced at the same time as deoxynivalenol, which, although without acute toxicity, is oestrogenic and it may be undesirable to expose the human population to a known oestrogen through the food chain. Indeed, the EC has set quite stringent limits for zearalenone (EC, 2006) for unprocessed cereals except maize 100 μg/kg; for cereal flours except maize 75 μg/kg; for bread, bakery products and breakfast cereals 50 μg/kg; and for processed cereal-based products for infants 20 μg/kg. These regulations were extended in 2007 to include maize products (EC, 2007).

Other *Penicillium* toxins include citrinin which, like ochratoxin, is nephrotoxic, but it is much less stable than ochratoxin and there is no evidence of it being a kidney carcinogen. A number of species of *Aspergillus* produce sterigmatocystin which is a precursor in the biosynthetic pathway to the aflatoxins as well as a metabolite in its own right. It is much less commonly found in human foods than aflatoxin and is very much less acutely toxic and carcinogenic. One species that secretes sterigmatocystin is *A. versicolor* which can grow at relatively low temperatures and can form part of the mould flora on the surfaces of cheese stored for long periods at low temperatures. Sterigmatocystin has been found in such cheeses, but it does not penetrate more than a few millimetres below the surface.

Toxic metabolites of moulds continue to be discovered although their significance in human health may not be apparent and the risks associated with their presence in foods may be low. It is important to maintain an awareness of the potential for moulds to produce toxic metabolites in foods without detracting from the long-standing and invaluable use of moulds in the production of a wide range of foods.

28.13 Detection and analysis

Apart from liquid commodities such as milk and fruit juices, mycotoxins are generally not evenly distributed throughout a consignment of food, feed or raw materials. Indeed sampling is an issue, and a sampling plan needs to be derived based on some understanding of the distribution. Studies such as those of Sharman *et al.* (1991) were referred to in Section 28.3.

Once a sample has been obtained it may need to be ground and thoroughly mixed to ensure homogeneity and an appropriate sub-sample prepared for analysis (see Whitaker *et al.*, 2002; Whitaker, 2004). The analyst has a wide range of tools available now, including thin layer chromatography, liquid chromatography, gas chromatography and a range of enzyme-linked immunosorbent assays (ELISAs). The availablity of antibodies specific to a range of mycotoxins is not only useful in the development of ELISA methodology but, by coupling such antibodies to a solid matrix, they can be used in immunoaffinity columns for the purification of toxins from complex extracts.

With the increasing introduction of legislation specifying quantitative limits for a range of mycotoxins which are chemically very diverse, there is a need for analytical methods which are sensitive, specific, quantitative and reproducible, as well as sampling procedures which are acceptable to both consumers and producers. Many of these methods are dealt with by appropriate experts in DeVries *et al.* (2002) and Magan and Olsen (2004).

28.14 Future trends

Although there will continue to be an extension of regulatory limits set by individual countries and groups of countries such as the EU, and analytical methods of surveillance will continue to be developed, there is also a need to understand the conditions and practices which lead to mycotoxin formation, such as the ecology and physiology of the fungi implicated, including their interactions with the crop as both endophytes and overt pathogens, so that their production can be reduced in the first place. There is also a need to assess the impact of regulations on the export of commodities from developing countries to ensure that indigenous populations are not exposed to high levels of mycotoxins from foods which they cannot export.

The Codex Alimentarius Commission is active in initiating Codes of Practice for the prevention and reduction of mycotoxin contamination. That there is a continuous need to involve producing countries was demonstrated when the Code of Practice for the Prevention and Reduction of Aflatoxin Contamination of Tree Nuts was extended to include Brazil nuts which are collected from the wild rather than from cultivated orchards. Recommended practices include clearing the area under Brazil nut trees, which can be 60 m high, removing residual pods and nuts from the former crop, collecting nuts as soon as most pods have fallen from the trees, and it is safe to do so, to minimise mould contamination, as well as removing broken and damaged pods.

While recognising that the complete elimination of mycotoxin-contaminated commodities is not achievable at present, the EC has published a detailed Code of Practice for the prevention and reduction of mycotoxin-contamination in cereals with special emphasis on ochratoxin A, zearalenone, fumonisins and trichothecenes. These are essentially based on details of GAP and GMP, from the time of planting the crop, management of the crop pre-harvest and at harvest, as well as subsequent storage. Although it is recognised that it is difficult or impossible to control such factors as weather and insect activity at the farm level, it is considered desirable to develop a HACCP approach to the control of mycotoxins in the future. Significant progress in this aspect of mycotoxin control was the publication in 2001 of a *Manual on the Application of the HACCP System in Mycotoxin Prevention and Control* (FAO Food and Nutrition Paper No 73, Rome).

Clearly, to ensure continued progress in mycotoxin control it is essential to continue to develop understanding of the conditions influencing their production, especially the factors involved in the endophytic relationships between fungi and plants.

28.15 Sources of further information and advice

The following journals are especially useful for details of mycotoxin studies:

- *Applied and Environmental Microbiology*
- *Food Additives and Contaminants*
- *Journal of the Association of Official Analytical Chemists*
- *Mycotoxin Research*
- *Mycotoxicology Newsletter* (www.mycotoxicology.org)

Reports of the Joint FAO/WHO Expert Committee on Food Additives (JECFA), the International Commission on Food Mycology and the International Agency for Research on Cancer (IARC), as well as publications from the European Food Safety Authority (www.efsa.europa.eu), and the following monographs are also invaluable sources of information:

Desjardins A E (2006) *Fusarium Mycotoxins Chemistry, Genetics, and Biology*, APS Press, St Paul, MN.

DeVries J W, Trucksess M W and Jackson L S (eds) (2002) *Mycotoxins and Food Safety*, Kluwer Academic, New York.

Hazel C and Walker R (eds) (2005) Ochratoxin A in food: recent developments and significance, *Food Additives & Contaminants* **22**(1), 1–107.

ICMSF (1996) *Microorganisms in Foods 5. Microbiological Specifications of Food Pathogens*, International Commission on Microbiological Specifications for Foods, Blackie Academic and Professional, London.

Jackson L S, DeVries J W and Bullerman L B (eds) (1996) *Fumonisins in Food*, Plenum, New York.

Magan N and Olsen M (eds) (2004) *Mycotoxins in Food, Detection and Control*, Woodhead, Cambridge.

Marasas W F O, Nelson P E and Toussoun T A (1984) *Toxigenic Fusarium Species, Identity and Mycotoxicology*, Pennsylvania State University Press, University Park, PA.

Sinha K K and Bhatnagar D (eds) (1998) *Mycotoxins in Agriculture and Food Safety*, Marcel Dekker, New York.

Smith J E and Henderson R S (eds) (1991) *Mycotoxins and Animal Foods*, CRC, Boca Raton, FL.

Van Egmond H P (ed.) (1989) *Mycotoxins in Dairy Products*, Elsevier Applied Science, London.

28.16 References

Aldred D and Magan N (2004) Prevention strategies for trichothecenes, *Toxicology Letters*, **153**, 165–71.

Bartoli A and Maggi O (1978) Four new species of *Aspergillus* from Ivory Coast soil, *Transactions of the British Mycological Society*, **71**, 383–94.

Battilani P and Pietri A (2002) Ochratoxin in grapes and wine, *European Journal of Plant Pathology*, **108**, 639–43.

Beretta B, Gaiaschi A, Galli C L and Restani P (2000) Patulin in apple-based foods: occurrence and safety evaluation, *Food Additives and Contaminants*, **17**, 399–406.

Boutrif E and Canet C (1998) Mycotoxin prevention and control: FAO programmes, *Review de Médicine Vétérinaire*, **149**, 681–94.

Castegnaro M and McGregor D (1998) Carcinogenic risk assessment of mycotoxins, *Review de Médicine Vétérinaire*, **149**, 671–8.

Connolly K (2004) Hungary in stew over cancer ban on paprika, *Daily Telegraph*, 29th October.

Cruikshank R H and Pitt J I (1990) Isoenzyme patterns in *Aspergillus flavus* and closely related species, in R A Samson and J I Pitt (eds), *Modern Concepts in Penicillium and Aspergillus Classification*, Plenum, New York, 259–65.

Dailey R E, Blaschka A M and Brouwer E A (1977) Absorption, distribution and excretion of [14C] patulin by rats, *Journal of Toxicological and Environmental Health*, **3**, 479–89.

DeVries J W, Trucksess M W and Jackson L S (eds) (2002) Mycotoxins and Food Safety, Kluwer Academic, New York.

EC (2006) Commission Regulation (EC) No. 1881/2006 of 19 December 2006 setting maximum levels for certain contaminants in foodstuffs, *Official Journal of the European Union*, **L364**, 20 December, 5–24.

EC (2007) Commission Regulation (EC) No 1126/2007 of 28 September 2007 amending Regulation (EC) No 1881/2006 setting maximum levels for certain contaminants in foodstuffs as regards *Fusarium* toxins in maize and maize products, *Official Journal of the European Union*, **L255**, 29 September, 14–17.

Edwards S G (2004) Influence of agricultural practices on fusarium infection of cereals and subsequent contamination of grain by trichothecene mycotoxins, *Toxicology Letters*, **153**, 29–35.

Frank J M (1999) HACCP and its mycotoxin control potential: an evaluation of ochratoxin A in coffee production, in *Proceedings, 3rd FAO/WHO/UNEP International Conference on Mycotoxins*, Tunis, March 3–6.

Frisvad J C, Frank J M, Houbraken J A M P, Kuipers A F A and Samson R A (2004) New ochratoxigenic A producing species of *Aspergillus* section *Circumdati*, *Studies in Mycology*, **50**, 23–43.

Hagler W M (1990) Aflatoxin in field crops in the South-eastern United States, in J F Robens (ed.), *A Perspective on Aflatoxins in Field Crops and Animal Food Products in the United States: a Symposium*, US Department of Agriculture, Springfield, VA, ARS-83, National Technical Information Service, 35–42.

Hazel C and Walker R (eds) (2005) Ochratoxin A in food: recent developments and significance, *Food Additives & Contaminants*, **22**(1), 1–107.

Ito Y, Peterson S W, Wicklow D T and Goto T (2001) *Aspergillus pseudotamarii*, a new aflatoxin producing species in *Aspergillus* section *Flavi*, *Mycological Research*, **103**, 233–9.

Jackson L and Jablonski J (2004) Fumonisins, in N Magan and M Olsen (eds), *Mycotoxins in Food, Detection and Control*, Woodhead, Cambridge, 367–405.

Jackson L S, DeVries J W and Bullerman L B (eds) (1996) *Fumonisins in Food*, Plenum, New York.

Kellerman T S, Marasas W F O, Thiel P G, Gelderblom W C A, Cawood M and Coetzer J A W, (1990) Leucoencephalomalacia in two horses induced by dosing of fumonisin B1, *Ondersterpoort Journal of Veterinary Research*, **57**, 269–75.

Klich M A, Mullaney E J, Daly C B and Cary J W (2000) Molecular and physiological aspects of aflatoxin and sterigmatocystin biosynthesis by *Aspergillus tamarii* and *A. ochraceoroseus*, *Applied Microbiological Biotechnology*, **53**, 605–9.

Krishnamachari K A V R, Bhat R V, Nagarajon V and Tilek T B G (1975) Hepatitis due to aflatoxicosis, *Lancet*, **i**, 1061–3.

Lewis L, Onsongo M, Njapau H *et al.* (2005) Aflatoxin contamination of commercial maize products during an outbreak of acute aflatoxicosis in Eastern and Central Kenya, *Environmental Health Perspectives*, **113**, 1763–7.

Lopez-Garcia R, Park D L and Phillips T D (1999) Integrated mycotoxin management systems, *Food, Nutrition and Agriculture*, **23**, 38–47.

Lovett J and Peeler J T (1973) Effect of pH on the thermal destruction kinetics of patulin in aqueous solutions, *Journal of Food Science*, **38**, 1094–5.

Magan N and Olsen M (eds) (2004), *Mycotoxins in Food, Detection and Control*, Woodhead, Cambridge.

Mantle P G (2000) Uptake of radiolabelled ochratoxin A from soil by coffee plants, *Phytochemistry*, **53**, 377–8.

Mantle P G (2002) Risk assessment and the importance of ochratoxins, *International Biodeterioration & Biodegradation*, **50**, 143–6.

Marasas W F O, Wehner F C, Van Rensburg S J and Van Schalkwyk D J (1981) Mycoflora of corn produced in human esophageal cancer areas in Transkei, Southern Africa, *Phytopathology*, **71**, 792–6.

Miller J D (2002) Aspects of the ecology of fusarium toxins in cereals, in J W DeVries, M W Trucksess and L S Jackson (eds), *Mycotoxins and Food Safety*, Kluwer Academic, New York, 19–27.

Moss M O (1998) Recent studies of mycotoxins, *Journal of Applied Microbiology*, Symposium Supplement **84**, 62S–76S.

Moss M O (2002) Risk assessment for aflatoxins in foodstuffs, *International Biodeterioration & Biodegradation*, **50**, 137–42.

Moss M O and Long M T (2002) Fate of patulin in the presence of the yeast *Saccharomyces cerevisiae*, *Food Additives & Contaminants*, **19**, 387–99.

Nichols T E (1983) Economic effects of aflatoxin in corn, in U L Diener, R L Asquith and J W Dickens (eds), *Aflatoxin and Aspergillus flavus in Corn*, Department of Research Information, Auburn University, Alabama, 67–71.

Page S W (ed.) (2004) Trichothecenes with a special focus on DON, Proceedings of a Workshop held on 10–12 September 2003 in Dublin, Ireland, *Toxicology Letters*, **153**, 1–189.

Pieters M N, Freijer J, Baars B J, Fiolet D C M, Van Klavaren J And Slob W (2002) Risk assessment of deoxynivalenol in food: concentration limits, exposure and effects, in J W DeVries, M W Trucksess and L S Jackson (eds), *Mycotoxins and Food Safety*, Kluwer Academic, New York, 235–48.

Pittet A (1998) Natural occurrence of mycotoxins in foods and feeds – an updated review, *Review de Médicine Vétérinaire*, **149**, 479–92.

Riley R T and Norred W P (1999) Mycotoxin prevention and decontamination – a case study on maize, *Food, Nutrition and Agriculture*, **23**, 25–30.

Rippon J W (1982) *Medical Mycology*, 2nd edn, W R Saunders, Philadelphia, PA, 714–20.

Sargeant K, Sheridan A, O'kelly J and Carnaghan R B A (1961) Toxicity associated with certain samples of groundnuts, *Nature*, **192**, 1096–7.

Schatzki T F (1995) Distribution of aflatoxin in pistachios. 1. Lot Distributions. 2. Distribution in freshly harvested pistachios, *Journal of Agricultural Food Chemistry*, **43**, 1561–9.

Schatzki T F and Pan J L (1996) Distribution of aflatoxin in pistachios. 3.Distribution in pistachio process streams, *Journal of Agricultural Food Chemistry*, **44**, 1076–84.

Scudamore K A (2004) Control of mycotoxins: secondary processing, in N Magan and M Olsen (eds), *Mycotoxins in Food, Detection and Control*, Woodhead, Cambridge, 224–43.

Shapira R (2004) Control of mycotoxins in storage and techniques for their decontamination, in N Magan and M Olsen (eds), *Mycotoxins in Food, Detection and Control*, Woodhead, Cambridge, 190–223.

Sharman M, Patey A L, Bloomfield D A and Gilbert J (1991) Surveillance and control of aflatoxin contamination of dried figs and fig paste imported into the United Kingdom, *Food Additives and Contamination*, **8**, 299–304.

Speijers G J A (2004) Patulin, in N Magan and M Olsen (eds), *Mycotoxins in Food, Detection and Control*, Woodhead, Cambridge, 339–52.

Stenske K A, Smith J R, Newman S J, Newman L B and Kirk C A (2006) Aflatoxicosis in dogs and dealing with suspected contaminated commercial foods, *Veterinary Medicine Today*, **228**, 1686–91.

Whitaker T B (2004) Sampling for mycotoxins, in N Magan and M Olsen (eds), *Mycotoxins in Food, Detection and Control*, Woodhead, Cambridge, 69–87.

Whitaker T B, Hagler W M, Giesbrecht F G and Johansson A S (2002) Sampling wheat for deoxynivalenol, in J W DeVries, M W Trucksess and L S Jackson (eds), *Mycotoxins and Food Safety*, Kluwer Academic, New York, 73–83.

29

Mycobacterium paratuberculosis

M. Griffiths, University of Guelph, Canada

Abstract: *Mycobacterium avium* spp. *paratuberculosis* (MAP) is the causative agent for Johne's disease, an incurable, chronic, infectious enteritis of ruminants, which results in diarrhoea, weight loss and, ultimately, death. Because of the similar pathology of Johne's and Crohn's disease in humans, MAP has also been proposed as the aetiological agent of Crohns' disease. The latter is a chronic inflammatory disease that most commonly affects the distal ileum and colon, but it can occur in any part of the gastrointestinal tract. The organism has been isolated from the large intestine of patients suffering from Crohn's disease and antibodies to MAP proteins can be isolated from sera of sufferers. However, evidence against MAP being the cause of Crohn's disease has also been amassed. These opponents to the role of MAP as the cause of Crohn's point to a large trial conducted in Australia, which concluded that there was no prolonged remission of disease when antimycobacterial drugs were administered. MAP has been isolated from raw and pasteurized milk and this led to the conclusion that the organism could survive pasteurization. There have been many studies to determine the accuracy of this statement but definitive experiments conducted in New Zealand indicate that the organism is unable to survive HTST pasteurization. Because the organism is difficult to cultivate, there has been an interest in the development of rapid diagnostic tests. These involve immunological and molecular approaches, both of which have advantages and disadvantages. Several approaches have been adopted to control the bacterium, including vaccination, but the most effective approach involves culling calves of infected cows, early culling of suspect cows and test reactors along with animals that have been in contact with these cattle. In conclusion, a causal link between MAP and Crohn's disease has yet to be established but the precautionary principle approach has been advocated until the role of the organism has been definitively recognized. However, elements of the disease appear to consist of genetic predisposition, immune-mediated tissue damage and enteric factors such as intestinal microbial flora and so it is unlikely that a single aetiological agent is responsible.

Key words: *Mycobacterium avium* spp. *paratuberculosis*, Johne's disease, Crohn's disease.

29.1 Introduction

'Johne's bacillus', the organism now known as *Mycobacterium avium* subsp. *paratuberculosis*, was first described by Johne and Frothingham in 1895 during an investigation of the cause of chronic diarrhoea in cattle. Based on its cell wall composition it has been classified as a member of the *M. avium* complex and as such *Mycobacterium avium* subsp. *paratuberculosis* (MAP) is a member of the family Mycobacteriaceae (Legrand *et al.*, 2000). Mycobacteriaceae are Gram-positive, strictly aerobic, non-motile, acid-fast rod-shaped bacteria with fastidious growth requirements. They are characterized by their slow growth rate and resistance to acid and alcohol; the latter is due to a strong cell wall containing a high lipid concentration. MAP requires the presence of mycobactins, which are iron-binding hydroxamate compounds, for growth. The organism will grow in the temperature range 25–45 °C with an optimum of 39 °C. It will grow at salt concentrations below 5 % and at a pH of 5.5 or greater (Collins *et al.*, 2001).

29.2 Johne's disease

MAP is the causative agent of Johne's disease, an incurable, chronic, infectious enteritis of ruminants, which results in diarrhoea, weight loss and, ultimately, death; although asymptomatic carriage may occur in cattle (Olsen *et al.*, 2002; Hayton 2007). Johne's disease is among the most common bacterial infections in domesticated animals worldwide, but is most common in cattle. It has been estimated that economic losses to the cattle industry in the USA are about $1.5 billion annually (Harris and Barletta, 2001). The organism has also been isolated from deer and wild boar (Alvarez *et al.*, 2005; Robino *et al.*, 2008), rabbits (Raizman *et al.*, 2005), racoons, opossums, and other small mammals as well as sparrows and starlings (Corn *et al.*, 2005), albeit at low frequency. MAP persists in rabbit populations, and this is now thought to be a significant route of transmission to livestock via the faecal–oral route (Judge *et al.*, 2005, 2007). A national survey for Johne's disease conducted in the USA in the 1980s showed that the infection was present in 1.6 % of all cattle and 2.9 % of cull cows (Merkal *et al.*, 1987). The United States Department of Agriculture has estimated that between 20 % and 40 % of US dairy herds are infected with MAP (Broxmeyer, 2005), and herd size has a major impact on herd prevalence (Manning and Collins, 2001). Forty percent of herds consisting of more than 300 head were infected with MAP. The prevalence of MAP in 4579 beef cattle from 115 ranches in Texas was determined using a commercial enzyme-linked immunosorbent assay (ELISA) (Roussel *et al.*, 2005). Serum samples were analyzed for antibodies and faecal matter from seropositive cattle was submitted for mycobacterial culture. Three percent of cattle were seropositive and 50 of the 115 (43.8 %) herds had at least one seropositive animal. Faeces of 73 % of the seropositive cattle were culture positive, and these animals were found in 18 % of the seropositive herds.

A survey conducted in south-west England concluded that 1 % of farms had cattle with Johne's disease and 2 % of the animals in these herds were infected (Cetinkaya *et al.*, 1994). The prevalence of culture-positive animals at slaughter was 2.6 % (Cetinkaya *et al.*, 1996). Similar prevalence rates have been found in other European countries (EC, 2000) and Canada (van Leeuwen *et al.*, 2005). However, higher prevalence rates (16.1 %) of MAP in culled dairy cattle have been reported for Eastern Canada and Maine based on a systematic random sample of 984 cattle at abattoirs. In Europe, herd prevalence ranges from 7–55 % and rates in Australia range from 9–22 % (Manning and Collins, 2001). A seasonal pattern of positive cows has also been reported, with the highest proportion of cows being positive in June (42.5 %). The prevalence rates also vary according to the method of detection. In a study of MAP in cattle in Alberta, Canada, it was determined that the true herd-level prevalence, as determined by ELISA, was 26.8 % +/– 9.6 %, whereas the rate ranged from 27.6 % +/– 6.5 % to 57.1 % +/– 8.3 % when MAP was detected by faecal culture (Sorensen *et al.*, 2003). Bannantine *et al.*, (2008a, b) experimentally infected cattle with MAP and were able to detect MAP proteins in cattle sera from as early as 70 d after exposure. They identified at least two antigens that may be useful in the early diagnosis of Johne's disease.

There is some evidence that strains are host-specific (Motiwala *et al.*, 2004; Marsh *et al.*, 2006; Sevilla *et al.*, 2007), with strains isolated from sheep and goats being more difficult to culture than the bovine strain (Juste *et al.*, 1991), and that infection depends on the MAP genotype (Gollnick *et al.*, 2007). Strains from bovine and ovine sources can be distinguished using the NLAP1506 gene locus (Griffiths *et al.*, 2008). There is evidence indicating that sheep can transmit MAP to cattle, albeit at low rates (Moloney and Whittington, 2008), and that horizontal transmission between calves can occur (van Roermund *et al.*, 2007).

Johne's disease usually occurs in young animals as a result of ingestion of feed contaminated with MAP. The ingested bacteria are transported across the epithelium through M-cells overlying the Peyer's patches and are subsequently released and taken up by macrophages (Tessema *et al.*, 2001). Unlike other intestinal epithelial cells, M-cells express integrins on their luminal faces. Efficient attachment and ingestion of MAP by cultured epithelial cells requires the expression of a fibronectin (FN) attachment protein homologue (FAP-P), which mediates FN binding by the bacterium. More recently, it has been postulated that the route of entry of MAP into intestinal epithelial cells involves the dystrophin–glycoprotein complex in a similar manner to that for Schwann cell invasion by *Mycobacterium tuberculosis* (Warth, 2008).

Targeting and invasion of M-cells by MAP *in vivo* is mediated primarily by the formation of a FN bridge formed between FAP-P and the integrins

(Secott *et al.*, 2001, 2002, 2004). Inside the macrophage, the mycobacteria are able to resist degradation and grow until the infected macrophage ruptures. The thick, lipid-rich cell envelope of MAP is mainly responsible for their resistance, as well as providing a physical barrier. The mycobacterial cell wall also contains several components that down-regulate the bactericidal function of macrophages. It has been shown that MAP can invade bovine epithelial cells more readily when the bacterium is suspended in milk than in broth or water (Patel *et al.*, 2006). The basic intracellular survival strategy of pathogenic mycobacteria requires (i) the use of entry pathways that do not trigger oxidative attack, (ii) modification of the intracellular trafficking of mycobacteria-containing phagosomes and (iii) modification of the link between the innate and specific immunity. These survival strategies, as well as their implications in the epidemiology, diagnosis and control of Johne's disease have been reviewed (Tessema *et al.*, 2001; Woo and Czuprynski, 2008).

29.3 Crohn's disease

Crohn's disease is a chronic inflammatory disease of humans that most commonly affects the distal ileum and colon, but it can occur in any part of the gastrointestinal tract (Hendrickson *et al.*, 2002). It is a life-long debilitating illness, the symptoms of which generally first appear in people aged 15–24. As yet, there is no known cure for Crohn's disease. However, management of the disease has become easier with the advent of drugs such as aminosalicylates, budesonide and immunosuppressive drugs (Achkar and Hanauer, 2000; Selby, 2000; Hendrickson *et al.*, 2002; Hermon-Taylor, 2002). Revolutionary treatments involving a genetically-engineered anti-TNFα antibody, infliximab, that targets a specific inflammatory mediator may mitigate against severe Crohn's disease (Bell and Kamm, 2000; Sandborn, 2005).

The illness is often cyclical with patients undergoing intermittent remission followed by recurrence. Surgical intervention is necessary in a large proportion of sufferers. It is primarily a disease of the industrialized world, which has led to several hypotheses for this phenomenon. These include the 'hygiene hypothesis' which concludes that the lack of enteric parasitic infections, all of which have been mainly eradicated in developed countries, causes a weakened systemic immune system, and this leads to the development of immunomediated diseases such as Crohn's disease (Wells and Blennerhassett, 2005) and the 'cold-chain hypothesis', which states that cold-chain development has paralleled the emergence of Crohn's disease during the 20th century and suggests that psychrotrophic bacteria found in food, such as *Yersinia* spp. and *Listeria* spp., contribute to the disease as they have been found in Crohn's disease lesions (Hugot *et al.*, 2003). Amre *et al.* (2006) have epidemiological evidence that exposure to infection at an early age is a risk factor for development of pediatric onset Crohn's disease, which they claim supports the hygiene hypothesis.

Several other causes of Crohn's disease have been postulated, such as infectious agents including bacteria and viruses, allergic and nutritionally related causes and microparticles, which forms part of the concept behind toothpaste being the causative agent (Korzenik, 2005). The underlying theory behind many of these ideas is that the central defect leading to Crohn's disease is an increased intestinal permeability, which may be the result of an innate immune deficiency (Korzenik, 2005; De Hertogh *et al.*, 2008).

29.4 *Mycobacterium paratuberculosis* and Crohn's disease

Because the pathological changes that occur in the small intestines of people with Crohn's disease are similar to those observed in cattle suffering from Johne's disease, it has been postulated that MAP is the aetiological agent responsible for Crohn's disease (Chiodini, 1989; Chiodini and Rossiter, 1996; Hermon-Taylor, 1998, 2000, 2001; Hermon-Taylor *et al.*, 2000; Hermon-Taylor and Bull, 2002). However, despite the similarities between the two conditions, detailed pathological comparisons reveal important differences, including important extraintestinal manifestations (Van Kruiningen, 1999). The clinical features of the two infections are shown in Table 29.1. It has also been pointed out that Crohn's disease may be more similar to inflammatory bowel disease (IBD) in dogs and cats (Qin, 2008).

Evidence linking MAP and Crohn's disease is far from conclusive; however, it is claimed that sufficient epidemiological evidence has been accumulated to point to an association between MAP and Crohn's disease (Uzoigwe *et al.*, 2007). Studies have shown that the organism can be cultured from about 7.5 % of patients with Crohn's disease but from only 1 % of healthy individuals (Chiodini and Rossiter, 1996) and can be detected in tissues from Crohn's disease patients by microscopy (Jeyanathan *et al.*, 2007). The results of several surveys are summarized in Table 29.2. A two-fold increase in the prevalence of the organism in Crohn's patients compared with control groups has been reported. However, the biggest differences were noted when a polymerase chain reaction (PCR) method based on the IS900 sequence was used to detect the organism (Moss *et al.*, 1992; Wall *et al.*, 1993; EC, 2000). However, concerns over the use of PCR have been raised because of the difficulty in isolating DNA from mycobacteria (Hermon-Taylor, 1998), the inability of PCR to differentiate between viable and dead cells and the fact that some IS900 primers may not be specific for MAP (Cousins *et al.*, 1999; Vansnick *et al.*, 2004; St Amand *et al.*, 2005; Semret *et al.*, 2006). Indeed, other mycobacteria have been recovered from tissues of Crohn's disease patients in a study that failed to detect MAP (Tzen *et al.*, 2006). The identification of MAP DNA in Crohn's disease tissue only reveals that MAP infection and Crohn's disease can exist at the same point in time. MAP infection may

Table 29.1 Clinical features of Crohn's disease and Johne's disease

Feature	Crohn's disease	Johne's disease
Pre-clinical stage		
Symptoms and signs	Not known	Decreased milk yield
Incubation period	Not known	Minimum 6 months
Clinical stage		
Presenting symptoms and signs	Chronic diarrhoea, abdominal pain, weight loss	Chronic diarrhoea, dull hair, weight loss, decrease in lactation
Gastrointestinal symptoms and signs		
Diarrhoea	Chronic (3 weeks+)	Chronic[a]
Blood in stool	Rare	Rare
Vomiting	Rare	No
Abdominal pain	Yes	Unknown
Obstruction	Yes	No
Extraintestinal manifestations		
Polyarthritis	Yes, but rare	No
Uveitis	Yes, but rare	No
Skin lesions	Yes	No
Amyloidosis	Yes, but rare	Yes[b]
Hepatic granulomatosis	Yes, but rare	Yes
Renal involvement	Yes, but rare	Yes[c]
Clinical course		
Remission and relapse	Yes	Yes

[a]Not seen in sheep.
[b]Goats primarily.
[c]Goats, deer, primates primarily, also camelids
Source: Rideout *et al.* (2003).

occur before or after the development of Crohn's disease, and in either case may have little bearing on the aetiology of the disease (Cook, 2000). Other studies have shown that the prevalence of MAP in Crohn's disease patients is no different from that in controls (Table 29.2). In a population-based case control study of seroprevalence of MAP in patients with Crohn's disease and ulcerative colitis, the authors could not identify a difference between IBD patients and healthy volunteers (Bernstein *et al.*, 2004). However, seropositive rates were high (approximately 35 %) for all study groups, and there was no difference in rates among Crohn's disease patients, ulcerative colitis patients, healthy controls and unaffected siblings.

Two systematic reviews of the literature linking the isolation of MAP from patients with Crohn's disease have concluded that the data confirm the observation that MAP is detected more frequently in Crohn's disease sufferers, but the role of the organism in disease remains inconclusive (Feller *et al.*, 2007; Abubakar *et al.*, 2008). A further systematic review of the literature regarding the zoonotic potential of MAP concluded that 'Evidence for the zoonotic potential of MAP is not strong, but should not be ignored'

Table 29.2 Isolation of MAP from patients with Crohn's disease

Methodology	Crohn's	Ulcerative colitis	Control	Reference
Serology				
Serum Ab by ELISA[a]	14/42[b] (33.3 %)	nd	3/34 (8.8 %)	Barta *et al.*, 2004
Serum Ab by ELISA	107/283 (37.8 %)	50/144 (34.7 %)	135/402 (33.6 %) 47/138 (34.1 %)[c]	Bernstein *et al.*, 2004
38kD Ag	16/28 (57.1 %)	nd	nd	Elsaghier *et al.*, 1992
24kD Ag	15/28 (53.6 %)	0/20 (0 %)	nd	
18kD Ag	15/28 (53.6 %)	2/20 (10 %)	nd	
p36 Ag	77/89 (86.5 %)	5/42 (11.9 %)	4/40 (10 %)	El-Zaatari *et al.*, 1999
p36 Ag	40/53 (75.5 %)	1/10 (10 %)	3/35 (8.6 %)	Naser *et al.*, 2000a
p35 Ag	79/89 (88.8 %)	4/27 (14.8 %)	5/50 (10 %)	
p35 + p36 Ag	39/53 (73.6 %)	1/10 (10 %)	0/35 (0 %)	
Serum IgA binding to mycobacterial HupB protein	9/10 (90 %)	0/10 (0 %)	0/10 (0 %)	Cohavy *et al.*, 1999
Culture				
Culture from blood	14/28 (50 %)	2/9 (22.2 %)	0/15 (0 %)	Naser *et al.*, 2004
Culture for 18 months	3/14 (21.4 %)	nd	nd	Chiodini *et al.*, 1984
Culture from colonic material	1/30 (3.3 %)	nd	nd	Coloe *et al.*, 1986
Culture from surgical tissue	4/82 (4.9 %)	nd	1/55 (1.8 %)	Gitnick *et al.*, 1989
Culture	1/66 (1.5 %)	nd	nd	Haagsma *et al.*, 1991
Culture from faeces	8/24 (33.3 %)	5/22 (22.7 %)	1/40 (2.5 %)	Del Prete *et al.*, 1998
Culture from surgical tissue	1/5 (20 %)	nd	nd	Pavlik *et al.*, 1994
Intestinal mucosal biopsies and surgical tissue	1/21 (4.8 %)	0/5 (0 %)	0/11 (0 %)	Clarkston *et al.*, 1998
Culture of whole blood	0/22 (0 %)	nd	nd	Freeman and Noble 2005
PCR[d] & related methods				
PCR of IS900 sequence from culture	6/18 (33.3 %)	nd	1/11 (9.1 %)	Moss *et al.*, 1992
PCR of IS900 sequence from culture	6/17 (35.3 %)	nd	0/13 (0 %)	Wall *et al.*, 1993

PCR of IS900 sequence from tissue	10/12 (83.3 %)	2/2 (100 %)	1/6 (16.7 %)	Romero *et al.*, 2005
FISH[e] using tissue	8/12 (66.7 %)	nd	0/6 (0 %)	
PCR of IS900 and IS1311 sequences from tissue	0/18 (0 %)	nd	nd	Baksh *et al.*, 2004
PCR in uncultured buffy coats	13/28 (46.4 %)	4/9 (44.4 %)	3/15 (20 %)	Naser *et al.*, 2004
Nested PCR based on IS900 sequence	0/13 (0 %)	0/14 (0 %)	0/13 (0 %)	Kanazawa *et al.*, 1999
PCR of IS900 sequence from surgical tissue	26/40 (65 %)	nd	6/63 (9.5 %)	Sanderson *et al.*, 1992
PCR of IS900 sequence from surgical tissue	11/24 (45.8 %)	nd	5/38 (13.2 %)	Lisby *et al.*, 1994
PCR of IS900 sequence from surgical tissue	13/18 (72.2 %)	nd	10/35 (28.6 %)	Dell'Isola *et al.*, 1994
PCR of IS900 sequence from surgical tissue	4/31 (12.9 %)	nd	0/30 (0 %)	Fidler *et al.*, 1994
PCR of IS900 sequence from colon biopsy	2/9 (22.2 %)	2/15 (13.3 %)	0/11 (0 %)	Murray *et al.*, 1995
PCR of IS900 sequence	10/10 (100 %)	11/18 (61.1 %)	14/16 (87.5 %)	Suenaga *et al.*, 1995
PCR of IS900 sequence	10/26 (38.5 %)	nd	4/35 (11.4 %)	Erasmus *et al.*, 1995
PCR of IS900 sequence	17/36 (47.2 %)	2/18 (11.1 %)	3/20 (15 %)	Gan *et al.*, 1997
PCR-DEIA[f] of IS900 sequence	15/17 (88.2 %)	9/18 (50 %)	1/40 (2.5 %)	Del Prete *et al.*, 1998
PCR of IS900 sequence from ileal mucosa	8/8 (100 %)	2/2 (100 %)	0/2 (0 %)	Mishina *et al.*, 1996
PCR	3/11 (27.3 %)	nd	0/11 (0 %)	Tiveljung *et al.*, 1999
PCR of IS900 sequence	0/68 (0 %)	0/49 (0 %)	1/26 (3.8 %)	Rowbotham *et al.*, 1995
PCR on intestinal biopsies	0/36 (0 %)	0/13 (0 %)	0/23 (0 %)	Dumonceau *et al.*, 1996
PCR of colon specimens	0/27 (100 %)	nd	0/11 (0 %)	Frank and Cook 1996
Multiplex PCR on ileal biopsy	0/10 (0 %)	nd	0/27 (0 %)	Al Shamali *et al.*, 1997
PCR on faeces, serum and intestinal tissue	0/21 (0 %)	0/5 (0 %)	0/11 (0 %)	Kallinowski *et al.*, 1998
PCR of IS900 sequence from biopsies and surgical tissue	0/34 (0 %)	nd	0/17 (0 %)	Chiba *et al.*, 1998
PCR	0/47 (0 %)	0/27 (0 %)	0/20 (0 %)	Cellier *et al.*, 1998
PCR of intestinal tissue	0/13 (0 %)	0/14 (0 %)	0/13 (0 %)	Kanazawa *et al.*, 1999
IS900 PCR of biopsies	20/23 (87 %)	nd	3/20 (15 %)	Scanu *et al.*, 2007
Microscopy of surgical resections using acid fast staining and *in situ* hybridization	10/17 (59 %)	nd	5/35 (14.3 %)	Jeyanathan *et al.*, 2007
PCR of IS900 sequences in blood	122/361 (33.8 %)	nd	42/200 (21.5 %)	Bentley *et al.*, 2008

Table 29.2 (Cont'd)

Methodology	Crohn's	Ulcerative colitis	Control	Reference
PCR of IS900 sequences of venous blood samples	0/73 (0 %)	nd	0/74 (0 %)	Lozano-Leon *et al.*, 2006
p89 & p92 PCR of mucosal biopsies	25/37 (67.6 %)	nd	7/34 (20.6 %)	Sechi *et al.*, 2005a
PCR of IS900 sequences of intestinal tissue	9/13 (69.2 %)	nd	2/14 (14.3 %)	Cheng *et al.*, 2005

[a] Enzyme-linked immunosorbent assay.
[b] No. of positive samples/total no. sampled.
[c] Unaffected siblings of Crohn's patients.
[d] Polymerase chain reaction.
[e] Fluorescence *in situ* hybridisation.
[f] Polymerase chain reaction – DNA enzyme immunoassay.
nd = not determined.

(Waddell *et al.*, 2008). Abubakar *et al.* (2007) conducted a case control study to assess the role of water and dairy products potentially contaminated with MAP in the aetiology of Crohn's disease. They concluded that there was no evidence to support the role of water and dairy products in acquisition of the disease; indeed there was a negative association between Crohn's disease and consumption of pasteurized milk but a positive association with consumption of meat. In a study of people suffering from Crohn's disease and irritable bowel syndrome in Italy, MAP with a conserved SNP at position 247 of IS900 was isolated, and this isolate was also found in seven of eight dairy sheep studied (Scanu *et al.*, 2007). This led to the conclusion that there was a significant association between consumption of hand-made sheep's milk cheese and development of Crohn's disease and irritable bowel syndrome. There is evidence linking cell-wall deficient forms of MAP with Crohn's disease (Hulten *et al.*, 2001; Sechi *et al.*, 2001, 2004), but the results produced by this group have been questioned (Roholl *et al.*, 2002).

Many other bacteria have been isolated from biopsies taken from Crohn's patients, including *Helicobacter* spp., *Listeria monocytogenes* and *Escherichia coli* (Tiveljung *et al.*, 1999). Indeed, Crohn's disease of the terminal ileum, especially if associated with a NOD2 mutation, is characterized by a reduced defensin (antimicrobial peptides) response and defective IL-8 response to bacterial peptidoglycan, which could lead to increased bacterial invasion of the intestinal mucosa (Subramanian *et al.*, 2006). Although it is uncertain that this deficient defensin response leads to a reduced antibacterial activity of the intestinal mucosa, it has been proposed that Crohn's disease is a defensin deficiency syndrome (Schmid *et al.*, 2004). The ileal mucosa-associated flora of patients with Crohn's disease involving the ileum, Crohn's disease restricted to the colon (CCD) and healthy individuals were examined by a combination of culture-independent analysis of bacterial diversity using 16S rDNA library analysis, quantitative PCR and fluorescence in situ hybridization as well as molecular characterization of cultured bacteria (Baumgart *et al.*, 2007). Analysis of 16S rDNA libraries of ileitis mucosa revealed that they were enriched in sequences for *E. coli*, but relatively depleted in a subset of Clostridiales. PCR of mucosal DNA failed to detect the presence of MAP, *Shigella* or *Listeria*. The number of *E. coli* present correlated with the severity of ileal disease, and invasive *E. coli* was restricted to inflamed mucosa. The majority of *E. coli* strains isolated from the ileum were novel in phylogeny, displayed pathogen-like behavior *in vitro* and harbored chromosomal and episomal elements similar to those described in extraintestinal pathogenic *E. coli* and pathogenic Enterobacteriaceae. These data suggest that a selective increase in a novel group of invasive *E. coli* is involved in Crohn's disease involving the ileum. A role for adherent-invasive *E. coli* in Crohn's disease has also been postulated by Meconi *et al.* (2007) who isolated the organism from patients and demonstrated that it could produce granulomas similar to the early stages of epithelioid granulomas found in Crohn's disease.

The presence of bacterial metabolites in the colonic lumen causing a specific breakdown of fatty acid oxidation in colonic epithelial cells may be an initiating event in IBD (Roediger, 1988). L-carnitine, a small highly polar zwitterion, plays an essential role in fatty acid oxidation and adenosine triphosphate (ATP) generation in the intestine, and the organic cation/carnitine transporters, OCTN1 and OCTN2, are primarily involved in the transport of carnitine and elimination of cationic drugs in the intestine. A region of about 250 kb in size, encompassing the OCTN1 and OCTN-2 genes, appears to confer susceptibility to Crohn's disease (Peltekova *et al.*, 2004). Lamhonwah *et al.* (2005) have shown that OCTN1 and OCTN2 are strongly expressed in target areas for IBD such as the ileum and colon. These researchers identified a nine amino acid epitope of this functional variant of OCTN1 that is shared by *Campylobacter jejuni* and MAP. They suggest that a specific antibody raised to this epitope during *C. jejuni* or MAP enterocolitis would cross-react with the intestinal epithelial cell functional variant of OCTN1, leading to an impairment of mitochondrial β-oxidation which, in turn, would trigger an initiating event in IBD.

El-Zaatari and colleagues (El-Zaatari *et al.*, 1994, 1995, 1997, 1999, 2001; Naser *et al.*, 2000a) produced further evidence implicating MAP as the aetiological agent in Crohn's disease. They screened the sera of Crohn's patients for the presence of antibodies to two MAP proteins, identified as p35 and p36, and found that 86 % of patients were seropositive for the p36 antibody and 74 % of patients were seropositive when challenged with both proteins. The corresponding values for sera from controls were 11 % and 0 %, respectively. However, 100 % of BCG vaccinated subjects and 89 % of subjects with tuberculosis or leprosy were also seropositive for the p36 antibody. Indeed, it has been proposed that the detection of p35 and p36 can form the basis of a diagnostic test for Crohn's disease (Shafran *et al.*, 2002b). An association between Crohn's disease and the presence of MAP was also established using antibodies against an IS900-glutathione S-transferase fusion protein (Nakase *et al.*, 2006), and a cytoadherent, immunogenic heparin binding haemagglutinin, analogous to a similar protein involved in the dissemination of *M. tuberculosis*, has been isolated from MAP (Sechi *et al.*, 2006a). It is worth noting that other research has failed to find a link between seroprevalence of MAP antibodies and Crohn's disease (Kobayashi *et al.*, 1989; Tanaka *et al.*, 1991; Walmsley *et al.*, 1996). Interestingly, Polymeros *et al.* (2006) showed that antibodies against MAP glucosyl transferase and human gastrointestinal glutathione peroxidase homologues cross-reacted which led the authors to conclude the existence of disease-specific mimics in patients with Crohn's disease. This may account for some of the correlations between Crohn's disease and seropositive microbial responses.

Data concerning seroprevalence are also difficult to interprete because many Crohn's patients take immunosuppressive drugs which can interfere with immunological assays, the assays themselves can be unreliable and Crohn's disease results in a 'leaky' intestine and so people with the disease readily form antibodies to intestinal and foodborne microorganisms (Blaser *et al.*, 1984). As a result, Ryan *et al.* (2004) were able to isolate *E. coli* DNA more frequently in granulomas from Crohn's patients than in other non-Crohn's bowel granulomas. They suggested that there was a tendency for lumenal bacteria to colonize inflamed tissue, or there may have been an increased uptake of bacterial DNA by gut antigen presenting cells. They concluded that the non-specific nature of the type of bacterial DNA present in granulomas was evidence against any one bacterium having a significant causative role in Crohn's disease. Difficulties in identification of MAP in Crohn's disease have led to the argument that infection with the organism in these patients is caused by low numbers, that is the infection is paucibacillary (Selby, 2004).

Other agents, including viruses, have been postulated to cause Crohn's disease (Smith and Wakefield, 1993; Huda-Faujan *et al.*, 2007). Indeed, it has been suggested that it is a pathogenic microbial community comprised of interactive groups of organisms, rather than a single microorganism, which may account for pathogenesis of Crohn's disease in genetically predisposed individuals (Bibiloni *et al.*, 2006; Eckburg and Relman, 2007). The microbial population attached to the ileocolonic mucosa of Crohn's disease patients is different from that of healthy subjects (Subramanian *et al.*, 2006). This led to the proposal that individuals who are predisposed to Crohn's disease are less able to regulate the microbial composition of their intestines, which leads to an unstable microbial population (Martinez-Medina *et al.*, 2006). Microbial mannans, including those of mycobacterial origin, can suppress mucosal phagocyte function and may contribute to the pathogenesis of Crohn's disease (Mpofu *et al.*, 2007).

Our understanding of the pathogenesis of Crohn's disease has progressed rapidly with the discovery of NOD-2/CARD15 variants associated with ileal involvement and better knowledge of the significance of various inflammatory mediators. It is now well-established that Crohn's disease is associated with polymorphisms of NOD2 (CARD15) (Girardin *et al.*, 2003; Sechi *et al.*, 2005a). Previous work has shown that NOD2 acts as an intracellular receptor for bacteria and bacterial breakdown products and, because it appears capable of both activating and inhibiting inflammatory responses, NOD2 plays an important role in the gastrointestinal tract's response to infectious organisms. NOD2 activation leads to the modification of NEMO (the nuclear factor-κB (NF-κB) essential modulator), a central component of the NF-κB signaling pathway controlling inflammatory responses. NOD2 mutations responsible for Crohn's disease cause polymorphisms that prevent the NOD2 protein from properly modifying NEMO (Abbott *et al.*, 2004). More than 30 different genetic variations have been reported so far in Crohn's disease patients, but three major mutations account for 82 % of the total NOD2 (CARD15) mutations (Girardin *et al.*, 2003).

Two groups, one consisting of Crohn's disease patients and the other consisting of uninfected blood donors, were screened for mutations in NOD2 associated with the disease and no association was found between carriage of one or two of the mutations and a significantly higher risk of Crohn's disease (Bentley *et al.*, 2008). However, when the incidence of MAP in blood of Crohn's disease patients was compared with the control group of blood donors, there was a significantly higher carriage of MAP (33.8 %) in the Crohn's disease group than the control (21.5 %). No association was found between NOD2 genotype and seroprevalence of MAP for patients suffering from Crohn's disease or ulcerative colitis (Bernstein *et al.*, 2007). However, other studies have reported such a link (Sechi *et al.*, 2005a), although the interpretation of the results of this study has been questioned (Sieswerda and Bannatyne, 2006).

Another genetic link with Crohn's disease was identified by Prescott *et al.* (2007), who observed a link between an autophagy-related gene, *ATG16L1*, and Crohn's in 727 patients which they claimed suggested a role for autophagy (a catabolic process involving the degradation of a cell's own components through the lysosomal machinery) in the pathogenesis of IBD. It was later shown that this mutation is not associated with the presence of MAP in the blood of patients (Valentine *et al.*, 2008). A further gene polymorphism, this time in the natural resistance-associated macrophage protein 1 (NRAMP1) at locus 823C/T, is associated with Crohn's disease, and this polymorphism was not associated with the presence of MAP in these patients (Sechi *et al.*, 2006b). As well as NOD2, Ferwerda *et al.* (2007) have proposed a role for two Toll-like receptors, TLR2 and TLR4, in the recognition of MAP by the innate immune system. Thus, it is clear that the etiology of Crohn's disease involves a complex interplay of genetic, infectious and immunologic factors.

Ren *et al.* (2008) examined whether MAP detection by nested PCR using an IS900 sequence in gut biopsies was associated with a different cytokine secretion profile. The latter was measured using an ELISA assay in subjects with Crohn's disease (n = 46), ulcerative colitis (n = 30), irritable bowel syndrome (n = 22) and normal controls (n = 18). MAP status was defined unique to MAP. Significantly higher levels of interleukin (IL)-4 and IL-2 were found in MAP-positive Crohn's patients compared to MAP-negative sufferers of the disease. Both normal and disease controls that were positive for MAP had similar levels of IL-4 and IL-2. IL-4 secretion was correlated with IL-2 production in blood cultures in Crohn's disease patients, consistent with a skewed Th2 immune response and provides the first evidence of altered T cell function linked to MAP infection in Crohn's disease, and the authors state that the results provide a link between detection of MAP and disease. They also claim that the pattern of cytokine shift in Crohn's disease is consistent with the concept that its increasing incidence is partly related to the hygiene theory. In previous work, these authors reported an increased secretion of another cytokine, tumour necrosis factor-α (TNFα), in MAP-positive Crohn's patients, which they claim was consistent with abnormal macrophage handling of the bacterium (Clancy *et al.*, 2007). The involvement of conserved *Mycobacterium* species antigen-specific T-cell responses in colitis is indicated by work using a mouse model (Singh *et al.*, 2007). Both bovine and human isolates of MAP, regardless of their short sequence repeat genotype, induced similar global gene expression patterns in a human monocytic cell line (Motiwala *et al.*, 2006a), and survival of MAP in human polymorphonuclear cells mimics that of *M. tuberculosis*; providing evidence that the organism is virulent in humans (Rumsey *et al.*, 2006).

Saleh Naser and colleagues have reported that, although MAP DNA can be detected in the circulation of patients with IBD and in controls without the disease, the viable organism could only be cultured from blood of patients with Crohn's disease or ulcerative colitis (Naser *et al.*, 2004). They have also reported the culture of MAP from intestinal tissue and breast milk in people with Crohn's disease (Naser *et al.*, 2000b). The existing evidence that MAP is one of the causes of Crohn's disease is inconclusive, despite arguments of protagonists (Hermon-Taylor, 2000, 2001; Chamberlin *et al.*, 2001; Quirke, 2001; Bull *et al.*, 2003; Greenstein, 2003; Greenstein and Collins, 2004). The identification of clusters of Crohn's disease cases indicates that an aetiological agent causes the disease, but such clusters are uncommon. For example, in Australia, Johne's disease is not uniformly distributed yet Crohn's disease occurs throughout the country.

The equal circulation of MAP DNA in patients with and without inflammatory bowel disease suggests that environmental exposure to MAP is widespread, possibly from water, milk or other sources (Greenstein and Collins, 2004). Nevertheless, Naser *et al.* (2004) were only able to culture the bacterium in patients with IBD. This may be due to a defect in the mucosal barrier in patients with inflammatory bowel disease that allows passage or persistence of viable organisms taken up by circulating monocytes, or impaired killing of MAP by macrophages in these patients. The presence of MAP in blood could thus be an effect of the disease rather than its cause. Naser and colleagues have subsequently concluded that there is no association between MAP and Crohn's disease (Boedeker *et al.*, 2007).

Strangely, no correlation was found between the presence of circulating MAP and immunosuppressive therapy. Such a correlation may be expected, as reactivation of tuberculosis is a significant complication of infliximab therapy. However, MAP infection has not been described with this treatment and there is only one report of Crohn's disease in a patient with AIDS (Richter *et al.*, 2002).

There have been reports of the efficacy of antimycobacterial drugs in treatment of Crohn's patients, but the antibiotics used are active against many other bacteria (Prantera *et al.*, 1989, 1994, 1996; Hermon-Taylor, 2002; Shafran *et al.*, 2002a). The remission in symptoms achieved with these drugs is generally short-lived, and similar results can be obtained with antibiotics not known to be effective against mycobacteria (Prantera *et al.*, 1996). However, Borody *et al.* (2002) have described reversal of severe Crohn's disease in six out of 12 patients using prolonged combination anti-

MAP therapy alone. Three patients achieved long-term remission with no detectable symptoms of Crohn's disease remaining when all therapy was removed, and Borody *et al.* (2007) have described healing of the mucosa of Crohn's patients following anti-mycobacterial therapy. More recently, Beckler *et al.* (2008) have shown that a mutation in the rpoB gene of MAP results in resistance to rifabutin and rifampicin, which have been used in the treatment of Crohn's disease. Shin and Collins (2008) determined that simultaneous exposure of MAP to the immunosuppressant drug 6-mercaptopurine and the antibiotic ciprofloxacin resulted in greater survival of the organism than obtained in the presence of the antibiotic alone, and they conclude that this may explain the paradoxical response of Crohn's disease patients infected with MAP to treatment with immunosuppressive thiopurine drugs. In these patients symptoms do not worsen with anti-inflammatory treatment as might be expected if the disease were caused by a microbial aetiological agent.

The results of an eagerly awaited Australian study that used a combination antibiotic therapy with clarithromycin, rifabutin and clofazimine for up to two years to treat 213 Crohn's disease sufferers did not find evidence of a sustained benefit (Selby *et al.*, 2007). The authors claim that their study does not support a significant role for MAP in the pathogenesis of Crohn's disease in the majority of patients. However, short-term improvement in symptoms was seen when this combination of antibiotics was added to corticosteroids, most likely because of non-specific antibacterial effects. Peyrin-Biroulet *et al.* (2007) reviewed the results of the Australian study, together with other published data related to antimycobacterial drugs to treat Crohn's disease, and concluded that 'it should help to refute arguments favoring MAP as an etiologic agent, although it does not definitively eradicate that hypothesis'. In response to the conclusions of the study it has been pointed out that sub-therapeutic doses were used and the authors did not attempt to isolate MAP from patients (Lipton and Barash, 2007) and alternative conclusions could be drawn as to the effectiveness of the treatments (Chamberlin, 2007; Gitlin and Biesecker, 2007; Kuenstner, 2007).

There is conclusive evidence that hereditary and environmental factors play an important role in the aetiology of Crohn's disease (Kornbluth *et al.*, 1993). This, together with the conflicting results of studies aimed at confirming a link between MAP and Crohn's disease, suggest that even if MAP is involved in the development of Crohn's disease it is not the sole cause. Also, if bacteria are involved in Crohn's disease then their action may be the result of a dysfunctional immune response and not due to the virulence of the organism *per se* (Griffiths, 2002, 2006; Braun and Targan, 2006).

Economou and Pappas (2008) sought to analyze the trends in prevalence of Crohn's disease worldwide over the last 25 years in an attempt to correlate them with genetic, environmental and socioeconomic factors. New Zealand, Canada, Scotland, France, the Netherlands and Scandinavia were the regions showing highest prevalence. They concluded that industrialization and affluence are the common denominators between endemic areas, but these terms are too broad to strongly indicate an etiological role. Although the incidence of Crohn's disease is increasing in Asia, the prevalence of the disease is still low and the increase may reflect an improved ability of local health systems to diagnose the illness. Genetic associations are variably reproduced worldwide, and this is inconsistent with a strong etiologic relationship. Data on paratuberculosis incidence are scarce, and the existing ones are inconclusive regarding even an indirect correlation between Crohn's disease and MAP infections in animals. Indeed, the prevalence of Crohn's disease is not any higher among dairy farmers than in the general population (Jones *et al.*, 2006).

Ghadiali *et al.* (2004) identified two alleles among short sequence repeats of MAP isolated from Crohn's disease patients, and these clustered with strains derived from animals with Johne's disease. The authors concluded that the identification of a limited number of genotypes among human strains suggests the existence of human disease-associated genotypes and strain sharing with animals. It has also been postulated that MAP is the causative agent of Type 1 diabetes mellitus in genetically prone individuals (Dow, 2006; Sechi *et al.*, 2008).

It has been established that MAP does not meet all of Koch's postulates when applied to Crohn's disease. In an attempt to bring Koch's postulates up to date, Fredericks and Relman (1996) introduced the concept of 'Sequence-Based Identification of Microbial Pathogens: A Reconsideration of Koch's Postulates'. These guidelines are to be used with newer technologies to establish causality, but strict adherence to these guidelines is not needed for a compelling case to be made regarding causation. Chamberlin *et al.* (2007a) addressed these points as they apply to MAP as the causative agent of Crohn's disease.

1. A nucleic acid sequence belonging to a putative pathogen should be present in most cases of an infectious disease. Microbial nucleic acids should be found preferentially in those organs known to be diseased.
   Response: MAP DNA is found more frequently in intestinal tissue, mesenteric lymph nodes and blood of Crohn's disease patients than in normal controls (Autschbach *et al.*, 2005; Sechi *et al.*, 2005b) and MAP is cultured from these same tissues at greater frequencies than controls (Naser *et al.*, 2004).
2. Hosts or tissues without disease should contain fewer, or no, copy numbers of pathogen-associated nucleic acid sequences.
   Response: MAP DNA is found and the organism is cultured less frequently in unaffected controls.
3. If the disease is resolved (for example, with clinically effective treatment), the copy number of pathogens should decrease or become undetectable. With clinical relapse, the opposite should occur.
   Response: Only one study has accomplished this. MAP was cultured from one patient who was then treated with antimycobacterial antibiotics,

resulting in complete clinical remission, during which he exhibited a normal colonic mucosa and his subsequent blood cultures were negative (Chamberlin *et al.*, 2007b).

4. When nucleic acid sequence detection predates disease, or sequence copy number correlates with severity of disease or pathology, the sequence–disease association is more likely to be a causal relationship.
   Response: When MAP was isolated from a Crohn's disease patient and given to a baby goat it produced intestinal infiammation (Chiodini, 1989; Greenstein, 2003).
5. The nature of the microorganism inferred from the available sequence should be consistent with the known biological characteristics of that group of organisms.
   Response: MAP naturally infects many animal species causing an infiammatory bowel disease similar to Crohn's disease.
6. Tissue–sequence correlations should be present at the cellular level and specific *in situ* hybridization of microbial sequence should occur in areas of tissue pathology.
   Response: *In situ* hybridization of MAP-specific IS900 sequence locates MAP intestinal granulomas in 83 % of Crohn's disease patients studied (Sechi *et al.*, 2001).
7. The evidence should be reproducible.
   Response: MAP has been consistently cultured from patient's blood and this has been reported by four different research groups. Chamberlin *et al.* (2007a) state that conflicting negative results are due to the use of different methods for detection of MAP.

Although these arguments appear convincing, they fail to take into account the many studies that do not support these principles of causation.

29.5 Prevalence of *Mycobacterium paratuberculosis* in foods

Because of the widespread occurrence of Johne's disease, it seems likely that MAP would be present in raw meats from ruminants as well as in raw vegetables and water due to environmental contamination by faeces. However, Jaravata *et al.* (2007) were unable to detect MAP in any of 200 retail, ground beef samples, either by IS900 PCR or by culture. Using a real-time PCR assay targeting the F57 sequence, Bosshard *et al.* (2006) detected MAP in 4.9 % of mesenteric lymph nodes, 0.9 % of ileum tissue and 2.9 % of diaphragmatic tissue of 101 healthy dairy cattle at slaughter. Few other data appear to exist in the literature on the prevalence of this organism from these sources, although MAP has been isolated from municipal potable water (Mishina *et al.*, 1996) where it appears to be resistant to chlorination (Greenstein, 2003). The organism was detected in 8 % of untreated water in Northern Ireland (Whan *et al.*, 2005b). Possible routes of transmission to humans are milk, undercooked beef, water and animal contact (Greenstein, 2003; Greenstein and Collins 2004). To ensure that produce is free from MAP, composting of manure at 55 °C is recommended as an efficient way of inactivating MAP (Grewal *et al.*, 2006).

Several studies have shown that MAP can be cultured from the milk of cows clinically infected with paratuberculosis (Doyle, 1954; Smith, 1960; Taylor *et al.*, 1981). MAP has been cultured from the faeces of 28.6 % of cows in a single herd with high prevalence of infection. Of the faecal culture-positive cows, MAP was isolated from the colostrum of 22.2 % and from the milk of 8.3 %. Cows that were heavy faecal shedders were more likely to shed the organism in the colostrum than were light faecal shedders (Sweeney *et al.*, 1992). Levels of the organism in nine culture-positive raw milks from clinically normal, faecal culture-positive cows were between 2 and 8 cfu/ml (Sweeney *et al.*, 1992). However, other studies have suggested that faecal contamination was the most important contributor to MAP contamination of milk and levels as high as 10^4 cfu/ml could be attained (Nauta and van der Giessen, 1998). Several surveys of pasteurized milk for the presence of MAP have now been carried out and these are summarized in Table 29.3.

Millar *et al.* (1996) examined cream, whey and pellet fractions of centrifuged whole cow's milk for MAP by IS900 PCR and found that the PCR assay gave the expected results for spiked milk and for native milk samples obtained directly from MAP-free, sub-clinically and clinically infected cows. These researchers also tested individual cartons and bottles of whole pasteurized cow's milk obtained from retail outlets throughout Central and Southern England and South Wales from September 1991–March 1993 by PCR. They found that 7.1 % (22 of 312) tested positive for MAP. After 13–40 months of culture, 50 % of the PCR-positive milk samples and 16.7 % of the PCR-negative milk samples contained MAP, but the plates were overgrown by other organisms. More positive samples were found in winter and autumn using the PCR assay.

During the period March 1999–July 2000, a survey was undertaken to determine the prevalence of MAP in raw and pasteurized milks in the UK (Grant *et al.*, 2002a). The organism was detected in milk using an initial rapid screening procedure involving immunomagnetic separation coupled to PCR (IMS-PCR). Conventional culture was used to confirm viability of the isolates and the organism was determined to be MAP if it met the following criteria: (i) acid fast, (ii) slow growth and typical colony morphology on Herrold's egg yolk medium, (iii) presence of IS900 insertion element confirmed by PCR and (iv) dependent on mycobactin J for growth. MAP DNA was detected by IMS-PCR in 7.8 % (95 % confidence interval, 4.3–10.8 %) and 11.8 % (95 % confidence interval, 9.0–14.2 %) of the raw and pasteurized milk samples, respectively. When culture, following chemical decontamination with 0.75 % cetylpyridinium chloride for 5 h, was used to confirm the presence of MAP, 1.6 % (95 % confidence interval, 0.04–3.1 %) and 1.8 % (95 % confidence interval, 0.7–2.8 %) of the raw and pasteurized milk samples, respectively,

Table 29.3 Prevalence of MAP in milk

Sample	Method	No. of samples tested	No. MAP +ve (% +ve)	Country	Reference
Commercially pasteurized milk	Culture for 32 weeks	244	4 (1.6)	Czech Republic	Ayele *et al.*, 2005
Raw milk	IMS-PCR[a] of IS900	389	50 (12.9)	Ireland	O'Reilly *et al.*, 2004
Commercially pasteurized milk		357	35 (9.8)		
Raw milk	Culture	389	1 (0.3)		
Commercially pasteurized milk		357	0 (0)		
Raw milk	PCR[b] of IS900	1384	273 (19.3)	Switzerland	Corti and Stephan, 2002
Raw milk	Culture	232	0 (0)	Switzerland	Hummerjohann *et al.*, 2008
Raw bulk tank milk	Real-time PCR of F57	100	3 (3)	Switzerland	Bosshard *et al.*, 2006
Raw bulk tank milk	IMS-PCR of IS900	244	19 (7.8)	United Kingdom	Grant *et al.*, 2002a
Commercially pasteurized milk		567	67 (11.8)		
Raw bulk tank milk	Culture	244	4 (1.6)		
Commercially pasteurized milk		567	10 (1.8)		
Commercially pasteurized milk	Nested PCR of IS900	710	110 (15.5)	Canada	Gao *et al.*, 2002
	Culture on Middlebrook medium	244 (44 PCR +ve)	0 (0)		
Commercially pasteurized whole milk	PCR of IS900 and/or *hspX*	702	452 (64.4)	USA	Ellingson, *et al.*, 2005
	Culture on HEYA[c] or ESP® II	702	20 (2.8)		
Pooled quarter milk	PCR of IS900	1493	201 (13.5)	USA	Jayarao *et al.*, 2004
	Culture on HEYM[d]		43 (2.8)		
Raw bulk tank milk	PCR of IS900	29	8 (27.5)		
	Culture on HEYM		2 (6.8)		

Raw milk from udder	PCR of IS900	87	18 (20.7)	Poland	Szteyn *et al.*, 2008
	Culture on HEYM	87	2 (2.3)		
Commercially pasteurized milk	Culture on HEYM	83	0 (0)	Venezuela	Mendez *et al.*, 2006
Raw bulk tank milk	Peptide-mediated capture and ISMAV2 PCR	423	23 (5.4)	Germany	Stratmann *et al.*, 2006

[a] Immuno-magnetic separation-polymerase chain reaction.
[b] Polymerase chain reaction.
[c] Herrold's egg yolk agar.
[d] Herrold's egg yolk medium.

tested positive for the organism. The 10 culture-positive pasteurized milk samples were from only eight of the 241 dairies that participated in the survey. Seven of these culture-positive pasteurized milks had been heat treated at 72–74 °C for 15 s, whereas the remaining three had been treated at 72–75 °C for an extended holding time of 25 s.

Gao *et al.* (2002) collected 710 retail milks from retail stores and dairy plants in southwest Ontario, and these were tested for the presence of MAP by nested IS900 PCR. The PCR reaction was positive for 110 samples (15.5 %); however, no survivors were isolated on Middlebrook 7H9 culture broth nor Middlebrook 7H11 agar slants from 44 PCR-positive and 200 PCR-negative retail milks. The absence of viable MAP in the retail milks tested may have been a true result or may have been due to the presence of low numbers of viable cells below the detection limit of the culture method (Gao *et al.*, 2002). A double-blind study involving two laboratories was undertaken to determine the prevalence of MAP in retail pasteurized whole milk in the USA. Two culture media, Herrold's egg yolk agar and the ESP culture system, were used and identity of isolates was confirmed by PCR of the IS900 sequence and/or the *hspX* gene (Ellingson *et al.*, 2005). A total of 702 retail whole milk samples were purchased from three states (233 from California, 234 from Minnesota and 235 from Wisconsin) over a 12-month period. Viable MAP was detected in 2.8 % of the retail whole milks tested, although about 64 % of samples were positive for MAP DNA when tested using PCR alone. The prevalence of MAP in milk was similar among states, but more MAP-positive samples were identified during late summer/early autumn (July–September).

As well as cow's milk, MAP DNA has been detected in goat's milk in Norway (Djonne *et al.*, 2003) and the UK (Grant *et al.*, 2001), but the UK study failed to detect the organism in a limited number (14) of sheep milk samples. A study carried out in the Czech Republic isolated MAP from 49 % and 35.3 % of 51 dried infant milk formulae from 10 manufacturers in seven countries using IS900 sequence PCR and F57 real-time PCR, respectively (Hruska *et al.*, 2005).

Stephan *et al.* (2007) conducted a survey of 143 raw milk cheese obtained from retail outlets in Switzerland. MAP was detected using immunomagnetic capture coupled with culture in 7H10-PANTA medium or supplemented BACTEC® 12 B medium, as well as by an F57-based real-time PCR assay. They were not able to culture MAP from any of the cheeses, but 4.2 % of the cheeses were positive when the PCR method was used. A similar study of retail cheeses from Greece and the Czech Republic showed the organism was present in 50 % (21 of 42 samples) of Greek cheeses and 12 % (five of 42 samples) of Czech cheeses when IS900 sequence PCR was used (Ikonomopoulos *et al.*, 2005). However, when culture on Herrold's egg yolk medium was used to detect the organism, only three (3.6 %) of the cheeses were shown to be contaminated with MAP. The Greek cheeses were all of the Feta variety made from a mixture of sheep and goat's milk.

29.6 Survival of *Mycobacterium paratuberculosis* in foods

Very little work has been published on the ability of MAP to survive in foods. To determine the ability of MAP to survive in cheese, Sung and Collins (2000) investigated the effect of pH, salt and heat treatment on the viability of the organism. They showed faster rates of inactivation of MAP at lower pH, but NaCl concentrations between 2 % and 6 % had little effect on the ability of the organism to survive regardless of pH. However, the inactivation rates were higher in acetate buffer (pH 6, 2 % NaCl) than in Queso Fresco cheese (pH 6.06, 2 % NaCl). They concluded that heat treatment of milk together with a 60-day curing period will reduce numbers of MAP in cheese by about 3 log cycles.

Donaghy *et al.* (2004) studied the survival of MAP over a 27-week ripening period in model Cheddar cheeses prepared from pasteurized milk artificially contaminated with high (10^4–10^5 cfu/ml) and low (10^1–10^2 cfu/ml) levels of three different MAP strains. The manufactured Cheddar cheeses were similar in pH, salt, moisture and fat composition to commercial Cheddar. A slow gradual decrease in the count of MAP in all cheeses was observed over the ripening period. In all cases where high levels (> 10^4 cfu/g) of MAP were present in one-day old cheeses, the organism could be recovered after the 27-week ripening period. At low levels of contamination, only one of the three strains of MAP used was recovered from the 27-week old cheese.

Similar research has been carried out in model hard (Swiss Emmentaler) and semi-hard (Swiss Tisliter) cheese made from raw milk artificially contaminated with declumped cells of two strains of MAP at a concentration of 10^4–10^5 cfu/ml (Spahr and Schafroth, 2001). As seen with the Cheddar cheeses, MAP counts decreased gradually in both the hard and the semi-hard cheeses during ripening. However, viable cells could still be detected in 120-day old cheese. The temperatures applied during cheese manufacture and the low pH at the early stages of cheese ripening were the factors contributing most to the death of MAP in cheese. The authors concluded that counts of MAP in these cheeses would be reduced by 10^3–10^4 cfu/g during the ripening period, which lasts at least 90–120 d. MAP has been shown to survive in cheese made from raw caprine and bovine milk for 60 and 45 d, respectively (Cirone *et al.*, 2006), but the organism was not detected in cheese made from milk heat-treated at 65 °C for 20 min. Although no viable MAP could be cultured from retail cheese curds bought in Wisconsin and Minnesota, 5 % of the 98 cheese curds were positive for MAP with both IS900 and *hspX* PCR (Clark *et al.*, 2006). It is interesting to note that the acid resistance of MAP has been shown to vary with growth conditions (Sung and Collins, 2003).

Since the revelation that MAP can be isolated from pasteurized milk, there have been numerous studies to ascertain its heat stability in milk. Using a holder method, Chiodini and Hermon-Taylor (1993) observed that heat treatments simulating a batch pasteurization (63 °C for 30 min) and a high temperature–short time (HTST) treatment (72 °C for 15 s) resulted in over 91 % and 95 % destruction of MAP, respectively. When MAP was heated in milk in a sealed vial, *D*-values of 229, 48, 22 and 12 s were reported at 62, 65, 68 and 71 °C, respectively (Sung and Collins, 1998; Collins *et al.*, 2001).

When a holder method (63.5 °C for up to 40 min) and a laboratory-scale pasteurizer (72 °C for 15 s) were used, numbers declined rapidly during heating, but approximately 1 % of the initial population remained after heating (Grant *et al.*, 1996). This has been attributed to clumping of the bacterial cells rather than due to the presence of a more heat-resistant sub-population of cells (Rowe *et al.*, 2000; Klijn *et al.*, 2001). Other experiments led to the conclusion that laboratory heat treatments simulating pasteurization did not effectively eliminate MAP from the milk unless initial numbers were below 10 cfu/ml (Grant *et al.*, 1998a) or the holding time at 72 °C was extended to 25 s (Grant *et al.*, 1999). Using similar methods to heat treat raw milk inoculated with 10^3–10^7 cfu/ml of MAP, Gao *et al.* (2002) found no survivors on culture of seven milks treated by a batch process at 63 °C for 30 min, but MAP cells were detected in two of the 11 HTST (72 °C for 15 s) treated milks. The positive samples were obtained from raw milks containing 10^5 cfu/ml and 10^7 cfu/ml MAP. Lund *et al.* (2002) pointed out that the laboratory pasteurizing apparatus used by Grant and her colleagues (Grant *et al.*, 1996, 1998a, 1999) may be liable to error due to condensate and splashed cells which may be able to reach the portions of the inlet and outlet tubes of the apparatus that are above the heating liquid. These cells would receive less than the full heat treatment and may drip back into the heating medium. Also there is a significant heat-up time of about 50 s for the milk to reach approximately 72 °C.

When information first showed the presence of MAP in pasteurised milk, most of major dairies in the UK changed their process conditions by increasing the pasteurisation time from 15 to 25 s, although the Food Standards Agency recommended that the recognized minimum HTST process of 72 °C for 15 s should not be changed. Similar minimum pasteurization standards are in place in Canada and the USA.

The importance of clumping of MAP cells on heat resistance was studied by Keswani and Frank (1998). These workers used clumped and de-clumped suspensions of cultures to determine the rate of heat inactivation and survival at pasteurization temperatures in sealed capillary tubes. At 55 °C, minimal thermal inactivation was observed for both clumped and declumped cells. At 58 °C, thermal inactivation ranging from 0.3–0.7 log cycles was observed for both clumped and de-clumped suspensions for all times tested. *D*-values at 60 °C ranged from 8.6–11 min and 8.2–14.1 min for clumped and declumped cells, respectively, and the respective values at 63 °C ranged from 2.7–2.9 and 1.6–2.5 min. Keswani and Frank (1998) also studied the survival of MAP at initial levels ranging from 44–10^5 cfu/ml at 63 °C for 30 min and 72 °C for 15 s. No survivors were observed after incubating plates for up to four months on Middlebrook 7H11 agar and up to two months on Herrold's egg yolk medium. This led to the conclusion that low levels of MAP, as might be found in raw milk, will not survive pasteurization treatments.

A laboratory method, in which milk inoculated with MAP was heated in sealed tubes at temperatures ranging from 65–72 °C for up to 30 min, was compared to results obtained using a small-scale pasteurizer designed to simulate HTST units used in processing plants (Stabel *et al.*, 1997). About 10 cfu/ml of MAP (from an original population of about 1×10^6 cfu/ml) survived heating for 30 min at 72 °C in the sealed tubes. However, results obtained using the laboratory-scale pasteurizer showed that there were no detectable cells of MAP after heating for 15 s at 65, 70 or 75 °C. Thus, results obtained using the holder method could not be extrapolated to a commercial HTST unit, and that continuous flow is essential for effective killing of MAP in milk.

Further studies using a continuous flow system were described by Hope *et al.* (1996). However, because a linear holding tube was used which failed to generate turbulent flow, this study did not simulate exactly the conditions that would exist in a commercial pasteurizer. Seventeen batches of raw milk were inoculated with 10^2–10^5 cfu/ml of MAP and pasteurized at temperatures ranging from 72–90 °C for 15–35 s. The organism could not be isolated from 96 % (275/286) of pasteurized milk samples, representing at least a 4 log cycle reduction in count. Viable mycobacteria were not recovered from the heat-treated milk when raw whole milk was loaded with less than 10^4 mycobacteria per ml, and were not cultured in any of five batches of milk pasteurized at 72–73 °C for 25–35 s, which are the minimum conditions applied to correct for laminar flow in the holding tube when this machine is used commercially. Adequate holding time appeared to be more effective in killing MAP than higher temperatures in the small number of batches treated, and this is similar to results reported by Grant *et al.* (1999).

Studies on the destruction of MAP using holder and HTST pasteurization standards were conducted using a plug-flow pasteurizer and a laboratory-scale pasteurizer, both of which were used to treat UHT milk inoculated with 10^5 and 10^8 cfu/ml of three different strains of MAP (Stabel *et al.*, 2004b; Stabel and Lambertz, 2004). Five different time–temperature combinations were evaluated: 62.7 °C for 30 min, 65.5 °C for 16 s, 71.7 °C for 15 s, 71.7 °C for 20 s, and 74.4 °C for 15 s. Regardless of bacterial strain or method of heating, the heat treatments resulted in an average 5.0 and 7.7 log cycle reduction in MAP count for milk inoculated with the low and high inoculum levels, respectively.

The survival of MAP in inoculated batches of milk in a small-scale commercial unit cannot be directly extrapolated to commercial pasteurization of naturally-infected milk because of the artificially high mycobacterial loads used in these experiments, possible differences between the thermoresistance of laboratory-cultured mycobacteria, features of the small-scale unit (Hope *et al.*, 1996) and the variation in heat resistance of MAP inocula grown under different culture conditions (Sung *et al.*, 2004). However, pasteurization in a continuous-flow small-scale unit was more efficient at killing MAP than batch experiments performed in the laboratory. The inaccuracies in heat resistance data produced by laboratory-scale experiments were pointed out by Cerf and Griffiths (2000) who stated that, considering the laws of thermodynamics, it was unfeasible to expect that extending the holding time would have a greater effect on survival of MAP than increasing the heating temperature as proposed by Grant *et al.* (1999). Cerf and colleagues (Cerf *et al.*, 2007) went on to produce a quantitative model describing the probability of finding MAP in pasteurized milk under the conditions prevailing in industrialized countries. The model suggests that the main reasons for reports that purport to find MAP in pasteurized milk include improper pasteurization and cross-contamination during analysis of the samples (Mendez *et al.*, 2006).

Several studies have now been completed to determine the effect of commercial pasteurization on the survival of MAP in milk with conflicting results. Pearce *et al.* (2001) used a pilot-scale pasteurizer operating under validated turbulent flow (Reynolds number, 11 050) to study the heat sensitivity of five strains of MAP (ATCC 19698 type strain, the human isolate designated Linda and three bovine isolates) in raw whole milk for 15 s at 63, 66, 69, and 72 °C. No strains survived at 72 °C for 15 s; and only one strain survived at 69 °C. Means of pooled *D*-values at 63 and 66 °C were 15.0 ± 2.8 s and 5.9 ± 0.7 s, respectively. The mean extrapolated $D_{72\,°C}$ was < 2.03 s, which was equivalent to a > 7 $\log_{10}$ kill at 72 °C for 15 s. The mean *z*-value (the temperature increase required for a 1 $\log_{10}$ cycle reduction in *D*-value) was 8.6 °C. The five strains behaved similarly when recovery was performed on Herrold's egg yolk medium containing mycobactin or by a radiometric culture method (BACTEC®).

In an additional experiment, milk was inoculated with fresh faecal material from a high-level faecal shedder with clinical Johne's disease and heated at 72 °C for 15 s (Pearce *et al.*, 2001). Under these conditions, a minimum inactivation of > 4 $\log_{10}$ cfu/ml of MAP kill was achieved, indicating that properly maintained and operated pasteurizers should ensure the absence of viable MAP in retail milk and other pasteurized dairy products. Similar values for $D_{72\,°C}$ (1.2 s) and z (7.7 °C) were determined by Rademaker *et al.* (2007) who used high concentrations of faeces from cows with clinical symptoms of Johne's disease to contaminate raw milk in an attempt to simulate real-life situations. Final MAP concentrations varied from 10^2–3.5×10^5 cfu/ml of raw milk which was then subjected to heat treatments on a pilot-scale HTST heat exchanger at temperatures between 60 °C and 90 °C and holding times of 6–15 s. Heating the milk at 72 °C for 6 s, 70 °C for 10 and 15 s, or under more stringent conditions, resulted in no viable MAP cells being recovered. This corresponded to log reductions of > 4.2 to > 7.1, depending on the original inoculum concentrations. The authors concluded that minimum recommended HTST pasteurization conditions of 72 °C for 15 s reduced MAP counts by greater than 7 log cycles.

In another study undertaken using a pasteurizer with a validated Reynolds number (62 112) and a flow rate of 3000 l/h, 20 batches of milk inoculated with 10^3–10^4 cfu/ml MAP were homogenized and then processed with combinations of three temperatures of 72, 75, and 78 °C and three time intervals of 15, 20 and 25 s (McDonald *et al.*, 2005). Surviving cells were assayed using a culture technique capable of detecting one organism per 10 ml of milk. In 17 of the 20 runs, no viable MAP cells were detected, whereas for three of the 20 runs of milk, pasteurized at 72 °C for 15 s, 75 °C for 25 s and 78 °C for 15 s, a small number of viable cells (corresponding to 0.002–0.004 cfu/ml) were detected using a most probable number technique. Pasteurization at all temperatures and holding times was found to be very effective in killing MAP, resulting in a reduction of > 6 $\log_{10}$ in 85 % of runs and > 4 $\log_{10}$ in 100 % of runs.

Both the studies by Pearce *et al.* (2001) and McDonald *et al.* (2005) involved milk which had been artificially contaminated with MAP. A similar study was undertaken using raw cow's milk naturally infected with MAP and pasteurized using an APV HXP commercial-scale pasteurizer (capacity 2000 l/h) (Grant *et al.*, 2002b). The milk was pasteurized at 73 °C for 15 s or 25 s, with and without prior homogenization (2500 psi in two stages) in an APV Manton Gaulin KF6 homogenizer. Raw and pasteurized milk samples were tested for MAP by IMS-PCR and culture after decontamination with 0.75 % (wt/vol) cetylpyridinium chloride for 5 h. On 10 of the 12 processing occasions, MAP was detectable in either raw or pasteurized milk by IMS-PCR, culture or both methods, and viable cells were cultured from four of 60 (6.7 %) raw and 10 of 144 (6.9 %) pasteurized milks. Results suggested that survival was related to the initial load of MAP in the raw milk and that homogenization increased the lethality of subsequent heat treatments.

Unlike previous work conducted using a laboratory pasteurizer (Grant *et al.*, 1999), extending the holding time at 73 °C to 25 s was no more effective at killing MAP than the 15 s holding time. In contrast to the work of Pearce *et al.* (2001) and McDonald *et al.* (2005), this study provides evidence that MAP present in naturally-contaminated milk is capable of surviving commercial HTST pasteurization if present in sufficient numbers in the raw milk (Grant *et al.*, 2002b). However, Lynch *et al.* (2007) also demonstrated the efficacy of HTST pasteurization (72.5 °C for 27 s) using turbulent flow for both inoculated milk (10^2–10^5 cfu/ml MAP) and naturally-contaminated milk. Following heat treatment they were unable to detect MAP in any milk sample. These authors also found that MAP levels in milk from herds containing faecal-positive animals was low (Lynch *et al.*, 2007). The combination of homogenization and HTST pasteurization was more effective at inactivating MAP than HTST pasteurization alone, possibly due to the disruption of clumps of the organism present in the raw milk (Grant *et al.*, 2005a). Lund *et al.* (2000, 2002) have reviewed the reasons for the differences reported in heat resistance studies of MAP.

An alternative processing treatment involving high-voltage electric pulses, or pulsed electric field (PEF), in combination with heat treatment has been explored to determine its effect on the viability of MAP cells suspended in peptone water and in sterilized cow's milk (Rowan *et al.*, 2001). PEF treatment at 50 °C (2500 pulses at 30 kV/cm) reduced the level of viable MAP cells by about 5–6 log cycles in peptone water and in cow's milk. Heating at 50 °C for 25 min or at 72 °C for 25 s (extended HTST pasteurization) resulted in reductions of MAP counts of approximately 0.01 and 2.4 log cycles, respectively. Electron microscopy revealed that exposure to PEF treatment caused substantial damage to MAP cell membranes. UV-treatment of MAP-contaminated milk was less effective than treatment of the organism in Middlebrook 7H9 broth; inactivation rates of 0.5–1 log cycle reductions per 1000 mJ/ml and 2.5–3.3 log cycle reductions per 1000 mJ/ml were achieved, respectively (Altic *et al.*, 2007). The group at Queen's University, Belfast has also shown that high-pressure processing of milk at 500 MPa for 10 min resulted in a > 6 log reduction in MAP counts (Donaghy *et al.*, 2007), confirming earlier work in which a 4 log reduction in count of MAP in sterilized milk was obtained after treatment with 500 MPa for 10 min (Lopez-Pedemonte *et al.*, 2006). Centrifugation of milk, pre-heated to 60 °C, at 7000 g for 10 s or microfiltration (pore size 1.2 μm) were both able to remove MAP from artificially-contaminated milk with efficiencies of 95–99.9 % (Grant *et al.*, 2005b).

29.7 Survival of *Mycobacterium paratuberculosis* in the environment

One possible route of transmission of MAP from cattle to humans is through contaminated water. MAP has been isolated from water run-off from cattle farms (Raizman *et al.*, 2004). A study of the River Taff in South Wales, UK, which runs into a populated coastal region from hill pastures grazed by livestock in which MAP is endemic, showed that the organism could be detected in 31 of 96 daily samples (32.3 %) by IS900 PCR-positive and 12 of the samples by culture (Pickup *et al.*, 2005). Sequencing of DNA obtained from cultured isolates and from river water extracts revealed that 16 of 19 sequences from river water DNA extracts had a single-nucleotide polymorphism at position 214 suggesting the presence of a different, unculturable strain of MAP in the river. The strains that were isolated remained culturable in lake water microcosms for 632 d and persisted to 841 d; however, other studies have shown that MAP can survive in dam water for about 350 d and in sediment for about 500 d (Whittington *et al.*, 2005). Of four reservoirs controlling the catchment area of the Taff, MAP was present in surface sediments from three and in sediment cores from two, consistent with deposition over at least 50 years. Similar studies on the river Tywi in South Wales revealed that 68.8 % of river water samples were contaminated with MAP and in the English Lake District nine of 10 major lakes were positive for MAP (Pickup *et al.*, 2006). The same study found evidence of MAP in one of 54 domestic cold water tanks. The authors concluded that there was potential for MAP to cycle within human populations through the water supply. Epidemiological studies have indicated a highly significant increase of Crohn's disease in districts bordering the river. Whan *et al.* (2005b) have also detected the organism in untreated water entering water treatment plants in Northern Ireland and have proposed that there may be possible contact with MAP during use of recreational waters.

MAP can also survive in faecal material for up to 55 weeks in a dry, fully shaded environment, but for much shorter periods in sunny locations (Whittington *et al.*, 2004) and can survive in pasture soil for more than 200 d (Cook and Britt, 2007). The organism survived for up to 24 weeks on grass that germinated through infected faecal material applied to the soil surface in completely shaded boxes and for up to nine weeks on grass in 70 % shade. The results obtained indicated that the cells were able to enter a dormant state and sequences homologous to genes involved in dormancy responses in other mycobacteria were present in the MAP genome sequence. MAP can also survive for long periods of time in protozoa (Mura *et al.*, 2006; Whan *et al.*, 2006).

When MAP was present in water at high numbers (10^6 cfu/ml), chlorine treatment at concentrations of 2 μg/ml for up to 30 min was insufficient to completely eradicate the organism and < 3 log cycle reduction in counts was observed (Whan *et al.*, 2001). However, radiofrequency heating to temperatures of 60–65 °C for less than 1 min was capable of reducing MAP populations in wastewater from dairy and animal farms by > 6 log cycles (Lagunas-Solar *et al.*, 2005). The strategies employed by MAP for survival and dissemination have been reviewed by Rowe and Grant (2006).

29.8 Detection, enumeration and typing

Diagnostic tests for MAP have been reviewed (Grant and Rowe, 2001; Nielsen *et al.*, 2001; Tiwari *et al.*, 2006). Culture methods are laborious and time-consuming, taking 8 to 16 weeks to obtain results. Because of this long incubation period, contamination is often a problem and samples have to be treated with selective agents to reduce numbers of non-mycobacterial organisms, although care must be taken in choosing the correct decontaminant and decontamination conditions to avoid a detrimental effect on the recovery rate in different sample matrices (Dundee *et al.*, 2001; Donaghy *et al.*, 2003; Grant and Rowe, 2004; Gao *et al.*, 2005; Gwozdz, 2006; Johansen *et al.*, 2006). The low numbers present in a sample often necessitate a concentration step, such as centrifugation, filtration, or, more recently, immunomagnetic separation (Grant *et al.*, 1998b, 2000; Djonne *et al.*, 2003) and peptide-mediated capture in which a peptide, aMptD, that binds with high affinity to the MptD protein of MAP is coated on paramagnetic beads (Stratmann *et al.*, 2006). Usually an enrichment step in liquid medium, such as Dubos broth, is used prior to plating onto a solid medium. Of these, one of the most commonly used is Herrold's egg yolk medium. Confirmation of the identity of the isolate involves meeting the following criteria: (i) it must be acid-fast; (ii) it must exhibit slow growth with typical colony morphology (i.e. colonies are 1–2 mm in diameter, entire and white); (iii) it must be IS900 PCR-positive; (iv) it must require mycobactin J for growth. To decrease the time to obtain results, spiral plating in combination with microscopic colony counting (Smith *et al.*, 2003) and a modified culture method using C-18-carboxypropylbetaine also in combination with microscopic screening (Ruzante *et al.*, 2006) have been investigated. These techniques allowed counts to be obtained in 8–45 d. De Juan *et al.* (2006) studied four different media for the isolation of MAP from tissue samples of cattle and goats. Because of strain differences, they recommended prolonging the incubation period to greater than six months.

A modification of the culture method has been used successfully in which radioisotope-labelled substrates are incorporated in the growth medium. The assimilation of these substrates can be detected using the BACTEC® system (Becton Dickinson Ltd). Radiometric culture is faster and more sensitive than traditional culture techniques (Nielsen *et al.*, 2001). The radiometric method has also been combined with PCR to rapidly confirm the MAP status of the sample. Other automated methods are available, such as the ESP® II Culture System (Trek Diagnostics) with *para*-JEM liquid culture reagents that uses pressure sensing instead of radioisotopes to detect growth (Kim *et al.*, 2004), the MB/BacT system in which CO_2 produced during growth in a modified Middlebrook 7H9 broth is monitored based on changes in reflection of light from a sensor in the bottom of the culture vessel (Stich *et al.*, 2004) and the Mycobacteria Growth Indicator Tube (MGIT) which has an O_2-sensitive fluorescent sensor embedded in its base that fluoresces when O_2 becomes depleted in the medium due to growth (Grant *et al.*, 2003). In a comparison of BACTEC and MGIT, it was reported that BACTEC® remained the method of choice when the genotype of the strain was unknown (Gumber and Whittington, 2007). However, another study concluded that quantification with the MGIT system was less expensive, faster, more accurate and more sensitive than BACTEC (Shin *et al.*, 2007).

Because of its slow growth there has been considerable interest in the development of more rapid detection tests. These have mainly focused on PCR of the IS900 insertion element, which is presumed to be unique to MAP and is present in multiple copies in the genome of the bacterium (Ikonomopoulos *et al.*, 2004). However, recent work has shown that this assay cross-reacts with environmental *Mycobacteria* spp. in ruminant faeces, which share 71–79 % (Cousins *et al.*, 1999) and 94 % in sequence homology (Englund *et al.*, 2002) to IS900. The limitations of the IS900 PCR assay have been discussed by Nielsen *et al.* (2001), but the main problem seems to be the high number of false-positive reactions generated by PCR as compared to conventional culture. Because of these limitations, work has also been carried out to develop PCR assays based on other genomic loci including 251 (Rajeev *et al.*, 2005); ISMAV2 (Shin *et al.*, 2004); *hspX* (Ellingson *et al.*, 2000); *dnaA* (Rodriguez-Lazaro *et al.*, 2004); ISMap02 (Stabel and Bannantine, 2005); and F57 (Vansnick *et al.*, 2004; Tasara and Stephan, 2005; Herthnek and Bolske, 2006). There has also been interest in multiplex assays targeting a combination of genes (Tasara *et al.*, 2005). A robust TaqMan® (Roche Molecular Systems) real-time PCR assay based on the amplification of F57 and IMav2 sequences for the detection of MAP in faeces has a relative accuracy of 99.9 % and a sensitivity of 94.4 % using samples containing <10 cfu/g of faeces (Schonenbrucher *et al.*, 2008). With the availability of the complete genome sequence of MAP, identification of unique open reading frames has been facilitated (Paustian *et al.*, 2004, 2005, 2008; Li *et al.*, 2005) and has opened new avenues to explore the possibility that MAP may be a foodborne pathogen (Bannantine *et al.*, 2004). However, Mobius *et al.* (2008) have indicated that there may be problems with single round and even nested PCR for the detection of MAP regardless of the target sequence.

To improve the sensitivity of PCR assays and to assist in the removal of inhibitors present in the sample that interfere with the polymerase enzyme, methods have also been developed that combine immunomagnetic capture of the organism with PCR (Grant *et al.*, 2000; Djonne *et al.*, 2003). Using this approach it was possible to detect 10 cfu/ml of MAP in water (Whan *et al.*, 2005a) and to automate the testing (Metzger-Boddien *et al.*, 2006). Other techniques have been investigated to capture MAP from suspension. For example, Halldorsdottir and colleagues (Halldorsdottir *et al.*, 2002) used buoyant density in a Percoll gradient to remove cells prior to IS900 sequence capture PCR and dot blot assay to detect MAP in faeces, whereas Stratmann *et al.* (2002) have developed a peptide-mediated magnetic separation technique based on phage display technology to aid selective isolation of MAP from milk. Nine recombinant bacteriophages binding to MAP were isolated from a commercial phage-peptide library encoding random 12-mer peptides. Paramagnetic beads coated with the phage or with a peptide, aMP3, allowed capture of MAP from milk, and when this was combined with an ISMav2-based PCR the bacterium could be detected at levels of 100 cfu/ml in artificially-spiked milk. Experiments using milk from naturally-infected cows and bulk milk samples from infected herds demonstrated that the peptide-mediated capture PCR was able to detect single strong shedders of MAP in pooled milk samples.

Several commercial PCR assays for MAP are now available, and the performance of three of these was evaluated by Taddei *et al.* (2004). There is also growing interest in the development of real-time PCR methods. Many approaches have been used including primers and fluorescent probes targeting the 251 genomic locus (Rajeev *et al.*, 2005); fluorescence energy transfer probes targeting IS900 (O'Mahony and Hill 2004); SYBR® Green

(Molecular Probes) in combination with the LightCycler® Roche Molecular Systems) (O'Mahony and Hill, 2002; Ravva and Stanker, 2005); molecular beacons (Fang *et al.*, 2002; Rodriguez-Lazaro *et al.*, 2004); TaqMan probes (Kim *et al.*, 2002; Ravva and Stanker, 2005; Rodriguez-Lazaro *et al.*, 2005). Studies comparing real-time PCR with molecular beacons to a commercial PCR/Southern blot technique, nested PCR and culture for the detection of MAP have shown that real-time PCR assays are valid alternatives to culture (Fang *et al.*, 2002; Christopher-Hennings *et al.*, 2003). Real-time PCR has also been combined with IMS to produce a method that was capable of detecting about 10 cells of MAP in 2 ml of milk or 200 mg of faeces (Khare *et al.*, 2004) and with peptide-mediated capture which is capable of detecting about 500 cells in 1 ml of milk (Stratmann *et al.*, 2006). It is essential to develop internal controls to validate real-time PCR assays, and Brey *et al.* (2006) have described a plasmid containing the *hspX* and IS900 sequences for use with MAP PCR assays. A robust triplex real-time PCR assay amplifying ISMAV2, F57 and an internal control, which the authors claim can be used for routine diagnostic analysis, has been developed (Schonenbrucher *et al.*, 2006).

PCR methods, when used alone, cannot distinguish between viable and non-viable states of the organism and, to this end, PCR has been combined with culture methods including the ESP II system (Ellingson *et al.*, 2004) and an agar culture enrichment step (Secott *et al.*, 1999). To overcome the problem of distinguishing live and dead cells, real-time PCR and an isothermal RNA amplication method, nucleic acid sequence-based amplification (NASBA), have also been investigated (Grant and Rowe, 2001; Rodriguez-Lazaro *et al.*, 2004). Another isothermal amplification method, loop-mediated isothermal amplication (LAMP), has been successfully used to detect MAP (Enosawa *et al.*, 2003). LAMP relies on a DNA polymerase with high strand displacement activity to catalyze auto-cycling strand displacement DNA synthesis. A specially designed set of two inner and two outer primers is used, but later during the cycling reaction only the inner primers are used for strand displacement DNA synthesis. The reaction is highly specific because the target sequence is recognized by six independent sequences in the initial stage and by four independent sequences during the later stages of the LAMP reaction. A large amount of DNA is synthesized during the LAMP reaction with the concomitant production of a large concentration of pyrophosphate ion, which can be detected as a white precipitate of magnesium pyrophosphate in the reaction mixture.

Work on optimization of enrichment protocols (Motiwala *et al.*, 2005) and DNA extraction protocols (Christopher-Hennings *et al.*, 2003; Chui *et al.*, 2004; Stabel *et al.*, 2004a; Sweeney *et al.*, 2006; Cook and Britt, 2007; Gao *et al.*, 2007) has also been carried out. O'Mahony and Hill (2004) found that a milk sample treatment involving a combination of centrifugation, harsh lysis, physical grinding, boiling, nucleic acid purification and real-time PCR allowed detection of 40 cfu/ml of milk, within 3 h. The method was also able to enumerate the initial titer of MAP in milk by using pre-determined standards. Detection of 10 cfu/ml of MAP in milk has been achieved using a bead beater (a device that breaks up bacterial cell wall mechanically by vibrating bacteria with microbeads at high speed) and lysis buffer to lyze MAP cells, followed by boiling and isopropanol precipitation to extract DNA, and inclusion of 0.0037 % bovine serum albumin in the PCR reaction mixtures. The improved assay was 10–10 000-fold more sensitive than PCR assays that used template DNA prepared by other lysis procedures including boiling alone, freeze–thaw plus boiling or use of commercial kits for lysis (Odumeru *et al.*, 2001). These researchers went on to show that MAP partitions in the cream layer of raw milk and used 0.75 % hexadecylpyridinium chloride to improve collection of the organism. When this new protocol was used a detection limit of 0.3–1 cfu/ml was achieved using IS900 PCR (Gao *et al.*, 2007). An interlaboratory ring trial determined that the Adiapure® DNA extraction kit (Adiagene) coupled with the Adiavet® PCR detection kit (Adiagene), allowed easy and sensitive detection of MAP in naturally-contaminated raw milk and that MGIT medium provides an effective method for culture-dependent detection of the organism (Donaghy *et al.*, 2008). Sequence-based capture methods have also been applied for the purification of DNA prior to PCR (Vansnick *et al.*, 2007).

Because detection is difficult when mycobacteria are intracellularly located or embedded within mammalian tissues, a culture-independent, *in situ* hybridization (ISH) assay for the detection of members of the *Mycobacterium avium* complex in culture, sputum and tissue has been developed based on a set of rRNA-based oligonucleotides (St Amand *et al.*, 2005).

Immunoassays are widely used for screening cattle for Johne's disease, but the use of ELISA assays for the detection of MAP in bulk tank milks has met with little success (Nielsen *et al.*, 2000). However, more recent work using a commercially-available ELISA for antibodies against MAP and preserved milk samples in an indirect ELISA assay format may be a convenient tool for the detection of the organism in dairy herds (Hendrick *et al.*, 2005). Other commercial assays have also been adapted for the detection of MAP in milk (Winterhoff *et al.*, 2002). Commercial ELISA and immunodiagnostic kits are available from a number of companies, although concern has been raised about variability between kit lots (Dargatz *et al.*, 2004).

As well as antibodies, aptamers, which are oligonucleotides that bind to specific protein sequences, against the hypothetical protein encoded by MAP0105c have been generated and tested for their ability to bind to MAP as well as other mycobacteria (Bannantine *et al.*, 2007b).

Other methods for detection have been investigated. The FASTPlaque TB assay (Biotic Laboratories) is an established diagnostic aid for the rapid detection of *Mycobacterium tuberculosis* from human sputum. Using the FASTPlaque TB assay reagents, viable MAP cells were detected as phage plaques in just 24 h (Stanley *et al.*, 2007). As the bacteriophage used in the assay infects other mycobacteria, a PCR-based identification method specific for MAP was included. Combining the plaque PCR technique with the phage-based detection assay allowed the rapid and specific detection of viable MAP in milk samples in just 48 h. Altic *et al.* (2007) used this method to assess the efficacy of UV irradiation to reduce MAP levels in milk. There is also a report of the development of a nanosensor based on magnetic relaxation of a superparamagnetic iron oxide nanoparticle functionalized with antibodies to MAP (Kaittanis *et al.*, 2007) and a solid phase cytometric method (D'Haese *et al.*, 2005). Methods for the detection of MAP have been reviewed by Stephan (2007).

To aid in epidemiological investigations, molecular typing methods have been studied to allow differentiation of MAP isolates from human and animal reservoirs (Motiwala *et al.*, 2006b). Among the methods studied are multilocus variable-number tandem-repeat analysis (MLVA) and IS900 restriction fragment length polymorphism (RFLP) typing. When the techniques were compared, MLVA typing sub-divided the most predominant RFLP type, R01, into six sub-types and, thus, showed improved discriminatory ability (Overduin *et al.*, 2004). Good discrimination has also been obtained using a multilocus short sequence repeat sequencing approach (Amonsin *et al.*, 2004; Harris *et al.*, 2006), IS900/ERIC-PCR using primers targeting the enterobacterial intergenic consensus (ERIC) sequence and the IS900 insertion sequence (Englund, 2003), a multiplex PCR of IS900 loci (MPIL) (Bull *et al.*, 2000) and randomly amplified polymorphic DNA (RAPD) using the OPE-20 primer (5′-AACGGTGACC-3′) (Pillai *et al.*, 2001). A highly discriminatory method for typing MAP, which involves PCR-based typing markers consisting of variable-number tandem repeats (VNTR) and mycobacterial interspersed repetitive units, placed 183 MAP isolates from bovine, caprine, ovine, cervine, leporine and human origins and from 10 different countries into 21 groups (Thibault *et al.*, 2007). It has been suggested that differences exist in genotyping results obtained from faecal isolates obtained following growth on liquid or solid media and that these differences need to be taken into account when evaluating diagnostic tests and interpreting data from molecular epidemiological studies (Cernicchiaro *et al.*, 2008).

29.9 Control

Guidelines for minimizing transmission of MAP have been published (Anon., 2004). Arguably the most effective way of controlling MAP is by adopting strategies that will eliminate Johne's disease from cattle on the farm. However, this is easier said than done because of the organism's ability to survive in the environment, the long incubation period of the disease and the lack of sensitive diagnostic tests to identify infected animals during this time (Kennedy and Benedictus, 2001). It is worth noting that test-and-cull strategies alone may not reduce the prevalence of paratuberculosis in cattle and are costly for producers to pursue, but improved calf-hygiene strategies were found to be critically important in every paratuberculosis control programme (Groenendaal and Galligan, 2003).

Risk factors for seropositivity included water source, use of dairy-type nurse cows, previous clinical signs of paratuberculosis, species of cattle and geographical location (Roussel *et al.*, 2005). Non-vertebrates, such as earthworms (Fischer *et al.*, 2003a), flies (Fischer *et al.*, 2004a, 2005), cockroaches (Fischer *et al.*, 2003b) and beetles (Fischer *et al.*, 2004b), as well as wild ruminants (Machackova *et al.*, 2004) and non-ruminant wildlife (Fischer *et al.*, 2000; Anderson *et al.*, 2007) can be vectors and potentially become a risk factor in the spread of MAP infection. Risk factors have also been identified for transmission of ovine Johne's disease (Dhand *et al.*, 2007).

The risk management strategies used to contain Johne's disease include prevention of infection of the herd through maintaining a closed herd whenever possible, and when new animals have to be introduced they should be from herds free from the disease. Cattle should also be kept from pastures and other environments that may be heavily contaminated due to prior exposure to animals with the disease. It has been demonstrated that MAP can survive in faeces kept outdoors for up to 246 d, depending on the conditions (Collins *et al.*, 2001), and composting of manure at 55 °C is an efficient way of inactivating MAP (Grewal *et al.*, 2006). High-temperature composting is more effective than pack storage or liquid storage of manure and, thus, thermophilic composting is recommended for treatment of manures destined for sensitive environments such as fields used for fruit and vegetable production or for application to rapidly draining fields.

Good management of water and effluent flows from neighbouring high-risk farms and manure management can also reduce environmental contamination. The survival of MAP on pasture after destocking of all cattle infected with paratuberculosis was found to be four months during winter. Dieguez *et al.* (2008) identified the use of colostrum from MAP seropositive cows and housing replacement calves with adult cattle before they were six months old as major risk factors for Johne's disease. MAP has also been isolated from vaginal secretion of goats suffering from Johne's disease, and this may be a potential source of neonatal contamination (Vohra *et al.*, 2008).

Vaccines have been developed that are effective in reducing the number of clinically infected animals, but vaccination does not appear to reduce the total number of infected animals (sub-clinical and clinical) in a herd. Current vaccines are not used in many countries because they interfere with subsequent tests for tuberculosis (Muskens *et al.*, 2002). Work is being carried out on new targets for vaccine development such as a recombinant heat shock protein, rHsp-70 (Langelaar *et al.*, 2002, 2005), a 74 kDa recombinant polyprotein Map74F (Chen *et al.*, 2008), a 35 kDa major membrane protein, NIMP (Bannantine *et al.*, 2007a) and the rMPT 85B antigen (Mullerad *et al.*, 2002).

Exposure of individual calves to paratuberculosis should be minimized by rearing in clean environments free from adult cattle, coupled with the adoption of good hygienic practices by farm personnel. The animals should be fed milk and water that are free from contamination, and ideally the calf should be provided with an adequate colostrum intake from a paratuberculosis-negative cow.

The most important source of MAP within the herd is the cow with advanced infection, as the rate of excretion of bacteria in the faeces increases as these animals approach clinical disease. Thus, infected animals should be identified as early as possible through testing and applying knowledge of the history of the disease in the herd. Control is best achieved by culling calves of infected cows, early culling of suspect cows and test reactors along with animals that have been in contact with these cattle. Recent work has suggested that calves infected during the first weeks of life can be a frequent and important source of infection and, in herds where poor control measures were in place, MAP was shed in faeces of 7.8–80 % of calves aged between four and six months (Pavlas, 2005). However, early separation of newborn calves from cows and grazing calves under 12 months of age in areas free of adult cattle were not found to be protective against Johne's disease by other workers (Ridge *et al.*, 2005).

Several countries, including Australia and Canada, have adopted national programmes to control paratuberculosis in dairy herds and/or to accredit MAP-negative herds as low risk sources of replacement cows (McKenna *et al.*, 2006). To reduce the level of MAP in milk, the UK has recommended increasing the holding time at the minimum pasteurization temperature of 72 °C from 15 to 25 s but, as yet, there have been no published data to show whether this strategy has been successful.

The control of MAP in animal populations has been reviewed (Kennedy and Benedictus, 2001; Whittington and Sergeant, 2001). A more recent report has suggested that the most effective interventions are the on-farm management programmes designed to reduce infection in cattle (Gould *et al.*, 2005).

29.10 Conclusions

A causal link between MAP and Crohn's disease has yet to be established, but the precautionary principle approach has been advocated until the role of the organism has been definitively recognized. However, elements of the disease appear to consist of genetic predisposition, immune-mediated tissue damage and enteric factors such as intestinal microbial flora and so it is unlikely that a single aetiological agent is responsible. The arguments for and against the role of MAP in Crohn's disease are summarized in Table 29.4.

Table 29.4 Evidence for and against role of MAP in Crohn's disease

For	Against
Clinical and pathological similarities between Johne's and Crohn's diseases	Differences in clinical and pathological responses in Johne's and Crohn's diseases
Presence in food and water supplies	Lack of convincing epidemiological support of transmissible infection
Increased detection of MAP in Crohn's disease tissues	No evidence of transmission to humans in contact with animals infected with MAP
Positive blood cultures of MAP in Crohn's disease patients	Genotypic variation in human and bovine MAP isolates
Increased serological responses to MAP in Crohn's disease patients	Variability in detection of MAP by PCR[a] and serological testing. Reliability of IS900 PCR.
Detection of MAP in human breast milk by culture and PCR	Limited evidence of mycobacterial cell wall by histochemical staining in Crohn's disease tissues
Progression of cervical lymphadenopathy to distal ileitis in a patient with MAP infection	No worsening of Crohn's disease with immunosuppressive agents or HIV infection
Therapeutic responses to combination antimycobacterial therapy	No documented cell mediated immune responses to MAP in patients with Crohn's disease Limited therapeutic response to traditional antimycobacterial drugs

[a]Polymerase chain reaction.
Source: adapted from Sartor, 2005.

29.11 Sources of further information and advice

Several reviews on the role of MAP in Crohn's disease have been published: (Hermon-Taylor *et al.*, 2000; Stabel, 2000; Manning, 2001; Manning and Collins, 2001; Griffiths, 2002; Hermon-Taylor and Bull, 2002; Rubery, 2002; Collins, 2003, 2004; Grant, 2003, 2005, 2006; Chacon *et al.*, 2004; Behr and Schurr, 2006; Braun, 2006; Chamberlin and Naser, 2006; Griffiths, 2006; Kuenstner, 2006; Ryan and Campbell, 2006; Smith and Shanahan, 2006; Behr and Kapur, 2008).

29.12 References

Abbott DW, Wilkins A, Asara JM and Cantley LC (2004) The Crohn's disease protein, NOD2, requires RIP2 in order to induce ubiquitinylation of a novel site on NEMO, *Current Biology*, **14**, 2217–27.

Abubakar I, Myhill DJ, Hart AR, Lake IR, Harvey I, Rhodes JM, Robinson R, Lobo AJ, Probert CS and Hunter PR (2007) A case-control study of drinking water and dairy products in Crohn's Disease–further investigation of the possible role of *Mycobacterium avium paratuberculosis*, *American Journal of Epidemiology*, **165**, 776–83.

Abubakar I, Myhill D, Aliyu SH and Hunter PR (2008) Detection of *Mycobacterium avium* subspecies *paratuberculosis* from patients with Crohn's disease using nucleic acid-based techniques: A systematic review and meta-analysis, *Inflammatory Bowel Diseases*, **14**, 401–10.

Achkar JP and Hanauer SB (2000) Medical therapy to reduce postoperative Crohn's disease recurrence, *American Journal of Gastroenterology*, **95**, 1139–46.

Al Shamali M, Khan I, Al Nakib B, Al Hassan F and Mustafa AS (1997) A multiplex polymerase chain reaction assay for the detection of *Mycobacterium paratuberculosis* DNA in Crohn's disease tissue, *Scandinavian Journal of Gastroenterology*, **32**, 819–23.

Altic LC, Rowe MT and Grant IR (2007) UV light inactivation of *Mycobacterium avium* subsp. *paratuberculosis* in milk as assessed by FASTPlaqueTB phage assay and culture, *Applied and Environmental Microbiology*, **73**, 3728–33.

Alvarez J, De Juan L, Briones V, Romero B, Aranaz A, Fernandez-Garayzabal JF and Mateos A (2005) *Mycobacterium avium* subspecies *paratuberculosis* in fallow deer and wild boar in Spain, *Veterinary Record*, **156**, 212–13.

Amonsin A, Li LL, Zhang O, Bannantine JP, Motiwala AS, Sreevatsan S and Kapur V (2004) Multilocus short sequence repeat sequencing approach for differentiating among *Mycobacterium avium* subsp. *paratuberculosis* strains, *Journal of Clinical Microbiology*, **42**, 1694–702.

Amre DK, Lambrette P, Law L, Krupoves A, Chotard V, Costea F, Grimard G, Israel D, Mack D and Seidman EG (2006) Investigating the hygiene hypothesis as a risk factor in pediatric onset Crohn's disease: A case control study, *American Journal of Gastroenterology*, **101**, 1005–11.

Anderson JL, Meece JK, Koziczkowski JJ, Clark DL Jr, Radcliff RP, Nolden CA, Samuel MD and Ellingson JL (2007) *Mycobacterium avium* subsp *paratuberculosis* in scavenging mammals in Wisconsin, *Journal of Wildlife Diseases*, **43**, 302–8.

Anon. (2004) *Guidance on control of Johne's disease in dairy herds*, Animal Disease Control, London, DEFRA.

Autschbach F, Eisold S, Hinz U, Zinser S, Linnebacher M, Giese T, Loffler T, Buchler MW and Schmidt J (2005) High prevalence of *Mycobacterium avium* subspecies *paratuberculosis* IS900 DNA in gut tissues from individuals with Crohn's disease, *Gut*, **54**, 944–9.

Ayele WY, Svastova P, Roubal P, Bartos M and Pavlik I (2005) *Mycobacterium avium* subspecies *paratuberculosis* cultured from locally and commercially pasteurized cow's milk in the Czech Republic, *Applied and Environmental Microbiology*, **71**, 1210–14.

Baksh FK, Finkelstein SD, Ariyanayagam-Baksh SM, Swalsky PA, Klein EC and Dunn JC (2004) Absence of *Mycobacterium avium* subsp. *paratuberculosis* in the microdissected granulomas of Crohn's disease, *Modern Pathology*, **17**, 1289–94.

Bannantine JP, Barletta RG, Stabel JR, Paustian ML and Kapur V (2004) Application of the genome sequence to address concerns that *Mycobacterium avium* subspecies *paratuberculosis* might be a foodborne pathogen, *Foodborne Pathogens and Disease*, **1**, 3–15.

Bannantine JP, Radosevich TJ, Stabel JR, Berger S, Griffin JFT and Paustian ML (2007a) Production and characterization of monoclonal antibodies against a major membrane protein of *Mycobactetium avium* subsp, *paratuberculosis*, *Clinical and Vaccine Immunology* **14**, 312–17.

Bannantine JP, Radosevich TJ, Stabel JR, Sreevatsan S, Kapur V and Paustian ML (2007b) Development and characterization of monoclonal antibodies and aptamers against major antigens of *Mycobacterium avium* subsp, *paratuberculosis*, *Clinical and Vaccine Immunology*, **14**, 518–26.

Bannantine JP, Bayles DO, Waters WR, Palmer MV, Stabel JR and Paustian ML (2008a) Early antibody response against *Mycobacterium avium* subspecies *paratuberculosis* antigens in subclinical cattle, *Proteome Science*, **6**, 5.

Bannantine JP, Paustian ML, Waters WR, Stabel JR, Palmer MV, Li L and Kapur V (2008b) Profiling bovine antibody responses to *Mycobacterium avium* subsp. *paratuberculosis* infection by using protein arrays, *Infection and Immunity*, **76**, 739–49.

Barta Z, Csipo I, Mekkel G, Zeher M and Majoros L (2004) Seroprevalence of *Mycobacterium paratuberculosis* in patients with Crohn's Disease, *Journal of Clinical Microbiology*, **42**, 5432; author reply 5432–3.

Baumgart M, Dogan B, Rishniw M, Weitzman G, Bosworth B, Yantiss R, Orsi RH, Wiedmann M, McDonough P, Kim SG, Berg D, Schukken Y, Scherl E and Simpson KW (2007) Culture independent analysis of ileal mucosa reveals a selective increase in invasive *Escherichia coli* of novel phylogeny relative to depletion of Clostridiales in Crohn's disease involving the ileum, *ISME Journal*, **1**, 403–18.

Beckler DR, Elwasila S, Ghobrial G, Valentine JF and Naser SA (2008) Correlation between *rpoB* gene mutation in *Mycobacterium avium* subspecies *paratuberculosis* and clinical rifabutin and rifampicin resistance for treatment of Crohn's disease, *World Journal of Gastroenterology*, **14**, 2723–30.

Behr MA and Kapur V (2008) The evidence for *Mycobacterium paratuberculosis* in Crohn's disease, *Current Opinion in Gastroenterology*, **24**, 17–21.

Behr MA and Schurr E (2006) Mycobacteria in Crohn's disease: A persistent hypothesis, *Inflammatory Bowel Diseases*, **12**, 1000–1004.

Bell SJ and Kamm MA (2000) Review article: the clinical role of anti-TNF alpha antibody treatment in Crohn's disease, *Alimentary Pharmacology & Therapeutics*, **14**, 501–14.

Bentley RW, Keenan JI, Gearry RB, Kennedy MA, Barclay ML and Roberts RL (2008) Incidence of *Mycobacterium avium* subspecies *paratuberculosis* in a population-based cohort of patients with Crohn's disease and control subjects, *American Journal of Gastroenterology*, **103**, 1168–72.

Bernstein CN, Blanchard JF, Rawsthorne P and Collins MT (2004) Population-based case control study of seroprevalence of *Mycobacterium paratuberculosis* in patients with Crohn's disease and ulcerative colitis, *Journal of Clinical Microbiology*, **42**, 1129–35.

Bernstein CN, Wang MH, Sargent M, Brant SR and Collins MT (2007) Testing the interaction between NOD-2 status and serological response to *Mycobacterium paratuberculosis* in cases of inflammatory bowel disease, *Journal of Clinical Microbiology*, **45**, 968–71.

Bibiloni R, Mangold M, Madsen KL, Fedorak RN and Tannock GW (2006) The bacteriology of biopsies differs between newly diagnosed, untreated, Crohn's disease and ulcerative colitis patients, *Journal of Medical Microbiology*, **55**, 1141–9.

Blaser MJ, Miller RA, Lacher J and Singleton JW (1984) Patients with active Crohn's disease have elevated serum antibodies to antigens of seven enteric bacteria, *Gastroenterology*, **87**, 888–94.

Boedeker E, Harpaz N, Naser S, Blackwelder W, Tacket C, Trucksis M, Frank D and Pace N (2007) *Mycobacterium avium* ssp. *paratuberculosis* (MAP): A blinded detection study (Epru 99-01) does not support an association of MAP with Crohn's disease (CD), *Gastroenterology*, **132**, A172–A173.

Borody TJ, Leis S, Warren EF and Surace R (2002) Treatment of severe Crohn's disease using antimyobacterial triple therapy – approaching a cure? *Digestive and Liver Disease*, **34**, 29–38.

Borody TJ, Bilkey S, Wettstein AR, Leis S, Pang G and Tye S (2007) Anti-mycobacterial therapy in Crohn's disease heals mucosa with longitudinal scars, *Digestive and Liver Disease*, **39**, 438–44.

Bosshard C, Stephan R and Tasara T (2006) Application of an F57 sequence-based real-

time PCR assay for *Mycobacterium paratuberculosis* detection in bulk tank raw milk and slaughtered healthy dairy cows, *Journal of Food Protection* **69**, 1662–7.
Braun J (2006) In search of desire: mimicry, MAP, and Crohn's disease, *Gastroenterology*, **131**, 312–14.
Braun J and Targan SR (2006) Multiparameter analysis of immunogenetic mechanisms in clinical diagnosis and management of inflammatory bowel disease, *Immune Mechanisms in Inflammatory Bowel Disease*, **579**, 209–18.
Brey BJ, Radcliff RP, Clark DL Jr and Ellingson JL (2006) Design and development of an internal control plasmid for the detection of *Mycobacterium avium* subsp. *paratuberculosis* using real-time PCR, *Molecular and Cellular Probes*, **20**, 51–9.
Broxmeyer L (2005) Thinking the unthinkable: Alzheimer's, Creutzfeldt-Jakob and Mad Cow disease: the age-related reemergence of virulent, foodborne, bovine tuberculosis or losing your mind for the sake of a shake or burger, *Medical Hypotheses*, **64**, 699–705.
Bull TJ, Hermon-Taylor J, Pavlik I, El-Zaatari F and Tizard M (2000) Characterization of IS900 loci in *Mycobacterium avium* subsp. *paratuberculosis* and development of multiplex PCR typing, *Microbiology*, **146**, 2185–97.
Bull TJ, McMinn EJ, Sidi-Boumedine K, Skull A, Durkin D, Neild P, Rhodes G, Pickup R and Hermon-Taylor J (2003) Detection and verification of *Mycobacterium avium* subsp. *paratuberculosis* in fresh ileocolonic mucosal biopsy specimens from individuals with and without Crohn's disease, *Journal of Clinical Microbiology*, **41**, 2915–23.
Cellier C, De Beenhouwer H, Berger A, Penna C, Carbonnel F, Parc R, Cugnenc PH, Le Quintrec Y, Gendre JP, Barbier JP and Portaels F (1998) *Mycobacterium paratuberculosis* and *Mycobacterium avium* subsp. *silvaticum* DNA cannot be detected by PCR in Crohn's disease tissue, *Gastroenterologie Clinique et Biologique*, **22**, 675–8.
Cerf O and Griffiths MW (2000) *Mycobacterium paratuberculosis* heat resistance, *Letters in Applied Microbiology*, **30**, 341–4.
Cerf O, Griffiths M and Aziza F (2007) Assessment of the prevalence of *Mycobacterium avium* subsp. *paratuberculosis* in commercially pasteurized milk, *Foodborne Pathogens and Disease*, **4**, 433–47.
Cernicchiaro N, Wells SJ, Janagama H and Sreevatsan S (2008) Influence of type of culture medium on characterization of *Mycobacterium avium* subsp. *paratuberculosis* subtypes, *Journal of Clinical Microbiology*, **46**, 145–9.
Cetinkaya B, Egan K and Morgan KL (1994) A Practice-Based Survey of the Frequency of Johnes-Disease in South-West England, *Veterinary Record*, **134**, 494–7.
Cetinkaya B, Egan K, Harbour DA and Morgan KL (1996) An abattoir-based study of the prevalence of subclinical Johne's disease in adult cattle in south west England, *Epidemiology and Infection*, **116**, 373–9.
Chacon O, Bermudez LE and Barletta RG (2004) Johne's disease, inflammatory bowel disease, and *Mycobacterium paratuberculosis*, *Annual Review of Microbiology*, **58**, 329–63.
Chamberlin W (2007) Importance of the Australian Crohn's disease antibiotic study, *Gastroenterology*, **133**, 1744–5; author reply 1745–6.
Chamberlin WM and Naser SA (2006) Integrating theories of the etiology of Crohn's disease – On the etiology of Crohn's disease: Questioning the Hypotheses, *Medical Science Monitor*, **12**, Ra27–Ra33.
Chamberlin W, Graham DY, Hulten K, El-Zimaity HMT, Schwartz MR Naser S, Shafran I and El-Zaatari FAK (2001) Review article: *Mycobacterium avium* subsp. *paratuberculosis* as one cause of Crohn's disease, *Alimentary Pharmacology & Therapeutics*, **15**, 337–46.
Chamberlin W, Borody T and Naser S (2007a) MAP-associated Crohn's disease MAP, Koch's postulates, causality and Crohn's disease, *Digestive and Liver Disease*, **39**, 792–4.
Chamberlin W, Ghobrial G, Chehtane M and Naser SA (2007b) Successful treatment of

a Crohn's disease patient infected with bacteremic *Mycobacterium paratuberculosis*, *American Journal of Gastroenterology*, **102**, 689–91.

Chen LH, Kathaperumal K, Huang CJ, McDonough SP, Stehman S, Akey B, Huntley J, Bannantine JP, Chang CF and Chang YF (2008) Immune responses in mice to *Mycobacterium avium* subsp. *paratuberculosis* following vaccination with a novel 74F recombinant polyprotein, *Vaccine*, **26**, 1253–62.

Cheng J, Bull TJ, Dalton P, Cen S, Finlayson C and Hermon-Taylor J (2005) *Mycobacterium avium* subspecies *paratuberculosis* in the inflamed gut tissues of patients with Crohn's disease in China and its potential relationship to the consumption of cow's milk: A preliminary study, *World Journal of Microbiology & Biotechnology*, **21**, 1175–9.

Chiba M, Fukushima T, Horie Y, Iizuka M and Masamune O (1998) No *Mycobacterium paratuberculosis* detected in intestinal tissue, including Peyer's patches and lymph follicles, of Crohn's disease, *Journal of Gastroenterology*, **33**, 482–7.

Chiodini RJ (1989) Crohn's disease and the mycobacterioses: a review and comparison of two disease entities, *Clinical Microbiology Reviews*, **2**, 90–117.

Chiodini RJ and Hermon-Taylor J (1993) The thermal resistance of *Mycobacterium paratuberculosis* in raw milk under conditions simulating pasteurization, *Journal of Veterinary Diagnostic Investigation*, **5**, 629–31.

Chiodini RJ and Rossiter CA (1996) Paratuberculosis: a potential zoonosis? *Veterinary Clinics of North America: Food Animal Practice*, **12**, 457–67.

Chiodini RJ, Van Kruiningen HJ, Merkal RS, Thayer WRJ and Coutu JA (1984) Characteristics of an unclassified *Mycobacterium* species isolated from patients with Crohn's disease, *Journal of Clinical Microbiology*, **20**, 966–71.

Christopher-Hennings J, Dammen MA, Weeks SR, Epperson WB, Singh SN, Steinlicht GL, Fang Y, Skaare JL, Larsen JL, Payeur JB and Nelson EA (2003) Comparison of two DNA extractions and nested PCR, real-time PCR, a new commercial PCR assay, and bacterial culture for detection of *Mycobacterium avium* subsp. *paratuberculosis* in bovine feces, *Journal of Veterinary Diagnosis and Investigation*, **15**, 87–93.

Chui LW, King R, Lu P, Manninen K and Sim J (2004) Evaluation of four DNA extraction methods for the detection of *Mycobacterium avium* subsp. *paratuberculosis* by polymerase chain reaction, *Diagnostic Microbiology and Infectious Disease*, **48**, 39–45.

Cirone KM, Morsella CG, Colombo D and Paolicchi FA (2006) Viability of *Mycobacterium avium* subsp. *paratuberculosis* in elaborated goat and bovine milk cheese maturity, *Acta Bioquimica Clinica Latinoamericana*, **40**, 507–13.

Clancy R, Ren Z, Turton J, Pang G and Wettstein A (2007) Molecular evidence for *Mycobacterium avium* subspecies *paratuberculosis* (MAP) in Crohn's disease correlates with enhanced TNF-alpha secretion, *Digestive and Liver Disease*, **39**, 445–51.

Clark DL, Anderson JL, Koziczkowski JJ and Ellingson JLE (2006) Detection of *Mycobacterium avium* subspecies *paratuberculosis* genetic components in retail cheese curds purchased in Wisconsin and Minnesota by PCR, *Molecular and Cellular Probes*, **20**, 197–202.

Clarkston WK, Presti ME, Petersen PF, Zachary PE, Fan WX, Leonardi CL, Vernava AM, Longo WE and Kreeger JM (1998) Role of *Mycobacterium paratuberculosis* in Crohn's disease – a prospective, controlled study using polymerase chain reaction, *Diseases of the Colon & Rectum*, **41**, 195–9.

Cohavy O, Harth G, Horwitz M, Eggena M, Landers C, Sutton C, Targan SR and Braun J (1999) Identification of a novel mycobacterial histone H1 homologue (HupB) as an antigenic target of pANCA monoclonal antibody and serum immunoglobulin A from patients with Crohn's disease, *Infection and Immunity*, **67**, 6510–17.

Collins MT (2003) Paratuberculosis: Review of present knowledge, *Acta Veterinaria Scandinavica*, **44**, 217–21.

Collins MT (2004) Update on paratuberculosis: 3. Control and zoonotic potential, *Irish Veterinary Journal*, **57**, 49–52.

Collins MT, Spahr U and Murphy PM (2001) Ecological characteristics of *M. paratuberculosis*, *Bulletin of the International Dairy Federation*, **362**, 32–40.

Coloe P, Wilkes CR, Lightfoot D and Tosolini FA (1986) Isolation of *Mycobacterium paratuberculosis* in Crohn's disease, *Australian Microbiology*, **7**, 188A.

Cook LS (2000) A causal role for *Mycobacterium avium* subspecies *paratuberculosis* in Crohn's disease? *Canadian Journal of Gastroenterology*, **14**, 479–80.

Cook KL and Britt JS (2007) Optimization of methods for detecting *Mycobacterium avium* subsp. *paratuberculosis* in environmental samples using quantitative, real-time PCR, *Journal of Microbiological Methods*, **69**, 154–60.

Corn JL, Manning EJ, Sreevatsan S and Fischer JR (2005) Isolation of *Mycobacterium avium* subsp. *paratuberculosis* from free-ranging birds and mammals on livestock premises, *Applied and Environmental Microbiology*, **71**, 6963–7.

Corti S and Stephan R (2002) Detection of *Mycobacterium avium* subspecies *paratuberculosis* specific IS900 insertion sequences in bulk-tank milk samples obtained from different regions throughout Switzerland, *BMC Microbiology*, **2**, 15.

Cousins DV, Whittington R, Marsh I, Masters A, Evans RJ and Kluver P (1999) Mycobacteria distinct from *Mycobacterium avium* subsp. *paratuberculosis* isolated from the faeces of ruminants possess IS900-like sequences detectable by IS900 polymerase chain reaction: implications for diagnosis, *Molecular and Cellular Probes* **13**, 431–42.

D'Haese E, Dumon I, Werbrouck H, De Jonghe V and Herman L (2005) Improved detection of *Mycobacterium paratuberculosis* in milk, *Journal of Dairy Research*, **72**, 125–8.

Dargatz DA, Byrum BA, Collins MT, Goyal SM, Hietala SK, Jacobson RH, Kopral CA, Martin BM, McCluskey BJ and Tewari D (2004) A multilaboratory evaluation of a commercial enzyme-linked immunosorbent assay test for the detection of antibodies against *Mycobacterium avium* subsp. *paratuberculosis* in cattle, *Journal of Veterinary Diagnosis and Investigation*, **16**, 509–14.

De Hertogh G, Aerssens J, Geboes KP and Geboes K (2008) Evidence for the involvement of infectious agents in the pathogenesis of Crohn's disease, *World Journal of Gastroenterology*, **14**, 845–52.

de Juan L, Alvarez J, Romero B, Bezos J, Castellanos E, Aranaz A, Mateos A and Dominguez L (2006) Comparison of four different culture media for isolation and growth of type II and type I/III *Mycobacterium avium* subsp. *paratuberculosis* strains isolated from cattle and goats, *Applied and Environmental Microbiology*, **72**, 5927–32.

Del Prete R, Quaranta M, Lippolis A, Giannuzzi V, Mosca A, Jirillo E and Miragliotta G (1998) Detection of *Mycobacterium paratuberculosis* in stool samples of patients with inflammatory bowel disease by IS900-based PCR and colorimetric detection of amplified DNA, *Journal of Microbiological Methods*, **33**, 105–14.

Dell'Isola B, Poyart C, Goulet O, Mougenot JF, Sadoun-Journo E, Brousse N, Schmitz J, Ricour C and Berche P (1994) Detection of *Mycobacterium paratuberculosis* by polymerase chain reaction in children with Crohn's disease, *Journal of Infectious Diseases*, **169**, 449–51.

Dhand NK, Eppleston J, Whittington RJ and Toribio JALML (2007) Risk factors for ovine Johne's disease in infected sheep flocks in Australia, *Preventive Veterinary Medicine*, **82**, 51–71.

Dieguez FJ, Arnaiz I, Sanjuan ML, Vilar MJ and Yus E (2008) Management practices associated with *Mycobacterium avium* subspecies *paratuberculosis* infection and the effects of the infection on dairy herds, *Veterinary Record*, **162**, 614–17.

Djonne B, Jensen MR, Grant IR and Holstad G (2003) Detection by immunomagnetic PCR of *Mycobacterium avium* subsp. *paratuberculosis* in milk from dairy goats in Norway, *Veterinary Microbiology*, **92**, 135–43.

Donaghy JA, Totton NL and Rowe MT (2003) Evaluation of culture media for the recovery of *Mycobacterium avium* subsp. *paratuberculosis* from Cheddar cheese, *Letters in Applied Microbiology*, **37**, 285–91.

Donaghy JA, Totton NL and Rowe MT (2004) Persistence of *Mycobacterium paratuberculosis* during manufacture and ripening of cheddar cheese, *Applied and Environmental Microbiology*, **70**, 4899–905.

Donaghy JA, Linton M, Patterson MF and Rowe MT (2007) Effect of high pressure and pasteurization on *Mycobacterium avium* ssp. *paratuberculosis* in milk, *Letters in Applied Microbiology*, **45**, 154–9.

Donaghy JA, Rowe MT, Rademaker JLW, Hammer P, Herman L, De Jonghe V, Blanchard B, Duhem K and Vindel E (2008) An inter-laboratory ring trial for the detection and isolation of *Mycobacterium avium* subsp. *paratuberculosis* from raw milk artificially contaminated with naturally infected faeces, *Food Microbiology*, **25**, 128–35.

Dow CT (2006) Paratuberculosis and Type I diabetes: is this the trigger? *Medical Hypotheses*, **67**, 782–5.

Doyle TM (1954) Isolation of Johne's bacilli from the udders of clinically infected cows, *British Veterinary Journal*, **110**, 218.

Dumonceau JM, VanGossum A, Adler M, Fonteyne PA, VanVooren JP, Deviere J and Portaels F (1996) No *Mycobacterium paratuberculosis* found in Crohn's disease using the polymerase chain reaction, *Digestive Diseases and Sciences*, **41**, 421–6.

Dundee L, Grant IR, Ball HJ and Rowe MT (2001) Comparative evaluation of four decontamination protocols for the isolation of *Mycobacterium avium* subsp. *paratuberculosis* from milk, *Letters in Applied Microbiology*, **33**, 173–7.

EC (2000) *Possible Links Between Crohn's Disease and Paratuberculosis*, Report of the Scientific Committee on Animal Health and Animal Welfare, 21 March, Brussels, European Commission DG Health and Consumer Protection.

Eckburg PB and Relman DA (2007) The role of microbes in Crohn's disease, *Clinical Infectious Diseases*, **44**, 256–62.

Economou M and Pappas G (2008) New global map of Crohn's disease: Genetic environmental and socioeconomic correlations, *Inflammatory Bowel Diseases*, **14**, 709–20.

El-Zaatari FAK, Naser SA, Engstrand L, Hachem CY and Graham DY (1994) Identification and characterization of *Mycobacterium paratuberculosis* recombinant proteins expressed in *E. coli.*, *Current Microbiology*, **29**, 177–84.

El-Zaatari FA, Naser SA, Engstrand L, Burch PE, Hachem CY, Whipple DL and Graham DY (1995) Nucleotide sequence analysis and seroreactivities of the 65K heat shock protein from *Mycobacterium paratuberculosis*, *Clinical and Diagnostic Laboratory Immunology*, **2**, 657–64.

El-Zaatari FA, Naser SA and Graham DY (1997) Characterization of a specific *Mycobacterium paratuberculosis* recombinant clone expressing 35,000-molecular-weight antigen and reactivity with sera from animals with clinical and subclinical Johne's disease, *Journal of Clinical Microbiology*, **35**, 1794–9.

El-Zaatari FAK, Naser SA, Hulten K, Burch P and Graham DY (1999) Characterization of *Mycobacterium paratuberculosis* p36 antigen and its seroreactivities in Crohn's disease, *Current Microbiology*, **39**, 115–19.

El-Zaatari FAK, Osato MS and Graham DY (2001) Etiology of Crohn's disease: the role of *Mycobacterium avium paratuberculosis*, *Trends in Molecular Medicine*, **7**, 247–52.

Ellingson JLE, Stabel JR, Bishai WR, Frothingham R and Miller JM (2000) Evaluation of the accuracy and reproducibility of a practical PCR panel assay for rapid detection and differentiation of *Mycobacterium avium* subspecies, *Molecular and Cellular Probes*, **14**, 153–61.

Ellingson JLE, Koziczkowski JJ and Anderson JL (2004) Comparison of PCR prescreening to two cultivation procedures with PCR confirmation for detection of *Mycobacterium avium* subsp. *paratuberculosis* in US Department of Agriculture fecal check test samples, *Journal of Food Protection*, **67**, 2310–14.

Ellingson JLE, Anderson JL, Koziczkowski JJ, Radcliff RP, Sloan SJ, Allen SE and Sullivan NM (2005) Detection of viable *Mycobacterium avium* subsp. *paratuberculosis* in retail pasteurized whole milk by two culture methods and PCR, *Journal of Food Protection*, **68**, 966–72.

Elsaghier A, Prantera C, Moreno C and Ivanyi J (1992) Antibodies to *Mycobacterium paratuberculosis*-specific protein antigens in Crohn's disease, *Clinical and Experimental Immunology*, **90**, 503–8.

Englund S (2003) IS900/ERIC-PCR as a tool to distinguish *Mycobacterium avium* subsp. *paratuberculosis* from closely related mycobacteria, *Veterinary Microbiology*, **96**, 277–87.

Englund S, Bolske G and Johansson KE (2002) An IS900-like sequence found in a *Mycobacterium* sp. other than *Mycobacterium avium* subsp. *paratuberculosis*, *FEMS Microbiology Letters*, **209**, 267–71.

Enosawa M, Kageyama S, Sawai K, Watanabe K, Notomi T, Onoe S, Mori Y and Yokomizo Y (2003) Use of loop-mediated isothermal amplification of the IS900 sequence for rapid detection of cultured *Mycobacterium avium* subsp. *paratuberculosis*, *Journal of Clinical Microbiology*, **41**, 4359–65.

Erasmus DL, Victor TC, Vaneeden PJ, Falck V and Vanhelden P (1995) *Mycobacterium paratuberculosis* and Crohn's disease, *Gut*, **36**, 942.

Fang Y, Wu WH, Pepper JL, Larsen JL, Marras SAE, Nelson EA, Epperson WB and Christopher-Hennings J (2002) Comparison of real-time, quantitative PCR with molecular beacons to nested PCR and culture methods for detection of *Mycobacterium avium* subsp. *paratuberculosis* in bovine fecal samples, *Journal of Clinical Microbiology*, **40**, 287–91.

Feller M, Huwiler K, Stephan R, Altpeter E, Shang A, Furrer H, Pfyffer GE, Jemmi T, Baumgartner A and Egger M (2007) *Mycobacterium avium* subspecies *paratuberculosis* and Crohn's disease: a systematic review and meta-analysis, *Lancet Infectious Diseases*, **7**, 607–13.

Ferwerda G, Kullberg BJ, de Jong DJ, Girardin SE, Langenberg DML, van Crevel R, Ottenhoff THM, van der Meer JWM and Netea MG (2007) *Mycobacterium paratuberculosis* is recognized by Toll-like receptors and NOD2, *Journal of Leukocyte Biology*, **82**, 1011–18.

Fidler HM, Thurrell W, Johnson NM, Rook GAW and McFadden JJ (1994) Specific detection of *Mycobacterium paratuberculosis* DNA associated with granulomatous tissue in Crohn's disease, *Gut*, **35**, 506–10.

Fischer O, Matlova L, Bartl J, Dvorska L, Melicharek I and Pavlik I (2000) Findings of mycobacteria in insectivores and small rodents, *Folia Microbiological*, **45**, 147–52.

Fischer OA, Matlova L, Bartl J, Dvorska L, Svastova P, du Maine R, Melicharek I, Bartos M and Pavlik I (2003a) Earthworms (*Oligochaeta*, Lumbricidae) and mycobacteria, *Veterinary Microbiology*, **91**, 325–38.

Fischer OA, Matlova L, Dvorska L, Svastova P and Pavlik I (2003b) Nymphs of the Oriental cockroach (*Blatta orientalis*) as passive vectors of causal agents of avian tuberculosis and paratuberculosis, *Medical and Veterinary Entomology*, **17**, 145–50.

Fischer OA, Matlova L, Dvorska L, Svastova P, Bartl J, Weston RT and Pavlik I (2004a) Blowflies *Calliphora vicina* and *Lucilia sericata* as passive vectors of *Mycobacterium avium* subsp. *avium*, *Mycobacterium avium paratuberculosis* and *Mycobacterium avium hominissuis*, *Medical and Veterinary Entomology*, **18**, 116–22.

Fischer OA, Matlova L, Dvorska L, Svastova P, Peral DL, Weston RT, Bartos M and Pavlik I (2004b) Beetles as possible vectors of infections caused by *Mycobacterium avium* species, *Veterinary Microbiology*, **102**, 247–55.

Fischer OA, Matlova L, Dvorska L, Svastova P, Bartos M, Weston RT, Kopecna M, Trcka I and Pavlik I (2005) Potential risk of *Mycobacterium avium* subspecies *paratuberculosis* spread by syrphid flies in infected cattle farms, *Medical and Veterinary Entomology*, **19**, 360–66.

Frank TS and Cook SM (1996) Analysis of paraffin sections of Crohn's disease for *Mycobacterium paratuberculosis* using polymerase chain reaction, *Modern Pathology*, **9**, 32–5.

Fredericks D and Relman D (1996) Sequence-based identification of microbial pathogens: a reconsideration of Koch's postulates, *Clinical Microbiology Reviews*, **9**, 18–3.
Freeman H and Noble M (2005) Lack of evidence for *Mycobacterium avium* subspecies *paratuberculosis* in Crohn's disease, *Inflammatory Bowel Diseases*, **11**, 782–83.
Gan H, Ouyang Q and Bu H (1997) *Mycobacterium paratuberculosis* in the intestine of patients with Crohn's disease, *Zhonghua Nei Ke Za Zhi*, **36**, 228–30.
Gao A, Mutharia L, Chen S, Rahn K and Odumeru J (2002) Effect of pasteurization on survival of *Mycobacterium paratuberculosis* in milk, *Journal of Dairy Science*, **85**, 3198–205.
Gao AL, Odumeru J, Raymond M and Mutharia L (2005) Development of improved method for isolation of *Mycobacterium avium* subsp. *paratuberculosis* from bulk tank milk: Effect of age of milk, centrifugation, and decontamination, *Canadian Journal of Veterinary Research*, **69**, 81–7.
Gao A, Mutharia L, Raymond M and Odumeru J (2007) Improved template DNA preparation procedure for detection of *Mycobacterium avium* subsp. *paratuberculosis* in milk by PCR, *Journal of Microbiological Methods*, **69**, 417–20.
Ghadiali AH, Strother M, Naser SA, Manning EJB and Sreevatsan S (2004) *Mycobacterium avium* subsp. *paratuberculosis* strains isolated from Crohn's disease patients and animal species exhibit similar polymorphic locus patterns, *Journal of Clinical Microbiology*, **42**, 5345–8.
Girardin SE, Hugot JP and Sansonetti PJ (2003) Lessons from Nod2 studies: towards a link between Crohn's disease and bacterial sensing, *Trends in Immunology*, **24**, 652–8.
Gitlin L and Biesecker J (2007) Australian Crohn's antibiotic study opens new horizons, *Gastroenterology*, **133**, 1743–4; author reply 1745–6.
Gitnick G, Collins J, Beaman B, Brooks D, Arthur M, Imaeda T and Palieschesky M (1989) Preliminary report on isolation of mycobacteria from patients with Crohn's disease, *Digestive Diseases and Sciences*, **34**, 925–32.
Gollnick NS, Mitchell RM, Baumgart M, Janagama HK, Sreevatsand S and Schukken YH (2007) Survival of *Mycobacterium avium* subsp. *paratuberculosis* in bovine monocyte-derived macrophages is not affected by host infection status but depends on the infecting bacterial genotype, *Veterinary Immunology and Immunopathology*, **120**, 93–105.
Gould G, Franken P, Hammer P, Mackey B and Shanahan F (2005) *Mycobacterium avium* subsp. *paratuberculosis* (MAP) and the food chain, *Food Protection Trends*, **25**, 268–97.
Grant IR (2003) *Mycobacterium paratuberculosis* and milk, *Acta Veterinaria Scandinavica*, **44**, 261–6.
Grant IR (2005) Zoonotic potential of *Mycobacterium avium* ssp. *paratuberculosis*: the current position, *Journal of Applied Microbiology*, **98**, 1282–93.
Grant IR (2006) *Mycobacterium avium* ssp. *paratuberculosis* in foods: current evidence and potential consequences, *International Journal of Dairy Technology*, **59**, 112–17.
Grant IR and Rowe MT (2001) Methods for detection and enumeration of viable *Mycobacterium paratuberculosis* from milk and milk products, *Bulletin of the International Dairy Federation*, **362**, 41–52.
Grant IR and Rowe MT (2004) Effect of chemical decontamination and refrigerated storage on the isolation of *Mycobacterium avium* subsp. *paratuberculosis* from heat-treated milk, *Letters in Applied Microbiology*, **38**, 283–8.
Grant IR, Ball HJ, Neill SD and Rowe MT (1996) Inactivation of *Mycobacterium paratuberculosis* in cow's milk at pasteurization temperatures, *Applied and Environmental Microbiology*, **62**, 631–6.
Grant IR, Ball HJ and Rowe MT (1998a) Effect of high-temperature, short-time (HTST) pasteurization on milk containing low numbers of *Mycobacterium paratuberculosis*, *Letters in Applied Microbiology*, **26**, 166–70.
Grant IR, Ball HJ and Rowe MT (1998b) Isolation of *Mycobacterium paratuberculosis*

from milk by immunomagnetic separation, *Applied and Environmental Microbiology*, **64**, 3153–8.

Grant IR, Ball HJ and Rowe MT (1999) Effect of higher pasteurization temperatures, and longer holding times at 72 °C, on the inactivation of *Mycobacterium paratuberculosis* in milk, *Letters in Applied Microbiology*, **28**, 461–5.

Grant IR, Pope CM, O'Riordan LM, Ball HJ and Rowe MT (2000) Improved detection of *Mycobacterium avium* subsp. *paratuberculosis* in milk by immunomagnetic PCR, *Veterinary Microbiology*, **77**, 369–78.

Grant IR, O'Riordan LM, Ball HJ and Rowe MT (2001) Incidence of *Mycobacterium paratuberculosis* in raw sheep and goats' milk in England, Wales and Northern Ireland, *Veterinary Microbiology*, **79**, 123–31.

Grant IR, Ball HJ and Rowe MT (2002a) Incidence of *Mycobacterium paratuberculosis* in bulk raw and commercially pasteurized cows' milk from approved dairy processing establishments in the United Kingdom, *Applied and Environmental Microbiology*, **68**, 2428–35.

Grant IR, Hitchings EI, McCartney A, Ferguson F and Rowe MT (2002b) Effect of commercial-scale high-temperature, short-time pasteurization on the viability of *Mycobacterium paratuberculosis* in naturally infected cows' milk, *Applied and Environmental Microbiology*, **68**, 602–7.

Grant IR, Kirk RB, Hitchings E and Rowe MT (2003) Comparative evaluation of the MGIT (TM) and BACTEC culture systems for the recovery of *Mycobacterium avium* subsp. *paratuberculosis* from milk, *Journal of Applied Microbiology*, **95**, 196–201.

Grant IR, Williams AG, Rowe MT and Muir DD (2005a) Efficacy of various pasteurization time-temperature conditions in combination with homogenization on inactivation of *Mycobacterium avium* subsp. *paratuberculosis* in milk, *Applied and Environmental Microbiology*, **71**, 2853–61.

Grant IR, Williams AG, Rowe MT and Muir DD (2005b) Investigation of the impact of simulated commercial centrifugation and microfiltration conditions on levels of *Mycobacterium avium* ssp. *paratuberculosis* in milk, *International Journal of Dairy Technology*, **58**, 138–49.

Greenstein RJ (2003) Is Crohn's disease caused by a mycobacterium? Comparisons with leprosy, tuberculosis, and Johne's disease, *Lancet Infectious Diseases*, **3**, 507–14.

Greenstein RJ and Collins MT (2004) Emerging pathogens: is *Mycobacterium avium* subspecies *paratuberculosis* zoonotic? *Lancet*, **364**, 396–7.

Grewal SK, Rajeev S, Sreevatsan S and Michel FC (2006) Persistence of *Mycobacterium avium* subsp. *paratuberculosis* and other zoonotic pathogens during simulated composting, manure packing, and liquid storage of dairy manure, *Applied and Environmental Microbiology*, **72**, 565–74.

Griffiths MW (2002) *Mycobacterium paratuberculosis*, in C de W Blackburn and PJ McClure (eds), *Foodborne Pathogens: Hazards, Risk Analysis and Control*, Cambridge, Woodhead, 489–500.

Griffiths MW (2006) *Mycobacterium paratuberculosis*, in Y Motarjemi and M Adams (eds), *Emerging Foodborne Pathogens*, Cambridge, Woodhead, 522–56.

Griffiths TA, Rioux K and De Buck J (2008) Sequence polymorphisms in a surface PPE protein distinguish types I, II, and III of *Mycobacterium avium* subsp. *paratuberculosis*, *Journal of Clinical Microbiology*, **46**, 1207–12.

Groenendaal H and Galligan DT (2003) Food Animal Economics – Economic consequences of control programs for paratuberculosis in midsize dairy farms in the United States, *Journal of the American Veterinary Medical Association*, **223**, 1757–63.

Gumber S and Whittington RJ (2007) Comparison of BACTEC 460 and MGIT 960 systems for the culture of *Mycobacterium avium* subsp. *paratuberculosis* S strain and observations on the effect of inclusion of ampicillin in culture media to reduce contamination, *Veterinary Microbiology*, **119**, 42–52.

Gwozdz JM (2006) Comparative evaluation of two decontamination methods for the isolation of *Mycobacterium avium* subspecies *paratuberculosis* from faecal slurry and sewage, *Veterinary Microbiology*, **115**, 358–63.

Haagsma J, Mulder CJJ, Egger A and Tytgat GNJ (1991) *Mycobacterium paratuberculosis* isolé chez des patients atteints de maladie de Crohn. Resultats préliminaires, *Acta Endoscopica*, **21**, 255–60.

Halldorsdottir S, Englund S, Nilsen SF and Olsaker I (2002) Detection of *Mycobacterium avium* subsp. *paratuberculosis* by buoyant density centrifugation, sequence capture PCR and dot blot hybridisation, *Veterinary Microbiology*, **87**, 327–40.

Harris NB and Barletta RG (2001) *Mycobacterium avium* subsp. *paratuberculosis* in veterinary medicine, *Clinical Microbiology Reviews*, **14**, 489–512.

Harris NB, Payeur JB, Kapur V and Sreevatsan S (2006) Short-sequence-repeat analysis of *Mycobacterium avium* subsp. *paratuberculosis* and *Mycobacterium avium* subsp. *avium* isolates collected from animals throughout the United States reveals both stability of loci and extensive diversity, *Journal of Clinical Microbiology*, **44**, 2970–73.

Hayton AJ (2007) Johne's disease, *Cattle Practice*, **15**, 79–87.

Hendrick SH, Duffield TE, Kelton DE, Leslie KE, Lissemore KD and Archambault M (2005) Evaluation of enzyme-linked immunosorbent assays performed on milk and serum samples for detection of paratuberculosis in lactating dairy cows, *Journal of the American Veterinary Medical Association*, **226**, 424–8.

Hendrickson BA, Gokhale R and Cho JH (2002) Clinical aspects and pathophysiology of Inflammatory Bowel Disease, *Clinical Microbiology Reviews*, **15**, 79–94.

Hermon-Taylor J (1998) The causation of Crohn's disease and treatment with antimicrobial drugs, *Italian Journal of Gastroenterology and Hepatology*, **30**, 607–10.

Hermon-Taylor J (2000) *Mycobacterium avium* subspecies *paratuberculosis* in the causation of Crohn's disease, *World Journal of Gastroenterology*, **6**, 630–32.

Hermon-Taylor J (2001) *Mycobacterium avium* subspecies *paratuberculosis* is a cause of Crohn's disease, *Gut*, **49**, 755–6.

Hermon-Taylor J (2002) Treatment with drugs active against *Mycobacterium avium* subspecies *paratuberculosis* can heal Crohn's disease: more evidence for a neglected public health tragedy, *Digestive and Liver Disease*, **34**, 9–12.

Hermon-Taylor J and Bull T (2002) Crohn's disease caused by *Mycobacterium avium* subspecies *paratuberculosis*: a public health tragedy whose resolution is long overdue, *Journal of Medical Microbiology*, **51**, 3–6.

Hermon-Taylor J, Bull TJ, Sheridan JM, Cheng J, Stellakis ML and Sumar N (2000) Causation of Crohn's disease by *Mycobacterium avium* subspecies *paratuberculosis*, *Canadian Journal of Gastroenterology*, **14**, 521–39.

Herthnek D and Bolske G (2006) New PCR systems to confirm real-time PCR detection of *Mycobacterium avium* subsp. *paratuberculosis*, *BMC Microbiology*, **6**, 87.

Hope AF, Tulk PA and Condron RJ (1996) Pasteurization of *Mycobacterium paratuberculosis* in whole milk, in RJ Chiodini, ME Hines II and MC Collins (eds), *Fifth International Colloquium on Paratuberculosis*, Madison, WI, International Association for Paratuberculosis, 377–82.

Hruska K, Bartos M, Kralik P and Pavlik I (2005) *Mycobacterium avium* subsp. *paratuberculosis* in powdered infant milk: paratuberculosis in cattle – the public health problem to be solved, *Veterinarni Medicina*, **50**, 327–35.

Huda-Faujan N, Mustafa S, Manaf MYA, Yee LY and Abu Bakar F (2007) The role of microbial agents in the pathogenesis of inflammatory bowel disease, *Reviews in Medical Microbiology*, **18**, 47–53.

Hugot JP, Alberti C, Berrebi D, Bingen E and Cezard JP (2003) Crohn's disease: the cold chain hypothesis, *Lancet*, **362**, 2012–15.

Hulten K, El-Zimaity HMT, Karttunen TJ, Almashhrawi A, Schwartz MR, Graham DY and El-Zaatari FAK (2001) Detection of *Mycobacterium avium* subspecies *paratuberculosis* in Crohn's diseased tissues by *in situ* hybridization, *American Journal of Gastroenterology*, **96**, 1529–35.

Hummerjohann J, Ulmann B and Schallibatun M (2008) Incidence of paratuberculosis pathogens in Swiss raw milk, *Agrarforschung* **15**, 104–7.

Ikonomopoulos J, Gazouli M, Pavlik I, Bartos M, Zacharatos P, Xylouri E, Papalambros E and Gorgoulis V (2004) Comparative evaluation of PCR assays for the robust molecular detection of *Mycobacterium avium* subsp. *paratuberculosis*, *Journal of Microbiological Methods*, **56**, 315–21.

Ikonomopoulos J, Pavlik I, Bartos M, Svastova P, Ayele WY, Roubal P, Lukas J, Cook N and Gazouli M (2005) Detection of *Mycobacterium avium* subsp. *paratuberculosis* in retail cheeses from Greece and the Czech republic, *Applied and Environmental Microbiology*, **71**, 8934–6.

Jaravata CV, Smith WL, Rensen GJ, Ruzante J and Cullor JS (2007) Survey of ground beef for the detection of *Mycobacterium avium paratuberculosis*, *Foodborne Pathogens and Disease*, **4**, 103–6.

Jayarao BM, Pillai SR, Wolfgang DR, Griswold DR, Rossiter CA, Tewari D, Burns CM and Hutchinson LJ (2004) Evaluation of IS900-PCR assay for detection of *Mycobacterium avium* subspecies *paratuberculosis* infection in cattle using quarter milk and bulk tank milk samples, *Foodborne Pathogens and Disease*, **1**, 17–26.

Jeyanathan M, Boutros-Tadros O, Radhi J, Semret M, Bitton A and Behr MA (2007) Visualization of *Mycobacterium avium* in Crohn's tissue by oil-immersion microscopy, *Microbes and Infection*, **9**, 1567–73.

Johansen KA, Hugen EE and Payeur JB (2006) Growth of *Mycobacterium avium* subsp. *paratuberculosis* in the presence of hexadecylpyridinium chloride, natamycin, and vancomycin, *Journal of Food Protection*, **69**, 878–83.

Jones PH, Farver TB, Beaman B, Cetinkaya B and Morgan KL (2006) Crohn's disease in people exposed to clinical cases of bovine paratuberculosis, *Epidemiology and Infection*, **134**, 49–56.

Judge J, Kyriazakis I, Greig A, Allcroft DJ and Hutchings MR (2005) Clustering of *Mycobacterium avium* subsp. *paratuberculosis* in rabbits and the environment: how hot is a hot spot? *Applied and Environmental Microbiology*, **71**, 6033–8.

Judge J, Davidson RS, Marion G, White PCL and Hutchings MR (2007) Persistence of *Mycobacterium avium* subspecies *paratuberculosis* in rabbits: the interplay between horizontal and vertical transmission, *Journal of Applied Ecology*, **44**, 302–11.

Juste RA, Marco JC, Deocariz CS and Aduriz JJ (1991) Comparison of different media for the isolation of small ruminant strains of *Mycobacterium paratuberculosis*, *Veterinary Microbiology*, **28**, 385–90.

Kaittanis C, Naser SA and Perez JM (2007) One-step, nanoparticle-mediated bacterial detection with magnetic relaxation, *Nano Letters*, **7**, 380–83.

Kallinowski F, Wassmer A, Hofmann M, Harmsen D, Heesemann J, Karch H, Herfarth C and Buhr H (1998) Prevalence of enteropathogenic bacteria in surgically treated chronic inflammatory bowel disease, *Hepato-Gastroenterology*, **45**, 1552–8.

Kanazawa K, Haga Y, Funakoshi O, Nakajima H, Munakata A and Yoshida Y (1999) Absence of *Mycobacterium paratuberculosis* DNA in intestinal tissues from Crohn's disease by nested polymerase chain reaction, *Journal of Gastroenterology*, **34**, 200–206.

Kennedy DJ and Benedictus G (2001) Control of *Mycobacterium avium* subsp. *paratuberculosis* infection in agricultural species, *Revue Scientifique et Technique de L'Office International des Epizooties*, **20**, 151–79.

Keswani J and Frank JF (1998) Thermal inactivation of *Mycobacterium paratuberculosis* in milk, *Journal of Food Protection*, **61**, 974–8.

Khare S, Ficht TA, Santos RL, Romano J, Ficht AR, Zhang SP, Grant IR, Libal M, Hunter D and Adams LG (2004) Rapid and sensitive detection of *Mycobacterium avium* subsp. *paratuberculosis* in bovine milk and feces by a combination of immunomagnetic bead separation-conventional PCR and real-time PCR, *Journal of Clinical Microbiology*, **42**, 1075–81.

Kim SG, Shin SJ, Jacobson RH, Miller LJ, Harpending PR, Stehman SM, Rossiter CA and Lein DA (2002) Development and application of quantitative polymerase chain reaction assay based on the ABI 7700 system (TaqMan) for detection and quantification of *Mycobacterium avium* subsp. *paratuberculosis*, *Journal of Veterinary Diagnostic Investigation*, **14**, 126–31.

Kim SG, Kim EH, Lafferty CJ, Miller LJ, Koo HJ, Stehman SM and Shin SJ (2004) Use of conventional and real-time polymerase chain reaction for confirmation of *Mycobacterium avium* subsp. *paratuberculosis* in a broth-based culture system ESP II, *Journal of Veterinary Diagnosis and Investigation*, **16**, 448–53.

Klijn N, Herrewegh AAPM and de Jong P (2001) Heat inactivation data for *Mycobacterium avium* subsp. *paratuberculosis*: implications for interpretation, *Journal of Applied Microbiology*, **91**, 697–704.

Kobayashi K, Blaser MJ and Brown WR (1989) Immunohistochemical examination for mycobacteria in intestinal tissues from patients with Crohn's disease, *Gastroenterology*, **96**, 1009–15.

Kornbluth K, Salomon P and Suchar DB (1993) Crohn's disease, in MH Sleisenger and JS Fordtran (eds), *Gastroenterological Disease*, Philadelphia, PA, W. B. Saunders, 127–304.

Korzenik JR (2005) Past and current theories of etiology of IBD–toothpaste, worms, and refrigerators, *Journal of Clinical Gastroenterology*, **39**, S59–S65.

Kuenstner JT (2006) *Mycobacterium avium* subspecies *paratuberculosis*: A human pathogen causing most cases of Crohn's disease, *American Journal of Gastroenterology*, **101**, 1157–8.

Kuenstner JT (2007) The Australian antibiotic trial in Crohn's disease: alternative conclusions from the same study, *Gastroenterology*, **133**, 1742–3; author reply 1745–6.

Lagunas-Solar MC, Cullor JS, Zeng NX, Truong TD, Essert TK, Smith WL and Pina C (2005) Disinfection of dairy and animal farm wastewater with radiofrequency power, *Journal of Dairy Science*, **88**, 4120–31.

Lamhonwah AM, Ackerley C, Onizuka R, Tilups A, Lamhonwah D, Chung C, Tao KS, Tellier R and Tein I (2005) Epitope shared by functional variant of organic cation/carnitine transporter, OCTN1, *Campylobacter jejuni* and *Mycobacterium paratuberculosis* may underlie susceptibility to Crohn's disease at 5q31, *Biochemical and Biophysical Research Communications*, **337**, 1165–75.

Langelaar M, Koets A, Muller K, van Eden W, Noordhuizen J, Howard C, Hope J and Rutten V (2002) *Mycobacterium paratuberculosis* heat shock protein 70 as a tool in control of paratuberculosis, *Veterinary Immunology and Immunopathology* **87**, 239–44.

Langelaar MFM, Hope JC, Rutten VPMG, Noordhuizen JPTM, van Eden W and Koets AP (2005) *Mycobacterium avium* ssp. *paratuberculosis* recombinant heat shock protein 70 interaction with different bovine antigen-presenting cells, *Scandinavian Journal of Immunology*, **61**, 242–50.

Legrand E, Sola C and Rastogi N (2000) *Mycobacterium avium intracellulare* complex: phenotypic and genotypic markers and molecular basis of inter-species transmission, *Bulletin De La Société De Pathologie Exotique*, **93**, 182–92.

Li LL, Bannantine JP, Zhang Q, Amonsin A, May BJ, Alt D, Banerji N, Kanjilal S and Kapur V (2005) The complete genome sequence of *Mycobacterium avium* subspecies *paratuberculosis*, *Proceedings of the National Academy of Sciences of the United States of America*, **102**, 12344–9.

Lipton JE and Barash DP (2007) Flawed Australian CD study does not end MAP controversy, *Gastroenterology*, **133**, 1742; author reply 1745–6.

Lisby G, Andersen J, Engbaek K and Binder V (1994) *Mycobacterium paratuberculosis* in intestinal tissue from patients with Crohn's disease demonstrated by a nested primer Polymerase Chain Reaction, *Scandinavian Journal of Gastroenterology*, **29**, 923–9.

Lopez-Pedemonte T, Sevilla I, Garrido JM, Aduriz G, Guamis B, Juste RA and Roig-Sagues AX (2006) Inactivation of *Mycobacterium avium* subsp. *paratuberculosis* in cow's milk by means of high hydrostatic pressure at mild temperatures, *Applied and Environmental Microbiology*, **72**, 4446–9.

Lozano-Leon A, Barreiro-de Acosta M and Dominguez-Munoz JE (2006) Absence of *Mycobacterium avium* subspecies *paratuberculosis* in Crohn's disease patients, *Inflammatory Bowel Diseases*, **12**, 1190–92.

Lund BM, Donnelly CW and Rampling A (2000) Heat resistance of *Mycobacterium paratuberculosis*, *Letters in Applied Microbiology*, **31**, 184–5.

Lund BM, Gould GW and Rampling AM (2002) Pasteurization of milk and the heat resistance of *Mycobacterium avium* subsp. *paratuberculosis*: a critical review of the data, *International Journal of Food Microbiology*, **77**, 135–45.

Lynch D, Jordan KN, Kelly PM, Freyne T and Murphy PM (2007) Heat sensitivity of *Mycobacterium avium* ssp. *paratuberculosis* in milk under pilot plant pasteurization conditions, *International Journal of Dairy Technology*, **60**, 98–104.

Machackova M, Svastova P, Lamka J, Parmova I, Liska V, Smolik J, Fischer OA and Pavlik I (2004) Paratuberculosis in farmed and free-living wild ruminants in the Czech Republic (1999–2001), *Veterinary Microbiology*, **101**, 225–34.

Manning EJB (2001) *Mycobacterium avium* subspecies *paratuberculosis*: a review of current knowledge, *Journal of Zoo and Wildlife Medicine*, **32**, 293–304.

Manning EJB and Collins MT (2001) *Mycobacterium avium* subsp. *paratuberculosis*: pathogen, pathogenesis and diagnosis, *Revue Scientifique et Technique de l'Office International des Epizooties*, **20**, 133–50.

Marsh IB, Bannantine JP, Paustian ML, Tizard ML, Kapur V and Whittington RJ (2006) Genomic comparison of *Mycobacterium avium* subsp. *paratuberculosis* sheep and cattle strains by microarray hybridization, *Journal of Bacteriology*, **188**, 2290–93.

Martinez-Medina M, Aldeguer X, Gonzalez-Huix F, Acero D and Garcia-Gil LJ (2006) Abnormal microbiota composition in the ileocolonic mucosa of Crohn's disease patients as revealed by polymerase chain reaction-denaturing gradient gel electrophoresis, *Inflammatory Bowel Diseases*, **12**, 1136–45.

McDonald WL, O'Riley KJ, Schroen CJ and Condron RJ (2005) Heat inactivation of *Mycobacterium avium* subsp. *paratuberculosis* in milk, *Applied and Environmental Microbiology*, **71**, 1785–9.

McKenna SL, Keefe GP, Tiwari A, VanLeeuwen J and Barkema HW (2006) Johne's disease in Canada part II: disease impacts, risk factors, and control programs for dairy producers, *Canadian Veterinary Journal*, **47**, 1089–99.

Meconi S, Vercellone A, Levillain F, Payre B, Al Saati T, Capilla F, Desreumaux P, Darfeuille-Michaud A and Altare F (2007) Adherent-invasive *Escherichia coli* isolated from Crohn's disease patients induce granulomas *in vitro*, *Cellular Microbiology*, **9**, 1252–61.

Mendez D, Gimenez F, Escalona A, Da Mata O, Gonzalez A, Takiff H and de Waard JH (2006) *Mycobacterium bovis* cultured from commercially pasteurized cows' milk: laboratory cross-contamination, *Veterinary Microbiology*, **116**, 325–8.

Merkal RS, Whipple DL, Sacks JM and Snyder GR (1987) Prevalence of *Mycobacterium paratuberculosis* in ileocecal lymph nodes of cattle culled in the United States, *Journal of the American Veterinary Medical Association*, **190**, 676–80.

Metzger-Boddien C, Khaschabi D, Schonbauer M, Boddien S, Schlederer T and Kehle J (2006) Automated high-throughput immunomagnetic separation-PCR for detection of *Mycobacterium avium* subsp. *paratuberculosis* in bovine milk, *International Journal of Food Microbiology*, **110**, 201–8.

Millar D, Ford J, Sanderson J, Withey S, Tizard M, Doran T and Hermon-Taylor J (1996) IS900 PCR to detect *Mycobacterium paratuberculosis* in retail supplies of whole pasteurized cow's milk in England and Wales, *Applied and Environmental Microbiology*, **62**, 3446–52.

Mishina D, Katsel P, Brown ST, Gilberts ECAM and Greenstein RJ (1996) On the etiology of Crohns disease, *Proceedings of the National Academy of Sciences of the United States of America*, **93**, 9816–20.

Mobius P, Hotzel H, Rassbach A and Kohler H (2008) Comparison of 13 single-round and nested PCR assays targeting IS900, ISMav2, f57 and locus 255 for detection of *Mycobacterium avium* subsp. *paratuberculosis*, *Veterinary Microbiology*, **126**, 324–33.

Moloney BJ and Whittington RJ (2008) Cross species transmission of ovine Johne's disease from sheep to cattle: an estimate of prevalence in exposed susceptible cattle, *Australian Veterinary Journal*, **86**, 117–23.

Moss MT, Sanderson JD, Tizard MLV, Hermon-Taylor J, Elzaatari FAK, Markesich DC and Graham DY (1992) Polymerase chain reaction detection of *Mycobacterium paratuberculosis* and *Mycobacterium avium* subsp. *silvaticum* in long-term cultures from Crohn's disease and control tissues, *Gut*, **33**, 1209–13.

Motiwala AS, Amonsin A, Strother M, Manning EJB, Kapur V and Sreevatsan S (2004) Molecular epidemiology of *Mycobacterium avium* subsp. *paratuberculosis* isolates recovered from wild animal species, *Journal of Clinical Microbiology*, **42**, 1703–12.

Motiwala AS, Strother M, Theus NE, Stich RW, Byrum B, Shulaw WP, Kapur V and Sreevatsan S (2005) Rapid detection and typing of strains of *Mycobacterium avium* subsp. *paratuberculosis* from broth cultures, *Journal of Clinical Microbiology*, **43**, 2111–17.

Motiwala AS, Janagama HK, Paustian ML, Zhu XC, Bannantine JP, Kapur V and Sreevatsan S (2006a) Comparative transcriptional analysis of human macrophages exposed to animal and human isolates of *Mycobacterium avium* subspecies *paratuberculosis* with diverse genotypes, *Infection and Immunity*, **74**, 6046–56.

Motiwala AS, Li L, Kapur V and Sreevatsan S (2006b) Current understanding of the genetic diversity of *Mycobacterium avium* subsp. *paratuberculosis*, *Microbes and Infection*, **8**, 1406–18.

Mpofu CM, Campbell BJ, Subramanian S, Marshall-Clarke S, Hart CA, Cross A, Roberts CL, Mcgoldrick A, Edwards SW and Rhodes JM (2007) Microbial mannan inhibits bacterial killing by macrophages: A possible pathogenic mechanism for Crohn's disease, *Gastroenterology*, **133**, 1487–98.

Mullerad J, Michal I, Fishman Y, Hovav AH, Barletta RG and Bercovier H (2002) The immunogenicity of *Mycobacterium paratuberculosis* 85B antigen, *Medical Microbiology and Immunology*, **190**, 179–87.

Mura M, Bull TJ, Evans H, Sidi-Boumedine K, McMinn L, Rhodes G, Pickup R and Hermon-Taylor J (2006) Replication and long-term persistence of bovine and human strains of *Mycobacterium avium* subsp. *paratuberculosis* within *Acanthamoeba polyphaga*, *Applied and Environmental Microbiology*, **72**, 854–9.

Murray A, Oliaro J, Schlup MMT and Chadwick VS (1995) *Mycobacterium paratuberculosis* and inflammatory bowel disease: frequency distribution in serial colonoscopic biopsies using the polymerase chain reaction, *Microbios*, **83**, 217–28.

Muskens J, van Zijderveld F, Eger A and Bakker D (2002) Evaluation of the long-term immune response in cattle after vaccination against paratuberculosis in two Dutch dairy herds, *Veterinary Microbiology*, **86**, 269–78.

Nakase H, Nishio A, Tamaki H, Matsuura M, Asada M, Chiba T and Okazaki K (2006) Specific antibodies against recombinant protein of insertion element 900 of *Mycobacterium avium* subspecies *paratuberculosis* in Japanese patients with Crohn's disease, *Inflammatory Bowel Diseases*, **12**, 62–9.

Naser SA, Hulten K, Shafran I, Graham DY and El-Zaatari FAK (2000a) Specific seroreactivity of Crohn's disease, patients against p35 and p36 antigens of *Mycobacterium avium* subsp. *paratuberculosis*, *Veterinary Microbiology*, **77**, 497–504.

Naser SA, Schwartz D and Shafran I (2000b) Isolation of *Mycobacterium avium* subsp. *paratuberculosis* from breast milk of Crohn's disease patient, *American Journal of Gastroenterology*, **95**, 1094–5.

Naser SA, Ghobrial G, Romero C and Valentine JF (2004) Culture of *Mycobacterium avium* subspecies *paratuberculosis* from the blood of patients with Crohn's disease, *Lancet*, **364**, 1039–44.

Nauta MJ and van der Giessen JWB (1998) Human exposure to *Mycobacterium paratuberculosis* via pasteurised milk: A modelling approach, *Veterinary Record*, **143**, 293–6.

Nielsen SS, Thamsborg SM, Houe H and Bitsch V (2000) Bulk-tank milk ELISA antibodies for estimating the prevalence of paratuberculosis in Danish dairy herds, *Preventive Veterinary Medicine*, **44**, 1–7.

Nielsen SS, Nielsen KK, Huda A, Condron R and Collins MT (2001) Diagnostic techniques for paratuberculosis, *Bulletin of the International Dairy Federation*, **362**, 5–17.

O'Mahony J and Hill C (2002) A real time PCR assay for the detection and quantitation of *Mycobacterium avium* subsp. *paratuberculosis* using SYBR Green and the Light Cycler, *Journal of Microbiological Methods*, **51**, 283–93.

O'Mahony J and Hill C (2004) Rapid real-time PCR assay for detection and quantitation of *Mycobacterium avium* subsp. *paratuberculosis* DNA in artificially contaminated milk, *Applied and Environmental Microbiology*, **70**, 4561–8.

O'Reilly CE, O'Connor L, Anderson W, Harvey P, Grant IR, Donaghy J, Rowe M and O'Mahony P (2004) Surveillance of bulk raw and commercially pasteurized cows' milk from approved Irish liquid-milk pasteurization plants to determine the incidence of *Mycobacterium paratuberculosis*, *Applied and Environmental Microbiology*, **70**, 5138–144.

Odumeru J, Gao A, Chen S, Raymond M and Mutharia L (2001) Use of the bead beater for preparation of *Mycobacterium paratuberculosis* template DNA in milk, *Canadian Journal of Veterinary Research*, **65**, 201–5.

Olsen I, Siguroardottir OG and Djonne B (2002) Paratuberculosis with special reference to cattle – a review, *Veterinary Quarterly*, **24**, 12–28.

Overduin P, Schouls L, Roholl P, van der Zanden A, Mahmmod N, Herrewegh A and van Soolingen D (2004) Use of multilocus variable-number tandem-repeat analysis for typing *Mycobacterium avium* subsp. *paratuberculosis*, *Journal of Clinical Microbiology*, **42**, 5022–8.

Patel D, Danelishvili L, Yamazaki Y, Alonso M, Paustian ML, Bannantine JP, Meunier-Goddik L and Bermudez LE (2006) The ability of *Mycobacterium avium* subsp. *paratuberculosis* to enter bovine epithelial cells is influenced by preexposure to a hyperosmolar environment and intracellular passage in bovine mammary epithelial cells, *Infection and Immunity*, **74**, 2849–55.

Paustian ML, Amonsin A, Kapur V and Bannantine JP (2004) Characterization of novel coding sequences specific to *Mycobacterium avium* subsp. *paratuberculosis*: implications for diagnosis of Johne's Disease, *Journal of Clinical Microbiology*, **42**, 2675–81.

Paustian ML, Kapur V and Bannantine JP (2005) Comparative genomic hybridizations reveal genetic regions within the *Mycobacterium avium* complex that are divergent from *Mycobacterium avium* subsp. *paratuberculosis* isolates, *Journal of Bacteriology*, **187**, 2406–15.

Paustian ML, Zhu XC, Sreevatsan S, Robbe-Austerman S, Kapur V and Bannantine JP (2008) Comparative genomic analysis of *Mycobacterium avium* subspecies obtained from multiple host species, *BMC Genomics*, **9**, 135.

Pavlas M (2005) New findings of pathogenesis, diagnostics and control of paratuberculosis in cattle, *Acta Veterinaria Brno*, **74**, 73–9.

Pavlik I, Bejckova L and Fixa B (1994) DNA fingerprinting as a tool for epidemiological studies of paratuberculosis in ruminants and Crohn's disease, in RJ Chiodini, MT Collins and E Bassey (eds), *Proceedings of the Fourth International Colloquium on Paratuberculosis*, Madison, WI, International Association for Paratuberculosis, Inc, 278–89.

Pearce LE, Truong HT, Crawford RA, Yates GF, Cavaignac S and de Lisle GW (2001) Effect of turbulent-flow pasteurization on survival of *Mycobacterium avium* subsp.

paratuberculosis added to raw milk, *Applied and Environmental Microbiology*, **67**, 3964–9.

Peltekova VD, Wintle RF, Rubin LA, Amos CI, Huang Q, Gu X, Newman B, Van Oene M, Cescon D, Greenberg G, Griffiths AM, St George-Hyslop PH and Siminovitch KA (2004) Functional variants of OCTN cation transporter genes are associated with Crohn's disease, *Nature Genetics*, **36**, 471–5.

Peyrin-Biroulet L, Neut C and Colombel JF (2007) Antimycobacterial therapy in Crohn's disease: Game over? *Gastroenterology*, **132**, 2594–8.

Pickup RW, Rhodes G, Arnott S, Sidi-Boumedine K, Bull TJ, Weightman A, Hurley M and Hermon-Taylor J (2005) *Mycobacterium avium* subsp. *paratuberculosis* in the catchment area and water of the river Taff in South Wales, United Kingdom, and its potential relationship to clustering of Crohn's disease cases in the city of Cardiff, *Applied and Environmental Microbiology*, **71**, 2130–39.

Pickup RW, Rhodes G, Bull TJ, Arnott S, Sidi-Boumedine K, Hurley M and Hermon-Taylor J (2006) *Mycobacterium avium* subsp. *paratuberculosis* in lake catchments, in river water abstracted for domestic use, and in effluent from domestic sewage treatment works: diverse opportunities for environmental cycling and human exposure, *Applied and Environmental Microbiology*, **72**, 4067–77.

Pillai SR, Jayarao BM, Gummo JD, Hue EC, Tiwari D, Stabel JR and Whitlock RH (2001) Identification and sub-typing of *Mycobacterium avium* subsp. *paratuberculosis* and *Mycobacterium avium* subsp. *avium* by randomly amplified polymorphic DNA, *Veterinary Microbiology*, **79**, 275–84.

Polymeros D, Bogdanos DP, Day R, Arioli D, Vergani D and Forbes A (2006) Does cross-reactivity between *Mycobacterium avium paratuberculosis* and human intestinal antigens characterize Crohn's disease? *Gastroenterology*, **131**, 85–96.

Prantera C, Bothamley G, Levenstein S, Mangiarotti R and Argentieri R (1989) Crohn's disease and mycobacteria: two cases of Crohn's disease with high anti-mycobacterial antibody levels cured by dapsone therapy, *Biomedicine and Pharmacotherapy*, **43**, 295–9.

Prantera C, Kohn A, Mangiarotti R, Andreoli A and Luzi C (1994) Antimycobacterial therapy in Crohn's disease: results of a controlled double blind trial with a multiple antibiotic regimen, *American Journal of Gastroenterology*, **89**, 513–18.

Prantera C, Zannoni F, Scribano MI, Berto E, Andreoli A, Kohn A and Luzi C (1996) An antibiotic regimen for the treatment of active Crohn's disease: a randomized, controlled trial of metronidazole and ciprofloxacin, *American Journal of Gastroenterology*, **91**, 328–32.

Prescott NJ, Fisher SA, Franke A, Hampe J, Onnie CM, Soars D, Bagnall R, Mirza MM, Sanderson J, Forbes A, Mansfield JC, Lewis CM, Schreiber S and Matthew CG (2007) A nonsynonymous SNP in ATG16L1 predisposes to ileal Crohn's disease and is independent of CARD15 and IBD5, *Gastroenterology*, **132**, 1665–71.

Qin X (2008) What is human inflammatory bowel disease (IBD) more like: Johne's disease in cattle or IBD in dogs and cats? *Inflammatory Bowel Diseases*, **14**, 138.

Quirke P (2001) Antagonist, *Mycobacterium avium* subspecies *paratuberculosis* is a cause of Crohn's disease, *Gut*, **49**, 757–60.

Rademaker JLW, Vissers MMM and Giffel MCT (2007) Effective heat inactivation of *Mycobacterium avium* subsp. *paratuberculosis* in raw milk contaminated with naturally infected feces, *Applied and Environmental Microbiology*, **73**, 4185–90.

Raizman EA, Wells SJ, Godden SM, Bey RF, Oakes MJ, Bentley DC and Olsen KE (2004) The distribution of *Mycobacterium avium* ssp. *paratuberculosis* in the environment surrounding Minnesota dairy farms, *Journal of Dairy Science*, **87**, 2959–66.

Raizman EA, Wells SJ, Jordan PA, DelGiudice GD and Bey RR (2005) *Mycobacterium avium* subsp. *paratuberculosis* from free-ranging deer and rabbits surrounding Minnesota dairy herds, *Canadian Journal of Veterinary Research*, **69**, 32–8.

Rajeev S, Zhang Y, Sreevatsan S, Motiwala AS and Byrum B (2005) Evaluation of multiple genomic targets for identification and confirmation of *Mycobacterium avium* subsp. *paratuberculosis* isolates using real-time PCR, *Veterinary Microbiology*, **105**, 215–21.

Ravva SV and Stanker LH (2005) Real-time quantitative PCR detection of *Mycobacterium avium* subsp. *paratuberculosis* and differentiation from other mycobacteria using SYBR Green and TaqMan assays, *Journal of Microbiological Methods*, **63**, 305–17.

Ren Z, Turton J, Borody T, Pang G and Clancy R (2008) Selective Th2 pattern of cytokine secretion in *Mycobacterium avium* subsp. *paratuberculosis* infected Crohn's disease, *Journal of Gastroenterology and Hepatology*, **23**, 310–14.

Richter E, Wessling J, Lugering N, Domschke W and Rusch-Gerdes S (2002) *Mycobacterium avium* subsp. *paratuberculosis* infection in a patient with HIV, Germany, *Emerging Infectious Diseases*, **8**, 729–31.

Rideout BA, Brown ST, Davis WC, Gay JM, Giannella RA, Hines II ME, Hueston WD and Hutchinson LJ (2003) *Diagnosis and Control of Johne's Disease*, Washington, DC, The National Academies Press.

Ridge SE, Baker IM and Hannah M (2005) Effect of compliance with recommended calf-rearing practices on control of bovine Johne's disease, *Australian Veterinary Journal*, **83**, 85–90.

Robino P, Nebbia P, Tramuta C, Martinet M, Ferroglio E and De Meneghi D (2008) Identification of *Mycobacterium avium* subsp. *paratuberculosis* in wild cervids (*Cervus elaphus hippelaphus* and *Capreolus capreolus*) from Northwestern Italy, *European Journal of Wildlife Research*, **54**, 357–60.

Rodriguez-Lazaro D, Lloyd J, Herrewegh A, Ikonomopoulos J, D'Agostino M, Pla M and Cook N (2004) A molecular beacon-based real-time NASBA assay for detection of *Mycobacterium avium* subsp. *paratuberculosis* in water and milk, *FEMS Microbiology Letters*, **237**, 119–26.

Rodriguez-Lazaro D, D'Agostino M, Herrewegh A, Pla M, Cook N and Ikonomopoulos J (2005) Real-time PCR-based methods for detection of *Mycobacterium avium* subsp. *paratuberculosis* in water and milk, *International Journal of Food Microbiology*, **101**, 93–104.

Roediger WE (1988) What sequence of pathogenic events leads to acute ulcerative colitis, *Diseases of the Colon & Rectum*, **31**, 482–7.

Roholl PJM, Herrewegh A and van Soolingen D (2002) Positive IS900 *in situ* hybridization signals as evidence for role of *Mycobacterium avium* subsp. *paratuberculosis* in etiology of Crohn's disease, *Journal of Clinical Microbiology*, **40**, 3112.

Romero C, Hamdi A, Valentine JF and Naser SA (2005) Evaluation of surgical tissue from patients with Crohn's disease for the presence of *Mycobacterium avium* subspecies *paratuberculosis* DNA by *in situ* hybridization and nested polymerase chain reaction, *Inflammatory Bowel Diseases*, **11**, 116–25.

Roussel AJ, Libal MC, Whitlock RL, Hairgrove TB, Barling KS and Thompson JA (2005) Prevalence of and risk factors for paratuberculosis in purebred beef cattle, *Journal of the American Veterinary Medicine Association*, **226**, 773–8.

Rowan NJ, MacGregor SJ, Anderson JG, Cameron D and Farish O (2001) Inactivation of *Mycobacterium paratuberculosis* by pulsed electric fields, *Applied and Environmental Microbiology*, **67**, 2833–6.

Rowbotham DS, Mapstone NP, Trejdosiewicz LK, Howdle PD and Quirke P (1995) *Mycobacterium paratuberculosis* DNA not detected in Crohn's disease tissue by fluorescent polymerase chain reaction, *Gut*, **37**, 660–67.

Rowe MT and Grant IR (2006) *Mycobacterium avium* ssp. *paratuberculosis* and its potential survival tactics, *Letters in Applied Microbiology*, **42**, 305–11.

Rowe MT, Grant IR, Dundee L and Ball HJ (2000) Heat resistance of *Mycobacterium avium* subsp. *paratuberculosis* in milk, *Irish Journal of Agricultural and Food Research*, **39**, 203–8.

Rubery E (2002) *A review of the evidence for a link between exposure to Mycobacterium paratuberculosis (MAP) and Crohn's disease in humans*, London, Food Standards Agency.

Rumsey JW, Valentine JF and Naser SA (2006) Inhibition of phagosome maturation and survival of *Mycobacterium avium* subspecies *paratuberculosis* in polymorphonuclear leukocytes from Crohn's disease patients, *Medical Science Monitor*, **12**, Br130–Br139.

Ruzante JM, Smith WL, Gardner IA, Thornton CG and Cullor JS (2006) Modified culture protocol for isolation of *Mycobacterium avium* subsp. *paratuberculosis* from raw milk, *Foodborne Pathogens and Disease*, **3**, 457–60.

Ryan T and Campbell D (2006) *Mycobacterium paratuberculosis – A public health issue?* Wellington, New Zealand Food Safety Authority.

Ryan P, Kelly RG, Lee G, Collins JK, O'Sullivan GC, O'Connell J and Shanahan F (2004) Bacterial DNA within granulomas of patients with Crohn's disease – detection by laser capture microdissection and PCR, *American Journal of Gastroenterology*, **99**, 1539–43.

Sandborn WJ (2005) New concepts in anti-tumor necrosis factor therapy for inflammatory bowel disease, *Reviews in Gastroenterological Disorders*, **5**, 10–18.

Sanderson JD, Moss MT, Tizard ML and Hermon-Taylor J (1992) *Mycobacterium paratuberculosis* DNA in Crohn's disease tissue, *Gut*, **33**, 890–96.

Sartor RB (2005) Does *Mycobacterium avium* subspecies *paratuberculosis* cause Crohn's disease? *Gut*, **54**, 896–8.

Scanu AM, Bull TJ, Cannas S, Sanderson JD, Sechi LA, Dettori G, Zanetti S and Hermon-Taylor J (2007) *Mycobacterium avium* subspecies *paratuberculosis* infection in cases of irritable bowel syndrome and comparison with Crohn's disease and Johne's disease: common neural and immune pathogenicities, *Journal of Clinical Microbiology*, **45**, 3883–90.

Schmid M, Fellermann K, Wehkamp J, Herrlinger K and Stange EF (2004) The role of defensins in the pathogenesis of inflammatory bowel disease, *Zeitschrift Fur Gastroenterologie*, **42**, 333–8.

Schonenbrucher H, Abdulmawjood A and Buelte M (2006) Real Time-PCR-assay for the detection of *Mycobacterium avium* ssp. *paratubercuiosis* – development and validation, *Fleischwirtschaft*, **86**, 123–5.

Schonenbrucher H, Abdurnawjood A, Failing K and Bulte M (2008) New triplex real-time PCR assay for detection of *Mycobacterium avium* subsp. *paratuberculosis* in bovine feces, *Applied and Environmental Microbiology*, **74**, 2751–8.

Sechi LA, Manuela M, Francesco T, Amelia L, Antonello S, Giovanni F and Stefania Z (2001) Identification of *Mycobacterium avium* subsp. *paratuberculosis* in biopsy specimens from patients with Crohn's disease identified by *in situ* hybridization, *Journal of Clinical Microbiology*, **39**, 4514–17.

Sechi LA, Mura M, Tanda F, Lissia A, Fadda G and Zanetti S (2004) *Mycobacterium avium* sub. *paratuberculosis* in tissue samples of Crohn's disease patients, *Microbiologica*, **27**, 75–7.

Sechi LA, Gazouli M, Ikonomopoulos J, Lukas JC, Scanu AM, Ahmed N, Fadda G and Zanetti S (2005a) *Mycobacterium avium* subsp. *paratuberculosis*, genetic susceptibility to Crohn's disease, and Sardinians: the way ahead, *Journal of Clinical Microbiology*, **43**, 5275–7.

Sechi LA, Scanu AM, Molicotti P, Cannas S, Mura M, Dettori G, Fadda G and Zanetti S (2005b) Detection and isolation of *Mycobacterium avium* subspecies *paratuberculosis* from intestinal mucosal biopsies of patients with and without Crohn's disease in Sardinia, *American Journal of Gastroenterology*, **100**, 1529–36.

Sechi LA, Ahmed N, Felis GE, Dupre I, Cannas S, Fadda G, Bua A and Zanetti S (2006a) Immunogenicity and cytoadherence of recombinant heparin binding haemagglutinin (HBHA) of *Mycobacterium avium* subsp. *paratuberculosis*: functional promiscuity or a role in virulence? *Vaccine*, **24**, 236–43.

Sechi LA, Gazouli M, Sieswerda LE, Molicotti P, Ahmed N, Ikonomopoulos J, Scanu AM, Paccagnini D and Zanetti S (2006b) Relationship between Crohn's disease, infection with *Mycobacterium avium* subspecies *paratuberculosis* and SLC11A1 gene polymorphisms in Sardinian patients, *World Journal of Gastroenterology*, **12**, 7161–4.

Sechi LA, Paccagnini D, Salza S, Pacifico A, Ahmed N and Zanetti S (2008) *Mycobacterium avium* subspecies *paratuberculosis* bacteremia in type 1 diabetes mellitus: an infectious trigger? *Clinical Infectious Diseases*, **46**, 148–9.

Secott TE, Ohme AM, Barton KS, Wu CC and Rommel FA (1999) *Mycobacterium paratuberculosis* detection in bovine feces is improved by coupling agar culture enrichment to an IS900-specific polymerase chain reaction assay, *Journal of Veterinary Diagnosis and Investigation*, **11**, 441–7.

Secott TE, Lin TL and Wu CC (2001) Fibronectin attachment protein homologue mediates fibronectin binding by *Mycobacterium avium* subsp. *paratuberculosis*, *Infection and Immunity*, **69**, 2075–82.

Secott TE, Lin TL and Wu CC (2002) Fibronectin attachment protein is necessary for efficient attachment and invasion of epithelial cells by *Mycobacterium avium* subsp. *paratuberculosis*, *Infection and Immunity*, **70**, 2670–75.

Secott TE, Lin TL and Wu CC (2004) *Mycobactetium avium* subsp. *paratuberculosis* fibronectin attachment protein facilitates M-cell targeting and invasion through a fibronectin bridge with host integrins, *Infection and Immunity*, **72**, 3724–32.

Selby W (2000) Pathogenesis and therapeutic aspects of Crohn's disease, *Veterinary Microbiology*, **77**, 505–11.

Selby WS (2004) *Mycobacterium avium* subspecies *paratuberculosis* bacteraemia in patients with inflammatory bowel disease, *Lancet*, **364**, 1013–14.

Selby W, Pavli P, Crotty B, Florin T, Radford-Smith G, Gibson P, Mitchell B, Connell W, Read R, Merrett M, Ee H, Hetzel D and Grp CsDS (2007) Two-year combination antibiotic therapy with clarithromycin, rifabutin, and clofazimine for Crohn's disease, *Gastroenterology*, **132**, 2313–19.

Semret M, Turenne CY and Behr MA (2006) Insertion sequence IS900 revisited, *Journal of Clinical Microbiology*, **44**, 1081–3.

Sevilla I, Garrido JM, Geijo M and Juste RA (2007) Pulsed-field gel electrophoresis profile homogeneity of *Mycobacterium avium* subsp. *paratuberculosis* isolates from cattle and heterogeneity of those from sheep and goats, *BMC Microbiology*, **7**, 18.

Shafran I, Kugler L, El-Zaatari FAK, Naser SA and Sandoval J (2002a) Open clinical trial of rifabutin and clarithromycin therapy in Crohn's disease, *Digestive and Liver Disease*, **34**, 22–8.

Shafran I, Piromalli C, Decker JW, Sandoval J, Naser SA and El-Zaatari FAK (2002b) Seroreactivities against *Saccharomyces cerevisiae* and *Mycobacterium avium* subsp. *paratuberculosis* p35 and p36 antigens in Crohn's disease patients, *Digestive Diseases and Sciences*, **47**, 2079–81.

Shin SJ and Collins MT (2008) Thiopurine drugs azathioprine and 6-mercaptopurine inhibit *Mycobacterium paratuberculosis* growth in vitro, *Antimicrobial Agents and Chemotherapy*, **52**, 418–26.

Shin SJ, Chang YF, Huang C, Zhu JQ, Huang L, Yoo HS, Shin KS, Stehman S, Shin SJ and Torres A (2004) Development of a polymerase chain reaction test to confirm *Mycobacterium avium* subsp. *paratuberculosis* in culture, *Journal of Veterinary Diagnostic Investigation*, **16**, 116–20.

Shin SJ, Han JH, Manning EJB and Collins MT (2007) Rapid and reliable method for quantification of *Mycobacterium paratuberculosis* by use of the BACTEC MGIT 960 system, *Journal of Clinical Microbiology*, **45**, 1941–8.

Sieswerda LE and Bannatyne RM (2006) Mapping the effects of genetic susceptibility and *Mycobacterium avium* subsp. *paratuberculosis* infection on Crohn's disease: strong but independent, *Journal of Clinical Microbiology*, **44**, 1204–5.

Singh UP, Singh S, Singh R, Karls RK, Quinn FD, Potter ME and Lillard JW, Jr (2007)

Influence of *Mycobacterium avium* subsp. *paratuberculosis* on colitis development and specific immune responses during disease, *Infection and Immunity*, **75**, 3722–8.
Smith HW (1960) The examination of milk for the presence of *Mycobacterium johnei*, *Journal of Pathology and Bacteriology*, **80**, 440–42.
Smith G and Shanahan F (2006) *Mycobacterium avium paratuberculosis* in Crohn's disease: Player or spectator? in P Irving, D Rampton and F Shanahan (eds), *Clinical Dilemmas in Inflammatory Bowel Disease*, Oxford, Blackwell, 36–9.
Smith MSH and Wakefield AJ (1993) Viral association with Crohn's disease, *Annals of Medicine*, **25**, 557–61.
Smith WL, McGarvey KL and Cullor JS (2003) The use of spiral plating and microscopic colony counting for the rapid quantitation of *Mycobacterium paratuberculosis*, *Letters in Applied Microbiology*, **36**, 293–6.
Sorensen O, Rawluk S, Wu J, Manninen K and Ollis G (2003) *Mycobacterium paratuberculosis* in dairy herds in Alberta, *Canadian Veterinary Journal*, **44**, 221–6.
Spahr U and Schafroth K (2001) Fate of *Mycobacterium avium* subsp. *paratuberculosis* in Swiss hard and semihard cheese manufactured from raw milk, *Applied and Environmental Microbiology*, **67**, 4199–205.
St Amand AL, Frank DN, De Groote MA and Pace NR (2005) Use of specific rRNA oligonucleotide probes for microscopic detection of *Mycobacterium avium* complex organisms in tissue, *Journal of Clinical Microbiology*, **43**, 1505–14.
Stabel JR (2000) Johne's disease and milk: do consumers need to worry? *Journal of Dairy Science*, **83**, 1659–63.
Stabel JR and Bannantine JP (2005) Development of a nested PCR method targeting a unique multicopy element, ISMap02, for detection of *Mycobacterium avium* subsp. *paratuberculosis* in fecal samples, *Journal of Clinical Microbiology*, **43**, 4744–50.
Stabel JR and Lambertz A (2004) Efficacy of pasteurization conditions for the inactivation of *Mycobacterium avium* subsp. *paratuberculosis* in milk, *Journal of Food Protection*, **67**, 2719–26.
Stabel JR, Steadham EM and Bolin CA (1997) Heat inactivation of *Mycobacterium paratuberculosis* in raw milk: Are current pasteurization conditions effective? *Applied and Environmental Microbiology*, **63**, 4975–7.
Stabel JR, Bosworth TL, Kirkbride TA, Forde RL and Whitlock, RH (2004a) A simple, rapid, and effective method for the extraction of *Mycobacterium paratuberculosis* DNA from fecal samples for polymerase chain reaction, *Journal of Veterinary Diagnosis and Investigation*, **16**, 22–30.
Stabel JR, Hurd S, Calvente L and Rosenbusch RF (2004b) Destruction of *Mycobacterium paratuberculosis*, *Salmonella* spp. and *Mycoplasma* spp. in raw milk by a commercial on-farm high-temperature, short-time pasteurizer, *Journal of Dairy Science*, **87**, 2177–83.
Stanley EC, Mole RJ, Smith RJ, Glenn SM, Barer MR, McGowan M and Rees CED (2007) Development of a new, combined rapid method using phage and PCR for detection and identification of viable *Mycobacterium paratuberculosis* bacteria within 48 hours, *Applied and Environmental Microbiology*, **73**, 1851–7.
Stephan R (2007) Diagnostic system for detection of *Mycobacterium avium* subsp. *paratuberculosis*, *Journal of Consumer Protection and Food Safety*, **2**, 222–7.
Stephan R, Schumacher S, Tasara T and Grant IR (2007) Prevalence of *Mycobacterium avium* subspecies *paratuberculosis* in swiss raw milk cheeses collected at the retail level, *Journal of Dairy Science*, **90**, 3590–95.
Stich RW, Byrum B, Love B, Theus N, Barbar L and Shulaw WP (2004) Evaluation of an automated system for non-radiometric detection of *Mycobacterium avium paratuberculosis* in bovine feces, *Journal of Microbiological Methods*, **56**, 267–75.
Stratmann J, Strommenger B, Stevenson K and Gerlach GF (2002) Development of a peptide-mediated capture PCR for detection of *Mycobacterium avium* subsp. *paratuberculosis* in milk, *Journal of Clinical Microbiology*, **40**, 4244–50.

Stratmann J, Dohmann K, Heinzmann J and Gerlach GF (2006) Peptide aMptD-mediated capture PCR for detection of *Mycobacterium avium* subsp. *paratuberculosis* in bulk milk samples, *Applied and Environmental Microbiology*, **72**, 5150–58.

Subramanian S, Campbell BJ and Rhodes JM (2006) Bacteria in the pathogenesis of inflammatory bowel disease, *Current Opinion in Infectious Diseases*, **19**, 475–84.

Suenaga K, Yokoyama Y, Okazaki K and Yamamoto Y (1995) Mycobacteria in the intestine of Japanese patients with Inflammatory Bowel Disease, *American Journal of Gastroenterology*, **90**, 76–80.

Sung N and Collins MT (1998) Thermal tolerance of *Mycobacterium paratuberculosis*, *Applied and Environmental Microbiology*, **64**, 999–1005.

Sung N and Collins MT (2000) Effect of three factors in cheese production (pH, salt, and heat) on *Mycobacterium avium* subsp. *paratuberculosis* viability, *Applied and Environmental Microbiology*, **66**, 1334–9.

Sung NM and Collins MT (2003) Variation in resistance of *Mycobacterium paratuberculosis* to acid environments as a function of culture medium, *Applied and Environmental Microbiology*, **69**, 6833–40.

Sung N, Takayama K and Collins MT (2004) Possible association of GroES and antigen 85 proteins with heat resistance of *Mycobacterium paratuberculosis*, *Applied and Environmental Microbiology*, **70**, 1688–97.

Sweeney RW, Whitlock RH and Rosenberger AE (1992) *Mycobacterium paratuberculosis* cultured from milk and supramammary lymph nodes of infected asymptomatic cows, *Journal of Clinical Microbiology*, **30**, 166–71.

Sweeney RW, Whitlock RH and McAdams SC (2006) Comparison of three DNA preparation methods for real-time polymerase chain reaction confirmation of *Mycobacterium avium* subsp. *paratuberculosis* growth in an automated broth culture system, *Journal of Veterinary Diagnostic Investigation*, **18**, 587–90.

Szteyn J, Wismiewska-Laszczych A and Rusmynska A (2008) Effectiveness of *Mycobacterium paratuberculosis* isolation from raw milk by means of direct isolation of DNA and classic culture, *Polish Journal of Veterinary Sciences*, **11**, 25–8.

Taddei S, Robbi C, Cesena C, Rossi I, Schiano E, Arrigoni N, Vicenzoni G and Cavirani S (2004) Detection of *Mycobacterium avium* subsp. *paratuberculosis* in bovine fecal samples: comparison of three polymerase chain reaction-based diagnostic tests with a conventional culture method, *Journal of Veterinary Diagnosis and Investigation*, **16**, 503–8.

Tanaka K, Wilks M, Coates PJ, Farthing MJ, Walker-Smith JA and Tabaqchali S (1991) *Mycobacterium paratuberculosis* and Crohn's disease, *Gut*, **32**, 43–5.

Tasara T and Stephan R (2005) Development of an F57 sequence-based real-time PCR assay for detection of *Mycobacterium avium* subsp. *paratuberculosis* in milk, *Applied and Environmental Microbiology*, **71**, 5957–68.

Tasara T, Hoelzle LE and Stephan R (2005) Development and evaluation of a *Mycobacterium avium* subspecies *paratuberculosis* (MAP) specific multiplex PCR assay, *International Journal of Food Microbiology*, **104**, 279–87.

Taylor TK, Wilks CR and Mcqueen DS (1981) Isolation of *Mycobacterium paratuberculosis* from the milk of a cow with Johne's disease, *Veterinary Record*, **109**, 532–3.

Tessema MZ, Koets AP, Rutten VPMG and Gruys E (2001) How does *Mycobacterium avium* subsp. *paratuberculosis* resist intracellular degradation? *Veterinary Quarterly*, **23**, 153–62.

Thibault VC, Grayon M, Boschiroli ML, Hubbans C, Overduin P, Stevenson K, Gutierrez MC, Supply P and Biet F (2007) New variable-number tandem-repeat markers for typing *Mycobacterium avium* subsp. *paratuberculosis* and *M. avium* strains: comparison with IS900 and IS1245 restriction fragment length polymorphism typing, *Journal of Clinical Microbiology*, **45**, 2404–10.

Tiveljung A, Soderholm JD, Olaison G, Jonasson J and Monstein HJ (1999) Presence of eubacteria in biopsies from Crohn's disease inflammatory lesions as determined by 16S rRNA gene-based PCR, *Journal of Medical Microbiology*, **48**, 263–8.

Tiwari A, VanLeeuwen JA, McKenna SLB, Keefe GP and Barkema HW (2006) Johne's disease in Canada – Part I: Clinical symptoms, pathophysiology, diagnosis, and prevalence in dairy herds, *Canadian Veterinary Journal*, **47**, 874–82.

Tzen CY, Wu TY and Tzen CY (2006) Detection of mycobacteria in Crohn's disease by a broad spectrum polymerase chain reaction, *Journal of the Formosan Medical Association*, **105**, 290–98.

Uzoigwe JC, Khaitsa ML and Gibbs PS (2007) Epidemiological evidence for *Mycobacterium avium* subspecies *paratuberculosis* as a cause of Crohn's disease, *Epidemiology and Infection*, **135**, 1057–68.

Valentine JF, Elam J and Naser SA (2008) ATG16L1 (Autophagy-related 16-like 1) Crohn's disease (CD) associated SNP is not associated with viable *Mycobacterium avium* subspecies *paratuberculosis* (MAP) in the blood of CD patients, *Gastroenterology*, **134**, A461.

Van Kruiningen HJ (1999) Lack of support for a common etiology in Johne's disease of animals and Crohn's disease in humans, *Inflammatory Bowel Diseases*, **5**, 183–91.

van Leeuwen JA, Forsythe LA, Tiwari A and Chartier R (2005) Seroprevalence of antibodies against bovine leukemia virus, bovine viral diarrhea virus, *Mycobacterium avium* subspecies *paratuberculosis*, and *Neospora caninum* in dairy cattle in Saskatchewan, *Canadian Veterinary Journal*, **46**, 56–8.

van Roermund HJW, Bakker D, Willernsen PTJ and de Jong MCM (2007) Horizontal transmission of *Mycobacterium avium* subsp. *paratuberculosis* in cattle in an experimental setting: calves can transmit the infection to other calves, *Veterinary Microbiology*, **122**, 270–79.

Vansnick E, de Rijk P, Vercammen F, Geysen D, Rigouts L and Portaels F (2004) Newly developed primers for the detection of *Mycobacterium avium* subspecies *paratuberculosis*, *Veterinary Microbiology*, **100**, 197–204.

Vansnick E, de Rijk P, Vercammen F, Rigouts L, Portaels F and Geysen D (2007) A DNA sequence capture extraction method for detection of *Mycobacterium avium* subspecies *paratuberculosis* in feces and tissue samples, *Veterinary Microbiology*, **122**, 166–71.

Vohra J, Singh SV, Singh AV, Singh PK and Sohal JS (2008) Detection and characterization of *Mycobacterium avium* subspecies *paratuberculosis* from vaginal secretions of post-parturient farm goats, using culture, IS 900 and ELISA kit, *Indian Journal of Animal Sciences*, **78**, 441–4.

Waddell LA, Rajic A, Sargeant J, Harris J, Amezcua R, Downey L, Read S and McEwen SA (2008) The zoonotic potential of *Mycobacterium avium* spp. *paratuberculosis* – a systematic review, *Canadian Journal of Public Health*, **99**, 145–55.

Wall S, Kunze ZM, Saboor S, Soufleri I, Seechurn P, Chiodini R and McFadden JJ (1993) Identification of spheroplast-like agents isolated from tissues of patients with Crohn's disease and control tissues by polymerase chain reaction, *Journal of Clinical Microbiology*, **31**, 1241–5.

Walmsley RS, Ibbotson JP, Chahal H and Allan RN (1996) Antibodies against *Mycobacterium paratuberculosis* in Crohn's disease, *Quarterly Journal of Medicine*, **89**, 217–21.

Warth A (2008) Is alpha-dystroglycan the missing link in the mechanism of enterocyte uptake and translocation of *Mycobacterium avium paratuberculosis*?, *Medical Hypotheses*, **70**, 369–74.

Wells RW and Blennerhassett MG (2005) The increasing prevalence of Crohn's disease in industrialized societies: the price of progress? *Canadian Journal of Gastroenterology*, **19**, 89–95.

Whan LB, Grant IR, Ball HJ, Scott R and Rowe MT (2001) Bactericidal effect of chlorine on *Mycobacterium paratuberculosis* in drinking water, *Letters in Applied Microbiology*, **33**, 227–31.

Whan L, Ball HJ, Grant IR and Rowe MT (2005a) Development of an IMS-PCR assay for the detection of *Mycobacterium avium* ssp. *paratuberculosis* in water, *Letters in Applied Microbiology*, **40**, 269–73.

Whan L, Ball HJ, Grant IR and Rowe MT (2005b) Occurrence of *Mycobacterium avium* subsp. *paratuberculosis* in untreated water in Northern Ireland, *Applied and Environmental Microbiology*, **71**, 7107–12.

Whan L, Grant IR and Rowe MT (2006) Interaction between *Mycobacterium avium* subsp. *paratuberculosis* and environmental protozoa, *BMC Microbiology*, **6**, 63.

Whittington RJ and Sergeant ESG (2001) Progress towards understanding the spread, detection and control of *Mycobacterium avium* subsp. *paratuberculosis* in animal populations, *Australian Veterinary Journal*, **79**, 267–78.

Whittington RJ, Marshall DJ, Nicholls PJ, Marsh AB and Reddacliff LA (2004) Survival and dormancy of *Mycobacterium avium* subsp. *paratuberculosis* in the environment, *Applied and Environmental Microbiology*, **70**, 2989–3004.

Whittington RJ, Marsh IB and Reddacliff LA (2005) Survival of *Mycobacterium avium* subsp. *paratuberculosis* in dam water and sediment, *Applied and Environmental Microbiology*, **71**, 5304–8.

Winterhoff C, Beyerbach M, Homuth M, Strutzberg K and Gerlach GF (2002) Evaluation of an ELISA for the detection of *Mycobacterium avium* subspecies *paratuberculosis*-specific antibodies in milk, *Deutsche Tierarztliche Wochenschrift*, **109**, 230–34.

Woo SR and Czuprynski CJ (2008) Tactics of *Mycobacterium avium* subsp. *paratuberculosis* for intracellular survival in mononuclear phagocytes, *Journal of Veterinary Science*, **9**, 1–8.

30

Transmissible spongiform encephalopathy (prion disease)

P. Brown, National Institutes of Health (retired), USA and Institute of Emerging Diseases, Commissariat à l'Energie Atomique, France

Abstract: A variant form of Creutzfeldt-Jakob disease (vCJD) resulted from human exposure to beef products contaminated with bovine spongiform encephalopathy (BSE). In this chapter, the chronology of both disease outbreaks is recounted, with special attention to pathogenesis, difficulty of decontamination, past and present risk from BSE and other animal prion diseases and, despite the waning number of new cases of both BSE and vCJD, the need for continuing surveillance and protective measures.

Key words: transmissible spongiform encephalopathy, prion disease, Creutzfeldt-Jakob disease, variant Creutzefeldt-Jakob disease, bovine spongiform encephalopathy, scrapie, pathogen decontamination, food safety.

30.1 Introduction

Transmissible spongiform encephalopathies (TSEs, or prion diseases) live in a world apart from all other infectious disorders in that they appear to be caused by a misfolded host protein rather than an environmental pathogen. The prototype disease is scrapie, recognized since at least the mid-18th century to be endemic in sheep (Brown and Bradley, 1998). A range of other animal and human diseases have since expanded the list to include transmissible mink encephalopathy (TME), chronic wasting disease (CWD) of deer and elk, bovine spongiform encephalopathy (BSE), and several different disease syndromes in humans, of which the most important is Creutzfeldt-Jakob disease (CJD) (Fig. 30.1). The relevance of prions to food safety is well illustrated by the recent epidemic of BSE caused by the recycling of rendered carcasses as cattle feed, and by the subsequent outbreak in humans

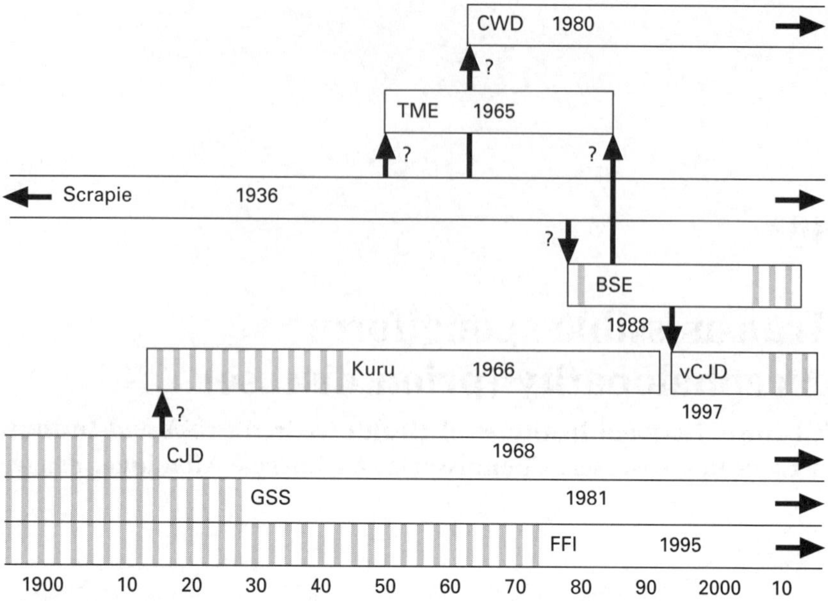

Fig. 30.1 Chronology of the transmissible spongiform encephalopathies. Bars represent known or presumed (striped) time periods of disease occurrence. Disease names are placed at time of first recognition (except for scrapie, first described in the early 18th century). Dates indicate year of first experimental transmission in the laboratory; vertical arrows are placed at times of known or possible (?) inter-species transmissions in nature. CWD = chronic wasting disease; TME = transmissible mink encephalopathy; BSE = bovine spongiform encephalopathy; vCJD = variant form of Creutzfeldt-Jakob disease; CJD = sporadic and familial forms of Creutzfeldt-Jakob disease; GSS = Gerstmann-Sträussler-Scheinker disease; FFI = fatal familial insomnia.

of a 'variant' form of CJD (vCJD) due to the consumption of infected bovine products in human food.

30.2 The pathogen

Thinking about the pathogenic agent has evolved from the original concept of a virus through successive stages of 'slow' virus, 'unconventional' virus, viroid, 'virino', and now 'prion', the name coined to indicate its proposed fully proteinaceous nature. Because of (i) the consistent failure to find any disease-specific foreign nucleic acid or protein and (ii) the almost equally consistent finding of a host-encoded but misfolded 'prion' protein in the brains of affected humans and animals, a consensus has arisen (but still not formally proven) that favors the protein as the principle if not sole cause of disease.

The proposed sequence of events is that the normal host protein (PrP^C), which is a soluble proteinase-sensitive and α-helix-rich glycoprotein of approximately 35 kD molecular weight, misfolds via one or more intermediate molecular species into a predominantly β-sheet amyloid configuration (PrP^{TSE}). This misfolded form of the protein 'seeds' a progressive conversion of normal protein causing functional derangement of living neurons, and is ultimately deposited in amyloid aggregates both inside and outside the central nervous system.

The chemical structure of the normal protein has been determined, together with the predicted structure of the misfolded conformer, which can be visualized in fibrillar form in detergent extracts of diseased tissue. PrP^{TSE} can have up to a third of its amino acids and all of its attached sugars removed and remain fully infectious. Recent fractionation experiments indicate that the most infectious particles have a molecular weight of 300–600 kD, equivalent to oligomers containing 14–28 molecules of PrP (Silveira *et al.*, 2005). The critical molecular species of the misfolding process may therefore lie upstream from the much larger amyloid aggregates, which could represent a 'dead-end' result of the body's effort to segregate the protein into a harmless extracellular deposit.

There are also recent indications that partially misfolded protein species are present in normal brain. Experimental studies on normal rodent brain tissue subjected to a newly developed protein amplification procedure have yielded detectable amounts of PrP^{TSE} and (in some cases) infectivity (Deleault *et al.*, 2007; U. Agrimi, personal communication). These surprising results are particularly interesting in light of a recent study in which a small amount of protein with some of the physicochemical characteristics of PrP^{TSE} was detected in normal human brains (Yuan *et al.*, 2006). It is therefore possible that 'seed' molecules are being continuously produced in normal brain tissue, and only require one or more as yet unidentified triggers to unleash the cascade of molecular events that terminate in symptomatic disease.

30.3 Epidemiology of BSE and vCJD

The first case of BSE was recognized in the UK in 1985 (Wells *et al.*, 1987), and was followed by a rapidly expanding, geographically widespread epidemic that by 1990 had reached an annual incidence of over 30 000 clinical cases (Fig. 30.2). The epidemic was magnified by the continued exportation of live cattle and of meat and bone meal feed to virtually every country in Europe and a number of non-European countries (no-one has ever been able to explain why, for several years after the cause was identified, the UK continued to export cattle and cattle feed – the most generous explanation would be a communications breakdown between scientists, government regulatory agencies, and commercial exporters). Sometimes, sick animals were identified before slaughter and did not result in disease outbreaks. However, in countries that imported large numbers of cattle, and thousands of tons

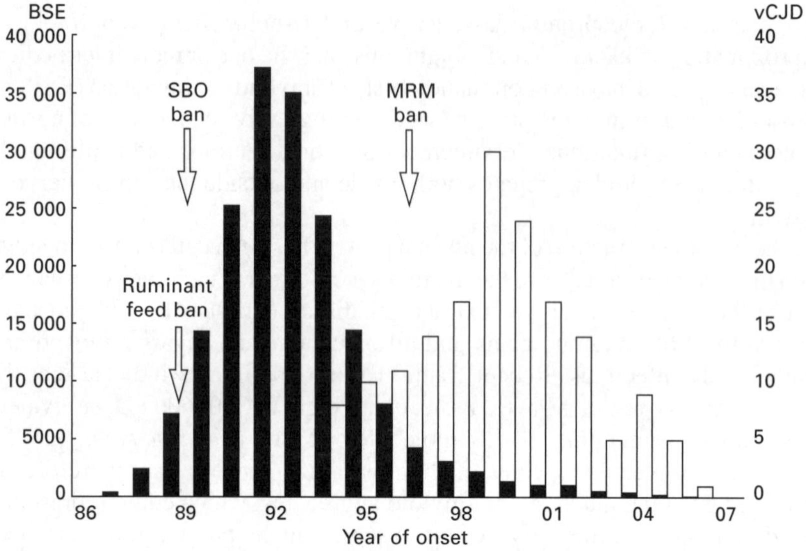

Fig. 30.2 Chronology of BSE and vCJD in the UK (as of September 2008). Solid bars are cases of BSE and open bars are cases of vCJD. Feed ban = prohibition of ruminant protein from use in ruminant feed; SBO ban = prohibition of specified bovine offals (thoracic and abdominal viscera and central nervous system tissue) from use in human food; MRM ban = prohibition of mechanically recovered meat from use in human food.

of meat and bone meal rendered from slaughtered carcasses, BSE was able to spread through carcass recycling of both imported and newly infected domestic cattle. Fortunately, these secondary outbreaks were substantially less severe than the original UK epidemic, which is estimated to have affected up to five million cattle.

In 1996, ten years after the first recognized case of BSE, several cases of CJD with an unusual clinical presentation were identified in UK adolescents and young adults (Will *et al.*, 1996). Systematic study of the first 10 cases revealed that this distinctive form of illness had not previously been seen in the UK (and was not occurring anywhere outside the country), and that the patients had no mutations or iatrogenic exposures to account for the disease. A suspicion that they might have resulted from infections with BSE was subsequently confirmed experimentally (see below), and increasing numbers of similar cases of what came to be known as variant CJD (vCJD) were identified in the next few years. The rise and fall of the annual incidence curve closely resembled that of BSE, but at a thousand-fold lower level (Fig. 30.2). After a delay of several years, cases also began to be recognized in other countries with 'imported' BSE outbreaks but, again, in much lower numbers than BSE. The global tally of vCJD currently stands at just over 200 cases, and the incidence of new cases is either in decline or has disappeared in every country in which it has occurred.

30.4 Origins of BSE

Any explanation of BSE must answer two questions: why did BSE begin in the mid-1980s, and why did it begin in the UK? Both of the competing hypotheses – a species-crossing leap of scrapie from sheep to cattle versus a spontaneous case of BSE in cattle – answer the question of timing by changes in livestock carcass rendering into meat and bone meal dietary supplements that took place around 1980.

At that time, continuous flow largely replaced single batch rendering (which could have created incomplete exposure to heat), and a final tallow extraction step consisting of exposure to hydrocarbon solvents under steam was eliminated. Neither of these changes had a very great effect on infectivity (Taylor *et al.*, 1995, 1997), but, if the level of infectivity entering the rendering process were near the threshold of transmissibility, small changes could have made the difference between survival and destruction. Whatever the reasons, it is a fact that the prohibition of meat and bone meal dietary supplements in 1987 was followed five years later by a downturn in BSE cases, and five years is the average incubation period between BSE infection and manifest illness.

The question of a scrapie versus cattle origin of BSE is more difficult to answer. Scrapie is prevalent in the UK (as in most other countries of the world), and the UK has a comparatively large proportion of sheep to cattle. Thus, a widespread potential source of infection was already in existence when the BSE epidemic began. The counter-argument that a spontaneous case of BSE initiated the epidemic does not explain why it began in the UK because, if spontaneous disease occurs, it should be a global phenomenon, yet no coincident epidemics occurred in other countries in which rendering processes were changed at about the same time as in the UK. Also, the distribution of early cases in the UK (and subsequently in other countries that imported BSE from the UK) is more consistent with multiple initiation points than with a single point source of infection. This would favor scrapie as the source of infection, because scrapie was widespread enough to have entered rendering plants throughout the UK, whereas it would have required simultaneous cases of sporadic BSE to achieve the same effect.

30.5 Origins of vCJD

The connection between BSE and vCJD, suspected on clinical grounds, was independently confirmed by two different experimental methods: (i) Western blots revealed that the PrP^{TSE} extracted from the brains of both BSE-affected cattle and vCJD-affected humans had an identical pattern that was different from the pattern seen in all other TSEs, including sporadic CJD; and (ii) inoculations of a panel of inbred mouse strains showed that brain tissue from cases of BSE and vCJD produced the same patterns of latency (incubation period) and brain lesion topography, which also were different from other TSEs, including sporadic CJD and scrapie.

The question of how humans were infected has been approached chiefly by a process of elimination, and lacks the kind of laboratory evidence that clinched the identification of BSE-infected cattle as the source of disease. Epidemiological study did not uncover any convincing disease clusters, or point to any regional peculiarities or vectors that might have linked BSE cattle to human disease, and investigation of 'high-risk' contact groups such as farmers, slaughterhouse workers, or butchers did not identify any cases of vCJD.

Attention then turned to exposure to cattle products. Consumption of meat and dairy products, and exposure to products containing either tallow or gelatin (or their derivatives) is virtually universal, and no correlations could be established between vCJD and exposure to any particular product. It nevertheless seemed probable, given its high concentration of infectivity, that nervous system tissue must somehow be involved, most likely in conjunction with the widespread consumption of meat products. Researchers (and the general public) were surprised to learn that mechanically recovered meat (MRM) consisted of a pressure-treated paste that included vertebral columns containing spinal cord fragments and paraspinal ganglia, and that MRM was routinely added to a variety of packaged meat products such as hot dogs, sausages, beef patties, luncheon meats, beef stews, etc., in proportions as high as 30 % (but usually in the range of 5–10 %). Thus, central nervous system tissues were entering the human food chain through these 'unadvertised' vehicles, and were surely the most likely cause of human infection.

Two questions remain unanswered: why, given the widespread exposure, have so few people been infected, and why have they been so young? It is possible that a fairly strong 'species barrier' and inefficient (oral) route of infection has protected all but a few exposures from producing disease; or it is possible that the contamination of MRM was exceedingly heterogeneous, such that infectivity was only rarely present in randomly distributed 'packets' within the MRM. Recent studies showing that the prion protein accumulates in association with inflammation (Heikenwalder *et al.*, 2005) suggest the further possibility that trivial (and thus unrecognized) foci of oral or gastrointestinal inflammation might have contributed to susceptibility to infection. As for the youthfulness of the patients, the age of cohorts in which maximum exposure to BSE is thought to have occurred (older adolescents) has been statistically correlated with both the developmental maximum of intestinal lymphatic tissue (Peyer's patches) (St Rose *et al.*, 2006) and the consumption of inexpensive meat products containing meat and bone meal (MBM) (Cooper and Bird, 2002). Inexplicably, however, only one instance of two members of a single family has been identified among the more than 200 families in which vCJD has occurred.

30.6 Symptoms of vCJD

The general public has been bewitched by confusion about the difference between sporadic and variant forms of CJD, to the point that for some years after the 'mad cow' scare, every new case of sporadic disease was believed to be due to eating contaminated beef. However, CJD has been occurring sporadically since well before the appearance of BSE (at least since 1920), with a global yearly incidence of approximately 1–2 cases per million population.

Patients with vCJD are typically between 15 and 30 years of age at the onset of illness, although a number of patients have been older, and a few have been younger (Fig. 30.3). They often present with either psychiatric or sensory symptoms (or both), neither of which can be traced to an organic origin by either physical or laboratory examination. For example, an 18 year-old girl might become uncharacteristically anxious and see a physician complaining of leg pains, but have no verifiable objective physical signs or abnormal laboratory tests, all of which would be more apt to be treated with a tranquillizer than raise a suspicion of vCJD. However, over a period of several months, mental decline becomes evident, and a variety of neurological deficits appear, including incoordination, tremor, visual deterioration, abnormal movements, and rigidity, and in the advanced state the illness resembles that of other forms of CJD, including sporadic disease. Several laboratory tests have proved extremely valuable as diagnostic aids, but they often do not become positive until the illness is well past its initial stages. Table 30.1 compares those clinical and laboratory features that are most useful in

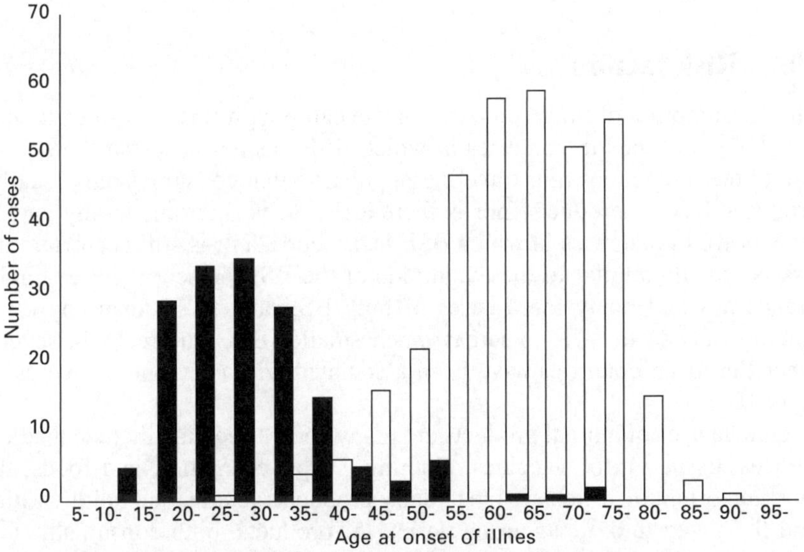

Fig. 30.3 Comparison of age at onset of illness in variant and sporadic forms of CJD. Solid bars are cases of vCJD; open bars are cases of sporadic CJD.

Table 30.1 Comparison of clinical and laboratory characteristics of sporadic and variant forms of CJD

	Variant CJD[a]	Sporadic CJD[b]
Average age at onset (years)	29	64
Clinical presentation (%)		
Cognitive	9	64
Psychiatric	63	29
Sensory symptoms	20	6
Incoordination	9	33
Visual abnormalities	0	19
Average duration of illness (months)	12	7
Laboratory tests		
Electroencephalogram (EEG) periodicity	Very rare	60 %
Cerebrospinal fluid (CSF) 14-3-3 protein	50 %	90 %
Magnetic resonance imaging (MRI) hyperintensity	Pulvinar	Striatum

[a]Will *et al.* (2000).
[b]Brown *et al.* (1994).

distinguishing vCJD from sporadic disease (Brown *et al.*, 1994; Will *et al.*, 2000).

30.7 Risk factors

The consumption of infectious tissue from cattle with BSE is a *sine qua non* for vCJD, and thus in countries in which BSE has not occurred there is no risk to the indigenous non-traveling population that consumes only locally-produced bovine products (nor is there a risk to vegetarians in any part of the world). In countries in which BSE has occurred, or is still occurring, the risk is broadly similar to the magnitude of the BSE presence: for example, France has had many more cases of both BSE and vCJD than any other country outside the UK, whereas much smaller BSE outbreaks in several other European countries have been associated with only one or two cases of vCJD.

Human uses of animal products are many and varied, and include medical devices, tissue grafts, vaccines, clothing, soaps, cosmetics, and foods, the most important of which are meat, dairy products, gelatin, and tallow, which find their way into a staggering variety of products, both consumable and non-consumable. Table 30.2 lists the tissue sources of consumable bovine products that have been a focus of attention since the appearance of BSE.

Table 30.2 Bovine tissues consumed by humans (not exhaustive)[a]

Used as such or after minimal processing
Meat on the bone[b] (T-bone, ox tail, rib, brisket)
De-boned meat
Offals[c] (e.g. liver, heart, kidney, thymus, pancreas, and brain)
Fat[d] from discrete adipose tissue (suet)
Bone (soup and gravy)
Skin (sausage casings, etc.)
Brain and endocrine powders in some health food supplements
Blood (black puddings, blood sausages)
Spinal cord, testicles, udder, colostrum[e]
Used after processing
Milk and milk products (e.g. butter, cheese, lactose and casein)
Rennet (chymosin and renin derived from abomasum) used in the production of cheese, whey and whey products such as lactose
Bone[b] and skin to make gelatin, gelatin derivatives, and collagen
Meat, including tongue, meat extracts used as flavoring, and 'mechanically recovered meat'[f]
Tripe (fore-stomachs, oesophagus, and abomasum)
Tallow[g] (used for frying and in food as shortening) and to make tallow derivatives[h]
Fat[d] (suet), beef drippings and fat derived by melting body fat
Blood and blood products (e.g. hemoglobin to clarify wine)
Intestines (duodenum to rectum) for natural sausage casings

[a]With the advent of BSE, safe sourcing has been practised by avoiding tissues that are known to harbor or replicate TSE agents, eliminating ruminant materials or, where there is no satisfactory alternative, sourcing them from countries, herds, and animals with an acceptably low geographical BSE risk assessment. This is therefore not a table of risk, but rather a reminder of the extent to which humans are exposed to products of bovine origin.

[b]Some bones like skulls and vertebral column (excluding the tail) from cattle over certain ages are classified as specified risk material (SRM) in the EU and some other countries.

[c]SRM includes skull, brain, ganglia, eyes, tonsils, intestines, and spinal cord that are compulsorily removed for destruction in the EU and some other countries.

[d]Some fat (e.g. mesenteric fat) is classified as SRM in the EU.

[e]Consumption limited to small population groups.

[f]Production of MRM from ruminant animals in the EU and USA is banned. Elsewhere, its inclusion in meat products usually does not exceed 5–10 % by weight.

[g]Tallow (fat from rendered bovine starting material is one of the two major end products from rendering; the other is meat and bone meal (MBM).

[h]Tallow derivatives include glycerol, fatty acids and their esters, stearates, polysorbates, and sorbitan esters. They are produced from tallow by hydrolysis at temperatures and pressures that have been shown to sterilize TSE infectivity (see Section 30.10 on prevention and control); saponification with alkali, and trans-esterification with alcohol.

To date, no cases of vCJD have been recognized as a consequence of non-consumable exposures to BSE.

The distribution of infectivity in the tissues of infected cows is surprisingly limited compared to other forms of TSE in other animal species (and to CJD in humans) (WHO, 2006). Almost all infectivity is confined to the central nervous system (CNS), which in the brain has been shown to have a titer of 10^6 mean lethal doses (LD_{50}) per gram of tissue. A few tissues closely associated with the CNS, including retina, optic nerve, and trigeminal and spinal ganglia, are also highly infectious. Much lower levels of infectivity have been detected outside the CNS in peripheral nerves, tonsils, and the terminal ileum. A single observation of infectivity in muscle was performed in transgenic mice that over-express bovine PrP, and may or may not equate to transmissibility under natural conditions.

Risk of infection is not only a function of tissue concentrations of the infectious agent, but also of the amount of tissue to which a person or animal is exposed, and the route by which infection is transmitted. For example, although the level of tissue infectivity is the most important factor in estimating the risk of transmission by instrument cross-contamination during surgical procedures (e.g. neurosurgery compared to general surgery), it will be only one determinant of the risk of transmission by blood transfusions, in which a large amount of low-infectivity blood is administered, or the risk of transmission by foodstuffs which, irrespective of high or low infectivity, involves a comparatively inefficient oral route of infection.

Host-associated risk factors include age at the time of exposure to infection, already discussed, and the host prion protein (*PRNP*) genotype. A polymorphism occurs in the *PRNP* gene at codon 129, which can encode either methionine or valine. In the general Caucasian population, approximately 50 % of people are methionine homozygotes, 10 % are valine homozygotes, and 40 % are heterozygotes. All of the nearly 200 tested cases of vCJD have occurred in methionine homozygotes, which is thus an important (and unexplained) human susceptibility factor.

30.8 Other potential sources of contamination

Cattle were not the only target for dietary supplements: other livestock species, such as sheep, pigs, and chickens, as well as several kinds of zoo animals, laboratory animals, and pets also received feed to which meat and bone meal had been added. Disease transmission to several non-livestock species has occurred in ungulates, felines, and lemurs in zoos, and in pet cats. The possibility of BSE-infected sheep has aroused concern because in the process of 'back-crossing' to sheep, the agent might carry its newly acquired ability to infect humans. Pigs are not susceptible to oral BSE infection, and chickens are resistant to infection by any route of administration. Dogs are also apparently resistant to TSE infection, as several animals inoculated intracerebrally with human strains of CJD and kuru did not become ill, and in countries with BSE no dogs have been identified with spongiform encephalopathy, despite the certainty that some dog food (like cat food) would have contained bovine tissue.

Quite apart from concerns about BSE, another member of the TSE family – chronic wasting disease (CWD) of deer and elk – has been the subject of increasing concern in North America, to which the disease has so far been restricted. CWD was first identified in the late 1960s in captive mule deer in a Colorado wildlife research facility, but was not recognized as a spongiform encephalopathy until 1978 (Williams and Young, 1980). Interestingly, its neuropathology included the hallmark 'florid plaque' of vCJD before that disease appeared in the UK. The known occurrence of CWD remained limited to captive mule deer until 1981, but within a decade was discovered in free ranging mule deer, white tail deer, and elk in Colorado and Wyoming.

By the year 2000, CWD had been identified in both farmed and free-ranging animals in several states near the original endemic area, as well as in contiguous regions of Canada. Intensified recent surveillance has identified what appears to be an expanding geographic range, thought to result more from commercial interstate transport of animals than from natural movement: cases have turned up in western Colorado, Wisconsin, New Mexico, in two deer farms in east central New York state, and in the wild in New York state, Pennsylvania, Maryland, and West Virginia. Elk exported from Canada have caused several endemic cases in Korean elk farms.

The pathogenesis of CWD has not yet received the same attention as scrapie or BSE, but what is known is consistent with the general outlines of TSE: early involvement of the lympho-reticular system, including gut-associated lymphoid tissue, and incubation periods ranging between 15 and 36 months, depending on the species and conditions of infection.

CWD can be experimentally transmitted to a variety of species, including at least one non-human primate species (the squirrel monkey) (Marsh *et al.*, 2005), with incubation periods similar to those seen in other types of TSE. CWD inoculation experiments in transgenic mice expressing either the cervid or human PrP gene indicate a substantial species barrier for transmission of elk CWD to humans (Kong *et al.*, 2005). To date, there is no documented instance of human infection, although infectivity has been detected in muscle from CWD deer (using cervidized transgenic mice) (Angers *et al.*, 2006); the few cases of young individuals who had regularly eaten venison and were suspected of having been infected with CWD were investigated and found to have died from sporadic CJD (Belay *et al.*, 2001).

In brief, CWD is spreading (or is now being recognized in areas distant from its original focus) and may have the potential to infect humans (Belay *et al.*, 2004). It is not known whether CWD exists outside North America, and programs of random testing of deer have recently been initiated in several European countries. Its unique and troubling feature is that unlike scrapie and BSE, it occurs in the wild, which poses enigmas both for understanding the means by which it is transmitted from animal to animal, and for devising strategies to prevent its spread. The durability of CWD infectivity in soil (Miller *et al.*, 2004), and a recent study in which infectivity was detected in the saliva of free-ranging deer (Mathiason *et al.*, 2006) have obvious implications for disease propagation in the wild.

30.9 Methods of detection

The gold standard of prion detection is the transmission of disease by inoculation of tissue from an affected animal or human to a normal host of the same or equally susceptible species. For animals, the species requirement is not a problem, and for humans the requirement can be met by using any one of several different non-human primate species. This method was used exclusively until the discovery in the early 1980s that TSE-specific fibrils in brain preparations can be visualized under the electron microscope (Merz *et al.*, 1984). Although much faster and less expensive than transmission bioassays, which can have pre-clinical incubation periods of months to years, the procedure is still labor-intensive, and so it came as a relief to discover that the fibrils consisted of PrP^{TSE}, which could be detected by Western blots of extracted brain tissue, using antibodies raised against PrP^{TSE} (McKinley *et al*, 1983; Brown *et al.*, 1986). Usually, the detection of PrP^{TSE} is equivalent to establishing the diagnosis of TSE, although there are rare exceptions in which the protein is not associated with infectivity. In recent years, many protein detection methods have been developed, including immuno-histochemistry for use on formalin-fixed tissues, and enzyme-linked immunosorbent assay (ELISA), which is the current method of choice for high throughput screening protocols (e.g. testing thousands of cattle brains for BSE).

Several rapid bioassay methods, involving either transgenic mice engineered to express PrP from the species under study or susceptible cell culture lines, are under development, as is an ultrasensitive protein misfolding cyclic amplification (PMCA) method. In fact, the sensitivity of PrP 'over-expressing' transgenic mice and PMCA assays is such as to call into question their relevance to transmissible levels of infectivity under natural conditions, and parallel titrations with standard bioassays are badly needed. It could even be argued that intracerebral inoculations in any type of mouse – transgenic or wild-type – do not necessarily imply a transmission risk in humans by less efficient routes of infection, and that all such experiments must ultimately be validated by oral dosing of susceptible non-human primates. The impracticality of this requirement, however, limits its use to only the most important public health questions, e.g. the risk of oral exposure to milk or meat, or intravenous exposure to blood.

30.10 Prevention and control

Within three years of its appearance in the UK, BSE was prevented from becoming a truly massive epidemic by a ruminant feed ban coupled with the prohibition of specified bovine offals (SBO) in both the animal and human food chains. This measure took virtually all of the high-risk tissues such as CNS and intestines off the market and, after a lag phase equal to the incubation period of the disease in bovines, led in the early 1990s to a sharp and continuing decline in the number of new cases. No subsequent precautionary measures (including the prohibition of MRM), designed to eliminate risk at the source and prevent the dispersion of contaminated material, have had quite the impact of these primary strategic bans.

Most countries followed the lead taken by the UK in a comprehensive effort to minimize source contamination, eliminate loopholes, and ensure compliance, but often not until after delays of several years – the European Union, for example, did not mandate similarly strict bans until the year 2000 (by which time the definition of SBO had been somewhat expanded and was renamed SRM, for specified risk material). Eventually, however, most of the world's affected countries put together prohibition packages that have been extremely successful in reducing the risk of BSE to both animals and humans.

All of the preventative measures to eliminate risk at the source have been immeasurably helped by the auxiliary strategy of identifying and destroying infected animals by testing cattle brains for the presence of PrP^{TSE}, the hallmark of disease. During the early years of the BSE epidemic, infected animals were identified by means of clinical observation and neuropathological confirmation of spongiform encephalopathy. The advent of rapid screening tests to detect PrP^{TSE} allowed systematic examination of all cattle coming to slaughter, independent of the presence of symptoms, and was invariably followed by a 'bump' upwards in the annual incidence curve of BSE in all countries in which testing programs were begun. The rises, however, were transient, and gave way to curves with a decreasing slope similar to the pre-testing era.

Coincident with the strategy of extinguishing risk at the source, a great deal of attention was also paid to the effect of protocols used to process bovine products and, when possible, the addition of steps that could reduce any infectivity that might be present while still maintaining the integrity of the product. An extensive literature dating back more than 50 years has documented the unique resistance of prions to all usual and many unusual disinfectants (Table 30.3); several multi-treatment studies, as well as reviews of primary references, have been published (Gibbs *et al.*, 1978; Brown *et al.*, 1982; Taylor, 2000; Fernie *et al.*, 2007). The only consistently sterilizing procedures are immersion in 1–2 N NaOH, 100 % formic acid, or undiluted bleach for at least 1 h, exposure to dry heat at a temperature > 600 °C for at least 15 min, or autoclaving at ≥ 134 °C for at least 20 min. The preferred method for general use in laboratories is a combination of exposure to NaOH or bleach followed by autoclaving.

It is fortunate that neither of the two major unprocessed foods – milk and meat (manually removed) – were among the small list of tissues that transmitted disease in the early pathogenesis study, because neither could be subjected to any of the fully effective sterilizing methods. Short exposures of beef products to 132 °C under ultra-high pressure was found to preserve their integrity and significantly reduce infectivity (Brown *et al.*, 2003), but

Table 30.3 Results of chemical and physical treatments to inactivate prion infectivity

Ineffective	Partially effective	Fully effective
Chemical methods[a]		
Alcohol	Chlorine dioxide	Hypochlorite (5 %)
Ammonia	Gluteraldehyde	Formic acid (100 %)
β-propiolactone	Iodophores	NaOH (1–2 N)
Detergents	Guanidinium thiocyanate	NaOH (0.75 N) +
Ethylene oxide	Sodium dichloroisocyanurate	Sodium dodecyl sulfate
Formaldehyde	Sodium metaperiodate	(SDS) (0.2 %)[b]
Hydrochloric acid	Urea (6–8 M)	
Hydrogen peroxide		
Peracetic acid		
Permangenate		
Phenolics		
Physical methods		
Boiling (100 °C/1 hr)	Steam heat (121 °C/20 min)	Steam heat (134 °C/20 min)
Microwave radiation	Dry heat (300 °C/15 min)	Dry heat (> 600 °C/15 min)
UV radiation	Ionizing radiation (≥ 50 kG)	
	UV-ozone (4–8 weeks)	
	UV-TiO_2 (2–12 h)	

[a]Exposure times from 15 min to 1 h.
[b]Fully effective in original report, but not yet independently validated.

was felt to require too large an expenditure for equipment to be commercially practical. Examples of processed bovine products that were subject to extensive infectivity reduction validation studies are MBM (still allowed as feed for animal species known to be resistant to BSE), which must be autoclaved, and gelatin and its derivatives, which must be exposed to both physical (heat) and chemical (NaOH) inactivating procedures.

Government regulatory agencies have also addressed the problem of possible 'secondary' disease transmission, that is, the transmission of vCJD from an individual during the unsuspected pre-clinical phase of disease to other humans via blood and organ donations, or from cross-contamination of instruments used in surgical or other invasive procedures. In particular, blood and blood products have received ongoing attention in many countries, and a variety of donor deferral policies have been implemented. The recent occurrence in the UK of four secondary infections due to blood donations has validated these concerns (Hewitt *et al.*, 2006; HPA 2007).

The situation with respect to deferral of organ donors is less well defined, depending as it does upon rules adopted by individual organ bank associations. In general, it can be said that these associations are aware of the potential risk posed by patients with vCJD, and of policies applied to blood donors, and that they tend to issue rules in line with these policies. Instrument cross-contamination is perhaps the most difficult issue to resolve, and it is likely that many if not most hospitals continue to employ routine decontamination procedures that are inadequate to guarantee the sterility of instruments used on patients with vCJD (or sporadic CJD).

As the science of TSE has advanced, and governments have implemented appropriate testing programs and prohibitory regulations, BSE and vCJD are today becoming historical events in nearly all countries in which they have occurred and, apart from continuing concerns about secondary vCJD transmissions, have largely disappeared from public consciousness.

30.11 Future trends

In spite of waning public concern, all aspects of the basic and applied science of TSE will continue to claim the interest of the prion research community. The precise characterization and mechanism of replication of the pathogen remain enigmatic, and, even with the disappearance of BSE and vCJD, the unaltered occurrence of sporadic disease will sustain research on pre-clinical diagnostic tests and preventive or curative strategies.

Moreover, every time it looks as though TSE is under control, something unexpected occurs: first it was the outbreak of contaminated growth hormone, then the outbreak of contaminated dura mater grafts, and then the outbreak of vCJD following on the heels of BSE. And now, when the primary outbreak of vCJD is ending, secondary infections are beginning to appear. Insofar as we can see ahead, poor track record notwithstanding, there are three principal causes for concern: atypical strains of BSE, atypical strains of scrapie, and CWD.

For several years after its identification, the pathogenic agent of BSE was monotonously uniform in both its neuropathology and molecular biological signature. In 2003, however, two different strains were independently discovered in Italy and France. In Italy, two of eight PrP^{TSE}-positive asymptomatic older cows were found to have a distinctive neuropathology that included amyloid plaques (never seen in typical BSE) and a Western blot PrP^{TSE} 'signature' subtly different from that of BSE (Fig 30.4) (Casalone

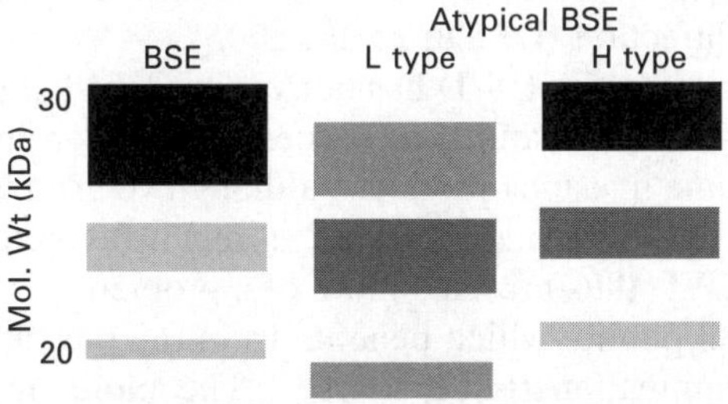

Fig. 30.4 Representation of Western blot patterns of PrP^{TSE} in typical bovine spongiform encephalopathy (BSE) and the two major types of atypical BSE; L type, atypical 'Light' pattern; H type, atypical 'Heavy' pattern.

et al., 2004). In France, three of five examined PrP^{TSE}-positive asymptomatic older cows were found to have a Western blot pattern different from both typical BSE and the Italian atypical strain (neuropathological features were not reported) (Biacabe *et al.*, 2004). Since these observations, both atypical strain patterns have been identified in a number of cattle in several other European countries, Canada, and the USA (Brown *et al.*, 2006). Except for France, it is not yet known what proportion of BSE-affected cattle identified in large-scale national screening test programs will be found to carry atypical agent strains: in France, approximately 2 % of all cases of BSE tested during the years 2001–2007 were found to be atypical (Biacabe *et al.*, 2008).

The principal cause for concern comes from studies showing a greater susceptibility to infection, and possibly also enhanced virulence, in experimentally-infected transgenic mice (P. Gambetti, personal communication), in cattle (Lombardi, *et al.*, 2007), and in a non-human primate (Comoy *et al.*, 2008). A further cause for concern is the finding that the Western blot protein pattern of the Italian strain is very similar to that of some patients having either the MV2 or MM2 sub-types of apparently sporadic CJD (Casalone *et al.*, 2004; Comoy *et al.*, 2008). It has been suggested that at least some CJD patients with these types of molecular 'signatures' might in fact have contracted disease from exposure to atypical BSE, a possibility all the more likely because such older asymptomatic cattle would not be identified without testing, and very few countries systematically employ the test to all animals coming to slaughter.

Similar but more speculative considerations apply to atypical strains of scrapie. All of the epidemiological evidence gathered over decades of research leads to the conclusion that scrapie is not pathogenic for humans, but we have no basis as yet to say the same thing about a newly identified strain of scrapie first observed in Norway in 1998 and designated Nor98 (Benestad *et al.*, 2003). It has since been identified in a UK sheep dying in 1989 (Bruce *et al.*, 2007) and is currently being found in varying, and sometimes large, proportions of infected sheep in many other European countries. Ironically, the Nor98 strain of scrapie, which seems to be increasing in numbers, can easily infect sheep that are being bred (based on PrP genotyping) to resist classical scrapie infection (Le Dur *et al.*, 2005).

Unlike BSE and scrapie, CWD has not yet produced any atypical strains, and its cervid hosts are the source of tissues that are consumed by far fewer people than consume livestock products, although venison is a staple in some restaurants and powdered antler 'velvet' is much favored in the Orient as an aphrodisiac. CWD thus presents more of a problem to individual hunters than to the general public, which benefits from the precautionary measures applied to all mammalian tissue sources. The more important potential for harm would arise if CWD were to find its way into non-cervid animal species, either wild or domesticated. We know that sheep and cattle can be experimentally infected with CWD by intracerebral inoculation, and tests are ongoing to determine if oral dosing with CWD brain tissue, or close contact with CWD-infected deer, can transmit disease to cattle.

Virtually all of the risk to human health posed by these diseases can be avoided by the continuation of systematic testing programs but, because of the lessening problems of BSE and vCJD, there is an increasingly strong inclination to begin relaxing precautionary measures, and scaling back or even eliminating testing. Given the present unknowns surrounding all three diseases – scrapie, BSE, and CWD – this course of action would seem to be premature, and short-term savings could evaporate before unanticipated long-term human costs.

30.12 Acknowledgment

The author is extremely grateful to Dr Raymond Bradley and Dr Linda Detwiler for critical reviews and helpful comments.

30.13 Sources of further information and advice

30.13.1 General reviews

Scrapie

Detwiler L A and Baylis M (2003) The epidemiology of scrapie, in *Risk Analysis of BSE and TSEs: Update on BSE and the Use of Alternative to MBM as Protein Supplements, Revue Scientifique et Technique*, Vol. 22, Office International des Epizooties (OIE), Paris, 121–43.

CWD

Sigurdson C J and Aguzzi A (2007) Chronic wasting disease, *Biochimica et Biophysica Acta*, **1772**, 610–18.

Williams E (2005) Chronic wasting disease, *Veterinary Pathology*, **42**, 530–49.

BSE and vCJD

Bradley R, Collee J G and Liberski P P (2006) Variant CJD (vCJD) and bovine spongiform encephalopathy (BSE): 10 and 20 years on (Part 1), *Folia Neuropathologica*, **44**, 93–101.

Brown P, Will R G, Bradley R, Asher D M and Detwiler L (2001) Bovine spongiform encephalopathy and variant Creutzfeldt-Jakob disease: background, evolution, and current concerns, *Emerging Infectious Diseases*, **7**, 6–16.

Collee J G, Bradley R and Liberski P P (2006) Variant CJD (vCJD) and bovine spongiform encephalopathy (BSE): 10 and 20 years on (Part 2), *Folia Neuropathologica*, **44**, 102–10.

Ward H J T, Head M W, Will R G and Ironside J W (2003) Variant Creutzfeldt-Jakob disease, *Clinical and Laboratory Medicine*, **23**, 87–108.

30.13.2 Websites

Scrapie

- USDA – http://www.aphis.usda.gov/animal_health/animal_diseases/scrapie

CWD

- CDC – http://www.cdc.gov/ncidod/dvrd/cwd
- USDA – http://www.aphis.usda.gov
- CWD Alliance – http://www.cwd-info.org
- EFSA – http://www.efsa.europa.eu/en/science.html

BSE

- OIE – http://ec.europa.eu/food/food/biosafety/bse/monitoring_en.htm; http://www.oie.int/eng/info/en_esbmonde.htm
- WHO – http://www.who.int/topics/encephalopathy_bovine_spongiform/en
- EFSA – http://www.efsa.europa.eu/en/science.html

vCJD

- UK Surveillance Unit – http://www.cjd.ed.ac.uk; http://www.eurocjd.ed.ac.uk/EUROINDEX.htm
- WHO – http://www.who.int/topics/creutzfeldtjakob_syndrome/en

30.14 References

Angers R C, Browning S R, Seward T S, Sigurdson C J, Miller M W, Hoover E A and Telling G C (2006) Prions in skeletal muscles of deer with chronic wasting disease, *Science*, **311**, 1117.

Belay E D, Gambetti P, Schonberger L B, Parchi P, Lyon D R, Capellari S, McQuiston J H, Bradley K, Dowdle G, Crutcher J M and Nichols C R (2001) Creutzfeldt-Jakob disease in unusually young patients who consumed venison, *Archives of Neurology*, **58**, 1673–8.

Belay E D, Maddox R A, Williams E S, Miller M W, Gambetti P and Schonberger L B (2004) Chronic wasting disease and potential transmission to humans, *Emerging Infectious Diseases*, **10**, 977–84.

Benestad S L, Sarradin P, Thu B, Schonheit J, Tranulis M A and Bratberg B (2003) Cases of scrapie with unusual features in Norway and designation of a new type, Nor98, *Veterinary Record*, **153**, 202–8.

Biacabe A-G, Laplanche J-L, Ryder S and Baron T (2004) Distinct molecular phenotypes in bovine prion diseases, *EMBO Reports*, **5**, 110–14.

Biacabe A G, Morignat E, Vulin J, Calavas D and Baron T (2008) Atypical bovine spongiform encephalopathies, France, 2001–2009, *Emerging Infectious Diseases*, **14**, 298–300.

Brown P and Bradley R (1998) 1755 and all that: a historical primer of transmissible spongiform encephalopathy, *British Medical Journal*, **317**, 1688–92.

Brown P, Gibbs C J Jr, Amyx H L, Kingsbury D T, Rohwer R G Sulima M P and Gajdusek D C (1982) Chemical disinfection of Creutzfeldt-Jakob disease virus, *New England Journal of Medicine*, **306**, 1279–82.

Brown P, Coker-Van M, Pomeroy K, Franko M, Asher D M, Gibbs C J Jr and Gajdusek D C (1986) Diagnosis of Creutzfeldt-Jakob disease by Western blot identification of marker protein in human brain tissue, *New England Journal of Medicine*, **314**, 547–51.

Brown P, Gibbs C J Jr, Rodgers-Johnson P, Asher D M, Sulima M P, Bacote A, Goldfarb L G and Gajudsek D C (1994) Human spongiform encephalopathy: the National Institutes of Health series of 300 cases of experimentally transmitted disease, *Annals of Neurology*, **35**, 513–29.

Brown P, Meyer R, Cardone F and Pocchiari M (2003) Ultra-high-pressure inactivation of prion infectivity in processed meat: a practical method to prevent human infection, *Proceedings of the National Academy of Science (USA)*, **100**, 6093–7.

Brown P, McShane L M, Zanusso G and Detwiler L (2006) On the question of sporadic or atypical bovine spongiform encephalopathy and Creutzfeldt-Jakob disease, *Emerging Infectious Diseases*, **12**, 1816–21.

Bruce M E, Nonno R, Foster J, Goldmann W, Di Bari M, Esposito E, Benestad S L and Agrimi U (2007) Nor98-like sheep scrapie in the United Kingdom in 1989, *Veterinary Record*, **160**, 665–6.

Casalone C, Zanusso G, Acutis P, Ferrari S, Capucci L, Tabliavini R, Monaco S and Caramelli M (2004) Identification of a second bovine amyloidotic spongiform encepahalopathy: molecular similarities with sporadic Creutzfeldt-Jakob disease, *Proceedings of the National Academy of Science (USA)*, **101**, 3065–70.

Comoy E E, Casalone C, Lescoutra-Etchegaray N, Zanusso G, Freire S, Marce D, Auvre F, Ruchoux M-M, Monaco S, Sales N, Caramelli M, Leboulch P, Brown P, Lasmezas C I and Deslys J-P (2008) Atypical BSE (BASE) transmitted from asymptomatic aging cattle to a primate, *PloS One*, **3**, e3017, available at: http://www.plosone.org/article/info%3Adoi%2F10.1371%2Fjournal.pone.0003017, accessed November 2008.

Cooper J D and Bird S M (2002) UK bovine carcass meat consumed as burgers, sausages and other meat products: by birth cohort and gender, *Journal of Cancer Epidemiology and Prevention*, **7**, 49–57.

Deleault N R, Harris B T, Rees J R and Supattpone S (2007) Formation of native prions from minimal components in vitro, *Proceedings of the National Academy of Science (USA)*, **104**, 9741–6.

Fernie K, Steele P J, Taylor D M and Somerville R A (2007) Comparative studies on the thermostability of five strains of transmissible-spongiform-encephalopathy agent, *Biotechnology and Applied Biochemistry*, **47**, 175–83.

Gibbs C J Jr, Gajdusek D C and Latarjet R (1978) Unusual resistance to ionizing radiation of the viruses of kuru, Creutzfeldt-Jakob disease, and scrapie, *Proceedings of the National Academy of Science (USA)*, **75**, 6268–70.

Heikenwalder M, Zeller N, Seeger H, Prinz M, Klohn P C, Schwarz P, Ruddle N H, Weissman C and Aguzzi A (2005) Chronic lymphocytic inflammation specifies the organ tropism of prions, *Science*, **307**, 1107–10.

Hewitt P E, Llewelyn C A, Mackenzie J and Will R G (2006) Creutzfeldt-Jakob disease and blood transfusion: results of the UK transfusion medicine epidemiological review study, *Vox Sanguinis*, **91**, 221–30.

HPA (2007) Fourth case of transfusion-associated variant-CJD infection, *Health Protection Report*, **1**(3), available at: http://www.hpa.org.uk/hpr/archives/2007/news2007/news0307.htm, accessed November 2008.

Kong Q, Huang S, Zou W, Vanegas D, Wang M, Wu D, Yuan J, Zheng M, Bai H, Deng

H, Chen K, Jenny A L, O'Rourke K, Belay E D, Schonberger L B, Petersen R B, Sy M S, Chen S G and Gambetti P (2005) Chronic wasting disease of elk: transmissibility to humans examined by transgenic mouse models, *Journal of Neuroscience*, **25**, 7944–9.

Le Dur A, Berinque V, Andreoletti O, Reine F, Lai T L, Baron T, Bratberg B, Vilotte J L, Sarradin P, Benestad S L and Laude H (2005) A newly identified type of scrapie agent can naturally infect sheep with resistant PrP genotypes, *Proceedings of the National Academy of Science (USA)*, **102**, 16031–6.

Lombardi G, Casalone C, D'Angelo A, Gelmetti D, Torcoli G, Barbieri I, Corona C, Fasoli E, Farinazzo A, Fiorini M, Gelati M, Iulini B, Tabliavini F, Ferrari S, Caramelli M, Monaco S, Cappucci L and Zanusso G (2007) Intraspecies transmission of BASE induces motor clinical dullness and amyotrophic changes, *PloS Pathogen*, **4**, e1000075, available at: http://www.plospathogens.org/article/info%3Adoi%2F10.1371%2Fjournal.ppat.1000075, accessed November 2008.

Marsh R F, Kincaid A E, Bessen R A and Bartz J C (2005) Interspecies transmission of chronic wasting disease prions to squirrel monkeys (*Saimiri sciureus*), *Journal of Virology*, **79**, 13794–6.

Mathiason C K, Powers J G, Dahmes S J, Osborn D A, Miller K V, Warren R J, Mason G L, Hays S A, Hayes-Klug J, Seelig D M, Wild M A, Wolfe L I, Spraker T R, Miller M W, Sigurdson C J, Telling G C and Hoover E A (2006) Infectious prions in the saliva and blood of deer with chronic wasting disease, *Science*, **314**, 133–6.

McKinley M P, Bolton D C and Prusiner S B (1983) A protease-resistant protein is a structural component of the scrapie prion, *Cell*, **35**, 57–62.

Merz P A, Rohwer R G, Kascsak R, Wisniewski H M, Sommerville R A, Gibbs C J Jr and Gajdusek D C (1984) Infection-specific particles from the unconventional slow-virus diseases, *Science*, **225**, 437–40.

Miller M W, Williams E S, Hobbs N T and Wolfe L L (2004) Environmental sources of prion transmission in mule deer, *Emerging Infectious Diseases*, **10**, 1003–6.

Silveira J R, Raymond G J, Hughson A G, Race R E, Sim V L and Caughey B (2005) The most infectious prion protein particles, *Nature*, **437**, 257–61.

St Rose S G, Hunter N, Matthews L, Foster J D, Chase-Topping M E, Kruuk L E B, Shaw D J, Rhind S M, Will R G and Woolhouse M E J (2006) Comparative evidence for a link between Peyer's patch development and susceptibility to transmissible spongiform encephalopathies, *BMC Infectious Diseases*, **6**, 5 available at: http://www.biomedcentral.com/1471-2334/6/5, accessed November 2008.

Taylor D M (2000) Inactivation of transmissible degeneratitve encephalopathy agents: a review, *Veterinary Journal*, **159**, 10–17.

Taylor D M, Woodgate S L and Atkinson M J (1995) Inactivation of the bovine Spongiform encephalopathy agent by rendering procedures, *Veterinary Record*, **137**, 605–10.

Taylor D M, Woodgate S L, Fleetwood A J and Cawthorne R J G (1997) Effect of rendering procedures on the scrapie agent, *Veterinary Record*, **141**, 643–9.

Wells G A H, Scott A C, Johnson C T, Gunning R F, Hancock R D, Jeffrey M, Dawson M and Bradley R (1987) A novel progressive spongiform encephalopathy in cattle, *Veterinary Record*, **121**, 419–20.

WHO (2006) Guidelines on Tissue Infectivity Distribution in Transmissible spongiform Encephalopathies, Report of a Consultation, World Health Organization, Geneva, Switzerland, 14–16 September 2005, available at: http://www.who.int/bloodproducts/TSEPUBLISHEDREPORT.pd, accessed November 2008.

Will R G, Ironside J W, Zeidler M, Cousens S N, Estibeiro K, Alperovitch A, Poser S, Pocchiari M, Hofman A and Smith P G (1996) A new variant of Creutzfeldt-Jakob disease in the UK, *Lancet*, **347**, 921–5.

Will R G, Zeidler M, Stewart G E, Macleod M A, Ironside J W, Cousens S N, Mackenzie J, Estibeiro K, Green A J E and Knight R S G (2000) Diagnosis of new variant Creutzfeldt-Jakob disease, *Annals of Neurology*, **47**, 575–82.

Williams E S and Young S (1980) Chronic wasting disease of captive mule deer: a spongiform encephalopathy, *Journal of Wildlife Diseases*, **16**, 89–98.
Yuan J, Xiao X, McGeehan J, Dong A, Cali I, Fujioka H, Kong Q, Kneale G, Gambetti P and Zou W Q (2006) Insoluble aggregates and protease-resistant conformers of prion protein in uninfected human brains, *Journal of Biological Chemistry*, **281**, 34848–58.

31

Histamine fish poisoning – new information to control a common seafood safety issue

P. Dalgaard and J. Emborg, Technical University of Denmark, Denmark

Abstract: Marine finfish products with more than 500–1000 mg of histamine/kg can cause histamine fish poisoning (HFP). With thousands of incidents reported worldwide during more than half a century, HFP is the most common seafood-borne disease. Nevertheless, characteristics, occurrence and quantitative activity of the specific bacteria responsible for histamine formation in seafood remain incompletely understood. Thus, in contrast to earlier reports, several recent incidents of HFP were caused by histamine formation due to the psychrotolerant bacteria *Morganella psychrotolerans* and *Photobacterium phosphoreum*. The present chapter discusses new data on HFP and describes how this information, including predictive models, can be used to control histamine formation in seafood and thereby HFP.

Key words: histamine, psychrotolerant bacteria, *Morganella*, *Photobacterium*, predictive microbiology.

31.1 Introduction

Histamine fish poisoning (HFP) is caused by marine finfish products where specific bacteria have been able to grow to high concentrations and form histamine by decarboxylation of histidine in the fish flesh. For seafood to cause HFP, more than 500–1000 mg of histamine/kg usually needs to be formed at some stage between catch and consumption. Figures 31.1 and 31.2 show the typical kinetics of histamine formation in fish products. Several studies have found mesophilic Enterobacteriaceae including *Morganella morganii*, *Raoultella planticola* (previously *Klebsiella pneumoniae*, see Section 31.3) and *Hafnia alvei* responsible for histamine formation in seafood. These bacteria do not form toxic concentrations of histamine in seafood at below

7–10 °C and chilling seems an obvious strategy to control the risk of HFP. However, once formed, histamine in seafood is rather stable and will not be inactivated by salting, freezing or heat treatments such as normal cooking, hot smoking or even canning. HFP, histamine formation and histamine-producing bacteria have been extensively studied, and various reviews are available (Kimata, 1961; Arnold and Brown, 1978; Taylor, 1986; Lehane and Olley, 2000; Flick *et al.*, 2001; Kim *et al.*, 2003; Dalgaard *et al.*, 2008). The available information has been used for example, to determine critical limits for histamine in seafood, maximum temperatures of storage, good manufacturing practices (GMP) and safety management relying on Hazard Analysis Critical Control Point (HACCP). Nevertheless, HFP remains one of the most common seafood-borne diseases world wide. It seems that available information on HFP and histamine formation in seafood is incomplete, or perhaps not used appropriately for management of this safety hazard. In fact, *Morganella psychrotolerans* and *Photobacterium phosphoreum* (previously N-group bacteria) were recently shown to cause HFP (Kanki *et al.*, 2004; Emborg *et al.*, 2005; Emborg & Dalgaard, 2006). The ability of these psychrotolerant bacteria to produce toxic concentrations of histamine in seafood at 0–5 °C seems a likely reason why HFP remains relatively common and is not efficiently controlled by chilling as used in practice.

The present chapter summarizes information on HFP, including occurrence, implicated products and their characteristics. More detailed information is provided on the relatively few types of bacteria that have been documented as responsible for histamine formation in seafood implicated in incidents of HFP. Options to control the growth and histamine formation by these bacteria are described including the use of predictive models.

31.2 Histamine fish poisoning (HFP)

31.2.1 Occurrence and implicated products

Outbreaks of HFP usually involve less than ten people but, due to a high frequency, this seafood-borne disease is quantitatively important (Table 31.1). During the 1990s, HFP caused 32–38 % of all reported seafood-borne incidents of human disease in the UK and USA. For the USA, HFP outbreaks were responsible for 18 % of all the people that became ill due to consumption of seafood and 1.2 % of all the people that became ill due to food consumption (Dewaal *et al.*, 2006; McLauchlin *et al.*, 2006; Hughes *et al.*, 2007). Data on the worldwide reported occurrence of HFP has been compiled recently (Table 31.1). This occurrence, however, is under-reported because (i) many countries do not collect data on incidents, (ii) symptoms can be mild and of short duration so that a physician is not necessarily contacted and (iii) HFP symptoms can be misdiagnosed and recorded, e.g. as food allergy.

Numerous species of marine finfish have caused HFP, but fish from freshwater do not seems to cause this disease. Amberjack, anchovies,

Table 31.1 Incidents and cases of histamine fish poisoning in various countries

Country	Year	Incidents or outbreaks	Cases	Cases/year/ million people[a]
Hawaii, USA	1990–2003	111	526	31.0
Denmark	1986–2005	64	489	4.9
New Zealand	2001–2005	11	62	3.1
Japan	1970–1980	42	4122	3.2
	1994–2005	68	1523	1.1
France	1987–2005	123	2635	2.5
Finland	1983–2005	41	162	1.3–2.1
Taiwan	1986–2001	8	535	1.5
Norway, UK, South Africa and Switzerland	1966–2004	608	1460	0.4–0.8
Australia, Canada, Netherlands, Philippines, Sweden and USA (states other than Hawaii)	1973–2005	603	3214	0.2–0.4

[a]To compare data between regions of different population size and for different recording periods cases/year/million people was calculated.
Source: Modified from Dalgaard *et al.* (2008).

bluefish, bonito, cape yellowtail, castor oil fish, escolar, garfish, kahawai, kingfish, mackerel, mahi-mahi, marlin, pilchard, sailfish, salmon, sardines, saury, swordfish and various tuna fishes have caused HFP, as shown by a summary of data from 142 outbreaks resulting in 1998 cases (Dalgaard *et al.*, 2008). Many of the implicated species are dark fleshed fish and have a relatively low pH of about 6.

Previously, HFP has been called scombroid fish poisoning, but this is misleading as many fish species that can cause the disease do not belong to Scombridae or Scomberesocidae. About 50 % of the reported incidents and cases of HFP are due to seafood with 1000–5000 mg histamine/kg, and about 90 % of the cases are caused by seafood with more than 500 mg/kg (Bartholomew *et al.*, 1987; Scoging, 1998; Dalgaard *et al.*, 2008). The concentrations of histamine that actually cause HFP are probably somewhat lower than just indicated. The reason for this is that histamine will continue to be formed in some implicated products between the time of an outbreak and until samples are analysed by a laboratory.

Histamine in seafood is formed by bacterial decarboxylation of the amino acid histidine, and more than 2000 mg of free histidine/kg is usual in flesh of fish species that cause HFP. However, above 10 000 mg of free histidine/kg is typically present in several species of tuna (*Thunnus*, *Euthynnus* and *Katuswonus*).

High concentrations of other biogenic amines (agmatine, cadaverine, β-phenethylamine, putrescine, spermine, spermidine, tryptamine and tyramine) may potentiate the oral toxicity of histamine (See Section 31.2.2). However, the concentration of histamine in seafood that actually caused HFP was shown to be about 10 times higher than the concentration of the sum of the other biogenic amines (Eq. 31.1). The species analysed included tuna, escolar, marlin, mackerel, mahi-mahi, swordfish and salmon. This suggests histamine rather than other biogenic amines is responsible for HFP, as further discussed in Section 31.2.2. For a typical meal size of 100 g seafood the concentrations of histamine that usually cause HFP corresponds to an intake of more than 50 mg of histamine with 100–500 mg being most common. The corresponding intake of other biogenic amines is above 5 mg with 10–50 mg being typical (Eq. 31.1).

$$\text{Sum of other biogenic amines (mg/kg)} = 0.1 * \text{Histamine (mg/kg)} \quad [31.1]$$

31.2.2 Symptoms and toxicology

Major HFP symptoms are cutaneous (rash/urticaria), neurological (flushing, headache) and gastrointestinal (diarrhoea, vomiting). Other symptoms include oral tingling, irregular or increased heart rate, i.e. palpitations, abdominal cramps and localized oedema. The symptoms are often mild but can be severe, and in rare situations, even life-threatening, particularly for individuals with asthma or heart disease. Each patient typically experiences several but rarely all of these symptoms. The sensitivity of consumers to HFP differs markedly, and it is unusual that everybody eating a seafood meal becomes ill even when it contains 1000–5000 mg of histamine/kg or more. The incubation time for HFP is short (2 min to 3 h), and people often develop symptoms while they are still eating. Symptoms last from a few hours up to about 1 d. The physiological effects of histamine include dilatation of the peripheral blood vessels and capillaries, increase in capillary permeability, contraction of intestinal smooth muscles as well as increased heart rate and increased force of heart contraction. These effects of histamine explain the symptoms of HFP (Taylor, 1986; Lehane and Olley, 2000; Glória, 2006; Parsons and Ganellin, 2006).

Histamine in small amounts is not toxic for humans as it is metabolized (detoxified) prior to reaching the blood. The enzymes histamine-N-methyltransferase (HMT), monoamine oxidase (MAO) and diamine oxidase (DAO or histaminase) transform histamine to less toxic metabolites that are excreted in urine and faeces. These enzymes are found primarily in the small intestine and liver of humans (Taylor, 1986; Glória, 2006). If the normal histamine metabolism is reduced or very large amounts of histamine are consumed then the concentration of histamine in the blood increases, resulting in the symptoms described above. These symptoms of HFP patients can be resolved by antihistaminic drugs (antihistamines) that block the binding of histamine to specific receptors in different tissues and thereby its effect (Glória, 2006; Parsons and Ganellin, 2006).

In some challenge studies with human volunteers, 67.5–300 mg of histamine administered in water, grapefruit juice or fish resulted in none or mild symptoms only. However, it has also been found that 180 mg histamine resulted in severe headache and flushing (Motil and Scrimshaw, 1979; van Gelderen *et al.*, 1992). Thus, available data from challenge studies with human volunteers suggest pure histamine cannot always explain the toxicity of histamine-containing seafood. This apparently low oral toxicity of pure histamine may, to some extent, be explained by variation in sensitivity of the few volunteers used in challenge studies and the relatively low amounts of histamine (< 100–500 mg) evaluated. Also, two different hypotheses to explain the apparently low oral toxicity of histamine have been extensively discussed in the scientific literature.

The histamine-potentiator hypothesis is based on numerous experiments with laboratory animals where various compounds (agmatine, cadaverine, ß-phenylethylamine, putrescine, trimethylamine, tyramine, combinations of these compounds and ethanol) inhibited normal histamine metabolizing enzymes (DAO, HMT and MAO) and thereby increased the oral toxicity of histamine (Taylor and Lieber, 1979; Lyons *et al.*, 1983; Hui and Taylor, 1985; Satter and Lorenz, 1990). However, data for humans are limited and do not clearly confirm the histamine-potentiator hypothesis (Taylor, 1986; van Gelderen *et al.*, 1992; Lehane and Olley, 2000).

The mast-cell-degranulation hypothesis suggests seafood that cause HFP should contain compounds that trigger a release of histamine from mast cells in the human intestinal tissue. HFP symptoms would then be due to indigenous histamine rather than due to histamine in seafood (Arnold and Brown, 1978; Taylor, 1986; Clifford *et al.*, 1991; Ijomah *et al.*, 1991; Lehane and Olley, 2000). Evidence to support this hypothesis is very limited. In fact, compounds including tryptase and prostaglandin D_2 are released from mast cells when degranulated. However, these compounds have not been detected in serum or urine from patients with HFP (Morrow *et al.*, 1991; Sanchez-Guerrero *et al.*, 1997).

When the observed concentrations of histamine and biogenic amines in seafood implicated in HFP are taken into account it seems this disease is caused primarily by histamine in seafood. Consequently, to reduce the occurrence of HFP the main objective should be to limit histamine formation in seafood.

31.3 Characteristics of histamine-producing bacteria

To efficiently detect, predict or inhibit the bacteria actually responsible for histamine formation in seafood their specific characteristics need to be known. Bacteria responsible for histamine formation in seafood implicated in HFP have been identified (Table 31.2). With thousands of HFP incidents (Table 31.1) it is surprising, however, that the histamine-producing bacteria have been identified for as few as ten incidents (Table 31.2). To determine

Table 31.2 Incidents of histamine fish poisoning where the bacteria responsible for histamine formation have been identified

Implicated seafood	Bacterium	Year reported
Mesophilic bacteria		
Fresh tuna	*Morganella morganii*	1956
Fresh tuna	*Hafnia alvei*	1967
Fresh tuna	*Morganella morganii*	1973
Fresh tuna	*Raoultella planticola*[a]	1978
Tuna heated in flexible film	*Morganella morganii*	2006
Psychrotolerant bacteria		
Dried sardines	*Photobacterium phosphoreum*	2004
Tuna in chilli-sauce	*Morganella psychrotolerans* and/or *Photobacterium phosphoreum*	2005
Cold-smoked tuna	*Photobacterium phosphoreum*	2006
Cold-smoked tuna	*Morganella psychrotolerans*	2006
Fresh tuna	*Photobacterium phosphoreum*	2006

[a] This isolate was initially identified as *Klebsiella pneumoniae* (Lerke *et al.*, 1978; Taylor *et al.*, 1979). However, more recent studies found the genus *Klebsiella* to be heterogeneous and showed histamine producing isolates belonged to *Raoultella* spp. (Drancourt *et al.*, 2001; Kanki *et al.*, 2002).
Source: Modified from Dalgaard *et al.* (2008).

the bacteria responsible for histamine formation in a fish product, part of the dominating microflora need to be isolated. Then, usually after some level of identification, the formation of histamine and possibly other biogenic amines is evaluated quantitatively for representative isolates (Taylor *et al.*, 1978; Emborg and Dalgaard, 2006). When leftovers are obtained, from a specific seafood sample that caused HFP, it is often possible to find the bacteria responsible for histamine formation, provided these bacteria have not been killed by processing or food preparation.

Numerous bacteria associated with seafood have the ability to produce histamine, but most species with strongly histamine-producing isolates/variants also include isolates/variants that produce little or no histamine. The genera *Morganella* and *Raoultella* may be the exceptions as all the studied seafood isolates of *M. morganii*, *M. psychrotolerans*, *R. planticola* and *R. ornithinolytica* seem to produce more than 1000 mg histamine/kg (Taylor *et al.*, 1978; Kanki *et al.*, 2002; Takahashi *et al.*, 2003; Dalgaard *et al.*, 2006; Emborg and Dalgaard, 2006). Formation of more than 1000 mg histamine/kg has also been reported for at least one isolate from several species of Enterobacteriaceae, two species of *Photobacterium* (*P. phosphoreum* and *P. damselae* subsp. *damselae,* previously named *P. histaminum*) and two Gram-positive bacteria, *Staphylococcus epidermidis* and *Tetragenococcus muriaticus* (Okuzumi *et al.*, 1994; Kim *et al.*, 2003). The Enterobacteriaceae included *Citrobacter freundii*, *Enterobacter aerogenes*, *Ent. agglomerans*, *Ent. cloacae*, *H. alvei*, *Klebsiella oxytoca*, *Kl. pneumoniae*, *Proteus mirabilis*, *Pr. vulgaris*, *Providencia stuartii*, *Rahnella aquatilis*, *R. ornithinolyticus*,

R. planticola and *Serratia marcescens*. Concerning the identification and histamine formation by *Klebsiella* it is noteworthy that Drancourt *et al.* (2001) devided species of the genus into three clusters and that cluster II was transferred to the genus *Raoultella*. In addition, Kanki *et al.* (2002) found *Kl. pneumoniae* (cluster I) and *Kl. oxytoca* (cluster III) unable to form histamine and suggested histamine-producing isolates similar to these species actually belonged to *R. ornithinolyticus* and *R. planticola*.

Prior to 2004, the general view was that HFP was caused by the activity of mesophilic histamine-producing bacteria in products that had not been properly chilled i.e. 'temperature abused' (Taylor, 1986; Kim *et al.*, 2003, 2004). Today, it is clear that both mesophilic bacteria (including *M. morganii*, *H. alvei* and *R. planticola*) and psychrotolerant bacteria (including *M. psychrotolerans* and *P. phosphoreum*) can produce toxic concentrations of histamine in seafood and thereby cause HFP (Table 31.2). It seems likely that bacteria other than the species shown in Table 31.2 can cause HFP due to their formation of high concentrations of histamine in seafood. However, additional studies of bacteria from seafood responsible for HFP are needed to determine if this is the case.

M. psychrotolerans and *P. phosphoreum* can grow at 0 °C (i.e. they are psychrotolerant) but not at 37 °C, and their temperature optimum is 20–30 °C. Interestingly, this corresponds to early studies from Japan where histamine formation in mackerel was faster and more pronounced at 17–23 °C compared to both 6–7 °C and 35 °C (Kimata, 1961). In contrast, when histamine is produced by mesophilic bacteria then the formation at 35–37 °C is markedly faster than at 17–23 °C (Frank *et al.*, 1981, 1983; Kim *et al.*, 2002).

Growth characteristics of species that include strongly histamine-producing isolates are shown in Table 31.3. It needs to be noted that environmental parameters corresponding to the limits of growth can differ markedly from the environmental conditions that allow an isolate to form toxic concentrations for histamine. For *H. alvei,* toxic histamine formation has been observed at temperatures of 7–19 °C or higher (Arnold *et al.*, 1980; Behling and Taylor, 1982), but the minimum growth temperature is ~2.5 °C (Table 31.3). Similarly, growth of at least some isolates of *M. morganii* and *R. planticola* can be detected at ~4 °C whereas toxic histamine formation typically is observed at 7–10 °C or above (Behling and Taylor, 1982; Wei *et al.*, 1990). Further studies are needed to determine the combined effect of several environmental parameters on limits for growth and histamine formation by the bacteria indicated in Table 31.3.

31.4 Detection of histamine-producing bacteria

For direct enumeration of histamine-producing bacteria in seafood various modifications of Nivens agar have been suggested, but false-positive colonies

Table 31.3 Growth characteristics of bacterial species that include strongly histamine producing isolates

Bacteria	Growth temperature (°C)		Maximum % NaCl for growth	Minimum pH for growth	References
	Minimum	Optimum			
Mesophilic bacteria					
Hafnia alvei[a]	2.5[a]	–	–	–	Ridell, 2000
Morganella morganii	~ 4	36–37	~ 9	~ 4.5	Janda and Abbott, 2005; Emborg *et al.*, 2006
Photobacterium damselae subsp. *damselae*[b]	> 4	30–35	6–8[c]	~ 4.5	Okuzumi *et al.*, 1994
Raoultella planticola	~ 4	–	–	–	Kanki *et al.*, 2002
Tetragenococcus myriaticus	~ 15	25–30	~ 25[d]	~ 5	Satomi *et al.*, 1997
Psychrotolerant bacteria					
Photobacterium phosphoreum	< 0	~ 20	~ 6[c]	4.3	Dalgaard, 1995; Dalgaard, 2002
Morganella psychrotolerans	< 0	26–27	6.5–8	4.5– > 5	Emborg *et al.*, 2006; Emborg and Dalgaard, 2008a,b

[a]*Hafnia alvei* includes two mesophilic variants with minimum growth temperatures of ~2.5 °C and ~11 °C, respectively.
[b]Previously identified as *Photobacterium histaminum* (Okuzumi *et al.*, 1994).
[c]~ 1 % and ~ 2 % NaCl are optimal for growth of *Photobacterium phosphoreum* and *Photobacterium damselae* subsp. *damselae*, respectively.
[d]5–7 % NaCl seems optimal for histamine formation by *Tetragenococcus myriaticus* (Kimura *et al.*, 2001).

remain problematic (Niven *et al.*, 1981; Marcobal *et al.*, 2006). Nivens agar relies on pour plating and incubation of plates at 35–37 °C. Thus *M. psychrotolerans* and *P. phosphoreum* will never be detected as they do not grow at 35–37 °C and the latter is killed by pour plating with 45 °C warm liquid agar. Incubation of Nivens agar at 25 °C allows *M. psychrotolerans* to form colonies with a purple halo but, as observed for other bacteria by Niven *et al.*, 1981, incubation at 25 °C is not optimal due to weak and diffuse purple halo formation.

M. psychrotolerans can be detected together with other Enterobacteriaceae using for example, Violet Red Bile Glucose agar and incubation at 25 °C. *P. phosphoreum* is more heat labile but can be detected by spread plating on Long and Hammer agar with 1 % NaCl and incubation at 15 °C (NMKL, 2006). Polymerase chain reaction (PCR) methods for detection of the histidine decarboxylase genes from both Gram-negative and Gram-positive bacteria have been suggested (Takahashi *et al.*, 2003; Rivas *et al.*, 2006; Landete *et al.*, 2007). It remains a challenge to develop sensitive methods for specific enumeration in seafood of the strongly histamine-producing variants of both psychrotolerant and mesophilic bacteria.

31.5 Management of histamine formation and HFP

Growth of histamine-producing bacteria, and the related formation of histamine, depends on several factors, including their presence in a specific seafood, storage conditions (temperature, atmosphere) and various product characteristics (pH, lactic acid, salt, smoke components and added antimicrobial agents). Quantitative information about the combined effect of these factors is important to manage histamine formation and thereby HFP. The information can be used for example, to determine safe shelf-life or to obtain a desired shelf-life by changing storage conditions or product characteristics.

31.5.1 Legislation, critical limits and methods to detect histamine

EU regulation EC 2073 (EC, 2005) includes sampling plans (n = 9 and c = 2) and limits for critical concentrations of histamine in 'fishery products from fish species associated with a high amount of histidine' where m = 100 mg/kg and M = 200 mg/kg. For 'fishery products which have undergone enzyme maturation treatment in brine, manufactured from fish species associated with a high amount of histidine' higher limits of m = 200 mg/kg and M = 400 mg/kg are applied. The EU regulation does not include critical limits for other biogenic amines (EU, 2005).

The USA uses a defect action level (m) of 50 mg of histamine/kg. Eighteen fish per lot should be analyzed individually or composited into, for example, six units of fish, but then the critical limit is reduced accordingly from 50 mg/kg to 17 mg/kg (FDA/CFSAN, 2001).

EU regulation EC 2073 (EC, 2005) indicates a high-performance liquid chromatography (HPLC) method (Malle *et al.*, 1996) for determination of histamine in seafood, whereas a less instrument-demanding fluometric AOAC method is used in the USA and several other countries (Rogers and Staruszkiewicz, 1997; AOAC, 2005). Many histamine test kits are available, and some perform well when compared to the AOAC method (Rogers and Staruszkiewicz, 2000).

31.5.2 Hygiene and disinfection

Information on the occurrence of histamine-producing bacteria in catch/harvesting water, live fish, fishing vessels and fish processing environments is important to understand routes of contamination and potentially limit or prevent contamination of seafood. However, the high frequency of HFP (Table 31.1) suggests contamination of seafood is common, and it may be difficult to prevent as explained below. In fact, histamine-producing bacteria are part of the natural microflora in seawater, intestines, gills and on the skin of marine fish. The flesh of live healthy fish is sterile but after slaughter histamine-producing bacteria from the live fish can directly or indirectly contaminate the fish flesh.

The intestinal content of many fish contains *P. phosphoreum* in high concentrations, often 10^6–10^8 cfu/g. Other strongly histamine-producing bacteria including *M. morganii, R. planticola, H. alvei, Pr. vulgaris and P. damselae* subsp. *damselae* are also found in fish intestines, but less is known about their typical concentrations (Omura *et al.*, 1978; Taylor *et al.*, 1979; Okuzumi and Awano, 1983; Okuzumi *et al.*, 1984b,c; Yoguchi *et al.*, 1990; Dalgaard, 2006). A high concentration of strongly histamine-producing bacteria in the intestinal content of at least some marine fish corresponds to the observation that histamine formation can be most pronounced in fish flesh close to the abdominal cavity (Lerke *et al.*, 1978; Frank *et al.*, 1981; Bédry *et al.*, 2000). Efficient gutting, removal of gills and rinsing is recommended (Frank *et al.*, 1981) but not possible in practice for smaller fish. Gills of different fish can also contain various strongly histamine-producing bacteria (Omura *et al.*, 1978; Kim *et al.*, 2001). However, quantitative information about their concentrations and transfer to fish flesh is limited (Kim *et al.*, 2001; Staruszkiewicz *et al.*, 2004). To improve the effect of hygiene and disinfection efforts, further studies are needed on occurrence and routes for contamination of seafood with strongly histamine-producing bacteria. At present it seems more realistic to manage histamine formation by reducing growth of the relevant bacteria.

31.5.3 Time and temperature

Chilling of seafood is highly important to reduce the rate of histamine formation (as described quantitatively in Section 31.5.5). Below 7–10 °C mesophilic and strongly histamine-producing bacteria do not form toxic concentrations of histamine in fish products. However, the psychrotolerant bacteria *M. psychrotolerans* and *P. phosphoreum* can produce toxic concentrations of histamine at 0–5 °C (Okuzumi *et al.*, 1982; van Spreekens, 1987; Kanki *et al.*, 2004; Emborg *et al.*, 2005; Dalgaard *et al.*, 2006). Recently, data for histamine formation were compiled from 124 storage trials with naturally contaminated seafood (Fig. 31.1). As shown by Fig. 31.1 toxic concentrations of histamine (> 500 mg/kg) can be formed both after short time at high temperature, e.g. from ~2 d at +11 °C to +15 °C as well as during more extended times of storage at below 5 °C, i.e. at temperatures in agreement with existing regulations within EU and USA.

EU regulation (EC 853/2004) states that 'Fresh fishery products, thawed unprocessed fishery products, and cooked and chilled products from crustaceans and molluscs, must be maintained at a temperature approaching that of melting ice' (EU, 2004). In some EU countries this is interpreted as temperatures between 0 °C and +2 °C. Lightly preserved seafood with less than 6 % salt and pH above 5, e.g. smoked and marinated products, should be at 5 °C or less. Regulations in the USA specifically indicate maximum times to reach critical chill storage temperatrures but the allowed chill storage temperature of 4.4 °C is relatively high (FDA/CFSAN, 2001). Storage at 2.0 °C or 4.4 °C can have a markedly different effect on growth and histamine formation as shown in Fig. 31.2(a) for *M. psychrotolerans*. In the USA, seafood in reduced oxygen packaging must be stored and distributed at below 3.3 °C. This is requested due to the risk of toxin formation by *Clostridium botulinum* type E (FDA/CFSAN, 2001) but, compared to storage at 4.4 °C, the risk of histamine formation in high concentrations is also considerably lower at 3.3 °C.

The variability of histamine formation shown in Fig. 31.1 for different naturally-contaminated fish products most likely results from (i) differences in contamination with strongly histamine-producing bacteria, (ii) differences in storage temperature and/or (iii) differences in product characteristics such as pH and the concentration of naturally-occurring lactic acid.

To manage histamine formation in seafood, the effect of temperature profiles from catch/harvest to consumption must be taken into account. This is a challenge, but mathematical models can be used to predict the effect of temperature profiles on histamine formation as discussed in Section 31.5.5. In some situations fish are not iced immediately after catch. The effect of this delayed or insufficient icing is a significant risk factor for histamine formation, and it has been the topic of much debate (Pan and James, 1985). Storage of fish during for example, 12-24 h at ~15 °C will not typically result in high concentrations of histamine (Fig. 31.1), but substantial growth of both psychrotolerant and mesophilic bacteria may occur. During subsequent chilled storage at 0–5 °C the psychrotolerant bacteria including both *M. psychrotolerans* and *P. phosphoreum* will continue to grow and, depending on the storage time, they will form toxic concentrations of histamine (Emborg *et*

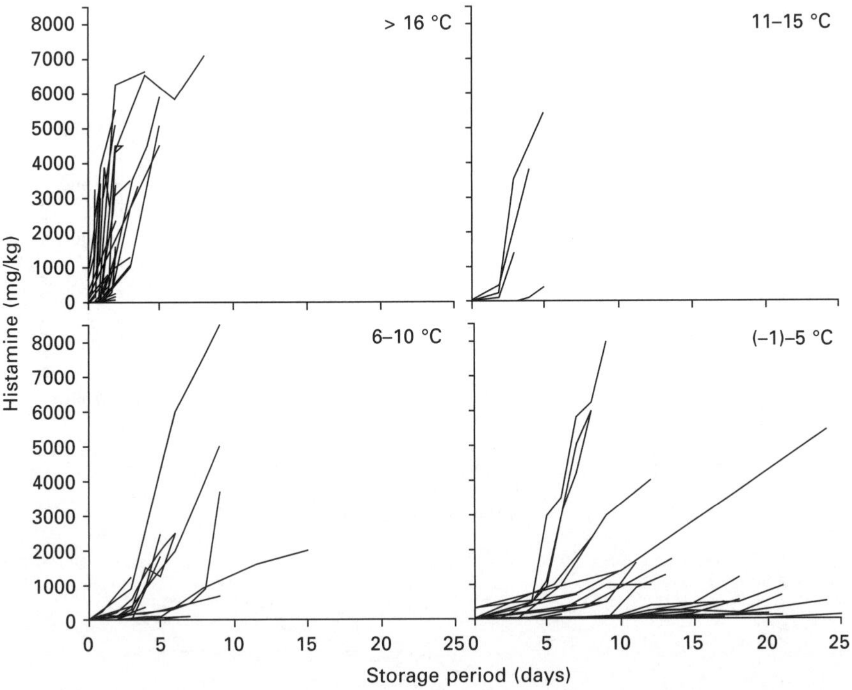

Fig. 31.1 Effect of temperature on histamine formation in seafood. Data from 124 storage trials with naturally-contaminated seafood (Dalgaard *et al.*, 2008).

al., 2005; Dalgaard *et al.*, 2006). It has also been suggested that the enzyme histidine decarboxylase produced prior to chilling by mesophilic bacteria remains active at below 5 °C and results in histamine formation, although the mesophilic bacteria grow very slowly or not at all at that temperature (Behling and Taylor, 1982). Histamine formation at 5 °C or below by the histidine decarboxylase enzyme in mesophilic bacteria has been documented, but very high concentrations of bacteria were required (Sakabe, 1973; Klausen and Huss, 1987; Kanki *et al.*, 2007). In relation to delayed icing of fish and seafood, studies are still lacking for the relative importance of psychrotolerant and mesophilic bacteria with respect to quantitative histamine formation.

31.5.4 Storage atmosphere, NaCl and pH

Vacuum-packed fresh fish as well as dried or smoked seafood with added NaCl have caused numerous incidents of HFP e.g. products of escolar, kahawai, mackerel and tuna (Scoging, 1998; Fletcher *et al.*, 1999; Kanki *et al.*, 2004; Emborg and Dalgaard, 2006). Information about the quantitative effect of storage atmosphere (vacuum or modified atmosphere packaging), NaCl and pH is clearly important to manage histamine formation in these seafoods.

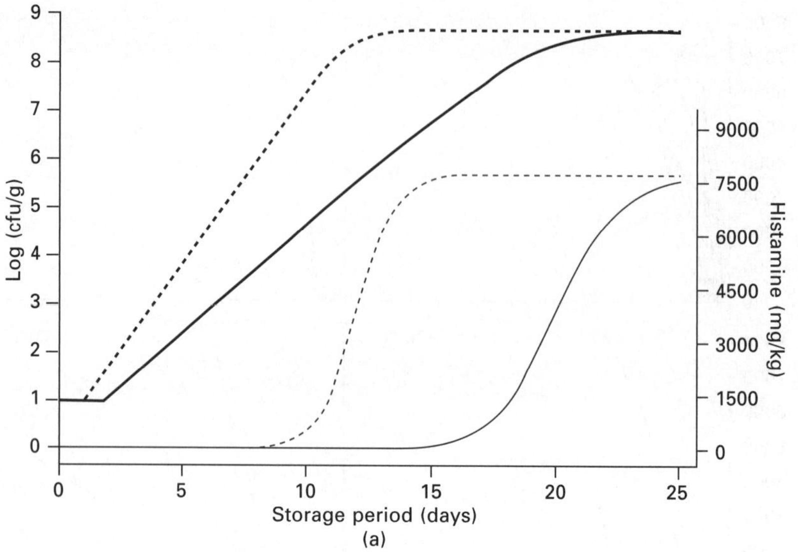

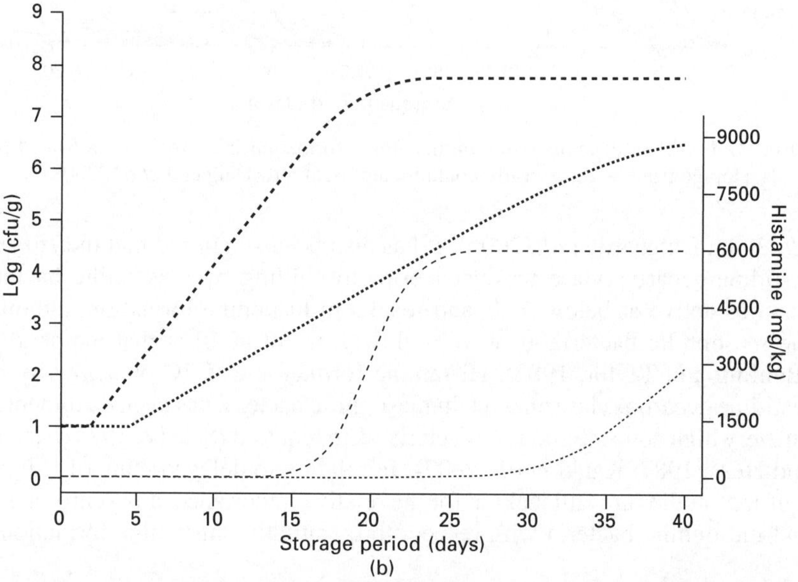

Fig. 31.2 Predicted growth (bold lines) and histamine formation (fine lines) by *Morganella psychrotolerans* at pH 5.9. (a) Predictions for 2.0 °C (solid lines) and 4.4 °C (dashed lines). (b) Predictions for 5.0 °C with 3.5 % NaCl (dashed lines) and 5.0 % NaCl (dotted lines) (Emborg and Dalgaard, 2008a).

Vacuum and modified atmosphere packaging (MAP) is used increasingly by the seafood sector. Vacuum packing reduces lipid oxidation in seafood but does not seem to delay histamine formation in fresh fish. MAP with gas mixtures containing carbon dioxide (CO_2) and nitrogen (N_2) can delay histamine formation when high CO_2 concentrations are used. Compared to fresh MAP fish, these atmospheres delay histamine formation more efficiently for frozen and thawed products where the highly CO_2–resistant bacteria *P. phosphoreum* has been inactivated due to the frozen storage (Dalgaard *et al.*, 2006). For lean fish, atmospheres with high concentrations of both CO_2 and oxygen (O_2) inhibit histamine formation markedly as shown, e.g. for chilled MAP tuna (Emborg *et al.*, 2005).

Concentrations of salt above 1–2 % NaCl reduce growth of the Gram-negative and strongly histamine-producing bacteria. In agreement with this the formation of histamine in mackerel at 20 °C was delayed from one to two days by addition of 3 % NaCl and from one to four days by addition of 4 % NaCl (Yamanaka *et al.*, 1985). For vacuum-packed cold-smoked tuna the potential histamine formation by *M. psychrotolerans* and *P. phosphoreum* can be controlled using 5 % water phase salt and a declared shelf-life of 3–4 weeks or less at 5 °C (Emborg and Dalgaard, 2006). As shown in Fig. 31.2(b) growth and histamine formation by *M. psychrotolerans* is delayed much more by 5.0 % water phase salt as compared to 3.5 % water phase salt. Gram-positive bacteria including *Staphylococcus epidermidis* and *Tetragenococcus muriaticus* can produce histamine at higher NaCl concentrations (Hernández-Herrero *et al.*, 1999; Kimura *et al.*, 2001). This may be important for fish sauce, fermented fish and salted-ripened fish, but the relative importance of these bacteria and of histidine decarboxylase produced by other bacteria prior to the mixing of fish and salt remains to be quantified.

Histamine formation in frigate tuna (*Auxis thazard*) was faster and much more pronounced at pH 6.2 as compared to both pH 5.6 and pH 6.7 (Kimata and Kawai, 1953). This corresponds to the effect of pH on histamine formation by isolated bacteria, e.g. *P. phosphoreum* (Okuzumi *et al.*, 1984a; Dalgaard *et al.*, 2006). Related to management of histamine formation in seafood, it is notable that a lowering of a product's pH to about 6 can contribute to increased risk of histamine formation.

31.5.5 Prediction of histamine formation

The kinetics of histamine formation during storage of seafood is characterized by a sometimes long phase with little or no histamine production followed by a second phase where the concentration can increase rapidly (Fig. 31.1 and 31.2). The first phase corresponds to the time needed for the specific histamine-producing bacteria to reach high concentrations, and the length of this phase depends primarily on the initial concentration of these bacteria, and their growth rate. The rate of histamine formation during the second phase corresponds to the activity of high concentrations of the histamine-producing bacteria and it is influenced by storage conditions and product characteristics (Fig. 31.2).

Simple empirical models to predict the kinetics of histamine formation have been suggested (Frank *et al.*, 1983; Frank, 1985; Pan and James, 1985;

Frank and Yoshinaga, 1987). The precision of these models is modest and they have not been widely adopted by the seafood sector (Dalgaard *et al.*, 2008). The most accurate of the empirical models has been developed by Frank (1985) for high-temperature storage (21.1 °C–37.8 °C) of Skipjack tuna (Eq. 31.2).

$$\text{Histamine (mg/kg)} = 2.031 \cdot 10^{-10} \cdot \text{Time(h)}^{3.523} \cdot (K-277.7)^{5.342} \quad [31.2]$$

where K is storage temperature in °Kelvin.

More recently, a predictive microbiology approach was used by Emborg and Dalgaard (2008a) to model and predict growth and histamine formation by *M. psychrotolerans*. A primary model, relying on a growth model and a constant yield factor for histamine formation, was used to predict the effect of the initial cell concentration and storage times on histamine formation. A secondary model described the effect of temperature (0–25 °C), atmosphere (0–100 % CO_2), NaCl (0–5 %) and pH (5.4–6.5) on growth and histamine formation. The deviation between measured and predicted times to formation of 500 mg of histamine/kg of seafood was about 10 % on average, but more variation was observed for individual experiments (Emborg and Dalgaard, 2008b). Importantly, this type of model has the potential to predict histamine formation depending on hygiene/initial cell concentrations as well as both storage conditions, including variable storage temperatures, and product characteristics (Fig. 31.2). Development of similar predictive microbiology models for other important histamine-producing bacteria is needed to predict histamine formation in various seafoods. We believe predictive models for histamine formation will become an important tool to manage HFP, and new models for *M. morganii* and *P. phosphoreum* are under development by our research group (http://sssp.dtuaqua.dk).

31.6 Conclusions

HFP remains a frequently occurring seafood-borne disease, and to improve this situation new controls are needed. Distribution of seafood at inappropriately high chill storage temperatures, e.g. above 7–10 °C, is one possible explanation for the high frequency of HFP. However, new information on psychrotolerant and strongly histamine-producing bacteria suggests the importance of histamine formation in seafood at below 7–10 °C has been under-estimated. Accepting the importance of histamine formation by psychrotolerant bacteria and using available information on *M. psychrotolerans* and *P. phosphoreum* provides new options to manage HFP:

- Shelf-life from catch to consumption of fresh marine fish should be a maximum of 14 d at 0 °C corresponding to 6–7 d at 4.4 °C when products are stored aerobically or vacuum-packed.
- Modified atmosphere packaging with high concentrations of both CO_2

and O_2 can be used to reduce histamine formation and extend the safe shelf-life, e.g. for lean MAP tuna steaks.
- New predictive models can facilitate the determination of safe shelf-life with respect to histamine formation depending on storage conditions and product characteristics – for example, determination of the NaCl concentrations required to obtain desired shelf-life of chilled cold-smoked tuna (Fig. 31.2(b)).

31.7 Future trends

To further improve the control of histamine formation in seafood several types of new information and activities are needed – a few are listed below.

- Specific and sensitive enumeration methods for species of strongly histamine-producing bacteria are needed to evaluate their concentration, for example, in fish raw material and newly processed seafood. For development of these methods the potential of real-time PCR procedures deserves to be evaluated.
- To reduce concentrations of strongly histamine-producing bacteria in seafood additional information is needed on their occurrence and routes of contamination. This information is particularly relevant for the newly identified species *M. psychrotolerans* where practically no information of this type is available.
- The relative importance of histamine formation in seafood by mesophilic and psychrotolerant bacteria deserves further attention to determine to what extent the new data on psychrotolerant bacteria obtained in Denmark and Japan are valid for other regions.
- To predict histamine formation, mathematical models for the effect of storage conditions and product characteristics on at least *P. phosphoreum* (psychrotolerant) and *M. morganii* (mesophilic) deserve further development.
- When sufficient information and predictive models become available, quantitative exposure and risk assessments will benefit decisions on improved ways to manage HFP.

31.8 Sources of further information and advice

An overview of the formation and importance of histamine and biogenic amines in seafood was recently presented by Dalgaard *et al.* (2008). That book chapter summarized results from a project within the EU SEAFOODplus programme (www.seafoodplus.org). New information was provided on occurrence of histamine fish poisoning (HFP), bacteria responsible for histamine formation and concentrations of histamine and biogenic amines in seafood implicated in incidents of HFP as well as on the importance of psychrotolerant and strongly histamine-producing bacteria.

Two of the mathematical models discussed in the present chapter to predict histamine formation have been included in the Seafood Spoilage and Safety Predictor (SSSP) software which is available at http://sssp.dtuaqua.dk/.

Lehane and Olley (2000) presented histamine fish poisoning in a risk-assessment framework. The extensive review included discussion of hypotheses on toxicology of histamine and HFP. Particular attention was given to the role of urocanic acid as a potential mast cell degranulating compound.

Earlier, Talyor (1986) presented an authoritative review on toxicology and occurrence of histamine food poisoning. That review also included a summary of early methods for detection of histamine.

31.9 References

AOAC (2005) Histamine in Seafood, Fluorometric method, 977.13, in P A Gunniff (ed.) *Official Methods of the Analysis of AOAC International*, 18th edn, Association of Official Analytical Chemists International, Gaithersburg, MD, 35.1.32.

Arnold S H and Brown W D (1978) Histamine (?) toxicity from fish products, *Advances in Food Research*, **24**, 113–54.

Arnold S H, Price R J and Brown W D (1980) Histamine formation by bacteria isolated from Skipjack tuna, *Katsuwonus pelamis*, *Bulletin of the Japanese Society of Scientific Fisheries*, **46**(8), 991–5.

Bartholomew B A, Berry P R, Rodhouse J C and Gilbert R J (1987) Scombrotoxic fish poisoning in Britain: feature of over 250 suspected incidents from 1976 to 1986, *Epidemiology and Infection*, **99**, 775–82.

Bédry R, Gabinski C and Paty M-C (2000) Diagnosis of scombroid poisoning by measurement of plasma histamine, *New England Journal of Medicine*, **342**(7), 520–1.

Behling A R and Taylor S L (1982) Bacterial histamine production as a function of temperature and time of incubation, *Journal of Food Science*, **47**, 1311–14, 1317.

Clifford M N, Walker R, Ijomah P, Wright J, Murray C K and Hardy R (1991) Is there a role for amines other than histamines in the etiology of scombrotoxicosis, *Food Additives and Contaminants*, **8**(5), 641–52.

Dalgaard P (1995) The effect of anaerobic conditions and carbon dioxide, in H H Huss (ed), *Quality and Quality Changes in Fresh Fish*, Food and Agriculture Organization of the United Nations, Rome, 78–83.

Dalgaard P (2002) Modelling and predicting the shelf-life of seafood, in H A Bremner (ed), *Safety and Quality Issues In Fish Processing*, Woodhead, Cambridge, 191–219.

Dalgaard P (2006) Microbiology of marine muscle foods, in Y H Hui (ed), *Handbook of Food Science, Technology and Engineering, Vol. I*, CRC, Boca Raton, FL, 53–1 to 53–20.

Dalgaard P, Madsen H L, Samieian N and Emborg J (2006) Biogenic amine formation and microbial spoilage in chilled garfish (*Belone belone belone*) – effect of modified atmosphere packaging and previous frozen storage, *Journal of Applied Microbiology*, **101**, 80–95.

Dalgaard P, Emborg J, Kjølby A, Sørensen N D and Ballin N Z (2008) Histamine and biogenic amines - formation and importance in seafood, in T Børresen (ed), *Improving Seafood Products for the Consumer*, Woodhead, Cambridge, 292–324.

Dewaal C S, Hicks G, Barlow K, Alderton L and Vegosen L (2006) Foods associated with food-borne illness outbreaks from 1990 through 2003, *Food Protection Trends*, **26**(7), 466–73.

Drancourt M, Bollet C, Carta A and Rousselier P (2001) Phylogenetic analysis of *Klebsiella* species delineate *Klebsiella* and *Raoultella* gen. nov., with description of *Raoultella ornithinolytica* comb. nov., *Raoultella terrigena* comb. nov. and *Raoultella planticola* comb. nov., *International Journal of Systematic and Evolutionary Microbiology*, **51**, 925–32.

EU (2004) Regulation (EC) No 853/2004 of the European Parliament and of the Council of 29 April 2004 laying down specific hygiene rules for food of animal origin, *Official Journal of the European Union*, **226**, 25 June, 22–82.

EU (2005) Commission regulation (EC) No 2073/2005 of 15 November 2005 on microbiological criteria for foodstuffs, *Official Journal of the European Union*, **338**, 22 December, 1–25.

Emborg J and Dalgaard P (2008a) Modelling the effect of temperature, carbon dioxide, water activity and pH on growth and histamine formation by *Morganella psychrotolerans*, *International Journal of Food Microbiology*, **128**, 226–233.

Emborg J and Dalgaard P (2008b) Growth, inactivation and histamine formation of *Morganella psychrotolerans* and *Morganella morganii* – development and evaluation of predictive models, *International Journal of Food Microbiology*, **128**, 234–243.

Emborg J, Dalgaard P and Ahrens P (2006) *Morganella psychrotolerans* sp. nov., a histamine producing bacterium isolated from various seafoods, *International Journal of Systematic and Evolutionary Microbiology*, **56**, 2473–9.

Emborg J, Laursen B G and Dalgaard P (2005) Significant histamine formation in tuna (*Thunnus albacares*) at 2 °C – effect of vacuum- and modified atmosphere-packaging on psychrotolerant bacteria, *International Journal of Food Microbiology*, **101**, 263–79.

FDA/CFSAN (2001) Scombrotoxin (histamine) formation, in *Fish and Fishery Products Hazards and Controls Guidance*, Washington, DC, Food and Drug Administration, Center for Food Safety and Applied Nutrition, Office of Seafood, 83–102.

Fletcher G C, Bremer P J, Osborne C M, Summers G, Speed S R and van Veghel P W C (1999) Controling bacterial histamine production in hot-smoked fish, in A C J Tuitelaars, R A Samson, F M Rombouts and S Notermans (eds), *Food Microbiology and Food Safety into the next Millennium,* Foundation Food Micro '99, Wageningen, 168–72.

Flick G J, Oria M P and Douglas L (2001) Potential hazards in cold-smoked fish: Biogenic amines, *Journal of Food Science*, **66** (Supplement) (7), S1088–S1099.

Frank H A (1985) Use of nomographs to estimate histamine formation in tuna, in B S Pan and D James (eds), *Histamine in Marine Products: Production by Bacteria, Measurement and Prediction of Formation*, Food and Agriculture Organization of the United Nations, Rome, 18–20.

Frank H A and Yoshinaga D H (1987) Table for estimating histamine formation in Skipjack tuna, *Katsuwonus pelamis*, at low nonfreezing temperatures, *Marine Fisheries Review*, **49**(4), 67–70.

Frank H A, Yoshinaga D H and Nip W K (1981) Histamine formation and honeycombing during decomposition of skipjack tuna (*Katsuwonus pelamis*), at elevated temperatures, *Marine Fisheries Review*, **43**(10), 9–14.

Frank H A, Yoshinaga D H and Wu I P (1983) Nomograph for estimating histamine formation in skipjack tuna at elevated temperatures, *Marine Fisheries Review*, **45**(4–6), 40–44.

Glória M B A (2006) Bioactive amines, in Y H Hui (ed), *Handbook of Food Science, Technology and Engineering, Vol. I*, CRC, Boca Raton, FL, 13–1 to 13–38.

Hernández-Herrero M M, Roig-Sagués A X, Rodigues-Lopez J and Mora-Ventura M T (1999) Halotolerant and halophilic histamine-forming bacteria isolated during the ripening of salted anchovies (*Engraulis encrasicholus*), *Journal of Food Protection*, **62**(5), 509–14.

Hughes C, Gillespie I A, O'Brien S J and The Breakdowns in Food Safety Group (2007) Foodborne transmission of infectious disease in England and Wales, 1992–2003, *Food Control*, **18**, 766–72.

Hui J Y and Taylor S L (1985) Inhibition of *in vivo* histamine-metabolism in rats by foodborne and pharmacologic inhibitors of diamine oxidase, histamine N-methyltransferase, and monoamine oxidase, *Toxicology and Applied Pharmacology*, **81**(2), 241–9.

Ijomah P, Clifford M N, Walker R, Wright J, Hardy R and Murray C K (1991) The importance of endogenous histamine relative to dietary histamine in the etiology of scombrotoxicosis, *Food Additives and Contaminants*, **8**(4), 531–42.

Janda J M and Abbott S L (2005) Genus XXl. Morganella *Fulton* 1943 81^{AL}, in O J Brenner, N R Kreig, J T Staley and G M Garrity (eds), *Bergye's Manual of Systematic Bacteriology*, 2nd edn, Vol. 2, part B, Springer, New York, 707–9.

Kanki M, Yoda T, Tsukamoto T and Shibata T (2002) *Klebsiella pneumoniae* produces no histamine: *Raoultella planticola* and *Raoultella ornithinolytica* strains are histamine producers, *Applied and Environmental Microbiology*, **68**(7), 3462–6.

Kanki M, Yoda T, Ishibashi M and Tsukamoto T (2004) *Photobacterium phosphoreum* caused a histamine fish poisoning incident, *International Journal of Food Microbiology*, **92**, 79–87.

Kanki M, Yoda T, Tsukamoto T and Baba E (2007) Histidine decarboxylase and their role in accumulation of histamine in tuna and dried saury, *Applied and Environmental Microbiology*, **73**(5), 1467–73.

Kim S-H, Field K G, Morrissey M T, Price R J, Wei C and An H (2001) Source and identification of histamine-producing bacteria from fresh and temperature-abused albacore, *Journal of Food Protection*, **64**(7) 1035–44.

Kim S H, Price R J, Morrissey M T, Field K.G, Wei C I and An, H J (2002) Histamine production by *Morganella morganii* in mackerel, albacore, mahi-mahi, and salmon at various storage temperatures, *Journal of Food Science*, **67**(4), 1522–8.

Kim S-H, Barros-Velázquez J, Ben-Gigirey B, Eun J-B, Jun S H, Wei C and An H (2003) Identification of the main bacteria contributing to histamine formation in seafood to ensure product safety, *Food Science and Biotechnology*, **12**(4), 451–60.

Kim S-H, Wei C, Clemens R A and An H (2004) Review: Histamine accumulation in seafoods and its control to prevent outbreaks of scombroid poisoning, *Journal of Aquatic Food Product Technology*, **13**(4), 81–100.

Kimata M (1961) The histamine problem, in G Borgstrøm (ed), *Fish as Food*, Academic Press, New York, 329–52.

Kimata M and Kawai A (1953) The freshness of fish and the amount of histamine presented in the meat. I., *Memoirs of the Research Institute for Food Science*, no. **3**, 25–54.

Kimura B, Konagaya Y and Fujii T (2001) Histamine formation by *Tetragenococcus muriaticus,* a halophilic lactic acid bacterium isolated from fish sauce, *International Journal of Food Microbiology*, **70**, 71–7.

Klausen N K and Huss H H (1987) Growth and histamine production by *Morganella morganii* under various temperature conditions, *International Journal of Food Microbiology*, **5**, 147–56.

Landete J M, De las Rivas B, Marcobal A and Munoz R (2007) Molecular methods for the detection of biogenic amine-producing bacteria in foods, *International Journal of Food Microbiology*, **117**, 258–69.

Lehane L and Olley J (2000) Histamine fish poisoning revisited, *International Journal of Food Microbiology*, **58**, 1–37.

Lerke P A, Werner S B, Taylor S L and Guthertz L S (1978) Scombroid poisoning, *Western Journal of Medicine*, **129**(5), 381–6.

Lyons D E, Beery J T, Lyons S A and Taylor S L (1983) Cadaverine and aminoguanidine potentiate the uptake of histamine *in vitro* in perfused intestinal segments of rats, *Toxicology and Applied Pharmacology*, **70**, 445–58.

Malle P, Valle M and Bouquelet S (1996) Assay of biogenic amines involved in fish decomposition, *Journal of AOAC International*, **79**, 43–9.

Marcobal A, De las Rivas B and Munoz R (2006) Methods for the detection of bacteria producing biogenic amines on foods: a survey, *Journal für Verbraucherschutz und Lebensmittelsicherheit*, **1**, 187–96.

McLauchlin J, Little C L, Grant K A and Mithani V (2006) Scombrotoxic fish poisoning, *Journal of Public Health*, **28**(1), 61–2.

Morrow J D, Margolies G R, Rowland J and Roberts L J (1991) Evidence that histamine is the causative toxin of scombroid fish poisoning, *New England Journal of Medicine*, **324**(11), 716–20.

Motil K J and Scrimshaw N S (1979) The role of exogenous histamine in scombroid poisoning, *Toxicology Letters*, **3**, 219–23.

Niven C F, Jeffrey M B and Corlett D A Jr (1981) Differential plating medium for quantitative detection of histamine-producing bacteria, *Applied and Environmental Microbiology*, **41**(1), 321–2.

NMKL (2006) *Aerobic Count and Specific Spoilage Organisms in Fish and Fish Products*, Nordic Committee on Food Analysis, NMKL no. 184.

Okuzumi M and Awano M (1983) Seasonal variations in numbers of psychrophilic and halophilic histamine-forming bacteria (N-group bacteria) in seawater and on marine fishes, *Bulletin of the Japanese Society of Scientific Fisheries*, **49**(8), 1285–91.

Okuzumi M, Okuda S and Awano M (1982) Occurrence of psychrophilic and halophilic histamine-forming bacteria (N-group bacteria) on/in red meat fish, *Bulletin of the Japanese Society of Scientific Fisheries*, **48**(6), 799–804.

Okuzumi M, Awano M and Ohki Y (1984a) Effect of temperature, pH value and NaCl concentration on histamine formation of N-group bacteria (psychrophilic and halophilic histamine-forming bacteria), *Bulletin of the Japanese Society of Scientific Fisheries*, **50**(10), 1757–62.

Okuzumi M, Yamanaka H and Kubaczka M (1984b) Occurrence of various histamine-forming bacteria on/in fresh fishes, *Bulletin of the Japanese Society of Scientific Fisheries*, **50**(1), 161–7.

Okuzumi M, Yamanaka H, Kubozuka T, Ozaki H and Matsubara K (1984c) Changes in numbers of histamine-forming bacteria on/in common mackerel stored at various temperatures, *Bulletin of the Japanese Society of Scientific Fisheries*, **50**(4), 653–7.

Okuzumi M, Hiraishi A, Kobayashi T and Fujii T (1994) *Photobacterium histaminum* sp. nov., a histamine-producing marine bacterium, *International Journal of Systematic Bacteriology*, **44**(4), 631–6.

Omura Y, Price R J and Olcott H S (1978) Histamine-forming bacteria isolated from spoiled skipjack tuna and jack mackerel, *Journal of Food Science*, **43**, 1779–81.

Pan B S and James D (1985) *Histamine in Marine Products: Production by Bacteria, Measurement and Prediction of Formation*, FAO Fisheries Technical Paper no. 252, Food and Agriculture Organization of the United Nations, Rome.

Parsons M & Ganellin C R (2006) Histamine and its receptors, *British Journal of Pharmacology*, **147**, S127–S135.

Ridell J (2000) *Hafnia alvei*, in R K Robinson, C A Batt and P D Patel (eds), *Encyclopedia of Food Microbiology*, Academic Press, San Diego, CA, 973–6.

Rivas B D L, Marcobal A, Carrascosa A V and Munoz R (2006) PCR detection of foodborne bacteria producing the biogenic amines histamine, tyramine, putrescine, and cadaverine, *Journal of Food Protection*, **69**(10), 2509–14.

Rogers P L and Staruszkiewicz W F (1997) Gas chromatographic method for putrescine and cadaverine in canned tuna and mahimahi and fluorometric method for histamine (minor modification of AOAC Official Method 977.13): Collaborative study, *Journal of AOAC International*, **80**(3), 591–602.

Rogers P L and Staruszkiewicz W F (2000) Histamine Test Kit Comparison, *Journal of Aquatic Food Product Technology*, 9(2), 5–17.

Sakabe Y (1973) Studies on allergy-like food poisoning. 2. Studies on the relation between storage temperature and histamine production, *Journal of Nara Medical Association*, **24**, 257–64.

Sanchez-Guerrero I M, Vidal J B and Escudero A I (1997) Scombroid fish poisoning: A potentially life-threatening allergic-like reaction, *Journal of Allergy and Clinical Immunology*, **100**(3), 433–4.

Satomi M, Kimura B, Mizoi M, Sato T and Fujii T (1997) *Tetragenococcus muriaticus* sp. nov., a new moderately halophilic lactic acid bacterium isolated from fermented squid liver sauce, *International Journal of Systematic Bacteriology*, **47**(3), 832–6.

Satter J and Lorenz W (1990) Intestinal diamine oxidases and enteral-induced histaminosis: studies on three prognostic variables in an epidemiological model, *Journal of Neural Transmission [Supplementum]*, **32**, 291–314.

Scoging A (1998) Scombrotoxic (histamine) fish poisoning in the United Kingdom: 1987 to 1996, *Communicable Disease and Public Health*, **1**(3), 204–5.

Staruszkiewicz W F, Barnett J D, Rogers P L, Brenner R A, Wong L L and Cook J (2004) Effect of on-board and dockside handling on the formation of biogenic amines in Mahimahi (*Coryphaena hippurus*), Skipjack tuna (*Katsuwonus pelamis*), and Yellowfin tuna (*Thunnus albacares*), *Journal of Food Protection*, **67**(1), 134–41.

Takahashi H, Kimura B, Yoshikawa M and Fujii T (2003) Cloning and sequencing of the histidine decarboxylase genes of gram-negative, histamine-producing bacteria and their application in detection and identification of these organisms in fish, *Applied and Environmental Microbiology*, **69**(5), 2568–79.

Taylor S L (1986) Histamine food poisoning: toxicology and clinical aspects, *CRC Critical Reviews in Toxicology*, **17**(2), 91–128.

Taylor S L, Guthertz L S, Leatherwood M, Tillman F and Lieber E R (1978) Histamine production by food-borne bacterial species, *Journal of Food Safety*, **1**, 173–187.

Taylor S L and Lieber E R (1979) In vitro inhibition of rat intestinal histamine-metabolizing enzymes, *Food and Cosmetics Toxicology*, **17**, 237–40.

Taylor S L, Guthertz L S, Leatherwood M and Lieber E R (1979) Histamine production by *Klebsiella pneumoniae* and an incident of scombroid fish poisoning, *Applied and Environmental Microbiology*, **37**(2), 274–8.

van Gelderen C E M, Savelkoul T J F, van Ginkel L A and van Dokkum W (1992) The effects of histamine administered in fish samples to healthy volunteers, *Clinical Toxicology*, **30**(4), 585–96.

van Spreekens K J A (1987) Histamine production by the psychrophilic flora, in D E Kramer and J Liston (eds), *Seafood Quality Determination*, Elsevier Science Amsterdam, 309–18.

Wei C I, Chen C-M, Koburger J A, Otwell W S and Marshall M R (1990) Bacterial growth and histamine production on vacuum packed tuna, *Journal of Food Science*, **55**(1), 59–63.

Yamanaka H, Itagaki K, Shiomi K, Kikuchi T and Okuzumi M (1985) Influence of the concentration of sodium chloride on the formation of histamine in the meat of mackerel, *Journal of the Tokyo University of Fisheries*, **72**(2), 51–6.

Yoguchi R, Okuzumi M and Fujii T (1990) Seasonal variation in number of halophilic histamine-forming bacteria on marine fish, *Nippon Suisan Gakkaishi*, **56**(9), 1473–9.

32

Gastroenteritis viruses

E. Duizer and M. Koopmans, National Institute for Public Health and the Environment (RIVM), The Netherlands

Abstract: Viruses causing gastroenteritis (GE) are among the most common causes of foodborne illness worldwide, with the noroviruses (NoV) ranking number one in many industrialized countries. The gastroenteritis viruses belong to several virus families and include NoVs and sapoviruses (family *Caliciviridae*), kobuvirus (Aichivirus, family: *Picornaviridae*) (Yamashita *et al.*, 1998), rotavirus (family: *Reoviridae*), adenoviruses (family: *Adenoviridae*) and astrovirus (family: *Astroviridae*).

Gastroenteritis viruses are transmitted by the faecal-oral route; they are shed in vomitus and faeces and enter their next victim orally. In persons with clinical illness, GE viruses typically are detected in stools at levels far exceeding 10^6 virus particles/ml. Even though the average attack rate of the GE viruses is high (exceeding 45 %), asymptomatic infections do occur and these individuals do excrete virus. Due to the lack of an envelope, these GE viruses are very stable and can remain infectious in the environment, in water or on foods for prolonged periods. Therefore, foods can be contaminated anywhere along the food chain: from farm to fork. High risk commodities are fresh produce, catered sandwiches and salads, and bivalve filterfeeding shellfish.

Prevention of gastroenteritis virus infection including foodborne transmission will rely on high standards for personal and environmental hygiene. The goal should be to prevent contamination of food items rather than to rely on treatment processes to inactivate the viruses once present in the foods or water. In daily practice this means that awareness among food-handlers, including seasonal workers, about the transmission of enteric viruses is needed (including the spread of viruses by vomiting), with special emphasis on the risk of transmission by asymptomatically infected persons and those continuing to shed virus following recovery from illness.

Key words: gastroenteritis, noroviruses, foodhandler, outbreaks.

32.1 Introduction

Viruses are increasingly recognised as causes of foodborne and waterborne illness in humans (Motarjemi *et al.*, 1995). Numerous viruses replicate in the human gut, but only a few of these are commonly recognised as important foodborne pathogens. The enteric viruses can be categorised into three main groups, according to the type of illness they produce: viruses that cause gastroenteritis, the enterically transmitted hepatitis viruses, and a third group of viruses that replicate in the human intestine, but cause illness after they migrate to other organs such as the central nervous system (Codex Alimentarius Committee on Food Hygiene, 1999). Foodborne illness has been documented for most of these viruses, but recent studies suggest that the noroviruses (NoVs, formerly known as Norwalk-like viruses or small round structured viruses) and hepatitis A virus (HAV) are the most common cause of illness by this mode of transmission (Mead *et al.*, 1999). Large foodborne outbreaks have also occurred with group B and C rotaviruses and waterborne outbreaks with hepatitis E virus and group B rotavirus (Hung *et al.*, 1984; Su *et al.*, 1986; Matsumoto *et al.*, 1989; Oishi *et al.*, 1993). This chapter will review foodborne gastroenteritis viruses with special emphasis on the NoVs.

32.2 Noroviruses and other gastroenteritis viruses

> Viruses are unique in nature. They are the smallest of all self-replicating organisms, historically characterised by their ability to pass through filters that retain even the smallest bacteria. In their most basic form, viruses consist solely of a small segment of nucleic acid encased in a simple protein shell. They have no metabolism of their own but rather are obliged to invade cells and parasitize subcellular machinery, subverting it to their own purposes. Many have argued that viruses are not even living, although to a seasoned virologist they exhibit a life as robust as any other creature. (Condit, R. C., Principles of virology, in *Fields Virology* (2001), with permission)

This definition naturally covers the gastroenteritis viruses and is a good starting point for consideration of viruses in matters of food safety. NoVs and sapoviruses (SaV) are human enteric caliciviruses. Other gastroenteritis viruses include human kobuvirus (Aichivirus, family: *Picornaviridae*), rotavirus (family: *Reoviridae*), adenoviruses (family: *Adenoviridae*) and astrovirus (family: *Astroviridae*) (Yamashita *et al.*, 1998).

The families *Caliciviridae*, *Picornaviridae* and *Astroviridae* all belong to the picornavirus-like superfamily: small, RNA viruses without an envelope. The calicivirus particles (virions) measure between 27 and 38 nm and consist of a spherical protein shell and a genome of a single strand of approximately

7.6 kb positive sense RNA. The caliciviruses known to infect humans belong to two genera: the genus *Norovirus* and the genus *Sapovirus* (SaV, formerly known as Sapporo-like viruses, or typical caliciviruses). In addition, the family holds two genera of animal viruses, named *Vesivirus* and *Lagovirus* (Green *et al.*, 2001). The genus NoV is further subdivided into an increasing number of poorly defined genogroups (Karst *et al.*, 2003) and each genogroup can be subdivided in several genotypes, defined by > 80 % amino acid homology across the capsid gene. Currently, five NoV genogroups are defined, the genogroups (G) I and II are most prevalent in humans and are subdivided into eight and 19 genotypes, respectively (Green, 2006). The NoVs are officially denoted by the following cryptogram: host species/genus abbreviation/species abbreviation (i.e. genogroup)/strain designation/year of detection/country of origin. For NoVs, the strain designation is the location where the strain was first detected, for example: Hu/NV/GI/Norwalk Virus/1968/US, Hu/NV/GI/Southampton virus/1991/UK, Hu/NV/GII/Hawaii virus/1971/US and Hu/NV/GII/Bristol/1993/UK (Green *et al.*, 2001).

Being such small and 'simple' organisms, with a small genome, viruses rely to a great extent on the enzymes in the host cell they invade to undergo their replicative cycle. Unlike most bacteria, viruses are obligate intracellular parasites. Many viruses, including the human enteric caliciviruses, have a narrow species- and tissue tropism, i.e. efficient replication of the human NoVs occurs only in the gastrointestinal tract of humans. Replication in other human tissues has not been detected and, even although NoV genotypes were found in cattle (e.g. bovine Jenavirus, GIII) and pigs (e.g. porcine enteric calicivirus, GII), none of these animal genotypes has ever been detected in human patients.

32.2.1 Symptoms caused in humans

People infected with NoV, SaV, astrovirus, human kobuvirus and rotavirus develop symptoms after a short incubation period, typically 24–48 h, whereas incubation times for adenoviruses are somewhat longer (average 7–14 d) (see Table 32.1).

Most illnesses associated with astrovirus, rotavirus, NoV, SaV and adenovirus occur in (young) children (Koopmans *et al.*, 2000; De Wit *et al.*, 2001a,b,c). The NoVs are the major causative agents of classical viral gastroenteritis in all age groups (de Wit *et al.*, 2001a, d). Explosive outbreaks are often seen in institutions such as nursing homes and hospitals (Vinjé and Koopmans, 1996; Vinjé *et al.*, 1997). After a short incubation period, the disease is characterised by an acute onset of symptoms including (non-bloody) diarrhoea, vomiting, (low-grade) fever, nausea, chills, weakness, myalgia, headache and abdominal pain. Dehydration may develop, particularly in rotavirus infections, causing most severe illness due to active secretion of fluids. In developing countries, rotavirus infections are considered to be a leading cause of childhood mortality, but foodborne transmission is considered to

Table 32.1 Charactreristics of (infections with) enteric viruses

Virus	Calicivirus	Rotavirus	Adenovirus	Astrovirus	Human kobuvirus
Size (nm)	30–38	70	80–110	28–30	22–30
Genome	ssRNA	dsRNA	dsDNA	ssRNA	ssRNA
Average incubation period	24–48 h	24–48 h	7–14 d	24–48 h	24–48 h
Symptoms	Vomiting and diarrhoea	Diarrhoea, vomiting, fever	Diarrhoea	Diarrhoea	Diarrhoea vomiting
Affected age groups	All	Children < 5 y	Children < 5 y	Children < 10 y	All
Duration of illness	Days	Days	Days–weeks	Days	4–6 d
Severity of illness	Mild	Major cause of death in developing countries	Mild	Mild	Mild (probably often asymptomatic)
Special risk groups	Institution-alised hospitalised				Unknown

be uncommon for rotaviruses (de Wit *et al.*, 2001c, 2003). NoV-associated disease on average is milder, but the high rate of secondary spread results in (large) outbreaks, termed 'gastric' or 'stomach flu'. The attack rate of NoV infection during outbreaks is typically around 45 % or higher, but this differs with virus genotype and is also dependent on the host's genetic susceptibility. Most notable is the usually sudden onset of projectile vomiting, resulting in efficient spreading of viruses. Although patients do in general feel acutely ill, and dehydration can be severe, the illness is rarely fatal and subsides on average after 2–6 d, with the longest duration of illness in children (Rockx *et al.*, 2002). Deaths associated with NoV outbreaks have been reported in the elderly (over 65 y), and long-term illness and shedding in patient groups with comorbidities (Jansen *et al.*, 2008).

No antiviral treatment has been found to be effective in treating NoV infection and no vaccine is available yet, although clinical trials with recombinant vaccines based on the capsid protein of NoV are in progress. For NoV and SaV no long-term sequelae have been reported, but the duration of the infectious period might have been under-estimated. It was found that shedding of NoV continued for three weeks in 26 % of the patients (Rockx *et al.*, 2002), but more recent reports in which highly sensitive molecular techniques were used for stool testing suggest even longer shedding (Atmar *et al.*, 2008; Siebenga *et al.*, 2008). In immunocompromised persons, chronic infection may develop with persistent diarrhoea and long-term shedding (Nilsson *et al.*, 2003; Gallimore *et al.*, 2004b, Siebenga *et al.*, 2008).

32.3 Epidemiology of viral gastroenteritis and examples of viral foodborne outbreaks

Gastroenteritis viruses are transmitted by the faecal–oral route; they are shed in vomitus and faeces and enter their next victim orally. In persons with clinical illness, NoVs are typically detected in stools at levels far exceeding 10^6 virus particles/ml (the titers for vomit are not known). Since the minimal infectious dose of the NoVs is believed to be very low (from 1–10 particles), and immunity short-lived, introduction of NoV in a population easily leads to an outbreak, affecting many people. Even though the attack rate is rather high, asymptomatic infections are quite common. For example, in a community study in the Netherlands, evidence of virus shedding was found in 5.2 % of controls, i.e. persons without gastrointestinal complaints (de Wit. *et al.*, 2001a) and in 19 % of people without gastrointestinal illness in an outbreak setting (Vinjé *et al.*, 1997). The efficient spread of viruses by (projectile) vomiting, the prolonged shedding in faeces after recovery from illness and the shedding of viruses from asymptomatic carriers are important factors contributing to the impressive numbers of NoV infections.

32.3.1 Burden of NoV disease studies

Few studies have looked at the incidence and health impact of NoV infection at the community level. The most extensive data are from the UK (Tompkins *et al.*, 1999; Wheeler *et al.*, 1999) and the Netherlands, where a randomised sample of the community participated in cohort studies of infectious intestinal disease (IID). The incidence of community-acquired IID was calculated as 190 per 1000 person years in the UK and 283 per 1000 person years in the Netherlands (Tompkins *et al.*, 1999; de Wit *et al.*, 2001a). Viruses were the most frequently identified causes of community-acquired gastroenteritis, with NoV detected in 11 % of cases in the Netherlands and 7 % in the UK. This difference may partly result from the different methods used for virus detection: de Wit *et al.* used reverse transcriptase-polymer chain reaction (RT-PCR) whereas the study in the UK employed the far less sensitive electron microscopy.

In both studies, the referral of patients to a general practitioner was also investigated. Approximately 5 % of cases in the Netherlands sought treatment, compared with 4 % of cases in Wales and 17 % in England. The physician-based patient group was younger, and had more severe symptoms and longer duration of illness (de Wit *et al.*, 2001b). Some 5 % of physician-based cases were NoV positive in the Netherlands and 6.5 % in the UK. The lower proportion of NoV disease in this study population compared with the community cases confirms that on average NoV illness is relatively mild. The financial consequences of this mild disease can, however, be serious: cost estimates related to NoV infections for the Netherlands showed that 13 % (i.e. € 46 million) of costs for gastroenteritis were due to NoV infections alone (Van den Brandhof *et al.*, 2004).

Smaller studies in selected patient populations have been conducted elsewhere, and show that NoV are known to occur as a prominent cause of illness in countries throughout Europe, the USA, Australia, Hong Kong and Japan (Fankhauser *et al.*, 1998, 2002; Iritani *et al.*, 2002; 2003; Lopman *et al.*, 2002, 2003, 2004; Marshall *et al.*, 2003; Lau *et al.*, 2004). Additionally, evidence is mounting that the disease may be common in countries with different degrees of development across the world, with studies from, for example, Hungary, Argentina, Brazil, Pakistan and India (Parks *et al.*, 1999; Farkas *et al.*, 2002; Girish *et al.*, 2002; Martinez *et al.*, 2002; Reuter *et al.*, 2003; Gallimore *et al.*, 2004a; Phan *et al.*, 2004). NoV infection is common in all age groups, but the incidence is highest in young children (< 5 y).

32.3.2 NoV outbreak studies

Probably the best known presentation of NoV is that of large outbreaks of vomiting and diarrhoea, that lend the disease the initial description of 'winter vomiting disease' (Zahorsky, 1929; Mounts *et al.*, 2000). Since the development of molecular detection methods, NoV have emerged as the most important cause of outbreaks of gastroenteritis in institutional settings (i.e. hospitals, nursing homes). The majority of NoV gastroenteritis cases result from direct person-to-person transmission. However, NoV-related outbreaks have been shown to be food- or waterborne, caused by, for example, contaminated shellfish (Kingsley *et al.*, 2002b; Le Guyader *et al.*, 2003; Doyle *et al.*, 2004), raspberries (Ponka *et al.*, 1999) or drinking water (Kukkula *et al.*, 1999; Carrique-Mas *et al.*, 2003; Parshionikar *et al.*, 2003). Additionally, environmental spread of NoV was found, for instance by contaminated carpets in hotels (Cheesbrough *et al.*, 2000), toilet seats and door handles in a rehabilitation centre (Kuusi *et al.*, 2002), and contaminated fomites on hard surfaces, carpets and soft furnishings in a concert hall (Evans *et al.*, 2002).

In the Netherlands, approximately 12–15 % of community cases of NoV gastroenteritis was attributed to foodborne transmission, based on analysis of questionnaire data. This makes NoV as common a cause of foodborne gastroenteritis as *Campylobacter*, and more common than *Salmonella* (de Wit *et al.*, 2003).

Outbreaks are quite often a result of a combination of several transmission routes, for example introduction of the virus in a sensitive population by food, water or an asymptomatic shedder, followed by efficient spread of the virus through the susceptible population by direct person-to-person transmission. Large food- or waterborne outbreaks due to a common (point) source introduction are less common than multiple transmission routes outbreaks, but they do occur (see for example: Cannon *et al.*, 1991; Kohn *et al.*, 1995; Daniels *et al.*, 2000; Girish *et al.*, 2002). Several waterborne outbreaks have been reported as a result of contaminated private wells or communal water systems in Sweden (Carrique-Mas *et al.*, 2003; Nygard *et al.*, 2003), municipal water in Finland (Kukkula *et al.*, 1999), and well water in Wyoming, USA (Anderson *et al.*, 2003). Interestingly, no signs of faecal contamination by testing for indicator bacteria were found by Carrique-Mas and co-workers (Carrique-Mas *et al.*, 2003).

Foods can be contaminated with NoV anywhere along the food chain: from farm to fork. Wherever a NoV carrier comes in contact with food, contamination might occur and, due to stability of these pathogens, they are likely to survive many food processes (see Section 32.6; Koopmans and Duizer, 2004). When viral contamination occurs through a person by touching food, the contamination will be localised in spots (focally). Infections caused by focally-contaminated foodstuffs are most likely to be recognised as foodborne when the contamination has occurred at the end of the food chain. Two reported examples show the characteristics of such outbreaks. One report from Sweden describes the large-scale outbreak detected in 30 day care centres in Stockholm area in March 1999. All centres obtained their lunch meals from one caterer, and approximately 25 % of all customers (n = 1500 customers) fell ill. The early cases had an average incubation time of 34 h, but the outbreak went on for at least 12 d, due to secondary infections. One of the food handlers fell ill in the same time window as the early cases and none of the other food handlers reported symptoms. This implies that the most likely route of food contamination was by a pre- or asymptomatic foodhandler (Gotz *et al.*, 2002).

Another large point-source foodborne outbreak occurred in the Netherlands in January 2001. A baker who had been sick with vomiting the previous day had prepared the rolls for a buffet lunch at a New Year's reception. Within 50 h after the reception, over 200 people had reported ill with gastrointestinal symptoms. Epidemiological and microbiological investigations indicated the rolls as point-source of this outbreak with a total of 231 people sick with diarrhoea and vomiting (de Wit *et al.*, 2007). It was noted that the baker was still ill during preparation of the rolls and that he vomited in the sink. Cleaning the sink with chlorine did not prevent the outbreak. It is probable that the spread of infectious viruses by vomiting was not restricted to the sink or the cleaning method used was not stringent enough to remove these resistant viruses from the sink.

In many outbreaks, however, the seeding event is not recognised (Verhoef *et al.*, 2009). This can be on two levels: (i) the route of introduction of the NoV to the affected population is unknown, or (ii) the vehicle of introduction is known, but it is unclear how and when that vehicle (vector) was contaminated. Resolving the transmission routes and vectors involved in such outbreaks is the challenge for viral food safety, now and in the near future.

32.3.3 Immune response and genetic susceptibility to NoV infection

The most novel finding of the past few years is that NoV evolve rapidly and thus escape what little immunity builds up in the population. Infected individuals do develop short-term immunity to homologous virus (up to 14 weeks), but the existence or development of long-term immunity to NoV is still elusive (Matsui and Greenberg, 2000). Pre-existing antibodies do not always correlate with protection from infection, and insight into the correlation between NoV genotypes and serotypes is only just beginning to develop. Serum IgG, IgA and IgM responses are observed after infection with NoV, and common epitopes within genogroups and between genogroups have been identified (Erdman *et al.*, 1989a, b; Treanor *et al.*, 1993; Gray *et al.*, 1994; Hale *et al.*, 2000; Yoda *et al.*, 2000; Harrington *et al.*, 2002; Kitamoto *et al.*, 2002). However, there are contradictory data on the level of cross-reactivity after NoV infections in humans. It has been shown that the specificity of the immune response varies greatly within genogroups but, nevertheless, other studies have observed cross-reactivity between viruses belonging to different genogroups (Madore *et al.*, 1990; Treanor *et al.*, 1993; Farkas *et al.*, 2003).

Recent reports show that mutations occurring in the region coding for the capsid protein (ORF2) do lead to mutant viruses that escape immunity developed against the prior strain (Siebenga *et al.*, 2007; Buesa *et al.*, 2008; Donaldson *et al.*, 2008).

Another recent finding is that sensitivity to NoV infection is dependent on ABH-histo-bloodgroup antigens (carbohydrates), Lewis antigens and secretor status (Hutson *et al.*, 2002; Marionneau *et al.*, 2002; Huang *et al.*, 2003; Harrington *et al.*, 2004; Rockx *et al.*, 2005a) and that the genetically determined susceptibility is different for different genotypes. For example, bloodgroup B and non-secretor status confer protection against infection by Norwalk virus, bloodgroup A confers protection against infection by Snow Mountain virus, while none of the bloodgroups offers protection to infection with GII.4 viruses (reviewed by Hutson *et al.*, 2004). These studies also suggest that the susceptibility to NoV infection is related to specific carbohydration of the host receptor, and that NoV binding can be inhibited by the specific carbohydrates in solution. These findings may lead to the development of (strain-specific) antiviral treatments. Research into the genetic susceptibility is booming at the moment and insight may develop and change over time.

32.4 Detection

32.4.1 Detection in patients

Infection with gastroenteritis viruses is typically diagnosed by the detection of the virus in stool samples from sick people, rather than by measuring immune response of the host to the virus. The virus capsids (protein) can be detected by enzyme immunoassay (EIA)-like tests and the virus genome by (RT) PCR assays.

Currently the method of choice for the diagnosis of gastroenteritis viruses is the molecular biological technique called PCR, for all RNA viruses preceded by a RT step to make cDNA out of the original RNA genome. This method is based on the detection of viral nucleic acid in faecal samples. The high sensitivity of this method (detection limit of 10–100 viral particles per gram stool) is one of its major advantages. Another advantage is that – theoretically – it can be applied to all kinds of substrates such as faeces, vomitus, serum, food-matrices and water. A bottleneck has been the high genetic diversity of these viruses. Several protocols are used which have been developed for the detection of a broad range of virus strains but, still, no single primer pair can detect all NoV strains (reviewed by Atmar and Estes, 2001). In food items, cross-reactivity of primers between human and animal strains is an additional issue requiring consideration (Wolf *et al.*, 2007). Additionally, generic tests are optimised for the detection of a broad range of viruses, not for the specific and more sensitive detection of one strain (Vennema *et al.*, 2002). While this may be appropriate for patient diagnostics (where high levels of virus are shed in the stool samples), virus detection in environmental samples, food or water requires highly optimised assays with low detection limits. This is not compatible with broad range detection.

Recent developments in molecular biology techniques have yielded several new diagnostic tests. One is the development of real-time RT-PCR methods, which are faster than the original RT-PCR protocols, mainly due to the fact that the RT-PCR and confirmation stages are incorporated in one step (Kageyama *et al.*, 2003; Richards *et al.*, 2004). Currently, multiplex real-time RT-PCR assays are being developed that allow for parallel testing of several groups of gastroenteritis viruses (Phan *et al.*, 2005; Kittigul *et al.*, 2009). This approach may provide a step forward in setting up syndrome surveillance for gastroenteritis. Another new method is the nucleic acid sequence based amplification (NASBA) technique which, in theory, should be simpler to perform and at least as sensitive as the RT-PCR (Jean *et al.*, 2003; Moore *et al.*, 2004). Whichever genome detection assay is used for virus detection, one should realise that false negative tests are relatively common, due to the great genetic diversity resulting in primer mismatches (Vinjé *et al.*, 2003). Therefore, back-up protocols should be used for unexplained outbreaks of suspected viral etiology.

32.4.2 Detection in foods and water

Detection of NoVs in foods and water also relies mostly on RT-PCR methods. Due to the enormous variety of food matrices, there is a considerable challenge for food microbiologists in the first stage of the RT-PCR protocol: the extraction of the viral RNA from the matrix. Many studies have focussed on method development to extract viral RNA from shellfish, which has resulted in a variety of methods, reviewed by Lees (2000). Common features of these are dissection of the digestive tract and hepatopancreas (Schwab *et al.*, 1998) and subsequent homogenisation, followed by variable (partial) purification and RNA extraction methods. However, the methods applied are laborious and specialised and are generally not available to most labs that carry out routine testing. Other matrices for which virus detection methods have been developed include fresh produce and fruits (Bidawid *et al.*, 2000; Leggitt and Jaykus, 2000; Dubois *et al.*, 2002; Sair *et al.*, 2002; Le Guyader *et al.*, 2004), and hamburgers, ham, turkey and roast beef (Leggitt and Jaykus, 2000; Schwab *et al.*, 2000; Sair *et al.*, 2002). Since foods are often contaminated by an infected food handler, the contamination of food items may be low level and focal.

Contamination with irrigation water may result in more diffuse presence of viruses, but primarily at the surface of the produce or product. Therefore most protocols are based on elution of the virus particles from the surface of the product, followed by a concentration step (mostly ultra-centrifugation or -filtration) and RNA extraction from the concentrate (Le Guyader *et al.*, 2004). Recently an immunomagnetic capture method was used to capture (concentrate and purify) NoV from a crude food suspension (Kobayashi *et al.*, 2004). This method at present cannot be generalised because antibodies to NoV are not broadly reactive, although some cross-reactivity exists between genotypes within the same genogroup. Further evaluation is needed to see if sufficiently broad reactivity exists.

For NoV detection in water a variety of methods is available (Gilgen *et al.*, 1997; Straub and Chandler, 2003; Haramoto *et al.*, 2004; Karim *et al.*, 2004). Here too, the challenge exists in obtaining concentrates with a detectable level of viral RNA, but a low level of RT-PCR inhibitors. Many protocols are variations on a general method which, in short, entails concentration by adsorption to a filter (e.g. a positively charged membrane, glasswool), subsequent elution from the filter and further concentration using a microconcentrator or precipitation step. However, immunoaffinity concentration and purification protocols have been developed for application in water too (Schwab *et al.*, 1996).

The difficulties in food safety issues regarding the cumbersome detection of NoVs (or HAV) in shellfish and water are further complicated by the recognition that potential indicators for those viral pathogens, such as bacteria or bacteriophages, may not be as appropriate as was assumed before. For example, the NoV may be present, or persistent, in water or shellfish whereas the indicator organisms are not (Lees, 2000; Formiga-Cruz *et al.*, 2002; Horman *et al.*, 2004; Myrmel *et al.*, 2004). In conclusion, the detection of NoVs in foods is still difficult, due largely to the low level, and focal route of contamination. Therefore, successful (pre-marketing) screening of foods is unlikely to be implemented soon, with the possible exception of shellfish. The generic RT-PCR protocols can, however, be optimised for sensitive detection of one strain when sequence data for the contaminating strain are already obtained from patient diagnostics (Fig. 32.1). This approach was found to be successful in establishing a contaminated recreational fountain as the source of a gastroenteritis outbreak in schoolchildren by the NoV Birmingham strain (GGI.3)(Hoebe *et al.*, 2004). Thus, for the detection of

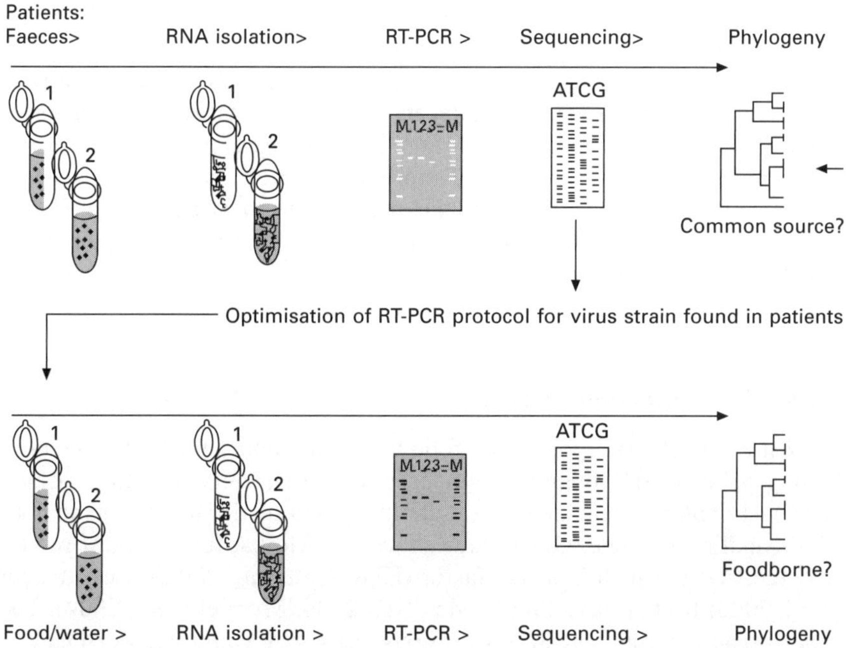

Fig. 32.1 Scheme for foodborne virus tracking. Virus detection in patients and possible vectors such as food or water, to determine common source outbreaks.

NoV in suspected foods (foods implicated by epidemiological data), optimised RT-PCR protocols may be used which might help to unambiguously link foods to outbreaks.

While methods have worked reasonably well with artificially-contaminated food items, assays applied to suspected food items collected in outbreaks have rarely yielded a positive result. Routine monitoring of foods for bacterial pathogens is itself acknowledged to be an ineffective method of assuring food safety (Motarjemi *et al.*, 1996). The additional practical difficulties in obtaining reliable data for viral contamination mean that routine viral monitoring of foods is very unlikely to become widespread.

Next to the problem of obtaining false-negative results due to practical difficulties in virus detection in foods and water, there is the problem of obtaining false-positive results using genome detection by PCR, i.e. the presence of viral RNA does not necessarily indicate the presence of infectious virus (Richards, 1999). Several authors have reported a poor correlation between conventional RT-PCR detection and virus infectivity in a cell culture assay: Slomka and Appleton (1998) and Duizer *et al.* (2004b) for FeCV after heat treatment or UV irradiation, Lewis *et al.* (2000) for poliovirus after UV irradiation, and Nuanualsuwan and Cliver (2002) for HAV, poliovirus and FeCV. On the other hand, for hypochlorite inactivation a good correlation was found between RNA detectability and infectivity (Nuanualsuwan and

Cliver, 2002; Duizer *et al.*, 2004b), and pretreatment (prior to RT-PCR) of heat or UV inactivated virus suspensions with RNase and proteinase K resulted in an improved correlation between PCR detectability and infectivity (Nuanualsuwan and Cliver, 2002). Recently it was shown that quantification of detectable RNA, by quantification of the RT-PCR signal or by application of quantitative RT-PCR methods, could improve the correlation too (Bhattacharya *et al.*, 2004; Duizer *et al.*, 2004b). However, one must bear in mind that neither PCR, nor electron microscopy (EM) nor enzyme-linked immunoassay (ELISA) are viability or infectivity tests.

32.5 Transmission routes

De Wit *et al* (2003) reported the following risk factors for NoV infection: having a household member with gastroenteritis, contact with a person with gastroenteritis outside the household (symptomatic shedding) and poor food handling hygiene. Other studies have shown that the eating of raw or undercooked shellfish is a risk factor (Kingsley *et al.*, 2002b; Le Guyader *et al.*, 2003; Butt *et al.*, 2004; Doyle *et al.*, 2004; Myrmel *et al.*, 2004). The efficient, aerosolised spread of infectious viruses by vomiting, worsened by the environmental survival of NoV, contributes to efficient transmission. Another factor in the spreading of NoV is the shedding of infectious viruses by asymptomatic (and pre- or postsymptomatic) carriers and the employment of NoV shedders in the food chain (Emerson and Purcell, 2006; Hollinger and Emerson, 2006; Pallansch and Roos, 2006).

32.5.1 Person-to-person transmission

As mentioned before, NoV are transmitted by the faecal–oral route and the most important way of spreading is from person to person. If the virus is introduced in a sensitive population by a symptomatic shedder, the origin of the virus is clear. However, asymptomatic, pre- or post-symptomatic shedders and food or water may also introduce the viruses. In addition, prolonged outbreaks may occur following environmental contamination and persistence of NoV. In these cases the origin of the virus is less clear.

Since viruses do not replicate in foods, and zoonotic transmission of NoV has not been found so far, all significant transmission routes are presumed to be a variation on person-to-person transmission. This means that NoV can be passed on from a shedder to a food item to the next victim, or water can be contaminated by a shedder and used for direct consumption or used for irrigation or washing of foods. It also implies that all foods handled by a food handler might get contaminated. However, even if the source of contamination will always be a person, infections with food or water as vector for NoV transmission are referred to as food- or waterborne infections.

32.5.2 Food- and waterborne transmission

The most notorious foods with respect to viral infections are the filter-feeding shellfish (oysters, mussels, clams). These sea animals actively accumulate viruses from water contaminated with human excrement. The viruses are concentrated in the intestinal tract of the shellfish and may remain infectious for weeks in the natural environment of the shellfish or when stored cooled. Moreover, depuration is not an effective method to reduce the viral load as it is for reducing bacterial contamination (Schwab *et al.*, 1998; Muniain-Mujika *et al.*, 2002; Kingsley and Richards, 2003), and the composition of the bodymass (proteins, fat, sugars) of shellfish contributes to increased virus stability in heat processing (Croci *et al.*, 1999). Matters are further complicated by the fact that viral contamination of shellfish is reported for batches that meet bacteriological safety standards and the absence of indicator organisms in growing waters or marketed batches is not always a guarantee for the absence of infectious NoV (Chalmers and McMillan, 1995; Schwab *et al.*, 1998; Le Guyader *et al.*, 2003; Nishida *et al.*, 2003). Besides shellfish, many other food items have been implicated in NoV outbreaks: ice (Khan *et al.*, 1994), raspberries (Ponka *et al.*, 1999), salad vegetables, poultry, red meat, fruit, soups, desserts, savoury snacks (Lopman *et al.*, 2003), sandwiches (Parashar *et al.*, 1998; Daniels *et al.*, 2000).

As mentioned before, all foods handled by an infectious food handler might be contaminated. Foods can be contaminated anywhere in the food chain, from farm to fork. The items posing the highest risk will be the products that are eaten raw or eaten without any (decontaminating) treatment. This explains why fruits and vegetables have been implicated in outbreaks, especially when harvested under conditions of poor sanitary hygiene or when irrigated with polluted water. The largest outbreak of NoV-associated gastroenteritis in Australia involved over 3000 individuals who had consumed orange juice (Fleet *et al.*, 2000). Although no virus could be detected in the orange juice, the outbreak terminated when the juice was no longer distributed. Several areas where contamination of the juice could have occurred were identified in the production facilities. A high-risk practice is catering. Catering products are by definition products that are manually handled by a food handler at the end of the food chain, and therefore might pose a risk.

32.5.3 Airborne transmission

Airborne transmission of NoV was also noted in several outbreaks (Sawyer *et al.*, 1988; Chadwick *et al.*, 1994; Marx *et al.*, 1999). For example, airborne spread of NoV and infection by inhalation with subsequent ingestion of virus particles was described for an outbreak in a hotel restaurant (Marks *et al.*, 2000). An outbreak in a school showed evidence of direct infection by aerosolised viruses after a case of vomiting (Marks *et al.*, 2003). In that same study it was found that cleaning with a quaternary ammonium preparation was ineffective. In all cases, viruses were spread by acts of vomiting.

32.5.4 Zoonotic transmission

The NoV that have been found in cattle (e.g. Jena and Newbury agent), pigs, a lion and a cat, so far are genetically distinct from the currently identified human NoVs (Sugieda *et al.*, 1998; Dastjerdi *et al.*, 1999; Liu *et al.*, 1999; van der Poel *et al.*, 2000, 2003; Sugieda and Nakajima, 2002; Deng *et al.*, 2003; Oliver *et al.*, 2003; Smiley *et al.*, 2003; Wise *et al.*, 2004, Martella *et al.*, 2007, 2008). In experimental settings, chimpanzees and macaques were found to seroconvert after inoculation with the prototype strain Norwalk virus (Green *et al.*, 2006; Subetki *et al.*, 2002; Rockx *et al.*, 2005b). Recent studies of 'natural' infections showed NoV antibody prevalence in mangabey, pigtail and rhesus monkeys and chimpanzees in the USA (Jiang *et al.*, 2004). The bovine enteric caliciviruses (BEC) have all been classified into a distinct genogroup (GIII), and some may even be in a distinct genus (Oliver *et al.*, 2003; Smiley *et al.*, 2003; Wise *et al.*, 2004). The viruses found in pigs, a lion and a dog are more closely related to human viruses. Moreover, one report mentions the detection of the most common human strain, the NoV GII.4, in cattle and swine in Canada, and GII.4 strains were able to infect gnotobiotic piglets as well (Mattison *et al.*, 2007; Souza *et al.*, 2008).

A recent study of the seroprevalence in the general population and a cohort of veterinarians specialising in bovines showed IgG reactivity to recombinant bovine NoV capsid protein in 20 % of controls and 28 % of the vets. Since cross-reactivity with human strains could not explain the difference in seroprevalence between these groups, these data indicate that infections of humans by bovine NoV have occurred (Widdowson *et al.*, 2005). Thus, recent reports suggest that animal NoV strains may be able to infect humans, and probably that human strains can infect animals, and thus that interspecies transmission cannot be excluded. However, these will be rare events and these results need further confirmation. Furthermore, the role of animal NoV strains in human disease needs to be studied more deeply. If interspecies transmission does occur, it is possible that animals (or humans) might play a role as 'mixing vessels' for recombination of NoV strains. While there is currently no proof of such a scenario, the genetic flexibility of RNA viruses is such that it should not be deemed impossible.

32.6 Conditions of growth and survival

Viruses, unlike most bacteria, are strict intacellular parasites and cannot replicate in food or water. Therefore, viral contamination of food will not increase during processing, and may actually decrease. Most food- or waterborne viruses are relatively resistant to heat, disinfection and pH (> pH 3) and it is no coincidence that most virus groups implicated in outbreaks are small, non-enveloped particles, rather than large, fragile, enveloped viruses. Abad *et al.* (1994) have shown that viruses persisted for extended periods on several types of materials commonly found in institutions and domestic environments. Overall, human rotavirus was more resistant than enteric adenovirus, but NoVs were not tested in this study (Abad *et al.*, 1994). Adenoviruses were found to survive in desiccated circumstances for up to 35 d on a plastic surface (Nauheim *et al.*, 1990). Rotaviruses were found to survive in aerosols at medium relative humidity (50 % ± 5 %), with a half-life of nearly 40 h. Three per cent of the infectious virus was detectable in the air after 9 d of aerosol age at 20 °C. Rotavirus could, therefore, survive in air for prolonged periods, thus making air a possible vehicle for their dissemination (Sattar *et al.*, 1984). Enteroviruses, though, behave quite differently, with most favourable conditions for survival at high relative humidity (Ijaz *et al.*, 1985). These findings stress the need for independent assessment of behaviour for different viruses. In water, viruses may survive for prolonged periods of time, with over 1 y survival of rotavirus in mineral water at 4 °C (Biziagos *et al.*, 1988). The inability to culture NoVs (Duizer *et al.*, 2004a) greatly limits the knowledge on stability and survival of these viruses. Currently available information is based on studies using model viruses (reviewed in Koopmans and Duizer, 2004) or quantitative RT-PCR methods to asses RNA disintegration of NoV (Nuanualsuwan and Cliver, 2002; Duizer *et al.*, 2004b).

Surrogates for human NoVs include feline calicivirus (FeCV) and murine NoV (MNV). Duizer *et al.* (2004a), and Lamhoujeb *et al.* (2008) do, however, report that FeCV is most likely less resistant than human NoV. Bae and Schwab (2008) concluded that MNV shows promise as a human NoV surrogate due to its genetic similarity and environmental stability. In another study using MNV, Cannon *et al.* (2006) reported that the strain tested survived for long periods on stainless steel coupons and also at room temperature, in solution, concluding that this was a better surrogate than FeCV. Baert *et al.* (2008) showed that MNV remained infectious for 6 d in vegetable wash water at room temperature and on frozen vegetables (spinach and onions) for six months. Lamhoujeb and co-workers found that NoV remained infectious for at least 10 d on refrigerated ready-to-eat foods, such as lettuce and turkey (Lamhoujeb *et al.*, 2008).

32.6.1 Inactivation of caliciviruses

The inability to culture NoVs has also hampered the development of reliable methods for testing their inactivation. Therefore, knowledge on efficient inactivation methods and effective intervention in transmission pathways is limited and mostly based on studies using model viruses (reviewed in Koopmans and Duizer, 2004) or quantitative RT-PCR methods to assess RNA disintegration of NoV (Nuanualsuwan and Cliver, 2002; Duizer *et al.*, 2004b). The most commonly used model for NoV in inactivation studies is the FeCV and this virus is proposed as surrogate virus for testing of virucidal activity of disinfectants (Steinmann, 2004). In several tests it was found that FeCV was not efficiently inactivated on environmental surfaces or in suspension by, for example, 1 % anionic detergents, quaternary ammonium (1:10), hypochlorite solutions with < 300 ppm free chlorine, or less than 50 % or more than 80 % alcohol preparations (ethanol or 1- and 2-propanol) (Scott, 1980; Gehrke *et al.*, 2004; Duizer *et al.*, 2004a). Varying efficacies of 65–75 % alcohol preparations are reported, however, short contact times (< 1 min) rarely resulted in more than 4 log inactivation. Moreover, the presence of faecal or other organic material reduces the virucidal efficacy of many chemicals tested.

The efficacy of seven commercial disinfectants for the inactivation of FeCV on strawberries and lettuce was tested by Gulati and co-workers and they found that none of the disinfectants was effective (defined as reducing the virus titer by at least 3-$\log_{10}$) when used at the FDA permitted concentration (Gulati *et al.*, 2001). These data were recently confirmed by Alwood and co-workers for sodium bicarbonate, chlorine bleach, peroxyacetic acid and hydrogen peroxide at FDA approved concentrations (Allwood *et al.*, 2004). More recently, Belliot *et al.* (2008) used MNV as a surrogate in disinfectant tests. They reported that MNV was sensitive to 60 % alcohol, alcohol hand rubs, bleach and povidone iodine-based disinfectant in suspension tests using very low-level interfering substances. They also used real-time RT-PCR and they confirmed that the presence of viral RNA did not correlate with the presence of infectious virus. In a study investigating virucidal effects of a new ethanol-based sanitiser, Macinga *et al.* (2008) used human rotavirus and MNV-1 and reported that the new sanitiser was more effective than a benchmark alcohol-based sanitiser, reducing infectivity by more than 3 logs of all viruses tested after 30 s exposure.

Recent studies using FeCV and canine caliciviruses (CaCV) as model for NoV showed that heat inactivation of these two animal caliciviruses was highly comparable. Based on the temperature-dependent inactivation profiles it was suggested that a greater inactivation of viruses may be expected from regular batch (63 °C for 30 min) or classical pasteurisation (70 °C for 2 min) than from high-temperature short-time (HTST) pasteurisation (72 °C for 15 s). Inactivation at 100 °C was, however, complete within seconds (Duizer *et al.*, 2004a). Likewise, blanching spinach (80 °C for 1 min) resulted in at least 2.44 log reductions of infectious MNV-1 (Baert *et al.*, 2008). A study investigating the effects of heat treatment on hepatitis A virus and norovirus in New Zealand Greenshell mussels (Hewitt and Greening, 2006) concluded that when mussels open on heating, their internal temperature may not be sufficient to reach the parameters required for virus inactivation and immersion for a minimum of 3 min in boiling water rather than steaming is recommended to reduce the risk of viral foodborne illness.

The resistance of FeCV (in suspension) to inactivation by UV 253.7 nm radiation was reported to be highly variable. Doses required to achieve 3-$\log_{10}$ reduction in FeCV infectivity ranged from 12–26 mJ/cm^2 (Thurston-Enriquez *et al.*, 2003; De Roda Husman *et al.*, 2004), while others found only 1-$\log_{10}$ reduction at 48 mJ/cm^2 (Nuanualsuwan *et al.*, 2002). These differences may result from differences in composition (turbidity) of the irradiated suspensions, although it was shown that at low protein levels (< 4 μg/ml) the effect of suspension composition was negligible.

From two studies comparing FeCV and CaCV it may be concluded that UV-B radiation is less effective than UV 253.7 nm in inactivating caliciviruses in suspension (De Roda Husman *et al.*, 2004; Duizer *et al.*, 2004a). The relative resistance of FeCV to UV radiation is intermediate, i.e. comparable to that for the enteroviruses (Gerba *et al.*, 2002), less effective than for vegetative bacteria, but more effective than for phage MS2 (De Roda Husman *et al.*, 2004) and for adenoviruses 2 and 40 (Gerba *et al.*, 2002; Thurston-Enriquez *et al.*, 2003) and *Bacillus subtilis* spores (Chang *et al.*, 1985). The inactivation of caliciviruses by ionising (gamma) radiation was found to be ineffective, requiring doses of 300–500 Gy to achieve 3 log reduction, i.e. more than twice as much as for phage MS2. Moreover, the low efficacy was even further reduced by increasing the protein content of the suspension to 3–4 μg/ml (De Roda Husman *et al.*, 2004). Other studies reported no apparent effect of ultrasonic energy (26 kHz) on FeCV (Scherba *et al.*, 1991), but complete inactivation by high hydrostatic pressure after 5 min treatments with 275 MPa or more (Kingsley *et al.*, 2002a). Based on comparative PCR data, it was concluded that FeCV was a valuable model for NoV in inactivation experiments. However, the enteric NoV was significantly less sensitive to low-pH treatments, indicating the need for a truly enteric virus as model for NoV or for an *in vitro* method for the detection of NoV viability (Duizer *et al.*, 2004b). A more recent study using high pressure (400 MPa for 5 min at 4 °C) with MNV-1 in oyster tissue reported inactivation of 4 log PFU (Kingsley *et al.*, 2007) and predictive models for effects of heat and high pressure are now becoming available (Buckow *et al.*, 2008).

Due to their high stability, the NoVs can survive in the environment, in water or on foods for prolonged periods. This means that the contamination can have occurred a long time ago. Additionally the NoVs are quite resistant to refrigeration and freezing, low pH and chemical disinfectants. This combination of high stability and resistance with a low minimal infectious dose has led to several remarkable outbreaks as described before. In the absence of further data on inactivation of NoV by milder disinfectants or treatments it is probably best to apply thorough heating (cooking) for water and food, and high hypochlorite concentrations (> 1 %) for disinfection of surfaces when contamination with NoV is suspected.

32.7 Prevention and control

Prevention of gastroenteritis virus infection including foodborne transmission will rely on high standards of personal and environmental hygiene. The goal should be to prevent contamination of food items rather than to rely on treatment processes to inactivate the viruses once present in the foods or water. The only way to achieve that will depend on increased awareness on viral food safety throughout the food chain, in particular implementation of good agricultural practice, good manufacturing practice and good hygienic practice during food preparation. Application of hazard analysis critical control point (HACCP) can further enhance preventive measures by identifying and reinforcing the implementation of specific control measures.

To reduce the number of viral foodborne outbreaks in the future, governments should consider viruses in the microbial food safety guidelines. In agriculture, primary products must be protected from contamination by human, animal and agricultural wastes. It should be noted that the currently operating sewage treatment systems do not provide effluents free from viruses (Lodder *et al.*, 1999; Kukavica-Ibrulj *et al.*, 2003; Le Cann *et al.*, 2004). Water used for the cultivation, preparation or packing of food should therefore be of suitable quality to prevent the introduction of virally-contaminated water into the food chain. Also, guidelines specifically aimed at reduction of viral contamination are needed, as it has become clear that the current indicators for, for example, water and shellfish quality are insufficient as predictors of viral contamination.

Furthermore, awareness among food handlers, including seasonal workers, on the transmission of enteric viruses is needed (including the spread of viruses by vomiting), with special emphasis on the risk of transmission by asymptomatically infected persons and those continuing to shed virus following recovery from illness. This implies that food handlers who have increased risk of being NoV shedders have to be excluded from contact with food. Strict personal hygiene will not only reduce the number of virus introductions but also reduce the size of the outbreaks by minimising secondary transmission (Koopmans and Duizer, 2004).

32.8 Future trends in viral food safety

Given the importance of foodborne viruses and the impact that different factors (globalisation of the market, increased international travel, consumer demands, changes in food-processing, pathogen evolution, etc.), may have on the emergence of disease, it is clear that priority needs to be given to expanding foodborne disease surveillance to cover foodborne viruses. The expansion should include three areas: (i) a more complete coverage of qualified laboratories per country, (ii) inclusion of more countries in international surveillance networks and (iii) development and implementation of detection methods of more classes of viruses in the surveillance programs.

It is also worthwhile paying attention to the impact of new mild preservation techniques used to inactivate bacteria in foods on viruses or, for example, alcohol-based hand-sanitisers used to inactivate bacteria on hands. Most of these methods have been introduced and put into use without proper evaluation of the efficiency against viruses, creating a window of opportunity for foodborne viruses to cause problems.

Increased awareness of food as a vector for viruses, combined with increased laboratory capabilities and the implementation of virus tracking methods and molecular epidemiology might result in detection of foodborne transmission of as yet unforeseen viruses. Hopefully the diagnostic gap in gastrointestinal illnesses such as gastroenteritis and hepatitis will be closed. It is not very likely, however, that the problems caused by foodborne virus transmission are going to be resolved quickly. Continued research is needed to assess efficacy of different control measures along food chains to be able to work towards virus safe food in the future.

32.9 References and further reading

Abad, F.X., Pinto, R.M. and Bosch, A. (1994) Survival of enteric viruses on environmental fomites, *Appl Environ Microbiol*, **60**(10), 3704–10.

Allwood, P.B., Malik, Y.S., Hedberg, C.W. and Goyal, S.M. (2004) Effect of temperature and sanitizers on the survival of feline calicivirus, *Escherichia coli*, and F-specific coliphage MS2 on leafy salad vegetables, *J Food Prot*, **67**, 1451–6.

Anderson, A.D., Heryford, A.G., Sarisky, J.P., Higgins, C., Monroe, S.S., Beard, R.S., Newport, C.M., Cashdollar, J.L., Fout, G.S., Robbins, D.E., Seys, S.A., Musgrave, K.J., Medus, C., Vinjé, J., Bresee, J.S., Mainzer, H.M. and Glass, R.I. (2003) A waterborne outbreak of Norwalk-like virus among snowmobilers–Wyoming, 2001, *J Infect Dis*, **187**, 303–6.

Atmar, R.L. and Estes, M.K. (2001) Diagnosis of noncultivatable gastroenteritis viruses, the human caliciviruses, *Clin Microbiol Rev*, **14**, 15–37.

Atmar, R.L., Opekun, A.R., Gilger, M.A., Estes, M.K., Crawford, S.E., Neill, F.H. and Graham, D.Y. (2008) Norwalk virus shedding after experimental human infection, *Emerg Infect Dis*, **14**(10), 1553–7.

Bae, J. and Schwab, K.J. (2008) Evaluation of murine norovirus, feline calicivirus, poliovirus, and MS2 as surrogates for human norovirus in a model of viral persistence in surface water and groundwater, *Appl Env Microbiol*, **74**, 477–84.

Baert, L., Uyttendaele, M., Vermeersch, M., Van Collie, E. and Debeverei (2008) Survival and transfer of murine norovirus 1, a surrogate for human noroviruses, during the production process of deep-frozen onions and spinach, *J Food Prot*, **71**, 1590–97.

Belliot, G., Laveran, H. and Monroe, S.S. (1997) Detection and genetic differentiation of human astroviruses: phylogenetic grouping varies by coding region, *Arch Virol*, **142**(7), 1323–34.

Belliot, G., Lavaux, A., Souihel, D., Agnello, D. and Pothier, P. (2008) Use of murine norovirus as a surrogate to evaluate resistance of human norovirus to disinfectants, *Appl Env Microbiol*, **74**, 3315–18.

Beuret, C., Kohler, D. and Luthi, T. (2000) Norwalk-like virus sequences detected by reverse transcription-polymerase chain reaction in mineral waters imported into or in Switzerland, *J Food Prot*, **63**(11), 1576–82.

Bhattacharya, S.S., Kulka, M., Lampel, K.A., Cebula, T.A. and Goswami, B.B. (2004) Use of reverse transcription and PCR to discriminate between infectious and non-infectious hepatitis A virus, *J Virol Methods*, **116**, 181–7.

Bidawid, S., Farber, J.M. and Sattar, S.A. (2000) Rapid concentration and detection of hepatitis A virus from lettuce and strawberries, *J Virol Methods*, **88**, 175–85.

Biziagos, E., Passagot, J., Crance, J.M. and Deloince, R. (1988) Long-term survival of hepatitis A virus and poliovirus type 1 in mineral water, *Appl Environ Microbiol*, **54**(11), 2705–10.
Brown, C.M., Cann, J.W., Simons, G., Fankhauser, R.L., Thomas, W., Parashar, U.D. and Lewis, M.J. (2001) Outbreak of Norwalk virus in a Caribbean island resort: application of molecular diagnostics to ascertain the vehicle of infection, *Epidemiol Infect*, **126**, 425–32.
Buckow, R., Isbarn, S., Knorr, D., Heinz, V., Lehmacher, A. (2008) Predictive model for inactivation of feline Calicivirus, a norovirus surrogate, by heat and high hydrostatic pressure. *Appl Env Microbiol*, **74**, 1030–38.
Buesa, J., Montava, R., Abu-Mallouh, R., Fos, M., Ribes, J.M., Bartolomé, R., Vanaclocha, H., Torner, N. and Domínguez, A.J. (2008) Sequential evolution of genotype GII.4 norovirus variants causing gastroenteritis outbreaks from 2001 to 2006 in Eastern Spain, *Med Virol*, **80**(7), 1288–95.
Burton-MacLeod, J.A., Kane, E.M., Beard, R.S., Hadley, L.A., Glass, R.I. and Ando, T. (2004) Evaluation and comparison of two commercial enzyme-linked immunosorbent assay kits for detection of antigenically diverse human noroviruses in stool samples, *J Clin Microbiol*, **42**, 2587–95.
Butot, S., Putallaz, T. and Sánchez, G. (2008) Effects of sanitation, freezing and frozen storage on enteric viruses in berries and herbs, *Int J Food Microbiol*, **126**, 30–35.
Butt, A.A., Aldridge, K.E. and Sanders, C.V. (2004) Infections related to the ingestion of seafood Part I: Viral and bacterial infections, *Lancet Infect Dis*, **4**, 201–12.
Cannon, R.O., Poliner, J.R., Hirschhorn, R.B., Rodeheaver, D.C., Silverman, P.R., Brown, E.A., Talbot, G.H., Stine, S.E., Monroe, S.S., Dennis, D.T. and others (1991) A multistate outbreak of Norwalk virus gastroenteritis associated with consumption of commercial ice, *J Infect Dis*, **164**, 860–63.
Cannon, J.L., Papafragkou, E., Park, G.W., Osborne, J., Jaykus, L.A., and Vinje, J. (2006) Surrogates for the study of norovirus stability and inactivation in the environment: a comparison of murine norovirus and feline calicivirus, *J Food Prot*, **69**, 2761–5.
Carrique-Mas, J., Andersson, Y., Petersen, B., Hedlund, K.O., Sjogren, N. and Giesecke, J. (2003) A norwalk-like virus waterborne community outbreak in a Swedish village during peak holiday season, *Epidemiol Infect*, **131**, 737–44.
Chadwick, P. R. and McCann, R (1994) Transmission of a small round structured virus by vomiting during a hospital outbreak of gastroenteritis, *J Hosp Infect*, **26**, 251–9.
Chadwick, P.R., Walker, M. and Rees, A.E. (1994) Airborne transmission of a small round structured virus, *Lancet*, **343**, 171. See also comments: *Lancet*, 1994, **343**(8897), 608–9; *Lancet*, 1994, **343**(8897), 609.
Chalmers, J.W. and McMillan, J.H. (1995) An outbreak of viral gastroenteritis associated with adequately prepared oysters, *Epidemiol Infect*, **115**, 163–7.
Chang, J.C., Ossoff, S.F., Lobe, D.C., Dorfman, M.H., Dumais, C.M., Qualls, R.G. and Johnson, J.D. (1985) UV inactivation of pathogenic and indicator microorganisms, *Appl Environ Microbiol*, **49**, 1361–5.
Cheesbrough, J.S., Green, J., Gallimore, C.I., Wright, P.A. and Brown, D.W. (2000) Widespread environmental contamination with Norwalk-like viruses (NLV) detected in a prolonged hotel outbreak of gastroenteritis, *Epidemiol Infect*, **125**, 93–8.
Codex Alimentarius Committee on Food Hygiene (1999) *Discussion Paper on Viruses in Food*, FAO/WHO document CX/FH pp/11, Rome/Geneva, Food and Agriculture Organization of the United Nations/World Health Organization.
Condit, R.C. (2006) Principles of virology, in DM Knipe, PM Howley, DE Griffin and RA Lamb (eds), *Fields Virology*, 5th edn, Philadelphia PA, Lippincott Williams and Wilkins, 19–52.
Croci, L., Ciccozzi, M., De Medici, D., Di Pasquale, S., Fiore, A., Mele, A. and Toti, L. (1999) Inactivation of hepatitis A virus in heat-treated mussels, *J Appl Microbiol*, **87**, 884–8.

Daniels, N.A., Bergmire-Sweat, D.A., Schwab, K.J., Hendricks, K.A., Reddy, S., Rowe, S.M., Fankhauser, R.L., Monroe, S.S., Atmar, R.L., Glass, R.I. and Mead, P. (2000) A foodborne outbreak of gastroenteritis associated with Norwalk-like viruses: first molecular traceback to deli sandwiches contaminated during preparation, *J Infect Dis*, **181**, 1467–70.

Dastjerdi, A.M., Green, J., Gallimore, C.I., Brown, D.W. and Bridger, J.C. (1999) The bovine Newbury agent-2 is genetically more closely related to human SRSVs than to animal caliciviruses, *Virology*, **254**, 1–5.

De Leon, R., Matsui, S.M., Baric, R.S., Herrmann, J.E., Blacklow, N.R., Greenberg, H.B. and Sobsey, M.D. (1992) Detection of Norwalk virus in stool specimens by reverse transcriptase-polymerase chain reaction and nonradioactive oligoprobes, *J Clin Microbiol*, **30**, 3151–7.

De Roda Husman, A.M., Bijkerk, P., Lodder, W., Van Den Berg, H., Pribil, W., Cabaj, A., Gehringer, P., Sommer, R. and Duizer, E. (2004) Calicivirus inactivation by nonionizing 253.7-Nanometer-Wavelength, *Appl Environ Microbiol*, **70**, 5089–93.

De Serres, G., Cromeans, T.L., Levesque, B., Brassard, N., Barthe, C., Dionne, M., Prud'Homme, H., Paradis, D., Sharpio, C.N., Nainan, O.V. and Margolis, H.S. (1999) Molecular confirmation of hepatitis A virus from well water: epidemiology and public health implications, *J Infect Dis*, **179**(1), 37–43.

de Wit, M.A., Hoogenboom-Verdegaal, A.M., Goosen, E.S., Sprenger, M.J. and Borgdorff, M.W. (2000) A population-based longitudinal study on the incidence and disease burden of gastroenteritis and *Campylobacter* and *Salmonella* infection in four regions of the Netherlands, *Eur J Epidemiol*, **16**, 713–18.

de Wit, M.A., Koopmans, M.P., Kortbeek, L.M., Wannet, W.J., Vinje, J., van Leusden, F., Bartelds, A.I. and van Duynhoven, Y.T. (2001a) Sensor, a population-based cohort study on gastroenteritis in the Netherlands: incidence and etiology, *Am J Epidemiol*, **154**, 666–74.

de Wit, M.A., Kortbeek, L.M., Koopmans, M.P., de Jager, C.J., Wannet, W.J., Bartelds, A.I. and van Duynhoven, Y.T. (2001b) A comparison of gastroenteritis in a general practice-based study and a community-based study, *Epidemiol Infect*, **127**, 389–97.

de Wit, M.A., Koopmans, M. P., Kortbeek, T., van Leeuwen, N., Bartelds, A. and van Duynhoven, Y. (2001c) Gastroenteritis in sentinel practices in the Netherlands, *Emerg Infect Dis*, **7**, 82–91.

de Wit, M.A., Koopmans, M.G., Kortbeek, T., van Leeuwen, N., Vinjé, J. and van Duynhoven, Y. (2001d) Etiology of gastroenteritis in sentinel general practices in the Netherlands, *Clin Infect Dis*, **33**, 280–88.

de Wit, M.A., Koopmans, M.P. and van Duynhoven, Y.T. (2003) Risk factors for norovirus, Sapporo-like virus, and group A rotavirus gastroenteritis, *Emerg Infect Dis*, **9**, 1563–70.

de Wit, M.A., Widdowson, M.A., Vennema, H., de Bruin, E., Fernandes, T. and Koopmans, M. (2007) Large outbreak of norovirus: the baker who should have known better, *J Infect*, **55**(2), 188–93.

Deng, Y., Batten, C.A., Liu, B.L., Lambden, P.R., Elschner, M., Gunther, H., Otto, P., Schnurch, P., Eichhorn, W., Herbst, W. and Clarke, I.N. (2003) Studies of epidemiology and seroprevalence of bovine noroviruses in Germany, *J Clin Microbiol*, **41**, 2300–305.

Donaldson, E.F., Lindesmith, L.C., Lobue, A.D. and Baric, R.S. (2008) Norovirus pathogenesis: mechanisms of persistence and immune evasion in human populations, *Immunol Rev*, **225**, 190–211.

Doyle, A., Barataud, D., Gallay, A., Thiolet, J.M., Le Guyaguer, S., Kohli, E. and Vaillant, V. (2004) Norovirus foodborne outbreaks associated with the consumption of oysters from the Etang de Thau, France, December 2002, *Euro Surveill*, **9**, 24–6.

Duan, S.M., Zhao, X.S., Wen, R.F., Huang, J.J., Pi, G.H., Zhang, S.X., Han, J., Bi, S.L., Ruan, L. and Dong, X.P. (2003) Stability of SARS coronavirus in human specimens and environment and its sensitivity to heating and UV irradiation, *Biomed Environ Sci*, **16**, 246–55.

Dubois, E., Agier, C., Traore, O., Hennechart, C., Merle, G., Cruciere, C. and Laveran, H. (2002) Modified concentration method for the detection of enteric viruses on fruits and vegetables by reverse transcriptase-polymerase chain reaction or cell culture, *J Food Prot*, **65**, 1962–9.

Duizer, E., Bijkerk, P., Rockx, B., De Groot, A., Twisk, F. and Koopmans, M. (2004a) Inactivation of caliciviruses, *Appl Environ Microbiol*, **70**, 4538–43.

Duizer, E., Schwab, K.J., Neill, F.H., Atmar, R.L., Koopmans, M.P. and Estes, M.K. (2004b) Laboratory efforts to cultivate noroviruses, *J Gen Virol*, **85**, 79–87.

Emerson, S.U. and Purcell, R.H. (2006) Hepatitis E virus, in D M Knipe, P M Howley, D E Griffin, R A Lamb, M A Martin, B Roizman and S E Straus (eds), *Fields Virology*, 5th edn, Philadelphia, PA, Lippincott Williams and Wilkins, 3047–58.

Erdman, D.D., Gary, G.W. and Anderson, L.J. (1989a) Development and evaluation of an IgM capture enzyme immunoassay for diagnosis of recent Norwalk virus infection, *J Virol Methods*, **24**, 57–66.

Erdman, D.D., Gary, G.W. and Anderson, L.J. (1989b) Serum immunoglobulin A response to Norwalk virus infection, *J Clin Microbiol*, **27**, 1417–18.

Evans, M.R., Meldrum, R., Lane, W., Gardner, D., Ribeiro, C.D., Gallimore, C.I. and Westmoreland, D. (2002) An outbreak of viral gastroenteritis following environmental contamination at a concert hall, *Epidemiol Infect*, **129**, 355–60.

Fankhauser, R.L., Noel, J.S., Monroe, S.S., Ando, T. and Glass, R.I. (1998) Molecular epidemiology of "Norwalk-like viruses" in outbreaks of gastroenteritis in the United States, *J Infect Dis*, **178**, 1571–8.

Fankhauser, R.L., Monroe, S.S., Noel, J.S., Humphrey, C.D., Bresee, J.S., Parashar, U.D., Ando, T. and Glass, R.I. (2002) Epidemiologic and molecular trends of "Norwalk-like viruses" associated with outbreaks of gastroenteritis in the United States, *J Infect Dis*, **186**, 1–7.

Farkas, T., Berke, T., Reuter, G., Szucs, G., Matson, D.O. and Jiang, X. (2002) Molecular detection and sequence analysis of human caliciviruses from acute gastroenteritis outbreaks in Hungary, *J Med Virol*, **67**, 567–73.

Farkas, T., Thornton, S.A., Wilton, N., Zhong, W., Altaye, M. and Jiang, X. (2003) Homologous versus heterologous immune responses to Norwalk-like viruses among crew members after acute gastroenteritis outbreaks on 2 US Navy vessels, *J Infect Dis*, **187**, 187–93.

Fleet, G.H., Heiskanen, P., Reid, I. and Buckle, K.A. (2000) Foodborne viral illness – status in Australia, *Int J Food Microbiol*, **59**, 127–36.

Formiga-Cruz, M., Tofino-Quesada, G., Bofill-Mas, S., Lees, D.N., Henshilwood, K., Allard, A.K., Conden-Hansson, A.C., Hernroth, B.E., Vantarakis, A., Tsibouxi, A., Papapetropoulou, M., Furones, M.D. and Girones, R. (2002) Distribution of human virus contamination in shellfish from different growing areas in Greece, Spain, Sweden, and the United Kingdom, *Appl Environ Microbiol*, **68**, 5990–98.

Gallimore, C.I., Barreiros, M.A., Brown, D.W., Nascimento, J.P. and Leite, J.P. (2004a) Noroviruses associated with acute gastroenteritis in a children's day care facility in Rio de Janeiro, Brazil, *Braz J Med Biol Res*, **37**, 321–6.

Gallimore, C.I., Lewis, D., Taylor, C., Cant, A., Gennery, A. and Gray, J.J. (2004b) Chronic excretion of a norovirus in a child with cartilage hair hypoplasia (CHH), *J Clin Virol*, **30**, 196–204.

Gehrke, C., Steinmann, J. and Goroncy-Bermes, P. (2004) Inactivation of feline calicivirus, a surrogate of norovirus (formerly Norwalk-like viruses), by different types of alcohol in vitro and in vivo, *J Hosp Infect*, **56**, 49–55.

Gentsch, J.R., Glass, R.I., Woods, P., Gouvea, V., Gorzigilia, M., Flores, J., Das, B.K. and Bhan, M.K. (1992) Identification of group A rotavirus gene 4 types by polymerase chain reaction, *J Clin Microbiol*, **30**(6), 1365–73.

Gerba, C.P., Gramos, D.M. and Nwachuku, N. (2002) Comparative inactivation of enteroviruses and adenovirus 2 by UV light, *Appl Environ Microbiol*, **68**, 5167–9.

Gilgen, M., Germann, D., Luthy, J. and Hubner, P. (1997) Three-step isolation method for sensitive detection of enterovirus, rotavirus, hepatitis A virus, and small round structured viruses in water samples, *Int J Food Microbiol*, **37**, 189–99.

Girish, R., Broor, S., Dar, L. and Ghosh, D. (2002) Foodborne outbreak caused by a Norwalk-like virus in India, *J Med Virol*, **67**, 603–7.

Glass, P.J., White, L.J., Ball, J.M., Leparc-Goffart, I., Hardy, M.E. and Estes, M.K. (2000) Norwalk virus open reading frame 3 encodes a minor structural protein, *J Virol*, **74**, 6581–91.

Gotz, H., De, J.B., Lindback, J., Parment, P.A., Hedlund, K.O., Torven, M. and Ekdahl, K. (2002) Epidemiological investigation of a food-borne gastroenteritis outbreak caused by Norwalk-like virus in 30 day-care centres, *Scand J Infect Dis*, **34**, 115–21.

Gouvea, V., Glass, R.I., Woods, P., Taniguchi, K., Clark, H.F., Forrester, B. and Fang, Z.Y. (1990) Polymerase chain reaction amplification and typing of rotavirus nucleic acid from stool specimens, *J Clin Microbiol*, **28**(2) 276–82.

Gouvea, V., Allen, J.R., Glass, R.I., Fang, Z.Y., Bremont, M., Cohen, H., McCrae, M.A., Saif L.J., Sinarachatanant, P. and Caul, E.O. (1991) Detection of group B and C rotaviruses by polymerase chain reaction, *J Clin Microbiol*, **28**(2), 276–82.

Gray, J.J., Cunliffe, C., Ball, J., Graham, D.Y., Desselberger, U. and Estes, M.K. (1994) Detection of immunoglobulin M (IgM), IgA, and IgG Norwalk virus-specific antibodies by indirect enzyme-linked immunosorbent assay with baculovirus-expressed Norwalk virus capsid antigen in adult volunteers challenged with Norwalk virus, *J Clin Microbiol*, **32**, 3059–63.

Green, K.Y. (2006) Caliciviridae: the noroviruses, in D M Knipe, P M Howley, D E Griffin, R A Lamb, M A Martin, B Roizman and S E Straus (eds), *Fields Virology*, 5th edn, Philadelphia, PA, Lippincott Williams and Wilkins, 949–80.

Green, K.Y., Lew, J.F., Jiang, X., Kapikian, A.Z. and Estes, M.K. (1993) Comparison of the reactivities of baculovirus-expressed recombinant Norwalk virus capsid antigen with those of the native Norwalk virus antigen in serologic assays and some epidemiologic observations, *J Clin Microbiol*, **31**, 2185–91.

Green, S.M., Lambden, P.R., Caul, E.O., Ashley, C.R. and Clarke, I.N. (1995) Capsid diversity in small round-structured viruses: molecular characterization of an antigenically distinct human enteric calicivirus, *Virus Res*, **37**, 271–83.

Gulati, B.R., Allwood, P.B., Hedberg, C.W. and Goyal, S.M. (2001) Efficacy of commonly used disinfectants for the inactivation of calicivirus on strawberry, lettuce, and a food-contact surface, *J Food Prot*, **64**, 1430–34.

Hale, A., Mattick, K., Lewis, D., Estes, M., Jiang, X., Green, J., Eglin, R. and Brown, D. (2000) Distinct epidemiological patterns of Norwalk-like virus infection, *J Med Virol*, **62**, 99–103.

Haramoto, E., Katayama, H. and Ohgaki, S. (2004) Detection of noroviruses in tap water in Japan by means of a new method for concentrating enteric viruses in large volumes of freshwater, *Appl Environ Microbiol*, **70**, 2154–60.

Harrington, P.R., Yount, B., Johnston, R.E., Davis, N., Moe, C. and Baric, R.S. (2002) Systemic, mucosal, and heterotypic immune induction in mice inoculated with Venezuelan equine encephalitis replicons expressing Norwalk virus-like particles, *J Virol*, **76**, 730–42.

Harrington, P.R., Vinje, J., Moé, C.L. and Baric, R.S. (2004) Norovirus capture with histo-blood group antigens reveals novel virus-ligand interactions, *J Virol*, **78**, 3035–45.

Hedberg, C.W. and Osterholm, M.T. (1993) Outbreaks of foodborne and waterborne viral gastroenteritis, *Clin Microbiol Rev*, **6**, 199–210.

Herrmann, J.E., Blacklow, N.R., Matsui, S.M., Lewis, T.L., Estes, M.K., Ball, J.M. and Brinker, J.P. (1995) Monoclonal antibodies for detection of Norwalk virus antigen in stools, *J Clin Microbiol*, **33**, 2511–13.

Hewitt, J. and Greening, G.E. (2006) Effect of heat treatment on hepatitis A virus and norovirus in New Zealand Greenshell mussels (*Perna canaliculus*) by quantitative real-time reverse transcription PCR and cell culture, *J Food Prot*, **69**, 2217–23.

Hoebe, C.J., Vennema, H., de Roda Husman, A.M. and van Duynhoven, Y.T. (2004) Norovirus outbreak among primary schoolchildren who had played in a recreational water fountain, *J Infect Dis*, **189**, 699–705.

Hollinger, F.B. and Emerson, S.U. (2006) Hepatitis A virus, in D M Knipe, P M Howley, D E Griffin, R A Lamb, M A Martin, B Roizman and S E Straus (eds), *Fields Virology*, 5th edn, Philadelphia, PA, Lippincott Williams and Wilkins, 911–48.

Horman, A., Rimhanen-Finne, R., Maunula, L., von Bonsdorff, C.H., Torvela, N., Heikinheimo, A. and Hanninen, M.L. (2004) *Campylobacter* spp., *Giardia* spp., *Cryptosporidium* spp., noroviruses, and indicator organisms in surface water in southwestern Finland, 2000–2001, *Appl Environ Microbiol*, **70**, 87–95.

Huang, P., Farkas, T., Marionneau, S., Zhong, W., Ruvoen-Clouet, N., Morrow, A.L., Altaye, M., Pickering, L.K., Newburg, D.S., LePendu, J. and Jiang, X. (2003) Noroviruses bind to human ABO, Lewis, and secretor histo-blood group antigens: identification of 4 distinct strain-specific patterns, *J Infect Dis*, **188**, 19–31.

Hung, T., Chen, G.M., Wang, C.G., Yao H.L., Fang, Z.Y., Chao, T.X., Chou, Z.Y., Ye, W., Chang, X.J., Den, S.S. *et al.* (1984) Waterborne outbreak of rotavirus diarrhoea in adults in China caused by a novel rotavirus, *Lancet*, **1**(8387), 1139–42.

Hutson, A.M., Atmar, R.L., Graham, D.Y. and Estes, M.K. (2002) Norwalk virus infection and disease is associated with ABO histo-blood group type, *J Infect Dis*, **185**, 1335–7.

Hutson, A.M., Atmar, R.L. and Estes, M.K. (2004) Norovirus disease: changing epidemiology and host susceptibility factors, *Trends Microbiol*, **12**, 279–87.

Ijaz, M.K., Sattar, S.A., Johnson-Lussenburg, C.M., Springthorpe, V.S. (1985) Comparison of the airborne survival of calf rotavirus and poliovirus type 1 (sabin) aerosolized as a mixture, *Appl Environ Microbiol*, **49**(2), 289–93.

Iritani, N., Seto, Y., Kubo, H., Haruki, K., Ayata, M. and Ogura, H. (2002) Prevalence of "Norwalk-like virus" infections in outbreaks of acute nonbacterial gastroenteritis observed during the 1999–2000 season in Osaka City, Japan, *J Med Virol*, **66**, 131–8.

Iritani, N., Seto, Y., Kubo, H., Murakami, T., Haruki, K., Ayata, M. and Ogura, H. (2003) Prevalence of Norwalk-like virus infections in cases of viral gastroenteritis among children in Osaka City, Japan, *J Clin Microbiol*, **41**, 1756–9.

Jansen, A., Stark, K., Kunkel, J., Schreier, E., Ignatius, R., Liesenfeld, O., Werber, D., Göbel, U.B., Zeitz, M., Schneider, T. (2008) Aetiology of community-acquired, acute gastroenteritis in hospitalised adults: a prospective cohort study, *BMC Infect Dis*, **22**(8), 143.

Jean, J., D'Souza, D. and Jaykus, L.A. (2003) Transcriptional enhancement of RT-PCR for rapid and sensitive detection of Noroviruses, *FEMS Microbiol Lett*, **226**, 339–45.

Jiang, X., Matson, D.O., Ruiz-Palacios, G.M., Hu, J., Treanor, J. and Pickering, L.K. (1995) Expression, self-assembly, and antigenicity of a snow mountain agent-like calicivirus capsid protein, *J Clin Microbiol*, **33**, 1452–5.

Jiang, B., McClure, H.M., Fankhauser, R.L., Monroe, S.S. and Glass, R.I. (2004) Prevalence of rotavirus and norovirus antibodies in non-human primates, *J Med Primatol*, **33**, 30–33.

Kageyama, T., Kojima, S., Shinohara, M., Uchida, K., Fukushi, S., Hoshino, F.B., Takeda, N. and Katayama, K. (2003) Broadly reactive and highly sensitive assay for Norwalk-like viruses based on real-time quantitative reverse transcription-PCR, *J Clin Microbiol*, **41**, 1548–57.

Kaplan, J.E., Feldman, R., Campbell, D.S., Lookabaugh, C. and Gary, G.W. (1982) The frequency of a Norwalk-like pattern of illness in outbreaks of acute gastroenteritis, *Am J Public Health*, **72**, 1329–32.

Karim, M.R., Pontius, F.W. and LeChevallier, M.W. (2004) Detection of noroviruses in water–summary of an international workshop, *J Infect Dis*, **189**, 21–8.

Karst, S.M., Wobus, C.E., Lay, M., Davidson, J. and Virgin, H.W. 4th (2003) STAT1-dependent innate immunity to a Norwalk-like virus, *Science*, **299**, 1575–8.

Khan, A.S., Moe, C.L., Glass, R.I., Monroe, S.S., Estes, M.K., Chapman, L.E., Jiang, X., Humphrey, C., Pon, E., Iskander, J.K. and others (1994) Norwalk virus-associated gastroenteritis traced to ice consumption aboard a cruise ship in Hawaii: comparison and application of molecular method-based assays, *J Clin Microbiol*, **32**, 318–22.

Kingsley, D.H. and Richards, G.P. (2003) Persistence of hepatitis A virus in oysters, *J Food Prot*, **66**, 331–4.

Kingsley, D.H., Hoover, D.G., Papafragkou, E. and Richards, G.P. (2002a) Inactivation of hepatitis A virus and a calicivirus by high hydrostatic pressure, *J Food Prot*, **65**, 1605–9.

Kingsley, D.H., Meade, G.K. and Richards, G.P. (2002b) Detection of both hepatitis A virus and Norwalk-like virus in imported clams associated with foodborne illness, *Appl Environ Microbiol*, **68**, 3914–18.

Kingsley, D.H., Hollinian, D.R., Calci, K.R., Chen, H.Q. and Flick, G.J. (2007) Inactivation of a norovirus by high-pressure processing, *Appl Env Microbiol*, **73**, 581–5.

Kitamoto, N., Tanaka, T., Natori, K., Takeda, N., Nakata, S., Jiang, X. and Estes, M.K. (2002) Cross-reactivity among several recombinant calicivirus virus-like particles (VLPs) with monoclonal antibodies obtained from mice immunized orally with one type of VLP, *J Clin Microbiol*, **40**, 2459–65.

Kittigul L, Pombubpa K, Taweekate Y, Yeephoo T, Khamrin P and Ushijima H. (2009) Molecular characterization of rotaviruses, noroviruses, sapovirus, and adenoviruses in patients with acute gastroenteritis in Thailand, *J Med Virol*, **81**(2), 345–53.

Knipe, D.M., Howley, P.M., Griffin, D.E., Lamb, R.A., Martin, M.A., Roizman, B. and Straus, S.E. (eds) (2006) *Fields Virology*, 5th edn, Philadelphia, PA, Lippincott Williams and Wilkins.

Kobayashi, S., Sakae, K., Natori, K., Takeda, N., Miyamura, T. and Suzuki, Y. (2000) Serotype-specific antigen ELISA for detection of chiba virus in stools, *J Med Virol*, **62**, 233–8.

Kobayashi, S., Natori, K., Takeda, N. and Sakae, K. (2004) Immunomagnetic capture RT-PCR for detection of norovirus from foods implicated in a foodborne outbreak, *Microbiol Immunol*, **48**, 201–4.

Kohn, M.A., Farley, T.A., Ando, T., Curtis, M., Wilson, S.A., Jin, Q., Monroe, S.S., Baron, R.C., McFarland, L.M. and Glass, R.I. (1995) An outbreak of Norwalk virus gastroenteritis associated with eating raw oysters. Implications for maintaining safe oyster beds, *JAMA*, **273**, 466–71.

Koopmans, M. and Duizer, E. (2004) Foodborne viruses: an emerging problem, *Int J Food Microbiol*, **90**, 23–41.

Koopmans, M., Vinjé, J., de Wit, M., Leenen, I., van der Poel, W. and van Duynhoven, Y. (2000) Molecular epidemiology of human enteric calciviruses in the Netherlands, *J Infect Dis*, **181**(6), S262–9.

Koopmans, M., von Bonsdorff, C.H., Vinje, J., de Medici, D. and Monroe, S. (2002) Foodborne viruses, *FEMS Microbiol Rev*, **26**, 187–205.

Koopmans, M., Vennema, H., Heersma, H., van Strien, E., van Duynhoven, Y., Brown, D., Reacher, M. and Lopman, B. (2003) Early identification of common-source foodborne virus outbreaks in Europe, *Emerg Infect Dis*, **9**, 1136–42.

Kukavica-Ibrulj, I., Darveau, A. and Fliss, I. (2003) Immunofluorescent detection and quantitation of hepatitis A virus in sewage treatment effluent and on agri-food surfaces using scanning confocal microscopy, *J Virol Methods*, **108**, 9–17.

Kukkula, M., Maunula, L., Silvennoinen, E. and von Bonsdorff, C.H. (1999) Outbreak of viral gastroenteritis due to drinking water contaminated by Norwalk-like viruses, *J Infect Dis*, **180**, 1771–6.

Kuusi, M., Nuorti, J.P., Maunula, L., Minh, N.N., Ratia, M., Karlsson, J. and von Bonsdorff, C.H. (2002) A prolonged outbreak of Norwalk-like calicivirus (NLV) gastroenteritis in a rehabilitation centre due to environmental contamination, *Epidemiol Infect*, **129**, 133–8.

Lamhoujeb, S., Fliss, I., Ngazoa, S.E. and Jean, J. (2008) Evaluation of the persistence of infectious human noroviruses on food surfaces by using real-time nucleic acid sequence-based amplification, *Appl Environ Microbiol*, **74**(11), 3349–55.

Lau, C.S., Wong, D.A., Tong, L.K., Lo, J.Y., Ma, A.M., Cheng, P.K. and Lim, W.W. (2004) High rate and changing molecular epidemiology pattern of norovirus infections in sporadic cases and outbreaks of gastroenteritis in Hong Kong, *J Med Virol*, **73**, 113–17.

Le Cann, P., Ranarijaona, S., Monpoeho, S., Le Guyader, F. and Ferre, V. (2004) Quantification of human astroviruses in sewage using real-time RT-PCR, *Res Microbiol*, **155**, 11–15.

Le Guyader, F.S., Neill, F.H., Dubois, E., Bon, F., Loisy, F., Kohli, E., Pommepuy, M. and Atmar, R.L. (2003) A semiquantitative approach to estimate Norwalk-like virus contamination of oysters implicated in an outbreak, *Int J Food Microbiol*, **87**, 107–12.

Le Guyader, F.S., Schultz, A.C., Haugarreau, L., Croci, L., Maunula, L., Duizer, E., Lodder-Verschoor, F., von Bonsdorff, C.H., Suffredini, E., van der Poel, W.M.M., Reymundo, R., Koopmans, M. (2004) Round-robin comparison of methods for the detection of human enteric viruses in lettuce, *J Food Prot*, **67**, 2315–19.

Lees, D. (2000) Viruses and bivalve shellfish, *Int J Food Microbiol*, **59**, 81–116.

Leggitt, P.R. and Jaykus, L.A. (2000) Detection methods for human enteric viruses in representative foods, *J Food Prot*, **63**, 1738–44.

Lewis, G.D., Molloy, S.L., Greening, G.E. and Dawson, J. (2000) Influence of environmental factors on virus detection by RT-PCR and cell culture, *J Appl Microbiol*, **88**, 633–40.

Liu, B.L., Lambden, P.R., Gunther, H., Otto, P., Elschner, M. and Clarke, I.N. (1999) Molecular characterization of a bovine enteric calicivirus: relationship to the Norwalk-like viruses, *J Virol*, **73**, 819–25.

Lodder, W.J., Vinje, J., van De Heide, R., de Roda Husman, A.M., Leenen, E.J. and Koopmans, M.P. (1999) Molecular detection of Norwalk-like caliciviruses in sewage, *Appl Environ Microbiol*, **65**, 5624–7.

Lopman, B.A., Brown, D.W. and Koopmans, M. (2002) Human caliciviruses in Europe, *J Clin Virol*, **24**, 137–60.

Lopman, B.A., Reacher, M.H., Van Duijnhoven, Y., Hanon, F.X., Brown, D. and Koopmans, M. (2003) Viral gastroenteritis outbreaks in Europe, 1995–2000, *Emerg Infect Dis*, **9**, 90–96.

Lopman, B., Vennema, H., Kohli, E., Pothier, P., Sanchez, A., Negredo, A., Buesa, J., Schreier, E., Reacher, M., Brown, D., Gray, J., Iturriza, M., Gallimore, C., Bottiger, B., Hedlund, K.O., Torvén, M., von Bonsdorff, C.H., Maunula, L., Poljsak-Prijatelj, M., Zimsek, J., Reuter, G., Szücs, G., Melegh, B., Svennson, L., van Duijnhoven, Y. and Koopmans, M. (2004) Increase in viral gastroenteritis outbreaks in Europe and epidemic spread of new norovirus variant, *Lancet*, **363**(9410), 682–8.

Macinga, D.R., Sattar, S.A., Jaykus, L.A., Arbogats, J.W. (2008) Improved inactivation of non-enveloped enteric viruses and their surrogates by novel alcohol-based hand sanitizer, *Appl Env Microbiol*, **74**, 5047–52.

Madore, H.P., Treanor, J.J., Buja, R. and Dolin, R. (1990) Antigenic relatedness among the Norwalk-like agents by serum antibody rises, *J Med Virol*, **32**, 96–101.

Marionneau, S., Ruvoen, N., Le Moullac-Vaidye, B., Clement, M., Cailleau-Thomas, A., Ruiz-Palacois, G., Huang, P., Jiang, X. and Le Pendu, J. (2002) Norwalk virus binds to histo-blood group antigens present on gastroduodenal epithelial cells of secretor individuals, *Gastroenterology*, **122**, 1967–77.

Marks, P.J., Vipond, I.B., Carlisle, D., Deakin, D., Fey, R.E. and Caul, E.O. (2000) Evidence for airborne transmission of Norwalk-like virus (NLV) in a hotel restaurant, *Epidemiol Infect*, **124**, 481–7.

Marks, P.J., Vipond, I.B., Regan, F.M., Wedgwood, K., Fey, R.E. and Caul, E.O. (2003) A school outbreak of Norwalk-like virus: evidence for airborne transmission, *Epidemiol Infect*, **131**, 727–36.

Marshall, J.A., Hellard, M.E., Sinclair, M.I., Fairley, C.K., Cox, B.J., Catton, M.G., Kelly, H. and Wright, P.J. (2003) Incidence and characteristics of endemic Norwalk-like virus-associated gastroenteritis, *J Med Virol*, **69**, 568–78.

Martella, V. Campolo, M. Lorusso, E., Cavicchio, P., Camero, M. Bellacicco, A.L., Decaro, N., Elia, G., Greco, G., Corrente, M., Desario, C., Arista, S., Banyai, K., Koopmans, M. and Buonavoglia, C. (2007) Norovirus in captive lion cub (*Panthera leo*), *Emerg Infect Dis*, **13**(7), 1071–3.

Martella, V., Lorusso, E., Decaro, N., Elia, G., Radogna, A., D'Abramo, M., Desario, C., Cavalli, A., Corrente, M., Camero, M., Germinario, C.A., Bányai, K., Di Martino, B., Marsilio, F., Carmichael, L.E. and Buonavoglia, C. (2008) Detection and molecular characterization of a canine norovirus, *Emerg Infect Dis*, **14**(8), 1306–8.

Martinez, N., Espul, C., Cuello, H., Zhong, W., Jiang, X., Matson, D.O. and Berke, T. (2002) Sequence diversity of human caliciviruses recovered from children with diarrhea in Mendoza, Argentina, 1995–1998, *J Med Virol*, **67**, 289–98.

Marx, A., Shay, D.K., Noel, J.S., Brage, C., Bresee, J.S., Lipsky, S., Monroe, S.S., Ando, T., Humphrey, C.D., Alexander, E.R. and Glass, R.I. (1999) An outbreak of acute gastroenteritis in a geriatric long-term-care facility: combined application of epidemiological and molecular diagnostic methods, *Infect Control Hosp Epidemiol*, **20**, 306–11.

Matsui, S.M. and Greenberg, H.B. (2000) Immunity to calicivirus infection, *J Infect Dis*, **181**, Supplement 2, S331–5.

Matsumoto, K., Hatano, M., Kobayashi, K., Hasegawa, A., Yamazaki, S., Nakata, S., Chiba, S. and Kimura, Y. (1989) An outbreak of gastroenteritis associated with acute rotaviral infection in schoolchildren, *J Infect Dis*, **160**(4), 611–15.

Mattison, K., Shukla, A., Cook, A., Pollari, F., Friendship, R., Kelton, D., Bidawid, S. and Farber, J.M. (2007) Human noroviruses in swine and cattle, *Emerg Infect Dis*, **13**(8), 1184–8.

Mead, P.S., Slutsker, L., Dietz, V., McCaig, L.F., Bresee, J.S., Saphior, C. *et al.* (1999) Food-related illness and death in the United States, *Emerg Infect Dis*, **5**(5), 840–42.

Moore, C., Clark, E.M., Gallimore, C.I., Corden, S.A., Gray, J.J. and Westmoreland, D. (2004) Evaluation of a broadly reactive nucleic acid sequence based amplification assay for the detection of noroviruses in faecal material, *J Clin Virol*, **29**, 290–96.

Motarjemi, Y., Käferstein, F., Moy, G., Miyagishima, K., Miyagawa, S. and Reilly, A., (1995) *Food technologies and Public Health*, WHO/FNU/FOS/95.12, Geneva, World Health Organization.

Motarjemi, Y., Káferstein, F., Moy, G., Niyagawa, S., Miyagishima, K. (1996) Importance of HACCP for public health and development, *Food Cont*, **7**, 77–85.

Mounts, A.W., Ando, T., Koopmans, M., Bresee, J.S., Noel, J. and Glass, R.I. (2000) Cold weather seasonality of gastroenteritis associated with Norwalk-like viruses, *J Infect Dis*, **181**, Supplement 2, S284–7.

Muniain-Mujika, I., Girones, R., Tofino-Quesada, G., Calvo, M. and Lucena, F. (2002) Depuration dynamics of viruses in shellfish, *Int J Food Microbiol*, **77**, 125–33.

Myrmel, M., Berg, E.M., Rimstad, E. and Grinde, B. (2004) Detection of enteric viruses in shellfish from the Norwegian coast, *Appl Environ Microbiol*, **70**, 2678–84.

Nauheim, R.C., Romanowski, E.G., Araullo-Cruz, T., Kowalski, R.P., Turgeon, P.W., Stopak, S.S. and Gordon, Y.J. (1990) Prolonged recoverability of desiccated adenovirus type 19 from various surfaces, *Ophthalmology*, **97**(11), 1450–53.

Nilsson, M., Hedlund, K.O., Thorhagen, M., Larson, G., Johansen, K., Ekspong, A. and

Svensson, L. (2003) Evolution of human calicivirus RNA in vivo: accumulation of mutations in the protruding P2 domain of the capsid leads to structural changes and possibly a new phenotype, *J Virol*, **77**, 13117–24.

Nishida, T., Kimura, H., Saitoh, M., Shinohara, M., Kato, M., Fukuda, S., Munemura, T., Mikami, T., Kawamoto, A., Akiyama, M., Kato, Y., Nishi, K., Kozawa, K. and Nishio, O. (2003) Detection, quantitation, and phylogenetic analysis of noroviruses in Japanese oysters, *Appl Environ Microbiol*, **69**, 5782–6.

Nuanualsuwan, S. and Cliver, D.O. (2002) Pretreatment to avoid positive RT-PCR results with inactivated viruses, *J Virol Methods*, **104**, 217–25.

Nuanualsuwan, S., Mariam, T., Himathongkham, S. and Cliver, D.O. (2002) Ultraviolet inactivation of feline calicivirus, human enteric viruses and coliphages, *Photochem Photobiol*, **76**, 406–10.

Nygard, K., Torven, M., Ancker, C., Knauth, S.B., Hedlund, K.O., Giesecke, J., Andersson, Y. and Svensson, L. (2003) Emerging genotype (GGIIb) of norovirus in drinking water, Sweden, *Emerg Infect Dis*, **9**, 1548–52.

Oishi, I., Yamazaki, K. and Minekawa, Y. (1993) An occurrence of diarrheal cases associated with group C rotavirus in adults, *Microbiol Immunol*, **37**(6), 505–9.

Oliver, S.L., Dastjerdi, A.M., Wong, S., El-Attar, L., Gallimore, C., Brown, D.W., Green, J. and Bridger, J.C. (2003) Molecular characterization of bovine enteric caliciviruses: a distinct third genogroup of noroviruses (Norwalk-like viruses) unlikely to be of risk to humans, *J Virol*, **77**, 2789–98.

Pallansch, M.A. and Roos, R.P. (2006) Enteroviruses : polioviruses, coxsackie viruses, echoviruses and newer enteroviruses, D M Knipe, P M Howley, D E Griffin, R A Lamb, M A Martin, B Roizman and S E Straus (eds), *Fields Virology*, 5th edn, Philadelphia, PA, Lippincott Williams and Wilkins, 839–94.

Parashar, U.D., Dow, L., Fankhauser, R.L., Humphrey, C.D., Miller, J., Ando, T., Williams, K.S., Eddy, C.R., Noel, J.S., Ingram, T., Bresee, J.S., Monroe, S.S. and Glass, R.I. (1998) An outbreak of viral gastroenteritis associated with consumption of sandwiches: implications for the control of transmission by food handlers, *Epidemiol Infect*, **121**, 615–21.

Parker, S., Cubitt, D., Jiang, J.X. and Estes, M. (1993) Efficacy of a recombinant Norwalk virus protein enzyme immunoassay for the diagnosis of infections with Norwalk virus and other human "candidate" caliciviruses, *J Med Virol*, **41**, 179–84.

Parks, C.G., Moe, C.L., Rhodes, D., Lima, A., Barrett, L., Tseng, F., Baric, R., Talal, A. and Guerrant, R. (1999) Genomic diversity of "Norwalk like viruses" (NLVs): pediatric infections in a Brazilian shantytown, *J Med Virol*, **58**, 426–34.

Parshionikar, S.U., Willian-True, S., Fout, G.S., Robbins, D.E., Seys, S.A., Cassady, J.D. and Harris, R. (2003) Waterborne outbreak of gastroenteritis associated with a norovirus, *Appl Environ Microbiol*, **69**, 5263–8.

Patterson, W., Haswell, P., Fryers, P.T. and Green, J. (1997) Outbreak of small round structured virus gastroenteritis arose after kitchen assistant vomited, *Comm Dis Rep Rev*, **7**, R101–3.

Phan, T.G., Okame, M., Nguyen, T.A., Maneekarn, N., Nishio, O., Okitsu, S. and Ushijima, H. (2004) Human astrovirus, norovirus (GI, GII), and sapovirus infections in Pakistani children with diarrhea, *J Med Virol*, **73**, 256–61.

Phan, T.G., Nguyen, T.A., Yan, H., Yagyu, F., Kozlov, V., Kozlov, A., Okitsu, S., Müller, W.E., Ushuijma, H. (2005) Development of a novel protocol for RT-multiplex PCR to detect diarrheal viruses among infants and children with acute gastroenteritis in Eastern Russia, *Clin Lab*, **51**(7–8), 429–35.

Ponka, A., Maunula, L., von Bonsdorff, C.H. and Lyytikainen, O. (1999) An outbreak of calicivirus associated with consumption of frozen raspberries, *Epidemiol Infect*, **123**, 469–74.

Reuter, G., Jiang, X. and Szucs, G. (2003) Noroviruses are the most common pathogens causing nosocomial gastroenteritis outbreaks in Hungarian hospitals, *Orv Hetil*, **144**, 1611–16.

Richards, G.P. (1999) Limitations of molecular biological techniques for assessing the virological safety of foods, *J Food Prot*, **62**, 691–7.

Richards, A.F., Lopman, B., Gunn, A., Curry, A., Ellis, D., Cotterill, H., Ratcliffe, S., Jenkins, M., Appleton, H., Gallimore, C.I., Gray, J.J. and Brown, D.W. (2003) Evaluation of a commercial ELISA for detecting Norwalk-like virus antigen in faeces, *J Clin Virol*, **26**, 109–15.

Richards, G.P., Watson, M.A. and Kingsley, D.H. (2004) A SYBR green, real-time RT-PCR method to detect and quantitate Norwalk virus in stools, *J Virol Methods*, **116**, 63–70.

Robertson, B.H., Jansen, R., Khanna, B., Totsuka, A., Nainan, O.V., Siegl, G., Widell, A., Margolis, H.S., Isomura, S., Ito, K. *et al.* (1992) Genetic relatedness of hepatitis A virus strains recovered from different geographical regions, *J Gen Virol*, **73**, 1365–77.

Robertson, B.H., Averhoff, F., Cromeans, T.L., Han, X.H., Khoprasert, B., Nainan, O.V., Rosenberg, J., Paikoff, L., Debess, E., Shapiro, C.N. and Margolis, H.S., (2000) Genetic relatedness of hepatitis A virus isolates during a community-wide outbreak, *J Med Virol*, **62**(2), 144–50.

Rockx, B., De Wit, M., Vennema, H., Vinjé, J., De Bruin, E., Van Duynhoven, Y. and Koopmans, M. (2002) Natural history of human calicivirus infection: a prospective cohort study, *Clin Infect Dis*, **35**, 246–53.

Rockx, B.H., Vennema, H., Hoebe, C.J., Duizer, E. and Koopmans, M.P. (2005) Association of histo-blood group antigens and susceptibility to norovirus infections, *J Infect Dis*, **191**(5), 749–54.

Rockx, B., Regeer, C., Bogers, W., van Amerongen, G., Heeney, J. and Koopmans, M. (2005b) Experimental norovirus infection in non-human primates, *J Med Virol*, **75**(2), 313–20.

Sair, A.I., D'Souza, D.H., Moe, C.L. and Jaykus, L.A. (2002) Improved detection of human enteric viruses in foods by RT-PCR, *J Virol Methods*, **100**, 57–69.

Sattar, S.A., Ijaz, M.K., Johnson-Lussenburg, C.M. and Springthorpe, V.S. (1984) Effect of relative humidity on the airborne survival of rotavirus SA11, *Appl Environ Microbiol*, **47**(4), 879–81.

Sawyer, L.A., Murphy, J.J., Kaplan, J.E., Pinsky, P.F., Chacon, D., Walmsley, S., Schonberger, L.B., Phillips, A., Forward, K., Goldman, C. and others (1988) 25- to 30-nm virus particle associated with a hospital outbreak of acute gastroenteritis with evidence for airborne transmission, *Am J Epidemiol*, **127**, 1261–71.

Scherba, G., Weigel, R.M. and O'Brien, W.D. Jr (1991) Quantitative assessment of the germicidal efficacy of ultrasonic energy, *Appl Environ Microbiol*, **57**, 2079–84.

Schwab, K.J., De Leon, R. and Sobsey, M.D. (1996) Immunoaffinity concentration and purification of waterborne enteric viruses for detection by reverse transcriptase PCR, *Appl Environ Microbiol*, **62**, 2086–94.

Schwab, K.J., Neill, F.H., Estes, M.K., Metcalf, T.G. and Atmar, R.L. (1998) Distribution of Norwalk virus within shellfish following bioaccumulation and subsequent depuration by detection using RT-PCR, *J Food Prot*, **61**, 1674–80.

Schwab, K.J., Neill, F.H., Fankhauser, R.L., Daniels, N.A., Monroe, S.S., Bergmire-Sweat, D.A., Estes, M.K. and Atmar, R.L. (2000) Development of methods to detect "Norwalk-like viruses" (NLVs) and hepatitis A virus in delicatessen foods: application to a food-borne NLV outbreak, *Appl Environ Microbiol*, **66**, 213–18.

Scott, F.W. (1980) Virucidal disinfectants and feline viruses, *Am J Vet Res*, **41**, 410–14.

Siebenga, J.J., Vennema, H., Renckens, B., de Bruin, E., van der Veer, B., Siezen, R.J. and Koopmans, M. (2007) Epochal evolution of GGII.4 norovirus capsid proteins from 1995 to 2006, *J Virol*, **81**(18), 9932–41.

Siebenga, J.J., Beersma, M.F., Vennema, H., van Biezen, P., Hartwig, N.J., Koopmans, M. (2008) High prevalence of prolonged norovirus shedding and illness among hospitalized patients: a model for in vivo molecular evolution, *J Infect Dis*, **198**(7), 994–1001. Erratum in: *J Infect Dis*, 2008, **198**(10), 1575.

Slomka, M.J. and Appleton, H. (1998) Feline calicivirus as a model system for heat inactivation studies of small round structured viruses in shellfish, *Epidemiol Infect*, **121**, 401–7.

Smiley, J.R., Chang, K.O., Hayes, J., Vinje, J. and Saif, L.J. (2002) Characterization of an enteropathogenic bovine calicivirus representing a potentially new calicivirus genus, *J Virol*, **76**, 10089–98.

Smiley, J.R., Hoet, A.E., Traven, M., Tsunemitsu, H. and Saif, L.J. (2003) Reverse transcription-PCR assays for detection of bovine enteric caliciviruses (BEC) and analysis of the genetic relationships among BEC and human caliciviruses, *J Clin Microbiol*, **41**, 3089–99.

Souza, M., Azevedo, M.S., Jung, K., Cheetham, S., Saif, L.J. (2008) Pathogenesis and immune responses in gnotobiotic calves after infection with the genogroup II.4-HS66 strain of human norovirus, *J Virol*, **82**(4), 1777–86.

Steinmann, J. (2004) Surrogate viruses for testing virucidal efficacy of chemical disinfectants, *J Hosp Infect*, **56**, Supplement 2, S49–54.

Straub, T.M. and Chandler, D.P. (2003) Towards a unified system for detecting waterborne pathogens, *J Microbiol Methods*, **53**, 185–97.

Su, C.Q., Wu, Y.L., Shen, H.K., Wang, D.B., Chen Y.H., Wu, D.M., He, L.N. and Yang, Z.L. (1986) An outbreak of epidemic diarrhoea in adults caused by a new rotavirus in Anjui Province of China in the summer of 1983, *J Med Virol*, **19**(2), 167–73.

Subekti, D.S., Tjaniadi, P., Lesmana, M., McArdle, J., Iskandriati, D., Budiarsa, I.N., Walujo, P., Suparto, I.H., Winoto, I., Campbell, J.R., Porter, K.R., Sajuthi, D., Ansari, A.A. and Oyofo, B.A. (2002) Experimental infection of Macaca nemestrina with a Toronto Norwalk-like virus of epidemic viral gastroenteritis, *J Med Virol*, **66**, 400–406.

Sugieda, M. and Nakajima, S. (2002) Viruses detected in the caecum contents of healthy pigs representing a new genetic cluster in genogroup II of the genus "Norwalk-like viruses", *Virus Res*, **87**, 165–72.

Sugieda, M., Nagaoka, H., Kakishima, Y., Ohshita, T., Nakamura, S. and Nakajima, S. (1998) Detection of Norwalk-like virus genes in the caecum contents of pigs, *Arch Virol*, **143**, 1215–21.

Tei, S., Kitajima, N., Takahashi, K. and Mishiro, S. (2003) Zoonotic transmission of hepatitis E virus from deer to human beings, *Lancet*, **362**, 371–3.

Thurston-Enriquez, J.A., Haas, C.N., Jacangelo, J., Riley, K. and Gerba, C.P. (2003) Inactivation of feline calicivirus and adenovirus type 40 by UV radiation, *Appl Environ Microbiol*, **69**, 577–82.

Tompkins, D.S., Hudson, M.J., Smith, H.R., Eglin, R.P., Wheeler, J.G., Brett, M.M., Owen, R.J., Brazier, J.S., Cumberland, P., King, V. and Cook, P.E. (1999) A study of infectious intestinal disease in England: microbiological findings in cases and controls, *Commun Dis Public Health*, **2**, 108–13.

Treanor, J.J., Jiang, X., Madore, H.P. and Estes, M.K. (1993) Subclass-specific serum antibody responses to recombinant Norwalk virus capsid antigen (rNV) in adults infected with Norwalk, Snow Mountain, or Hawaii virus, *J Clin Microbiol*, **31**, 1630–34.

van Duynhoven, Y.T.H.P, de Jager, C.M., Kortbeek, L.M., Vennema, H., Koopmans, M.P.G., van Leusden, F., van der Poel, W.H.M., van den Broek, M.J.M. (2005) A one-year intensified study of outbreaks of gastroenteritis in the Netherlands, *Epidemiol Infect*, **133**, 9–21.

van den Brandhof, W.E., De Wit, G.A., de Wit, M.A. and van Duynhoven, Y.T. (2004) Costs of gastroenteritis in the Netherlands, *Epidemiol Infect*, **132**, 211–21.

van der Poel, W.H., Vinje, J., van Der Heide, R., Herrera, M.I., Vivo, A. and Koopmans, M.P. (2000) Norwalk-like calicivirus genes in farm animals, *Emerg Infect Dis*, **6**, 36–41.

van der Poel, W.H., van der Heide, R., Verschoor, F., Gelderblom, H., Vinjé, J. and Koopmans, M.P. (2003) Epidemiology of Norwalk-like virus infections in cattle in the Netherlands, *Vet Microbiol*, **92**, 297–309.

Vennema, H., de Bruin, E. and Koopmans, M. (2002) Rational optimization of generic primers used for Norwalk-like virus detection by reverse transcriptase polymerase chain reaction, *J Clin Virol*, **25**, 233–5.

Verhoef, L.P., Kroneman, A., van Duynhoven, Y., Boshuizen, H., van Pelt, W., Koopmans, M. *et al.* (2009) Selection tool for foodborne norovirus outbreaks, *Emerg Infect Dis*, **15**(1), 31–8.

Vinjé, J. and Koopmans, M. (1996) Molecular detection and epidemiology of NLV in outbreaks of gastroenteritis in the Netherlands, *J Infect Dis*, **174**, 610–15.

Vinjé, J., Altena, S.A. and Koopmans, M.P. (1997) The incidence and genetic variability of small round-structured viruses in outbreaks of gastroenteritis in the Netherlands, *J Infect Dis*, **176**, 1374–8.

Vinjé, J., Deijl, H., Van de Heide, R., Lewis, D., Hedlund, K.-O., Svensson, L. and Koopmans, M. (2000) Molecular detection and epidemiology of typical human calciviruses, *J Clin Microbiol*, **38**(2), 530–36.

Vinjé, J., Vennema, H., Maunula, L., von Bonsdorff, C.H., Hoehne, M., Schreier, E., Richards, A., Green, J., Brown, D., Beard, S.S., Monroe, S.S., de Bruin, E., Svensson, L. and Koopmans, M.P. (2003) International collaborative study to compare reverse transcriptase PCR assays for detection and genotyping of noroviruses, *J Clin Microbiol*, **41**, 1423–33.

Vinjé, J., Hamidjaja, R.A. and Sobsey, M.D. (2004) Development and application of a capsid VP1 (region D) based reverse transcription PCR assay for genotyping of genogroup I and II noroviruses, *J Virol Methods*, **116**, 109–17.

Vipond, I.B., Pelosi, E., Williams, J., Ashley, C., Lambden, P., Clarke, I. and Caul, O. (2000) A diagnostic EIA for detection of the prevalent SRSV strain in UL outbreaks of gastroenteritis, *J Med Virol*, **61**, 132–7.

Wheeler, J.G., Sethi, D., Cowden, J.M., Wall, P.G., Rodrigues, L.C., Tompkins, D.S., Hudson, M.J. and Roderick, P.J. (1999) Study of infectious intestinal disease in England: rates in the community, presenting to general practice, and reported to national surveillance. The Infectious Intestinal Disease Study Executive, *BMJ*, **318**, 1046–50.

Widdowson, M.A., Cramer, E.H., Hadley, L., Bresee, J.S., Beard, R.S., Bulens, S.N., Charles, M., Chege, W., Isakbaeva, E., Wright, J.G., Mintz, E., Forney, D., Massey, J., Glass, R.I. and Monroe, S.S. (2004) Outbreaks of acute gastroenteritis on cruise ships and on land: identification of a predominant circulating strain of norovirus–United States, 2002, *J Infect Dis*, **190**, 27–36.

Widdowson, M.A., Rockx, B. , Schepp, R., Vinjé, J., van Duynhoven, Y. T. and Koopmans, M.P. (2005) Detection of serum antibodies to bovine norovirus in veterinarians and the general population in the Netherlands, *J Med Virol*, **1**, 119–28.

Wise, A.G., Monroe, S.S., Hanson, L.E., Grooms, D.L., Sockett, D. and Maes, R.K. (2004) Molecular characterization of noroviruses detected in diarrheic stools of Michigan and Wisconsin dairy calves: circulation of two distinct subgroups, *Virus Res*, **100**, 165–77.

Wolf, S., Williamson, W.M., Hewitt, J., Rivera-Aban, M., Lin, S., Ball, A., Scholes, P. and Greening, G.E. (2007) Sensitive multiplex real-time reverse transcription – PCR assay for the detection of human and animal noroviruses in clinical and environmental samples, *Appl Environ Microbiol*, **73**(17), 5464–70.

Yamashita, T., Sakae, K., Tsuzuki, H., Suzuki, Y., Ishikawa, N., Takeda, N., Miyamura, T and Yamazaki, S. (1998) Complete nucleotide sequence and genetic organization of Aichi virus, a distinct member of the Picornaviridae associated with acute gastroenteritis in humans, *J Virol*, **72**(10), 8408–12.

Yazaki, Y., Mizuo, H., Takahashi, M., Nishizawa, T., Sasaki, N., Gotanda, Y. and Okamoto, H. (2003) Sporadic acute or fulminant hepatitis E in Hokkaido, Japan, may be foodborne, as suggested by the presence of hepatitis E virus in pig liver as food, *J Gen Virol*, **84**, 2351–7.

Yoda, T., Terano, Y., Suzuki, Y., Yamazaki, K., Oishi, I., Utagawa, E., Shimada, A.,

Matsuura, S., Nakajima, M. and Shibata, T. (2000) Characterization of monoclonal antibodies generated against Norwalk virus GII capsid protein expressed in *Escherichia coli*, *Microbiol Immunol*, **44**, 905–14.
Zahorsky, J. (1929) Hyperemesis hiemis or the winter vomiting disease, *Arch Pediatr*, **46**, 391–5.

Index